June 28, 2013

Dear Customer:

Thank you for your purchase of *Webb's Physics of Medical Imaging, Second Edition*, edited by M.A. Flower.

The following corrections were not made prior to printing:

p. xv, para 4, line 1: 'Clarie Skinner' should be 'Claire Skinner'.

p. 183, equation 5.10: Change '$d_p \cos^3\theta$' to '$d_p^2 \cos^3\theta$'

p. 237, para 2, line 3: Change '(see Section 14.3.3)'to '(see Section 14.3.2)'

p. 245, Figure 5.66, caption, line 3: Change 'greater/less than 11 cm, respectively' to 'greater/less than the normalisation distance (11 cm in this case)'

p. 258, caption for Figure 5.85: Change 'the kidney' to 'the transplanted kidney'.

p. 294: Table 5.17, column 3, row starting '^{13}N': the target reaction should be '^{16}O (p,α) ^{13}N'

p. 294: Table 5.17, column 3, row starting '^{11}C': the target reaction should be '^{14}N (p,α) ^{11}C'

p. 370, last para of section 6.2.2, line 6: Change 'radio frequency point spread function' to 'radio-frequency point spread function'

p. 485, Further Reading, Dunn & O'Brien 1977 reference: Change '*Ultrasonic Biophysics Benchmark Papers in Acoustics*' to 'Ultrasonic Biophysics *Benchmark Papers in Acoustics* '

p. 520, Section 7.6.2, line 2: Start the italics earlier so that it reads 'The key idea in MRI is that *if one makes B vary with position, then*'

p. 598, Figure 7.63, caption, line 3: Change '(Data from Poole and Bowtell, 2007.)' to '(ROU, region of uniformity; ROS, region of shielding. Data from Poole and Bowtell, 2007.)'

p. 611, Beckman Institute 1998 reference, line 2: Change 'accessed on May 3, 2007' to 'accessed on February 27, 2012'

p. 776, Figure 15.4 caption, line 2: Change 'NaI' to '^{131}I-NaI'

Colour Images, p. 13, Figure 7.63, caption, line 3: Change '(Data from Poole and Bowtell, 2007.)' to '(ROU, region of uniformity; ROS, region of shielding. Data from Poole and Bowtell, 2007.)'

Colour Images, p.16, Figure 15.4 caption, line 2: Change 'NaI' to '^{131}I-NaI'

We sincerely regret any inconvenience this may have caused you. Please let us know if we can be of any assistance regarding this title or any other titles that Taylor & Francis publishes.

Best regards,
Taylor & Francis

ISBN 978-0-7503-0573-0/IP743

Webb's Physics of Medical Imaging

Second Edition

Series in Medical Physics and Biomedical Engineering

Series Editors: John G Webster, E Russell Ritenour, Slavik Tabakov, Kwan-Hoong Ng

Other recent books in the series:

Correction Techniques in Emission Tomography
Mohammad Dawood, Xiaoyi Jiang, and Klaus Schäfers (Eds)

Physiology, Biophysics, and Biomedical Engineering
Andrew Wood (Ed)

Proton Therapy Physics
Harald Paganetti (Ed)

Correction Techniques in Emission Tomography
Mohammad Dawood, Xiaoyi Jiang, Klaus Schäfers (Eds)

Practical Biomedical Signal Analysis Using MATLAB®
K J Blinowska and J Żygierewicz (Ed)

Physics for Diagnostic Radiology, Third Edition
P P Dendy and B Heaton (Eds)

Nuclear Medicine Physics
J J Pedroso de Lima (Ed)

Handbook of Photonics for Boimedical Science
Valery V Tuchin (Ed)

Handbook of Anatomical Models for Radiation Dosimetry
Xie George Xu and Keith F Eckerman (Eds)

Fundamentals of MRI: An Interactive Learning Approach
Elizabeth Berry and Andrew J Bulpitt

Handbook of Optical Sensing of Glucose in Biological Fluids and Tissues
Valery V Tuchin (Ed)

Intelligent and Adaptive Systems in Medicine
Oliver C L Haas and Keith J Burnham

A Introduction to Radiation Protection in Medicine
Jamie V Trapp and Tomas Kron (Eds)

A Practical Approach to Medical Image Processing
Elizabeth Berry

Biomolecular Action of Ionizing Radiation
Shirley Lehnert

An Introduction to Rehabilitation Engineering
R A Cooper, H Ohnabe, and D A Hobson

Series in Medical Physics and Biomedical Engineering

Webb's Physics of Medical Imaging

Second Edition

Edited by

M A Flower

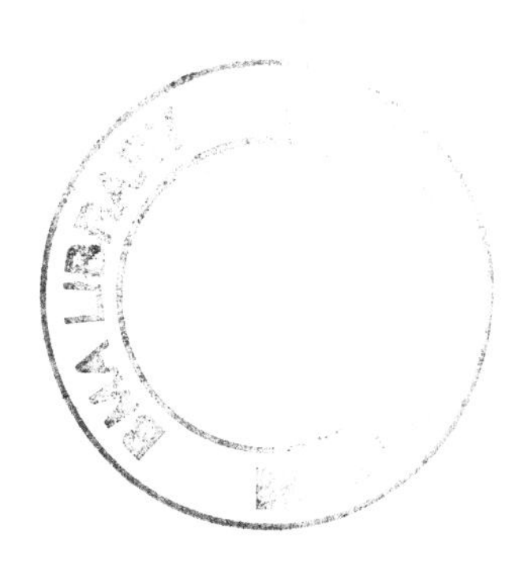

CRC Press is an imprint of the
Taylor & Francis Group, an **informa** business
A TAYLOR & FRANCIS BOOK

CRC Press
Taylor & Francis Group
6000 Broken Sound Parkway NW, Suite 300
Boca Raton, FL 33487-2742

Printed in the United States of America on acid-free paper
Version Date: 20120515

International Standard Book Number: 978-0-7503-0573-0 (Hardback)

Library of Congress Cataloging-in-Publication Data

Webb's physics of medical imaging / editor, M.A. Flower. -- 2nd ed.
p. ; cm. -- (Series in medical physics and biomedical engineering)
Physics of medical imaging
Rev. ed. of: Physics of medical imaging / edited by Steve Webb. c1988.
Includes bibliographical references and index.
Summary: "This updated text explains the physical principles governing medical imaging. It places different techniques in context with each other to provide historical perspective and insight on future developments. It also illustrates techniques through examples of clinical imaging applications. The book covers the use of x-rays in diagnostic radiology and computed tomography in radiotherapy planning. It also discusses the tools and techniques of radioisotope, ultrasound, magnetic resonance, infrared, and optical imaging, along with the mathematics involved in image formation and processing, image interpretation, and computational requirements of imaging systems"--Provided by publisher.
ISBN 978-0-7503-0573-0 (hardback : alk. paper)
I. Webb, Steve, 1948- II. Flower, M. A. (Maggie A.) III. Physics of medical imaging. IV. Title: Physics of medical imaging. V. Series: Series in medical physics and biomedical engineering.
[DNLM: 1. Diagnostic Imaging. 2. Physics. WN 180]

616.07'54--dc23

2012008918

Visit the Taylor & Francis Web site at
http://www.taylorandfrancis.com

and the CRC Press Web site at
http://www.crcpress.com

Series Preface

The Series in Medical Physics and Biomedical Engineering describes the applications of physical sciences, engineering and mathematics in medicine and clinical research.

The series seeks (but is not restricted to) publications in the following topics:

- Artificial organs
- Assistive technology
- Bioinformatics
- Bioinstrumentation
- Biomaterials
- Biomechanics
- Biomedical engineering
- Clinical engineering
- Imaging
- Implants
- Medical computing and mathematics
- Medical/surgical devices
- Patient monitoring
- Physiological measurement
- Prosthetics
- Radiation protection, health physics and dosimetry
- Regulatory issues
- Rehabilitation engineering
- Sports medicine
- Systems physiology
- Telemedicine
- Tissue engineering
- Treatment

The Series in Medical Physics and Biomedical Engineering is an international series that meets the need for up-to-date texts in this rapidly developing field. Books in the series range in level from introductory graduate textbooks and practical handbooks to more advanced expositions of current research. This series is the official book series of the International Organization for Medical Physics.

The International Organization for Medical Physics

The International Organization for Medical Physics (IOMP), founded in 1963, is a scientific, educational and professional organisation of 76 national adhering organisations, more than 16,500 individual members, several corporate members and 4 international regional organisations.

IOMP is administered by a council, which includes delegates from each of the adhering national organisations. Regular meetings of the council are held electronically as well as every three years at the World Congress on Medical Physics and Biomedical Engineering. The president and other officers form the executive committee, and there are also committees covering the main areas of activity, including education and training, scientific, professional relations and publications.

Objectives

- To contribute to the advancement of medical physics in all its aspects
- To organise international cooperation in medical physics, especially in developing countries
- To encourage and advise on the formation of national organisations of medical physics in those countries which lack such organisations

Activities

The official journals of the IOMP are *Physics in Medicine and Biology* and *Medical Physics and Physiological Measurement*. The IOMP publishes a bulletin, *Medical Physics World*, twice a year, which is distributed to all members.

A World Congress on Medical Physics and Biomedical Engineering is held every three years in cooperation with International Federation for Medical and Biological Engineering (IFMBE) through the International Union for Physics and Engineering Sciences in Medicine (IUPESM). A regionally based international conference on medical physics is held between world congresses. IOMP also sponsors international conferences, workshops and courses. IOMP representatives contribute to various international committees and working groups.

The IOMP has several programmes to assist medical physicists in developing countries. The joint IOMP Library Programme supports 69 active libraries in 42 developing countries, and the Used Equipment Programme coordinates equipment donations. The Travel Assistance Programme provides a limited number of grants to enable physicists to attend the world congresses. The IOMP website is being developed to include a scientific database of international standards in medical physics and a virtual education and resource centre.

Information on the activities of the IOMP can be found on its website at www.iomp.org.

Quotation

'What is the use of a book,' thought Alice, 'without pictures?'

Lewis Carroll
Alice's Adventures in Wonderland

Contents

Contributors.....xi
Editor.....xvii
Acknowledgement to the First Edition.....xix
Acknowledgement to the Second Edition.....xxi
Abbreviations.....xxiii
Introduction and Some Challenging Questions.....xxxiii

1. **In the Beginning: the Origins of Medical Imaging**.....1
S. Webb

2. **Diagnostic Radiology with X-Rays**.....13
D. R. Dance, S. H. Evans, C. L. Skinner and A. G. Bradley

3. **X-Ray Transmission Computed Tomography**.....97
I. A. Castellano and S. Webb

4. **Clinical Applications of X-Ray Computed Tomography in Radiotherapy Planning**.....153
H. J. Dobbs and S. Webb

5. **Radioisotope Imaging**.....165
R. J. Ott, M. A. Flower, A. D. Hall, P. K. Marsden and J. W. Babich

6. **Diagnostic Ultrasound**.....351
J. C. Bamber, N. R. Miller and M. Tristam

7. **Spatially Localised Magnetic Resonance**.....487
S. J. Doran and M. O. Leach

8. **Physical Aspects of Infrared Imaging**.....623
C. H. Jones

9. **Imaging of Tissue Electrical Impedance**.....647
B. H. Brown and S. Webb

10. **Optical Imaging**.....665
J. C. Hebden

11. **Mathematics of Image Formation and Image Processing**.....687
S. Webb

12. Medical Image Processing 713
J. Suckling

13. Perception and Interpretation of Images 739
C. R. Hill

14. Computer Requirements of Imaging Systems 755
G. D. Flux, S. Sassi and R. E. Bentley

15. Multimodality Imaging 767
G. D. Flux

16. Epilogue 783
S. Webb and C. R. Hill

Index 791

Contributors

The 15 contributors to the first edition of this book (1988) were at that time almost all on the staff of the Joint Department of Physics at the Institute of Cancer Research (ICR) and (then called) Royal Marsden Hospital (RMH) at either Chelsea, London, or Sutton, Surrey, UK. At the time of writing the second edition (2012), some are still in this department and the (now called) Royal Marsden National Health Service Foundation (NHSF) Trust, but many have since retired. They have been joined by 11 physicists from here and from other institutions to write this second edition.

John Babich is the president and chief scientific officer at Molecular Insight Pharmaceuticals in Cambridge, Massachusetts. He was a founder member of this new radiopharmaceutical company in 1997. Prior to this, he was an assistant professor of radiology at Harvard Medical School and an assistant radiopharmaceutical chemist at Massachusetts General Hospital, Boston, Massachusetts, for eight years. Prior to that, he was the principal radiopharmaceutical scientist at ICR/RMH (Sutton) for six years. John's current research focuses on the discovery and clinical development of small-molecule-based radiopharmaceuticals for applications in cancer, including targeting of cellular transporters and receptors, metabolic and enzyme targets, integrins, and hypoxia-inducible targets. His recent research has included the development of unique radiolabelling platforms for radiohalogens as applied to therapy and novel chelate systems for radiometals in diagnostic applications.

Jeff Bamber is a reader in physics applied to medicine. His main research interests are in ultrasound and optical imaging, where he has pioneered new techniques in speckle reduction, the use of microbubbles, elastography and photoacoustics. His work has included tissue characterisation, ultrasonic investigation of the breast, imaging of skin tumours, intraoperative imaging in neurosurgery, radiotherapy and ultrasound therapy guidance, ultrasound gene transfection and mathematical modelling of the interaction of ultrasound waves with tissue. He has published 14 book chapters, 5 patents and over 180 research papers. He is team leader in medical ultrasound and optics and senior tutor at ICR, Sutton, and medical physicist at the RMH. He has worked for the Hewlett-Packard Medical Products Group and acts as advisor to several medical instrument manufacturers. He has taught medical ultrasound at masters level on several courses, the first since 1979.

Roy Bentley obtained his PhD at the University of Birmingham in 1955, then spent the greater part of his working life at the ICR and the associated RMH. After working on the assessment of radioactive nuclides in the environment, he switched to working on the application of computers in medicine, and in particular medical physics. Starting with computers in a number of applications in radiotherapy, he subsequently turned his attention to nuclear medicine and methods of presentation of the images produced. In 1967–1968 he spent a year's sabbatical in the Biomedical Computer Department at Washington University, St. Louis, where he was one of the first people to link a computer online to a gamma camera. He was also instrumental in the introduction of a computerised information system for the RMH. Before retiring in 1995, he returned to an interest in radiotherapy and worked with an advanced system of imaging and planning developed at DKFZ in Heidelberg.

Andy Bradley is a clinical physicist at the Manchester Royal Infirmary working in nuclear medicine, with a special interest in radiation protection. Prior to his move north, he was a member of the Radiological Physics Group at the RMH (Chelsea) with an interest in radiation dosimetry.

Brian Brown is an emeritus professor of medical physics and clinical engineering at the University of Sheffield/Central Sheffield University Hospitals. His major interest is in medical electronics for physiological measurement. He pioneered electrical impedance tomography and spectroscopy as routine clinical tools.

Elly Castellano is a consultant clinical scientist working at The Royal Marsden NHSF Trust as head of the Diagnostic Radiology Physics Group. In 2006, she completed a part-time PhD. Elly's main research interests are in CT dosimetry and CT optimisation, although she has also published papers in the fields of mammography and fluoroscopy. She runs the X-ray and CT imaging module of the MSc in medical engineering and physics at King's College London, and lectures extensively in the UK and abroad. She is a contributing author to several textbooks and handbooks on CT and patient radiation dosimetry. She was chair of the CT Users Group between 2004 and 2007 and is chair of the IPEM's Diagnostic Radiology Special Interest Group.

David Dance was a consultant clinical scientist and head of the Department of Physics at The Royal Marsden NHSF Trust (Chelsea) before he retired in 2005. He led research in the physics of X-ray imaging, where he published widely. The results of David's Monte Carlo calculations form the basis of the United Kingdom, European and IAEA protocols for mammographic dosimetry. For many years, David collaborated with the Department of Radiation Physics at Linkoping University in Sweden, and he has an MD *honoris causa* from that university. David was a founder member of Symposium Mammographicum and UKMPG and has been chairman of both organisations. He is a consultant to the IAEA and a co-editor of the agency's *Handbook of Diagnostic Radiology Physics*. The lure of research into the physics of mammography proved too great after David's retirement. He is presently a consultant physicist at the Royal Surrey County Hospital and a visiting professor at the University of Surrey.

Jane Dobbs is an emeritus consultant in clinical oncology at Guy's & St. Thomas' Hospital and, formerly, lecturer in radiotherapy at ICR/RMH. Her research experience has included assessing the role of X-ray computed tomography in radiotherapy treatment planning. She has published a textbook on radiotherapy planning and external beam techniques entitled *Practical Radiotherapy Planning* (fourth edition, 2009). She was course director of the ESTRO Teaching Course Imaging for Target Volume Determination in Radiotherapy.

Simon Doran graduated from Cambridge University in 1988 with a degree in physics and theoretical physics. A final year project in MRI with the late professor Laurie Hall sparked his interest in 3D imaging, and he has remained in the same field ever since. After a 'gap year', he returned to do a PhD in Cambridge on the subject of quantitative T1 and T2 imaging and went on to do a post-doc, performing research in ultra-rapid MRI in Grenoble, France. On returning to England, he took up a lecturing post in the Department of Physics at the University of Surrey, where he stayed for 11 years, teaching at all levels and in a variety of subjects from special relativity to industrial applications of ultrasound. His research at the University of Surrey focussed on quantitative MRI and pioneering developments in the new field of optical computed tomography for the verification of complex treatment plans in radiotherapy. In 2006, he moved to the ICR, where, as senior staff scientist, he is based in the MRI section of the CR-UK and EPSRC Cancer Imaging Centre and leads the effort in advanced imaging informatics.

Stephen Evans is head of medical physics, Northampton General Hospital, and is responsible for 40 scientific/technical/administrative staff. The services provided by the medical physics department include radiotherapy physics, imaging (diagnostic X-ray) physics, radiation protection and nuclear medicine. He is the radiation protection advisor (RPA) to Northampton General Hospital, Milton Keynes NHSF Trust, Kettering General Hospital and various private clinics and dental surgeries. Since 1979, he has worked in the field of radiation protection. He has published texts on radiological protection and scientific papers on X-ray imaging. He has research interests in digital tomographic imaging, small angle X-ray scattering and diffraction-enhanced X-ray imaging. Stephen's professional roles include director of the IPEM's Science Board.

Maggie Flower was a senior lecturer in physics as applied to medicine at RMH/ICR (Sutton) before she retired in 2004. She received her PhD from the ICR in 1981. Her research interests have included the evaluation and development of quantitative radioisotope imaging techniques using SPECT and PET, and the application of quantitative imaging methods to a variety of clinical research projects, in particular new dosimetry techniques for patients undergoing targeted radionuclide therapy. Maggie has published over 90 peer-reviewed papers and has contributed to 15 chapters in textbooks on medical imaging and radionuclide therapy. She is a fellow of the Institute of Physics and Engineering in Medicine (FIPEM).

Glenn Flux is head of radioisotope physics at The Royal Marsden NHSF Trust and ICR in Sutton. His main research focus is on molecular imaging and internal dosimetry for molecular radiotherapy, with particular emphasis on clinical applications. He currently chairs the European Association of Nuclear Medicine Dosimetry Committee and the UK CTRad Group on Molecular Radiotherapy and was chair of a British Institute of Radiology (BIR) working party that reported on the status of molecular radiotherapy in the UK. He has authored over 60 papers and several book chapters and is the holder of several research grants.

Adrian Hall has been head of radiopharmacy at The Royal Marsden NHSF Trust, Sutton, since 1991 and director of education for the Joint Department of Physics since 2000. He received his PhD in medicinal chemistry from the University of Essex before being appointed as a lecturer in pharmaceutical chemistry at the Department of Pharmacy, King's College, London, in 1987. His research interests include metal chelation chemistry, the radiolabelling of peptides and antibodies for radioisotope imaging, and development of systems for the semi-automated synthesis of Ga-68 radiopharmaceuticals. He is currently a member of the UK Radiopharmacy Group Committee.

Jem Hebden is a professor of biomedical optics and has been head of the Department of Medical Physics and Bioengineering at University College London (UCL) since 2008. After receiving his PhD in astronomy and spending two years in Arizona exploring high-resolution methods for mapping stellar atmospheres, he spent five years at the University of Utah investigating new optical imaging techniques for functional imaging of human tissues. He pioneered the experimental development of time-resolved methods which overcome the blurring effects of scatter. A Wellcome Trust Senior Fellowship enabled him to establish a group at UCL devoted to the development of clinical prototypes for optical imaging of human subjects, with particular emphasis on the study of the premature infant brain at risk of damage resulting from hypoxia-ischaemia. His group has developed time-resolved instruments for 3D optical tomography, utilising unique source and detector technology. These have been used to produce the first whole-brain images of evoked functional activity in the newborn infant, and this work is now focussed on the study of infant seizures. His group has also built systems for mapping the haemodynamic response in the cortex to sensory stimulation and other cognitive activity, and to acquire EEG measurements simultaneously.

Kit Hill is an emeritus professor of physics at the University of London and a former head of the Joint Department of Physics at ICR/RMH. Following a period in industrial engineering and subsequent work in the physics of environmental radiation, he took up a long-term interest in the potential value of ultrasound in cancer management. This covered two areas, both of which combined practical applications with studies of the underlying biophysics: the use of strongly focussed beams for non-invasive surgery and pulse-echo technology with its extensions into tissue characterisation and parametric imaging.

Colin Jones was head of department at Fulham Road for many years. He was a senior lecturer in physics as applied to medicine and retired in 1997. He has a long history of interest in thermal imaging and has published widely. Among other interests, he has published papers on the physics of radiotherapy and dosimetry, radiation protection and breast imaging. After retirement, he served as visiting expert in brachytherapy for the International Atomic Energy Authority on several assignments and throughout the period 2000–2006 was visiting lecturer in brachytherapy at the European School for Medical Physics, Archamps.

Martin Leach is a professor of physics as applied to medicine and codirector of the CR-UK and EPSRC Cancer Imaging Centre and director of the Wellcome/MRC/NIHR Clinical Research Facility at The Institute of Cancer Research and The Royal Marsden NHSF Trust, Sutton. He is deputy head of the division of radiotherapy and imaging and was previously co-chair of the section of magnetic resonance. His current research includes magnetic resonance imaging and spectroscopy, and imaging in radiotherapy. He has also contributed to imaging using SPECT, PET and CT. He received his PhD in physics from the University of Birmingham for research on *in vivo* activation analysis and inert-gas metabolism and also worked on cyclotron isotope production. He has published over 270 peer-reviewed articles and is a fellow of the Academy of Medical Sciences, the Institute of Physics, the IPEM, the Society of Biology and the International Society of Magnetic Resonance in Medicine. He was awarded the Silvanus Thompson Medal of the BIR in 2010 and the BIR Barclay Medal in 2009 and appointed an NIHR senior investigator in 2008. He was honorary editor of the international journal *Physics in Medicine and Biology* from 1996 to 1999.

Paul Marsden received his PhD from the ICR in 1987, working on the MUP-PET positron emission tomography system. He is currently professor of PET physics and scientific director of the PET Imaging Centre at King's College London, where his research interests include PET methodology, clinical and research applications of PET, and the development of simultaneous PET and MR scanning.

Naomi Miller received her PhD from the ICR in 2000, working in the field of image perception. She then worked for nine years as a clinical and research physicist at ICR and The Royal Marsden NHSF Trust, Sutton, specialising in medical ultrasound.

Bob Ott is an emeritus professor of radiation physics at the University of London, and a past head of the Joint Department of Physics and academic dean of the ICR. He was in charge of research and development in radioisotope imaging at ICR before he retired in 2005. His research interests include the development of position-sensitive detectors (including positron cameras and CCD-based high-resolution systems) for radioisotope imaging, methods for quantification of radioisotope imaging using SPECT and PET, and the applications of these techniques to clinical oncology. He is a fellow of the Institute of Physics and an honorary fellow of the BIR.

Salem Sassi was a principal PET physicist at The Royal Marsden NHSF Trust in Sutton for nearly 5 years. Prior to this, he worked at St. George's NHS Trust as a principal medical physicist. In 2011, Salem took up the position of senior consultant at the Riyadh Military Hospital in Saudi Arabia.

Clarie Skinner is a consultant clinical scientist and head of radiological physics and radiation safety at the Royal Free Hampstead NHS Trust in London. Prior to this, she worked at The Royal Marsden NHSF Trust in Fulham Road, specialising in diagnostic radiology physics. Claire has a special interest in mammography and has contributed to a number of papers in this field. She is the regional coordinator for physics services to the NHS breast screening programme in the East of England and lectures on a number of courses in the UK.

John Suckling is professor of psychiatric neuroimaging and co-director of the brain mapping unit in the Department of Psychiatry at the University of Cambridge. Beginning his career in positron emission tomography at the ICR, he subsequently worked on image processing applications in digital mammography, computed tomography, and structural and functional magnetic resonance imaging. John has also published over 150 peer-reviewed articles. His current interests include imaging as primary outcome variables in randomised controlled trials of treatments for mental health disorders.

Maria Tristam contributed to the first edition when she was a research fellow at the Joint Department of Physics. She went on to become a clinical physicist at Southampton General Hospital until her retirement in 2009. Her research interests included the development of new imaging techniques to assess tissue movement using ultrasound. Her early research experience was in cosmic-ray physics detecting high-energy muons deep underwater and underground.

Steve Webb is an emeritus professor of radiological physics and was head of the Joint Department of Physics at the ICR/RMH from 1998 until he retired in 2011. He was a team leader in radiotherapy physics, with a special interest in conformal radiation therapy and intensity-modulated radiation therapy. He has a PhD and DSc and is a fellow of the Institute of Physics (FInstP), the Institute of Physics and Engineering in Medicine (FIPEM) and the Royal Society of Arts (FRSA). He is also a chartered physicist (CPhys) and a chartered clinical scientist (CSci). Steve has published some 200 peer-reviewed papers in medical imaging and the physics of radiation therapy, as well as 5 single-author textbooks. From 2006 until

2011, he was editor in chief of the international journal *Physics in Medicine and Biology*. He was awarded the Silvanus Thompson Medal of the BIR in 2004, the BIR Barclay Medal in 2006 and the EFOMP Medal in 2011. He has been visiting professor at DKFZ Heidelberg, the University of Michigan at Ann Arbor, Memorial Sloan Kettering Cancer Centre, New York, and Harvard Mass General Hospital, Boston.

Photos of contributors, from left to right, starting at top row: Simon Doran, Jeff Bamber, John Babich, Kit Hill, Martin Leach, Stephen Evans, Salem Sassi, Colin Jones, Paul Marsden, Steve Webb, David Dance, Bob Ott, John Suckling, Glenn Flux, Roy Bentley, Jane Dobbs, Maggie Flower, Maria Tristam, Adrian Hall, Elly Castellano, Brian Brown, Claire Skinner, Jem Hebden, Naomi Miller and Andy Bradley.

Editor

Maggie Flower graduated from St Hugh's College, Oxford, with a BA in physics in 1971. Immediately after her graduation, she embarked on her medical physics career by taking up an NHS basic grade post in radiotherapy physics at King's College Hospital, Denmark Hill, London. During her three years in this post, Maggie studied part time for her MSc in radiation physics at St Bartholomew's Hospital Medical School. In 1974, she was appointed to a university post at the ICR, working in the radioisotope section of the physics department, and was based at the Sutton branch of the RMH. She continued working at what became the ICR/RMH Joint Department of Physics for 30 years. During that time, she studied part time and received her PhD in medical physics from the University of London in 1981. Her PhD thesis was on the quantitative estimation of radioactive sources *in vivo* using transaxial emission tomography and conventional scanning. She was promoted to a lecturer's post in 1982 and to senior lecturer in 1997. A year later, she became a faculty member. By the time Maggie took early retirement in 2004, she had become joint team leader for the ICR's Radioisotope Research Team and head of Radioisotope Imaging in the RMH's physics department. As joint team leader, she helped to win several research grants. Amongst these were grants from the Neuroblastoma Society for the development of new dosimetry techniques for unsealed source therapy and from the EPSRC for the development of a novel PET camera using multiwire proportional chambers. Maggie's teaching responsibilities included lecturing to radiographers, radiologists, nuclear medicine physicians, as well as clinical scientists, and MSc and PhD students in the field of medical physics. She was in at the beginning of the course of lectures at Chelsea College from which the first edition of this book arose, and she continued to lecture every year on this course, which became the radioisotope imaging module of the MSc in medical engineering and physics at King's College, London, until she retired.

Acknowledgement to the First Edition

This book has been compiled by members of the Department of Physics at the Institute of Cancer Research and Royal Marsden Hospital, and our principal debt of gratitude is to those, too numerous to mention by name, whose wisdom foresaw the importance of medical imaging for cancer and other diseases and whose perseverance established imaging as a central component of our work. We particularly thank three past professors, Val Mayneord, Jack Boag and Ged Adams, who brought most of us here and whose influence still governs what we do and how we do it. We would also like to acknowledge the late Professor Roy Parker, whose own serious illnesses did not weaken his research initiative in imaging. The direction of the department owes much to the wisdom of Dr. Nigel Trott, who has also provided help and support for the first edition of this book.

The book arose from a course of lectures given initially at Chelsea College, University of London, as part of the MSc in biophysics and bioengineering organised by Dennis Rosen. We would like to acknowledge our long association with Dr. Rosen and his past and present colleagues.

Joint-author papers in the reference lists give some indication of specific credit due to our own past and present colleagues for some of the material included in this book. We are very grateful to the large number of people who have lent us figures and images and who are mentioned by name at the appropriate places. Our thanks also go to the publishers who have allowed this material to be freely used.

Several illustrations have been drawn from work at The Royal Marsden Hospital, and we are grateful to clinical colleagues who have allowed us to use this material. To say that Sheila Dunstan, Rosemary Atkins and Marion Barrell typed the manuscript gives no indication of the organisational effort they have put into this project. They have been typing at times of the day and night when ordinary mortals have been enjoying themselves or asleep. We cannot thank them enough. We should also like to thank Sue Sugden for help with references and Ray Stuckey for help with the production of figures.

Our thanks go to Neville Hankins (commissioning editor) and to Sarah James (desk editor) at Adam Hilger Publishers for their most professional work.

Steve Webb, 1988

Acknowledgement to the Second Edition

It is more than 20 years since we wrote the first edition. *The Physics of Medical Imaging* has been widely adopted throughout the world as a course book for taught courses as well as in the form of a support vehicle for research in medical imaging. It has been translated into Russian. Hence our first acknowledgement is to those who have promoted the work and led to the demand for a second edition. The teaching of medical imaging for the MSc in biophysics and bioengineering at Chelsea College, London (now run by King's College London and renamed MSc in Medical Engineering and Physics), continues, albeit in a revised form which is constantly being updated.

Professors Mayneord, Adams and Boag have all now died, as have Nigel Trott and Dennis Rosen. Ten of the 15 contributing authors in 1987 have now retired, though some are still scientifically active. For this second edition, they have been joined by 11 more authors, and the book is no longer written by just scientists at The Royal Marsden NHSF Trust and ICR.

Much of the essential physics of medical imaging never changes, so readers should not be surprised to find much of the first edition material surviving. We have trimmed this where needed, but, more importantly, built on this to update the material. A guiding principle, given that whole volumes have appeared on many of the topics we have covered, has been to retain student accessibility to this subject and keep the contents in one manageable book.

All publishers and copyright holders were contacted with request for permission to reproduce copyright material. We thank all the authors and publishers who have given permission for their material to be used here for illustration. We acknowledge that most of the clinical images in the second edition are from clinical colleagues at The Royal Marsden NHSF trust, but additional images have been obtained from other London hospitals. These include St George's Hospital, The PET Imaging Centre at St Thomas' Hospital and Moorfields Eye Hospital.

Our thanks go to John Navas (commissioning editor), to Amber Donley (project coordinator) and to Randy Burling (project editor) at Taylor & Francis Publishers as well as to Vinithan Sethumadhavan (senior project manager) at SPi Global.

Maggie Flower, 2012

Note from the Editor of the Second Edition

Many readers know that the second edition of this book has been planned and underway for many years. When I joined Steve as co-editor in 2006, I had no idea that it would take another six years to get the second edition published. With the edited text at the publisher and following a serious illness, Steve resigned as co-editor in March 2011 and I took over as sole editor. I hope that readers will find that the second edition of *Webb's Physics of Medical Imaging* is well worth the long wait.

Abbreviations

[Hb]	Concentration of haemoglobin (Hb)
[HbO_2]	Concentration of oxygenated haemoglobin (HbO_2)
1D	One-dimensional
2D	Two-dimensional
3D	Three-dimensional
4D	Four-dimensional
A/D	Analogue-to-digital
AAPM	American Association of Physicists in Medicine
ABC	Automatic brightness control
ACD	Annihilation coincidence detector
ACR	American College of Radiologists
ADC	Analogue-to-digital converter
ADC	Apparent diffusion coefficient
ADP	Adenosine diphosphate
AEC	Automatic exposure control
AERE	Atomic Energy Research Establishment
AIUM	American Institute of Ultrasound in Medicine
Al	Alanine
ALARA	As low as reasonably achieved
AMA	Active matrix array
AMD	Advanced micro devices
AMP	Adenosine monophosphate
AMPR	Adaptive multiple plane reconstruction
amu	Atomic mass unit
ANSI	American National Standards Institute
AP	Antero-posterior (projection)
APD	Avalanche photodiode
APT	Applied potential tomography
ARSAC	Administration of Radioactive Substances Advisory Committee
ART	Algebraic reconstruction technique
aw	Acoustic-working
a-Se	Amorphous selenium
a-SiH	Hydrogenated amorphous silicon
Asp	Aspartate
ASS	Aberdeen section scanner
ASSR	Advanced single-slice rebinning
ATM	Asynchronous transfer mode
atm	Atmosphere
ATP	Adenosine triphosphate
bash	Bourne-again shell
BB-SSRB	Cone-beam single-slice rebinning
BEM	Boundary element method
BGO	Bismuth germanate
BIR	British Institute of Radiology

B-ISDN	Broadband Integrated Series Digital Network
BLAST	Broad-use linear acquisition speed-up technique
BMUS	British Medical Ultrasound Society
BOLD	Blood oxygenation level dependant
BPH	Benign prostatic hyperplasia
BSI	British Standards Institute
Ca	Carcinoma
CBP	Convolution and back projection
CC	Correlation coefficient
CCD	Charge coupled device
CD	Compact disk
CDRH	Centre for Devices and Radiological Health
CDROM	Compact disk: read only memory
CE	Contrast enhanced
CEP	Centre for Evidence-based Purchasing
CF	Correction factor
CFOV	Central field of view
CHESS	Chemical-shift selective
Cho	Choline
CISC	Complex instruction set computer
Cit	Citrate
C-mode	Constant-depth scan
CMOS	Complementary metal-oxide semiconductor
CMT	Cadmium mercury telluride
CMUT	Capacitive micromachined ultrasound transducer
COR	Centre of rotation
COSY	Correlation spectroscopy
CP	Carr–Purcell
CPMG	Carr–Purcell sequence modified by Meiboon and Gill
cps	Counts per second
CPU	Central processing unit
CR	Computed radiography
Cr	Creatine
CRO	Cathode ray oscilloscope
CRT	Cathode ray tube
CR-UK	Cancer Research-UK
CSE	Correlated signal enforcement
CSF	Cerebrospinal fluid
CSI	Chemical shift imaging
CT	Computed tomography
CTDI	CT dose index
CtOx	Oxidized cytochrome oxidase
CVS	Chorionic villus sampling
CW	Continuous-wave
DAT	Digital audio tape
dB	Decibels
DC	Direct current
DCE-MRI	Dynamic contrast-enhanced MRI
DDI	Detector dose indicator

DDPR	Depth (or distance) dependent point response
DDR	Direct digital radiography
DEC	Digital Equipment Corporation
DFT	Discrete Fourier transform
DHGC	Dual headed gamma camera
DHSS	Department of Health and Social Security
DICOM	Digital Imaging and Communications in Medicine
DLP	Dose–length product
DMF	Demand modulation function
DOI	Diffuse optical imaging
DQE	Detective quantum efficiency
DR	Digital radiography
DSA	Digital subtraction angiography
DSF	Digital subtraction fluorography
DSI	Digital spot imager
dsv	Diameter spherical volume
DT	Distance transform
DTGS	Deuterated triglycerine sulphate
DTI	Diffusion tensor imaging
DTPA	Diethylenetriamine pentaacetic acid
DVD	Digital versatile disk
DVT	Deep vein thrombosis
DW	Diffusion weighted
DWI	Diffusion-weighted imaging
EBUS	Endobronchial ultrasound
ECD	Ethyl cysteine dimer
ECG	Electrocardiogram or electrocardiography
ECT	Emission computed tomography
EDTA	Ethylenediaminetetraacetic acid
EDV	End diastole volume
EEG	Electroencephalography
EFOMP	European Federation of Organisations for Medical Physics
e-h	Electron-hole
EIT	Electrical impedance tomography
ELD	Electroluminescent display
EMF	Electromagnetic field
EMF	Electromotive force
EMI	Electric and Musical Industries
EPI	Echo-planar imaging
EPR	Electronic patient records
EPSRC	Engineering and Physical Sciences Research Council
ESR	Electron spin resonance
ESV	End systole volume
ETL	Echo-train length
EU	European Union
EUS	Endoscopic ultrasound
FBP	Filtered back projection
FDA	Food and Drug Administration
FDG	Fluorodeoxyglucose

FDK	Feldhamp, Davis and Kress
FDV	Finite difference method
FEM	Finite element method
FFE	Fast field echo
FFT	Fast Fourier transform
FID	Free induction decay
FIESTA	Fast imaging employing steady-state acquisition
FISP	Fast imaging with steady-state free precession
FLASH	Fast low-angle shot (MRI)
fMRI	Functional magnetic resonance imaging
FN	False negative
FNA	Fine-needle aspiration
FONAR	Field-focussed NMR
FOV	Field of view
FP	False positive
FPA	Focal-plane array
FPD	Flat panel display
FPF	False positive fraction
FREE	Fitted retention and excretion equation
FSE	Fast spin echo
FT	Fourier transform
ftp	File transfer protocol
FVM	Finite volume method
FWHM	Full width at half maximum
FWTM	Full width at tenth maximum
GABA	Gamma-aminobutyric acid
GB	Gigabyte
GE	Gradient echo
GFR	Glomerular filtration rate
GI	Gastrointestinal
GLEEP	Graphite low-energy experimental pole
Glu	Glutamine
GMP	Good manufacturing practice
GPC	Glycerophosphoethanolamine choline
GPE	Glycerophosphoethanolamine
GRASE	Gradient echo and spin echo
GRASS	Gradient recalled acquisition in steady state
GSO	Gadolinium oxyorthosilicate
GTV	Gross tumour volume
GUI	Graphical user interface
Gy	Gray
HASTE	Half-Fourier acquisition single-shot turbo spine echo
Hb	Haemoglobin
HbO_2	Oxygenated haemoglobin
HH	Half height
HIFU	High-intensity focussed ultrasound
HIS	Hospital information systems
HL7	Health Level 7
HMPAO	Hexamethylpropylene amine oxime

HO	Half O Gauge
HPA	Hospital Physicists' Association
HPA	Health Protection Agency
HPGe	High purity germanium
HRRT	High resolution research tomograph
HT	High tension
HV	High voltage
HVL	Half-value layer
Hz	Hertz
IA	Intel architecture
IAUGC	Initial area under the gadolinium curve
IBM	International Business Machines
IC	Integrated circuit
ICG	Indocyanide green
ICNIRP	International Commission on Non-Ionising Radiation Protection
ICR	Institute of Cancer Research
ICRP	International Commission on Radiation Protection
ICRU	International Commission on Radiation Units and Measurements
IEC	International Electrotechnical Commission
I-EPI	Interleaved EPI
IGRT	Image-guided radiation therapy
IHE	Integrating the healthcare enterprise
ILST	Iterative least squares technique
IMRT	Intensity-modulated radiotherapy
InSb	Indium antimonide
IOMP	International Organization for Medical Physics
IOPP	Institute of Physics Publishing
IPEM	Institute of Physics and Engineering in Medicine
IPEMB	Institution of Physics and Engineering in Medicine and Biology
IPSM	Institute of Physical Sciences in Medicine
IR	Infrared
IR	Inversion-recovery
ISDN	Integrated series digital network
ISIS	Image-selected *in vivo* spectroscopy
ISO	International Standards Organisation
ITO	Indium tin oxide
IUPAP	International Union for Pure and Applied Physics
IVU	Intravenous urography
keV	Kilo electron volt
ksh	Korn shell
kV	Kilovoltage
kVCT	Kilovoltage computed tomography
kVp	Peak potential difference, or peak kilovoltage
Lac	Lactate
LAL	Limulus amoebocyte lysate (assay)
LAN	Local area network
LaOBr	Lanthanum oxybromide
LC	Lumped constant
LCD	Liquid crystal display

LCT	Liquid crystal thermography
LEGP	Low-energy general-purpose
LEHR	Low-energy high-resolution
LEHS	Low-energy high-sensitivity
LET	Linear energy transfer
LHS	Left-hand side
LI	Linear interpolation
LOIS	Laser optoacoustic imaging system
LOR	Line of response
LROC	Location ROC
LSF	Line spread function
LSO	Lutetium oxyorthosilicate
LSRF	Line source response function
LV	Left ventricle
LVEF	Left ventricular ejection fraction
LYSO	Lutetium yttrium oxyorthosilicate
Mac	Macintosh (Apple computer)
MAG3	Mercaptoacetyltriglycine
MAS	Magic-angle spinning
mAs	Milliampere seconds
mAs_{eff}	Effective tube-current scan-time product
MB	Megabyte
MC	Megacycle pulsed ultrasound
MCPMT	Multichannel photo-multiplier tube
MDA	Medical Devices Agency
MEHS	Medium-energy high-sensitivity
MFI	Multi-slice filtered interpolation
MHRA	Medicine and Healthcare Products Regulation Agency
MHz	Megahertz
mI	Myo-inositol
MI	Mechanical index
MIP	Maximum intensity projection
MIPS	Million instructions per second
MIT	Magnetic induction tomography
ML	Maximum likelihood
MLC	Multileaf collimators
ML-EM	Maximum likelihood expected maximisation
MLI	Multi-slice linear interpolation
MMI	Multimodality imaging
M-mode	Motion mode
MMX	Multimedia extension/matrix math extension
MOS	Metal-oxide semiconductor
MPR	Multi-planar reformatting
MR	Magnetic resonance
MRC	Medical Research Council
MRI	Magnetic resonance imaging
MRS	Magnetic resonance spectroscopy
MRTD	Minimum resolvable temperature difference
MS	Microsoft

MS	Multiple sclerosis
MS-DOS	Microsoft disk operating system
MSPTS	Multicrystal single-photon tomographic scanner
MTF	Modulation transfer function
MTFA	Modulation transfer function area
MUGA	Multigated acquisition
MVCT	Megavoltage computed tomography
MWPC	Multiwire proportional chamber
NAA	*N*-acetyl aspartate
NaI(Tl)	Thallium-activated sodium iodide
NDCT	Non-diagnostic computed tomography
NEC	Noise equivalent counts
NECR	Noise equivalent count rate
NEMA	National Electrical Manufacturers Association
NEP	Noise equivalent power
NEQ	Noise equivalent quanta
NETD	Noise equivalent temperature difference
NF	Noise figure
NHSF	National Health Service Foundation
NIHR	National Institute for Health Research
NIR	Near-infrared
NMR	Nuclear magnetic resonance
NOESY	Spectroscopy using the nuclear Overhauser effect
NPS	Noise power spectrum
NRM	Noise required modulation
NRPB	National Radiological Protection Board
NT	New technology
OB	Oblique
OCT	Optical coherence tomography
OD	Optical density
ODS	Output display standard
OS	Operating system
OSA	Optical Society of America
OSEM	Ordered subject expectation maximisation
OTF	Optical transfer function
PA	Posterior-anterior (projection)
PA	Performance assessment
Pa	Pascal
PACS	Patient archiving and communication system
PAS	Patient administrative system
PC	Personal computer
PC	Phosphocholine
PCr	Phosphocreatine
PD	Photodiode
PDE	Phosphodiester
PDP	Programmed data processor
PE	Phase-encoding
PE	Phosphoethanolamine
PEDRI	Proton-electron double-resonance imaging

PEEP	Positive end-expiratory pressure
PET	Positron emission tomography
PET/CT	Positron emission tomography/computed tomography
PGSE	Pulsed (field) gradient spin-echo
PHA	Pulse height analyser
Pi	Inorganic phosphate
PI	Perfusion index
PME	Phosphomonoester
PMMA	Polymethylmethacrylate
PMT	Photomultiplier tube
ppm	Parts per million
PPSI	Pulse-pressure-squared integral
PR	Projection reconstruction
PRESS	Point-resolved spectroscopy
PRF	Pulse repetition frequency
PSF	Point spread function
PSIF	Reversed FISP
PSRF	Point source response function
PTV	Planning target volume
PVDF	Polyvinylidine difluoride
PVE	Partial-volume effect
PZT	Lead zirconate titanate
QA	Quality assurance
QC	Quality control
QWIP	Quantum well infrared photodetector
RAGE	Rapid acquisition with gradient echo
RAL	Rutherford Appleton Laboratory
RAM	Random-access memory
RARE	Rapid acquisition with relaxation enhancement
RF	Radio frequency
RFOV	Reconstruction field-of-view
RGBA	Red green blue alpha
RHS	Right-hand side
RIS	Radiology information system
RISC	Reduced instruction set computer
RMH	Royal Marsden Hospital
RMS	Root mean square
ROC	Receiver operating characteristic
ROI	Region of interest
RP	Radiation protection
RPM	Revolutions per minute
RTE	Radiative transport equation
RV	Right ventricle
SAD	Sum of absolute differences
SAFT	Synthetic-aperture focussing technique
SAR	Specific absorption rate
SARS	Severe acute respiratory syndrome
SATA	Spatial-average temporal-average
SCA	Single channel analyser

SCP	Service class providers
SCU	Service class users
SCV	Superior vena cava
SE	Spin echo
SENSE	Sensitivity encoding
SI	Spectroscopic imaging
SIMD	Single instruction multiple datastream
SiPM	Silicon photomultiplier
SIPSRF	Space-invariant point source response function
SIRT	Simultaneous iterative reconstruction technique
SL	Spin locking
SMASH	Simultaneous acquisition of spatial harmonics
SMPTE	Society of Motion Picture and Television Engineers
SMR	Society of Magnetic Resonance
SNR	Signal-to-noise ratio
SOP	Service-object pairs
SPARC	Scalable processor architecture
SPEC	Standard Performance Evaluation Corporation
SPECT	Single photon emission tomography
SPECT/CT	Single photon emission tomography/computed tomography
SPGR	Spoiled gradient echo
SPIE	Society of Photo-Optical Instrumentation Engineers
SPRITE	Signal processing in the element
SPTA	Spatial-peak pulse-average
SPTP	Spatial-peak temporal-peak
SR	Saturation-recovery
SS	Steady state
SSE	Streaming single instruction multiple datastream extensions
SSFP	Steady-state free precession
STEAM	Stimulated echo acquisition mode
STEP	Simultaneous transmission/emission protocols
STP	Signal transfer property
STRAFI	Stray-field imaging
SUN	Stanford University Network
SUV	Standard uptake value
Sv	Sievert
SVPSRF	Space-variant point source response function
TA	Temporal average
TB	Terabyte
TCP/IP	Transmission control protocol/Internet protocol
TDC	Time-domain correlation
TDI	Tissue Doppler imaging
TEW	Triple energy window
TFT	Thin-film transistor
TGC	Time gain control
TGS	Triglycerine sulphate
THI	Tissue harmonic imaging
TI	Thermal index
TIB	Bone-at-focus thermal index

TIC	Cranial bone thermal index
TIS	Soft-tissue thermal index
TLD	Thermoluminescent dosimeter
TM	Time motion
TMAE	Tetrakis(dimethylamino)ethylene
TMM	Tissue-mimicking material
TN	True negative
TNF	True negative fraction
TO	Test object
TOCSY	Total correlation spectroscopy
TOE	Transoesophageal
TOF	Time of flight
TP	True positive
TPC	Transaction Processing Performance Council
TPF	True positive fraction
tsch	Tenex C shell
TSE	Turbo spin echo
TV	Television
UFOV	Useful field of view
UK	United Kingdom
UNIX	Uniplexed information and computing system
UPA	Ultra port architecture
US	United States
UTE	Ultrashort T_E
VAX	Virtual address extension
VDU	Visual display unit
VOI	Volume of interest
WAN	Wide area network
WHO	World Health Organization

Introduction and Some Challenging Questions

The role of accurate investigation and diagnosis in the management of all diseases is unquestionable. This is especially true for cancer medicine, and central to the diagnostic process are physical imaging techniques. Medical imaging not only provides for diagnosis but also serves to assist with planning and monitoring the treatment of malignant disease. Additionally, the role of imaging in cancer screening is widening to complement the imaging of symptomatic disease, as the appreciation of the importance of early diagnosis has become established.

When in its infancy, imaging with ionising x-radiation could be, and was, carried out by general practitioners and other medical specialists, but it has since become the raison d'être of diagnostic radiographers and radiologists and is rightly regarded as a speciality in its own right. In the mid-1940s, the only satisfactory medical images available were radiographs of several kinds. Within the last 70 years or so, the situation has changed so dramatically that diagnostic imaging is rarely contained within a single hospital department, there being many imaging modalities, some important enough to be regarded as medical specialities in their own right. The creation of these imaging techniques called on the special skills of physicists, engineers, biologists, pharmacists, mathematicians and chemists, which collectively complement medical expertise. Diagnostic imaging has become a team activity: it is to this group of paramedical scientists that this book is mainly directed. The intention has been to explain the physical principles governing medical imaging in a manner that should be accessible to the postgraduate or final-year undergraduate physical scientist. We have attempted to set different techniques in context with each other to provide some historical perspective and to speculate on future developments. As clinical imaging is the end point for such physical developments, clinical applications are discussed but only to the level of clarifying intentions and providing examples. The medical specialist should therefore find this a useful adjunct to the several comprehensive volumes that have appeared describing state-of-the-art clinical imaging. The practising or training radiographer may also find something of interest.

Fifteen members from one department, the Joint Department of Physics at the Institute of Cancer Research and Royal Marsden NHS Foundation Trust, created the first edition of this book. This team had conducted research together in medical imaging for a number of years, and the diversity of medical imaging serves to explain why so much individual expertise is required even within one department. It was hoped that the advantage of choosing authors in this way had produced a readable text by contributors who were aware of what their colleagues were providing. An attempt was made substantially to reframe contributions editorially to simulate a single-author text for convenient reading while retaining the separate expertise rarely available from one author. Mistakes introduced by this process were entirely the responsibility of the editor. For the second edition, we asked 11 other well-known scientists to contribute, reflecting the fact that (i) the scientific interests of some of the original contributors had changed in the intervening 20 years and that (ii) some imaging specialties are not part of the portfolio of research at the Joint Department of Physics. Also, despite being in retirement, 7 of the original 15 contributors willingly revised their sections.

The framework of the text arose from a course of lectures on medical imaging given as part of the MSc in biophysics and bioengineering at Chelsea College, London (now the Intercollegiate London MSc run from King's College London). This course (and hence this book) was designed to introduce the physics of medical imaging to students from widely differing undergraduate science courses. Introduced in 1976, this course is still being taught some 36 years later, albeit in a somewhat different format, and now on our own campus.

It is today difficult to imagine how diagnosis was carried out without images. There is, of course, no one alive whose experience pre-dates the use of X-rays to form images. There are, however, plenty of people who can remember how things were done before the use of ultrasound, radionuclide and nuclear–magnetic–resonance imaging and other techniques became available, and they have witnessed how these imaging methods were conceived, developed and put into practice. They have observed the inevitable questioning of roles and the way the methods compare with each other in clinical utility. They have lived in a unique and exciting period in which the science of diagnostic medicine has matured and diversified.

Against this background, it is reasonable to raise a number of challenging questions, some of which this book hopes to answer. Why are there so many imaging modalities and does each centre need all of them? (This is a favourite question of hostile critics and some financial providers!) Almost, without exception, new techniques come to be regarded as complementary rather than replacing existing ones, and resources of finance, staff and space are correspondingly stretched. The essence of an answer to this question is not difficult to find. We shall see how different imaging methods are based on separate physical interactions of energy with biological tissue and thus provide measurements of different physical properties of biological structures. By some fortunate provision of nature, two (or more) tissues that are similar in one physical property may well differ widely in another. It does, of course, become necessary to interpret what these physical properties may reflect in terms of normal-tissue function or abnormal pathology, and for some of the new imaging modalities, these matters are by no means completely resolved.

The reverse of the question concerning the multiplicity of imaging methods is to ask whether one should expect the number of (classes of) imaging techniques to remain finite and in this sense small. Stated differently, will the writer of a text on medical imaging in 10 years' time or 100 years' time be expected to be able to include grossly different material from that of today? Certainly one may expect the hardware of existing classes of imaging to become less expensive, more compact, computationally faster and more widely available. With these changes, procedures that can be perfectly well specified today but cannot yet be reasonably done routinely will be achieved and included in the armament of practising radiologists. None of this, however, really amounts to crystal-ball gazing for quite new methodology. It would be a foolish person who predicts that no major developments will arise, but one must, of course, observe the baseline principle that governs the answer to this question. All imaging rests on the physics of the interaction of energy and matter. It is necessary for the energy to penetrate the body and be partially absorbed or scattered. The body must be semitransparent and there is a limited (and dare one say finite?) number of interactions for which such a specification exists. This requirement for semi-opacity becomes obvious from considering two extremes. The body is completely opaque to long-wavelength optical electromagnetic radiation, which therefore cannot be used as an internal probe. Equally, neutrinos, to which the body is totally transparent, are hardly likely to be the useful basis for imaging! The windows currently available for *in vivo* probing of biological tissue are all described in this book. External probes lead to either resonant or non-resonant interaction between matter and energy. When the wavelength of the energy of the probe matches a scale size of a typical length in the tissue, a resonant interaction leads to inelastic scattering, energy absorption and re-emission; attenuation of the energetic probe is the key to forming images from the transmitted intensity. Such is the case, for example, for X-rays interacting with inner and outer electron shells or γ-rays interacting with atomic nuclei. Infrared and optical radiation similarly interacts with outer electron shells. When the frequency of matter and energy differ widely, elastic scattering, which is isotropic in a homogeneous material, leads to the classical Huygens-optics behaviour. The scattering becomes anisotropic at tissue boundaries and in inhomogeneous matter, and the reflections and refractions generated form the basis of imaging. This is, for example, the essence of how ultrasound waves may be used to form images. With this in mind, the reader should consider whether the interactions described in this book comprise a complete set of those likely to form the basis of imaging techniques and whether the energy probes described are the only ones to which the body is semi-opaque. If readers reach this conclusion affirmatively, then they must also conclude that the present time has its own unique historical place in the evolution of medical imaging.

As readers proceed, they should also enquire what combinations of factors led to each imaging modality appearing when it did? In some cases, the answer lies simply in the discovery of the underlying physical principle itself. In others, however, developments were predicted years before they were achieved. Tomography using X-rays, internal radionuclides or nuclear magnetic resonance all have a requirement for fast digital computing and, although simple analogue reconstruction pre-dated the widespread use of the digital computer, the technique was impracticable. (The reader may like to work out (after Chapter 3) how long it would take to form a computed tomography scan 'by hand'.) Inevitably, some techniques owe their rapid development to parallel military research in wartime (e.g. the development of imaging with ultrasound following sonar research, particularly in the Second World War) or to the by-products of nuclear reactor technology (e.g. radionuclide imaging), or to research in high-energy nuclear physics (e.g. particle and photon detector development).

Even if one were to conclude that the fundamental fields of imaging have all been identified, there is no reason for complacency or a sense of completion. Let us immediately recognise an area in which virtually nothing has been achieved in image production. Most of the images we shall encounter demonstrate macroscopic structure with spatial scales of 1 mm or so being described as 'good' or 'high' resolution. Diagnosticians are accustomed to viewing images of organs or groups of organs, even the whole body, and requesting digital images with pixels of this magnitude. Yet cancer biology is proceeding at the cellular level and it is at this spatial scale that one would like to perform investigative science. It should be appreciated that before abnormal pathology can be viewed by any of the existing modalities, one of two events must occur. Either some 10^6 cells must in concert demonstrate a resolvable physical difference from their neighbours or a smaller number of abnormal cells must give rise to a sufficiently large signal that they swamp the cancelling effect of their 'normal' neighbours within the smallest resolvable voxel. An analogy for the latter might be the detection of a single plastic coin in a million metal coins by a measurement technique that cannot register less than a million coins but possesses a sensitivity better than 1 in 10^6.

Two further questions challenge any false complacency. First, we may enquire whether the time will come when all investigative imaging, which currently carries some small risk or is otherwise unpleasant, will be replaced by hazard-free alternative procedures that are more acceptable to the patient? A number of contenders are certainly known to carry far less risk than those based on the use of ionising radiation but currently by no means make X-ray imaging redundant. In the past decade or so, we have possibly seen the opposite as members of the public and scientists have noted potentially adverse effects even from hitherto-thought harmless modalities (such as ultrasound and MR). Secondly, will medical diagnosis ever become a complete science or must it remain partly an art?

An enormous number of questions concerning the interpretation of images require to be resolved before the imprecision is removed; the randomness, introduced by the biological component of the process, may make this quite impossible. The importance of this question is enormous, not only for the management of the patient but also regarding the training and financing of clinical staff.

The human body is in a sense remarkably uncooperative as an active component of imaging. It emits infrared photons; it generates surface electrical potentials and some acoustic energy in the thorax relating to cardiac blood flow and pulmonary air movement. All these natural emissions have been used in diagnosis. It is, however, the paucity of natural signals that calls for the use of external probes or artificial internal emissions. Indeed, two- or three-dimensional sectional images with good spatial resolution are difficult or impossible to form with these natural emissions, and they are generally disregarded in any description of medical imaging. Also disregarded is the most primitive (but very important) form of imaging, namely, visual inspection. Indeed, medical imaging is generally restricted to imply the imaging of internal structure, and we shall here adopt this convention. Finally, as is customary, the reader is informed what this book does not attempt to do. In keeping with the intention to overview the whole field of medical imaging in one teaching text, it should go without saying that each chapter cannot be regarded as a complete review of its subject. Whole volumes have already appeared on aspects of each imaging modality. This is even more true now than when the first edition of this book appeared over 20 years ago. I hope, however, to have tempted the student to the delights of the physics of medical imaging and to have honed that natural inquisitiveness which is the basis for all significant progress.

S. Webb

1

In the Beginning: the Origins of Medical Imaging

S. Webb

CONTENT

References..9

In the chapters that follow, an attempt is made to describe the physical principles underlying a number of imaging techniques that prove useful in diagnostic medicine and some of which simultaneously underpin the physical basis of determining structure volumes for radiotherapy planning. For brevity, we have concentrated largely on describing state-of-the-art imaging, with a view to looking to the future physical developments and applications. For the student anxious to come to grips with today's technology and perhaps about to embark on a research career, this is in a sense the most realistic and practical approach. It does, however, lead to a false sense that the use of physics in medicine has always been much as it is today and gives no impression of the immense efforts of early workers whose labours underpin present developments. In this chapter, we take a short backwards glance to the earliest days of some aspects of medical imaging.

Even a casual glance at review articles in the literature tells us that historical perspective is necessarily distorted by the experience and maybe prejudices of the writer. In a sense, history is best written by those who were involved in its making. During the 75th anniversary celebrations of the *British Journal of Radiology* in 1973, an anniversary issue (*Br. J. Radiol.* **46** 737–931) brought together a number of distinguished people to take stock, and although the brief was wider than to cover imaging, the impact of physics in medical imaging was a strong theme in their reviews. Since that time, of course, several new imaging modalities have burst into hospital practice and, with the excusable preoccupation with new ideas and methods, it is these which occupy most of current research effort, feature most widely in the literature and are looked to for new hope particularly in the diagnosis, staging and management of cancer and other diseases. At the time of the Röntgen centenary (1995), attention was again focused on the origins of medical imaging (e.g. Mould 1993, Aldrich and Lentle 1995, Thomas et al. 1995).

Against this background, what features of the landscape do we see in our retrospective glance? Within any one area of imaging, there exists a detailed and tortuous path of development, with just the strongest ideas surviving the passage of time. Numerous reviews chart these developments and are referred to in subsequent chapters. Many of us are, however, fascinated to know what was the *first* reported use of a technique or announcement of a piece of equipment, and might then be content to make the giant leap from this first report to how the situation stands today, with cavalier disregard for what lies between. So, by way of introduction, this is what we shall do here. Even so, a further difficulty arises. In a sense each variation on a theme is new and would certainly be so claimed by its originators, and yet it is not all these novelties that history requires us to remember. The passage of time acts as a filter to perform a natural selection. Perhaps what was regarded as important at its discovery has paled and some apparently unimportant announcements have blossomed beyond expectation. The organised manner in which research is required to be documented also veils the untidy methods by which it is necessarily performed. Most of the imaging modalities that are in common use were subject to a period of laboratory development and it is useful to distinguish clearly between the first reported experimental laboratory equipment and the first truly clinical implementation. Physicists might be tempted to rest content that the potential had been demonstrated in a laboratory, but were they patients they would take a different view! In most of the topics we shall meet in later chapters, there has been (or still is!) a lengthy intermediate time between these two 'firsts'.

Wilhelm Conrad Röntgen's laboratory discovery of X-rays, when he was professor of physics at Wurzburg, is perhaps the only 'first' that can probably be pinned down to the approximate time and exact day! – the late evening of 8 November 1895. This date was reported in *McClure's Magazine* by the journalist H J W Dam (1896): Röntgen himself never apparently stated a date. The discovery must also rank as one of the fastest ever published – submitted on 28 December 1895 and made known to the world on 5 January 1896. The prospects for X-ray diagnosis were immediately recognised. Röntgen refused, however, to enter into any commercial contract to exploit his discovery. He was of the opinion that his discovery belonged to humanity and should not be the subject of patents, licences and contracts. The result was undoubtedly the wide availability of low-cost X-ray units. A portable set in the United States cost $15 in 1896. Since it is believed that the first X-radiograph taken with clinical intent was on 13 January 1896 by two Birmingham (UK) doctors to show a needle in a woman's hand, the 'clinical first' followed the 'experimental first' with a time lapse also surely the shortest by comparison with any subsequent development. A bromide print of the image was given by Ratcliffe and Hall-Edwards to the woman, who next morning took it to the General and Queen's Hospital where the casualty surgeon J H Clayton removed the needle – the first X-ray-guided operation. The single discovery of X-rays has clearly proved so important that it has already been the subject of many reviews (see, e.g. Brailsford 1946, Mould 1980, 1993, Burrows 1986). It is amusing to note that many newspaper reports of the discovery were anything but enthusiastic. As equipment was readily available to the general public, there was at the time an abundant number of advertisements for X-ray sets. Mould (1980, 1993) has gathered together a plethora of these and other photographs of clinical procedures with X-rays. These are with hindsight now known to have been risky for both patient and radiologist, and many early radiation workers became casualties of their trades.

In 1995, the radiological community celebrated the centenary of the discovery of the X-ray. Congresses were held throughout the world and many encyclopaedic books appeared documenting the century (e.g. Webb 1990, Mould 1993, Aldrich and Lentle 1995, Thomas et al. 1995).

Digital subtraction angiography involves the subtraction of two X-rays, precisely registered, one using contrast material, to eliminate the unwanted clutter of common structures. Historically, we have a precedent. Galton (1900) wrote (in the context of photography):

> If a faint transparent positive plate is held face to face with a negative plate, they will neutralise one another and produce a uniform grey. But if the positive is a photograph of individual, A, and the negative a photograph of individual, B, then they will only cancel one another out where they are identical and a representation of their differences

> will appear on a grey background. Take a negative composite photograph and superimpose it on a positive portrait of one of the constituents of that composite and one should abstract the group peculiarities and leave the individuality.

We shall not attempt to document the first application of X-radiography to each separate body site, but it is worth noting (Wolfe 1974) that Salomon (1913) is reported to have made the first mammogram. Thereafter, the technique was almost completely abandoned until the early 1950s.

The announcement of a machine used to perform X-ray computed tomography (CT) in a clinical environment, by Hounsfield at the 1972 British Institute of Radiology annual conference, has been described as the greatest step forward in radiology since Röntgen's discovery. The relevant abstract (Ambrose and Hounsfield 1972) together with the announcement entitled 'X ray diagnosis peers inside the brain' in the *New Scientist* (27 April 1972) can be regarded as the foundation of clinical X-ray CT. The classic papers that subsequently appeared (Ambrose and Hounsfield 1973, Hounsfield 1973) left the scientific community in no doubt as to the importance of this discovery. Hounsfield shared the 1979 Nobel Prize for Physiology and Medicine with Cormack. The Nobel lectures (Cormack 1980, Hounsfield 1980) were delivered on 8 December 1979 in Stockholm.

It was made quite clear, however, by Hounsfield that he never claimed to have 'invented CT'. The importance of what was announced in 1972 was the first practical realisation of the technique, which led to the explosion of clinical interest in the subsequent years. Who really did 'invent CT' has been much debated since. The history of CT has been written in detail by Webb (1990). The original concept of reconstructing from projections is usually credited to Radon (1917), whilst Oldendorf (1961) is often quoted as having published the first laboratory X-ray CT images of a 'head' phantom. In fact, he did not do this but devised a sensitive point method in which signals were preferentially recorded from certain points in the body whilst others were blurred out. What Oldendorf actually did was to rotate a head phantom (comprising a bed of nails) on a gramophone turntable and provide simultaneous translation by having an HO-gauge railway track on the turntable and the phantom on a flat truck, which was pulled slowly through a beam of X-rays falling on a detector. He showed how the internal structures in the phantom gave rise under such conditions to characteristic signals in the projections as the centre of rotation traversed the space relative to the fixed beam and detector. He was well aware of the medical implications of his experiment, but he did not actually generate a CT image. In his paper, he referred to the work of Cassen and also Howry, who appear elsewhere in this chapter in other contexts. It is certainly not difficult to find papers throughout the 1960s describing the potential of reconstruction tomography in medicine, suggesting methods and testing them by both simulation and experiment. Cormack, in particular, was performing laboratory experiments in CT in 1963 (Cormack 1980), work for which he later shared the Nobel Prize for Medicine in 1979 with Hounsfield. Kuhl had made a CT image of a crude kind by adapting his emission tomographic scanner as early as 1965. It is perhaps less well known that a CT scanner was built in (formerly) Russia in 1958. Korenblyum et al. (1958) published the mathematics of reconstruction from projections together with experimental details and wrote: 'At the present time at Kiev Polytechnic Institute, we are constructing the first experimental apparatus for getting X-ray images of thin sections by the scheme described in this article'. This was an analogue reconstruction method, based on a television detector and a fan-beam source of X-rays. Earlier reports from Russia have also been found (e.g. Tetel'Baum 1957).

It might also be argued that the origins of CT can be traced back even further to the work of Takahashi in the 1940s and even to that of Frank in 1940. However, a careful study of these developments shows that they did not quite get the technique right (they did not understand about removing the point spread function) and, whatever view one takes of the way history determines later events, the key point is that it was Hounsfield and EMI who made the method a practical reality (Webb 1990). Hounsfield died on 12 August 2004 and Alan Cormack died on 7 May 1998.

The history of the detection of gamma-ray photons emitted from the body after injection of a radionuclide is a fascinating mix of contemporary detector physics and the development of radiopharmaceuticals. Detection techniques are almost as old as the discovery of radioactivity itself. The Crookes' spinthariscope (1903), the Wilson cloud chamber (1895), the gold-leaf electroscope and the Geiger counter (1929) were all used to detect, although not image, radiation. Artificial radionuclides did not arrive until Lawrence invented the cyclotron in 1931. Interestingly, $^{99}Tc^{m}$ was first produced in the 37 in. cyclotron at Berkeley in 1938. Following the first experimental nuclear reactor in 1942, several reactors, notably at Oak Ridge and at the Brookhaven National Laboratory, produced medically useful radionuclides. Nuclear medicine's modern era began with the announcement in the 14 June 1946 issue of *Science* that radioactive isotopes were available for public distribution (Myers and Wagner 1975). A famous one-sentence letter from Sir J D Cockcroft at the Ministry of Supply, AERE, Harwell, to Sir E Mellenby at the MRC, dated 25 November 1946, said: 'I have now heard that the supply of radioactive isotopes to the UK by the US Atomic Energy Project is approved'. In September 1947, UK hospitals were receiving the first shipments from the GLEEP (graphite low-energy experimental pile) reactor at AERE, Harwell, Europe's first nuclear reactor (N G Trott, private communication).

Mallard and Trott (1979) reviewed the development in the United Kingdom of what is today known as nuclear medicine. The *imaging* of radiopharmaceuticals was a logical extension of counting techniques for detecting ionising radiation, which go back to the invention of the Geiger–Müller tube in 1929. The development of Geiger–Müller counting was largely laboratory based even in the early 1940s (notably in the United Kingdom at the National Physical Laboratory), and commercial detectors did not appear until after the Second World War (McAlister 1973). It was not, however, until 1948 that the first point-by-point image (of a thyroid gland) was constructed by Ansell and Rotblat (1948), which might be regarded as the first clinical nuclear medicine scan. (Rotblat later also received a Nobel Prize but for Peace, not Physics (Hill 2008).) The advantages of employing automatic scanning were recognised by several early workers. Cassen et al. (1950) developed a scintillation (inorganic calcium tungstate) detector and wrote of their desire to mount it in an automatic scanning gantry, which they later achieved (Cassen et al. 1951). Cassen et al.'s paper appeared in August 1951, whilst in July 1951 Mayneord et al. (1951a, 1951b) introduced an automatic gamma-ray scanner based on a Geiger detector. Mayneord et al. (1955) reported an improved scanner, which used a coincident pair of scintillators and storage tube display. Whilst these two developments are usually cited as the origin of rectilinear scanning in actuality Ziedses des Plantes (1950) had built a rectilinear scanner 1 year earlier.

The concept of the gamma camera might be credited to Copeland and Benjamin (1949), who used a photographic plate in a pinhole camera. Their invention was made in the context of replacing autoradiographs, and long exposure times of the order of days were required. It is interesting to note that they came to criticise their own instrument's usefulness because 'many of the tracers used in biological work have little gamma activity', a situation rather different from what we know today! Anger (1952) first announced an electronic gamma camera with a crystal acting as an image intensifier for a film (also with a pinhole collimator), which used a sodium iodide crystal of size $2 \times 4 \times 5/16\,\text{in.}^3$. This was regarded as a large crystal at the time. The first electronic gamma camera with multiple photomultiplier tubes (PMTs) was reported in 1957 (Anger 1957, 1958). It had a 4 in. diameter crystal of thickness 0.25 in. and just seven 1.5 in. diameter PMTs. Commercial cameras followed soon afterwards, amongst the first in the United Kingdom being the prototype Ekco Electronics camera evaluated by Mallard and Myers (1963a, 1963b) at London's Hammersmith Hospital. In 1968, Anger (1968) was also the first to report how a gamma camera could be used in a rectilinear scanning mode to perform multiplane longitudinal tomography. The new machine made redundant the need to perform several rectilinear scans with different focal-length collimators on single or double detectors.

Single-photon emission computed tomography (SPECT) stands in relation to planar Anger camera imaging as X-ray CT stands to planar X-radiology. Its importance in diagnostic nuclear medicine is now clearly established. Who invented SPECT and when? Once again, the honours are disputable, largely because what was suspected to be possible did not become a clinical reality for some while. It is also possible to identify several 'firsts' since SPECT has been achieved in several widely different ways. Kuhl and Edwards (1963) published the first laboratory single-photon emission tomography (SPET) (note: no 'C') images based on a rotate–translate arrangement and collimated crystal detectors (Webb 1990). What we would now call a transverse section tomogram was generated entirely by analogue means without the need for a computer. The angular increment for rotation was a coarse 15°. Kuhl and Edwards (1964) published a photograph of the first SPET scanner (which was also capable of other scanning modes such as rectilinear and cylindrical scanning) and, in their explanation of how the analogue image is built up on film, provided what is possibly the first description of windowed tomography by which potentially overbright values were 'top-cut'. They also refined the image formation method to produce an image on paper tape whose contrast could be adjusted *a posteriori*. One of the first tomographic images visualised a malignant astrocytoma using an intravenous injection of ^{197}Hg chlormerodrin. Interestingly, in an addendum to the 1964 paper, they wrote: 'we have had good results with technetium 99 m (^{99}Tc^{m}) in pertechnetate form for brain scanning since March 1964'. This represents one of the earliest reports of the use of this isotope, which was to have such an important impact thereafter. The technetium generator was one of the first 'radioactive cows' and was conceived at the Brookhaven National Laboratory by Green, Tucker and Richards around the mid-1950s, being first reported in 1958 (see Tucker et al. 1958, Tucker 1960, Richards et al. 1982, Ketchum 1986), although the original identification of technetium as an impurity in the aluminium oxide generator of ^{132}I was reportedly made by a customer for one such generator. Generators were also available in the United Kingdom from AERE in the 1950s, supplied by G B Cook, and were in use at the (then known as) Royal Cancer Hospital (N G Trott, private communication).

Anger himself showed how SPECT could be achieved with a gamma camera as early as 1967, rotating the patient in a chair in front of the stationary camera and coupling the line of scintillation corresponding to a single slice at each orientation to an optical camera (Anger et al. 1967). Tumours were satisfactorily delineated and the technique was established, but it was far from today's clinical situation. These results were reported on 23 June 1967 at the *14th Annual Meeting of the Society of Nuclear Medicine* in Seattle. In 1971 Muehllehner and Wetzel (1971) produced some laboratory images by computer, but the lack of a clinical SPECT system based on a rotating camera was still being lamented as late as 1977 when Jaszczak et al. (1977) and Keyes et al. (1977) were reporting clinical results obtained with a home-made camera gantry. The first commercial gamma-camera-based SPECT systems appeared in 1978 and at much the same time single-slice high-resolution SPECT systems were also marketed (Stoddart and Stoddart 1979). Almost a decade earlier, however, SPECT tomograms had been obtained in the clinic by Bowley et al. (1973) using the Mark 1 Aberdeen Section

Scanner, whose principles were largely similar to those of Kuhl and Edwards' laboratory scanner. We see, therefore, that with regard to SPECT imaging there was no clear date separating the impossible from the possible.

Logically complementing imaging with single photons is the detection of the annihilation gammas from positron emitters in order to form images of the distribution of a positron-labelled radiopharmaceutical in the body. The technology to achieve positron emission tomography (PET) is now established commercially but the market initially grew much slower than for gamma-camera SPECT. However, a number of specialist centres started conducting clinical PET in the early 1960s. Perhaps it is surprising, therefore, to find that the technique of counting gammas from positron annihilation was discussed as early as 1951 by Wrenn et al. (1951). They were able to take data from a source of ^{64}Cu in a fixed brain enclosed within its skull using thallium-activated sodium iodide detectors. Images as such were not presented, but certainly by 1953 simple scanning arrangements had been engineered for the creation of images. Brownell and Sweet (1953) showed the *in vivo* imaging of a recurrent tumour using ^{74}As. In this paper they wrote: 'we have been working independently on this [i.e. PET imaging] problem for a period of approximately 2 years'. It would be reasonable then to assign the beginnings of PET imaging to the year 1951. A lengthy text on the history of nuclear medicine tomographic imaging has been written by Webb (1990).

The discovery of the phenomenon of nuclear magnetic resonance (NMR) was announced simultaneously and independently in 1946 by groups headed by Bloch and by Purcell, who shared a Nobel Prize. Thereafter, there was a steady development of NMR spectroscopy in chemistry, biology and medicine. NMR imaging followed much later and several 'firsts' are worth recording. In a letter to *Nature* in 1973, Lauterbur (1973) published the first NMR image of a heterogeneous object comprising two tubes of water, but the date of publication is preceded by a patent filed in 1972 by Damadian (1972), who proposed without detail that the body might be scanned for clinical purposes by NMR. The first human image of a live finger was reported by Mansfield and Maudsley (1976) and there followed the first NMR image of a hand (Andrew et al. 1977) and of a thorax (Damadian et al. 1977) in 1977. An article in the *New Scientist* in 1978 amusingly entitled 'Britain's brains produce first NMR scans' (Clow and Young 1978) was the first NMR image produced by a truly planar technique. In the same year, the first abdominal NMR scan was reported by Mansfield et al. (1978). This (1978) was also the year in which the first commercial NMR scanner became available, and the first demonstration of abnormal human pathology was reported in 1980 by Hawkes et al. (1980). The beginnings of NMR imaging clearly require the specification of a large number of 'firsts'!

It is believed that after X-radiology the use of ultrasound in medical diagnosis is the second-most frequent investigative imaging technique. The earliest attempts to make use of ultrasound date from the late 1930s, but these mimicked the transmission method of X-rays and cannot really be recorded as the beginnings of ultrasound imaging as we know it today. Ultrasonic imaging based on the pulse-echo principle, which is also the basis of radar, became possible after the development of fast electronic pulse technology during the Second World War. The use of ultrasound to detect internal defects in metal structures preceded its use in medicine and was embodied in a patent taken out by Firestone in 1940. The first 2D ultrasound scan was obtained using a simple sector scanner and showed echo patterns from a myoblastoma of the leg in a living subject (Wild and Reid 1952). This paper was received for publication on 25 October 1951. Prior to this, ultrasonic echo traces from human tissue had been demonstrated as early as 1950 by Wild (1950), but 2D images had not been constructed. Very shortly afterwards, on 2 June 1952, Howry and Bliss (1952) published the results of their work, which had been in progress since 1947, and their paper included a 2D image of a human wrist.

Wild and Reid (1957) went on to develop the first 2D ultrasound scanner and used it to image the structure of the breast and rectum. It was not until 1958 that the prototype of the first commercial 2D ultrasonic scanner was described by Donald et al. (1958) as a development of a 1D industrial flaw detector made by Kelvin and Hughes. This machine was used to carry out the first investigations of the pregnant abdomen (Donald and Brown 1961). For a more detailed history of ultrasonic imaging, one might consult Hill (1973), White (1976) and Wild (1978). A number of other groups were actively investigating the use of ultrasound in medicine in the early 1950s, including Leksell in Sweden and a group led by Mayneord in what was then known as the Royal Cancer Hospital in London (now The Royal Marsden NHSF Trust). Wild visited the Royal Cancer Hospital in 1954 and concluded that the group was quite familiar with the high-amplitude echo from the cerebral midline and the connection between its displacement and cerebral disease, which has subsequently become the basis of cerebral encephalography. According to Kit Hill (previously Head of Ultrasound Physics at The Marsden) the early UK work was documented only in the Annual Reports of the British Empire Cancer Campaign but may have provided a basis for the subsequent work of Donald's group in obstetrics.

In the following chapters, we shall encounter the attempts that have been made to image a wide variety of different physical properties of biological tissue. Some of these are the basis of well-established diagnostic techniques whose origins have been mentioned earlier. Other imaging methods are less widely applied or are still to reach the clinic. Some are still subject to controversy over their usefulness. For example, we shall find that the measurement of the light transmission of tissue may be used for early diagnosis of breast disease. Commercially available equipment appeared for the first time in the late 1970s and yet Cutler (1929) reported the first attempts at transillumination some 50 years earlier in New York. Perhaps in contrast to the time lapse between the discovery and clinical use of X-rays, this ranks as the longest delay! Thermometric methods for showing the pattern of breast disease were first reported by Lloyd-Williams et al. (1961). The use of xeroradiographic techniques for mammography were pioneered in the late 1960s and early 1970s when the image quality of the electrostatic technique became comparable with film imaging and the usefulness of the extra information offered was appreciated (Boag 1973). The first medical xeroradiographic image (of a hand) was, however, published in 1907 by Righi (1907), and was reproduced by Kossel (1967). (Righi had been working on the method since 1896.) The process was patented in 1937 by Carlson. This long delay was largely due to the inadequacies of the recording process, which were to be dramatically improved by the development of the Xerox copying process in the 1940s and 1950s. Again this technique has fallen out of favour. In 1955 Roach and Hilleboe (1955) described their feelings that xeroradiography was a logical replacement for film radiography particularly for mobile work. Their paper begins with an extraordinary justification for the work, namely the preservation of the lives of U.S. casualties in the aftermath of a nuclear attack. They wrote:

> In the event of the explosion of an atomic bomb over one of the major cities in the United States, the number of casualties produced and requiring emergency medical care would be tremendous. ... xeroradiography offers a simple, safe and inexpensive medium for the recording of röntgen images. No darkrooms or solutions of any type are needed. No lead lined storage vaults are required and there is no film deterioration problem. No transport of large supplies is involved.

Hills et al. (1955) were already comparing xeroradiography and screen-film radiography.

Imaging the electrical impedance of the body is a new technique whose description in Chapter 9 also serves for its history. The first *in vivo* cross-sectional clinical images were recorded early in the 1980s and there is now commercially available equipment for electrical impedance tomography (EIT).

In Figure 1.1 some of the firsts discussed in this chapter have been plotted on a non-linear clock. The period between 1895 and 1987 has been divided up such that the 92 years correspond to 12 h. The 'hour hand' represents in a non-linear way the fraction of the 92 year period elapsed since 1895. This figure has not been updated because, with respect to history, no major new imaging modalities have emerged since 1988. Of course, the technological details, uses and applications have certainly changed.

In the introduction, it was tentatively suggested that the period between the mid-1940s and the present may have 'completed the set' of all physical probes to which the patient is semi-opaque and which are, therefore, available as the basis of imaging modalities.

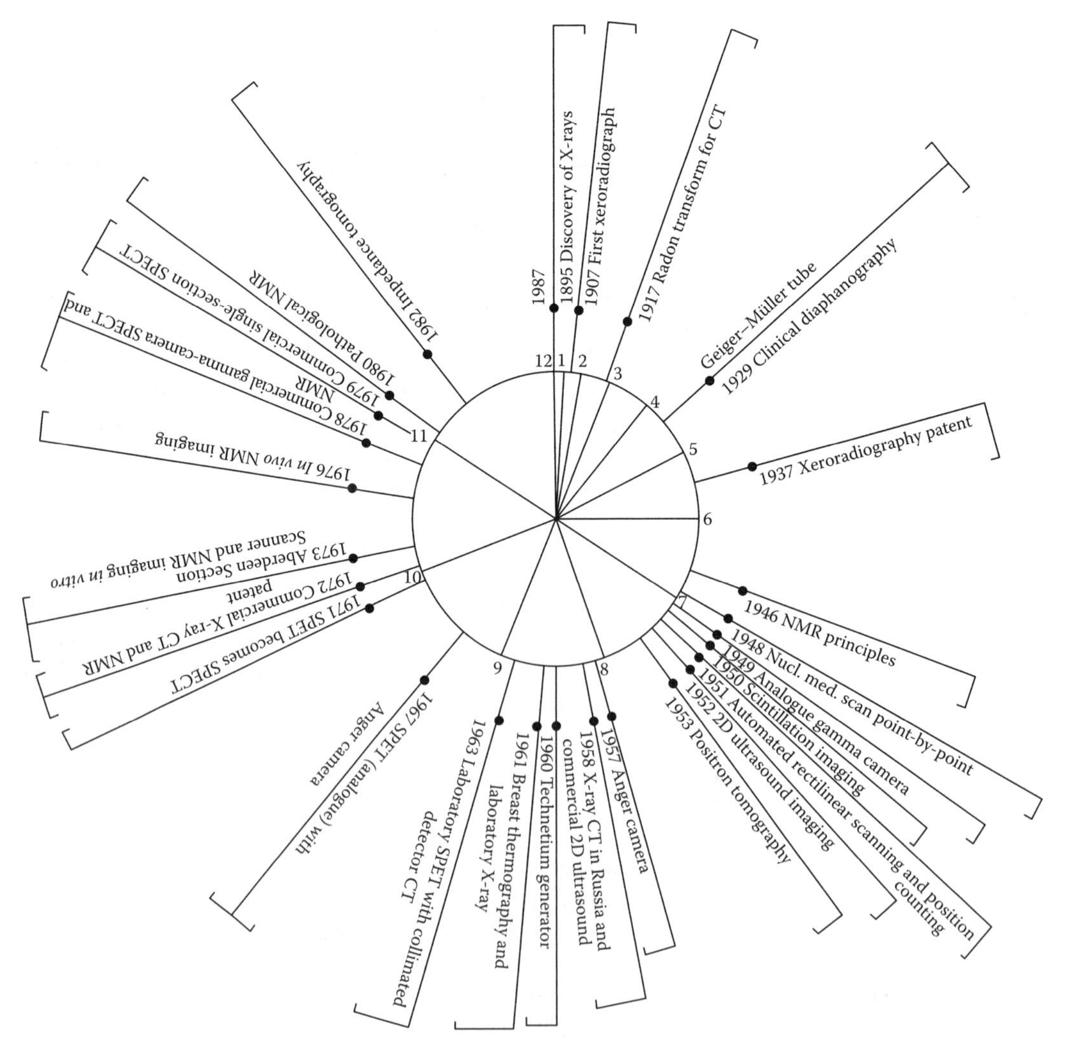

FIGURE 1.1
Non-linear clock of important 'firsts'.

This is graphically illustrated in the figure by the crowding of events in the latter half of the time span.

As time passes by and those who were the pioneers of medical imaging grow old and eventually die, it is perhaps pertinent to record that a number of texts have appeared documenting the contributions of individuals. Among these we may note the following: The history of the UK (then called) Hospital Physicists' Association (now Institute of Physics and Engineering in Medicine) was written in 1983 (HPA 1983). A retrospective on a more genteel age of Medical Physics has been written by Roberts (1999). An ultrasound pioneer (and one of the authors in this book) Kit Hill recently wrote a personal retrospective (Hill 2006). Recently an 'oral history' day in which many distinguished pioneers contributed was arranged by the Wellcome Foundation and the transcript published (Christie and Tansey 2006).

With this brief historical scene setting, let us proceed to examine the physical basis of imaging.

References

Aldrich J E and Lentle B C 1995 A *New Kind of Ray* (Vancouver, British Columbia, Canada: Canadian Association of Radiologists).

Ambrose J and Hounsfield G 1972 Computerised transverse axial tomography *Br. J. Radiol.* **46** 148–149.

Ambrose J and Hounsfield G 1973 Computerised transverse axial scanning (tomography). Part 2: Clinical applications *Br. J. Radiol.* **46** 1023–1047.

Andrew E R, Bottomley P A, Hinshaw W S, Holland G N, Moore W S and Simaroj C 1977 NMR images by the multiple sensitive point method: Application to larger biological systems *Phys. Med. Biol.* **22** 971–974.

Anger H O 1952 Use of a gamma-ray pinhole camera for in-vivo studies *Nature* **170** 200–201.

Anger H O 1957 A new instrument for mapping gamma-ray emitters. *Biol. Med. Quart. Rep.* UCRL-3653, January 1957 p. 38.

Anger H O 1958 Scintillation camera *Rev. Sci. Instrum.* **29** 27–33.

Anger H O 1968 Multiplane tomographic gamma camera scanner *Medical Radioisotope Scintigraphy* STI Publ. 193 (Vienna, Austria: IAEA) pp. 203–216.

Anger H O, Price D C and Yost P E 1967 Transverse section tomography with the gamma camera *J. Nucl. Med.* **8** 314.

Ansell G and Rotblat J 1948 Radioactive iodine as a diagnostic aid for intrathoracic goitre *Br. J. Radiol.* **21** 552–558.

Boag J W 1973 Xeroradiography *Modern Trends in Oncology I* Part 2 *Clinical Progress* ed. R W Raven (London, UK: Butterworths).

Bowley A R, Taylor C G, Causer D A, Barber D C, Keyes W I, Undrill P E, Corfield J R and Mallard J R 1973 A radioisotope scanner for rectilinear, arc, transverse section and longitudinal section scanning (ASS—the Aberdeen Section Scanner) *Br. J. Radiol.* **46** 262–271.

Brailsford J F 1946 Röntgen's discovery of X-rays *Br. J. Radiol.* **19** 453–461.

Brownell G L and Sweet W H 1953 Localisation of brain tumours with positron emitters *Nucleonics* **11** 40–45.

Burrows E H 1986 *Pioneers and Early Years: A History of British Radiology* (Alderney: Colophon).

Cassen B, Curtis L and Reed C W 1950 A sensitive directional gamma ray detector *Nucleonics* **6** 78–80.

Cassen B, Curtis L, Reed C W and Libby R 1951 Instrumentation for ^{131}I use in medical studies *Nucleonics* **9** (2) 46–50.

Christie D A and Tansey E M 2006 *Development of Physics Applied to Medicine in the UK, 1945–1990* (London, UK: The Trustee of the Wellcome Trust).
Clow H and Young I R 1978 Britain's brains produce first NMR scans *New Scientist* **80** 588.
Copeland D E and Benjamin E W 1949 Pinhole camera for gamma ray sources *Nucleonics* **5** 44–49.
Cormack A M 1980 Early two-dimensional reconstruction and recent topics stemming from it (Nobel Prize lecture) *Science* **209** 1482–1486.
Cutler M 1929 Transillumination as an aid in the diagnosis of breast lesions *Surg. Gynaecol. Obstet.* **48** 721–729.
Dam H J W 1896 The new marvel in photography *McClure's Mag.* **6** 403 *et seq.*
Damadian R V 1972 Apparatus and method for detecting cancer in tissue *US Patent* 3789832, filed 17 March 1972.
Damadian R, Goldsmith M and Minkoff L 1977 NMR in cancer: FONAR image of the live human body *Physiol. Chem. Phys.* **9** 97–108.
Donald I and Brown T G 1961 Demonstration of tissue interfaces within the body by ultrasonic echo sounding *Br. J. Radiol.* **34** 539–546.
Donald I, McVicar J and Brown T G 1958 Investigation of abdominal masses by pulsed ultrasound *Lancet* **271** 1188–1195.
Galton F 1900 Analytic portraiture *Nature* **62** 320.
Hawkes R C, Holland G N, Moore W S and Worthington B S 1980 NMR tomography of the brain: A preliminary clinical assessment with demonstration of pathology *J. Comput. Assist. Tomogr.* **4** 577–586.
Hill C R 1973 Medical ultrasonics: An historical review *Br. J. Radiol.* **47** 899–905.
Hill C R 2006 Ultrasound for cancer investigation: An anecdotal history *Ultrasound* **14** 78–86.
Hill C R 2008 Professor Pugwash, The man who fought nukes: The life of Sir Joseph Rotblat (Wellington, UK: Halsgrove).
Hills T H, Stanford R W and Moore R D 1955 Xeroradiography—The present medical applications *Br. J. Radiol.* **28** 545–551.
Hounsfield G N 1973 Computerised transverse axial scanning (tomography). Part 1: Description of system *Br. J. Radiol.* **46** 1016–1022.
Hounsfield G N 1980 Computed medical imaging (Nobel Prize lecture) *Science* **210** 22–28.
Howry D H and Bliss W R 1952 Ultrasonic visualisation of soft tissue structures of the body *J. Lab. Clin. Med.* **40** 579–592.
HPA 1983 History of the Hospital Physicists' Association 1943–1983 (Newcastle upon Tyne, UK: HPA).
Jaszczak R J, Murphy P H, Huard D and Burdine J A 1977 Radionuclide emission computed tomography of the head with $^{99}Tc^{m}$ and a scintillation camera *J. Nucl. Med.* **18** 373–380.
Ketchum L E 1986 Brookhaven, origin of $^{99}Tc^{m}$ and ^{18}F FDG, opens new frontiers for nuclear medicine *J. Nucl. Med.* **27** 1507–1515.
Keyes J W, Orlandea N, Heetderks W J, Leonard P F and Rogers W L 1977 The Humongotron—A scintillation camera transaxial tomograph *J. Nucl. Med.* **18** 381–387.
Korenblyum B I, Tetel'Baum S I and Tyutin A A 1958 About one scheme of tomography *Bull. Inst. Higher Educ. Radiophys.* **1** 151–157 (translated from the Russian by H H Barrett, University of Arizona, Tucson, AZ).
Kossel F 1967 Physical aspects of xeroradiography as a tool in cancer diagnosis and tumour localisation *Progress in Clinical Cancer* ed. I M Ariel Vol. 3 (New York: Grune and Stratton) pp. 176–185.
Kuhl D E and Edwards R Q 1963 Image separation radioisotope scanning *Radiology* **80** 653–662.
Kuhl D E and Edwards R Q 1964 Cylindrical and section radioisotope scanning of the liver and brain *Radiology* **83** 926–936.
Lauterbur P C 1973 Image formation by induced local interactions: Examples employing nuclear magnetic resonance *Nature* **242** 190–191.

Lloyd-Williams K, Lloyd-Williams F J and Handley R S 1961 Infra-red thermometry in the diagnosis of breast disease *Lancet* **2** 1378–1381.

Mallard J R and Myers M J 1963a The performance of a gamma camera for the visualisation of radioactive isotopes in vivo *Phys. Med. Biol.* **8** 165–182.

Mallard J R and Myers M J 1963b Clinical applications of a gamma camera *Phys. Med. Biol.* **8** 183–192.

Mallard J R and Trott N G 1979 Some aspects of the history of nuclear medicine in the United Kingdom *Semin. Nucl. Med.* **9** 203–217.

Mansfield P and Maudsley A A 1976 Planar and line scan spin imaging by NMR in *Magnetic Resonance and Related Phenomena Proc. 19th Congr. Ampere,* Heidelberg, Germany (IUPAP) pp. 247–252.

Mansfield P, Pykett I L, Morris P G and Coupland R E 1978 Human whole-body line-scan imaging by NMR *Br. J. Radiol.* **51** 921–922.

Mayneord W V, Evans H D and Newbery S P 1955 An instrument for the formation of visual images of ionising radiations *J. Sci. Instrum.* **32** 45–50.

Mayneord W V, Turner R C, Newbery S P and Hodt H J 1951a A method of making visible the distribution of activity in a source of ionising radiation *Nature* **168** 762–765.

Mayneord W V, Turner R C, Newbery S P and Hodt H J 1951b A method of making visible the distribution of activity in a source of ionising radiation *Radioisotope Techniques* Vol. 1 (*Proc. Isotope Techniques Conf. Oxford* July 1951) (London, UK: HMSO).

McAlister J 1973 The development of radioisotope scanning techniques *Br. J. Radiol.* **46** 889–898.

Mould R F 1980 *A History of X-Rays and Radium* (London, UK: IPC Business Press Ltd).

Mould R F 1993 *A Century of X-Rays and Radioactivity in Medicine* (Bristol, UK: IOP Publishing).

Muehllehner G and Wetzel R A 1971 Section imaging by computer calculation *J. Nucl. Med.* **12** 76–84.

Myers W G and Wagner H N 1975 Nuclear medicine: How it began *Nuclear Medicine* ed. H N Wagner (New York: H P Publishing).

Oldendorf W H 1961 Isolated flying spot detection of radiodensity discontinuities: Displaying the internal structural pattern of a complex object *IRE Trans. Biomed. Electron.* **BME-8** 68–72.

Radon J 1917 Uber die Bestimmung von Funktionen durch ihre Integralwerte langs gewisser Mannigfaltigkeiten *Ber. Verh. Sachs. Akad. Wiss. Leipzig Math. Phys.* K1 **69** 262–277.

Richards P, Tucker W D and Shrivastava S C 1982 Technetium-99m: An historical perspective *Int. J. Appl. Radiat. Isot.* **33** 793–799.

Righi A 1907 *Die Bewegung der Ionen bei der elektrischen Entladung* (Leipzig, Germany: J Barth).

Roach J F and Hilleboe H E 1955 Xeroradiography *Am. J. Roentgenol.* **73** 5–9.

Roberts J E 1999 *Meandering in Medical Physics—A Personal Account of Hospital Physics* (Bristol, UK: IOPP).

Salomon A 1913 Beitrage zue Pathologie und Klinik der Mammacarcinoma *Arch. Klin. Chir.* **101** 573–668.

Stoddart H F and Stoddart H A 1979 A new development in single gamma transaxial tomography: Union Carbide focussed collimator scanner *IEEE Trans. Nucl. Sci.* **NS-26** 2710–2712.

Tetel'Baum S I 1957 About a method of obtaining volume images with the help of X-rays *Bull. Kiev Polytechnic Inst.* **22** 154–160 (translated from the Russian by J W Boag, Institute of Cancer Research, London, UK).

Thomas A M K, Isherwood I and Wells P N T 1995 *The Invisible Light: 100 Years of Medical Radiology* (Oxford, UK: Blackwell Science).

Tucker W D 1960 Radioisotopic cows *J. Nucl. Med.* **1** 60.

Tucker W D, Greene M W, Weiss A J and Murrenhoff A P 1958 BNL 3746 *American Nuclear Society Annual Meeting, Los Angeles* June 1958 *Trans. Am. Nucl. Soc.* **1** 160.

Webb S 1990 *From The Watching of Shadows: The Origins of Radiological Tomography* (Bristol, UK: IOPP).

White D N 1976 *Ultrasound in Medical Diagnosis* (Kingston, Ontario, Canada: Ultramedison).

Wild J J 1950 The use of ultrasonic pulses for the measurement of biological tissues and the detection of tissue density changes *Surgery* **27** 183–188.

Wild J J 1978 The use of pulse-echo ultrasound for early tumour detection: History and prospects *Ultrasound in Tumour Diagnosis* eds. C R Hill, V R McCready and D O Cosgrove (London, UK: Pitman Medical).

Wild J J and Reid J M 1952 The application of echo-ranging techniques to the determination of structure of biological tissues *Science* **115** 226–230.

Wild J J and Reid J M 1957 Progress in the techniques of soft tissue examination by 15 MC pulsed ultrasound *Ultrasound in Biology and Medicine* ed. E Kelly (Washington, DC: American Institute of Biological Sciences) pp. 30–48.

Wolfe J N 1974 Mammography *Radiol. Clin. N. Am.* **12** 189–203.

Wrenn F R, Good M L and Handler P 1951 The use of positron emitting radioisotopes in nuclear medicine imaging *Science* **113** 525–527.

Ziedses des Plantes B G 1950 Direct and indirect radiography (reprint of Zeidses des Plantes 1973) *Selected works of B G Ziedses des Plantes* (Amsterdam, the Netherlands: Excerpta Medica).

2

Diagnostic Radiology with X-Rays

D. R. Dance, S. H. Evans, C. L. Skinner and A. G. Bradley

CONTENTS

2.1 Introduction ... 14
2.2 Imaging System and Image Formation ... 16
2.3 Photon Interactions ... 18
2.4 Important Physical Parameters ... 20
 2.4.1 Contrast ... 20
 2.4.2 Unsharpness ... 22
 2.4.3 Noise and Dose ... 23
 2.4.4 Dynamic Range ... 27
2.5 X-Ray Tubes ... 27
 2.5.1 Cathode ... 28
 2.5.2 Anode ... 28
 2.5.3 X-Ray Spectra ... 30
 2.5.4 Generation of High Voltage ... 32
 2.5.5 Geometric Unsharpness ... 34
2.6 Scatter Removal ... 36
 2.6.1 Magnitude of the Scatter-to-Primary Ratio ... 36
 2.6.2 Anti-Scatter Grids ... 37
 2.6.3 Air-Gap Techniques ... 40
 2.6.4 Scanning Beams ... 41
2.7 Analogue Image Receptors ... 42
 2.7.1 Direct-Exposure X-Ray Film and Screen–Film ... 42
  2.7.1.1 Construction and Image Formation ... 42
  2.7.1.2 Characteristic Curve ... 43
 2.7.2 Screen–Film Combinations ... 46
  2.7.2.1 Construction and Image Formation ... 46
  2.7.2.2 Unsharpness ... 46
  2.7.2.3 Sensitivity ... 48
  2.7.2.4 Contrast ... 50
  2.7.2.5 Noise ... 50
 2.7.3 Image Intensifiers ... 54
  2.7.3.1 Construction and Image Formation ... 54
  2.7.3.2 Viewing and Recording the Image ... 57
2.8 Digital Image Receptors ... 58
 2.8.1 Digital Fluoroscopy ... 59
 2.8.2 Computed Radiography ... 60
 2.8.3 Receptors Using Phosphor Plates and Charge-Coupled Devices ... 64

2.8.3.1 Phosphor Plates 64
2.8.3.2 Charge-Coupled Devices 65
2.8.4 Flat-Panel Receptors Using Active-Matrix Arrays 66
2.8.4.1 Systems Using Fluorescent Screens 66
2.8.4.2 Systems Using Selenium Photoconductors 66
2.8.5 Photon Counting Detectors 68
2.8.6 Advantages of Digital Imaging Systems 68
2.9 Automatic Exposure Control 69
2.9.1 Automatic Brightness Control 70
2.9.2 Automatic Control of X-Ray Spectrum in Mammography 70
2.10 Contrast Media 70
2.11 Applications 72
2.11.1 Digital Subtraction Imaging 72
2.11.1.1 Digital Subtraction Angiography 72
2.11.1.2 Dual-Energy Imaging 73
2.11.2 Digital Mammography 73
2.11.3 Stereotactic Biopsy Control and Spot Imaging in Mammography 75
2.11.4 Digital Chest Imaging 75
2.11.5 Digital Tomosynthesis 75
2.11.6 Equalisation Radiography 76
2.12 Quality Control 76
2.12.1 Definitions 77
2.12.2 Quality Control Tests 77
2.12.3 Tube and Generator Tests 77
2.12.3.1 Tube Voltage 78
2.12.3.2 Exposure Time 79
2.12.3.3 Tube Output 79
2.12.3.4 Half-Value Layer and Filtration 80
2.12.3.5 Focal-Spot Size 80
2.12.3.6 Light-Beam Diaphragm/X-Ray Field Alignment and Beam Perpendicularity 82
2.12.3.7 Practical Considerations 82
2.12.4 Automatic Exposure Control Tests 83
2.12.5 Conventional Tomography 83
2.12.6 Fluoroscopy Systems 84
2.12.7 Fluorography Systems 86
2.12.8 Mammographic Systems 87
2.12.9 Digital Radiographic Systems 89
2.12.10 Image Display Monitors 91
References 91

2.1 Introduction

X-rays have been used to produce medical images ever since their discovery by Wilhelm Röntgen in 1895 (see Chapter 1). In the United Kingdom, it has been estimated that there are 705 medical and dental radiographic examinations per 1000 population per year (Tanner et al. 2001), so that the technique is of major importance in medical imaging.

In view of this place in history as well as the vast application of X-rays, it is appropriate to begin the discussion of the physics of medical imaging by considering diagnostic radiology with X-rays. In later chapters, we shall see how many of the concepts formed for describing radiography with X-rays are also useful for other modalities. In a sense, the language of imaging was framed for X-radiology, including concepts such as image contrast and noise and spatial resolution, and it has subsequently been taken across to describe these other techniques for imaging the human body. This chapter covers both the essential physics of the design of X-ray imaging equipment and the quality control of the equipment. Quality control is an important component of modern radiographic practice, facilitating the maintenance of the quality of the image and the radiation protection of patients and staff.

The radiographic image is formed by the interaction of X-ray photons with a photon detector and is therefore a distribution of those photons, which are transmitted through the patient and are recorded by the detector. These photons can be either primary photons, which have passed through the patient without interacting, or secondary photons, which result from an interaction in the patient (Figure 2.1). The secondary photons will in general be deflected from their original direction and, for our purposes, can be considered as carrying no useful information.* The primary photons do carry useful information. They give a measure of the probability that a photon will pass through the patient without interacting and this probability will itself depend upon the sum of the X-ray attenuating properties of all the tissues the photon traverses. The image is therefore a *projection* of the attenuating properties of all the tissues along the paths of the X-rays. It is a two-dimensional projection of the three-dimensional distribution of the X-ray attenuating properties of tissue.

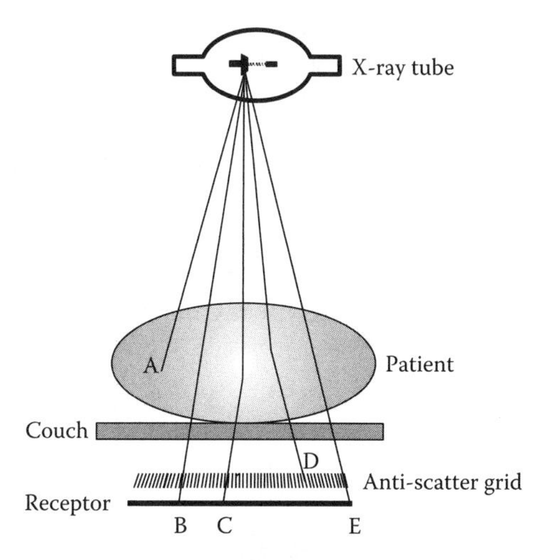

FIGURE 2.1
The components of the X-ray imaging system and the formation of the radiographic image. B and E represent photons that have passed through the patient without interacting. C and D are scattered photons. D has been stopped by an anti-scatter grid. Photon A has been absorbed.

* The development of X-ray imaging systems that make use of the information contained in scattered photons is (at the time of writing) an active research area.

It is possible to obtain an image of a two-dimensional slice through the three-dimensional distribution of attenuating properties using the techniques of conventional tomography, digital tomosynthesis, or computed tomography. These techniques are dealt with in Sections 2.12.5 and 2.11.5 and Chapter 3, respectively.

2.2 Imaging System and Image Formation

The components of a typical X-ray imaging system are shown in Figure 2.1. The photons emitted by the X-ray tube enter the patient, where they may be scattered, absorbed or transmitted without interaction. The primary photons recorded by the image receptor form the image, but the scattered photons create a background signal, which degrades contrast. In most cases, the majority of the scattered photons can be removed by placing an anti-scatter device between the patient and the image receptor (see Section 2.6). This device can simply be an air gap or a grid formed from a series of lead strips, which will transmit most of the primary radiation but reject most of the scatter. Alternatively, a smaller radiation field can be used, which scans the patient to build up a projection image. Nevertheless, even with a good scatter-rejection technique, the radiographic contrast between different tissues may not be sufficient for normal or abnormal anatomy to be adequately visualised. In some situations the contrast can be enhanced by introducing contrast material into the patient, with a different attenuation coefficient to that of normal tissue. This is discussed in Section 2.10.

The image recorded by the receptor is processed (e.g. an X-ray film is developed, a computed radiography (CR) image plate is read out) and can then be viewed by the radiologist. It is important that the radiograph is presented at the correct illumination and is viewed using a distance and magnification appropriate to the detail in the image and the angular-frequency response of the eye. The interpretation of the radiograph is a very skilled task and involves both the perception of small differences in contrast and detail as well as the recognition of abnormal patterns. The mechanisms of perception are discussed in Chapter 13 and our treatment here is limited to the production of the radiographic image.

It will be useful in what is to follow to have a simple mathematical model of the radiographic imaging process. We start by considering a monochromatic X-ray source that emits photons of energy E (Figure 2.2). The central ray of the X-ray beam is in the z-direction

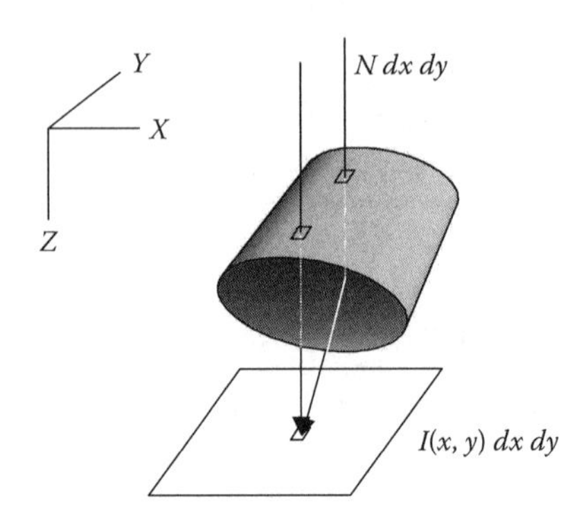

FIGURE 2.2
Simple model of the formation of the radiographic image showing both a primary and a secondary photon.

and the image is recorded in the xy plane. We assume that each photon interacting in the image receptor is locally absorbed and that the response of the receptor is linear, so that the image may be considered as a distribution of absorbed energy. If there are N photons per unit area incident on the receptor in the absence of the patient and $I(x, y)\,dx\,dy$ is the energy absorbed in area $dx\,dy$ of the receptor, then

$$I(x,y) = primary + secondary \tag{2.1}$$

$$I(x,y) = N\varepsilon(E,0)E\exp\left[-\int \mu\,dl\right] + \int \varepsilon(E_s,\theta)E_s S(x,y,E_s,\Omega)\,dE_s\,d\Omega \tag{2.2}$$

where the line integral is over all tissues along the path of the primary photons reaching the point (x, y) and μ is the linear attenuation coefficient, which will vary throughout the path length. The scatter distribution function S is defined so that $S(x, y, E_s, \Omega)\,dE_s\,d\Omega\,dx\,dy$ gives the number of scattered photons in the energy range E_s to $E_s + dE_s$ and the solid angle range Ω to $\Omega + d\Omega$, which pass through area $dx\,dy$ of the receptor. The energy absorption efficiency ε of the receptor is a function of both the photon energy and the angle θ between the photon direction and the z-axis (which, for simplicity, has been taken as zero for the primary photons). The effects of the anti-scatter device can be easily added to this equation if required. Equation 2.2 assumes that none of the energy deposited in the image receptor arises from electrons generated in the patient or the other components of the imaging system. For the X-ray energies used in diagnostic radiology, this is a good approximation.

In most applications, the receptor will not have an efficiency close to unity and the path length of the photon through the receptor will have an important effect on efficiency. Scattered photons will usually be absorbed more efficiently than primary photons, so that inefficient receptors will enhance the effects of scatter on the image.

The scatter function S has a complicated dependence on position and the distribution of tissues within the patient. For many applications, it is sufficient to treat it as a slowly varying function and to replace the very general integral in Equation 2.2 with the value at the centre of the image. As the scatter will decrease away from the centre, this will give a maximal estimate of the contrast-degrading effects of the scatter. Equation 2.2 then simplifies to

$$I(x,y) = N\varepsilon(E,0)E\exp\left[-\int \mu\,dl\right] + \left(\overline{\varepsilon_s E_s}\right)\bar{S} \tag{2.3}$$

where

the function $\bar{S}$ is the number of scattered photons incident per unit area at the centre of the image, and

$\left(\overline{\varepsilon_s E_s}\right)$ is the average energy absorbed per scattered photon.

In practice, it is the ratio of the energies absorbed per unit area from scattered and primary radiation that is either measured or calculated. An appropriate form of Equation 2.3 is then

$$I(x,y) = N\varepsilon(E,0)E\exp\left[-\int \mu\,dl\right](1+R). \tag{2.4}$$

The quantity R is known as the scatter-to-primary ratio.

2.3 Photon Interactions

To understand the X-ray imaging system, we need to understand the interactions of photons with matter. It is not our intention here to give a full treatment of these interactions, and the reader is referred to Attix (1986) for a more detailed discussion. There are, however, several aspects of these interactions that are important to consider.

We start by establishing the appropriate photon energy range for diagnostic radiology. Figure 2.3 shows how the transmission of monoenergetic photons through tissue varies with photon energy and tissue thickness. If the transmission is very low, then very few photons will reach the image receptor and the radiation dose to the tissue will be very high. If the transmission is close to unity, then there will be very little difference in transmission through different types of tissue and the contrast in the image will be poor. The choice of energy will, therefore, be a compromise between the requirements of low dose and high contrast. This is an example of the general principle of requiring semi-opacity for imaging discussed in the introduction. The photon energy range 15–150 keV encompasses current clinical practice, with the higher energies generally being more appropriate for imaging the thicker body sections. In this energy range, the important photon interactions are the photoelectric effect and scattering. Figure 2.4 shows the variation of the linear attenuation coefficient with photon energy for these two processes. Scattering interactions do not result in absorption of the photon and the coefficient for the energy-absorptive component of the scatter, which is due to the transfer of energy to recoil electrons, is also shown. It will be seen that, for soft tissue, the photoelectric cross section is larger than the scatter cross section for energies up to about 25 keV.

Figure 2.5 shows the variation with energy of the linear attenuation coefficients for both bone and soft tissue. The difference between the two coefficients is largely due to differing photoelectric cross sections and densities, and explains why X-rays are so good at imaging broken bones. The difference between the two curves, and hence the contrast between soft tissue and bone, decreases with increasing photon energy.

The contribution of photoelectric interactions to the linear attenuation coefficient varies approximately as the third power of the atomic number and inversely as the third power

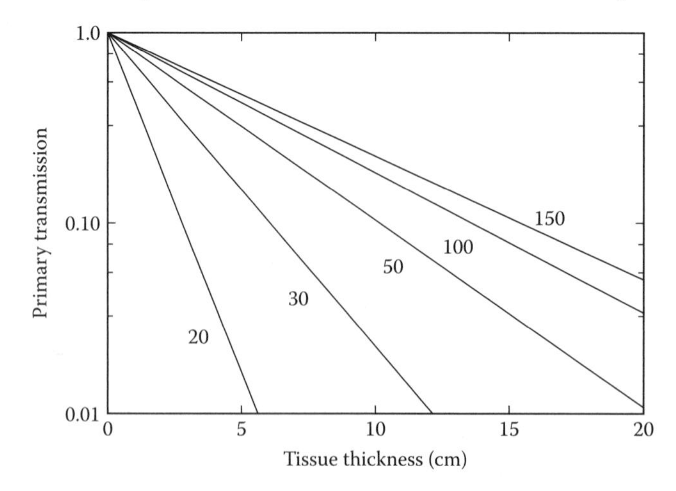

FIGURE 2.3
Transmission of monoenergetic photons through soft tissue. Curves are shown for photon beams with energies 20, 30, 50, 100 and 150 keV.

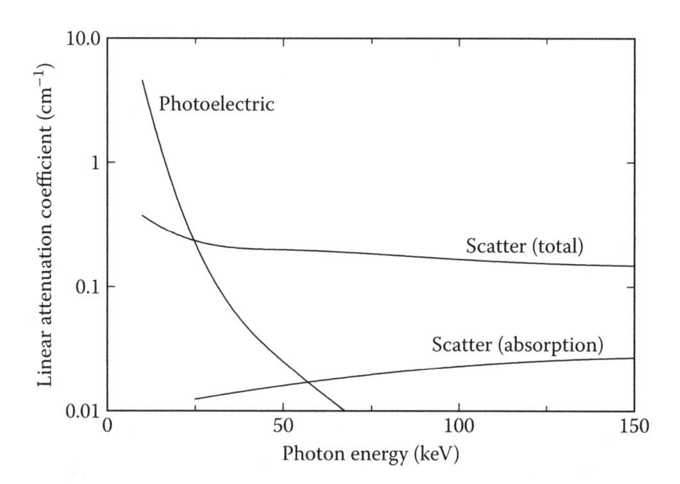

FIGURE 2.4
Variation of the linear attenuation coefficient with photon energy for soft tissue. The coefficient is shown for photoelectric absorption, scatter and the absorptive part of the scatter process.

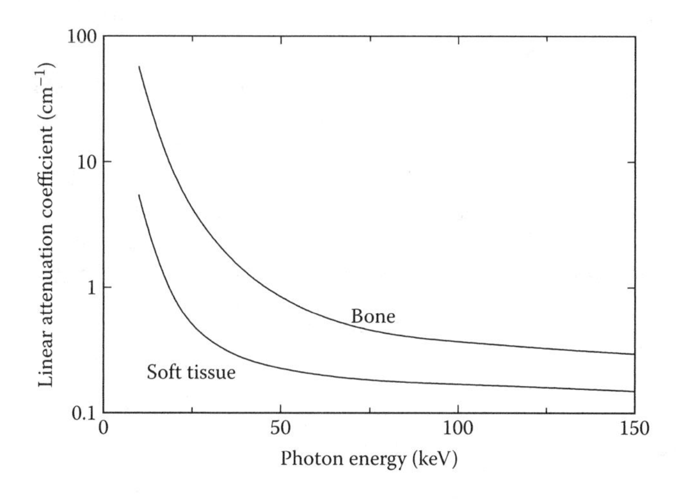

FIGURE 2.5
Variation with energy of the linear attenuation coefficients for soft tissue and cortical bone.

of the photon energy, and shows discontinuities at absorption edges where there is an increase in the interaction cross section due to new processes becoming energetically possible (Attix 1986). The existence of an absorption edge can affect the performance of the image receptor, which can have different resolution and efficiency immediately below and above the edge (see, e.g. Section 2.7.2.3).

A photoelectric interaction is followed by the ejection of a photoelectron and by one or more characteristic X-rays and Auger electrons. The highest electron energy likely to be produced in this way in diagnostic radiology is 150 keV, and such an electron would have a range in water of only 0.3 mm. For most purposes, therefore, the electrons can be considered as being locally absorbed and will greatly contribute to patient dose and to the energy absorbed in the image receptor. Characteristic X-rays, however, cannot always be considered as being locally absorbed. This is illustrated in Table 2.1, which gives the energies,

TABLE 2.1

Properties of K-shell characteristic X-rays.

Element	Mean Energy (keV)	Fluorescent Yield	Mean Free Path in Water (cm)
O	0.5	0.01	<0.001
Ca	3.7	0.16	0.01
Ag	22.6	0.83	1.8
Cs	31.7	0.89	2.9
Gd	43.9	0.93	4.0

mean free paths and fluorescent yields for selected K-shell characteristic X-rays. (The fluorescent yield gives the probability of a fluorescent X-ray being emitted after the ejection of a photoelectron.) Elements are tabulated to illustrate characteristic X-ray production both in tissue (oxygen and calcium) and in image receptors (silver, caesium and gadolinium). It will be seen that the spread of the characteristic X-rays is important for the elements of higher atomic number that are used in image receptors, where it can lead to both a loss of efficiency and to image blurring.

The contribution of interactions involving scatter to the linear attenuation coefficient only varies slowly with energy and is largely independent of atomic number. They are, therefore, less important for providing contrast between tissues with differing average atomic numbers than the photoelectric effect, except at the higher photon energies where the photoelectric cross sections for tissue become small. The energy of the scattered photon is similar to that of the incident photon (for incident photon energies of 25 and 100 keV, the energies of backscattered photons are 22.8 and 71.9 keV, respectively) so that, as illustrated in Figure 2.4, the absorptive component of the scatter interaction is small. Because of this and because of the high atomic numbers generally used, most of the energy deposited in the image receptor is due to photoelectric interactions.

2.4 Important Physical Parameters

The performance of the diagnostic radiology imaging system and the performance of its components can be assessed and specified in terms of just a few physical parameters, and in this section we discuss the most important of these, namely contrast, unsharpness, absorbed (radiation) dose, noise and dynamic range. This discussion will form a basis for the remainder of the chapter.

2.4.1 Contrast

In Section 2.2 we derived an equation for the radiographic image. We now use this result to obtain an expression for the radiographic contrast. Consider the simple model shown in Figure 2.6. The patient is represented by a uniform block of tissue of thickness t and linear attenuation coefficient μ_1, containing an embedded block of 'target' tissue of thickness x and linear attenuation coefficient μ_2. The 'target' tissue is the volume that it is required to image in the projection radiograph. The contrast C of the 'target' tissue is defined in terms of the image

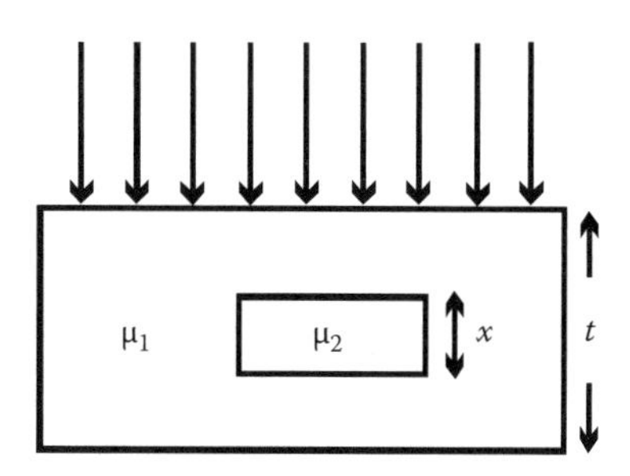

FIGURE 2.6
Simple model used for the estimation of contrast.

distribution functions I_1 and I_2, which give the energy absorbed per unit area of the receptor outside and inside the (shadow) image of the 'target' tissue, respectively. It is given by

$$C = \frac{I_1 - I_2}{I_1}. \tag{2.5}$$

The 'target' is considered to be in the centre of the image and the region with which it is compared is close by, so that the scatter fields in the two regions are very similar. For this to be a reasonable assumption, the lateral extent of the target tissue is considered to be sufficiently small. The functions I_1 and I_2 can then be obtained from Equation 2.3 and are given by

$$I_1 = N\varepsilon(E,0)E\exp[-\mu_1 t] + \left(\overline{\varepsilon_s E_s}\right)\overline{S} \tag{2.6}$$

$$I_2 = N\varepsilon(E,0)E\exp\left[-\mu_1(t-x) - \mu_2 x\right] + \left(\overline{\varepsilon_s E_s}\right)\overline{S}. \tag{2.7}$$

So the contrast is given by

$$C = \frac{N\varepsilon(E,0)E\exp[-\mu_1 t]\left(1-\exp[(\mu_1-\mu_2)x]\right)}{I_1} \tag{2.8}$$

which can be simplified using Equation 2.4 to

$$C = \frac{(1-\exp[(\mu_1-\mu_2)x])}{1+R}. \tag{2.9}$$

The factors that affect the contrast are, therefore, seen to be the thickness of the 'target', the difference in linear attenuation coefficients and the scatter-to-primary ratio R. Figure 2.7 shows the dependence of contrast on photon energy for two objects that are important in mammography (the X-ray examination of the breast). It will be seen that the contrast decreases rapidly with increasing photon energy, so that for the best contrast we should use a low photon energy. However, as we have seen already, low energy means high patient dose, and a compromise must be reached between these two quantities. The magnitude of the scatter-to-primary ratio and its effect on contrast are considered in Section 2.6.

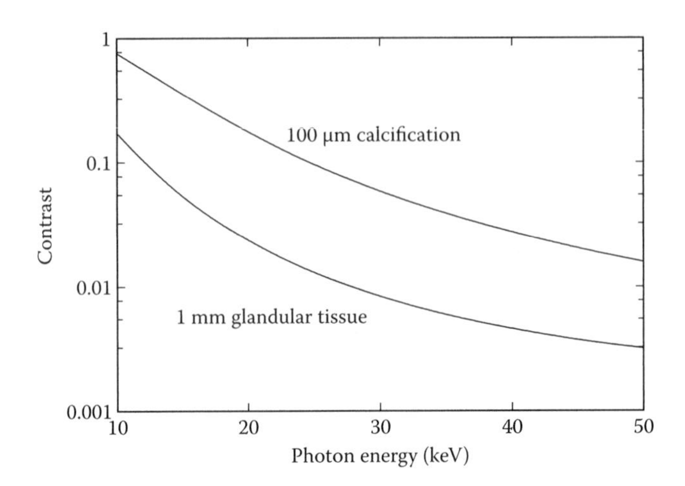

FIGURE 2.7
Variation of contrast with photon energy for two objects of importance in mammography. The upper curve is for a 100 μm calcification (calcium hydroxyapatite) and the lower curve is for 1 mm of glandular tissue. Contrast degradation due to scatter and image unsharpness has been ignored.

In deriving Equation 2.9, we have assumed that the response of the receptor is linear with the energy absorbed. In practice, however, this is often not the case and the contrast given by the equation then needs to be modified (see, e.g. Section 2.7.1.2).

2.4.2 Unsharpness

The unsharpness of the radiological imaging system is another very important parameter and can be expressed in a variety of ways. One simple approach is to use a measure of blurring, for example, the full width at half maximum (FWHM) of the point- or line-spread function. Another approach is to use the concept of resolution (as opposed to unsharpness). For example, a bar pattern can be imaged and the highest bar frequency in line pairs per mm that can be visualised by the system determined. However, the most general and useful approach is to use the modulation transfer function (MTF) (see, e.g. ICRU 1986 or Barrett and Swindell 1981), which characterises how well the system can image information at any spatial frequency. The MTF is much utilised for comparing different imaging systems. It is also employed to combine together the contributions to the total unsharpness from the various components in the imaging system. In our case, the contributions to the unsharpness arise from the focal spot of the X-ray tube, which gives a penumbra to the image (geometric unsharpness – see Section 2.5.5), from the receptor itself and from patient movement during the exposure. In some cases, movement unsharpness can be neglected but it is very important when imaging parts of the body that are in motion, such as the heart and its associated blood vessels. Figures 2.25, 2.34 and 2.45 will give examples of the use of the MTF.

The MTF of radiological image receptors may be determined from measurements of the line-spread response function, the edge-response function or the square-wave response. In the first case, the MTF is calculated from the Fourier transform of the line-spread response, and in the second case, the line-spread response is obtained as the derivative of the edge response. In the third case, the sine-wave response is calculated from the square-wave response. The line-spread response can be obtained by imaging a narrow slit placed in front of the receptor and the edge response by imaging a translucent or opaque edge test device. For digital systems, account must be taken of the Nyquist sampling theorem

(Barrett and Swindell 1981) and adequate sampling must be employed (see Section 11.6). The MTF thus obtained is referred to as the 'pre-sampled MTF'. The reader is referred to Dobbins (1995) and Samei et al. (2006) for more information. An example of a pre-sampled MTF will be shown in Figure 2.38.

2.4.3 Noise and Dose

Although the imaging system may have high contrast and good resolution, the radiologist will fail to identify even a large object if the noise level in the image is very high. There are two major contributions to the noise in a radiographic image: statistical fluctuations in the number of X-ray photons detected per unit area (quantum noise) and fluctuations due to the properties of the image receptor and display system. The appearance of the noise in the image will be affected by the spatial-frequency response of the system. A poor high-frequency response will give rise to a somewhat blotchy image and this has led to the use of the terms 'quantum mottle' and 'radiographic mottle'. Because of the frequency variation, it is sometimes convenient to express the noise in terms of its Wiener spectrum or noise power spectrum and this is considered further in Section 2.7.2.5 (also see Barrett and Swindell 1981 or ICRU 1996 for a detailed discussion).

The noise due to quantum mottle can be reduced by increasing the number of photons used to form the image. This will also increase the dose to the patient and it is instructive to explore the relationship between these two quantities. We use the model shown in Figure 2.6 and answer the following question: 'What is the surface dose required to be able to see a contrast C over an area A against a background noise arising purely from quantum mottle?' We first compare the signal we are trying to observe with the background noise to obtain a signal-to-noise ratio (SNR). The signal ΔIA that we are trying to detect may be found from Equations 2.4 and 2.5 by putting $\Delta I = I_1 - I_2$ and $I = I_1$. We then have

$$Signal = \Delta I\ A = C\ I\ A = C\ N\ \varepsilon\ A\ E \exp[-\mu_1 t](1+R). \tag{2.10}$$

The quantum noise in the image arises from fluctuations in the energy absorbed by the receptor. For simplicity, we assume that each photon that interacts with the receptor is completely absorbed and that the receptor efficiency ε is the same for both primary and secondary photons. The number of photons detected in area A of the receptor is then simply $(IA)/E$. As this quantity is a Poisson variate, the noise associated with an area A adjacent to our target area is $E(IA/E)^{1/2}$, or

$$Noise = E(N\ \varepsilon\ A \exp[-\mu_1 t](1+R))^{1/2}. \tag{2.11}$$

The SNR is, therefore,

$$SNR = C(N\ \varepsilon\ A \exp[-\mu_1 t](1+R))^{1/2}. \tag{2.12}$$

Substituting Equation 2.9 for C gives

$$SNR = \{1-\exp[(\mu_1-\mu_2)x]\}\left(\frac{N\ \varepsilon\ A \exp[-\mu_1 t]}{1+R}\right)^{1/2}. \tag{2.13}$$

According to Rose (1973), an object becomes detectable when its SNR exceeds a certain minimum or threshold value. Rose suggested that this threshold was a ratio of 5, but for the moment we shall use the symbol k to denote this quantity. The minimum patient dose will occur at this threshold and, equating Equation 2.13 to k and solving, we obtain a value for the number of photons incident on the patient per unit area as

$$N = \frac{k^2(1+R)\exp[\mu_1 t]}{\varepsilon(\Delta\mu x)^2 x^2}. \tag{2.14}$$

In deriving this equation we have assumed that the contrast is small, expanded the first exponential in Equation 2.13 to the second term and used $\Delta\mu = \mu_1 - \mu_2$. We have also assumed that the object of interest is a cube of side x and have substituted the appropriate value for the area A. The surface dose is then simply obtained as the product of the number of photons per unit area (N), the mass energy absorption coefficient for tissue (μ_{En}/ρ) and the photon energy (E):

$$Dose = \left(\frac{\mu_{En}}{\rho}\right) E \frac{k^2(1+R)\exp[\mu_1 t]}{\varepsilon(\Delta\mu)^2 x^4}. \tag{2.15}$$

An important result follows from this equation: the minimum dose required to visualise an object increases as the inverse fourth power of the size of the object. For fixed dose and contrast, there will be a minimum object size that can be visualised and the low-contrast resolution of the system will vary with object size. This subject is taken up again in some detail in Chapter 13. It should be pointed out, however, that our model is very idealised and the real problem of viewing abnormality against a background with complicated architecture is considerably more difficult. It should also be noted that film-based receptors constrain the dose to within certain limits because a minimum dose is required to achieve any appreciable blackening and the film itself will saturate above a maximum dose. The speed of the image receptor will have a critical effect on the noise in the image and, hence, on the resolution attainable at low contrasts.

It is instructive to substitute numerical values into Equations 2.14 and 2.15 and to calculate the number of incident photons per unit area and the surface dose for imaging 1 mm^3 of tissue with 1% contrast. We use the values $E = 50$ keV, $\varepsilon = 0.3$, $x = 1$ mm, $k = 5$, $\mu_1 = 22.6\,\text{m}^{-1}$, $(\mu_{En}/\rho) = 0.004\,\text{m}^2/\text{kg}$, $\Delta\mu x = 0.03$, $(1 + R) = 3$ and $t = 0.2$ m, which gives $N = 2.6 \times 10^{13}$ photons/m^2 and a surface dose of 0.8 mGy.

The aforementioned treatment considers the SNR for the task of detecting a particular object. For the study of detector performance, it is necessary to consider the transfer of the SNR of the pattern of photons incident on the detector (SNR_{in}) to the SNR in the image (SNR_{out}). For this purpose an alternative SNR is used, based on the comparison of a signal from a uniform field with the fluctuations in this field. So that, if there are N_{in} photons incident per unit area on a photon counting detector, SNR_{in} is given by

$$(SNR_{in})^2 = N_{in}. \tag{2.16}$$

For a detector which has a photon detection efficiency ε and which gives the same signal per interacting photon, SNR_{out} is given by

$$(SNR_{out})^2 = N_{in}\varepsilon = N_{out} \tag{2.17}$$

so that

$$\frac{(SNR_{out})^2}{(SNR_{in})^2} = \varepsilon. \tag{2.18}$$

When the signal per interacting photon varies, which is generally the case in practice, Equation 2.18 is no longer valid. Nevertheless, the square of the ratio of the two SNR values gives a very useful measure of the transfer of SNR by the detector, and hence of detector performance. By analogy with Equation 2.18, it is known as the *detective quantum efficiency* (DQE) and is given by

$$DQE = \frac{(SNR_{out})^2}{(SNR_{in})^2}. \tag{2.19}$$

From Equation 2.19, it can be seen that for an imaging system consisting of cascaded processes, the overall DQE is the product of the DQE values for each stage in the cascade.

When studying image receptors, it is also important to consider the actual value of the final SNR. By analogy with Equation 2.17, the quantity used is called the *noise equivalent quanta* (NEQ), and is given by

$$NEQ = (SNR_{out})^2. \tag{2.20}$$

We now turn our attention to the magnitude of the radiation dose and the associated risk of radiation detriment. Table 2.2 gives some typical doses obtained in clinical practice and based on UK surveys (Young and Burch 2000, Hart et al. 2002). The doses in the table are values of the entrance surface dose apart from that for mammography, where the mean glandular dose (Dance et al. 1999) is given. It will be seen that the dose values are comparable in magnitude to our earlier dose estimate and, therefore, that the radiographs are often taken fairly close to noise-limited conditions.

It is well known that there is a risk associated with the use of ionising radiation (see Chapter 16). It is important, therefore, to discuss how this risk is related to the radiation dose for X-ray examinations. The risks to an individual arising from medical irradiation can be classified into deterministic effects and stochastic effects (ICRP 2008).

TABLE 2.2

Doses for some common adult radiological examinations.

Examination[a]	Dose (mGy)
OB breast	2.1
PA chest	0.12
AP lumbar spine	4.3
AP pelvis	3.2
AP abdomen	4.1

[a] OB, medio-lateral oblique projection; PA, postero-anterior projection; AP, antero-posterior projection.

For deterministic effects, the probability of harm will be zero up to a certain dose and will then increase rapidly to 100% above a dose threshold. In diagnostic and interventional radiology, the dose is usually well below such thresholds, with the exception of certain high-dose interventional procedures (ICRP 2000), when the skin dose can approach or exceed the threshold for transient erythema (2 Gy). Contrary to deterministic effects, stochastic effects occur with a certain probability, which is related to the dose level and the organ or tissues irradiated. The stochastic risk itself can be split into somatic and genetic components with the former corresponding mainly to the risk of carcinogenesis and the latter to the risk of hereditary effects in progeny. The cancer risks are the sum of the different risks for the various organs and tissues in the body. In most situations the irradiation is non-uniform and the ICRP has introduced the concept of *effective dose* for radiation protection purposes. The effective dose is a weighted average dose for the organs and tissues within the body. It takes into account the varying risks for the different organs and tissues and is expressed in units of sievert (Sv). The weighting is chosen so that the radiation detriment for an effective dose of a particular value is the same as that for a uniform whole body dose of the same value. The ICRP (2008) gives the stochastic risks of radiation detriment as 0.041 and 0.001/Sv for cancer induction and severe hereditable effects, respectively. These figures are for a particular population of adult workers. The concept of effective dose also finds utility in diagnostic radiology, although it should always be used with caution because the ICRP weighting factors used in its calculation may not be applicable to a patient of a particular age and sex. Notwithstanding this difficulty, it is a very useful quantity for comparing risks associated with different imaging techniques, for combining together the doses for a procedure where the body parts imaged vary during the examination, and as the basis for optimisation of imaging system design.

Table 2.3 gives typical values of the effective dose for various examinations as determined in the national survey of Hart and Wall (2002). The effective dose was obtained by applying the results of Monte-Carlo-based model calculations to measurements of entrance dose or related quantities. An alternative approach is to use as a measure of risk the energy imparted to the patient by the radiation field. This is simpler to estimate and is reasonably well correlated to the effective dose (Alm Carlsson et al. 1999).

Using the ICRP risk factors given previously, the total lifetime risks of radiation detriment for chest PA and barium enema examinations are estimated as 0.0001% and 0.04%,

TABLE 2.3

Values of the effective dose (mSv) for selected radiological examinations.

Examination[a]	Effective Dose (mSv)
Barium swallow	1.5
Barium enema	7.2
IVU	2.4
PA chest	0.016
AP + lateral lumbar spine	1.0
AP pelvis	0.67
AP abdomen	0.76
Intraoral dental	0.005

[a] PA, postero-anterior projection; AP, antero-posterior projection; IVU, intravenous urography.

respectively, which can usually be considered small when compared with the potential benefit of the examination. The values of the effective dose may also be viewed in the context of the average *per capita* radiation dose to the UK population per year, arising from all sources. This is 2.5 mSv per annum, of which 0.3 mSv arises from the medical uses of ionising radiation (Hughes et al. 1989).

The risks from the use of ionising radiation in diagnostic radiology are further discussed and set in the context of risks from other imaging techniques in Chapter 16.

2.4.4 Dynamic Range

In many applications in medical X-ray imaging, the pattern of X-ray photons incident on the image receptor will vary considerably over the image because of associated variations in photon attenuation for passage through neighbouring regions of the body. The incident photon pattern is said to have a *dynamic range*, which can, for example, be defined as the ratio of the maximum to minimum values of the photon intensity distribution. On the other hand, image receptors may perform well over a limited range of intensities, for which the DQE (or transfer of SNR) can be considered adequate. For higher or lower intensities, the DQE may be poor and the information displayed inadequate. The dynamic range of the *image receptor* can thus be considered as the range of incident intensities for which the displayed information is adequate. Ideally, the exposure level and the dynamic range of the image receptor should be such that the whole range of intensities in the incident photon pattern is adequately imaged.

2.5 X-Ray Tubes

The X-ray tube used in diagnostic radiology consists of an evacuated, heat resistant, borosilicate glass envelope containing an anode and cathode (Figure 2.8). Electrons emitted from the cathode are accelerated across the vacuum within the tube by an electric field and strike the anode (also called the target). Bremsstrahlung and characteristic X-rays are produced as these high-energy electrons interact with the atoms in the target. The glass

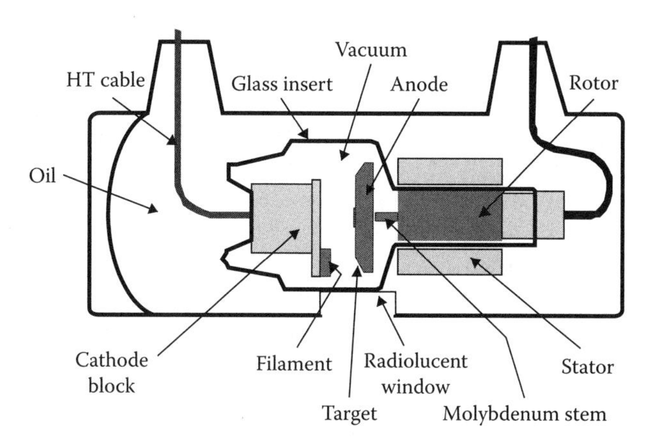

FIGURE 2.8
The construction of a rotating-anode X-ray tube.

envelope is surrounded by oil to help with the electrical insulation and to absorb the heat radiated from the anode. As the oil expands when heated, the housing has bellowed ends to allow for this expansion. The housing has a small radio-lucent window close to the anode to allow X-rays to leave the tube in the required direction. X-rays that are emitted in other directions are absorbed by the housing. The X-ray beam that leaves the exit window of the tube passes through collimators that allow the size of the X-ray field to be varied. The complete tube assembly is held in a suitable mounting so that the beam position and direction can be adjusted as necessary.

2.5.1 Cathode

The cathode of an X-ray tube comprises a thin tungsten wire wound into a helical filament that is set in a nickel block. Tungsten is used because it has a high and stable thermionic emission, a high melting point (3410°C), is strong and has a low tendency to vaporise. The filament is heated by the passage of an electrical current of the order of a few amps. Electrons are released from the surface of the filament by thermionic emission and are accelerated towards the anode by a potential difference of between 25 and 150 kV. The nickel block supports the filament and is shaped to create an electric field that focuses the electrons into a slit beam. The design of the filament and the electron optics that guide the electrons to the anode is very important. The unsharpness in the image may be limited by the size of the X-ray source (see Section 2.5.5), whereas the output of the tube is related to the magnitude of the electron current striking the anode. Many tubes contain a cathode with two filaments. One filament is smaller than the other and this gives the operator the option of two focal-spot sizes. The larger filament provides greater output than the smaller filament but this is at the expense of increased unsharpness in the image.

In some special tubes, called grid-controlled or grid-biased X-ray tubes, the electron optics are insulated from the cathode and can be biased to a more negative potential than the cathode. If this biasing is high enough, the flow of electrons from the cathode to the anode is stopped. By removing the negative biasing, the flow of electrons can recommence. This allows rapid switching of the X-ray beam and is used in applications such as pulsed fluoroscopy. This is preferable to switching the high-voltage supply on and off. The voltage would then have to ramp up to the desired value, resulting in higher patient dose from the increased number of low-energy X-rays.

2.5.2 Anode

When the electrons interact with the anode of the X-ray tube, they slow down and stop. Most of the energy absorbed by the anode from the electrons appears in the form of heat, but a small amount (less than 1%) appears in the form of X-rays. The anode material, therefore, needs to be able to withstand very high temperatures and, like the filament, is required to have a low tendency to vaporise. Vaporised metal can interfere with the passage of the electrons and when deposited on the glass surface can affect the insulation. This can reduce the lifetime of the tube because glass is particularly vulnerable to electron bombardment (hence, some heavy-duty tubes are made from ceramic material or metal instead of glass). The anode is usually constructed from tungsten although molybdenum or rhodium is used for special applications where a low-energy X-ray beam is required (see Section 2.5.3). Tungsten has an atomic number of 74, acceptable thermal conductivity and thermal capacity, and a high melting point. The high

TABLE 2.4

Properties of molybdenum, rhodium and tungsten.

	Mo	Rh	W
Atomic number	42	45	74
K X-ray energies (keV)	17.4–19.8	20.2–22.8	58.0–67.7
Relative density	10.2	12.41	19.3
Melting point (°C)	2617	1966	3410
Specific heat (J/(kg °C))	250	242	125

atomic number is important because the Bremsstrahlung yield from the target increases with the atomic number. Improved tube lifetime can be obtained by using a 90/10 tungsten/rhenium alloy. This reduces crazing of the anode surface caused by the continual heating and cooling processes to which it is subjected.

It is important that the anode has a high thermal capacity. A larger anode will give a higher rating and a shorter exposure time, and the greater thermal capacity associated with an increase in anode volume will allow the possibility of a shorter time interval between exposures. For heavier-duty applications, the thermal capacity can be increased by using a molybdenum backing to the anode. Molybdenum has a higher specific heat than tungsten (Table 2.4) and the heat capacity for an anode of this type would typically be 250,000 J. A further increase in heat capacity can be made by adding a graphite layer to the tungsten/molybdenum anode. This can increase the heat capacity to 1,000,000 J or more.

One of the problems faced by the tube designer is how to limit the heat deposited in the target area and how to remove it from that area as quickly as possible. It is the heat deposited in the anode that limits the possible X-ray output from the tube. The use of a slit source of electrons helps by spreading out the target area, and this idea can be extended by using a rotating anode. The electron beam impinges on the bevelled edge of a rotating disc and the target area is spread out around the periphery of the disc. A rotation speed of about 3000 RPM and an anode diameter of around 10 cm are used in general-purpose units. It is possible to use a stationary target and these are sometimes found in simple, low-power equipment, such as mobile and dental X-ray units. In this case the target is mounted in a large block of copper. The copper absorbs the heat generated in the target and radiates it to the surrounding oil.

The disc of a rotating anode is mounted on a thin molybdenum stem. Molybdenum is a poor conductor of heat so the thin stem reduces the heat flow backwards and prevents the rotor bearings, which are made from copper, from overheating. The heat loss from the rotating anode is, therefore, mainly radiative. In some tubes, where a high output is required for long periods of time, for example, for use in CT or cardiology, the ball bearings have been replaced by a liquid metal bearing system where the rotor 'floats' on a film of liquid metal. These tubes can cope with much higher working temperatures without causing the bearings to fail.

It has been mentioned that the size of the focal spot is important in determining the geometrical unsharpness in the image. Using a long filament wire will lead to a large focal spot on the anode. However, the apparent size of the focal spot, as viewed from the image receptor, will be smaller. As noted earlier, an anode with a bevelled edge is used. This bevel is at a steep angle to the direction of the electron beam and the tube exit window accepts X-rays that are approximately at right angles to the electron beam. Thus the X-ray source as viewed from the receptor appears approximately square

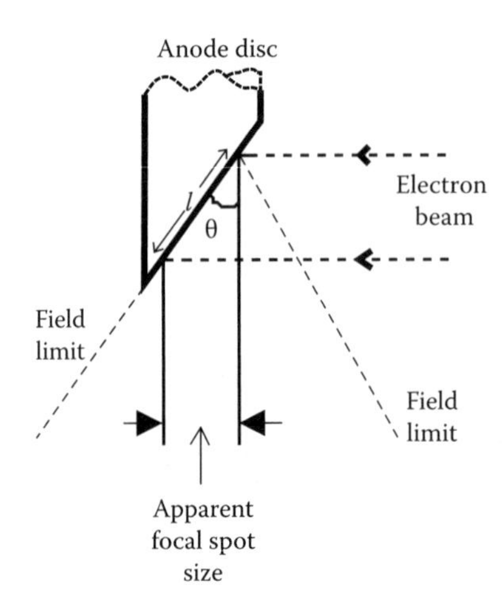

FIGURE 2.9
The use of a bevelled anode to reduce the effective focal-spot size. The width of the electron beam is $l \cos \theta$ whereas the focal-spot size as viewed from the central axis of the X-ray field is $l \sin \theta$.

even though the electron beam impinging on the target is slit shaped (Figure 2.9). The choice of the anode angle θ will depend upon the application, with the angle being varied according to the requirements of field- and focal-spot sizes and tube output. For general-purpose units, an angle of about 12°–16° is appropriate whilst units used with image intensifiers or for skull work, where a smaller field size is acceptable, may have angles more like 10°–12°. Decreasing the anode angle increases the focal area for the same focal-spot size and allows the anode to withstand a greater load.

There is a loss of intensity in the X-ray field at the anode side of the tube. This is due to absorption of the X-rays in the anode. For a 35 × 43 cm cassette, the difference in intensity across the area of the cassette can be greater than 20% at a distance of 1 m from the focus, and drops to around 10% at 2 m. This loss of intensity across the X-ray field is known as the heel effect.

2.5.3 X-Ray Spectra

The shape of the X-ray spectrum will depend upon the target material, the peak voltage and voltage waveform applied to the tube and the effects of any filters placed in the X-ray beam. Table 2.4 shows the properties of the three commonly used target materials and Figures 2.10 through 2.12 show three typical X-ray spectra. The first two spectra are for tubes with tungsten targets and the third spectrum is for a tube with a molybdenum target. Also shown are the spectra after modification by transmission through the body. Characteristic X-rays will only be produced when the energy of the bombarding electron is greater than the binding energy of the atomic electrons. For tungsten, the K-shell binding energy (K-edge) is 69.5 keV and for molybdenum 20 keV. Significant levels of characteristic X-rays will only be produced once the tube voltage has been raised sufficiently far above the K-edge.

The tungsten spectra are well suited to imaging thicker body sections because the energies of the tungsten K-shell characteristic X-rays are sufficiently high (average K_α energy 58.8 keV, average K_β energy 67.6 keV). Molybdenum has lower-energy K-shell characteristic X-rays (average K_α energy 17.4 keV, average K_β energy 19.6 keV), which are more appropriate

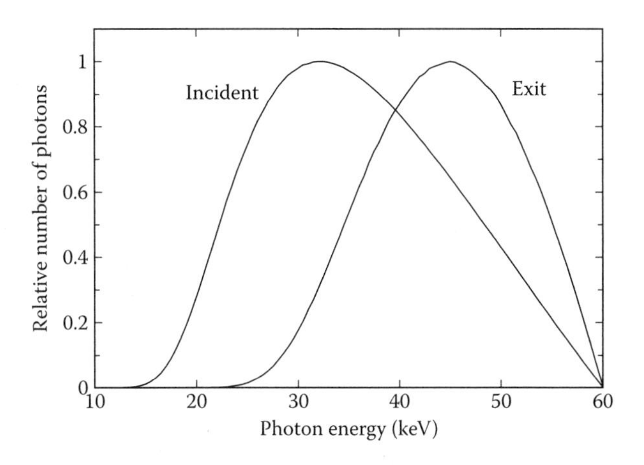

FIGURE 2.10
X-ray spectra for an X-ray tube with a tungsten target; 60 kV constant potential with 2.5 mm aluminium added. The spectra are shown both before and after attenuation by 9.5 cm soft tissue plus 0.5 cm bone. (The spectra have been calculated using data from Cranley et al., 1997.)

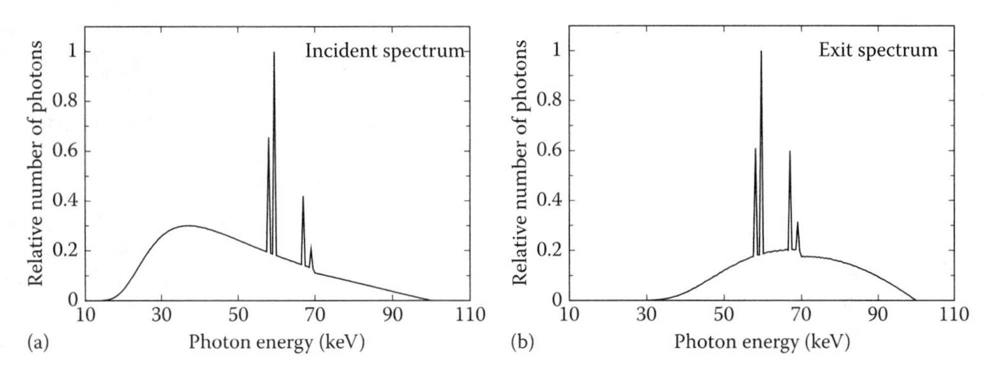

FIGURE 2.11
X-ray spectra for an X-ray tube with a tungsten target; 100 kV constant potential with 2.5 mm aluminium added. The spectra are shown both before (a) and after (b) attenuation by 18.5 cm soft tissue plus 1.5 cm bone. (The spectra have been calculated using data from Cranley et al., 1997.)

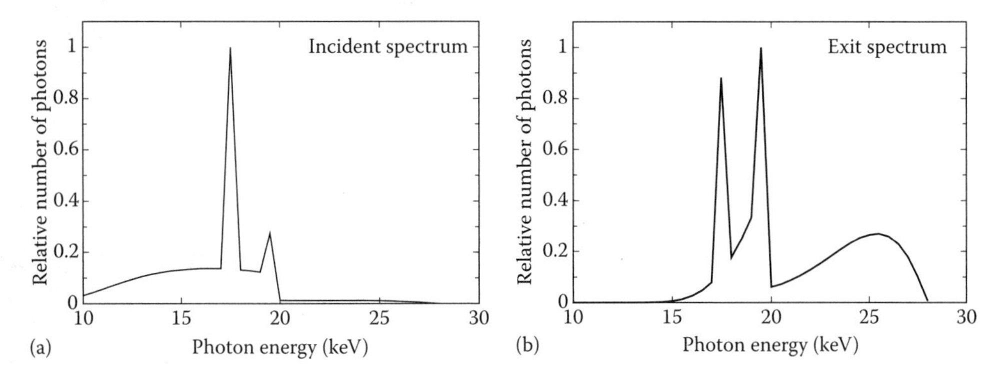

FIGURE 2.12
X-ray spectra for an X-ray tube with a molybdenum target; 28 kV constant potential with 0.03 mm molybdenum filter. The spectra are shown both before (a) and after (b) attenuation by 5 cm tissue. (The spectra have been calculated using data from Cranley et al., 1997.)

for imaging thinner body sections at high contrast. They are used in X-ray units specifically designed for mammography. An X-ray tube with a rhodium target may also be used for this purpose. The characteristic X-rays from rhodium have energies about 3 keV higher than those from molybdenum. They are, therefore, better at penetrating thicker or denser breasts. In Figure 2.12 the X-ray spectrum from the molybdenum target has been modified by the use of a molybdenum filter, which heavily attenuates any X-rays that have energies above the molybdenum K-edge. If rhodium is used as the target, a rhodium K-edge filter is used. Such a filter may also be used with a molybdenum target. It is also possible to use a tungsten target for mammography and, depending upon the imaging requirements, a rhodium, silver or aluminium filter may then be used.

There is a large difference between the X-ray spectra before and after passage through the patient. Knowledge of the X-ray spectrum that leaves the patient is important as it is these X-rays that interact with the image receptor. The difference between the incident and exit spectra is due to the photons that interact in the patient and are scattered or absorbed. These photons deliver the radiation dose to the patient. Low-energy photons are more likely to be absorbed inside the patient because the probability of photoelectric absorption increases with decreasing photon energy. If the spectrum incident on the patient is too soft, that is, it has a relatively high proportion of low-energy photons, then these low-energy photons will only contribute to patient dose and not to image contrast. It is, therefore, important that they are removed from the X-ray beam before it reaches the patient. For general radiography, this can be achieved by the use of aluminium or sometimes copper filters, with the amount of filtration often being increased for higher tube voltages. The spectra in Figures 2.10 and 2.11 have been filtered by 2.5 mm of aluminium. In the case of mammography, the molybdenum filter used to remove photons above 20 keV will also filter out the very-low-energy photons. The filtration of the beam exiting the X-ray tube will increase slightly with the age of the tube because metal is slowly vaporised from the anode and deposited on the surface of the glass envelope and exit window. This results in a gradual hardening of the X-ray spectrum.

2.5.4 Generation of High Voltage

The high potential difference, used to accelerate the electrons between the cathode and the anode, is produced using a high-voltage transformer and rectifying circuits (Forster 1985, Ammann 1990). The stability of the applied voltage will vary depending on the design of the rectification in the high-voltage generator. The oscillating variation of the applied voltage produced by most types of generator is called the voltage ripple. A medium-/high-frequency inverter generator, for example, produces a voltage ripple of 10% or less of the set voltage. The ripple in the applied voltage causes the maximum energy of the X-rays to vary with time. Hence, for the same set peak voltage, a unit with a large voltage ripple will produce a softer X-ray spectrum than a unit with a small voltage ripple.

The ideal high-voltage generator would produce no voltage ripple, but in practice this is very difficult to achieve. The high-voltage generator that produces the lowest voltage ripple is the 'constant-potential generator'. This uses a three-phase AC input and triodes on the high-voltage side of the transformer to produce a high-voltage output with less than 2% voltage ripple. The triodes control the high-voltage applied to the X-ray tube and the exposure time. Such generators are bulky and expensive.

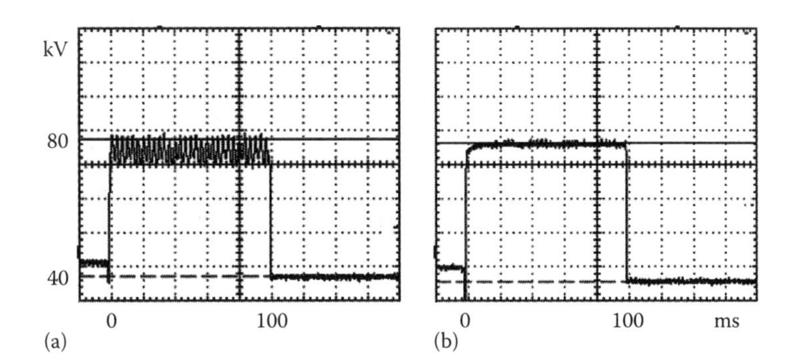

FIGURE 2.13
Peak voltage measurements for (a) a three-phase, six-pulse generator and (b) a medium-frequency generator. The displayed voltage waveforms demonstrate the voltage ripple associated with these types of generator and allow measurements of tube voltage and exposure time to be made (see Section 2.12.3). Both exposures were set at 80 kV, 100 ms. (Courtesy of A. Shawarby.)

The simplest high-voltage generator takes a single-phase AC input to a transformer and uses the X-ray tube itself to rectify the high-voltage applied across the anode and cathode. This produces a 100% voltage ripple and only generates X-rays during half of the voltage cycle. To prevent the polarity of the anode and cathode changing where higher power is required, the high-voltage can be half-wave rectified before it is applied across the X-ray tube. A higher output can be achieved using full-wave rectification, which produces output over the entire voltage cycle. The voltage ripple is still 100% but at a frequency of 100 Hz. If a smoothing circuit is used, the voltage ripple is much reduced and a 'sawtooth' output is produced. Single-phase generators can be used in low-output tubes such as those required for dental radiography or for fluoroscopy at low tube current.

Where higher outputs and harder X-ray spectra are required, three-phase high-voltage generators may be employed. Using a three-phase transformer and rectification significantly reduces the voltage ripple. The transformer configuration can be either delta or wye, or a combination of both, producing either '3-phase, 6-pulse' (300 Hz peak-to-peak) or '3-phase, 12-pulse' (600 Hz peak-to-peak) generators. The six-pulse generators can produce voltage ripple of between 13% and 25% (Figure 2.13) and the 12-pulse generator voltage ripple less than 10%. The capacitance of the high-voltage cables produces further smoothing of the voltage waveform. The voltage applied to the X-ray tube is varied using an autotransformer to adjust the low-voltage input to the high-voltage transformer.

Three-phase generators are still to be found on older equipment but modern generators tend to be the 'medium' or 'high frequency' type. These still use three-phase AC input but this is rectified and smoothed to produce a constant voltage waveform. This in turn is converted to a low-voltage high-frequency square wave using an inverter. The high-frequency square wave is applied to the input of a high-frequency transformer and the resulting high-voltage high-frequency waveform is rectified. Altering the frequency of the low-voltage input to the transformer varies the voltage applied to the X-ray tube. These generators produce a voltage ripple of less than 10% with a frequency that varies between generators, but can be of the order of several kHz. They are much smaller than the older three-phase, 12-pulse generators because the voltage produced by a transformer is proportional to the product of the frequency, the cross-sectional area of the core and the number of windings. If the frequency is higher, the cross section of the core and/or the number of windings can be reduced.

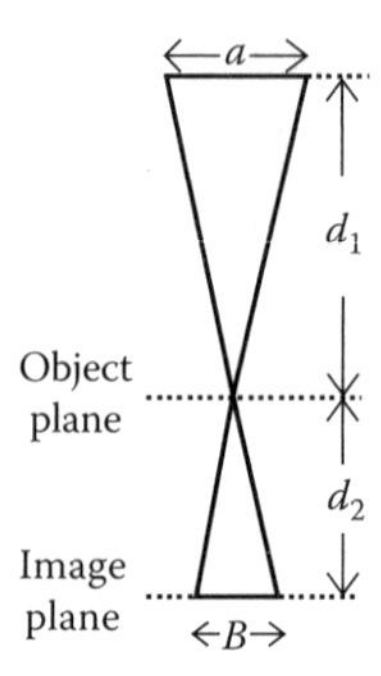

FIGURE 2.14
Geometry and notation used for the calculation of geometric unsharpness (Equation 2.21) (not to scale).

2.5.5 Geometric Unsharpness

Geometric unsharpness is the unsharpness U_g in the image due to penumbra from the finite size of the X-ray source. Using the geometry and notation shown in Figure 2.14, the blurring B in the image is given by

$$B = a\frac{d_2}{d_1} = a(m-1) \tag{2.21}$$

where
a is the effective size of the focal spot of the X-ray tube, and
m is the image magnification.

It is convenient in practice to correct for the image magnification by dividing by m, so that the geometric unsharpness becomes

$$U_g = a\left(1 - \frac{1}{m}\right). \tag{2.22}$$

In order to assess the importance of the geometric unsharpness, we must combine it with the other components of the image unsharpness (Section 2.4.2). If we neglect movement unsharpness and if U_r is the unsharpness due to the receptor, then the overall unsharpness U in the image is given by

$$U = \left(U_g^2 + U_r^2\right)^{1/2} \tag{2.23}$$

or

$$U = \frac{F}{m}\left(1 + [m-1]^2\frac{a^2}{F^2}\right)^{1/2} \tag{2.24}$$

where we have assumed for simplicity that the two types of unsharpness can be added in quadrature and have used the relation

$$U_r = \frac{F}{m} \tag{2.25}$$

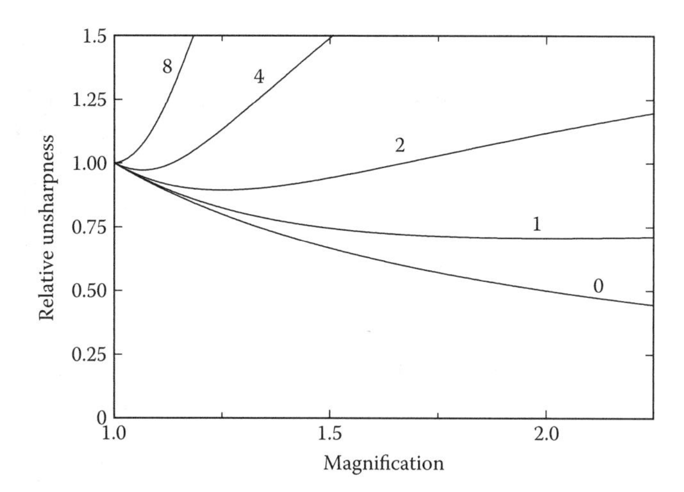

FIGURE 2.15
Variation of overall unsharpness with image magnification. The unsharpness has been normalised to the receptor unsharpness *F*. The curves are labelled 0, 1, 2, 4, 8 for focal-spot sizes of 0, *F*, 2*F*, 4*F* and 8*F*, respectively.

in deriving Equation 2.24. Here *F* is the intrinsic receptor unsharpness which would be obtained if the object had zero thickness and was in contact with the receptor.

We can see from Equation 2.24 that if the geometric unsharpness is much smaller than the receptor unsharpness, then the overall unsharpness will be inversely proportional to the magnification and magnification radiography will reduce the overall image unsharpness. If, however, the receptor unsharpness is much smaller than the geometric unsharpness, then the overall unsharpness will increase with increasing magnification. This is illustrated in Figure 2.15, which shows how U/F varies with magnification. Each curve is for a different value of the ratio a/F. A value of less than about 2 for a/F indicates that magnification radiography will reduce unsharpness and a value of more than about 4 indicates that magnification radiography will increase the unsharpness. The following conclusions can be drawn from these curves:

1. For general radiography with $a/F > 4$, the overall unsharpness will increase with increasing magnification; the patient should be as close as possible to the image receptor and the focus–receptor distance should be as large as possible.
2. For magnification radiography with $a/F < 2$, the overall unsharpness will decrease with increasing magnification at small-to-moderate magnifications and will then increase; a significant reduction in unsharpness is only possible if an appropriate magnification is used.

The focal-spot sizes of a general radiography unit with twin foci are typically 0.6 and 1.0 mm, whereas the receptor unsharpness for a screen–film receptor might be 0.1–0.2 mm. The condition $a/F > 4$ holds for the larger focus and the overall unsharpness will be limited by the size of the focal spot and the geometry used. Magnification radiography uses a very small focal spot and exposure times will need to be increased to compensate for the reduced tube output. In addition, it is necessary to increase dose or use a noisier receptor to maintain the same density on the X-ray film. The image will also increase in size and may become too large for the receptor. For magnification mammography, a focal-spot size of 0.1–0.15 mm and a magnification of 1.5–2.0 are often used but the increase in dose and the image size mean that the technique is used to provide an additional view of a suspicious area rather than as a standard part of each examination.

It should be noted that the aforementioned analysis is rather simplified. For a more realistic consideration of the detection of small objects, it is also necessary to take account of the contrast and noise in the image.

2.6 Scatter Removal

The most important factors that allow different features to be distinguished in an image are the difference in density or contrast between a feature and its background. In Section 2.4.1, we derived an expression for the contrast (Equation 2.9), which we now rewrite as follows:

$$C = \frac{(1-\exp[(\mu_1-\mu_2)x])}{1+R} = C_p D_s. \tag{2.26}$$

In rewriting this expression, we have recognised that the overall contrast is simply the product of the contrast produced by the primary photons, C_p, and a factor, D_s, which represents the degradation of contrast by the presence of scatter. The contrast degradation factor is thus given in terms of the scatter-to-primary ratio R by

$$D_s = \frac{1}{(1+R)}. \tag{2.27}$$

The effect of scatter on the image contrast can thus be understood by considering the magnitude of R.

2.6.1 Magnitude of the Scatter-to-Primary Ratio

Table 2.5 shows the scatter-to-primary ratio for five different radiographic examinations. The values correspond to imaging without any attempt to remove scatter. By reference to Equation 2.26, it will be seen that there is a very significant loss of contrast due to scatter, with the contrast degradation factor varying between 0.65 and 0.15. For small field sizes, the

TABLE 2.5

Values of the scatter-to-primary ratio, R, for various examinations.

Examination[a]	R	Field Size (cm^2)	Patient Thickness (cm)
AP paediatric pelvis	2.0	22 × 27	13
AP lumbar spine	4.4	29 × 42	20
LAT lumbar spine	5.7	16 × 42	35
PA chest	2.0	35 × 43	20
Mammography	0.54	100	5

Sources: Data taken from Sandborg et al., 1993; Dance et al., 1992.

Note: Apart from the pelvis examination, the data are for adult examinations.

[a] PA, postero-anterior projection; AP, antero-posterior projection; LAT, lateral examination.

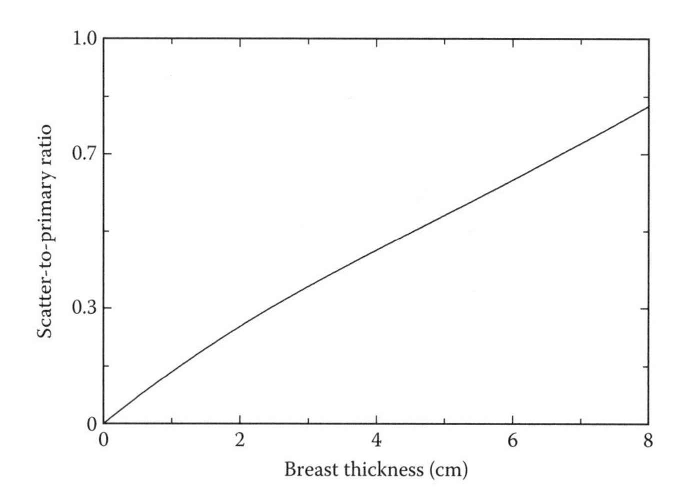

FIGURE 2.16
Dependence of the scatter-to-primary ratio for mammography on breast thickness. (Data taken from Dance et al., 1992.)

ratio R increases roughly linearly with the field size, and then gradually flattens out with further increase in field size. The magnitude of the scatter also varies across the field, falling off by up to 50% at the centre of a field edge. As the thickness of the body section increases, the transmissions of scattered and primary photons decrease by roughly the same factor, whereas more scattered radiation is produced, so that R is approximately proportional to the thickness of the tissue imaged. This is illustrated in Figure 2.16, which shows the thickness dependence of the scatter-to-primary ratio for mammography. The energy dependence of R is more complicated. The probability of a single scatter decreases with increasing energy, but the scatter distribution gradually becomes more forward peaked and the probability of multiply scattered photons reaching the image receptor increases.

It is important to note that the energy absorption efficiency of the image receptor may be different for primary and scattered photons. This is for two reasons. First, the path length in the receptor for scattered and primary photons will be different. This effect is important for inefficient detectors, which will enhance the value of R. Second, the average energy of the scattered photons will be less than that of the primary photons. This can either increase or decrease the value of R, depending upon the location of the photoelectric absorption edge of the receptor.

In view of the loss in contrast due to scatter that we have demonstrated, it is important in many situations to remove as much of the scatter as practicable. This is usually achieved using an anti-scatter grid although in some situations an air gap is used for this purpose. In addition, some specialised X-ray imaging systems employing a scanning X-ray beam can achieve excellent scatter rejection.

2.6.2 Anti-Scatter Grids

An anti-scatter grid is a device placed between the patient and the image receptor, which has different absorption properties for scattered and primary photons, so that it reduces the value of the scatter-to-primary ratio R. It is constructed from strips of a material with a high absorption, which are intended to absorb scattered radiation, and from material with a low absorption, which is intended to allow the passage of primary photons. Most anti-scatter grids are linear grids (IEC 2001), which may be parallel (Figure 2.17) or focused (Figure 2.1).

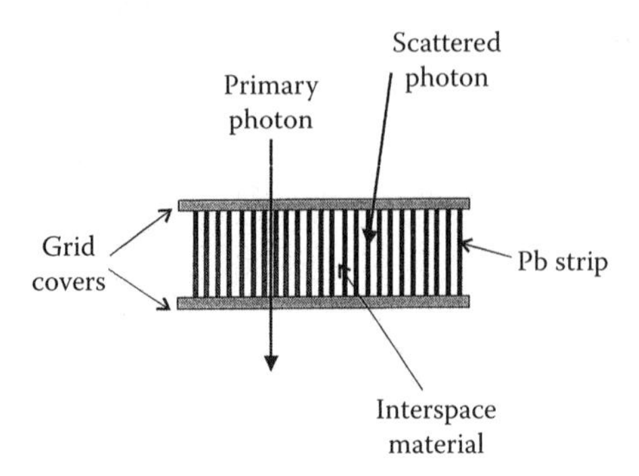

FIGURE 2.17
Construction and function of a linear parallel anti-scatter grid. The figure shows the absorption of a scattered photon and the transmission of a primary photon.

They are usually constructed from thin strips of lead, separated by an interspace material. For a parallel grid, the lead strips are parallel. For a focused grid, the lead strips point back to a line focus passing through the focal spot of the X-ray tube. With this geometry, if the lead strips are thin enough, most of the primary photons will be incident on the interspace material, and will traverse the grid without hitting a lead strip. They will, thus, have a high transmission probability. The interspace is not completely transparent to primary photons, and it is desirable that both it and the protective covers of the grid are constructed from materials with a low atomic number. Sandborg et al. (1993) have shown that the patient dose can be a factor of 1.3 higher if a grid with aluminium covers and interspace is used compared with a grid which uses lower atomic number materials (e.g. paper or cotton fibre for the interspace and carbon fibre composite for the covers).

The scattered photons incident on the grid will, in general, have a high probability of hitting a lead strip. If the lead strips are thick enough, or enough lead strips are traversed, such photons will be absorbed and, thus, will not reach the image receptor. The probability of a scattered photon hitting a lead strip will depend upon the ratio of the height of the strip to the thickness of the interspace material. This quantity is known as the grid ratio. In general, the greater the grid ratio, the better the scatter rejection.

As for many aspects of imaging, the design of the grid is a compromise. As the thickness of the lead strips increases, the secondary transmission decreases, but so does the primary transmission. As the grid ratio increases for a fixed interspace thickness, the secondary transmission decreases, but so does the primary transmission, this time because of increased absorption in the interspace material. Alternatively, it is possible to increase the grid ratio by keeping the strip height fixed and decreasing the width of the interspace material. In this case, the primary transmission will decrease because the lead strips then occupy a greater fraction of the grid volume. In addition, the strip density (number of grid lines per cm) will increase and the grid will be more difficult to fabricate. Grids with a high grid ratio tend to be used for situations where the scatter-to-primary ratio is high, such as the imaging of thicker body parts in adults, whereas a lower grid ratio is used to image thinner body parts such as the female breast. Thicker lead strips are required for imaging higher energy radiation. For example, for adult lumbar spine radiography a grid with ratio 12, lead strip thickness 36 μm and 36 lines/cm may be used, whereas for mammography typical values of the corresponding parameters are 5, 18 μm and 31 lines/cm, respectively. Fabrication difficulties presently limit the grid ratio to about 15 or 18 and the strip density to about 80 lines/cm. In practice, general radiographic units will only have one grid, and some compromise in the choice of grid parameters will be necessary.

The performance of an anti-scatter grid may be measured in terms of the contrast improvement factor, K, and the dose increase factor B (also known as the Bucky Factor, IEC 2001). The contrast improvement factor is given by

$$K = \frac{(1+R)}{(1+R_g)}. \tag{2.28}$$

It is, thus, calculated as the ratio of the contrast degradation factors (Equation 2.27) with and without the grid. The quantity R_g is the scatter-to-primary ratio with the grid and is given by

$$R_g = \frac{(T_s S)}{(T_p P)} \tag{2.29}$$

where T_s and T_p are the transmissions through the grid of secondary and primary photons, respectively. Equation 2.28 ignores a small loss of contrast due to hardening of the primary beam by the interspace material, which usually amounts to just a few per cent and Equation 2.29 ignores any scatter generated within the grid itself.

The increase in dose when the grid is used arises from the need to maintain the same energy imparted per unit area of the receptor. Since some of the primary photons and the majority of the secondary photons are removed from the beam, it is necessary to increase the dose to maintain this quantity. The dose increase factor is given by

$$B = \frac{K}{T_p}. \tag{2.30}$$

Table 2.6 shows values of the contrast improvement factor and dose increase factors when grids are used for the X-ray examinations illustrated in Table 2.5. It will be seen that although

TABLE 2.6

Values of the contrast improvement factor, K, and the dose increase factor, B, for various examinations when a grid is used to remove scatter.

Examination[a]	K	B
AP paediatric pelvis	2.4	3.2
AP lumbar spine	3.7	4.7
LAT lumbar spine	3.8	4.8
PA chest	2.3	2.9
Mammography	1.4	1.9

Sources: Data taken from Sandborg et al., 1993; Dance et al., 1992.

Note: Field sizes and patient thickness are as for Table 2.5. Apart from the pelvis examination, the data are for adult examinations. For general radiography, the results are for a grid with ratio 12, lead strip thickness 36 μm and 36 lines/cm. For mammography, the corresponding grid parameters are 5, 18 μm and 31 lines/cm, respectively. Low-atomic-number interspace and grid covers were used in all cases.

[a] PA, postero-anterior projection; AP, antero-posterior projection; LAT, lateral examination.

a substantial improvement in contrast can be achieved, this is at the cost of a significant increase in dose, with the largest dose increases occurring for situations where the scatter is highest. Examination of the values of K and R (Table 2.5) shows that even with a high grid ratio, there may still be a substantial amount of scatter recorded by the image receptor. This is because the grid is poor at rejecting scattered radiation travelling in some directions. For example, for the lateral lumbar spine examination illustrated, the scatter-to-primary ratio with the grid is 0.8.

There are several important practical requirements for the use of a linear focused grid. First, the lead strips may be visible on the radiograph and can partially obscure small image detail. In such circumstances, the grid is usually moved throughout the exposure. The mechanism used to move the grid is referred to as a Bucky or a Potter–Bucky mechanism. Uniform movement of the grid is rarely used. There are three main types of grid movement: reciprocating movement, where the grid moves fast in one direction and slowly in another; oscillating movement, where the grid vibrates back and forth across the film during exposure; and catapult movement, where the grid is accelerated quickly and then decelerated. Where movement is not possible, the grid is known as a stationary grid and higher strip densities and thinner lead strips may be used. The alignment of a focused grid is very important, particularly at higher strip densities. The grid must also be used at the correct focusing distance. Inadequate alignment or positioning will give rise to a reduction in the primary transmission at the edge of the radiation field, known as grid cut-off.

In addition to the linear focused grid, a number of other grid designs are in use. These include the aforementioned parallel linear grids, crossed grids (constructed from two linear grids) and grids with a cellular structure (Rezentes et al. 1999).

2.6.3 Air-Gap Techniques

The introduction of an air gap between the patient and the image receptor leads to a decrease in the number of scattered and primary photons reaching the image receptor per unit time. Both of these quantities will decrease in accordance with the inverse square law, and because the source of the secondary photons is closer to the receptor than the source of the primary photons, the fractional decrease in the number of secondary photons reaching the receptor will be greater than that of the primary photons. As a consequence, the scatter-to-primary ratio R will decrease with increasing separation between the patient and the receptor. This is illustrated in Figure 2.18 for a patient thickness of 20 cm, a field 30 cm in diameter and a 130 kV X-ray spectrum (Persliden and Alm Carlsson 1997).

As well as lowering the ratio R, an air gap will produce a magnification effect, so that the technique can be particularly useful when a fine focal spot and magnification are used to reduce geometrical unsharpness (Section 2.5.5). A good example is magnification mammography. However, magnification imaging is not always a practical option because of the increased size needed for the image receptor. In some situations, the magnification can be reduced by maintaining the air gap and increasing the focus–skin distance. However, this is at the cost of an increased exposure time and tube output requirement. In addition, the scatter rejection achieved with the air-gap technique may not be as good as that achieved using a grid. As a consequence, the air-gap technique at low magnification is only used for scatter rejection in special situations. It finds its main application for chest radiography at high kV (Sandborg et al. 2001, Ullman et al. 2006).

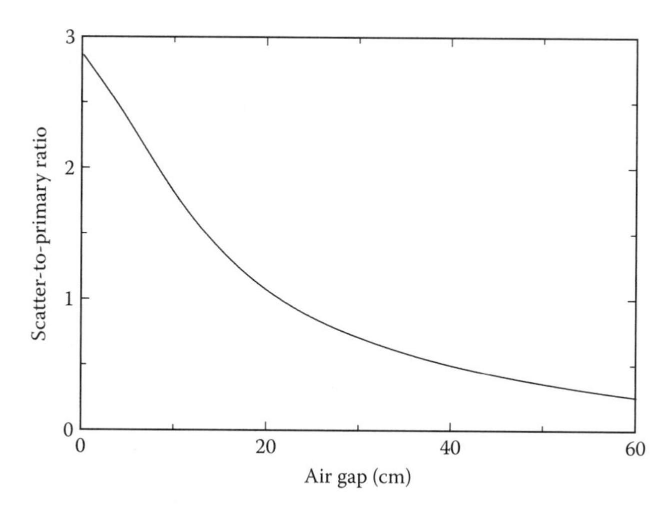

FIGURE 2.18
Dependence of the scatter-to-primary ratio on the size of the air gap between the patient and image receptor. Data are for a patient thickness of 20 cm, a field diameter of 30 cm and an X-ray spectrum at 130 kV with 4 mm aluminium filtration. (Data taken from Persliden and Alm Carlsson, 1997.)

2.6.4 Scanning Beams

Imaging systems that use a scanning X-ray beam and receptor have the important advantage that they have excellent scatter rejection. A collimator is used before the patient to restrict the primary beam to that area necessary to irradiate the receptor and another collimator is used after the patient to reject scatter. This is illustrated in Figure 2.19, which shows the slit-scanning system developed by Tesic et al. (1983) for digital imaging of the chest. The same principle is also used in some systems for digital mammography (Tesic et al. 1997). Projection radiographs produced by CT scanners as an aid to CT slice selection also make use of the scanning-slit principle.

An important disadvantage of scanning systems is that much of the useful output of the X-ray tube is wasted and long exposure times may be necessary. For the method developed

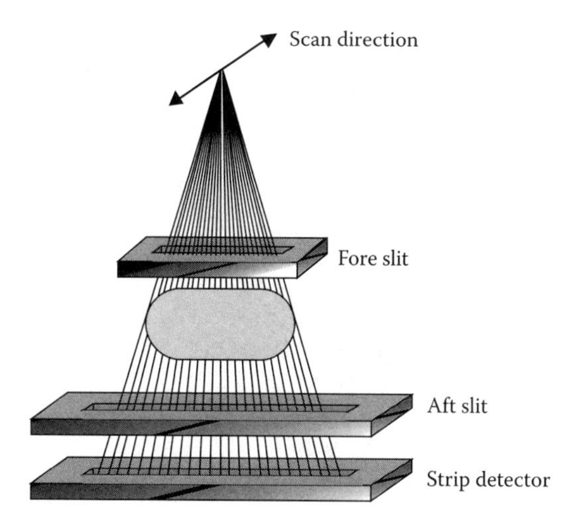

FIGURE 2.19
Slit-scanning system for digital chest radiography.

by Tesic et al. (1983), the exposure time was about 10 s, which decreases the useful working life of the X-ray tube and may create problems with movement. However, it should be noted that, although the total exposure time is long, the time spent in any region of the image is quite short, so that movement losses are much less important than those for a conventional radiograph with the same exposure time.

2.7 Analogue Image Receptors

Many different types of image receptor are used in modern diagnostic radiology. They all form an image by the absorption of energy from the X-ray beam but a variety of techniques are used to convert the resulting energy distribution into something that can be visualised by eye. This section discusses the various analogue image receptors and the physical factors that determine and limit their performance. Digital image receptors, which are now much more widely used than analogue receptors, are described in Section 2.8.

2.7.1 Direct-Exposure X-Ray Film and Screen–Film

Direct-exposure X-ray film is only used in special applications in radiology because it has a low absorption efficiency for photons in the diagnostic energy range. However, film is used as the final display medium for many types of imaging system (in particular, for screen–film imaging where it detects the light emitted following the absorption of X-rays by a fluorescent screen, Section 2.7.2) and we, therefore, study its properties in some detail.

2.7.1.1 Construction and Image Formation

Figure 2.20 shows the construction of a typical direct-exposure (or non-screen) film. The film consists of two photographic emulsions deposited on either side of a transparent polyester or acetate sheet, which is known as the film base. The emulsions are attached to the base by a subbing layer and have a thin surface coating to give protection against abrasion. Each emulsion consists of grains of silver bromide and 0%–10% of silver iodide suspended in gelatin, which prevents the clumping together of the grains, and aids development. The size of the grains depends upon the application but may be in the range 0.2–10 μm.

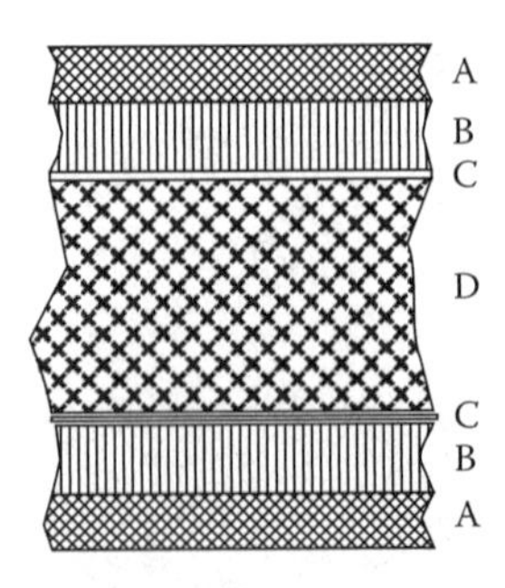

FIGURE 2.20
Construction of direct-exposure X-ray film: A, protective coating; B, film emulsion (20 μm) of silver halide grains in gelatin; C, subbing layer; D, film base (200 μm). (Modified from Barrett and Swindell, 1981.)

The shape and orientation of the grains can have an important effect on the performance of the emulsion. Emulsions with tabular grains are used in some screen–film systems. Such grains present an approximately hexagonal cross section to the incident radiation, but are thin in the direction of the beam. They enable a greater number of grains to be packed into a given emulsion thickness and increase the surface-to-volume ratio. The latter can be important if absorptive dyes are used to improve the response of the emulsion to the wavelengths emitted by the fluorescent screen. The use of two emulsions reduces the development time for a given total thickness of emulsion and is necessary when the film is used in conjunction with a double fluorescent screen (Section 2.7.2.2).

The photographic process is rather complicated and the description given here has been considerably simplified. The reader is referred to Mees and James (1966) for a more detailed treatment. It is the silver halide grains within the emulsion which are sensitive to the light photons or X-rays. This is due to the presence of one or more sensitivity specks, which are formed within each grain during manufacture. For simplicity, we consider the case where the emulsion is exposed with light photons. Following the interaction of a light photon with a silver bromide grain, a bromide ion is neutralised and a mobile electron produced, which can migrate within the grain and become trapped at a sensitivity speck, where it attracts and neutralises a silver ion. This is a reversible reaction, but if the process is repeated and (about) four silver atoms are captured at the sensitivity speck, a stable configuration is reached and the grain is said to be sensitised. A latent image of the pattern of light or X-ray photons is built up of such grains. When the image is developed and fixed, the sensitised grains are converted to silver and the unsensitised grains are removed. It is important to note that, unlike exposure with light photons, one or more grains will be sensitised by the absorption of a single X-ray photon, because of the much larger energy deposition and the release of many low-energy electrons.

2.7.1.2 Characteristic Curve

The blackening on film is usually expressed in terms of the optical density, D, which is defined using the transmission of light through the film by

$$D = \log_{10}\left(\frac{I_0}{I}\right) \tag{2.31}$$

where I_0 and I are the intensities of a light beam before and after passage through the film. The optical density depends upon the number of developed silver grains per unit area, g, and the average cross-sectional area, σ, of a developed grain:

$$D = 0.434\, g\, \sigma. \tag{2.32}$$

This relationship is known as Nutting's law (Dainty and Shaw 1974, p. 41). If we substitute $g = G$, where G is the number of grains present per unit area of the unexposed emulsion, we obtain an expression for the maximum possible density on the film, that is,

$$D_{\max} = 0.434\, G\, \sigma. \tag{2.33}$$

A developed grain diameter of 1.0 μm and a maximum optical density of 3.0, thus, correspond to 9×10^8 grains/cm^2.

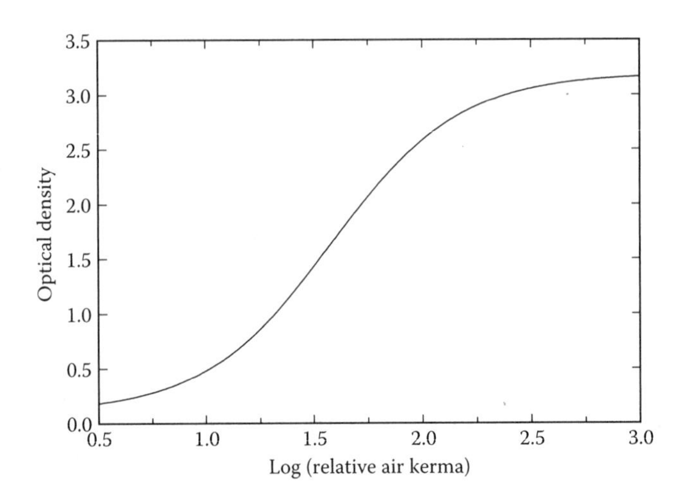

FIGURE 2.21
Typical characteristic curve for a screen–film.

For direct exposure film at low optical densities, the number of sensitised grains per unit area is proportional to the exposure. This is not the case, however, for screen–film where more than one interaction is required to sensitise a grain. For such films the relationship between the optical density and logarithm of the exposure is of the form shown in Figure 2.21.

Curves that relate optical density to film exposure are known as characteristic curves or H and D curves (after Hurter and Driffield). For screen–films, their important features are as follows:

- A background region at zero and low exposures corresponding to the optical density due to the film base and any underlying fog level;
- A toe region for which there is only a small change of optical density with increasing exposure;
- A central region for which the relationship between optical density and the logarithm of exposure (X) is approximately linear

$$D = \gamma \log_{10}\left(\frac{X}{X_0}\right) \tag{2.34}$$

where X_0 is a scaling factor;
- A shoulder region for which there is only a small change in optical density with increasing exposure; and
- A region of saturation where the optical density reaches D_{max} (Equation 2.33).

The shape of the toe region is due to the multi-hit nature of the grain sensitisation process. That of the shoulder region is due to the fact that most of the grains are already sensitised at this high exposure level.

For a small contrast $\Delta X/X$ in the input exposure, we can expand the logarithm in Equation 2.34, and obtain the associated change in optical density ΔD as

$$\Delta D = 0.434\gamma\left(\frac{\Delta X}{X}\right). \tag{2.35}$$

The quantity γ is known as the film gamma and when Equation 2.34 holds, the difference in optical density is proportional to the product of the gamma and the input contrast. Even when Equation 2.34 no longer holds, Equation 2.35 still gives the change in optical density associated with the small contrast $\Delta X/X$, provided the definition

$$\gamma = \frac{dD}{d\log_{10} X} \tag{2.36}$$

is used.

Figure 2.22 shows the dependence of the film gamma (Equation 2.36) on optical density for a typical screen–film. The peak of this curve corresponds to the position of maximum contrast for the film and, at this position, gamma typically takes values of 2–4, depending upon the properties of the film. The film gamma can be increased by increasing the grain size (Equation 2.32), so that the optical density increases more quickly with the number of sensitised grains.

Figures 2.21 and 2.22 both demonstrate that the contrast will be very poor for exposures that result in a low or high optical density. Film, thus, has quite a limited dynamic range. It is important, therefore, that it is exposed so that the regions of interest in the patient produce optical densities for which the film gamma is sufficiently large. The range of exposures for which this is possible is known as the film latitude. High-contrast films with a high gamma will have a narrow latitude, and low-contrast films with a low gamma will have a wide latitude. As far as possible, the contrast and latitude of the film must be matched to the application.

It is clear that the correct exposure of the radiographic film is very important. In most cases, however, it is difficult to judge the correct exposure from the size of the patient and it is advantageous to use automatic exposure control. This is discussed in Section 2.9.

In Section 2.8 it will be seen that with digital receptors it is possible to overcome the limitations imposed by the dynamic range associated with the H and D curve.

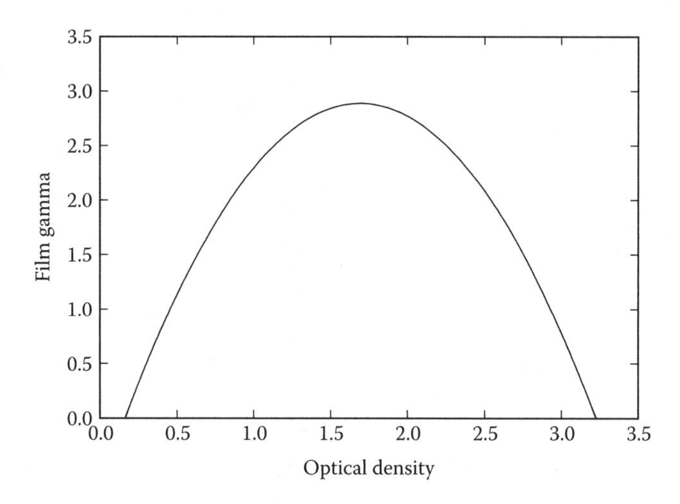

FIGURE 2.22
Dependence of the film gamma on optical density. (Same film as for Figure 2.21.)

2.7.2 Screen–Film Combinations

Screen–film receptors are faster than direct-exposure films but do not have such good inherent resolution. However, when digital receptors are not available, they are the receptor of choice for many radiographic examinations where the dose saving is more important than the loss of very fine detail.

2.7.2.1 Construction and Image Formation

Screen–film receptors produce an image in a four-stage process. An X-ray photon is absorbed by the screen and some of its energy is then re-emitted by the phosphor in the form of light-fluorescent photons. These photons expose the emulsion of a film in contact with the screen and the film is then developed and viewed in the normal way. Figure 2.23 shows the construction of a typical fluorescent screen. The phosphor layer consists of active phosphor particles in a binding material. The average size of the phosphor particles can vary between 2–10 μm, depending upon the application. For general radiography, the phosphor is typically some 70–200 μm thick (Barrett and Swindell 1981) with a total coating of 70–200 mg/cm^2 (based on data in Ginzburg and Dick 1993). The packing fraction of the phosphor within the binder is about 50% and the binder itself is a synthetic polymer (e.g. cellulose acetate or polyurethane).

The active layer of the screen is supported by a sheet of cardboard or plastic and, depending upon the application, the backing layer between the phosphor and this sheet may be reflective to increase light output. Conversely, absorptive dyes may be used to reduce the light output (but improve unsharpness).

2.7.2.2 Unsharpness

The resolution of a screen–film system is limited by the construction of the screen and in particular by the lateral spread of the light-fluorescent photons as they pass from screen to film. This spread increases with increasing separation between the point of emission of the light and the emulsion; good contact between screen and film is essential. This is achieved by exposing the film in a specially designed light-tight cassette.

Fluorescent screens can be used in pairs, with a screen on either side of a double-emulsion film, or singly, with the screen placed behind the film but in contact with the emulsion (Figure 2.24). More X-ray photons are absorbed in the front half of a screen than in the back half and this configuration for a single screen brings the production point for the fluorescent light photons as close as possible to the emulsion. Single screens

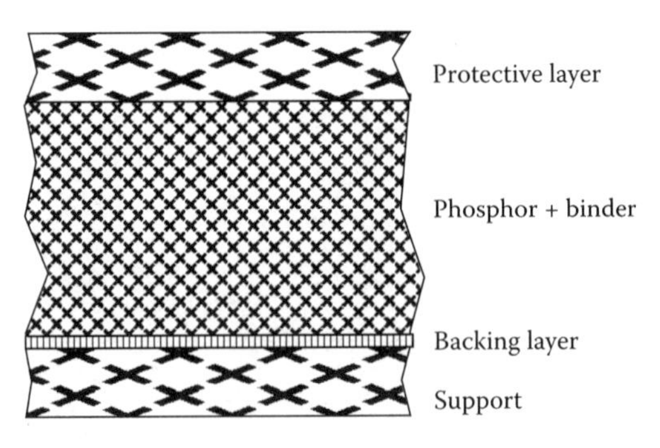

FIGURE 2.23
Construction of a fluorescent screen.

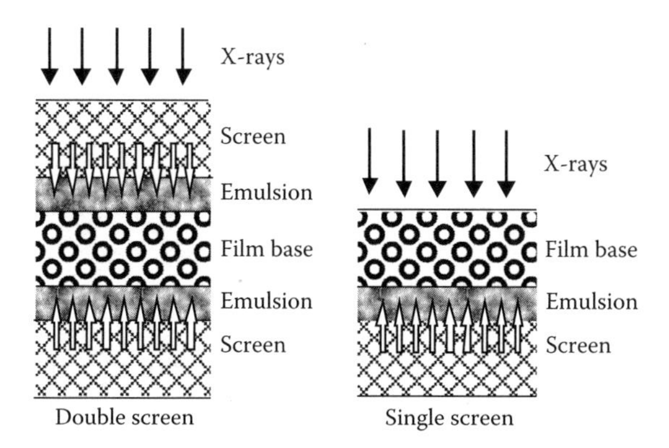

FIGURE 2.24
The use of double and single screens. The arrows at the top of each figure represent the incident X-rays, which interact in the screens, and the arrows originating in the screens represent the light-fluorescent photons, which expose the film emulsion. The lateral spread of the light-fluorescent photons has been omitted for clarity.

are used for applications where the improved resolution is more important than the dose reduction associated with the use of two screens (e.g. mammography).

The unsharpness of the screen–film receptor can be controlled by varying the thickness of the phosphor and by the use of absorptive dyes that favour those light-fluorescent photons produced closest to the emulsion. The effect of phosphor thickness is illustrated in Figure 2.25. This shows calculations of MTF for a screen 90 μm thick and the contributions to the MTF for light emitted at distances of 9 and 81 μm from the emulsion. The data are taken from the work of Van Metter and Rabbani (1990).

There are two further factors which may influence the unsharpness of the screen–film system, blurring from characteristic X-rays and crossover of light-fluorescent photons. Characteristic X-rays may be emitted by a heavy atom in the screen following the photoelectric absorption of the initial X-ray, and may be subsequently absorbed by the screen at some distance from the original interaction point. In such a case, the unsharpness of the receptor may be different below and above the position of the absorption edge of the atom concerned (Arnold and Bjarngard 1979). Crossover of light-fluorescent photons can

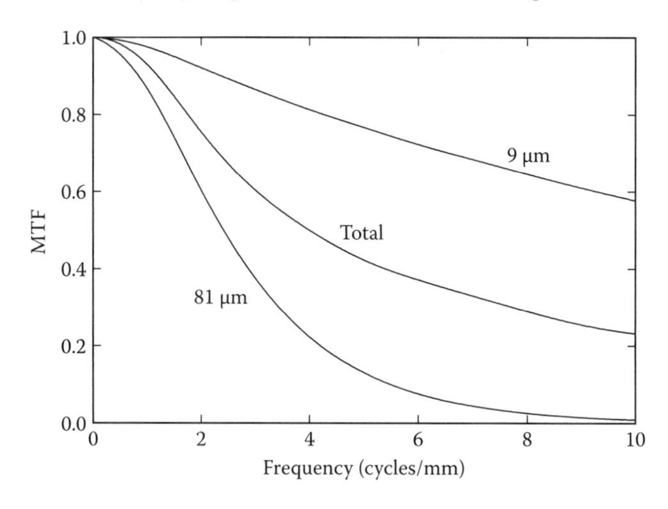

FIGURE 2.25
Effect of phosphor thickness on MTF.

be important for double-screen double-emulsion systems. Light-fluorescent photons produced in one screen pass through the nearest emulsion and expose the emulsion of the other side of the film base. Because of the thickness of the film base, the lateral spread of such photons can be substantial and the reduction in the MTF significant.

2.7.2.3 Sensitivity

The sensitivity of a screen–film receptor will depend upon the X-ray absorption efficiency of the screen(s), the efficiency with which the energy deposited in the screen(s) is converted to light-fluorescent photons, the probability of a light-fluorescent photon reaching the emulsion, the sensitivity of the emulsion and the development of the film. Absorption of the X-rays by the film emulsion contributes only a few per cent of the total film blackening.

Various phosphors can be used in the construction of fluorescent screens, but for simplicity we limit our discussion to calcium tungstate, gadolinium oxysulphide (an example of a rare earth oxysulphide), lanthanum oxybromide and yttrium tantalate. All of these phosphors contain atoms of high atomic number and have a high energy-absorption coefficient. Table 2.7 lists some of their important properties and Figure 2.26 shows the mass attenuation coefficients for three of them. The mass attenuation coefficient for yttrium tantalate is not shown as it is similar to calcium tungstate (similar atomic numbers). The figure demonstrates that the phosphors with the highest atomic number give the highest probability of photon interaction

TABLE 2.7

Properties of phosphor materials.

Phosphor	Activator	K-Edge keV	Light Colour	Energy-to-Light Conversion Efficiency
Gd_2O_2S	Tb	50.2	Green	0.18
LaOBr	Tm	38.9	Blue	0.18
$CaWO_4$		69.5	Blue	0.05
$YTaO_4$	Nb	67.4	UV/blue	0.11

Source: Data for light-conversion efficiencies taken from Curry et al., 1990.

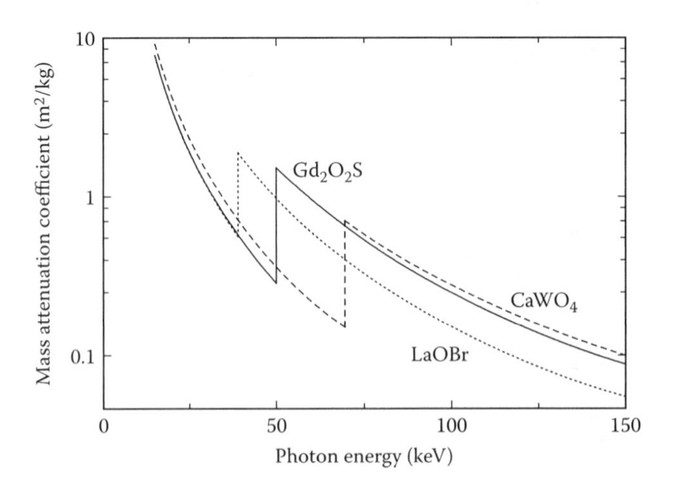

FIGURE 2.26
Variation with photon energy of the mass attenuation coefficient for calcium tungstate (dashed curve), gadolinium oxysulphide (solid curve) and lanthanum oxybromide (dotted curve).

per unit mass, and that the location of the K-absorption edge will have a strong influence on performance. The gadolinium oxysulphide and lanthanum oxybromide phosphors have better X-ray absorption properties in the important energy range of 40–70 keV than the tungstate or tantalate phosphors. However, the advantage is not as large as that suggested by the figure. This is because photoelectric interactions can be followed by the emission of a K-fluorescent photon and such photons may escape from the screen without further energy loss.

The fluorescent properties of the phosphors listed in Table 2.7, apart from calcium tungstate, rest on the addition of a small amount of an activator. For example, for gadolinium oxysulphide about 0.3% of a terbium activator is used. The table shows that the efficiency for conversion of energy absorbed in a screen to light-fluorescent photons is worse for the tungstate screens and this is why they are now little used. The efficiency for the light-fluorescent photons to reach the emulsion varies with screen construction. For the purposes of illustration, we use a value of 0.3 to calculate the number of light photons reaching the emulsion per incident X-ray photon absorbed. If we take the incident X-ray energy to be 50 keV and use a gadolinium oxysulphide screen that emits green light with a wavelength of say 0.5 μm (2.5 eV), then the number of light photons reaching the emulsion is

$$\frac{(50 \times 10^3 \times 0.15 \times 0.3)}{2.5} = 900.$$

The final stage of the image production process is the exposure and development of the film, and it is important here that the screen and receptor be matched. Conventional screen film is sensitive to blue light and for rare earth screens it is necessary to use special film that has molecules of a green-sensitive dye adsorbed on the surface of the silver halide grains.

For most exposures, the density on the film depends only on dose and shows little dependence upon exposure time. However, for very long or for short exposures, some dose-rate dependence is evident. This phenomenon is known as reciprocity law failure and it can be important when long exposures are used. In this situation, a small clump of silver atoms at a sensitivity speck (Section 2.7.1.1) is more likely to be broken up before it becomes large enough to be stable.

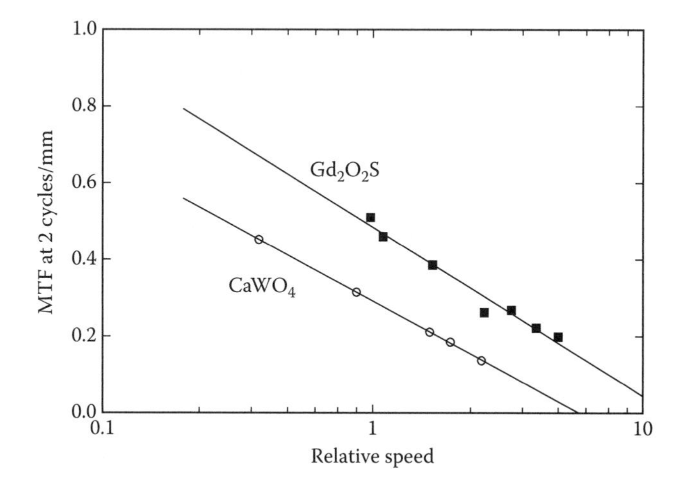

FIGURE 2.27
Relationship between relative sensitivity and resolution for screen–film systems. The MTF at a frequency of 2 cycles/mm is chosen as a measure of resolution. The sensitivity of the screen–film systems is relative to that of the Par/XRP system measured at 80 kV. The solid squares are for systems using Gd_2O_2S screens and the open circles for systems using $CaWO_4$ screens. (Data taken from Doi et al., 1982, 1986.)

We have seen that there are many factors that influence the sensitivity of a screen–film detector and that there is a compromise between speed and unsharpness. For example, the thicker the screen, the greater the speed, but the worse the resolution because of the increased lateral spread of light-fluorescent photons. If a dye is added to reduce the effect of this lateral spread, the speed will decrease. Figure 2.27 quantifies these compromises and shows the relative sensitivity and MTF (at a frequency of 2 cycles/mm) for a range of screen–film systems. The data are taken from Doi et al. (1982, 1986) and clearly demonstrate the dose advantage of using a rare earth screen.

2.7.2.4 Contrast

Screen–film receptors can either enhance or reduce the contrast in the radiographic image. Two factors are at play. Most obviously, the contrast will depend upon the slope or gamma of the film characteristic curve, which can be selected to give wide latitude and low contrast, or narrow latitude and high contrast, depending upon the requirements of the object. This has been discussed in Section 2.7.1.2. Additionally, the contrast will be affected by the efficiency of the screen, which in general decreases with increasing energy but will contain a large discontinuity at the position of an absorption edge (see previous section). As a consequence, the contrast can either be enhanced or reduced by the energy response of the receptor.

2.7.2.5 Noise

The noise in the image produced by a screen–film receptor arises from five principal sources:

(i) fluctuations in the number of X-ray photons absorbed per unit area of the screen(s) (quantum mottle);
(ii) fluctuations in the energy absorbed per interacting photon;
(iii) fluctuations in the screen absorption associated with inhomogeneities in the phosphor coating (structure mottle);
(iv) fluctuations in the number of light-fluorescent photons emitted per unit energy absorbed; and
(v) fluctuations in the number of silver halide grains per unit area of the emulsion (film granularity).

The most important of these are the quantum mottle and the film granularity. Structure mottle contributes typically 10% to the total noise. Noise contributions (ii) and (iv) may be regarded as making modifications to the magnitude of the quantum mottle.

In Section 2.7.1.2 we derived an expression that related optical density to contrast. We now use this expression to calculate the quantum mottle. We average over area A of the image and substitute $(A\varepsilon N)^{1/2}/(A\varepsilon N)$ for $\Delta X/X$ in Equation 2.35. Here ε is the probability that an X-ray photon will interact in the screen, N is the number of photons incident per unit area of the screen and we have used the fact that $(A\varepsilon N)$ has a Poisson distribution. The resulting expression for ΔD_Q, the noise due to quantum mottle, is

$$\Delta D_Q = 0.434\gamma \frac{(A\varepsilon N)^{1/2}}{(A\varepsilon N)} = 0.434\gamma (A\varepsilon N)^{-1/2}. \quad (2.37)$$

This expression neglects noise sources (ii) and (iv). For gadolinium oxysulphide phosphors, these contributions together reduce the DQE by 10%–50%, depending upon photon energy and screen construction (Ginzburg and Dick 1993).

The film granularity can also be estimated using the results of Section 2.7.1.2. We use Equation 2.32 to calculate the fluctuation ΔD_G in the average optical density in area A arising from a fluctuation $(gA)^{1/2}$ in the number of developed grains (gA). After some algebra, the result is

$$\Delta D_G = \left(0.434D\frac{\sigma}{A}\right)^{1/2}. \tag{2.38}$$

This expression neglects fluctuations arising from variation in grain size and sensitivity but gives reasonable agreement with measurement (Barnes 1982).

Figures 2.28a and 2.28b show an example of measurements of the characteristic curve and the gamma and calculations of the quantum mottle and film granularity for a particular screen–film combination. The quantum mottle is proportional to the slope γ of the film characteristic curve (Equation 2.37) and inversely proportional to the square root of the exposure.

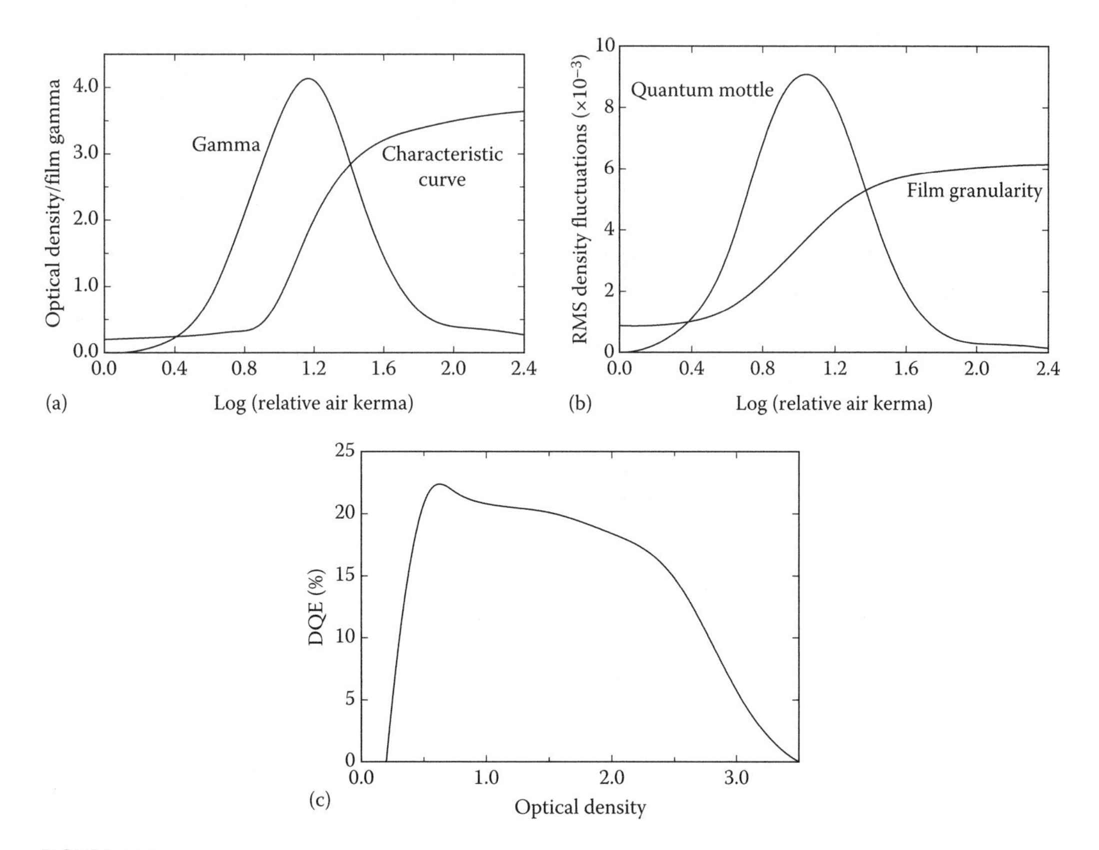

FIGURE 2.28
Measurements and calculations for the DuPont Cronex HiPlus/Kodak X-Omatic RP screen–film combination: (a) film characteristic curve and gamma, (b) calculations of the variation of quantum mottle and film granularity with exposure and (c) variation of detective quantum efficiency with optical density. Noise estimates are for a circular sampling area of diameter 0.5 mm. (Data taken from Barnes, 1982.)

It therefore shows a maximum value in the central region of the characteristic curve and approaches zero in the toe and shoulder regions. The film granularity is proportional to the square root of the optical density (Equation 2.38) and therefore increases with increasing density. As a consequence of these effects, quantum mottle is the dominant effect for a correctly exposed film but is overtaken by the film granularity for regions where the film is either under- or overexposed.

The radiographic noise first increases with density and then decreases, and we must look at the DQE (Section 2.4.3) to see its significance. Figure 2.28c shows the DQE for the screen–film system used in Figures 2.28a and b. The DQE is at most ~20% and is quite close to this maximum for the density range 0.5–2.5. Outside of this range, it falls off rapidly, and this once again illustrates the limited latitude of screen–film combinations and the importance of correct film exposure.

The aforementioned treatment has neglected the effects of image unsharpness on the noise. This is important for viewing small objects at low contrast and is best treated by considering the noise power or Wiener spectrum $W_{tot}(\omega)$ (as mentioned in Section 2.4.3). For fluctuations in the optical density recorded on the film, this is made up of the three noise components we have already identified: quantum mottle, film granularity and structure mottle. Figure 2.29 shows the dependence of the quantum mottle, film granularity and the total noise power spectrum on spatial frequency for a mammographic screen–film. It will be seen that the quantum mottle is dominant at lower frequencies, but the film granularity becomes dominant at higher frequencies, when the higher-frequency information in the incident pattern of X-ray photons is not resolved by the receptor.

If we fix the optical density in the image, and increase the speed of the screen–film system by increasing the useful light output per X-ray photon absorbed or the speed of the film, fewer X-ray photons will be required and the noise and noise power spectrum will increase in magnitude (Equation 2.37). On the other hand, if the light output per X-ray photon and the speed of the film remain fixed and speed is increased by increasing the energy absorption efficiency of the phosphor (e.g. by changing its thickness), the increase in speed will not be accompanied by an increase in noise as the product εN in Equation 2.37 will remain fixed. In practice, there is an overall tendency for noise associated with screen–film

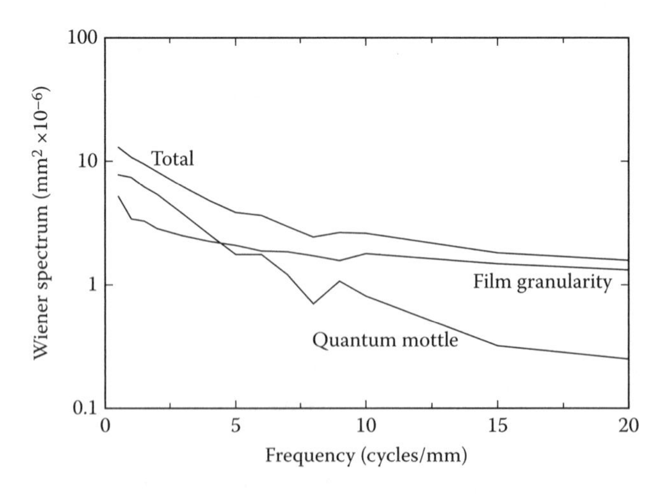

FIGURE 2.29
Contribution of quantum mottle and film granularity to the total noise power spectrum for a mammographic screen–film combination. (Data taken from Nishikawa and Yaffe, 1990.)

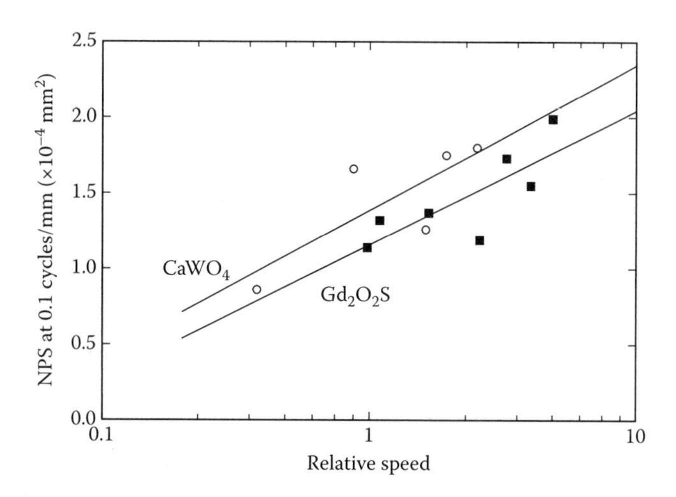

FIGURE 2.30
Dependence of the noise power spectrum (NPS) at 0.1 cycles/mm on the speed of the screen–film system. The solid squares are for systems using Gd_2O_2S screens and the open circles for systems using $CaWO_4$ screens. (Data taken from Doi et al., 1982, 1986.)

systems to increase with speed. This is illustrated in Figure 2.30, which shows the Wiener spectrum at a spatial frequency of 0.1 cycles/mm as a function of speed for a selection of tungstate and rare earth screen–film systems.

It is possible to express the DQE and the NEQ as a function of frequency (ω). They are evaluated as (Sandrik and Wagner 1982, ICRU 1996)

$$DQE(\omega) = \frac{(0.434\,\gamma)^2\,|MTF(\omega)|^2}{W_{tot}(\omega)\,N_{in}}$$

$$NEQ(\omega) = \frac{(0.434\,\gamma)^2\,|MTF(\omega)|^2}{W_{tot}(\omega)} \tag{2.39}$$

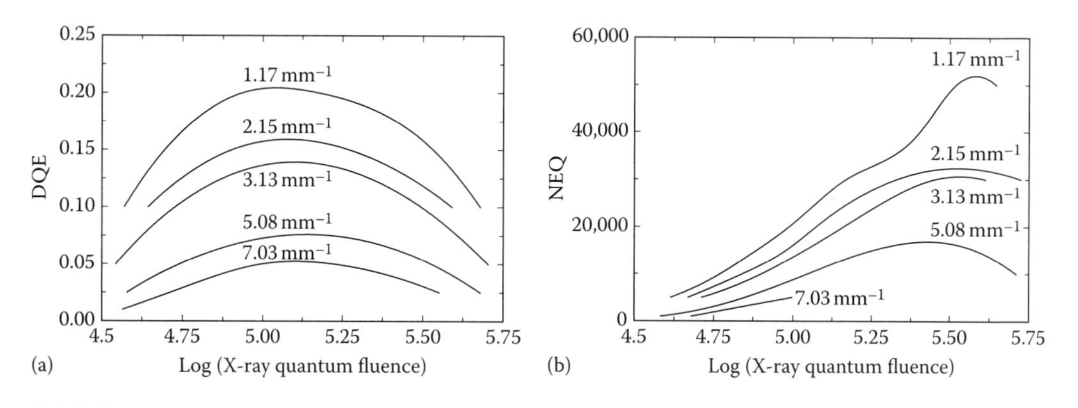

FIGURE 2.31
Dependence of DQE (a) and NEQ (b) on spatial frequency and exposure for a 200 speed class gadolinium oxysulphide screen–film system. Data are shown for spatial frequencies of 1.17, 2.15, 3.13, 5.08 and 7.03 cycles/mm. (Based on data in Bunch, 1994.)

where N_{in} is the number of X-ray photons incident per unit area (see Equations 2.16, 2.19, and 2.20). Determination of these quantities as a function of exposure gives valuable information about the useful exposure range for screen–film systems. Figure 2.31 shows the DQE and NEQ for a particular screen–film combination as a function of frequency and exposure (expressed here as the logarithm of the incident photon fluence). It can be seen that the highest DQE is achieved at a limited range of exposures and only at low frequencies. As the frequency increases, so the DQE falls, thus reflecting the increasing importance of noise contributions other than quantum mottle. It should be noted that the NEQ peaks at a higher exposure than the DQE. This is because the NEQ, being the square of the SNR in the image, is the product of the DQE and N_{in}.

2.7.3 Image Intensifiers

The radiographic image intensifier is a high-gain device for imaging X-ray photons. The radiation dose associated with its use for simple procedures can be low but the corresponding unsharpness and noise are inferior to those associated with screen–film systems. The image intensifier can produce single and serial radiographs with low radiation dose (fluorography) and can also operate in a fluoroscopy mode where the X-ray tube runs continuously, but at very low current. It is particularly valuable for the study of processes that involve movement, flow or filling, for intra-operative control during surgery and for the fluoroscopic control of the insertion of cannulae and catheters. It is used for investigations employing radiographic contrast media and can be used as the source of the image for digital radiology systems. It can also inform four-dimensional radiotherapy planning of moving tissues.

2.7.3.1 Construction and Image Formation

Figure 2.32 shows the construction of a typical image intensifier. It consists of an evacuated tube with an intensifying screen at either end, a photocathode and some electron optics (Curry et al. 1990, pp. 165–174, de Groot 1991). The entry window of the tube should have a high X-ray transmission. It can be constructed from glass but thin metal windows

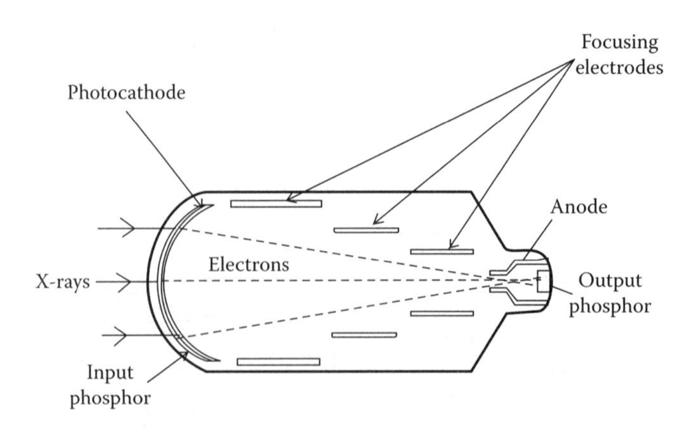

FIGURE 2.32
Construction of an image intensifier showing the conversion of X-rays to an electron stream, which is focused to the output phosphor.

(e.g. aluminium, with a typical transmission of 92%) are now most often used. After passage through the entry window, the X-rays strike a fluorescent screen deposited on the inner surface of the window. The light-fluorescent photons emitted by this screen then hit a photocathode, where they eject photoelectrons. The photoelectrons in turn are accelerated through a potential difference of some 20–30 kV and are guided by electron optics to strike the exit fluorescent screen and produce further fluorescent photons, which can be viewed at the exit window in various ways. The diameter of the input screen will limit the field of view of the intensifier and is typically in the range 12.5–35 cm, although image intensifiers with a 57 cm field of view are available. The diameter of the output screen is typically 2.5 cm and this demagnification of the image coupled with the electron acceleration means that the image intensifier has a very high photon gain. In fact, the increase in the light output intensity of the image intensifier compared with that from a standard fluorescent screen can be as high as a factor of 10,000 due to demagnification and electron optics. The number of photons emitted from the output phosphor is about 50 times the number produced by the input phosphor. This is referred to as the flux gain.

For analogue systems, the image produced at the exit window is recorded on film or viewed with a television (TV) camera (also known as a pickup tube). Digital images may be obtained by digitising the signal from a TV camera or charge-coupled device (CCD) camera connected to the exit window (Webster 1992) (see Section 2.8.3.2). Direct viewing by eye using appropriate optics is also possible but is no longer used. In some applications, the patient is screened* or monitored using the image intensifier and a final image is then captured using analogue or digital means.

Most image intensifiers have caesium iodide input screens. This phosphor has the major advantage that it can be deposited with a packing density of 100%. No binding material is required and a high energy-absorption efficiency is possible. Figure 2.33 shows how the mass attenuation coefficient for the caesium iodide screen varies with photon energy. This coefficient has discontinuities at the iodine and caesium K-edges at 33.2 and 36.0 keV but gives rise to an energy absorption efficiency that can be as high as 0.6. The caesium iodide

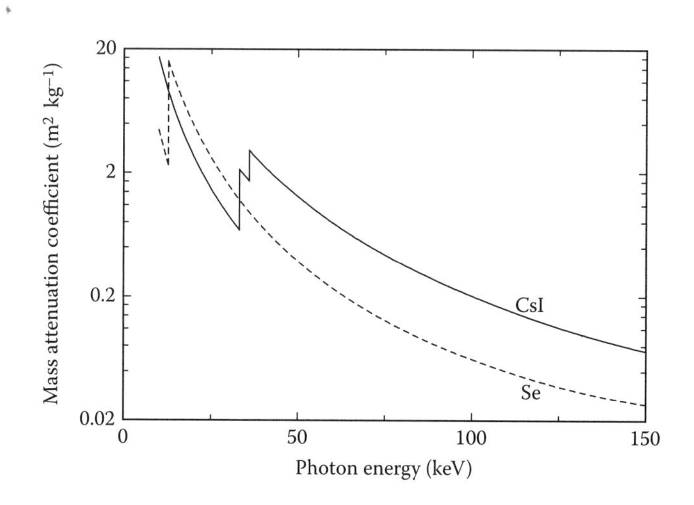

FIGURE 2.33
Variation of the mass attenuation coefficients for caesium iodide and selenium with photon energy.

* The use of this term should not be confused with screening a population to detect disease at an early stage.

screen has the further advantage that its crystals can be aligned with the direction of the X-ray beam, which reduces the lateral spread of the light-fluorescent photons produced within the screen. The spread is also minimised by ensuring close contact between the input screen and the photocathode.

The caesium iodide screen produces about 2000 light-fluorescent photons per incident X-ray absorbed (Nudelman et al. 1982). Their wavelength is matched to the response of the photocathode by the use of sodium doping. The photocathode efficiency is about 0.1, so that 200 photoelectrons are ejected from the photocathode per incident X-ray photon absorbed by the screen. These photoelectrons are then focused and accelerated towards the output screen. The focusing electrodes are designed to reduce the image size with as little distortion as possible but, because of the shape of the input windows, the images may still suffer from distortion.

The output phosphor should have high resolution but will still contribute significantly to the image unsharpness because of the minification of the image. In addition, there will be loss of contrast or glare due to the scatter and reflection of the fluorescent photons produced in the output phosphor, known as veiling glare. This can be reduced by using a tinted-glass exit window. Direct coupling of the exit window to the TV camera using fibre optics can also be used to reduce glare, and this has the added advantage of a greater photon collection efficiency than a lens system.

The unsharpness of the image intensifier will depend upon the degree of minification used. Some tubes are designed to operate at more than one magnification and their response can then be varied depending upon the application, with the unsharpness improving as the input field size is reduced. This is illustrated in Figure 2.34, which shows the MTF for an image intensifier with a large field of view operating at field sizes of 34, 47 and 57 cm. The MTF curves for two screen–film systems have also been included to show the increase of unsharpness associated with the use of the image intensifier.

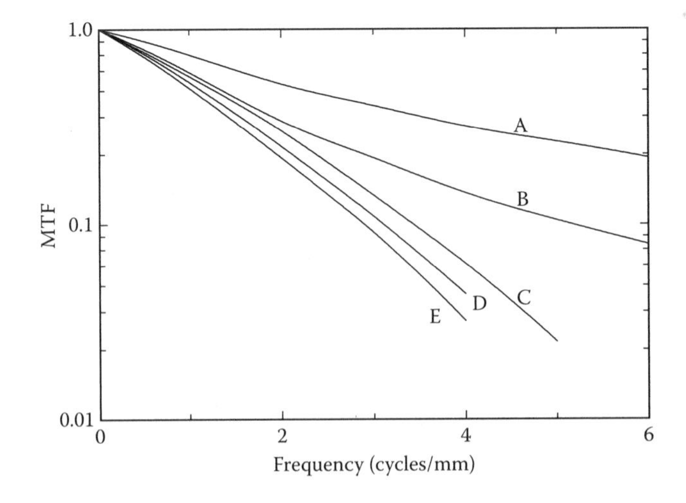

FIGURE 2.34
MTFs for an image-intensifier and for two screen–film systems. Curves A and B are for DuPont Detail and Cronex Par Speed screens, respectively. (Taken from the tabulation of Doi et al., 1982.) Curves C, D and E are for a 57 cm image intensifier at field diameters of 34, 47 and 57 cm, respectively, and were taken from a Siemens data sheet.

2.7.3.2 Viewing and Recording the Image

In this section we discuss the viewing and recording of the image from analogue image intensifier systems. In many situations these have been replaced by digital systems, which will be discussed in some detail in Section 2.8.1. We have seen in the previous section that for analogue imaging, the output from the image intensifier may be recorded on film or may be viewed with a TV camera. Both of these methods can in fact be employed for the same image intensifier by using a partially silvered mirror, which will allow an image to reach both film and TV camera. (In addition, the image may be captured using a screen–film system positioned before the image intensifier. This is known as a 'spot film').

Still photography uses 100 mm cut film and the image size will depend upon the focal length of the camera. The unsharpness will be slightly worse than that of the output screen but there is a dose saving of typically a factor of 5 compared with a conventional screen–film receptor. The dose saving is limited by the quantum mottle, which is large because of the high photon gain of the system, and the film granularity, which becomes more important because of the smaller size of the image. Such films also have some more practical advantages: they are cheaper than conventional film because they contain less silver, and they are easier to store.

A particular advantage of the high photon gain of the image intensifier over the screen–film receptor is the fact that images can be taken with very short exposure times. This is important for imaging in the presence of motion.

The TV camera has the advantage over a film camera in that it produces an image that can be viewed in real time on a TV screen. It performs best when it is directly coupled to the output screen of the image intensifier using fibre optics. The image produced by the TV system can be stored on video disc or tape or used as input to a digital radiology system. The use of a frame store offers the possibility of further dose reduction because the X-ray beam can be switched off whilst the image on the screen is studied.

The choice of TV camera will depend upon the application. There are several types of camera available, but we restrict discussion here to the vidicon and the plumbicon (Curry et al. 1990). The vidicon camera has an antimony trisulphide photoconductor, which has a significant dark current. Its output current is proportional to the 0.7 power of the input light intensity, so it has a gamma of 0.7 and will reduce contrast. Its redeeming feature is the fact that its response is slow. It will, therefore, integrate the light output from the intensifier and reduce quantum mottle. It is well suited to imaging stationary organs but would not be suitable for cardiac studies. The plumbicon camera has a lead monoxide photoconductor in the form of a semiconducting p–n junction. It has a very small dark current, a gamma of 1 and negligible lag. It is well suited to imaging moving organs.

The noise level in the video signal produced by the TV camera and associated preamplifier will depend upon the bandwidth used (or the read-out time per pixel). Nudelman et al. (1982) quoted a noise current of about 2 nA for a bandwidth of 5 MHz and a signal or operating current of 1.6 μA for a plumbicon camera. This gives an idealised SNR or dynamic range of 800:1 and would mean that the noise introduced by the TV camera is negligible compared with the noise in the image due to quantum mottle. In practice, the SNR may be less than this and there may be a transition from quantum to video noise domination with increasing dose rate.

The gain of the system and the dose rate and/or X-ray voltage must be adjusted so that the video-tube current is sufficiently high that electronic noise is unimportant and is sufficiently low that the tube does not saturate. The overall gain of the system can be controlled at the design stage and during the exposure (by varying the aperture of the optical coupling between intensifier and TV camera), but the design of a system that can accommodate both fluoroscopic and radiographic dose rates can be difficult.

Although the TV camera can reduce contrast, the TV monitor will enhance the contrast in the image because it has a gamma of about 2 (Curry et al. 1990), but the use of the TV system will result in a loss of resolution because of the line-scanned nature of the image. The resolution of the TV system has to be considered in terms of its vertical and horizontal components. The vertical resolution V (lines per image) can be expressed in terms of the actual number of image scan lines N via the relation

$$V = kN. \tag{2.40}$$

The parameter k is known as the Kell factor and is usually taken to be 0.7 (Moores 1984). For an input diameter of 20 cm and a 625-line TV system, the smallest vertically resolved element would be $200/(625 \times 0.7) = 0.45$ mm. For a 1250-line system the resolution would be 0.22 mm. The horizontal resolution of the system is limited by the bandwidth. The number H of horizontal resolution elements across the picture is given by

$$H = \text{bandwidth}/(\text{frames/s} \times \text{lines/frame}). \tag{2.41}$$

If the bandwidth is selected to make H and V equal and we use 625 lines with 25 interlaced frames per second, we find that we need a bandwidth of 6.8 MHz. (In an interlaced TV, the image is scanned in two passes with alternate lines being skipped on each pass.) Doubling the number of lines would increase the bandwidth by a factor of 4 and would also increase the video noise in the image, as noted previously.

2.8 Digital Image Receptors

Screen–film-based radiographic imaging systems record and display data in an analogue fashion with a continuous range of possible optical densities up to some limiting value. They have strict exposure requirements because of the narrow exposure latitude of film and offer little possibility of image processing. By comparison, imaging systems that capture radiographic information in a digital form often have a very wide exposure latitude and allow the image to be processed and displayed in a variety of ways. Such systems are more expensive to purchase than screen–film systems but may not have the consumable overheads associated with the use of film.

Digital systems can enhance the displayed image using techniques such as windowing, pan/zoom, contrast enhancement, histogram equalisation and edge enhancement (see Chapter 12). Digital image processing techniques are generally applied to the acquired images before they are displayed (e.g. logarithmic transformation to correct for exponential X-ray attenuation). Other digital processing techniques are also necessary to overcome limitations associated with the manufacturing process for individual image receptors. These include flat-field corrections, to overcome spatial-sensitivity variations, and median filtering, to remove the effects of 'bad' pixels.

The use of digital systems is now common as hospitals realise the value of digital image capture and storage. Such systems also have found particular advantage where automatic exposure control (AEC) is not available, for instance, mobile radiography, and where low dose is of particular importance such as paediatric imaging. As well as systems suitable for general

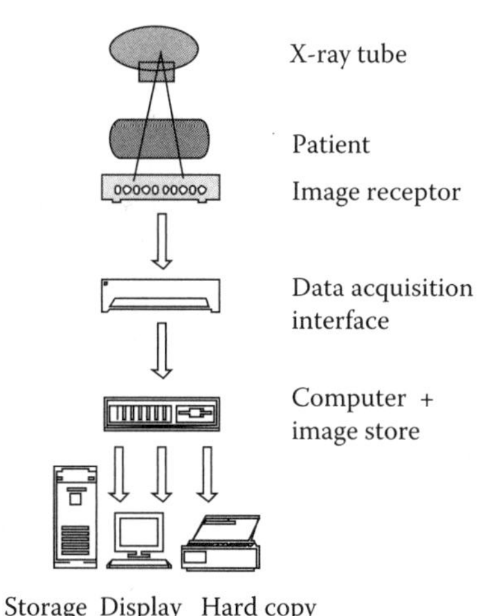

FIGURE 2.35
Typical components of a digital radiology system.

radiology, dedicated digital imaging systems are available for specialised applications such as chest radiography (Section 2.11.4), mammography (Section 2.11.2) and cardiology.

Figure 2.35 shows the components of a typical digital radiographic system. The image receptor captures information that is digitally stored on a computer. The processed image (soft copy) is displayed on a TV monitor at the operator's control console and on the radiologist's reporting workstation. The digital image is usually transferred to the hospital PACS for storage. If required, a film-writing device can be used to make a permanent analogue copy (hard copy) of the image. Information from the digital image can be used to provide feedback to the X-ray system, and hence to control the exposure. This is possible for both single exposures (where a brief pre-exposure may be possible) and series of fluoroscopic or fluorographic images.

In general, digital-imaging systems will have inferior resolution to the equivalent analogue system, such as the screen–film receptor or the image intensifier with analogue image capture, but, provided the matrix size and image receptor are matched to the application, this will not be of clinical significance. For example, a 512 × 512 image may be sufficient for digital fluoroscopy whereas a digital system for planar imaging might require a 2048 × 2048 matrix and a pixel size of 0.1–0.2 mm or smaller. The pixel depth in the image will also depend upon the application. An 8 bit analogue-to-digital converter (ADC), giving a precision of 0.4%, will be adequate for noisy images or a large image matrix (smaller pixels are noisier), but 10 bit conversion (0.1% precision) or higher is appropriate for most applications. In addition, the pixel depth and system gain must be balanced to ensure that the required precision is achieved over the whole grey-scale range of the image. One effect of this is to increase the pixel depth needed.

The size and depth of the image matrix will affect the data storage requirements, and image compression may be necessary for storage of large numbers of high-resolution digital images.

2.8.1 Digital Fluoroscopy

The use of an image intensifier for analogue imaging has been described in Section 2.7.3. For digital fluoroscopic imaging, the output of the pickup tube can be suitably amplified and sampled using an ADC. Alternatively a CCD camera (Section 2.8.3.2) can be used instead of the pickup tube. Both approaches are well established for digital fluoroscopy and fluorography. CCDs are generally smaller and more robust than pickup tubes, they have almost no lag and generally suffer from higher noise (Gould 1999).

The use of image-intensifier systems for digital imaging has led to important opportunities for dose reduction as well as the instant playback of fluorographic runs and digital subtraction angiography (DSA) (Section 2.11.1.1). The system may also be capable of producing hard film copy. A dose saving can be achieved using slower frame rates for fluoroscopy and by making use of frame freezing or frame grabbing to reduce the exposure time. A very important feature of digital fluorography is that the images are available immediately.

Digital image-intensifier systems are often designed for specific applications. For example, for cardiac systems, the field of view of the image intensifier is typically 17 cm, giving a pixel size of 0.17 mm for a 1024-line TV system. A powerful X-ray tube will be required to produce sufficient output to acquire adequate quality images for later viewing. Larger field sizes (e.g. 40 cm field of view) are required for peripheral angiography. In this case, a 1024-line TV system will give a pixel size of 0.4 mm.

An advantage of image intensifiers is that they are quantum-limited devices (Yaffe and Rowlands 1997), but the images suffer from non-uniformity and distortion, as well as the presence of veiling glare (Section 2.7.3.1). Image intensifiers are also very bulky. There is, thus, great interest in using real-time digital detectors such as flat-panel detectors (Section 2.8.4), which can be designed to have rapid read-out of the image. Such systems do not require any optical coupling and, thus, avoid the losses in image quality associated with this stage in the production of the image from an image intensifier.

2.8.2 Computed Radiography

The receptors we have studied so far fall into two categories: those producing an image that is immediately transferred to another medium and those with a storage capacity themselves. The system, known as computed radiography (CR), makes use of a photostimulable phosphor and falls into the latter category. It was developed by the Fuji Photo Film Company (Sonoda et al. 1983) and is still a widely used technique for digital radiography.

CR systems have a wide dynamic exposure range approximately from 0.1 μGy to 1 mGy (10,000:1) and do not suffer from reciprocity law failure. This improvement in dynamic range is an important advantage over screen–film receptors and is particularly valuable in mobile or bedside radiography, where AEC may not be available.

The CR image receptor is a flexible plate coated with a photostimulable phosphor. Such phosphors can store the energy absorbed from the incident X-ray beam in quasi-stable states and emit this energy in the form of light photons when stimulated by visible or infrared radiation of an appropriate frequency. The requirements for this phosphor are that it should have a high X-ray absorption efficiency and a high light output per unit energy absorbed. The read-out and emitted light must be of different frequencies. The read-out light must be able to penetrate the full depth of the phosphor and the emitted light must be of a suitable frequency for monitoring with a photomultiplier tube. The response time of the phosphor should be sufficiently short to allow rapid read-out of the image but the lifetime of the latent image should be sufficiently long to allow read-out after a delay of tens of minutes. Europium-activated barium fluorohalide compounds ($BaFX:Eu^{2+}$ where X is either Cl, Br or I and Eu is the dopant) match these requirements well and are the basis of the stimulated luminescence receptors in present use.

The mechanism for stimulated fluorescence in the barium fluorohalide phosphor is as follows (Figure 2.36 and Rowlands 2002). Each phosphor crystal has colour centres

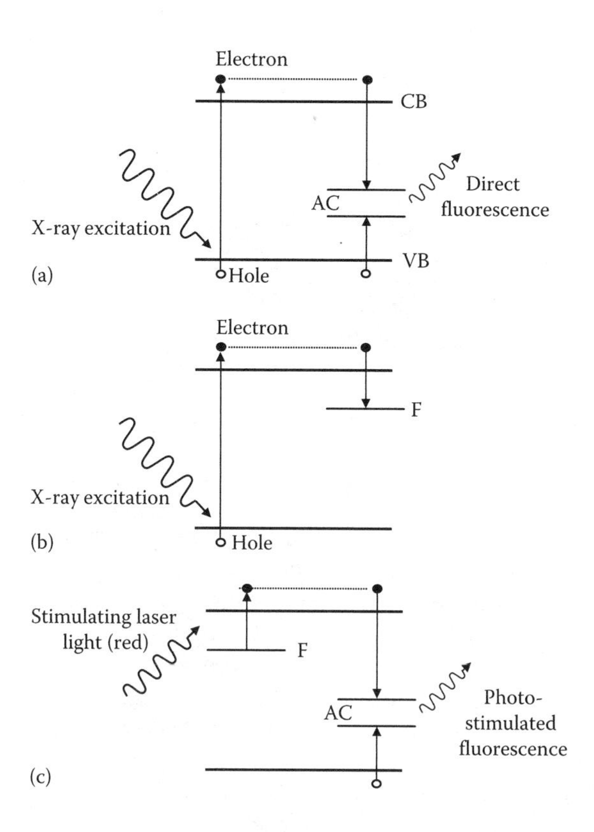

FIGURE 2.36
Processes involved in the exposure and read-out of a CR plate: (a) X-ray excitation (charging) of the storage phosphor. X-rays lift electrons from the valency band (VB) into the conduction band (CB) from where about half fall into the deeper recombination centres (activators or luminescence centres, AC) and spontaneously emit light. (b) The rest of the electrons are caught in electron traps (F or colour centres, F). (c) Optical readout (discharging) of the storage phosphor. The phosphor is irradiated with red laser light, which raises the trapped electrons back into the conduction band. About half these electrons are re-trapped and the rest fall into lower energy levels emitting photo-stimulated light at the same time.

(F centres) (McKinlay 1981) and luminescence centres (Eu ions). When the plate is irradiated, the energy absorbed from the incident X-rays produces electron-hole pairs. The holes can become trapped at the luminescence centres and the electrons and holes can then recombine resulting in spontaneous luminescence. The phosphor is chosen so that this recombination occurs with a probability of about 50%. If the recombination were lower, the probability of retrapping during read-out would be higher. The remaining 50% of electrons are trapped at F-centres in the valence band. In this way a latent image of trapped electrons is built up. This image is semi-stable with a loss of typically 25% in 8 h. When the plate is irradiated with light of the appropriate frequency, the trapped electrons are raised to the conduction band and can recombine with holes producing light emission. The optimum light output is achieved when the probabilities of the excited electrons being retrapped or stimulating fluorescence are balanced. This results in a 25% conversion efficiency compared to the same phosphor without traps (Yaffe and Rowlands 1997).

The image plate is constructed from phosphor particles in an inorganic binder. The phosphor particles are 5–10 μm in size. Larger particles have a greater light-emission efficiency, but scatter the read-out light more and will lead to greater structure mottle.

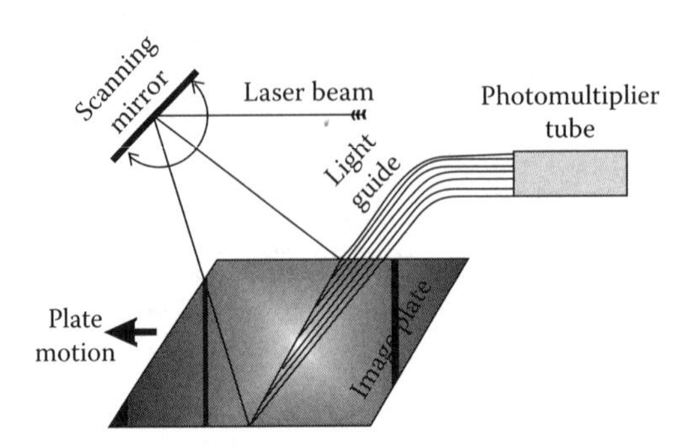

FIGURE 2.37
Schematic diagram showing the basic components of a CR system for reading a photostimulable image plate. (Modified from Sonoda et al., 1983.)

Typical image-plate thicknesses are in the range 140–210 μm. A backing layer protects the plate from damage during transfer and storage.

Figure 2.37 shows the technique used to read out the image plate. The phosphor is scanned using a laser beam, which is deflected by an oscillating mirror so as to pass back and forth across the plate. The plate is advanced after each traverse of the laser beam so that the complete image is read out in a raster pattern. The laser beam has a typical spot size of 0.1 mm and the image is sampled at 5–10 pixels/mm. The radiation emitted by the plate is collected by a light guide, which is coupled to a photomultiplier tube. The resulting signal is digitised, using an ADC, and stored. The plate is read out in about 1 min, with a typical read-out time per pixel of 14 μs. For early system designs using 10 bit ADCs, the plate was first pre-scanned at 256 × 256 and low intensity. The resulting image was then segmented into unattenuated beam, collimated beam and useful image. The distribution of the useful image readings could then be used to select the sensitivity and latitude of the digitisation system to allow optimisation of the electronic circuitry. In modern designs, typically using 12 bit ADCs, the range of amplification is decided by the amount of photostimulated luminescence expected from the clinical study under investigation. The resulting images can be processed to adjust the latitude and gamma of the displayed image, and may also be processed to aid interpretation (e.g. unsharp masking). Hard copy is typically produced by writing the image to a fine-grain single-emulsion film using a laser printer at a typical pixel size of 100 μm. Laser printers that use a dry process are also available. Large images may be reduced in size first.

The factors that limit the resolution of the image plate are quite different to those that limit the performance of screen–film systems. Scatter of emitted light does not affect resolution, which is determined by the size of the read-out light beam, the sampling interval and the scatter of the read-out light. The more light emitted from each read-out volume collected, the less the image noise and the greater the receptor DQE. However, scattering of the read-out light will increase the read-out volume for a given pixel location and, hence, will increase image unsharpness. Because of this, a black support layer is used to reduce light reflections. Spread of read-out light can also be decreased by reducing the thickness of the phosphor or using dyes to attenuate the scattered light. Both of these approaches will make the system less sensitive. A typical CR system requires about two times the exposure compared with a 400-speed screen–film system. Figure 2.38 shows examples of MTFs for a CR system and for slow and medium speed screen–film systems. It can be seen that the MTF for the CR system is comparable to that of the medium speed screen–film system and is poorer than that of the slower speed system designed for high detail use. In practice, the MTF for CR systems

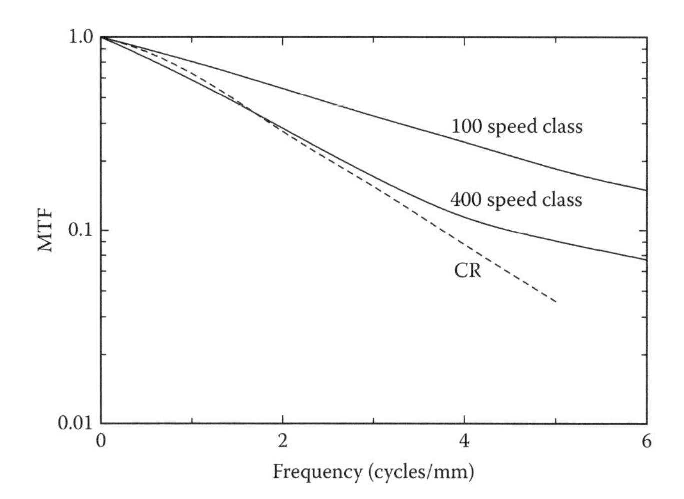

FIGURE 2.38
MTFs for the Fuji ST-V CR plate and for slow (100 speed class) and medium speed (400 speed class) screen–film systems manufactured by Konica. (Data taken from Medical Devices Agency (MDA), 1997, 1998.)

measured in directions along and across the image plate will show small differences due to the asymmetry of the read-out configuration. It should be noted that the MTF for the CR system shown stops at 5 cycles per mm, which is the inverse of twice the sampling interval (the Nyquist frequency) of 100 μm. No information is available above the Nyquist frequency and the MTF is artificially enhanced close to this frequency if under-sampled.* For an image plate with 200 μm sampling, the Nyquist frequency is 2.5 cycles per mm.

Developments in CR systems include the replacement of the black support layer with an optically clear backing material. This means that the fluorescent light emitted in both the forward and backward directions can escape the image plate. The use of a dual-sided read-out system and the detection of the light emitted from both sides of the receptor results in two images, which can then be combined. Such a system offers a reduction in noise and, hence, an improved DQE and reduced patient dose (Fetterly and Schueler 2006, Monnin et al. 2006). In addition, a thicker phosphor layer can be used, resulting in increased X-ray absorption.

At present, there is no standard way of expressing the speed of digital X-ray image receptors. (For information on the relationship between exposure and pixel values for a number of CR manufacturers see Samei et al. 2001). The light output of the phosphor per unit incident air-kerma will depend upon the photon energy, the plate thickness and composition, the depth distribution of the latent image and the image read-out parameters. The associated image noise will receive contributions from quantum mottle; image plate structure mottle; reader noise; digitisation noise and, when appropriate, film granularity. The relative importance of the various noise sources depends upon exposure conditions. At low dose quantum mottle is important (the low-frequency Wiener spectrum is then proportional to incident air-kerma) whilst at high dose other noise sources become significant. Of great importance in the comparison with screen–film imaging will be the DQE and its variation with exposure level and spatial frequency. Figure 2.39 shows the DQE for the same CR system as Figure 2.38 for three values of the air-kerma incident on the cassette. The DQE for

* The digital MTF illustrated is a pre-sampling MTF, which does not show the effects of aliasing (Dobbins, 1995). See also Section 2.4.2.

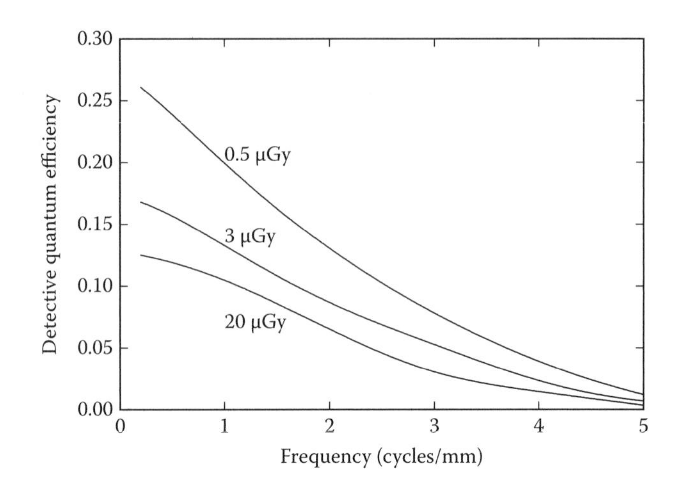

FIGURE 2.39
DQE for a Fuji ST-V CR plate digitised at a pixel size of 100 μm and incident air-kermas at the cassette of 0.5, 3 and 20 μGy. Data were measured using a 75 kV X-ray spectrum. (Data taken from Medical Devices Agency (MDA), 1997.)

CR can be better than for screen–film at low frequencies; screen–film DQE can be better at high frequencies, but only when the receptor is correctly exposed. For an experimental comparison of detector performance for CR systems see Samei and Flynn (2002).

2.8.3 Receptors Using Phosphor Plates and Charge-Coupled Devices

2.8.3.1 Phosphor Plates

We saw in our discussion on screen–film systems (Section 2.7.2) that fluorescent screens can be used as the basis of an analogue imaging system with good resolution and reasonable DQE. They can also form the basis of digital receptors with similar or better performance by coupling their light output to suitable digital detectors. Two types of detector are used: a CCD, which is described here, and an active-matrix amorphous-silicon semiconductor, which is described as part of Section 2.8.4.

Figure 2.40 shows two examples of phosphors that can be used with CCDs. Caesium iodide will give better resolution at the same phosphor thickness because its 'cracked' structure channels the fluorescent photons emitted by the phosphor towards the CCD and

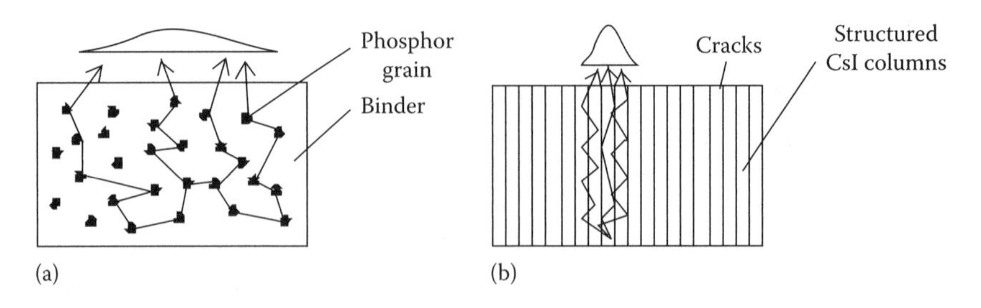

FIGURE 2.40
Schematic diagrams of two phosphors suitable for use with CCDs and in other digital detectors: (a) shows a fluorescent screen comprising phosphor grains (for example Gd_2O_2S) embedded in a binder, (b) shows a CsI phosphor, which has a 'cracked' structure which limits the spread of emitted light. (Based on Yaffe and Rowlands, 1997. With permission from IOPP.)

reduces the lateral spread which is possible with a fluorescent screen such as gadolinium oxysulphide. The number of light photons produced per X-ray absorbed in gadolinium oxysulphide is greater than that for caesium iodide. The two phosphors require on average 13 and 19 eV, respectively, per light photon emitted (Yaffe and Rowlands 1997). There will be a loss in DQE due to the energy distribution of the light-fluorescent photons, which is given by M_1^2/M_0M_2, where M_i is the ith moment of the light photon-energy distribution. This is known as the 'Swank factor' (Swank 1973).

2.8.3.2 Charge-Coupled Devices

A CCD is an integrated circuit consisting of a series of electrodes or 'gates' deposited on a semiconductor substrate. These form an array of metal-oxide-semiconductor (MOS) capacitors. When a voltage is applied to the gates, the semiconductor immediately below is depleted to form charge storage wells or 'buckets'. The CCDs form a mosaic structure varying from less than 256 × 256 up to 2048 × 2048 elements. The charge liberated by the photoelectric absorption of the incident optical quanta is captured in the 'buckets' and the distribution of electrons in the buckets forms the image. The image is read out electronically by passing the charge from 'bucket' to 'bucket' along lines of the array and digitising the signals reaching the ends of the lines. The overall transfer efficiency is related to the nth power of the bucket-to-bucket transfer efficiency, where n is the number of elements in each line. The efficiency of each transfer is necessarily extremely high (of the order of 0.999999). Read-out times of about 30 frames/s are possible with devices up to 1000 × 1000 pixels (Yaffe and Rowlands 1997).

The size of a CCD array is limited by practical engineering considerations to a maximum dimension of about 5 cm. Separate arrays of CCDs may be 'patched' together to provide larger imaging areas, but the 'stitching' between the patches may be problematical due to differences in the electronics between each array and 'patch' lines may be visible in the image. An alternative method of imaging areas larger than the area of a single CCD array is to employ demagnification techniques. This can be achieved using a lens or a fibre-optic taper between the phosphor and the array. However, the light loss associated with a lens coupling is much greater than that for coupling with a fibre-optic taper at the same demagnification (Figure 2.41). The DQE of systems using the latter approach can, therefore, be expected to be better.

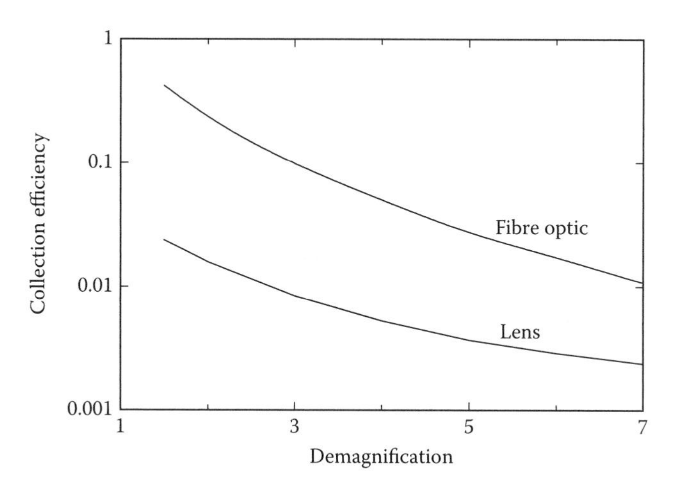

FIGURE 2.41
Light collection efficiency for optical and fibre-optic coupling. (Based on data in Hezaji and Trauernicht, 1996.)

2.8.4 Flat-Panel Receptors Using Active-Matrix Arrays

Flat-panel receptors contain large numbers of thin-film field-effect transistors (TFTs) arranged in a large matrix. The TFTs are 'active' devices and the arrangement is, therefore, known an active-matrix array (AMA) (Rowlands and Kasap 1997). The potential of AMAs was first realised in the early 1990s when the use of hydrogenated amorphous silicon (a-Si:H) for large-area devices became technologically possible. Active-matrix liquid-crystal displays (LCDs) constructed from amorphous silicon are widely used in notebook computer displays. The same technology is used to provide the image capture and read-out function for digital X-ray image receptors (Rowlands and Yorkston 2000, Yaffe and Rowlands 1997). Amorphous-silicon active-matrix arrays can be configured as an alternative to a CCD for the detection of light emitted by a fluorescent screen (Section 2.8.4.1) or to measure the charge liberated by X-rays interacting in a photoconductor (Section 2.8.4.2). This technology offers improved static and real-time imaging.

2.8.4.1 Systems Using Fluorescent Screens

Figure 2.42 shows a cross section and plan view of a receptor using a caesium iodide phosphor and an AMA. The layer of caesium iodide is evaporated directly onto the active matrix. Each pixel in the AMA is configured as a photodiode, which converts the light-fluorescent photons emitted by the phosphor to electrical charge (Rowlands and Yorkston 2000). Part of each pixel area is configured as a TFT for read-out purposes. All TFTs in rows have gates connected and all TFTs in columns have their sources connected. When gate line i is activated, all TFTs in that row are turned on. The N source lines from $j = 1$ to N then read the charges on pixels in row i. These parallel data are multiplexed into serial data, digitised and fed into a computer. The scanning control then activates the next row, $i + 1$, and so on.

As the area of each pixel must accommodate the TFT and read-out control circuitry, there is some loss of efficiency. The fraction of the area of each pixel that is available for signal collection is known as the 'fill-factor'.

2.8.4.2 Systems Using Selenium Photoconductors

The use of a photoconductor in conjunction with an AMA allows the photoconductor to directly produce electrical charge following the interaction of an X-ray photon, and

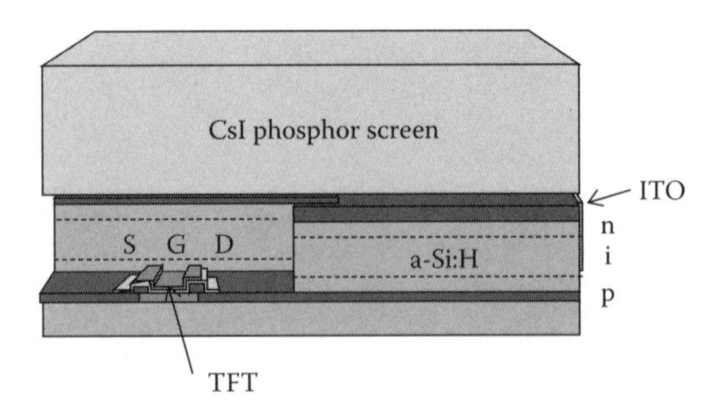

FIGURE 2.42
Schematic diagram of an amorphous silicon detector showing the thin-film transistor (TFT) and the n-i-p a-Si:H photodiode. The photodiode is connected to a transparent conductive indium tin oxide (ITO) electrode at a common bias of ~5 V and to the drain (D) of the TFT. The potential on the gate (G) acts as a switch to connect the photodiode to the source (S). The amorphous silicon is coupled to a CsI phosphor. (Modified from Yaffe and Rowlands, 1997.)

without the need to produce fluorescent-light photons (Rowlands and Yorkston 2000). For this reason, such systems are referred to as direct systems as opposed to indirect systems using a fluorescent screen as described previously.

The general requirements of an X-ray photoconductor are high X-ray absorption efficiency; small amount of energy (W) to create a single free electron-hole pair; and negligible dark current (which conflicts with low W). Electrons and holes need to be able to travel farther than the thickness of the detector; otherwise they could be trapped before reaching the electrode. The mean distance travelled by a charge carrier along the applied electric field before being trapped is referred to as the 'Schubweg'.

The need for a high X-ray absorption efficiency implies that a detector with a high Z is required (the photoelectric contribution to the linear attenuation coefficient is proportional to Z^3, see Section 2.3). In order that most of the X-rays interact with the detector, its thickness should be sufficiently greater than the inverse of the linear attenuation coefficient (known as the absorption depth). There may, thus, be a conflict between the requirements of high photon absorption and of the Schubweg.

Amorphous selenium (a-Se) is a photoconductor, which provides a suitable compromise between these conflicting requirements. Being amorphous (i.e. non-crystalline), it can be spread over large areas. It has a low melting point (220°C) and a low glass transition point compared with other photoconductors. These properties make it suitable for the deployment with an AMA. The a-Se can be easily deposited on the AMA by thermal evaporation in a vacuum providing a coat 200 μm thick in under an hour. The Schubweg is 3–6 mm for holes and 1–2 mm for electrons at an operating field strength of 5 V/μm (Rowlands and Kasap 1997). The efficiency with which charge is produced depends upon the electric field strength in the selenium, but a typical value is 50 eV per electron-hole pair. The absorption depth is 0.05 mm at 20 keV, and 1 mm at 60 keV. Figure 2.33 compares the mass attenuation coefficient for selenium with that for CsI. Selenium has a relatively low atomic number (34) and its mass attenuation coefficient is much lower than that for CsI above 36 keV.

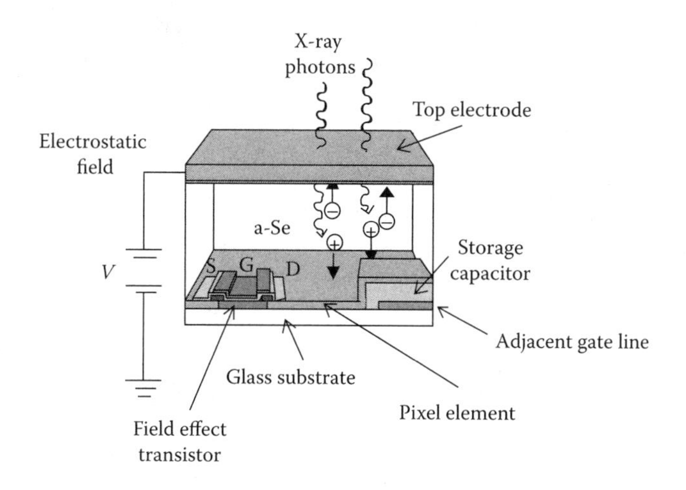

FIGURE 2.43
Schematic diagram of a pixel element of an amorphous selenium (a-Se) area detector. A layer of a-Se is evaporated directly onto an active-matrix array. The charge created by irradiating the detector with X-rays is collected by the capacitor. S, G and D are explained in the caption to Figure 2.42. (Modified from Yaffe and Rowlands, 1997.)

In the production of the flat-panel detector, an electrode is deposited on top of the a-Se layer to provide a biasing potential (Figure 2.43). Electron-hole pairs travel along the field lines and charge reaching the amorphous Se is stored in capacitors within each pixel. Because the movement of the electrical charge within the selenium is along the direction of the electric field, there is very little lateral spread of image information as the charge is moved from its production point to its measurement point. There is thus potential for both excellent spatial resolution and high DQE for lower energy spectra.

2.8.5 Photon Counting Detectors

All of the image receptors described so far record a signal whose magnitude depends upon the amount of energy deposited by each photon that interacts. This will be different for each individual photon because of variations in three quantities: the incident X-ray photon energy, the nature of the photon interaction in the receptor and the efficiency with which the energy initially absorbed is converted to an electrical signal. As a consequence, the DQE will always be less than the probability of interaction in the receptor. An exciting innovation is the development of digital image receptors, which have the capability of counting the individual photons incident on each pixel in the image. Such receptors have zero or very small electronic noise and a DQE which is higher than that of an equivalent receptor which produces a signal which is energy dependent. Such a system is available for mammography and offers a significant improvement in DQE (see Section 2.11.2).

2.8.6 Advantages of Digital Imaging Systems

The advantages of digital radiographic systems may be divided into four classes: image display, dose reduction, image processing and image storage and retrieval (see Section 14.3.2).

We first consider advantages associated with the image display. Probably, the most important is the fact that the brightness and contrast of a digital image on a TV screen or the density on a film can be controlled. For example, every film produced from a digitally recorded image can be properly exposed and can have a characteristic that is properly matched to the range of pixel values in the image. Alternatively, the full range of density or brightness can be used to display just part of the range of pixel values, thereby enhancing the contrast in a region of interest (this technique is known as windowing). Algorithms are available for controlling the displayed (analogue) image so that best use is made of the capabilities of the display. For example, the histogram equalisation technique maps the digital image onto the display in such a way that there are equal numbers of pixels in each brightness or density level in the analogue image (see Chapter 12). The image may also be filtered in various ways to improve visualisation. For example, edge enhancement or noise reduction may be applied to individual images. Noise in a series of ciné images can be reduced by using motion-dependent smoothing filters.

The possibility of dose reduction is another important advantage of digital radiology. In conventional radiology, the dose is determined by the sensitivity of the image receptor and the film latitude. In digital radiology, dose savings can be achieved by adjusting the dose to give the required noise level in the image, often without regard for the absolute magnitude of the signal. (Although the plates used for CR that match the resolution characteristics of modern screen–film systems do not normally produce a dose saving). Further reductions in dose may be possible by using the X-ray spectrum that gives the lowest dose for a given SNR and by recovering any losses in contrast by adjusting the displayed image.

Certain imaging techniques or image applications require the image to be digitally processed. Digital systems have an important advantage over analogue systems where this is necessary. A very important example is digital subtraction imaging (Section 2.11.1). Radiologists have to identify abnormal structures from a complicated background of normal tissue architecture. They may miss a small object that the system has resolved, or a low-contrast object that is visible above the noise level in the image, simply because of the complexity of the surrounding and overlying architecture. Subtraction radiology can remove much of the unwanted background architecture and can thus improve the visualisation of the important features of the radiograph. Another important application of image processing and analysis is computer-aided detection and diagnosis. Here pattern-recognition techniques are used to automatically detect or classify abnormalities in the radiograph.

2.9 Automatic Exposure Control

The input dose range to imaging receptors can be large. For example, the range of X-ray intensities in mammography is of the order of 20:1. In order to ensure that the amount of radiation incident on an imaging receptor produces an adequate signal (e.g. film blackening for screen–film systems), use is made of automatic exposure control (AEC). There are three main types of detector used with AECs: photomultiplier detectors (phototimers), ionisation chambers and solid-state detectors.

A phototimer or photomultiplier detector consists of a plastic sheet coated with a thin luminescent (phosphor) layer placed between the grid and cassette. Light output from the phosphor is sampled using a photomultiplier tube (see Section 5.2.1) with an electronic feedback to the X-ray console. Exposure is terminated when a predetermined output from the phosphor is reached.

An ionisation chamber detector (see Section 5.2.2) consists of two thin films of foil filled with a gas. The chamber is pre-charged before exposure and any ionisation within the gas volume reduces the voltage across the two foils. A voltage monitor on the ionisation chamber provides electronic feedback to the exposure controls and terminates the exposure when a preset voltage is reached.

AECs using solid-state detectors (see Section 5.2.3) are also available based on PN junction techniques or photoconductivity. They can be made small, fast in response and without any significant X-ray absorption.

To produce an image with optimum exposure in the area(s) of interest, three detectors are often available. Any one of these detectors or a combination of them may be selected by the operator, depending upon the examination protocol.

It should be noted that in mammography it is necessary to place the AEC detector behind the cassette. The low energies deployed with this technique would otherwise result in the detector being imaged and unwanted absorption of the beam by the detector would result in an increase in patient exposure.

The response of the image receptor and the AEC detector may vary with the energy or incident angle of the radiation, leading to problems as the tube voltage or patient size change. This difficulty may be overcome by using a measure of beam quality as the basis for compensation. For screen–film systems, a further correction may also be required to account for reciprocity law failure of the film.

To avoid unnecessarily high patient doses, exposure is terminated automatically using a guard timer if the AEC does not detect any exposure.

2.9.1 Automatic Brightness Control

Automatic brightness control (ABC) is a special case of automatic exposure control. ABCs are used with image intensifiers to adjust the light output from the image intensifier or the strength of the video signal from the TV camera to a preset value. The X-ray tube current and/or voltage are controlled so that sufficient signal is received to provide suitable image quality. The ABC will reduce output to avoid sudden surges of brightness on the monitor. The changes made by the ABC will also result at times in higher tube outputs. It is, therefore, important for the operator to be aware of the exposure settings.

It should also be noted that it is possible for monitor glare to be avoided by adjusting the gain of the TV camera without affecting patient dose. This is known as automatic gain control (Curry et al. 1990, p. 185).

2.9.2 Automatic Control of X-Ray Spectrum in Mammography

The optimum spectrum for use in mammography depends upon breast size and composition. Some mammographic systems have an X-ray tube with a dual target, and can select the target and filtration for each exposure automatically. They make use of the dose rate recorded after the receptor and/or the breast thickness to act as an indicator of the transmission through the breast. The 'optimal' target/filter combination can then be determined. Systems which use the dose rate must first make a brief trial exposure. The result from this trial then determines the target material, beam filter and tube voltage used for the remainder of the exposure.

2.10 Contrast Media

Some anatomical features can be so similar in density and atomic number to their surroundings that they cannot be radiographically distinguished from their background. In Section 2.4.1 we derived an expression for the contrast C between objects of differing compositions in terms of the difference between their linear attenuation coefficients μ_1 and μ_2:

$$C = \frac{(1 - \exp[(\mu_1 - \mu_2)x])}{1 + R} \tag{2.42}$$

where

R is the scatter-to-primary ratio, and
x is the object thickness.

From this equation we can see that contrast will be small for small differences in μ. It is sometimes possible, however, to introduce a contrast medium into the patient, which has a different attenuation coefficient to normal tissue and, hence, enhances the contrast. The medium needs to be either of lower density than normal tissue (negative contrast), for example, air, which is used for imaging brain ventricles (now used very little due to the development of CT), or of higher density (positive contrast). Higher-density contrast media include barium compounds (which are sticky) for examination of the digestive tract (e.g. oesophagus, stomach, colon) and iodine compounds (which are less sticky) for blood vessels, heart, kidney and ureter investigations. A third method for contrast enhancement is to use a combination of positive (barium or iodine) and negative (gas) contrast enhancement (e.g. used for colon, bladder imaging).

The contrast medium, therefore, needs to be of a suitable atomic number Z to provide an adequate change in attenuation, and, because it is introduced into the body, it must be non-toxic and possess a viscosity suitable for either injection or ingestion. It must also have a suitable persistence and be economic. As noted previously, barium and iodine are commonly used contrast media. Barium has an atomic number of 56 and a K-edge at 37.4 keV. It is normally used in a barium sulphate mix, for example, 450 g barium sulphate in 2.5 L of water with a density of 1.2 g/cm^3. The contrast variation between muscle and barium at different peak tube voltages is shown in Table 2.8. Due to the increased contrast above the barium K-edge, the contrast at 80 and 90 kV is greater than that at 70 kV.

Iodine has a K-edge at 33.2 keV and the optimum tube voltage (Table 2.9) will be less than that for barium. Iodine is normally injected into the patient at body temperature (37°C). For example, when used for femoral angiography, the iodine may be injected at a concentration of 300 mg/mL (density 1.4 g/cm^3) and at a rate of 8 mL/s for 10 s. Once injected, it becomes diluted as it passes through the body. The CsI phosphor (as used in image intensifiers and some flat-panel detectors) is particularly well suited to imaging iodine contrast media. This is because of its enhanced absorption efficiency just above the iodine K-edge.

Because of their more invasive nature, examinations requiring contrast media involve an increased risk compared with non-contrast examinations. Digital radiography employing

TABLE 2.8

Variation of contrast, *C*, between barium and muscle with peak tube voltage.

	70 kV	80 kV	90 kV
μ muscle cm^{-1}	0.26	0.24	0.23
μ Ba mix cm^{-1}	0.88	2.8	2.4
C for 1 cm	0.46	0.92	0.88
C for 0.1 cm	0.06	0.23	0.20

Note: The table shows that better contrast is achieved when the proportion of X-ray energies above the energy of the barium K-edge increases significantly.

TABLE 2.9

Variation of contrast, *C*, between a small blood vessel and iodine with peak tube voltage.

	70 kV	80 kV	90 kV
μ muscle cm^{-1}	0.26	0.24	0.23
μ I contrast cm^{-1}	11.5	9.0	7.5
C for 0.75 mm vessel	0.57	0.48	0.42

contrast enhancement or subtraction techniques (Section 2.11.1) has the potential to reduce the concentration of the contrast media required for some of these examinations.

2.11 Applications

2.11.1 Digital Subtraction Imaging

2.11.1.1 Digital Subtraction Angiography

Digital subtraction angiography (DSA) is a subtraction technique that is used for the visualisation of blood vessels following the intravenous or intra-arterial injection of contrast media. An image of the region of interest is obtained before the arrival of the iodinated contrast agent and is used as a mask to subtract from the images showing the passage of the contrast medium through the blood vessels. The SNR in the subtracted image can be improved by using spatial and temporal filters on the complete image set, and corrections can also be applied for patient movement. For intravenous injection, the contrast medium may be much diluted by the time it reaches the region of interest, so that the high-contrast resolution achievable with digital techniques is very important (a contrast of 1% can be seen using DSA). For peripheral imaging, the transit time for passage of the contrast through the region of interest will be several seconds and a moderate frame rate (1 per second) is used with a pulsed X-ray beam. For intra-arterial injection, there is greater contrast in a shorter time interval and this allows the visualisation of smaller vessels. A 1024 × 1024 matrix together with a short X-ray beam pulse (4–33 ms) and a frame rate of four per second may be appropriate. For cardiac imaging, a frame rate of 12.5 or 25 per second may be necessary with continuous or pulsed X-ray exposure. Figure 2.44 shows two examples of DSA images.

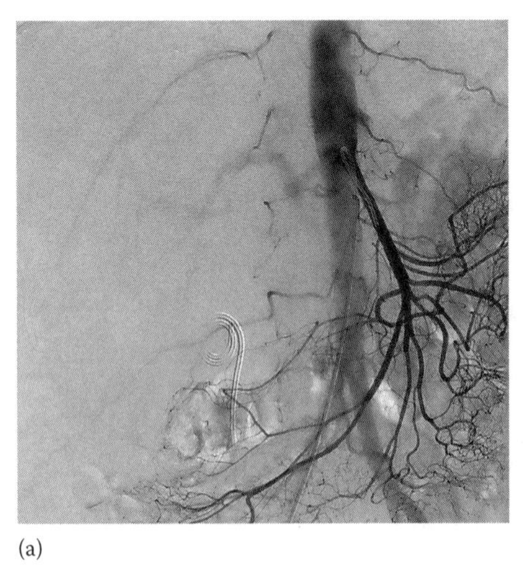
(a)

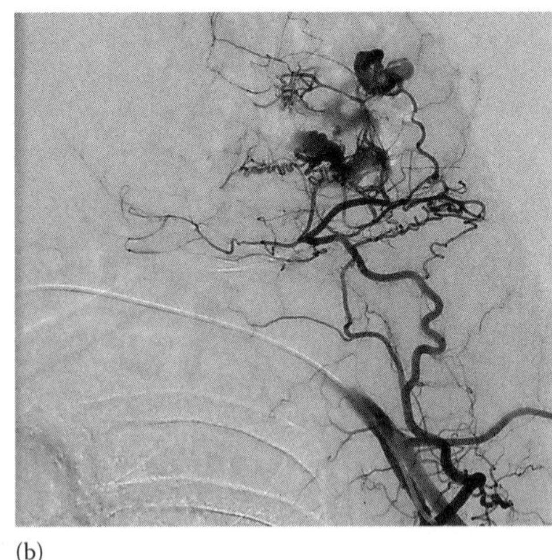
(b)

FIGURE 2.44
Examples of digital subtraction angiograms. (a) DSA image of the superior mesenteric artery feeding the intestines. The projection is AP over the upper abdomen; the line catheter is in the aorta. (b) DSA image of one of the arteries feeding a haemangioma in the left trapezius muscle in the shoulder. The projection is AP over the left lung and shoulder; the catheter is in the left subclavian artery. (Images courtesy of the Radiology Dept, The Royal Marsden Hospital, Chelsea, UK.)

2.11.1.2 Dual-Energy Imaging

Dual-energy radiology is another form of subtraction imaging. In this technique two images are obtained using different X-ray spectra. The signal recorded in each pixel gives an indication of the attenuation caused by the overlying tissue and the attenuation itself can be divided into components due to photoelectric absorption and to photon scatter. As these two components have different energy dependence, it is possible to extract a photoelectric and a scatter component from the two initial images. Equally, since different materials will have different amounts of scatter and photoelectric absorption, it is possible to produce images of the thicknesses of two selected materials, which, in combination, will reproduce the attenuation maps recorded in the two original images (Lehmann et al. 1981). Suitable selection of these materials can produce images that show or exclude any desired tissue type. For example, it is possible to choose soft tissue and bone (with some nominal composition) as basis materials and to view the soft tissue or bony images separately.

2.11.2 Digital Mammography

In mammography, it is important that the images have good resolution to detect the presence of microcalcifications, high image contrast in order to see small differences in soft-tissue architecture, and wide dynamic range with a good SNR (NEQ) at low dose.

The use of a single-screen single-emulsion film receptor satisfies two of the aforementioned requirements for mammography. It has a resolution typically of 13 and up to 20 line pairs/mm is possible for a good system. A reasonably low breast dose of 2 mGy (breast-size dependent) at high contrast can also be achieved. Screen–film systems do, however, have some important disadvantages. Probably, the most important of these is the limited dynamic range of the film, its low SNR at low dose and its relatively poor (in comparison with digital detectors) DQE at low spatial frequencies.

In comparison, digital receptors do not suffer from dynamic range limitations. Furthermore, it may be possible to achieve a dose saving with a digital system by using a higher-energy X-ray spectrum (Dance et al. 2000).

The MTFs of current commercially available digital receptors for mammography are inferior to those of typical mammographic screen–film systems. This may not be too important, however, since the DQE of a system better characterises the detectability of small objects than the maximum number of line pairs/mm that can be visualised in a high-contrast bar pattern.

A brief discussion of the characteristics of examples of digital receptors that have been or are currently (at the time of writing) used for mammography is given in the following. For each technique, hard copies, at a resolution matched to the pixel size, can be produced using a laser printer. Soft copies can also be viewed using high-resolution (2000 line) monitors.

A 3 × 4 array of CCD detectors coupled to a caesium iodide phosphor using fibre optics was developed for full-field mammography by Lorad (Danbury, Conn.). The coupling produced a 2:1 demagnification that achieved a pixel size of 40 μm (Nyquist frequency 12.5 line pairs/mm) in the image plane (Kimme-Smith et al. 1998).

An amorphous silicon-based flat-panel detector employing a CsI phosphor has been developed for full-field mammographic imaging by General Electric. The detector has a pixel size of 87 μm with a pixel spacing of 100 μm (Nyquist frequency 5 line pairs/mm) and a fill factor of 75% (Muller 1999).

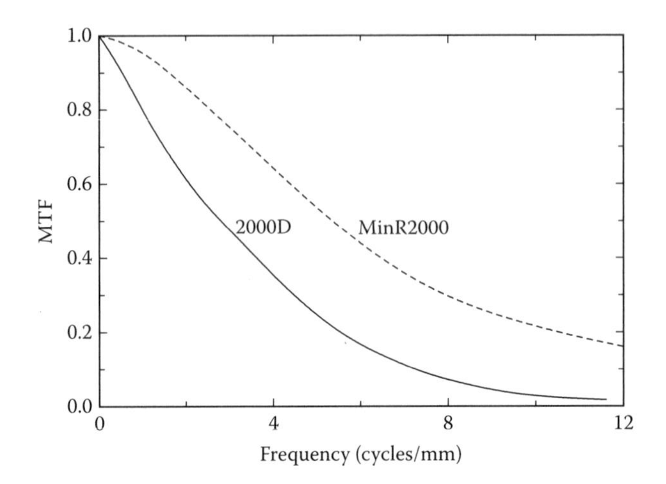

FIGURE 2.45
MTFs for examples of screen–film (dotted curve, Kodak MinR2000) and digital (solid curve, GE2000D system) mammography systems. The Nyquist frequency for the 2000D is 5 cycles/mm. (Data for dotted curve based on Bunch, 1999; data for solid curve based on Medical Devices Agency (MDA), 2001.)

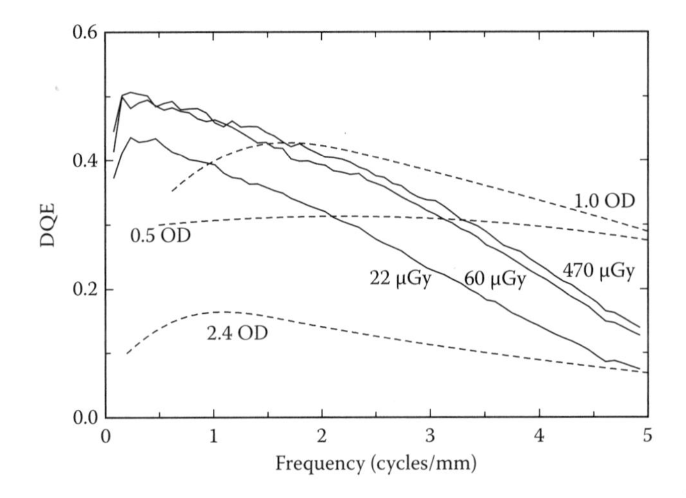

FIGURE 2.46
Detective quantum efficiency for examples of screen–film (dotted curves at optical densities of 0.5, 1.0 and 2.4 for Kodak MinR2000) and digital (solid curves for incident air-kermas at the receptor of 22, 60 and 470 μGy for GE2000D system) mammography systems. The Nyquist frequency for the 2000D is 5 cycles/mm. (Data for dotted curves based on Bunch, 1999; data for solid curves based on MDA, 2001.)

Figures 2.45 and 2.46 compare the MTFs and DQEs for examples of screen–film and flat-panel receptors for mammography. For the examples shown, the digital receptor has an inferior MTF but better DQE for low spatial frequencies and for exposures which give low or high optical densities for the screen–film system.

CR systems were first used for digital mammography in 1994 (Brettle et al. 1994). Thin CR plates were used with a pixel size of 100 μm. The plates were kept thin because of the scatter of the read-out light, which reduces resolution and this, together with the absorption of light by the opaque back support of the plate, reduced the DQE. Nowadays, CR plates are available for mammography, which are read out from both sides (see Section 2.8.2).

Such plates have improved DQE compared with that for single-sided read-out, but reduced MTF (Fetterly and Schueler 2003, Monnin et al. 2007).

The resolution of CR plates is limited by their read-out process, whereas for flat-panel systems it is limited by the inherent size of the electronic components associated with the fill factor. Neither flat-panel detectors nor CR plates need to be 'patched' or suffer from demagnification losses associated with CCDs (see Section 2.8.3.2).

An interesting development is the use of a photon counting receptor (Section 2.8.5) for digital mammography. A system is available which uses a photon counting Si-strip detector. Because of the linear geometry, a scanned X-ray beam is used to obtain the image (Section 2.6.4) with the benefit of improved scatter rejection. A DQE higher than that for flat-panel and CR systems has been reported (Monnin et al. 2007).

2.11.3 Stereotactic Biopsy Control and Spot Imaging in Mammography

Mammographic X-ray sets often have an attachment for stereotactic localisation and biopsy. The procedure involves making exposures from two different tube positions with the breast remaining fixed in position. The three-dimensional location of the region of interest can be calculated from the two images, and a biopsy taken or a localisation marker inserted into the breast. Screen–film receptors are not well suited to such procedures because the patient's breast has to remain compressed whilst the film is processed and the stereotactic localisation calculations made. In such a situation the advantage of a digital system with very fast image availability is clear.

The field size required for stereotactic localisation is smaller than that for full-field mammography. Consequently, a fibre-optic taper with a small or unit demagnification can be used to link a phosphor plate to a CCD. A system developed by Siemens (the Opdima) used a 100 μm thick caesium iodide phosphor with an active CCD area of 49 × 85 mm and achieved a pixel size of 12 μm. The images needed to be corrected for dark current, flat-field variations and for pixel defects. The latter was achieved by binning pixels to 24 or 48 μm, which still provided high resolution.

2.11.4 Digital Chest Imaging

As discussed previously, dynamic range is a difficult challenge for a screen–film receptor for chest radiography since it is desirable to image regions such as the lungs and the retro-cardiac area on the same film. For this examination the large dynamic range of digital systems is an advantage, combined with the usual ability of digital systems for improved image display.

Large-area (e.g. 41 × 41 cm) flat-panel detectors are available for dedicated chest imaging. Detectors based on caesium iodide and amorphous silicon show excellent uniformity, repeatability and linearity (Floyd et al. 2001). Such detectors can produce a considerable dose reduction (33%) compared with 400-speed screen–film systems (Strotzer et al. 2002). Bacher et al. (2006) have compared their performance with that of an amorphous-selenium flat-panel detector and found that they could be operated at a lower dose with equal or even improved image quality. CR plates with dual-sided read-out, incorporated into a cassette-free chest imaging system were available at the time of writing for chest radiography (Riccardi et al. 2007).

2.11.5 Digital Tomosynthesis

The superimposition of overlaying tissues in a two-dimensional image can hinder the detection of lesions and can also result in false positive results. Classical tomography (Section 2.12.5) aims to visualise tissue in a selected plane of interest using analogue methods, but overlying structures are still present in the image in a blurred form and a separate exposure must be made for each image plane required. CT provides the ultimate in sectional imaging by removing the blurred structures, but has an associated detriment in terms of high patient dose. CT units are also very expensive to purchase. Tomosynthesis offers an improvement over classical tomography with much simpler equipment, and a patient dose similar to that for single-plane radiographs. Using the technique it is possible to calculate any number of tomographic planes. This is done retrospectively following the exposure of a series of digital projection radiographs obtained during a single trajectory of the X-ray tube and digital detector (Dobbins and Godfrey 2003). During the exposure, the digital detector may remain stationary or may move in synchrony with the X-ray tube (e.g. on a C-arm).

The technique finds application in mammography, angiography, chest and dental radiography.

2.11.6 Equalisation Radiography

Screen–film radiographic images often exhibit a large density range. Because film has a limited dynamic range and poor DQE at low and high optical densities, the exposure latitude required is often greater than that available. Quantum noise is also higher than desirable in some parts of the image and lower than it needs to be elsewhere. One solution to reduce the dynamic range requirement and to improve the quantum noise would be to increase the exposure when the transmission through the subject is low and to decrease the exposure when this transmission is high.

Chest imaging is an important example of this problem. One possibility here is to use a high-voltage (120–150 kV) technique to reduce the relative absorption between low- and high-density structures. Another simple solution for producing broad equalisation of lung, spine and heart in a PA chest radiograph is to use a trough filter positioned at the light-beam diaphragm. Such a filter may be constructed from aluminium or translucent material of sufficiently high density. It is used to preferentially attenuate the X-ray beam in the regions of the lung, whilst allowing greater irradiation in the region of the spine (Manninen et al. 1986).

A limitation of the simple construction of trough filters is that they cannot be customised to the equalisation needs of a particular image. An alternative, much more advanced approach is to use a scanned X-ray beam whose intensity can be modulated in accordance with the transmission through the patient (Plewes 1994).

2.12 Quality Control

In this chapter we have so far discussed the physical principles of diagnostic X-ray imaging systems. It is evident that the different components of the imaging system must be well designed in order to produce images of good quality whilst minimising the radiation dose to the patient. However, good design alone will not ensure optimum performance of an imaging system throughout its useful working life. Regular performance monitoring as part of a robust quality-assurance programme is essential. In this section we discuss some of the tests that should be carried out on the various components of the imaging chain as part of such a programme. The need for performing each test is discussed, and a brief description of the equipment required and the test procedure is given where appropriate. A complete description of all the tests required for different types of diagnostic X-ray equipment, including suggested testing frequencies and tolerances is not appropriate here. Suitable references are, therefore, provided.

2.12.1 Definitions

The terms 'quality assurance' and 'quality control' are often used interchangeably, but it should be recognised that they do not have identical meaning. Quality assurance may be defined as all those planned and systematic actions necessary to provide adequate confidence that a structure, system, component or procedure will perform satisfactorily in compliance with agreed standards (EU 1997). The need for establishing a good quality-assurance programme in diagnostic radiology has become increasingly important in recent years and Council Directive 97/43/Euratom (EU 1997) makes several direct references to the implementation of such a programme. Quality assurance has two components, quality management and quality control. Quality management may be defined as that aspect of the overall management function that determines and implements quality policy. Quality control may be defined (EU 1997) as the set of operations (programming, coordinating, implementing) intended to maintain or to improve quality. It covers monitoring, evaluation and maintenance at required levels of all characteristics of performance of equipment that can be defined, measured and controlled. The tests presented in this section are, therefore, best described as quality control tests.

The quality control tests performed on a diagnostic X-ray imaging system fulfil different functions according to the status of the system. When a new piece of equipment is installed, tests should be carried out to verify that the equipment complies with its specification agreed as part of the procurement contract. This is called *acceptance testing*. Tests must also be carried out in order to establish baseline measurements and to ensure that the equipment is ready for clinical use. This is known as *commissioning*. In practice, acceptance and commissioning often comprise similar tests and are, therefore, performed concurrently, but it should be recognised that they fulfil different functions. Quality control tests should also be carried out on a routine basis throughout the life of the equipment, and following any modification or repair which might affect performance.

2.12.2 Quality Control Tests

Technological advances have resulted in diagnostic X-ray equipment becoming increasingly complex. The greater the functionality, the greater the number of quality-control tests required to ensure its optimum performance. It is useful to devise distinct sets of tests for the different equipment functions and/or components encountered. In this way, the relevant set of tests can be selected for any particular piece of equipment according to its functionality. This modular approach is particularly useful when dealing with large numbers of different X-ray units.

2.12.3 Tube and Generator Tests

The performance of the tube and generator is fundamental to the satisfactory performance of the complete X-ray imaging system. Details of suitable tube and generator tests, including equipment required and descriptions of how to carry out the tests, are given in IPEMB (1996) Part I and BIR (2001). Recommended test frequencies and tolerances are given in IPEM (2005b). In this section we discuss the rationale for the tests and the underlying physical principles where appropriate.

2.12.3.1 Tube Voltage

The excitation potential across an X-ray tube affects the number of photons emitted from the tube and the photon-energy distribution, both of which influence the radiation dose to the patient. The photon-energy distribution also affects the contrast in the resulting image. The potential applied across an X-ray tube during an X-ray exposure will be affected by the type of X-ray generator. Single-phase, three-phase and medium-frequency generators have characteristic voltage ripples which are described in Section 2.5.4.

The voltage indicated on the X-ray control unit normally refers to the peak potential applied across the X-ray tube. It is expressed in units of kV and is often referred to as 'kilovoltage' or kVp. A wide range of tube voltages are encountered in diagnostic radiology. In mammography, a low voltage of around 25–32 kVp is required in order to achieve good contrast between similar soft tissues. Most general radiography of the trunk is carried out between 70 and 90 kVp. Chest radiography is often performed at around 120–140 kVp. It is important that the voltage indicated on the X-ray control unit is in good agreement with that actually delivered. The service engineer will sometimes measure voltage directly across the X-ray tube (invasive kVp measurement). This technique is not appropriate for routine quality control and a non-invasive technique should be adopted, where the voltage is estimated from measurements on the emerging X-ray beam. The most common technique is the use of a non-invasive potential divider (Figure 2.47). This instrument may contain two or more solid-state detectors. Filters with differing composition and/or thickness are placed over each detector. The instrument is placed in the X-ray beam such that all detectors are irradiated. The filters will attenuate the X-ray beam to different extents, the degree of attenuation by each filter depending on the photon-energy distribution and, hence, on the applied voltage. If I_1 and I_2 are the values of photon flux emerging from two filters, and the filters are of thickness x_1 and x_2 and are composed of materials of linear attenuation coefficient μ_1 and μ_2, then the ratio I_1/I_2 of the photon fluxes emerging from the filters is approximately given by

$$\frac{I_1}{I_2} = \exp(\mu_2 x_2 - \mu_1 x_1). \tag{2.43}$$

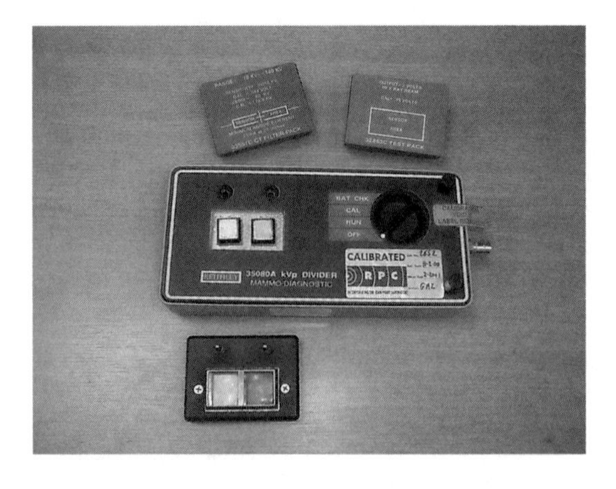

FIGURE 2.47
Potential divider containing two solid-state detectors. Three filter packs are also shown. (The potential divider and filter packs were manufactured by Keithley Instruments Inc., Cleveland, OH.)

Since the linear attenuation coefficients depend upon photon energy, the ratio I_1/I_2 may be used to estimate the peak tube voltage. This works best when heavy filtration is used in the instrument so that the low-energy portions of the X-ray spectrum are removed.

Using the potential divider alone provides a means of assessing the maximum tube voltage for an X-ray exposure, but provides no information about the variation of the tube voltage during the exposure. This information can be obtained by connecting the kV divider to a storage oscilloscope or a computer running suitable waveform-analysis software and displaying the voltage waveform. The voltage waveform contains much useful information and its analysis is fundamental to the complete assessment of an X-ray tube and generator. The peak tube voltage at any stage of the X-ray exposure can be measured, and the type and frequency of the generator can be assessed. The voltage ripple (required for the calculation of filtration from half-value layer, Section 2.12.3.4) can also be inferred. Problems with the generator such as over- or undershooting of the tube voltage at the beginning of an exposure, or loss of a phase, are clearly identified from the waveform. The waveform also provides a convenient means of measuring the X-ray exposure time. Figure 2.13 shows typical voltage waveforms obtained from a three-phase, six-pulse generator and a medium-frequency generator, both at 80 kVp. It is evident that measurements of peak tube voltage, voltage ripple, generator frequency and exposure time can all be made from these waveforms.

2.12.3.2 Exposure Time

The number of photons produced during an X-ray exposure will depend not only on the beam quality but also on the tube loading and the exposure time. These latter two parameters are often quoted together as the tube-current exposure-time product or mAs. Some X-ray generators set tube current and exposure time separately; others operate using an mAs integrator to ensure that the correct total mAs is delivered for each exposure. In the former case, a change in either tube current or exposure time will result in a change in tube output. In the latter case, if the tube current is incorrect the exposure time will change to compensate. If exposure times are too long, image artefacts due to patient motion may become evident. It is, therefore, important that the exposure time set on an X-ray control unit or selected under AEC is in good agreement with the true exposure time.

The most convenient method of measuring the exposure time is from the voltage waveform as described in the previous section (see also Figure 2.13). The waveform will also give an indication of the time taken for the maximum voltage to be achieved. An alternative method of measuring exposure time is to use an electronic timer. Such timers incorporate a photodiode that switches on an electronic timing circuit during the exposure. They are convenient to use but do not provide an indication of the time taken for the voltage to reach maximum.

2.12.3.3 Tube Output

We have already seen that the number and distribution of photons emerging from the X-ray tube will depend on a combination of the tube voltage, tube filtration and tube-current exposure-time product. If any one of these parameters changes, the tube output will change. It is, therefore, useful to measure this quantity, especially since direct measurement of the tube current is inappropriate for routine measurement as it involves an invasive measurement within the X-ray generator. Tube output is also a fundamental property of the X-ray tube. If measurements are made for two X-ray units with identical tube filtration and at identical tube voltage and tube-current exposure-time product settings, it is likely that tube output would still be different for the two units. This is because tube output also relates to the thermionic emission from the filament and the properties of the anode, which are unique to each X-ray tube.

Tube output is expressed in air-kerma and can be measured using a variety of instruments including ionisation chambers and solid-state detectors. Instruments must be suitable for primary beam measurements and calibrated at diagnostic energies. The tube output should be proportional to the tube current and to the exposure time. The relationship between tube output and tube voltage is more complex, but for general X-ray units, tube output is approximately proportional to the square of the voltage. If there has been a change in tube output but no change in tube voltage, exposure time, tube current or filtration, this could be indicative of a more serious problem such as damage to the anode.

Many multi-functional instruments are available, which measure tube voltage, exposure time and tube output in a single X-ray exposure, facilitating quick and simple assessment of tube and generator performance.

2.12.3.4 Half-Value Layer and Filtration

The measurement of tube voltage and assessment of the voltage waveform have been discussed in Section 2.12.3.1. In order to fully describe the quality of an X-ray beam, the total filtration of the beam must also be known. Too little filtration increases the number of low-energy X-rays emerging from the X-ray tube. These soft X-rays will be absorbed by the patient without contributing to the final image, thus unnecessarily increasing patient dose (Section 2.5.3). On the other hand, too much filtration increases the loading on the X-ray tube, which may shorten its life. Filtration is most conveniently estimated by measurement of the half-value layer of the X-ray beam using an ionisation chamber or solid-state detector under narrow-beam conditions. This is accomplished by introducing increasing thicknesses of aluminium into the beam and determining the thickness required to reduce tube output to one half of its value with no aluminium filter present. The experimental configuration is shown in Figure 2.48. The half-value layer depends on the peak tube voltage, the total filtration, the ripple in the voltage waveform, and the anode angle. The total filtration can be estimated from tables published by Cranley et al. (1991) with knowledge of the other parameters.

2.12.3.5 Focal-Spot Size

The focal spot has a finite size, which results in geometric unsharpness in the image. The magnitude of the focal spot will thus affect the resolution of the imaging system. Changes in its value can give early indication of tube failure. Most X-ray tubes have more than one filament and consequently more than one focal-spot size. A broad focus is used for general work and a fine focus is used where greater detail is required. During fluoroscopy a fine focus is usually preferred.

The size of the focal spot can be measured using two basic techniques. The first employs a pinhole or slit in a gold/platinum alloy plate or other suitable material, mounted in a larger plate made from radio-opaque material and positioned on the central ray of the X-ray field. A magnified image of the pinhole or slit is taken for each focus. Two exposures are required for each focus when using the slit; the width and length of the focus are determined by aligning the slit parallel and perpendicular to the anode–cathode axis, respectively. A typical image of a focal spot produced using the pinhole technique in shown in Figure 2.49. The actual size of the focal spot can be determined from the image with knowledge of the magnification. The measured dimensions of the focal spot should be compared with the X-ray tube manufacturer's specification.

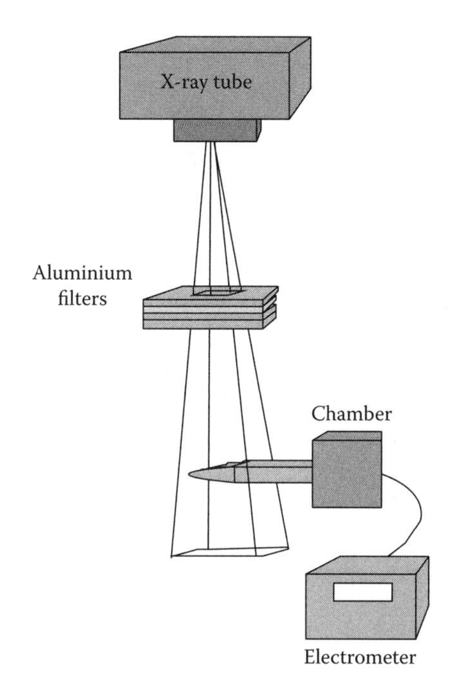

FIGURE 2.48
Experimental configuration for the measurement of half-value layer.

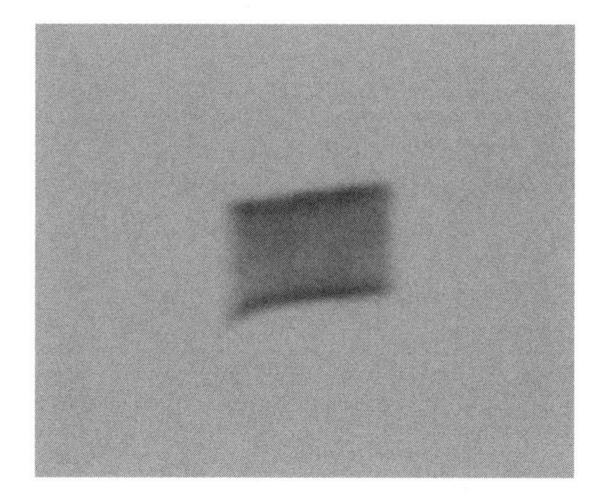

FIGURE 2.49
Image of a radiographic focal spot obtained with a pinhole.

The pinhole and slit techniques can be time consuming because of the difficulty in locating the central ray of the X-ray beam. However, they provide the best assessment of the focal-spot size since they produce an image of the focal spot that can be evaluated qualitatively as well as quantitatively. These techniques are often only used at the acceptance of a new X-ray tube. For routine measurements it may be easier to measure the focal-spot size indirectly using a resolution pattern such as the star. Several such test patterns are commercially available but all employ the same basic principle. The star consists of a pattern of lead-foil spokes sandwiched between two plastic plates, arranged such that the spatial frequency of the spokes increases with decreasing distance from the centre of the test object. The pattern is positioned on the exit port of the X-ray tube and imaged. Close to the centre

FIGURE 2.50
Image of a star test pattern used for inferring focal-spot size.

of the image, the pattern is of high frequency and will be blurred. Further out, the pattern is visualised, but is phase reversed, and finally the pattern is properly resolved with the correct phase. The focal-spot size can be calculated from the dimensions of the resolved region with knowledge of the magnification of the test pattern image. A typical image of a star pattern is shown in Figure 2.50.

2.12.3.6 Light-Beam Diaphragm/X-Ray Field Alignment and Beam Perpendicularity

For X-ray units which employ a light beam to indicate the position of the X-ray field, it is important to ensure that there is good coincidence between the light and X-ray fields. If the X-ray field is larger than that indicated by the light field, the patient will receive an unnecessary radiation dose; if it is smaller, patient anatomy crucial to the examination may be missed, resulting in a repeat exposure and more unnecessary radiation. The position of the X-ray field will depend on whether the central beam axis is perpendicular to the image plane. Both light/X-ray field alignment and perpendicularity can be tested using simple test tools such as those described in IPEMB (1996) Part I and BIR (2001).

2.12.3.7 Practical Considerations

In this section we have presented a brief discussion of the basic tests necessary to check tube and generator performance. It should be noted that the techniques for carrying out these tests have to be adjusted, sometimes quite radically, for different types of X-ray equipment. For X-ray units capable of both radiography and fluoroscopy, measurements should be made in radiographic/fluorographic and fluoroscopic modes as appropriate. Not all tests will be required in each mode, but measurement of tube voltage and output, for example, may be necessary in both modes since tube current and circuitry may be different for radiography and fluoroscopy. An increasing number of fluoroscopy systems have no means of manually selecting exposure factors, and the voltage and current are selected automatically by the system. This can make measurements of parameters such as half-value layer difficult. In these cases the manufacturer should be consulted, since there may be special modes of operation available to the service engineer, which are suitable for such measurements. Other specialist equipment, which may require adaptation of the usual test protocols, include panoramic dental equipment (where stopping the tube rotation may be helpful in order to carry out the tests) and equipment which scans the patient using a narrow slit beam.

2.12.4 Automatic Exposure Control Tests

Much general radiography is carried out using AEC devices, where the X-ray exposure is automatically terminated when the image receptor has received sufficient radiation for an adequate diagnostic image (Section 2.9). It is important that an AEC system operates correctly so that consistent image quality at minimum patient dose is ensured. The requirements for AEC devices are different for screen–film and digital radiography but the basic tests are similar. For screen–film radiography, consistency in the resultant optical density is assessed. For digital radiography the analogous quantity is the detector dose indicator (DDI), which indicates the absorbed dose to the detector and the sensitivity of the read-out system as appropriate for either CR or direct digital radiography (DDR). For digital radiography image quality is less sensitive to changes in exposure than screen–film radiography and these tests are consequently more concerned with patient dose than image quality.

Descriptions of suitable AEC tests for screen–film and digital radiography are given in IPEM (1997) Part IV and BIR (2001). Recommended test frequencies and tolerances for many of these tests are given in IPEM (2005b). For most tests the requirement is that a consistent optical density or DDI is achieved when different parameters are varied. Most modern AEC devices give an indication of the final tube-current exposure-time product delivered by the system for each exposure. Recording this parameter will monitor X-ray tube output and give an indication of patient dose. This is particularly important in digital radiography where adequate image quality will be achieved over a much wider dynamic range than in screen–film radiography. The following tests should be made: consistency of a single radiation-monitoring AEC device; consistency between the different monitoring devices; consistency with changing beam load; and consistency with changing tube voltage. In addition to this, the increments in optical density or DDI provided by the fine control may be assessed. The input dose to the image receptor should also be assessed. For all these measurements, suitable materials should be placed in the X-ray beam to simulate a typical patient. Water-equivalent material or polymethyl methacrylate (PMMA) make good soft-tissue substitutes. However, they can be awkward and cumbersome in practice and many AEC devices can be tested using suitable thicknesses of copper as beam load. The guard timer should also be tested for each radiation monitoring device. This ensures that an exposure is terminated at a predetermined exposure time or tube-current exposure-time product if the radiation monitor detects no radiation during exposure.

2.12.5 Conventional Tomography

The use of conventional tomography has declined in recent years, largely due to the increased use of computed tomography. However, the technique still finds application in some diagnostic X-ray procedures such as intravenous urography and it is important that where such a facility exists, it is functioning correctly.

Linear tomography makes use of a standard radiographic X-ray set where the X-ray tube and cassette holder are connected such that during exposure the X-ray tube and cassette move in opposite directions about a pivot axis that is adjusted to the height of the desired image plane (Figure 2.51). The resulting image shows sharp detail in the imaging (fulcrum) plane and progressive blurring of detail outside this plane. The angle through

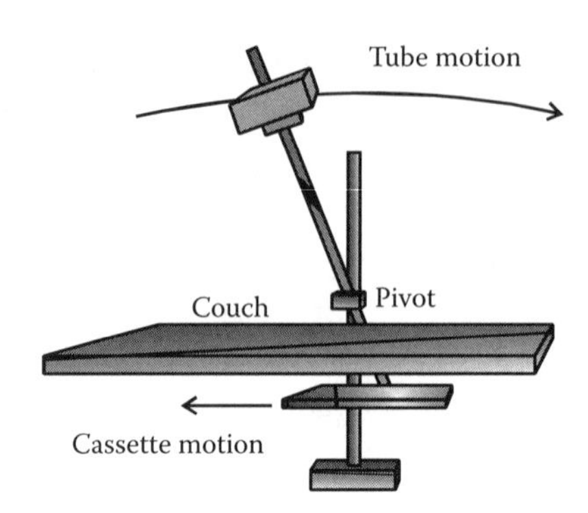

FIGURE 2.51
Configuration used for conventional tomography.

which the X-ray tube housing and cassette move is adjusted to provide the amount of blurring required. To avoid image artefacts being formed due to the aforementioned linear motion, more complex motions may be employed during the exposure such as circular, elliptical, spiral and hypocycloidal (Curry et al. 1990, p. 251).

Full descriptions of suitable conventional tomography tests are given in IPEMB (1996) Part V and BIR (2001). Recommended test frequencies and tolerances for some of these tests are given in IPEM (2005b). Suitable tests include assessment of the cut height calibration and cut plane thickness; measurement of swing angle and asymmetry; and measurement of spatial resolution in the cut plane.

2.12.6 Fluoroscopy Systems

As discussed in Section 2.7.3, fluoroscopy involves the production of dynamic X-ray images which are displayed in real time on a TV monitor. X-ray units capable of fluoroscopy may employ image-intensifier TV systems or, more recently, flat-panel detectors. The tests described in this section apply to both types of system; however, for flat-panel detectors some of the DDR tests described later in the chapter may also be applicable. The procedures followed when testing a fluoroscopy system will depend on its functional capabilities. The different modes of operation should be fully understood before attempting to perform any quality control checks.

We have already discussed the tests necessary to check the performance of the X-ray tube and generator and noted that these tests will need to be adapted when testing a fluoroscopy system. This section describes the additional tests required for fluoroscopy systems. Full descriptions of these tests are provided in IPEMB (1996) Part II and BIR (2001). Test frequencies and tolerances are given in IPEM (2005b).

Fluoroscopy systems operate using AEC and may also be capable of operating under manual control. It is important to ensure that the AEC is operating correctly in order to maintain good image quality and minimise patient dose. For image-intensifier TV systems, the automatic control system is designed to maintain a given output signal irrespective of patient thickness or anatomy, usually by control of the fluoroscopic exposure factors. This is achieved by controlling the air-kerma rate at the input surface of the detector. For flat-panel detectors the same principle applies, but changes in detector air-kerma rate with changing field size may be different since the requirement for image intensifiers that the TV camera receives the same amount of light irrespective of field size is no longer valid. A fundamental test of a fluoroscopy system is, therefore, the measurement of air-kerma rate at the input surface of the detector in the different modes of operation of the unit. A suitable measurement instrument (ionisation chamber or solid-state detector) is placed on the input surface of the detector, and appropriate beam-load materials as specified by the equipment supplier are placed between the X-ray tube and measurement instrument. The input air-kerma rate may vary with selected field size, and there may be a choice of levels for any given field size. The equipment supplier should be able to provide nominal values measured under pre-defined conditions. It is important that input air-kerma rates are measured under the conditions specified by the supplier.

It is also useful to measure the air-kerma rate at the input surface of phantoms representing typical patients. Measurements should be made in those modes which are in routine clinical use. The patient is best simulated using water-equivalent material or PMMA. However, such materials can be cumbersome and blocks of aluminium may provide a reliable and more manageable substitute. It should be noted that copper is unsuitable for these measurements. Its composition is unrepresentative of human tissue due to its high atomic number and it may produce a different response from the detector.

The most usual means of assessing image quality is by making subjective measurements using suitable test objects. Although it can be difficult to make absolute measurements, subjective assessment of this nature is useful in determining how the performance of the imaging system is maintained throughout its operating life. A set of test objects were originally designed by Hay et al. (1985). These test objects form the basis of many of the tests recommended in IPEMB (1996) Part II and are available from Leeds Test Objects Ltd. The test objects enable subjective measurements to be made of brightness and contrast, greyscale rendition, threshold contrast, contrast-detail detectability, limiting resolution and homogeneity of resolution. Figure 2.52 shows an image of a composite test object (TOR 18FG) used for the assessment of brightness and contrast, threshold contrast and limiting resolution. Figure 2.53 shows an image of test object TO.10, which is used for assessing contrast-detail detectability. It should be noted that these figures have been derived from contact radiographs and are for illustrative purposes only; the detail visible may not be

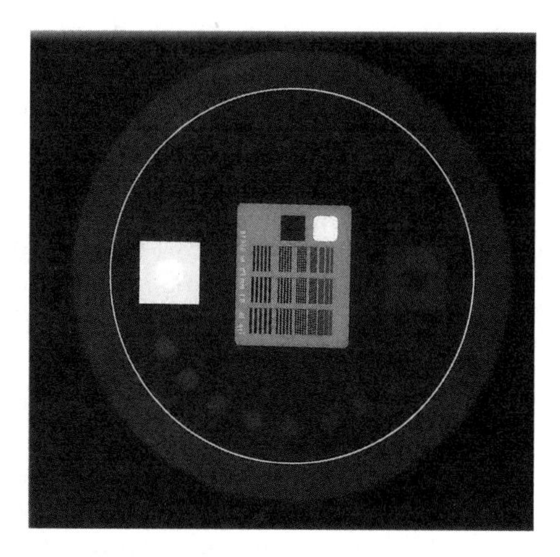

FIGURE 2.52
Radiograph of Leeds test object TOR 18FG, which is used for assessing the performance of fluoroscopy systems. The following features are visible: (a) two discs of maximum and minimum transmissions within squares of slightly different transmissions to facilitate correct adjustment of brightness and contrast, (b) a circular arrangement of low-contrast details for the assessment of large-detail detectability, (c) a high-contrast resolution grating and (d) a high-contrast circle for the assessment of geometrical distortion.

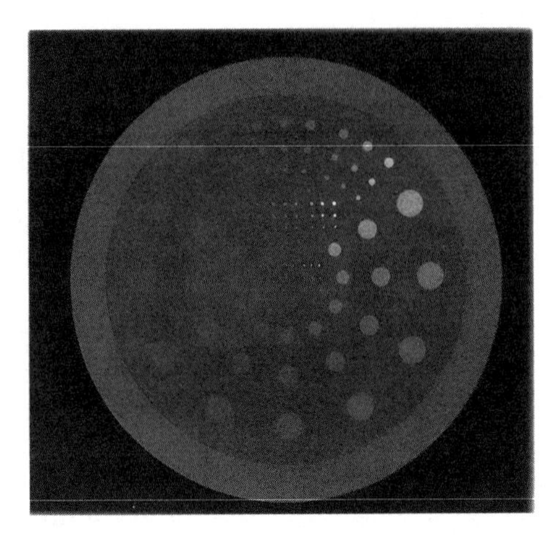

FIGURE 2.53
Radiograph of Leeds test object TO 10, which is used for assessing the performance of fluoroscopy systems. The test object comprises an array of circular details of various contrasts and sizes, facilitating an assessment of threshold-contrast-detail detectability.

representative of that achieved for fluoroscopy. Test objects are also available for the measurement of the video-signal-output voltage and assessment of the visible field size on the TV monitor. An assessment should also be made of the actual area irradiated.

2.12.7 Fluorography Systems

Many fluoroscopy units will also have a facility for acquiring images. For analogue systems (Section 2.7.3.2), the acquisition facility may be a spot-film device positioned between the X-ray tube and image intensifier or a small-format film camera, which uses the output signal from the image intensifier. Analogue systems used for dynamic imaging employ a ciné camera for the recording of images. These types of acquisition systems have been largely superseded by digital fluorography, where the acquired signal from the image-intensifier TV system is digitised and stored. Flat-panel fluoroscopy systems will also have facilities for digital acquisition. Different tests are required for analogue and digital fluorography. A spot-film device may be tested in the same way as a conventional screen–film AEC device although adaptations to the test procedure may be required and additional checks of the automatic film collimation should be made. Analogue fluorographic techniques such as small-format film or ciné film acquisition involve the image intensifier and this should be reflected in the test procedures followed. A description of the tests necessary for analogue fluorographic systems is given in IPEMB (1996) Part VI.

For modern systems incorporating digital acquisition, the basic tests are similar to those performed for fluoroscopy. However, different test objects and additional tests may be required depending on the sophistication of the system. Some systems incorporate a simple digital-acquisition feature, which utilises similar dose rates as a high-dose fluoroscopy mode. In this instance, the tests required may be very similar to those performed for fluoroscopy and the same test objects and test tolerances may be adopted. For more sophisticated systems, true digital acquisition is performed at a much higher dose rate than for fluoroscopy, resulting in high-quality images comparable to those obtained using conventional analogue fluorography. In this instance, the test objects used for fluoroscopic image quality may not be adequate. Digital subtraction fluorography (DSF) is another high-dose acquisition technique, which is used for contrast enhancement, and again special test objects will be required.

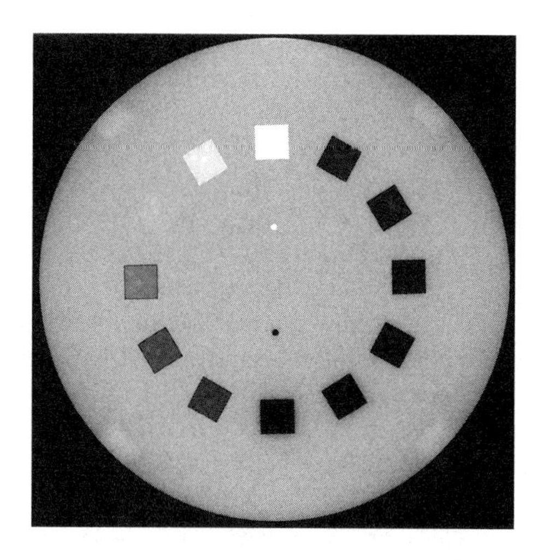

FIGURE 2.54
Radiograph of Leeds test objects TO.DR, which are used for assessing the performance of digital fluorography systems. Two test objects have been used to form this image. The first comprises a circular array of 12 square filters of varying contrast. The second comprises a circular array of 12 identical sets of 5 contrasting details, which are then superimposed on the filters in the image. The test objects are used for the assessment of dynamic range.

Examples of test objects suitable for testing these types of system are discussed by Cowen (1994) and are available from Leeds Test Objects Ltd. Test objects have been constructed for assessing contrast-detail detectability and dynamic range for digital non-subtractive and subtractive fluorographic systems. Figure 2.54 shows an image derived from a contact radiograph of test object TO.DR, which is used to assess dynamic range. A test object is also available for testing pixel mis-registration in DSF. It should be noted that tests relating to dosimetry and beam geometry are the same as those made for fluoroscopy. Recommended tests and tolerances for digital fluorography are summarised in IPEM (2005b); however, this publication acknowledges that the testing of complex digital fluorography and DSF systems is still evolving and some flexibility may have to be applied to these procedures and tolerances.

2.12.8 Mammographic Systems

Many of the basic quality-control tests required to check the performance of a mammography X-ray unit are similar to those for general radiography. However, different equipment, techniques and test tolerances are generally required due to the special nature of mammography. Tolerances tend to be much tighter than for general radiography since small changes in equipment performance can have a significant effect on image quality and dose. Furthermore, since mammography is currently the technique of choice for screening asymptomatic women as part of national breast screening programmes, it is essential that image quality and dose are optimised. Tests need to be carried out more frequently, some on a daily basis, in order to ensure performance consistency.

When assessing tube and generator performance, the instruments used for measuring tube voltage and tube output should be specifically designed and calibrated for the low energies used in mammography. Dedicated mammography ionisation chambers incorporate a very thin entrance window. Measured values of peak tube voltage, tube output and half-value layer will be different from those encountered in general radiography and there may be different relationships between the various parameters. For example, for a molybdenum target the tube output is approximately proportional to the cube of the tube voltage.

Focal-spot sizes are smaller than those used in general radiography in order to achieve the requirements for excellent spatial resolution and dedicated mammography test tools will be required. Tighter tolerances are needed for the alignment of the X-ray beam with the image receptor since it is particularly important that the entire breast is imaged (including the region closest to the chest wall, which is particularly susceptible to malignancy) whilst ensuring that other tissues are not irradiated. Tube and generator checks will be required for the different available combinations of target and filter materials as appropriate.

In screen–film mammography it is crucial that optimum film density is achieved in order to ensure that all breast tissues are adequately imaged. Frequent assessment of AEC performance and film-processor sensitometry is, therefore, required. Different thicknesses of PMMA are used to simulate breasts of different compressed breast thickness and glandular tissue content. Film density and post-exposure tube-current exposure-time product may be assessed for different combinations of breast thickness, target/filter material and tube voltage. The exposure parameters can also be used to calculate the mean glandular dose to the breast.

A variety of phantoms are available for assessing image quality in mammography, such as those manufactured by Leeds Test Objects Ltd. These phantoms provide a means of assessing parameters such as high-contrast resolution and threshold contrast of different types of detail. An image of one such phantom (TOR MAM, IPEM 2005a) is shown in Figure 2.55. This phantom contains filament structures, clusters of small high-contrast particles and larger low-contrast details in order to simulate different pathological features often seen in an abnormal breast.

Tests should be made in all modes of operation including all image formats, magnification mode and small-field digital mammography mode where available. Other tests specific to mammography include assessment of the force of the compression paddle and accuracy of the breast thickness display. Suitable tests for monitoring the performance of mammography units are given in IPEM (2005a) and test tolerances and frequencies are given in IPEM (2005b). Techniques and methods for testing small-field digital mammography systems are available in NHSBSP (2001).

The use of full-field digital mammography is increasing and gradually replacing conventional screen–film techniques. Greater dynamic range and the ability to manipulate images remove some of the constraints associated with screen–film mammography. However, there is still a requirement to optimise image quality and patient dose, and optimal beam qualities may be significantly different for digital mammography. Test protocols will need to be adapted; for example, whilst tests on the film processor become redundant,

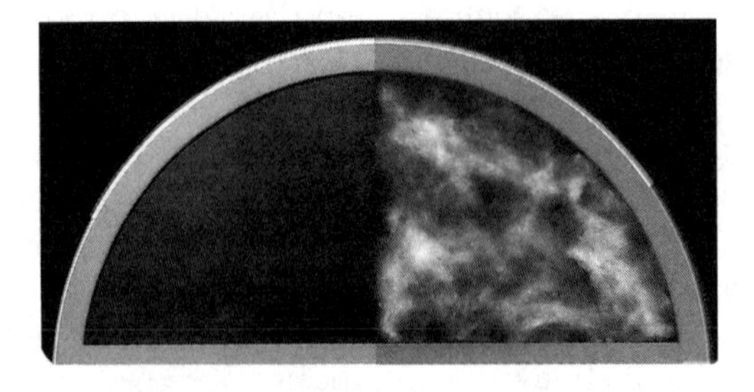

FIGURE 2.55
Radiograph of Leeds test object TOR MAM, which is used for the assessment of mammography X-ray units. One half of this phantom contains an array of filaments, particles and circular details, which represent pathological features often seen in an abnormal breast. The other half simulates breast tissue and contains microcalcifications, fibrous details and nodular details. The contrast on the left side of the image has been adjusted in an attempt to make some of the fine detail in the phantom more visible.

the use of display monitors for the clinical evaluation of images introduces new requirements for quality-control testing. Techniques and methods for testing full-field digital-mammography systems are available in NHSBSP (2009).

2.12.9 Digital Radiographic Systems

The rapid evolution of digital technology has presented new challenges for quality control. We have already considered how tests of AEC devices need to be adapted for digital radiography, and discussed methods for testing digital fluorography and DSF systems. We now turn our attention to the additional tests required for testing digital radiography systems. Quality control is particularly important when digital radiographic techniques are employed. With a screen–film combination it is obvious if the wrong exposure has been given; the film will appear too light or too dark. With digital techniques however, faults with the X-ray unit or AEC system may go unnoticed because an adequate image may still be achieved over a wide range of exposures due to the increased dynamic range and the ability to manipulate the image following exposure. A careful programme of quality control for the tube, generator and AEC will therefore be required, although these tests may require some adaptation. Tests for systems which employ photostimulable phosphor plates (CR) will differ slightly from tests for those types of system where images are directly read out from the image receptor without the user moving it from the exposure position (referred to in this section as DDR systems) and these distinctions are made where appropriate.

We have already considered the use of the DDI as a means of monitoring the performance of digital radiography systems with respect to AEC devices. This indicator may also be used for monitoring other aspects of digital radiography performance. However, the relationship between the DDI and exposure is not linear and is highly dependent on beam quality, and indicator values for CR may also be affected by latent image decay of the signal between exposure and read-out. The DDI should be calibrated according to the supplier's method, which may specify particular beam qualities and exposure parameters to be used and, for CR, may specify the time elapsed between exposure and read-out.

Protocols available for clinical imaging may not be suitable for quality control purposes due to the amount of image processing applied. Suppliers of digital radiography systems generally provide test protocols suitable for quality control, which apply minimal processing and ideally maintain a direct relationship between mean pixel value and dose. This relationship may not be linear and is known as the signal transfer property (STP) of the system. The STP must be measured in order that measurements of pixel value obtained for certain tests may be converted to the dose delivered to the detector. We will now consider some of the tests required for digital radiography systems.

The performance of a digital radiography system may deteriorate over time. For CR, this may be due to a drift in the gain of the photomultiplier tube in the reader, dirt in the reader's optical assembly or a deterioration of the plates themselves. A decline in system performance may be monitored by assessing the reproducibility of the DDI. This may be achieved by exposing the CR cassette or direct digital detector to a known exposure, for example, 10 μGy, using a specific tube voltage, beam filtration, test protocol and (for CR) reading the cassette after a specified time period. The resultant DDI is recorded. The specific conditions used may be those specified by the supplier – in which case the measured DDI may be compared with a nominal value – or, for routine comparisons, a standard set of test conditions may be used for all types of detector. Tolerances should be based on the actual dose delivered since the DDI is not linear with exposure. The repeatability of the indicator should also be assessed by repeating the test a number of times. For CR, the DDI obtained from a known exposure should be assessed for all new cassettes to ensure consistency of performance. Cassettes and image plates will also need to be cleaned and erased regularly according to the supplier's recommendations.

The STP of the system may be established by performing this test at different dose levels within the dynamic range of the system, for example, at 1, 10 and 30 μGy, and measuring the resultant mean pixel values using region-of-interest analysis in addition to recording the DDI. Image uniformity may be assessed by measuring the mean pixel value at different positions in the image and converting these pixel values to detector dose using the system STP. Images should also be inspected for artefacts using a narrow window width. Uniformity may be influenced by the anode heel effect; this effect can be minimised either by using a large focus–detector distance, or by making two exposures with the detector rotated by 180° with respect to the tube focus between exposures.

The presence of noise in the system and, for CR, any light leakage in the cassettes or reader, may be assessed by measuring the system dark noise. For CR, cassettes should first be erased, left to rest for 5 min and then read under standard conditions. For direct digital detectors an image should be obtained without exposing or, if this is not possible, using a very low exposure. The DDI should be recorded and the mean pixel value measured using region-of-interest analysis and converted to dose using the system STP. Suppliers often provide tolerances on the DDI obtained for this test. Images should also be inspected for structural noise.

Most digital radiography systems incorporate electronic callipers for distance measurement. Calliper accuracy and non-linearity of the detector or CR reader may be assessed by imaging a grid or other object of known dimensions using all sampling rates available. Measurements should be made in orthogonal directions; for CR it is particularly important that measurements are made at the edges of images in the direction of scanning since faults may occur at the beginning or end of a scan.

For CR the efficiency of the erasure cycle should be assessed. This may be achieved by exposing a cassette to a high dose with the cassette partially occluded by a high attenuation material such as lead. The cassette is read, re-exposed at a low exposure without the high attenuation material and then read again. The second image should be examined for any evidence of retention of the first image. Region-of-interest or line profile analysis may be helpful in establishing whether any residual image remains when visual analysis is inconclusive.

The image quality of digital radiography systems may be assessed using test objects similar in design to those used for assessing digital fluorographic systems. For example, test object TO.20 manufactured by Leeds Test Objects Ltd. is useful for the assessment of threshold-contrast-detail detectability. Baseline values should be established at acceptance to facilitate comparisons during routine testing. The test conditions should be kept constant to ensure measurement consistency. The visibility of test details will be influenced by the contrast and brightness settings; in general, it is useful to adjust these settings such that image noise is just visible in the background in order to optimise detail visibility.

Limiting spatial resolution may be assessed by imaging a lead-grating resolution bar pattern under low-noise and high-contrast conditions. For CR tests should be made for all available sampling rates. The limiting resolution should be expected to approach the Nyquist limit. Positioning the grating at 45° relative to the detector gives more consistent results since this avoids interference effects; however, care should be taken when comparing measured values with the nominal pixel pitch, and the Nyquist frequency at 45° must be calculated. Uniformity of resolution may be assessed using a fine wire mesh and checking the image for blurring and discontinuities.

The testing of digital radiography systems is still evolving. Suitable tests and test methodologies are given in IPEM (2010). Other useful publications include BIR (2001), Samei et al. (2001) and SCAR (2002). Test tolerances and frequencies are given in IPEM (2005b).

2.12.10 Image Display Monitors

The growth of digital imaging technology and its implementation in the clinical environment have resulted in the displacement of film by display monitors for medical image evaluation. It is important that these monitors perform optimally and a suitable quality-assurance programme is, therefore, required. Display monitors may be used either for the interpretation of medical images; or for review of these images for other purposes, usually after a report has already been made available. Both types of display must meet minimum standards of performance, but the requirements for displays used for image interpretation are more stringent.

When using display monitors for interpreting medical images it is essential that the ambient viewing conditions are suitable. Low ambient lighting conditions are required and monitors should be free from reflections. The ambient light level may be assessed by measuring the amount of light incident on the surface of the monitor (illuminance) with the monitor switched off, using a suitable photometer. All monitors should be kept clean and free from dust and fingerprints.

Most display monitor tests involve the assessment of electronic test pattern images such as the Society of Motion Picture and Television Engineers (1991) test pattern or those recommended in AAPM (2005). These test patterns incorporate areas of varying grey level, details indicating contrast at the extremes of grey level, and high-and low-contrast resolution patterns. Monitor grey scale can be characterised by measuring the luminance of areas of varying grey level from black (0%) to white (100%) using a photometer. The photometer should have a suitable photopic spectral response and should be calibrated. Contrast may be measured by assessing the visibility of 5% and 95% details superimposed on 0% and 100% backgrounds, respectively. Uniformity may be assessed by displaying an area of 50% grey level over the entire monitor and measuring luminance at the centre and four corners of the screen. The variation in luminance between monitors should also be assessed, particularly where multiple monitors are used at the same workstation. Other tests include the assessment of low- and high-contrast resolution at the centre and periphery of the monitor; accuracy of distance and angle measurement tools; and monitor distortion. A more detailed description of suitable tests and test methodologies is given in AAPM (2005). Test tolerances and frequencies are given in IPEM (2005b).

References

Alm Carlsson G, Dance D R, Sandborg M and Persliden J 1999 Use of the concept of energy imparted in diagnostic radiology. In: *Subject Dose in Radiological Imaging* K-H Ng, D A Bradley and H M Warren-Forward (Eds.) (Amsterdam, the Netherland: Elsevier) pp. 39–62.

American Association of Physicists in Medicine (AAPM) 2005 *Assessment of display performance for medical imaging systems AAPM Task Group 18 On-line Report OR-03* [URL http://www.aapm.org/pubs/reports/].

Ammann E 1990 X-ray generators and control circuits. In: *Imaging Systems for Medical Diagnostics* E Krestel (Ed.) (Berlin and Munich, Germany: Siemens Aktiengesellschaft) pp. 284–317.

Arnold B A and Bjarngard B E 1979 Effect of phosphor K X-rays on MTF of rare earth screens *Med. Phys.* **6** 500–503.

Attix F H 1986 *Introduction to Radiological Physics and Radiation Dosimetry* (New York: Wiley-Interscience).

Bacher K, Smeets P, Vereecken L, De Hauwere A, Duyck P, De Man R, Verstraete K and Thierens H 2006 Image quality and radiation dose on digital chest imaging: Comparison of amorphous silicon and amorphous selenium flat-panel systems *Am. J. Roentgenol.* **187** 630–637.

Barnes G T 1982 Radiographic mottle: A comprehensive theory *Med. Phys.* **9** 656–667.

Barrett H H and Swindell W 1981 *Radiological Imaging* vols I and II (London, UK: Academic Press).

Brettle D S, Ward S C, Parkin G J S, Cowen A R and Sumsion H J 1994 A clinical comparison between conventional and digital mammography utilising computed radiography *Br. J. Radiol.* **67** 464–468.

British Institute of Radiology (BIR) 2001 *Assurance of Quality in the Diagnostic X-ray Department* (2nd edn.) (London, UK: British Institute of Radiology).

Bunch P C 1994 Performance characteristics of high-MTF screen-film systems *SPIE* **2163** 14–34.

Bunch P C 1999 Advances in high-speed mammographic image quality *SPIE* **3659** 120–130.

Cowen A R 1994 The physical evaluation of the imaging performance of television fluoroscopy and digital fluorography systems using the Leeds x-ray test objects: A UK approach to quality-assurance in the diagnostic radiology department. In: *Specification, Acceptance Testing and Quality Control of Diagnostic X-Ray Imaging Equipment* J A Seibert, G T Barnes and R G Gould (Eds.) Medical Physics Monograph No. 20 (Woodbury, NY: AAPM) pp. 499–568.

Cranley K, Gilmore B J and Fogarty G W A 1991 Data for estimating X-ray tube total filtration *Institute of Physical Sciences in Medicine Report 64* (York, UK: IPEM).

Cranley K, Gilmore B J, Fogarty G W A, Desponds L and Sutton D 1997 *Catalogue of Diagnostic X-ray Spectra and Other Data* IPEM78 (York, UK: IPEM).

Curry S, Dowdey J E and Murry C 1990 *Christensen's Physics of Diagnostic Radiology* (4th edn.) (Philadelphia, PA: Lea & Febiger).

Dainty J C and Shaw R 1974 *Image Science* (London, UK: Academic Press).

Dance D R, Persliden J and Alm Carlsson G 1992. Monte Carlo calculation of the properties of mammographic anti-scatter grids *Phys. Med. Biol.* **37** 235–248.

Dance D R, Skinner C L and Alm Carlsson G 1999. Breast dosimetry. In: *Subject Dose in Radiological Imaging* K-H Ng, D A Bradley and H M Warren-Forward (Eds.) (Amsterdam, the Netherlands: Elsevier) pp. 185–203.

Dance D R, Thilander Klang A, Sandborg M, Skinner C L, Castellano Smith I A and Alm Carlsson G 2000 Influence of anode/filter material and tube potential on contrast, signal-to-noise ratio and average absorbed dose in mammography: A Monte Carlo study *Br. J. Radiol.* **73** 1056–1067.

de Groot P M 1991 Image intensifier design specifications. In: *Specification, Acceptance Testing and Quality Control of Diagnostic X-ray Imaging Equipment*, J A Seibert, G T Barnes and R G Gould (Eds.) Medical Physics Monograph No. 20 (Woodbury, NY: AAPM) pp. 429–460.

Dobbins J T 1995 Effects of undersampling on the proper interpretation of modulation transfer function, noise power spectra, and noise equivalent quanta of digital x-ray imaging systems *Med. Phys.* **22** 171–181.

Dobbins J T and Godfrey D J 2003 Digital x-ray tomosynthesis: Current state of the art and clinical potential *Phys. Med. Biol.* **48** R65–R106.

Doi K, Holje G, Loo L-N, Chan H-P, Sandrik J M, Jennings R J and Wagner R F 1982 *MTFs and Wiener Spectra of Radiographic Screen-Film Systems* FDA 82–8187 (Rockville, MD: Bureau of Radiological Health).

Doi K, Yoshie K, Loo L-N, Chan H-P, Yoshiharu H and Jennings R J 1986 *MTFs and Wiener Spectra of Radiographic Screen-film Systems* vol. II FDA 86–8257 (Rockville, MD: Centre for Devices and Radiological Health).

European Union (EU) 1997 Council Directive 97/43/EURATOM 1997 on health protection of individuals against the dangers of ionizing radiation in relation to medical exposure *Offi. J. Eur. Commun.* No L 180.

Fetterly K A and Schueler B 2003 Performance evaluation of a 'dual-side read' dedicated mammography computed radiography system *Med. Phys.* **30** 1843–1854.

Fetterly K A and Schueler B 2006 Performance evaluation of a computed radiography imaging device using a typical 'front side' and novel 'dual side' readout storage phosphors *Med. Phys.* **33** 290–296.

Floyd C E, Warp R J, Dobbins J T, Chotas H G, Baydush A H, Vargas-Voracek R and Ravin C E 2001 Imaging characteristics of an amorphous silicon flat-panel detector for digital chest radiography *Radiology* **218**:683–688.

Forster E 1985 *Equipment for Diagnostic Radiography* (Lancaster, UK: MTP Press).

Ginzburg A and Dick C E 1993 Image information transfer properties of x-ray intensifying screens in the energy range from 17 to 320 keV *Med. Phys.* **20** 1011–1021.

Gould R G 1999 Digital angiography and fluoroscopy: An overview. In: *Practical Digital Imaging and PACS* Seibert J A, Filipow L J and Andriole K P (Eds.), Medical Physics Monograph No. 25 (Madison, WI: American Association of Physicists in Medicine) pp. 91–106.

Hart D, Hillier M C and Wall B F 2002 *Doses to Patients from Medical X-ray Examinations in the UK – 2000 Review* NRPB-W14 (Chilton, UK: National Radiological Protection Board).

Hart D and Wall B F 2002 *Radiation Exposure of the UK population from Medical and Dental X-ray Examinations* NRPB-W4 (Chilton, UK: National Radiological Protection Board).

Hay G A, Clarke O F, Coleman N J and Cowen A R 1985 A set of X-ray test objects for quality control in television fluoroscopy *Br. J. Radiol.* **58**, 335–344.

Hezaji S and Trauernicht D P 1996 Potential image quality in scintillator CCD-based imaging systems for digital radiography and digital mammography *SPIE* **924** 253–261.

Hughes J S, Shaw K B and O'Riordan M C 1989 *Radiation Exposure to the UK Population – 1988 Review* NRPB-R227 (Chilton, UK: National Radiological Protection Board).

ICRP (International Commission on Radiological Protection) 2000 *Avoidance of Radiation Injuries from Medical Interventional Procedures* ICRP Publication 85, Annals ICRP 30(2) (Oxford, UK: Pergamon).

ICRP (International Commission on Radiological Protection) 2008 *The 2007 Recommendations of the International Commission on Radiological Protection* ICRP Publication 103, Annals ICRP 36(2–4) (Oxford, UK: Pergamon).

ICRU (International Commission on Radiation Units and Measurements) 1986 *Modulation Transfer Function of Screen Film Systems* ICRU Report 41 (Bethesda, MD: ICRU).

ICRU (International Commission on Radiation Units and Measurements) 1996 *Medical Imaging – The Assessment of Image Quality* ICRU Report 54 (Bethesda, MD: ICRU).

Institute of Physics and Engineering in Medicine (IPEM) 1997 *Measurement of the Performance Characteristics of Diagnostic X-ray Systems used in Medicine* Report 32 (2nd edn.) *Part IV: X-ray Intensifying Screens, Films, Processors and Automatic Exposure Control Systems* (York, UK: IPEM).

Institute of Physics and Engineering in Medicine (IPEM) 2005a *The Commissioning and Routine Testing of Mammographic X-Ray Systems* Report 89 (York, UK: IPEM).

Institute of Physics and Engineering in Medicine (IPEM) 2005b *Recommended Standards for the Routine Performance Testing of Diagnostic X-ray Imaging Systems* Report 91 (York, UK: IPEM).

Institute of Physics and Engineering in Medicine (IPEM) 2010 *Measurement of the Performance Characteristics of Diagnostic X-ray Systems* Report 32 *Part VII: Digital Imaging Systems* (York, UK: IPEM).

Institution of Physics and Engineering in Medicine And Biology (IPEMB) 1996 *Measurement of the Performance Characteristics of Diagnostic X-ray Systems used in Medicine* Report 32 (2nd edn.) *Part I: X-ray Tubes and Generators; Part II: X-ray Image Intensifier Television Systems; Part V: Conventional Tomographic Equipment; Part VI: X-ray Image Intensifier Fluorography Systems* (York, UK: IPEMB).

International Electrotechnical Commission (IEC) 2001 *Diagnostic X-ray imaging equipment – Characteristics of General Purpose and Mammographic Anti-Scatter Grids* IEC Publication 60627 (2nd edn.) (Geneva, Switzerland: IEC).

Kimme-Smith C, Lewis C, Beifuss M, Williams M B and Bassett L W 1998 Establishing minimum performance standards, calibration intervals and optimal exposure values for a whole breast digital mammography unit *Med. Phys.* **25** 2410–2416.

Lehmann L A, Alvarez R E, Macovski A and Brody W R 1981 Generalized image combinations in dual kVp digital radiography *Med. Phys.* **8** 659–667.

Manninen H, Rytkönen H, Soimakallio S, Terho E O and Hentunen J 1986 Evaluation of an anatomical filter for chest radiography *Br. J. Radiol.* **59** 1087–1092.

McKinlay A F 1981 *Thermo-Luminescent Dosimetry* (Bristol, UK: Adam Hilger) p 12.

Medical Devices Agency (MDA) 1997 A technical evaluation of the Fuji ST-VA computed radiography image plates under standard radiographic conditions Medical Devices Agency Evaluation Report MDA/97/12 (London, UK: MDA).

Medical Devices Agency (MDA) 1998 Evaluation of Konica radiographic screen and films. Part 2: Image quality measurements Medical Devices Agency Evaluation Report MDA/98/10 (London, UK: MDA).

Medical Devices Agency (MDA) 2001 Evaluation of the IGE medical systems senographe 2000D full field digital mammography unit Medical Devices Agency Evaluation Report MDA 01041 (London, UK: MDA).

Mees C E K and James T H 1966 *The Theory of the Photographic Process* (3rd edn.) (New York: Macmillan).

Monnin P, Gutierrez D, Bulling S, Guntern D and Verdun F R 2007 A comparison of the performance of digital mammography systems *Med. Phys.* **34** 906–914.

Monnin P, Holzer Z, Wolf R, Neitzel U, Vock P, Gudinchet F and Verdun F R 2006 An image quality comparison of standard and dual-side read CR systems for pediatric radiology *Med. Phys.* **33** 411–420.

Moores B M 1984 Physical Aspects of Digital Fluorography *Digital Radiology-Physical and Clinical Aspects* R M Harrison and I Isherwood (Eds.) IPSM (London, UK: Hospital Physicists' Association) pp. 45–57.

Muller S 1999 Full-field digital mammography designed as a complete system *Eur. J. Radiol.* **31** 25–34.

National Health Service Breast Screening Programme (NHSBSP) 2001 *Commissioning and Routine Testing of Small Field Digital Mammography Systems* Report 01/09 (Sheffield, UK: NHS Cancer Screening Programmes).

National Health Service Breast Screening Programme (NHSBSP) 2009 Commissioning and routine testing of full field digital mammography systems Report 06/04 Version 3 (Sheffield, UK: NHS Cancer Screening Programmes).

Nishikawa, R M and Yaffe M J 1990 SNR properties of mammographic film-screen systems *Med. Phys.* **12** 32–39.

Nudelman S, Roehrig H and Capp M P 1982 A study of photoelectronic-digital radiology – Part III: Image acquisition components and system design *Proc. IEEE* **70** 715–727.

Persliden J and Alm Carlsson G 1997 Scatter rejection by air gaps in diagnostic radiology *Phys. Med. Biol.* **42** 155–175.

Plewes D B 1994 Scanning equalisation radiography. In: *Specification, Acceptance Testing and Quality Control of Diagnostic X-ray Imaging Equipment* Seibert J A, Barnes G T and Gould R G (Eds.) Medical Physics Monograph No. 20 (New York: AAPM) pp. 359–381.

Rezentes P S, de Almeida A and Barnes G T 1999 Mammography grid performance *Radiology* **228** 227–232.

Riccardi L, Cauzzo M C, Fabbris R, Tonini E and Righetto R 2007 Comparison between a 'dual side' chest imaging device and a standard 'single side' CR *Med. Phys.* **34** 119–126.

Rose A 1973 *Vision: Human and Electronic* (New York: Plenum) pp. 21–23.

Rowlands J A 2002 The physics of computed radiography *Phys. Med. Biol.* **47** R123–R166.

Rowlands J and Kasap S 1997 Amorphous semiconductors usher in digital x-ray imaging *Physics Today* **50** 24–30.

Rowlands J A and Yorkston J 2000 Flat-panel detectors for digital radiography. In: *Handbook of Medical Imaging* Beutal J, Kundel H L, Van Metter R L (Eds.) (Bellingham, WA: SPIE Press) pp. 331–367.

Samei E and Flynn M J 2002 An experimental comparison of detector performance for computed radiography systems *Med. Phys.* **29** 447–459.

Samei E, Ranger N T, Dobbins J T and Chen Y 2006 Intercomparison of methods for image quality characterization. I. Modulation transfer function *Med. Phys.* **33** 1454–1465.

Samei E, Seibert J A, Willis C E, Flynn M J, Mah E and Junck K L 2001 Performance evaluation of computed radiography systems *Med. Phys.* **28** 361–371.

Sandborg M, Dance D R, Alm Carlsson G and Persliden J 1993 Selection of anti-scatter grids for different imaging tasks: The advantage of low atomic number cover and interspace material *Br. J. Radiol.* **66** 1151–1163.

Sandborg M, McVey G, Dance D R and Alm Carlsson G 2001 Schemes for the optimization of chest radiography using a computer model of the patient and x-ray imaging system. *Med. Phys.* **28** 2007–2019.

Sandrik J M and Wagner R F 1982 Absolute measure of physical image quality: Measurement and application to radiographic magnification *Med. Phys.* **9** 540–549.

Society for Computer Applications in Radiology (SCAR) 2002 *Quality-Assurance: Meeting the Challenge in the Digital Medical Enterprise* (Great Falls, VA: SCAR).

Society of Motion Picture & Television Engineers (SMPTE) 1991 *Specifications for Medical Diagnostic Imaging Test Pattern for Television Monitors and Hardcopy Recording Cameras SMPTE RP133* (New York: SMPTE).

Sonoda M, Takano M, Miyahara J and Kato H 1983 Computed radiography utilising scanning laser stimulated luminescence *Radiology* **148** 833–838.

Strotzer M, Völk M, Fründ R, Hamer O, Zorger N and Feuerbach S 2002 Routine chest radiography using a flat panel detector *Am. J. Roentgenol.* **178** 169–71.

Swank R K 1973 Absorption and noise in x-ray phosphors *J. Appl. Phys.* **44** 4199–203.

Tanner R J, Wall B F, Shrimpton P C, Hart D and Bungay D R 2001 *Frequency of Medical and Dental X-Ray Examinations in the UK – 1997/98* NRPB-R320 (Chilton, UK: National Radiological Protection Board).

Tesic M M, Fisher Picarro M and Munier B 1997 Full field digital mammography scanner *Eur. J. Radiol.* **31** 2–17.

Tesic M M, Mattson R A, Barnes G T, Sones R A and Stickney J B 1983 Digital radiology of the chest: Design features and considerations for a prototype unit *Radiology* **148** 259–264.

Ullman G, Sandborg M, Dance D R, Hunt R A and Alm Carlsson G 2006. Towards optimization in digital chest radiography using Monte Carlo modelling *Phys. Med. Biol.* **51** 2729–2743.

Van Metter R and Rabbani M 1990 An application of moment generating functions to the analysis of signal and noise propagation in radiographic screen-film systems *Med. Phys.* **17** 65–71.

Webster J G 1992 *Medical Instrumentation* (Boston, MA: Houghton Mifflin) p. 648.

Yaffe M and Rowlands J 1997 X-ray detectors for digital radiography *Phys. Med. Biol.* **42** 1–39.

Young K C and Burch A 2000 Radiation doses received in the UK breast screening programme in 1997 and 1998 *Br. J. Radiol.* **73** 278–287.

3

X-Ray Transmission Computed Tomography

I. A. Castellano and S. Webb

CONTENTS

3.1 Need for Sectional Images 98
3.2 Principles of Sectional Imaging 100
 3.2.1 Source–Detector Geometries 100
 3.2.2 Line Integrals 103
 3.2.3 Projection Sets 105
 3.2.4 Information Content in Projections and the Central-Section Theorem 106
 3.2.5 Reconstruction by 2D Fourier Methods 108
 3.2.6 CT Image 108
3.3 Developments in Scanner Design 109
 3.3.1 Electron-Beam CT 109
 3.3.2 Spiral CT 109
 3.3.3 Multi-Slice CT 110
 3.3.4 Dual-Source CT 112
3.4 Scanner Operation and Components 112
 3.4.1 Gantry 112
 3.4.2 X-Ray Generation 113
 3.4.3 Beam Filtration and Collimation 115
 3.4.4 Data Acquisition System 115
 3.4.5 Patient Couch 117
 3.4.6 Computer System 117
 3.4.7 Image Reconstruction and Display 118
3.5 Data Interpolation in Spiral CT 119
 3.5.1 Interpolation Techniques for Single-Slice Scanners 119
 3.5.2 Interpolation Techniques for Multi-Slice Scanners 120
3.6 2D Image Reconstruction 121
 3.6.1 2D Filtered Backprojection 122
 3.6.2 Practical Implementation 124
3.7 Iterative Methods of Reconstruction 128
3.8 Cone-Beam Reconstruction 130
 3.8.1 Exact Cone-Beam Reconstruction Algorithms 131
 3.8.2 Approximate Cone-Beam Reconstruction Algorithms 131
3.9 Image Quality Considerations 134
 3.9.1 Image Quality Parameters 134
 3.9.2 Angular Sampling Requirements 136

3.9.3 Relation between Image Quality Parameters 136
3.9.4 Partial-Volume Artefacts 137
3.9.5 Beam-Hardening Artefacts 138
3.9.6 Aliasing Artefacts 138
3.9.7 Motion Artefacts 139
3.9.8 Equipment-Related Artefacts 139
3.9.9 Effect of Spiral Interpolation Algorithms on Image Quality 140
3.9.10 Effect of Iterative Reconstruction Algorithms on Image Quality 141
3.9.11 Effect of Cone-Beam Reconstruction Algorithms on Image Quality 141
3.10 Recent Developments in CT Scanning 142
3.11 Performance Assessment of CT Scanners 143
3.11.1 Tube and Generator Tests 144
3.11.2 Scan Localisation 144
3.11.3 CT Dosimetry 145
3.11.4 Image Quality 145
3.11.5 Helical Scanning 146
3.A Appendix 148
References 148

3.1 Need for Sectional Images

When we look at a chest X-ray (see Figure 3.1), certain anatomical features are immediately apparent. The ribs, for example, show up as a light structure because they attenuate the X-ray beam more strongly than the surrounding soft tissue, so the film receives less exposure in the shadow of the bone. Correspondingly, the air-filled lungs show up as darker regions.

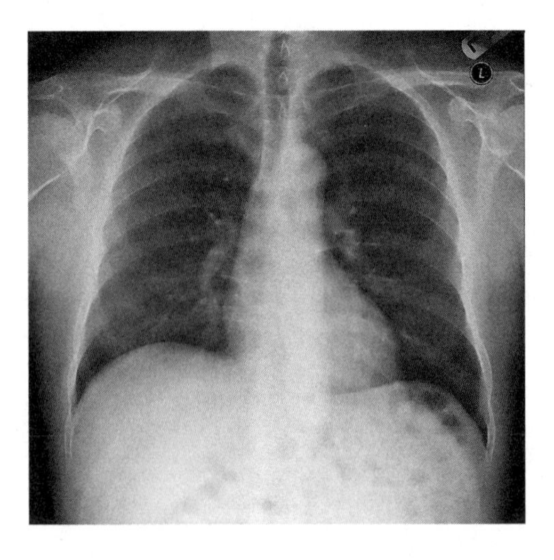

FIGURE 3.1
Typical chest X-radiograph.

A simple calculation illustrates the type of structure that one could expect to see with this sort of conventional transmission radiograph. The linear attenuation coefficients for air, bone, muscle and blood are

$$\mu_{\text{air}} = 0$$

$$\mu_{\text{bone}} = 0.48\,\text{cm}^{-1}$$

$$\mu_{\text{muscle}} = 0.180\,\text{cm}^{-1}$$

$$\mu_{\text{blood}} = 0.178\,\text{cm}^{-1}$$

for the energy spectrum of a typical diagnostic X-ray beam. Thus, for a slab of soft tissue with a 1 cm cavity in it, the results of Table 3.1 follow at once using Beer's expression for the attenuation of the primary beam, namely,

$$I(x) = I_0 \exp(-\mu x) \tag{3.1}$$

where
I_0 is the fluence entering the slab, and
$I(x)$ is the fluence exiting the slab.

X-ray films usually allow contrasts of the order of 2% to be seen easily, so a 1 cm thick rib or a 1 cm diameter air-filled trachea can be visualised. However, the blood in the blood vessels and other soft-tissue details, such as details of the heart anatomy, cannot be seen on a conventional radiograph. In fact, to make the blood vessels visible, the blood has to be infiltrated with a liquid contrast medium containing iodine compounds; the iodine temporarily increases the linear attenuation coefficient of the fluid medium to the point where visual contrast is generated (see Section 2.10, where contrast media are discussed in detail). Photon scatter further degrades contrast (see Sections 2.4.1 and 2.6.1).

Another problem with the conventional radiograph is the loss of depth information. The 3D structure of the body has been collapsed, or projected, onto a 2D film and, while this is not always a problem, sometimes other techniques such as tomosynthesis or conventional tomography (see Section 2.12.5) are needed to retrieve the depth information.

It is apparent that conventional radiographs are inadequate in these two respects, namely, the inability to distinguish soft tissue and the inability to spatially resolve structures along the direction of X-ray propagation.

TABLE 3.1

Contrast in a transmission radiograph.

Material in Cavity	$I(x)/I_0$ (x = 1 cm)	Difference (%) with respect to Muscle
Air	1.0	+20
Blood	0.837	+0.2
Muscle	0.835	0
Bone	0.619	−26

3.2 Principles of Sectional Imaging

With computed tomography (CT), a planar slice of the body is defined and X-rays are passed through it only in directions that are contained within, and are parallel to, the plane of the slice (see Figure 3.2). No part of the body that is outside of the slice is interrogated by the X-ray beam, and this eliminates the problem of 'depth scrambling'. The CT image is as though the slice (which is usually a few millimetres thick) had been physically removed from the body and then radiographed by passing X-rays through it in a direction perpendicular to its plane. The resulting images show the human anatomy in section with a spatial resolution of about 1 mm and a density (linear attenuation coefficient) discrimination of better than 1% (see Figure 3.3). This chapter is about the method of converting the X-ray measurements of Figure 3.2 into the images shown in Figure 3.3.

There are several seminal papers on the principles of CT; see, for example, Brooks and Di Chiro (1975, 1976a). There are also several excellent publications which deal extensively with this imaging technique, such as Kalender (2005) and Hsieh (2009). The commercial development of CT has been described as the most important breakthrough in diagnostic radiology since the development of the first planar radiograph.

3.2.1 Source–Detector Geometries

As far as the patient is concerned, the CT scanner is a machine with a large hole in it. The body or the head is placed inside the hole in order to have the pictures taken (see Figure 3.4). The covers of the machine hide a complicated piece of machinery, which has evolved through several versions since its inception (Hounsfield 1973). A short description of the development follows.

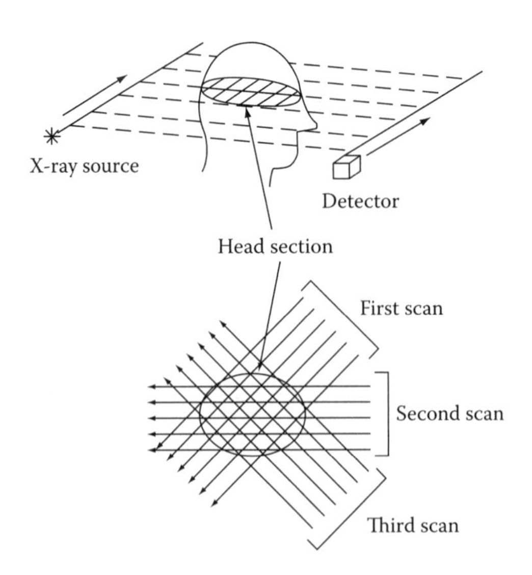

FIGURE 3.2
Simple scanning system for transaxial tomography. A pencil beam of X-rays passes through the object and is detected on the far side. The source–detector assembly is scanned sideways to generate one projection. This is repeated at many viewing angles and the required set of projection data is obtained. (Reproduced from Barrett and Swindell, 1981.)

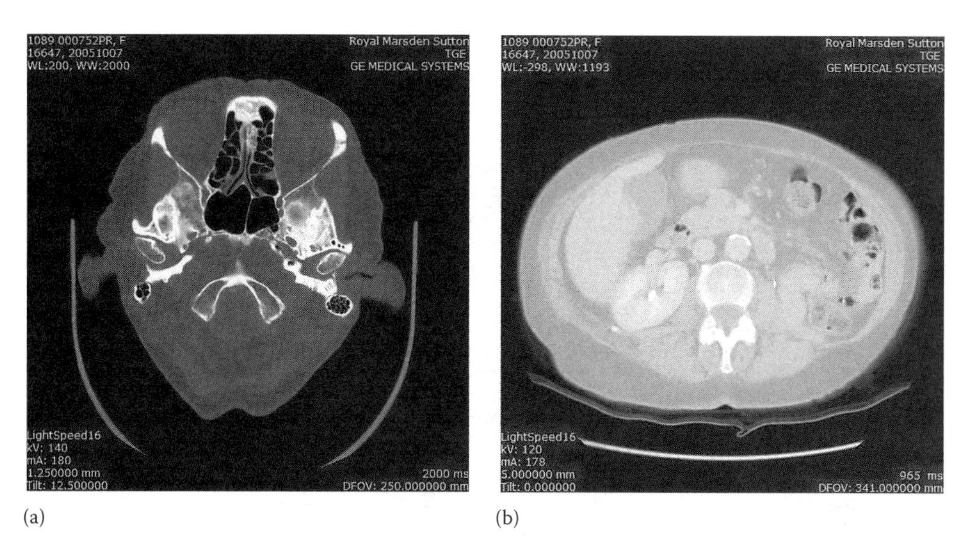

FIGURE 3.3
(a) CT image of a head taken at an angle through the eyes. (b) Abdominal section through the kidneys.

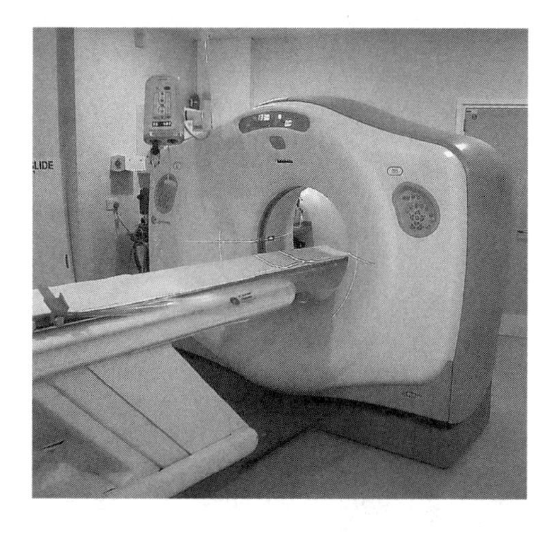

FIGURE 3.4
Typical CT scanner.

In a first-generation scanner, a finely collimated source defined a pencil beam of X-rays, which was then measured by a well-collimated detector. This source–detector combination measured parallel projections, one sample at a time, by stepping linearly across the patient. After each projection, the gantry rotated to a new position for the next projection (see Figure 3.5). Since there was only one detector, calibration was easy and there was no problem with having to balance multiple detectors; also costs were minimised. The scatter rejection of this first-generation system was higher than that of any other generation because of the 2D collimation at both source and detector. The system was slow, however, with typical acquisition times of 4 min per section, even for relatively low-resolution images.

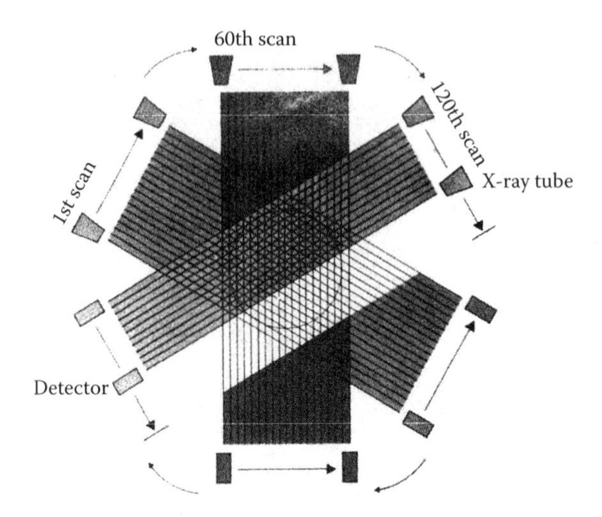

FIGURE 3.5
Schematic representation of a first-generation CT scanner. It utilises a single pencil beam and single detector for each scan slice. The X-ray source and detector are passed transversely across the object being scanned, with incremental rotations of the system at the end of each transverse motion. (Reproduced from Maravilla and Pastel, 1978.)

Data gathering was speeded up considerably in the second generation, which was introduced in 1974. Here a single source illuminated an array of detectors with a narrow (~10°) fan beam of X-rays (see Figure 3.6). This assembly traversed the patient and measured N parallel projections simultaneously (N being the number of detectors). The gantry angle was incremented by an angle equal to the fan angle between consecutive traverses. These machines could complete the data gathering in about 20 s. If the patient could suspend

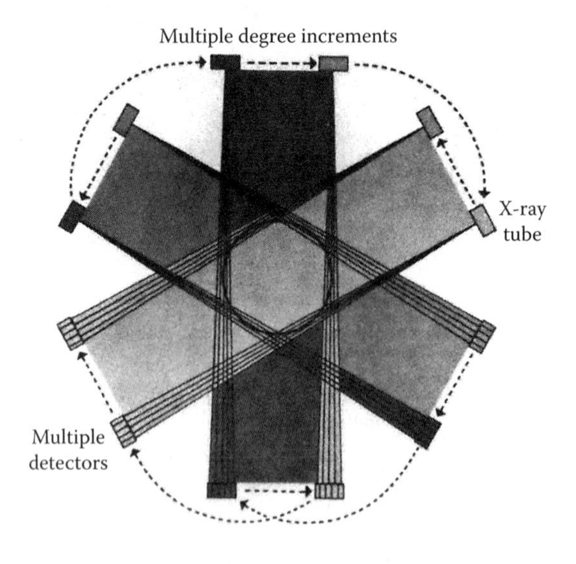

FIGURE 3.6
Schematic representation of a second-generation CT scanner. A narrow-angle fan beam of X-rays and multiple detectors record several pencil beams simultaneously. As the diverging pencil beams pass through the patient at different angles, this enables the gantry to rotate in increments of several degrees and results in markedly decreased scan times of 20 s or less. (Reproduced from Maravilla and Pastel, 1978.)

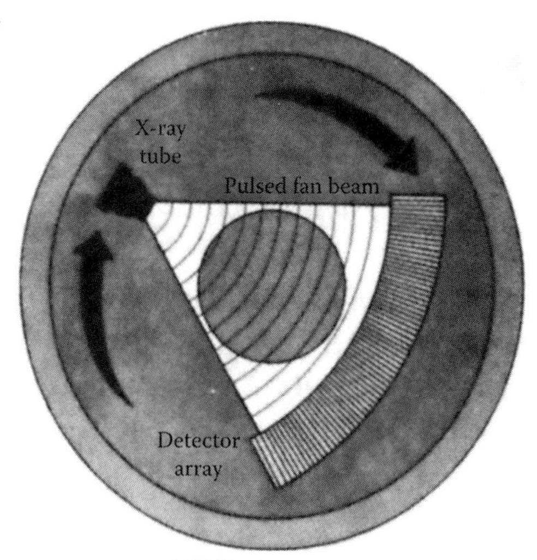

FIGURE 3.7
Schematic representation of a third-generation CT scanner in which a wide-range fan beam of X-rays encompasses the entire scanned object. Several hundred measurements were recorded with each pulse of the X-ray tube (the X-ray tube does not pulse in modern CT scanners). (Reproduced from Maravilla and Pastel, 1978.)

breathing for this period, the images would not be degraded by motion blur which would otherwise be present in chest and abdominal images.

The third generation of scanner geometry became available in 1975. In these systems, the fan beam was enlarged to cover the whole field of view (see Figure 3.7), typically 50 cm. Consequently, the gantry needed only to rotate, which it could do without stopping, and the data gathering could be done in less than 5 s. It is relatively easy for a patient to remain still for this length of time.

Fourth-generation systems became available a year or so after the third-generation geometry was introduced. In this design, a stationary detector ring was used and only the source rotated along a circular path contained within the detector ring (see Figure 3.8). Scan speeds were comparable to that of third-generation scanners.

For the present-day CT scanner, manufacturers have almost exclusively adopted third-generation geometry. For over two decades, third- and fourth-generation systems competed for technical supremacy in the clinic. The relative merits of the two geometries will be discussed in Section 3.4.

3.2.2 Line Integrals

The data needed to reconstruct the image are transmission measurements through the patient. Assuming, for simplicity, that we have (1) a very narrow pencil beam of X-rays, (2) monochromatic radiation and (3) no scattered radiation reaching the detector, then the transmitted intensity is given by

$$I_\phi(x') = I_\phi^0(x')\exp\left(-\int_{AB} \mu[x, y]\,dy'\right) \tag{3.2}$$

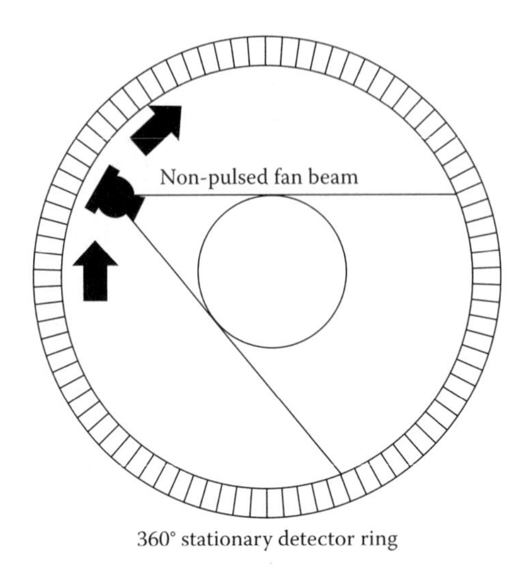

FIGURE 3.8
Schematic representation of a fourth-generation CT scanner. There is a rotating X-ray source and a continuous 360° ring of detectors, which are stationary. Leading and trailing edges of the fan beam pass outside the patient and are used to calibrate the detectors. (Reproduced from Maravilla and Pastel, 1978.)

where
 $\mu[x, y]$ is the 2D distribution of the linear attenuation coefficient,
 ϕ and x' define the position of the measurement, and
 $I_\phi^0(x')$ is the unattenuated intensity (see Figure 3.9).

The x' y' frame rotates with the X-ray source position such that the source is on the y' axis. Equation 3.2 is simply an extension of Beer's law (Equation 3.1) to take the spatial variation of μ into account.

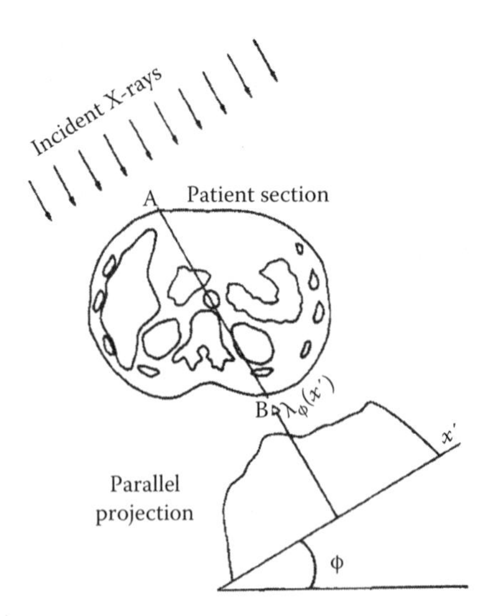

FIGURE 3.9
Projections are defined as the negative logarithm of the fractional X-ray transmittance of the object, $\lambda_\phi(x') = -\ln[I_\phi(x') / I_\phi^0(x')]$; ϕ is the angle at which the projection data are recorded.

A single projection of the object $\lambda_\phi(x')$ is defined as

$$\lambda_\phi(x') = -\ln\left[\frac{I_\phi(x')}{I_\phi^0(x')}\right]$$

$$= \int_{-\infty}^{\infty}\int_{-\infty}^{\infty} \mu[x,y]\delta(x\cos\phi + y\sin\phi - x')\,dx\,dy \quad (3.3)$$

where, now, the Dirac delta function δ picks out the path of the line integral, since the equation of AB is $x' = x \cos \phi + y \sin \phi$.

Equation 3.3 expresses the linear relationship between the object function $\mu[x, y]$ and the measured projection data λ_ϕ. The problem of reconstructing is precisely that of inverting Equation 3.3, that is, recovering $\mu[x, y]$ from a set of projections $\lambda_\phi(x')$.

3.2.3 Projection Sets

The quantity $\lambda_\phi(x')$ in Equation 3.3 may be interpreted as the 1D function λ_ϕ of a single variable x' with ϕ as a parameter, and, with the arrangement of Figure 3.9, this $\lambda_\phi(x')$ is referred to as a parallel projection. To gather this sort of data, a single source and detector are translated across the object at an angle ϕ_1, producing $\lambda_{\phi_1}(x')$. The gantry is then rotated to ϕ_2 and $\lambda_{\phi_2}(x')$ is obtained, and so on for many other angles. As we mentioned in the previous section, the inefficiencies of this first-generation scanning are no longer tolerated in commercial systems, and the projection data are measured using a fan beam. In this case, the distance x' is measured in a curvilinear fashion around the detector from the centre of the array and ϕ is the angle of the central axis of the projection (see Figure 3.10). In what follows we analyse the case for parallel projections simply because it is the easier case to study. The added complexity of fan-beam geometry obscures the basic solution method, while adding but little to the intellectual content (see also end of Section 3.6.2).

In practice, the X-ray source and X-ray detector are of finite size. The projection data are better described as volume integrals over long, thin 'tubes' rather than as line integrals. One effect of this is to average over any detail within the object that is small compared to

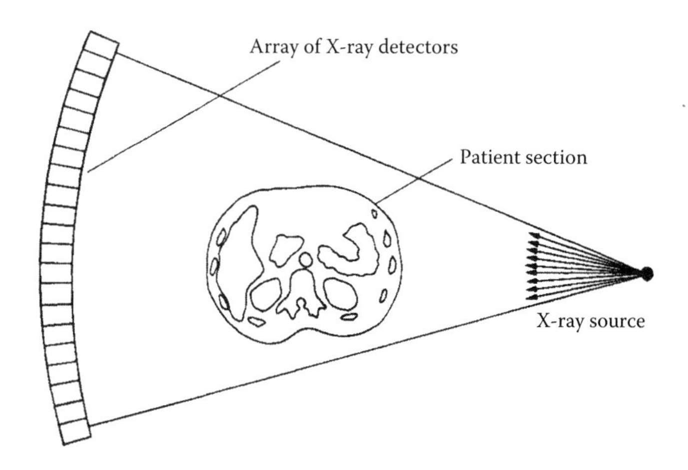

FIGURE 3.10
For second- and higher-generation systems, data are collected using the fan-beam geometry as shown here.

the lateral dimensions of the tube. The highest spatial frequencies that would be present in a 'perfect' projection are, thus, not measurable and the reconstructed image is band limited because of this low-pass filtering by the measuring system. This has important consequences, which will be discussed later (Section 3.6).

3.2.4 Information Content in Projections and the Central-Section Theorem

Up to this point, we have assumed that Equation 3.3 has a solution. We shall now show that a complete set of projection data do indeed have enough information contained to permit a solution. In doing so, we shall point the way to the method of solution that is most commonly used in X-ray CT scanners.

First we specify the notation. The Fourier transform of the density distribution $\mu[x, y]$ is $M[\zeta, \eta]$. The square brackets serve to remind us that the coordinates are Cartesian. In polar coordinates, the corresponding quantities are $\mu^P(r, \theta)$ and $M^P(\rho, \phi)$. (M is upper-case Greek 'mu'.) The various quantities defined in the x' y' frame are $\mu'[x', y']$, $M'[\zeta', \eta']$, etc. It is not necessary to use the prime on $\lambda_\phi(\,)$, etc., since the different functional form of λ for each ϕ value is implicitly denoted by the subscript ϕ.

The angular orientation of the $[x, y]$ reference frame is arbitrary, so without loss of generality we can discuss the projection at $\phi = 0$. From Equation 3.3 we have

$$\lambda_0(x') = \int_{-\infty}^{\infty}\int_{-\infty}^{\infty} \mu[x, y]\delta(x - x')\,dx\,dy. \tag{3.4}$$

The integration over x is trivial, that is

$$\lambda_0(x) = \int_{-\infty}^{\infty} \mu[x, y]\,dy \tag{3.5}$$

which is an obvious result anyway.

The next step is to take the 1D Fourier transform of both sides of Equation 3.5. Readers unfamiliar with the basic concepts of the Fourier transform may care to study the appendix to Chapter 11 (Section 11.A.1). Writing the transformed quantity as $\Lambda_0(\zeta)$, we have

$$\begin{aligned}\Lambda_0(\zeta) &\equiv \int_{-\infty}^{+\infty} \lambda_0(x)\exp(-2\pi i\zeta x)\,dx \\ &= \int_{-\infty}^{\infty}\int_{-\infty}^{\infty} \mu[x, y]\exp[-2\pi i(\zeta x + \eta y)]\,dy\,dx\big|_{\eta=0}.\end{aligned} \tag{3.6}$$

An extra term $\exp(-2\pi i\eta y)$ has been slipped into the Fourier kernel on the right-hand side, but the requirement that the integral be evaluated for $\eta = 0$ makes this a null operation.

However, in this form the RHS of Equation 3.6 is recognisable as the 2D Fourier transform $M[\zeta, \eta]$ evaluated at $\eta = 0$, so Equation 3.6 can be rewritten as

$$\Lambda_0(\zeta) = M[\zeta, 0]. \tag{3.7}$$

Because the Cartesian ζ axis (i.e. $\eta = 0$) coincides with the polar coordinate ρ at the same orientation, Equation 3.7 can be written as

$$\Lambda_0(\zeta) = M^P(\rho, 0]. \tag{3.8}$$

This is an important result. In otherwords, it says that if we take the 1D Fourier transform of the projection λ_0, the result Λ_0 is also the value of the 2D transform of μ along a particular line. This line is the central section that is oriented along the direction $\phi = 0$. Now we can restore the arbitrary angular origin of the reference frames and state the general result, namely,

$$\begin{aligned}\Lambda_\phi(\zeta') &= M'[\zeta', \eta'] \\ &= M^P(\rho, \phi).\end{aligned} \tag{3.9}$$

This important result is known as the central-section or central-slice theorem. To illustrate the theorem, consider a general, bounded object. This object can always be synthesised from a linear superposition of all of its 2D spatial frequency components. Now, consider just one of those cosinusoidal frequency components (see Figure 3.11a). Only when the projection direction is parallel to the wave crests does the projection differ from zero.

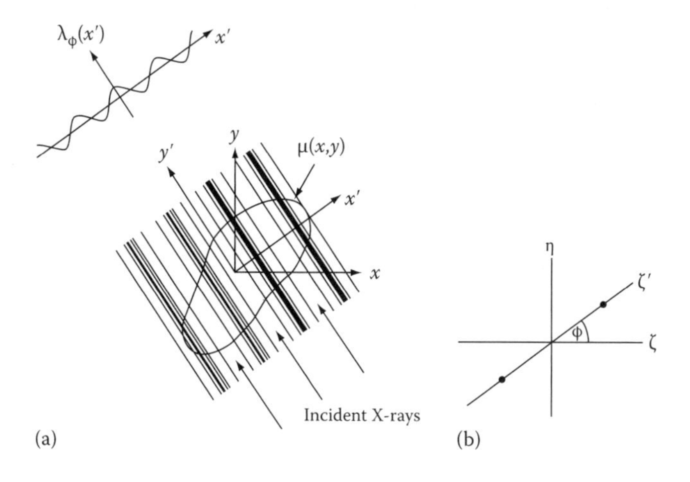

FIGURE 3.11
(a) A general object distribution $\mu(r)$ can be decomposed into Fourier components of the form sin $(2\pi\rho r)$ or cos $(2\pi\rho r)$. One of the latter is depicted here. There is only one direction ϕ for which the projection of this component is nonzero, and at this particular ϕ the component is fully mapped onto the projection. (b) The Fourier transform of this component is a pair of δ functions (shown here by dots) located on the ζ' axis. (Reproduced from Barrett and Swindell, 1981.)

However, for that particular direction, the full cosine distribution is projected onto the x' axis. The Fourier transform of this one component is shown in Figure 3.11b. The original object is a superposition of many component waves of various phases, periods and directions, and it follows that only those waves that are parallel to the first one will have their transforms located on the ζ' axis, and that these are the only waves that will change the form of $\lambda_\phi(x')$.

3.2.5 Reconstruction by 2D Fourier Methods

It now follows that a complete set of projections contains the information that is needed to reconstruct an image of μ. This can be seen by considering a large number of projections at evenly spaced angles ϕ_n.

The value of $M^P(\rho, \phi)$ can then be determined along the radial spokes of the same orientations. If M^P is thus defined on a sufficiently well-sampled basis (more about this later – Section 3.9.2), then $\mu[x, y]$ can be obtained by a straightforward 2D transformation of M, which can be obtained from M^P by means of interpolation from the polar to the Cartesian coordinate systems.

It is worth noting that the projections must be taken over a full 180° rotation without any large gap in angle. If there are large gaps, there will be corresponding sectors in Fourier space that will be void of data. The object μ cannot faithfully be constructed from its transform M if this latter is incompletely defined. We shall see in Chapter 5 that certain classes of PET scanners suffer the problem of limited-angle projection data. In the case of fan beams, projections are acquired over 180° plus the fan angle. The fan projection data are then rebinned to generate a complete set of parallel projections.

The solution method just outlined is not a very practicable one for a number of reasons, but the discussion demonstrates that, in principle, an object can be reconstructed from a sufficiently complete set of its projections. The commonly used 'filtered backprojection' method is described in Section 3.6. It should be noted that if the X-ray flux or the detector response change whilst acquiring the projection set, or the object moves, then inconsistencies will be introduced into the projection data. These in turn will give rise to reconstruction errors that will appear as artefacts in the image (see Section 3.9).

3.2.6 CT Image

The reconstruction of μ is usually made on a rectangular array, where each element or pixel has a value μ_i ascribed to it $(1 \leq i \leq I)$. Before these data are displayed on a monitor, it is conventional to rescale them in terms of a 'CT number', which is typically defined as

$$\text{CT number} = \frac{(\mu_{\text{tissue}} - \mu_{\text{water}})}{\mu_{\text{water}}} \times 1000 \tag{3.10}$$

and is given in terms of Hounsfield units (HU). Thus, the CT number of any particular tissue is the fractional difference of its linear attenuation coefficient relative to water, measured in units of 0.001, that is, tenths of a per cent. The CT numbers of different soft tissues are relatively close to each other and relatively close to zero. However, provided that the projection data are recorded with sufficient accuracy, different soft tissues can be differentiated with a high degree of statistical confidence. Similar tissues, which could not be resolved on conventional transmission X-radiographs, can be seen on CT reconstructions.

3.3 Developments in Scanner Design

Ever since the inception of clinical CT, there has been an overwhelming drive towards reducing scan times. Long scan times are uncomfortable for the patient, and they give rise to situations in which image quality and diagnostic efficacy may be impaired by motion. For example, in the lung region a separate breath-hold is required for each slice; if the breath-holds are not identical, a detail located just below the scan plane in one acquisition could rise to just above it in the next, and hence be missed. This section gives an overview of the major technical developments that have taken place since 1980.

3.3.1 Electron-Beam CT

The electron-beam CT scanner was introduced in 1983, designed specifically for the purpose of imaging the heart. It operates with sweep times of 50 or 100 ms, and produces X-rays by steering an electron beam over up to four tungsten targets shaped as a 210° circular arc (see Figure 3.12); it is the lack of moving parts that enables these short scan times to be achieved. It was heralded as a fifth-generation device (Peschmann et al. 1985) but its design is based on fourth-generation geometry, with two contiguous 216° detector rings. In the single-slice mode, the scanner utilises one target, one detector ring and 100 ms sweep time. The requisite photon statistics are achieved by performing multiple sweeps at each slice location. In this mode, the lungs can be rapidly imaged. In the multiple-slice mode, four targets are used in combination with the two detector rings to obtain eight contiguous slices with no table feed. If cardiac gating is used to synchronise data acquisition to the cardiac cycle, image sequences can be obtained to demonstrate cardiac blood flow and cardiac motion.

3.3.2 Spiral CT

In 1989, the first spiral (or helical) CT scanner was introduced (Kalender et al. 1990). When used in spiral mode, the X-ray tube performs a continuous rotation about the isocentre whilst the patient is translated through the scan plane (Figure 3.13). The name of the

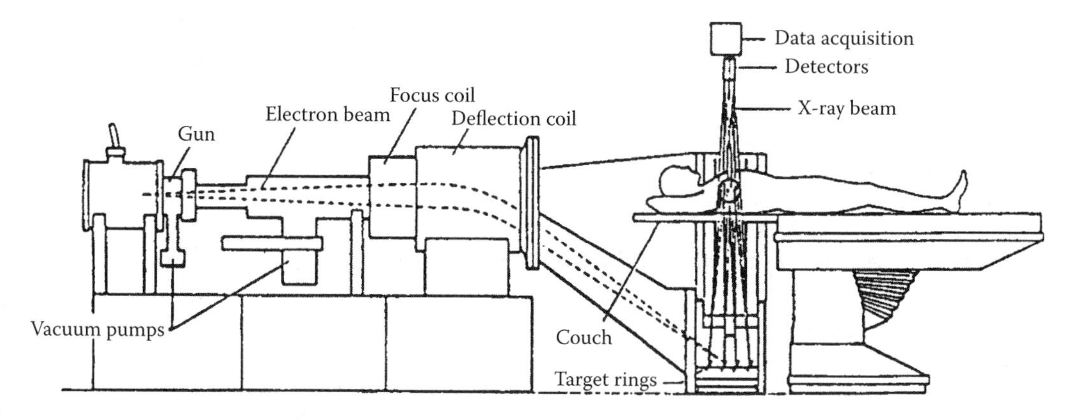

FIGURE 3.12
Imatron CT-100 ciné CT scanner; longitudinal view. Note the use of four target rings for multi-slice examination. (Courtesy of Imatron, Saskatoon, Saskatchewan, Canada.)

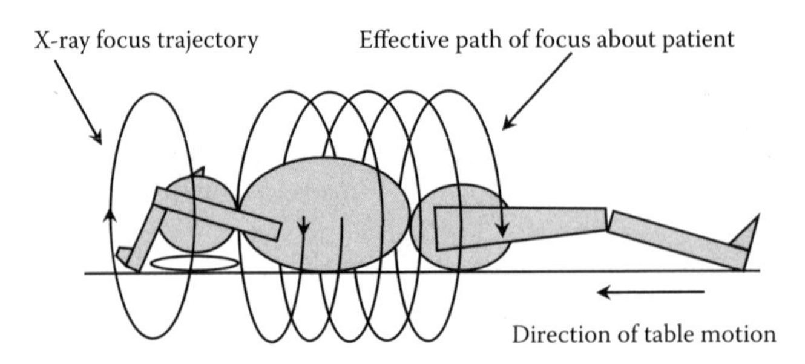

FIGURE 3.13
Principle of helical CT scanning.

technique arises from the shape of the path of the X-ray focus about the patient – a helix with a pitch q, defined as

$$q = \frac{T}{b} \tag{3.11}$$

where
T is the table increment in one revolution, and
b is the nominal slice thickness.

Helical scanning can be achieved in both third- and fourth-generation scanners. It is not a new idea, but turning theory into practice had to wait for two important developments: the application of slip-ring technology to high voltage (HV) and electronic signal transfer and the increase in computer power. These breakthroughs are described in Section 3.4. The continuous motion of the patient through the scan plane gives rise to inconsistencies in the projection data, which unless corrected for would produce artefacts in the reconstructed image. The techniques that have been developed to make these corrections are described in Section 3.5.

3.3.3 Multi-Slice CT

There has been concern for some time that in CT acquisition so much of the useful X-ray beam is wasted. In 1992, Elscint introduced the CT-Twin, a CT scanner with two detector arrays, placed side-by-side along the rotation axis. In 1998, several major manufacturers extended this principle by introducing 2D detector arrays (Figure 3.14). The benefit of this design is that contiguous slices can be acquired simultaneously (up to 320 at the time of writing), thus making better use of the X-ray beam, and reducing the total investigation time. The maximum possible number of simultaneous slices is constrained whilst it is assumed for the purpose of reconstruction that the X-ray fan beam is bounded by parallel planes along the axis of rotation. When this assumption breaks down, that is when the X-ray fan beam becomes a cone beam with a significant divergence in this direction, artefacts are introduced. At this point cone-beam reconstruction techniques must be used (see Section 3.8).

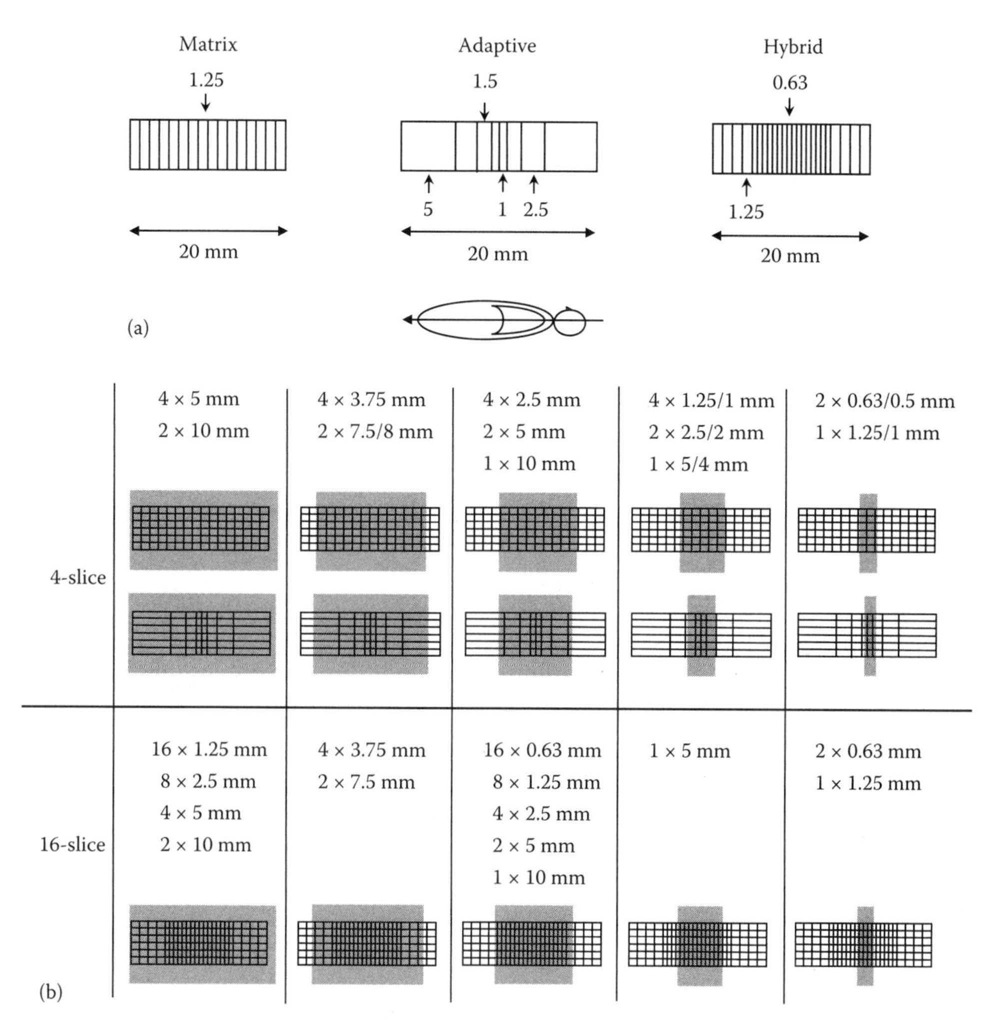

FIGURE 3.14
(a) Schematic representation of detectors used for multi-slice CT. The matrix detector shown is used by GE Medical Systems. The adaptive detector shown is used by Siemens Medical Solutions and Philips Medical Systems. The hybrid detector shown is used by GE Medical Systems. The size of the detector elements is given in mm. (b) Slice selection schemes for axial scanning using 4-slice (matrix and adaptive detectors) and 16-slice (hybrid detector) scanners. A section of the detector array is shown; the axis of rotation lies along the horizontal direction. The grey boxes indicate the extent of the X-ray beam along the axis of rotation. In four-slice CT, where the slice selection available differs according to the type of detector, the option for the matrix detector is given first, followed by that for the adaptive, separated by a slash.

The definition of pitch given by Equation 3.11 was initially adopted for multi-slice CT scanners, but gave values typically much greater than those found in single-slice systems. An alternative definition has therefore been formulated:

$$q = \frac{T}{n \cdot b} \tag{3.12}$$

where $n \cdot b$ is the nominal X-ray beam width.

This equation will give rise to values of pitch directly comparable to those obtained with a single-slice scanner. Note that n is generally, but not always, the number of slices reconstructed simultaneously.

3.3.4 Dual-Source CT

In 2005, Siemens released the first dual-source CT scanner model. It has two imaging chains mounted orthogonally to each other in the same plane. Chain A comprises a conventional X-ray tube and multi-slice detector array. Chain B comprises an X-ray tube and narrow-arc multi-slice detector array. Chain A offers full imaging capabilities and can be selected on its own. Chain B offers a reduced FOV (34 cm at the time of writing), and can be used only in conjunction with chain A. When both imaging chains are used together, the time taken to acquire the required projection set is reduced. This scanner is, therefore, suited to applications where short investigation times are essential.

3.4 Scanner Operation and Components

The typical modern CT scanner is based on third-generation geometry, and can be of single-slice or, more commonly, multi-slice design. In order to ascertain where to start and stop the scan sequence, a scan projection radiograph, known as a preview, surview scout view or topogram, is first obtained by keeping the X-ray tube stationary whilst the table is translated through the scan plane. The scan sequence can be acquired over a volume by either introducing table feed between slices (axial scanning) or by moving the table through the scan plane continuously during data acquisition (spiral or helical scanning). The scan sequence can also be acquired at a single location to detect changes over time (dynamic scanning).

Slices are reconstructed from full (360°) or partial (180° plus the fan angle) rotation of the X-ray source. Full rotation is generally selected for axial scanning as the redundancy in projection data can be used to correct for inconsistencies that might give rise to artefacts (see Section 3.9). Typically 1000 fan-beam projections per 360° rotation are acquired. Single-slice scanners generate one slice per rotation; multi-slice scanners can generate a variable number of slices ranging from one up to the maximum allowed for the particular detector configuration.

CT equipment can be divided into three major components: gantry, patient couch and computer system. The first two are illustrated in Figure 3.4. The gantry houses the imaging components shown schematically in Figure 3.15. These are similar to those that would be encountered in a conventional radiographic system: X-ray source, additional filters, collimators, scatter removal device and detectors. However, tomography imposes special requirements on these components, which are discussed in the sections that follow.

3.4.1 Gantry

The gantry houses and provides support for the rotation motors, HV generator, X-ray tube (one or two), detector array and preamplifiers (one or two), temperature control system and the slip rings. Slip rings enable the X-ray tube (and detectors in a third-generation system) to rotate continuously. The HV cables have been replaced with conductive tracks

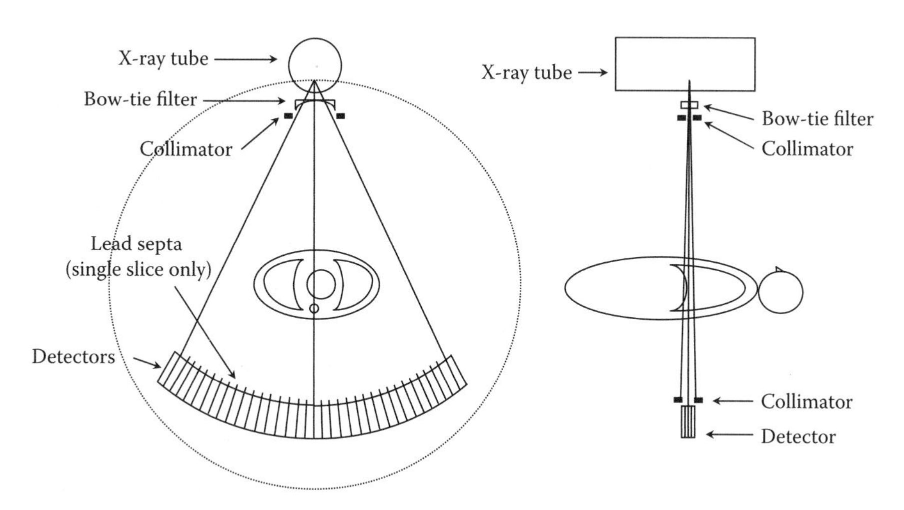

FIGURE 3.15
Schematic representation of the components of a CT scanner system, shown as viewed from the patient's feet and the patient's side.

on the slip rings that maintain continuous contact with the voltage supply via graphite or silver brushes. Two slip-ring designs were initially used in commercial scanners: (1) low-voltage slip rings, in which the HV transformer was mounted on the rotating part of the gantry with the X-ray tube, and only low voltages were present on the stationary and moving parts and (2) high-voltage slip rings, in which the HV was generated in the stationary part of the gantry, thus reducing the inertia of the moving parts. The low-voltage design is now universally adopted. The detector cables have been replaced with either conductive tracks on the slip rings that maintain continuous contact with the detector electronics via graphite or silver brushes, or a radio-frequency wireless transmission system.

The scan plane is usually vertical; other inclinations away from the vertical are possible by tilting the entire gantry through angles no greater than 30°. The position of the scan plane is indicated by lights or lasers within the aperture. The centre of tube rotation is referred to as the isocentre of the scanner. Rotation speeds range typically from 360° in 0.3–2 s. A selection of scan times is always available, based on partial and 360° rotation.

3.4.2 X-Ray Generation

Most CT scanners operate at three to four tube voltages – 80, 100, 120 and 140 kV are typical – and at a range of tube currents up to about 800 mA. It is crucial that the X-ray output is stable. Short-term fluctuations will give rise to inconsistent projection data. Long-term drift will affect the calibration of the scanner. The HV generator is, therefore, required to fulfil two primary tasks: to supply an extremely stable HV across the X-ray tube and to monitor and correct any changes in X-ray tube performance, such as the reduction in filament emissivity which can occur with ageing. Modern high-frequency generators use microprocessor control systems for tube-voltage and tube-current regulation. In addition, information about X-ray flux is supplied to the microprocessor by radiation monitors in the primary beam. X-ray generation is continuous during data acquisition; in older scanners, the X-ray beam was pulsed to gate data collection, and, hence, the pulsing determined the angular sampling rate.

The tube current can be prescribed manually by the operator, or automatically according to image quality preferences specified by the operator. Manufacturers have developed automatic exposure algorithms which modulate the prescribed tube current according to the radio-opacity of the patient, whilst maintaining image noise constant at a preset value, or varying slowly, with patient size. Tube current modulation can be achieved in three ways: first, by adjusting the tube current to the size of the patient; second, by adjusting the tube current with angular position of the X-ray tube (xy modulation); and third, by adjusting the tube current along the axis of rotation (z-axis modulation). Thus, higher tube currents will be automatically prescribed for thicker body sections, such as lateral projections or projections through the abdomen, whereas lower tube currents will be used for thinner sections, such as antero-posterior projections or projections through the neck. The practical implementation of tube current modulation is manufacturer specific; it is important that it is understood by the user. A description of individual algorithms is beyond the scope of this chapter. The reader is referred to Kalra et al. (2004) and McCollough et al. (2006) for comprehensive reviews of these techniques.

The X-ray tube must withstand much higher tube loadings than in conventional radiography (see Table 3.2). Improvements in tube rating and heat dissipation over those of a conventional X-ray tube can be achieved by increasing the diameter of the anode disk, and circulating the insulating oil through a heat exchanger. Higher-specification X-ray tubes have a metal/graphite anode, which has a greater heat storage capacity, and 'liquid metal' bearings which increase conductive loss through the rotor by increasing the contact area with the outer casing. The Straton X-ray tube (Siemens Medical Solutions, Erlangen) offers a novel solution to the tube loading problem. The tube is mounted on a spindle so that the whole of it rotates, rather than just the anode. As there is no rotor, the conductive loss from the anode is very efficient. The electron beam is steered to maintain the point of impact on the anode fixed in space (Schardt et al. 2004).

X-ray tubes designed for CT offer at least two foci with sizes typically in the range 0.5–1.0 mm for the fine focus, and 0.8–1.6 mm for the broad focus. The fine focus is used for high-resolution scanning; geometric unsharpness, which arises from the large separation between patient and detector array, is reduced by selecting this focus. Interestingly, geometrical unsharpness was perceived as a disadvantage of fourth-generation geometry where the separation between patient and detector array is larger.

TABLE 3.2

Typical properties of an X-ray tube used for CT compared to those of a conventional radiographic tube.

	Conventional X-Ray Tube	CT X-Ray Tube
Typical exposure parameters	70 kV, 40 mAs	120 kV, 10,000 mAs
Energy requirements	2,800 J	1,200,000 J
Anode diameter	100 mm	160 mm
Anode heat storage capacity	450,000 J	3,200,000 J
Maximum anode heat dissipation	120,000 J/min	540,000 J/min
Maximum continuous power rating	450 W	4000 W
Cooling method	Fan	Circulating oil

3.4.3 Beam Filtration and Collimation

Filtration additional to that inherent in the X-ray tube is found in all CT scanners. A beam-shaping ('bow-tie') filter is ubiquitous; some manufacturers also add a flat filter. The purpose of the flat filter is to remove the low-energy (soft) radiation incident on the patient and, thus, reduce the beam-hardening artefacts that arise from differential X-ray absorption in the patient's tissues (see Section 3.9.5). The bow-tie filter modulates the primary X-ray beam intensity, so that it is highest along the central beam axis and lowest along the edge of the beam. This modulation ensures that the dynamic range of the X-ray intensity exiting the patient is reduced, and, hence, the performance specification of the detectors can be relaxed.

A selection of X-ray beam apertures is always available. In single-slice systems, the slice thickness is determined by the beam collimation, and ranges from 10 to 1 or 2 mm, defined at the isocentre. The desired slice thickness is usually achieved by setting the pre-patient collimation. However, the smallest slice thickness is often obtained by also using post-patient collimators to reduce the width of the X-ray beam impinging on the detectors. The post-patient collimators fulfil the additional function of scatter removal; grids or lead septa are usually also present. Poor scatter removal was considered to be another disadvantage of fourth-generation geometry as the septa cannot be aligned with the X-ray focus in this configuration.

In multi-slice systems, the X-ray aperture determines the number of detectors fully or partially irradiated in the direction parallel to the axis of rotation. The slice thickness and the number of slices per rotation are chosen according to the manufacturer's scheme. An example is given in Figure 3.14b. The 'slice collimation' is often quoted instead of the nominal X-ray beam width to indicate explicitly the number and width of the active detector elements in the selected scan protocol. For example, a slice collimation of $32 \times 0.6\,\text{mm}$ corresponds to a nominal X-ray beam width of 19.2 mm.

3.4.4 Data Acquisition System

The ideal detector for CT imaging has a high detection efficiency and short afterglow. The detection efficiency is a combination of the detective quantum efficiency (DQE) and the geometric efficiency (expressed as the percentage of the X-ray beam incident on the detector elements, bearing in mind the existence of spaces or septa between them). Matching between detector elements in an array must be ensured or corrected for across all X-ray energies and intensities in order to avoid inconsistent projection data. Two types of detectors have been used in commercial scanners: xenon ionisation chambers and solid-state scintillators. The latter include crystal materials such as cadmium tungstate and ceramics such as gadolinium oxysulphide. The properties of these detectors are given in Table 3.3.

Before the advent of ceramic scintillators the choice of detector depended on the scanner geometry. Fourth-generation systems universally adopted crystal scintillators, as the single-chamber structure of xenon detectors did not lend itself to a full-ring configuration (a full detector ring contains typically 2000–4800 detectors). Both types of detector were found in third-generation systems. As scan times shortened, the long afterglow associated with crystal scintillators became troublesome. Manufacturers that had favoured these detectors due to their higher detection efficiency were forced to replace them with xenon detectors. The DQE of the latter could be improved to some extent by increasing the gas pressure and the interaction depth of the detector chamber

TABLE 3.3

Properties of detectors in common use in CT scanning.

	Xenon Detectors	Crystal Scintillator	Ceramic Scintillator
Detector	High pressure (8–25 atm) Xe ionisation chamber	$CaWO_4$ + silicon photodiode	Gd_2O_2S + silicon photodiode
Detector array	Single chamber, divided into elements by septa	Discrete detectors	Discrete detectors
Signal	Proportional to ionisation intensity	Proportional to light intensity	Proportional to light intensity
Detector efficiency	40%–70%	95%–100%	90%–100%
Geometric efficiency (in fan direction)	>90%	>80%	>80%
Afterglow limitations	No	Yes	No
Detector matching	No	Yes	Yes

(typically 3–10 cm). Ceramic materials possess the advantages of both xenon and crystal scintillator detectors, and exhibit an increased light yield per incident X-ray photon; they are now the detectors of choice.

Third-generation detector arrays contain typically 600–1000 detector elements in the plane of the X-ray fan (the detector row). Single-slice CT scanners use a single-row detector array. Two designs are used to combine detector rows to form the 2D array for four-slice scanners: adaptive and matrix arrays, shown in Figure 3.14a. Adaptive arrays have been developed to minimise the number of septa between detector rows and, hence, improve geometric efficiency; matrix arrays are more versatile and can be upgraded more easily. Sixteen-slice detector arrays are a hybrid between these two designs: 16 narrow detector rows in the centre of the array (typically 0.5–0.8 mm), and eight detector rows of twice the width straddling these, four on each side. The total width of the array along the axis of rotation is comparable between 16-slice and four-slice arrays, and is approximately 20–32 mm. 32- and 40-slice CT scanners have detector arrays of similar design which range in width between 28 and 40 mm. Matrix detector arrays reappear in 64-slice scanners as the challenges in manufacturing large-area detectors with small elements are mostly resolved; the detector width for these scanners ranges between 32 and 40 mm. Higher-slice scanners offer matrix detector arrays up to 160 mm in width for 320 slices. Multi-slice CT heralded the demise of fourth-generation geometry due to the prohibitive cost of manufacturing a 2D full detector ring.

The electronic signal from the detectors is collected through a set of electronic channels. The number of channels along the detector row is equal to the number of detector elements; the number of channels parallel to the axis of rotation is one for a single-slice scanner and, at the time of writing, 2, 4, 8, 10, 16, 32, 40, 64, 128 and 320 for multi-slice scanners. For 2- to 40-slice scanners, when more detector rows are activated than channels available, the detector signals are summed into the available channels as illustrated in Figure 3.14b. For 64- and higher-slice scanners, the number of detector rows equals the number of channels parallel to the axis of rotation and such summing is not necessary.

Some manufacturers describe their multi-slice scanners as offering twice as many slices per rotation as the number of channels available along the axis of rotation. This is because the X-ray focal spot is toggled between two positions on the anode during data acquisition, yielding two sets of X-ray projections for each channel row. In this chapter, scanners are described throughout in terms of the number of detector-channel rows available; for example, a 64-slice scanner is a scanner with 64 detector-channel rows.

Electronic gating is used to specify the sampling rate, typically in the region of 1 kHz, and sampling window. The detector signals are amplified and digitised before leaving the gantry, so that the signal transmitted to the computer system is less susceptible to interference. The electronics are required to have a high dynamic range, typically 1:1,000,000, and low noise. A 20-bit word would be required to cover the dynamic range of the electronic signal, so auto-ranging techniques are adopted in which a 14 bit word with a 2 bit range code is used to cover the range by decreasing the precision with which high-intensity signals are read.

3.4.5 Patient Couch

The patient couch must be able not only to support the weight of the patient (which may exceed 150 kg), but also translate the patient into the gantry aperture without flexing whilst achieving a positional accuracy of the order of 1 mm. In addition, it must be radiolucent, safe for the patient and easy to clean. Most modern couches are made of a carbon fibre composite which can provide the required rigidity and radiolucency.

3.4.6 Computer System

During the course of a CT scan, a multitude of tasks take place: interfacing with the operator, gantry and couch movement control; acquiring, correcting, reconstructing and storing the data; image display and archive; generating hard copies, and pushing images to PACS (see Section 14.3.2) and networked workstations. For a clinical scanner to work efficiently, as many as possible of these tasks must take place concurrently. This was originally achieved with a multiprocessor system and parallel architecture, but more recently with a multitasking workstation. A typical multiprocessor system consisted of an operator interface computer (which also acted as host), scanner interface computer, an array processor for data processing in a 'pipeline' fashion and a display computer. In a multitasking system, the workstation also acts as host to an image reconstruction engine and a graphics engine (the engines may be boards within the computer, or additional processors).

Figure 3.16 shows a typical axial scan sequence undertaken on a scanner from the 1990s. In this example, the computer performs a maximum of two tasks concurrently. It is noticeable that by far the most time-consuming task is that of image reconstruction and display. During this phase, the acquired raw data are amplified, digitised, corrected (for X-ray output changes, detector mismatches, detector dark current and coherent noise sources), reconstructed and displayed.

Spiral scanning imposes additional requirements on the computer system, not least in the size of the memory. For example, in axial scanning about 700,000 samples are collected per slice and processed immediately to generate the image. In single-slice helical scanning, as many as 42,000,000 samples may be collected, and over 100,000,000 samples for multislice helical scanning. This magnitude of data acquisition has become feasible only with the advent of multitasking systems.

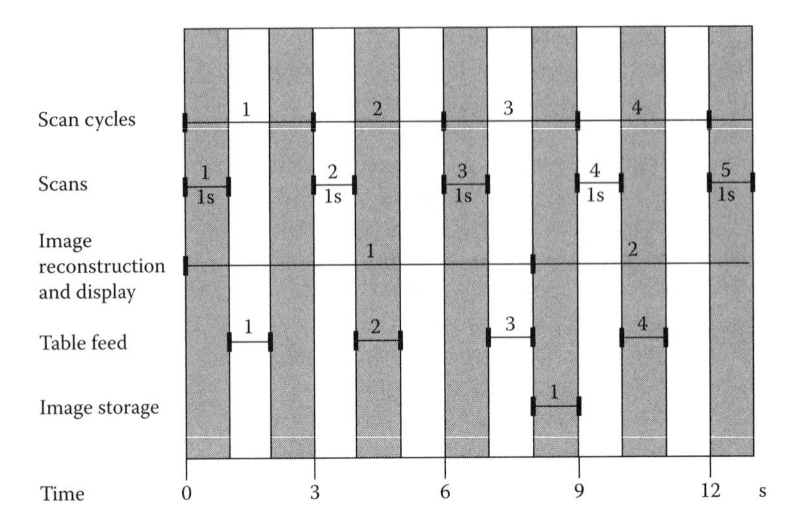

FIGURE 3.16
Schematic representation of a typical scanning sequence for a single-slice spiral CT scanner c. 1990. A CT image can now be reconstructed in a fraction of a second. (Courtesy of Siemens, Erlangen, Germany.)

3.4.7 Image Reconstruction and Display

The CT image is reconstructed using 2D filtered backprojection, or, in the case of 16- and higher-slice scanners, cone-beam reconstruction. These techniques are described in Sections 3.6 and 3.8, respectively. The user can specify the reconstruction filter from a selection offered by the manufacturer; for example, a smoothing filter suppresses noise and a sharp filter enhances high frequencies. Examples of such filters are also given in Section 3.6. The image is typically available as a 512 × 512 matrix, where each pixel has a depth of 12 bits. This means that CT numbers between –1000 and 3095 are possible. Some manufacturers provide an extended CT number scale, which increases the range to approximately 20,000 and is useful when scanning areas containing metal implants.

The display monitor typically provides 256 grey levels; hence it is not possible to assign a grey level to each CT number in the image. Windowing techniques are used to map a selected range of CT numbers (the window width) onto the grey scale (see Figure 3.17)

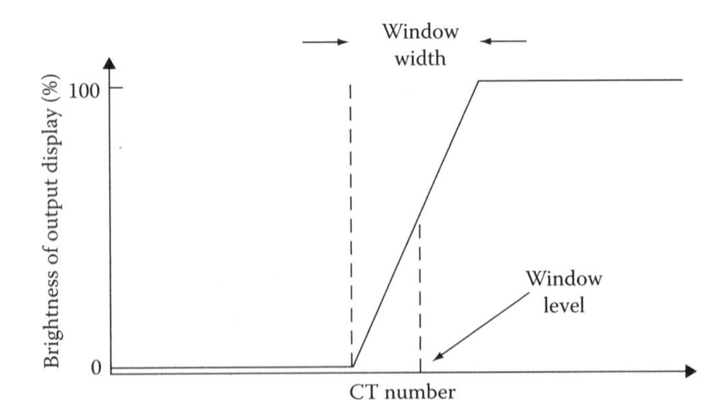

FIGURE 3.17
The 'windowing' facility allows the display brightness range to be fully modulated by any desired range of CT values as determined by adjusting the window 'level' and 'width'.

and, thus, enhance the contrast between the tissues of interest. Windowing is applied automatically or interactively. In some anatomical areas two windows may be necessary; for example, the thorax is best visualised using a lung window (for lung tissue) and a soft tissue window (for the mediastinum and bone).

The anatomy displayed in the image will depend on the reconstruction field of view (RFOV) and the image centre selected by the operator. In the case where the RFOV corresponds to the scan field and the latter is, say, 50 cm, each pixel in the matrix represents approximately 1.0 mm of patient anatomy. If the RFOV is 12.5 cm, then each pixel will represent 0.25 mm of patient anatomy, and so on. It should be noted that the operator can change the displayed FOV ('zoom the image') after reconstruction; in this instance, the size of the pixel on the monitor will change, but the size of patient anatomy represented by each pixel will not.

3.5 Data Interpolation in Spiral CT

In spiral scanning the fan projection data are acquired along a helix, rather than within a 2D plane. Thus, only one (single-slice scanners) or a few (multi-slice scanners) projections are available in a given reconstruction plane perpendicular to the direction of motion. Before image reconstruction can take place, a full set of projections needs to be obtained at the desired slice position. This section describes the techniques developed to achieve this in single-slice and four-slice spiral scanning.

3.5.1 Interpolation Techniques for Single-Slice Scanners

Consider Figure 3.18a. The locus of the X-ray tube as viewed from the side of the patient is shown by the solid helix. We want to obtain a full projection set at position z_0. The image plane at this location is indicated by the solid ellipse. Only one fan projection is available at z_0 – when the X-ray source is located as marked by the dot. In order to obtain the projection $L_\phi(z_0, i)$ at an angular position ϕ and for detector channel i, we must make use of the real projections at this angle and channel, $L_\phi(z_1, i)$ and $L_\phi(z_2, i)$, which occur at positions z_1 and z_2, one before z_0, one after. Note that $z_2 - z_1$ corresponds to the distance covered in one tube rotation. The central ray path through the isocentre $L_\phi(z_0, 0)$ is indicated by the solid radial line in the figure.

In the first instance, we can estimate the required projection by interpolating linearly between $L_\phi(z_1, i)$ and $L_\phi(z_2, i)$. Thus,

$$L_\phi(z_0, i) = (1 - v) \cdot L_\phi(z_1, i) + v \cdot L_\phi(z_2, i) \tag{3.13}$$

where $v = (z_0 - z_1)/(z_2 - z_1)$.

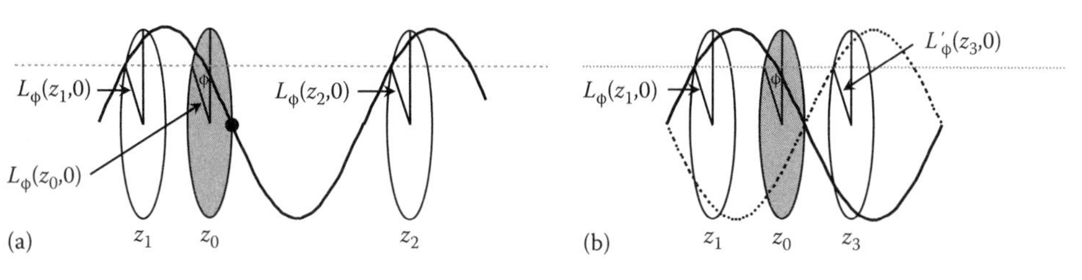

FIGURE 3.18
Illustration of interpolation techniques in helical CT scanning. (a) 360° LI and (b) 180° LI.

This operation is simply repeated for all detector channels, and all projection angles to generate the full set of projection data. For one slice to be reconstructed, a 2 × 360° range of data is, therefore, required. This linear interpolation technique is called 360° LI, and is more fully described in Kalender et al. (1990), Crawford and King (1990) and Kalender and Polancin (1991).

The estimate can be improved by making use of the equivalence of opposed line integrals. For example, the central ray $L_\phi(z,0)$ can be rebinned as $L_{\phi+\pi}(z,0)$, and thus a new sample is synthesised. This process is applied to the measured samples to synthesise fan projections, the source of which lies along a helix with a phase shift of 180° to the real helix (Figure 3.18b). The synthesised projection $L'_\phi(z_3,i)$ is closer to the required plane z than $L_\phi(z_2,i)$ and will, therefore, yield a better estimate of $L_\phi(z_0,i)$ than the latter. Only a 2 × (180° + fan angle) range of data is now used. This linear interpolation technique is called 180° LI. Crawford and King (1990) and Polancin et al. (1992) discuss it in greater detail.

More sophisticated interpolation techniques are also possible, for example, cubic spline fitting to $L_\phi(z_1,i)$, $L'_\phi(z_3,i)$ and $L_\phi(z_2,i)$, or motion correction algorithms originally used in axial scanning. Further details can be found in Wang and Vannier (1993), Hsieh (1996), Besson (1998), Hu and Shen (1998), amongst others. Regardless of the interpolation technique adopted, the *nominal* slice width of the reconstructed slice is equivalent to the beam collimation, as in axial scanning.

3.5.2 Interpolation Techniques for Multi-Slice Scanners

Whilst the X-ray beam cone angle can be neglected, the interpolation techniques described in the preceding section can be readily extended to multi-slice spiral CT. This is true for multi-slice scanners with up to four data acquisition channels. The number of helices available is equivalent to the number of data acquisition channels in the direction parallel to the axis of rotation. Helices offset by 180° are synthesised in an analogous manner (see Figure 3.19). The two projections required for interpolation are chosen from the nearest real or synthesised helices straddling the plane of interest. This technique is referred to as 180° MLI (multi-slice linear interpolation).

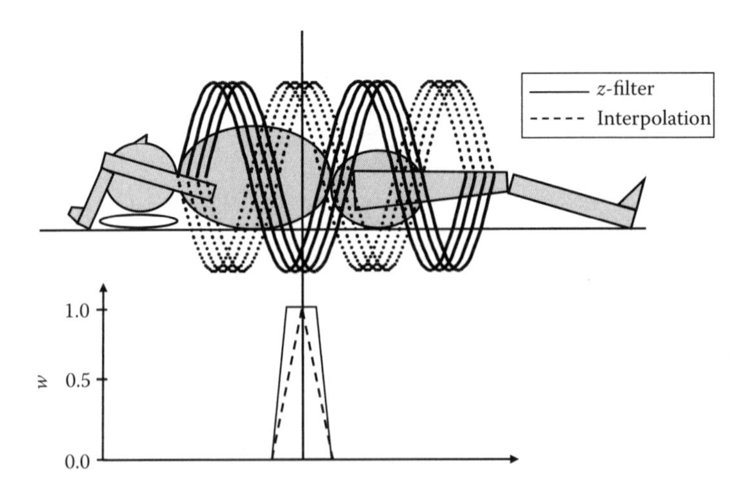

FIGURE 3.19
Illustration of interpolation techniques in helical CT scanning for multi-slice CT. The synthesised helices are shown with a dotted line. w is a weighting factor applied to the projection data at the same angular position as the required projection in the scan plane.

An alternative interpolation strategy makes more efficient use of the multiple data available at a given angular position in four-slice scanners. A z-filter is applied to the projection data together with a multi-point interpolation weighting factor (see Figure 3.19). The shape and width of the z-filter determines the width of the reconstructed slice. This technique is described as 180° MFI (multi-slice filtered interpolation). It should be noted that the reconstructed slice width can never be less than the individual slice collimation. An advantage of this technique is that slices of different widths can be reconstructed from one helical acquisition simply by changing the width of the z-filter. The slice collimation and the thickness of the reconstructed slice are now (almost) independent variables.

Further details of the interpolation techniques described earlier can be found in Hu (1999) and Kalender (2005).

3.6 2D Image Reconstruction

The mathematics of transmission CT, or the theory of reconstruction from projections, has itself acquired the status of attracting attention as an independent research area. The literature is enormous and many excellent books reviewing the field have existed since its inception. Some, such as Herman (1980), review the subject from the viewpoint of the theoretically minded physicist or engineer, whereas others, for example, Natterer (1986), are really only accessible to persons who first and foremost regard themselves as mathematicians. Optionally, for a practical discussion, the books by Barrett and Swindell (1981) and Kalender (2005) may be consulted. Against the backcloth of this formidable weight of literature, the purpose of this section is to provide a simplified view of theory applicable to the most elementary scanning geometry in order to come to grips with some basic principles. From this beginning, it should be possible to go on to view the analogous developments in single-photon emission CT (Chapter 5) as well as reconstruction techniques using ultrasound (Chapter 6) and nuclear magnetic resonance (Chapter 7). The serious research student will find the treatment in this chapter over-simple and could do no better than branch out starting with one of the aforementioned books.

The theory of reconstruction from projections pre-dates the construction of any practical scanner for CT. It is often stated, although even this is disputed (Webb 1990), that the problem was first analysed by Radon (1917). The theory has been 'rediscovered' in several distinct fields of physics including radioastronomy (Bracewell and Riddle 1967) and electron microscopy (Gilbert 1972). An account of a method and the first system for reconstructing X-ray medical images probably originated from Russia (Tetel'baum 1956, 1957, Korenblyum et al. 1958) (see Chapter 1), although it is the work of Hounsfield (1973) that led to the first commercially developed system. This was a head scanner marketed by EMI. Many different techniques for solving the reconstruction problem have been devised and, in turn, each of these has received a great deal of attention regarding subtle but important additional features that improve image quality. It would be true to say that, just as scanner hardware has developed rapidly, so parallel developments in reconstruction theory have kept pace to service new designs and to predict optimal scanning conditions and the design criteria for new scanners.

Reconstruction techniques can be largely classified into two distinct groups, namely, the convolution and backprojection methods (or equivalent Fourier techniques) and iterative methods. For a long time, there was much debate as to the relative superiority of one algorithm or another, and, in particular, whether one of the two classes was in some way superior. Today, this debate has been rekindled by the advent of new iterative reconstruction algorithms. The inevitable conclusion is that each method has its advantages, it being important to tailor the reconstruction technique to the physics of the imaging modality. For example, iterative techniques have found some important applications in emission CT (Section 5.8) where photon statistics are poorer.

3.6.1 2D Filtered Backprojection

Next we derive the algorithm that is used in all CT scanners up to four-slice systems, and to a more limited extent in axial scanning in 16- and higher-slice systems. It is the method of 'filtered backprojection' or 'convolution and backprojection'. A formal statement of the 2D inverse polar Fourier transform yielding $\mu[x, y]$ is given by

$$\begin{aligned}\mu^{P}(r,\theta) &= \mu[x,y] \\ &= \int_{0}^{\pi}\int_{-\infty}^{\infty} M^{P}(\rho,\phi)\exp[2\pi i\rho(x\cos\phi + y\sin\phi)\,|\,\rho\,|\,d\rho\,d\phi\end{aligned} \tag{3.14}$$

where $x(=r\cos\theta)$ and $y(=r\sin\theta)$ denote the general object point.

If Equation 3.14 is broken into two parts, the method of solution becomes immediately apparent (indeed, these two equations give the method its name):

$$\mu[x,y] = \int_{0}^{\pi} \lambda_{\phi}^{\dagger}(x')\,d\phi\,|_{x'=x\cos\phi+y\sin\phi} \tag{3.15}$$

where

$$\lambda_{\phi}^{\dagger}(x') = \int_{-\infty}^{\infty} M^{P}(\rho,\phi)\,|\,\rho\,|\exp(2\pi i\rho x')\,d\rho. \tag{3.16}$$

Consider Equation 3.16, which defines an intermediate quantity $\lambda^{\dagger}$. For reasons that will become obvious, $\lambda^{\dagger}$ is called the filtered projection. The first point to notice is that Equation 3.16 is the 1D Fourier transform of the product of M^P and $|\rho|$. As such, it is possible to write it as the convolution of the Fourier transforms of M^P and $|\rho|$ (see Section 11.A.2). Taking M^P first, its transform is known from the central-slice theorem. It is just the projection data, $\lambda_{\phi}(x')$, that is,

$$\lambda_{\phi}(x') = \int_{-\infty}^{\infty} \Lambda_{\phi}(\zeta')\exp(2\pi i\zeta' x')\,d\zeta' \tag{3.17}$$

where, from Equation 3.9,

$$\Lambda_{\phi}(\zeta') = M^{P}(\rho,\phi).$$

Now consider $|\rho|$. This is not a sufficiently well-behaved function for its transform to exist. In practice, however, we have seen in Section 3.2.3 that $M(\rho, \phi)$ is band limited by the measuring system, so if the maximum frequency component of $M(\rho, \phi)$ is ρ_{max} then $|\rho|$ can be similarly truncated. Thus, we need the transform $p(x')$ of $P(\rho)$, where

$$\begin{aligned} P(\rho) &= 0 \qquad |\rho| \geq \rho_{max} \\ P(\rho) &= |\rho| \qquad |\rho| < \rho_{max} \end{aligned} \tag{3.18}$$

that is,

$$p(x') = \int_0^{\rho_{max}} \rho \exp(2\pi i \rho x')\, d\rho - \int_{-\rho_{max}}^{0} \rho \exp(2\pi i \rho x')\, d\rho. \tag{3.19}$$

Equation 3.19 is straightforward to evaluate (see Appendix, Section 3.A), with the result

$$p(x') = \rho_{max}^2[2\,\mathrm{sinc}(2\rho_{max}x') - \mathrm{sinc}^2(\rho_{max}x')] \tag{3.20}$$

which is perfectly well behaved.

Using the convolution theorem, Equation 3.16 can now be written as

$$\lambda_\phi^\dagger(x') = \int_{-\infty}^{\infty} \lambda_\phi(x) p(x' - x)\, dx \tag{3.21}$$

or in the conventional shorthand notation (* denoting convolution)

$$\lambda_\phi^\dagger(x') = \lambda_\phi(x') * p(x'). \tag{3.22}$$

The dagger (†) indicates a *filtered* projection because the original projection is convolved with $p(x')$, which constitutes a filtering operation.

Now we look at Equation 3.15. This represents the process of *backprojection* in which a given filtered projection $\lambda_\phi^\dagger$ is distributed over the $[x, y]$ space. For any point x, y and projection angle ϕ, there is a value for x' given by

$$x' = x \cos\phi + y \sin\phi.$$

This is the equation of a straight line (parallel to the y' axis), so the resulting distribution has no variation along the y' direction. A simple analogy is to think of dragging a rake, with a tooth profile given by $\lambda_\phi^\dagger(x')$, through gravel in the y' direction. The 1D tooth profile is transferred to the 2D bed of gravel. Backprojection is not the inverse of projection. If it were, the reconstruction-from-projection problem would be trivial! It is very important to be clear that pure backprojection of unfiltered projections will not suffice as a reconstruction technique. Equation 3.15 also contains an integration over ϕ, which represents the summation of the backprojections of each filtered projection, each along its own particular direction. It is like raking the gravel from each projection direction with a different tooth profile for each filtered projection. The analogy breaks down, however, since each raking operation would destroy the previous distribution rather than add to it, as required by the integration process.

The total solution is now expressed by Equations 3.22 and 3.15. In otherwords, each projection $\lambda_\phi(x')$ is convolved (filtered) with $p(x')$ (Equation 3.20). The filtered projections are each backprojected into $[x, y]$ and the individual backprojections (for each projection angle) are summed to create the image $\mu[x, y]$.

3.6.2 Practical Implementation

In practice, the data are discretely sampled values of $\lambda_\phi(x')$. Thus, the continuous convolution of Equation 3.21 must be replaced by a discrete summation, as must also the angular integration of Equation 3.15. We deal first with the convolution. The Whittaker–Shannon sampling theorem states that a band-limited function with maximum frequency component ρ_{max} can be completely represented by, and reconstructed from, a set of uniform samples spaced s apart, where $s \le (2\rho_{max})^{-1}$. This requirement corresponds to adjacent samples being taken approximately $w/2$ apart, where w is the width of a detector. Provided that the data are band limited in this manner, the continuous convolution can be replaced by a discrete convolution. Grossly misleading results can occur if the sampling is too wide to satisfy this Nyquist condition.

From Equation 3.20 and using $s = (2\rho_{max})^{-1}$, it is seen that

$$\begin{aligned} p(ms) &= 0 && m \text{ even}, m \neq 0 \\ p(ms) &= -(\pi ms)^{-2} && m \text{ odd} \\ p(ms) &= (2s)^{-2} && m = 0 \end{aligned} \tag{3.23}$$

where ms denotes the positions along x' at which the discrete filter is defined. The projection data λ_ϕ are sampled at the same intervals, so that Equation 3.21 can be replaced by its discrete counterpart

$$\lambda_\phi^\dagger(ms) = \frac{1}{4s}\lambda_\phi(ms) - \frac{1}{\pi^2 s}\sum_{\substack{n \\ (m-n)\text{odd}}} \frac{\lambda_\phi(ns)}{(m-n)^2} \tag{3.24}$$

where m and n are integers. Figure 3.20 shows the continuous and sampled versions of $p(x')$.

Equation 3.24 is the result obtained in a quite different way in the classic paper by Ramachandran and Lakshminarayanan (1971) and is the discrete version of the result obtained by Bracewell and Riddle (1967). Note that, although the Fourier transform has featured in its derivation, the reconstruction technique is entirely a real-space operation. The convolution function (the transform of the bounded $|\rho|$) is the same for all the projections and can, therefore, be computed, stored and reused for each projection. Equations 3.15 and 3.24 show that the contributions to the reconstruction $\mu[x, y]$ can be computed from the projections one by one, as they arrive from the scanner, and once 'used' the projection may be discarded from computer memory. This is a distinct advantage over the 2D Fourier transform method (Section 3.2.5) of recovering $\mu[x, y]$, when all the transformed projections are in use simultaneously. One can even view the reconstruction 'taking shape' as the number of contributing projections increases.

The discrete backprojection is shown in Figure 3.21. It is necessary to assign and then sum to each element in the image array μ_i the appropriate value of $\lambda_\phi^\dagger$. This can be done on a nearest-neighbour basis, but it is better to interpolate between the two nearest sampled values of $\lambda_\phi^\dagger$. Formally, the process is described by

$$\mu_i = \sum_{n=1}^{N} \lambda_{\phi_n}^\dagger(m^* s) \tag{3.25}$$

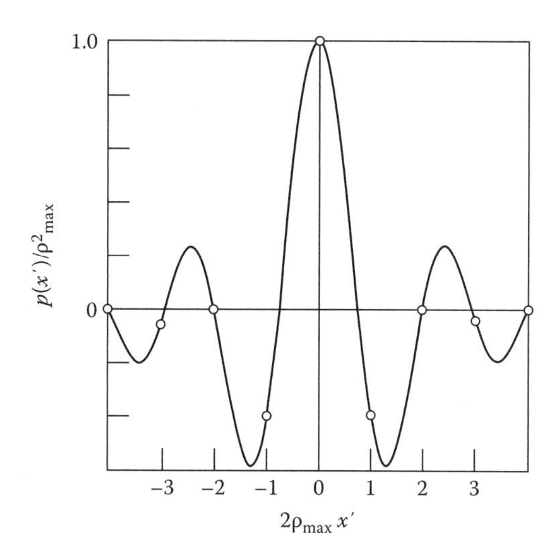

FIGURE 3.20
The full curve shows continuous form of the Ramachandran–Lakshminarayanan filter. The open circles show the points at which the filter is sampled for digital filtering methods. (Reproduced from Barrett and Swindell, 1981. With permission.)

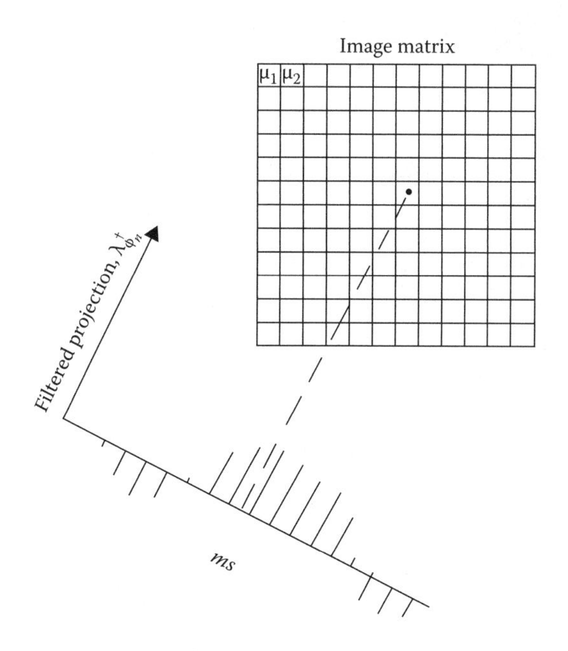

FIGURE 3.21
Object is reconstructed into an array of pixels $\mu_1, \mu_2, \ldots$, by backprojecting each filtered projection onto the array and summing the results for each projection angle.

where subscript n denotes the nth projection and m^* denotes an interpolated point within m for which the interpolated value of $\lambda^\dagger$ is calculated. The backprojection through m^* passes through the centre of the ith pixel.

The total process is not as daunting as it seems. For parallel projection data, the whole process can be coded into less than 25 lines of FORTRAN code (see Figure 3.22).

```
      DIMENSION P(65), PSTAR(65), F(4225)
      DATA W/.3333333/, M/50/ , F/4225*0./
C  FOLLOWING STEPS ARE DONE FOR EACH PROJECTION
      DO 50 K=1 ,M
C READ ONE SET PROJECTION DATA AND ANGLE
      READ (1 ,100 ) P , PHI
100   FORMAT (66F6.2)
      SINE = SIN(PHI)
      COSINE = COS(PHI)
C CALCULATE FILTERED PROJECTION PSTAR
      DO 30 I=1 ,65
      Q = P(I)*2.467401
      JC = 1 + MOD(I ,2)
      DO 20 J=JC ,65 ,2
20    Q = Q - P(J)/(I-J)**2
30    PSTAR(I) = Q/(3.141593*M*W)
C BACK PROJECT FILTERED PROJECTION ONTO IMAGE ARRAY
      DO 50 J=1 ,65
      IMIN = J*65-32-INT(SQRT(1024.-(33-J)**2)
      IMAX = (2*J-1)*65 - IMIN + 1
      X = 33 + (33-J)*SINE + (IMIN-J*65+31)*COSINE
      DO 50 I=IMIN,IMAX
      X = X + COSINE
      IX = X
50    F(I) = F(I) + PSTAR(IX) + (X-IX)*(PSTAR(IX+1)-PSTAR(IX))
C DENSITY VALUES ARE NOW STORED IN F ARRAY READY FOR PRINTOUT
      STOP
      END
```

FIGURE 3.22
Despite the apparent complexity, the reconstruction process of filtering the projections and backprojecting into the image array can be coded into just a few lines of FORTRAN. This code is for parallel-beam reconstruction. (After Brooks and Di Chiro, 1975.)

Returning briefly to the filtering operation, it is sometimes advantageous to reduce the emphasis given to the higher-frequency components in the image for the purpose of reducing the effects of noise. One widely used filter due to Shepp and Logan (1974) replaces $|\rho|$ with

$$\left|\left(\frac{2\rho_{\max}}{\pi}\right)\sin\left(\frac{\pi\rho}{2\rho_{\max}}\right)\right|$$

(see Figure 3.23). The digital filter has the form

$$p(ms) = -2(\pi s)^{-2}(4m^2 - 1)^{-1} \quad m = 0, \pm 1, \pm 2, \ldots \tag{3.26}$$

Another widely used filter, the Hanning window, uses an apodising factor $A(\rho)$ given by

$$A(\rho) = \alpha + (1-\alpha)\cos\left(\frac{\pi\rho}{\rho_{\max}}\right)$$

which multiplies into $|\rho|$. The quantity α is a variable parameter that gives $A(0) = 1$ for all α and $A(\rho_{\max}) = 2\alpha - 1$, varying from -1 for $\alpha = 0$ to 1 for $\alpha = 1$. In practical terms, when $A(\rho)$ is included on the RHS of Equation 3.16, this becomes equivalent to convolving the projection

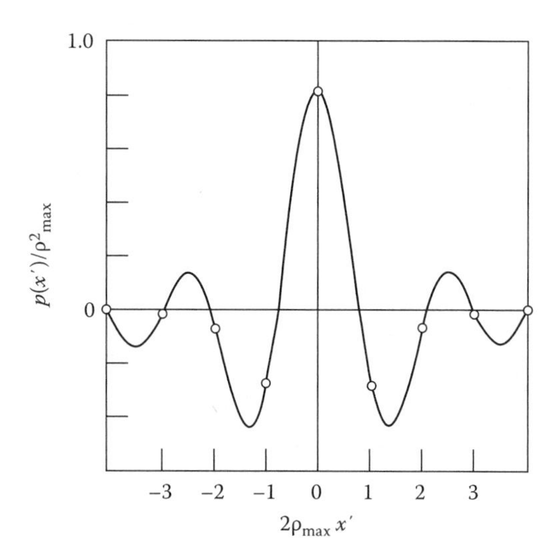

FIGURE 3.23
Continuous and discrete versions (full curve and open circles, respectively) of the Shepp and Logan filter. (Reproduced from Barrett and Swindell, 1981.)

data $\lambda_\phi(x')$ with a modified kernel $p(x')$ whose form is different from that in Equation 3.20 (or discretely, Equation 3.23). A fuller discussion of windowing is given in the section on SPECT (Section 5.8).

In this section, the filtered backprojection method has been described. Since both the filtering and the backprojection are linear and shift-invariant operations, it does not matter which is performed first. When backprojection is performed first, however, the filtering becomes a 2D operation. Backprojecting unfiltered projections would yield a result $\mu_B[x, y]$, where

$$\mu_B[x,y] = \int_0^\pi \lambda_\phi(x')\,\mathrm{d}\phi\,|_{x'=x\cos\phi+y\sin\phi}. \tag{3.27}$$

Rewriting Equation 3.17 for the central-slice theorem, we have

$$\lambda_\phi(x') = \int_{-\infty}^{\infty} \left(\frac{M^P(\rho,\phi)}{|\rho|}\right)|\rho|\exp(2\pi i\rho x')\,\mathrm{d}\rho. \tag{3.28}$$

Comparing the pairs of Equations 3.27 and 3.28 with 3.15 and 3.16, it is quite clear that $\mu_B[x, y]$ is related to $\mu[x, y]$ by a function that compensates for the denominator in the integral (3.28). Deconvolution of this function from $\mu_B[x, y]$ to yield $\mu[x, y]$ is possible, but is a very clumsy way of tackling the reconstruction problem. If it is also remembered that filtering can take place in real space (by convolution) or Fourier space, it is clear that there are many equivalent ways of actually performing the reconstruction process (Barrett and Swindell 1981).

After the first generation of transmission CT scanners, the technique of rotate-translate scanning was largely abandoned in favour of faster scanning techniques involving fan-beam geometry. Viewed at the primitive level, however, these scanning geometries merely in-fill Fourier space in different ways and a reconstruction of some kind will always result.

Indeed, it is perfectly possible to imagine merging projection data for the same object taken in quite different geometries. Once this is realised, it is soon apparent that the multitude of reconstruction methods that exist are in a sense mere conveniences for coping with less-simple geometry. The methods do, however, possess some elegance and many of the derivations are quite tricky! Without wishing to be overdismissive of a very important practical subject, we shall make no further mention of the mathematics of more complex geometries for reconstructing 2D tomograms from 1D projections.

3.7 Iterative Methods of Reconstruction

In the early days of CT, iterative methods were popular. Various techniques with names such as algebraic reconstruction technique (ART), simultaneous iterative reconstruction technique (SIRT) and iterative least squares technique (ILST) were proposed and implemented. Such iterative techniques are no longer used for X-ray CT but still find application where the data sets are very noisy or incomplete, as they often are in emission CT (see Section 5.8).

For several decades iterative methods were not used at all in X-ray CT. However, the last few years have seen a resurgence in ILSTs. A grasp of the basic principles is, therefore, a prerequisite for understanding the new methods. To this end, iterative reconstruction in its simplest form is discussed in this section.

The principle of the method is described in Figure 3.24. The image (the estimate of the object) is composed of I 2D square pixels with densities μ_i, $1 \leq i \leq I$. The projections $\hat{\lambda}(\phi, x')$ that would occur if this were the real object are readily calculated using

$$\hat{\lambda}(\phi, x') = \sum_{i=1}^{I} \alpha_i(\phi, x')\mu_i \tag{3.29}$$

where $\alpha_i(\phi, x')$ is the average path length traversed by the (ϕ, x') projections through the ith cell. These coefficients need only be calculated once; they can then be stored for future use. For a typical data set, Equation 3.29 represents 10^5 simultaneous equations. The solution method is to adjust the values of the μ_i iteratively until the computed projections $\hat{\lambda}$ most closely resemble the measured projections λ. These final values μ_i are then taken to be the solution, that is, the image. Equation 3.29 is not soluble using direct matrix inversion for a variety of reasons that relate not only to the size (α is typically a $10^5 \times 10^5$ square matrix, albeit a very sparse one) but also to the conditioning of the data.

Because of measurement noise, and the approximations engendered by the model, there will not be a single exact solution. The arguments are very similar to those in Chapter 11 explaining why image deconvolution is difficult. Furthermore, there are usually far more equations than there are unknowns, so a multiplicity of solutions may exist. Part of the difficulty of implementing the solution is in deciding upon the correct criteria for testing the convergence of the intermediate steps and knowing when to stop. This is but one example of a whole class of *ill-posed* problems, which are dealt with in a branch of mathematics bearing this same name.

The many iterative algorithms differ in the manner in which the corrections are calculated and reapplied during each iteration. They may be applied additively or multiplicatively; they may be applied immediately after being calculated; optionally, they may be stored and applied only at the end of each round of iteration. The order in which the projection data are taken into consideration may differ as well.

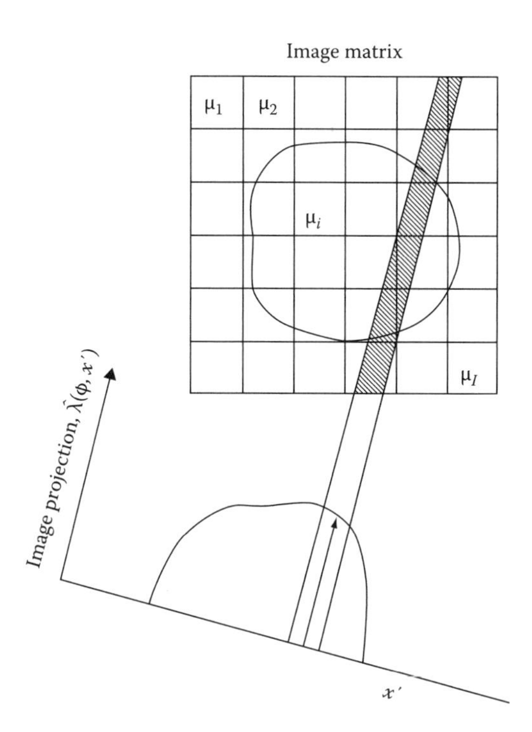

FIGURE 3.24
In iterative reconstruction methods, a matrix of I cells represents the object. The line integrals for the projection data are then replaced by a set of linear equations (3.29). (Reproduced from Barrett and Swindell, 1981.)

The simple example shown in Figure 3.25 illustrates additive immediate correction. Four three-point projections are taken through a nine-point object O, giving rise to projection datasets P_1 through P_4. Taken in order, these are used successively to calculate estimates E_1 through E_4 of the original object.

The initial estimate is obtained by allocating, with equal likelihood, the projection data P_1 into the rows of E_1. Subsequent corrections are made by calculating the difference between the projection of the previous estimate and the true projection data and equally distributing the difference over the elements in the appropriate row of the new estimate. For example, the difference between the projection of the first column of E_1 shown in parentheses (15), and the true measured value (16) is 1. In creating the first column of E_2, one-third of this difference (1/3) is added to each element of the first column of E_1. The first iteration is completed with the calculation of E_4. That the process converges in this numerical example is demonstrated by calculating the root-mean-square (RMS) deviation of elements of E_1 through E_4 from the true values in O. As the figure shows, these RMS errors decrease monotonically.

All the CT manufacturers have introduced iterative reconstruction in the last few years. The algorithms differ between the manufacturers, but can be broadly split into two classes: algorithms where the image is reconstructed iteratively from the projection data and algorithms where the image is post-processed iteratively to decrease image noise whilst maintaining spatial resolution. The second class of algorithms is not, strictly speaking, iterative reconstruction. In practice, iterative reconstruction algorithms initialise the image matrix with the filtered backprojected image rather than a blank image in order to reduce the number of iterations needed and, hence, the computing time.

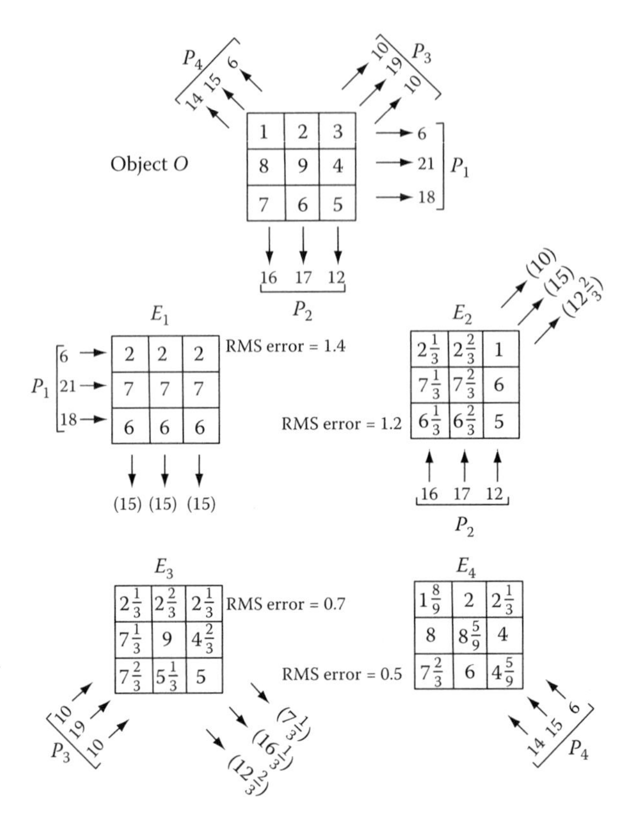

FIGURE 3.25
Simple example of iterative reconstruction. See text for explanation.

Reconstruction times are significantly longer than for filtered backprojection techniques; iterative methods are, therefore, not always clinically useful. Further details can be found in Hsieh (2009) and Fleischmann and Boas (2011).

3.8 Cone-Beam Reconstruction

The central-slice theorem can be extended from 2D to 3D to yield the result that the full set of 2D Fourier transforms of the plane integrals through an object $\mu[x, y, z]$ is identical to the 3D Fourier transform of μ. As in the 2D case, it is assumed that the object is bounded by the X-ray beam. It is, therefore, theoretically feasible to obtain $\mu[x, y, z]$ by 3D Fourier inversion or by using 3D filtered backprojection. However, difficulties are encountered in practice: the measurement process yields line integrals of μ, not plane integrals; a divergent X-ray cone beam is used which does not readily lend itself to summing line integrals to obtain plane integrals; the human body is a long object which cannot be easily bound by the X-ray beam (nor would it be desirable to do so, if, for example, only a scan of the chest is needed); and only truncated projection data may be available due to limitations in the X-ray focus trajectory and the shape of the human body. The following sections briefly describe the solutions that have been proposed to date to tackle these difficulties. More detailed accounts can be found in Flohr et al. (2005), Kalender (2005), Wang et al. (2007) and Hsieh (2009).

3.8.1 Exact Cone-Beam Reconstruction Algorithms

Exact reconstruction algorithms convert the measured 1D cone-beam projection data into plane integrals and then use 3D Fourier techniques or filtered backprojection to reconstruct the image. The formulae used contain no approximations – hence the name. Image reconstruction from cone-beam projection data is possible if each plane through the object also intersects the X-ray focus trajectory at least once. This is called the Tuy condition (Tuy 1983).

The 3D Fourier inversion method uses Grangeat's formula (Grangeat 1991). Filtered backprojection algorithms have been developed by Kudo, Defrise and Tam, with increasing emphasis on solving the long object problem (see, for example, Defrise and Clack 1994, Kudo and Saito 1994, Tam 1995). Algorithms have been formulated specifically for spiral trajectories of the X-ray focus over a limited range along the long axis of the patient (e.g. Kudo et al. 1998, Tam et al. 1998, Defrise et al. 2000, Schaller et al. 2000, Tam 2000).

In 2002, Katsevich published a seminal filtered backprojection algorithm for spiral CT (Katsevich 2002) for use with cone-beam projection data. The following year he published a generalised filtered backprojection scheme for other X-ray focus trajectories (Katsevich 2003). Katsevich proved that it is possible to reconstruct an image exactly from only a subset of cone-beam projection data, rather than the full set of plane integrals. Katsevich's work is a landmark in image reconstruction research. Recent activity has focused on deriving suitable formulae for the filter function and producing computational algorithms which implement Katsevich's scheme.

The primary drawbacks of exact reconstruction techniques are: the computational complexity of the reconstruction; the difficulties in defining the backprojection filter function; and the existence of data redundancy, that is, projection data are acquired that are not used for reconstruction, leading to unnecessary patient dose. For these reasons, exact reconstruction techniques are currently considered impractical for medical applications.

3.8.2 Approximate Cone-Beam Reconstruction Algorithms

Approximate cone-beam reconstruction techniques avoid calculating the full set of plane integrals through the object. They are computationally simpler to implement than the exact algorithms, but may yield artefacts when the cone angle increases. They divide into two classes described as follows.

The first class comprises algorithms based on the work of Feldkamp et al. (1984), often referred to as FDK-type algorithms. Such algorithms are available on all current CT scanners; however, the exact nature of the algorithm varies from CT manufacturer to CT manufacturer.

The original approach by Feldkamp, Davis and Kress was formulated for a circular focus trajectory. A set of planar projections of the object, akin to conventional X-ray images, was obtained using an area detector. First, the detector plane was transposed to a plane containing the rotation axis. The divergence of the X-ray beam was neglected so that the X-rays impinging on the detector were assumed to be normal to the detector; a geometric correction was made to the detected signal to account for the increased pathlength through the object. The corrected projection data were then convolved with a 1D ramp filter on a row-by-row basis (a row being parallel to the plane of the focus trajectory). Finally, the filtered projections were backprojected onto the image volume, this time taking account of the cone angle (3D backprojection).

The 1990s saw a rapid development in FDK-type algorithms. Reconstruction formulae were generalised to deal with a variety of focus trajectories, including spiral (Yan and Leahy 1992, Wang et al. 1993). Activity in the 2000s centred on focus trajectories with clinical potential. The hybrid Tent-FDK (Grass et al. 2000) and the angular weighted hybrid

Tent-FDK (Grass et al. 2001) algorithms have been developed for use with single and sequential circular focus trajectories. They allow reconstruction to take place using projection data acquired over 180° as well as 360°, so that a greater extent of the object along the axis of rotation can be reconstructed. In an alternative approach, the 2π sequential method (Köhler et al. 2001), the table increment between rotations is chosen so that a given point in the object not illuminated by the X-ray beam over a full 360° in one focus orbit will be illuminated in the next for the required angular range. The π-method algorithm has been formulated for helical trajectories (Danielsson et al. 1999). As its name suggests, it is based on 180° reconstruction techniques. Again, this algorithm has undergone a number of revisions in order to increase the usage of the projection data. In the $n\pi$ method (Proksa et al. 2000), a given region of interest (ROI) in the object is illuminated over $n\pi$ radians, yielding improved reconstruction characteristics and image quality (see Section 3.9.11). Several workers have also found benefits in varying the direction along which the filtering function is applied, for example, tangentially to the spiral trajectory (π-slant technique, described in Köhler et al. 2002).

More recent FDK-type algorithms implement an initial rebinning step to generate parallel cone-beam projection data, so that the geometric correction described previously need only be applied in the direction of the cone angle (Kachelrieß et al. 2004). Other algorithms use weighting schemes to reduce the influence of the projection data acquired at the larger cone angles (Tang et al. 2005, 2006), or to minimise the amount of redundant data (and thereby improve dose efficiency) (Stierstorfer et al. 2004, Taguchi et al. 2004, Schöndube et al. 2009).

The second class of approximate cone-beam reconstruction algorithms encompasses algorithms based on single-slice rebinning (SSRB) techniques. They are not, strictly speaking, cone-beam reconstruction algorithms, as the reconstruction step is taken in two dimensions. Their use is limited to helical focus trajectories in multi-slice CT scanners yielding no more than 32 slices per rotation.

One of the first approaches in this class is the CB-SSRB algorithm developed by Noo et al. (1999). Their method rebins the cone-beam projection data into a stack of fan-beam projections parallel to the rotation axis. This is accomplished by first defining virtual transaxial reconstruction planes, and then estimating the fan-beam projection data in the reconstruction plane from cone-beam projection data. A geometric correction is made for the increased pathlength through the object. The images are then reconstructed from the virtual fan-beam projection data using conventional 2D techniques.

A natural extension of this approach is to define reconstruction planes tilted to the transaxial plane (see Figure 3.26a). The benefit of this scheme is that it minimises the errors in the rebinning step identified previously, as the tilted planes will match the helical trajectory of the focal spot better. Reconstruction then takes place to generate a set of nutating slices. A final step is needed to obtain the required transaxial images by using a triangular weighting function to interpolate between the tilted slices. This algorithm is called advanced single-slice rebinning (ASSR), and was developed by Kachelrieß et al. (2000). Typically, two to five slices contribute to each transaxial image (Köhler et al. 2002). Two disadvantages have been identified with this technique, with significant consequences for medical applications: (1) the algorithm has been optimised for a pitch of 1.5, and (2) about 30% of the cone-beam projection data are not used for reconstruction, giving rise to unnecessary radiation dose to the patient (Flohr et al. 2003). Decreasing the pitch increases the angular range over which the object is illuminated, so that it is possible to increase the angular range of parallel projection data used to reconstruct the image and thus improve image quality. However, the ASSR algorithm cannot take advantage of this because it is impossible to optimise

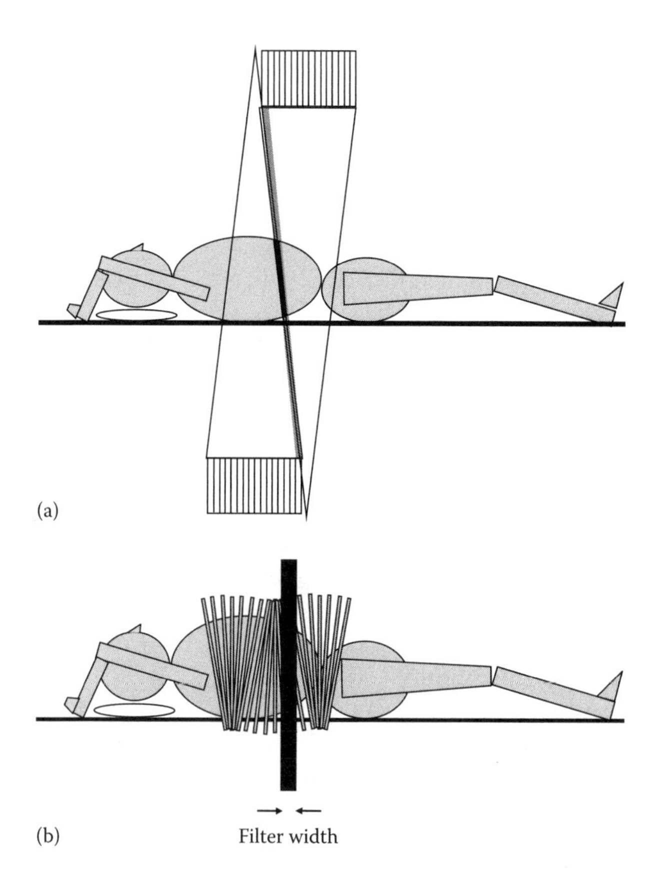

FIGURE 3.26
Figure illustrating AMPR. (a) Tilted reconstruction planes can be defined which are optimised to the helical path of the focus. Projection data that lie closest to the tilted plane are selected from different parts of the detector; in this case, data are taken from opposite ends of the detector. (b) 'Booklets' of tilted images are reconstructed using this technique. Here only the booklets with spines in the anterior and posterior positions are shown. A z-filter is then applied to generate the transaxial slices with the desired width.

the fit of the image plane to the spiral path. The adaptive multiple plane reconstruction algorithm (AMPR), described in Flohr et al. (2003), solves the first problem. Here, the spiral segment which illuminates a given transaxial plane is identified. It is then divided into overlapping subsegments, each one with an angular range of at least π radians. An image is reconstructed from the projection data in each sub-segment in a plane tilted optimally to the spiral trajectory. The number of tilted images reconstructed at any transaxial plane depends on the pitch, and they fan out from the point where the spiral trajectory intersects the transaxial plane like the pages of a book (Figure 3.26b). A z-filter in then used to interpolate between these images and yield a transverse image of the required thickness. The shape of the filter is chosen so as to keep the full width at half maximum (FWHM) of the slice sensitivity profile constant, regardless of the pitch.

On a general note, the reader should be aware that developments in approximate reconstruction algorithms take place alongside and with due regard to the advances in exact reconstruction techniques. For the time being, approximate techniques remain more efficient in terms of computational effort and radiation doses and are the reconstruction algorithms of choice for all CT manufacturers.

3.9 Image Quality Considerations

The quality of the CT image can be described in a manner similar to that found in planar radiography. The parameters used may be different, but they still reflect the physical quantities of noise, resolution and signal-to-noise ratio (SNR) in an image formed at a given dose level. In Section 3.9.1, the parameters used to describe the quality of the image are introduced, and the manner in which they are interrelated is discussed.

CT images are occasionally marred by artefacts, which can take the form of streaks, shading, banding or rings. It is often difficult to ascertain the exact cause of artefacts present in an image. In Sections 3.9.4 through 3.9.8 the processes that can create artefacts are described, and possible solutions presented. A comprehensive treatise of the causes and correction of image artefacts can be found in Hsieh (1995) and Hsieh (2009).

3.9.1 Image Quality Parameters

Image quality in CT is determined by the physical processes that govern the acquisition of the projection data and the reconstruction algorithm selected, which might, for example, suppress noise or enhance high frequencies. In the following paragraphs the image quality parameters in general use are introduced, and the principal physical processes underlying their behaviour identified.

The *spatial resolution in the scan plane* is described by the modulation transfer function (MTF), or the limiting spatial (high-contrast) resolution. It is determined by the operating bandwidth imposed by the measuring system, the sampling efficacy and the reconstruction algorithm. Sampling requirements are discussed in Section 3.9.2.

The *spatial resolution perpendicular to the scan plane*, the z-axis resolution, is also of interest. The FWHM of the system response function parallel to the rotation axis (the sensitivity profile) is often used to quantify this parameter. The z-axis resolution is determined primarily by the slice collimation.

The *noise* in a CT image is quantified by the standard deviation of a sample of CT numbers obtained by placing an ROI on the image of a uniform phantom, preferably a water phantom. The area of the chosen ROI is typically 1% of the cross-sectional area of the object. The image noise is predominantly influenced by the Poisson statistics governing the measurement of projection data (see Section 3.9.3), the reconstruction algorithm and, to a lesser extent, electronic noise. Image noise is not affected by the pixel size, unlike in other imaging modalities. This is because the detection aperture is determined by the effective ray width, not the reconstructed pixel size (Judy 1995).

Low-contrast resolution is used to describe the imaging efficacy of the scanner. It is quoted as the diameter of the smallest low-contrast detail visible at a given contrast level and surface dose, for example, 5 mm at 3 HU contrast for a surface dose of 18 mGy on a 20 cm diameter Catphan phantom (Nuclear Associates, New York). The low-contrast resolution is influenced by the SNR, the spatial resolution and the reconstruction algorithm.

The *radiation dose* required to form a single image can be expressed in terms of the surface dose, or the average dose within the scan plane to a dosimetry phantom. The average dose is quoted as the weighted CT dose index ($CTDI_w$). The CT dose index was first defined by Shope et al. (1981) as follows:

$$CTDI = \left(\frac{1}{n \cdot b}\right) \int_{-\infty}^{\infty} D(z)\, \mathrm{d}z \tag{3.30}$$

where
n is the number of slices acquired simultaneously,
b is the nominal slice thickness, and
D is the dose profile along the axis of rotation z.

For practical purposes, the limits of the *CTDI* integral are taken as ±50 mm. The term $CTDI_{100}$ is used to make this simplification explicit.

$CTDI_w$ is given by

$$CTDI_w = \frac{1}{3}CTDI_{c,100} + \frac{2}{3}CTDI_{p,100} \tag{3.31}$$

where
$CTDI_{c,100}$ is the CT dose index at the centre of the phantom, and
$CTDI_{p,100}$ is the average of the CT dose index measured 1 cm under the surface of the phantom at the four cardinal points (Leitz et al. 1995, CEC 1999).

Two dosimetry phantoms are available. They are cylindrical Perspex phantoms of 16 and 32 cm diameter, designed by the Centre for Devices and Radiological Health (CDRH) (Shope et al. 1982, FDA 2000) to represent the head and body, respectively.

The dose delivered to the dosimetry phantoms, for a complete scan sequence is indicated in the first instance by refining $CTDI_w$ to include the packing factor (the ratio between the table increment between rotations and beam collimation in axial scanning) or pitch (in helical scanning). The new quantity is called $CTDI_{vol}$ and is defined as

$$CTDI_{vol} = \frac{CTDI_w}{q} \tag{3.32}$$

where q is the packing factor or pitch, as appropriate (IEC 2002). $CTDI_{vol}$ indicates the average dose in the volume irradiated.

The dose to the dosimetry phantoms can also be expressed in terms of the dose-length product (*DLP*), given by

$$DLP = CTDI_{vol} \cdot R \tag{3.33}$$

where R is the range of the scanned volume along the axis of rotation (CEC 1999).

For multi-slice scanners with X-ray beams wider than 40 mm, these quantities significantly underestimate the radiation dose delivered to the dosimetry phantoms because the limits of the *CTDI* integral are not sufficiently far apart. At the time of writing, two solutions have been proposed to this problem, the first involving a new dose paradigm for CT, and the second a correction factor for the *CTDI* quantities. The reader is referred to AAPM (2010) and IAEA (2011) for further details.

The radiation dose delivered to a patient is a complex function of the X-ray beam quality, scanner geometry, characteristics of the beam-shaping filter, the size of the patient, and the acquisition parameters. The quantities $CTDI_w$, $CTDI_{vol}$ and *DLP*, albeit defined for dosimetry phantoms, are used to indicate the radiation dose delivered to a standard adult patient. Modern CT scanners will display $CTDI_w$ or $CTDI_{vol}$ and DLP for the prescribed scan parameters. An empirical relationship exists between DLP and effective dose for different body regions that facilitates the estimation of the latter (Shrimpton et al. 2005).

3.9.2 Angular Sampling Requirements

Assuming a point-like source of X rays, the effect of having a rectangular detector profile of width w in the direction of the projection is to modulate the frequency spectrum of the projection with a $\sin(\pi\rho w)/\pi\rho w$ apodisation. The first zero of this function is at $\rho = 1/w$. If we equate this to the maximum frequency component, that is, $\rho_{max} = 1/w$, then the sampling interval s along the projection must be $s \leq w/2$, as required by the sampling theorem (see also Section 11.6). Frequencies higher than ρ_{max} will, of course, persist in the sidelobes of the apodising function but at greatly reduced amplitudes, so the sampling requirement is only approximately fulfilled. Additional high-frequency attenuation will take place, however, owing to the finite source size (and possibly patient motion), and this $w/2$ criterion is found to be an acceptable compromise between generating aliasing artefacts and processing massive amounts of data. The question regarding the number N_ϕ of angular samples remains. The number N_ϕ is taken to be the number of projections in the range $0 \leq \phi < 180°$.

If the final image is to have equal resolution in all directions, then the highest spatial-frequency components must be equally sampled in the radial and azimuthal directions in the neighbourhood of $\rho = \rho_{max}$ in the (ρ, ϕ) Fourier space.

For an object space of diameter D and projection data that are sampled with an interval d, the number of samples per projection is $N_s = D/d$ and the radial sampling interval in Fourier space is thus $2\rho_{max}d/D$. The azimuthal interval at $\rho = \rho_{max}$ is $\rho_{max}\Delta\phi$ where $\Delta\phi = \pi/N_\phi$. Equating these sampling intervals yields the result

$$N_\phi = \left(\frac{\pi}{2}\right)N_s. \tag{3.34}$$

In practice, projections are usually taken over 360° to reduce partial-volume and other artefacts, so $2N_\phi$ projections are usually taken. Equivalently, the angular increment in projection angle is

$$\Delta\phi = \frac{2}{N_s} \tag{3.35}$$

for a uniformly sampled image data set.

3.9.3 Relation between Image Quality Parameters

The projection data are subject to measurement noise. If a particular measurement were repeated many times, yielding an average measured value of n detected X-ray photons, then the random noise associated with a single reading would be $\sqrt{n}$. These fluctuations result from the Poisson statistics of the photon beam, and cannot be eliminated.

These measurement fluctuations propagate through the reconstruction algorithms, with the result that a perfectly uniform object of density μ will appear to have a mottled appearance. An SNR can be defined as

$$SNR = \frac{\mu}{\Delta\mu} \tag{3.36}$$

where $\Delta\mu$ is the RMS fluctuation in the reconstructed value of μ about its mean value.

On the assumption that this photon noise is the only source of noise in the image, several authors (see, for example, Brooks and Di Chiro 1976b) determined an expression relating the X-ray dose U delivered to the centre of a cylindrical object to the spatial resolution ε and the SNR in first-generation CT scanners. The expression has the form

$$\eta U = \frac{k_1 (SNR)^2}{\varepsilon^3 b} \tag{3.37}$$

where

- b is the thickness of the slice,
- η is the DQE of the detector, and
- k_1 is a constant that depends on beam energy, the diameter of the object and the precise manner in which ε is defined.

The points to note are that the dose depends on the second power of the SNR and, to all intents and purposes, on the inverse fourth power of the resolution. This latter claim is made because in any reasonable system the slice width will be scaled in proportion to the resolution required: thickness $b = k_2\varepsilon$ where k_2 is typically 2–5, that is, the reconstruction voxel is a rather skinny, rectangular parallelepiped. If k_2 becomes too large, partial-volume effects, as described in Section 3.9.4, will become obtrusive. One of the assumptions used in deriving Equation 3.37 is that the exit fluence is at least an order of magnitude smaller than the incident fluence. For objects of size $d \leq \mu^{-1}$, Equation 3.37 does not apply.

The relationship between the X-ray dose U, the SNR and the slice thickness b derived above holds for multi-slice CT scanners. However, the relationship between U and the resolution ε is not as straightforward because in modern CT scanners data acquisition and the image pixel size are not as intimately related as in first-generation systems.

3.9.4 Partial-Volume Artefacts

Because the X-ray beam diverges in a direction perpendicular to the scan plane (i.e. it is a cone beam with a finite albeit small cone angle), a projection measured in one direction may be slightly different from the projection taken along the same path but in the opposite direction. For example, a dense object at the periphery of the scan field might protrude into the X-ray beam when the detector array is close, but miss it when the gantry rotates to the opposite side. An inconsistency will be introduced into the projection data that will give rise to streak and shading artefacts. A full 360° scan of the patient can be used to compensate for the inconsistencies in the data by combining data from opposite directions. These artefacts become more pronounced with increasing X-ray beam cone angle.

A different but related partial-volume effect arises from the observation that anatomical structures do not in general intersect the section at right angles. A long, thin voxel could well have one end in soft tissue and the other end in bone. As a result, the reconstructed μ would have an intermediate value that does not correspond to any real tissue at all. This is the main reason for scaling d with ε and not letting k_2 get too large (see Section 3.9.3).

Narrow slices should, therefore, always be used in preference to wide ones in anatomical regions where there is danger of partial-volume artefacts occurring. In multi-slice CT it is possible to acquire and reconstruct thin slices, and then fuse them together to form a set of wider slices.

3.9.5 Beam-Hardening Artefacts

As the X-ray beam passes through tissue, the lower-energy components are attenuated more rapidly than the higher-energy components. The average photon energy increases; the beam becomes *harder*. As a result, the exponential law of attenuation no longer holds.

With no absorber in the beam, the detector output is

$$I_0 = k_3 \int_0^{E_{max}} S(E)\,\mathrm{d}E \tag{3.38}$$

where the source spectrum $S(E)$ is defined such that $S(E)$ dE is the energy fluence in the energy range E to E + dE.

With an object in the beam, Beer's law must be weighted by the energy spectrum and integrated, to give

$$I = k_3 \int_0^{E_{max}} S(E)\exp\left(-\int_{AB} \mu_E[x,y]\,\mathrm{d}y'\right)\mathrm{d}E. \tag{3.39}$$

and the projection λ is thus

$$\lambda = -\ln\left(\frac{\int_0^{E_{max}} S(E)\exp\left(-\int_{AB} \mu_E[x,y]\,\mathrm{d}y'\right)\mathrm{d}E}{\int_0^{E_{max}} S(E)\,\mathrm{d}E}\right). \tag{3.40}$$

It is this *non*-linear relationship between λ and μ that causes the problem. The principal effects of this artefact show up as a false reduction in density in the centre of a uniform object (referred to as 'cupping') and the creation of false detail such as light streaks and dark bands in the neighbourhood of bone/soft tissue interfaces. These artefacts are particularly troublesome in the skull. There are several ways that the problems can be overcome, primarily by applying software corrections. At this point, the reader is referred to the appropriate literature, for example, Hsieh (1995).

3.9.6 Aliasing Artefacts

If the sampling requirements discussed in Section 3.9.2 are not met, aliasing artefacts may occur. These take the form of streak artefacts through the under-sampled object or at some distance from it, depending on whether the cause was insufficient samples in a projection, or insufficient projections (Hsieh 1995). Satisfying the sampling requirements was a problem initially with third-generation geometry, as the sampling interval in the projection fan is only a detector width. Fourth-generation geometry was proposed as a solution to this problem. In this instance, the number of samples in a projection fan (which has the detector as focus) is determined by the number of times the detector is sampled during the transit of the X-ray tube.

Aliasing artefacts can be reduced in third-generation systems by adopting longer scan times, which enable more projections to be acquired if the detector sampling rate remains constant. Alternatively the sampling rate can be increased for shorter scan times. Manufacturers have also developed a number of schemes to decrease the sampling interval. They are most likely to be utilised when scanning small, high-contrast objects. An example of such a scheme is the quarter-detector offset. In this technique, the centre of the detector array is displaced by one-quarter of a detector width with respect to the central ray. Opposing projections are now no longer equivalent, but interlaced, and hence the sampling interval is halved. A further example is the 'flying focal spot', where the electron beam is steered between two locations on the anode. A second set of projection fans is produced which is interleaved with the original set; the sampling interval is again halved.

3.9.7 Motion Artefacts

If the patient moves during data acquisition, there will be inconsistencies in the projections which will be most pronounced at the beginning and at the end of the acquisition. This discontinuity will appear as a sudden change in attenuation properties; if a high-density object is involved, the result will be streak artefacts in the direction of the start angle. These artefacts can be reduced by overscanning, say to 400°, and merging or 'feathering' the first and last 40° worth of data. However, this method increases scan time and patient dose (Gould 1991). An alternative technique is described as underscanning, where the first and last few projections are weighted so that their contribution to the reconstructed image is reduced, the weighting factor being smallest for those at the start angle. To compensate for this, weights greater than 1 are given to the parallel-opposed projections, where there will be no discontinuity in the data.

3.9.8 Equipment-Related Artefacts

Changes in the operating performance of the scanner may result in artefacts. The exact nature of the artefacts depends on the faulty component and the scanner generation. Drifts in detector performance manifest themselves as ring artefacts in third-generation systems and divergent streak artefacts in fourth-generation systems. In the latter case, an entire projection fan will be erroneous. In the third-generation geometry the X-ray focus and the detector describe a line which is at a constant distance from the isocentre. The erroneous transmission measurement is backprojected along this line. As the focus and detector rotate around the object, these lines become superimposed and a circle is formed. Changes in tube output will cause artefacts, but it is difficult to predict whether they will appear as rings or streaks as the effect on the image will be determined on the type and length of the change. If the selected exposure parameters are insufficient to provide adequate photon statistics, streak artefacts will result due to the large fluctuations in the projection data; the image is described as 'photon starved'. Foreign bodies in the scan plane, for example, a drop of contrast on the gantry, will cause streaks in the image, which appear to converge on the location of the foreign body.

3.9.9 Effect of Spiral Interpolation Algorithms on Image Quality

The interpolation algorithms used in single-slice spiral scanning will introduce some degree of blurring into the image. Figure 3.27 shows the sensitivity profiles for a 5 mm slice obtained in a single-slice scanner using 360° LI and 180° LI (Figure 3.27a), and two different pitches (Figure 3.27b). It is evident that the FWHM depends on the interpolation algorithm adopted, and hence so will the z-axis resolution. Increasing the pitch results in an increase in the FWHM, indicating that the z-axis resolution deteriorates with increasing pitch. This effect will be most pronounced for small spherical details, which will appear elongated, and linear structures such as blood vessels lying at a shallow angle to the scan plane. This deterioration in image quality would appear to be undesirable, but it is balanced by a very appealing feature of helical scanning: the ability to reconstruct an image at any chosen scan plane. Thus, by careful selection of the scan plane, and the use of overlapping slices, optimum viewing of small features can be achieved, and only a slight loss in z-axis resolution is perceived (Brink et al. 1992).

For single-slice interpolation algorithms the noise depends on the interpolation technique, and is independent of pitch. This is due to the total number of samples used for interpolation, which is algorithm dependent. For example, the noise at the object centre is reduced by a factor of $\sqrt{(2/3)}$ over the equivalent axial scan when using 360° LI, as interpolation over two full cycles has a smoothing effect on the data (Kalender and Polancin 1991). However, for 180° LI noise increases by a factor of $\sqrt{(4/3)}$ over the equivalent axial scan, and $\sqrt{2}$ over 360° LI due to the reduced number of samples used (Polancin et al. 1992). Higher-order interpolation techniques are expected to increase image noise.

Spatial resolution in helical images is comparable to that in images acquired by axial scanning using the same acquisition parameters. The radiation dose delivered to a scan volume remains approximately the same as that delivered by axial scanning if the acquisition parameters are the same and the table feed equals the slice thickness (pitch = 1). Looked at in greater detail, there is a small increase in dose in helical scanning, as an extra half or whole scan, depending on interpolation algorithm, is required at each end of the volume to enable the first and last slices to be reconstructed (helical z-overscan). If the pitch is increased, the dose is reduced following an inverse relationship, as expected.

The interpolation algorithm 180° MLI used in multi-slice scanning influences the sensitivity profile and noise in the same manner as described previously. z-filtering interpolation

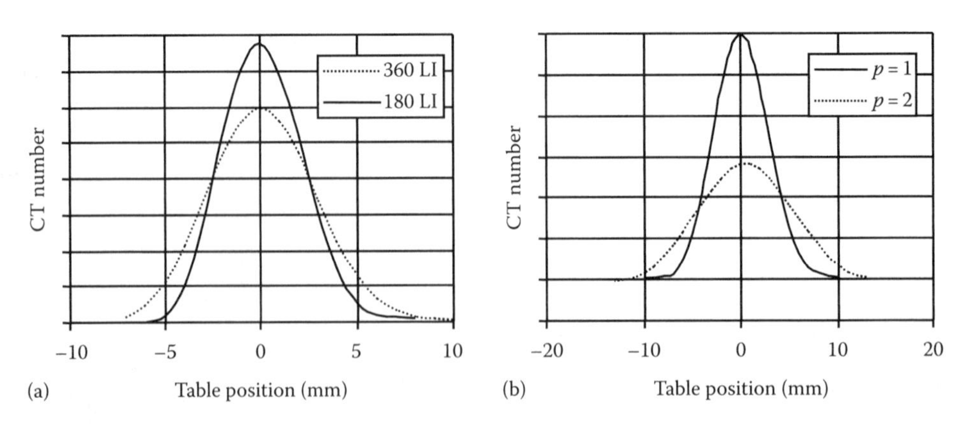

FIGURE 3.27
Graphs showing the dependence of the sensitivity profile on (a) the interpolation algorithm and (b) the pitch, p, in helical CT scanning.

(180° MFI) is more often encountered clinically. In this case, the number of samples within the filter window is reduced as the pitch increases. Consequently, the z-axis resolution remains unchanged with increasing pitch, but the noise will increase. This situation is diametrically opposite to that found with single-slice scanning. All manufacturers have chosen to compensate for the loss of samples by increasing the tube current linearly with pitch; in this instance, the image noise remains constant. However, it has been postulated that maxima and minima might exist at particular pitch values that are associated with specific ratios of the table increment per rotation and the separation between slices (Hu 1999).

In multi-slice CT, radiation dose follows an inverse relationship with pitch when the tube current is held constant. However, as all manufacturers automatically increase the tube current linearly with pitch, the radiation dose appears to be independent of pitch. The term 'effective tube current scan time product' (mAs_{eff}) has been introduced as an exposure parameter to emphasize that images acquired with the same mAs_{eff} will have equivalent image quality and patient dose, regardless of pitch. The mAs_{eff} equals the actual mAs for a pitch of 1. The helical z-overscan is considerably greater in multi-slice than in single-slice CT. A helical z-overscan between 40 and 60 mm, depending on the slice thickness and the helical pitch, is typical for 16-slice scanners.

3.9.10 Effect of Iterative Reconstruction Algorithms on Image Quality

Noise in CT images depends on the selected reconstruction filter chosen (Section 3.9.1). This is because the process of filtering the projection data enhances its inherent noise properties to a greater or lesser extent. Iterative reconstruction algorithms do not use a reconstruction filter; images reconstructed iteratively are, therefore, not prone to noise to the same extent as filtered backprojected images. In addition, the process of iteration itself is likely to have a smoothing effect on the image. Iterative reconstruction, therefore, has two major impacts on CT image quality: first, image noise is lower, and second, the texture of the noise is different. Lower image noise opens the possibility of reducing the exposure parameters and, hence, patient dose. The change in noise texture poses a challenge to the radiologist; at least one manufacturer offers images which are a mix of filtered backprojected and iteratively reconstructed images to retain the image texture familiar to clinicians. Iterative reconstruction has also been shown to reduce beam-hardening artefacts. Further details on the impact of iterative reconstruction on image quality can be found in Fleischmann and Boas (2011).

3.9.11 Effect of Cone-Beam Reconstruction Algorithms on Image Quality

Objective performance assessments of cone-beam reconstruction algorithms are based on simulations using virtual phantoms. Of particular interest is the behaviour of the algorithms with increasing cone angle and increasing sampling along the z-axis (i.e. increasing the number of rows in an area detector of given width). Köhler et al. (2002) and Sourbelle et al. (2002) have demonstrated the presence of 'wave' and 'windmill' artefacts emanating from high-contrast features when FDK-type algorithms and ASSR are used. The artefacts are reduced when the number of rows in the detector is increased, indicating that they are caused by aliasing along the z-axis (Köhler et al. 2002). As the cone angle is increased, the 'tilted' FDK-type algorithm performs better than the ASSR algorithm, but the most robust behaviour is demonstrated by exact algorithms, as expected (Sourbelle et al. 2002). Within the family of FDK-type algorithms, the most robust appears to be the n-π type (Köhler et al. 2002).

The AMPR algorithm has been implemented clinically. Flohr et al. (2003) report that image noise is maintained within 10% for a pitch range 0.5–1.5 if the tube current is increased in line with pitch. In addition, the slice sensitivity profile in the centre of the image corresponds closely to the nominal slice thickness, regardless of pitch. This is achieved by choosing the weighting function judiciously in the z-reformation step. However, a tendency towards broadening of the slice-sensitivity profile has been observed.

3.10 Recent Developments in CT Scanning

The increasing availability of multi-slice CT has encouraged the use of narrow slices to acquire projection data. This, coupled with rapid developments in computer power, facilitates the reconstruction of CT images with near isotropic voxels (i.e. with the same dimensions along the orthogonal axes). It is, thus, now possible to display images routinely in planes other than the transverse plane without detrimental loss in image quality. This facility is described as multi-planar reformatting (MPR). Sophisticated image graphics techniques such as surface and volume rendering are increasingly used to produce 2D images from 3D data sets that convey an impression of depth. It is also feasible to obtain quantified data from the scan by using automatic analysis tools. The reader is referred to Kalender (2005) for descriptions of these techniques.

CT angiography is a clinical application that relies heavily on the aforementioned image display tools. In this technique, a contrast bolus is injected and a rapid acquisition is performed over the volume of interest during the arterial flow phase. The projection set is reconstructed as a 3D array. The arterial tree can be displayed in a variety of ways. If volume rendering is used, a CT threshold is set by the user, and the voxels along a line of sight which exceed it are selected and displayed. An angiographic examination is illustrated in Figure 3.28. The time taken to perform such an investigation is now comparable to conventional

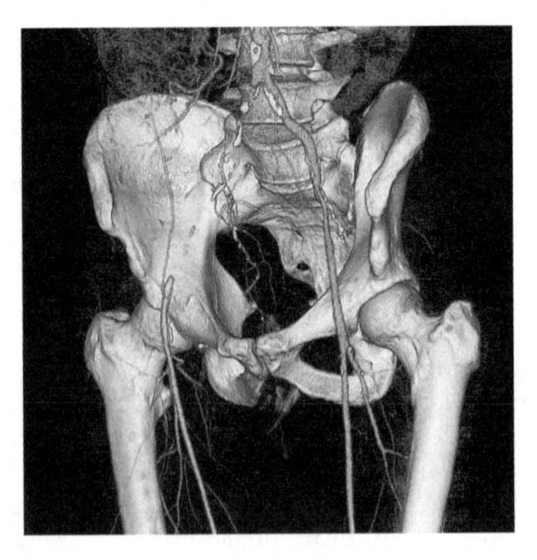

FIGURE 3.28
Clinical example of CT angiography. (Courtesy of D. Mears.)

X-ray angiography, making CT a viable alternative. Further details can be found in Rankin (1999), and Remy-Jardin and Remy (1999), and Fishman (2001).

CT scanning has also found an application as a virtual endoscope, since surface rendering can be applied to features internal to the scan volume to obtain an image of inner surfaces such as the walls of the colon. Animation techniques are then used to allow the operator to travel virtually through the patient. Although less invasive than conventional endoscopy, this tool has limited use due to the associated radiation risk and the difficulties in obtaining examination time on the CT scanner. Applications in the bronchi are described in Grenier et al. (2002), and in the colon in Rogalla et al. (2000).

In 1993, CT fluoroscopy was introduced to aid biopsies and other localisations performed under CT guidance. Conventionally, the radiologist would place a localising needle within the target volume and its position would be checked by performing a CT scan of the area. During CT fluoroscopy, a live image of the requisite slice is provided and refreshed several times per second. This is possible by scanning, using a very low tube current, 1 s or sub-second rotation speeds and smaller reconstruction matrices. For example, if a frame rate of 8 frames per second is used, and 1 s rotation, every 1/8 of a second the latest 45° sector of data is reconstructed and added to the image, and the contribution from the earliest sector is removed. Thus, it is not necessary to carry out a full reconstruction, and the image is updated quickly. The facility is available upon pressing a foot switch. This technique is attractive, but has potentially serious radiation protection implications for the patient and the attendant staff (Keat 2001). It is described in greater detail in Kalender (2005).

An area of increasing interest is cardiac imaging. Data acquisition can be triggered using ECG gating to obtain images of the thorax with minimal cardiac motion; this technique is described as prospective gating. It is also possible to acquire a spiral scan over the heart together with the ECG signal. The projection data can then be tagged with the point of the cardiac cycle at which they were acquired (retrospective gating). Images are reconstructed using single-phase or multi-phase selection of the projection data. In single-phase reconstruction, all the projection set is selected from within one heartbeat at the requisite location. In multi-phase reconstruction, different segments of the projection set are selected from different heartbeats at the requisite location. The latter technique results in improved temporal resolution because it is determined by the temporal window defined by the segment, rather than the temporal window defined by a partial scan. Dual-source CT and scanners with wide detector arrays come into their own in cardiac imaging. Fuller descriptions of these techniques and the dedicated interpolation algorithms that have been developed can be found in the literature (e.g. Kachelrieß and Kalender 1998, 2000, Kalender 2005).

Patient dose is a serious issue in CT scanning: it was estimated that by the year 2000 this modality delivered 47% of the collective radiation dose to the population arising from medical use (Hart and Wall 2004). The increasing popularity of the advanced scanning techniques described previously can only result in an increased radiation burden to the population as well as to the individual.

3.11 Performance Assessment of CT Scanners

A CT scanner is in the first instance an X-ray unit, and as such, the quality control tests that would be performed on a conventional radiographic X-ray unit also apply to it. However, the beam geometry and rotational movement impose an additional level of difficulty in the performance of these tests which often require ingenuity and cooperation from the manufacturer if the difficulties are to be overcome.

The quality control tests that need to be performed can be divided into four modules: tube and generator, scan localisation, dosimetry and image quality. The tests are described briefly in the following subsections. More details can be found in McCollough and Zink (1995), and in IPEM report 32 Part III (2003), and IPEM report 91 (2005). The tests required for helical scanning are described in Section 3.11.5.

3.11.1 Tube and Generator Tests

In a manner analogous to a conventional radiographic X-ray unit, the calibration of the following parameters needs to be performed: tube voltage, exposure time and the tube output; all three for a range of tube voltage, tube current and exposure times. In addition, the half-value layer and the focal spot sizes need to be assessed. A facility for producing X-rays with the tube stationary is often available to the service engineer; with their cooperation, these tests can be performed relatively easily. If the tester does not have access to this service mode, then tests can still be performed in the preview mode.

The kVp meter used in conventional radiography can also be used in CT to measure tube voltage and exposure time calibration. The tester should be aware that, unlike in conventional radiography, the X-ray beam may be finely collimated, especially in single-slice scanners, and may not cover the whole of the sensitive area on the kVp meter. Some instrument manufacturers provide a dedicated filter pack; otherwise, the response of the meter to a collimated beam will need to be ascertained. In all events, the largest slice available with each focus (it will not necessarily be the same for both foci) should be used. If the preview mode is used, only the tube voltage calibration can be assessed.

The tube output cannot usually be measured with a standard small-volume chamber, as the X-ray fan width will generally be small. Instead, a pencil ionisation chamber with an active length of 100 mm is used. This chamber integrates the dose profile along its length, and thus avoids positioning difficulties which might be experienced with a small chamber in a rapidly changing dose distribution. It is most convenient to measure the tube output at the isocentre.

The half-value layer can be measured using the pencil ionisation chamber placed at the isocentre; high purity aluminium is placed directly on the gantry. It should be measured at 80 kVp (so that the inherent filtration can be inferred) and at the tube voltages used clinically. Typical values at 120 kVp range from 6 to 10 mm of aluminium. It should be noted that when a bow-tie filter is present, the HVL will vary between the central axis and the edge of the beam, and, therefore, correct positioning is important. The focal spot sizes can be assessed only with a pinhole or a slit camera, again placed directly on the gantry.

These tests are usually performed when the scanner is accepted, as the availability of the service mode cannot be guaranteed during routine testing.

3.11.2 Scan Localisation

The accuracy with which the positioning laser or light localises the scan plane can be determined in a variety of ways. One method uses a direct exposure film placed at the isocentre and a fine needle to prick the film at the position of the laser. An exposure is made using a fine slice with the tube stationary. The slice location is compared with that of the pin pricks, visible on the film as fogging. A second more common method uses a dedicated phantom, where the lasers are aligned with markings on the phantom surface that coincide with details in the test objects that will give an indication of the accuracy of alignment.

The X-ray beam collimation can also be measured using direct exposure film. The film is placed at the isocentre and irradiated in service or clinical mode. When possible, the fine focus should be used to reduce the geometrical unsharpness in the image. The X-ray beam collimation is given by the FWHM of the dose profile thus obtained. This is only true at the centre of the film if it was exposed in clinical mode.

The accuracy of gantry tilt can be confirmed using the laser light projected onto a white vertical surface, and measuring the angular deviation from the vertical; alternatively, a direct exposure film placed vertically can be irradiated with the tube stationary in a lateral position. It is important to ensure that the zero tilt position is accurate; this can be done most easily by inspecting a lateral preview or a test object placed on the table.

The accuracy of the table movement can be measured using the laser light projected onto graph paper that has been placed on the table. It is important that the table is loaded to simulate the clinical situation. An alternative technique utilises direct exposure film which is irradiated either at the isocentre with the X-ray tube stationary or wrapped round a phantom in conventional axial mode. Table movements directly under operator control, and table feeds during automatic scan sequences should be assessed.

3.11.3 CT Dosimetry

The CT dose index can be measured in practice using the pencil ionisation chamber, or an array of thermoluminescent dosimeters (TLDs). However, it should be remembered that the chamber will only integrate over 100 mm, and the TLDs over the length of the array. For practical quality control, this is not a problem, as long as the same technique is used each time. Measuring the $CTDI_{100}$ in air at the isocentre provides a convenient means of monitoring the performance of the X-ray tube without having to stop the rotation of the tube.

$CTDI_w$ is obtained by measuring the $CTDI_{100}$ in the centre and at the cardinal points of the CT dosimetry phantoms described in Section 3.9.1. The phantoms are centred on the isocentre of the scanner, and a range of exposure conditions is tested. These measurements can be compared with the values specified by manufacturers as long as their standard conditions have been duplicated. Corrections may be needed to convert from air-kerma in air to air-kerma in Perspex, and to correct for the limited range of the integral. The phantoms are cumbersome and heavy and, hence, the absorbed dose distribution will often only be measured during acceptance.

3.11.4 Image Quality

The image quality parameters of interest are CT number calibration, image noise, image uniformity, high and low-contrast resolution in the image, and *z*-axis resolution. All these parameters are evaluated using phantoms in axial scanning mode. CT number calibration, image noise and uniformity, and *z*-axis resolution are assessed for each of the detector rows available. It is customary for the manufacturer to provide a set of phantoms with the scanner, and these are normally sufficient to ensure that image quality is being maintained.

A water phantom, typically 20 cm in diameter, is used to assess the CT number of water, the image noise and the uniformity of the CT number for a number of clinical protocols. An ROI of predetermined size is placed on the image at the centre and at the four cardinal points; the mean CT number and the standard deviation, which is representative of the noise, are recorded.

The CT number of other materials of known composition, such as air, and plastics simulating fat, muscle and bone, can be measured using a phantom that contains suitable inserts.

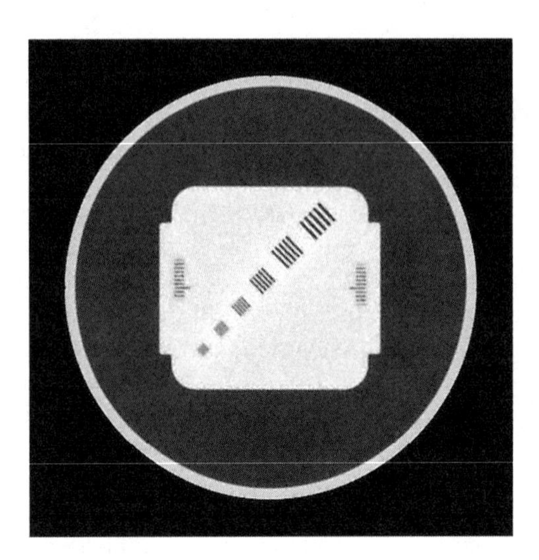

FIGURE 3.29
Test object used for the assessment of high-contrast resolution.

The high-contrast resolution can be measured in a variety of ways. A thin, high-contrast wire is used to obtain the point spread function and, hence, the MTF. Alternatively, a high-contrast edge will yield the edge response function, which can be differentiated to obtain the line spread function and, hence, the MTF along one axis. A high-contrast grating, comprising line pairs of increasing spatial frequency (Figure 3.29), provides a third method. An ROI is placed within each group of line pairs in turn, and the standard deviation is recorded. The standard deviation is related to the modulation in the signal, which decreases as the limiting resolution of the system is approached (Droege and Morin 1982). Finally, high-contrast details of diminishing size and spatial frequency can be assessed visually to obtain a subjective measure of limiting high-contrast resolution.

The low-contrast resolution is invariably always assessed using a suitable phantom. The most widely accepted are the Catphan phantom, a 20 cm diameter phantom with a low-contrast insert, and the ACR accreditation phantom. Manufacturers quote their low-contrast resolution in terms of these phantoms. Unfortunately, the phantoms are cumbersome and expensive; for routine assessment the manufacturer's phantom should provide some means of assessing low-contrast resolution.

Z-axis resolution is usually quoted as the FWHM of the sensitivity profile. Sensitivity profiles are obtained by scanning a thin, high-contrast strip, for example, aluminium, which is inclined at a known angle to the scan plane. The apparent width of the strip on the image is determined by the slice thickness and the inclination of the strip (Figure 3.30). The FWHM can be obtained from a plot of CT number against distance across the image of the strip. Alternatively, the window level can be set to halfway between the maximum CT number in the strip and the background, the window width to 1, and the apparent width of the strip can be evaluated using the measurement tool on the scanner console.

3.11.5 Helical Scanning

All the aforementioned tests are performed in axial scanning mode; additional tests are required to evaluate the helical scanning mode. They are described briefly in the following; more details can be found in Suess et al. (1995).

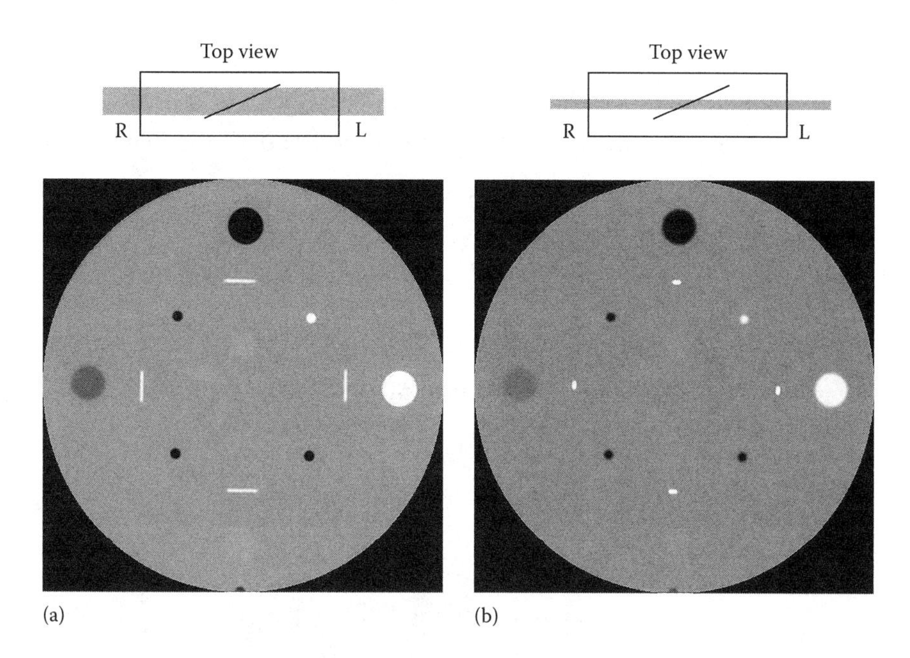

FIGURE 3.30
Schematics and images of the Catphan phantom illustrating how pairs of horizontal and vertical aluminium strips are used for the assessment of sensitivity profiles. The strips are angled to the scan plane, so that when imaged, their apparent width reflects the selected slice thickness (shown as grey in the schematics): (a) wide slice thickness and (b) narrow slice thickness. Pairs of strips are used to eliminate the effect of any skewness in positioning the sensitivity phantom. The images also show four cylinders constructed from a range of materials which are used to check the CT number calibration and markers which are used to evaluate spatial linearity.

The table motion during helical acquisition can best be assessed using direct exposure film wrapped round a phantom; the pitch and uniformity of table motion can then be determined from the separation along the direction of travel between consecutive irradiations at a given location on the phantom.

The absorbed dose distribution is not expected to vary between axial and helical modes for a pitch of 1. Comparisons between these two modes can be made by carrying out multiple scans which comfortably cover the length of the 16 and 32 cm dosimetry phantoms. Changes in the absorbed dose distribution with pitch can be assessed using the same method.

The majority of additional tests are concerned with image quality. Noise and uniformity should be assessed for a range of pitches, slice thicknesses and X-ray beam collimations. Phantoms used for low-contrast resolution that rely on the partial volume effect cannot quantify the low-contrast resolution along the z-axis. There is considerable interest in the development of true 3D test objects to carry out this assessment. Sensitivity profiles in helical scanning cannot be evaluated using the angled aluminium strip method, as artefacts are caused by the interpolation algorithm. Instead a thin high-contrast bead or foil – the latter placed parallel to the scan plane – is used; the scan images are reconstructed with a very small table increment between them. The CT number will change as shown in Figure 3.27. This technique is described fully in Polancin and Kalender (1994). Sensitivity profiles will need to be assessed for the range of pitches, slice thickness and X-ray beam collimations used clinically.

3.A Appendix

3.A.1 Evaluation of the Fourier Transform $p(x')$ of the Band-Limited Function $|\rho|$

We start from Equation 3.19, which is integrated as follows:

$$p(x') = \int_0^{\rho_{max}} \rho \exp(2\pi i\rho x')\,d\rho - \int_{-\rho_{max}}^{0} \rho \exp(2\pi i \rho x')\,d\rho$$

$$= \left[\frac{\rho \exp(2\pi i\rho x')}{2\pi i x'}\right]_0^{\rho_{max}} - \int_0^{\rho_{max}} \frac{\exp(2\pi i \rho x')\,d\rho}{2\pi i x'}$$

$$- \left[\frac{\rho \exp(2\pi i\rho x')}{2\pi i x'}\right]_{-\rho_{max}}^{0} + \int_{-\rho_{max}}^{0} \frac{\exp(2\pi i \rho x')\,d\rho'}{2\pi i x'}$$

$$= \frac{\rho_{max}\sin(2\pi\rho_{max}x')}{\pi x'} - \left[\frac{\exp(2\pi i\rho x')}{(2\pi i x')^2}\right]_0^{\rho_{max}} + \left[\frac{\exp(2\pi i\rho x')}{(2\pi i x')^2}\right]_{-\rho_{max}}^{0}$$

$$= 2\rho_{max}^2 \operatorname{sinc}(2\rho_{max}x') + \frac{\cos(2\pi\rho_{max}x') - 1}{2\pi^2 x'^2}$$

$$= 2\rho_{max}^2 \operatorname{sinc}(2\rho_{max}x') - \rho_{max}^2 \operatorname{sinc}^2(\rho_{max}x')$$

We thus obtain the result given by Equation 3.20, that is,

$$p(x') = \rho_{max}^2\left[2\operatorname{sinc}(2\rho_{max}x') - \operatorname{sinc}^2(\rho_{max}x')\right]$$

References

AAPM Report No 111 2010 *Comprehensive Methodology for the Evaluation of Radiation Dose in X-Ray Computed Tomography: Report of AAPM Task Group 111: The Future of CT Dosimetry* AAPM Task Group 111 (College Park, MD: AAPM).

Barrett H H and Swindell W 1981 *Radiological Imaging: The Theory of Image Formation, Detection and Processing* vol. II (New York: Academic Press).

Besson G 1998 New classes of helical weighting algorithms with applications to fast CT reconstruction *Med. Phys.* **25** 1521–1532.

Bracewell R N and Riddle A C 1967 Inversion of fan-beam scans in radio astronomy *Astrophys. J.* **150** 427–434.

Brink J A, Heiken J P, Balfe D M, Sagel S S, Dicroce J and Vannier M W 1992 Spiral CT: Decreased spatial resolution in vivo due to broadening of section-sensitivity profile *Radiology* **185** 469–474.

Brooks R A and Di Chiro G 1975 Theory of image reconstruction in computed tomography *Radiology* **117** 561–572.

Brooks R A and Di Chiro G 1976a Principles of computer assisted tomography (CAT) in radiographic and radioisotope imaging *Phys. Med. Biol.* **21** 689–732.

Brooks R A and Di Chiro G 1976b Statistical limitations in x-ray reconstructive tomography *Med. Phys.* **3** 237–240.

CEC 1999 Quality criteria for computed tomography, EUR 16262 EN (Brussels, Belgium: CEC).

Crawford C R and King K F 1990 Computed tomography scanning with simultaneous patient translation *Med. Phys.* **17** 967–982.

Danielsson P E, Edholm P, Eriksson J, Magnusson S M and Turbell H 1999 The original PI-method for helical cone-beam CT *Proceedings of International Meeting on Fully 3D Image Reconstruction* Egmond aan Zee, Holland, June 3–6.

Defrise M and Clack R 1994 A cone-beam reconstruction algorithm using shift variant filtering and cone-beam backprojection *IEEE Trans. Med. Imag.* **MI-13** 186–195.

Defrise M, Noo F and Kudo H 2000 A solution to the long object problem in helical cone-beam tomography *Phys. Med. Biol.* **45** 623–643.

Droege R T and Morin R L 1982 A practical method to measure the MTF of CT scanners *Med. Phys.* **9** 758–760.

FDA Center for Devices and Radiological Health 2000 21 CFR Part 1020.33 *Performance Standard for Diagnostic X-ray Systems. Computed Tomography Equipment, Federal Register Rules and Regulations* Ed. 1/4/00 (Silver Spring, MD: FDA—Food and Drug Administration).

Feldkamp L A, Davis L C and Kress J W 1984 Practical cone-beam algorithm *J. Opt. Sci. Am.* A **1** 612–619.

Fishman E K 2001 From the RSNA refresher courses: CT angiography: Clinical applications in the abdomen *Radiographics* **21** Spec no S3–S16.

Fleischmann D and Boas F 2011 Computed tomography—Old ideas and new technology *Eur. Radiol.* **21** 510–517.

Flohr T G, Schaller S, Stierstorfer K, Bruder H, Ohnesorge B M and Schoepf U J 2005 Multi-detector row CT systems and image-reconstruction techniques *Radiology* **235** 756–773.

Flohr Th, Stierstorfer K, Bruder H, Simon J, Polancin A and Schaller S 2003 Image reconstruction and image quality evaluation for a 16-slice CT scanner *Med. Phys.* **30** 832–844.

Gilbert P 1972 Iterative methods for the three-dimensional reconstruction of an object from projections *J. Theor. Biol.* **36** 105–117.

Gould R G 1991 CT overview and basics In *Specification, Acceptance Testing and Quality Control of Diagnostic X-ray Imaging Equipment 1991 AAPM Summer School* eds. Seibert J A, Barnes G T and Gould R G (Woodbury, NJ: American Institute of Physics).

Grangeat P 1991 Mathematical framework of cone beam 3D reconstruction via the first derivative of the Radon transform In *Mathematical Methods in Tomography* (Lecture Notes in Mathematics) Vol. 1497, eds. Herman G T, Louis A K and Natterer F (Berlin, Germany: Springer) pp. 66–97.

Grass M, Köhler Th and Proksa R 2000 3D cone-beam reconstruction for circular trajectories *Phys. Med. Biol.* **45** 329–347.

Grass M, Köhler Th and Proksa R 2001 Angular weighted hybrid cone-beam CT reconstruction for circular trajectories *Phys. Med. Biol.* **46** 1595–1610.

Grenier P A, Beigelman-Aubry C, Fetita C, Preteux F, Brauner M W and Lenoir S 2002 New frontiers in CT imaging of airway disease *Eur. Radiol.* **12** 1022–1044.

Hart D and Wall B F 2004 UK population dose from medical x-ray examinations *Eur. J. Radiol.* **50** 285–291.

Hermann 1980 *Image Reconstruction from Projections—The Fundamentals of Computed Tomography* (New York: Academic Press).

Hounsfield G N 1973 Computerised transverse axial scanning (tomography). Part 1: Description of system *Br. J. Radiol.* **46** 1016–1022.

Hsieh J 1995 Image artefacts, causes, and correction In *Medical CT and Ultrasound: Current Technology and Applications* eds. Goldman L W and Fowlkes J B (Madison, WI: Advanced Medical Publishing).

Hsieh J 1996 A general approach to the reconstruction of x-ray helical computed tomography *Med. Phys.* **23** 221–229.

Hsieh J 2009 *Computed Tomography: Principles, Design, Artifacts, and Recent Advances* 2nd edn (Bellingham, WA: SPIE Press).

Hu H 1999 Multi-slice helical CT: Scan and reconstruction *Med. Phys.* **26** 5–18.
Hu H and Shen Y 1998 Helical CT reconstruction with longitudinal filtration *Med. Phys.* **25** 2130–2138.
IAEA Human Health Reports 5 2011 *Status of Computed Tomography Dosimetry for Wide Cone Beam Scanners* (Vienna, Austria: IAEA).
IEC 2002 Amendment 1 to Medical Electrical equipment—Part 2–44: Particular requirements for the safety of x-ray equipment for computed tomography IEC 60601-2-44-am1 (2002-09) (IEC).
IPEM report 91 2005 *Recommended Standards for the Routine Performance Testing of Diagnostic X-ray Imaging Systems* IPEM, CoR and NRPB Working Party (York, UK: IPEM).
IPEM report 32 Part III 2003 *Measurement of the Performance Characteristics of Diagnostic X-ray Systems Used in Medicine, Part III: Computed Tomography X-ray Scanners* eds. Edyvean S, Lewis M A, Keat N and Jones A P (York, UK: IPEM).
Judy P F 1995 Evaluating computed tomography image quality In *Medical CT and Ultrasound: Current Technology and Applications* eds. Goldman L W and Fowlkes J B (Madison, WI: Advanced Medical Publishing).
Kachelrieß M and Kalender W A 1998 Electrocardiogram-correlated image reconstruction from subsecond spiral computed tomography scans of the heart *Med. Phys.* **25** 2417–2431.
Kachelrieß M and Kalender W A 2000 ECG-correlated image reconstruction from subsecond multislice spiral CT of the heart *Med. Phys.* **27** 1881–1902.
Kachelrieß M, Knaup M and Kalender W A 2004 Extended parallel backprojection for standard three-dimensional and phase-correlated 4D axial and spiral cone-beam CT with arbitrary pitch, arbitrary cone-angle, and 100% dose usage *Med. Phys.* **31** 1623–1641.
Kachelrieß M, Schaller S and Kalender W A 2000 Advanced single-slice rebinning in cone-beam spiral CT *Med. Phys.* **27** 754–772.
Kalender W A 2005 *Computed Tomography: Fundamentals, System Technology, Image Quality, Applications* 2nd edn (Erlangen, Germany: Publicis Corporate Publishing).
Kalender W A and Polancin A 1991 Physical performance characteristics of spiral CT scanning *Med. Phys.* **18** 910–915.
Kalender W A, Seissler W, Klotz E and Vock P 1990 Spiral volumetric CT with single-breath-hold technique, continuous transport, and continuous scanner rotation *Radiology* **176** 181–183.
Kalra M K, Maher M M, Toth T L, Schmidt B, Westerman B L, Morgan H T and Saini S 2004 Techniques and applications of automatic tube current modulation for CT *Radiology* **233** 649–657.
Katsevich A 2002 Analysis of an exact inversion algorithm for spiral cone-beam CT *Phys. Med. Biol.* **47** 2583–2597.
Katsevich A 2003 A general scheme for constructing inversion algorithms for conebeam CT *Int. J. Math. Math. Sci.* **21** 1305–1321.
Keat N 2001 Real-time CT and CT fluoroscopy *Br. J. Radiol.* **74** 1088–1090.
Köhler T, Proksa R, Bontus C and Grass M 2002 Artefact analysis of approximate helical cone-beam CT reconstruction algorithms *Med. Phys.* **29** 51–64.
Köhler T, Proksa R and Grass M 2001 A fast and efficient method for sequential cone-beam tomography *Med. Phys.* **28** 2318–2327.
Korenblyum B I, Tetel'baum S I and Tyutin A A 1958 About one scheme of tomography *Bull. Inst. Higher Educ. Radiophys.* **1** 151–157 (translated from the Russian by H H Barrett, University of Arizona, Tucson).
Kudo H, Noo F and Defrise M 1998 Cone-beam filtered-backprojection algorithm for truncated helical data *Phys. Med. Biol.* **43** 2885–2909.
Kudo H and Saito T 1994 Derivation and implementation of a cone-beam reconstruction algorithm for non-planar orbits *IEEE Trans. Med. Imag.* **MI-13** 186–195.
Leitz W, Axelsson B and Szendrö G 1995 Computed tomography dose assessment—A practical approach *Radiat. Protect. Dosim.* **57** 377–380.
Maravilla K R and Pastel M S 1978 Technical aspects of CT scanning *Comput. Tomogr.* **2** 137–144.
McCollough C H, Bruesewitz M R and Kofler J M Jr 2006 CT dose reduction and dose management tools: Overview of available options *Radiographics* **26** 503–512.

McCollough C H and Zink F E 1995 Quality control and acceptance testing of CT systems In *Medical CT and Ultrasound: Current Technology and Applications* eds. Goldman L W and Fowlkes J B (Madison, WI: Advanced Medical Publishing).

Natterer S 1986 *The Mathematics of Computerised Tomography* (New York: Wiley).

Noo F, Defrise M and Clackdoyle R 1999 Single slice rebinning method for helical cone beam CT *Phys. Med. Biol.* **44** 561–570.

Peschmann K R, Napel S, Couch J L, Rand R E, Alei R, Ackelsberg S M, Gould R and Boyd D P 1985 High speed computer tomography: Systems and performance *Appl. Opt.* **24** 4052–4060.

Polancin A and Kalender W A 1994 Measurement of slice sensitivity profiles in spiral CT *Med. Phys.* **21** 133–140.

Polancin A, Kalender W A and Marchal G 1992 Evaluation of section sensitivity profiles and image noise in spiral CT *Radiology* **185** 29–35.

Proksa R, Köhler Th, Grass M and Timmer J 2000 The n-PI method for helical cone-beam CT *IEEE Trans. Med. Imag.* **19** 848–863.

Radon J 1917 Uber die Bestimmung von Funktionen durch ihre Integralwerte langs gewisser Manningfaltigkeiten *Ber. Verh. Sachs. Akad. Wiss. Leipzig Math. Phys. Kl.* **69** 262–277.

Ramachandran G N and Lakshminarayanan A V 1971 Three-dimensional reconstruction from radiographs and electron micrographs: Applications of convolutions instead of Fourier transforms *Proc. Natl. Acad. Sci. USA* **68** 2236–2240.

Rankin S C 1999 CT angiography *Eur. Radiol.* **9** 297–310.

Remy-Jardin M and Remy J 1999 Spiral CT angiography of the pulmonary circulation *Radiology* **212** 615–636.

Rogalla P, Meiri N, Ruckert J C and Hamm B 2000 Colonography using multislice CT *Eur. J. Radiol.* **36** 81–85.

Schaller S, Noo F, Sauer F, Tam K C, Lauritsch G and Flohr T 2000 Exact Radon rebinning algorithm for the long object problem in helical cone-beam CT *IEEE Trans. Med. Imag.* **19** 361–375.

Schardt P, Deuringer J, Freudenberger J, Hell E, Knupfer W, Mattern D and Schild M 2004 New x-ray tube performance in computed tomography by introducing the rotating envelope tube technology *Med. Phys.* **31** 2699–2706.

Schöndube H, Stierstorfer K and Noo F 2009 Accurate helical cone-beam CT reconstruction with redundant data *Phys. Med. Biol.* **54** 4625–4644.

Shepp L A and Logan B F 1974 The Fourier reconstruction of a head section *IEEE Trans. Nucl. Sci.* **NS–21** 21–43.

Shope T B, Gagne R M and Johnson G C 1981 A method for describing the doses delivered by transmission x-ray computed tomography *Med. Phys.* **8** 488–495.

Shope T B, Morgan T S, Showalter C K, Pentlow K S, Rothenberg L N, White D R and Speller R D 1982 Radiation dosimetry survey of computed tomography systems from ten manufacturers *Br. J. Radiol.* **55** 60–69.

Shrimpton P C, Hillier M C, Lewis M A and Dunn M 2005 Doses from computed tomography (CT) examinations in the UK – 2003 review Report NRPB-W67 (Chilton, UK: National Radiological Protection Board (NRBP)).

Sourbelle K, Lauritsch G, Tam K C, Kachelrieß M and Kalender W A 2002 Comparison of cone-beam (CB) reconstruction algorithms in spiral computed tomography (CT) *Radiology* **225(P)** 253.

Stierstorfer K, Rauscher A, Boese J, Bruder H, Schaller S and Flohr T 2004 Weighted FBP- a simple approximate 3D FBP algorithm for multislice spiral CT with good dose usage for arbitrary pitch *Phys. Med. Biol.* **49** 2209–2218.

Suess C, Kalender W and Polancin A 1995 Performance evaluation and quality control in spiral CT In *Medical CT and Ultrasound: Current Technology and Applications* eds. Goldman L W and Fowlkes J B (Madison, WI: Advanced Medical Publishing).

Taguchi K, Chiang B S and Silver M D 2004 A new weighting scheme for cone-beam helical CT to reduce the image noise *Phys. Med. Biol.* **49** 2351–2364.

Tam K C 1995 Method and apparatus for converting cone beam x-ray projection data to planar integral and reconstructing a three-dimensional computerized tomography (CT) image of an object US Patent No. 5257183.

Tam K C 2000 Exact local regions-of-interest reconstruction in spiral cone-beam filtered-backprojection CT *Proc. SPIE* **3979** 520–532.

Tam K C, Samarasekera S and Sauer F 1998 Exact cone-beam CT with a spiral scan *Phys. Med. Biol.* **43** 1015–1024.

Tang X, Hsieh J, Hagiwara A, Nilsen R A, Thibault J B and Drapkin E 2005 A three-dimensional weighted cone beam filtered backprojection (CB-FBP) algorithm for image reconstruction in volumetric CT under a circular source trajectory *Phys. Med. Biol.* **50** 3889–3905.

Tang X, Hsieh J, Nilsen R A, Dutta S, Samsonov D and Hagiwara A 2006 A three-dimensional-weighted cone beam filtered backprojection (CB-FBP) algorithm for image reconstruction in volumetric CT-helical scanning *Phys. Med. Biol.* **51** 855–874.

Tetel'baum S I 1956 About the problem of improvement of images obtained with the help of optical and analog instruments *Bull. Kiev Polytechnic Inst.* **21** 222.

Tetel'baum S I 1957 About a method of obtaining volume images with the help of x-rays *Bull. Kiev Polytechnic Inst.* **22** 154–160 (translated from the Russian by J W Boag, Institute of Cancer Research, London, UK).

Tuy H K 1983 An inversion formula for cone-beam reconstruction *SIAM J. Appl. Math.* **43** 546–552.

Wang G, Lin T-H, Cheng P and Shinozaki D M 1993 A general cone-beam reconstruction algorithm *IEEE Trans. Med. Imag.* **12** 486–496.

Wang G and Vannier M W 1993 Helical CT image noise—Analytical results *Med. Phys.* **20** 1635–1640.

Wang G, Ye Y and Yu H 2007 Approximate and exact cone-beam reconstruction with standard and non-standard spiral scanning *Phys. Med. Biol.* **52** R1–R13.

Webb S 1990 *From the Watching of Shadows—The Origins of Radiological Tomography* (Bristol, UK: IOPP).

Yan X and Leahy R M 1992 Cone-beam tomography with circular, elliptical and spiral orbits *Phys. Med. Biol.* **37** 493–506.

4

Clinical Applications of X-Ray Computed Tomography in Radiotherapy Planning

H. J. Dobbs and S. Webb

CONTENTS

4.1 Computed Tomography and Its Role in Planning 153
4.2 Non-Standard Computed Tomography Scanners 159
4.2.1 CT Imaging on a Radiotherapy Simulator 159
4.2.1.1 Cone-Beam X-Ray CT on a Simulator 159
4.3 Megavoltage Computed Tomography 161
4.4 Concluding Remarks 162
References 162

4.1 Computed Tomography and Its Role in Planning

Cancer is a major cause of mortality in the Western world, ranking second only to cardiovascular disease. Overall, 30% of patients with cancer are cured and return to a normal life. Chemotherapy has had a dramatic improvement on the survival rate of patients treated for some of the less common cancers, such as testicular and paediatric tumours and the lymphomas, which present with disseminated disease. However, the large majority of patients who are cured have localised disease at presentation and are treated by surgery or radiotherapy or by a combination of the two modalities.

The success of radiotherapy is dependent upon ensuring that a tumour and its microscopic extensions are treated accurately to a tumouricidal dose. It is, therefore, essential to define the precise location and limits of a tumour with clinical examination and by using the optimum imaging methods available for a particular tumour site. The normal organs that surround a tumour limit the radiation dose that can be given because of their inherent and organ-dependent radiosensitivity. If this radiation tolerance is not taken into account when planning treatment, permanent damage to normal tissues would result.

Conventionally, and in early pre-CT days, localisation of a tumour and adjacent sensitive organs within the body was carried out using orthogonal radiographs with contrast media where appropriate, for example, introduced into the bladder via a catheter. All available patient data, i.e. clinical, surgical and radiological, were taken into account when determining the volume for treatment, in addition to a knowledge of the natural history of the tumour and likely routes of microscopic spread. Radiotherapy plans were produced in the cross-sectional plane with the use of a planning computer (Figure 4.1). Outlines were often obtained by very crude techniques using simple lead wire. The limitations of the conventional process were due to the inability of plane radiography

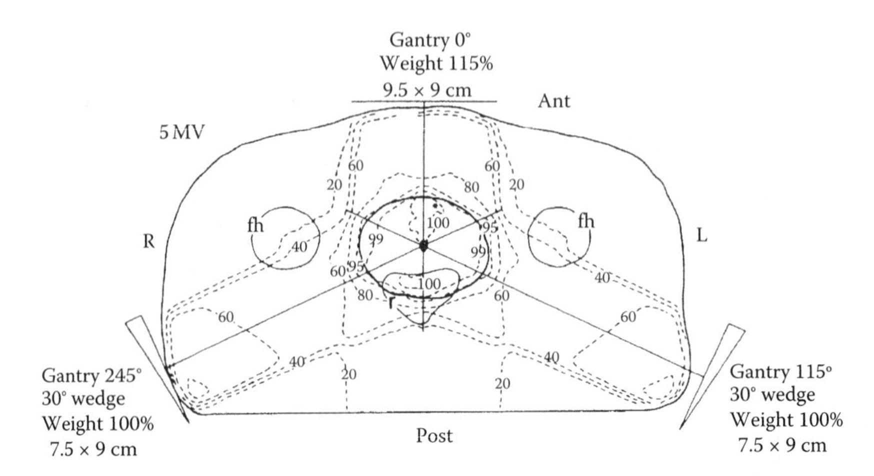

FIGURE 4.1
Conventional radiotherapy plan for treatment of Ca prostate. Bold line = PTV, r = rectum, fh = femoral head. (From Dobbs et al., 1999. Reproduced by permission of Hodder Education.)

to visualise the tumour and the difficulty in transcribing data into the transverse plane required for dosimetry. These days now seem very distant and virtually no radiography planning in the developed world is based on 2D techniques and non-CT data. This paragraph simply documents and grossly simplifies about 75 years of radiotherapy planning from 1896 to 1971.

Computed tomography (CT), available commercially since April 1972 and going through a rapid development in its first 3 years of life (see Chapter 3), has become the mainstay of cancer imaging and staging of primary tumours. In some situations, magnetic resonance imaging (MRI) provides superior information and both modalities have a major role in defining the gross tumour volume (GTV) for radiotherapy (Webb 2000, 2004). For instance, for bladder cancer, MRI is better than CT for staging early tumours but CT is accurate for advanced disease (Figure 4.2), and they provide equivalent data on lymph-node involvement

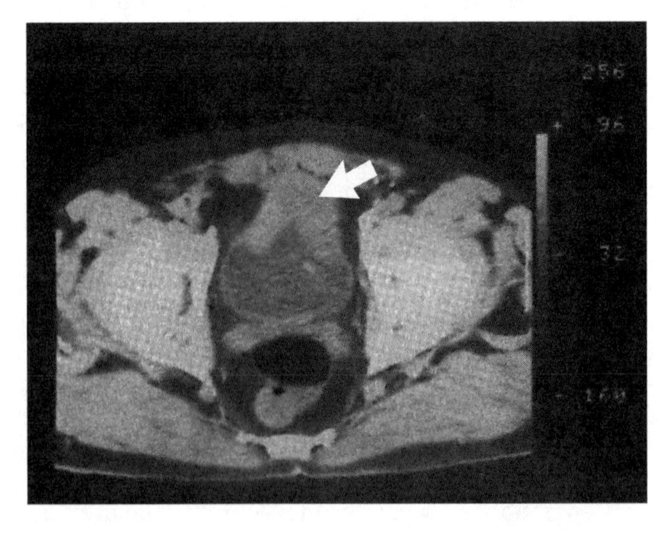

FIGURE 4.2
CT section showing extensive bladder tumour with extra-vesicle spread anteriorly (arrowed).

(Husband 1998). For prostate cancer, MRI is considered the optimum diagnostic modality to show the anatomy of the prostate, capsular penetration and involvement of the seminal vesicles (Schiebler et al. 1993).

CT and MRI provide imaging with high spatial resolution of both the tumour and anatomical structures providing good anatomical geometric data for 3D radiotherapy planning. However, more recently, techniques such as positron emission tomography (PET), single-photon emission computed tomography (SPECT) and magnetic resonance spectroscopy (MRS) give additional information about physiology, metabolism and molecular biology of the tumours, creating a new class of biological imaging (Ling et al. 2000). For lung tumours, PET scanning gives superior information in staging mediastinal lymph nodes and the combined PET/CT scan, see Chapter 5, gives optimum imaging for defining the GTV for conformal radiotherapy of lung cancer (Chapman et al. 2003). Imaging of tumour hypoxia, angiogenesis and proliferation leads to the identification of tumour cells within an inhomogeneous tumour mass that can then be individually targeted. Clinical trials are ongoing to evaluate the role of biological imaging in radiotherapy treatment planning (van Baardwijk et al. 2006). Although the term CT can apply to medical tomographic imaging in general, most of this chapter will deal specifically with X-ray CT. The subject of image-guided radiation therapy (IGRT) is now a huge development area and not greatly covered in this book. One of the authors has produced a quartet of books on technical radiation-therapy physics and all these contain large sections on the application of imaging to radiation therapy (Webb 1993, 1997, 2000, 2004). Another excellent review is Evans (2008).

CT images are ideal for planning radiotherapy treatment because they can be obtained in the transverse plane (as well as reconstructed in the sagittal and coronal planes), provide detailed visualisation of the tumour and adjacent organs, depict the complete body contour and provide electron densities necessary for dosimetry (Figure 4.3). However, as MRI, PET, ultrasound or SPECT may provide the optimum tumour information, it has become necessary to find methods of combining these diagnostic tumour data with CT imaging for treatment planning. This correlation has to take into account variables such

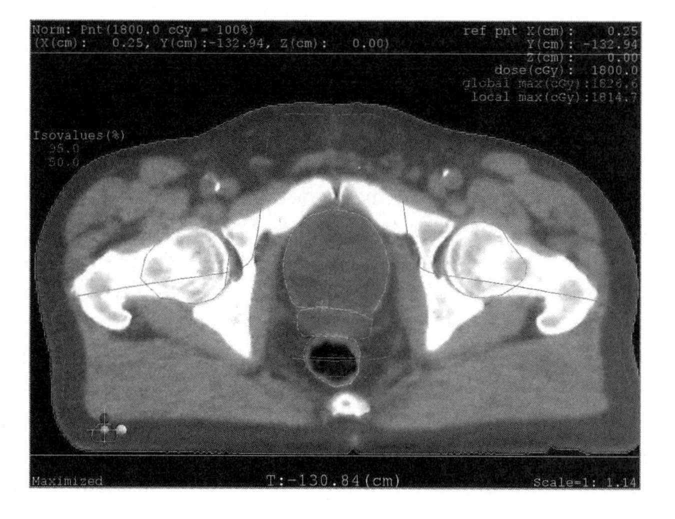

FIGURE 4.3
(See colour insert.) CT radiotherapy plan for treatment of Ca prostate. Red = PTV, blue = bladder, yellow = rectum.

as patient positioning, MRI geometric distortions, varying anatomical boundaries and differences in spatial resolution. Image registration methods are being developed to permit tumour data from a variety of diagnostic modalities to be used with CT data for treatment planning. See, for example, figure 2C in Gandhe et al. (1994). The subject of multimodality imaging and image registration for radiotherapy planning has been reviewed in detail by Webb (1993, 1997) and is also reviewed in Chapter 15.

CT scans taken for radiotherapy treatment planning are different from those taken for diagnostic purposes, as their aim and use are different. The radiotherapy CT planning scan is used to define the exact position of a tumour, its 3D extent, the position of adjacent normal structures and the relationship between the tumour, normal organs and external skin landmarks used to locate treatment beams. Ideally, CT scans are taken with the help of a therapy radiographer familiar with the position of the patient during radiotherapy, as this must be reproduced on the CT scanner. The couch must be flat as on the treatment machine and immobilisation devices used on the scanner exactly as for radiotherapy. Skin tattoos are made over adjacent bony landmarks and marked with radio-opaque catheters or barium paste so that they are visible on the CT scan and topogram (see Section 3.4). Midline and lateral laser lights must be available in the CT scanning room to align the patient using skin tattoos to ensure correct patient positioning.

A topogram recording the positions of the CT scans is used as a record, but is not divergent in the cranio-caudal direction. Measurements of the tumour in the cranio-caudal dimension must therefore be made using the CT couch positions which must have a strict quality control system to ensure accuracy. Small slice thickness and interval (2–5 mm) are essential for 3D reconstructions and dose calculations in order to ensure an adequate quantity of data. If oral or rectal contrast media are used, then care must be taken with inhomogeneity corrections in the pelvis as the presence of contrast media will affect pixel-by-pixel corrections.

Studies comparing conventional planning with CT planning for localisation of the target volume were performed in the late 1970s and 1980s (Goitein 1982, Dobbs et al. 1983). These showed that in approximately 30% of cases, a change in site and the margins of the target volume was necessary in order to adequately cover the tumour. Rothwell et al. (1985) attempted to study whether this improvement in accuracy of localising the target using CT translated into an improvement in survival, but they did not study local control. They made a retrospective comparison of the survival rate of patients whose bladder tumours were or were not adequately treated within a 90% isodose as judged by CT. They found that, if all bladder tumours had been included in the target volume, a 4%–5% increase in survival of the whole group of patients would have been predicted.

CT scanning is used routinely for many tumour sites and initially was only used as a 2D or 2½D localisation tool. This restricts the use of CT to planning at the central slice in the target volume with off-central-axis verification to check that all tumour is included within that depicted on the central slice. Dose distributions can be computed, displayed and viewed at other levels but doses do not incorporate 3D calculations using this method because calculations did not truly follow the trajectories of photons in 3D.

With the advent of 3D image handling of CT scans, 3D tumour localisation, 3D dose calculations and treatment planning, conformal radiotherapy has now become routine in most centres (Webb 1993, 1997, 2000, 2004). This technique aims at conforming the shape of the high-dose volume to that of the target volume so minimising the dose to normal structures by shaping the beams using multileaf collimators (MLC) or shaped cerrobend blocks. The use of conformal therapy is standard now for patients undergoing radical treatment. It is particularly useful at tumour sites where (a) escalating the total dose may be advantageous (e.g. prostate, lung cancer), (b) the tumour is very close to sensitive normal structures (e.g. brain), (c) the target volume is an irregular shape (e.g. thyroid, prostate tumours) and (d) the macroscopic tumour is being treated prior to surgery (e.g. rectum). In addition, intensity-modulated radiotherapy (IMRT) has been developed to vary the intensity as well as the geometry of the beam (Webb 2000, 2004). The use of IMRT is being evaluated at a number of tumour sites such as breast, prostate, head and neck in clinical trials (Bentzen 2008).

CT scans are essential for conformal therapy planning. The electron-density numbers of the X-ray CT scans, unlike for MRI scans, can be used for dose calculations. CT examinations must of course be made under geometrical conditions identical to those during radiotherapy in order to ensure accurate reproduction of subsequent treatments. Basic requirements such as the immobilisation of the patient have now become of prime importance, because if the patient or organ bearing the tumour moves either intra- or interfractionally then a geographical miss can occur. Complex immobilisation devices such as Perspex shells, vacuum bags and Alpha cradles have been developed as well as conventional pads and poles to secure limbs. A system of skin tattoos and laser lights to align the patient on a daily basis is necessary to ensure accurate positioning (Figure 4.4). Verification studies are now considered mandatory to check both geometric reproducibility using portal imaging and dosimetry using diodes or thermoluminescent dosimetry (Evans 2008).

Tumours of the prostate, bladder, lung and rectum are treated with small volumes confined to the primary tumour with an adequate margin to allow for both organ movement and set-up variations. Data on organ movement are especially important in conformal therapy where margins around the tumour are as small as possible in order to escalate dose (Padhani et al. 1999). In addition, each centre has to quantify their own set-up variations by measuring systematic and random errors for a particular tumour site and technique. Many set-up verification studies are available in the literature and summarised in a review by Hurkmans et al. (2001).

A variety of immobilisation techniques are used for patients undergoing breast radiotherapy designed to abduct the arm out of the beam pathway. If localisation is to be performed in a CT scanner, then the arms need to be abducted above the head with the

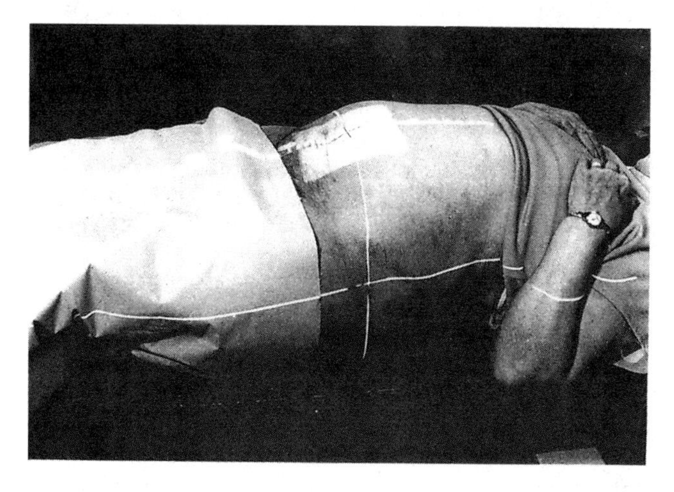

FIGURE 4.4
Patient lying supine on treatment couch, position aligned using a system of skin tattoos and laser lights as in the CT scanner.

patient supine. Gagliardi et al. (1992) showed that with their technique for patients with breast cancer using 3D CT treatment planning, the arm position greatly affected breast target-volume position. Patients were immobilised in a vacuum-moulded polystyrene bag in the supine position with one arm above the head. A horizontal grid was drawn on the patient's breast and axilla and any movement of the grid with change in position of the ipsilateral arm was analysed. Up to 4 cm displacement of the grid was found in the axilla when the arm was abducted to 130° to go through the scanner. This study showed that variation in the arm position between the CT scanner and therapy unit could lead to an error of set-up of breast radiotherapy due to movement of skin tattoos used as landmarks.

Multislice CT scanning has led to much shorter scan times of a few seconds so that all the data for a tumour in the chest can be taken in a single breath hold. However, radiotherapy treatment takes several minutes. CT may, therefore, not show the movement of the tumour in the cranio-caudal direction which will occur with respiration during radiotherapy. It may be necessary to use slower CT scans in the chest for treatment planning to include the full dimensions of the tumour with respiration. Alternatively, pulsed treatment could be gated to a particular phase of the respiratory cycle and matched with the same phase during diagnostic CT scanning. Respiration-correlated CT scanning and treatment using 4D multislice CT scans to incorporate respiratory movement is under development.

The whole subject of measuring intrafraction tumour motion and strategies for correcting it is now at the forefront of research in radiation-therapy physics. There is an enormous number of technologies dedicated to the task (see e.g. Webb 2006, Evans 2008).

CT scanning is used for planning radical radiotherapy for lung tumours. It is anticipated that conformal radiotherapy for lung cancer will enable escalation of dose whilst keeping doses to adjacent sensitive organs, such as the normal lung, heart and spinal cord, to acceptable levels.

In the head and neck region, primary tumours with their adjacent lymph-node areas are now defined using new target volume definition protocols (Grégoire et al. 2000). Conformal therapy and IMRT have a role in improving the therapeutic ratio by reducing doses to adjacent normal tissues and delivering a concomitant high-dose small-volume boost dose to the tumour.

3D localisation and dose calculations are increasingly being used in departments which have the infrastructure for complex immobilisation, adequate time for 3D localisation and 3D planning, treatment units with MLC for beam shaping and facilities for verification both of geometry and dose to the target. All these facilities have to be developed and staff training put into place before 3D conformal therapy can be given.

Intensity-modulated radiotherapy (IMRT) is the most advanced form of conformal radiotherapy and can create treatment dose volumes which have concave surfaces in which may lie organs at risk. For these problems, it is essential to feed the planning process (generally inverse planning) with detailed 3D maps of targets and normal structures (Webb 1997, 2000, 2004). A thorough development and implementation programme is needed to introduce this technique safely and appropriately into clinical practice in a department.

ICRU Reports 50 (1993) and 62 (1999) have defined a series of tumour and target volumes to enable an international consensus of volume definition. The recent explosion of technological advances has given the potential for major improvements in the results of radical radiotherapy treatments. Many prospective studies of conformal therapy are now in progress to assess these new techniques and the early results suggest an improvement in local control of tumours such as prostate cancer (Zelefsky et al. 1998).

4.2 Non-Standard Computed Tomography Scanners

The first part of this chapter has discussed how CT images may be used to assist the planning of radiotherapy and illustrated how the role of X-ray CT is much wider than providing for diagnosis and staging. The machines utilised to produce such images are referred to as commercial diagnostic CT scanners, since they generate images that have the required resolution and contrast for diagnosis. It should, however, be recognised that there are applications in radiotherapy for which diagnostic-quality images are not required and for which a number of special-purpose CT scanners have been constructed. In view of their different purposes, it is not surprising that these CT scanners differ in their technological bases. At the time of writing the first edition of this book, these machines were the subject of intense interest and development. They have now been almost entirely superseded by the regular use of CT devices with suitable-sized apertures to accommodate even the largest of patients. For this reason, a detailed report on these developments is excluded here. It can be found if required in Webb (1993, 1997).

4.2.1 CT Imaging on a Radiotherapy Simulator

A number of groups worldwide have investigated the extent to which the imaging requirements for radiotherapy planning may be met by the development of non-standard CT scanners. Some of these groups have constructed apparatus based on a radiotherapy simulator gantry. Others have built special-purpose CT devices with radioactive sources. The major use of simulator-based machines has been in planning radiotherapy in the upper thorax where tissue-inhomogeneity corrections are important.

Full 3D diagnostic-quality CT datasets are essential for planning most conformal radiotherapy. However, a small number of people developed simpler CT machines to assist with *specific* treatment-planning problems. Planning to conform the high-dose volume to the breast avoiding irradiating too much lung is one such application. A full review was written by Webb (1987, 1990).

The term non-diagnostic CT (NDCT) is here used to mean all those *single-slice* computed tomography machines whose aim is to produce CT cross sections of lower quality – i.e. poorer spatial and electron-density resolution – than obtained by state-of-the-art commercial CT scanners. The latter were of course developed for diagnostic purposes, their use in treatment planning being inspired lateral thinking afterwards. An abundance of terms appears in the literature for NDCT machines, including 'simple CT scanners', 'alternative CT scanners', 'non-commercial CT scanners', etc. The term NDCT is preferred because it immediately differentiates the purpose of such machines and is a valid generic term. It removes the sense that NDCT scanners are in some way inferior. They can and did provide enough data for the purpose for which they were designed and engineered.

Most NDCT machines were designed and built in university and hospital departments rather than by commercial companies. Hence, in regarding their construction and capabilities, we are faced with a miscellany of technology. No two machines were alike although some used common detector technology.

4.2.1.1 Cone-Beam X-Ray CT on a Simulator

Another limitation of most of the early commercial simulator-based CT systems was that they produced only one slice per rotation. Since rotations were also quite slow (of the order a minute), multiple-rotation studies are impractical. Yet volumetric imaging is required for 3D treatment planning. A number of authors suggested that the solution is to perform *cone-beam* CT using an area detector (Silver et al. 1992, Cho and Griffin 1993). Cho et al. (1995) describe the use of a Philips medical systems digital spot imager (DSI) which consists of an image-intensifier digital processor and image display and transfer facilities. The DSI can acquire frames with a maximum rate of eight frames per second. The DSI had automatic adjustment of X-ray flux with gantry position for non-circularly-symmetric objects, variable dynamic range compression and automatic unsharp-masking contrast variation, and these non-linear transformations limited the utility of the projection data for quantitative CT reconstruction. However, Cho et al. (1995) showed very good qualitative reconstructions.

The area detector was offset to create 'half-projections' since the image intensifier was not wide enough to span the whole patient. Since the detector is circular, there is a trade-off between the diameter of reconstruction and the axial height of the reconstructed volume. The reconstructed volume was 32 cm in diameter when 12 cm in axial length. If only the central slice was reconstructed (axial length = 0), the reconstructed diameter increased to 48 cm. Conversely, an axial length of 35 cm was available with a reduced reconstruction diameter. Cho et al. (1995) described a method of splicing the half-projections in which artefacts due to the sudden edges of each of the half-projections are avoided. Basically, the half-projections were made to overlap slightly and each was padded out prior to convolution with data from nearest opposing rays. After convolution, the data were only retained for reconstruction at filtered-projection pixel sites corresponding to measurement positions. Reconstruction made use of the well-known Feldkamp algorithm (Feldkamp et al. 1984). Strictly, the Tuy-completeness condition (the condition that all reconstructed planes must intersect the source trajectory at least once) was violated but this was shown to be unimportant for data with a long source-to-isocentre distance.

From a series of measurements with phantoms, it was established that (using 'width-truncated data') the geometric fidelity of the reconstruction was 1.08 mm at worst. A 1% contrast enhancement was visible in a 1.25 cm diameter tube and the spatial resolution was established, using crow's foot and line-set phantoms, as a visible 5 cycles cm^{-1}. Cho et al. (1995) successfully imaged both head and chest anthropomorphic phantoms, and whilst some ring artefacts were present in the images (Figures 4.5 and 4.6), the quality was clearly

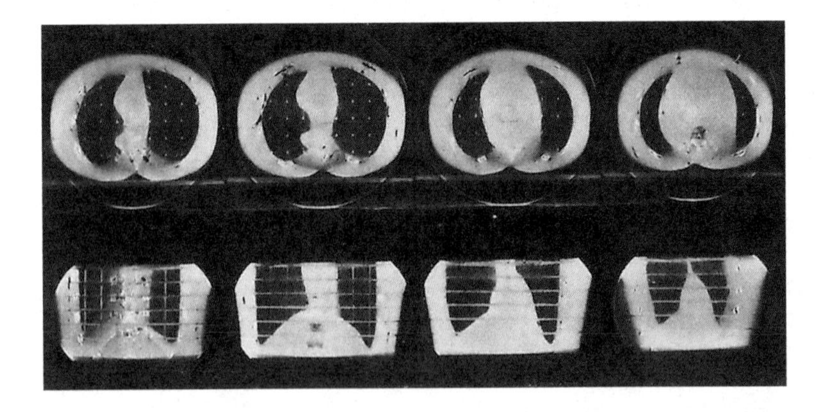

FIGURE 4.5
Cone-beam reconstruction of a Rando chest phantom using 360° width-truncated projections. Top row: trans-axial planes. From left to right the planes are at distances $z = -3.5, -1.5, 1.5$ and 3.5 cm axially from the $z = 0$ plane in which the source rotates. Bottom row: coronal planes. From left to right $x = -3.4, -1.7, 1.7$ and 3.4 cm with the plane $x = 0$ containing the rotation axis of the scanner. (From Cho et al., 1995.)

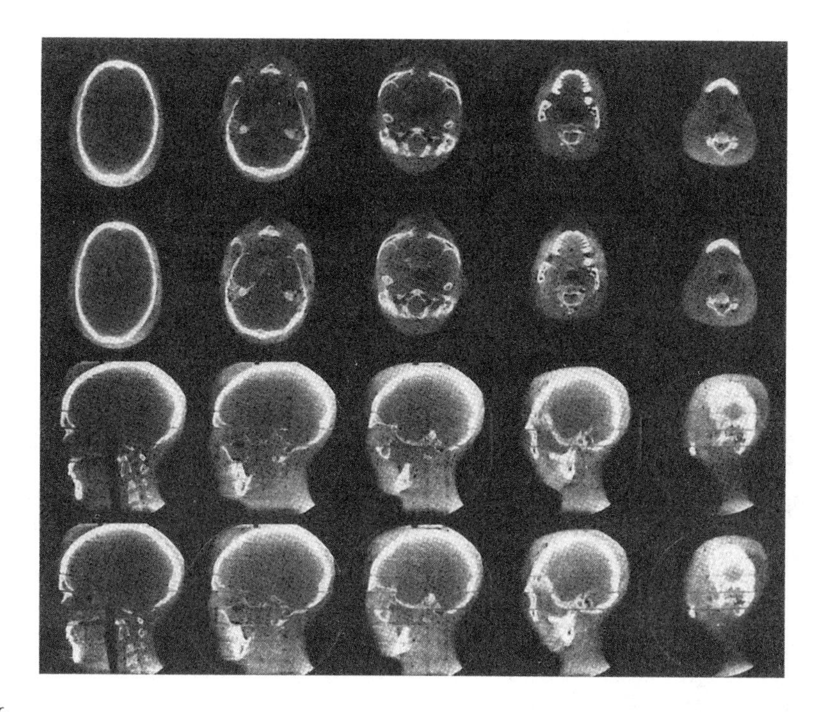

FIGURE 4.6
Cone-beam reconstruction of a Rando head phantom from 337 views. Cylindrical holes used for TLD placement are visible in some of the planes. Top row: transaxial planes of reconstruction using full-width data (455 × 455 pixel^2). From left to right $z = -5.5$, 0, 1.6, 4.7 and 8.6 cm. Second row: transaxial planes of reconstruction using width-truncated data (248 × 455 pixel^2). Third row: sagittal planes of reconstruction using full-width data. From left to right $y = 0$, 2.3, 4.7, 7 and 11.7 cm. Bottom row: sagittal planes of reconstruction using width-truncated data. (From Cho et al., 1995.)

acceptable for a treatment-planning system which uses segmented regions of interest with assigned density values. No attempt to generate truly quantitative data was made because of the non-linearities described earlier in the projection-data-taking chain.

Cone-beam CT has been included here because the latest embodiment is for kVCT actually on a linac. All the manufacturers now provide this and the origins go long back in history. Its main use is to correct for interfraction variation.

4.3 Megavoltage Computed Tomography

In Tucson, Arizona, and later at The Royal Marsden NHSF Trust, Sutton, Swindell and colleagues first developed megavoltage computed tomography (MVCT) (Simpson et al. 1982, Swindell et al. 1983, Lewis and Swindell 1987, Lewis et al. 1988, 1992). A Swedish development was also reported (Brahme et al. 1987, Källman et al. 1989) as well as a replication of Swindell's system in Japan (Nakagawa et al. 1991). Megavoltage CT scanning has recently been developed by one of the major linac manufacturers, Siemens (Evans 2008).

The development of MVCT achieves several goals. Firstly, it may be noted that X-ray linear attenuation coefficients derived from a machine operating at megavoltage energy would be immediately applicable to making tissue-inhomogeneity corrections in planning (whereas for a diagnostic machine a conversion is needed). Secondly, images taken on a linac would help to ensure the patient was in the treatment position. The aim is to take images just prior to treatment to verify the positioning of the patient by comparison with CT images at treatment simulation. This provides a connection with CT imaging on a simulator since these latter images might well be obtained for certain treatments (e.g. breast radiotherapy) with a simulator NDCT scanner (Webb 1990). Periodic checks of the patient anatomy (e.g. to determine whether a tumour has changed its size) become feasible during the course of fractionated radiotherapy.

We conclude this section by emphasising that, as conformal radiotherapy and IMRT have become more sophisticated, the emphasis is returning to the question of whether the target has been correctly identified and is located correctly at the time of treatment. 4D multislice spiral CT is now possible so that techniques can be developed to plan for the moving tumour (see review in Webb 2004). At the time of writing, most of the main accelerator manufacturers are offering kVCT systems for reconstructing a volume of patient data at the time of treatment and using it to reposition the patient and maybe compute *a posteriori* the dose delivered. This is a vast topic too large for this textbook and there are many other texts that can be consulted by those whose specialty is radiation therapy rather than imaging.

4.4 Concluding Remarks

The position today is that the use of X-ray CT and MRI CT is completely standard for 3D and some 4D treatment planning. The use of SPECT and PET is more patchy and the debate still rages on how to use integrated imaging technology. kVCT and MVCT are routinely used in the clinic for correcting interfraction rotation. The wide topic of measuring and correcting for intrafraction motion is hardly yet started in the clinic. Special purpose-built CT scanners based on simulators are little used.

References

Bentzen S M 2008 Radiation oncology health technology assessment – The best is the enemy of the good *Nat. Clin. Pract. Oncol.* **5** 563.

Brahme A, Lind B and Näfstadius P 1987 Radiotherapeutic computed tomography with scanned photon beams *Int. J. Radiat. Oncol. Biol. Phys.* **13** 95–101.

Chapman J D, Bradley J D, Eary J F, Haubner R, Larson S M, Michalski J M, Okunieff P G, Strauss H W, Ung Y C and Welch M J 2003 Molecular (functional) imaging for radiotherapy application: An RTOG symposium *Int. J. Radiat. Oncol. Biol. Phys.* **55** 294–301.

Cho P S and Griffin T W 1993 Single-scan volume CT with a radiotherapy simulator *Med. Phys.* **20** 1292.

Cho P S, Johnson R H and Griffin T W 1995 Cone-beam CT for radiotherapy applications *Phys. Med. Biol.* **40** 1863–1883.

Dobbs H J, Barrett A and Ash D 1999 *Practical Radiotherapy Planning* (3rd edn.) (London, UK: Arnold).

Dobbs H J, Parker R P, Hodson N J, Hobday P and Husband J E 1983 The use of CT in radiotherapy treatment planning *Radiother. Oncol.* **1** 133–141.

Evans P 2008 Anatomical imaging for radiotherapy *Phys. Med. Biol.* **53** R151–R191.

Feldkamp L A, Davis L C and Kress J W 1984 Practical cone-beam algorithm *J. Opt. Soc. Am.* **A1** 612–619.

Gagliardi G, Lax I and Rutqvist L E 1992 Radiation therapy of Stage I breast cancer: Analysis of treatment technique accuracy using three-dimensional treatment planning tools *Radiother. Oncol.* **24** 94–101.

Gandhe A J, Hill D L G, Studholme C, Hawkes D J, Ruff C F, Cox T C S, Gleeson M J and Strong A J 1994 Combined and three-dimensional rendered multimodal data for planning cranial base surgery: a prospective evaluation *Neurosurgery* **35** 463–471.

Goitein M 1982 Applications of computed tomography in radiotherapy treatment planning In: *Progress in Medical Radiation Physics* ed. Orton C G (New York: Plenum Press) pp. 195–287.

Grégoire V, Coche E, Cosnard G, Hamoir M and Reychler H 2000 Selection and delineation of lymph node target volumes in head and neck conformal radiotherapy. Proposal for standardising terminology and procedure based on surgical experience *Radiother. Oncol.* **56** 136–150.

Hurkmans C W, Remeijer P, Lebesque J V and Mijnheer B J 2001 Set up verification using portal imaging: Review of current clinical practice *Radiother. Oncol.* **58** 105–120.

Husband J 1998 Bladder cancer In: *Imaging in Oncology* eds. Husband J E S and Reznek R H (Oxford, UK: Isis Medical Media Ltd).

ICRU Report 50 1993 Prescribing, recording and reporting photon beam therapy (Bethesda, MD: ICRU).

ICRU Report 62 1999 (Supplement to Report 50) Prescribing, recording and reporting photon beam therapy (Bethesda, MD: ICRU).

Källman P, Lind B, Iacobeus C and Brahme A 1989 A new detector for radiotherapy computed tomography verification and transit dosimetry *Proc 17th Int. Congress Radiol.* Paris, France, p. 83.

Lewis D and Swindell W 1987 A MV CT scanner for radiotherapy verification *Proc. 9th Int. Conf. on Computers in Radiotherapy, Holland in The Use of Computers in Radiation Therapy* eds. I A D Bruinvis, P H van der Giessenn, H J van Kleffens and F W Wittkämper (Amsterdam, North-Holland: Elsevier) pp. 339–340.

Lewis D G, Morton E J and Swindell W 1988 A linear-accelerator-based CT system in *Megavoltage Radiotherapy 1937–1987, Brit. J. Radiol.* (Suppl 22), p. 24.

Lewis D G, Swindell W, Morton E J, Evans P and Xiao Z R 1992 A megavoltage CT scanner for radiotherapy verification *Phys. Med. Biol.* **37** 1985–1999.

Ling C C, Humm J, Larson S, Amols H, Fuks Z, Leibel S and Koutcher J A 2000 Towards multidimensional radiotherapy (MD-CRT): Biological imaging and biological conformality *Int. J. Radiat. Oncol. Biol. Phys.* **47** 551–560.

Nakagawa K, Aoki Y, Akanuma A, Onogi Y, Karasawa K, Terahara A, Hasezawa K and Sasaki Y 1991 Development of a megavoltage CT scanner using linear accelerator treatment beam *J. Japan Soc. Therap. Radiol. Oncol.* **3** 265–276.

Padhani A R, Khoo V S, Suckling J, Husband J E, Leach M O and Dearnaley D P 1999 Evaluating the effect of rectal distension and movement on prostate gland position using cine MRI *Int. J. Radiat. Oncol. Biol. Phys.* **44** 525–533.

Rothwell R I, Ash D V and Thorogood J 1985 An analysis of the contribution of computed tomography to the treatment outcome in bladder cancer *Clin. Radiol.* **36** 369–372.

Schiebler M L, Schnall M D, Pollack H M, Lenkinski R E, Tomaszewski J E, Wein A J, Whittington R, Rauschning W and Kressel H Y 1993 Current role of MR imaging in the staging of adenocarcinoma of the prostate *Radiology* **189** 339–352.

Silver M D, Yahata M, Saito Y, Sivers E A, Huang S R, Drawert B M and Judd T C 1992 Volume CT of anthropomorphic phantoms using a radiation therapy simulator *Medical Imaging – 6: Instrumentation (SPIE 1651)* (Bellingham, WA: SPIE) pp. 197–211.

Simpson R G, Chen C T, Grubbs E A and Swindell W 1982 A 4 MV CT scanner for radiation therapy: The prototype system *Med. Phys.* **9** 574–579.

Swindell W, Simpson R G, Oleson J R, Chen C T and Grubb E A 1983 Computed tomography with a linear accelerator with radiotherapy applications *Med. Phys.* **10** 416–420.

Van Baardwijk A, Baumert B G, Bosmans G, van Kroonenburgh M, Stroobants S, Gregoire V, Lambin P and De Ruysscher D 2006 The current status of FDG-PET in tumour volume definition in radiotherapy treatment planning *Cancer Treat. Rev.* **32** 245–260.

Webb S 1987 A review of physical aspects of x-ray transmission computed tomography *IEEE Proc. A* **134** 126–135.

Webb S 1990 Non-standard CT scanners: Their role in radiotherapy *Int. J. Radiat. Oncol. Biol. Phys.* **19** 1589–1607.

Webb S 1993 *The Physics of Three-Dimensional Radiation Therapy: Conformal Radiotherapy, Radiosurgery and Treatment Planning* (Bristol, UK: IOP Publishing).

Webb S 1997 *The Physics of Conformal Radiotherapy: Advances in Technology* (Bristol, UK: IOP Publishing).

Webb S 2000 *Intensity Modulated Radiation Therapy* (Bristol, UK: IOP Publishing).

Webb S 2004 *Contemporary IMRT – Developing Physics and Clinical Implementation* (Bristol, UK: IOP Publishing).

Webb S 2006 Motion effects in (intensity modulated) radiation therapy: A review *Phys. Med. Biol.* **51** R403–R426.

Zelefsky M J, Leibel S A, Gaudin P B, Kutcher G J, Fleshner N E, Venkatramen E S, Reuter V E, Fair W R, Ling C C and Fuks Z 1998 Dose escalation with three dimensional conformal radiation therapy affects the outcome in prostate cancer *Int. J. Radiat. Oncol Biol. Phys.* **41** 491–500.

5

Radioisotope Imaging

R. J. Ott, M. A. Flower, A. D. Hall, P. K. Marsden and J. W. Babich

CONTENTS

5.1 Introduction 168
5.2 Radiation Detectors 170
5.2.1 Scintillation Detectors 171
5.2.2 Gas Detectors 175
5.2.3 Semiconductor Detectors 176
5.2.4 Summary 176
5.3 Radioisotope Imaging Equipment 177
5.3.1 History 177
5.3.2 Gamma Camera: Basic Principles 178
5.3.3 Gamma-Camera Developments 188
5.3.4 Multi-Crystal Scanners for Single-Photon and PET Imaging 191
5.3.5 Multi-Wire Proportional Chamber Detectors 193
5.3.6 Semiconductor Detectors 197
5.3.7 Summary 200
5.4 Radionuclide Production and Radiopharmaceuticals 200
5.4.1 Radioactive Decay 201
5.4.2 Production of Radionuclides 202
5.4.3 Radioactive Decay Modes 209
5.4.4 Choice of Radionuclides for Imaging 212
5.4.5 Fundamentals of Radiopharmaceutical Chemistry 213
5.4.6 Biological Distribution of Radiopharmaceuticals 219
5.4.7 Radiopharmaceuticals Using Single-Photon Emitters 221
5.4.8 Quality Control of Radiopharmaceuticals 223
5.4.9 Summary 227
5.5 Role of Computers in Radioisotope Imaging 227
5.5.1 Data Acquisition 228
5.5.2 Online Data Correction 230
5.5.3 Data Processing and Image Reconstruction 230
5.5.4 Image Display and Manipulation 235
5.5.5 Data Storage 236
5.5.6 System Control 237
5.5.7 Summary 238
5.6 Static Planar Scintigraphy 238
5.6.1 Basic Requirements 238
5.6.2 Important Data-Acquisition Parameters 240
5.6.3 Dual-Isotope Imaging 243

5.6.4 Quantification 244
5.6.5 Whole-Body Imaging 248
5.6.6 High-Energy Photon Emitters 249
5.6.7 Conclusions 250
5.7 Dynamic Planar Scintigraphy 250
5.7.1 General Principles 250
5.7.2 Cardiac Imaging 251
5.7.2.1 Multigated Acquisition Studies 251
5.7.2.2 First-Pass Studies 255
5.7.3 Renal Imaging 256
5.7.4 Summary 261
5.8 Single-Photon Emission Computed Tomography 262
5.8.1 Limited-Angle Emission Tomography 264
5.8.2 Basic Principles of Transaxial SPECT 266
5.8.2.1 Data Acquisition 267
5.8.2.2 Specific Problems Associated with Emission Tomography 267
5.8.2.3 Data Sampling Requirements 268
5.8.2.4 Image Reconstruction 268
5.8.3 Instrumentation for Transaxial SPECT 268
5.8.3.1 Gamma-Camera-Based Transaxial SPECT 269
5.8.3.2 Hardware Developments 271
5.8.3.3 Fan- and Cone-Beam SPECT 274
5.8.3.4 Special-Purpose SPECT Systems 274
5.8.3.5 Dual-Modality Devices (SPECT/PET and SPECT/CT) 274
5.8.3.6 SPECT/PET/CT 277
5.8.4 Image Reconstruction Methods for SPECT 277
5.8.4.1 Filtered Backprojection 277
5.8.4.2 Iterative Reconstruction Techniques for SPECT 278
5.8.5 Data Correction Methods for SPECT 281
5.8.5.1 Correction for Nonuniformity 281
5.8.5.2 Correction for Misaligned Projection Data 281
5.8.5.3 Correction for Photon Attenuation 281
5.8.5.4 Correction for Photon Scatter 283
5.8.5.5 Compensation for Distance-Dependent PRF 284
5.8.5.6 Motion Correction 284
5.8.5.7 Compensation for Statistical Noise 285
5.8.5.8 Correction for the Partial-Volume Effect 286
5.8.6 Optimisation of SPECT Data Processing and Display 287
5.8.7 Clinical Applications of SPECT and SPECT/CT 288
5.8.8 Summary 293
5.9 Positron Emission Tomography 293
5.9.1 Introduction 293
5.9.2 Basic Principles of PET 294
5.9.2.1 Coincidence Detection of Gamma Rays from Positron-Electron Annihilation 294
5.9.2.2 Spatial Resolution in PET 295
5.9.2.3 Gamma-Ray Detection Efficiency 296
5.9.2.4 Types of Coincidence Event 297
5.9.2.5 Attenuation and Scatter 297

5.9.2.6 True and Random Coincidences 298
5.9.2.7 2D and 3D Sensitivity 299
5.9.2.8 Dead Time and Count-Rate Losses 300
5.9.2.9 Noise Equivalent Counts 300
5.9.3 Detectors and Cameras 301
5.9.3.1 Detectors Used in a Positron Camera 301
5.9.3.2 Block Detector 302
5.9.3.3 Standard Positron Camera Configuration 303
5.9.3.4 Dedicated PET NaI Camera 304
5.9.3.5 Modified Dual-Head Gamma Camera 305
5.9.3.6 Wire-Chamber Systems 305
5.9.3.7 Research Tomographs 305
5.9.3.8 PET/SPECT Cameras 305
5.9.3.9 PET/CT Systems 306
5.9.4 Recent Developments in PET Camera Technology 306
5.9.4.1 New Scintillators 306
5.9.4.2 New Light Detectors 307
5.9.4.3 Time of Flight 307
5.9.4.4 Small Animal PET Systems 308
5.9.4.5 PET/MR 309
5.9.5 Clinical Data Acquisition 309
5.9.5.1 Standard Acquisition Protocols Using ^{18}FDG 309
5.9.5.2 Dynamic PET Acquisition Modes 309
5.9.6 Image Reconstruction Algorithms 310
5.9.6.1 2D Image Reconstruction 310
5.9.6.2 3D Image Reconstruction Algorithms 310
5.9.6.3 Iterative and Model-Based Image Reconstruction Algorithms 312
5.9.7 Data Correction and Image Quantification 313
5.9.7.1 Normalisation 313
5.9.7.2 Attenuation Correction 313
5.9.7.3 Randoms Correction 314
5.9.7.4 Scatter Correction 314
5.9.7.5 Dead-Time Correction 315
5.9.7.6 Quantification/Calibration 315
5.9.7.7 Partial-Volume Effect 315
5.9.8 Image Quantification 316
5.9.8.1 Standard Uptake Values 316
5.9.8.2 Compartmental Modelling 316
5.9.8.3 Other Analysis Techniques 317
5.9.9 PET Radiopharmaceuticals 317
5.9.9.1 General Properties 317
5.9.9.2 Production of PET Radioisotopes and Radiopharmaceuticals 318
5.9.9.3 ^{18}F-Labelled Fluorodeoxyglucose 318
5.9.9.4 Other PET Radiopharmaceuticals 319
5.9.10 Diagnostic and Research Applications of PET 319
5.9.10.1 Staging of Cancer 321
5.9.10.2 Assessing Response to Therapy 321
5.9.10.3 PET in Radiotherapy Treatment Planning 322
5.9.10.4 Applications in Cardiology 322

5.9.10.5 Applications in Neurology 322
5.9.10.6 Applications in Clinical Research 322
5.10 Performance Assessment and Quality Control of Radioisotope Imaging Equipment 323
5.10.1 Flood-Field Uniformity 325
5.10.2 Spatial Resolution 326
5.10.3 Energy Resolution 329
5.10.4 Spatial Distortion 329
5.10.5 Plane Sensitivity 330
5.10.6 Count-Rate Performance 330
5.10.7 Multiple-Energy-Window Spatial Registration 331
5.10.8 Shield Leakage 331
5.10.9 Performance Assessment and Quality Control for SPECT 332
5.10.10 Performance Assessment and Quality Control for PET 333
5.10.11 Summary 335
5.11 Concluding Remarks 335
References 336

5.1 Introduction

From methods of imaging human anatomy with X-rays, attention is now turned to imaging physiological function using radiation emitted by radioactive isotopes inside the human body. The techniques discussed in this chapter contrast with most other medical imaging techniques, which generally provide anatomical details of the body organs. The use of radioisotopes in tracer quantities for the clinical diagnosis of human disease grew rapidly in the 1970s and 1980s and became the recognised medical speciality known as nuclear medicine (Wagner 1975, Maisey 1980, Sharp et al. 1998, Ell and Gambhir 2004). In the 1980s, radioisotope imaging split into two major areas, the most common being *single-photon imaging* in which single gamma rays emitted from the radionuclide are detected (Figure 5.1). The less common alternative is *positron emission tomography* (PET) (Phelps 1986) in which the two annihilation photons emitted from a positron-emitting radionuclide are detected simultaneously (Figure 5.2). In both techniques, images of the biodistribution of radionuclide-labelled agents in the body are formed. These agents, known as radiopharmaceuticals, are designed to determine the physiological function of individual tissues or organs in the body. The distribution of these agents within the body is determined by route of administration and by such factors as blood flow, blood volume and a variety of metabolic processes.

The first use of a radioisotope, ^{131}I, to investigate thyroid disease was carried out in the late 1930s. Early developments in imaging equipment include the production of the rectilinear scanner and the scintillation camera during the 1950s. Both these devices became widely available in the mid-1960s. The rectilinear scanner and similar devices are no longer widely used since the Anger/gamma camera has become readily available – this is now the instrument of choice for single-photon imaging. Also in the early 1950s, the first devices for the detection of annihilation photons from positron emitters were used clinically. This technology has developed rapidly in the 1990s and, with the increased availability of positron-emitting radionuclides from cyclotrons, there is now a steady growth in the number of hospitals performing PET. Unlike X-ray imaging, where both the emission and detection

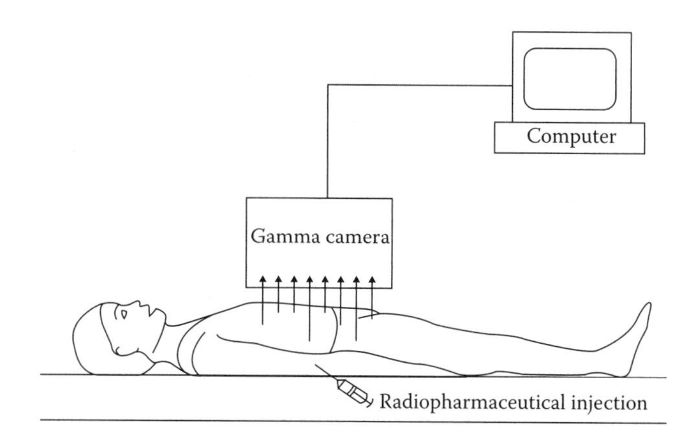

FIGURE 5.1
Illustration of the radioisotope imaging process carried out using a single-photon-emitting radionuclide. The direction of the incident photon is determined by a collimator mounted on the gamma camera.

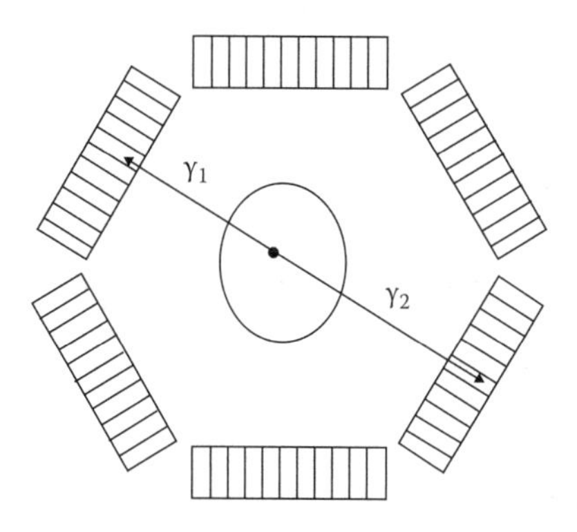

FIGURE 5.2
Illustration of radioisotope imaging using a positron-emitting radionuclide. The two annihilation photons are detected by a pair of the many detectors surrounding the patient defining the line of response (LOR) through the point of radionuclide decay.

position of each X-ray is known, only the γ-ray detection position is determined for a radioisotope source within the body. To produce an image it is, hence, necessary to provide some form of collimation, which defines the photon direction. This takes the form of mechanical (i.e. lead) collimation in the gamma camera or electronic collimation in the positron camera. In this chapter, we will explain the function and development of the modern gamma camera and positron camera, including the most recent improvements brought about by the use of improved scintillation crystals, faster electronics and advanced microprocessor technology.

Different modes of radioisotope imaging are possible with planar static imaging with a gamma camera (planar scintigraphy) being the most common form. These single-view images consist of two-dimensional (2D) projections of the three-dimensional (3D) activity distributions in the detector's field of view (FOV). Temporal changes in the spatial distribution of radiopharmaceuticals can be studied by taking multiple planar gamma-camera images over periods of time that may vary from milliseconds to hundreds of seconds. This form of imaging (dynamic scintigraphy) is fundamental to the use of radioisotopes in showing the temporal function of the organ/system being examined as compared to the 'snap-shot' of function obtained in planar scintigraphy.

Since the planar image contains information from a three-dimensional object, it is often difficult to determine clearly the function of tissues deep in the body. Tomographic images derived from multiview acquisitions of the object overcome most problems caused by superposition of information in a single planar view. The technique of emission computed tomography (ECT), which encompasses single-photon emission computed tomography (SPECT) and PET, has parallels with X-ray computed tomography (CT) but also some important differences. X-ray CT is based on the determination of photon attenuation in body tissues, whereas in ECT it is necessary to correct for the effect of photon attenuation in order to accurately determine the distribution of radioactivity within the body. In addition, modest count densities in radioisotope studies, which are limited by the total quantity of radioactivity that can be safely injected into the body, limit the statistical accuracy of radioisotope images in comparison with X-ray CT. However, as ECT is a functional-imaging technique by nature, it is not sensible to compare image quality directly with those obtained from essentially anatomical modalities which use X-rays, ultrasound or magnetic resonance.

5.2 Radiation Detectors

The radioactive decay of radioisotopes leads to the emission of α-, β-, γ- and X-radiation, depending on the radionuclide involved (see Section 5.4). The range of α- and β-particles in tissue is too small to allow *in vivo* imaging using detectors external to the body, but X- and γ-rays penetrate tissue, attenuating exponentially dependent upon both the photon energy and the tissue being traversed (see Section 2.3). Figure 5.3 shows the mass attenuation coefficient for photons in muscle and bone as a function of photon energy and illustrates that, at energies below 50 keV, X- and γ-rays are greatly attenuated by body tissues due to photoelectric absorption. Above 50 keV, photon interactions in tissue are dominated by Compton scattering. At energies above 511 keV, it is increasingly more difficult to detect photons efficiently due to the high penetration of likely detector materials. Absorption due to pair production does not become important until well above 1000 keV. Hence,

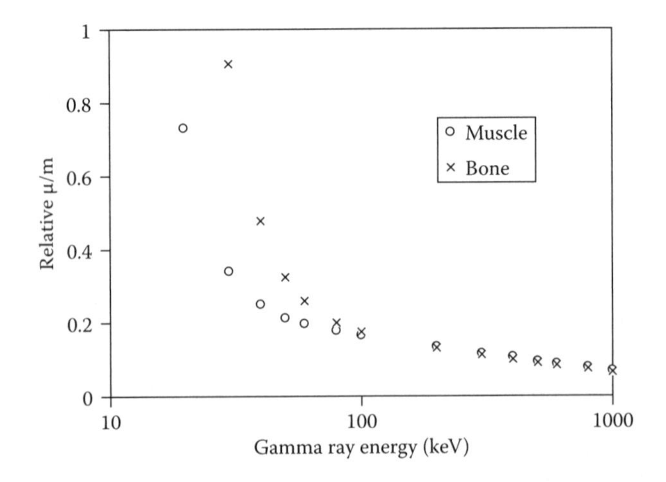

FIGURE 5.3
Relative values of γ-ray mass attenuation coefficients (μ/m) in muscle and bone as a function of photon energy.

radioisotope imaging is restricted to the use of radionuclides emitting photons with energies in the range 50–511 keV. The photoelectric and Compton interaction processes transfer energy from the emitted X- or γ-photons to a detector via atomic excitation and ionisation. Excitation is followed by the emission of low-energy photons often in the optical region that are detectable by a suitable optically sensitive device such as a photomultiplier or silicon diode. Ionisation causes the removal of electrons from their atomic orbits producing quasi-free particles that can be detected directly via ionisation or by the production of excitation photons. We will now consider the main types of radiation detectors that have been used to image X- and γ-rays emitted from the body.

5.2.1 Scintillation Detectors

Materials that emit visible or near-visible light when energy is absorbed from ionising radiation have been utilised both to count and to image radioisotopes (Birks 1964). Table 5.1 shows a range of inorganic scintillators that have a high Z and, hence, good X- and γ-photon stopping power. The light-emission characteristics of most of these scintillators match the spectral sensitivity of a photomultiplier tube (PMT) (Flyckt and Marmonier 2002) or a photodiode (PD) (Gowar 1993). As scintillators are relatively transparent to their own light, ionising radiation detectors using a scintillator-PMT/PD combination (a scintillation counter) will have high X- and γ- ray detection sensitivity. The signal output of a scintillation counter detecting a mono-energetic gamma ray (Figure 5.4) shows a total energy absorption peak (the photopeak) and a continuum produced by absorption of part of the photon energy following a Compton scatter. Light emission by inorganic scintillators is proportional to energy deposited in the material and, hence, it is possible to measure photon energy and to discriminate between photons of different energy detected simultaneously. The width of the photopeak is determined by Poisson noise in the system and is essentially proportional to the square root of the number of photoelectrons produced when the light enters the PMT or PD. Energy resolution is defined as the full width at half maximum (FWHM) of the photopeak and is typically of the order of 10%–12% at 100–200 keV for NaI(Tl), for example. For a Gaussian-shaped photopeak,

$$\mathrm{FWHM} = 2.35\sigma \tag{5.1}$$

where σ is the standard deviation of the Gaussian peak.

TABLE 5.1

Physical properties of inorganic scintillators.

	NaI (Tl)	BaF_2	BGO	LSO	GSO	LYSO	$LaBr_3$
Density (g cc^{-1})	3.67	4.89	7.13	7.4	6.71	7.3	5.1
Effective atomic number	51	54	75	66	58	66	~50
Refractive index	1.85	1.49	2.15	1.82	1.85	1.82	~1.9
Attenuation length[a] (mm)	25.6	22.7	11.2	11.4	15.0	11.6	21.3
Emission wavelength (nm)	410	195/220 310	480	420	430	428	380
Relative light yield	100	5 16	15	75	20	75	166
Light decay time (ns)	230	0.6 620	300	~40	~60	40	16–30

[a] Distance in which a 511 keV photon beam is attenuated by a factor of $1/e$.

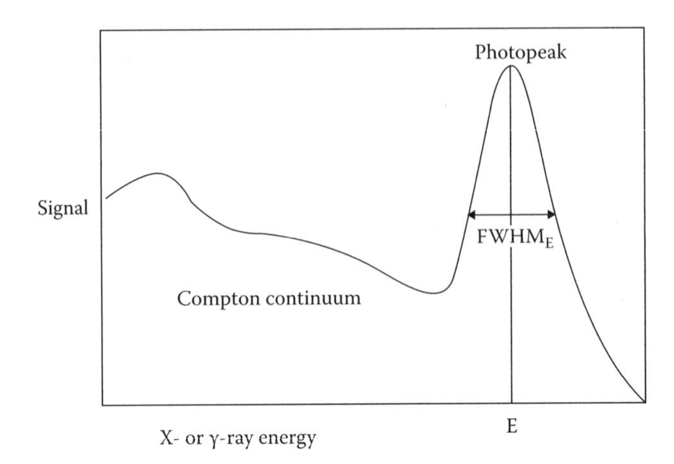

FIGURE 5.4
Signal output from a scintillation counter detecting mono-energetic photons showing the effects of photoelectric absorption and Compton scattering in the detector.

Energy resolution determines the ability of the scintillation counter to discriminate, for instance, between γ-rays emitted from the body unscattered and those which have scattered and lost energy in the process. This is an important factor in reducing the effect of scatter in images produced by a scintillation detector. Positional information is usually obtained by using either multiple small crystal/PMT detectors or a single large-area scintillation crystal coupled to a large number of PMTs. The physical size of the scintillation crystals and the light detectors determine the application as imaging detectors. Small crystals a few millimetres across and 10–30 mm long are available in most materials and are suitable for multidetector-based imaging systems such as positron cameras. The limitations here are usually the size of the PMT, although the use of 'block detectors' (see Section 5.9.3.2) has allowed large volume imaging systems to be made with small crystals. Larger (typically 400 × 500 mm^2) single crystals of thickness ~6–15 mm suitable for gamma cameras are difficult to grow except with NaI(Tl). Figure 5.5 shows the relative absorbed fraction of photon energy in a 13 mm thick NaI(Tl) crystal as a function of photon energy. When combined with Figure 5.3, we can see that there is a window between 50 and 500 keV

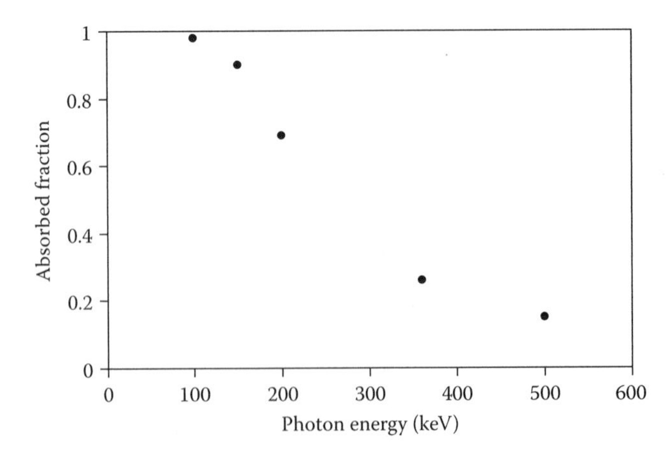

FIGURE 5.5
Relative photon-energy absorbed fraction in a 13 mm thick NaI(Tl) crystal as a function of photon energy.

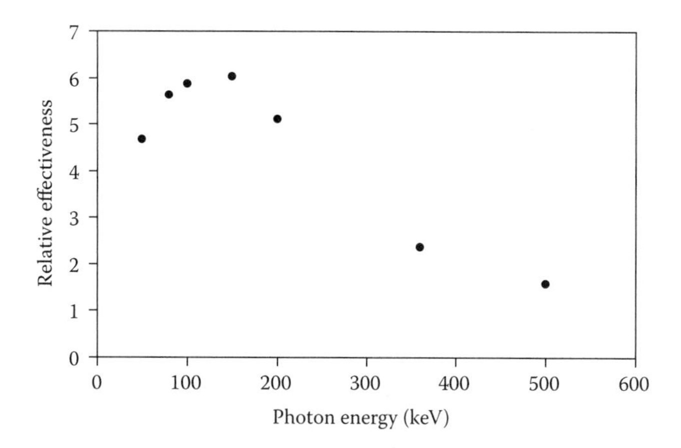

FIGURE 5.6
Relative effectiveness of NaI(Tl)-based scintillation counters for the detection of gamma rays as a function of photon energy.

(Figure 5.6), in which scintillation counters can be used effectively as radioisotope imaging detectors. The density, atomic number and gamma-ray stopping power of these crystals (Table 5.1) vary significantly, making them suitable for the detection of different energy gamma rays. NaI(Tl) has been the crystal of choice for cameras used to detect 50–200 keV photons, whereas BGO was initially the best option for annihilation photons (511 keV). New scintillation materials such as lutetium oxyorthosilicate (LSO) (see Table 5.1) are now replacing BGO due to their enhanced light output and shorter optical decay time.

The PMT forms the traditional light-sensitive device for a scintillation counter (Figure 5.7). It consists of a glass vacuum vessel with the inner front face coated with a thin layer of photosensitive material, commonly known as the photocathode, which can convert optical photons to photoelectrons. Typical conversion efficiencies for photons of wavelength 200–800 nm are between 10% and 40%, depending upon the material used. Inside the vacuum vessel are a series of electron-multiplying dynodes which are held at increasingly greater positive voltages with respect to the photocathode. An electron produced in the photocathode will be accelerated down the vessel and will undergo electron multiplication by collision with the dynodes. The electron gain per dynode, known as the secondary emission yield, is typically 5 so that a substantial gain of 5^n in signal can be achieved for a PMT with n dynodes. After the last dynode, an anode collects the total charge produced and generates an output signal. The amplitude of this signal is proportional to the quantity

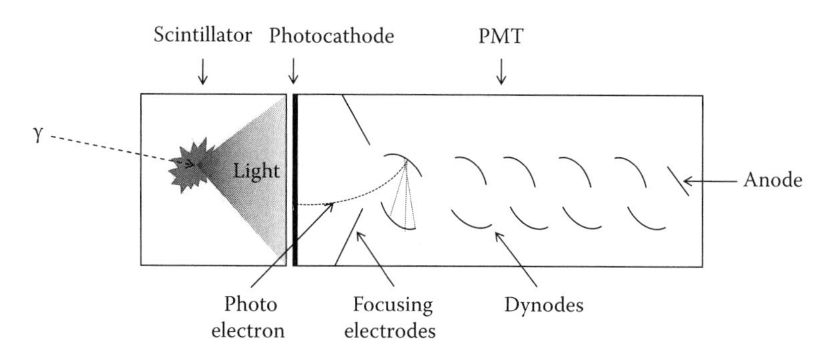

FIGURE 5.7
Scintillation counter consisting of a scintillation crystal coupled to a PMT.

of light hitting the photocathode and hence the signal is proportional to the energy deposited in the crystal. The structure of the dynodes allows fast output signals of a few ns rise times. A range of PMT diameters (1 cm to >10 cm) and cross sections (circular, hexagonal, square) are available to suit the type of detector.

Position-sensitive PMTs have also found uses in the development of small-scale imaging devices especially for PET animal imagers (Cherry et al. 1996, 1999). An example of such a device has multiple anode wires to read out the position of the signal generated by the photocathode and amplified by the dynode chain. Typically, the anode may have up to 16 × 16 wires over an area of ~50 mm and, when coupled to a thin phosphor, can provide a spatial resolution of 1–2 mm.

PDs (Figure 5.8) have also been used to detect light from scintillating crystals. In this case the signal is produced by generating multiple electron-hole pairs in the junctions formed between *p* and *n* type semiconducting (known as *pin* junctions) when illuminated by light. Most diodes are better suited to the detection of the red/green light produced by CsI(Tl) rather than the blue light from the other crystals (see Table 5.1). As the intrinsic signal produced by a PD is small, the signals require substantial amplification which produces significant noise especially if fast signals are desired. Hence PDs work best where large numbers of optical photons are produced. PDs are cheap, fast, require low voltages to power them and are usually small (a few mm across) which makes them ideally suited for close packing with small crystals. Like a PMT, the gain of a PD is quite sensitive to temperature changes. The problem of amplifier noise can be overcome by the use of avalanche photodiodes (APDs) (Kagawa et al. 1981) that have substantial signal gain due to the internal avalanche multiplication effect. In this case, a gain of up to 100 compared to conventional PDs is possible using 100–200 V. It is possible to obtain a gain of 1000 when using special doping techniques and for this ~1500 V is needed across the diode. However APDs are more expensive than PDs. Overall the application of PDs and APDs in nuclear medicine have been limited to small experimental systems (Bergeron et al. 2008). A new device which will almost certainly find applications in X-/γ-ray imaging is the silicon photomultiplier (SiPM). These devices are formed of arrays of close-packed APDs and perform in a similar way. Typical devices (Moehrs et al. 2006) consist of a 4 × 4 array of SiPMs. Each SiPM, of size 2.85 × 2.85 mm^2, contains 3640 APD microcells running at about 30 V with a gain ~10^6. The light-detection efficiency of most of these devices is at green/red wavelengths so that the quantum efficiency for the blue light from most phosphors is ~6%. Nevertheless, when

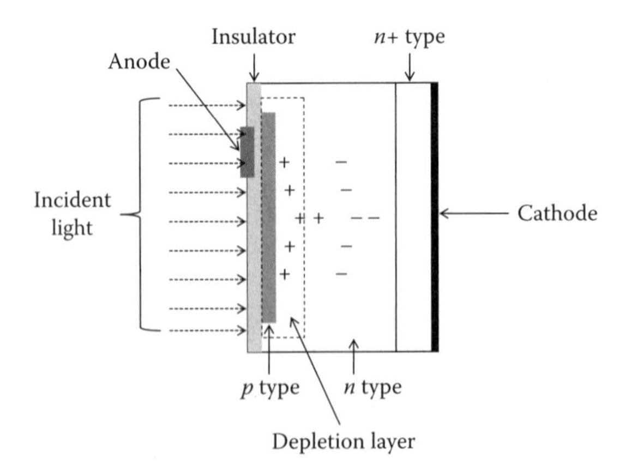

FIGURE 5.8
Silicon photodiode used for detecting light from a scintillation crystal.

coupled to small LSO crystals SiPMs can achieve a spatial resolution of 1–2 mm, an energy resolution of ~12% and a timing resolution of ~1 ns. This kind of device could allow the manufacture of a small PET scanner for use inside an MRI scanner (see Section 5.9.4.5).

5.2.2 Gas Detectors

The traditional gas-filled ionisation chamber (Attix and Roesch 1966) is a very sensitive and accurate detector of directly ionising radiation. In its simplest form (Figure 5.9) the gas detector can both count the amount of ionising radiation and differentiate between radiation types or energies. In this kind of detector all the ion-pairs produced in the initial ionisation process are collected (Parker et al. 1978). As the voltage across the electrodes is increased the detector enters the proportional counter region where the signal produced by the detection of ionising radiation undergoes gas amplification (up to $\times 10^8$) due to secondary ionisation of the counter gas. This produces a large signal that is proportional to the applied chamber voltage. This technology has been used in the multi-wire proportional chamber (MWPC) for radioisotope imaging. Figure 5.10 shows an example of how

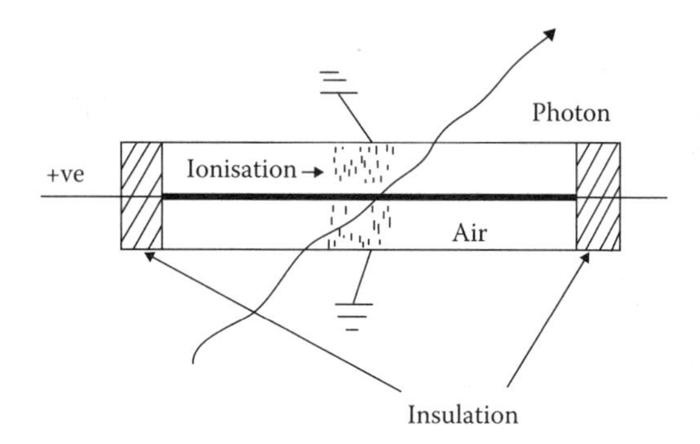

FIGURE 5.9
Schematic of a simple ionisation chamber used to detect an incident gamma photon.

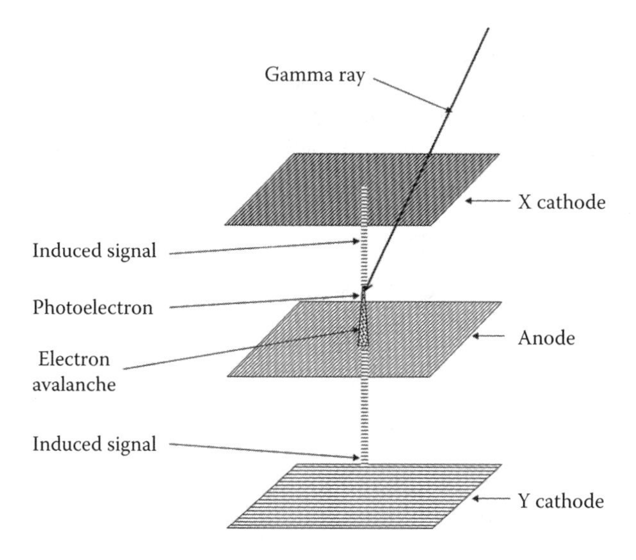

FIGURE 5.10
Schematic of a multi-wire proportional chamber showing how secondary ionisation from a gamma photon is detected via electron avalanche effects.

a MWPC can detect gamma rays that enter the chamber gas. Interactions via the photoelectric or Compton scatter processes, produce ionisation in the chamber. Electrons are accelerated rapidly towards the anode wires, whilst the positive ions travel more slowly towards the cathode planes. The very high field close to the anode wires causes local gas amplification that produces a large signal close to the nearest anode wire. This in turn induces a signal on the adjacent cathode planes that can be read out, using delay lines for instance, to provide the spatial localisation of the ionisation. Examples of MWPC gamma-ray detectors will be discussed in Section 5.3.5. This form of detector is, however, very transparent to X- or γ-rays with energies above a few 10s of keV unless high-Z, high-pressure gases such as xenon are used or gamma-ray converters are inserted into the chamber gas. At still higher applied voltages, the Geiger counter has been used as a γ-ray detector, particularly in radiation protection, but it has little application in radioisotope imaging.

5.2.3 Semiconductor Detectors

In the last 10 years, the development of low-cost silicon detectors has provided a potentially useful alternative to scintillators. It is possible to make small-area silicon detectors $5 \times 5\,cm^2$ and about 1 mm thick, and to obtain a spatial resolution down to a few μm using micro-strip or pixel technology (Gerber et al. 1977, Gatti et al. 1985, Acton et al. 1991). This, allied with the excellent energy resolution (a few keV at room temperature), is outweighed by the low Z of silicon. The semiconductor, in the form of a charge-coupled device (CCD) (Eastman Kodak 2001) or a CMOS active pixel sensor, may have useful applications in autoradiography, particularly the imaging of thin tissue sections, as a replacement for film emulsion, (Ott et al. 2000, Cabello et al. 2007) and in small-area radioisotope imagers (Ott et al. 2009).

Germanium, either in a high-purity form or doped with lithium, has a much higher Z than silicon and would appear to have some advantages over silicon at the photon energies used for radioisotope imaging. However, the high-energy resolution of better than 1 keV is obtained only at liquid-nitrogen temperatures. Developmental gamma cameras have been produced that show the potential advantages of the semiconductor over scintillators, but the cost of the raw material and cryogenics is still prohibitive. Other semiconductor materials such as cadmium telluride (CdTe) and cadmium zinc telluride (CdZnTe) have an acceptable stopping power for gamma rays of energy ~100 keV (Barber et al. 1997, Eisen et al. 1999) and have been used in gamma probes and, more recently, in small-area gamma cameras (Verger et al. 2004, Gu et al. 2011). This material is still quite difficult to manufacture consistently in large wafers needed to make them viable for gamma-ray imaging devices.

5.2.4 Summary

From the previous discussion, it is clear that scintillation materials have had the greatest impact on radioisotope imaging, although developments in gas proportional chambers and semiconductors may become more competitive in the next decade. It is worth noting here the application of film emulsion and small-scale digital devices for the detection of β emission and X-rays from radioisotopes used in autoradiography, that is, the imaging of thin tissue sections containing radiolabelled tracers (Rogers 1979, Lear 1986). MWPC and semiconductor (CCD/CMOS) autoradiography systems have been developed recently and may have a significant role in this area in the near future.

5.3 Radioisotope Imaging Equipment

5.3.1 History

The earliest radioisotope images were produced by the use of a scintillation counter connected to a focused lead collimator. The collimator allowed only those photons emitted from a small region in the object near the collimator focus to be detected efficiently by the counter. By scanning this counter in a rectilinear motion across the body (Figure 5.11), it was possible to build up an image of the radiopharmaceutical distribution. In the 1950s, rectilinear scanners were developed to include one or two detectors with large-volume NaI(Tl) crystals and heavy focused lead collimators, which could image the distribution of radionuclides such as ^{18}F, ^{51}Cr, ^{131}I, ^{198}Au and ^{59}Fe (Mayneord and Newbery 1952, Mallard and Peachey 1959, Beck 1964). Although detecting high-energy photons with reasonable sensitivity, rectilinear scanners spend very little time imaging any particular part of the body and, hence, had a very low efficiency for imaging large areas of the body. Subsequently, other imaging devices based on multiple crystal/collimator combinations were developed specifically as linear scanners for whole-body imaging and for emission computed tomographs.

Dedicated whole-body scanners were a direct extension of the rectilinear scanner where the rectilinear motion was replaced by a single longitudinal scan with a laterally extended scanning head. In one design (Figure 5.12) the system contained one or two scanning heads consisting of single slabs of NaI(Tl), 50 × 3.2 cm^2 in area and 2 cm thick, with a single lead collimator specially designed to give a spatial resolution of 0.7–1.2 cm at 8 cm in tissue and a depth of focus in air of 18–21 cm. These special-purpose imaging systems are no longer commercially available, being of little use other than for skeletal imaging and, for this application, they have been replaced by scanning gamma cameras (see Section 5.6.5).

The production of single large-area crystals of NaI(Tl) in the late 1950s led to the development of what is currently the most important radioisotope imaging device with the invention of the Anger or gamma camera (Anger 1958, 1964). In this device, a single, large-area lead collimator is placed in front of the scintillating crystal. An array of PMTs attached to the rear surface of the scintillator detects the scintillation light produced by the detected gamma rays. Electronics coupled to the PMTs are used to determine the position of the photon interaction in the scintillator and the energy of the photon. Large-area 6–10 mm thick NaI(Tl) crystals have a high sensitivity at photon energies between 80 and 200 keV but become rapidly less sensitive

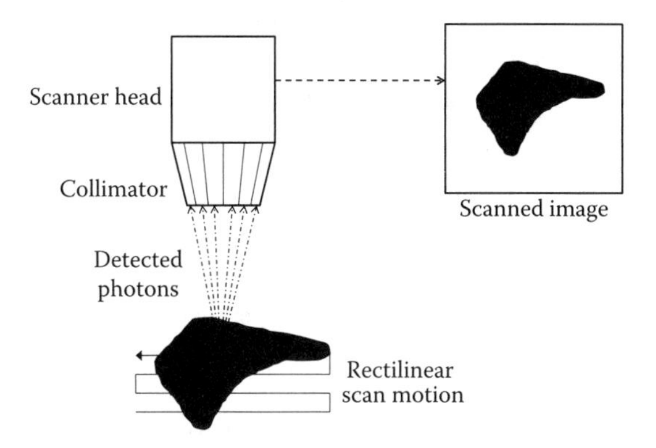

FIGURE 5.11
Schematic of radioisotope image production using a rectilinear scanner. The rectilinear motion of the collimator focus is in a plane that is perpendicular to the central axis of the scanner head.

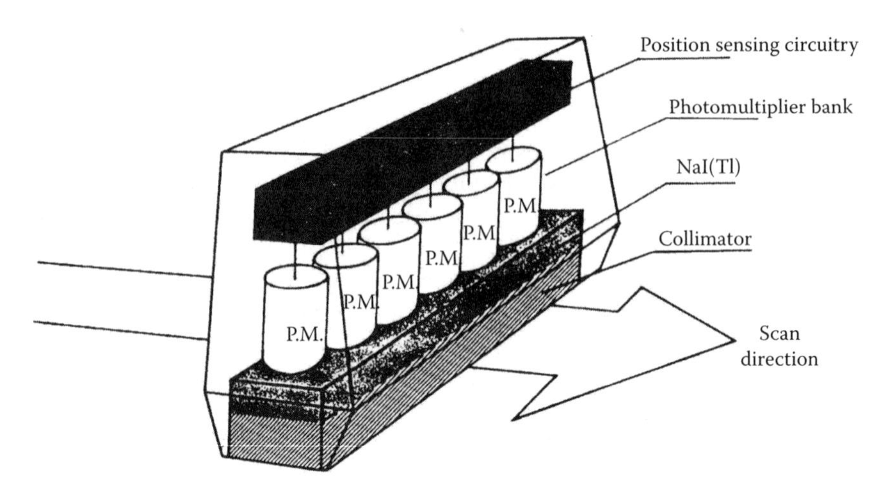

FIGURE 5.12
Head of a linear radionuclide scanner.

to higher-energy photons. With the interest in using double-headed gamma cameras for annihilation photon imaging of positron emitters, thicker crystals of 15–20 mm have been used to increase the detection efficiency for 511 keV gamma rays.

From the modest beginning of a device using 7 PMTs, a modern gamma camera may have a sensitive area of up to $500 \times 400\,\text{mm}^2$ which is ideally suited to large-organ and whole-body imaging. Dynamic scintigraphy involving the production of a sequence of images to determine the temporal changes in the radiopharmaceutical distribution became possible once the camera was attached to a digital computer. Similarly rotation of the camera around the radioactive object to acquire multiple 2D projections allowed 3D images of the radionuclide distribution to be made using SPECT. Further details of the gamma camera are outlined later (in Sections 5.3.2 and 5.3.3).

Radioisotope imaging systems based on gas-filled detectors have been a more recent development. Detectors based on MWPCs have been developed for autoradiography (Englert et al. 1995), single-photon imaging (Lacy et al. 1984, Barr et al. 2002) and for PET (Duxbury et al. 1999, Jeavons et al. 1999, Divoli et al. 2005). A number of MWPC-based autoradiography systems have been used for imaging electrophoresis plates but the gamma and positron camera developments are still at the research stage and have yet to make a significant impact on clinical nuclear medicine. Very much the same applies for semiconductor-based detectors, although materials which can be used at or near room temperature, such as silicon, gallium arsenide and cadmium zinc telluride, may begin to impact on the field in the coming decade.

5.3.2 Gamma Camera: Basic Principles

In this section we will consider the basic principles of the gamma camera with analogue electronics to provide the positional and energy information used to create an image (Short 1984). The individual components of the camera (Figure 5.13) including the collimator, the scintillating crystal, the PMTs, the analogue position and energy logic, the camera shield and image display are discussed together with basic performance parameters such as spatial resolution, sensitivity, count-rate capacity and dead time. Definitions and methods of measurement of these performance parameters are provided in Section 5.10. Some of the limitations of the analogue gamma camera have been overcome using digital techniques, which will be discussed in Section 5.3.3.

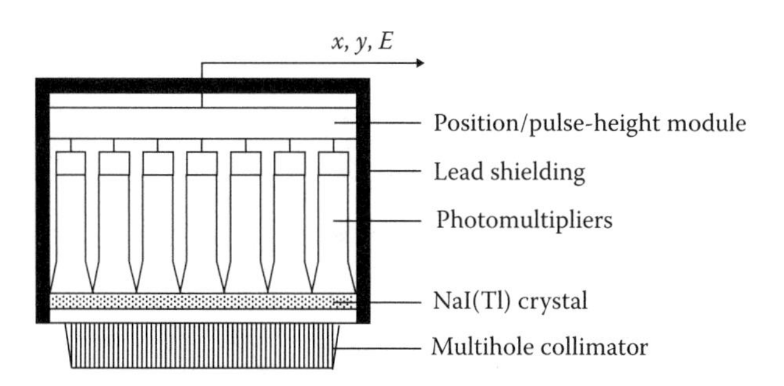

FIGURE 5.13
Schematic showing the basic components of a gamma-camera head.

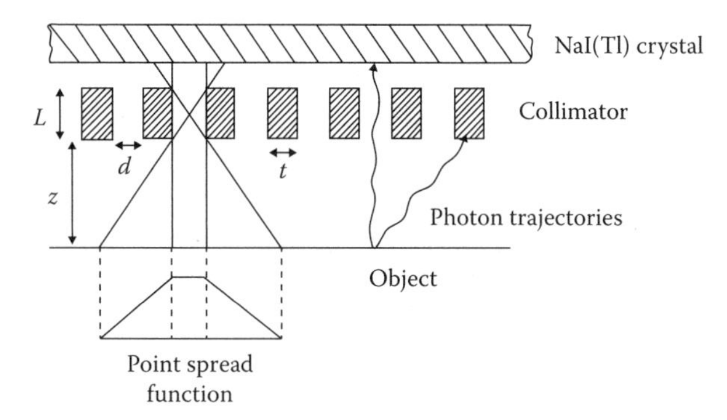

FIGURE 5.14
Schematic of a parallel-hole collimator showing how near-normal incident photons are selected preferentially for detection by the camera, and how the point spread function (PSF) is determined by the collimator-hole dimensions. The sensitivity is proportional to the area under the PSF, whilst the spatial resolution is often defined as the FWHM of the PSF.

The *collimator* is a critical part of the system as it determines the direction of the photons incident on the camera. Essentially, a collimator is a sheet of lead penetrated by many fine holes. The holes are separated by thin lead septa and it is the combination of septal width and collimator thickness which determines the ability of the collimator to reject gamma rays which do not pass through the holes. This is particularly important when imaging high-energy photons where septal penetration degrades image quality. In the case of a parallel-hole collimator (Figure 5.14), only photons incident near-normal to the collimator surface reach the scintillator. The collimator also defines the geometrical FOV of the camera and essentially determines both the spatial resolution and sensitivity of the system. A range of collimators can be used to image different photon energies and to achieve sufficient compromise between spatial resolution and sensitivity (Table 5.2). As well as needing several parallel-hole collimators, it is useful to have a pinhole collimator for imaging small superficial organs such as the thyroid or skeletal joints. For brain SPECT, a fan-beam collimator can be used to optimise the use of the whole scintillator area.

For a *parallel-hole collimator* it is possible to express the spatial resolution and geometrical efficiency in terms of the collimator dimensions. If L is the hole length, d the hole diameter and z the source-to-collimator distance (Figure 5.14), then the collimator geometrical spatial resolution (R_c) is given by

$$R_c \approx \frac{d(L+z)}{L}. \tag{5.2}$$

TABLE 5.2

Examples of parallel-hole collimator types and dimensions for use with a 400 mm diameter gamma camera.

Description	Hole Size[a] (mm)	Number of Holes	Septal Thickness (mm)
Low-energy, high-resolution (LEHR)	1.8	30,000	0.3
Low-energy, general-purpose (LEGP)	2.5	18,000	0.3
Low-energy, high-sensitivity (LEHS)	3.4	9,000	0.3
Medium-energy, high-sensitivity (MEHS)	3.4	6,000	1.4

[a] Diameter of round holes, or distance across flats of hexagon.

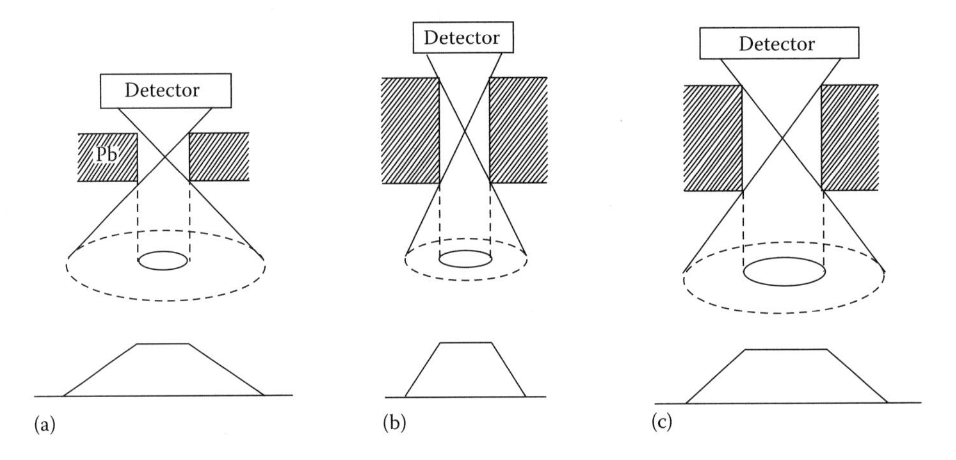

FIGURE 5.15
Effect of collimator-hole diameter and length for a parallel-hole collimator. A short hole (a) provides a wide response function, a longer hole (b) a narrower response function, whilst widening the collimator hole (c) broadens the response function again.

From Equation 5.2 it can be seen that spatial resolution is improved (i.e. lower R_c) by increasing the hole length or by reducing the hole diameter with a consequent increase in the number of holes per unit area (whilst maintaining adequate septal thickness). Figure 5.15 shows schematically how the spatial resolution varies with the diameter and length of a hole.

The intrinsic spatial resolution (R_i) of the gamma camera, as determined by the information from the PMTs, is typically 3–4 mm, but when combined with the geometrical spatial resolution of the collimator (R_c), this gives a *system spatial resolution* (R_s) of

$$R_s = (R_i^2 + R_c^2)^{1/2}. \tag{5.3}$$

The collimator spatial resolution can be improved by minimising the distance between the source and collimator surface. Figure 5.16 shows that the value of R_s varies with distance from the collimator surface for a parallel-hole collimator and that the system resolution is dominated by the collimator at all distances. Hence, there is little to be gained from improving the intrinsic spatial resolution of a gamma camera beyond present values.

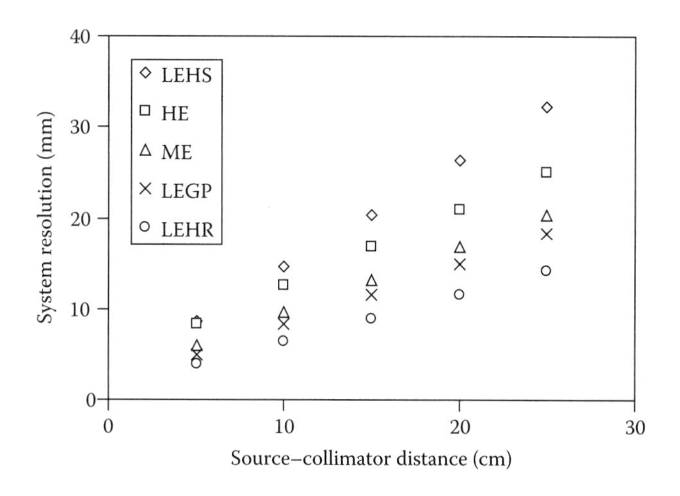

FIGURE 5.16
Variation of system spatial resolution with source–collimator distance for typical low-energy high-sensitivity (LEHS), high-energy (HE), medium-energy (ME), low-energy general-purpose (LEGP) and low-energy high-resolution (LEHR) collimators.

The geometrical efficiency g of the collimator is given by:

$$g \approx \left[\frac{Kd^2}{L(d+t)}\right]^2 \tag{5.4}$$

where
t is the thickness of the lead septa between the holes, and
K is a constant dependent upon hole shape (e.g. $K = 0.26$ for hexagonal holes in a hexagonal array).

Note that g is independent of source–collimator distance for a point source imaged in air since the inverse-square-law relationship is compensated by the detector area exposed. Performance characteristics of some commercially available parallel-hole collimators are shown in Table 5.3.

For source locations many multiples of the collimator hole length from its surface (i.e. $z >> L$) and for holes whose diameter is much greater than the septal thickness (i.e. $d >> t$) the relationship (from Equations 5.2 and 5.4) becomes

$$g \propto R_c^2. \tag{5.5}$$

TABLE 5.3

Performance characteristics of some typical commercially manufactured parallel-hole collimators.

Type[a]	Maximum Energy (keV)	g (%)	R_c (mm) at 10 cm
LEHR	150	0.0184	7.4
LEGP	150	0.0268	9.1
LEHS	150	0.0574	13.2
MEHS	400	0.0172	13.4

Source: Data from Sorenson and Phelps, 1980.
[a] See Table 5.2.

This shows that collimator spatial resolution can be improved only at the expense of reduced collimator geometric efficiency for a given septal thickness. In order to correct for septal penetration, L may be replaced by the effective length of the hole

$$L' = L - 2\mu^{-1} \tag{5.6}$$

where μ is the attenuation coefficient of the collimator material at the energy of the photons being imaged.

It is possible to calculate a value for the septal thickness in terms of the length of the septa (and hole), the hole diameter and the linear attenuation coefficient μ of lead at the appropriate photon energy (Sorenson and Phelps 1980). If w is the shortest path in lead traversed by photons passing between adjacent holes, then

$$t \approx \frac{2dw}{(L-w)}. \tag{5.7}$$

If 5% septal penetration is allowed, then $\mu w \geq 3$ and

$$t \approx \frac{6d}{\mu}\left[L-\left(\frac{3}{\mu}\right)\right]^{-1}. \tag{5.8}$$

Typical values for septal thickness vary from 0.3 mm for photons with energy less than 150 keV to 4.7 mm for ~400 keV photons.

The parallel-hole collimator is the workhorse of nuclear medicine imaging and, with advances in the control of other factors affecting camera performance, is the limiting component of the gamma camera in terms of sensitivity and spatial resolution. Only by using special-purpose focused collimators is it possible to improve image quality when imaging small parts of the body such as the brain, thyroid, skeletal joints and for paediatric imaging.

Figure 5.17 shows schematically how a *pinhole collimator* can be used to create an image of a small object. The collimator consists of a single small aperture a few mm in diameter at the end of a conical lead shield containing sufficient attenuating material to minimise photon penetration for energies up to about 500 keV. The collimator projects an inverted image onto the scintillator and, by suitable positioning of the object, significant image magnification can be achieved, although at the expense of some distortion particularly at the image edge.

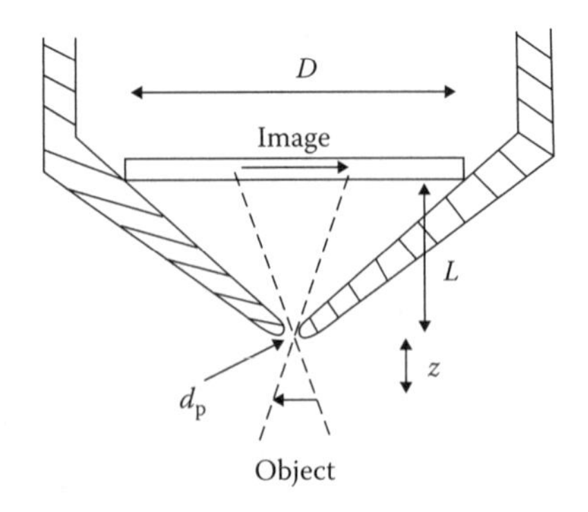

FIGURE 5.17
Creation of a magnified and inverted image using a pinhole collimator.

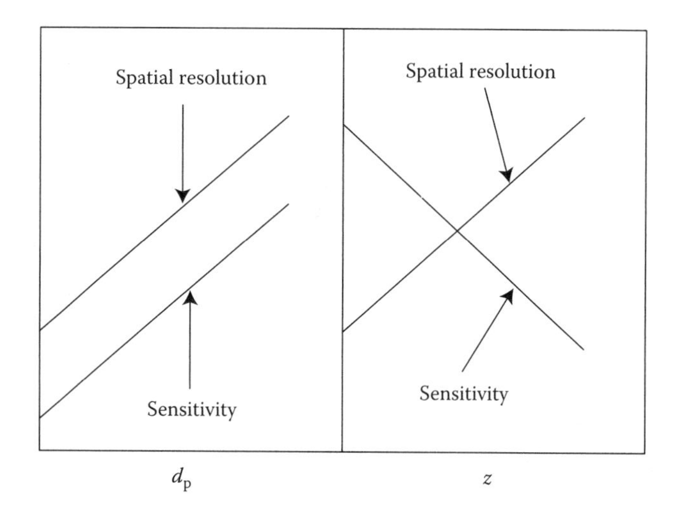

FIGURE 5.18
Relative variation of spatial resolution (i.e. FWHM) and sensitivity with the parameters d_p and z for a pinhole collimator.

The spatial resolution (R_c) and efficiency (g) of a pinhole collimator are determined by the pinhole diameter (d_p), the scintillator diameter (D), the collimator length (L) and the source-to-collimator aperture distance (z) (Figure 5.18):

$$R_c = \frac{d_p(L+z)}{L} \tag{5.9}$$

$$g = \frac{d_p \cos^3\theta}{16z^2} \tag{5.10}$$

where θ is the cone angle of the collimator ($\tan\theta/2 = D/(2L)$).

For a given value of d_p then, by reducing z for a given object size, the sensitivity and spatial resolution can be improved at the expense of some image distortion. A pinhole collimator is most suitable for small-organ imaging, where the object can be placed close to the collimator, for example, thyroid gland and skeletal joints.

Converging multi-hole collimators have been used to provide the best combination of high resolution and sensitivity (Figure 5.19) at the expense of reduced FOV (Murphy et al. 1975) and, as with the pinhole collimator, some image distortion. The spatial resolution and efficiency relationships now depend upon the focal length (f) as well as the other aforementioned parameters:

$$R_c = \frac{d(L+z)}{L} \cdot \frac{L}{\cos\theta} \cdot \left(1 - \frac{L}{2(f+L)}\right) \tag{5.11}$$

$$g = \left(\frac{Kd}{L}\right)^2 \cdot \left(\frac{d}{(d+t)}\right)^2 \cdot \left(\frac{f}{(f-z)}\right)^2. \tag{5.12}$$

Modern versions of this collimator are the fan- or cone-beam collimators which are used to produce high-resolution images of the brain, for instance. Figure 5.20 shows, in 1D, how a

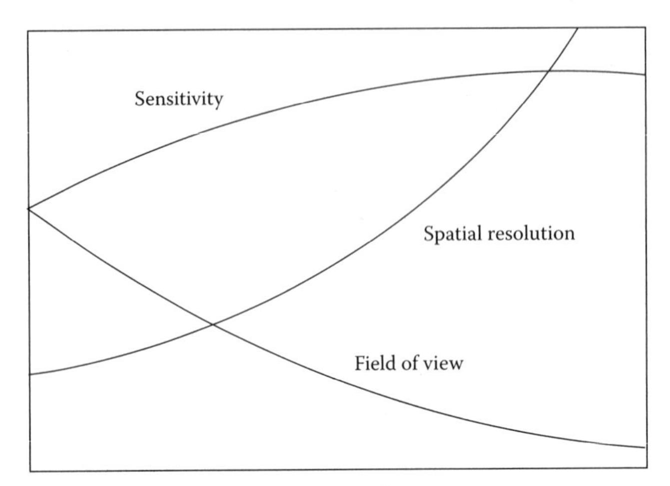

FIGURE 5.19
Relative variation of spatial resolution (i.e. FWHM), sensitivity and FOV with source-to-collimator distance for a converging collimator.

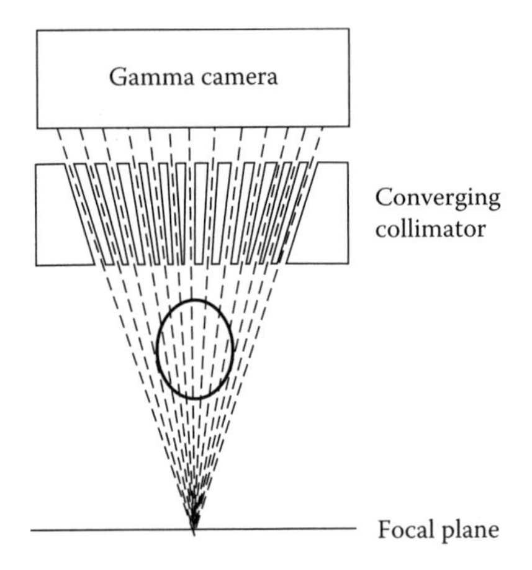

FIGURE 5.20
Use of a converging collimator to optimise brain imaging with a gamma camera.

fan- or cone-beam collimator can be used to image the brain onto the large area of a gamma camera to maximise spatial resolution and sensitivity. These special-purpose collimators have been developed specifically for SPECT with a gamma camera (see Section 5.8.3).

The majority of gamma cameras use a large-area (typically 50 cm × 40 cm rectangular or 40 cm diameter), thin (6–10 mm), single *scintillation crystal* of thallium-activated sodium iodide, NaI(Tl). This scintillator emits blue-green light with a spectral maximum at 415 nm that is matched reasonably well to the light-photon response characteristics of standard bialkali PMTs (Figure 5.21). The crystal has a reasonably high atomic number and density (Table 5.1), and the linear attenuation coefficient at 150 keV is 2.22 cm^{-1} giving a 90% absorption in ~10 mm at this energy. The light-emission decay constant of 230 ns means that event rates of many tens of thousands of counts per second (kcps) can be attained without serious degradation of performance. The light output is the highest of all the readily available inorganic scintillators (Table 5.1) and the material is highly

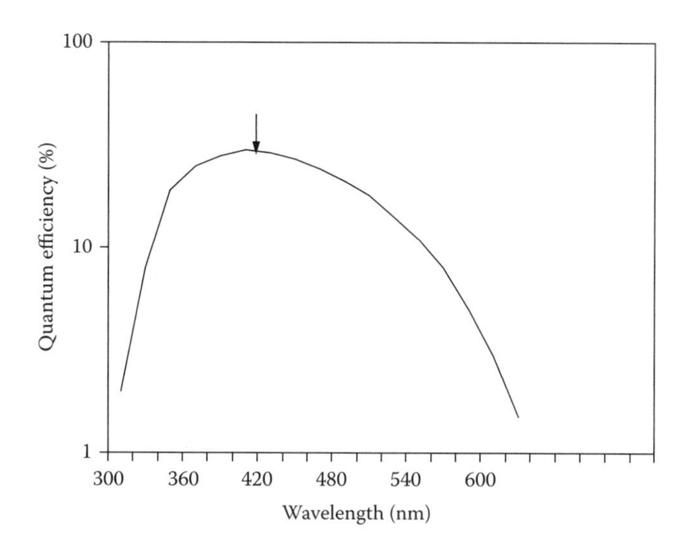

FIGURE 5.21
Spectral response of a bialkali photomultiplier compared with the maximum emission wavelength (arrow) of the light output from NaI(Tl) scintillator.

transparent to its own light emission. Although hygroscopic and, hence, requiring hermetic encapsulation, NaI(Tl) is still unsurpassed for the detection of photon energies close to 100 keV. Because of the high light output, the energy resolution (width of the photopeak) of a thin NaI(Tl) crystal is typically 10%–12% FWHM at 150 keV.

As the refractive index of NaI(Tl) is 1.85, there is the potential for substantial non-uniform light collection across the face of the crystal. This can be minimised by using a *light guide* to interface the scintillator to the PMT. This reduces light losses in the transport of light to the PMT as the guides are made of a transparent plastic with a refractive index close to 1.85 and are carefully shaped to match the shape of the photomultiplier cathode. In most modern cameras, the light guide has been dispensed with in favour of microprocessor-based correction procedures to account for the sensitivity variation across the FOV.

The optimal arrangement of *PMTs* (with circular or hexagonal cross sections) closely packed onto the surface of a circular scintillation crystal is a hexagonal array (Figure 5.22) containing 7, 19, 37, 61, etc. PMTs. In the 1970s, 50 mm diameter PMTs were available, limiting the number used on a 40 cm diameter camera to 37. From about 1980 onwards, gamma cameras

FIGURE 5.22
Hexagonal arrangement of PMTs on the circular crystal of gamma cameras containing 7, 19, 37 or 61 photomultiplier tubes.

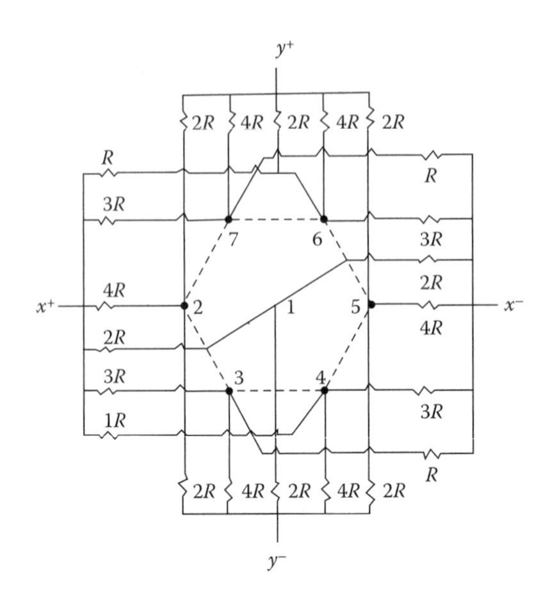

FIGURE 5.23
Example of a resistor network used to provide positional information from the analogue outputs of the gamma-camera PMTs.

utilising smaller-diameter (25–30 mm) PMTs enabled cameras with 61 or 75 tubes (by adding extra tubes around the edges) to become available. With the use of rectangular scintillation crystals, PMTs with square cross sections are used by some camera manufacturers.

The PMTs are chosen from batches having closely matched gain response characteristics so that the application of high voltage and the gain adjustment to provide uniform response over the scintillator surface area is simplified.

Traditionally a capacitor network or resistive-coupled network (Figure 5.23) is used to provide *positional information* from the analogue outputs of the PMTs. The relative intensity of these outputs determines the position (x,y) and energy (E) of the scintillation event and provides signals to produce an image on a cathode ray oscilloscope (CRO), storage oscilloscope or for computer storage after digitisation. The energy signal is determined by

$$E = \Sigma S_i \tag{5.13}$$

and the x and y positions by

$$x = \frac{\Sigma S_i W_x}{\Sigma S_i} \tag{5.14}$$

$$y = \frac{\Sigma S_i W_y}{\Sigma S_i} \tag{5.15}$$

where

S_i is the signal from the ith PMT, and

W_x and W_y are the weighting factors calculated to give optimum spatial linearity.

The denominator ensures that x and y are independent of the photon energy. This methodology is sometimes referred to as 'Anger' or analogue logic. For events near the edge, information is only provided from one side of the scintillation position and this leads to

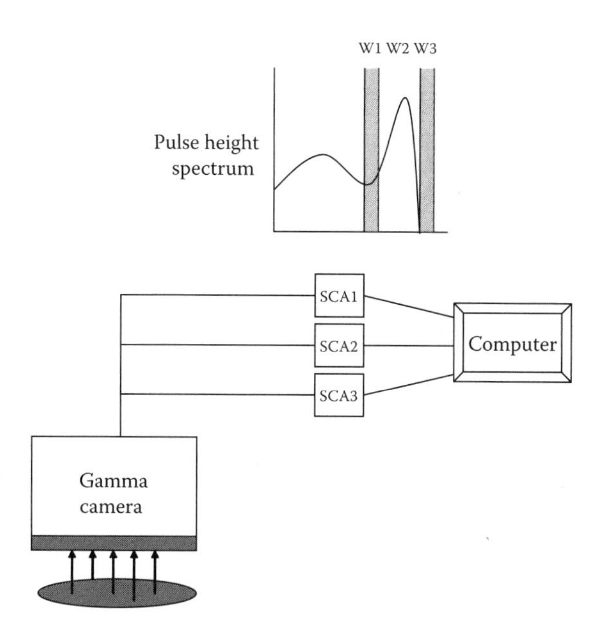

FIGURE 5.24
Use of single-channel pulse-height analysers (SCAs) to select widths of the photopeak window (W2) and scatter-correction energy windows (W1 and W3) for a gamma camera.

distortion or 'edge effects'. This part of the camera is usually excluded from the useful FOV, thus reducing the effective size of the camera when imaging.

The photon energy signal is sent to a pulse-height analyser (PHA), with up to six *upper and lower energy discriminator levels* to determine whether the E signal corresponds to that expected from the gamma photon(s) emitted from the radionuclide being imaged. Figure 5.24 shows how the PHA (or a set of single-channel analysers (SCAs)) is used to choose the 'photopeak(s)' (or total-absorption peak(s)) of the radionuclide. For a single-photon emitter, such as $^{99}Tc^{m}$, a single energy window is set to reject the majority of low-energy scattered photons, which would degrade the spatial accuracy of the image. When imaging a radionuclide that emits more than one photon per decay, such as ^{111}In, several photopeak windows (in this case two) are used to maximise the unscattered component of the image.

The scintillation crystal and electronics are surrounded by a large *lead shield* of sufficient thickness to minimise the detection of unwanted radiation from outside the collimator FOV. With the development of rotating gantries for gamma cameras to allow ECT to be performed, the weight of this shielding must be minimised, although modern camera gantries are robust enough to support and rotate up to three camera heads.

Analogue gamma cameras were historically equipped with *image display facilities* to provide image hard copy film. Generally, an image-formatting system was provided to allow multiple grey-scale images to be recorded on an X-ray film. The quality of images from an analogue camera is highly dependent on matching the video display output intensity to the film response and requires the information density (counts per unit area) to be sufficient to expose the film adequately. The dynamic range of information available can sometimes be too wide to be stored on a single film exposure and computer-stored images are required to optimise image quality and minimise information loss (see Section 5.5).

At high count rates, the analogue signal circuitry can become overloaded due to *pulse pile-up* of scintillation signals from the detector (Figure 5.25). This happens when events from several photons are collected within the time of the pulse width produced by the electronics.

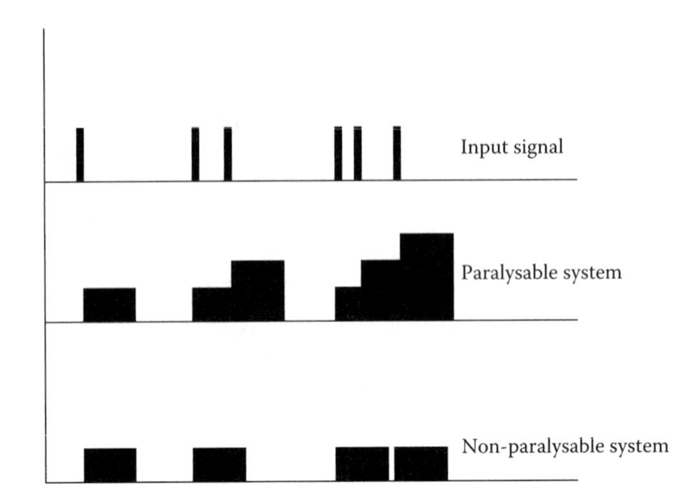

FIGURE 5.25
Schematic of the output signals from paralysable and non-paralysable detectors when counting randomly distributed signals. For the paralysable system the pulse pile-up effect is shown.

This will affect both the positional and energy information and, hence, image quality from the camera. In addition, the system will begin to lose counts because of the inherent *dead time* generated in the camera electronics. The detected count rate (n_d) of the system is related to the true count rate (n_t) by

$$n_t = \frac{n_d}{(1 - n_d \tau)} \tag{5.16}$$

where τ is the electronic dead time, typically about 4 μs for an analogue camera.

This relationship assumes that the electronics of the camera is nonparalysable so that no new signals can be processed until after the dead time of the electronics produced by the previous signal. However, some parts of the system (the phosphor, for example) will be paralysable in which case,

$$n_d = n_t \exp(-n_t \tau). \tag{5.17}$$

Here, as n_t gets very large, n_d becomes very small until the whole system paralyses. This type of electronics also leads to pulse pile-up.

5.3.3 Gamma-Camera Developments

Since the mid-1980s there have been major developments, particularly in the use of microchip technology, that have led to a significant improvement in the performance of the gamma camera. Using this technology most gamma-camera development has been aimed at improved uniformity and stability to meet the demands of emission tomography (see Section 5.8). Each PMT is matched as best as possible to the others in the camera head but it is impossible to obtain identical gain response as the voltage or temperature varies. The response of a PMT varies with the amount of light and the position where the light hits the photocathode and, hence, the signal produced by a PMT varies substantially across its surface area. This is a major cause of spatial non-linearity. In addition, PMT gain is sensitive to changes in temperature, magnetic field and ageing and components in the PMT

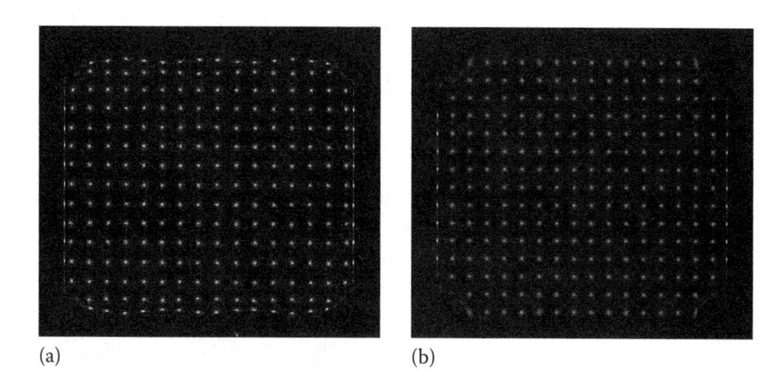

FIGURE 5.26
(See colour insert.) Images of a linearity phantom comprising an array of $^{99}Tc^{m}$ sources: (a) no linearity correction and (b) with linearity correction. (Images courtesy of S. Heard.)

electronics suffer similar problems. Further camera nonuniformity is due to variations in sensitivity caused by variable energy response across the camera FOV.

Correction for PMT gain drift can be minimised by exposing the tubes to a fixed light output such as that generated by a light-emitting diode, for instance. The output signal from each PMT can be adjusted by varying either the high voltage or the signal amplifier gain to make the PMT output constant. This can be done continuously to ensure PMT gain stability over an extended time period, although it does not ensure that each PMT responds in the same way across the photocathode surface. Once the PMT gains are stabilised, it is possible to apply corrections to reduce spatial and energy response non-linearities. The former is achieved by generating a distortion map using a linearity phantom, and correcting the measured x, y positions accordingly. Figure 5.26 illustrates the effect of applying a uniformity correction on an image of an array of small $^{99}Tc^{m}$ sources. The distortion map is usually generated at the factory or before the camera is installed. The variable energy response is minimised by using an energy-correction map generated by the user. Using an appropriate uniform radiation source, for instance, images are produced using two narrow-energy windows placed at equal energies above and below the true photopeak energy. If the photopeak is symmetric, then the two images should have equal counts in them on a pixel-by-pixel basis. A correction matrix is generated and applied iteratively until the two images are 'identical'. Once the energy and linearity corrections have been applied, the uniformity of the camera is greatly improved. Further improvements can be gained by uniformity-correction procedures to remove small spatially dependent variations in collimator and crystal sensitivity. This involves exposing the camera with an appropriate collimator and energy window to a uniform flood source to generate a uniformity-correction matrix. Since this matrix is determined by the inverse of the flood image, very high counts (>10,000 counts per pixel) are required.

The use of large square PMTs in a modern gamma camera matches the rectangular geometry of the crystal that in turn matches the FOV required for whole body and tomographic imaging. A camera with a rectangular crystal of $520 \times 370\,\text{mm}^2$ can be covered using 48 square PMTs, for example. The use of square PMTs makes it difficult to use conventional Anger logic to get a uniform image because the distance between the centres of neighbouring PMTs is no longer the same in all directions. The four PMTs touching the sides of the centre PMT are significantly closer than the four touching the corners and if a point source is moved from one PMT centre to the next, the ratio of signal strengths from different PMTs changes by different amounts depending on whether the source is moved diagonally or parallel to an edge of the PMT. Although modern correction techniques can

	1	2	5	8	5	2	1
	2	8	25	31	25	10	4
	4	24	134	521	130	20	4
	4	31	511	2570	499	30	4
	2	25	127	500	133	18	2
	2	14	20	30	22	12	1
Vertical sum	15	104	822	3660	814	92	16
Threshold (100) then sum	0	0	472	3291	462	0	0
Sum then threshold (200)	0	0	622	3460	614	0	0
%age extra photons	0	0	32%	5%	33%	0	0

FIGURE 5.27
Illustration of the effect of correlated signal enhancement on the signals used to define the position of the photon interaction in a gamma camera.

be used to create a uniform image from even highly distorted images, the presence of rapidly changing distortions implies that even very small changes in PMT characteristics (such as gain) over time may result in corrections becoming invalid.

One way of overcoming this problem is the use of *correlated signal enhancement* (CSE).* The basis of this method (Figure 5.27) is the summing of rows and columns of PMT signals without prior signal processing. Here, the sums of every row and column are treated as a 1D array, since the signal contained on a column of PMTs is approximately independent of where a source is located along the length of the column (except near the edge of the image). This signal independence is only achieved with square PMTs (for hexagonal or round PMTs, the strength of signal along any row depends on the distance between the source and the PMT centres). A second benefit of using CSE is the increase in the number of light photons detected per event, important for improved spatial and energy resolution. The number of photons detected by any PMT decreases as the scintillation source–PMT distance increases. Anger logic makes use of signals from all PMTs for the position determination, and small signals will have a poor signal-to-noise degrading the position calculation. By excluding small signals using thresholding it is possible to improve spatial resolution. Figure 5.27 shows the average number of photons seen by each square PMT when an event occurs over the centre of a PMT near the middle of the array. Apart from the four PMTs that touch the sides of the centre tube, all other PMTs contain 'small signals' which would be rejected using the Anger thresholding method. Using the CSE method of first summing the signals along a column, and then applying a threshold, the total number of photons used to determine the position is increased by ~30%.

In order for a camera to perform well at high count rates, it is necessary to detect the initial part of each pulse accurately. This is especially important if pulse pile-up occurs when two events are detected close together in time. In this case, the energy signal never returns to the baseline, and simple energy-windowing schemes miss the start of the second event, thus causing errors in positioning. Together with the CSE system, very-high-speed digitisation of the energy signal processed in real-time can determine the exact timing of events, and this information is used to trigger event integration correctly at all count rates. Accurate integration of signal is performed with high-speed (flash) digitisation of detector signals followed by a digital integration stage. This allows shorter integration times at higher count rates to be used to reduce dead time or increased integration time for better spatial and energy resolution at lower rates.

* See http://www.gehealthcare.com/usen/fun_img/docs/digitalcse.pdf, accessed February 20, 2012.

Examples of other improvements that have been implemented on modern gamma cameras are as follows:

- At every point in the circuit where analogue-to-digital conversion takes place, circuitry is included which verifies continuously that a zero-volt input signal is translated correctly into a digital zero output ensuring stability of the image geometry with time and environmental variations.
- Auto-tune circuits stabilise the PMTs through their ageing process, thermal fluctuations and magnetic fields. Electronics is included in the detector that allows the gain of each PMT to be measured without disturbing normal operation of the camera. This makes it possible to diagnose PMT problems before they seriously affect the detector performance.
- Special stabilisation of the dynode chain in the PMT voltage supply is employed to ensure that sudden changes in local count rate do not affect PMT gain.
- Most of the light output from NaI has a decay time of 230 ns. However, a fraction of the energy deposited in the crystal is trapped in states from which emission of photons takes place over a large fraction of a second. At high count rates, this slow light component gives rise to a shift in the energy and a corresponding shift in the photopeak of the energy spectrum. It is possible to 'detect' the presence of the slow component of the crystal output, and to make appropriate corrections for it.

Overall, these technical modifications to the gamma camera have transformed it into a device with greatly improved stability and uniformity of response so that tomographic imaging can be performed without substantial artefact generation. Quality control (QC) is no longer the continuously demanding task required with older cameras and many problems are corrected by routine maintenance with minimum disruption to clinical use.

5.3.4 Multi-Crystal Scanners for Single-Photon and PET Imaging

Very early transaxial emission tomographic scans were obtained using detectors containing several scintillating crystals, each with a focused collimator (Kuhl and Edwards 1963). The crystal/collimator systems were translated and rotated to provide sufficient views of the object for tomographic imaging. A typical example of this type of detector, built by Aberdeen University (the Aberdeen Section Scanner, ASS), consisted of two small-area detectors equipped with long focal-length collimators (Bowley et al. 1973). The detector heads were scanned in tandem at one azimuthal angle to the object and then rotated to a new azimuthal angle, where scanning was repeated. The system was designed to produce single transaxial sections with high sensitivity and depth-independent resolution. An improved version of this system was the Aberdeen Section Scanner Mark II (Evans et al. 1986) which had 24 detectors arranged along the sides of a square, 6 detectors on each side, each with a 20 cm focal-length collimator. The detectors were moved tangentially up to 64 mm and rotated through an angular range of 95°. The spatial resolution achieved in the single-section image plane was 9 mm and the plane thickness was 14 mm. The large increase in scintillation detector volume gave a greatly increased sensitivity (300 cps kBq^{-1} mL within a single slice). Few of these multi-crystal, single-photon tomographic scanners (MSPTSs) were produced commercially, mainly because they were limited to imaging single slices. Examples were the Cleon systems of Union Carbide Imaging Systems Inc. (Stoddart and Stoddart 1979), one of which was marketed

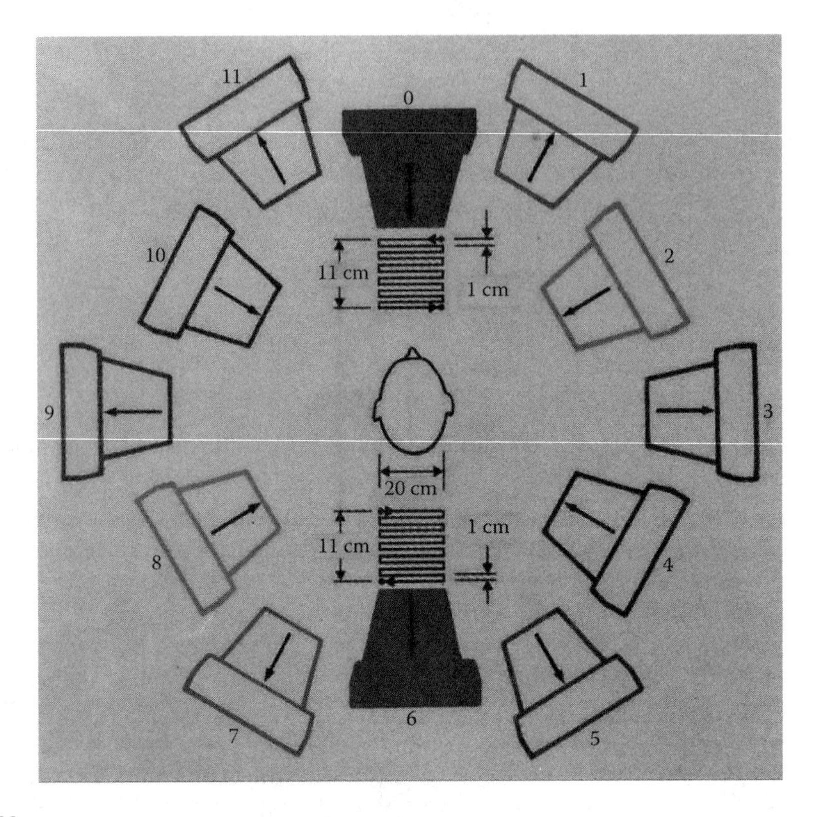

FIGURE 5.28
Schematic diagram showing the configuration and movement of the detectors for the multi-crystal Cleon 710 single-photon tomographic brain imager. The brain imager had 12 large-area detectors, each mounted on a separate scanning frame at 30° intervals around a central axis. Each detector required a 9 cm diameter PMT and a 15 cm focal-length collimator, which subtended a 30° angle at the focal plane. Each detector underwent a rectilinear scan within the plane of interest so that the focal point scanned half the FOV. All detectors moved either clockwise or anticlockwise when scanning tangentially but opposing pairs of detectors moved radially together, either towards or away from the central axis, at the end of each tangential scan.

later as the 810 Radionuclide Brain Imager by Strichman Medical Equipment Inc. In contrast to the Aberdeen scanners, these used large-volume, large-area scintillators fitted with short focal-length collimators and a unique scanning geometry (as illustrated in Figure 5.28). The brain imager achieved a transaxial resolution of about 1 cm in a single section with an effective slice thickness of about 2 cm and a sensitivity of approximately 420 cps kBq^{-1} mL (Flower et al. 1979). Another example of a MSPTS was the Tomomatic 64 (Stokely et al. 1980), which was designed specifically for regional cerebral blood-flow (CBF) studies using ^{133}Xe, and was able to image a few slices per scan. This scanner had four banks of 16 detectors and 25 PMTs, each bank with the PMTs arranged in three rows (8 + 9 + 8), so that each scintillator was viewed by three PMTs, one in each row making the detectors position sensitive in the axial direction. With the use of multi-slice collimators the scanner produced three sections simultaneously.

The multi-crystal SPECT systems were all designed to provide images of a single (or a few) section(s) with higher sensitivity and spatial resolution than a rotating gamma camera. However, since the gamma camera can be used for planar (Section 5.6), dynamic (Section 5.7) and multi-slice tomographic scintigraphy (Section 5.8), this versatility has overshadowed the development of dedicated SPECT scanners, and their contribution to clinical nuclear medicine was, consequently, very limited.

In PET, the decay of the radionuclide produces two approximately antiparallel γ-rays, each of 511 keV. The most efficient method of imaging the *in vivo* distribution of the positron-emitting radiopharmaceutical is to surround the patient with a large number of individual scintillation detectors, with each detector in electronic coincidence with those on the opposite side of the patient. Multi-crystal detectors provide the state-of-the-art technology for PET, although the high cost of such systems still limits their use in clinical nuclear medicine. A more complete description of the detectors used for clinical PET is deferred to Section 5.9.3.

5.3.5 Multi-Wire Proportional Chamber Detectors

Whilst gas-filled detectors have not played a major part in radioisotope imaging, there are a few applications of note. One area of development has been in *low-cost PET*. Although MWPCs have a relatively low detection efficiency at 511 keV, they can be built in sufficiently large areas (i.e. 60 × 40 cm^2) to obtain adequate sensitivity for PET. The intrinsic resolution is in general very good, that is, ~5 mm or less, and the resolving time can be <10 ns. Imaging is usually performed by placing detectors on either side of the object and rotating azimuthally in order to acquire the complete set of data necessary to reconstruct a 3D image.

The detectors used vary between different research groups, but the first systems made use of lead inserts to convert incident photons to electrons, which are subsequently detected via ionisation of the chamber gas. The lead inserts increase the detection efficiency from a fraction of 1% to approximately 10%. Two designs that have been used clinically were those developed at CERN in Switzerland (Jeavons et al. 1983, Townsend et al. 1984) and at the UK Rutherford-Appleton Laboratory (RAL) (Marsden et al. 1986). The latter design used a stack of interleaved wire planes and thin lead foil converters (Figure 5.29). The clinical version of this system (MUP-PET) was 60 × 30 cm in area, had a spatial resolution of 6 mm and a sensitivity of 490 cps kBq^{-1} mL for a 20 cm diameter cylinder of activity 12 cm high. The CERN design used 30 cm × 30 cm × 5 mm converters drilled with a fine matrix of holes (0.8 mm in diameter on a 1 mm pitch) (Figure 5.30). Photoelectrons produced in the lead escaped into

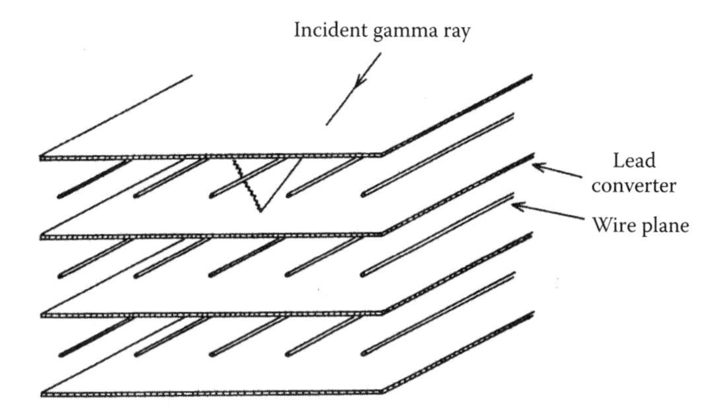

FIGURE 5.29
Stack of multi-wire proportional chambers used for detecting annihilation photons from a positron emitter, showing interleaved wire-plane anodes and thin lead-foil cathodes (i.e. which act as photon converters).

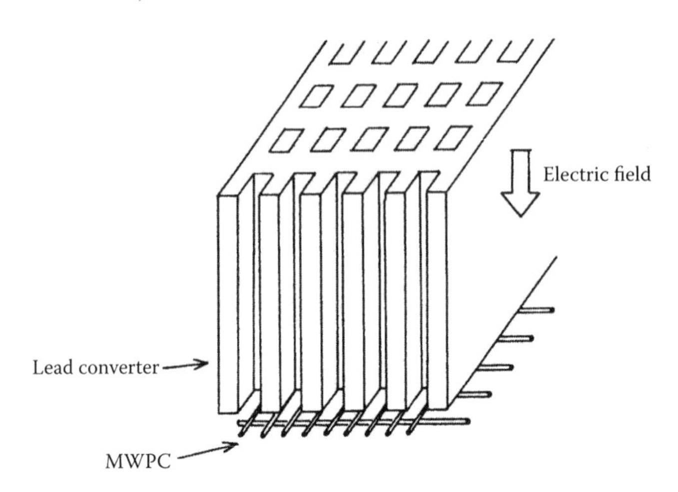

FIGURE 5.30
Lead channel-plate photon converter for use with a multi-wire proportional chamber positron camera.

the holes of the converter. A 10 kV cm^{-1} electric field pulled the electrons out of the converter onto a wire plane, where they were detected. For a ^{22}Na line source in a plastic holder, a line spread function with a FWHM of 2 mm was obtained. The temporal resolution was poor due to the time taken to extract the electrons from the holes but the sensitivity was similar to that of the RAL system. Owing to the low sensitivity of MWPC systems, the high spatial resolution was only realised in practice when imaging small organs (e.g. the thyroid gland) containing high levels of radioactivity compared to surrounding tissues. Images of larger organs required smoothing in order to reduce statistical fluctuations reducing the spatial resolution to 15–20 mm. In a typical study, about two million coincidences were recorded in 30 min. A system based on the CERN design has been manufactured for laboratory imaging and achieves a spatial resolution close to 1 mm, (Jeavons et al. 1999). As coincidence events collected by MWPC positron cameras (and indeed any other positron camera using large-area detectors) are not confined to a set of planes through the object, image reconstruction has not usually been performed by the standard 2D filtered backprojection (FBP) method (see Chapter 3). Instead, the acquired data are first backprojected into a 3D array, which is subsequently filtered in 3D (Townsend et al. 1983, Clack et al. 1984, Erlandsson et al. 1998).

One development of gas detectors to PET imaging is a *hybrid system* in which a scintillation crystal is coupled to an MWPC (Anderson et al. 1983, Mine et al. 1988, Schotanus et al. 1988). Barium fluoride (BaF_2) has three light emission components with 80% emitted with a decay constant of 620 ns at a wavelength of ~310 nm and 20% emitted with a decay constant of 0.7 ns at wavelengths of 195 and 220 nm. The material has a density of 4.89 g cm^{-3} (compared with 3.67 g cm^{-3} for NaI (see Table 5.1) but the light output is only 5% that of NaI. Tetrakis(dimethylamino)ethylene (TMAE) is a vapour which can be photoionised by photons of wavelengths below 200 nm. A camera (PETRRA) consisting of two detectors incorporating BaF_2 crystals and a MWPC filled with TMAE mounted on a rotating gantry has been built and evaluated to determine its application as a PET detector (Visvikis et al. 1997, Duxbury et al. 1999, Ott et al. 2002). Gamma-ray photons are detected efficiently by the BaF_2, and the short-wavelength light is converted by the TMAE, producing a few photoelectrons per event. The photoelectrons are amplified by the attached MWPC (Figure 5.31) sufficiently for a fast coincidence pulse to be generated and for positional information to be produced. The camera has a spatial resolution of

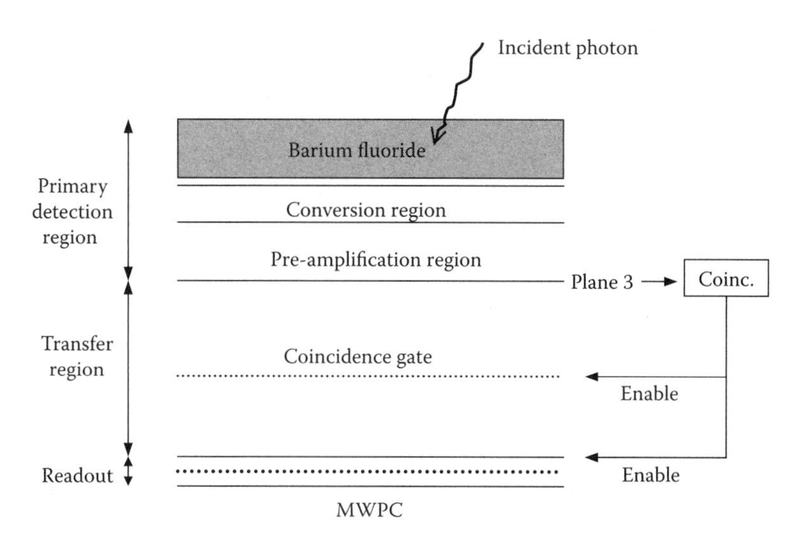

FIGURE 5.31
Multi-wire proportional-chamber/BaF_2 annihilation-photon detector system which is part of the novel positron camera (PETRRA).

~6 mm, a timing resolution of 3.5 ns and count-rate performance of 50–100 kcps. Figure 5.32 shows that the image quality for PETRRA is comparable to that of a commercial PET/CT system. A recent upgrade of PETRRA has increased the camera detection efficiency by ×5 and the spatial resolution to ~4 mm. Similar BaF_2-TMAE technology was adopted in a high-resolution PET scanner for laboratory studies (Bruyndonckx et al. 1997).

A second application of MWPC technology to radioisotope imaging was the development of an *MWPC gamma camera* for imaging low-energy photons (<100 keV) (Lacy et al. 1984). The detector (Figure 5.33) consisted of a drift region and a detection region encased in an aluminium pressure vessel, with a thin entrance window. Photons entering the window interacted with the high-pressure gas (90% xenon, 10% methane at 3–5 Torr) and the resulting ionisation was 'drifted' into the detection region by an electric field of 1 kV cm^{-1}. The detection region contained three parallel wire-planes, one anode and two cathodes that detected the gas avalanche. The two cathodes, containing wires oriented orthogonal to each other, provide the spatial information of the detection position. The detector-sensitive

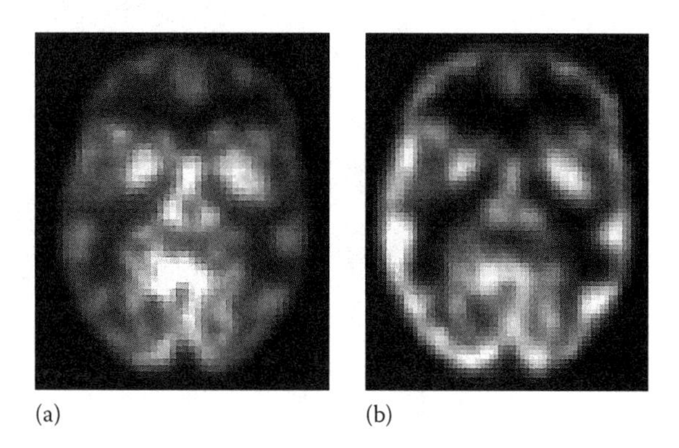

FIGURE 5.32
Images of the Hoffman brain phantom taken with (a) PETRRA and (b) a commercial PET/CT system.

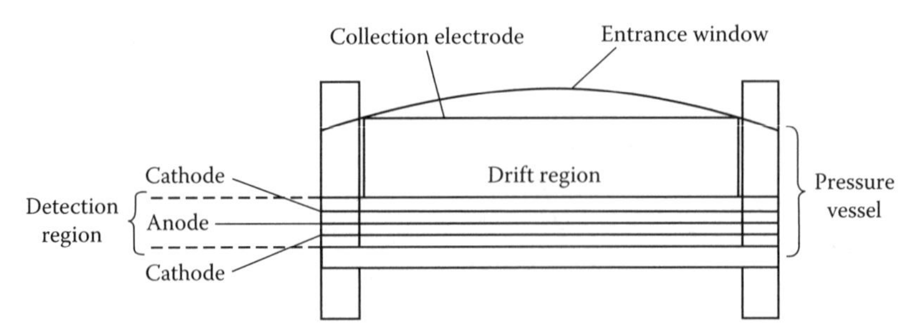

FIGURE 5.33
High-pressure multi-wire proportional-chamber gamma camera for imaging γ-ray energies below 100 keV. (After Lacy et al., 1984.)

area was 25 cm in diameter and the detection efficiency reached a maximum of about 50%–60% at roughly 40 keV, falling to 10%–15% at 100 keV. The intrinsic energy resolution at 60 keV was 33%, the count-rate capacity was 850 kcps for a 50% dead-time loss and the intrinsic spatial resolution was ~2.5 mm FWHM and ~5 mm FWTM. The detector had a very limited application to medicine, as few useful imaging radioisotopes emit photons in the energy window 50–100 keV. ^{133}Xe and ^{201}Tl are potentially suitable and ^{178}Ta from a ^{178}W/^{178}Ta generator has been used for first-pass cardiac imaging. A similar detector has been evaluated (Barr et al. 2002) which uses up to 10 bar of Xenon and has a sensitive area of 25 cm × 25 cm. This camera has a useful sensitivity at energies up to 120 keV.

A third important application of MWPC technology is the development of *quantitative macroscopic autoradiography* using agents labelled with β emitters, such as ^{14}C, ^{32}P, ^{35}S, ^{3}H as well as the X-ray emitter, ^{125}I (Bateman et al. 1985). This type of detector has been used to image electrophoresis plates for protein and DNA sequencing. The Instant Imager (Englert et al. 1995) is based on the CERN MWPC design, shown in Figure 5.34. The system contains a 240 mm × 240 mm array of 200,000 microchannels, each 0.4 mm viewed by an MWPC. Using ^{14}C-labelled agents a spatial resolution of ~0.5 mm has been achieved, sufficient to resolve DNA sequence bands in electrophoresis plates. The notable advantage of the MWPC detector over film autoradiography is the high

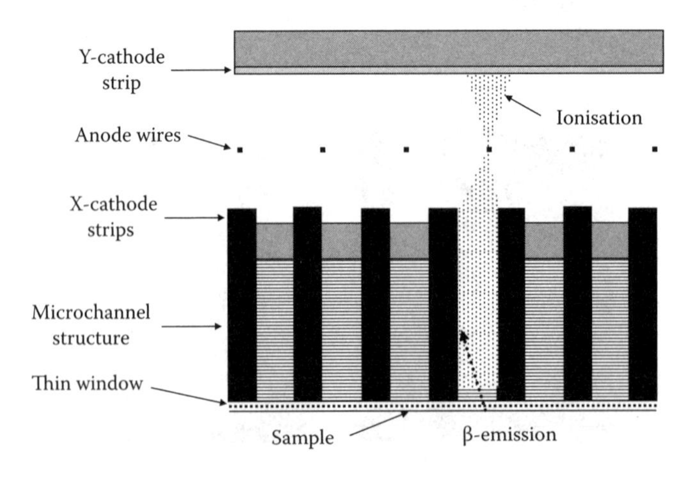

FIGURE 5.34
Multi-wire proportional-chamber autoradiographic camera for imaging beta emitters such as ^{3}H, ^{14}C, ^{32}P and ^{35}S.

sensitivity, giving exposure times of 100 times less than conventional film techniques (a few hours instead of a few weeks).

5.3.6 Semiconductor Detectors

It takes only a few electron volts of energy deposited in a silicon or germanium semiconductor to produce an electron–hole (e–h) pair and, if the material is fully depleted, it functions like a solid-state ionisation chamber, giving very high detection efficiency for each e–h pair. Hence, it is possible to obtain energy resolutions of about 0.6 keV for high-purity germanium (HPGe) at liquid-nitrogen temperature and a few keV for silicon at room temperature.

The high atomic number (Z) and excellent energy resolution of germanium attracted the attention of several groups of workers in the 1970s who constructed small-scale gamma cameras using lithium-drifted germanium crystals. A *germanium gamma camera* was produced in the United Kingdom in the early 1970s (McCready et al. 1971). The detector used orthogonal electrical contact strips on a single slice of lithium-drifted germanium to form a *p*-type/intrinsic/*n*-type (pin) structure. The detector was 44 × 44 mm^2 in area and 10 mm thick. The energy resolution was a few keV and the intrinsic spatial resolution was 3 mm. A similar detector developed in the mid-1970s (Gerber et al. 1977) used a charge-splitting resistor network to determine the spatial position of the photon conversion point (Figure 5.35). This detector was 30 × 30 mm^2 in area, 5 mm thick and had 14 electrode strips on each side spaced on 3 mm centres. The detector had a 5 keV energy resolution and 1.7 mm intrinsic spatial resolution. The potential of these types of gamma camera have never been fully realised because of overall cost, the need for cryogenics and specifically the difficulty of making large-area Ge crystals.

Several attempts to overcome this have been made using 'coded-aperture' systems. Perhaps the most interesting of these is the *Compton gamma camera*. The idea was first proposed in the mid-1970s (Everett et al. 1977), in which a segmented semiconductor (silicon) was suggested as the scattering material in coincidence with a conventional gamma

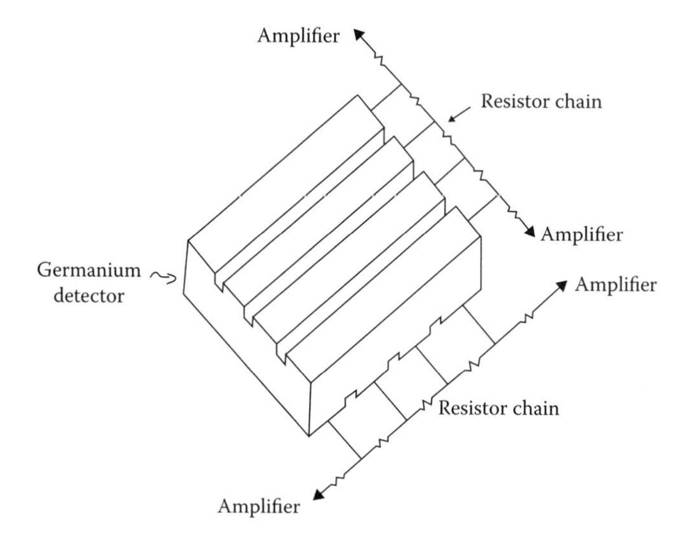

FIGURE 5.35
Charge-splitting system used to read out positional information from a semiconductor gamma camera. (Adapted from Gerber et al., 1969.)

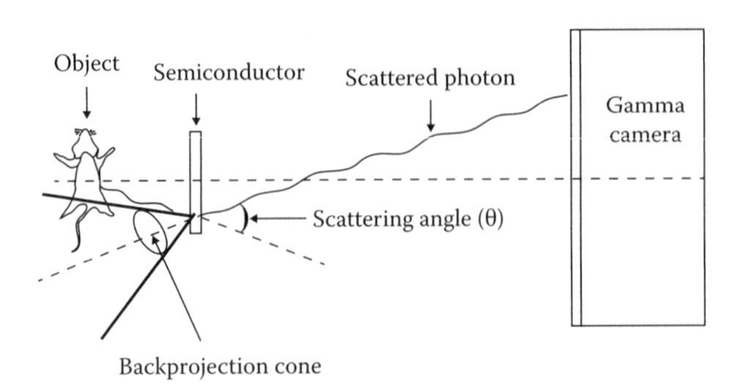

FIGURE 5.36
Schematic diagram of a semiconductor/scintillation counter Compton camera configuration, showing the backprojection cone of reconstruction determined using the coordinates of the scattered photon as measured by the semiconductor and the gamma-camera absorber.

camera (minus collimator) as the absorber (Figure 5.36). The Compton scatter of photons is governed by the relationship

$$\cos\theta = \frac{E - k\delta E}{E - \delta E} \tag{5.18}$$

where
$k = 1 + m_0c^2/E$,
E is the photon energy,
m_0c^2 is the rest mass of the electron,
δE is the energy loss in the scatter process, and
θ is the scatter angle of the photon.

It is possible to determine the scatter angle θ by measuring the coordinates of the detection positions in the semiconductor and camera and values for the energy loss, δE, and the scattered photon energy, $(E - \delta E)$. When each event is backprojected, the incident photon vector is positioned on a conical surface. The distribution of the photon source can, in principle, be obtained by appropriate iterative reconstruction techniques. A Compton camera using small germanium detectors as the scattering material provided high-sensitivity whole-body measurements (Singh and Doria 1983). The camera contained a 33 × 33 array of HPGe detectors coupled to an uncollimated conventional gamma camera. Each HPGe element was 5 × 5 mm^2 in area and 6 mm thick. It was predicted that a spatial resolution of 12 mm at 10 cm from the detector was achievable, and the sensitivity should be 15 times higher than that of a conventional gamma camera/collimator system. However, the reconstruction of images from a Compton camera produces substantial noise in the images and as a rule ~10× the number of events is required to provide image quality similar to that using a conventional gamma camera. Hence, the gain is not as much as might be expected.

The development of silicon wafers, with spatial read-out using either microstrip or pixel techniques, has motivated work on devices for autoradiography and other small radioisotope imaging systems. The recent development in spatially resolving silicon detectors raises the possibility of a small-scale Compton camera in which the scatter position is known to better than 1 mm and the energy loss in the scatter process to a few keV. The spatial accuracy in the image is a function of the accuracy of both the energy and spatial measurements in the scattering material and absorber. Although the spatial resolution of the scintillation camera is no better than 3 mm (at best), the use of appropriate geometry can provide a spatial resolution of ~1 mm in an object close to the scattering material. If scatter in the object is ignored, the energy of the photon entering the scattering material is known and, hence, if the energy loss due to scatter is accurately measured, the poor energy resolution in the NaI is of little importance. It should be possible to image small objects (a few square centimetres) with a spatial resolution of better than 1 mm, ideal for measurements of the distribution of radiolabelled agents (antibodies, etc.) in small human tumours and animal models.*

A further and important application of semiconductors is the use of silicon microstrip or pixel detectors to tissue autoradiography. In a double-sided *microstrip detector*, where the conducting strips are spaced ~50 μm apart, each strip can be read out using edge-connected amplifier-discriminator circuits and spatial accuracy of 50–100 μm achieved. A *pixel* detector has the advantage of giving the exact coordinates of the ionising radiation and in the case of a CCD the read-out system is cheap and simple (Figure 5.37). Such a system has been developed (Ott et al. 2000) and tested by imaging tissue samples labelled with ^{14}C (Figure 5.38), ^{125}I, ^{35}S, ^{32}P and ^{18}F and has been shown to have a resolution of 35 μm and a sensitivity of 10–100 times that of film to these radionuclides. At the time of writing, this is comparable or better than any other digital autoradiography device. More recently, CMOS-based pixel detectors have begun to show promise for radioisotope imaging. Small devices with pixel sizes of 25 μm and similar in size to CCDs have intelligence built into each pixel allowing high-speed read-out. When coupled to a CsI(Tl) phosphor this kind

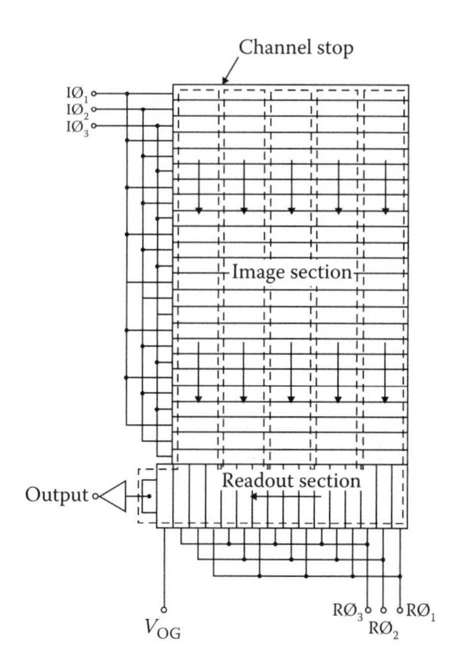

FIGURE 5.37
Schematic of a charged coupled device (CCD) showing the imaging and read-out sections. In this example, there are 40 pixels in a 5 × 8 array. Pulsing of the IØ electrodes causes the charge to move through the image section in the vertical direction. Once this charge reaches the read-out section, similar pulsing of the RØ electrodes transfers the charge to the output node. The output charge-detection amplifier is shielded from the read-out register by the output gate that is held at a fixed bias of V_{OG}. (After EEV, 1987. Reproduced with permission from e2v Technologies, Chelmsford, UK.)

* Up-to-date information on Compton cameras can be found at the following website: http://www.mpb.ucl.ac.uk/research/acadradphys/researchactivities/comptoncameras.htm (accessed February 20, 2012).

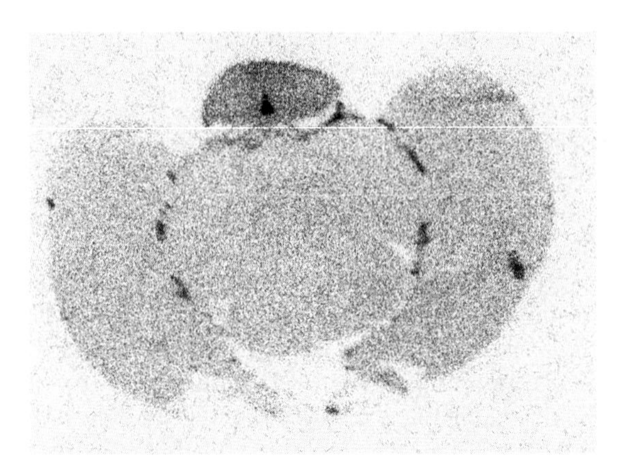

FIGURE 5.38
Autoradiograph of a section through mouse brain using a CCD autoradiography system showing the biodistribution of ^{14}C-glucose. Twenty-four-hour image consisting of 1440, 60 s frames. The spatial resolution in this image is ~22 μm. (From Ott et al., 2000. Reproduced with permission from IOP Publishing Limited.)

of device has been shown to give a spatial resolution of ~80 μm suitable for imaging lymph nodes with $^{99}Tc^m$, for example (Ott et al. 2006). Further applications of semiconductors to radionuclide imaging can be expected in the future with the development of position-sensitive devices based on materials such as gallium arsenide and cadmium zinc telluride which have a higher photon-stopping power than silicon and can be used at or near room temperature unlike germanium.

5.3.7 Summary

We can conclude from this section that the detection of γ-photons has been dominated for the last 50 years by the use of inorganic scintillators, in particular NaI(Tl), coupled to PMTs. The scintillator is an efficient γ-ray absorber and produces plenty of light photons. The extended temporal nature of the light emission is often a problem in high-count-rate imaging. The PMT limits the performance of scintillator-based detectors as its analogue stability inhibits the uniformity of a conventional gamma camera. This is now largely overcome using sophisticated correction processes using microprocessor technology, see Section 5.5. The use of semiconductor devices, especially APDs, instead of PMTs may significantly improve the properties of future scintillating-crystal-based imaging systems. The advances in gas and semiconductor detectors has been slow with the result that the scintillator–PMT combination still dominates radioisotope imaging and seems likely to do so for some time to come.

5.4 Radionuclide Production and Radiopharmaceuticals

One of the primary advantages associated with the use of radionuclides in medicine is the large signal (in this case, the emitted radiation) obtained from the relatively small mass of radionuclide employed for a given study. Nuclear medicine takes advantage of this physical characteristic by using a variety of compounds, labelled with radionuclides (radiopharmaceuticals), in order to 'trace' various functions of the body (Sandler et al. 2002). The minute

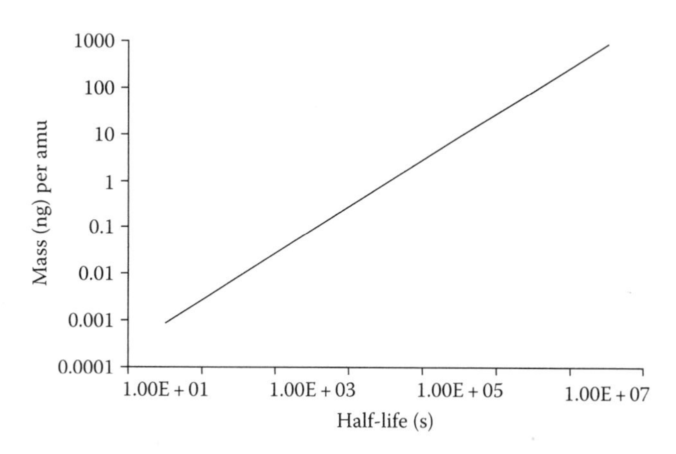

FIGURE 5.39
Relationship between the mass of 37 GBq of a radionuclide per atomic mass unit and the radionuclide half-life.

mass of radiolabelled material allows for non-invasive observation without disturbance of the system under study through pharmacological or toxicological effect. For most nuclear medicine studies, the mass of tracer used is in the range of nanograms (Figure 5.39) and no other physical technique could be employed to measure mass at these levels. Therefore, the sensitive measurement of biochemical and physiological processes through the use of radioactivity and its detection comprise the fundamental basis of nuclear medicine and are the key to its continued growth.

5.4.1 Radioactive Decay

The radioactivity Q of a number (N) of nuclei is given by

$$Q = -\lambda N = \frac{dN}{dt} \tag{5.19}$$

where λ is defined as the decay constant for the radioisotope. We can see that the rate of decay of nuclei depends only upon λ and the number of nuclei, N. The solution to Equation 5.19 is

$$N = N_0 \exp(-\lambda t) \tag{5.20}$$

where N_0 is the number of nuclei at some reference time $t = 0$. If $T_{1/2}$ is the time for half the nuclei to decay, the so-called half-life, then

$$T_{1/2} = \frac{(\ln 2)}{\lambda}. \tag{5.21}$$

An alternative form of Equation 5.20 is

$$N = N_0 \left(\frac{1}{2}\right)^m \tag{5.22}$$

where m is the number of half-lives since the reference time $t = 0$.

TABLE 5.4

Approximate mass (g) of 37 GBq (1 Ci) of radionuclide for a given half-life and atomic weight.

	Atomic Weight of Atom (amu)		
$T_{½}$	18	99	201
15 s	2.4×10^{-11}	1.3×10^{-10}	2.7×10^{-10}
15 min	1.4×10^{-9}	7.7×10^{-9}	1.6×10^{-8}
6 h	3.5×10^{-8}	1.9×10^{-7}	3.8×10^{-7}
8 days	1.1×10^{-6}	6.3×10^{-6}	1.2×10^{-5}
15 a	7.7×10^{-4}	4.2×10^{-3}	8.3×10^{-3}

Using these simple equations, it is possible to calculate the radioactivity of any mass of nuclei at any time subsequent to a measurement at a reference time. By rearranging Equation 5.19, the number of atoms of a given radionuclide is given by

$$N = \frac{Q}{\lambda}. \tag{5.23}$$

The mass of radionuclide present in this number of atoms is given by

$$Mass = \frac{Q}{\lambda} \times \frac{m}{\mathrm{Av}} \tag{5.24}$$

where

m is the atomic weight of the radionuclide, and
Av is Avogadro's number (the number of atoms in 1 mol = 6.02×10^{23}).

Table 5.4 shows the approximate mass of 37 GBq of a radionuclide as a function of half-life and atomic weight.

5.4.2 Production of Radionuclides

The fundamental property of all radioactive elements is the imbalance of the proton-to-neutron ratio of the nucleus. A proper balance of protons and neutrons is essential for maintaining a stable atomic nucleus. The balance must be maintained to overcome electrostatic repulsion of the charged protons, and Figure 5.40 shows how the neutron-to-proton (n/p) ratio changes with increasing mass. There are four ways by which radionuclides are produced:

- neutron capture (also known as neutron activation);
- nuclear fission;
- charged-particle bombardment; and
- radionuclide generators.

Each method affords useful isotopes for nuclear medicine imaging. A description of each of the methods, and examples of some of the useful radioisotopes produced, is presented in the following paragraphs.

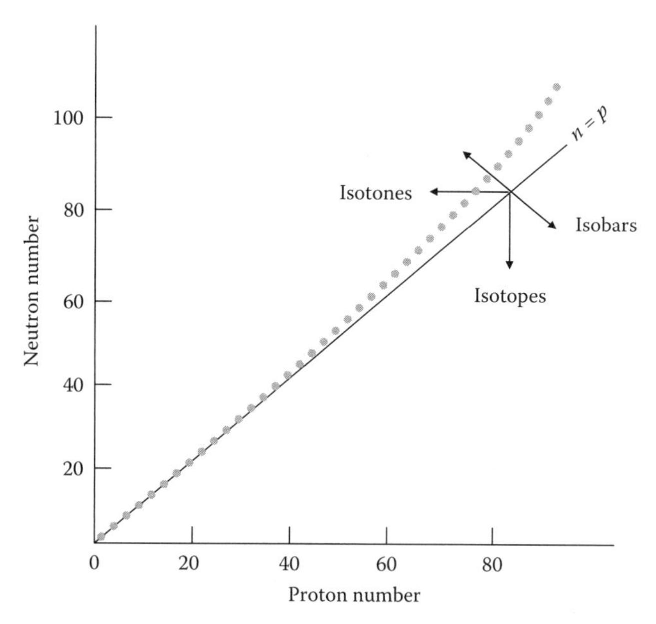

FIGURE 5.40
Graph of neutron number n versus proton number p, showing the increase in the n/p ratio with atomic number.

Neutron capture is the absorption of a neutron by an atomic nucleus, and the production of a new radionuclide via reactions such as

$$n + {}^{98}\text{Mo} \rightarrow {}^{99}\text{Mo} + \gamma \tag{5.25}$$

$$n + {}^{32}\text{S} \rightarrow {}^{32}\text{P} + \text{p}. \tag{5.26}$$

To produce radioactive elements through neutron capture, neutrons must have a mean energy of 0.03–100 eV. These 'thermal' neutrons are best suited for interaction with, and absorption into, the atomic nucleus. The most efficient means of producing radioisotopes by this method is through the use of a nuclear reactor. To produce a radioactive species, a sample of a target element is placed in a field of thermal neutrons. The reaction yield depends on the flux density of incident particles, ϕ ($cm^{-2}\ s^{-1}$), the number of accessible target nuclei (n_t) and the likelihood or cross section of the reaction, σ (barn). The yield (N_y) is given by

$$N_y = \frac{n_t \phi \sigma}{\lambda}\left[1 - \exp(-\lambda t)\right]. \tag{5.27}$$

The radionuclide produced via the (n, γ) reaction is an isotope of the target material, that is, the two nuclei have the same number of protons. This means that radionuclides produced via the (n, γ) reaction are not carrier free (compare with nuclear fission), and, thus, both the ratio of radioactive atoms to stable atoms and the specific activity are relatively low. Separation of the radionuclide from other target radionuclides is possible by physical and chemical techniques. Useful tracers produced by neutron absorption are shown in Table 5.5.

TABLE 5.5

Radionuclides produced by neutron absorption.

Radionuclide	Production Reaction	Gamma-Ray Energy (keV)	Half-Life	σ (Barn)
^{51}Cr	$^{50}Cr(n, \gamma)^{51}Cr$	320	27.7 days	15.8
^{59}Fe	$^{58}Fe(n, \gamma)^{59}Fe$	1099	44.5 days	1.28
^{99}Mo	$^{98}Mo(n, \gamma)^{99}Mo$	740	66.02 h	0.13
^{131}I	$^{130}Te(n, \gamma)\ ^{131}Te \rightarrow\ ^{131}I$	364	8.04 days	0.29

Source: From Mughabghab et al., 1981.

Nuclear fission is the process whereby heavy nuclei (^{235}U, ^{239}Pu, ^{237}U, ^{232}Th and others with atomic numbers greater than 92) are rendered unstable when irradiated with thermal neutrons, due to absorption of these neutrons. Consequently, these unstable nuclei undergo 'fission', the breaking up of the heavy nuclei into two lighter nuclei of approximately similar atomic weight, for example,

$$^{235}_{92}U + ^{1}_{0}n \rightarrow ^{236}_{92}U \rightarrow ^{99}_{42}Mo + ^{133}_{50}Sn + 4^{1}_{0}n. \tag{5.28}$$

Nuclides produced by fission range in atomic number between about 28 and 65. Figure 5.41 shows the distribution of fission products by atomic mass number produced following fission of ^{235}U. As seen from Equation 5.28, this reaction produces four more neutrons, which may be absorbed by other heavy nuclei, and the fission process can continue until the nuclear fuel is exhausted. An interaction such as that in Equation 5.28 must, of course, conserve Z and A.

Nuclides produced by fission of heavy nuclei must undergo extensive purification in order to harvest one particular radionuclide from the mixture of fission products. The fission

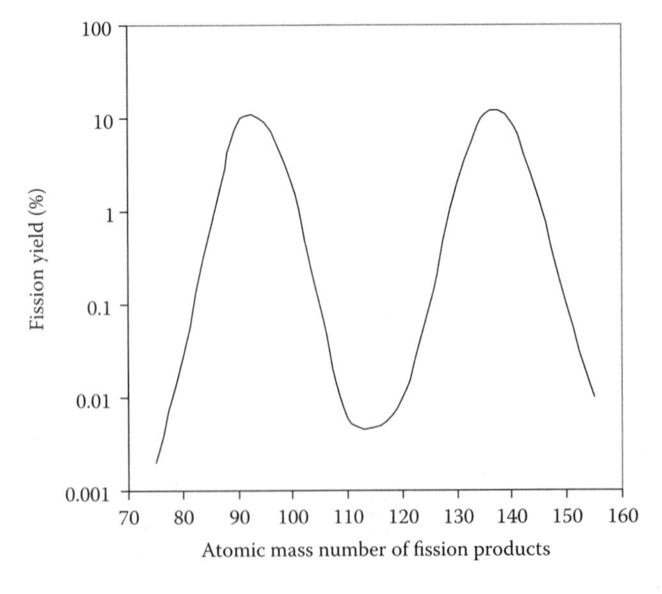

FIGURE 5.41
Fission product yield for ^{235}U irradiated by thermal neutrons.

TABLE 5.6

Radionuclides produced by nuclear fission.

Isotope	Gamma-Ray Energy (keV)	Half-Life	Fission Yield (%)
^{99}Mo	740	66.02 h	6.1
^{131}I	364	8.05 days	2.9
^{133}Xe	81	5.27 days	6.5
^{137}Cs	662	30 a	5.9

Source: From BRH, 1970.

process affords high specific activity due to the absence of carrier material (non-radioactive isotope of the same element). However, fission products are usually rich in neutrons and therefore decay principally via β^- emission (see Section 5.4.3), a physical characteristic that is undesirable for medical imaging, but of interest in radionuclide therapy. Useful nuclides produced by nuclear fission are shown in Table 5.6.

Charged-particle bombardment is the process of production of radionuclides through the interaction of charged particles ($H^{\pm}$, D^+, $^3He^{2+}$, $^4He^{2+}$) with the nuclei of stable atoms. The particles must have enough kinetic energy to overcome the electrostatic repulsion of the positively charged nucleus. Two basic types of accelerator are used for this purpose, the linear accelerator and the cyclotron. In both systems, charged particles are accelerated over a finite distance by the application of alternating electromagnetic potentials (Figure 5.42). In both types of machine, particles can usually be accelerated to a range of energies. Examples of typical reactions in a target are

$$p + {}^{68}Zn \rightarrow {}^{67}Ga + 2n \tag{5.29}$$

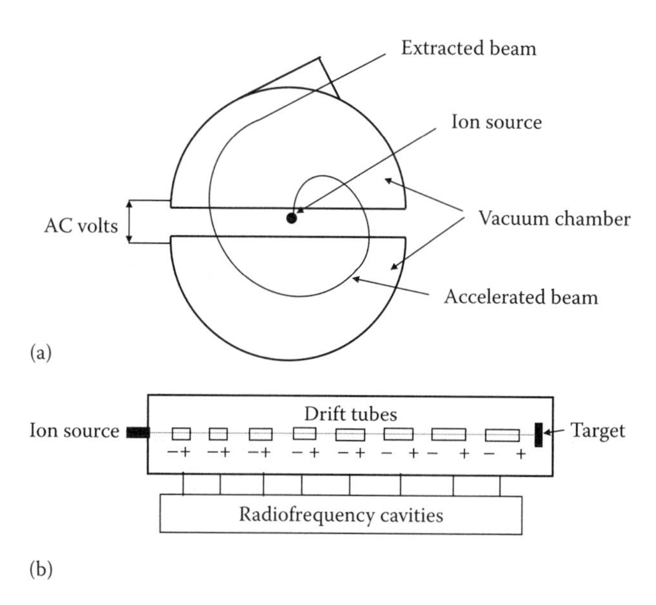

FIGURE 5.42
Schematic diagrams of (a) a cyclotron and (b) a linear accelerator for radionuclide production.

and

$$\alpha + {}^{16}\text{O} \rightarrow {}^{18}\text{F} + \text{p} + \text{n}. \tag{5.30}$$

For the production of medically useful radionuclides, particle energies per nucleon in the range 1–100 MeV are commonly used. One major advantage of producing isotopes through charged-particle bombardment is that the desired isotope is almost always of different atomic number to the target material. This theoretically allows for the production of radionuclides with very high specific activity and high radionuclidic purity. However, the actual activity and purity obtained is related to the isotopic and nuclidic purity of the target material, the cross section of the desired reaction and the cross section of any secondary reaction.

Charged-particle reactions yield radionuclides that are predominantly neutron deficient and, therefore, decay by β^+ emission or electron capture (see Section 5.4.3). The latter radioisotopes are particularly useful for clinical imaging due to the lack of particulate emission. Examples of accelerator-produced radionuclides routinely used in nuclear medicine are shown in Table 5.7.

Radioactive decay can lead to the generation of either a stable or a radioactive nuclide. In either case, the new nuclide may have the same or different atomic number depending on the type of decay (see Section 5.4.3). Radioactive decay leading to the production of a radioactive daughter with a different Z allows for the possibility of simple chemical separation of the parent–daughter combination. If the daughter radionuclide has good physical characteristics compatible with medical imaging and the parent has a sufficiently long half-life to allow for production, processing and shipment, then remote parent–daughter separation provides a potentially convenient source of a medically useful, short-lived radionuclide. This type of radionuclide production system is known as a *radionuclide generator.*

A radionuclide generator provides a method for simple chemical separation of a daughter radionuclide from the corresponding parent nuclide, giving a ready supply of short-lived radionuclide. This can be accomplished through the use of chromatographic techniques, distillation or phase partitioning. However, chromatographic techniques have been the most widely explored and form the basis of most

TABLE 5.7

Radionuclides produced by charged-particle bombardment.

Radionuclide	Principal Gamma-Ray Energy (keV)	Half-Life	Production Reaction
^{11}C	511 (β^+)	20.4 min	^{14}N(p, α)^{11}C
^{13}N	511 (β^+)	9.96 min	^{13}C(p, n)^{13}N
^{15}O	511 (β^+)	2.07 min	^{15}N(p, n)^{15}O
^{18}F	511 (β^+)	109.7 min	^{18}O(p, n)^{18}F
^{67}Ga	93, 184, 300	78.3 h	^{68}Zn(p, 2n)^{67}Ga
^{111}In	171, 245	67.9 h	^{112}Cd(p, 2n)^{111}In
^{120}I	511 (β^+)	81 min	^{127}I(p, 8n)^{120}Xe $\rightarrow$ ^{120}I
^{123}I	159	13.2 h	^{124}Te(p, 2n)^{123}I ^{127}I(p, 5n)^{123}Xe $\rightarrow$ ^{123}I
^{124}I	511 (β^+)	4.2 days	^{124}Te(p, n)^{124}I
^{201}Tl	68–80.3	73 h	^{203}Tl(p, 3n)^{201}Pb $\rightarrow$ ^{201}Tl

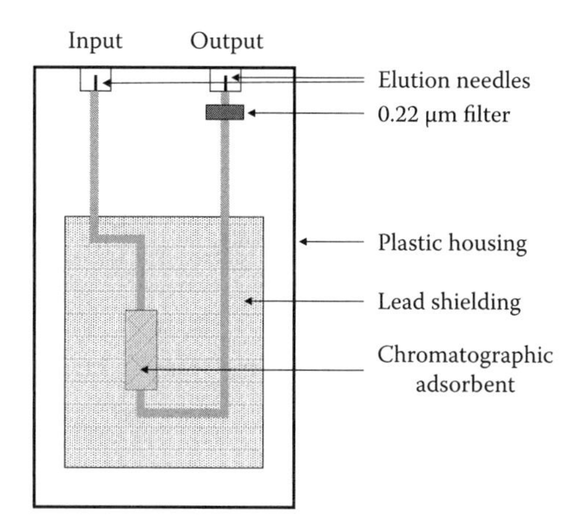

FIGURE 5.43
Schematic of a radioisotope generator.

radionuclide generators in use today (Babich 1989, Knapp and Mirzadeh 1994, Welch and McCarthy 2000) (Figure 5.43).

The equations governing generator systems stem from the formula

$$A_2 = \frac{\lambda_2}{\lambda_2 - \lambda_1} A_1^0 \left[\exp(-\lambda_1 t) - \exp(-\lambda_2 t)\right] \tag{5.31}$$

where
- A_1^0 is the parent activity at time $t = 0$,
- t is the time since the last elution of the generator,
- A_2 is the activity of the daughter product ($A_2^0 = 0$), and
- λ_1 and λ_2 are the decay constants of parent and daughter radioisotopes, respectively.

For the special case of *secular equilibrium*, defined by $\lambda_2 >> \lambda_1$, we have

$$A_2 = A_1^0[\exp(-\lambda_1 t) - \exp(-\lambda_2 t)]. \tag{5.32}$$

If t is much less than the half-life of the parent, $\ln(2)/\lambda_1$, and greater than approximately seven times the daughter half-life, $\ln(2)/\lambda_2$, then

$$A_2 \approx A_1^0. \tag{5.33}$$

This is the equilibrium condition. The growth of the daughter here is given by

$$A_2 = A_1^0[1 - \exp(-\lambda_2 t)]. \tag{5.34}$$

An example of a radionuclide generator system that follows this decay/growth pattern is the $^{68}Ge/^{68}Ga$ generator. The parent radionuclide, ^{68}Ge, has a half-life of 288 days, whilst the daughter radionuclide, ^{68}Ga, has a half-life of only 68 min. The rate of ^{68}Ga activity build-up and decay following elution of the generator is shown in Figure 5.44.

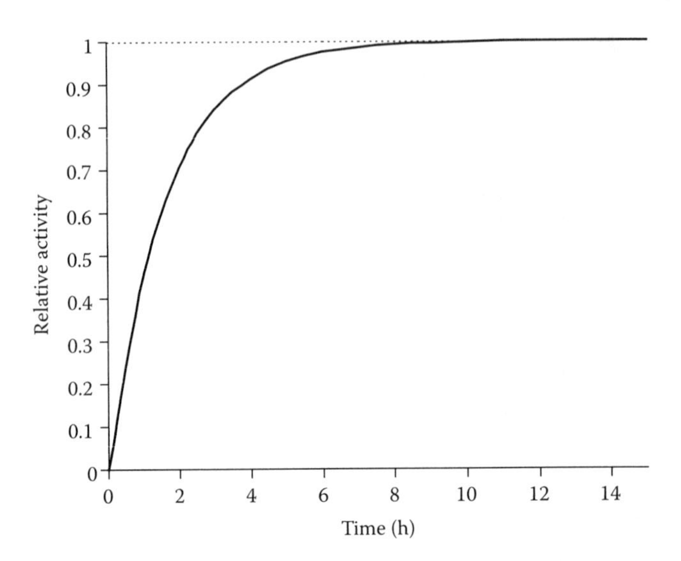

FIGURE 5.44
Secular equilibrium: build-up of ^{68}Ga activity on a $^{68}Ge/^{68}Ga$ generator following generator elution.

For *transient equilibrium*, defined by $\lambda_2 > \lambda_1$ but λ_2 not *very* much greater than λ_1, the growth of the daughter is given by

$$A_2 = \frac{\lambda_2 A_1^0}{(\lambda_2 - \lambda_1)}. \tag{5.35}$$

Figure 5.45 illustrates the transient equilibrium situation, using the $^{99}Mo/^{99}Tc^m$ generator as an example.

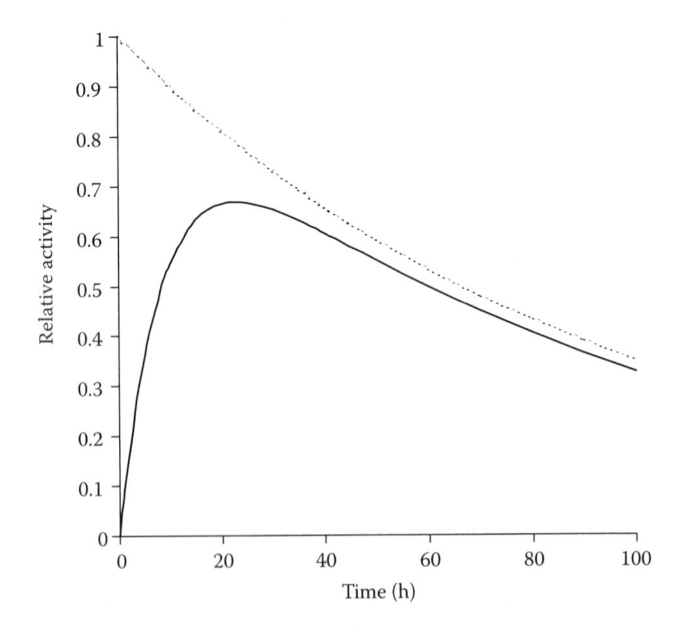

FIGURE 5.45
Transient equilibrium: build-up of $^{99}Tc^m$ activity on a $^{99}Mo/^{99}Tc^m$ generator following elution.

TABLE 5.8

Generator-produced radionuclides.

Parent P	Parent Half-Life	Mode of Decay P → D	Daughter D	Mode of Decay of D	Daughter Half-Life	Daughter γ Energy (keV)
^{62}Zn	9.1 h	β^+	^{62}Cu	β^+	9.8 min	511
		EC		EC		1173
^{68}Ge	280 days	EC	^{68}Ga	β^+	68 min	511
				EC		1080
^{81}Rb	4.7 h	EC	$^{81}Kr^m$	IT	13 s	190
^{82}Sr	25 days	EC	^{82}Rb	EC	76 s	777
				β^+		511
^{99}Mo	66.02 h	β^-	$^{99}Tc^m$	IT	6.02 h	140
^{113}Sn	115.1 days	EC	$^{113}In^m$	IT	1.66 h	392
$^{195}Hg^m$	40 h	IT	$^{195}Au^m$	IT	30.6 s	262
		EC				

The most widely used generator-produced radionuclide in nuclear medicine is technetium-99m ($^{99}Tc^m$). The parent, ^{99}Mo, which has a half-life of about 66 h, can be produced through neutron activation or as a fission product, can be chemically adsorbed onto an $A1_2O_3$ (alumina) column and decays to $^{99}Tc^m$ (85%) and ^{99}Tc (15%).

$^{99}Tc^m$ has a half-life of 6.02 h and decays to ^{99}Tc (half-life 2.1×10^5 years) by isomeric transition (IT), emitting a 140 keV γ-ray (98%) with no associated particulate radiation:

$$^{99}\mathrm{Mo} \xrightarrow[\beta^-]{} {}^{99}\mathrm{Tc}^m \xrightarrow[\mathrm{IT}]{} {}^{99}\mathrm{Tc} + \gamma. \tag{5.36}$$

The $^{99}Tc^m$ is eluted from the alumina chromatographic column by passing a solution of isotonic saline (0.9% NaCl) through the column. This saline solution and the solid phase of $A1_2O_3$ allow for efficient separation of $^{99}Tc^m$ from the ^{99}Mo with only minute amounts of ^{99}Mo breakthrough (normally significantly less than 0.1%). The eluted $^{99}Tc^m$ can be chemically manipulated so that it binds to a variety of compounds, which will then determine its fate *in vivo* (see Section 5.4.6). Other generator systems exist, producing both single-photon and positron-emitting radionuclides, examples of which are given in Table 5.8.

5.4.3 Radioactive Decay Modes

All radionuclides used in nuclear medicine are produced via one of the aforementioned four production methods. Each of these radionuclides has a unique process by which it decays. The *decay scheme* describes the type of decay, the energy associated with it and the probability for each type of decay. These decay schemes can be very complex since many radionuclides decay via multiple nuclear processes (Figure 5.46).

Alpha decay often occurs during the decay of high-mass-number radionuclides, and involves the spontaneous emission of an α-particle (a helium nucleus) with a discrete energy in the range 4–8 MeV. If α decay leaves the nucleus in an excited state, the de-excitation

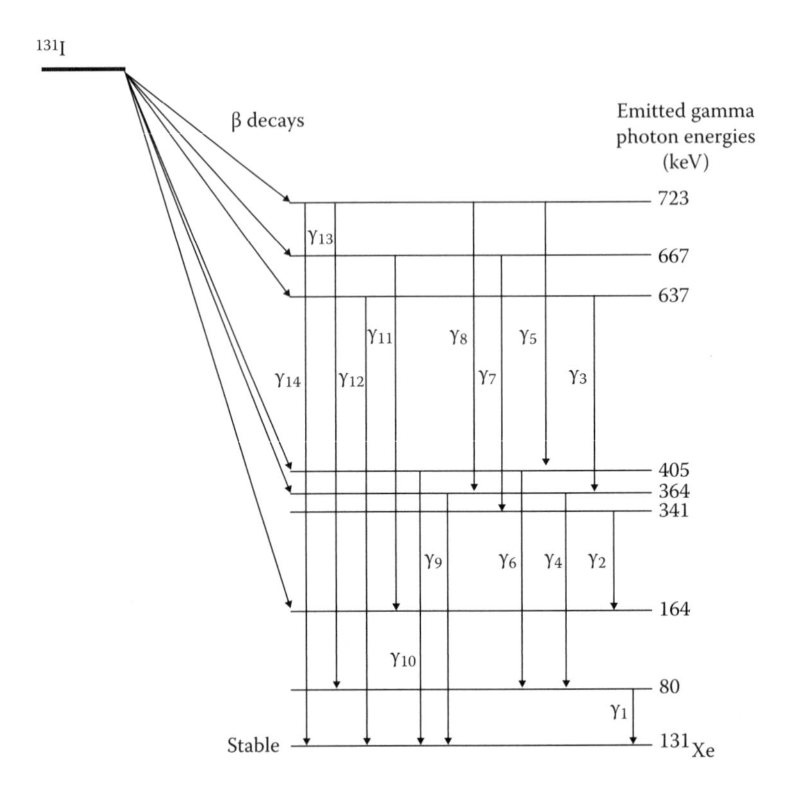

FIGURE 5.46
Nuclear decay scheme for ^{131}I. (Derived from Dillman and Von der Lage, 1975.)

will be via the emission of γ-radiation. Most of the energy released in the transition is distributed between a daughter nucleus (as recoil energy) and the α-particle (as kinetic energy). In the α-decay process, the parent nucleus loses four units of mass and two units of charge. An example of alpha decay is

$$^{226}_{88}\mathrm{Ra} \xrightarrow{\alpha \text{ decay}} {}^{222}_{86}\mathrm{Rn} + {}^{4}_{2}\mathrm{He}. \qquad (5.37)$$

Although α-emitting nuclides have no use in medical imaging (since the α-particles travel only micrometer distances in tissue), encapsulated α-emitting radionuclides have been used as sealed sources emitting X- or γ-rays for therapy, and there has been a renewed interest in their use for targeted radionuclide therapy because of the very high LET in tissue (Hoefnagel 1991, Volkert et al. 1991, Vaidyanathan and Zalutsky 1992, Couturier et al. 2005, Boyd et al. 2006, Bruland et al. 2006, Vandenbulcke et al. 2006).

As discussed previously, many radionuclides are unstable due to the neutron/proton imbalance within the nucleus. The decay of neutron-rich radionuclides involves the ejection of a β^--particle (e^-), resulting in the conversion of a neutron into a proton. Decay via β^- *emission* results in the atomic number of the atom changing but the atomic weight remaining the same. The energy of the emitted β^--particles is not discrete but a continuum (i.e. varies from zero to a maximum). Since the total energy lost by the nucleus during disintegration must be discrete, energy conservation is maintained by the emission of a third particle – the anti-neutrino ($\bar{\nu}$). The anti-neutrino has neither measurable mass nor charge and interacts only weakly with matter. An example of β^- decay is

$$ {}^{99}_{42}\text{Mo} \xrightarrow{\beta\ \text{decay}} {}^{99}_{43}\text{Tc}^{m} + e^{-} + \bar{\nu}. \tag{5.38} $$

Beta decay may be accompanied by γ-ray emission if the daughter nuclide is produced in an excited state.

Nuclei that are rich in protons or are neutron-deficient may decay by *positron emission* from the nucleus. This decay is also accompanied by the emission of an uncharged particle, the neutrino (ν). After positron decay, the daughter nuclide has an atomic number that is one less than that of the parent, but again the atomic weight is the same. The range of a positron (e^+ or β^+) is short (of the order of 1 mm in tissue) and, when the particle comes to rest, it combines with an atomic electron from a nearby atom, and is annihilated. Annihilation (the transformation of these two particles into pure energy) gives rise to two photons, both of energy 511 keV, which are emitted approximately antiparallel to each other. These photons are referred to as annihilation radiation. Positron emission only takes place when the energy difference between the parent and daughter nuclides is larger than 1.02 MeV. An example of positron decay is

$$ {}^{68}_{32}\text{Ge} \xrightarrow{\text{EC}} {}^{68}_{31}\text{Ga} \xrightarrow{\beta^{+}\ \text{decay}} {}^{68}_{30}\text{Zn} + e^{+} + \nu. \tag{5.39} $$

An alternative to positron emission for nuclides with a proton-rich nucleus is *electron capture*. Electron capture involves the capture within the nucleus of an inner orbital electron, transforming a proton into a neutron. For this process to occur, the energy difference between the parent and the daughter nuclides can be small, unlike positron emission. Hence, this process dominates over positron emission in high-mass radionuclides. Usually, the K-shell electrons are captured because of their proximity to the nucleus. Electrons from the outer orbitals fill the vacancy created in the inner electron orbitals. The difference in energy between these electron shells appears as an X-ray that is characteristic of the daughter radionuclide. The probability of electron capture increases with increasing atomic number because electron shells in these nuclei are closer to the nucleus. An example of electron capture is

$$ {}^{51}_{24}\text{Cr} + e^{-} \xrightarrow{\text{EC}} {}^{51}_{23}\text{V} + \gamma + \nu. \tag{5.40} $$

A nucleus produced in a radioactive decay can remain in an excited state for some time. Such states are referred to as *isomeric states*, and decay to the ground state can take from fractions of a second to many years. A transition from the isomeric or excited state to the ground state is accompanied by *γ emission*. When the isomeric state is long-lived, this state is often referred to as a metastable state. $^{99}\text{Tc}^{m}$ is the most common example of a metastable isotope encountered in nuclear medicine (see Equation 5.38).

Internal conversion is the process that can occur during γ-ray emission when a photon emitted from a nucleus may interact with one of the atomic electrons from the atom, causing it to be ejected. This particularly affects K-shell electrons, as they are the nearest to the nucleus. The ejected electron is referred to as the *conversion electron* and will have a kinetic energy equal to the energy of the γ-ray minus the electron-binding energy. The probability of internal conversion is highest for low-energy photon emission. Vacancies in the inner orbitals are filled by electrons from the outer shells, leading to the emission of *characteristic X-rays*. Furthermore, characteristic X-rays produced during internal conversion may themselves interact with other outer orbital electrons, causing their ejection, provided that the X-rays have an energy greater than the binding energy of the electron with which they interact. This emitted electron is then referred to as an *Auger electron*. Vacancies in the electron shells resulting from Auger emission are filled by electrons from outer orbitals, leading to further X-ray emission.

5.4.4 Choice of Radionuclides for Imaging

The physical characteristics of radionuclides that are desirable for nuclear medicine imaging include:

- a physical half-life long enough to allow preparation of the radiopharmaceutical and completion of the imaging procedure, but short enough to ensure that delivered radiation doses to patients and staff are minimised;
- decay via either isomeric transition or electron capture to produce photon emission;
- a sufficiently high photon abundance to ensure good counting statistics;
- associated photon energy high enough to penetrate the body tissue with minimal tissue attenuation but low enough for minimal thickness of collimator septa; and
- absence of particulate emission that would deliver unnecessary radiation dose to the patient.

The effective half-life T_E of a radiopharmaceutical is a combination of the physical half-life, T_P, and the biological half-life, T_B, that is,

$$\frac{1}{T_E} = \frac{1}{T_B} + \frac{1}{T_P}. \tag{5.41}$$

Close matching of the effective half-life with the duration of the study is an important dosimetric, as well as practical, consideration in terms of availability and radiopharmaceutical synthesis.

The emitted photon energy is critical for various reasons. The photon must be able to escape from the body efficiently, and it is desirable that the photopeak should be easily separated from any scattered radiation. These two characteristics favour high-energy photons. However, at very high energies, detection efficiency using a conventional gamma camera is poor (Figure 5.6) and the increased septal thickness required for collimators decreases the sensitivity further. In addition, high-energy photons are difficult to shield and present practical handling problems.

The radionuclide that fulfils most of the aforementioned criteria is $^{99}Tc^m$, which is used in more than 80% of all nuclear medicine studies. It has a physical half-life of 6.02 h, is produced via decay of a long-lived ($T_{1/2}$ = 66 h) parent, ^{99}Mo, and decays via isomeric transition to ^{99}Tc emitting a 140 keV γ-ray. The short half-life and absence of $\beta^{\pm}$ emission result in a low radiation dose to the patient. The 140 keV γ emission has a half-value tissue thickness of 4.6 cm but is easily collimated by lead. Most importantly, the radionuclide can be produced from a generator that has a useful life of approximately 1 week.

Other single-photon-emitting radioisotopes in common use in nuclear medicine include ^{67}Ga, $^{81}Kr^m$, ^{111}In ^{123}I and ^{201}Tl. ^{123}I has proved a valuable imaging replacement for ^{131}I as it decays via electron capture, emitting a γ-ray of energy 59 keV and has a 13.2 h half-life.

TABLE 5.9

Main gamma emissions of ^{67}Ga and ^{111}In.

Radionuclide	Gamma-Ray Emission	Gamma-Ray Energy (keV)	Mean Number per Disintegration
^{67}Ga	γ_2	93	0.38
	γ_3	185	0.24
	γ_5	300	0.16
	γ_6	394	0.043
^{111}In	γ_1	171	0.90
	γ_2	245	0.94

It is easily incorporated into a wide range of proteins and other pharmaceuticals capable of being iodinated. Like most of the other radionuclides in the previous list, it is cyclotron produced (Table 5.7) and, therefore, quite expensive.

^{111}In and ^{67}Ga have similar chemical properties (see Section 5.4.5), and both decay via electron capture (Table 5.9). ^{111}In is the superior imaging isotope, having a much higher photon yield and better photon energies for gamma-camera studies, but it is expensive as it is produced by charged-particle bombardment. ^{67}Ga has long been used as a tumour localising agent in the form of gallium citrate and has also proved useful in the same form in the detection of infection. ^{201}Tl is utilised by the cardiac muscle in a similar fashion to potassium and has been extensively used for imaging myocardial perfusion and for assessing myocardial viability. However, the photon emissions used in ^{201}Tl myocardial imaging are the 80 keV X-rays, which are close in energy to lead X-rays produced by the collimator, and this, together with the long half-life (73 h), results in poor image quality. In an attempt to overcome these disadvantages with the use of ^{201}Tl for myocardial imaging, a number of lipophilic ^{99}Tc^{m} complexes have been developed, which act as cationic tracers and are good myocardial perfusion agents (Jain 1999).

5.4.5 Fundamentals of Radiopharmaceutical Chemistry

The strength of nuclear medicine is its ability to image, qualitatively and quantitatively, dynamic physiological processes within the body. This ability is important because anatomical or morphological changes due to pathological conditions are often preceded by physiological or biochemical alterations, and to obtain such physiological or biochemical information one must be able to assess function. By using radiolabelled compounds, it is possible to study various biological functions such as blood flow, glucose metabolism, the functional state of an organ, the fate of a drug and neurotransmitter receptor binding. In order to accomplish these tasks, an appropriate radiopharmaceutical must be employed, along with a suitable imaging system. This section will highlight some aspects of radiopharmaceutical chemistry, including design rationale.

A radiopharmaceutical is a radioactive compound (biomolecule or drug) of a suitable quality to be safely administered to humans for the purpose of diagnosis, therapy or research (see Section 5.4.8 for details on QC of radiopharmaceuticals). A radiopharmaceutical is usually made up of two components, the radionuclide and the pharmaceutical compound which confers the desired biodistribution on the radiopharmaceutical. However, some radiopharmaceuticals have no associated pharmaceutical component (^{133}Xe gas, for instance), whilst in others, the pharmaceutical component is a simple counter ion (e.g. sodium iodide, NaI, or thallous chloride, TlCl).

The development of radiopharmaceuticals has followed various strategies (Burns 1978, Heindel et al. 1981, Boudreau and Efange 1992, Schibli and Schubiger 2002, Wong and Pomper 2003, Aloj and Morelli 2004) and the majority of radiopharmaceutical development is based on the design of agents that will successfully incorporate readily available inexpensive radionuclides with good imaging characteristics.

As with the selection of suitable radionuclides, there are certain properties of radiopharmaceuticals (or the pharmaceutical component thereof) that are either desirable, or essential, for application as clinical imaging agents.

Since radiopharmaceuticals are often used to study organ function, it is important that they impart no effect on the system under study, through pharmacological or toxicological effects. In most cases, the concentration of radiopharmaceutical is subpharmacological, usually between micromolar (10^{-6} M) and nanomolar (10^{-9} M), and at this concentration imparts no measurable effect. The radiopharmaceutical should also demonstrate a high uptake in target tissue compared with non-target tissue (specificity) and have an effective half-life appropriate to the duration of the study. From a practical viewpoint, the radiopharmaceutical should be easily synthesised or labelled, and should have a sufficiently long shelf life, both before and after radiolabelling to make it cost effective. Finally, the radiopharmaceutical should be of the necessary pharmaceutical quality once labelled (see Section 5.4.8).

Reviewing the radionuclides discussed in Section 5.4.4, it can be seen that the majority of clinically useful radionuclides are either metals or halogens.

Technetium is a group VIIB transition metal, which can exist in a variety of oxidation states, from +7 to 0, and possibly even –1. $^{99}Tc^{m}$ is eluted from the $^{99}Mo/^{99}Tc^{m}$ generator (Section 5.4.2) in the form of sodium pertechnetate, $Na^{+}TcO^{-}_{4}$ (Tc(VII)). Technetium in this +7 oxidation state is very stable and will not react with pharmaceutical components unless it is reduced (i.e. the oxidation number lowered). Reduction is usually accomplished by using a large excess of stannous ions (often more than 10^{6} × the concentration of technetium) in the presence of both pertechnetate and the molecule to be labelled:

$$2TcO_{4^-} + 16H^{+} + 3Sn^{2+} \rightleftharpoons 2Tc^{4+} + 3Sn^{4+} + 8H_2O. \tag{5.42}$$

The stannous ions are added in vast excess to ensure that the reduction is performed as completely as possible. Failure to achieve complete reduction of the technetium will result in the radiopharmaceutical preparation being contaminated with unreduced technetium in the +7 oxidation state, as pertechnetate. This is one of the reasons why it is important to perform QC assessments on technetium radiopharmaceuticals.

As with other transition metals, the chemistry of technetium is dominated by its ability to form coordination complexes with compounds containing *donor atoms*. Such complexes involve the formation of coordinate covalent bonds, in which both electrons in the bond are donated by an electron-rich donor atom such as nitrogen, oxygen, phosphorus or sulphur. Extensive reviews of technetium coordination chemistry may be found in Srivastava and Richards (1983), Deutsch et al. (1983), Clarke and Podbielski (1987), Jurisson et al. (1993), Banerjee et al. (2001) and Arano (2002).

The oxidation states which dominate the chemistry of technetium are the +5 (e.g. Tc(V) in $^{99}Tc^{m}$-MAG3 and $^{99}Tc^{m}$-1,1-ECD), +3 (e.g. Tc(III) in $^{99}Tc^{m}$-IDAs) and +1 (e.g. Tc(I) in $^{99}Tc^{m}$-sestamibi) states (Owunwanne et al. 1998). Whilst technetium can exist in the form of simple complexes, in which the metal atom is bound to a single donor atom on a ligand, such as in $^{99}Tc^m$-sestamibi, maximum complex stability is achieved via chelation (Cotton and Wilkinson 1980). A *chelate* is formed when a metal atom is bound to more than one donor atom of a complexing molecule or ligand, and generally confers high stability on the resulting complex. Numerous technetium complexes have been developed that demonstrate accumulation in various organs, and these and other agents will be discussed further in Section 5.4.6.

In general, the *in vivo* distribution of chelate complexes is governed by their characteristic physicochemical properties (including molecular weight and lipid solubility). However, the low concentration of radiopharmaceutical in typical preparations ($\sim 10^{-9}$M) makes it extremely difficult, if not impossible, to definitively determine the structure of a technetium radiopharmaceutical. In addition, the large excess of tin present in the preparation may lead to inconsistent incorporation of stannous ions in the radiopharmaceuticals. Indeed, there is evidence to suggest that within a single preparation, there may be several different chemical forms of the same technetium radiopharmaceutical, which have different biological behaviours (Tanabe et al. 1983). The net result is that, despite being used very widely, the detailed structures of many technetium radiopharmaceuticals are still open to some debate.

An important characteristic of some technetium chelates is the 'essential' nature of the technetium atom in determining their biodistribution. Burns et al. (1978) divided technetium radiopharmaceuticals into two categories – *technetium-tagged* and *technetium-essential* radiopharmaceuticals. The latter class includes those technetium complexes whose biodistribution is dependent upon the chemical properties of the final complex, rather than the properties of the uncomplexed ligand. In such cases, the biological behaviour of the metal chelate will be governed by a combination of the stability of the complex, the final charge of the chelate complex and the types of chemical group interacting with the surrounding environment.

An example of a technetium-essential radiopharmaceutical is $^{99}Tc^m$-HMPAO, which crosses the blood–brain barrier and localises in brain tissue in proportion to blood flow (Neirinckx et al. 1987). However, the unlabelled ligand (HMPAO) does not cross the blood–brain barrier (McKenzie et al. 1985). In contrast, technetium-tagged radiopharmaceuticals often show only minor alterations of biological behaviour after the addition of the technetium atom. Most of the materials in this category are high-molecular-weight substances such as colloids, particles, proteins and cells (Burns et al. 1978), and the addition of the relatively small $^{99}Tc^m$ atom to these larger substances results in minimal alteration. However, the procedures used for labelling may be harsh (i.e. extremes of pH, the presence of oxidising or reducing agents) and can adversely affect the substance to be labelled.

Similarly, some biological molecules, such as peptides and proteins, have critical parts of their structure that may be responsible for the specific nature of their biochemistry. One such example is the antibody molecule or immunoglobulin, which is comprised of amino-acid chains held together by disulphide bonds (see Figure 5.47). There are areas known as hypervariable regions, in which the structural sequences of the chain varies between antibodies, which gives the antibody its unique ability to identify, and bind to, a particular antigen or foreign substance (Ritter 1995). Radiolabelling procedures and/or the addition of a chelate-group to the hypervariable region can affect the ability of the antibody to bind to the antigen for which it is specific (Paik et al. 1983).

Ligands that are to be attached to other carrier molecules, such as proteins or peptides, are known as *bifunctional ligands*. These are ligands that have one or more reactive groups

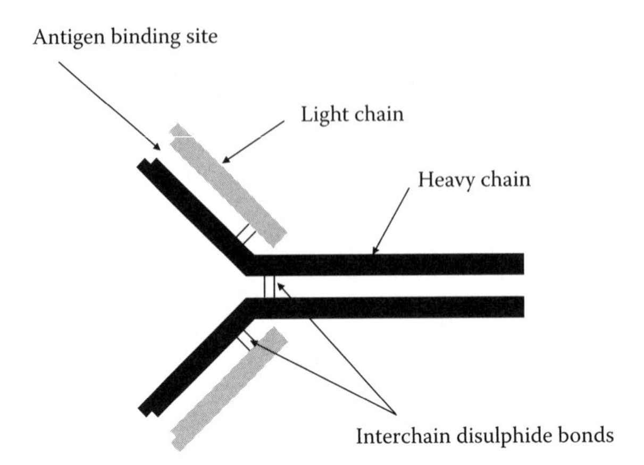

FIGURE 5.47
Generalised structure of an IgG immunoglobulin molecule of molecular weight ~150 kDa.

capable of being attached to the carrier molecule, leaving the correct arrangement of donor atoms to bind the radiometal. The chemical conjugation of bifunctional chelating agents such as diethylenetriaminepentaacetic acid (DTPA), desferrioxamine (DFO) and ligands such as 1,4,7-triazacyclononane-1,4,7-triacetic acid (NOTA) and 1,4,7,10-tetraazacyclododecane-N,N′,N″,N‴-tetraacetic acid (DOTA) to peptides and proteins is now well established (Roselli et al. 1991, Lewis et al. 1994, Meijs et al. 1996, Lee et al. 1997, Liu and Edwards 2001, Chappell et al. 2003, Banerjee et al. 2005). While most of the bifunctional chelate work has involved labelling with radionuclides of indium and gallium, this methodology has also been used to label proteins with technetium (Arano et al. 1987, Purohit et al. 2004). However, it is also possible to use an alternative method for labelling of proteins in general, and immunoglobulins in particular, with technetium. This so-called direct-labelling methodology does not use an external ligand, but simply relies on the affinity of reduced technetium for donor atoms on the surface of the protein (Mather and Ellison 1990). However, the same groups required for technetium binding are often essential for maintaining the 3D shape of the immunoglobulin, and alteration of some of these groups may result in a reduction in the ability of the antibody to bind to an antigen.

Bifunctional chelates have also been used to label biomolecules, such as fatty acids, which show little or no native affinity for technetium (Eckelman et al. 1975). Such molecules are often fairly low molecular-weight compounds, and the addition of a bifunctional chelate may sufficiently alter the structure of the molecule that it is unable to perform its biochemical function. Such changes may be due to factors including the increase in molecular weight (or steric bulk) or altered lipid solubility or polarity, although it may be expected that the overall effect would be due to a combination of factors.

Whilst many technetium chelates have been prepared and are widely used in nuclear medicine, these agents, as yet, do not lend themselves to the investigation of more intricate biochemical processes, such as glucose or fatty-acid metabolism, steroid and protein synthesis and neurotransmitter receptor studies. This is largely due to the problems associated with incorporating $^{99}Tc^{m}$ and/or a chelating group into the required tracer without significant structural, electronic or hydrophobic alterations (Eckelman and Volkert 1982).

An attractive alternative for labelling such biochemical tracers is to use radionuclides of the *halogens*. The coordination number of the common halogens (fluorine, chlorine, bromine and iodine) is 1, in the −1 oxidation state. As the halogens (group VIIA) need only one electron to reach the noble-gas configuration, they easily form the −1 anion (X^-) or a single covalent bond (Cotton and Wilkinson 1980). This latter property is a distinct advantage over technetium when labelling low-molecular- weight biologically active molecules. While extensive research effort has gone into the radiochemistry of the halogens (see Palmer and Taylor 1986), this section will use iodine to illustrate many common properties of the halogens (Coenen et al. 2006).

The radionuclides of *iodine* (Tables 5.5 through 5.7) are of considerable interest in nuclear medicine due to the range of physical half-lives and decay modes available. The ability to use essentially identical labelling methodologies to incorporate any radionuclide of iodine into a radiopharmaceutical allows the radiopharmaceutical to be tailored to particular applications. The iodine atom occupies a similar volume to that of a methyl ($-CH_3$) or ethyl ($-C_2H_5$) group and the electro-negativity of iodine is similar to that of carbon. Therefore, iodine can be introduced into an organic molecule without changing the molecule's polar or steric configuration excessively (Coenen et al. 1983). Whilst the addition of a halogen is not without effect, halogenation of pharmacologically active or biochemically active molecules is an accepted method of 'fine-tuning' *in vivo* behaviour or potency. Furthermore, in contrast to technetium radiopharmaceuticals, iodine-containing molecules can be synthesised on a macroscopic scale, allowing for chemical, pharmacological and toxicological evaluation.

Iodine radiopharmaceuticals may be produced by either direct or indirect iodination methods (Baldwin 1986). The *direct iodination* method involves the generation of I^+, followed by exchange of the iodine with a group in the molecule to be iodinated. However, this mechanism of iodination often involves the production of weak carbon–iodine bonds, which can lead to extensive deiodination of the radiopharmaceutical, particularly *in vivo*. The *indirect iodination* method involves conjugation of an iodinated compound to the molecule of interest, or less commonly, the process of recoil labelling. These processes will be outlined in the following paragraphs.

Direct iodination of molecules may be accomplished using an oxidising agent capable of producing I^+, such as chloramine-T, iodogen and iodobeads, which will all give iodination efficiencies in excess of 80% under most conditions, although this may be at the expense of some loss in biological activity. In contrast, alternative methods of oxidising iodine, including electrolytic and enzymatic methods, give lower overall iodination efficiencies, but in many cases, biological molecules retain more of their native biological activity using such methods.

It is also possible to iodinate molecules that do not have any moieties capable of being directly iodinated, using one of the many conjugation reagents available which may subsequently be coupled to the molecule in question (Wilbur 1992). This methodology is often used when iodinating peptides or proteins in which the amino acid residues that would be used for incorporation of the iodine are vital for retention of the biological activity of the molecule.

Recoil iodination involves the reaction between positively charged iodine species (produced as a result of the decay of the corresponding xenon radionuclide by electron capture) with the compound to be iodinated. The positively charged iodine species is produced from the xenon atom as a result of Auger-electron emission following electron capture decay processes (see Section 5.4.3). This method has been used on a commercial scale for the production of ^{123}I IBZM, a radiopharmaceutical that targets dopamine D2 receptors (Cygne 1993).

Biological molecules, such as proteins or peptides, may be iodinated either through the use of prelabelled reagents or through the direct iodination of amino acids. It is possible to iodinate a number of different amino acids, including tyrosine, histidine, phenylalanine, tryptophan and cysteine.

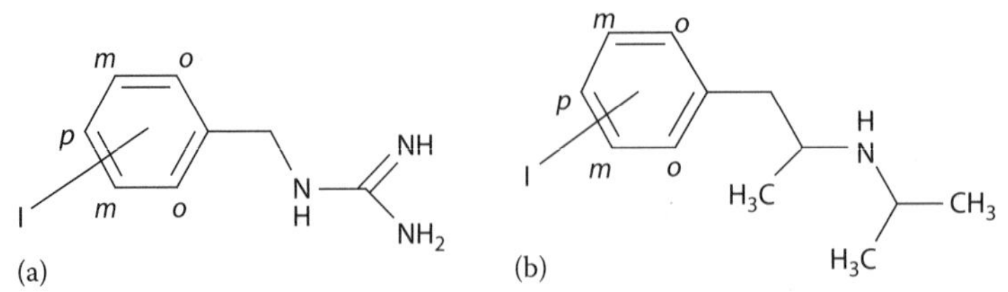

FIGURE 5.48
Molecular structures of (a) iodobenzylguanidine and (b) iodopropylamphetamine (*o*, *ortho*; *m*, *meta*; *p*, *para* position on aromatic rings).

When selecting an iodination method, it must be remembered that the position of the iodine atom often has an effect on the chemical and biological behaviour of the radiotracer (Coates 1981). Two iodinated compounds used in the clinical setting, *m*-iodobenzylguanidine (*m*IBG) and *p*-iodoamphetamine (Figure 5.48) are such examples. Winchell et al. (1980) demonstrated that iodoamphetamine compounds iodinated in the *para* position had the highest brain uptake and brain-to-blood activity ratios of the isomers studied. Similarly, Wieland et al. (1980) demonstrated that the *para* and *meta* isomers of iodobenzylguanidine exhibit greater uptake in the target tissue (adrenal medulla) than the *ortho* isomer, and that the *meta* isomer is more resistant to *in vivo* dehalogenation than the *para* or *ortho* isomers.

Indium is a group IIIA metal whose solution chemistry is dominated by the +3 oxidation state. Indium (and gallium) often mimics the behaviour of iron in a biological system, due to the similarities in their hydrated ionic radii, coordination numbers and their preference for donor atoms such as oxygen (Hider and Hall 1991). As with technetium, indium can form complexes with single donor atoms, or with several donor atoms, from a ligand to form chelate structures. The major difference between indium and technetium lies in the preference of technetium for the softer donor atoms, such as sulphur and phosphorus, whereas indium has a preference for hard donor atoms, such as oxygen and nitrogen. However, there is an overlap in selectivity, and certain ligands, such as DTPA, will bind both technetium and indium.

As with technetium, there is a wide range of small molecule ligands for indium, and a range of bifunctional ligands for conjugation of indium to peptides or proteins. However, in contrast to technetium, there is no equivalent of the direct-labelling approach for incorporating indium into proteins without the use of bifunctional chelates.

In order to demonstrate the differences that radiolabelling with metallic radionuclides makes in comparison with radiohalogens, the specific example of the peptide radiopharmaceutical, *octreotide*, will be considered. Octreotide is a synthetic peptide analogue of the neuropeptide somatostatin, which inhibits the secretion of growth hormone and is used to inhibit the growth of various types of tumour. Somatostatin is inactivated by enzymes in the plasma, with a biological half-life of less than 2 min and, as a result, a range of synthetic analogues have been synthesised which retain the receptor-binding properties of somatostatin, but which have significantly longer biological half-lives. Octreotide is arguably the most successful of the analogues, with a plasma half-life of approximately 90 min, but unfortunately there is no site for radiolabelling. Octreotide derivatives have been produced which contain sites for radiolabelling with iodine ([Tyr^3]-octreotide), technetium ([CpTT-D-Phe^1]-octreotide), indium ([DTPA-D-Phe^1]-octreotide) and yttrium ([DOTA-D-Phe^1-Tyr^3]-octreotide) (Paganelli et al. 1999). Despite the fact that each of these derivatives still recognises the somatostatin receptor (i.e. they all retain biological activity) the pharmacokinetics of each is different, with different tissue distribution and clearance profiles. For example, the iodinated derivative, ^{123}I [Tyr^3]-octreotide, shows a very rapid hepatic clearance of the radiopharmaceutical, with 28% of the injected dose present in the liver 4 h after administration (Bakker et al. 1991). Whilst hepatic clearance results in intact radiolabelled peptide being excreted, peptide remaining in the circulation is deiodinated *in vivo*, with subsequent excretion of free iodide via the kidneys. This metabolic profile, particularly the high hepatic uptake, makes it difficult to visualise sites of uptake in the abdomen.

As a result of these difficulties, an indium-labelled octreotide derivative was developed ([DTPA-D-Phe^1]-octreotide), which demonstrated a completely different pharmacokinetic profile to the iodinated derivative (Krenning et al. 1993). Following administration, rapid renal excretion is observed, although the rate of plasma clearance is rather slower than with the iodinated derivative. The hepatic clearance profile is also significantly different, with less than 2% of the injected dose being present in the liver 4 h after administration. This change in biodistribution means that it is significantly easier to visualise sites of uptake in the abdomen with the indium-labelled derivative than with the iodinated derivative.

Octreotide has also been labelled with technetium, using both the direct-labelling approach (Kolan et al. 1996) and the indirect-labelling approach (Maina et al. 1994, Spradau et al. 1999). Use of the direct-labelling technique resulted in almost total loss of specific receptor-mediated radiopharmaceutical uptake, indicating the importance of the disulphide bond in maintaining the structure of the peptide. In contrast, when using the indirect-labelling methodology with a bifunctional chelate, the octreotide derivative retained much of its biological activity (i.e. the ability to bind to somatostatin receptors) but, as with ^{123}I [Tyr^3]-octreotide, displayed a very high hepatic uptake and retention. From these examples, it is clear that differing labelling methodologies can affect the biological activity to radically different extents, and that even using a similar type of labelling technique (bifunctional chelate), major differences in biodistribution can be observed (cf. [DTPA-D-Phe^1]-octreotide and ^{99}Tc^m [CpTT-D-Phe^1]-octreotide).

The aforementioned comparison of radiopharmaceuticals incorporating metallic and halogen radionuclides has been intended to indicate how the use of different radionuclides and/or labelling methodologies allows radiopharmaceuticals to be tailored to the study of particular physiological functions, and how biological properties may be totally altered by changing the radiolabelling methodology.

5.4.6 Biological Distribution of Radiopharmaceuticals

A number of chemical factors are responsible for the mechanism by which a radiopharmaceutical is distributed within a living system (Jambhekar 1995). These factors include, but are not limited to, the 3D structure of the molecule, the size of the molecule (which correlates with the molecular weight of the compound) and the nature of the functional groups incorporated into the structure.

Given that the majority of radiopharmaceuticals are administered intravenously, distribution and elimination are the two major mechanisms determining their *in vivo* biodistribution (see Figure 5.49). Absorption properties are important only for the limited number of radiopharmaceuticals administered orally, intradermally, intrathecally or by inhalation. Distribution and elimination are both affected by the same parameters, namely blood flow, capillary permeability and interaction with both intracellular and extracellular components.

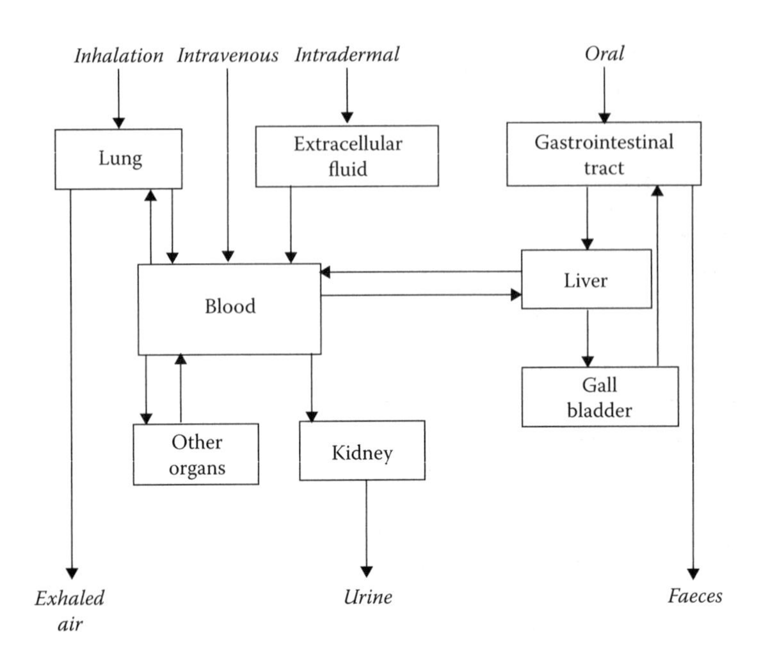

FIGURE 5.49
Routes of administration, distribution and excretion of radiopharmaceuticals.

In order for radiopharmaceuticals to localise at target sites, they need to traverse one or more biological membranes. The ability of a radiopharmaceutical to effectively diffuse across biological membranes is related to its hydrophobicity, or lipid solubility. Cell membranes are primarily composed of phospholipid bilayers, which are assembled with water-soluble phosphate groups exposed on the outer surfaces of the membrane, whilst the lipid portion of the membrane is enclosed between these two water-soluble surfaces. In order for a drug to cross a membrane, it must possess a hydrophobicity such that it is soluble in the lipid portion of the membrane. The hydrophobicity of a compound is controlled in part by its pK_a (which controls the extent to which the molecule is ionised, or electrically charged, at physiological pH (pH = 7.4)). The mechanisms controlling membrane transport, on which the mechanism of uptake of any radiopharmaceutical is dependent, include both passive transport systems (encompassing simple diffusion and filtration) and more specialised transport systems (active transport, facilitated transport and phagocytosis). Passive membrane transport requires no energy expenditure, and simply involves diffusion of the radiopharmaceutical down an electrochemical (concentration) gradient, either through aqueous membrane channels in the case of hydrophilic (or ionised) radiopharmaceuticals, or directly across the membrane lipid bilayer in the case of more hydrophobic radiopharmaceuticals. The more specialised transport systems often either involve energy expenditure to accumulate radiopharmaceuticals against a concentration gradient (active transport) or utilise a non-energy-dependent carrier-mediated transport system to move radiopharmaceuticals down a concentration gradient.

The distribution of the radiopharmaceutical is also governed by such factors as blood flow and the drug's availability to tissue, and also depends on the proportion of the tracer that is bound to proteins in the blood. Once the radiopharmaceutical has been delivered to the cells at the target site, the final distribution depends on continued retention in those cells. Such retention is entirely dependent upon cellular interaction, either at the cell surface or, more commonly, some form of intracellular interaction following membrane transport. Such interaction may involve biochemical processing of the radiopharmaceutical into a product that is unable to leave the cell, by trapping the radiopharmaceutical either physically or biochemically or by physical adsorption onto the surface of intracellular components.

Elimination of intravenously administered radiopharmaceuticals occurs primarily via either the urine or the bile. Renal excretion can occur via passive glomerular filtration, tubular diffusion or energy-requiring tubular secretion, whilst biliary excretion follows three major pathways dependent on whether the molecule in question is positively charged, electrically neutral or negatively charged. It is still not entirely clear as to which factors determine whether a particular radiopharmaceutical will follow a renal or biliary clearance pathway. However, it is known that hydrophilic compounds with a molecular weight below about 325 tend to follow a urinary excretion pattern, whilst more lipophilic and higher molecular-weight compounds tend to follow a biliary excretion pathway. However, in practice, a combination of urinary and biliary excretion is often observed, given that the parent compound and its metabolites may well behave differently.

In order to design a radiopharmaceutical rationally with a specific mechanism of localisation, one must consider all the aforementioned factors, as well as selecting a radionuclide whose incorporation does not change the properties of the pharmaceutical component.

5.4.7 Radiopharmaceuticals Using Single-Photon Emitters

Whilst the biological fate of most radiopharmaceuticals is controlled by the pharmaceutical component, the properties of the radiopharmaceutical are dependent to a large extent on the nature of the radionuclide incorporated. It is possible to classify radionuclides into three broad categories when considering radiopharmaceutical behaviour, namely biologically relevant radionuclides, halogen radionuclides and metallic radionuclides.

Biologically relevant radionuclides are generally considered to be nuclides of elements which occur naturally in biological molecules, including hydrogen, carbon, nitrogen and oxygen. Of these, hydrogen is the element that does not have a radionuclide of use in medical imaging, although fluorine, by virtue of its similarity in size to the hydrogen atom, can often mimic hydrogen without significantly affecting the behaviour of biological molecules. The most useful radionuclide of each of these elements is a positron-emitting radionuclide (^{11}C, ^{13}N, ^{15}O and ^{18}F), and these elements will, therefore, be discussed in more detail in Section 5.9.

The *halogens* form a single group in the periodic table, and the chemistry of each of the elements (including fluorine, chlorine, bromine, iodine and astatine) in the group is similar. Each of the elements, with the exception of astatine, has radionuclides that are of potential interest in medical imaging, although astatine (as the α-emitter ^{211}At) is of interest for targeted radionuclide therapy applications. Fluorine, by virtue of its similarity to hydrogen, is often considered to be a pseudo-biologically relevant radionuclide rather than with the rest of the halogens.

With few exceptions, the rest of the radionuclides which find use in medical imaging applications are *metallic radionuclides*, with many of them being classifiable as either true transition metals, or having chemical behaviour allowing them to be thought of as pseudo-transition metals. For this reason, many metallic radionuclides share common features with respect to the nature of the interaction of the radionuclide with the pharmaceutical component of the radiopharmaceutical.

Details of some common radiopharmaceuticals used for single-photon imaging applications are given in Table 5.10. This table lists the radiopharmaceuticals in order of

TABLE 5.10

Examples of radiopharmaceuticals used for planar and SPECT applications.

Radionuclide	Pharmaceutical	Indication/Use	Typical Administered Activity (MBq)
^{67}Ga	Citrate	Tumour imaging, infection/inflammation imaging	150[a]
^{81}Kr^{m}	Krypton gas	Lung ventilation imaging	6000[a]
^{99}Tc^{m}	Albumin	Cardiac blood-pool imaging, peripheral vascular imaging	800[a]
^{99}Tc^{m}	Colloids, including tin colloid and sulphur colloid	Oesophageal transit and reflux	40[a]
		Liver imaging	80[a], 200 (SPECT)[a]
		Bone marrow imaging, GI bleeding	400[a]
^{99}Tc^{m}	DTPA	Lung ventilation imaging (aerosol)	80[a]
		Renal imaging/renography	300[a]
		Brain imaging (static)	500[a], 800 (SPECT)[a]
		First-pass blood-flow studies	800[a]
^{99}Tc^{m}	DMSA	Renal imaging (DMSA(III))	80[a]
		Tumour imaging (DMSA(V))	400[a]
^{99}Tc^{m}	ECD	Brain imaging	500[a]
^{99}Tc^{m}	Erythrocytes (normal)	GI bleeding	400[a]
		Cardiac blood-pool imaging or peripheral vascular imaging	800[a]
^{99}Tc^{m}	Erythrocytes (heat denatured)	Spleen imaging	100[a]
^{99}Tc^{m}	Exametazime	Cerebral blood-flow imaging (SPECT)	500[a]
^{99}Tc^{m}	Iminodiacetates (IDAs)	Functional biliary system imaging	150[a]
^{99}Tc^{m}	Leucocytes	Infection/inflammation imaging	200[a]
^{99}Tc^{m}	Macroaggregated albumin	Lung perfusion imaging	100[a], 200 (SPECT)[a]
^{99}Tc^{m}	MAG3	Renal imaging/renography	100[a]
		First-pass blood-flow imaging	200[a]
^{99}Tc^{m}	Nanocolloids	Lacrimal drainage	4[a]
		Sentinel node or lymph node imaging	20[a]
^{99}Tc^{m}	Pertechnetate	Micturating cystogram	25[a]
		Thyroid uptake	40[a]
		Thyroid imaging, salivary gland imaging	80[a]
		Ectopic gastric mucosa imaging (Meckel's)	400[a]
		First-pass blood-flow imaging	800[a]
^{99}Tc^{m}	Phosphonate and phosphate compounds	Bone imaging	600[a], 800 (SPECT)[a]
^{99}Tc^{m}	Sestamibi	Myocardial imaging	300[a], 400 (SPECT)[a]
		Tumour imaging, breast imaging	900[a]
^{99}Tc^{m}	Sulesomab	Infection/inflammation imaging	750[a]
^{99}Tc^{m}	Technegas	Lung ventilation imaging	40[a]
^{99}Tc^{m}	Tetrofosmin	Myocardial imaging	300[a], 400 (SPECT)[a]
		Parathyroid imaging	900[a]
^{111}In	Capromab Pendetide	Biopsy-proven prostate carcinoma imaging	185[b]
^{111}In	DTPA	GI transit	10[a]
		Cisternography	30[a]

TABLE 5.10 (continued)

Examples of radiopharmaceuticals used for planar and SPECT applications.

Radionuclide	Pharmaceutical	Indication/Use	Typical Administered Activity (MBq)
^{111}In	Leucocytes	Infection/inflammation imaging	20[a]
^{111}In	Pentetreotide	Somatostatin receptor imaging	110[a], 220 (SPECT)[a]
^{111}In	Platelets	Thrombus imaging	20[a]
^{123}I	Iodide	Thyroid uptake	2[a]
		Thyroid imaging	20[a]
		Thyroid metastases imaging	400[a]
^{123}I	Ioflupane	Striatal dopamine transporter visualisation	185[a]
^{123}I	*m*IBG	Neuroectodermal tumour imaging	400[a]
^{131}I	*m*IBG	Neuroectodermal tumour imaging	20[a]
^{131}I	Iodide	Thyroid uptake	0.2[a]
		Thyroid metastases imaging	400[a]
^{133}Xe	Xenon gas	Lung ventilation studies	400[a]
^{201}Tl	Thallous chloride	Myocardial imaging	80–120[a]
		Parathyroid imaging	80[a]
		Tumour imaging	150[a]

Source: Compiled from two sources.

[a] *Notes for Guidance on the Clinical Administration of Radiopharmaceuticals and Use of Sealed Radioactive Sources*, Administration of Radioactive Substances Advisory Committee, March 2006, revised 2006, 2007 (twice) and 2011 (London, UK: Health Protection Agency).

[b] Prostascint (Capromab pendetide) product information leaflet, EUSA Pharma (USA) Ltd., Langhorne, PA.

increasing atomic number of the radionuclide and, in addition to their clinical indications or use, typical administered activities are quoted. The size of the table reflects the large number of radiopharmaceuticals that is available for planar scintigraphy and SPECT. In the United Kingdom, the administered activities for all nuclear medicine studies are recommended by ARSAC. For more information about internal dosimetry and radiation safety for clinical nuclear medicine, the reader is referred to other textbooks, for example, Cormack et al. (2004).

5.4.8 Quality Control of Radiopharmaceuticals

Previous sections have highlighted some of the constraints placed on the development of radiopharmaceuticals. While the development of clinically useful radiotracers is not accomplished without great effort, a variety of tracers exist that make nuclear medicine imaging a viable diagnostic modality. In order to guarantee that the radiopharmaceutical formulation is safe and will yield diagnostically useful information, a number of QC measures are required. These processes will help to determine the biological and radiopharmaceutical purity of the product.

With the majority of commercially available radiopharmaceuticals, such QC measures are a legal requirement, with the various specifications being laid down in the relevant pharmacopoeia. The three most widely quoted pharmacopoeias are the British Pharmacopoeia (BP), the European Pharmacopoeia (EP) and the United States Pharmacopoeia (USP). Each pharmacopoeia contains both general monographs, giving outline information about the type of tests which should be performed on radiopharmaceuticals and advice on the general test methods to be employed, as well as specific monographs for individual radiopharmaceuticals, which give details of specific tests and limits which must be achieved if the material is to be used in a clinical setting. Whilst it is not essential to perform the required QC tests using the methods specified in the pharmacopoeia, any alternative method used must be demonstrably at least as good as the pharmacopoeial test method. It should be appreciated that the pharmacopoeia is a legally enforceable document in most countries, and that there is a legal obligation to ensure that all radiopharmaceuticals comply with the relevant monographs.

Biological purity testing is required to assure that the radiopharmaceutical formulation and/or the components that make up the final preparation are safe for human administration. The range of tests which may be required by the pharmacopoeia are those which determine *sterility* (absence of live micro-organisms), *apyrogenicity* (generally taken to indicate the absence of components which may cause a pyrogenic response or fever in patients, but often specifically considered to indicate the absence of cellular debris of micro-organisms) and *toxicity*.

For most radiopharmaceuticals in use today, the constituent components in the final product are produced to Good Manufacturing Practice (GMP) standards, as required by the national licensing authorities, such as the Medicines and Healthcare products Regulatory Agency (MHRA) in the United Kingdom, or the Food and Drug Administration (FDA) in the United States. The manufacturer must assure that the components of any kit produced explicitly for human administration (especially non-oral formulations) are sterile and pyrogen free.

Sterility may be tested by incubating appropriate samples of the component or radiopharmaceutical in approved culture medium (i.e. thioglycollate or soybean-casein digest) for 7–14 days at a specific temperature (see relevant pharmacopoeia for further details). If any growth occurs, the product fails sterility testing and cannot be used. Radiopharmaceuticals incorporating short-lived isotopes cannot be tested in this manner, due to the long incubation period required for confirmation of sterility, and pharmacopoeial specifications, therefore, allow these materials to be used before the results are obtained.

Apyrogenicity may be tested by injecting three rabbits with the test material and then monitoring the rectal temperature of the animals over a 3 h period. If the total temperature rise in all three animals is less than 1.4°C, and no single animal shows a temperature rise greater than 0.6°C, then the material is considered to be nonpyrogenic. If one or all of the animals demonstrate temperature increases greater than that stated, the sample must be tested with five more rabbits. A temperature rise of 0.6°C in more than three of the eight animals, or a total temperature rise of more than 3.7°C in all eight animals, is taken to indicate the presence of pyrogens. This is still the assay for pyrogenic compounds that is required by the British Pharmacopoeia, although few of the included radiopharmaceuticals, with the exception of $^{99}Tc^{m}$ labelled macro-aggregated albumin, are required to have this test performed. However, it is recognised that a pyrogenic response may be caused by the presence of radioactivity in the preparation, and that it may be necessary to allow the activity to decay before performing the test. In addition, this test is not considered sensitive enough for radiopharmaceuticals that are to be administered intrathecally, that is, directly into the cerebrospinal fluid. In such cases, a test that is specific for bacterial endotoxins (breakdown products of bacterial cell walls) is recommended.

The limulus amoebocyte lysate (LAL) assay provides a sensitive and rapid method for the detection of *endotoxins*. The material that is responsible for the detection of endotoxins using this method is the lysate of amoebocytes from the blood of the horseshoe crab

TABLE 5.11

British Pharmacopoeia (2012) radionuclidic purity specifications for $^{99}Tc^m$ sodium pertechnetate produced from fission-derived ^{99}Mo and from non-fission-derived ^{99}Mo.

Maximum Permissible Radionuclidic Impurity Levels in $^{99}Tc^m$ Sodium Pertechnetate (Fission)		Maximum Permissible Radionuclidic Impurity Levels in $^{99}Tc^m$ Sodium Pertechnetate (Non-Fission)	
^{99}Mo	1×10^{-1} %	^{99}Mo	1×10^{-1} %
^{131}I	5×10^{-3} %	All other radionuclides	1×10^{-2} %
^{103}Ru	5×10^{-3} %		
^{89}Sr	6×10^{-5} %		
^{90}Sr	6×10^{-6} %		
All α-emitters	1×10^{-7} %		

(*Limulus polyphemus*). A number of test methods have been developed using this approach, using gel-clot, turbidimetric or chromogenic assays. The choice of test method depends largely upon the type of material to be tested, its potential to interfere with the assay, and the required lower limit of endotoxin detection.

Radiopharmaceutical purity can be divided into three areas: *radionuclidic purity, radiochemical purity* and *chemical purity.*

No sample of radionuclide is completely free of any other radioactive components, the amount and type of which will depend on the production method used for the desired radionuclide. Whilst it is essentially impossible to remove all of the contaminating species, safeguards must be applied to ensure that the contaminants fall within specified limits, as determined by the radionuclidic purity of the sample in question. *Radionuclidic purity* refers to the percentage of the total radioactivity of a sample that is in the desired radionuclidic form (Table 5.11). To assess radionuclidic purity, γ-ray spectroscopy using a multichannel analyser is most often used. As each radionuclide has an emission spectrum that is unique, spectral data can allow for the determination of impurities. Other means of impurity detection include half-life determination and γ-ray filtration using attenuators known to block certain energies with high efficiency. This approach is commonly adopted in the so-called molybdenum breakthrough assay (a technique commonly used for assessment of ^{99}Mo contamination in $^{99}Tc^m$ samples produced from a $^{99}Mo/^{99}Tc^m$ generator). In this assay, a 6 mm thick lead shield is used as a selective filter to absorb virtually all of the 140 keV γ photons emitted by the $^{99}Tc^m$, whilst absorbing only 50% of the higher-energy 740 keV γ photons emitted by ^{99}Mo. It is then possible to use this information to calculate the total level of ^{99}Mo contamination with a sample of $^{99}Tc^m$. Potential consequences of radionuclidic impurities include increased radiation dose to the patient and image degradation from collimator penetration and scattered photons. Many pharmacopoeial monographs use generic specifications which simply state that the γ-ray spectrum should not differ significantly from a standardised source of the same radionuclide, whilst other monographs set specific limits for individual radionuclidic impurities. This is particularly true of radionuclides that have been prepared from fission products, where there may be expected to be a more significant range of contaminating radionuclides. The pharmacopoeial limits for contaminating radionuclides are often set on the basis of the radiological hazard that such radionuclides pose to the patient.

Radiochemical purity refers to the percentage of total activity in a sample that is present in the desired chemical form. Many conventional analytical techniques have been modified to detect the presence of radiochemical compounds, although chromatographic techniques (thin-layer, paper, liquid, high-performance liquid, etc.) are by far the most widely used. These systems separate various components of a sample due to the varying affinity/solubility that these components display for the solid and liquid phases used. Other methods include electrophoresis and precipitation. For further discussion on radiochromatography, see Wieland et al. (1986). The biodistribution and elimination of radiopharmaceuticals, following the pathways discussed earlier, is dependent upon the chemical integrity of the radiopharmaceutical. The presence of multiple chemical forms of the radionuclide will inevitably result in altered biodistribution of the radiopharmaceutical from that expected. For this reason, it is vital that the radiochemical purity of all radiopharmaceuticals is assessed prior to use. As with radionuclidic purity, there are both general and specific pharmacopoeial monographs relating to techniques to be used for radiochemical purity determination, and to minimum radiochemical purity levels for specific radiopharmaceuticals. Typical pharmacopoeial radiochemical purity limits range between 90% and 98% of the total radioactivity being present in the desired chemical form.

Chemical impurities are those compounds that may be inadvertently added to a radiopharmaceutical preparation that (1) are not specified by definition or (2) pose potential formulation or toxicity problems. These impurities are rare with manufacturers' kits, but may be potentially noteworthy when radiopharmaceuticals are prepared from generator eluants. Contamination of the pharmaceutical with breakdown products of the inorganic absorbent of the generator column may result in formulation problems. For example, aluminium (from the aluminium oxide support matrix in a ^{99}Mo/^{99}Tc^{m} generator) is a potential chemical impurity in the eluant from such generators, and can result in altered biodistribution of ^{99}Tc^{m} radiopharmaceuticals. This is particularly true of colloidal radiopharmaceuticals, where increased aluminium concentration can lead to flocculation of the colloidal particles, increasing their particle size, and leading to a significantly different biodistribution profile to that expected. Measurement of aluminium concentration may be readily performed using a colorimetric assay, in which aluminium (as Al^{3+} ions) in a sample of ^{99}Mo/^{99}Tc^{m} generator eluate interact specifically with a compound such as chrome azurol S, to produce a coloured complex. The colour of the complex may then be compared with the colour produced using a standard solution of aluminium ions at the maximum permissible pharmacopoeial concentration. Providing that the intensity of the colour produced with the generator eluate is less than the colour intensity of the standard solution, the radiopharmaceutical is deemed to have passed the test. In addition to measuring aluminium content of ^{99}Tc^{m} sodium pertechnetate obtained from a generator, there are requirements for measuring the tin content of many commercial kits for preparation of ^{99}Tc^{m} radiopharmaceuticals. As mentioned in Section 5.4.5, tin is added in many chemical forms to kits for reconstitution with sodium pertechnetate, to ensure that the technetium is reduced to the correct oxidation state to interact with the pharmaceutical component. The pharmacopoeial assay for tin is a colorimetric assay, using a standard tin solution for comparison.

In addition to biological and radiopharmaceutical purity, there may be additional pharmacopoeial requirements for *other QC tests* to be performed. Such tests may include measurement of pH, particle-size distribution, quantification of non-filterable radioactivity and specific assays for pharmaceutical components.

As indicated in Section 5.4.5, many radiopharmaceuticals labelled with metallic radionuclides are coordination complexes. The stability of such complexes (i.e. the strength of interaction of the metallic radionuclide with the pharmaceutical component) may be critically dependent upon the pH of the medium. It is, therefore, vital that the pH of radiopharmaceuticals is measured prior to administration, to ensure that the stability of the radiopharmaceutical will not be compromised, which would be likely to lead to significantly altered biodistribution. Pharmacopoeial monographs often suggest the use of a glass electrode for pH measurement although, in practice, it is sufficient to use narrow range pH paper for pH assessment, only resorting to a glass electrode if the pH appears to be outside the set limits.

Several common radiopharmaceuticals are based on particulate materials of a discrete size range, and these may be broadly classified as colloidal radiopharmaceuticals or particulate radiopharmaceuticals. The biodistribution of these radiopharmaceuticals is critically dependent upon the size distribution of the particles being within certain ranges. Whilst there are no pharmacopoeial requirements to measure particle sizes or size distribution of colloidal radiopharmaceuticals, it is necessary to determine the range of particle sizes present in a particulate radiopharmaceutical preparation. In contrast to many other tests, there are no general pharmacopoeial monographs for particle sizing techniques, with methodological recommendations and limits being specified in individual monographs.

Many individual radiopharmaceuticals also have pharmacopoeial monographs, which call for a specific assay to be performed, to ensure that the identity of the pharmaceutical component is correct. Such assays are disparate in nature, and due to the complexity of many of the suggested techniques, are often beyond the scope of many end users to perform in-house. Such assays are, therefore, often performed by specialist chemical-analysis contract laboratories.

One important aspect of QC testing of radiopharmaceuticals to pharmacopoeial specifications is that the material is not considered to be of pharmacopoeial quality unless it meets the minimum specifications for the duration of its specified shelf life. Pharmacopoeial testing is often performed shortly after reconstitution of radiopharmaceutical kits, but rarely checked throughout the life of the radiopharmaceutical.

5.4.9 Summary

The properties of radiopharmaceuticals are an essential feature of nuclear medicine imaging. The biodistribution of these tracers provide the images with biochemical and physiological information not available in other imaging techniques. Whilst few of these tracers are specific to diseases or even particular organs in the body, their properties allow sensitivity not available with any other technique. The range of pharmaceutical/radioisotope combinations allows nuclear medicine to make an unparalleled range of measurements relating to tissue function. The growth of tracers with greater specificity through the radiolabelling of molecules identified as biologically essential to tissue function will undoubtedly result in radioisotope imaging continuing to play a major role in the diagnosis and treatment of disease.

5.5 Role of Computers in Radioisotope Imaging

With the development of minicomputers in the mid-1960s, it became possible to store large quantities of data and to perform complex data manipulation well beyond the range of previous human experience. However, not until the mid-1970s were computers attached to a gamma camera (Lieberman 1977), partly because of the complexity of interface electronics and partly through lack of suitable computer software. In the 1980s it became possible to have powerful computers on a single microchip and the cost of memory, data storage and input/output facilities was dramatically reduced. In the 1990s with the development of UNIX workstations and PCs, the cost of interfacing a powerful computer system to a gamma camera has become less significant than the cost of the software. Thus, technology has provided a revolution in the application of computers to medical imaging and, in particular, to nuclear medicine (see e.g. Todd-Pokropek 1982, Ell and Gambhir 2004, Cook et al. 2006), where the need for tomographic and dynamic information has become essential. There are multiple applications for computers in radioisotope imaging, including:

- data acquisition;
- online data correction;
- data processing and image reconstruction;
- multimodality image registration;
- image display and manipulation;
- data storage; and
- system control.

We will discuss each of these in the following subsections and/or refer the reader to other chapters where appropriate. In many cases separate computers dedicated to, for example, acquisition, processing and system control are provided in the overall system.

5.5.1 Data Acquisition

In nuclear medicine, data can be acquired for planar (2D) scintigraphy, whole-body 2D/3D scanning, dynamic 2D and 3D imaging, and tomography such as SPECT and PET. Data from devices such as the gamma camera or a positron camera can be acquired and stored on a computer in list mode, frame mode or, in the case of PET and SPECT, as 1D/2D projections. In *list mode*, information is recorded as serial spatial and temporal data, which can later be either reordered into dynamic data files (frame mode) or in some special applications directly reconstructed into tomographic images. For example, in a high data-rate acquisition with a gamma camera, such as a first-pass cardiac study, we can store the spatial information (x, y) for each detected event and a timing parameter (t), which can be used in subsequent reordering of the data into a series of time-dependent frames. Another example is the storage of spatial information in a positron-camera system, where the reconstruction process will produce a backprojected image directly without the need for framing the raw data into projections.

In *frame mode* the spatial values (x, y) are stored directly into a 2D matrix (or frame) in the computer using, for a gamma camera, analogue-to-digital converters (ADCs). The analogue information from a gamma camera is digitised in such a way that the storage location for the detected event is determined (Figure 5.50). The data are then stored in the computer memory if a suitable E pulse is available. In the so-called digital gamma camera, each PMT output is digitised and the signals used to allocate the event to the appropriate matrix in the computer. The advantage of this is that local digitisation reduces sensitivity to noise compared to the use of analogue systems.

A single frame or image is usually stored in a matrix where the pixel size (picture element size) is smaller than the desired image resolution (see Chapters 3 and 11 on sampling).

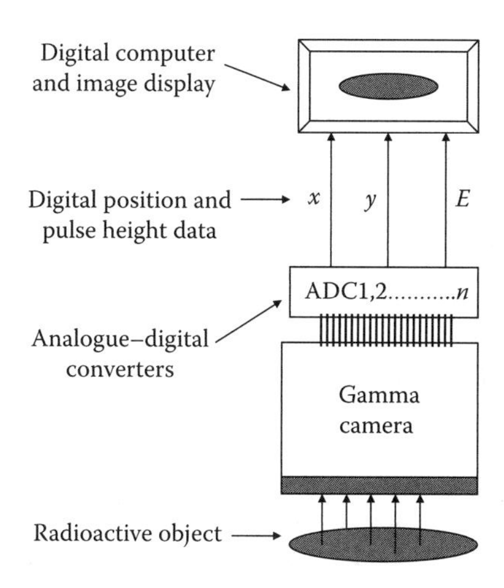

FIGURE 5.50
Schematic of the image production process for a gamma camera using analogue-to-digital converters to read positional (*x*,*y*) and energy (*E*) information into a digital computer.

For instance, if a 40 cm diameter gamma camera is used to image the body, a 64 × 64 matrix will give a linear sampling size of 6.25 mm. This is adequate for data storage from a detector with a spatial resolution of 15 mm (2–3 times this value). For imaging systems with a higher resolution of about 5–6 mm, the image should be stored in a 128 × 128 or 256 × 256 array. For small organs, such as the heart or thyroid, a 32 × 32 matrix may be used if the number of image counts is low. In determining the size of the array, it should be remembered that, for a multi-frame data acquisition, more memory is needed for storage. For example, a typical tomographic acquisition may have 64–128 frames each containing 128 × 128 pixels, needing several Mbytes of computer memory in a 16-bit computer (Table 5.12).

Images obtained from data acquired using a multi-ring positron camera, as described in Section 5.9, are usually reconstructed from either sinograms (see Section 5.8.2.1) or parallel projections and for this histogramming is a simple and rapid way of storing the data.

The data acquisition hardware must be supplemented by sophisticated acquisition software that will allow a range of parameters to be varied to suit the study being performed.

TABLE 5.12

Computer memory/storage requirements for typical studies performed in nuclear medicine.

Image Type	Window Number	Frame Size	Frame Number	Mbytes
Static	1	256 × 256	1	0.065
Static[a]	3	256 × 256	3	0.2
Dynamic	1	128 × 128	100	1.6
Whole body	1	512 × 1024	1	0.5
SPECT	1	128 × 128	128	2.0
SPECT[a]	3	128 × 128	128 × 3	6.0
List mode	1	10^7 × 4 byte events	—	40

Notes: All data assume single-byte-mode acquisition unless otherwise stated.

[a] With triple-energy-window scatter correction.

5.5.2 Online Data Correction

With the advent of cheap, powerful microprocessor chips, it has become possible to correct acquired data online to remove some of the inadequacies of gamma-camera performance. In particular, the performance of PMTs used in both gamma cameras and positron cameras requires rapid, automatic monitoring and subsequent correction to 'iron out' errors in image production caused by the instability of these devices. The most obvious of these are corrections for variations in pulse-height gain of the PMT with time or temperature and the differences between even well-matched PMT pulse-height response to scintillation light.

Hence, the most important feature is the ability to monitor PMT pulse-height drift and to correct for this drift on a time scale much shorter than that over which the image is acquired. Once PMT gains have been stabilised, it is possible to correct for spatial non-linearities by imaging specially designed phantoms (see Section 5.10.4) and determining a correction matrix, which is then applied to each acquired event. Another example is the use of a pre-acquired uniform flood image to correct gamma-camera images for nonuniformities caused by collimator, crystal and light-guide imperfections. This uniformity correction process may be performed during or after data acquisition using the appropriate microprocessor software. Typical data corrections required for single-photon planar and dynamic imaging include:

- PMT gain drift which can be corrected by adjustment of the voltages on ms time scale by monitoring PMT pulse heights;
- spatial non-linearities via a correction matrix using a phantom;
- energy-response variations via energy-dependent spatial shifts using a range of radioactive sources emitting different gamma-ray energies; and
- crystal/collimator nonuniformities via illumination with a flood source.

Additionally SPECT/PET require corrections for:

- attenuation;
- scatter;
- randoms (PET only);
- dead time; and
- centre of rotation (SPECT and rotating positron cameras).

These corrections will be described in more detail in Sections 5.8.5 and 5.9.7.

5.5.3 Data Processing and Image Reconstruction

Examples of processing procedures provided by commercially available nuclear medicine computer systems are shown in Table 5.13. This area of computer application can be subdivided into several categories. One example is the production of images in SPECT or PET by applying complex *data reconstruction* techniques such as FBP or iterative reconstructions (see Chapter 3 and Section 5.8.4), which require substantial computer-based processing power. A second example is the processing of images either to remove statistical noise or to enhance a particular image feature such as the edge of a structure. These are usually known respectively as *smoothing* and *filtering*. In addition images from both single-frame and multi-frame acquisitions can, subsequently, be processed using appropriate software to provide *numerical* or *curve data*, which can add to the functional diagnostic features of the images themselves.

TABLE 5.13

Types of processing procedures for a gamma-camera/computer system.

Type	Example
Data reconstruction	Filtered backprojection or iterative reconstruction
Image enhancement	Smoothing or filtering
Image analysis	ROIs and activity–time curves
Image manipulation	Rotation, minification, subtraction, addition
Image-based modelling	Numerical analysis and curve fitting

Considering these particular examples of data processing in terms of the computer needs, we will see that they all have requirements for high-speed, complex mathematical processes that make full use of the processing power of modern microprocessor systems.

The processes involved in FBP require the repeated use of algorithms to apply 2D Fourier transforms and various other mathematical manipulations to large data files. A successful *reconstruction program* will take acquired multiple 2D projection data and produce multi-slice tomographic images in a matter of minutes. Often the requirement for operator interaction can extend the image-production process more than the computer processing. The production of 3D volume images using iterative reconstruction methods may require longer reconstruction times with conventional computer techniques, but algorithms such as OSEM (Hutton et al. 1997) make it possible to reconstruct images for clinical use rather than just for quantification.

If we now consider the need for image processing to remove statistical noise or to enhance image features, the computational requirements are directly related again to the volume of data being handled. The simplest *image filtering techniques* in one or two dimensions require little sophisticated software or hardware if matrices in real space are used. Spatial smoothing with a nine-point filter is performed in real space by convolving each image pixel with weighted values of each pixel's nearest eight neighbours (Figure 5.51). If a very smooth image is required, the filter elements are made more equal and vice versa. A 1D filter can be used for temporal smoothing of three adjacent frames (Figure 5.52). The relative values of the filter elements again decide the smoothness. The nine-point 2D and three-point 1D filters can, of course, be extended to cover a larger number of pixels or frames, such as 25-point 2D or 5-point 1D, for instance, at the expense of computing time. *Edge enhancement*, the opposite of

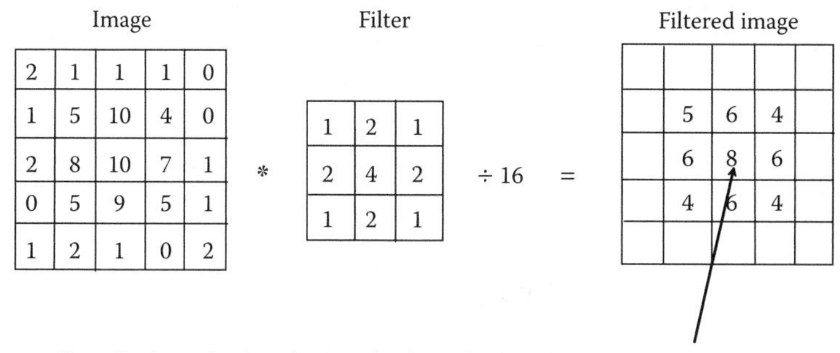

FIGURE 5.51
Illustration of how a nine-point spatial smoothing filter is applied to image data. Here the central pixel value is convolved with the eight surrounding values and replaced by this number divided by the total value of the convolution matrix (16).

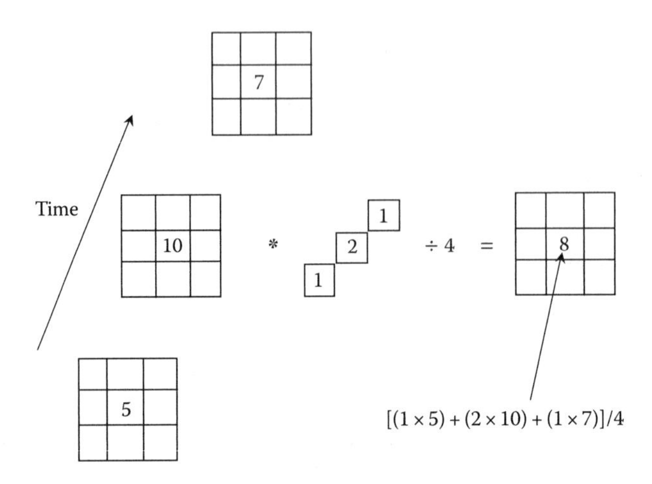

FIGURE 5.52
Illustration of how a 1D three-point temporal smoothing filter is applied to sequential frames.

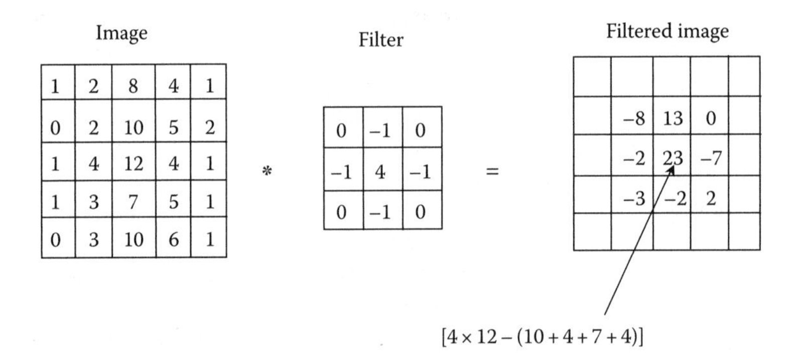

FIGURE 5.53
Illustration of a spatial edge-enhancement filter applied to image data.

smoothing, can be achieved by having negative values in the filtering matrix (Figure 5.53). This filter is useful for delineating organ boundaries in images.

An alternative method of filtering images is to perform the operation in Fourier space. This is more appropriate for *complex filtering operations* where real-space convolution is very time consuming. Here a range of algorithms can be used to remove or damp out unwanted spatial frequencies from an image. Whereas real-space filters discussed earlier are applied by convolution, frequency-space filters require simple multiplication (Brigham 1974) (see also Chapter 11). Two spatial-frequency filters that are commonly used in nuclear medicine are the Hanning and Butterworth filters.

The Hanning filter is defined by

$$A(f) = 0.5\left[1+\frac{\cos\left(\pi|f|\right)}{F}\right] \qquad |f| \le F$$

$$= 0 \qquad |f| > F \tag{5.43}$$

where

$A(f)$ is the amplitude of the filter at a spatial frequency f, and
F is called the cut-off frequency.

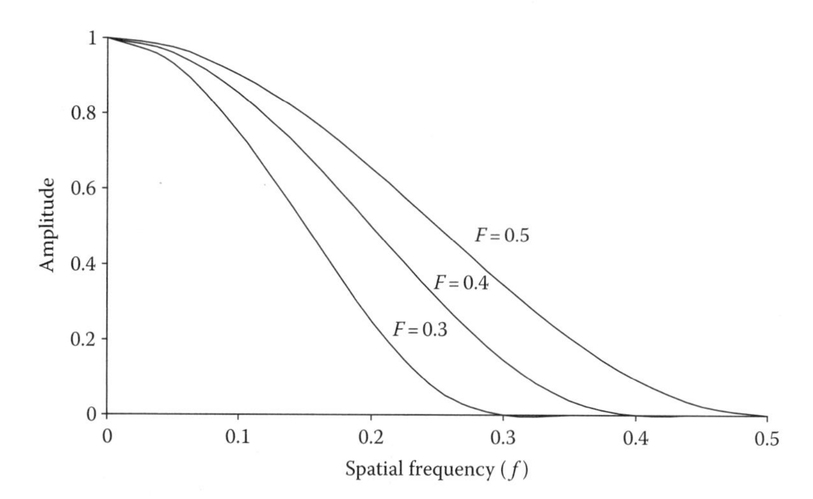

FIGURE 5.54
Graph showing variation of amplitude with spatial frequency (f) of Hanning filters with cut-off frequencies (F) of 0.3, 0.4 and 0.5 pixel^{-1}.

Examples of the Hanning filter are shown in Figure 5.54 for different values of F. The Butterworth filter is defined by

$$A(f) = \left[1 + \left(\frac{|f|}{Q}\right)^p\right]^{-1/2} \tag{5.44}$$

where values of the parameters Q and p are selected by the operator. Q is a frequency that controls roll-off and p is the power factor, which together with Q determines the filter shape. If F or Q (for the Hanning or Butterworth filters respectively) are high, the filters are 'sharp', allowing high-frequency noise and sharp edges in the image: for low values of F or Q, the image becomes smooth.

The application of frequency-space filters requires the image data to be transformed to and from Fourier space before and after the filtering operation – fast Fourier transform algorithms are necessary for these operations (see Chapter 11). An example of the application of a Fourier-space filter is shown in Figure 5.55 which shows transaxial cross-sectional images through a radioactive cylinder containing spheres with different levels of radioactivity. This figure shows how a frequency cut-off of 0.5 pixels^{-1} produces noisy images

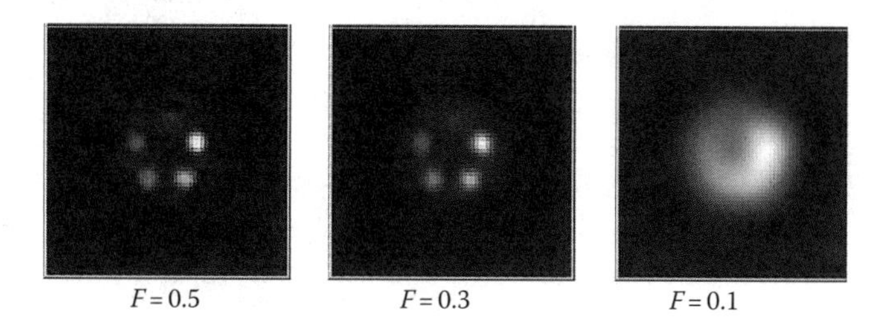

FIGURE 5.55
Illustration of the application of the Hanning filter on a transaxial cross section through a radionuclide-filled cylindrical phantom with five spherical inserts containing different amounts of radioactivity. The images show the effect on signal/noise and spatial resolution of three different values of the cut-off frequency (F, in pixel^{-1}).

with sharp features. As the cut-off is reduced, the image becomes smoother and spatial resolution is degraded.

In most cases, although the intrinsic spatial resolution of an imaging system may be adequate, poor image statistics require the image to be smoothed and the spatial resolution is necessarily degraded at the expense of noise removal. More sophisticated image-processing techniques, for instance, to remove the point-source response function (PSRF) from 2D and 3D images (Webb et al. 1985b, Yanch et al. 1988) are discussed elsewhere (Chapter 11).

Although frequency-space operations are time consuming for large datasets, even they are quick in comparison with real-space operations that filter images with large filter kernels. Real-space operations of this kind have been carried out to produce images, the significance of whose pixel values is precisely known in relation to their neighbours (Webb 1987). In these non-linear image-processing methods, the value of a pixel is compared with the local average of pixels in neighbouring regions and may be replaced by such an average if its departure from the mean is not significant by some preset criterion. It can be shown that such images have improved perceptual properties (see also Chapter 13).

A third category of data-processing features to be found on a nuclear medicine image-processing system involves the use of regions of interest (ROIs) to produce *numerical and curve data*. If a sequence of frames is acquired during a dynamic study, compound images can be formed by addition (reframing) of groups of frames. ROIs can then be drawn around the images of organs, background, etc. (Figure 5.56a) and the ROI counts in each frame can be used to produce 'activity–time' curves (Figure 5.56b), which provide temporal information about the radiopharmaceutical distribution. Curve subtraction and fitting options provide for accurate parametric determination as used in both kidney and heart studies (see Section 5.7). Parametric images may be formed by putting in each pixel the value of a parameter determined from the temporal variation of that pixel. Several parametric images can compress a large amount of information into a relatively acceptable format. An example of a parameter that can be used to form an image is the time to reach a peak value in each pixel of a renal study.

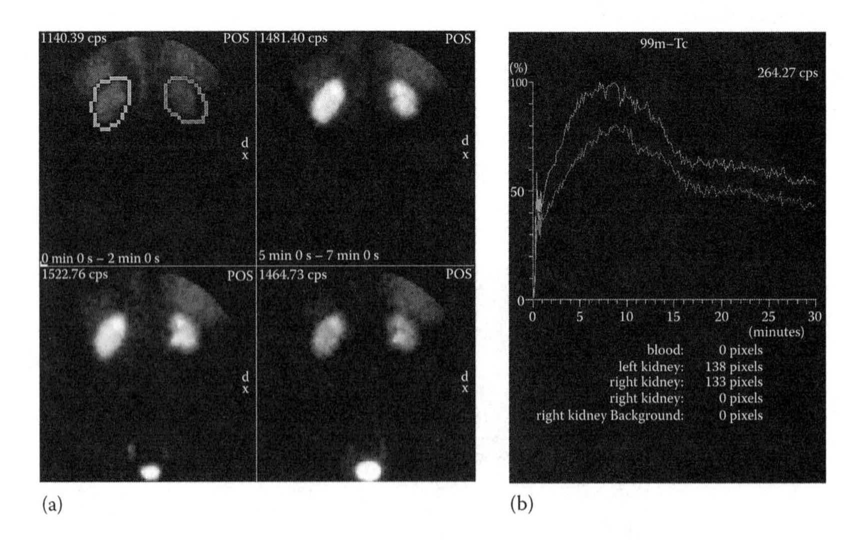

FIGURE 5.56
(See colour insert.) (a) ROIs positioned on a 2 min summed image of a dynamic renal study using $^{99}Tc^{m}$ MAG3. (b) Kidney activity–time curves produced from these ROIs. (Courtesy of S. Chittenden.)

This type of image could identify any parts of the kidney that have a time-to-peak value much different from the norm which may indicate abnormal function.

In this way dynamic datasets can be used to extract physiological parameters. For example, a model of tissue/organ function can be used to generate fits to the activity–time curves. In this case the values of the parameters obtained from the fits can be actual measures of physiological function such as blood flow or rates of metabolism. These types of data analysis will be discussed further in Section 5.7.

5.5.4 Image Display and Manipulation

The end product of data acquisition, correction and processing is the image, which must be displayed in a manner that best highlights the diagnostic information required. An immediate advantage of digital over analogue image display is the facility to alter the display (or grey) levels to optimise the viewing of the full dynamic range of information in the image. Figure 5.57 shows how an image displayed with all its grey levels can be enhanced by windowing to pick out different grey-level regions. This helps to highlight high-and low-contrast areas separately so that abnormal uptake in tissues can be seen over a wide dynamic range. Similar windowing procedures were discussed in the context of X-ray CT in Section 3.4.7.

Further enhancements can be achieved by careful selection of colour scales. If adjacent grey levels are now represented by high-contrast colours, a contouring effect can be achieved, which will both highlight areas of low/high uptake and may be used to give a first-order (roughly 10%) quantification of an image directly from the display.

Important display features allowing optimal image viewing are overall image size and the number of pixels used for display. Although a 64 × 64 pixel image (with ~6 mm pixel size) adequately describes a typical SPECT image (where spatial resolution is 15–20 mm), the viewer may find the pixel edges disturbing and this may detract from the diagnostic value. In some cases for PET where the spatial resolution is generally better (~6 mm), the use of larger matrices with smaller pixels is essential to optimise the image quality. Figure 5.58 shows an image of the single-slice Hoffman brain phantom imaged with a positron camera – the detail in the image is only seen when smaller pixel sizes (3 mm) are used. Use of 128 × 128 or 256 × 256 arrays provides images that are qualitatively as good as the best analogue images produced in radioisotope imaging.

The ability to rotate images, to provide multi-image display and to add text is an accepted and important feature of image display. For example, the viewing of dynamic or tomographic images can be enhanced by a 'snaking' multi-slice display (Figure 5.59), or by a ciné display, where images are shown sequentially at varying display rates to simulate temporal changes or

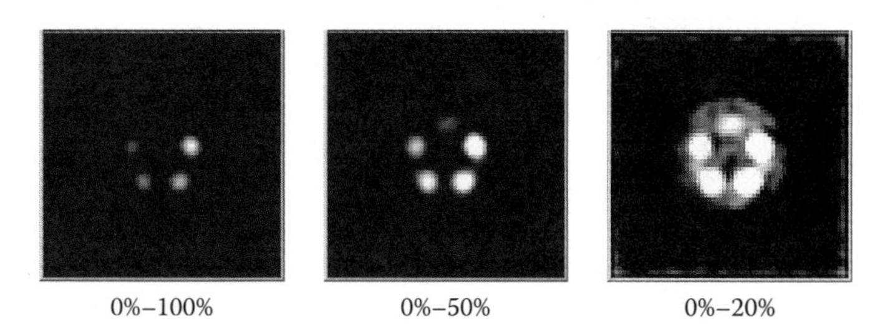

FIGURE 5.57
Effect on image display of using upper and lower thresholds (grey-level windowing). The three images show the phantom illustrated in Figure 5.55 ($F = 0.3$) with no window (0%–100%), a 'top cut' of 50% and a 'top cut' of 80%.

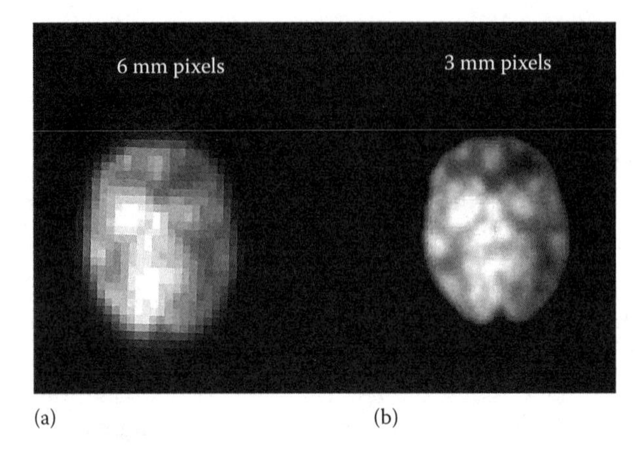

FIGURE 5.58
Effect of the number and size of pixels on an image of the single-slice Hoffman brain phantom: (a) 64 × 64, 6 mm pixels and (b) 128 × 128, 3 mm pixels.

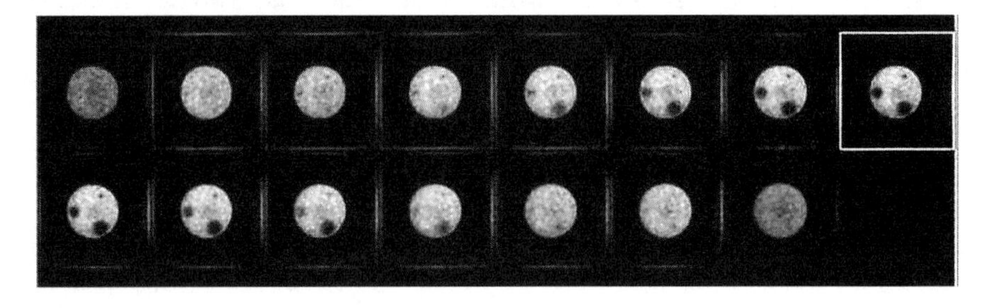

FIGURE 5.59
Multi-slice display of images, which are part of a tomographic PET study of a cylindrical phantom filled with ^{18}F containing three water-filled spherical inserts of different sizes.

spatial rotation, respectively. This latter form of image display is particularly useful in SPECT, where the planar projections can be viewed from many angles highlighting the 3D nature of the images and reducing the effect of image noise via the effect of persistence of vision.

With the increased interest in multimodality imaging, the overlay of images from, for example, SPECT and CT, can add substantially to the diagnostic value of radioisotope images as the CT provides a registered anatomical reference for the functional SPECT data (see Chapter 15). Similar displays using image overlay or linked cursors are essential for the analysis of radionuclide images taken over several days for radiotherapy dosimetry purposes.

The most important application of image display and manipulation is in clinical diagnosis, where the techniques of image subtraction, image overlay, and reduction of image noise, and the extraction of physiological parameters, aid the diagnostic procedure beyond that based on pure visual image interpretation.

5.5.5 Data Storage

The large volumes of data produced during dynamic and tomographic gamma camera acquisitions require modest computer memory for short-term storage, but considerable hard-disc space for semi-permanent storage (see Table 5.12). In addition, processing will produce further large data files, requiring more computer memory and disc space. A standard SPECT acquisition produces several Mbyte of data and probably similar-sized processed multi-slice data files. A multi-frame dynamic acquisition can be even more demanding of storage space depending upon the size and number of frames. A list-mode PET acquisition could easily produce a file of ~1 Gbyte. A typical computer system used for nuclear medicine imaging will have up to 4 Gbyte random-access memory (RAM) and a hard-disc system with Gbytes to Tbytes of fast-access storage. For permanent storage of small volumes of data, high-capacity (up to 900 Mbytes) CD-ROMs seem to be the favoured storage media. DVDs (up to 18 Gbyte) are used for archiving large volumes of data, and external hard drives (Tbytes) have become a cost-effective alternative for long-term storage.

The development of picture archiving and communication systems (PACS) allows the storage of all types of medical imaging data on a common architecture such as a networked central server that stores a database containing the images (see Section 14.3.3). These systems are connected to either a local- or wide-area network allowing all image data relating to a particular patient to be accessed remotely for improved diagnosis and patient management.

5.5.6 System Control

The availability of low-cost, powerful microprocessors has increased the role of computers in system control. In radioisotope imaging, this falls into several categories.

Gamma-camera gantry control has always been necessary for the provision of multiview data acquisition and the gantry control provides rotation needed for transaxial SPECT imaging and translation for planar and whole-body scanning. Other gantry motions may be necessary to allow automatic or semi-automatic collimator loading and to position the camera head(s) to allow QC procedures to be performed. In all cases, positional information from the gantry must be monitored and stored by the computer. This information is used to control the camera position during data acquisition and to provide correct spatial and angular information required for data processing. In most modern gamma camera systems the gantry can be controlled either using the computer terminal or by a mobile control system that can be moved close to the camera for local use.

Patient couch control is necessary for scanning camera systems and is also important for tomographic imaging. If circular gamma-camera orbits are used in SPECT, accurate patient positioning using the couch is necessary to optimise the multiview information and to ensure that at no point does the ROI move out of the camera's FOV (Figure 5.60a). Elliptical gamma-camera orbits (Figure 5.60b) allow the camera head to be placed optimally close to the patient at all angles and hence maximise spatial resolution. This process is best achieved by computer control of the patient couch during acquisition, so that the lateral and vertical couch motion is closely synchronised with the gantry rotation. Synchronised couch and camera head control is also essential for contoured whole-body imaging (Figure 5.60c), where the upper camera head is contoured as close to the body as possible throughout the scan.

Total *gamma-camera control* to provide a totally 'digital' gamma-camera system is now the default for most commercial manufacturers. Here, several microprocessors are used to provide acquisition set-up, image display, data processing, gantry/couch control and data storage. The analogue data from the gamma-camera PMTs are digitised right at the camera head to minimise noise effects. No analogue images are available, but are replaced by high-resolution (256×256 or 512×512) digital images. This total control package has several other advantages, including the provision of multiple correction procedures for a range of radioisotopes, collimators and energy window combinations, and the optimisation of image display and processing features by choice of dedicated processors that communicate via a common database.

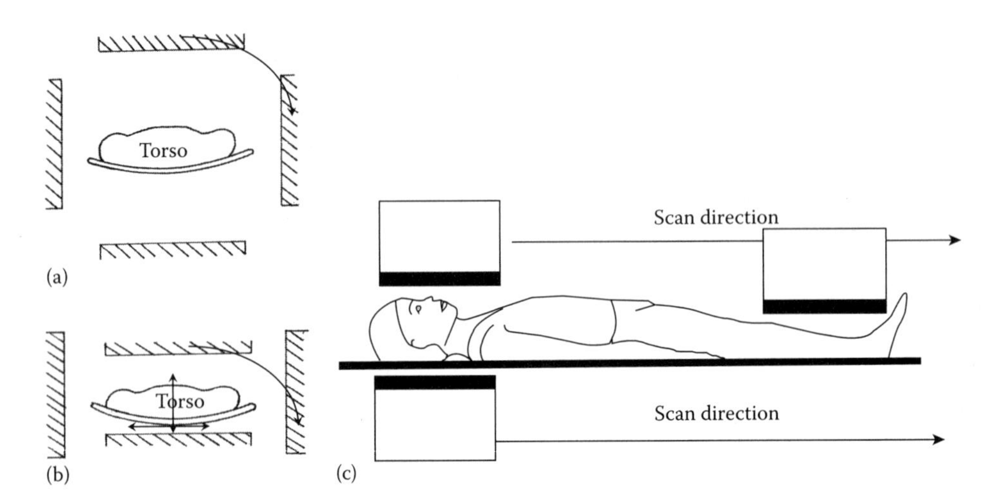

FIGURE 5.60
Schematic diagrams of patient and camera positions for (a) a tomographic study with a 'circularly' orbiting camera, (b) a tomographic study with an 'elliptically' orbiting camera and (c) a whole-body scan with the upper camera head contouring the patient.

5.5.7 Summary

The use of computers has both improved image quality and enabled the extraction of quantitative functional information from radionuclide imaging. Further aspects are addressed in Chapter 14. Computers are an essential part of clinical nuclear medicine both from the image display and manipulation viewpoint and for enhancing clinical diagnosis.

5.6 Static Planar Scintigraphy

Conventionally, the most common radioisotope image used in clinical diagnosis is the planar (2D) scintigram. Since modern gamma cameras come with an attached digital computer, these images are now stored in digital form and have gradually replaced the production of analogue film hard copy. In this section we discuss the basic principles of planar scintigraphy and give simple examples of its use in nuclear medicine.

5.6.1 Basic Requirements

The requirements for the production of this simplest form of radioisotope image include the following: a modern large-FOV gamma camera mounted on a flexible electromechanical gantry, a range of collimators to cover all gamma-ray energies to be used and a high-quality digital image production/storage system. In addition, as the gamma camera is now used extensively for whole-body imaging a motorised patient couch is essential. In many cases double-headed camera systems are now used for whole-body imaging as it provides simultaneous anterior and posterior views. This equipment, allied to a range of readily available radiopharmaceuticals, will provide the basic radioisotope image – 'the X-ray film' of nuclear medicine. The properties of the gamma camera and collimators have been described in Sections 5.3.2 and 5.3.3, so here we will comment only on the additional features mentioned earlier.

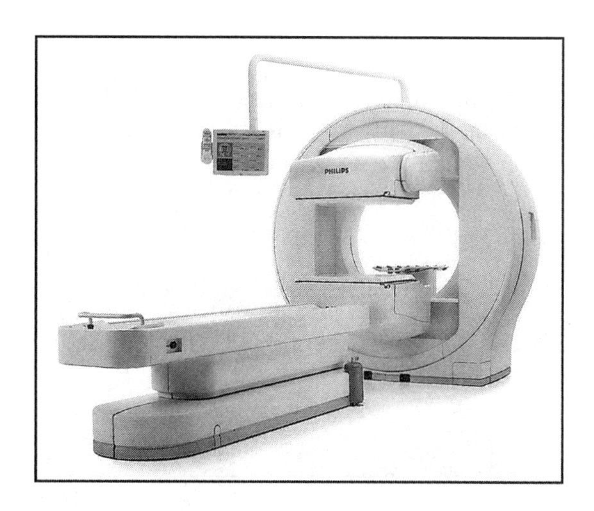

FIGURE 5.61
Photograph of a double-headed gamma-camera system showing the two camera heads mounted on an electro-mechanical gantry capable of 360° rotation for tomography. The patient couch (suitable for whole-body scanning as well as tomography), and the local control panel can also be seen. (Photograph courtesy of Philips Healthcare, Guildford, Surrey, UK.)

The *electro-mechanical gantry* must be capable of rapid and accurate movements to allow the gamma camera(s) to be raised, lowered or tilted in any direction and rotated for SPECT imaging. This will allow images of any part of the body to be obtained without physical discomfort to the patient. Additionally, aspects of safety must be included in gantry design to make them 'fail-safe' – they are, after all, supporting typically 1–2000 kg. An example of a commercial gantry for a double-headed gamma camera is shown in Figure 5.61. The design of the gantry must take into account the needs of collimator loading and unloading which have now become automatic or semiautomatic. Gantries are usually controlled by a local control panel that provides a digital display of position/orientation information as well as remotely via the attached computer system.

The *patient couch* used for whole-body and dynamic studies must be sufficiently flexible to allow height adjustment for easy patient access plus lateral and longitudinal motion to allow the optimal patient/camera position for data acquisition. Accessories that attach to the patient couch include head and arm rests for patient comfort and to ensure the correct positioning for the study. Laser systems are usually provided to simplify patient positioning with respect to the camera. These systems also aid reproducibility if repeat studies are necessary.

The production of high-quality *analogue images* was, historically, the central feature of the scintigraphic system. Gamma cameras were supplied originally with a formatting photographic system enabling many images to be produced on a single X-ray film (4–16 typically) (see Figure 5.62a and b). One of the most difficult parts of analogue imaging was always the reproducible production of good-quality hard copy. The problem lay in selecting correct intensity settings, so that the wide count-density range in the image was satisfactorily transferred to film. This is one reason why analogue imaging has been superseded by digital storage allowing, for example, post-imaging grey-scale manipulation to optimise the information on the film and to take advantage of the full dynamic range. Computer storage can be used to make planar scintigraphy more quantitative, whereas in the past the modality has been qualitative. More details of the facilities available for image manipulation using the online computer system has been given in Section 5.5.

5.6.2 Important Data-Acquisition Parameters

The optimisation of image quality in static scintigraphy depends on several parameters, and errors in any of these can seriously degrade the information content and, hence, the diagnostic value of the image. For instance, the choice of a *collimator* and *photopeak window* to match the γ-ray emission from the radioisotope used is obviously important in image production.

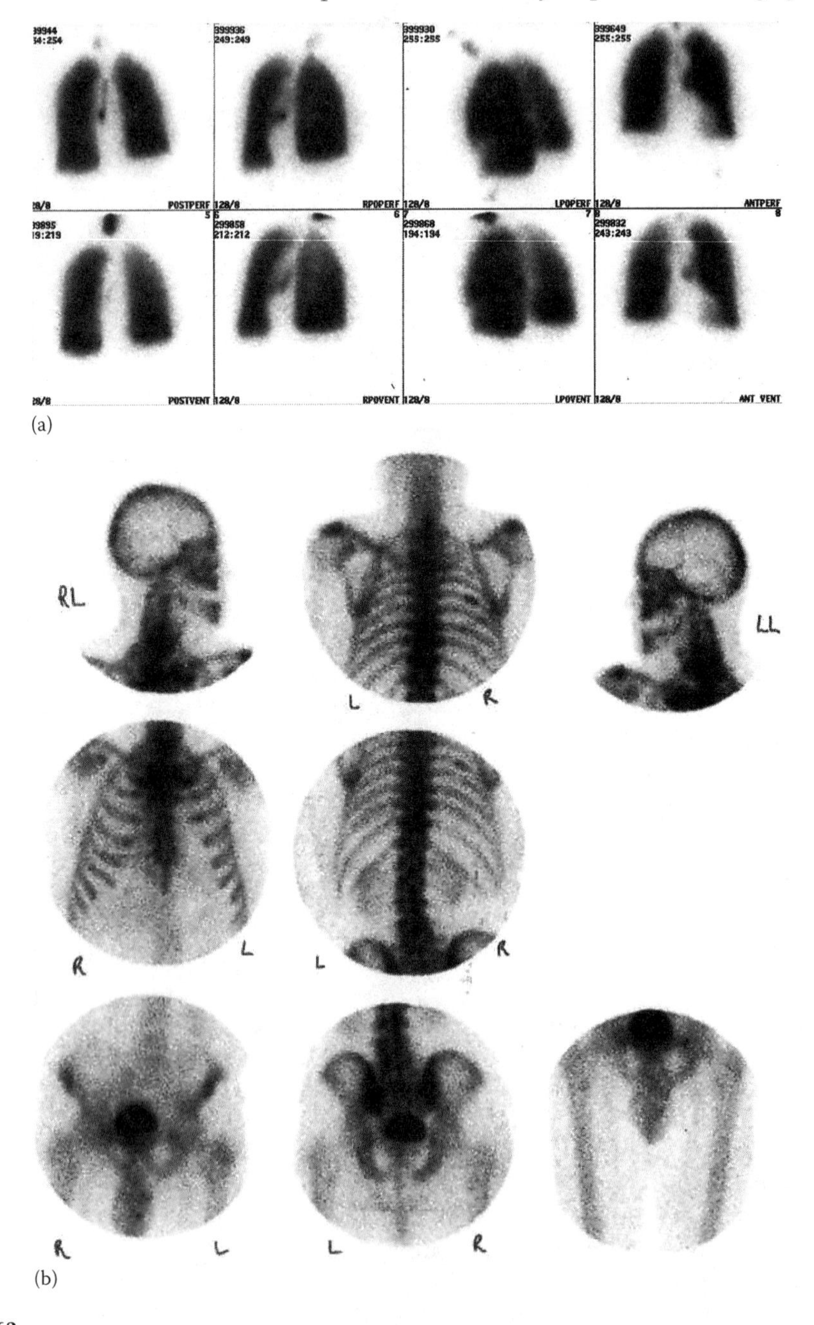

FIGURE 5.62
(a) Multi-formatted storage on an X-ray film showing lung perfusion scans (top) and matching ventilation scans (bottom), (b) multi-formatted images of the skeleton showing uptake of $^{99}Tc^{m}$ MDP.

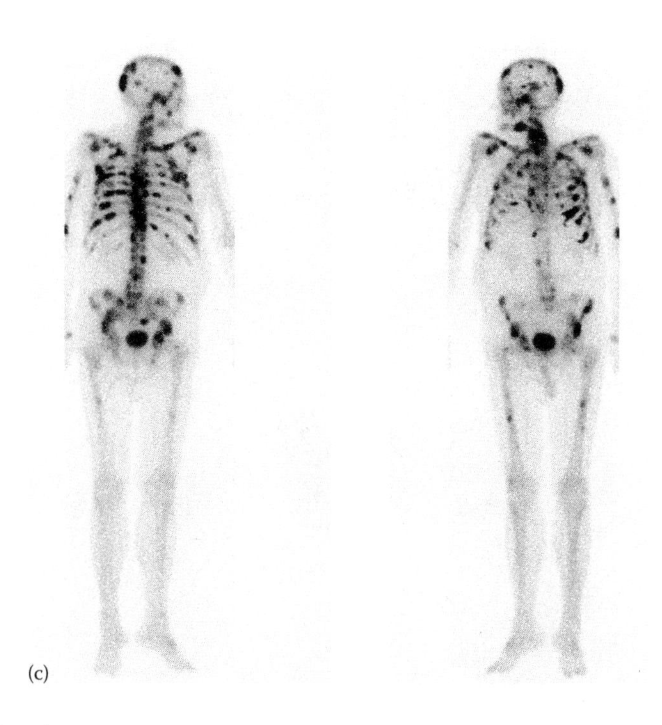

FIGURE 5.62 (continued)
(c) Whole-body double-headed gamma camera scans (posterior left, anterior right) of the skeleton with the same tracer showing metastatic disease from prostate cancer.

The collimator FOV should match the size of the object to be studied and the collimator efficiency should be chosen to provide a high enough count rate (possibly at the cost of spatial resolution) to give satisfactory information density in the image.

The *time after injection* can be an important factor if the radiopharmaceutical is required to clear from some tissues before the target organ can be satisfactorily imaged. For example, a few minutes wait time is sufficient for lung ventilation or perfusion imaging, whereas 2–3 h is necessary for a polyphosphate to clear the soft tissues sufficiently to allow skeletal imaging and even longer is desirable for static renal imaging with DMSA.

In order to obtain sufficient information about the physical location of abnormalities using planar scintigraphy, it is usually necessary to take *several views* of the object from different angles. Figure 5.62a shows images of the lungs from four different angles and with two different tracers. The four views (posterior, right posterior oblique, left posterior oblique and anterior) have been chosen to highlight lungs in a way that maximises the detection of abnormal function, whereas the two tracers show lung blood perfusion (top with $^{99}Tc^{m}$-macroagregates) and lung airway ventilation (bottom with $^{99}Tc^{m}$-technigas). This form of display provides a powerful tool to evaluate lung function. In contrast, a single-view image may be sufficient to determine the function of the thyroid or the existence of abnormal skeletal function in the feet/ankles as shown in Figures 5.63 and 5.64, respectively. The former is an anterior image taken with a pinhole collimator showing the uptake of $^{99}Tc^{m}$-pertechnetate in a normal thyroid gland and illustrates that this tracer also gets taken up into salivary glands. In this case, the smaller FOV is traded for a higher-spatial resolution via image magnification. The latter is an anterior image taken with a parallel-hole collimator showing the uptake of $^{99}Tc^{m}$-MDP in the bones of the ankle and feet of a patient illustrating hypermetabolic areas which in this case are due to a stress

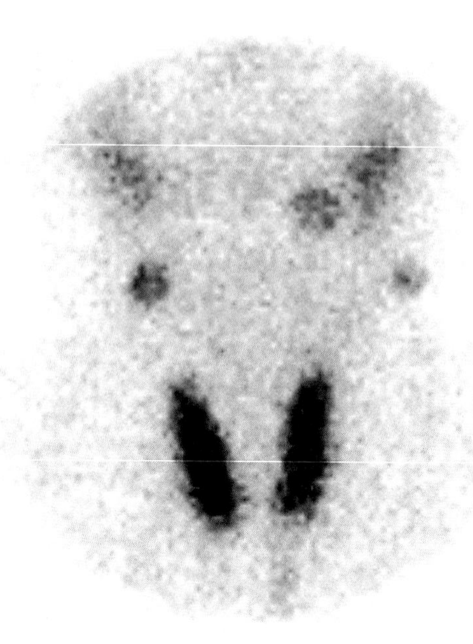

FIGURE 5.63
Image of the neck of a patient using a pinhole collimator mounted on a gamma camera. The collimator geometry magnifies the image onto the full FOV of the camera and shows detail of tracer uptake in the thyroid gland and salivary glands.

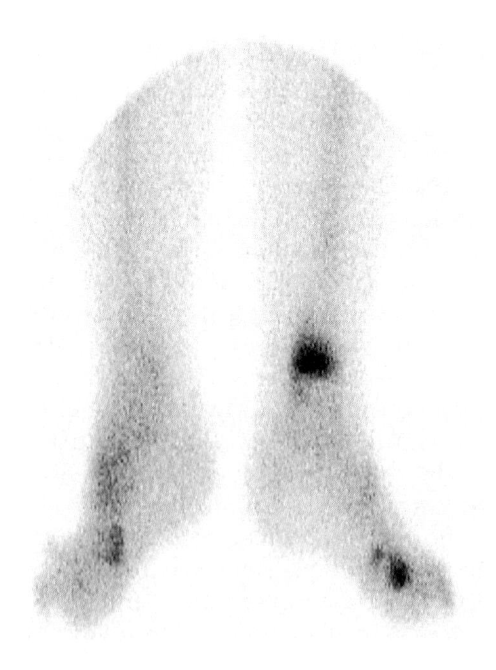

FIGURE 5.64
Anterior images of the legs of a patient taken with a parallel-hole collimator showing uptake of $^{99}Tc^{m}$-MDP in a stress fracture in the ankle. Increased uptake in other bones is due to degenerative disease.

fracture in the ankle and degenerative disease in the bones of the foot. Here, the larger FOV required limits the spatial resolution that can be achieved. To acquire this form of data, accurate *patient positioning* is an important consideration. The *duration time* of the study is usually determined by a compromise between the need to keep the patient immobile and gathering enough counts to give a satisfactory image. As a patient will have difficulty remaining still for more than a few minutes when sitting or standing, many studies are now done with the patient prone or supine on the couch. Figure 5.62c shows how whole-body imaging of $^{99}Tc^m$-MDP can achieve satisfactory skeletal imaging with minimal patient movement. This form of imaging may be a problem, however, for patients with poor lung function and breathing difficulties in the prone or supine positions and such factors have to be considered when clinical imaging is performed. It is highly likely that patient motion will be the dominant factor affecting the spatial resolution of most nuclear medicine studies.

The film used to record analogue images has a characteristic relationship between the count density and the film exposure (the *characteristic curve* of the film (see Chapter 2)), which determined the minimum total counts required per study. These total counts, even for digital images, will also depend on the size of the object and the distribution of the radiopharmaceutical. Typically a single-view image could take 5 min and contain 100–500,000 counts, but variations from organ to organ, patient to patient and with radiopharmaceutical can extend outside this range.

Clearly an improper choice of *exposure time* or count density can negate the value of the analogue image, problems which, as we have seen in Section 5.5, are overcome by the use of a computer to create digital images.

5.6.3 Dual-Isotope Imaging

One of the limitations of planar scintigraphy as a diagnostic technique is the lack of specificity of radiopharmaceuticals, which can lead to poor definition of the function of the target organ. Occasionally, two radiopharmaceuticals can be used to image different functions of the same organ. This may make it possible to improve image definition by the use of *dual-isotope imaging*. In its simplest form, images are obtained of the distribution of two radiopharmaceuticals labelled with different radioisotopes and by a comparison or subtraction of the two images a clearer image of target-organ function is obtained. The philosophy is not unlike that underlying digital subtraction techniques in X-radiology (Section 2.11.1). Historically, there are several clinical examples of the use of this technique. One example is shown in Figure 5.65, where images of the thyroid and parathyroid are

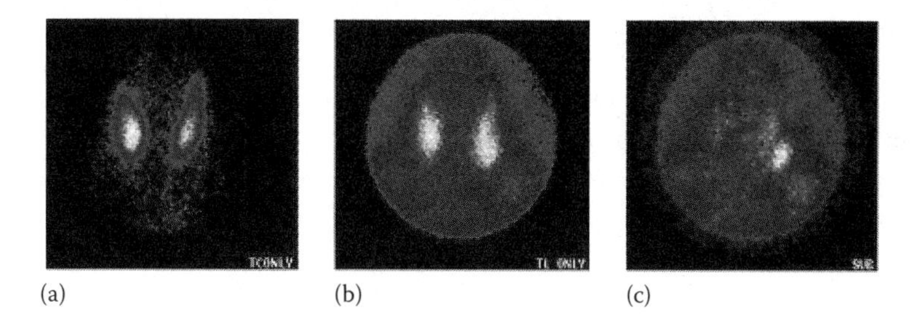

FIGURE 5.65
Dual tracer imaging of the neck with $^{99}Tc^m$ pertechnetate (a) showing uptake in the thyroid and ^{201}Tl chloride (b) showing uptake in the thyroid and parathyroid glands, and the result of image subtraction (c) showing the parathyroid clearly delineated.

taken with two tracers. The left image shows the uptake of $^{99}Tc^{m}$-pertechnetate, the central image shows the uptake of ^{201}Tl-chloride and the right image is the difference between the central and left images. The thallium chloride is taken up in normal thyroid and abnormal parathyroid whilst the pertechnetate is taken up by the thyroid only. The subtraction shows the uptake in an adenoma in the parathyroid gland. This method works in this case where the tracers emit photons with similar energies and in tissues where attenuation is low (Ferlin et al. 1983).

This type of study is more problematical when the photon energies of the tracers are quite different and in highly attenuating parts of the body. For example, if an ^{131}I-monoclonal antibody is used to localise tumour there will be a high non-specific residue in surrounding vascular tissues because the antibody is a large protein. $^{99}Tc^{m}$-macroagregates can be used to image the vascular tissue and a subtraction of two images, taken simultaneously to minimise movement artefact, should show the tumour more clearly (Ott et al. 1983b). Subtraction requires normalisation using count densities obtained from regions of normal uptake in the liver. If A and B are the counts per pixel in the ^{131}I and $^{99}Tc^{m}$ images, respectively, then the difference image pixel count C is given by

$$C = A - kB \tag{5.45}$$

where

A is made up of the pixel count due to normal vascular tissue (A_{NT}), to the tumour (A_T) and to over/underlying tissues (A_B), and

B is similarly made up of B_{NT}, B_T and B_B.

If we assume $B_T = 0$ (i.e. no macroaggregate uptake in tumour) and k is chosen so that

$$A_{NT} + A_B \approx k(B_{NT} + B_B) \tag{5.46}$$

then

$$C \approx A_T \tag{5.47}$$

Unfortunately the assumptions made are incorrect both physiologically and physically. The normal tissue/tumour distributions of the agents are different as in this case, images would be taken 24 h or more after the antibody administration but only 15–20 min after the $^{99}Tc^{m}$ macroaggregate administration. Also the physical attenuation of the photons emitted from two isotopes is quite different and this means that k is depth dependent. Figure 5.66 shows that the rate of attenuation as a function of tissue thickness is less for photons emitted by ^{131}I than those from $^{99}Tc^{m}$. Hence, whatever value of k is chosen will lead to the subtraction only being correct at one depth in the body, implying over/under subtraction at other depths. This can lead to the production of serious image artefacts (Figure 5.67), negating the objective of the technique. The problems associated with subtraction imaging are overcome with SPECT (see Section 5.8).

5.6.4 Quantification

Single planar views of an object represent a 2D superposition of 3D information, and the image of the target organ may be obscured and confused by under/overlying tissue activity. Furthermore, the spatial resolution of a gamma camera changes rapidly with distance

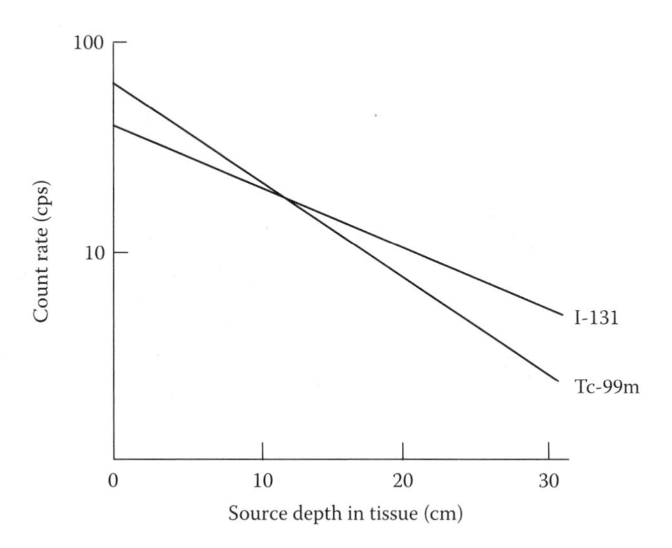

FIGURE 5.66
Relative transmission of γ-rays through different thicknesses of tissue for photons from ^{131}I and ^{99}Tc^{m}, showing that, if these two radioisotopes are used for subtraction imaging, over-/under-subtraction will occur at distances greater/less than 11 cm, respectively.

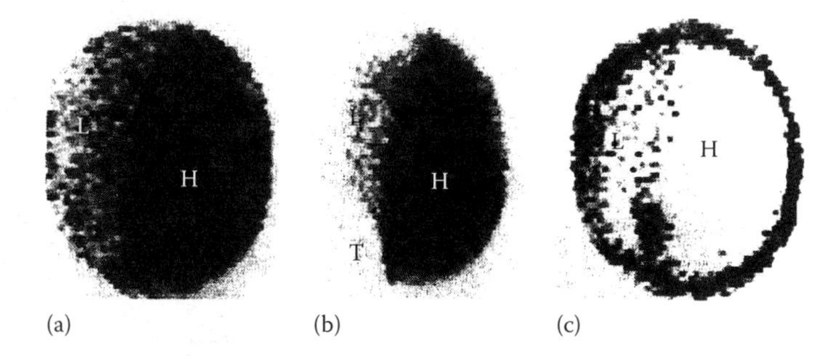

FIGURE 5.67
Planar images of the distribution of (a) ^{131}I-CEA (carcinoembryonic antigen) and (b) ^{99}Tc^{m}-HSA (human serum albumin), and (c) a difference image (a)–(b) showing the production of artefacts by dual-isotope subtraction. H, heart; L, limb; T, tumour.

from the collimator, and, hence, any region on a planar image is a complex function of the object radioactivity distribution. The counts in an image from any ROI will be depth dependent, as shown in Figure 5.66, and any quantitative information will be totally dominated by this effect. As attenuation correction of a single view is impossible (unless the depth of the organ is known), quantification is difficult unless attenuation is negligible or small. An example of this is the measurement of the uptake in the thyroid (Figure 5.63) where the organ is covered by only a few centimetres of tissue and the attenuation correction is ~0.8–0.9.

In the case where relative organ measurements are needed, such as divided kidney function (Figure 5.68), depth corrections are possible by measuring the position of the organs in the body using lateral views, as shown in Figure 5.69. Correction errors can be minimised by ensuring that the organs are as close to the collimator face as possible – hence, kidney images are made using the posterior view. For larger organs, such as the liver, depth correction is of little help as the organ thickness itself makes a substantial contribution

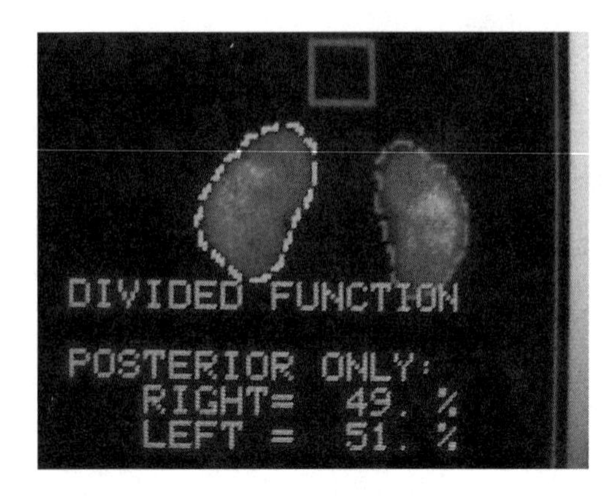

FIGURE 5.68
(See colour insert.) Example of the results for a divided renal function study. (Courtesy of S. Sassi.)

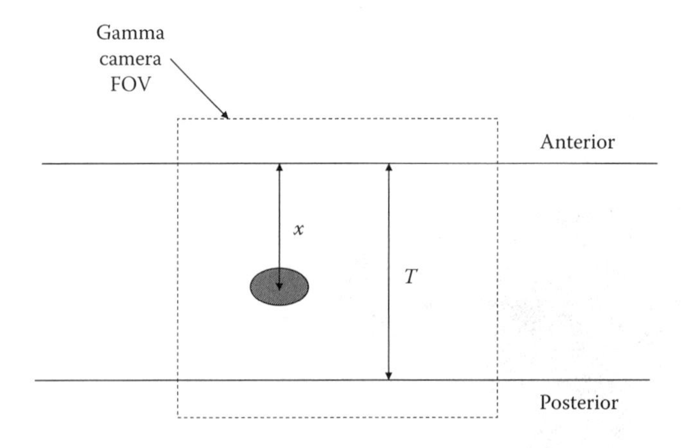

FIGURE 5.69
Use of the lateral view for depth corrections to reduce the effect of photon attenuation. Here x is the mean organ depth and T is patient thickness. The depth corrections are proportional to $\exp(\mu x)$ for anterior image and to $\exp[\mu(T - x)]$ for posterior image.

to photon attenuation. In the latter case, if it is possible to obtain anterior and posterior images without moving the patient using either a rotating gantry or a dual-headed camera, then attenuation problems may be minimised and crude quantitative measurements of organ function can be made.

If we consider a point source of activity A at a depth x below the anterior surface of the body, which is of thickness T (Figure 5.70), then the counts recorded in the anterior and posterior views will be given by

$$C_A = kA \exp(-\mu x) \tag{5.48}$$

and

$$C_P = kA \exp[-\mu(T - x)] \tag{5.49}$$

respectively, where

k represents a calibration factor for the gamma camera, and
μ represents the linear attenuation coefficient for photons in the body.

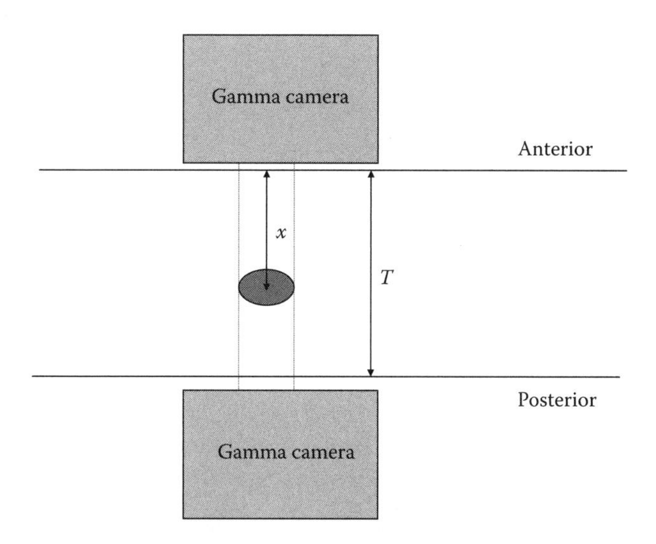

FIGURE 5.70
Illustration of how the combination of two opposed gamma-camera views of a source of activity (A) reduces the effect of photon attenuation. Here x and T are the same as in Figure 5.69. If the geometric mean is used in the combined view, then the combined counts, $C = \{kA \exp(-\mu x) \times kA \exp[-\mu(T-x)]\}^{1/2}$, i.e. $C = kA \exp(-\mu T/2)$.

If the anterior and posterior images are summed pixel by pixel, forming an arithmetic mean (AM) image, then the counts recorded in the combined image will be given by

$$C_{AM} = \tfrac{1}{2} kA \,\{\exp(-\mu x) + \exp[-\mu(T-x)]\} \tag{5.50}$$

which is less dependent upon source depth than C_A or C_P, but does not, of course, totally remove the dependence on x as shown in Figure 5.71a. If, however, the geometric mean (GM) is used when combining the images, then

$$C_{GM} = (C_A C_P)^{1/2} = kA \exp\left(\frac{-\mu T}{2}\right) \tag{5.51}$$

which is totally independent of source depth, as shown in Figure 5.71a. It is important, however, to remember that this result only holds for the situation where the attenuation coefficient is held constant. The technique of combining opposed views to achieve a depth-independent response is known as isosensitive or quantitative scanning. It was used originally with double-headed rectilinear scanners (Hisada et al. 1967) but was subsequently used with the gamma camera (Graham and Neil 1974, Fleming 1979). A further property of an image formed by combining opposing views is that the spatial resolution in the image can be made to be almost depth independent (Figure 5.71b). This is an important feature, which is utilised to good advantage in SPECT (Section 5.8).

The geometric mean is also dependent only on the total path length T for a uniform source of activity. For nonuniform source distributions, the geometric mean is *not* depth independent. In addition, attenuation is not constant throughout the body, especially in the thorax, scatter is present in both anterior and posterior images and this contribution varies with the depth of the activity in the tissue. Also the superposition of information may still obscure the target organ or ROI. Correction for scatter has been attempted using 'build-up factors' as reported by Wu and Siegel (1984) or by applying a scatter correction technique such as the triple window technique.

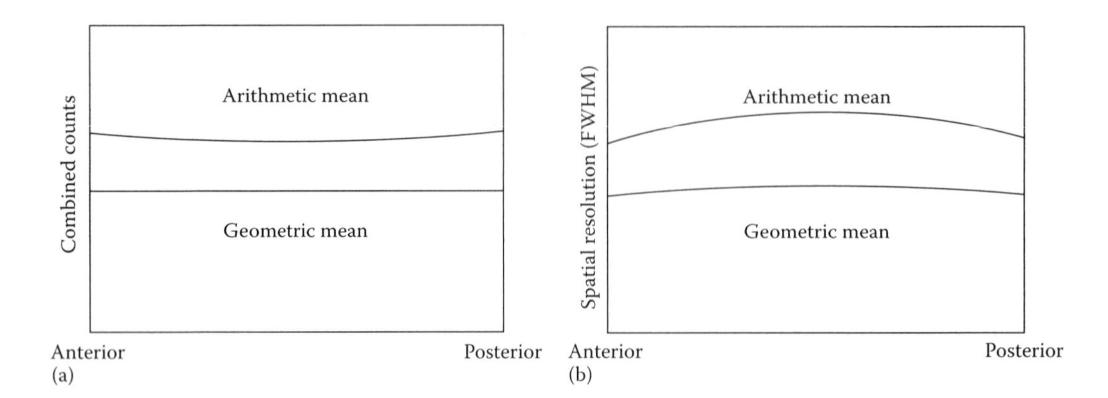

FIGURE 5.71
(a) Graphs of the combined counts from opposed views using arithmetic/geometric means, as a function of depth in tissue for a point source. The arithmetic mean curve is proportional to $\exp(-\mu x) + \exp[-\mu(T - x)]$, whereas the geometric mean curve is proportional to $\exp(-\mu T/2)$. (b) The variation of spatial resolution for the same combinations of data.

In the build-up method, build-up factors are generated using point-source measurements at different depths in a phantom to determine the effect of increased scatter with depth:

$$B(d) = \left(\frac{C_A}{C_0}\right) \exp(\mu d) \tag{5.52}$$

where
C_0 are the unattenuated counts per second,
C_A are the attenuated anterior counts per second, and
d is the depth in the object.

If the build-up factors $B(d)$ are included in the anterior and posterior images, we obtain

$$C_A = C_0[\exp(-\mu d)]\ B(d)\left[\frac{\sinh(\mu x/2)}{(\mu x/2)}\right] \tag{5.53}$$

$$C_P = C_0[\exp(-\mu(T-d))]\ B(T-d)\left[\frac{\sinh(\mu x/2)}{(\mu x/2)}\right]. \tag{5.54}$$

From these equations, we can generate a new depth-independent relationship:

$$\frac{C_A}{C_P} = \frac{[\exp(-\mu d)]\ B(d)}{[\exp(-\mu(T-d))]\ B(T-d)}. \tag{5.55}$$

Experience has shown that the use of the geometric mean with a triple energy window scatter correction gives very similar results to the use of build-up factors without the need to produce the build-up factor data. However, for more accurate quantification, the removal of counts from over- and underlying structures via ECT is required.

5.6.5 Whole-Body Imaging

With the development of the scanning gamma camera, whole-body imaging became an important addition to the nuclear medicine portfolio. Historically, skeletal imaging was performed by the acquisition of multiple views to enable the whole skeleton to be examined (Figure 5.62b). This was seen as ideal as it allowed the best possible spatial resolution to be achieved with the camera placed as close to the skeleton as possible. As discussed earlier (Section 5.3.1), dedicated whole-body scanners were developed in the 1980s specifically to perform this sort of imaging, although it was argued that the spatial resolution obtained was poorer than using multiple gamma-camera images because the scanning heads were not as close to the patient. Now most scanning cameras are equipped with a contouring facility so that the camera head imaging the anterior body could be kept at a minimum distance from the patient as the scan is performed. This allows the sort of images shown in Figure 5.62c to be acquired without any serious loss of image quality. This figure shows posterior (left) and anterior (right) whole-body scans of a patient with prostate cancer, showing the high uptake of $^{99}Tc^{m}$-MDP in the skeletal metastases and bladder (through excretion).

5.6.6 High-Energy Photon Emitters

The majority of radioisotope images are obtained using $^{99}Tc^{m}$-labelled radiopharmaceuticals but there are a number of studies using radioisotopes that emit higher-energy photons. It is important to consider whether the use of isotopes such as ^{131}I will be carried out when purchasing a new camera as a system with a 10 mm crystal, which has been optimised for imaging $^{99}Tc^{m}$ labelled agents, will have a poor detection efficiency for higher-energy photons. The most important of these radioisotopes are ^{131}I, ^{67}Ga and ^{111}In. Although the major role of ^{131}I is as a therapy agent, imaging of this isotope is essential if dosimetry for radionuclide therapy is to be carried out. Additionally, iodine labelling of proteins and peptides is increasingly important, with an enormous potential for nuclear medicine in oncology. With the development of techniques for chelating ^{67}Ga and ^{111}In to antibodies and peptides, these radionuclides are also being used as specific disease-localising agents. These applications, as well as the more conventional uses of these radioisotopes, indicate a continuing demand for images obtained with medium- and high-energy photons.

An examination of the physical properties of these radioisotopes reveals potential problems for the production of images with a gamma camera. In particular, the energies of the photons emitted (Tables 5.5 and 5.9) are, in general, well above the optimal sensitivity range of the modern camera (100–200 keV). Hence, detection efficiency is low (about 25% at 300 keV) and is made worse by the increasing difficulty of collimating photons of energy greater than 200 keV. The penetration capability of photons of energy 364 keV can be estimated from the half-value layer which, for lead, is 2.4 mm compared with 0.4 mm at 140 keV. Taking the criterion of 5% for the level of acceptable penetration (Anger 1964), the septa of a collimator suitable for ^{131}I imaging should be 10 mm thick. This is quite impracticable and, even if the collimator depth is increased, photons can still readily penetrate the 3–4 mm septa actually used. The result of this septal penetration is a considerable degradation of spatial resolution (20–25 mm typically), with associated image blurring. Finally, the decay characteristics of these radioisotopes (especially their long half-life and, in the case of ^{131}I, the existence of β emission) are such that the radiation dose per unit injected activity is much higher than for equivalent levels of $^{99}Tc^{m}$. This usually limits the administered activity levels, and the count rates obtained, consequently, lead to poor image quality. Hence, the information content of images made with high-energy photons is both statistically poor and lacks spatial definition as shown in, for example, a whole-body image of a patient with thyroid cancer (Figure 5.72). The image shows the uptake of $Na^{131}I$ in metastases in the lungs, skeleton and residual thyroid tissue as well as the bladder. This type of image can be used to provide dosimetry for

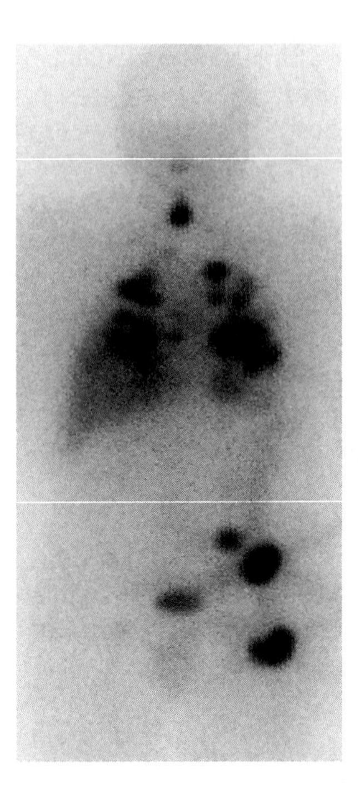

FIGURE 5.72
Whole-body scan of a patient with metastatic disease from thyroid cancer. The image shows uptake of ^{131}I in tumour in the lungs and the pelvis as well as excretion through the bladder. Image quality obtained when imaging with this high-energy (364 keV) photon emitter is inferior to that taken with $^{99}Tc^m$ because of the effects of scatter, lower detection efficiency and collimator penetration.

radionuclide therapy using a tracer amount of the same radiopharmaceutical. Attempts to improve their diagnostic quality using ECT (Section 5.8) have proved difficult.

5.6.7 Conclusions

Planar scintigraphy, especially whole-body scanning, is still the most commonly used method for diagnosis using radioisotopes. However, there will be many cases where this form of imaging is inadequate to give sufficient contrast to provide an accurate clinical diagnosis. There are several methods for providing semi-quantitative measurement of tracer uptake from planar images but these methods are all confounded by the effects of photon attenuation and scatter and are rarely a substitute for tomographic imaging.

5.7 Dynamic Planar Scintigraphy

5.7.1 General Principles

The technique of dynamic scintigraphy became a major part of radioisotope imaging with the development of low-cost, high-power microprocessors. Prior to this, the production of a sequence of images spaced throughout a 20–30 min study could supply only qualitative information about organ function on this timescale. Now, with a modern nuclear medicine computer, quantitative dynamic information is readily obtainable to determine the function of organs such as the heart, the lungs and the kidneys and indeed any organ in which a suitable radiotracer is available.

The basic technique involves the acquisition of data in frame mode (stored directly in a series of x, y matrices) or list mode (stored sequentially as x, y, t values and framed off-line) to describe the temporal changes in radiopharmaceutical distribution *in vivo*.

For cardiac imaging, relatively fast dynamic acquisition is required to produce multiple images of the heart during a heartbeat. By using the signal from an electrocardiogram (ECG) as a trigger to sum up the dynamic images from multiple beats, high-count images can be obtained of the heart at different times during the cardiac cycle. For renal function measurements, relatively slow dynamic acquisition is sufficient, as the renal uptake/excretion phase takes typically 30 min. In general, external respiratory gating is not used, as the movement artefact from respiration is small.

Acquired data are processed to produce 'activity–time' curves describing organ uptake/excretion rates. Images of extracted parameters (parametric images) can be formed to condense the information from the large amount of acquired data. The two examples mentioned earlier will be described in more detail to highlight the processes used and the parameters determined (Peters 1998).

5.7.2 Cardiac Imaging

Cardiac function measurements can be obtained from either a multigated acquisition (MUGA) or from a first-pass study (Berger et al. 1979, Ell and Gambhir 2004, Cook et al. 2006). *MUGA studies* require ECG triggering, where the R-wave signal, indicating the beginning of the systole phase (contraction of the left ventricle of the heart), is used to align the acquisition data from different cardiac cycles. Hence, a MUGA study will supply information about the cardiac function averaged over many heartbeats during a period when the radiopharmaceutical (e.g. $^{99}Tc^{m}$-labelled red blood cells) is uniformly distributed in the blood pool. By contrast, *first-pass studies* require very rapid acquisition of data from the left and right ventricles during the initial period when the injected activity first passes through the heart. The total passage of the bolus injection through the heart requires several heartbeats and *individual measurements* are made of the ejection fraction during each beat. As the two processes are entirely different, they are considered separately.

5.7.2.1 Multigated Acquisition Studies

If we consider the function of the heart as a two-stroke, four-chamber pump (Figure 5.73) then, in the diastole phase, the right and left atria (RA and LA) contract and pump blood into the left and right ventricles (RV and LV) whilst, in the systole phase, the ventricles contract and pump blood to the lungs (from RV) and body (from LV) and the atria expand taking blood from the body (to RA) and the lungs (to LA). The most important measurement obtained from a MUGA study is left-ventricular ejection fraction (LVEF), which determines the fraction of the blood in the LV that enters the body arterial system averaged over many heartbeats. The ECG signal (Figure 5.74) can supply a clear trigger from the time at end diastole using the R-wave signal, and this is used to align the acquired data into a series of frames during each heartbeat. The camera must be carefully positioned (usually left anterior oblique at 35°–45° with up to 15° caudal tilt) to optimise the

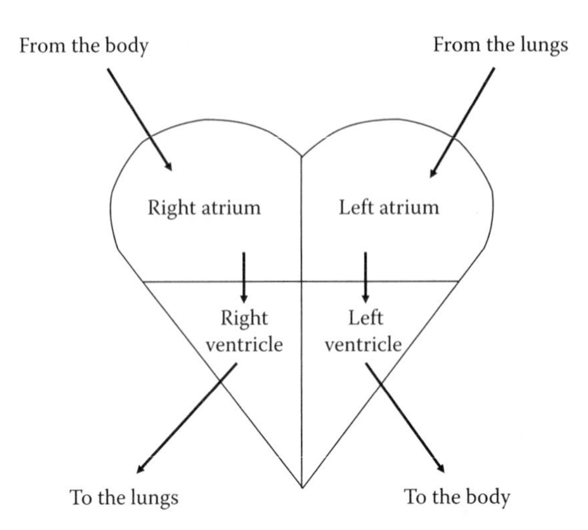

FIGURE 5.73
Schematic illustrating the function of the heart chambers.

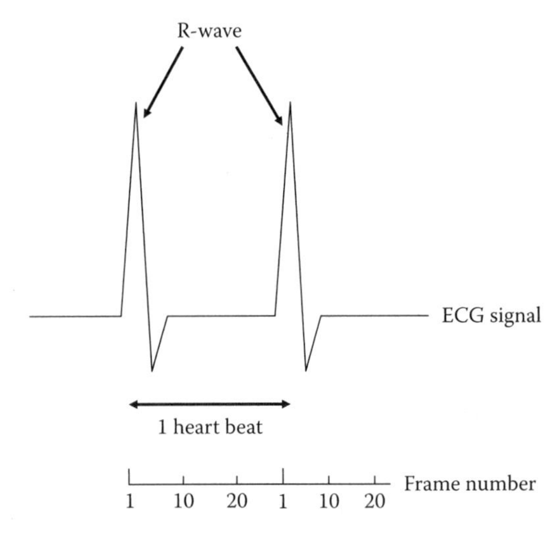

FIGURE 5.74
Use of the R-wave from an ECG to provide gating for the multiple-frame acquisition in a MUGA study.

image of the left ventricle and minimise information coming from other chambers shown in Figure 5.75. This is made easier by the fact that the LV has the largest muscle to cope with the task of pumping blood around the whole body.

If high count rates prevent data from being stored immediately into frames, a list-mode acquisition is used and the data are subsequently framed after the study. Each image frame is stored in a matrix of typically 64 × 64 pixels. Acquisition continues until an acceptable count density is reached in each frame, typically 4–5 min at 15,000 cps although longer may be needed for rest studies. Some patients have arrhythmia which causes a significant number of heart beats of abnormal length and it is possible to reject such beats automatically by comparing the R–R interval with the mean of 10 average-length beats measured before the start of the study. The images are then usually smoothed both spatially and temporally.

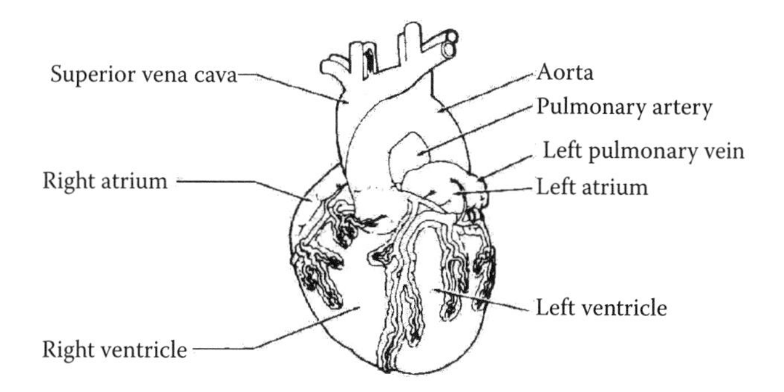

FIGURE 5.75
Schematic of the heart as seen from the left anterior oblique view used for a MUGA study. (Courtesy of GE Healthcare, Slough, Berkshire, UK.)

After this, ROI analysis can be carried out manually, automatically or semiautomatically to determine the outline of the left ventricle in each frame (Figure 5.76). Finding the left ventricle will depend very much on the quality of the data in each frame. Usually the method involves determining the centre of the ventricle and then, with some form of edge-detection algorithm, determining the outline of the ventricle. The methods also involve the generation of a background region around the ventricle that is used to remove activity from other parts of the heart. If the data are of good quality then an automatic method will work well, but operator intervention will be required in cases where the definition of the ventricle edge is difficult. Under the assumption that counts in the LV ROIs are proportional to LV volume, curves are generated of LV volume versus time during each heartbeat (Figure 5.76c) and the LVEF is calculated as

$$\text{LVEF} = \frac{(\text{EDV} - \text{ESV})}{\text{EDV}} \times 100\% \tag{5.56}$$

where EDV and ESV are the end-diastole and end-systole volumes, respectively, corrected for background ROI counts. It is possible to make this calculation using only the

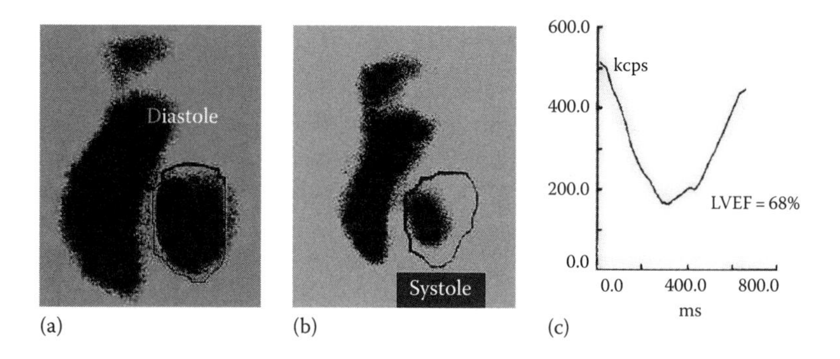

FIGURE 5.76
ROIs used to outline the left ventricle (LV) of the heart at (a) end diastole and (b) end systole. A background ROI is also shown on the end-systole image. The variation of LV counts (assumed to be proportional to LV volume) as a function of time (c) is used to determine the left-ventricular ejection fraction (LVEF).

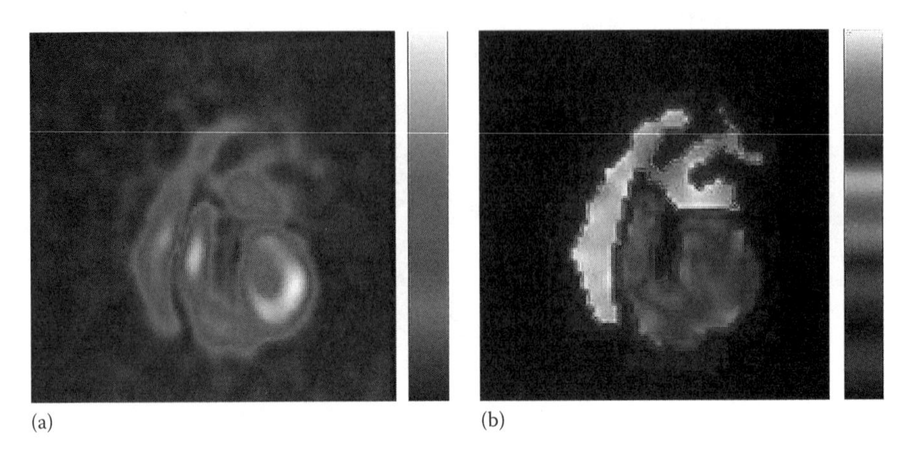

FIGURE 5.77
(See colour insert.) Parametric images of the amplitude (a) and phase (b) of the ejection from the left ventricle. (Courtesy of S. Chittenden.)

two frames corresponding to EDV and ESV to speed up the process but this assumes that the position of these frames is known without the production of the whole ventricular volume curve.

Other parameters can be obtained from a MUGA study and, in particular, parametric images of the amplitude and phase of the ejection from the LV can be made by fitting the LV activity–time curve with a single harmonic. If this is performed on a pixel basis, the phase of the fitted curve gives an indication of the contraction of each pixel, whilst the amplitude will indicate the stroke volume for each pixel (Figure 5.77). These parametric images may show malfunction in the cardiac cycle caused by muscle, valve or septal problems. Some idea of regional wall motion can be achieved by measuring the distance in pixels along several radii connecting the ventricular centre and the 25% isocontour (Figure 5.78a and b). These numbers represent the 2D movement of the silhouette of the ventricle and may identify regions of the wall that have poor motion.

In some cases, abnormal cardiac function is only seen if the heart is stressed and, hence, a repeat measurement of LVEF is made after stressing the patient (carefully!).

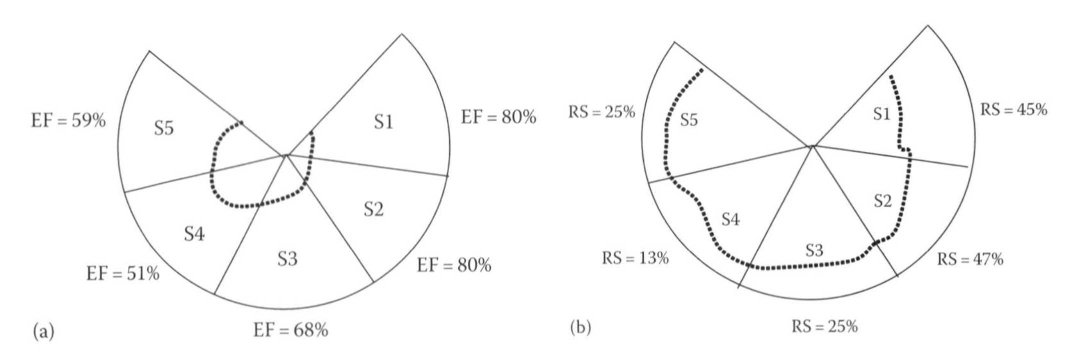

FIGURE 5.78
Measurement of the regional wall motion along different radii connecting the ventricular centre and the 25% contour of the ventricle: (a) shows the variation of ejection fraction (EF) for the different segments (S1–S5) of the left ventricle and (b) shows the apparent radial shortening (RS) of the heart beat for the same segments.

5.7.2.2 First-Pass Studies

Such studies usually require list-mode acquisition since very high initial count rates are needed to give acceptable statistics. The passage of the injected activity is tracked through the right and left ventricles (RV and LV) during the first few heartbeats. Data are acquired with the camera positioned to optimise the view of both ventricles simultaneously as well as the superior vena cava (SVC) (Figure 5.79). The data are reordered into frames using the timing information and then the frames are summed to provide a composite image from which the positions of SVC, RV and LV can be determined (Figure 5.80a). Functional curves of the passage of the radiopharmaceutical through these regions are then derived from the dynamic data (Figure 5.80b). The SVC curve determines the input function of the activity entering the heart and this should be less than 2.5 s wide to produce acceptable results (i.e. the injection must be a small bolus injected quickly). Frame markers are used to produce curves of the function of the two ventricles during the first pass of the radioactivity before recirculation spreads out the temporal information obtained. By careful positioning of frame markers (Figure 5.81a)

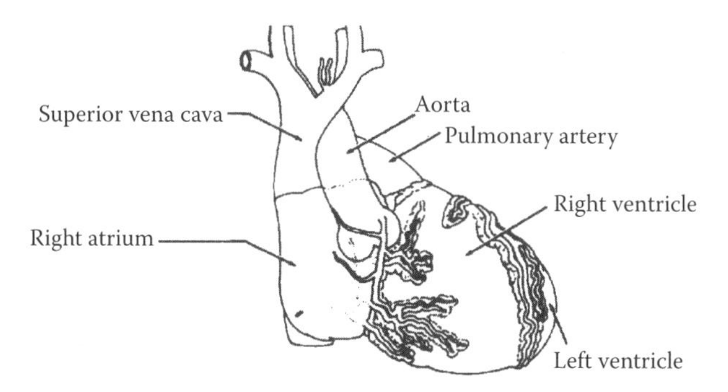

FIGURE 5.79
Schematic of the heart as seen from the right anterior oblique view used for a first-pass cardiac study. (Courtesy of GE Healthcare, Slough, Berkshire, UK.)

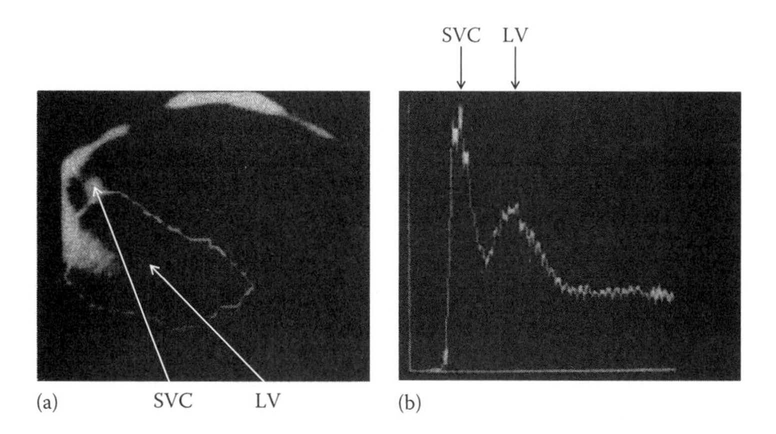

FIGURE 5.80
(a) Summed image from a first-pass study showing the contribution of the superior vena cava (SVC) and the left ventricle (LV), and an ROI used to produce (b) the activity–time curve.

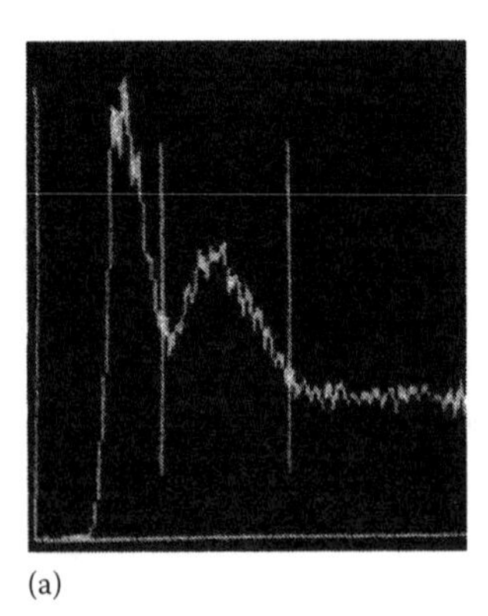
(a)

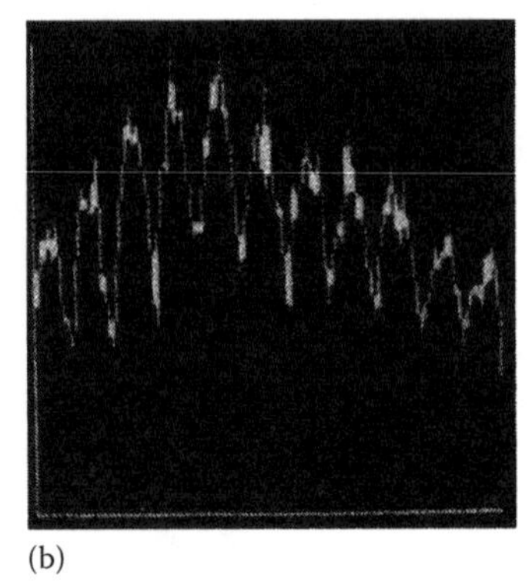
(b)

FIGURE 5.81
(a) Frame markers used to define the left-ventricular portion of the study shown in Figure 5.80 and (b) a left-ventricular volume versus time curve for this study showing the ejection fraction for a series of heart beats.

the LVEF for each heartbeat can be obtained (Figure 5.81b) free from effects of background, cardiac arrhythmia, etc. In comparison with a MUGA study, this process is more difficult to do well and statistically the data are often less than adequate. A first-pass acquisition may require 1000–1200 64 × 64 byte-mode images, a total of 4 Gbytes per study.

5.7.3 Renal Imaging

An example of a 'slow' dynamic study is the use of multiple temporal images of the kidney obtained with an appropriate radiopharmaceutical to produce renograms – activity–time curves representing kidney function (Britton and Brown 1971). The kidney extracts urine (containing low-molecular-weight waste products) from the blood in the renal arteries and excretes it via ureters to the bladder. The urine is then passed out of the body from the bladder during micturation (Figure 5.82). Dynamic renal function imaging can be carried out with any tracer that has a low molecular weight, high water solubility and low protein binding. Traditionally, the best tracers were the chelating agents DTPA and EDTA but these have been largely replaced by new tracers such as ^{99}Tc^{m}-MAG3 because this has a very high clearance rate from normal tissues, reducing background in a renogram.

The three major phases of kidney function (Figure 5.83) seen in a renogram are perfusion (0–30 s after injection), when the kidney is perfused by the radiopharmaceutical; filtration (1–5 min after injection), during which the radiopharmaceutical is filtered from the blood supply and excretion (more than 5 min after injection), when the radiopharmaceutical in the urine is passed to the bladder.

Perfusion is an important indication of the function of a transplanted kidney, when a sequential improvement of perfusion after the transplant can provide evidence of success – poor perfusion often indicates tissue rejection. The glomerular filtration rate (GFR), which is the rate of extraction of urine from blood, can be a measure of the intrinsic health of the kidneys. Abnormal function in the excretion phase can indicate problems (Figure 5.84) in the collection system, obstruction of the ureters or reflux (return of urine from bladder to kidney during micturation), any of which may lead to loss of kidney function, pain, infection and ultimately the need for transplantation.

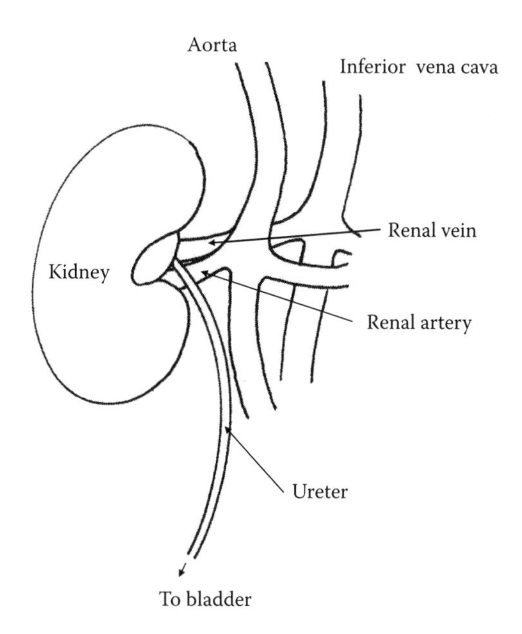

FIGURE 5.82
Schematic diagram of the anatomy of a kidney, its blood supply and ureter. (Courtesy of GE Healthcare, Slough, Berkshire, UK.)

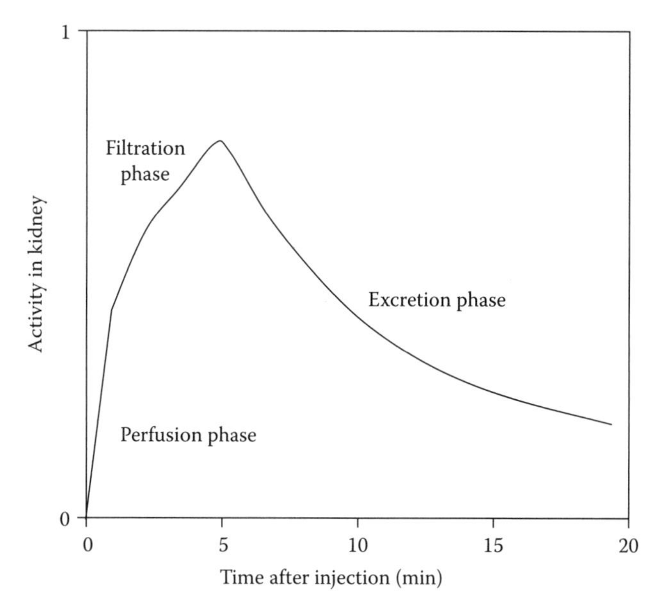

FIGURE 5.83
Typical activity–time curve for a normal kidney showing the three major phases of a dynamic renogram study: perfusion, filtration and excretion.

The *perfusion study* requires a rapid injection (bolus) of a suitable radiopharmaceutical and the kidney is imaged anteriorly (as the new kidney is usually placed anteriorly) providing typically 90 one second frames, each containing 64 × 64 pixels. ROIs placed over the kidney, iliac artery and a suitable background area (Figure 5.85) provide kidney and arterial function curves (Figure 5.86) which can be corrected for background counts.

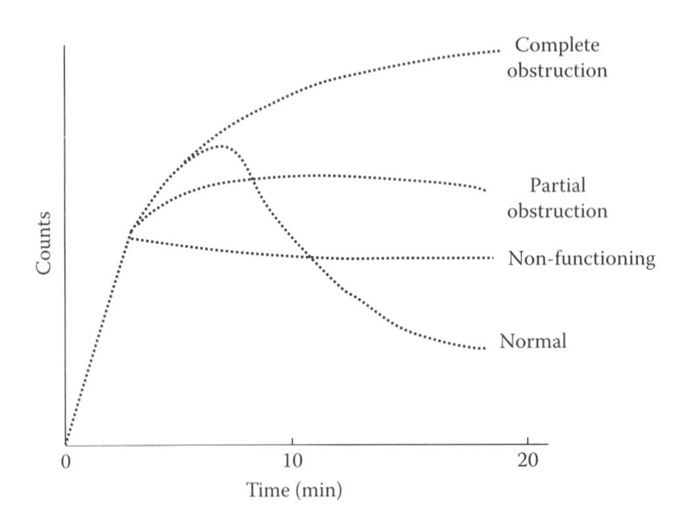

FIGURE 5.84
Activity–time curves of renal function showing the effects of various disorders compared to normal function.

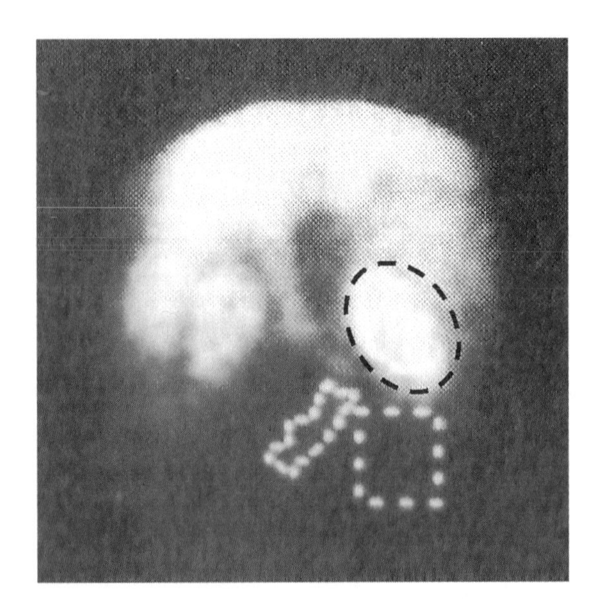

FIGURE 5.85
Summed perfusion images and ROIs to outline the kidney, the iliac artery and the background.

The ratio of the integrated arterial activity to the integrated kidney perfusion gives a *perfusion index* (PI) (Hilson et al. 1978) and sequential measurements of the PI can indicate success or failure of a transplant.

A *renogram study* again requires a bolus of injected activity to be imaged, but this time both kidneys and the LV of the heart should be in the camera FOV. The posterior view is selected in order to minimise attenuation of photons coming from the kidneys. Typically 60–90 twenty second frames, each of 64 × 64 pixels, are acquired with a total acquisition time of 20–30 min. An additional study of the micturation phase may be required, again with a number of 20 s frames. ROI analysis of the activity in both kidneys, the LV and suitable background region(s) is carried out (Figure 5.87) to produce activity–time curves

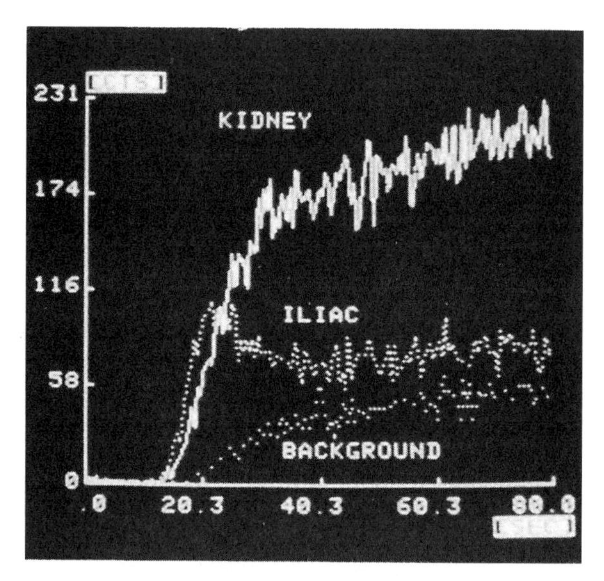

FIGURE 5.86
Activity–time curves for the ROIs shown in Figure 5.85.

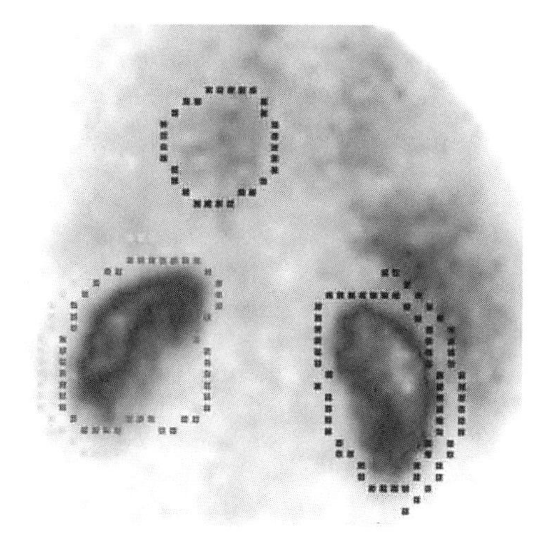

FIGURE 5.87
ROIs drawn on a summed $^{99}Tc^m$ MAG3 renogram image. The ROIs are placed over the left and right kidneys, the left ventricle and background (around each kidney ROI). The ROI counts per frame are used to generate activity–time data for renal analysis.

(Figure 5.88). The LV activity–time curve represents the spreading of the bolus by recirculation and provides the input function to the kidneys. This function must be removed from the kidney curves before estimations of GFR and excretion rates are made. The removal of the impulse curve may be performed by deconvolution (Diffey et al. 1976) or an appropriate model analysis such as fitted retention and excretion equations (FREE) (Hyde et al. 1988) or the graphical analysis technique leading to the production of Patlak plots (Patlak et al. 1983, Peters 1998).

Graphical analysis allows the determination of various parameters that describe renal function by using the dynamic data acquired with the gamma camera. The assumption

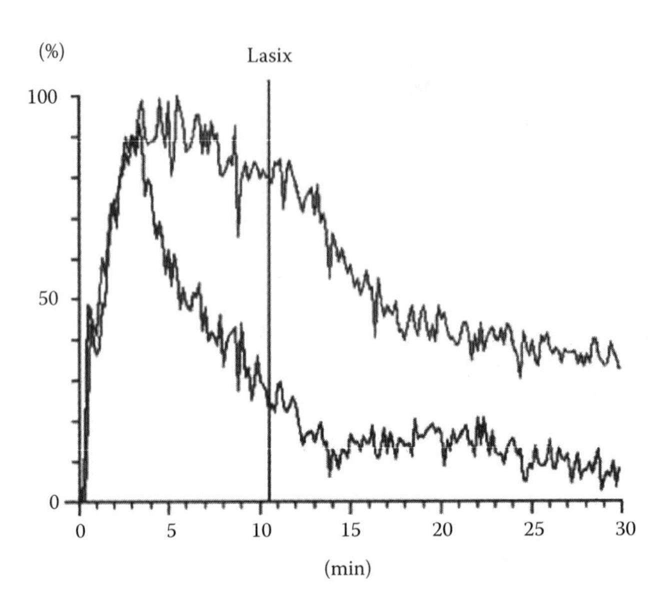

FIGURE 5.88
Activity–time curves for the two kidneys in Figure 5.87. The difference in the two curves is caused by poor excretion in one kidney. This is partly corrected when a diuretic (Lasix) is given.

is made that a three-compartmental model of renal clearance can be applied (Figure 5.89). The relationships between the renal clearance rate, β, the net tracer retention in the renal compartment as a function of time, $R(t)$, and the quantity of tracer in the plasma compartment as a function of time, $P(t)$, is given by

$$R(t) = \beta \cdot \int P(t)\mathrm{d}t. \tag{5.57}$$

However, when an ROI is placed over the kidney, the values obtained will include counts from vascular and extravascular compartments. The latter is removed by subtracting counts obtained in an ROI placed adjacent to the kidney. The vascular component must be accounted for in the model by modifying Equation 5.57 to give

$$R(t) = \alpha \cdot P(t) + \beta \cdot \int P(t)\mathrm{d}t \tag{5.58}$$

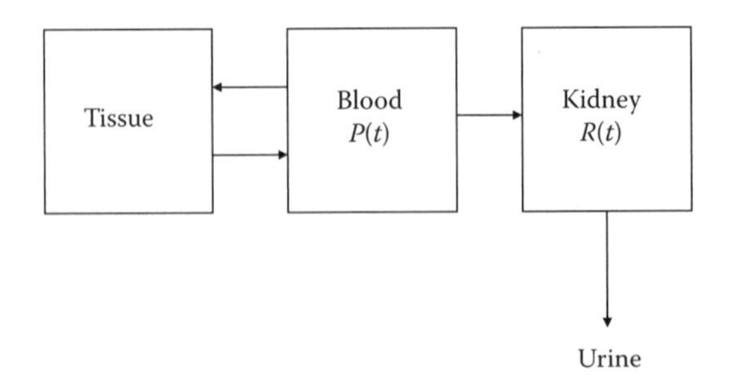

FIGURE 5.89
Compartments used for analysis of renogram data to extract parameters related to renal clearance rates and the fraction of vascular compartment in the kidney ROI.

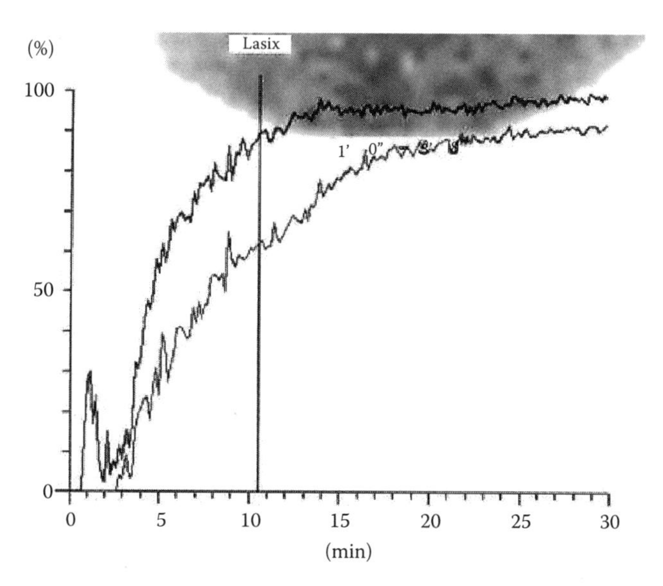

FIGURE 5.90
Percentage outflow curves for the two kidneys in Figure 5.87.

where α represents the fraction of vascular compartment in the kidney ROI. In order to determine $P(t)$ an ROI is placed over the left ventricle as representative of the vascular activity. Plotting $R(t)/P(t)$ against $\int P(t)\mathrm{d}t/P(t)$ produces a Patlak plot (Patlak et al. 1983) which, for a trapped tracer, yields a straight line which typically occurs between 1.5 and 3 min post injection for a kidney. The value of β is obtained from the slope of the curve during this period and α is the intercept at time zero. Several other parameters that might be useful in diagnosis can be derived from these data such as the retention function and percentage outflow as a function of time for each kidney (Figure 5.90). The retention function curves for each kidney are calculated by deconvolving the smoothed renal curves with the input blood function. The mean transit time of activity through each kidney is determined by integrating the retention function curve up to the point at which the outflow curve first reaches a plateau using the relationship

$$\text{Mean transit time} = \frac{\int t \cdot \mathrm{ret}(t)\,\mathrm{d}t}{\int \mathrm{ret}(t)\,\mathrm{d}t} \tag{5.59}$$

where ret(t) is the retention function curve. Abnormalities such as renal obstruction, dilated pelvis or ureters can often be diagnosed from the curves. More accurate information is gained from parametric images used to separate out the function of the uptake and excretion parts of the kidney. As with cardiac studies, these parametric images of, for instance, time-to-peak values or mean transit times for each pixel contain a wealth of information, which can often add considerably to diagnosis.

5.7.4 Summary

Dynamic scintigraphy is a technique that illustrates the applications of radioisotope imaging at its most powerful. The production of multiple time-frame images allows the performance of tissues to be measured over the period of time that they normally function. The application of mathematical models to these images allows the extraction of parameters that can be used to describe the function of the tissues on a pixel-by-pixel basis. The images and the extracted data are a substantial aid to the diagnosis of disease and the monitoring of treatment. The success of these techniques depends both on the ability of the radiopharmaceutical to trace the function of the tissue as accurately as possible and on the quality of the images produced.

5.8 Single-Photon Emission Computed Tomography

In Section 5.6 it was shown that planar scintigraphy, which involves the production of a 2D projection of a 3D radioisotope distribution, is seriously affected by the superposition of non-target or background radioactivity, which restricts the measurement of organ function and prohibits accurate quantification of that function. ECT is a technique whereby multiple cross-sectional images of tissue function can be produced. In these sectional images, the information from radioactivity in overlying and underlying sections is removed. Hence, the advantages of ECT over planar scintigraphy are improvement in image contrast, better spatial localisation of lesions, improved detection of abnormal function and greatly improved quantitative capabilities (since theoretically the count-rate per voxel should be proportional to the radioisotope concentration, in MBq mL^{-1}). Figures 5.91 and 5.92 illustrate some of these advantages. However, the spatial resolution of tomographic images is usually worse than that of planar scintigrams since the detectors are often much further away from the organ of interest, and the images are often smoothed heavily to reduce the inherent statistical noise.

The technique of ECT is usually classified as two separate modalities, namely, SPECT and PET. SPECT involves the use of pharmaceuticals labelled with radioisotopes such as $^{99}Tc^{m}$, where generally a single γ-ray is detected per nuclear disintegration (see Section 5.4.7). PET makes use of radioisotopes such as ^{18}F, where two co-linear γ-rays, each of 511 keV, are emitted simultaneously when a positron from a nuclear disintegration annihilates in tissue (see Section 5.9).

The technique of ECT can also be classified into limited-angle or transaxial tomography (Figure 5.93). In *limited angle* (or *longitudinal*) *ECT*, photons are detected within a limited

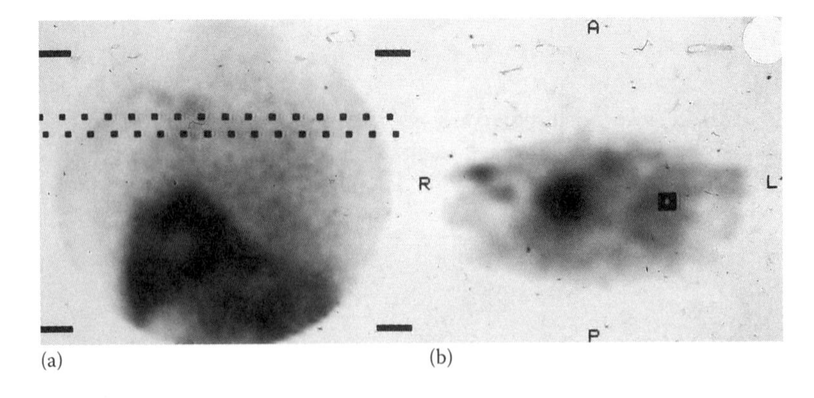

FIGURE 5.91
$^{99}Tc^{m}$-HMPAO images showing lesion in thorax with increased blood flow: (a) single projection and (b) tomographic slice (at level indicated on planar projection) illustrating improved spatial localisation in SPECT slice.

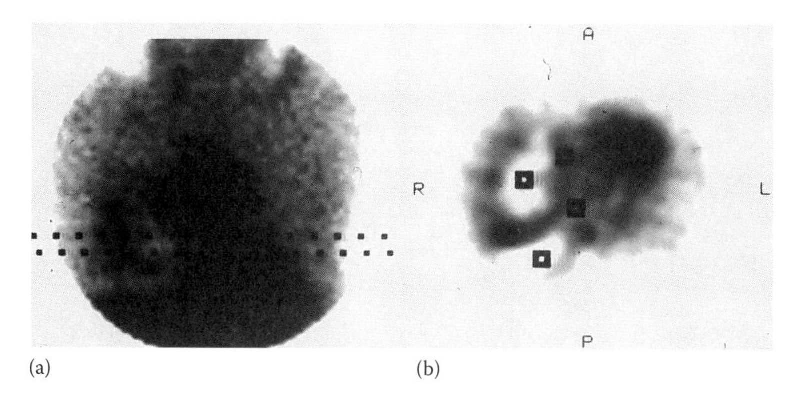

FIGURE 5.92
$^{99}Tc^{m}$-HMPAO images showing liver lesion with decreased blood flow: (a) single projection and (b) tomographic slice (at level indicated on planar projection) illustrating improved contrast in SPECT slice.

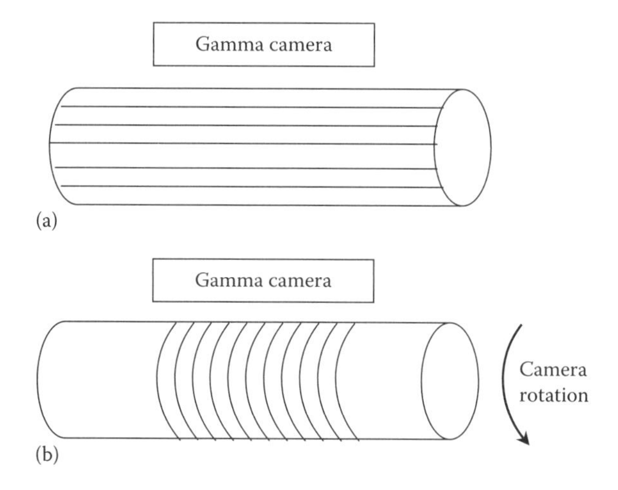

FIGURE 5.93
Schematics showing the orientation of the imaging planes for (a) longitudinal and (b) transaxial tomography with respect to the imaging system.

angular range from several sections of the body simultaneously. In this case, the image planes that are reconstructed are parallel to the face of the detector(s) (Figure 5.93a). In *transaxial* (or *transverse-section*) *ECT*, the detectors move around, or surround, the body to achieve complete 360° angular sampling of photons from either single or (more commonly) multiple sections of the body. Here the reconstructed image planes are perpendicular to the face of the detector(s) (Figure 5.93b). In both cases, when multiple sections are imaged, the reconstructed data may be redisplayed in other orthogonal planes (Figure 5.94). Oblique planes can also be obtained from the 3D dataset if necessary for easier image interpretation.

In this section, after a brief overview of longitudinal SPECT, we will concentrate in more detail on transaxial SPECT. Although there are some areas of overlap between SPECT and PET, the latter will be covered in more detail in Section 5.9. The technique of SPECT has had a long period of development spanning many decades and this section will cover some of this interesting historical development.

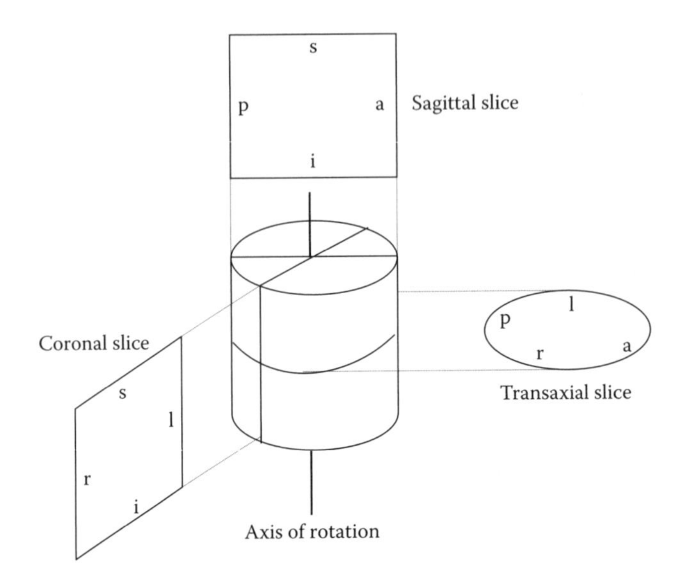

FIGURE 5.94
Orientation of transaxial, sagittal and coronal slices produced in tomographic imaging. s, superior; i, inferior; a, anterior; p, posterior; l, left; r, right.

5.8.1 Limited-Angle Emission Tomography

Several different types of equipment were developed during the 1970s for limited-angle or longitudinal-section emission tomography. The basic principle of these different systems is the same: the radioactive distribution is viewed from several different angles, so that information about the depth of the radioactive sources can be obtained. One of the most sophisticated systems was Anger's multiplane tomographic scanner (Anger 1969), which was later marketed by Searle as the Pho/Con Imager (Figure 5.95). This device combined the tomographic features of large-detector rectilinear scanners with the Anger camera electronics, so that six longitudinal tomograms were produced from a single scan of the object. In order to bring each plane into focus, the images were repositioned electronically. Figure 5.96 shows multiple longitudinal sections through the skeleton produced by a Pho/Con Imager.

The stereoscopic information required for limited-angle ECT was also acquired using a gamma camera with special (moving or stationary) collimators (e.g. Freedman 1970, Muehllehner 1971, Rogers et al. 1972, Vogel et al. 1978). A variety of moving (usually rotating) collimators were designed and built. These included slant-hole, pinhole, diverging and converging multi-hole collimators. The stationary collimators fell into two categories: coded apertures (such as the Fresnel zone-plate, stochastic multi-pinhole and random pinhole), which produced complex images that needed to be decoded, and segmented apertures (such as the multiple pinhole and the quadrant slant hole), which produced separated images, each providing a different angular view of the object. The reconstruction (decoding) processes for coded-aperture imaging were fairly complicated and time consuming. Segmented-aperture reconstruction was simpler.

Since insufficient angular sampling of the 3D object is obtained in limited-angle ECT, the tomographic images include blurred information from over- and underlying planes of activity. Methods of deblurring limited-angle tomograms were developed using either deconvolution or iterative techniques (e.g. Webb et al. 1978), but the depth resolution (effective slice thickness) was still poor compared with the in-slice resolution and worsened

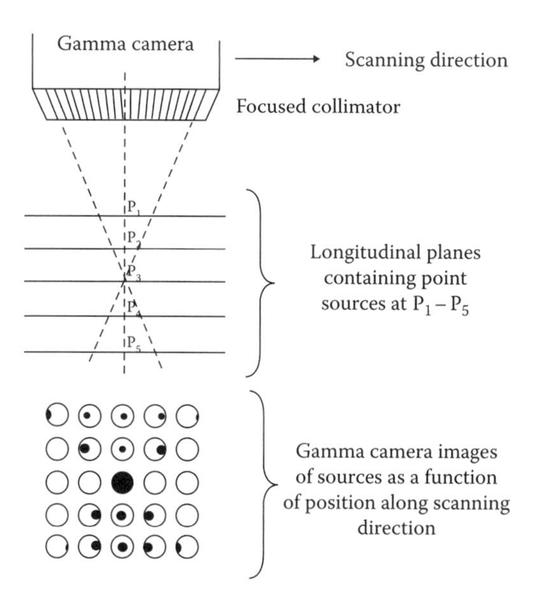

FIGURE 5.95
Multiplane longitudinal tomography using a Pho/Con Imager. Images of activity at five different planes in front of the focused collimator move at different rates across the detector. By compensating for these differences in motion for each plane, several longitudinal tomograms (i.e. in-focus images) are produced. (Adapted from Anger, 1969.)

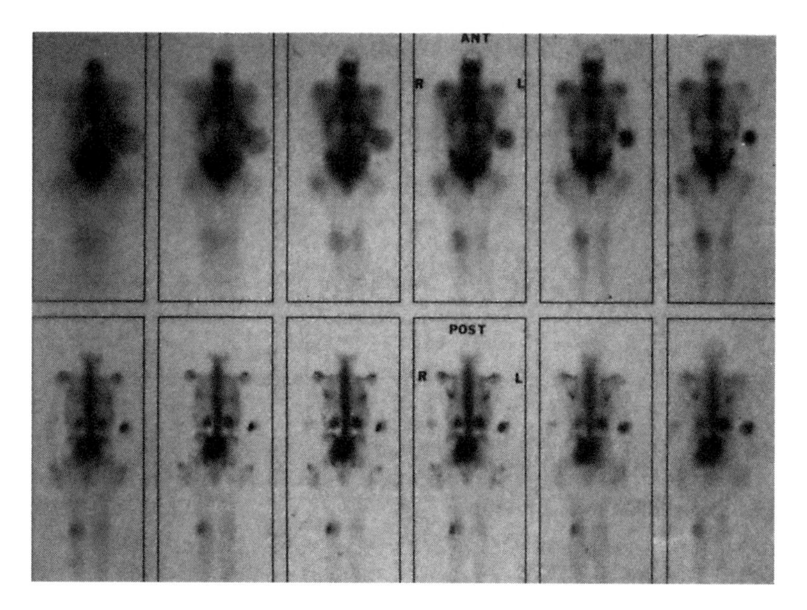

FIGURE 5.96
Twelve longitudinal tomograms of the skeleton recorded using a Pho/Con Imager.

with increasing depth. There are several other problems associated with limited-angle tomography: noise propagation occurs during the decoding process; compensation for photon attenuation is often ignored; the decoding process is computationally costly; and distortions can occur, especially with pinhole systems. As a result of these problems, limited-angle tomograms are essentially qualitative unless orthogonal views are used

(Bizais et al. 1983), and the technique of limited-angle tomography was gradually replaced by transaxial systems which, although not without their own problems, do offer the potential for quantitative imaging. However, one advantage for the stationary-detector limited-angle devices was the possibility of performing fast dynamic ECT.

5.8.2 Basic Principles of Transaxial SPECT

The basic principles of transaxial ECT are similar to those of X-ray transmission computed tomography (CT) (Chapter 3). However, there are several important differences between ECT and X-ray CT. In comparison with X-ray CT, where the *in vivo* distribution of X-ray linear attenuation coefficients μ is determined, the reconstruction of ECT images is more complicated, since an attempt is usually made to determine the distribution of activity A in the presence of an unknown distribution of μ. In SPECT, the projection data (see Figure 5.97) can be written in the form of an attenuated Radon transform:

$$p(x',\theta) = \int ds\, f(x,y) \exp\left(-\int dl\, \mu(u,v) \right) \tag{5.60}$$

where

$p(x',\theta)$ is the value of the projection at angle θ and position x',
$f(x,y)$ is the 2D distributions of radioactivity,
$\mu(u,v)$ is the 2D distribution of linear attenuation coefficients, and
the integrals over s and l are along the LOR as defined by x' and θ.

The integral over s is along the whole length of the LOR, whereas the integral over l is along the LOR from each point (x,y) in the object to the detector at angle θ.

ECT provides images of physiological and metabolic processes with a spatial resolution that is of the order of 10 mm. However, this can vary from <5 mm for PET to >20 mm for SPECT, depending on the instrumentation and the radionuclide used, and is much poorer than the anatomical resolution of X-ray CT (<1 mm). The radiation dose associated with ECT is distributed throughout much of the body for the effective lifetime of the

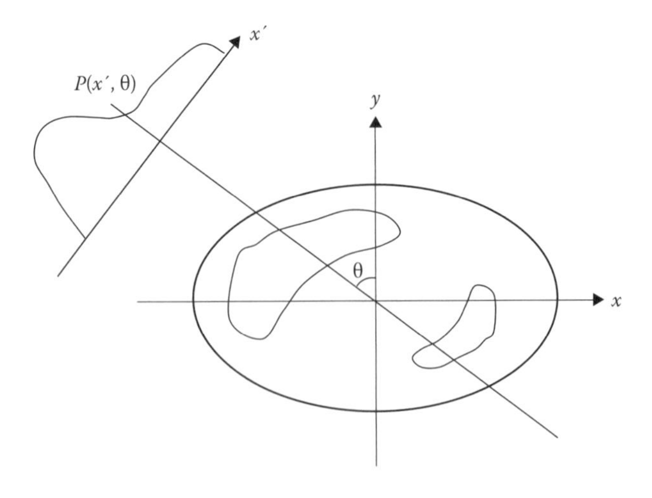

FIGURE 5.97
Schematic of an image profile (or the 1D projection) of radioactivity in a single transaxial slice for a camera with a parallel-hole collimator.

radiopharmaceutical, whereas in X-ray CT the radiation dose is limited to the duration of the X-ray exposure and to the section of the body irradiated. Hence, if identical radiation doses were given to a patient, the photon utilisation in SPECT is typically 10^4 times lower when compared with X-ray CT.

5.8.2.1 Data Acquisition

If suitable detectors are attached to a computer-controlled gantry (Figure 5.61), which allows rotation about a supine (or prone) patient, multiple views (or 2D projections) of the 3D radiopharmaceutical distribution can be acquired. For example, a gamma camera coupled to a parallel-hole collimator provides a set of 2D images consisting of multiple profiles, each profile representing a 1D projection of the radioactivity in a single slice of the patient (Figure 5.97 and Equation 5.60). Notice that the 3D imaging problem is first reduced to a 2D problem via the use of lead collimation, so that the 3D object is essentially divided up into multiple 2D sections. The 2D imaging problem is further reduced to 1D by taking linear measurements, that is, each 2D cross section is represented by a set of discrete 1D profiles. Ideally, each point on the profile should represent the linear sum of the activity elements along the line of view of the detector as determined by the collimator. However, in practice, due mainly to photon attenuation, this is not true. Nonetheless, if data sampling is adequate in both the linear and angular directions, then it is possible to use well-established mathematical techniques (known as image reconstruction from projections, and described in Section 3.6) to reconstruct cross-sectional images that represent the radiopharmaceutical distribution in the body. If photon attenuation and other image-degrading factors can be corrected for accurately, then quantitative SPECT images can be produced (Tsui et al. 1994a, 1994b, Rosenthal et al. 1995).

The 1D profiles relating to a single section can be displayed as a function of gantry angle (Figure 5.98). This form of display is referred to as a sinogram, since a point source offset from the centre of rotation (COR) would appear as a sine wave.

5.8.2.2 Specific Problems Associated with Emission Tomography

There are several problems to be solved in order to achieve quantitative SPECT. These are summarised in Table 5.14 and more details of the solutions to these problems will be provided in subsequent sections, especially Section 5.8.5.

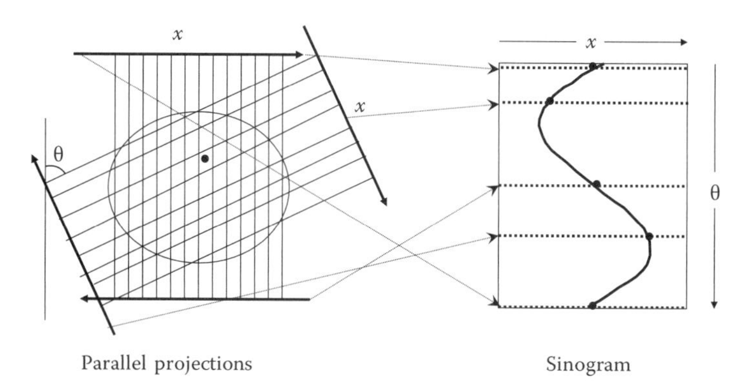

FIGURE 5.98
Diagram illustrating how a sinogram is formed from a set of 1D profiles.

TABLE 5.14

Problems and solutions in SPECT.

Problems	Solutions
Geometry, and spatially variant PSF	Special collimators, combined views, iterative reconstruction
Detector nonuniformity	Normalisation, uniformity scan and correction
Photon absorption	Attenuation correction
Scattered photons	Scatter correction
Statistical noise	High-sensitivity systems, smoothing filters, regularisation
Patient and organ motion	Realignment of projection data

5.8.2.3 Data Sampling Requirements

The sampling requirements of SPECT need to be carefully determined in order to minimise statistical fluctuations. The number of projections required for ECT is $\geq \pi D/2r$, where D is the object diameter and r is the image pixel size (note that this equation was also discussed (see Equation 3.34 in the context of X-ray CT in Section 3.9.2). In addition, the linear sampling interval should be (0.4–0.7)r (Huesman 1977). For a uniform disc of activity, the image-count standard deviation (σ) is given by (Budinger et al. 1977):

$$\sigma(\%) = 120 \times (\text{No. of pixels in object})^{3/4} \times (\text{No. of photons detected})^{-1/2}. \quad (5.61)$$

This equation can be used to give an indication of the uncertainty on the reconstructed image counts for a given number of acquired events and matrix size. It can also indicate the magnitude of the propagation of errors during the image reconstruction process, since the value of σ is much greater (typically 8–10 times) than that predicted from a Poisson distribution of counts. For a more realistic situation where there are one or two regions of uniform activity surrounded by a uniform background, the uncertainties are reduced and given by the empirical relationship (Budinger et al. 1978):

$$\sigma(\%) = 120 \times (\text{No. of photons detected})^{1/4} \times (\text{No. of counts per target pixel})^{-3/4}. \quad (5.62)$$

A typical gamma-camera SPECT acquisition may consist of 128 planar views (or projections), each containing 128 × 128 image pixels and acquired at 64–128 discrete angles covering 360° around the patient. Whereas 180° acquisition is adequate for single-slice X-ray CT, 360° is almost always required for SPECT in order to overcome both the effects of photon attenuation and the degradation of spatial resolution with distance from the collimator (referred to later (in Section 5.8.5.5) as depth-dependent point response, DDPR).

5.8.2.4 Image Reconstruction

The algorithms used for reconstruction of transaxial ECT images are similar to those already described for X-ray CT (Chapter 3). Image reconstruction algorithms used in SPECT and PET will be described in more detail in Sections 5.8.4 and 5.9.6, respectively.

5.8.3 Instrumentation for Transaxial SPECT

The most common hardware used for transaxial SPECT is one or more gamma-camera heads mounted on a rotating gantry. Hence, the same camera head can be used for the full range of nuclear medicine imaging, namely, planar, dynamic, whole-body and tomographic scanning. However, special-purpose SPECT devices have also been built, mainly in academic medical imaging centres.

System design involves the usual compromises between:

- spatial resolution, system sensitivity and image noise;
- tomographic efficiency and conventional capabilities;
- single slice and multiple slice; and
- data acquisition time and image quality.

Advantages of camera-based systems over single-slice machines are:

- better photon utilisation;
- simultaneous acquisition of many sections so that dynamic scanning of whole organ becomes possible, especially with multiple-headed systems;
- variable section thickness;
- oblique section scans; and
- selection of section through centre of area of interest.

5.8.3.1 Gamma-Camera-Based Transaxial SPECT

In order to use a gamma camera (see Section 5.3.2) for SPECT in clinical nuclear medicine, there are several basic requirements. First, the gamma camera must be extremely stable (both electronically and mechanically) and have excellent uniformity and spatial linearity. Using the most modern digital camera head fitted with microprocessors for real-time energy and linearity correction, it is possible to minimise the production of circular artefacts (Figure 5.99), which result from poor gamma-camera uniformity. Second, a selection of collimators that should include at least low-energy, high-resolution (LEHR) and low-energy, general-purpose (LEGP) collimators is necessary. Effort has been applied to the manufacture of special-purpose SPECT collimators (Moore et al. 1992). However, these have had less impact than that made by the introduction of multiple-headed camera systems. Third, the mechanical and electrical design of the camera gantry and patient couch is important. Poor-quality design hindered the early development of SPECT, but much more robust systems are now available. Finally, to complete the package, a state-of-the-art fast computer system with a special-purpose SPECT acquisition and processing software package is required. The latter includes control of gantry and couch motion during data acquisition, QC procedures, data correction, image reconstruction, image processing and display facilities.

There is a wide range of data acquisition parameters, the choice of which will affect the quality of the final image. The most important *acquisition parameters/variables* are:

- geometrical properties of collimator;
- radius of rotation;
- position and width of energy window(s);
- number of acquired views;
- angular range of these views;

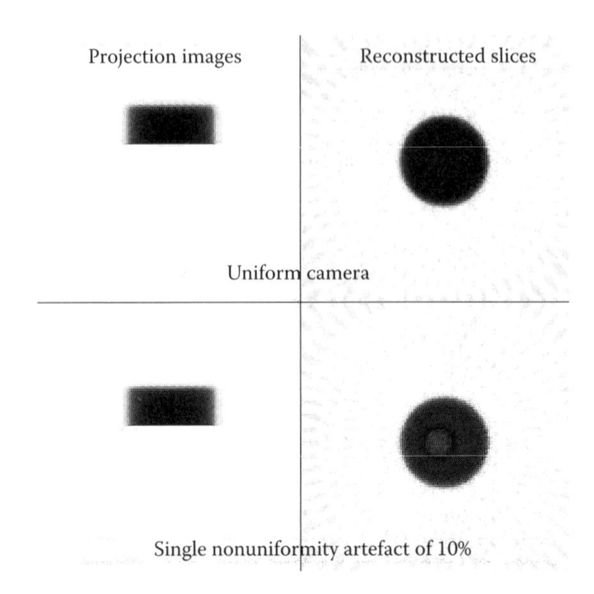

FIGURE 5.99
Illustration of camera nonuniformity, causing a circular artefact in a SPECT image. The artefact is centred on the axis of rotation. (Courtesy of J.S. Fleming.)

- pixel and matrix size of projections;
- pixel and matrix size of reconstructed image;
- time per view; and
- administered activity levels.

The choice of collimator depends on the radionuclide being used and on the organ being studied. For a high count-rate study (e.g. ^{99}Tc^{m}-HMPAO for cerebral perfusion studies), a LEHR collimator is recommended, but for a low count-rate study (e.g. ^{99}Tc^{m}-pertechnetate for evaluation of the blood–brain barrier), a higher-sensitivity collimator may be justified. In general, the thickest collimator should be chosen since, although the spatial resolution falls off as a linear function of distance from a parallel-hole collimator (see Equation 5.2), the rate of fall-off is lower with thicker collimators. The radius of rotation, defined as the distance between the outer surface of the collimator and the central axis of rotation, must be set to a minimum value consistent with patient size. The width and position of the photopeak energy window need to be carefully chosen, as discussed in Section 5.3.2. Additional energy windows may be needed, for example, for scatter correction (see Section 5.8.5.4) or for imaging radionuclides that emit several photons of different energy per disintegration. Although continuous rotation offers better photon utilisation, since the camera is constantly acquiring data, this can introduce blurring of the acquired projection data due to camera motion. For systems with resolution better than 10 mm, 'step-and-shoot' acquisition is preferable compared with continuous rotation. This then leads to the choice of the number of 'steps' or views. Improved angular sampling can be achieved by increasing the number of views but at the expense of reduced counts per view, unless the study time is extended. Streaking artefacts, caused be inadequate angular sampling, are most pronounced for small objects at the edge of the FOV, and 'disappear' as the number of views rises above about 90. 180°-angle acquisitions are used routinely for cardiac imaging (see Section 5.8.7) but 360° is required for larger and deeper organs of interest, to avoid image distortion. The acquisition matrix and pixels sizes are often set to the same values for the reconstructed images. However, while 64 × 64 matrices with 6mm pixels were satisfactory for most reconstructed images using the older, single-head SPECT systems, 128 × 128 matrices with 3 mm pixels are required for the multiple-headed systems with improved (better than 10 mm) spatial resolution. The larger matrix size requires longer reconstruction times and more storage space, but a smaller matrix would reduce image quality if it undersamples the data. If the object to be imaged is much smaller than the FOV, zooming the projections can optimise (i.e. reduce) both pixel and matrix sizes, but care in positioning the object is required to prevent movement out of the FOV during camera rotation. Finally, the longer the acquisition time per view, the more events are obtained but at the expense of possible artefacts from patient movement. A better solution is to increase the administered activity, but this obviously has to be weighed against the consequent increase in radiation dose to the patient.

5.8.3.2 Hardware Developments

Developments in the hardware used for gamma-camera-based SPECT have been aimed at improved electronic stability and better spatial resolution with the aim of achieving a reconstructed spatial resolution of <10 mm for $^{99}Tc^{m}$ SPECT so that it is much more comparable to PET. One of the most important ways of achieving improved resolution is by reducing the distance between the camera and the patient. There are several ways of achieving this for brain imaging. Examples are as follows:

- A special head support can be attached to the end of the patient couch.
- A slant-hole collimator and an angled camera head can be used (Esser et al. 1984).
- Fan-beam (Jaszczak et al. 1979, Webb et al. 1985a, Tsui et al. 1986) or cone-beam collimators, with either a symmetrical or asymmetrical geometry with respect to the axis of rotation (Jaszczak et al. 1986a, 1986b, 1988), have been proposed for both improved resolution and sensitivity.
- Astigmatic collimators (Hawman and Hsieh 1986), in which the holes converge to two orthogonal lines at different focal lengths, have been used to avoid the truncation artefacts that can occur with cone-beam collimators.
- A cut-away camera head can be used to prevent the shoulders from limiting the radius of rotation (Larsson et al. 1984), although higher-density material such as tungsten is required to provide adequate shielding.

For body imaging:

- An elliptical or non-circular orbit (see Section 5.5.6) can result in noticeably improved resolution and uniformity (Gottschalk et al. 1983, Todd-Pokropek 1983) since collimator-to-patient distance is minimised throughout detector rotation.
- Cone-beam collimators have been used for cardiac SPECT (Datz et al. 1994).
- A variable-focus collimator (Hawman and Haines 1994) was developed to avoid truncation artefacts in cardiac cone-beam SPECT. The focusing is strongest at the centre and is weakest (i.e. nearly parallel-hole collimation) at the edge of the collimator.

Improved sensitivity and resolution for both brain and body imaging have also been achieved via multiple (double, triple and quadruple)-headed gamma-camera systems. The sensitivity of these systems is in direct proportion to the number of heads. By increasing the sensitivity in this way, it is possible to achieve improved resolution simply by incorporating less smoothing in the reconstruction process. These improvements have naturally only been achieved at increased cost, but the improvements far outweigh those achieved by the special collimators described earlier. With the increased use of the gamma camera for SPECT, and the introduction of additional heads on the rotating gantry, came the introduction of rectangular rather than circular camera heads, and square rather than circular PM tubes (Kimura et al. 1990), the rectangular FOV of the heads allowing a larger axial FOV for SPECT.

One interesting development was the provision of hardware (and associated software) for simultaneous *transmission/emission* protocols (STEP), with the transmission data providing attenuation maps for attenuation correction (see Section 5.8.5.3). The simultaneous acquisition of transmission data from an external radioisotope source and the emission data is desirable since emission imaging times are long and it avoids problems with either patient movement between two separate acquisitions or image registration if the attenuation map is obtained from, for example, a CT scanner. Various geometric configurations (Figure 5.100 and Table 5.15) became available commercially, especially with the multiple-headed systems. Even with a single-headed system, provided the energy of the transmission source is less than that of the emission source, STEP can be achieved. Gadolinium-153 ($T_{1/2}$ = 242 days, $E\gamma$ = 97.4, 103.2 keV) is considered to be the radionuclide of choice for the transmission source for ^{99}Tc^{m} SPECT, but it is expensive. The fourth entry in Table 5.15 requires further explanation. A sliding electronic (spatial) window is required to separate the emission from the transmission signals. Figure 5.101 shows that appropriate parts of

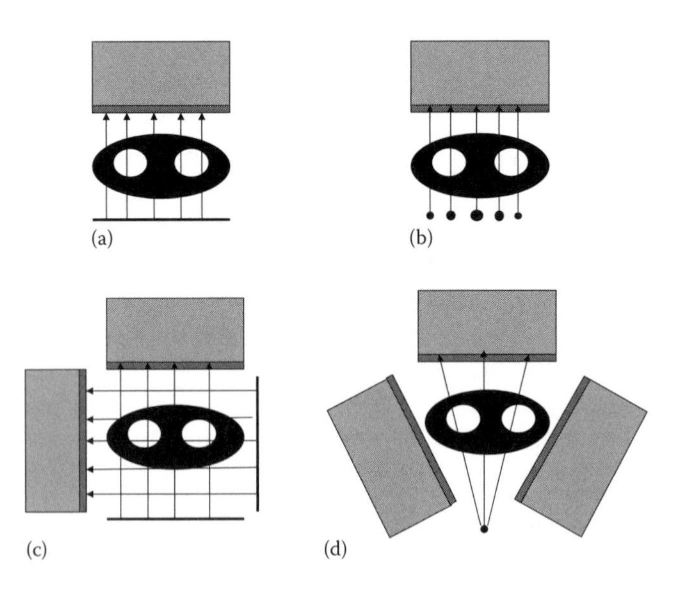

FIGURE 5.100
Diagram illustrating geometric configurations used for transmission imaging using external sources: (a) flood or moving line source; (b) multiple line sources, with size of source indicating its relative activity; (c) moving line sources on a dual-headed camera; (d) line source on focal line of fan-beam collimator for one of the three heads of a triple-headed camera. (Adapted from Bailey, 1998.)

TABLE 5.15

Hardware configurations used for transmission/emission protocols.

Collimator	Radioactive Source	References
Parallel	Flood	Bailey et al. (1987)
Fan-beam	Line on focal line	Welch et al. (1994)
Cone-beam	Point at focus	Welch et al. (1993)
Parallel	Scanning line source	Tan et al. (1993)
Parallel	Multiple line sources	Celler et al. (1998)

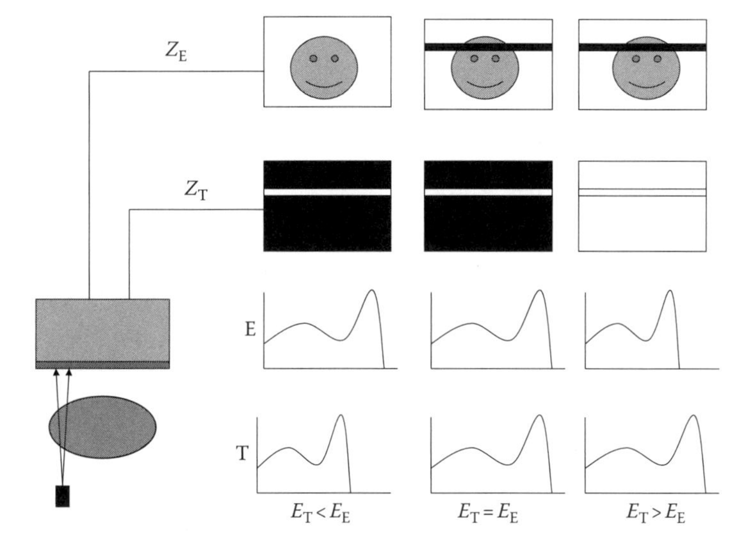

FIGURE 5.101
Illustration of the method of using a scanning line source for simultaneous transmission-emission tomography. Top rows: 2D projections for the emission (Z_E) and transmission (Z_T) data. Bottom rows: emission (E) and transmission (T) energy spectra. Different areas of the images are blanked off as shown, depending on the relative energies of the photons emitted by the emission and transmission radionuclides. (Adapted from Tan et al., 1993.)

the transmission and emission projections can be blanked out to avoid crosstalk between the two. The main problems associated with STEP are:

- crosstalk or 'spillover' between the transmission and emission photons in their respective energy windows;
- low SNR for the transmission data;
- practical implementation and ease of use;
- cost of initial system and ongoing source replacement;
- increased radiation doses to patient and staff; and
- possible truncation of transmission projection data.

The introduction of STEP added a level of complexity to SPECT, but this was justified by the substantial improvements in image quantification. It also paved the way for the development of SPECT/CT systems that are now commercially available. SPECT/CT will be discussed further under the headings of dual-modality devices (Section 5.8.3.5) and optimisation of SPECT display (Section 5.8.6).

5.8.3.3 Fan- and Cone-Beam SPECT

As with planar imaging, using a converging collimator with a gamma camera for SPECT imaging results in improved sensitivity and spatial resolution, and is particularly useful for heart and brain imaging, when the organs are much smaller than the size of the crystal, and for animal imaging. An approximate method of image reconstruction for cone-beam geometry was proposed by Feldkamp et al. (1984). For exact reconstruction of an object when using cone-beam geometry, it is required that every plane that intersects the object also intersects the focal-point path (Tuy 1983, Smith 1985). A single circular orbit of the camera head does not satisfy this condition. However, there are many different paths for the focal point which enable the aforementioned sufficiency condition to be fulfilled, for example, a circle and two lines, a helix, a circle and a one-turn helix, a circular sinusoid, a circular sinusoid and a circle. The first two options have been the most commonly employed.

5.8.3.4 Special-Purpose SPECT Systems

In contrast to gamma-camera SPECT, special-purpose devices have been designed usually for tomographic imaging alone, but occasionally with the option of performing planar imaging as well. The early special-purpose SPECT systems (some of which were described in Section 5.3.4) produced single-section tomograms and were based on scanning and rotating detectors comprising scintillating crystals with focused collimators (Kuhl and Edwards 1963, Bowley et al. 1973, Kuhl et al. 1976, Stoddart and Stoddart 1979, Stokely et al. 1980, Moore et al. 1984, Evans et al. 1986).

There were three basic advantages of the multi-crystal SPECT scanners over the rotating gamma camera. First, an increased sensitivity was achieved, since a larger crystal area was exposed for each slice imaged (Flower et al. 1979, 1980). Second, an improved resolution was achieved by the use of focused collimators. Third, the variation of resolution across the reconstructed FOV was minimised by the careful design of the focused collimator (the longer the focal length the better) (Figure 5.102).

Special-purpose SPECT systems dedicated to brain imaging have been designed with stationary cylindrical detector geometry. HEADTOME was comprised of a cylindrical array of NaI(Tl) crystal rods, while CERASPECT consisted of a single annular NaI(Tl) crystal (Zito et al. 1993). Both systems need special rotating collimators, as shown in Figure 5.103. The spatial resolution of these systems was reported to be <8.5 mm FWHM on the central axis. The excellent performance in terms of both sensitivity and spatial resolution was reflected in the quality of the images (Holman et al. 1990).

5.8.3.5 Dual-Modality Devices (SPECT/PET and SPECT/CT)

One area of instrumentation development has been the combination in hardware of SPECT with another imaging modality, namely SPECT/PET and SPECT/CT. SPECT/PET can be achieved via the modification of the electronics of a dual-head gamma camera (DHGC). This enables the user to switch between SPECT imaging (with collimators on) and PET imaging (with collimators removed, and using annihilation coincidence detection (ACD)). The idea was extended to triple-headed cameras, which provided increased angular sampling. The performance of DHGC-PET systems (Lewellen et al. 1999) gradually improved, but never matched that of the hardware described in Sections 5.9.3.1 through 5.9.3.3.

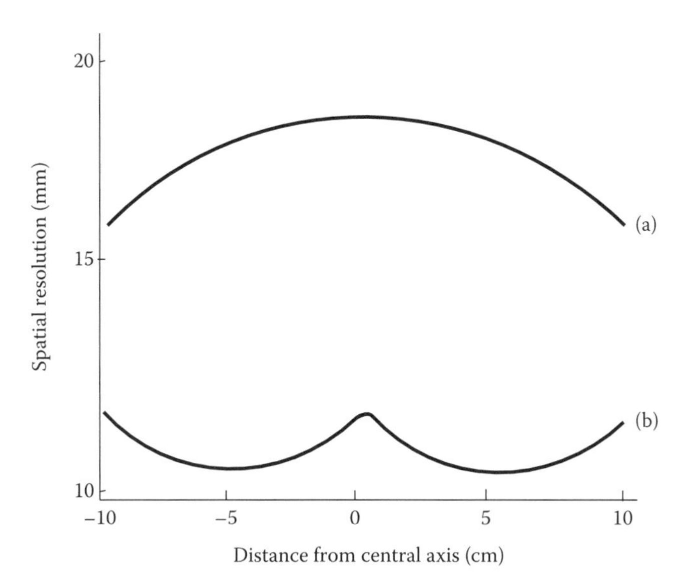

FIGURE 5.102
Variation of reconstructed spatial resolution with distance from the central axis of rotation for SPECT systems using (a) a parallel-hole collimator and (b) focused collimators.

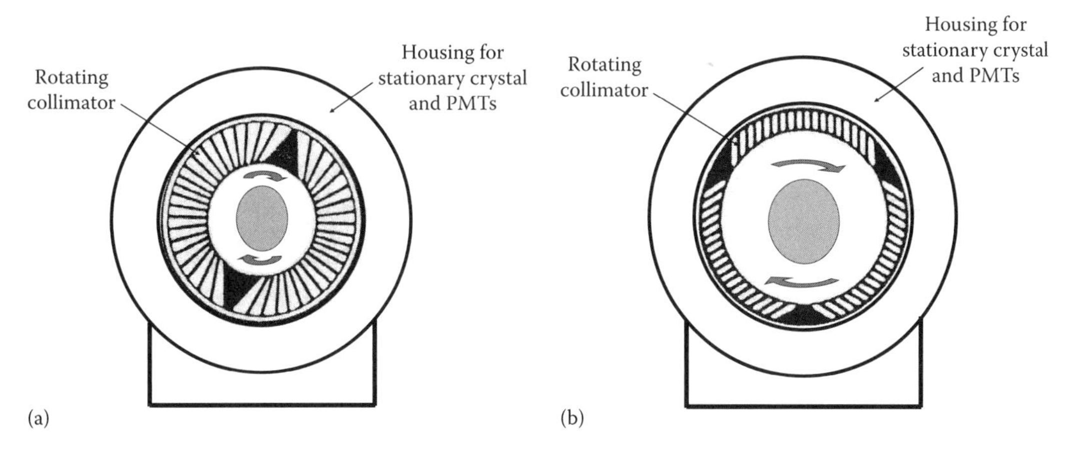

FIGURE 5.103
Stationary detector SPECT systems: (a) HEADTOME (Shimadzu) and (b) CERASPECT (Digital Scintigraphic). (Adapted from Tanaka, 1996.)

The main limitation is the count-rate capability that limits the amount of activity in the FOV. However gamma-camera PET was an improvement over the initial attempts to build 511-keV collimators for SPECT imaging of positron-emitting radionuclides (Jarritt and Acton 1996). A much more promising development, using a phoswich of LSO and NaI(Tl), for SPECT/PET is described in Section 5.9.3.8.

Dual-modality clinical SPECT/CT devices provide automatic co-registration of functional SPECT and anatomical CT images. They were built to overcome some of the problems associated with the transmission/emission systems and with registration of images from different imaging modalities. The early SPECT/CT systems (see Figure 5.104) used low-dose CT (Blankespoor et al. 1996, Bocher et al. 2000) since the main aim was to provide

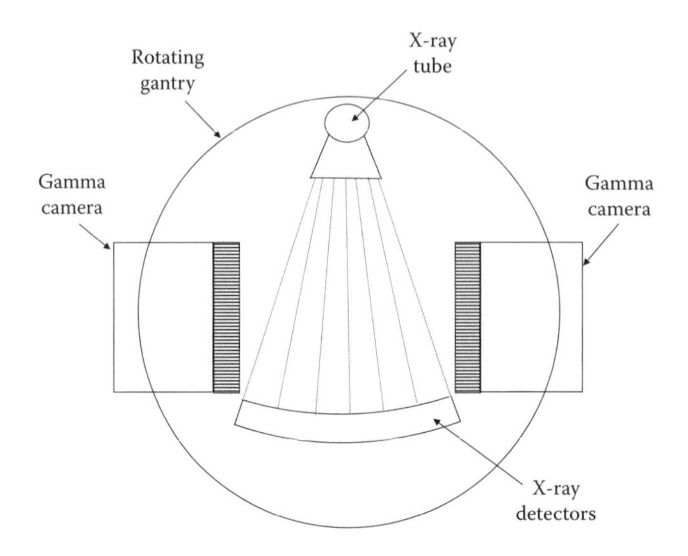

FIGURE 5.104
Schematic of an early SPECT/CT system.

X-ray electron density maps from which attenuation and scatter corrections factors could be derived. However, following the success and clinical impact of PET/CT systems in which state-of-the-art high-resolution CT was used, SPECT/CT systems now include similar high-performance CT so that radiopharmaceutical uptake can be both quantified and spatially localised on high-resolution anatomical images (Daube-Witherspoon et al. 2003, Buck et al. 2008). Figure 5.105 shows the quality of the low-dose transmission images obtained from an early hybrid gamma camera/CT scanner (Bocher et al. 2000). Other examples of SPECT/CT images, with both low-dose and state-of-the-art CT, can be seen in Belhocine et al. (2007) and Strobel et al. (2007), respectively. Despite the popularity of SPECT/CT and PET/CT they are not without their problems. These are mainly associated with patient movement and the conversion of the CT values to linear attenuation coefficients for the appropriate photon energy. Also the CT FOV is often truncated compared to the emission FOV. See Buck et al. (2008) for a review of SPECT/CT and Section 5.9.3.9 for more details on PET/CT.

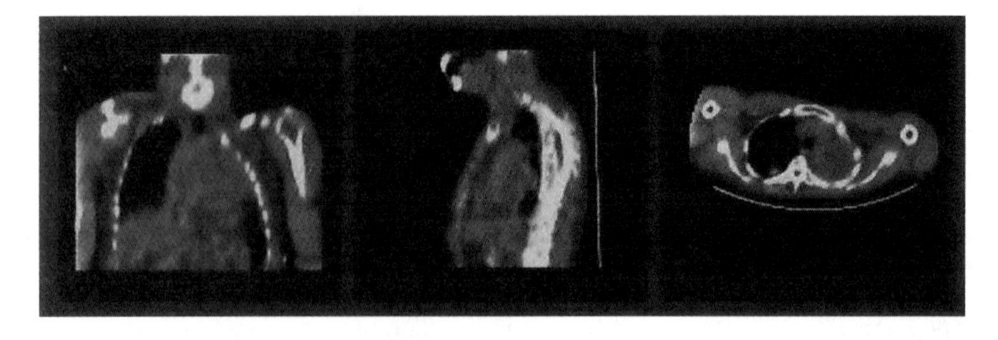

FIGURE 5.105
Coronal, sagittal and transaxial transmission images of a patient with a left pneumectomy, taken with a gamma camera mounted on an anatomical X-ray tomograph. A 2cm nodule in the posterior right apex is clearly seen. (Courtesy of GE Healthcare, Slough, Berkshire, UK.)

In addition to the clinical SPECT/CT systems, the NanoSPECT/CT system (marketed by Bioscan Inc., Washington, DC) provides very-high-resolution, high-efficiency, dual-modality *in vivo* imaging of small animals. The sub-millimetre resolution is achieved via multiplexed multi-pinhole SPECT technology and helical scanning.*

5.8.3.6 SPECT/PET/CT

Tri-modality SPECT/PET/CT systems became available for pre-clinical *in vivo* imaging in 2010. One example is the Siemens Inveon™ scanner which has the potential to be a powerful tool in the study of disease biology and the evaluation of novel therapeutic options in small animal models.

5.8.4 Image Reconstruction Methods for SPECT

The reconstruction algorithms, described in Chapter 3 for X-ray CT, require adaptation for use in SPECT in order to overcome the problems listed in Table 5.14 and especially if the aim is to achieve accurate image quantification.

5.8.4.1 Filtered Backprojection

The projections or profiles are acquired via a process often referred to as forward projection (Figure 5.106a). Unfiltered backprojection (BP) provides a blurred image, crudely representative of the object. This effect is a result of the lack of information about the actual emission point so that the backprojection is convolved with a $1/r$ function where r is the distance along the LOR from the centre of the image. In addition to the blurring, if there is insufficient angular sampling, spoke or star artefacts will be present in the image (see Section 3.9.6 and Figure 5.106b). By the use of FBP, which involves modifying the information in each profile to account for the $1/r$ function prior to backprojection, it is possible to remove most of these artefacts (Figure 5.106c). The important step of filtering each linear profile was considered

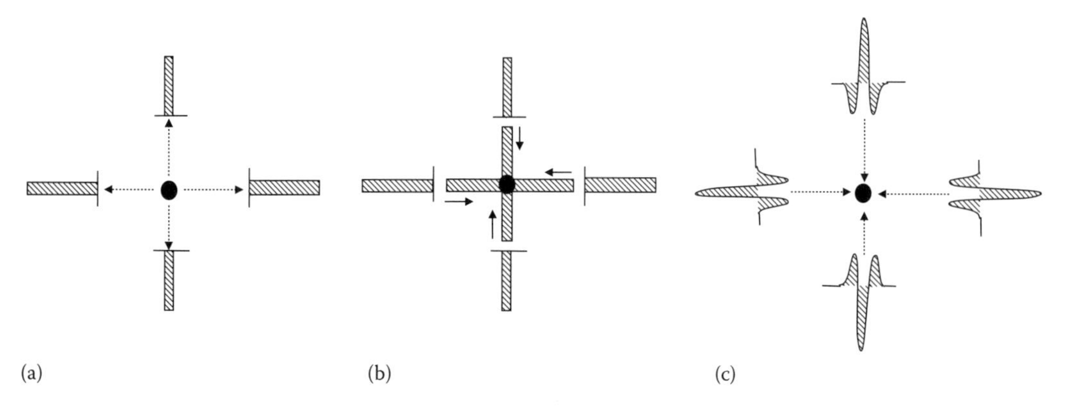

FIGURE 5.106
Schematic illustrating, for a small object and only four orthogonal projections, the processes of (a) forward projection (data acquisition), (b) backprojection (and the production of unwanted blurring and star artefacts) and (c) filtered backprojection (and the reduction of both blurring and star artefacts). (From Ott, 2003.)

* Examples of these high-quality dual-modality images can be seen at http://new.spect-ct.com/gallery/entries/ (accessed February 20, 2012).

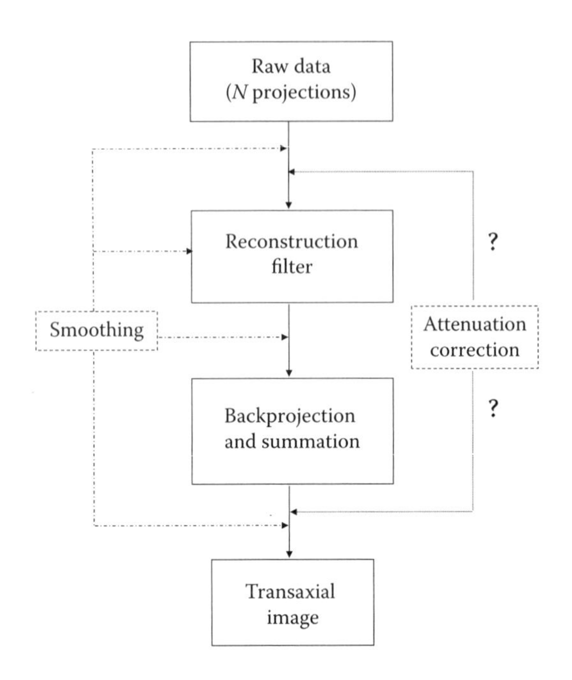

FIGURE 5.107
Flowchart illustrating the FBP process for SPECT and the options (indicated by the two question marks) for applying attenuation correction either before the application of the reconstruction filter or after the summation of the filtered backprojections.

carefully in Chapter 3, to which the reader is referred for details of the convolution kernels involved. FBP is based on Fourier theory. The filtering step can take place in Fourier or frequency space, but can also be applied in real space, when the reconstruction is referred to more accurately as convolution and backprojection (CBP). FBP or CBP is easy to implement since there is no need to store all the raw projection data prior to reconstruction. The reconstructed images (in frequency or real space) can be built up as each projection makes its contribution. However, the theory of image reconstruction from projections ignores photon attenuation and scattering which are important effects in radioisotope imaging.

Almost all methods of attenuation correction with FBP reconstruction are approximate. Attenuation correction can be applied before or after the FBP reconstruction process as illustrated in Figure 5.107. An analytical solution for the general form of Equation 5.60 has now been found (Natterer 2000) so that attenuation correction can in theory be applied during the FBP reconstruction process. Another disadvantage of FBP is the requirement that the PSF is space invariant which is not satisfied due to the depth-dependent resolution for gamma cameras. Also negative values can occur in FBP reconstructed images. Despite these disadvantages, FBP is still commonly used in SPECT.

5.8.4.2 Iterative Reconstruction Techniques for SPECT

Alternative methods of image reconstruction are those based on iterative reconstruction techniques (Gilbert 1972, Herman et al. 1973, Wallis and Miller 1993, Tsui and Zhao 1994). These have already been introduced in the context of X-ray CT in Section 3.7.

The basic method, illustrated in Figure 5.108, is to make an initial estimated image by, for example, setting each image pixel count to the mean pixel count (total counts/total number of pixels) and then to use an iterative procedure to alter this initial image gradually by

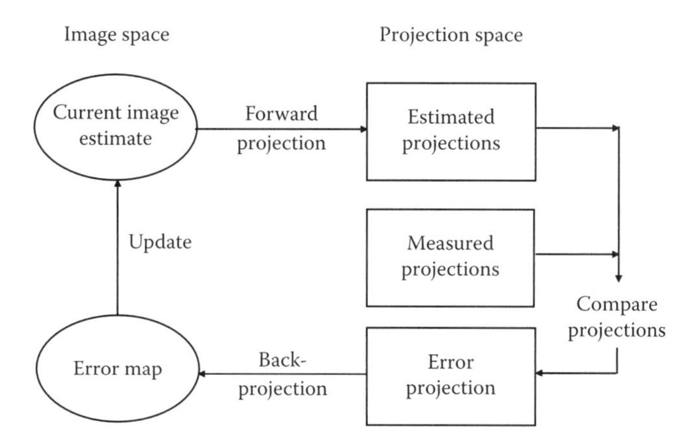

FIGURE 5.108
Flowchart illustrating the iterative reconstruction process. (From Ott, 2003.)

comparing the resultant estimated projections at each iteration with the measured projections. In forming the estimated projections from each interim image, attenuation and other effects can be taken into account. By minimising the difference between the estimated and measured projections, it is possible to re-create an accurate image of the activity distribution. Some workers minimised the number of iterations required by using, as the initial image, the FBP reconstruction with no correction for attenuation. Obviously, the estimated projections (which allow for attenuation) formed from the initial image (which does not) will be different from the measured projections, but the difference will be smaller than if a uniform initial image had been used. This technique was known as iterative convolution (Walters et al. 1976). The reconstruction time was further reduced by a two-step procedure proposed by Chang (1978).

Although iterative methods require storage of large quantities of data and can be time consuming, their main advantage for ECT is the easy incorporation of correction for attenuation and depth-dependent resolution into what is referred to as the system model. The latter can also account for non-standard geometries such as those used with fan- and cone-beam collimators. Other *a priori* information can be included in the model such as non-negativity constraints. Iterative techniques usually require optimisation of more than one free parameter. Images reconstructed using iterative algorithms have a very different appearance to those reconstructed using FBP, particularly in the structure of noise in the image. Since iterative reconstruction algorithms are non-linear, the SNR and spatial resolution may vary at different points within an image.

Many different iterative reconstruction methods have been applied to SPECT (and PET) data. These are characterised by the objective function that is minimised or maximised at each iteration (e.g. least squares or maximum likelihood (ML)). The aim is to achieve the best consistency of the image estimate with the measured data. There are two categories of algorithms: unregularised, which aims at agreement between the projection data and the image estimate, and regularised, for which the objective function becomes a compromise between consistency and extended properties of the image (e.g. not too noisy). An example of a regularised algorithm using a penalised function is the penalised weighted least squares (Fessler 1994). The regularisation is equivalent to smoothing in analytic methods of reconstruction. There has been a lot of effort to accelerate the convergence of the iterative procedure. One of the most popular methods of acceleration is via the use of ordered subsets of projections, as first suggested by Holte et al. (1990), and implemented by Hudson and Larkin (1994).

The mathematical formulae for a few of the commonly used iterative reconstruction algorithms are briefly discussed here. The forward-projection process, or the first step in Figure 5.108, aims to simulate the measurement process and can be written as

$$p_j = \sum_i M_{ji} a_i \tag{5.63}$$

where
p_j represents the projection data (or number of photons detected) in pixel j,
a_i is the activity in voxel i, and
M_{ji} is the probability that a photon emitted in voxel i is detected in pixel j.

M_{ji} can be referred to as the system model or the system transfer matrix. The more realistic the system model, the more accurate is the reconstructed activity distribution. A more rigorous version of Equation 5.63 is

$$p_j = \sum_i M_{ji} a_i + b_j + n_j \tag{5.64}$$

where
b_j is the background in pixel j due to scatter, and
n_j is the noise in pixel j.

Using the *iterative least-squares technique* (ILST) (Budinger and Gullberg 1974), the activity distribution can be determined by minimising the objective function:

$$O(a^k) = \sum_j \frac{\left(p_j - \sum_i M_{ji} a_i^k\right)^2}{\sigma_j^2} \tag{5.65}$$

where
k is the iteration number, and
σ_j is the standard deviation of p_j.

As the iterative process progresses, a better estimate of the activity distribution should be achieved and the estimated projections $\left(\sum_i M_{ji} a_i^k\right)$ should get closer to the measured projections (p_j).

One of the most widely used algorithms is the ML method which aims to provide the activity distribution for which the most likely raw dataset is the one that was actually collected. ML–EM or *maximum likelihood–expectation maximisation* is a particular implementation for obtaining the ML solution. The ML–EM algorithm is given by

$$a_i^{k+1} = \frac{a_i^k}{\sum_j M_{ji}} \sum_j \frac{M_{ji} p_j}{\sum_i M_{ji} a_i^k}. \tag{5.66}$$

Here the $k+1$ iteration estimate is equal to the kth estimate for projection data p. The algorithm assumes that the raw data are Poisson distributed. One variation on the ML–EM algorithm is the very fast OSEM (*ordered-subsets expectation maximisation*) algorithm:

$$a_i^{S+1} = \frac{a_i^S}{\sum_{j\in S} M_{ji}} \sum_{j\in S} \frac{M_{ji} p_j}{\sum_i M_{ji} a_i^S} \quad (5.67)$$

in which a subset S of projections is used. After addressing all subsets once, an iteration of OSEM is defined as done. Simplified versions of the aforementioned equations can be found in a useful review of OSEM (Hutton et al. 1997).

Iterative reconstruction methods are often referred to as statistical reconstruction techniques and the inclusion of a statistical noise model results in images with improved noise characteristics. Iterative algorithms are now available within most commercial SPECT software packages and being used in a range of clinical applications.

5.8.5 Data Correction Methods for SPECT

5.8.5.1 Correction for Nonuniformity

Nonuniformity of system response, whether from the camera or collimator, will show up as circular artefacts in the reconstructed image, unless removed by a uniformity correction using a high-count (typically 3×10^7) flood source acquisition. These flood data must be acquired with the same parameters (such as photopeak window, zoom, collimator) as the clinical study. The sinogram (see Figure 5.98) can be used to highlight the presence of camera nonuniformity or inadequate uniformity correction, since this will show up as a vertical streak in the sinogram superimposed on the sine-wave patterns of real structure.

5.8.5.2 Correction for Misaligned Projection Data

The *centre of rotation* (COR) used in the reconstruction process must be correct to avoid data misalignment and subsequent artefact production. A routine QC procedure to establish the correct COR coordinates is essential (see Section 5.10.9). The projection data are shifted in the transverse direction by the difference between the COR and the centre of the projection matrix prior to the reconstruction process.

5.8.5.3 Correction for Photon Attenuation

The use of an appropriate *attenuation-correction* algorithm is very important if quantitative SPECT images are required. The most accurate methods of correcting for attenuation involve the use of iterative reconstruction techniques (Section 5.8.4.2) and this requires an accurate 2D attenuation map (i.e. $\mu(x,y)$ for each slice). This map can be obtained either from a separate X-ray CT image (although this will require multimodality image registration which is not trivial (see Chapter 15)) or from a transmission scan using a radioisotope source or attached X-ray CT system (see Section 5.8.3.2). The latter is at the expense of increased imaging time, unless simultaneous transmission/emission imaging is feasible. The problem of attenuation correction in PET is simpler (see Section 5.9.7.2) since two photons are detected in coincidence, and thus it is easier to achieve accurate quantification with PET than with SPECT. The way in which iterative reconstruction techniques can correct for attenuation correction in SPECT is to include this effect in the forward-projection step of the iterative algorithm.

If iterative algorithms and attenuation maps are unavailable, simple *approximate methods of attenuation correction* are used in conjunction with FBP. Most FBP software packages allow user-defined variables, which may include a value for μ and a suitable choice of parameters to define a body outline for each slice. As discussed previously (Section 5.3.2), the PSRF or spatial resolution of a gamma camera varies with the distance of the source from the collimator, and one simple solution to minimise this *geometrical effect* is to combine opposed projections. This also enables the photon absorption problem to be solved approximately. We have already seen (in Section 5.6.4) that, for a point source in a uniform attenuating medium, the combined response from two opposed detectors is independent of source position and has a simple dependence on the total attenuating path length between the two detectors. Hence, provided an outline of the body can be obtained, a simple correction can be applied to the combined opposed projections. The correction factor (CF) is given by

$$\mathrm{CF} = \frac{(\mu T/2)\exp(\mu T/2)}{\sinh(\mu T/2)} \tag{5.68}$$

where

μ is the linear attenuation coefficient, and
T is the total attenuation path length, determined from the patient outline.

Equation 5.68 is derived by considering the attenuation within a uniformly distributed radioactive source, and by assuming that μ is constant within the body outline. However, the correction factor is dependent on how the radioisotope is distributed in the object. Although the combination of opposed views helps to produce an approximately space-invariant PSRF, the application of Equation 5.68 to correct for photon attenuation provides only a simple approximation for a non-uniform radioisotope distribution and a non-uniform attenuation map. In contrast to using the combination of opposed projections (i.e. a pre-reconstruction method), a *post-reconstruction correction* matrix can be applied, in which the mean attenuation factor (averaged over all angles) is used for each pixel in the reconstructed image. This is often referred to as the Chang method (Chang 1978), although his paper actually describes a two-step iterative technique. Another popular method is one that was described by Bellini et al. (1979). Bellini's method does not assume that the activity distribution is uniformly distributed and, therefore, works better than the combined projections method.

The approximate (constant μ) SPECT attenuation correction methods require the determination of the *body outline*. In some situations (e.g. in abdomen), an ellipse is used for all sections, while in others (e.g. brain) the elliptical outline varies for each section. A body outline can be determined either by imaging point sources at the ends of the major and minor axes of the body, or by use of Compton-scatter data using single- or dual-energy windows. If the attenuation is not uniform, such as in the thorax, then the use of an approximate (constant μ) method of attenuation correction should be used with caution, since it can introduce as much error as it remedies. Another idea for performing attenuation correction without using a transmission scan made use of the consistency conditions of the attenuated Radon Transform (Welch et al. 1997). Examples of attenuation correction maps using the Chang method are shown in Figure 5.109.

There have been many attempts to compare the different methods of attenuation correction. One example was the paper by Murase et al. (1987), in which the effects of non-uniform μ, errors in body contour, and statistical noise on reconstruction accuracy

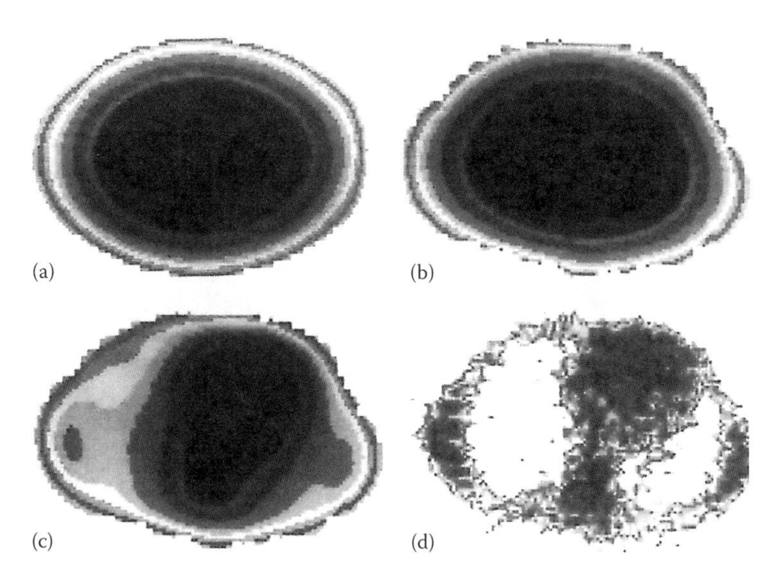

FIGURE 5.109
Attenuation correction factor maps using the Chang method are shown for (a) an elliptical outline and constant μ, (b) an accurate body outline and constant μ and (c) measured values of $\mu(x,y)$ derived from the attenuation map shown in (d). (From Cook et al., 2006. Reproduced by permission from Edward Arnold Limited.)

and computation time were studied. These authors showed that each algorithm had some disadvantages as well as advantages.

5.8.5.4 Correction for Photon Scatter

Photon scatter in both the patient and the collimator gives rise to image blurring and can lead to inaccuracies in quantification. With a standard 20% photopeak window, ~30% of counts in a $^{99}Tc^{m}$ image come from scattered photons (Webb et al. 1986). These events represent a serious problem in SPECT, particularly if accurate quantification is required. The simplest way of correcting for scatter is to use an empirical value of μ in the correction for photon attenuation (typically 0.12 cm^{-1} instead of 0.15 cm^{-1} for 140 keV photons of $^{99}Tc^{m}$), thus preventing overcorrection of attenuation. This technique is, however, an oversimplified approach to the problem. More sophisticated methods for removal of the scatter component from SPECT images can be applied either in the energy or the spatial domain. An example of an energy-domain method is the subtraction of a scatter SPECT dataset taken with a 'scatter window' placed over the Compton part of the energy spectrum (Jaszczak et al. 1984). This approach is usually applied to projection rather than reconstructed data. The main problem with this method is the uncertainty in the value of the subtraction factor (typically 0.35–0.6) which is geometry and energy-window dependent. A more sophisticated multi-window approach is the triple-energy window (TEW) subtraction method (Ichihara et al. 1993). Typical energy windows used for the TEW method are shown in Figure 5.24. The method assumes that the scattered counts form a trapezoid region in the photopeak window and is most useful for radionuclides which emit higher-energy photons in addition to those selected for imaging, or for dual-isotope studies. Examples of spatial-domain methods are 2D or 3D deconvolution of a PSRF which can be applied either pre- or postreconstruction (Yanch et al. 1988). Compared with the scatter subtraction method, 3D deconvolution of a PSRF looked by far the more promising (Figure 5.110), but this method would require a dedicated microprocessor for routine clinical use. A more sophisticated

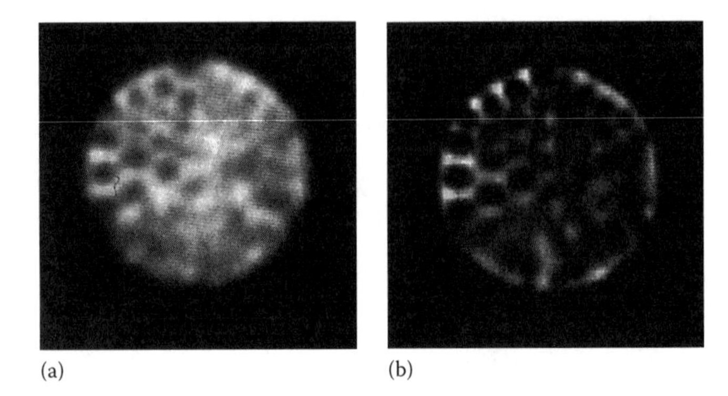

FIGURE 5.110
Use of 3D deconvolution of a PSRF to remove the effects of scattered photons from tomographic images: (a) original tomogram and (b) after 3D deconvolution.

subtraction technique makes use of Monte-Carlo simulations to determine the full spatial distribution and energy spectrum (Ljungberg and Strand 1990a, 1990b). The most accurate scatter correction techniques, especially for body regions like the thorax for which the uniform μ assumption is not valid, are those that are based on transmission data (Meikle et al. 1994, Welch et al. 1995, Welch and Gullberg 1997). For a review of scatter correction methods in SPECT, see Buvat et al. (1994) and for a comparison of two different methods, see Narita et al. (1996).

5.8.5.5 Compensation for Distance-Dependent PRF

It has been shown that, in addition to correction for attenuation and scatter, *compensation for distance-dependent point response (DDPR)* is very important (Liang et al. 1992). The variation of the PSRF with distance from the imaging system arises from various components. Although the major component is the collimator geometric response, the detector intrinsic response, septal penetration and collimator scatter also contribute to the DDPR. The geometric component can be modelled analytically, and Monte Carlo techniques can be used to model the other components. Alternatively, the overall response can be measured, but this involves much tedious work. There are different ways of implementing DDPR compensation. Like the other corrections discussed earlier, there are two approaches: pretreatment of the projections prior to FBP, and direct incorporation into an iterative reconstruction algorithm. In the first approach, the projections can be modified using stationary or non-stationary filters (King et al. 1988a, 1988b, Van Elmbt and Walrand 1993), or the frequency–distance relationship can be used (Xia et al. 1995). In the second approach, ray-tracing methods are accurate, but computationally expensive, whereas pixel-based methods are easier to implement. Rotation-based methods (Wallis and Miller 1997) are the most efficient (Figure 5.111), since the DDPR can be modelled by using row-by-row convolution in the spatial domain or filtering in the Fourier domain. The convolution kernel increases in size as the distance from the collimator increases, but a more efficient approach is to replace the large kernel with consecutive convolutions with a few small kernels (as illustrated in Figure 5.112).

5.8.5.6 Motion Correction

Patient and organ motion is a problem in ECT as a result of the relatively long (typically 20–30 min) acquisition times. *Body motion* can be tracked using fiducial radioactive markers. Any change in patient position will show up well in the sinogram data of the markers.

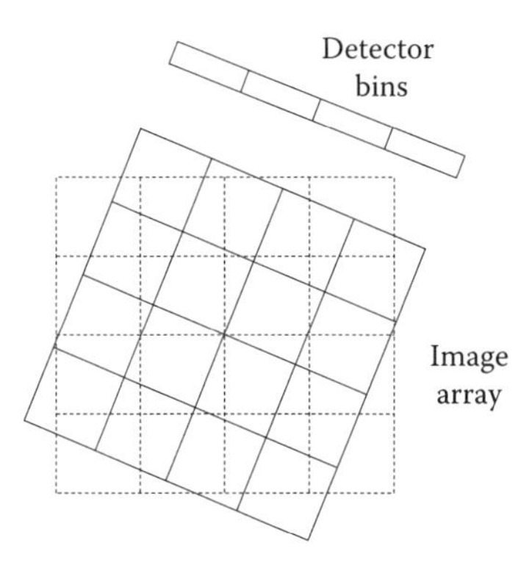

FIGURE 5.111
Diagram to illustrate a rotation-based projector, in which the image array is resampled to align with the detector bins at each projection angle. The distance-dependent PRF is obtained more easily using the rotated matrix.

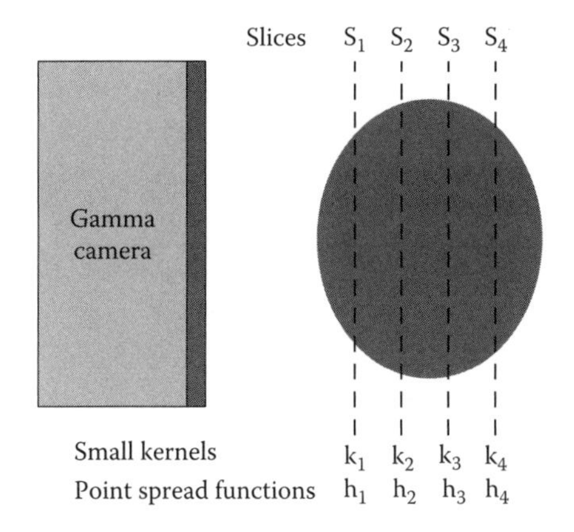

FIGURE 5.112
Diagram to illustrate an efficient approach to slice-by-slice blurring in order to model depth-dependent point response (DDPR). For the 4 slices in this simple example, the PSRFs (which get larger with increasing distance from the detector) can be equated to: $h_1 = k_1$; $h_2 = h_1*k_2$; $h_3 = h_2*k_3$; and $h_4 = h_3*k_4$; using the smaller depth-dependent kernels k_i (for $i = 1, 2, 3, 4$). The forward-projection data can then be calculated as $(((S_4*k_4) + S_3)*k_3 + S_2)*k_2 + S_1)*k_1$.

The projection data are shifted as appropriate in order to correct for body motion. Fiducial markers cannot correct for internal organ motion. A method of correcting for *organ motion* is as follows:

- find the position of the centre of the organ in each projection;
- fit a sine wave to these positions in the transverse direction and a linear function in the longitudinal direction; and
- perform a 2D translation of the projection data in order to centre the organ on a perfect sine wave in the transverse direction and a constant axial position.

5.8.5.7 Compensation for Statistical Noise

The *statistical noise* inherent in all radioisotope imaging poses a particular problem in ECT since the ramp filter (equivalent to the $1/r$ filter) used in FBP automatically amplifies the noise, which is predominantly at the high-frequency end of the spatial-frequency spectrum. Hence, smoothing filters (Section 5.5.3) are used to reduce image noise at the

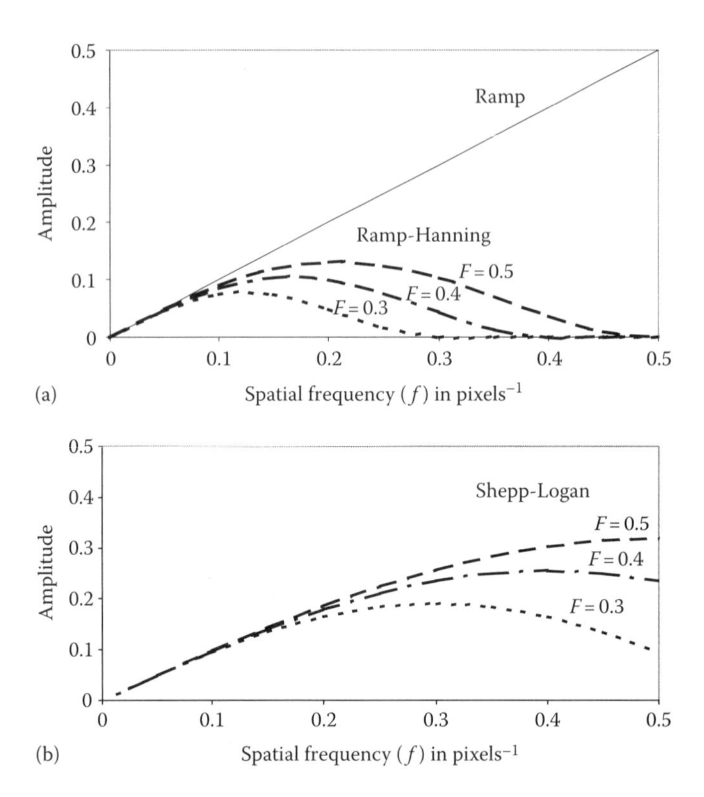

FIGURE 5.113
Filters used in FBP for SPECT with various values of the cut-off frequency, F: (a) Ramp and Ramp-Hanning and (b) Shepp-Logan.

expense of spatial resolution. A range of 2D/3D filters are usually provided within a SPECT software package and these allow pre- and/or post-reconstruction filtering as well as a choice of the smoothing process within FBP. The filter chosen is study dependent – a high (frequency)-pass filter being used for a high-statistics study but producing noisy images where acquired count rates are low. A 'too smooth' filter will degrade overall image quality and resolution, whereas a 'too sharp' filter will produce pseudostructuring of the noise that can be mistaken for real structure. Examples of Ramp-Hanning filters and Shepp-Logan filters, which incorporate the smoothing filter into the reconstruction filter (Shepp and Logan 1974), are shown in Figure 5.113. It should be noted that prefiltering of 2D projections introduces smoothing in both the axial and transaxial directions, whereas incorporation of smoothing within FBP only introduces smoothing in the transverse plane.

5.8.5.8 Correction for the Partial-Volume Effect

Partial-volume artefacts have already been mentioned in the context of CT imaging in Chapter 3. Small objects appear to be larger than they are as a result of the spatial resolution of the imaging system (King et al. 1991). The partial volume refers to that fraction of the image voxel or resolution volume that is occupied by the object. Although the reconstructed counts are smeared over an area that is larger than the object size, the total reconstructed counts should be preserved. This will only be true if all the other corrections described earlier are in place. The correction for the partial-volume effect (PVE) can be achieved by dividing the reconstructed counts by the *recovery coefficient* which is defined (by Hoffman et al. 1979) as the ratio of the apparent isotope concentration in the image to the true isotope concentration. The recovery coefficient is determined by imaging different sized objects filled with the same radioactive concentration. Other imaging modalities with better spatial resolution (such as X-ray CT, MRI and ultrasound) can be used to provide an estimate of the object size when selecting the appropriate recovery coefficient, although this relies upon the functional size of the object matching the anatomical size.

5.8.6 Optimisation of SPECT Data Processing and Display

In addition to the choice of acquisition parameters discussed in Section 5.8.3.1, there are also several parameters which require careful consideration during data correction and image reconstruction. The most important *processing parameters/variables* are:

- number of counts in uniformity flood image;
- slice thickness/number of slices;
- angular range;
- smoothing filter (pre-/post- and/or that used within FBP);
- reconstruction algorithm; and
- attenuation (and scatter) correction.

Some of these parameters have already been mentioned in previous sections, so only those that have not been covered previously are discussed here.

Large slice widths (in pixels) may be chosen to improve image statistics at the expense of PVEs (see end of previous section) and the subsequent loss of contrast of small abnormalities. The number of slices is usually limited to cover just the organ of interest, thus reducing the reconstruction time.

Commercial software packages all offer standard FBP and a few of these allow limited-angle reconstructions (e.g. 180°, 270°), which may be valuable in enhancing lesion contrast in some studies (Ott et al. 1983a). Most commercial companies now offer the OS-EM algorithm as well, but users need to experiment with and understand how the choice of free parameters affects the SPECT images.

Display options need the highest-quality hardware and software to maximise the visual impact of tomographic data. The ability to produce single- or multiple-slice images, to interrelate transaxial, sagittal and coronal sections, and to produce 3D shaded-surface display is fundamental to the optimal use of SPECT and PET. The use of cine mode in the display of projections or reconstructed images can be very useful, especially if the individual images are noisy. Cine display of projection data is an excellent way of demonstrating whether or not correction for patient movement is required. Maximum intensity projections, in which projections are formed from the pixels in the reconstructed image with maximum intensity for the corresponding LOR, can also be useful. Finally, fused images, in which the SPECT data (often in colour) are fused with registered CT data or attenuation maps derived from simultaneous transmission scans, can greatly improve the ease with which SPECT images can be interpreted. The viewing of fused images is usually much easier than viewing separate images from the two modalities. An example of image fusion and 3D display for SPECT/CT angiography can be seen in Strobel et al. (2007).

5.8.7 Clinical Applications of SPECT and SPECT/CT

Although SPECT can be performed using any of the radiopharmaceuticals listed in Table 5.10, the majority of clinical SPECT studies are of the heart, skeleton and brain, with the remainder being mainly tumour imaging. Figures 5.114 through 5.119 illustrate some of these clinical applications. In general SPECT provides increased sensitivity (i.e. percentage of patients with the disease of interest that will return a positive result) and specificity (i.e. percentage of patients without the disease that will return a negative result) when compared with planar scintigraphy.

The main application of SPECT in cardiology is the study of regional myocardial blood flow (Figures 5.114 and 5.115). Sensitivity for the detection of coronary artery disease is higher with SPECT than with planar imaging. However, the specificity is still limited by artefacts due mainly to inadequate correction for attenuation, scatter, DDPR and patient motion. Examples of the effects of breast attenuation, diaphragmatic attenuation and gut uptake are shown in Figure 5.116. Physicists play an important role in helping to identify these artefacts and in devising ways of preventing them. For example, new 3D iterative reconstruction methods have been developed to reduce the attenuation artefacts. Also, simple

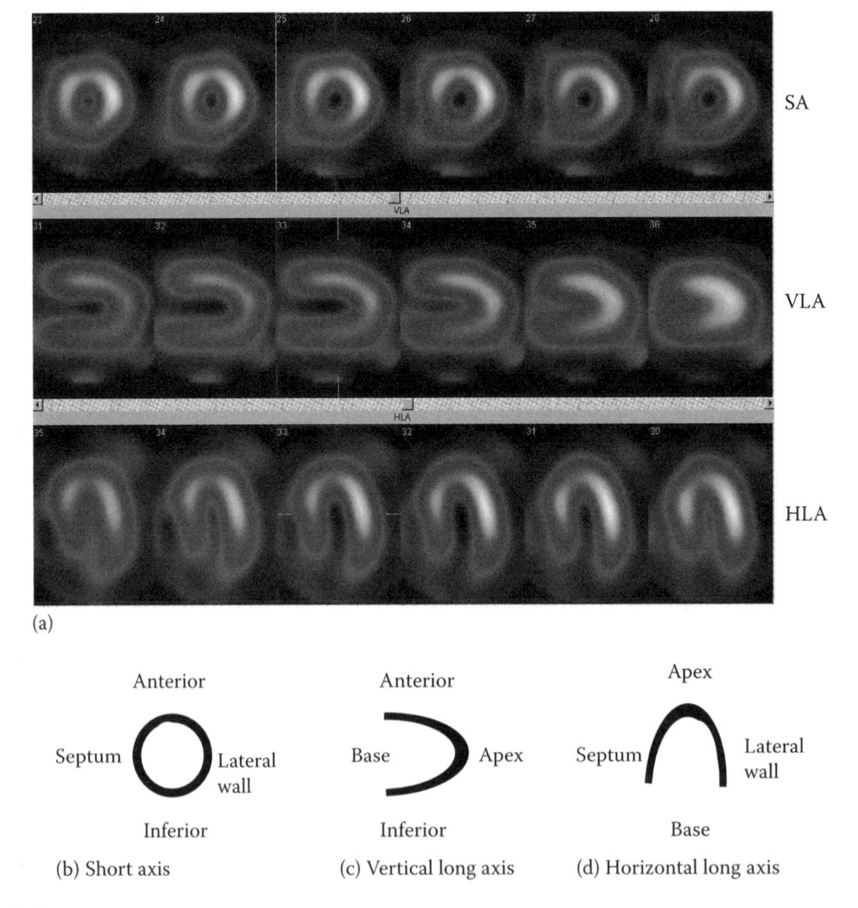

FIGURE 5.114
(See colour insert.) (a) ^{99}Tc^{m}-sestamibi SPECT scans showing myocardial perfusion in a normal male. Each row shows a different view: SA, short axis; VLA, vertical long axis; HLA, horizontal long axis. (b)–(d) Schematics showing the orientation of each view. (Courtesy of S. Sassi.)

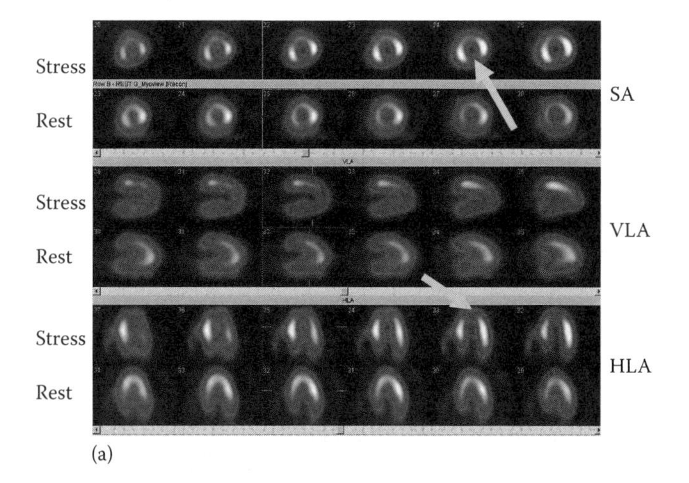

(a)

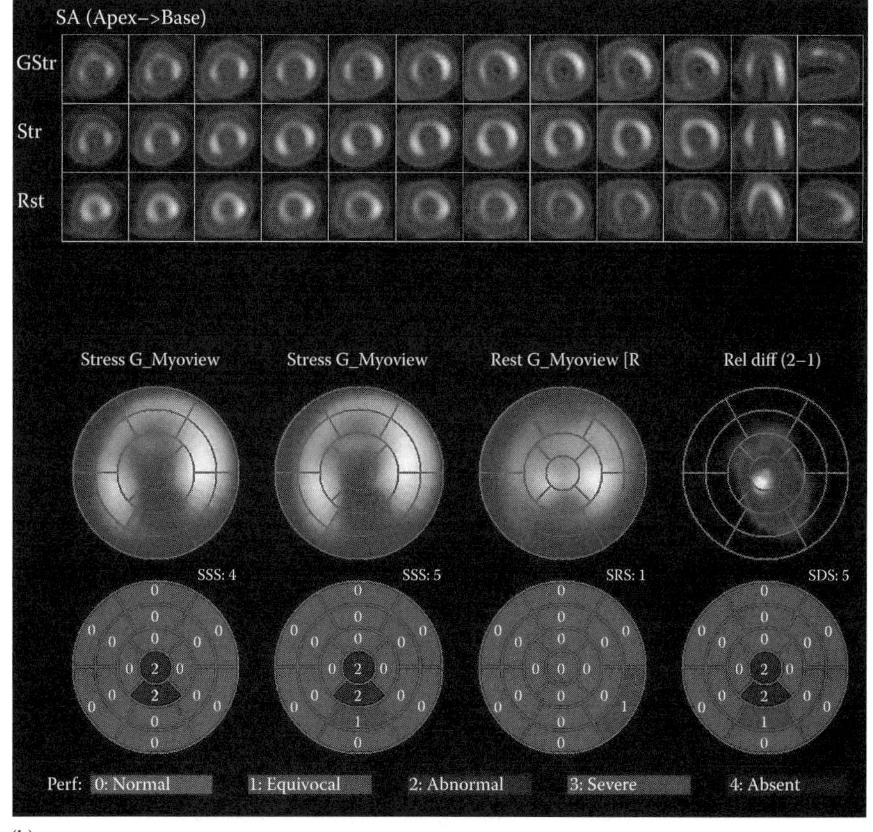

(b)

FIGURE 5.115
(See colour insert.) Myocardial perfusion images for a patient with ischaemia. (a) Images taken under different conditions: stress and rest. The abnormality is indicated by the yellow arrows. (b) Polar plots (or bull's eye display) that are derived from the short axis views and that include a detriment map (i.e. rest minus stress) and statistical analysis results.

(*continued*)

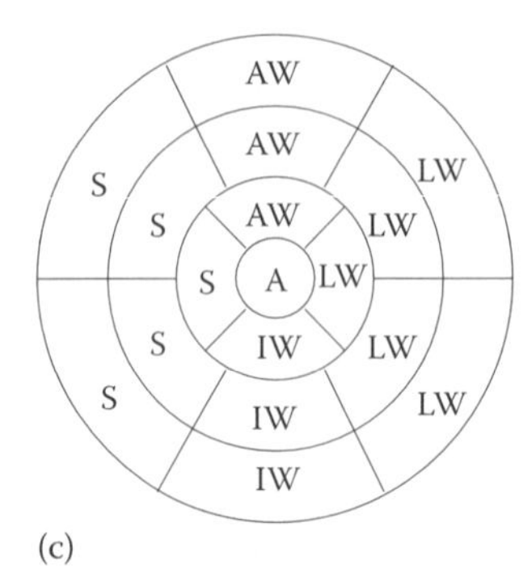

FIGURE 5.115 (continued)
(See colour insert.) (c) Orientation of the segments of the polar plots: A, apex; AW, anterior wall; S, septum; IW, inferior wall; LW, lateral wall. Statistical analysis of the polar plots is used to classify the blood flow in the segments as normal, equivocal, abnormal, severe, or absent. (Images in (a) and (b) are courtesy of S. Sassi.)

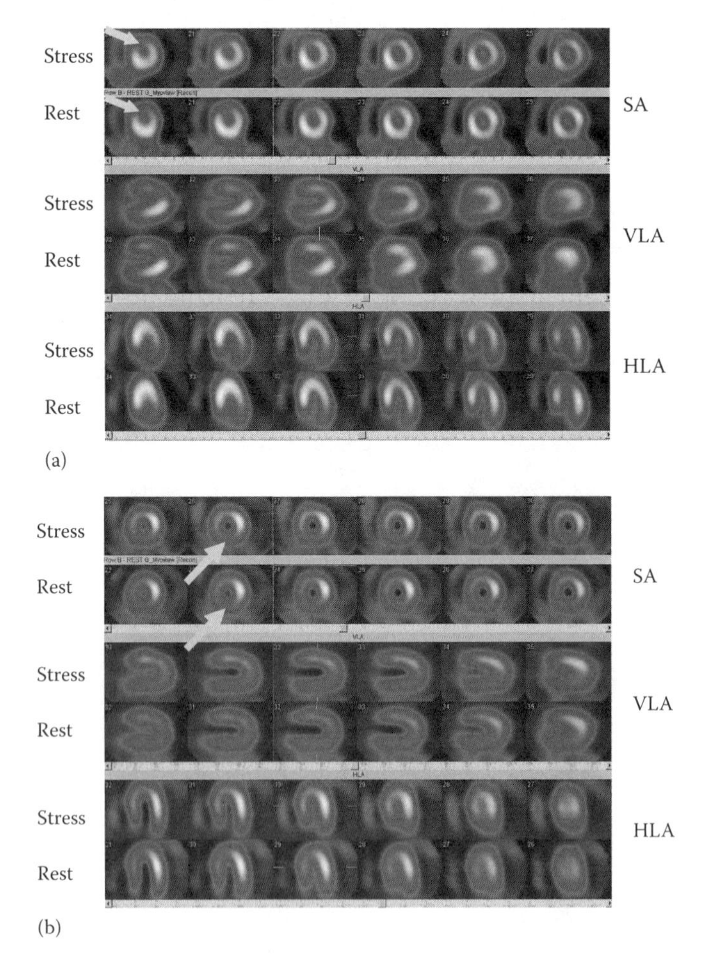

FIGURE 5.116
(See colour insert.) Examples of artefacts in cardiac SPECT images due to (a) breast attenuation, (b) diaphragmatic attenuation.

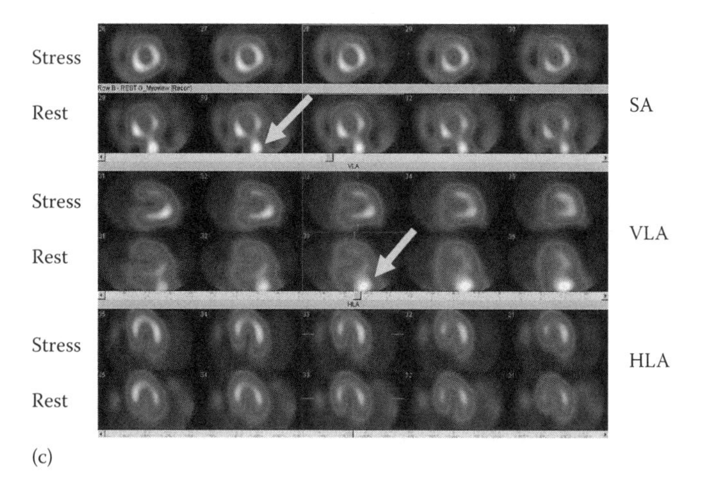

FIGURE 5.116 (continued)
(See colour insert.) (c) gut uptake. The yellow arrows indicate the position of the artefacts. Note that the attenuation artefacts appear in both the rest and stress images, whereas the ischaemic area is only seen on the stress images. (Courtesy of S. Sassi.)

techniques to reduce artefacts have been proposed (Iqbal et al. 2004). Gated myocardial perfusion SPECT can improve specificity by helping to identify artefacts in patients with suspected coronary artery disease, and also plays a key role in identifying patients with non-ischaemic cardiomyopathies. Gated blood-pool SPECT has the potential to replace planar equilibrium radionuclide angiography for analysis of LVEF, regional wall motion and right heart function.

SPECT has also had a major impact on the diagnostic accuracy of skeletal scintigraphy. The markedly improved visualisation of abnormalities arises in sites where bony structures would be superimposed in planar scintigraphy. Examples are the spine (Figure 5.117), knees and femoral head. The main clinical application of bone SPECT is in determining the causes of back pain. Since this is such a common problem, dual and triple-headed cameras are useful in maintaining patient throughput.

There are many diagnostic benefits associated with SPECT brain studies, as illustrated by the following examples. Ischaemia can be demonstrated during the early phase of acute cerebral infarction when CT and MRI are insensitive. Alzheimer's can be delineated

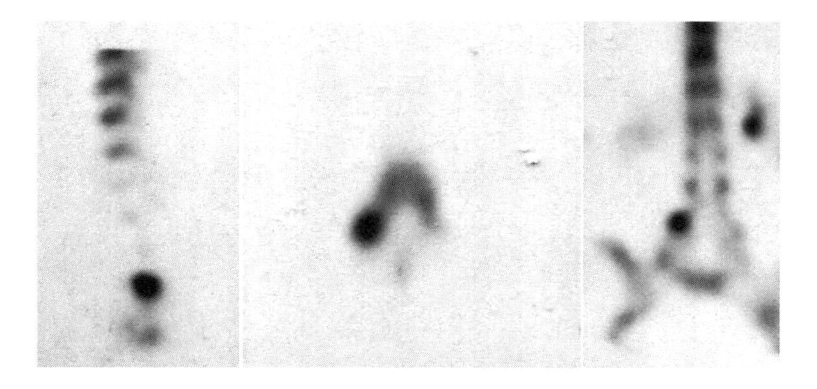

FIGURE 5.117
Sagittal, transaxial and coronal sections of the lower spine showing the enhanced uptake of $^{99}Tc^m$-MDP in part of a vertebra corresponding to a stress fracture.

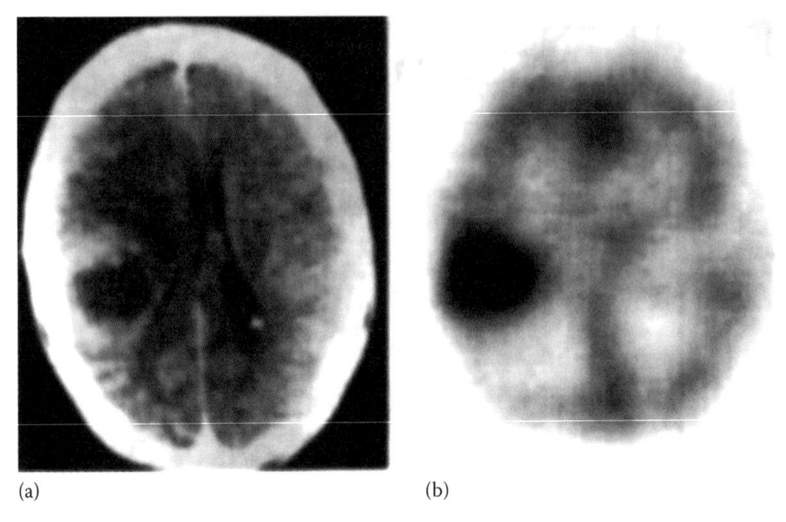

FIGURE 5.118
Transaxial sections through the brain of a patient with a glioma, showing an ill-defined lesion with peripheral enhancement on the X-ray CT scan (a) and intense $^{99}Tc^{m}$-HMPAO uptake in the corresponding SPECT image (b). (From Irvine et al., 1990. Reproduced with kind permission from Springer Science+Business Media.)

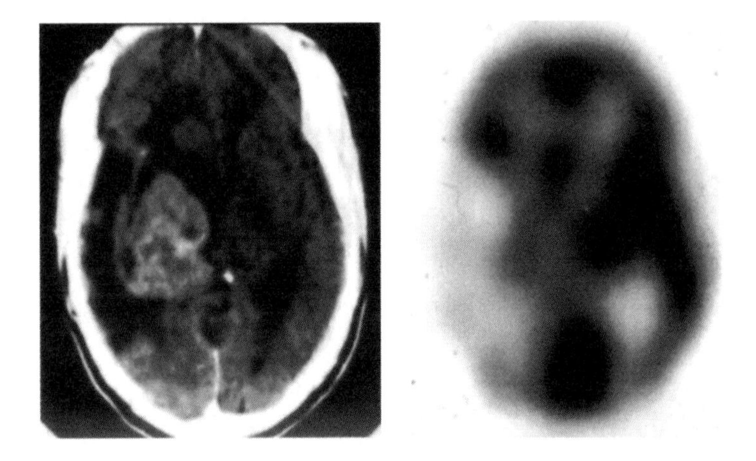

FIGURE 5.119
As for Figure 5.118, but for a large, well-defined contrast-enhancing glioma in the right parietal region. The $^{99}Tc^{m}$-HMPAO image (on the right) shows reduced $^{99}Tc^{m}$-HMPAO uptake in the tumour and surrounding oedema. (From Irvine et al., 1990. Reproduced with kind permission from Springer Science+Business Media.)

accurately in its early phase, and distinguished from treatable and benign conditions. The focus of epileptic disease can be identified compared with normal MRI. Receptor studies can play an important role in the diagnosis and management of psychiatric disease. Detection of tumour (primary, secondary and recurrence) can be demonstrated with perfusion SPECT (Figures 5.118 and 5.119), but PET imaging is preferred if available. Other examples of SPECT brain images can be found on the Harvard University website.*

Abdominal SPECT is used in the evaluation of hepatic haemangioma, diffuse hepatocellular disease, infective processes, renal disease and oncology. SPECT also has an important role to play in radionuclide therapy, where accurate quantification is required

* See the whole brain atlas at http://www.med.harvard.edu/AANLIB/ (accessed February 20, 2012).

TABLE 5.16

Advantages of SPECT and PET.

SPECT	PET
$^{99}Tc^{m}$ radiopharmaceuticals readily available	More useful choice of biologically significant chemical elements for labelling (^{15}O, ^{13}N, ^{11}C)
Less expensive imaging equipment	Higher and more uniform spatial resolution (limited ultimately by positron range)
Simple tracer labelling techniques	Higher sensitivity since no collimators required
Lower running costs	Easier to correct for photon attenuation
Simultaneous dual isotope examinations possible using energy discrimination	Short half-lives enable repeat studies with different compounds
Delayed scans can be accommodated using long-lived radionuclides	Dynamic 3D imaging, better spatial resolution and sensitivity enabling more accurate application of biological models

in order to produce 3D dose distributions for dose predictions prior to targeted radionuclide therapy using a suitable tracer, and/or for dose delivered following radionuclide therapy (Flux et al. 2006).

For more details of clinical applications of SPECT, the reader is referred to other textbooks (e.g. Ell and Gambhir 2004, Cook et al. 2006). A review of clinical SPECT/CT studies can be found in Buck et al. (2008).

5.8.8 Summary

Over the last couple of decades SPECT has progressed from being a research tool to a widely accepted clinical technique. Improvements in both SPECT instrumentation and the development of new SPECT radiopharmaceuticals have contributed to the expansion in the clinical role of SPECT.

SPECT provides functional images with improved contrast when compared with planar imaging but usually at the expense of spatial resolution. As usual, there is a compromise between resolution and total image counts, but there is no doubt that SPECT provides information which can be missed in planar scintigraphy. For the detection of low-contrast variations in radiopharmaceutical distributions, it is presently the technique of choice.

Most of the improvements in SPECT instrumentation, computer hardware and reconstruction algorithms and data correction methods have resulted from a cross-fertilisation of ideas between the SPECT and PET communities. Section 5.9 is dedicated to PET, and by way of introduction to this special branch of ECT, a comparison of the advantages of SPECT and PET is given in Table 5.16.

5.9 Positron Emission Tomography

5.9.1 Introduction

In PET, radiotracers labelled with positron-emitting radionuclides are used to investigate normal and abnormal body processes. The use of positron emitters as opposed to the single-photon emitters used in 'conventional' nuclear medicine has two important consequences. First, the most commonly used positron-emitting radionuclides, ^{15}O, ^{11}C, ^{13}N and ^{18}F

TABLE 5.17

Radionuclides commonly used in PET studies.

Radionuclide	Half-Life (min)	Target Reaction	Common Compounds	Usage
^{15}O	2.0	^{14}N (d,n) ^{15}O	$^{15}O_2$	Oxygen metabolism
			$C^{15}O_2$, $C^{15}O$, $H_2{}^{15}O$	Blood flow, blood volume
^{13}N	10.0	^{16}O (p,a) ^{13}N	$^{13}NH_3$	Myocardial perfusion
^{11}C	20.4	^{14}N (p,a) ^{11}C	$^{11}CO_2$, ^{11}CO, $^{11}CH_3I$	Radiolabelling a large range of compounds
^{18}F	110	^{18}O (p,n) ^{18}F	^{18}FDG	Glucose metabolism
^{124}I	6048	^{124}Te (d,2n) ^{124}I	$Na^{124}I$	Thyroid imaging, labelling proteins
^{82}Rb	1.3	^{82}Sr decay ($T_{1/2}$; 25 d)	$^{82}RbCl$	Myocardial perfusion

(see Table 5.17), are isotopes of biogenic elements that can be readily incorporated into many biological molecules. All have radioactive half-lives of <2 h and require a cyclotron for their production. Second, due to the physics of the imaging process, PET images are generally superior, in terms of both image quality and quantitative accuracy, to those obtained with single-photon emitters. These two aspects are exploited in a wide variety of applications. PET was used for many years as a tool for basic and clinical research in the brain, heart and for oncology, using a wide range of radiotracers. More recently, PET has also become firmly established in an increasing number of routine clinical roles based, almost exclusively, on the use of the tracer ^{18}F-labelled fluorodeoxyglucose (^{18}FDG). The need for both anatomical and functional information when making accurate diagnoses has led to the development of the combined PET-CT scanner, whereby the two modalities are mounted in-line around the same patient couch. This format allows rapid sequential acquisition of accurately aligned images. In this section, the basic principles and technology of PET, along with a brief outline of current clinical and research applications, are presented.

5.9.2 Basic Principles of PET

5.9.2.1 Coincidence Detection of Gamma Rays from Positron-Electron Annihilation

Many proton-rich low-atomic-number nuclei decay by the emission of a positron (β^+) which is the anti-particle of the electron with similar properties except for the positive charge. Positrons are emitted with a range of energies up to a maximum, E_{max}, which depends on the nuclide, and interact with surrounding atoms via multiple coulomb interactions. Depending on its energy, a positron will travel between a fraction of a millimetre and several millimetres from its point of production in tissue and, at the end of its path, will interact with a nearby atomic electron. The two particles annihilate and their masses convert into energy in the form of two 511 keV gamma rays according to Einstein's $E = mc^2$. In order to satisfy the rules governing conservation of energy and momentum, these gamma rays are emitted at ~180° to each other. If both gamma rays can be detected within a very short time of one another (typically within 10 ns), then it is assumed that they arose from the same annihilation event and that the original disintegration must have occurred somewhere along, or very close to, the line joining the two detection points as shown in Figures 5.2 and 5.120. A positron camera usually consists of a ring of radiation detector elements, arranged around the subject. The detector read-out electronics allows a coincidence to be

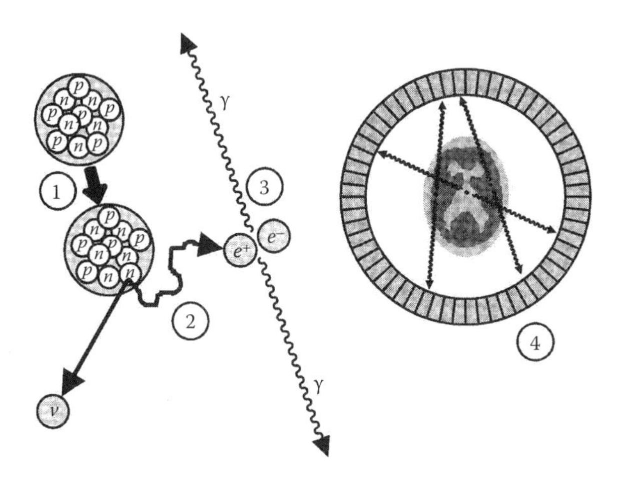

FIGURE 5.120
Annihilation coincidence detection. A positron and a neutrino are emitted by a proton-rich nucleus (1). The positron follows a convoluted path (2) of a millimetre or so. At the end of its path the positron interacts with an electron to produce two anti-parallel 511 keV photons (3), which are detected in coincidence (4) by the PET camera. (From Marsden, 2006. Reproduced by permission of Edward Arnold Limited.)

measured between detectors on opposing sides of the patient. Large numbers of annihilation photon pairs are collected and an image of the distribution of the radionuclide is reconstructed using a reconstruction algorithm. The use of 'annihilation coincidence detection' for radionuclide imaging was first described in the early 1950s (Brownell and Sweet 1953).

5.9.2.2 Spatial Resolution in PET

Fundamental limitations on the *spatial resolution* in PET are imposed by the positron range and the acollinearity of the two 511 keV photons. The positron range distribution in tissue is isotropic with a narrow peak and long tails and depends on E_{max} for the radionuclide. For ^{18}F (E_{max} = 0.94 MeV), the FWHM and FWTM of this distribution are 0.13 and 0.38 mm, respectively, whilst for ^{82}Rb (E_{max} = 3.35 MeV) the FWHM and FWTM are 0.42 and 1.9 mm (Derenzo et al. 1993). This contribution to the overall camera spatial resolution in a human scanner is small compared to other factors. A much more significant contribution comes from the acollinearity of the two annihilation photons. The magnitude of this error increases linearly with the diameter of the positron camera – for a detector separation of 100 cm the contribution of acollinearity to the spatial resolution is about 2 mm. For a small-animal scanner, Stickel and Cherry (2005) gave the absolute limit to the system spatial resolution that could be obtained with ^{18}F as 0.5 mm FWHM using a pixel detector size of 250 μm or smaller.

From the scanner design point of view, the most important parameter determining the system spatial resolution is usually the size of the individual detector elements that are often small scintillation crystals. The contribution of the crystal width, d, to the spatial resolution is $\sim d/2$ with d usually around 4–5 mm for current human scanners. This is the effect at the centre of the FOV, whereas, away from the centre, the resolution degrades due to crystal penetration.

The positron range, photon acollinearity and crystal size all determine the 'system spatial resolution' which is measured using a point or line source (see Section 5.10.10). However, clinical images, acquired in a limited time, contain a high level of noise and need to be

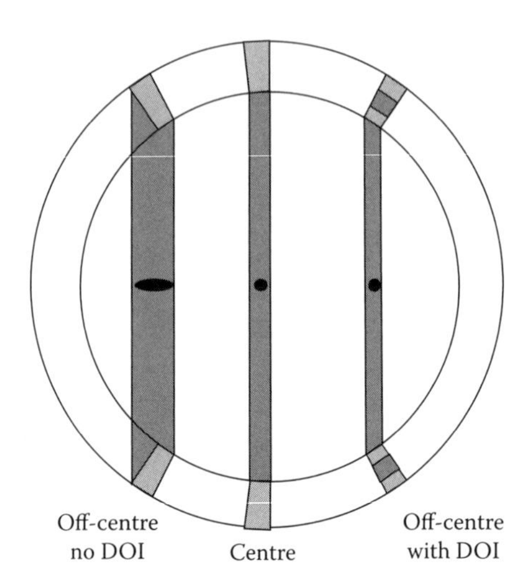

FIGURE 5.121
Depth of interaction (DOI) effect. The radial resolution at the centre of the scanner is better than that obtained off-centre. If the depth of interaction within the crystal is known, then the off-centre resolution is improved.

smoothed, usually during the image reconstruction process, in order to obtain an acceptable SNR. As a result of this smoothing, the final 'image spatial resolution' is invariably worse than the system spatial resolution. In practice, therefore, the scanner sensitivity and the reconstruction algorithm both play a major part in determining spatial resolution. The spatial resolution achieved in a clinical whole-body PET image is typically between 5 and 10 mm FWHM. In some specialised brain imaging systems, the spatial resolution may be as low as 2–3 mm FWHM, whilst in small animal systems it can approach 1 mm.

The radial component of the spatial resolution away from the centre of the FOV degrades rapidly with distance from the centre as gamma rays are incident on the crystal sides (Figure 5.121). This effect accounts for the increase in the radial resolution of the ECAT HR+ scanner from 4.3 mm FWHM at the centre to 8.3 mm FWHM at 20 cm. With the aim of maximising sensitivity, cameras have a small diameter in order to cover a large solid angle for minimum cost. This minimises the effects of photon acollinearity but exacerbates the degradation of off-centre radial resolution. To correct for this effect designs have been developed to measure the depth of interaction position within the scintillation crystal. These either use photosensors at each end of the crystal, determining the interaction position by comparing the amount of light measured at the two ends, or use pulse shape discrimination from a stack of two or more scintillators (a 'phoswich') to identify their different scintillation decay times – this is employed in the PET/SPECT system described in Section 5.9.3.8.

5.9.2.3 Gamma-Ray Detection Efficiency

The efficiency, ε, with which the positron camera detects 511 keV photons depends mainly on the crystal. The higher the density and Z value of the crystal, the greater the chance of photon detection. Hence, NaI(Tl) is not ideally suited to detect the higher-energy photons produced in PET. Bismuth germanate (BGO) was the crystal of choice for PET for many years, and has now been largely superseded by LSO. The crystals need to be several cm thick (cf. 6–9 mm in SPECT) in order to stop enough annihilation photons to make positron cameras clinically viable. Table 5.1 in Section 5.2.1 shows the physical properties of the most commonly used scintillators in PET.

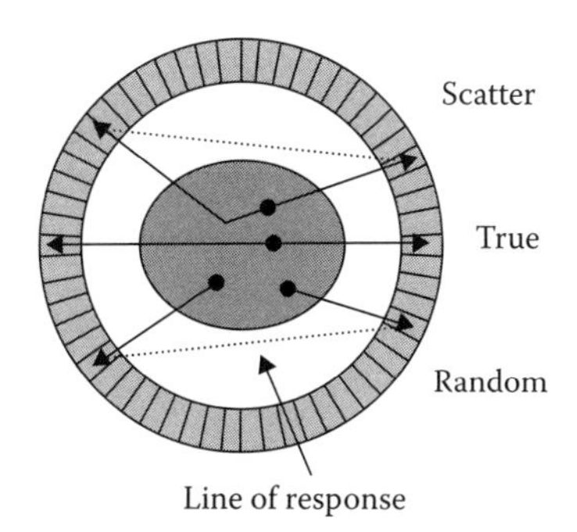

FIGURE 5.122
Types of coincidence event. (From Marsden, 2006. Reproduced by permission of Edward Arnold Limited.)

5.9.2.4 Types of Coincidence Event

Figure 5.122 illustrates the three different types of coincidence events: trues, scatters and randoms. The line that joins the two detection points of a coincidence event pair is called a line-of-response or LOR and, for a true coincidence event, the initial disintegration point lies along or very close to the LOR. The figure shows that, for scattered coincidences, where one or both of the gamma rays has scattered before being detected, and for random coincidences, where the two gamma rays come from different disintegrations, this is not the case. These events can comprise up to ~75% of the total number recorded in a clinical PET study, resulting in reduced image SNR and degraded quantification accuracy. Minimising the number of scattered and random coincidence events is, therefore, one of the main goals of scanner design.

5.9.2.5 Attenuation and Scatter

The total linear attenuation coefficient of 511 keV gamma rays in tissue is $0.096\,\text{cm}^{-1}$ – this corresponds to a half-value layer of ~7 cm. Therefore, for a positron emitter situated at the centre of the brain, less than one in four disintegrations results in both photons exiting the head unattenuated, and this can fall to a few percent in the abdomen. At 511 keV almost all first interactions with tissue are by Compton scattering, there being negligible photoelectric absorption.

When imaging 140 keV photons with a gamma camera, 75% of scattered photons can be removed using energy discrimination, but this approach is much less effective in PET for two reasons. First, at 511 keV a photon scattering through 30° loses only ~10% of its energy, which is less than the energy resolution of most PET detectors. Second, as much as 50% of the photon interactions in the detector involve Compton scattering which results in the photon depositing only a fraction of its energy in the scintillator. Such events are detected in the correct position and so, ideally, should not be rejected. The setting of a lower-energy discriminator level is, therefore, a trade-off between accepting good events that have Compton scattered in the detector and rejecting bad events that have Compton scattered in the subject. In practice, these discriminators are usually set between 250 and 450 keV.

For a given LOR, the attenuation factor is independent of the distribution of positron-emitting sources along the line joining the detectors, as shown in Figure 5.123. An accurate correction for attenuation can, therefore, be made by measuring the attenuation factor directly with, for example, an external source. This is one of the reasons why image quality and quantification is superior in PET compared to SPECT.

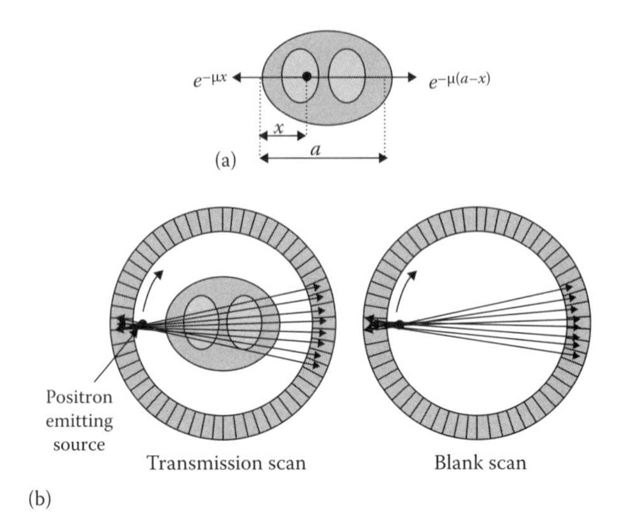

FIGURE 5.123
Attenuation and attenuation correction. (a) The total attenuation factor for a positron-emitting source at position x along the LOR, in a medium with linear attenuation coefficient, μ, is equal to $e^{-\mu x} \times e^{-\mu(a-x)} = e^{-\mu a}$, that is, it is independent of the position of the source along the line joining the detectors and in general is independent of the source distribution along the line. (b) The attenuation factor can be measured by comparing the count rates obtained for transmission and blank scans acquired with and without the patient in the scanner. The attenuation correction factor, equal to the inverse of the attenuation factor, is given by the ratio of the blank scan count rate to the transmission scan count rate for each detector pair.

5.9.2.6 True and Random Coincidences

The detectors used in positron cameras cannot determine the arrival time of a 511 keV gamma ray exactly. This uncertainty is characterised by the camera's temporal resolution, which is usually in the range 1–10 ns. A coincidence is deemed to occur when signals from two detectors occur within a predefined time of each other. In practice, an electronic gate of duration of the order of the temporal resolution is opened whenever the first detector produces a signal. When gates from any two detectors overlap a coincidence is said to have occurred between those two detectors, as shown in Figure 5.124. The true coincidence rate, C, is dependent upon the detection efficiency, ε, of the detectors and the singles rates of the two detectors involved in the coincidence. If the singles rates in each of a pair

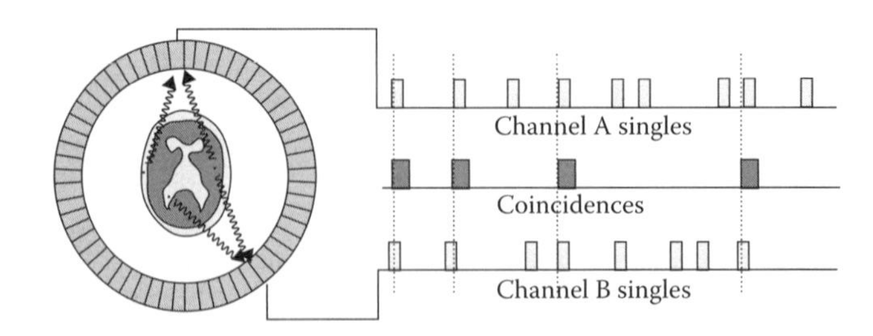

FIGURE 5.124
True and random coincidences. Single events are measured for each detector element. A true coincidence occurs when two singles from the same positron annihilation are registered on different channels within the coincidence resolving time 2τ. The singles rates are much higher than the coincidence rate due to the high fraction of unpaired singles. Unpaired singles may overlap purely by chance and produce random coincidences – there is no way of distinguishing between true and random coincidences.

of detectors is S then, for the ideal case of a central point source of a pure positron emitter with activity A (Bq), C is given by

$$C = \varepsilon S \quad (5.69)$$

where S is given by $A\Omega\varepsilon$ and Ω is the solid angle subtended by both detectors. It can be seen that, in order to maximise C, it is essential to have high detector efficiency. For an extended source less than 25% of single photons detected by any given detector will typically have a coincidence partner. This is either because one photon of the pair is not detected due to scattering or low detector efficiency, or because the detected photon arises from activity outside the FOV. Accidental (or 'random') coincidences can occur between unpaired photons purely by chance. The randoms rate, R, for a given detector pair is

$$R = 2\tau S^2 \quad (5.70)$$

where 2τ is the width of the coincidence time window. Assuming that the singles rates are similar for all detectors, the random rate is, therefore, seen to be proportional to the temporal resolution and to the square of the total singles rate.

5.9.2.7 2D and 3D Sensitivity

The sensitivity of a positron camera can be defined as the fraction of all the pairs of annihilation photons generated within the FOV that are detected and contribute to the formation of an image. The sensitivity depends primarily on the solid angle, Ω, subtended by the camera at the centre of the FOV in addition to the square of the individual detector efficiency, ε, because both photons need to be detected to register a coincidence. For modern cameras the sensitivity is typically 0.5%–5%. Multi-slice cameras usually consist of a stack of detector rings as shown in Figure 5.125. In 2D mode, data are collected and reconstructed from each ring independently to create a set of 2D images which are then reorganised into a 3D dataset (Figure 5.126). The detector rings are separated by lead or tungsten septa which reduce the fractions of scatter and random coincidences. The septa are not like the collimators of a gamma camera but function only in a plane. An image can be formed without septa, but it would have a lower SNR.

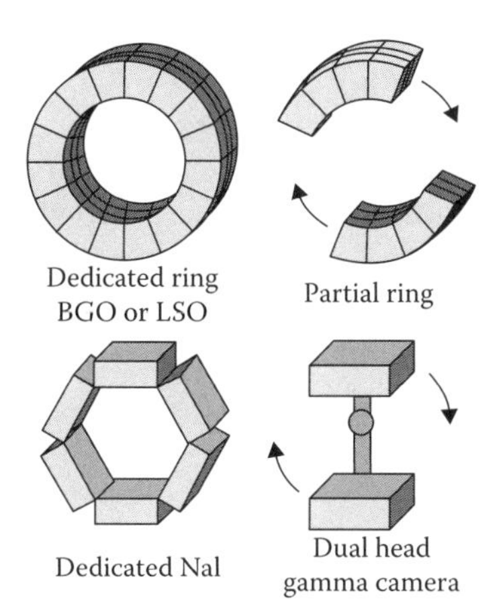

FIGURE 5.125
PET camera configurations.

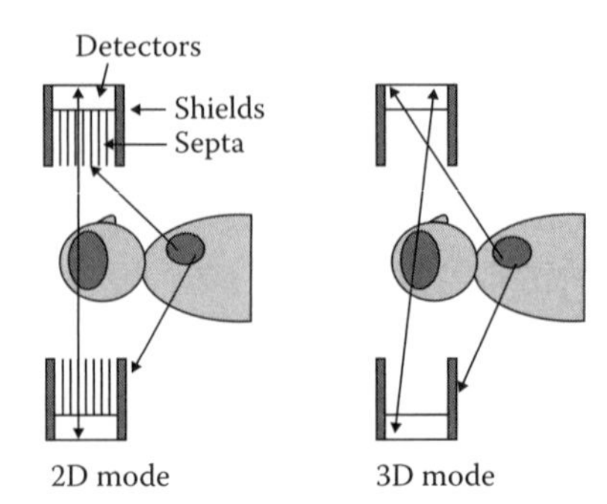

FIGURE 5.126
2D and 3D acquisition. In 3D mode, the larger solid angle subtended by the detectors results in a higher sensitivity to true coincidences, but also a higher sensitivity to scattered and random coincidences, particularly those originating from outside the FOV. (From Marsden, 2006. Reproduced by permission of Edward Arnold Limited.)

However, the improved SNR for 2D-PET is achieved at a great cost in sensitivity, as only coincidences where both annihilation photons are detected within the same ring are accepted. In 3D mode (see Figure 5.126), there are no septa (or they are retracted into the camera gantry) and additional coincidences are accepted between detectors in different rings (Townsend et al. 1991, Cherry et al. 1992). The fraction of all true coincidence pairs that are detected is much greater in 3D than in 2D mode (by ×5 to ×10), but at the expense of a greater sensitivity to scattered and random coincidences, particularly from activity outside of the FOV.

Whilst it has been shown that there may be advantages in retaining 2D-mode acquisition for large patients, where scatter and random fractions are high, most commercial scanners now only operate in 3D mode. As detectors with better scatter and random rejection capabilities become available, 2D mode is likely to become redundant.

5.9.2.8 Dead Time and Count-Rate Losses

The dead time (see Equations 5.16 and 5.17) of positron cameras is a key factor in their performance as the detectors generally see much higher count rates than is normal in SPECT. The dead time in most systems is usually a combination of nonparalysable and paralysable, but the dominant factors are usually related to the read-out electronics. Singles rates are particularly high if a short-lived tracer is used as the activity levels in the FOV are very large at the start of the study. Dead times which are large fractions of a μs lead to high levels of event losses. In these cases correction for dead-time losses (see Section 5.9.7.5) is essential if quantification is required, especially in dynamic studies.

5.9.2.9 Noise Equivalent Counts

The performance of a positron camera depends on a combination of factors. A useful parameter of camera performance is the noise equivalent count (NEC) rate (Strother et al. 1990) which is a function of the true, (T) scatter (S) and randoms (R) event rates:

$$\text{NEC} = \frac{T^2}{(T + S + fR)} \tag{5.71}$$

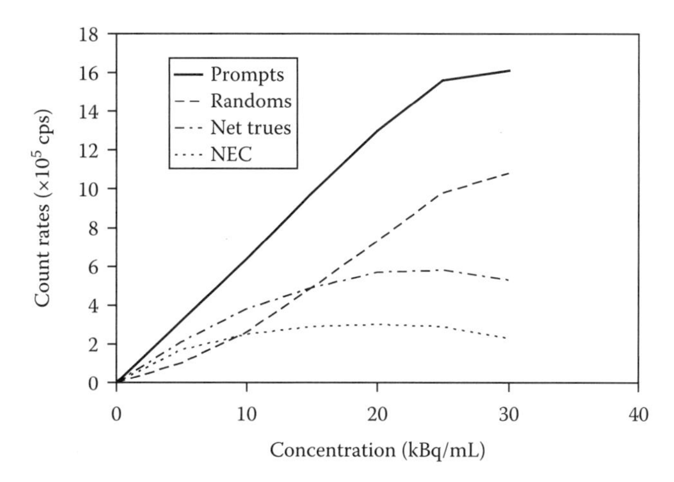

FIGURE 5.127
Example of the count rates as a function of the radioactive concentration produced by a modern positron camera when imaging a 20 cm diameter phantom. The 'net trues' are a sum of the true and scatter count rates.

where f is the fraction of randoms reconstructed within the patient boundary. The value of f is typically 0.25 for body imaging. A useful guide to the performance of a camera is the peak NEC rate measured when imaging a standard phantom. Figure 5.127 shows typical camera count-rate curves for trues, randoms, scatter and NEC.

For any pair of detectors, both T and S are roughly proportional to the amount of radioactivity in the FOV. However, because the randoms rate is proportional to S^2, the random/true ratio increases as the activity increases. In some cases, the randoms rate may become so large that it dominates the signal obtained from the camera. There are straightforward ways of correcting for the systematic effects of randoms, which contribute a broad background to the image (see Section 5.9.7.3), but there is no way to reduce the statistical noise that they introduce. The best solution is to minimise them either by shielding the detectors from single events originating from activity outside the FOV or by using detectors with a short temporal resolution.

5.9.3 Detectors and Cameras

5.9.3.1 Detectors Used in a Positron Camera

The most important part of a positron camera is the 511 keV photon detector. Most of these detectors are based on scintillation crystals coupled to PMTs (see Sections 5.2.1 and 5.3.4). The main requirements of a good detector for PET are as follows:

- A high detection efficiency, ε, for 511 keV gamma rays, preferably with a large photoelectric component to provide full energy deposition in the detector. This requires a high density and atomic number (Z).
- A high spatial resolution. If the scintillation light is divided between several PMTs to locate the interaction origin a high light yield is required. Additionally, a high-Z material ensures that the fraction of 511 keV photons that Compton scatters in the detector leading to mispositioning is minimised.
- A high temporal resolution. This depends on both the light yield and the scintillation decay time. A good temporal resolution minimises the random coincidence rate.

A short scintillation decay time also reduces the detector dead time, allowing data to be acquired at high rates.

- A good energy resolution. This depends primarily on the light yield of the scintillator. Whilst energy level discriminators are usually set quite low for PET imaging, good energy resolution is still desirable to maximise the rejection of scatter events.

Properties of BGO are shown in Table 5.1. Until the late 1990s most dedicated PET cameras used BGO because of its high detection efficiency at 511 keV. However, BGO has a relatively slow light decay time reducing the ability to set a narrow timing window and low light yield leading to poor energy resolution. Commercial cameras now make use of several other crystals, namely, cerium-doped LSO, gadolinium-oxyortho-silicate (GSO), cerium-doped lutetium yttrium orthosilicate (LYSO) and lanthanum bromide ($LaBr_3$). The characteristics of these newer crystals are also included in Table 5.1 and will be discussed in Section 5.9.4.1. BGO-based cameras produce a relatively small improvement in image quality for 3D imaging relative to 2D imaging in the body due to high random and scatter fractions. The use of crystals such as LSO makes 3D whole-body scanning much more effective. The newer crystals can also make it possible to provide time-of-flight (TOF) information (see Section 5.9.4.3), to perform simultaneous emission and transmission scanning and to help reject scatter.

5.9.3.2 Block Detector

For many years the detector design commonly used in a positron camera was the BGO block detector (Figure 5.128) (Casey and Nutt 1986, Dahlbom and Hoffman 1988). For example, the Siemens/CTI EXACT HR+ camera (Bendriem et al. 1996) contains blocks of scintillator cut into an 8 × 8 array of crystal elements, each of dimensions 4.39 (axial) × 4.05 (transaxial) × 30 (radial) mm. The cuts go part way through the whole depth of the crystal and distribute the light between the four square PMTs attached to the block. The interaction segment of crystal is identified by comparing the light collected by the four PMTs. Whilst the performance of the block detector is inferior to that of an array of individually coupled crystals, it is a cost-effective and practical solution to the problem of reading out a large number of crystal elements with a small number of PMTs. An image of ^{18}FDG uptake in a brain taken with the

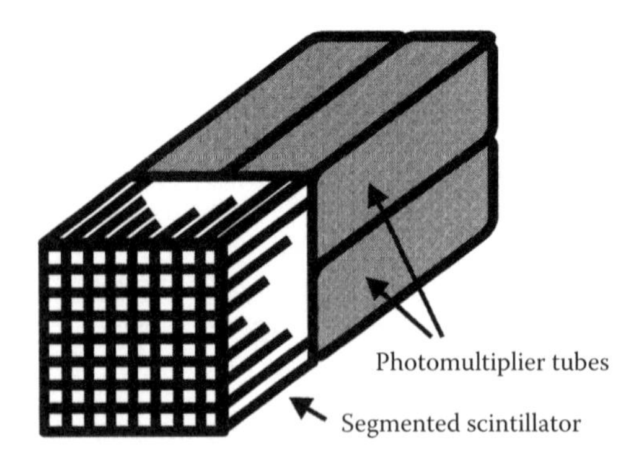

FIGURE 5.128
Block detector. A 511 keV gamma ray interacts in one of the segments of the crystal block. The interaction position is determined by comparing the amount of scintillation light that reaches each of the four photomultiplier tubes. The total amount of light collected reflects the energy deposited in the crystal. (From Marsden, 2006. Reproduced by permission of Edward Arnold Limited.)

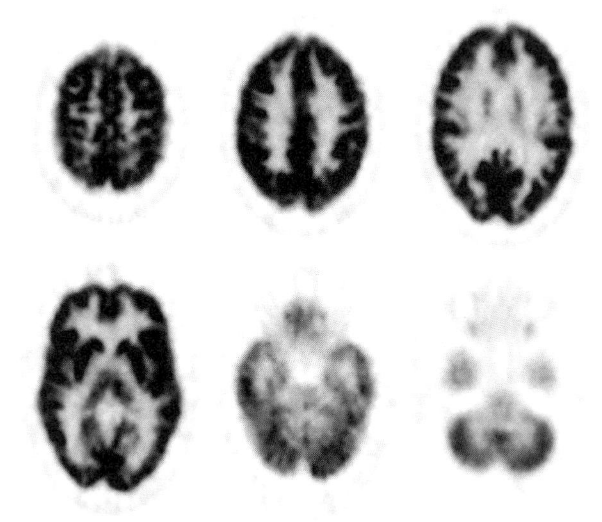

FIGURE 5.129
Six selected transaxial slices through an FDG brain image obtained with the EXACT HR+. A total of 63 contiguous slices are acquired. (Courtesy of C. La Fougère and University of Munich, Munich, Germany.)

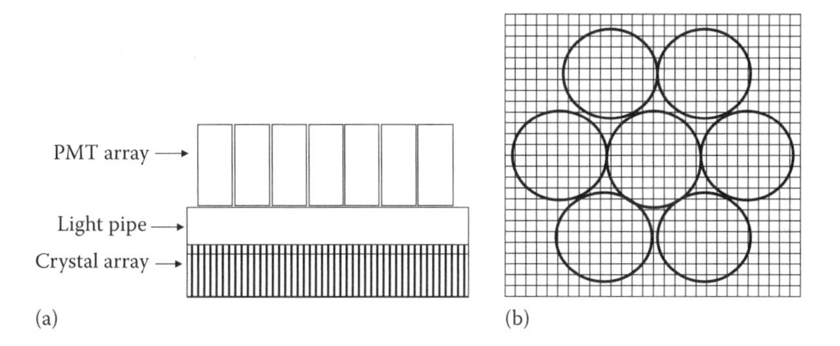

FIGURE 5.130
Schematic of an Anger-logic detection system for a modern positron camera. (a) Cross section showing how the PMT array is coupled to the crystal array using a partially slotted light guide. The light guide allows sharing of the crystal light between the local PMT cluster that produces pulses used to generate pulse-height and positional information. (b) The end-on view shows the arrangement of the PMT and crystal arrays.

HR+ is shown in Figure 5.129. Other designs involving the use of continuous arrays of small crystals coupled to large PMTs are more common now (see Figure 5.130).

5.9.3.3 Standard Positron Camera Configuration

Various geometries for a positron camera have been designed as shown in Figure 5.125. However, the vast majority of systems have been based on the multi-ring format. The configuration for the whole-body BGO camera, the ECAT EXACT HR+, mentioned in Section 5.9.3.2, is shown schematically in Figure 5.131. The system consists of four adjacent rings of 72 BGO block detectors. The diameter of the ring is 82.7 cm and the axial FOV is 15.5 cm which is adequate to image the whole of the brain or the heart. The intrinsic spatial resolution is 4.3 mm FWHM at the centre of the FOV and degrades to 4.7 (tangential)

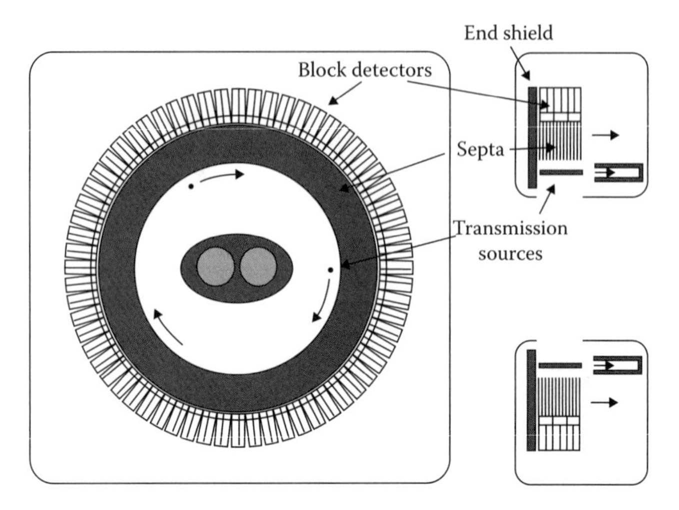

FIGURE 5.131
Schematic diagram of a typical dedicated PET BGO ring camera. The septa and transmission rod sources can be retracted out of the FOV when not required.

and 8.3 (radial) at 20 cm from the centre. The interslice tungsten septa (0.8 mm thick and 66.5 cm long) are retractable to allow 3D-mode acquisition, and there is substantial lead shielding at the ends of the camera to reject photons that originate from outside the FOV. The detector electronics and data acquisition system permit true coincidence count rates of up to 200–300 kcps. Data can be acquired in list mode, whereby each detected event is stored individually with an associated time flag, permitting greater flexibility in processing dynamic and gated studies. Transmission scans for attenuation correction are acquired using three rotating ^{68}Ge rod sources of activity 220 MBq each which retract into a shielded housing within the gantry when not in use. The overall result is a high true event sensitivity, maximum acquisition rate and SNR which leads to excellent image quality.

A variation on the BGO ring design is the rotating partial ring (Bailey et al. 1997). Here the cost of the system is reduced by mounting two opposing arrays of block detectors on a rotating gantry (Figure 5.125). This system operates in 3D mode only.

5.9.3.4 Dedicated PET NaI Camera

An example of a dedicated NaI positron camera was the system described by Karp et al. (1998). It consisted of six curve-plate Anger cameras arranged in a ring of 92 cm diameter and 25 cm axially. The crystal thickness of 25 mm was a trade-off between the low detection efficiency of NaI(Tl) and the spatial resolution, which deteriorates with crystal thickness. A version of this design was marketed as C-PET by ADAC Laboratories (Adam et al. 2001). NaI is cheap and its low efficiency can be offset by building large-area detectors operating in 3D mode. However, NaI detectors operate with very high singles rates and modest coincidence rates. The singles-to-true ratio for a detector plane may be ~100:1, so for a true coincidence rate of 10 kcps, the camera must handle singles rates of >1 Mcps. Pulse clipping and local centroiding were implemented to allow the camera to tolerate these very high rates.

The good energy resolution achievable with NaI allows emission data, at 511 keV, to be acquired simultaneously with transmission data acquired at 662 keV from a ^{137}Cs point or line source (Karp et al. 1995). This resulted in rapid scanning protocols and perfect registration of the emission and transmission scans.

5.9.3.5 Modified Dual-Head Gamma Camera

The advent of clinical PET applications with ^{18}FDG stimulated the production of modified DHGCs that could operate in coincidence mode (Lewellen et al. 1999). The advantage of this approach was that the system could be used for SPECT and PET studies (see Section 5.8.3.5). However, the PET performance suffered from the same limitations as (and to an even greater extent than) the dedicated NaI(Tl) PET systems (Section 5.9.3.4). The very-low, true-to-singles count-rate ratio was exacerbated by the open structure of the DHGC making it very susceptible to random and scatter events originating from activity outside the FOV. True coincidence rates were very low, and randoms and scatter fractions very high. In order to keep dead time and randoms to an acceptable level, whole-body ^{18}FDG studies were usually performed with ~⅓ of the injected activity that would be used with a BGO system. Gamma cameras for use in coincidence imaging systems had 16 or 19 mm thick NaI crystals compared with the 6–9 mm used for single-photon imaging. This resulted in an increase in efficiency at 511 keV without unduly compromising the spatial resolution at 140 keV. Dual-head coincidence systems were popular for a short period but are no longer readily available.

5.9.3.6 Wire-Chamber Systems

These have been described in detail in Section 5.3.5. Only a small number of clinical studies have been performed using wire-chamber positron cameras. However development continues due to the potential for the production of a high-performance camera at lower cost than would be possible with conventional systems.

5.9.3.7 Research Tomographs

High-resolution tomographs for research have included the use of 2 mm wide crystals in systems dedicated to imaging the human brain (Schmand et al. 1999), whilst crystals of width ~1 mm have been used in cameras for imaging small animals (Cherry et al. 1999).

The CTI high resolution research tomograph (HRRT) is an example of a high-performance research system that incorporated many of the innovations described earlier, including the use of LSO, very small crystal size and depth-of-interaction determination (Schmand et al. 1998b, 1999). It had a small transaxial FOV for brain-only imaging and consisted of eight flat detector units arranged in an octagon. The axial extent was 25.2 cm and the distance between opposing detectors was 46.9 cm. Each plane consisted of an array of crystal segments read out by PMTs. The LSO was arranged in two 7.5 mm thick layers to provide depth-of-interaction information – LSO with different decay times was used for the two layers. The small diameter and large axial extent resulted in an extremely high 3D sensitivity, and the very small size of the detector crystal segments ($2.1 \times 2.1 \times 7.5\,mm^3$) resulted in a reconstructed spatial resolution of <2.5 mm over a 20 cm diameter FOV. The HRRT also incorporated sophisticated hardware to acquire and process coincidence data at rates approaching 1000 kcps.

5.9.3.8 PET/SPECT Cameras

A PET/SPECT device capable of imaging both PET and single-photon tracers at different times was developed by Schmand et al. (1998a). This device used purpose-built block detectors mounted in two opposing arrays which rotated on a gamma-camera gantry. Each block

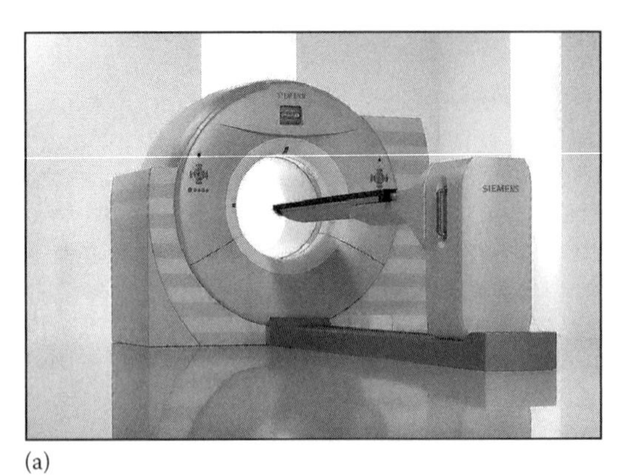

(a)

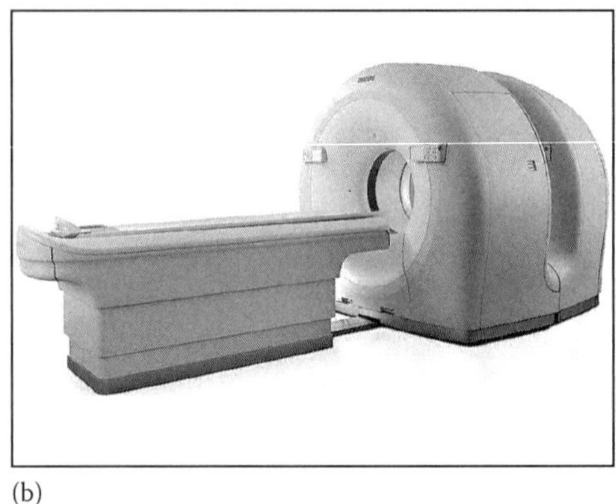

(b)

FIGURE 5.132
(a) Biograph™ mCT system. (Courtesy of Siemens Healthcare, Frimley, Surrey, UK.) (b) GEMINI™ TF PET/CT system. (Courtesy of Philips Healthcare, Guildford, Surrey, UK.)

detector consisted of a sandwich of LSO and NaI(Tl). Single-photon imaging was performed with NaI(Tl), whereas PET imaging used both the NaI and LSO components with the different optical decay times (230 and 40 ns, respectively) in order to obtain depth of interaction information.

5.9.3.9 PET/CT Systems

The first prototype combined PET/CT camera was constructed using a BGO rotating partial ring system mounted in the same gantry as a standard CT scanner allowing PET and CT studies to be performed in rapid succession (Beyer et al. 2000). This led to a major development in PET/CT systems by all the major manufacturers since they provide accurate PET and CT image registration as well as attenuation correction for the PET scans without the need for transmission sources. At the time of writing, PET/CT is the standard configuration for nearly all commercial PET scanners. Figure 5.132 shows two designs, one in which the two devices are closely coupled with a single patient port and a second where the PET and CT scanners are separated to allow patient access between the two scanners. Whole-body images produced by a PET/CT scanner are shown in Figure 5.133.

5.9.4 Recent Developments in PET Camera Technology

PET technology has changed rapidly since the early 1990s. Developments have been aimed both at low-cost high-volume clinical systems and at very-high-performance systems for research.

5.9.4.1 New Scintillators

Modern cameras are taking advantage of newer scintillators with improved performance – these new crystals combine high detection efficiency with a high light yield and a fast decay time. For example, as shown in Table 5.1, LSO has a slightly lower detection efficiency than BGO at 511 keV but it has a light output approaching that of NaI and a short scintillation decay time. A temporal resolution of <1 ns has been reported for LSO (Melcher and Schweitzer 1992, Daghighian et al. 1993). Other crystals used in commercial positron cameras are LYSO (Kimble et al. 2002) and GSO

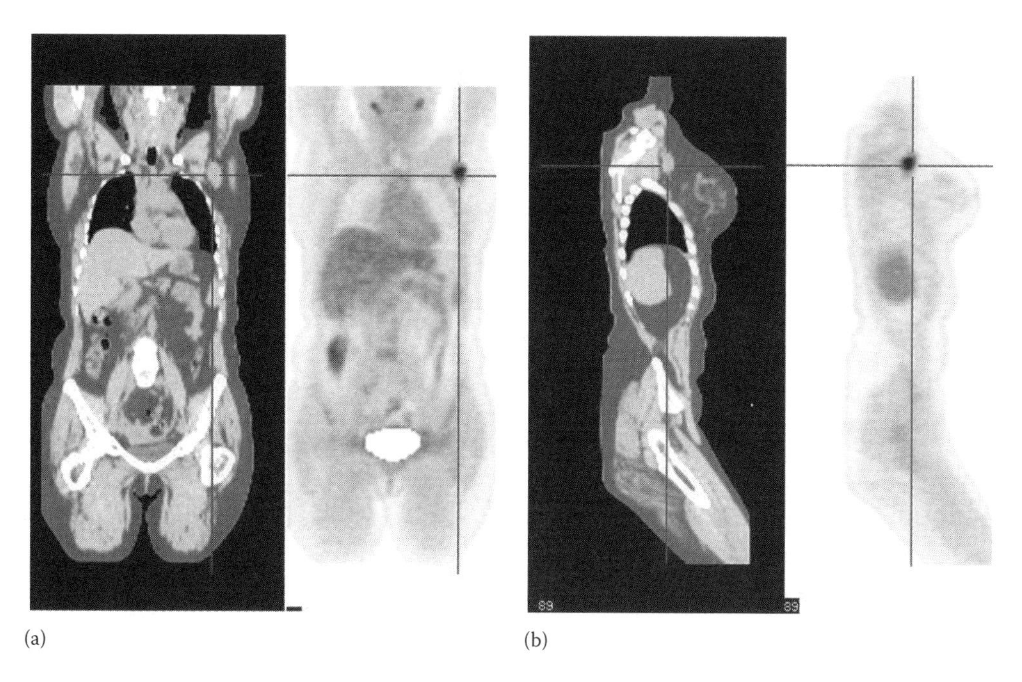

FIGURE 5.133
(See colour insert.) Whole-body images from a combined PET/CT scanner: (a) coronal images; (b) sagittal images. In each pair of images, CT is on the left, and PET is on the right. Linked cursors (red lines) have been used to indicate a lesion in the patient's left axilla. (Courtesy of GE Healthcare, Slough, Berkshire, UK.)

(Surti et al. 2000, 2007). In addition, a new 'bright' crystal $LaBr_3(Ce)$ has been developed and is under investigation (Kuhn et al. 2006).

5.9.4.2 New Light Detectors

PMTs are bulky, expensive and are subject to drift over time. Small cameras for animal imaging have been designed using position-sensitive PMTs (see Section 5.9.4.4) but these devices are not really practical for use in a large-volume camera for human imaging. Reliable solid-state devices to detect scintillation light are available with the most successful so far being the APD (see Section 5.2.1). These can be fabricated in pixel arrays that can be bonded directly onto a crystal array to form a compact robust unit (Pichler et al. 1998). APDs are still more expensive per unit area compared to PMTs and in some cases require cooling. Small-scale APD-based scanners have been made for animal imaging (Bergeron et al. 2008). The individual crystal/diode coupling results in simple crystal identification and uniformity of response compared to a block detector. A more recent development is the silicon photomultiplier (SiPM) which is essentially an array of tiny APDs (Section 5.2.1). These devices might also provide a cost-effective way to develop a simpler positron camera once they have been validated and the cost per unit reduced.

5.9.4.3 Time of Flight

Annihilation coincidence detection (ACD) locates a positron annihilation event to somewhere along or close to a line connecting the two photon detection positions. If the precise time of detection of each gamma ray could be determined, then it would be possible to locate

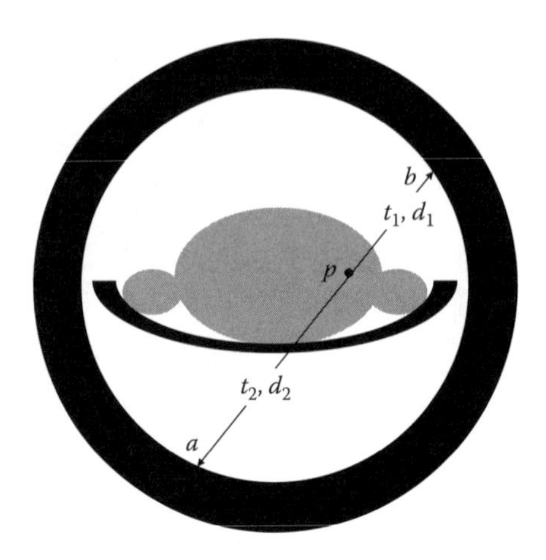

FIGURE 5.134
TOF principle. Measurement of the difference in time of arrival between the two coincidence pulses at a and b can improve the localisation of the annihilation point p. Here $t_1 = d_1/c$ and $t_2 = d_2/c$, where c is the velocity of light, which is ~30 cm ns^{-1} in air. Measurement of the time difference $t_1 - t_2$ can, in principle, provide a value for $d_1 - d_2$, from which the location of c can be determined. However, the errors on the measurement of $t_1 - t_2$ limit the accuracy of this estimation. As $d_1 - d_2 = (t_1 - t_2) \times c$ then, if the error on $t_1 - t_2$ is 0.5 ns, the uncertainty on $d_1 - d_2$ can be 15 cm!

the exact position of the annihilation along the line, and image reconstruction would not be required. In practice, as illustrated in Figure 5.134, the interaction position can only be localised with an accuracy of $c\tau/2$ where τ is the temporal resolution of the detector pair and c is the speed of light. TOF cameras with $\tau \sim 600$ ps have been constructed using detectors based on the fast scintillators BaF_2 and CsF (Lewellen 1998). The resulting position uncertainty of ~9 cm provides sufficient information to improve the SNR in an image when incorporated into the reconstruction algorithm. However, the noise reduction obtained using BaF_2 or CsF is outweighed by the higher overall sensitivity that can be obtained by using BGO which has a higher-detection efficiency (Surti et al. 2000, 2007). TOF cameras using denser crystals such as LSO and LYSO are now commercially available and can produce improved image quality for objects with high levels of scatter. Cameras that use very fast scintillators also have very-low dead time so that they can acquire at very high rates without losing data.

5.9.4.4 Small Animal PET Systems

The potential for PET in developing and assessing new drugs and therapies has resulted in the manufacture of very-high-resolution systems for imaging small animals. An example is the UCLA microPET system. The microPET detector unit consists of an 8×8 array of $2 \times 2 \times 10\,mm^3$ individual LSO crystals which is read out by a multichannel photomultiplier tube (MC-PMT). Due to their large size, the MC-PMTs are set back from the crystal array, and scintillation light is transferred to the MC-PMT by sixty four 24 cm long optical fibres. The original camera design consisted of a 17 cm diameter ring of 30 of these detector units and had a reconstructed spatial resolution of 2.0 mm (FWHM) (Cherry et al. 1996). A second-generation microPET camera with smaller crystals has a spatial resolution approaching ~1 mm (Cherry et al. 1999) This research group, now at UC Davis, reported a spatial resolution of 0.83 mm FWHM at the CFOV (Tai et al. 2003). The performance of small-animal PET systems are now being characterised using NEMA standards (NEMA 2008). See for example Prasad et al. (2011).

5.9.4.5 PET/MR

Small prototype NMR-compatible PET systems with ~5 cm diameter have been constructed (Shao et al. 1997). These produced accurate registered PET/MR images and allowed correlation of PET images with functional MR data (images or spectra) over time (Garlick et al. 1997). The main technical issues are that the positron camera must be able to operate within the high field of the NMR magnet, and that the operation of the positron camera must not interfere with the NMR acquisition in any way. PMTs will not operate in a strong magnetic field so, in the prototype systems, optical fibres were used to transfer scintillation light from the camera to MC-PMTs situated several meters away from the magnet. Prototype systems have been demonstrated to operate successfully in a variety of NMR imaging and spectroscopy systems with field strengths up to 9.4 T. A secondary aspect of performing PET in a magnetic field is that the motion of the positrons is constrained by the high magnetic field and the average positron range is reduced (Iida et al. 1986). In principle this could lead to improvements in PET spatial resolution; however, significant improvements are only expected for high-energy positron emitters (e.g. ^{68}Ga, ^{82}Rb) in high-field magnets (>5 T). In 2007 Siemens produced the first PET/MR prototype dedicated to human brain imaging and, 3 years later, a whole-body device. These PET scanners use APD technology (see Section 5.2.1) which can function well in a 3 T MRI system. Philips has also produced a whole-body PET/MR device in which the two gantries are separated but share a common patient-handling system. See Pichler et al. (2010) for more details on the technical evolution and clinical potential of PET/MRI.

5.9.5 Clinical Data Acquisition

5.9.5.1 Standard Acquisition Protocols Using ^{18}FDG

The simplest PET imaging protocol involves the injection of ~250–370 MBq ^{18}FDG intravenously. Following an uptake period of 30–90 min, a single emission scan is acquired lasting typically up to 20 min. Prior to the advent of PET/CT, a transmission scan to correct the data for photon attenuation was achieved either through a separate scan, by interleaving emission and transmission scans for successive bed positions (thus reducing artefacts due to patient movement between the emission and transmission scans) or by performing the transmission scan simultaneously with the emission scan as described by Meikle et al. (1995). PET/CT allows a CT scan to be performed just prior to the PET scan for attenuation correction and localisation. Most clinical studies require the whole body to be surveyed, for example, for the presence of tumour. The axial FOV of most dedicated PET cameras is limited to 15–20 cm, so the survey is performed in a series of discrete steps, each lasting a few minutes and covering the body from head to toe (or head to pelvis for a 'half-body' scan) in ~20 min.

Whole-body scans taken in 2D acquisition mode are usually reconstructed using iterative reconstruction algorithms such as OSEM (Section 5.8.4.2) to reduce the streak artefacts that occur in images reconstructed using FBP (Section 5.8.4.1). Whole-body scans are performed in 3D mode with overlapping bed positions in order to provide a more uniform sensitivity in the axial direction. 3D images are produced using either fully 3D algorithms or by rebinning the data into 2D so that 2D algorithms can be used. Reconstruction methods for PET are described in Section 5.9.6.

5.9.5.2 Dynamic PET Acquisition Modes

Conventionally, static imaging data are stored either as 2D projections or as sinograms (Section 5.8.2.1). For the higher data rates that are more typical of dynamic imaging, list-mode storage is used and the data are then framed up later. Dynamic acquisitions are usually required for absolute quantification, mostly for research protocols. In dynamic imaging, data acquisition starts at the same time as tracer injection and may continue for anything from several minutes to several hours depending on the tracer. The subject must remain still for the whole duration of the study so that activity–time curves for small regions can be traced through the whole dataset. Because the tracer distribution changes most rapidly immediately after injection, time frames may be as short as 5 s or less at the start of the study. A large amount of data (~1 Gbyte) is acquired very rapidly and in 3D mode this becomes larger still. For these types of studies, it is important to correct for camera dead time (Section 5.9.7.5) as the data rates can change substantially over time.

5.9.6 Image Reconstruction Algorithms

The aim of PET image reconstruction is usually to create an image in which the grey-scale values reflect the radiotracer concentration in the object as accurately as possible. However, in some applications, for example, the detection of small lesions or when looking for subtle changes between successive studies, the objective may be to maximise the SNR in the image at the expense of quantitative accuracy.

5.9.6.1 2D Image Reconstruction

2D-PET data can be reconstructed using the standard FBP algorithm used in X-ray CT (see Section 3.6.1) and SPECT (see Section 5.8.4.1). Raw data are acquired in the form of a sinogram (one for each transaxial plane of data), and corrections for detector uniformity ('normalisation'), photon attenuation, random coincidences and scatter are usually applied directly to the sinogram prior to reconstruction. In order to boost the sensitivity of 2D multi-slice cameras, coincidence data are often acquired between adjacent or nearly adjacent rings, as shown in Figure 5.135, and all the LORs that do not lie in a transaxial plane are simply assigned to the sinogram corresponding to the nearest transaxial plane. The errors arising from this procedure are negligible over the patient FOV. Because of the inadequate linear sampling inherent in multi-crystal ring cameras, sinogram data are reorganised prior to reconstruction. This improves the inadequate linear sampling by a factor of two at the expense of angular sampling; however, the linear sampling distance (now equal to half the crystal width) is still only roughly equal to the spatial resolution and does not fulfil the requirement of the sampling theorem, that sampling should equal 0.3–0.5 of the spatial resolution.

5.9.6.2 3D Image Reconstruction Algorithms

A reconstruction algorithm for 3D-PET data must be able to incorporate the 'cross-plane' rays shown in Figure 5.136 as well as the standard 2D direct-plane data. It is, in principle,

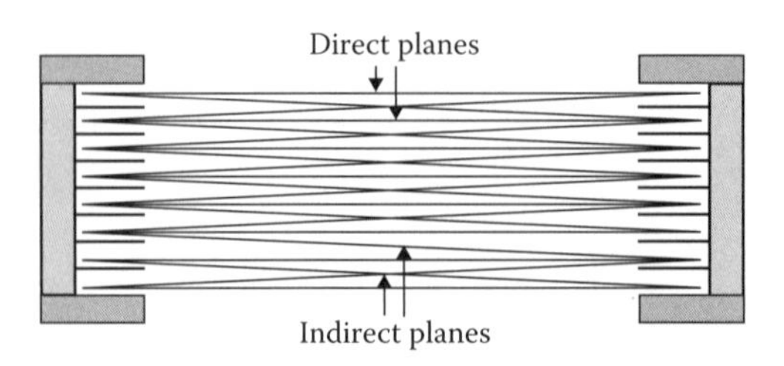

FIGURE 5.135
2D direct and indirect slices.

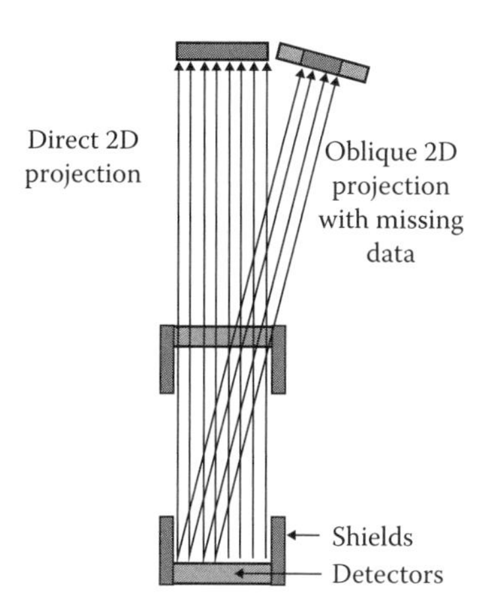

FIGURE 5.136
2D projections obtained in 3D PET acquisition.

fairly straightforward to develop a 3D equivalent of the 2D FBP – the 3D dataset is ordered into a set of 2D projections which are then filtered using a 2D analogue of the 1D ramp filter. An example of such a filter was described by Colsher (1980). The problem with this approach is that for the most common camera configuration (i.e. a cylindrical detector), most of the 2D projections are incomplete as shown in Figure 5.136. Several different algorithms have been proposed that address the problem of the 'missing data'.

The first widely used algorithm was the reprojection algorithm described by Kinahan and Rogers (1989). First, a reconstruction of the 2D direct planes only is performed and this is forward projected to estimate the missing parts of the 2D projections. The completed 2D projection can then be reconstructed using the Colsher filter. A speed advantage of about 50% is obtained by the 'FaVoR' algorithm (Defrise et al. 1992), in which the forward-projection step required to calculate the missing projection data is circumvented.

In order to overcome the computing burden of the 3D algorithms there has been a lot of interest in rebinning algorithms. In these algorithms, the data from several oblique planes are added together in a similar way to that for a multi-slice 2D camera. The advantage of this technique is that the summed sinograms can then be reconstructed using a fast 2D reconstruction algorithm such as 2D FBP or OSEM. The simplest rebinning approach is 'single-slice rebinning' (Figure 5.137). This approach is similar to the way in which indirect planes are combined in a multi-slice 2D system – each LOR is rebinned into the direct plane located halfway between the two detectors in coincidence. This very

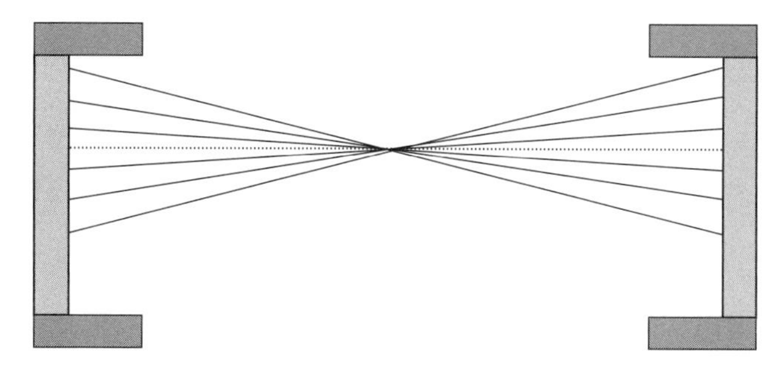

FIGURE 5.137
Single-slice rebinning. All the oblique rays are assigned to the central (dotted) plane.

crude approach obviously involves some gross approximations and can result in significant distortions for objects that extend some distance in the radial direction. However, for imaging structures close to the z-axis (e.g. structures deep within the brain), acceptable reconstructions can be obtained. An improved method is the 'Fourier rebinning' algorithm ('FORE') developed by Defrise et al. (1997) which exploits properties of the Fourier transform of a sinogram to obtain a highly accurate reconstruction. FORE is an order of magnitude faster than the 3D reprojection algorithm.

Whilst the 3D algorithms allow the incorporation of all the data collected by the camera into an artefact-free image with an overall high SNR (subject to the limitations described earlier), this may vary widely in the axial direction due to the triangular-shaped axial sensitivity profile. The image quality may be significantly degraded at the edge of the axial FOV particularly for systems with a large axial extent. For some algorithms, the axial variation in SNR is offset by an axial variation in the reconstructed spatial resolution.

5.9.6.3 Iterative and Model-Based Image Reconstruction Algorithms

The analytic reconstruction algorithms described earlier assume that the raw data contain no statistical noise, whereas statistical algorithms account for statistical noise in the data explicitly. Because these algorithms involve the process of iteration they are generally referred to as iterative algorithms. They can be applied to both 2D and 3D datasets with the aim of achieving the optimum image signal-to-noise properties without degrading the spatial resolution that occurs with simple smoothing procedures. Substantial effort has gone into the development of statistical algorithms, and they are now in widespread use. Images reconstructed using iterative techniques often have a very different appearance, particularly in the structure of noise in the image, to those reconstructed using standard methods, and reconstruction times can be long. Because iterative reconstruction algorithms are non-linear, the SNR and spatial resolution may vary at different points within an image.

One of the most widely used algorithms is ML-EM, described in Section 5.8.4.2. There are many variations of the ML-EM algorithm including the very fast OSEM which is available on most commercial systems (often in conjunction with the FORE 3D rebinning algorithm). The OSEM algorithm does not necessarily converge to the ML solution, but due to its speed, it is used widely, particularly for whole-body PET, where it appears to be able to reduce artefacts in these noisy images without any reduction in spatial resolution, as shown in Figure 5.138.

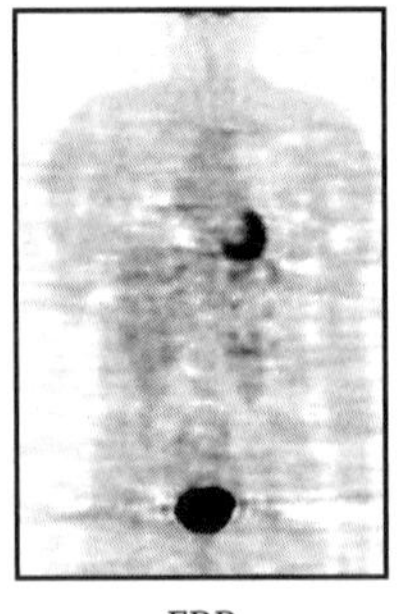

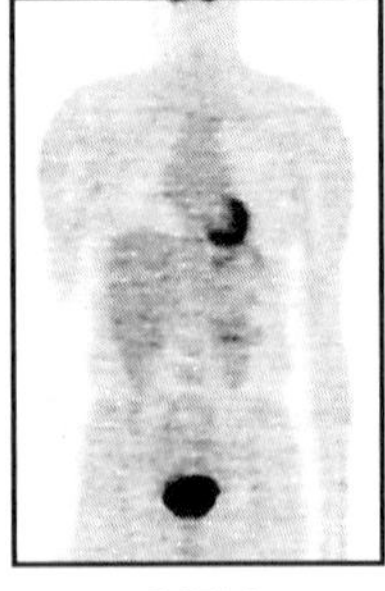

FBP OSEM

FIGURE 5.138
FBP and OSEM reconstructions for a whole-body FDG image in which uptake of FDG can be seen in the heart and the bladder. The OSEM reconstruction shows reduction of streak artefacts compared with FBP. (Courtesy of G. Cook.)

Whereas FBP implicitly assumes that the imaging system has perfect spatial resolution, iterative algorithms can be formulated to incorporate a model of the detector response. The model might include a spatially varying point spread function to account for the blurring introduced by the positron range and depth of interaction effects. Incorporating these effects into the model can only improve image quality to the extent that there is sufficient statistical information in the data to support it. However, in situations where the statistical quality of the data is good, very impressive results have been obtained (Qi et al. 1998a, 1998b).

5.9.7 Data Correction and Image Quantification

5.9.7.1 Normalisation

Random and systematic variations in detector efficiencies can lead to an increase in noise and artefacts in reconstructed images. In order to overcome these effects a correction factor is applied to every LOR in the emission sinogram, so that each ends up with the same sensitivity. A blank scan (a transmission scan with nothing in the FOV) or a scan of a uniformly active cylinder is performed. The correction factors are then proportional to the inverse of the count rates obtained for each LOR. A correction must be made for variations in the incident flux along each LOR and there are various methods of improving the statistical accuracy of the normalisation dataset. The normalisation procedure for 3D mode is essentially the same as that for 2D and is reviewed in Badawi et al. (1998).

For cameras with rotating detectors it is also necessary to apply a set of 'rotational' weights that account for the fall-off in efficiency for LORs at the edge of the transverse FOV as these are only detected by the rotating detectors for a relatively short time (Clack et al. 1984). For rotating systems, the detector efficiency normalisation factors depend only on the radial offset of the LOR.

5.9.7.2 Attenuation Correction

In principle attenuation correction in PET is straightforward, consisting simply of multiplying the emission sinogram by attenuation correction factors derived from the transmission scan. However, in practice the transmission data need to be processed in various ways to enhance the statistical quality of the data. This is because transmission scans are often acquired very rapidly, with a resulting reduction in SNR which will propagate into the final reconstructed image. The transmission sinogram can be smoothed to regain SNR; however, this results in a mismatch in the resolution of the transmission and emission data and results in artefacts, for example, at the lung boundary (Chatziioannou and Dahlbom 1996). A better solution for whole-body studies is to reconstruct the transmission images and then apply a segmentation algorithm using known values of μ (or deriving average values from the image) (Xu et al. 1996). This high-quality segmented image, showing areas of soft tissue, lung and bone, can then be forward projected to obtain the required attenuation correction factors with much improved SNR. Segmentation can also be used on transmission images acquired using a single-photon transmission source (permitting transmission scan times of <1 min) (Yu and Nahmias 1995). In this case, the values of μ are wrong due to the use of a ^{137}Cs source (gamma-ray energy 662 keV) and because of the large amount of scatter in the image. However, a high-quality image is still obtained and image segmentation is used to assign predetermined μ values to the relevant areas. Typical emission images with and without attenuation correction are shown in Figure 5.139.

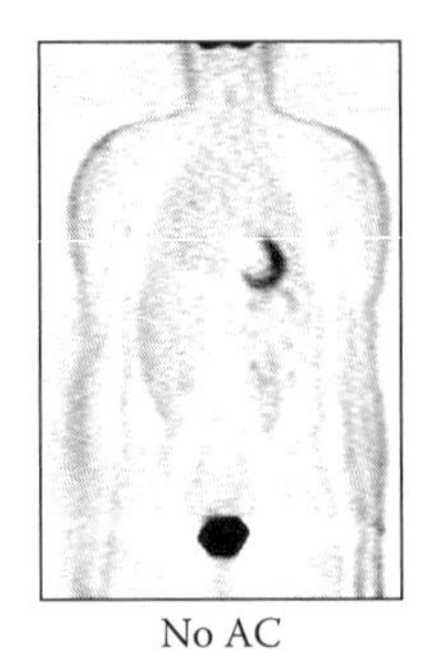

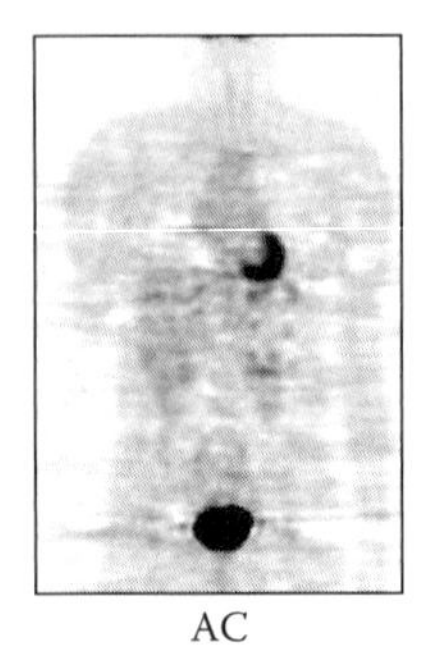

FIGURE 5.139
Whole-body coronal images before and after attenuation correction (AC). Note the apparent high uptake in the lungs and increased activity at the skin surface in the uncorrected (No AC) emission images. These artefacts are removed when the attenuation correction is applied. (Courtesy of G. Cook.)

For clinical brain scanning an attenuation correction can be performed using a simple segmentation procedure on the emission sinogram, thus removing the requirement for a transmission scan (Bergstrom et al. 1982).

Current whole-body scanners are invariably PET/CT systems and the CT data are used to provide an attenuation correction for the PET scan. Usually the CT scan precedes the PET scan and the quality of the CT image required (for attenuation only or for diagnosis as well) depends on the current used in the X-ray system and the time taken for the scan. It should be noted that the radiation dose to the patient from the CT scan may be comparable to that from the PET scan. The effect of the energy difference between the photons from the CT scanner and the positron emitters is corrected for prior to applying the CT-derived attenuation correction.

5.9.7.3 Randoms Correction

The most effective way of dealing with random coincidence events is to design the PET camera so that these events are minimised. This can be achieved either by shielding the detectors from single events originating outside the FOV, or by using detectors with a short temporal resolution.

In clinical whole-body imaging the randoms rates can easily be greater than or equal to the trues rates so, for quantification, randoms subtraction is essential. There are straightforward ways of correcting for the systematic effects of randoms (Hoffman et al. 1981). An estimate of the randoms contribution to each LOR in the raw data is usually acquired simultaneously with the real data (using a delayed-coincidence channel) and subtracted online so no further processing is require to correct for randoms. Randoms can also be corrected for by using the coincidence data itself – images can be reconstructed by choosing photon coordinates from different events (Divoli et al. 2004). This provides a randoms image with very high statistics unlike most other methods.

5.9.7.4 Scatter Correction

The total fraction of scatter in an image depends on the object being imaged but can be up to 30% for a 2D whole-body image, and larger still in 3D. Because the distribution of scatter in the reconstructed image is very broad (due to the large angles of scatter at 511 keV), scatter is usually ignored for clinical imaging in 2D mode as it has little effect on the appearance of the image other than to add a small background ~10%. If accurate quantification is required then various methods of scatter correction are possible. Deconvolution of a non-stationary scatter response function has been validated for brain studies (Endo and Iinuma 1984) but has not been widely used in the body. In 3D mode, scatter fractions can be high enough to make visual image interpretation difficult. In the brain, simple techniques, such as fitting smooth functions to the edge of the sinogram and extrapolating these across the FOV are adequate (Stearns 1995). For the body, more sophisticated approaches are required as the scatter distribution can show some structure. One approach is to model the scatter distribution, although this does not account for scatter arising from activity outside the FOV (Ollinger 1996).

5.9.7.5 Dead-Time Correction

As indicated earlier (Section 5.9.2.8) the dead time of a positron camera can be long enough to produce substantial data losses when high activity levels are being imaged. Correction for this effect is particularly important if image quantification is the objective following dynamic imaging of short-lived tracers. Corrections are usually determined by using the detector singles rates and a model of the camera dead time.

5.9.7.6 Quantification/Calibration

The calibration of image voxel values in units of Bq mL^{-1} can be derived from first principles, but is normally achieved by scanning a uniform cylinder with a known radioactivity concentration. Clinically, this is required for the evaluation of standard uptake values (SUVs) (see Section 5.9.8.1) which are an index of tracer uptake that can be compared between subjects. It is often necessary to cross-calibrate the camera against a well counter used to measure blood samples acquired during a study. This can be accomplished without reference to an absolute radioactivity measurement by imaging a uniform cylinder of activity and then measuring in the well counter a small aliquot taken from the cylinder. The image value is related to the well-counter count rate (per volume of the sample) by the factor, *K*:

$$K = \frac{\mathrm{SC} \times \mathrm{VA}}{\mathrm{WCA}} \tag{5.72}$$

where

SC is the average image voxel value for the calibration cylinder,
VA is the volume of the aliquot, and
WCA is the well-counter count rate for the aliquot.

5.9.7.7 Partial-Volume Effect

All regional measurements of tracer uptake determined from PET (and SPECT) images are subject to the limitations of the PVE, as discussed in Section 5.8.5.8. Objects that have a diameter smaller than two or three times the image spatial resolution (FWHM) have a reduced peak value relative to large objects as shown in Figure 5.140, and the apparent activity may even be reduced to a level below the threshold for visual detection (Hoffman et al. 1979). The ratio of the apparent activity to the true activity is called the recovery coefficient, which for a tumour of diameter equal to the spatial resolution may be as little as 0.1. In principle, recovery coefficients can be determined from phantom measurements; however, this is only of use if the dimensions of the functional volume of the object are known.

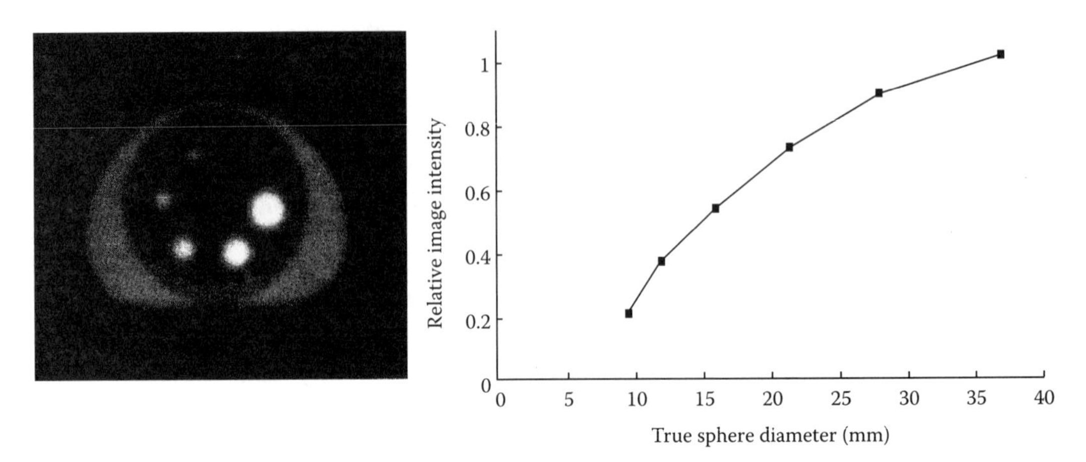

FIGURE 5.140
Partial-volume effect. Each sphere in the phantom contains the same activity concentration. The image on the left and the graph on the right show that the measured activity is reduced for the smaller spheres. (From Marsden, 2006. Reproduced by permission of Edward Arnold Limited.)

Reasonable assumption may be made in some situations – for example, various algorithms have been proposed for partial volume correction in the brain using anatomical information derived from MRI scans, although such methods are very sensitive to misregistration between the functional and anatomical images. For clinical applications, it may be possible to determine the dimensions of a tumour from CT or MRI but again there is no guarantee that the anatomical and functional volumes are the same. Overall these effects make quantification of uptake in small volumes in a radioactive background very difficult.

5.9.8 Image Quantification

5.9.8.1 Standard Uptake Values

Absolute quantification of PET images is rarely performed clinically partly due to the complexity of the procedures, but primarily because there are currently very few clinical situations where absolute quantification has a proven utility. A semi-quantitative index that is widely used in FDG-PET (Keyes 1995) is the standard uptake value (SUV) which is defined as

$$\mathrm{SUV} = \frac{\text{Tumour uptake (MBq mL}^{-1})}{(\text{Injected dose (MBq)/Patient weight (g))}}. \tag{5.73}$$

If the tracer were distributed uniformly throughout the body, the SUV would be equal to unity everywhere. Tumour SUVs may range from <1 to >20 and have been demonstrated to be related to grade and malignancy for certain tumours.

5.9.8.2 Compartmental Modelling

Some PET research studies attempt to measure physiological parameters in absolute units, for example, blood flow in mL min^{-1} g^{-1} or glucose utilisation rate in mol min^{-1} g^{-1}. These studies are complex to perform, requiring rapid dynamic imaging and measurement of the time course of the tracer concentration in arterial blood. Values for kinetic rate constants (k)

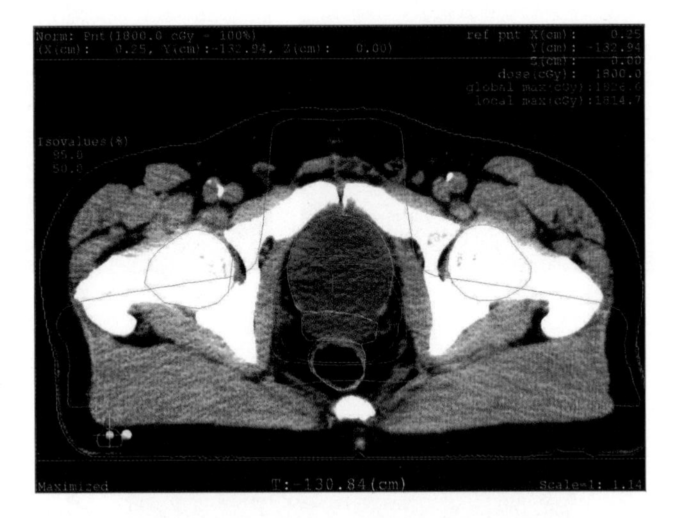

FIGURE 4.3
CT radiotherapy plan for treatment of Ca prostate. Red = PTV, blue = bladder, yellow = rectum. The greyscale in the CT radiotherapy plan has been windowed in an attempt to make the coloured lines more visible.

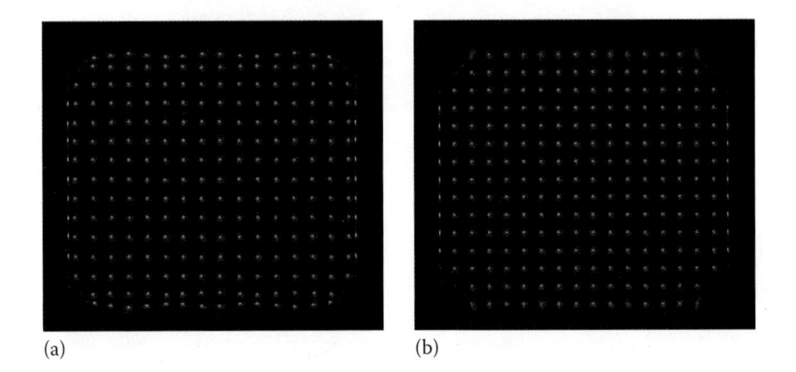

FIGURE 5.26
Images of a linearity phantom comprising an array of $^{99}Tc^{m}$ sources: (a) no linearity correction and (b) with linearity correction. (Images courtesy of S. Heard.)

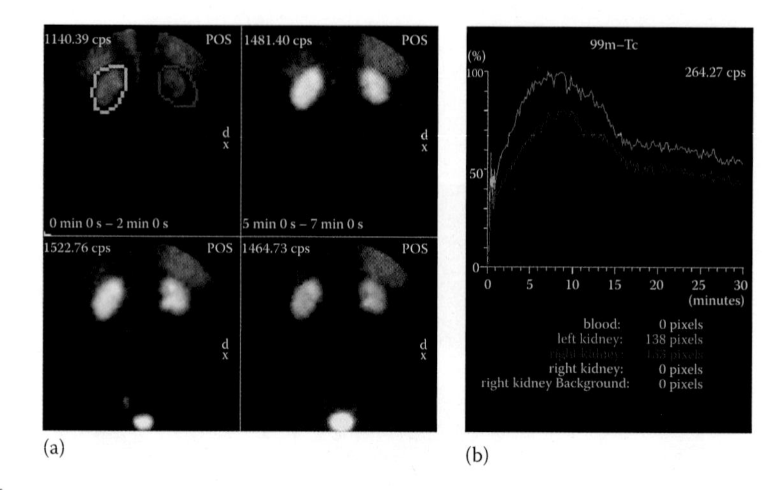

FIGURE 5.56
(a) ROIs positioned on a 2 min summed image of a dynamic renal study using $^{99}Tc^{m}$ MAG3. (b) Kidney activity-time curves produced from these ROIs. (Courtesy of S. Chittenden.)

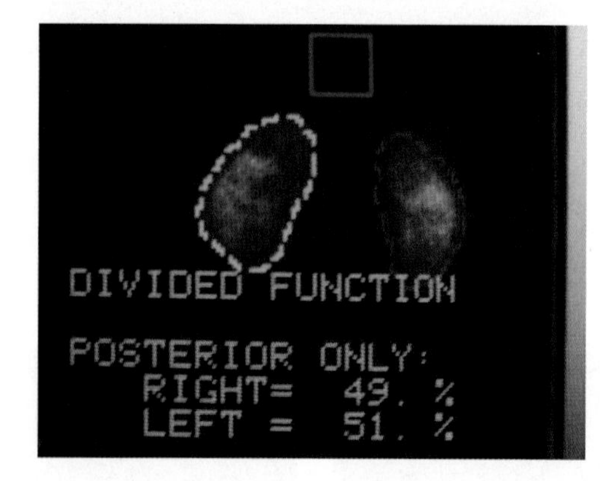

FIGURE 5.68
Example of the results for a divided renal function study. (Courtesy of S. Sassi.)

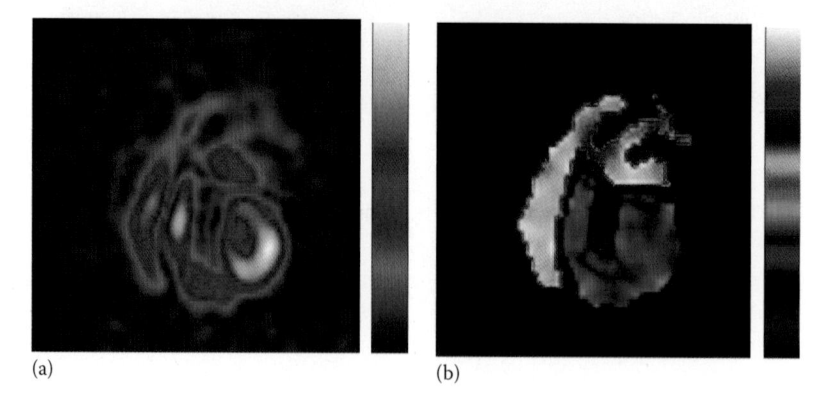

FIGURE 5.77
Parametric images of the amplitude (a) and phase (b) of the ejection from the left ventricle. (Courtesy of S. Chittenden.)

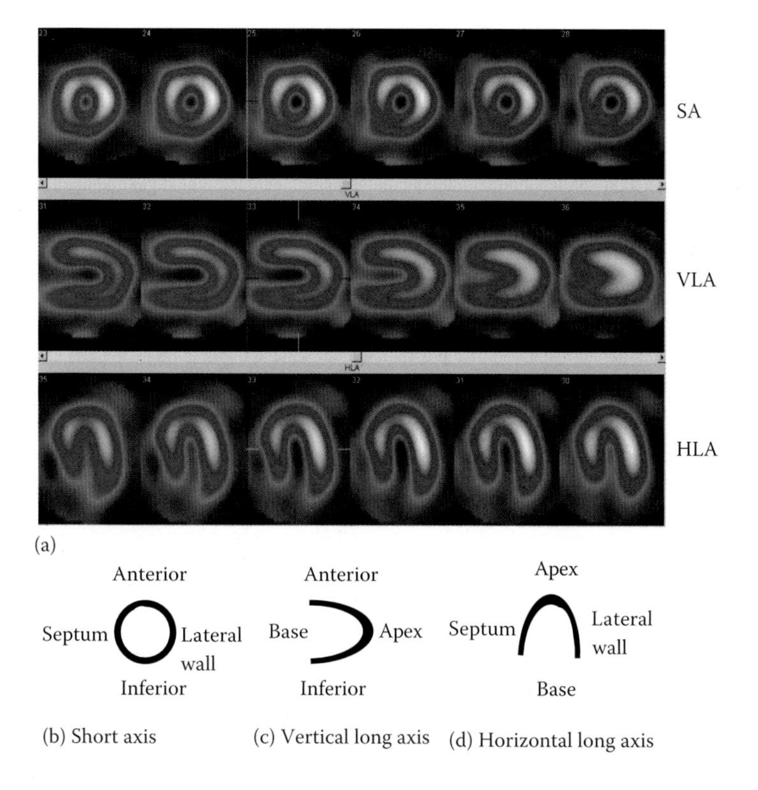

FIGURE 5.114
(a) $^{99}Tc^{m}$-sestamibi SPECT scans showing myocardial perfusion in a normal male. Each row shows a different view: SA, short axis; VLA, vertical long axis; HLA, horizontal long axis. (b)–(d) Schematics showing the orientation of each view. (Courtesy of S. Sassi.)

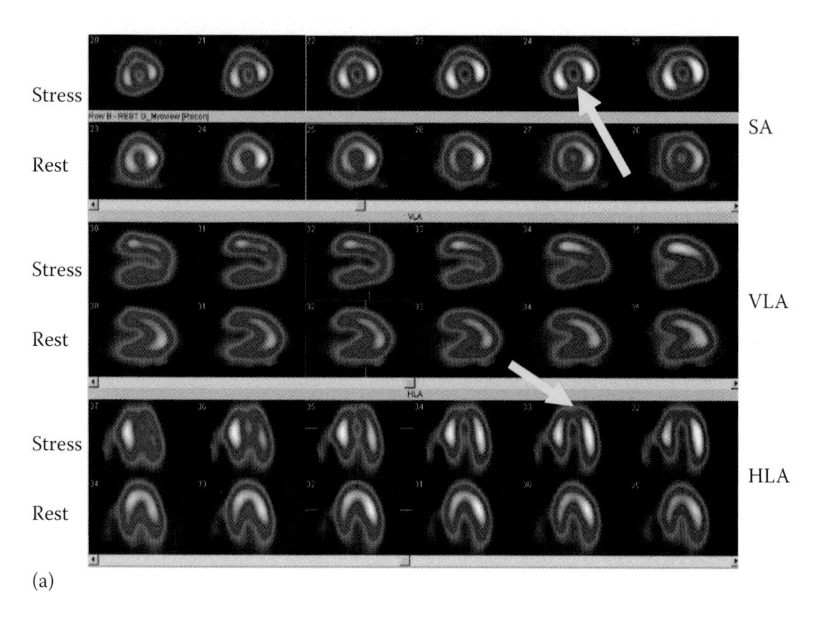

FIGURE 5.115
Myocardial perfusion images for a patient with ischaemia. (a) Images taken under different conditions: stress and rest. The abnormality is indicated by the yellow arrows.

(continued)

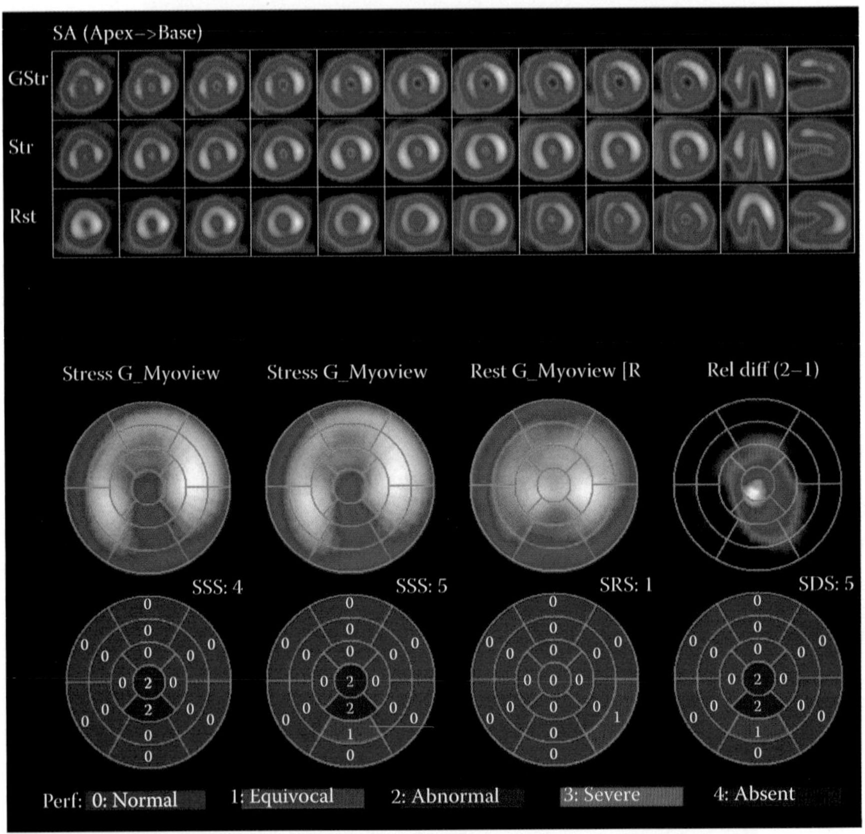

(b)

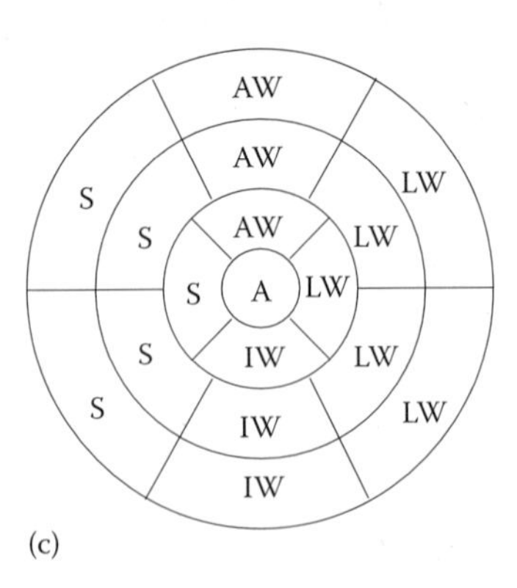

(c)

FIGURE 5.115 (continued)
(b) Polar plots (or bull's eye display) that are derived from the short axis views and that include a detriment map (i.e. rest minus stress) and statistical analysis results. (c) Orientation of the segments of the polar plots: A, apex; AW, anterior wall; S, septum; IW, inferior wall; LW, lateral wall. Statistical analysis of the polar plots is used to classify the blood flow in the segments as normal, equivocal, abnormal, severe, or absent. (Images in (a) and (b) are courtesy of S. Sassi.)

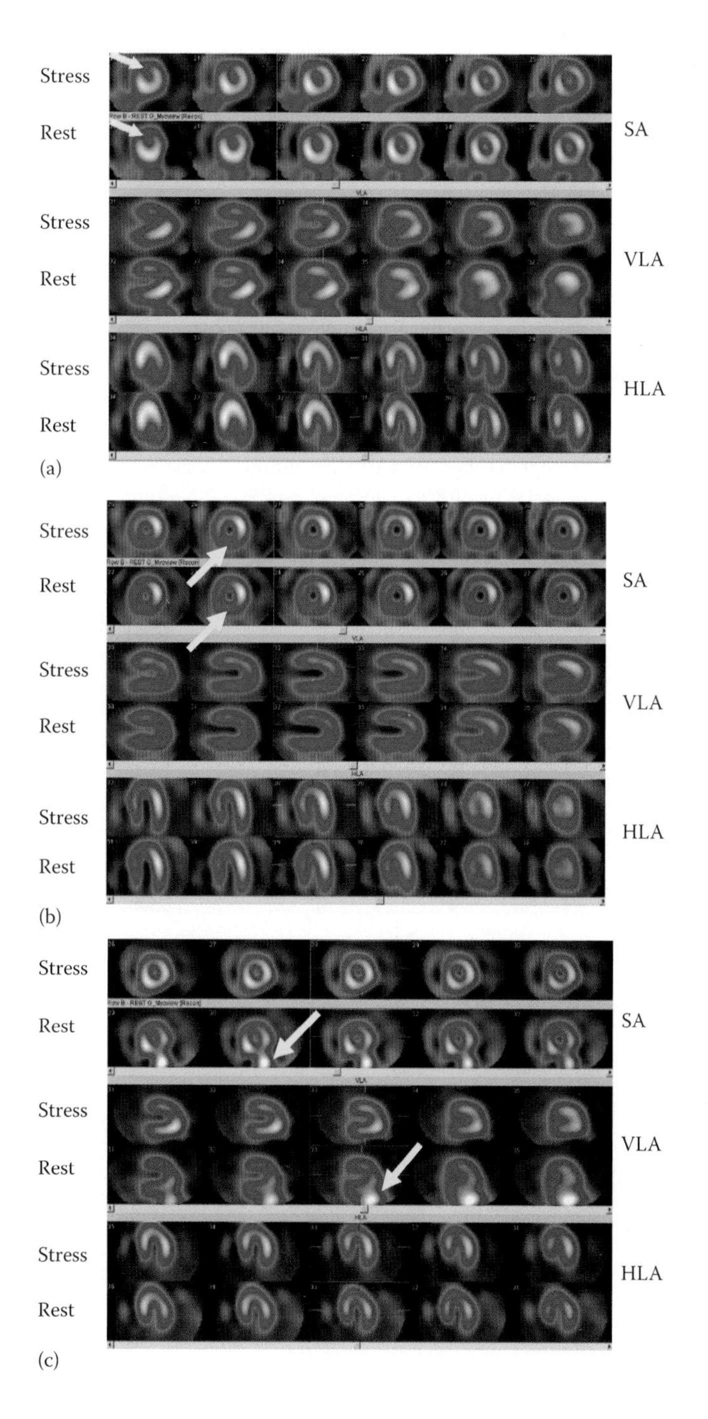

FIGURE 5.116
Examples of artefacts in cardiac SPECT images due to (a) breast attenuation, (b) diaphragmatic attenuation. (c) gut uptake. The yellow arrows indicate the position of the artefacts. Note that the attenuation artefacts appear in both the rest and stress images, whereas the ischaemic area is only seen on the stress images. (Courtesy of S. Sassi.)

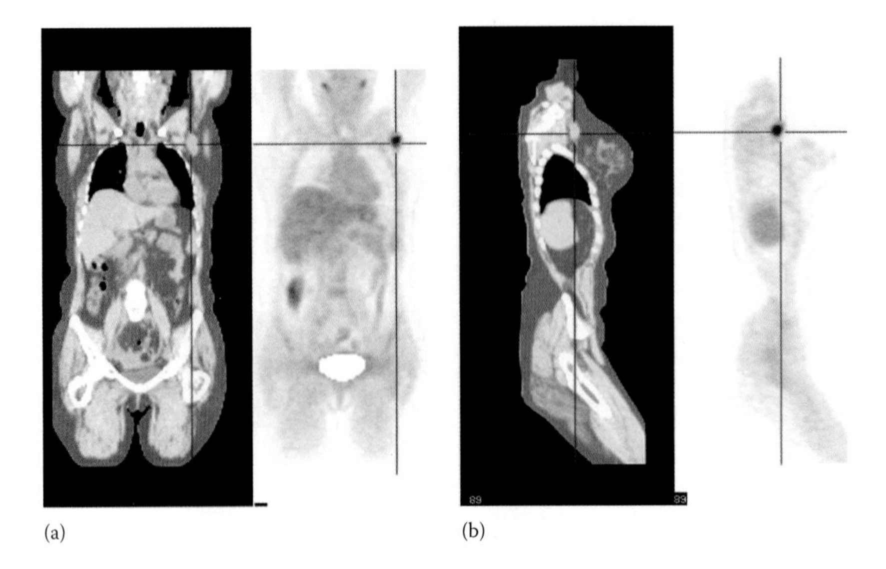

FIGURE 5.133
Whole-body images from a combined PET/CT scanner: (a) coronal images; (b) sagittal images. In each pair of images, CT is on the left, and PET is on the right. Linked cursors (red lines) have been used to indicate a lesion in the patient's left axilla. (Courtesy of GE Healthcare, Slough, Berkshire, UK.)

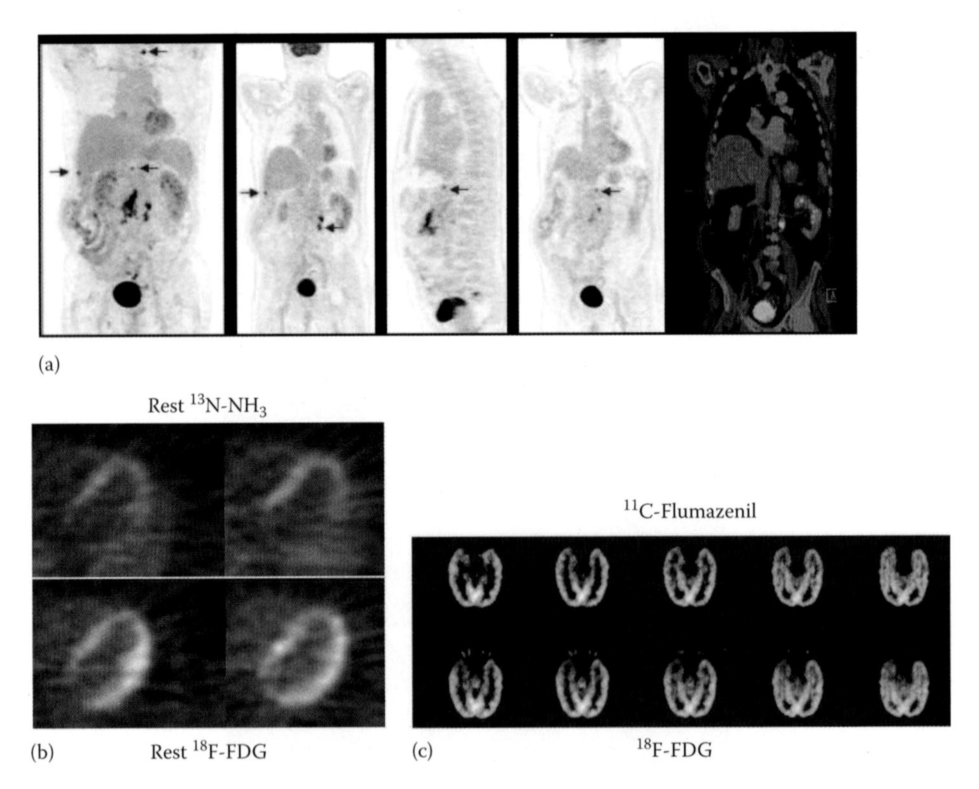

FIGURE 5.142
Examples of clinical PET images. (a) Clinical FDG-PET/CT images for primary staging of a patient with non-Hodgkin's lymphoma. A mesenteric nodal mass with increased ^{18}F-FDG uptake along with hypermetabolic para-aortic and celiac lymph nodes are visualised. A solitary ^{18}F-FDG avid left supraclavicular nodal lesion and a small lesion in the margin of the lower part of right lobe of liver suggest advanced disease. (Courtesy of University of Tennessee, Knoxville, TN.) (b) ^{13}N-NH$_3$ and ^{18}F-FDG images of a hibernating myocardium. (c) ^{11}C-Flumazenil and ^{18}F-FDG images of a patient with epilepsy.

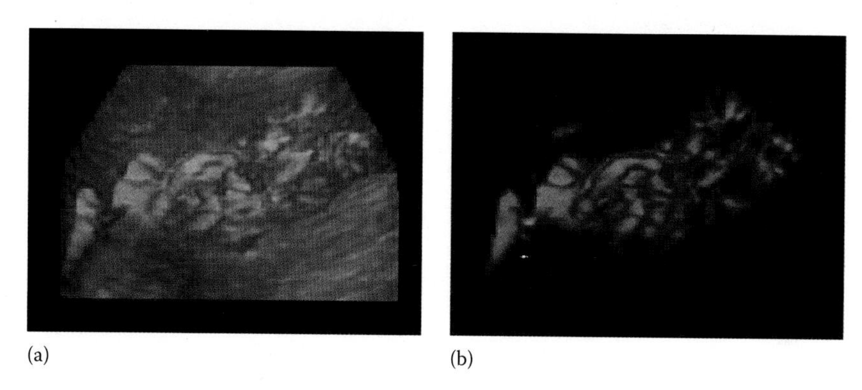

FIGURE 6.40
Example of volume rendering (of a 3D power Doppler scan of a normal kidney): (a) maximum intensity projection volume-rendering of grey-scale data, with blood flow superimposed using surface rendering and (b) blood-flow component of the image only.

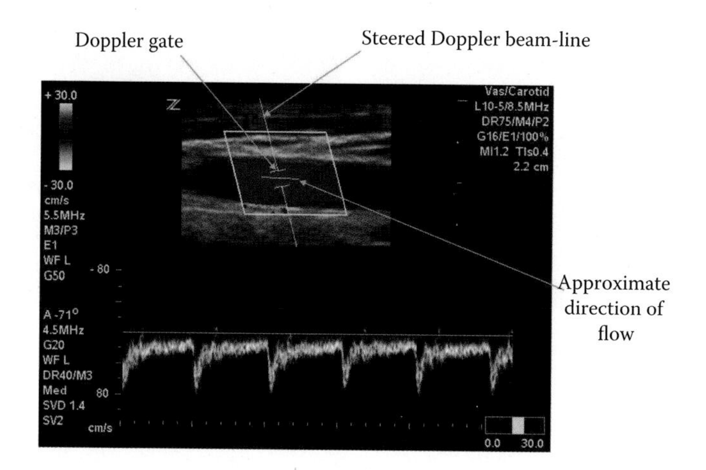

FIGURE 6.53
Triplex display of blood flow in the carotid artery. Above: the position and size of the Doppler gate, the steered Doppler beam line and the approximate direction of flow as indicated by the user are superimposed on the combined B-mode and colour Doppler image. Below: the spectral Doppler information is displayed as a sonogram, which depicts spectral power (brightness) as a function of flow velocity and direction (vertical) and time during the cardiac cycle (horizontal). When the user changes the scanner's direction-of-flow indicator, the scanner alters $\cos\theta$ in Equation 6.21 to change the vertical scale of the sonogram.

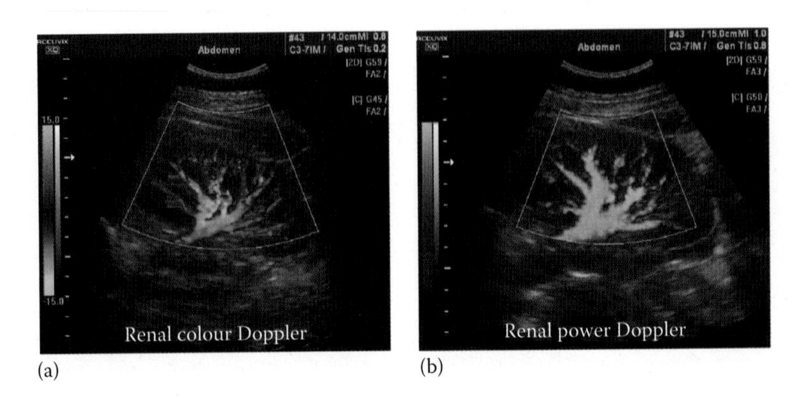

FIGURE 6.54
(a) Colour Doppler and (b) power Doppler images of the same view of a normal human kidney. (Courtesy of Medison Inc., Miami, FL.)

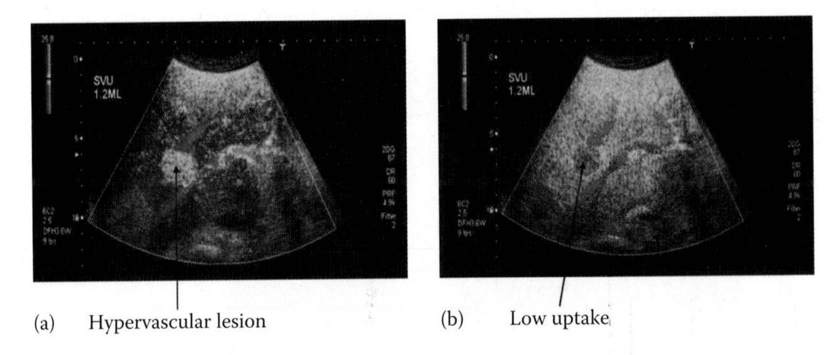

FIGURE 6.56
Illustration of bubble-specific imaging; liver metastasis from colo-rectal cancer. After Sonovue™ microbubble injection, the early 'arterial phase' (a) shows the lesion to be hypervascular, but in 'late phase' (b) the same lesion shows low uptake. The type of image display illustrated combines pulse-inversion bubble-specific imaging with colour Doppler. Green is used when the mean Doppler shift is below a threshold (i.e. 'stationary' bubbles) and red or blue (according to flow direction) when the Doppler shift is above the threshold (i.e. 'moving' bubbles). (Images courtesy of D.O. Cosgrove and M. Blomley, and reproduced with permission from Toshiba Medical Systems Visions Magazine.)

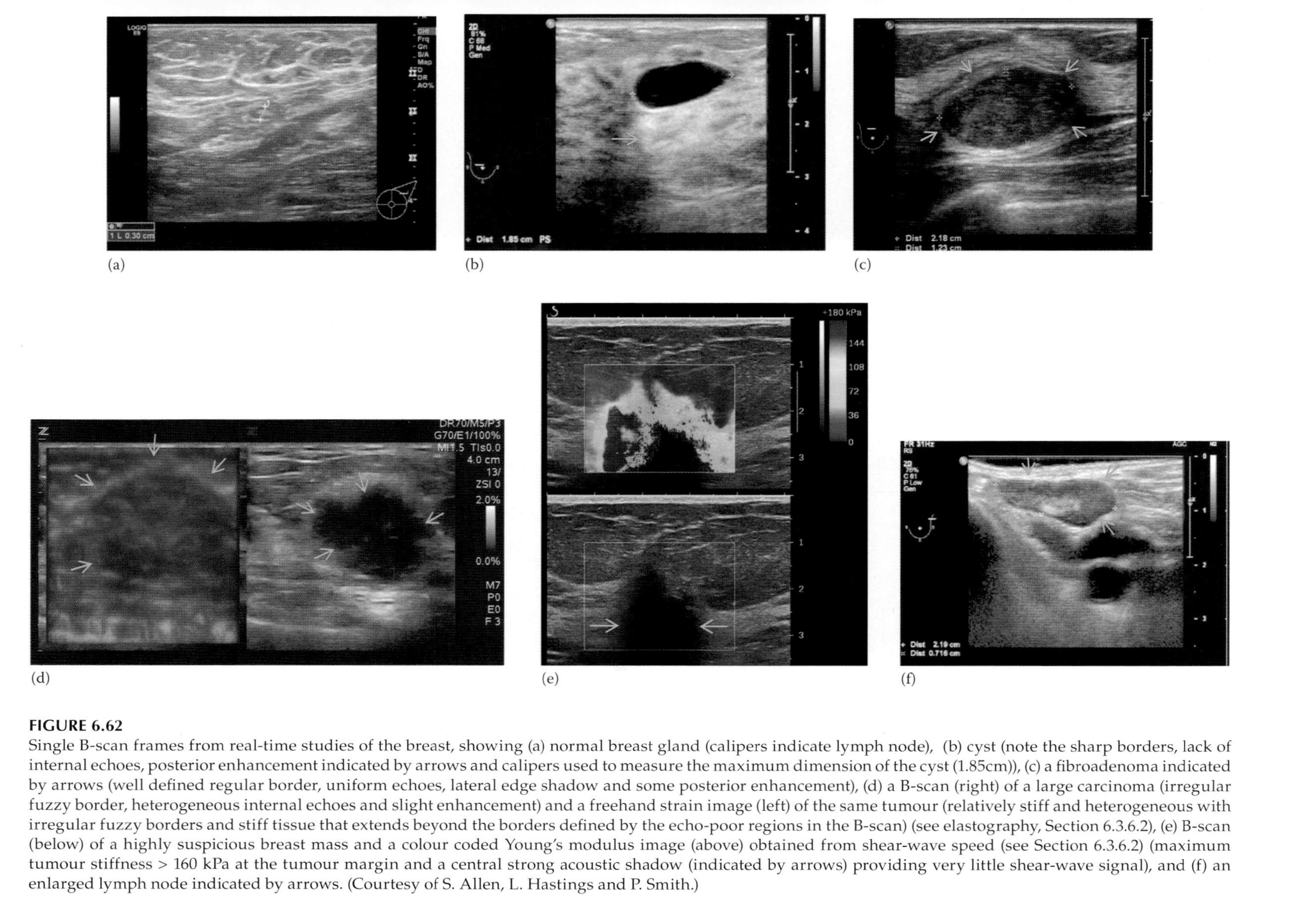

FIGURE 6.62
Single B-scan frames from real-time studies of the breast, showing (a) normal breast gland (calipers indicate lymph node), (b) cyst (note the sharp borders, lack of internal echoes, posterior enhancement indicated by arrows and calipers used to measure the maximum dimension of the cyst (1.85cm)), (c) a fibroadenoma indicated by arrows (well defined regular border, uniform echoes, lateral edge shadow and some posterior enhancement), (d) a B-scan (right) of a large carcinoma (irregular fuzzy border, heterogeneous internal echoes and slight enhancement) and a freehand strain image (left) of the same tumour (relatively stiff and heterogeneous with irregular fuzzy borders and stiff tissue that extends beyond the borders defined by the echo-poor regions in the B-scan) (see elastography, Section 6.3.6.2), (e) B-scan (below) of a highly suspicious breast mass and a colour coded Young's modulus image (above) obtained from shear-wave speed (see Section 6.3.6.2) (maximum tumour stiffness > 160 kPa at the tumour margin and a central strong acoustic shadow (indicated by arrows) providing very little shear-wave signal), and (f) an enlarged lymph node indicated by arrows. (Courtesy of S. Allen, L. Hastings and P. Smith.)

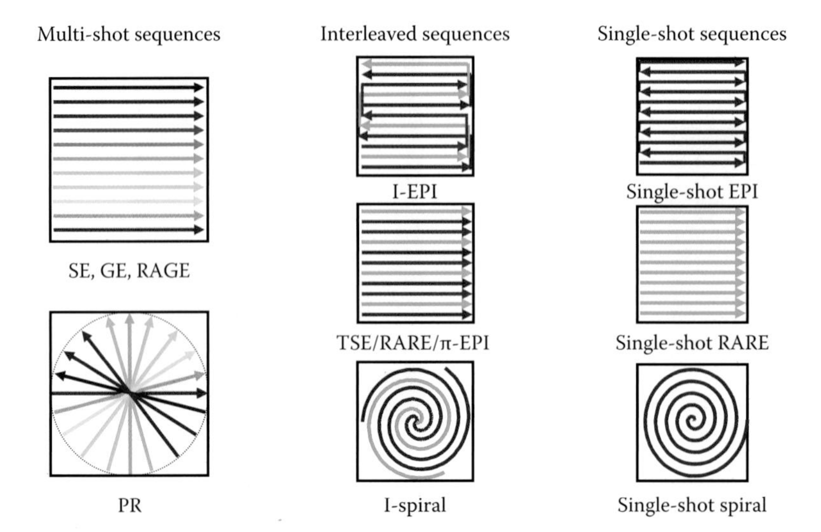

FIGURE 7.30
Schematic comparison of the *k*-space trajectories in multi-shot, interleaved and single-shot sequences. Different coloured lines represent data acquired on different shots.

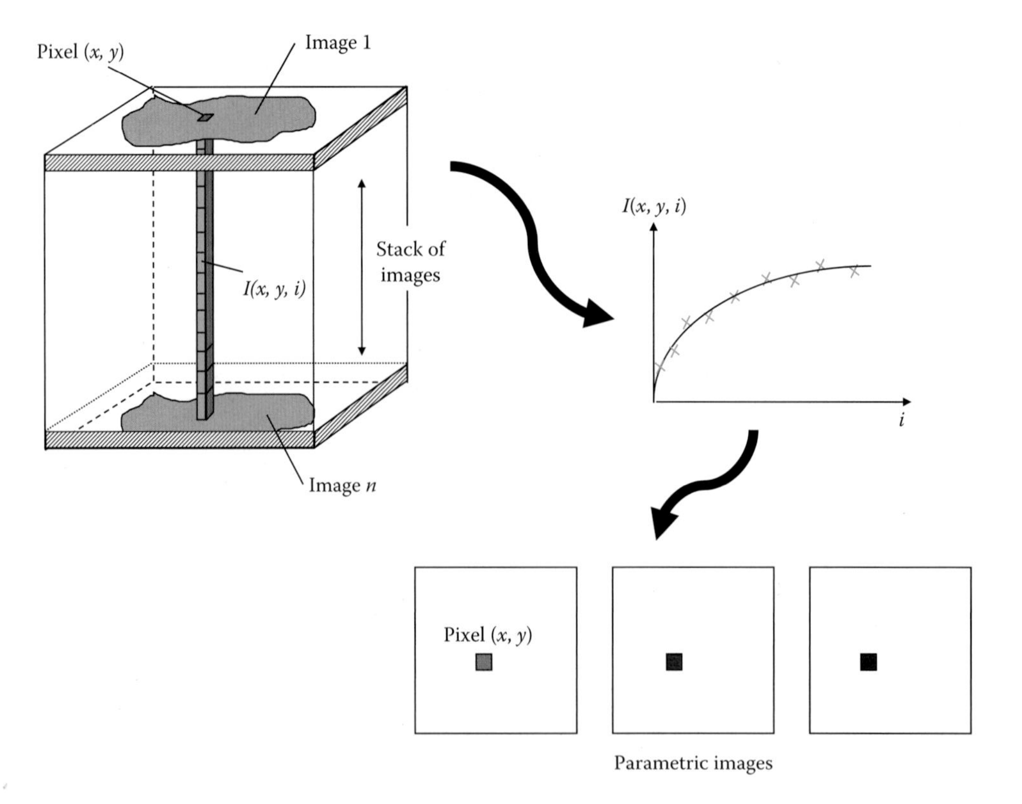

FIGURE 7.35
Principle of quantitative imaging: A stack of images is acquired, each weighted differently by the parameter to be measured. A model fit is performed for each pixel (x, y), and the results are placed in the output parametric images.

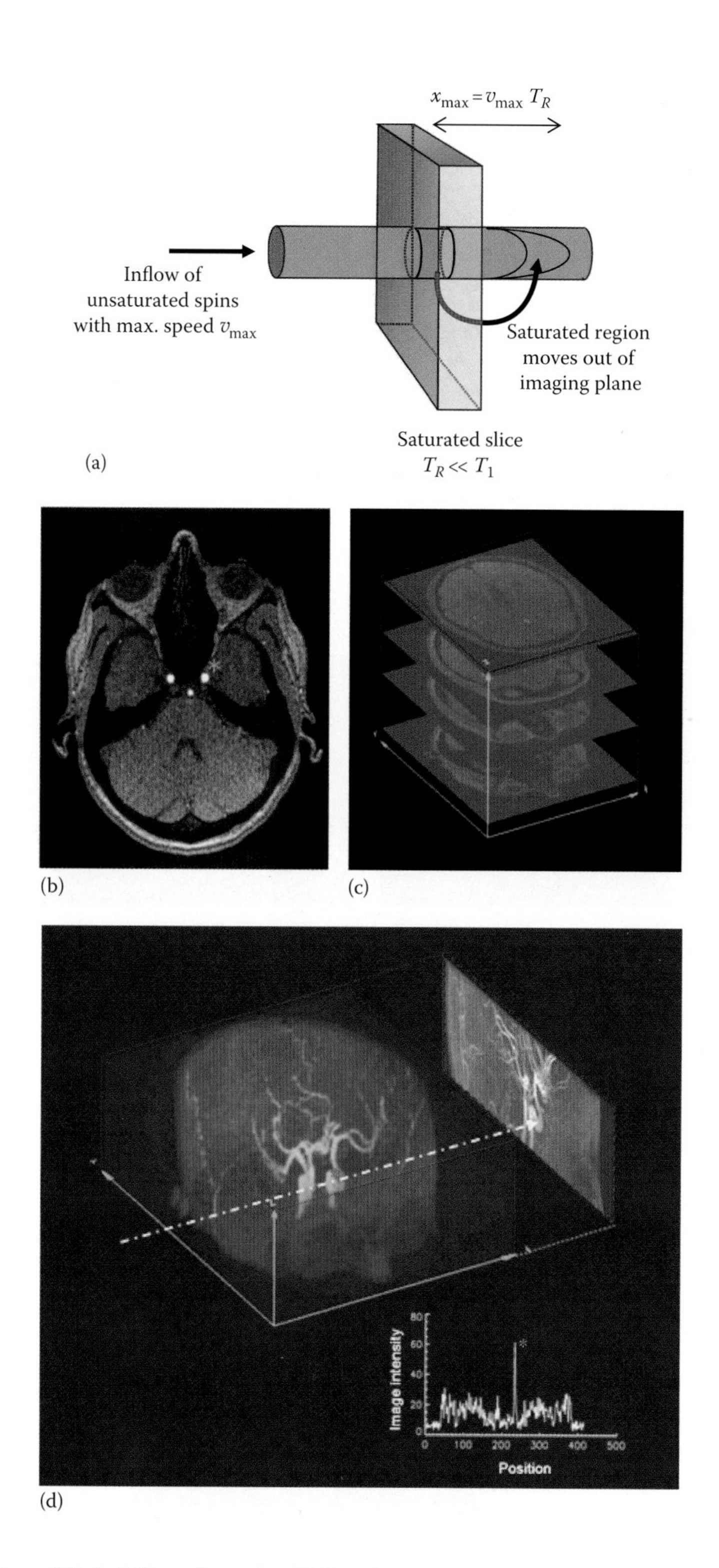

FIGURE 7.36
(a) Principle of time-of-flight MR angiography. (b) Sample angiography image of a cross section in the head. High vessel contrast is achieved by saturation of stationary spins and inflow of fully relaxed spins. (c) A 3D image is acquired consisting of a stack of highly contrasted slices. (d) A maximum intensity projection (MIP) image is formed from projections taken through the 3D data set, with the inset graph showing the signal intensity along the dotted line. The orange asterisks shown in (b) and (d) represent corresponding points in the different representations of the data.

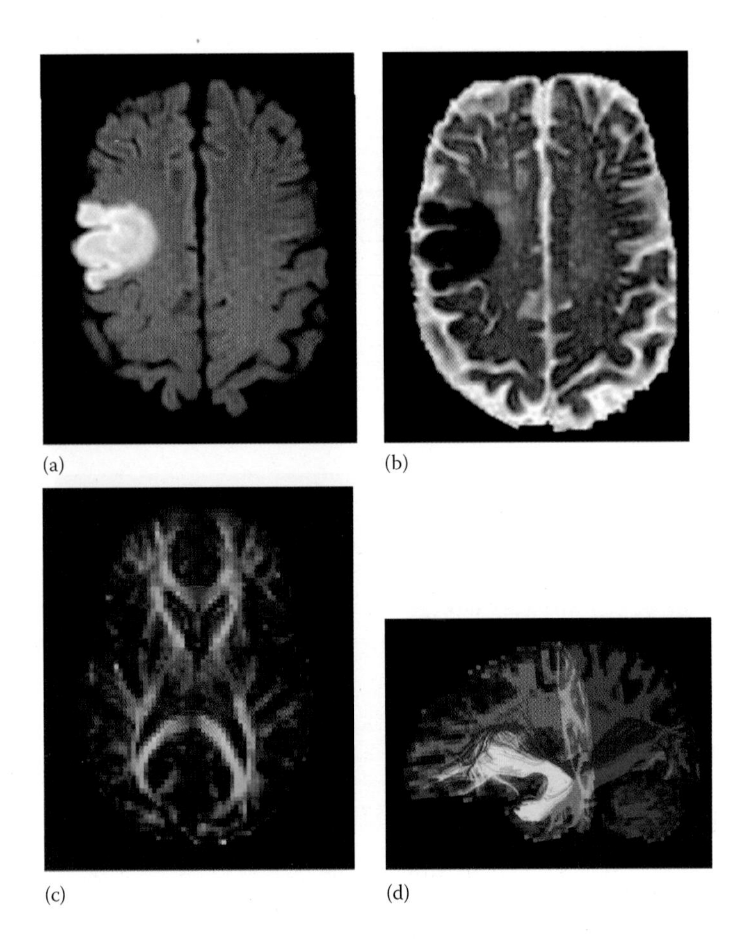

FIGURE 7.39
Examples of MR diffusion imaging. (a) Diffusion-weighted image of a patient with an acute stroke. (b) Calculated ADC image. (Data for (a) and (b) from Pipe and Zwart, 2006.) (c) Transverse DTI colour map: red, green and blue components of each colour represent fibres running along right-left, anterior-posterior and superior-inferior axes, respectively. (d) White matter tracts identified via 'diffusion streamlines'. (Data for (c) and (d) are Courtesy of the Autism Research Centre, University of Cambridge, Cambridge, UK.)

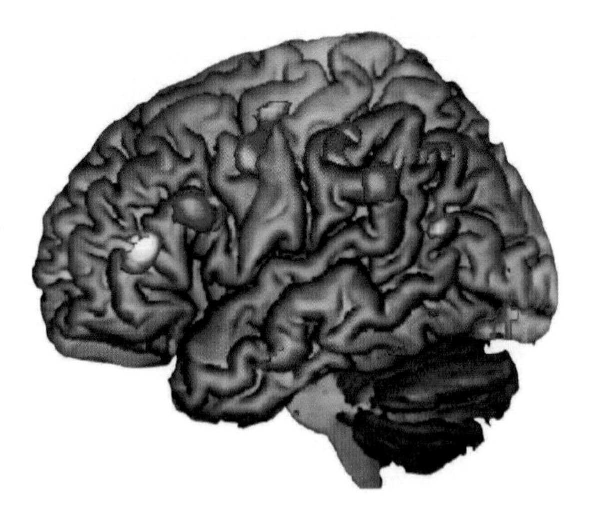

FIGURE 7.40
Example of a fMRI data set superimposed on a 3D surface rendering of a high-resolution anatomical data set. (Data Courtesy of the Autism Research Centre, University of Cambridge, Cambridge, UK.)

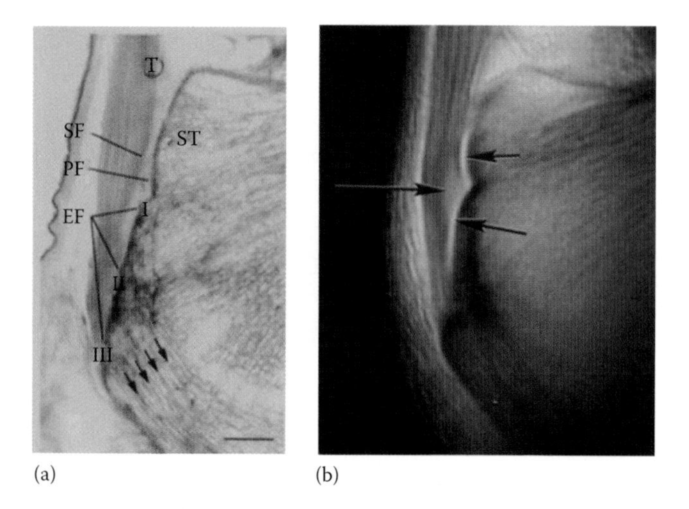

FIGURE 7.41
Achilles tendon. Anatomical specimen (a) and UTE image (T_R/T_E 500/0.08 ms) (b). The fibrocartilage in the tendon has a high signal (see arrows in (b)). The enthesis fibrocartilage (EF), periosteal fibrocartilage (PF) and sesomoid fibrocartilage (SF) are shown with the tendon (T) opposite the superior tuberosity (ST). Arrows in (a) show the direction of the trabecular bone. (Data from Robson and Bydder, 2006.)

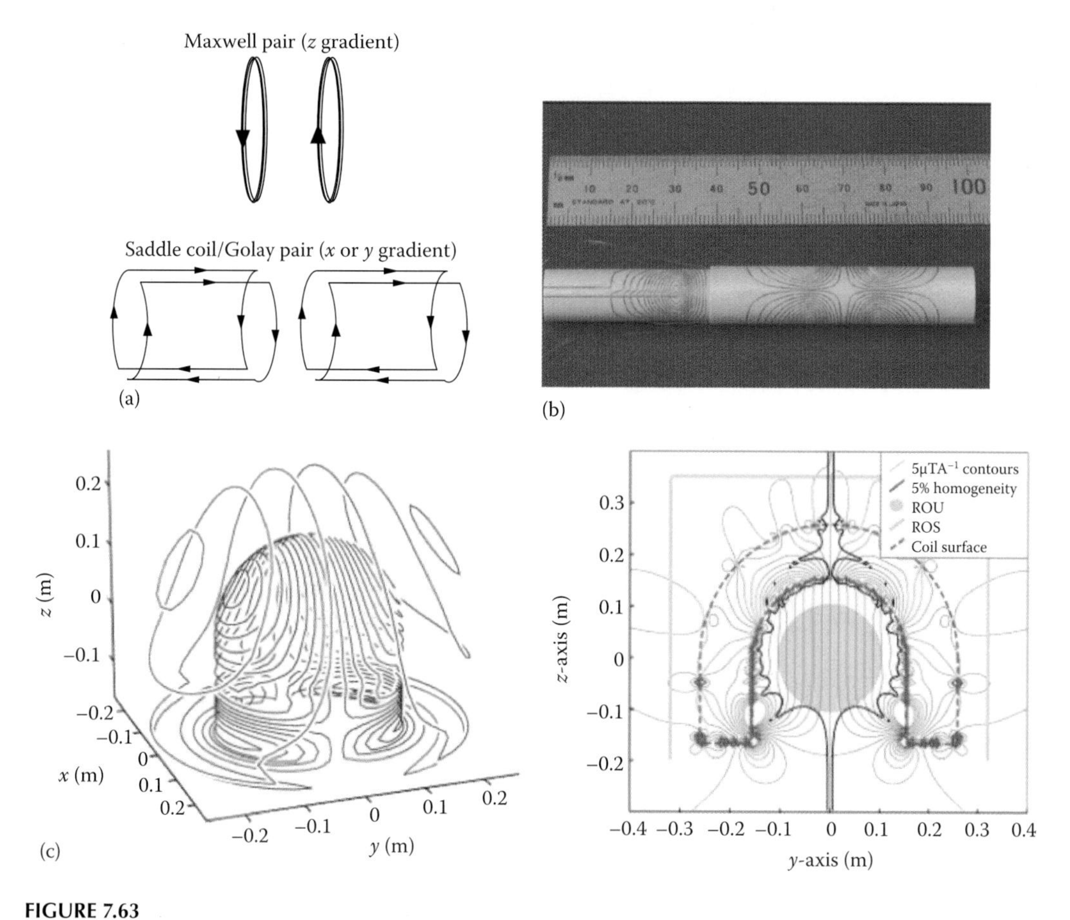

FIGURE 7.63
(a) Schematic of wire paths in an elementary Maxwell–Golay gradient set. (b) More complex wire paths in a dual-layer gradient set. (Data for (b) from Leggett et al., 2003.) (c) Design for *y*-gradient of ultra-efficient, highly asymmetrical head gradient coil. (Data for (c) from Poole and Bowtell, 2007.)

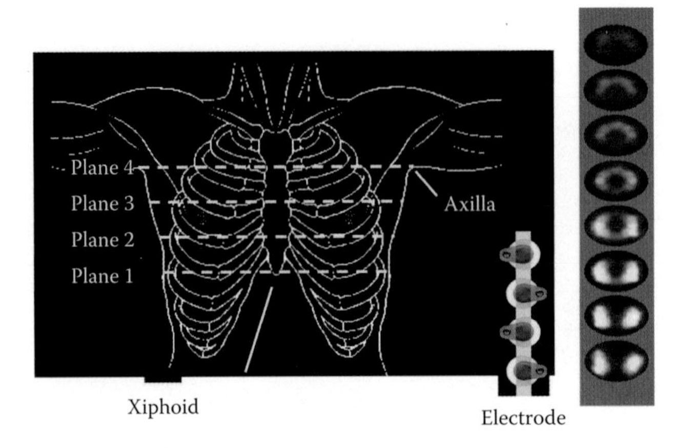

FIGURE 9.5
Four planes of 16 electrodes were used to reconstruct dynamic images at eight levels showing lung resistivity at inspiration with respect to expiration. (Courtesy of P. Metherall.)

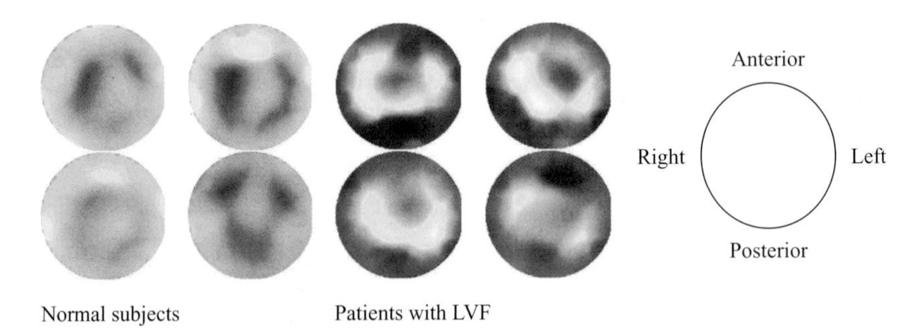

FIGURE 9.6
EIT images of the thorax for four normal subjects and four patients admitted to coronary care with lung water associated with left ventricular failure (LVF). The images show conductivity with respect to a mean normal thorax. The increased conductivity in the lungs of the patients with pulmonary oedema is clear.

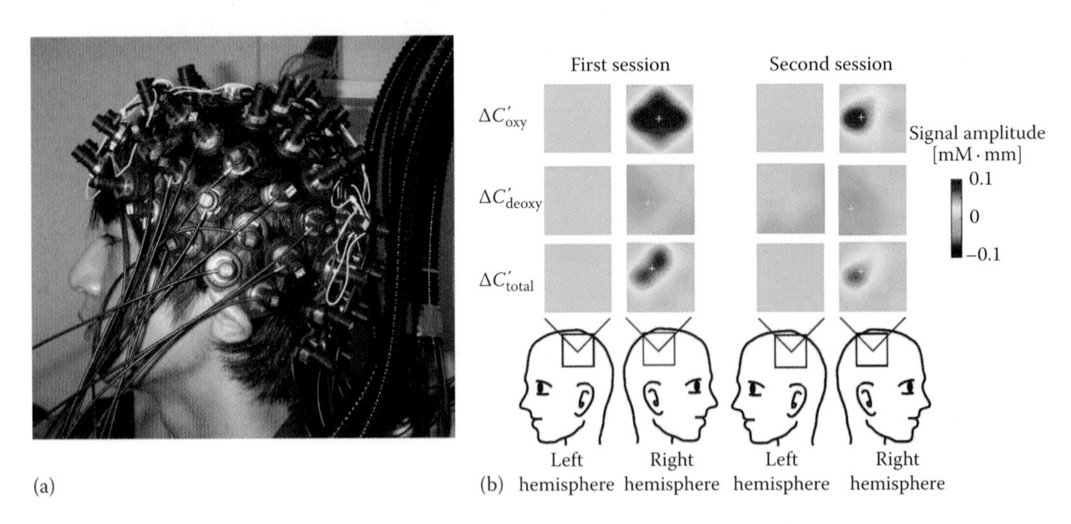

FIGURE 10.5
(a) The Hitachi ETG-7000 whole-head optical topography system. (Reproduced with permission from Koizumi et al., 2003.) (b) Topographic maps of activation on the left and right sides of the adult brain during a left-hand finger-tapping exercise, obtained using the Hitachi ETG-100 system. The maps show changes in concentrations of oxy-($\Delta C'_{oxy}$), deoxy-($\Delta C'_{deoxy}$), and total haemoglobin ($\Delta C'_{total}$). All changes occur on the contralateral side of the brain. (Reproduced with permission from Sato et al., 2006.)

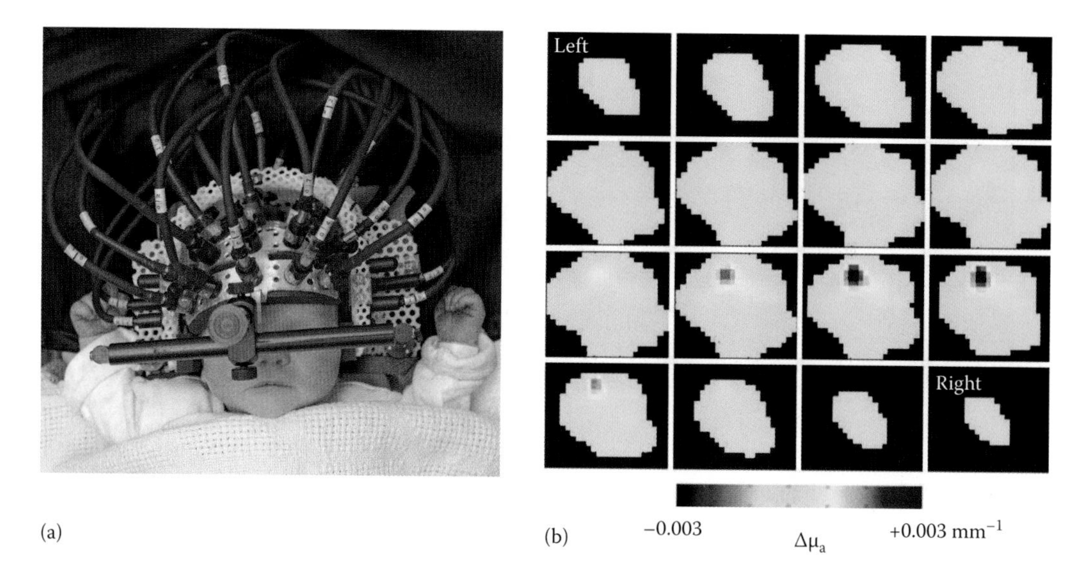

FIGURE 10.6
3D optical imaging of evoked response in the newborn infant brain: (a) a helmet attaching an array of optical fibres to the infant head; (b) sagittal slices across a 3D image of the absorption change occurring on the right side of the brain produced by passive movement of the left arm.

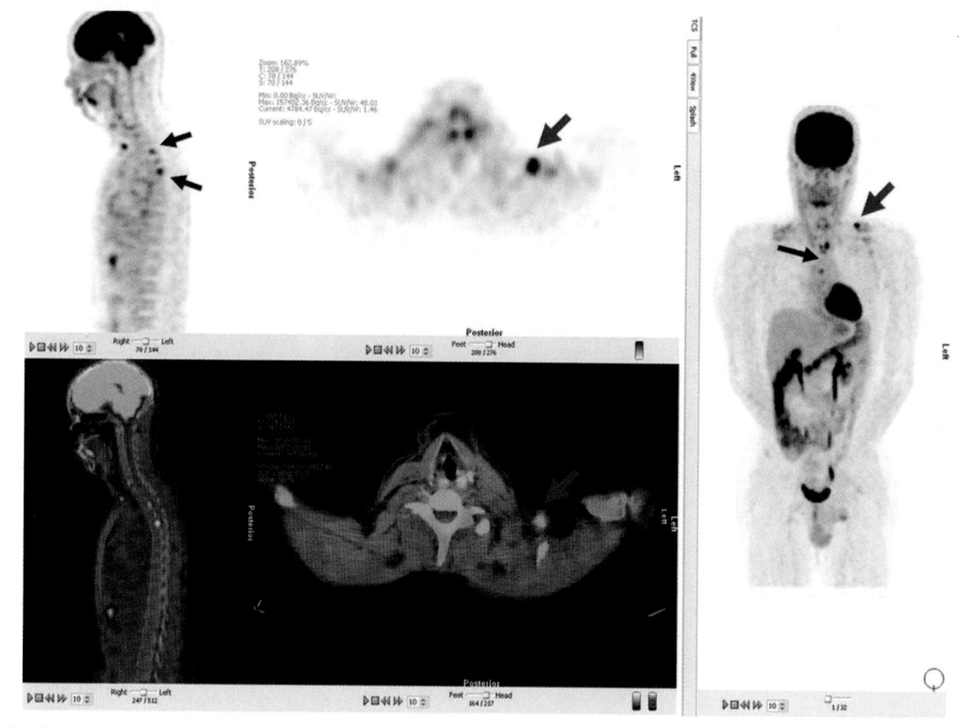

FIGURE 15.2
A 27 year old male had a total thyroidectomy and left-sided neck dissection for a papillary thyroid carcinoma. A few years later he presented with abnormally raised thyroglobulin level, but negative whole-body planar imaging I-131 (not shown). PET/CT showed evidence of thoracic bony metastases (black arrows) and a focal area of intense FDG uptake within the left supraclavicular fossa (red arrows), corresponding to a nodal lesion on CT, consistent with a nodal metastasis instead of a left clavicular bony deposit. (Courtesy of S. Chua.)

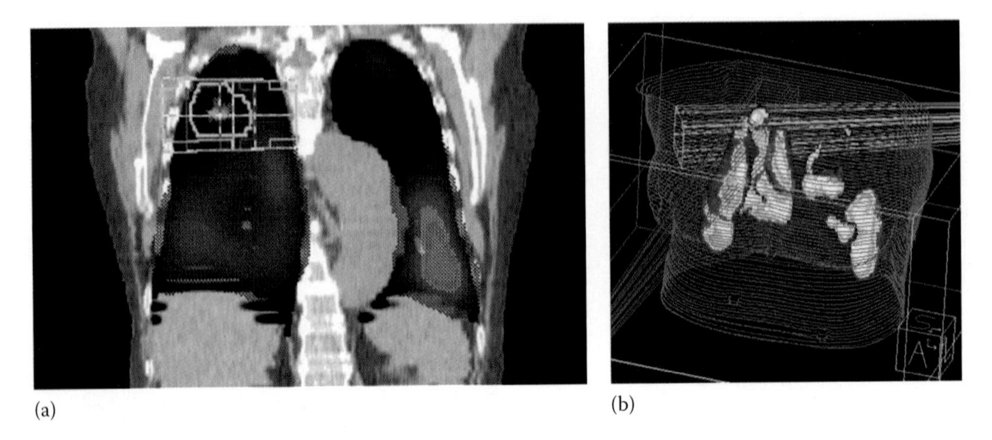

FIGURE 15.3
(a) A coronal CT slice with a planning target volume outlined in blue. (b) The registered and overlaid lung SPECT image, which indicates that the top of the contralateral lung is non-functioning and therefore can be irradiated if necessary. (Images courtesy of M. Partridge.)

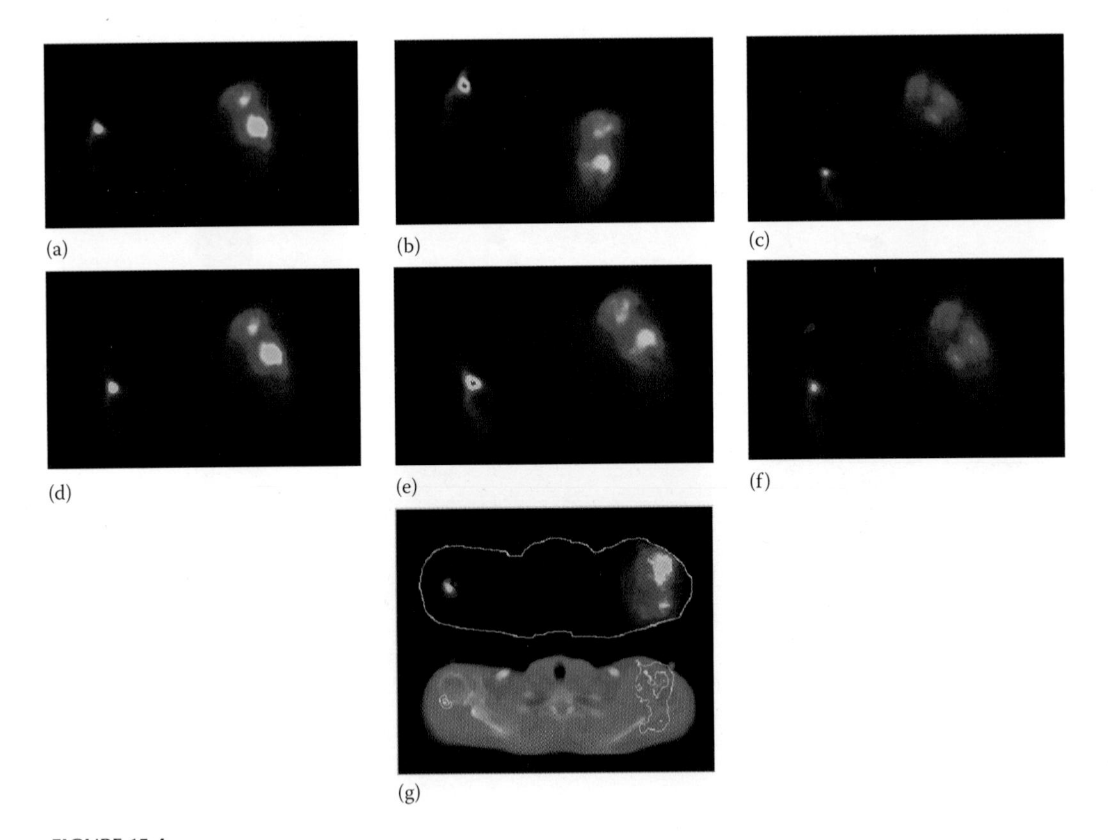

FIGURE 15.4
The application of image registration to dosimetry for targeted radionuclide therapy: (a)–(c) show successive SPECT scans taken on days 2, 3 and 6 following administration of 3000 MBq NaI for thyroid metastases. (d)–(f) show these same scans following registration using the mutual information algorithm; and (g) shows (at the top) the resultant absorbed dose distribution and (at the bottom) the isodose contours (ranging from 33 to 70 Gy) overlaid onto the CT scan that has been registered using external markers.

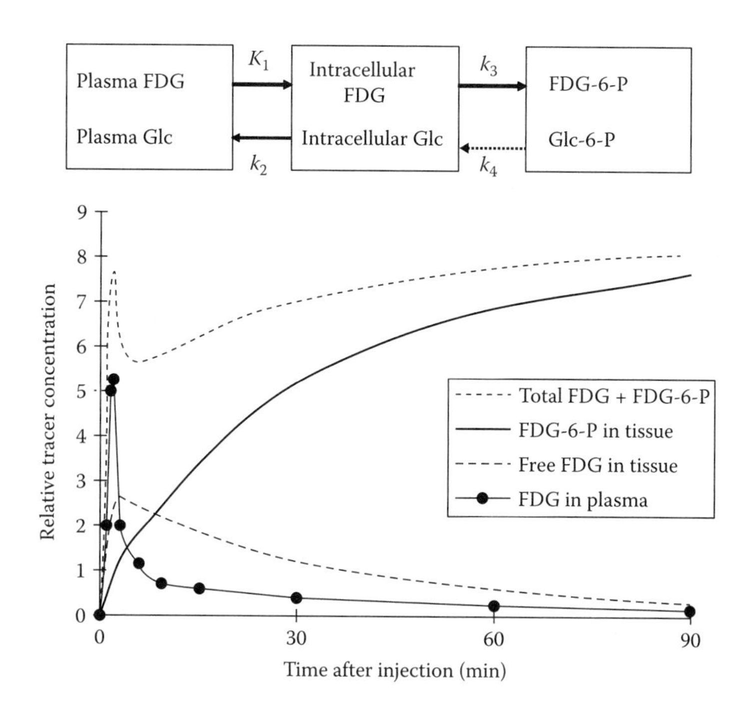

FIGURE 5.141
Three-compartmental model shown is used to describe the delivery and metabolism of FDG (and glucose (Glc)) in tissues. The flow rate constants K_1 k_2 and k_3 are large, whereas k_4 is small. This means that most of the tracer (FDG) is trapped in the tissue compartment as FDG-6-P. The general equations defining the rate of change of concentration in the tissue compartment are

$$dC_e(t) = K_1{}^*C_p(t) - k_2{}^*C_e(t) - k_3{}^*C_e(t) + k_4{}^*C_m(t) \quad \text{and} \quad dC_m(t) = k_3{}^*C_e(t) - k_4{}^*C_m(t)$$

where $C_e(t)$ and $C_m(t)$ are the concentrations of non-metabolised and metabolised tracer in the tissue and $C_p(t)$ is the concentration in the plasma. The activity–time data are used to calculate the flow constants using these equations.

can be extracted from the dynamic datasets by analysing the activity–time curves using compartmental models either for predefined ROIs or on a pixel-by-pixel basis. By using a model (Watabe et al. 2006) of the utilisation of the tracer and its metabolites (Figure 5.141), values of various parameters can be estimated and used to evaluate quantitatively the function of the tissues under study.

5.9.8.3 Other Analysis Techniques

A range of statistical and other data-led techniques have been used for the analysis of dynamic PET data. A review of the compartmental model approach to the analysis of PET data can be found in Koeppe (1996), and descriptions of data-led approaches such as Statistical Parametric Mapping and Spectral Analysis can be found in Friston et al. (1991) and Meikle et al. (1998).

5.9.9 PET Radiopharmaceuticals

5.9.9.1 General Properties

The most significant characteristic of PET in relation to the study of physiology, biochemistry, pathology and drug development is the availability of positron-emitting radioisotopes of ^{11}C, ^{13}N and ^{15}O. Stable forms of these elements are found in almost all biomolecules and drugs and, by using these radiotracers in minute quantities, information about organ function, disease processes and therapeutic intervention in living humans can be obtained without affecting the physiological processes involved. In addition, fluorine can act as a bio-isostere of hydrogen or the hydroxyl group which are commonly encountered in a variety of drugs. ^{18}F has, therefore, also found widespread application in PET chemistry where it has been used to label biological probes and drugs. PET radiotracers can be used to study the function in tissues of enzymes, receptors and other metabolically important compounds and their associated reactions. Furthermore, labelling a drug molecule by substituting a stable isotope with its positron-emitting isotope yields a tracer that is chemically and biologically indistinguishable from the stable or non-radioactive drug. Pharmacokinetic studies with PET tracers can in general be performed over a time frame of approximately four times the physical half-life of the tracer.

5.9.9.2 Production of PET Radioisotopes and Radiopharmaceuticals

The half-lives and means of production for the positron-emitting radionuclides in common use are given in Table 5.17. By far, the most commonly used are the low-Z nuclides ^{11}C, ^{13}N, ^{15}O, and ^{18}F. These short-half-life proton-rich nuclei are produced artificially by bombarding stable nuclei with protons [^{1}H] or deuterons [^{2}H] in a cyclotron (Figure 5.42a). Usually negative-ion cyclotrons are used as they can produce several different radionuclides simultaneously. In addition, as there is no long-lived radioactive activation of the cyclotron, maintenance can be performed within several hours of bombardment. 11 MeV protons can produce adequate yields of these four radionuclides – 18 MeV protons give better yields for some radionuclides.

The direct products of most of the nuclear reactions currently used to prepare positron-emitting radionuclides are very simple molecules such as $^{13}NH_3$, $^{11}CO_2$ or $H^{18}F$. These radiochemicals allow the synthesis of an array of physiologically active compounds as well as most drugs. Although chemical incorporation of these building blocks into drugs may appear straightforward, the constraints imposed by the short, radioactive half-lives often make radiotracer synthesis a formidable problem. In general, radiopharmaceutical preparation must be rapid enough to allow the target drug to be synthesised, isolated, purified and formulated as a sterile, pyrogen-free, isotonic solution within two to three half-lives. Furthermore, due to the short physical half-lives, large quantities of radioactive material (usually gigabecquerels) have to be handled and radiation exposure to personnel is an extremely important consideration. These issues have led to the design and synthesis of drug precursors that can be radiolabelled in a single step. This is often done using rapid, remote-controlled and robotics-based chemistry systems and commercial synthesisers (Phelps 2004).

PET radiopharmaceuticals are usually prepared using an 'on-site' cyclotron and radiochemistry facility. This limits the general availability of many of these compounds to specialised facilities at large hospitals and universities. However, the most widely used clinical PET tracer, ^{18}F-FDG ($T_{1/2}$ = 109.8 min), can be transported up to 2 h travelling time away from a regional radiotracer distribution facility on a daily basis – in principle, this would also apply to other ^{18}F-labelled tracers.

5.9.9.3 ^{18}F-Labelled Fluorodeoxyglucose

Despite the large number of PET radiopharmaceuticals that have been investigated, the tracer most commonly used clinically is ^{18}F-labelled fluorodeoxyglucose (^{18}FDG). The use of ^{18}FDG-PET stems from work done with ^{14}CDG autoradiography (Sokoloff et al. 1977), which was subsequently adapted for human use (Phelps et al. 1979, Reivich et al. 1985). FDG is an analogue of glucose that is transported into the cell and phosphorylated at the same rate as glucose. Unlike real glucose, the metabolite remains trapped in the cell and an image of the distribution of trapped FDG is, thus, related to the regional rate of glucose utilisation, often referred to as 'metabolic rate'. As the rate of glucose utilisation is a basic metabolic parameter in tissue it can often be used as a diagnostic tool. In many cases the rate of glucose utilisation in diseased tissue is different to that of normal tissue – for example, highly malignant tumours exhibit very high ^{18}FDG uptake rates. In principle, the absolute rate of glucose utilisation can be measured using ^{18}FDG, but in practice there are complications. The so-called lumped constant (LC) accounts for small differences in the affinity of glucose transporters and the hexokinase pathway between glucose and FDG. Although the LC has a well-known value in normal brain (~0.52 for grey matter), its value can vary when the balance between glucose transport and phosphorylation is disturbed by pathology. For example, the value of the LC in tumours is not known and is likely to be very variable.

Extraction of FDG into tissue is rapid over the first 10 min, and then continues less rapidly over the next hour or so (Figure 5.141). Free FDG in tissue peaks a few minutes after injection when it reaches equilibrium with the plasma concentration and then falls; 30 min after injection, the ratio between bound and free FDG is ~4:1. A PET image registers total labelled FDG, including contributions from FDG and FDG-6-P in tissue and FDG in plasma (a few per cent of the total activity after 30 min). Clinical oncology protocols usually consist of a 60 min uptake period after injection, followed by a whole-body scan over a period of between 10 and 20 min. This results in an image where the grey-scale level is roughly proportional to the rate of glucose utilisation.

5.9.9.4 Other PET Radiopharmaceuticals

An important limitation of ^{18}FDG imaging is its nonspecificity, since the tracer can be taken up by inflammatory responses or infection, and enhanced tumour uptake can often be seen as an early response to therapy. Several other imaging agents have already found their way into clinical use. Table 5.17 includes some of the radiopharmaceuticals and their usage for clinical applications. Table 5.18 gives examples of radioligands that have been investigated for PET brain and heart imaging, whereas Table 5.19 gives examples of some PET tracers that are used for cancer management, for example, FLT may be useful in cases where the use of FDG produces equivocal diagnostic information. Many tracers are still under investigation to determine if they have useful clinical applications. An extended review of many of these agents is described in Mercer (2007).

5.9.10 Diagnostic and Research Applications of PET

The most common acquisition protocol in oncology is a whole-body ^{18}FDG PET-CT scan, whereby much of the body is scanned by moving the patient through the camera in a series of discrete steps, the total scan time being around 20 min. Most of the current clinical applications of PET fall into the areas of oncology, cardiology, neurology and psychiatry (Maisey et al. 1999). The majority of clinical oncology scans are performed in order to determine the spread of the disease prior to therapy and to monitor response to therapy. PET has been shown to be cost effective in the management of patients with lung cancer and is also used for lymphoma, germ cell, gut and head-and-neck tumours. There are also clinical PET applications in cardiology and to a lesser extent in neurology and psychiatry. PET is widely used as a research tool in these same areas, particularly oncology, neurology and psychiatry.

TABLE 5.18

Examples of radioligands that have been used for PET imaging.

Receptor	Ligand	Specificity
Dopamine	^{11}C – SCH-23390	D_1
	^{11}C – SCH-39166	D_1
	^{11}C – Methylspiperone	D_2, ($5HT_2$)
	^{76}Br – Bromospiperone	D_2, ($5HT_2$)
	^{11}C – Raclopride	D_2
	^{76}Br – Bromolisuride	D_2
Serotonin	^{18}F – Setoperone	$5HT_2$
	^{11}C – Methylbromo LSD	$5HT_2$
Benzodiazepine	^{11}C – Flumazenil	BZ_1, BZ_2
	^{11}C – PK11195	Peripheral
Opioid	^{11}C – Carfentanil	μ
	^{11}C – Diprenorphine	μ, k
	^{18}F – Acetylcyclofoxy	μ
Acetylcholine (muscarinic)	^{11}C – MQNB	M_1, M_2
	^{11}C – Levetimide	M_1, M_2
	^{11}C – Scopolamine	M_1, M_2
Adrenergic	^{11}C – CGP 12177	β_1, β_2

TABLE 5.19

Examples of PET tracers and their applications in oncology.

Radiotracer	Target	Use in Cancer Management
^{18}FDG	Hexokinase activity, glucose metabolism	Tumour detection, grading, response to treatment
^{11}C-methionine ^{18}F-fluoroethyltyrosene	Amino acid transport and protein synthesis	Tumour detection, grading, therapy response
^{18}F-FLT ^{11}C-thymidine ^{124}I-UdR	Nucleoside transport and phosphorylation, DNA synthesis, cellular proliferation	Measurement of tumour growth rate and changes in cellular proliferation due to therapy
^{11}C-choline ^{18}F-choline ^{11}C-acetate	Choline transport and kinase activity	Phospholipid synthesis related to cell growth/division, prostate cancer therapy response
^{64}Cu-ATSM ^{18}F-FMISO	Tissue hypoxia	Resistance to chemotherapy and radiotherapy due to poor tissue oxygenation
^{18}F-annexin V	Cellular apoptosis	Detection of early cell death during treatment
^{18}F-gancyclovir/HSV-1tk reporter gene system	Detection of products associated with gene expression	Gene expression marker and gene therapy monitoring
Integrin VEGF/tk	Detection of angiogenesis in tissues	Angiogenesis effects of drugs to reduce blood flow to tissues
Antigens/antibodies	Specific localisation of tumours	Detection and treatment of tumours
Hormone receptor ligands	Detection of hormone expression levels	Localisation and treatment of hormone expressing tumours in the breast and prostate, for example

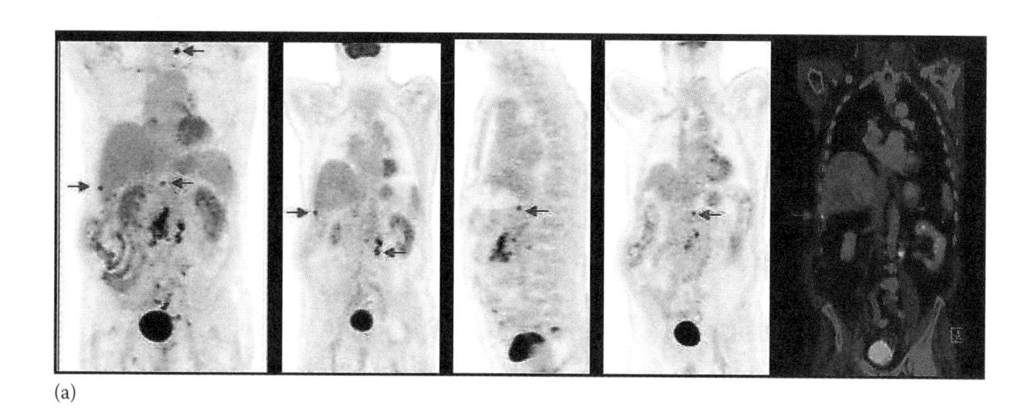

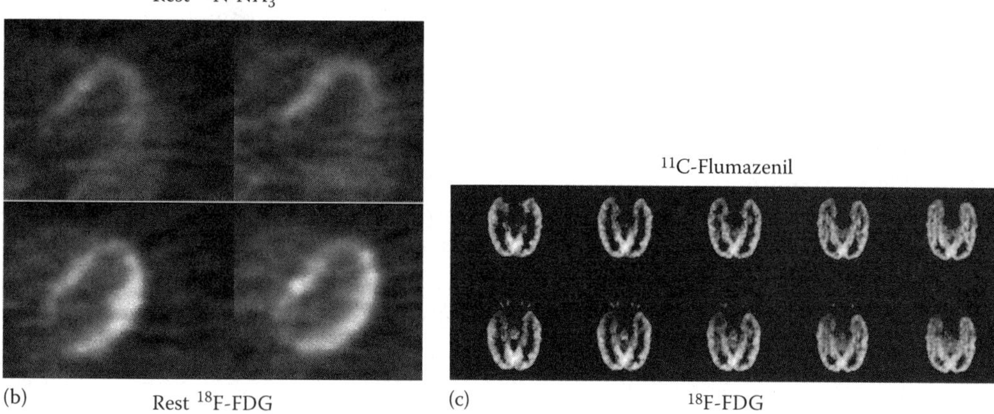

FIGURE 5.142
(See colour insert.) Examples of clinical PET images. (a) Clinical FDG-PET/CT images for primary staging of a patient with non-Hodgkin's lymphoma. A mesenteric nodal mass with increased ^{18}F-FDG uptake along with hypermetabolic para-aortic and celiac lymph nodes are visualised. A solitary ^{18}F-FDG avid left supraclavicular nodal lesion and a small lesion in the margin of the lower part of right lobe of liver suggest advanced disease. (Courtesy of University of Tennessee, Knoxville, TN.) (b) $^{13}N\text{-}NH_3$ and ^{18}F-FDG images of a hibernating myocardium. (c) ^{11}C-Flumazenil and ^{18}F-FDG images of a patient with epilepsy.

5.9.10.1 Staging of Cancer

PET cancer imaging has had a significant impact on staging accuracy which has led to changes in the treatment of patients. For example, in lung cancer, PET imaging has changed the staging information significantly, in most cases by the detection of extra disease sites not seen by other imaging modalities or, in some cases, confirming that the disease is localised solely in a single site (Farrell et al. 2000, Devaraj et al. 2007). This important contribution is based on the ability of PET to discriminate between malignant and benign disease more accurately than other modalities. An example of oncology staging scans is shown in Figure 5.142a.

5.9.10.2 Assessing Response to Therapy

Another important area that PET has contributed to is the investigation of the response of cancers to treatment. In many cases tumours respond at different rates in the same patient and the malignancy status of 'residual masses' seen by other modalities can be more accurately determined using PET – this makes a significant contribution to the management of individual patients by determining whether more or a different treatment is required. This is very important as radiotherapy can cause damage to normal tissues and an evaluation of the effect on tumour growth/regression can determine if further doses are likely to be useful (Shankar et al. 2006).

5.9.10.3 PET in Radiotherapy Treatment Planning

At present, most radiotherapy treatment plans (RTP) are based on X-ray CT images. However, it is clear that these images do not always accurately define the active tumour area. PET scans taken of patients undergoing radiotherapy have shown that the malignant part of tumours is often different to the anatomical mass seen by X-rays – in some extreme cases it appears that active tumour can be missed by the radiotherapy treatment. Many studies have been undertaken to evaluate the role of PET imaging in RTP (De Ruysscher and Kirsch 2010, Grosu and Weber 2010, Haie-Meder et al. 2010, Lambrecht and Haustermans 2010, Picchio et al. 2010, Troost et al. 2010).

5.9.10.4 Applications in Cardiology

For cardiac PET imaging, ^{13}N ammonia, ^{15}O-labelled water or ^{82}Rb and ^{18}FDG are used, the first three to measure myocardial perfusion and the fourth to measure metabolism. A region of low perfusion that nevertheless shows metabolism of glucose indicates viable or 'hibernating' myocardium that is likely to benefit from a revascularisation procedure (see Figure 5.142b). Regions of low perfusion and poor metabolism are indicative of infarcted tissue for which revascularisation will not work (Grandin et al. 1995). Gated ^{18}FDG scans can also be used to measure wall motion and wall thickening to assess the presence of viability in areas of hypo/dyskinesis.

5.9.10.5 Applications in Neurology

^{18}FDG and ^{11}C-flumazenil are used clinically to locate epileptic foci (see Figure 5.142c) (Van Paesschen et al. 2007). Although there are several ^{11}C-labelled tracers that can be used to investigate neurotransmitter systems, to date these have not yet found a significant role in clinical practice. ^{18}FDG and/or ^{11}C-methionine are also used to assess the grade and extent of tumours in the brain, and recurrence following therapy (Chen 2007).

5.9.10.6 Applications in Clinical Research

Research applications often involve acquiring a dynamic sequence of images (<5 s per image for some applications) and may last from several minutes to several hours. Very high count rates may be encountered when using short-half-life tracers, and the emphasis is on obtaining the maximum image quality by maximising the resolution, sensitivity and SNR of the system. In research, the overall cost is often less of a constraint. Table 5.20 shows some examples of the main research areas in oncology, cardiology and neurology/psychiatry. Much of this work is aimed at measuring *in vivo* the pharmacokinetics and pharmacodynamics of drugs using both radiolabelled drugs and markers of response. Whilst producing and using radiolabelled drugs is very complex, requiring sophisticated radiochemistry and modelling of tracer metabolism, the development and evaluation of generic radiolabelled therapy response markers has become a very active area.

TABLE 5.20
PET research application areas.

Oncology	• Monitoring of response to treatment – glucose metabolism, protein synthesis, DNA synthesis, perfusion, hypoxia • As an adjunct to CT in radiotherapy planning • Labelled therapy agents – dose ranging and pharmacokinetics • Receptor markers – new blood vessels and metastasising tumours
Cardiology	• Hibernation and blood flow/perfusion • Fatty acid metabolism • Atheroma formation • Cardiac beta receptors • Assessment of new local therapies to plaques
Neurology and psychiatry	• Monitoring of therapy for dementias using glucose metabolism, and specific receptor ligands • Movement disorders with dopamine and other neurotransmitter systems, monitoring the effects of treatment, assessment of new therapies • Stroke/vascular – assessment of new therapies, monitoring of interventions, glucose and oxygen consumption, neurotransmitter systems • Huntington's chorea and other inherited diseases • Psycho-pharmacology – labelled drugs, pharmacokinetics, dose ranging, monitoring of effects, mechanisms of action • Drug misuse – labelled drugs and metabolites, mechanisms of action, location of abnormalities, monitoring of treatment • Functional brain mapping – perfusion, metabolism and neurotransmitter systems

Here, the idea is to use PET to measure endpoints of a therapy which might be aimed at reduced cellular proliferation, reduced metabolism, overcoming drug resistance or tissue hypoxia, with the overall aim of improving therapy and bringing new drugs into the clinic more rapidly. This subject has a profusion of publications which can be accessed by review articles such as Rohren et al. (2004).

5.10 Performance Assessment and Quality Control of Radioisotope Imaging Equipment

The basic requirements of a QC procedure are that the tests must be easy to perform, the measurements should be sensitive to important changes and the results easily recorded for reference. Various recommendations for performance-assessment (PA) and QC procedures, particularly for gamma cameras, have come from a range of professional bodies (Table 5.21). Some of these recommendations also cover electrical and mechanical safety requirements. The issue of Nuclear Medicine software QC has also been addressed (Cosgriff and Sharp 1989, IPSM 1992a, 1992b, IPEM 2003). The aim of this section is to give a brief overview of PA and QC procedures for radioisotope imaging equipment.

There are two major classes of PA/QC tests performed. The first are those carried out immediately after installation of equipment, which provide both acceptance testing and baseline measurements for reference. The major questions here relate to

TABLE 5.21

Recommendations for performance assessment and QC of gamma cameras for planar imaging.

For the manufacturer	
British Standards	BSI (1998)
International Electrotechnical Commission	IEC (1998)
National Electrical Manufacturers' Association	NEMA (1986, 1994, 2001a)
For routine hospital use	
Hospital Physicists' Association	HPA (1978, 1983)
Department of Health and Social Security	DHSS (1980, 1982)
Institute of Physical Sciences in Medicine	IPSM (1992a, 1992b)
Institute of Physics and Engineering in Medicine	IPEM (2003)

Note: Some of the more recent publications have replaced the earlier ones.

whether the manufacturer's specifications have been met. Once the baseline measurements have proved to be satisfactory, the second category of tests is required. These include both routine daily and long-term tests to determine the overall performance and stability of the system for comparison with the baseline measurements. A series of routine checks at regular intervals have been suggested by NEMA (Hines et al. 1999) and IPEM (2003). The range of normal variations in the values of the performance parameters should be determined and appropriate action thresholds should be set. Action may be required if there is a sudden or a slow systematic change, or if there is an increase in the size of random variations. The importance of doing daily QC is illustrated in Figure 5.143.

The main PA/QC measurements for gamma-camera systems operating in planar imaging mode are:

- flood-field uniformity;
- spatial resolution;
- energy resolution;
- spatial distortion or linearity;

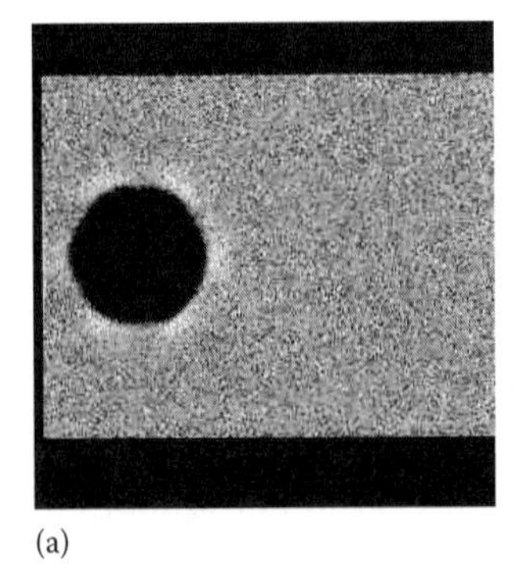

(a)

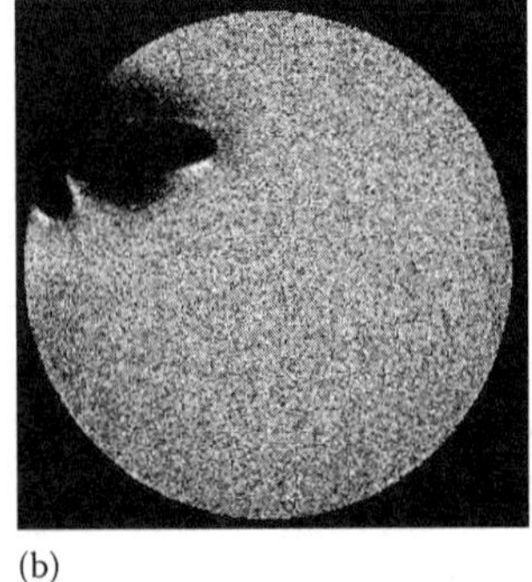

(b)

FIGURE 5.143
Flood-field uniformity images showing the effects of (a) PMT failure and (b) a cracked crystal. (Courtesy of K. Pomeroy.)

- plane sensitivity;
- count-rate performance;
- multiple-energy-window spatial registration; and
- shield leakage.

Some of the tests can be made with the collimator removed and a lead mask positioned on the crystal surface to limit the area of the detector tested to that normally defined by the collimator – these are known as *intrinsic measurements*. Alternatively, if the collimator is in place when the tests are made, then they are referred to as *system* or *extrinsic measurements*. In most cases, an online computer system is required for quantitative measurements. Each of the measurements listed earlier will now be considered in turn. The methods described for quantitative measurements are mainly those recommended by NEMA (1986, 1994, 2001a).

5.10.1 Flood-Field Uniformity

The uniformity of response of the gamma camera for a uniform flux of radiation can be determined by both global (integral) and local (differential) uniformity measurements.

Integral uniformity is an indication of the extreme range of variation in regional sensitivity over a defined FOV and is given by

$$U_{\mathrm{I}}(\%) = 100 \times \frac{(\max - \min)}{(\max + \min)} \tag{5.74}$$

where
'max' is the maximum pixel count, and
'min' is the minimum pixel count.

Differential uniformity indicates a maximum rate of change of sensitivity over a predetermined short distance within the FOV and is given by

$$U_{\mathrm{D}}(\%) = 100 \times \frac{(\mathrm{hi} - \mathrm{low})}{(\mathrm{hi} + \mathrm{low})} \tag{5.75}$$

where 'hi' and 'low' are the maximum and minimum pixel counts for all rows and columns within a localised line of five contiguous pixels for which hi – low is the largest deviation.

The *coefficient of variation* gives an alternative measure of the overall variation in image uniformity and is given by

$$\mathrm{COV}\ (\%) = 100 \times \text{standard deviation} \div \text{mean} \tag{5.76}$$

where the standard deviation and the mean are both expressed in counts pixel^{-1}.

Uniformity measurements are usually obtained for two FOVs of the camera. The useful field of view (UFOV) is the area used for imaging as specified by the manufacturer, and the central field of view (CFOV) is that defined by scaling all linear dimensions of the UFOV by 75%. Both intrinsic and system nonuniformity measurements can be made. Intrinsic nonuniformity is measured by exposing the camera (with the collimator removed) to a $^{99}\mathrm{Tc}^{\mathrm{m}}$ 'point' source. The minimum source-detector distance should be five times the maximum dimension of the UFOV and the count rate should be below 30 kcps with a

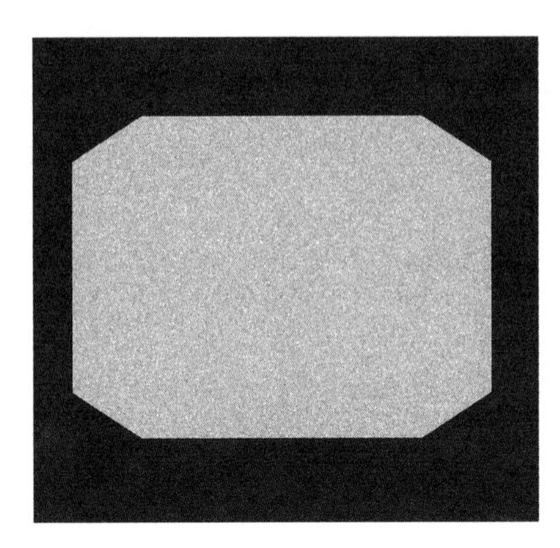

FIGURE 5.144
Typical intrinsic uniformity image.

photopeak window of 20%. System uniformity can be measured with a flood source containing ~80 MBq of $^{99}Tc^m$. The source diameter should be greater than the geometric FOV. High-count images (containing typically 30 million counts) are recommended. Figure 5.144 shows an example of an intrinsic uniformity image obtained from a modern digital gamma-camera system. Typical values of integral and differential uniformity are 3.0% and 2.2%, respectively.

5.10.2 Spatial Resolution

Spatial resolution characterises the ability of the camera to accurately resolve spatially separated radioactive sources.

Qualitative assessments of spatial resolution can be made using either transmission or emission phantoms. The former consist of lead foils in Perspex sheets and are uniformly irradiated by an appropriate source of γ-rays to provide images suitable for resolution measurements. Transmission phantoms can be used for qualitative assessment of either intrinsic or system spatial resolution. The different geometries used for the intrinsic and system measurements are shown in Figure 5.145. Images obtained with some transmission phantoms are illustrated in Figure 5.146. Emission phantoms are made of Perspex containers filled with a radioactive liquid. They are designed to have 'hot' and 'cold' regions of varying size in a uniform background. Figure 5.147 illustrates some images obtained with a selection of these phantoms. As no collimation is provided within emission phantoms, they can only be used for assessment of system spatial resolution.

Quantitative measurements of spatial resolution are made by determining the response of the detector to a radioactive line source. In real space this is referred to as the line-spread function (LSF), and in frequency space this is called the modulation transfer function (MTF) (see Chapters 2 and 11). In attempting to quote a single figure for comparative measurements of spatial resolution, the FWHM of the LSF or the 50% level of the MTF are often used. However, there are limitations in trying to represent the full LSF or MTF curves by a single parameter, and often the FWTM is quoted in addition to the FWHM as a compromise (see Figure 5.148).

Quantitative measurements of *intrinsic* spatial resolution are made with a collimated radioactive line source placed typically 50 mm from the crystal surface. The source,

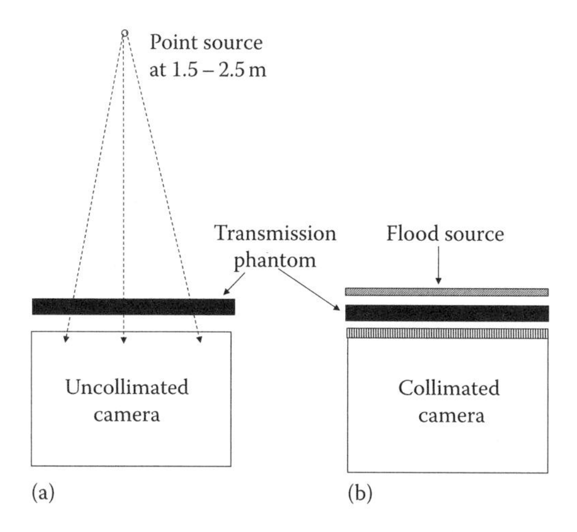

FIGURE 5.145
Schematic diagram showing geometry used for qualitative assessment of spatial resolution: (a) intrinsic measurement with a point source at a distance from the transmission phantom, which has been placed carefully on the uncollimated camera head; (b) system measurement with a flood source immediately above the transmission phantom, which has been placed on the collimated camera head.

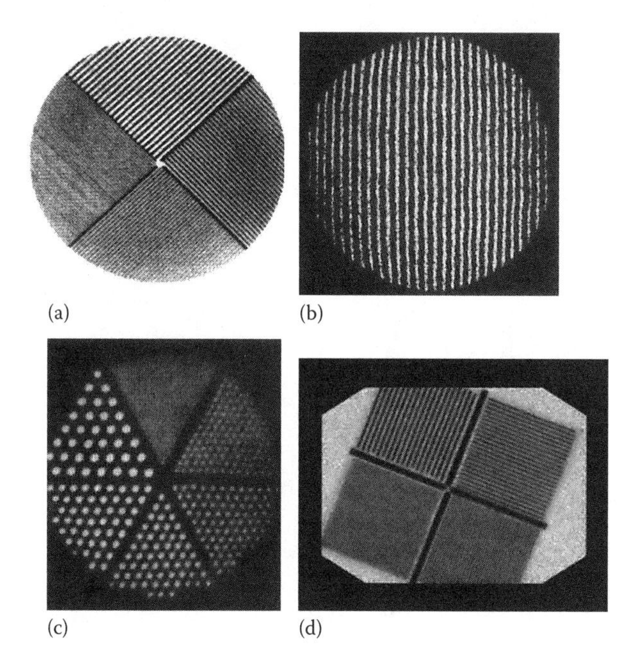

FIGURE 5.146
Spatial resolution images using transmission phantoms: (a) a quadrant bar phantom on a circular camera head; (b) parallel-line equal-spacing bar phantom; (c) an Anger pie phantom; and (d) a quadrant bar phantom on a rectangular camera head.

containing ~40 MBq of $^{99}Tc^{m}$, is aligned with the x or y axis and the rest of the detector shielded with lead sheets. Peak counts of approximately 10^4 are obtained, and LSF measurements repeated for different positions over the camera surface. The spatial resolution is given by the FWHM of the LSF. Special computing facilities are needed for this measurement since it is recommended that the pixel size must be small enough so that the FWHM

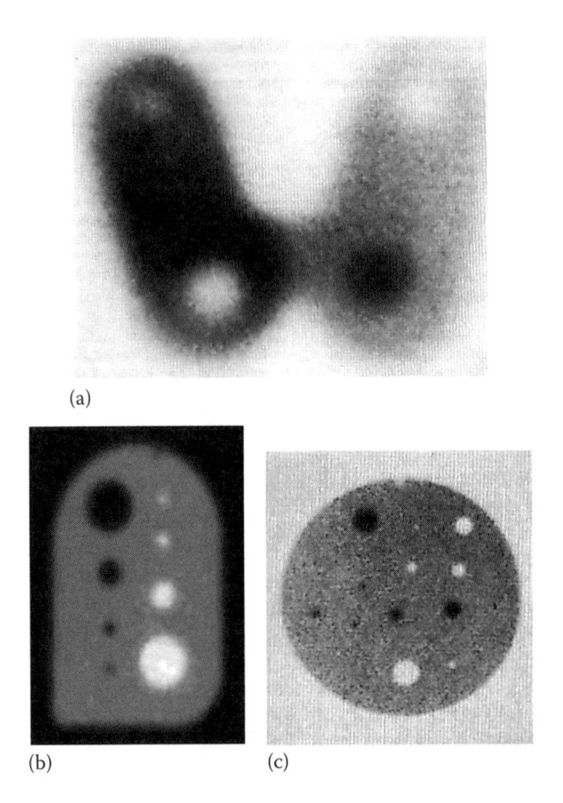

FIGURE 5.147
System spatial resolution images using emission phantoms: (a) Picker thyroid phantom; (b) Williams' liver-slice phantom; and (c) Shell phantom.

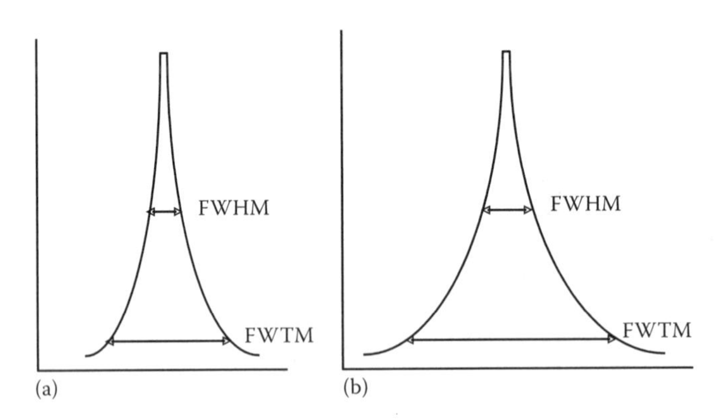

FIGURE 5.148
Schematics of line-spread functions for (a) $^{99}Tc^{m}$ and (b) ^{131}I showing greater differences between the FWTM than the FWHM.

contains at least 10 pixels. Typical values of intrinsic spatial resolution are 3–4 mm FWHM and 6–7 mm FWTM.

Quantitative measurements of *system* spatial resolution are usually made with an uncollimated radioactive line source. The source, ~40 mm long, is embedded in the face of a Perspex sheet ~240 × 240 mm^2 and 10 mm thick. Tissue-equivalent material can be used between the source and collimator to simulate the effects of scatter. The FWHM of the LSF is measured under a variety of conditions, namely:

- at different positions in each x, y plane;
- at different distances from the collimator;
- with different absorber thicknesses between detector and source;
- with different collimators; and
- with different radioisotopes.

5.10.3 Energy Resolution

Energy resolution is a measure of the ability to distinguish between photons of different energies. Quantitative measurements of the energy resolution of a gamma camera require the use of a multichannel analyser. The analyser is calibrated using standard γ-ray sources so that the energy spectrum can be measured in keV. Measurements are usually made of the intrinsic energy resolution using a point source. The analyser is usually set up so that there are at least 10 channels within the photopeak and 10^4 counts acquired in the peak channel. The energy resolution is given by

$$\Delta E(\%) = \left(\frac{\mathrm{FWHM}_E}{E}\right) \times 100 \tag{5.77}$$

where
FWHM_E is the width of the photopeak, and
E is the photopeak mean energy (see Figure 5.4).

5.10.4 Spatial Distortion

This term is used to describe the deviation of the image of a line source from linearity. *Qualitative assessment* can be made using a variety of linearity phantoms. For instance, a special linearity phantom, in the form of a set of line sources in an orthogonal pattern, can be used. Alternatively, transmission phantoms (Figure 5.146) can be used for simultaneous qualitative assessment of both spatial resolution and spatial distortion.

Quantitative measurements of spatial distortion require computer programs to determine the deviation of the image of a line source from a true straight line. These measurements should be carried out in both the x and y directions. Figure 5.149 shows the geometry that can be used for measuring the spatial distortion in the y direction of the camera head. A least-squares fit of a straight line to the imaged line can determine the minimum and maximum deviations $(\Delta y)_{\mathrm{min,max}}$ from the fitted line. The spatial distortion in, for example, the y direction (D_y) is expressed as

$$(D_y)_{\mathrm{min,max}}(\%) = \frac{(\Delta y)_{\mathrm{min,max}}}{L} \times 100 \tag{5.78}$$

where L is the length of the line source covering the UFOV. This measurement requires a large number of pixels to be included in the direction perpendicular to the line to give an accurate determination of the deviations.

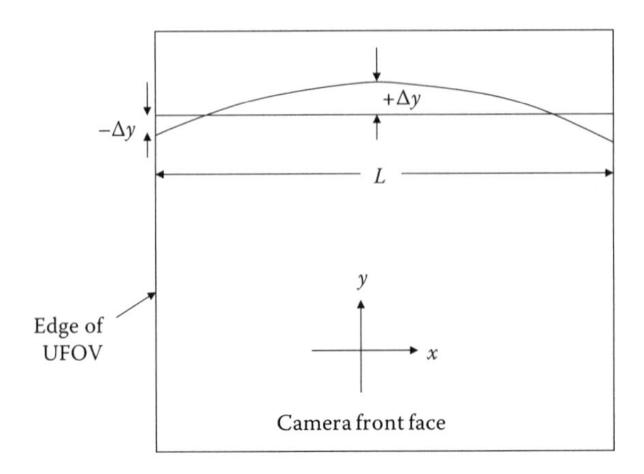

FIGURE 5.149
Schematic showing geometry and position of the line source for the measurement of spatial distortion in the y direction.

5.10.5 Plane Sensitivity

This is a measurement of the count rate obtained when imaging a given radioactive plane source and is expressed as counts per second per megabecquerel (cps MBq^{-1}) for an area source of dimensions $100 \times 100\,mm^2$. The source contains ~30 MBq of $^{99}Tc^m$ and is placed 100 mm from the collimator surface. The activity should be known to better than ±5%. Values for plane sensitivity can be determined for a range of radionuclide–collimator combinations.

5.10.6 Count-Rate Performance

Several measurements can be made to determine the maximum count-rate capacity and count-rate losses (see Section 5.3.2) of a gamma-camera system. These measurements are usually made with a standard phantom (Figure 5.150), which provides suitable scattering material.

Examples of count-rate performance parameters are:

- the observed count rate at which the value measured by the gamma camera is 90% of that expected from calculations;
- maximum observed count rate; and
- observed count rates at which losses are 10%, 20% and 30% of expected rates.

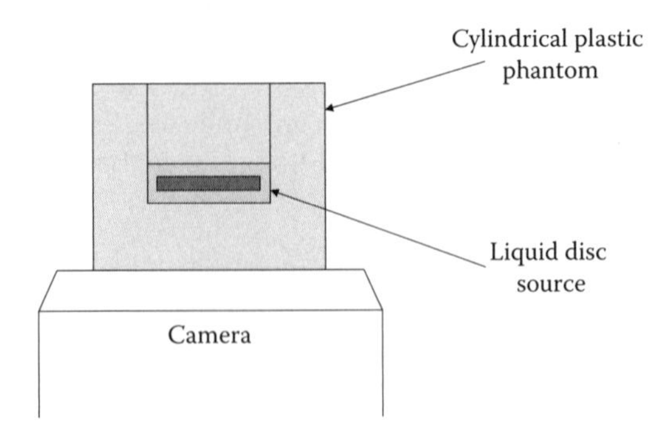

FIGURE 5.150
IEC source for measurement of count-rate capability.

5.10.7 Multiple-Energy-Window Spatial Registration

This test measures the difference in the detected positions of gamma rays of two different energies which emanate from the same point. It is important if the gamma camera is to be used for imaging radionuclides that emit photons with multiple energies and/or for dual radionuclide studies. ^{67}Ga, which emits photons with a wide range of energies (93, 184 and 300 keV), is recommended for this test. A well-collimated ^{67}Ga source should be placed at nine positions on the camera face and imaged using three windows centred on each photopeak. The centroid of the counts in the x and y directions for each image is determined from the count profiles. The maximum difference in position of the centroids for each photopeak should be reported. For any one of the nine positions, the maximum displacement is simply the largest value of D_{ij} between energy windows i and j, where

$$D_{ij} = |L_i - L_j| \tag{5.79}$$

where
$i = 1, 2$ or 3, and
$j = 1, 2$ or 3.

A typical value of the maximum displacement is 0.75 mm. Figure 5.151 shows typical images acquired for this test using the 93 and 300 keV windows.

5.10.8 Shield Leakage

The objective of this test is to determine the gamma-camera count rates when exposed to a radioactive source outside the collimator FOV. A small cubic source of approximately 10 mm sides is used to check count rates at various positions around the shield. These count rates are compared with the count rate recorded when the source is placed 100 mm from the collimator face on the axis of the collimator. Leakage count rates (L) are expressed (for each source position) as

$$L(\%) = \frac{\text{observed count rate}}{\text{reference count rate}} \times 100. \tag{5.80}$$

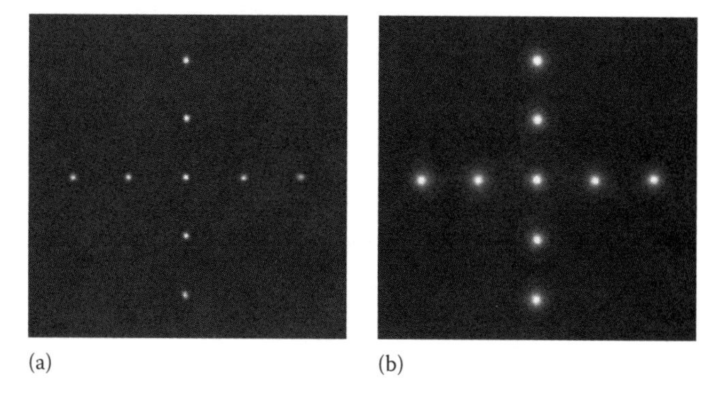

FIGURE 5.151
Images taken at (a) 93 keV and (b) 300 keV for the multiple-energy-window spatial registration test. (Courtesy of K. Pomeroy.)

Common regions containing shield weaknesses are cable entry points, bolt holes and collimator–detector junctions. Many modern gamma cameras have very limited shielding to photons above 300 keV to minimise overall camera weight for tomography. The test should, therefore, be performed with $^{99}Tc^m$ as well as the highest energy radionuclide for which the camera will be used.

5.10.9 Performance Assessment and Quality Control for SPECT

In addition to the PA/QC procedures discussed earlier, extra tests and measurements are required if the gamma camera is to be used in SPECT mode (HPA 1983, IPSM 1992a, 1992b, Hines et al. 1999).

Mechanical tests include checking that the gamma camera rotates about a single well-defined axis that is parallel to the front face of the collimator. In addition, measurements of angular precision and rotation speed are required. Errors in these parameters will produce gross SPECT image distortion. Further simple checks of the imaging couch alignment are necessary to minimise patient-positioning errors.

Electronic tests include measurements of the alignment of the x and y axes of the gamma camera with the axis of rotation, determination of any changes in energy spectrum as a function of camera angle, and estimation of image pixel size. The latter is particularly important, as the pixel size can be used to monitor the stability of the analogue-to-digital converters used to form the digital images. Variations in pixel size can affect SPECT image quality and quantification.

QC procedures include a tomographic acquisition of a point source, which is used to determine the COR in the x direction (Figure 5.152). The COR acquisition data are used to align the subsequent projection data with the reconstructed image. Also included is the acquisition of a high-count (3×10^7) flood-source image used for subsequent uniformity correction. Both tests must be carried out prior to data reconstruction and under the same conditions as are subsequently used for the SPECT acquisition, for example, using the same collimator, photon energy and pixel size.

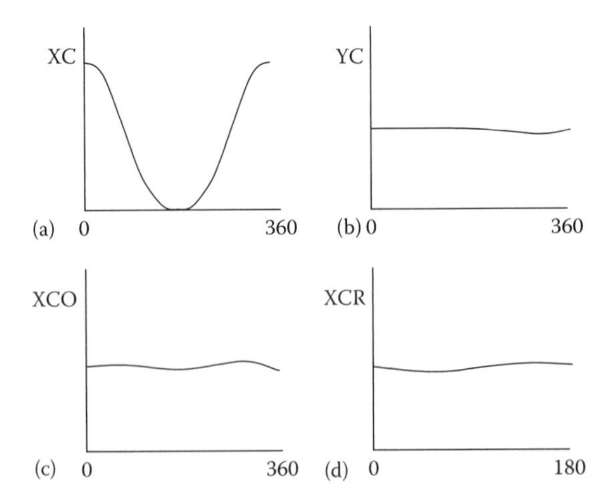

FIGURE 5.152
Schematic of data processed from a centre-of-rotation study, using a rotating gamma camera. The x-axis on all graphs represents the angular position of the camera in degrees. (a) Position of centre of point source in x direction (XC). (b) Position of centre of source in y direction (YC). (c) XCO = XC − sine wave fitted to curve for XC. (d) XCR = average of XC(θ) and XC (180 + θ). XC should be sinusoidal. YC should be constant if source is imaged in single slice. XCO and XCR should be within 0.25 pixel.

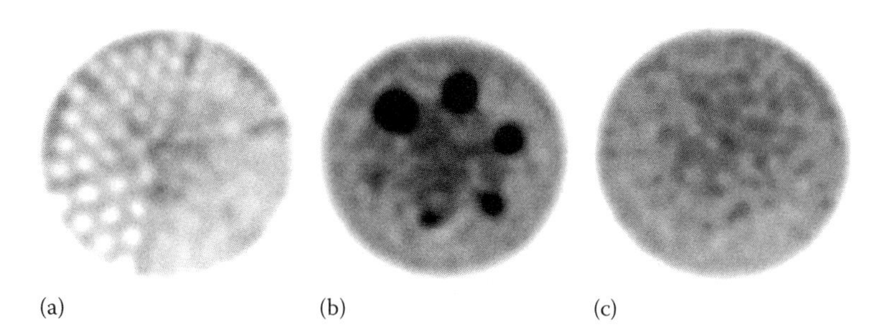

FIGURE 5.153
SPECT QC images using a Jaszczak phantom: (a) through section with cold rods for assessment of spatial resolution and linearity, (b) through section of hot spheres for assessment of image contrast, and (c) through uniform section for assessment of uniformity. (Courtesy of K. Pomeroy.)

Additional *performance assessment* for SPECT may include the following:

- Measurement of the uniformity of a reconstructed section through a cylindrical phantom. The reconstructed data should have twice the mean counts per voxel as compared to clinical studies. Visual inspection of the reconstructed images of a uniform phantom are also recommended since, if the uniformity correction is inadequate this will be revealed by circular artefacts (centred on the axis of rotation) as illustrated in Figure 5.99.
- Measurement of transaxial (in-slice) and axial (slice thickness) spatial resolution. The transaxial resolution is usually measured in both the radial and the tangential directions. This test can be performed at the same time as the COR test, that is, using the same point source, or can be performed using a line source either in air or in a cylinder of water. These measurements should be repeated at different positions in the tomographic FOV.
- Determination of system volume sensitivity in cps $MBq^{-1}mL$ using a Perspex cylinder uniformly filled with a radioactive liquid.
- Estimation of contrast in the reconstructed image, which determines what count-rate difference is required between a small ROI and surrounding areas to allow the ROI to be resolved from the background radioactivity.
- Determination of the recovery coefficient (i.e. the ratio of the apparent isotope concentration in the image divided by the true isotope concentration) as a function of object size and spatial resolution. Multiple performance characteristics of camera-based SPECT systems can be evaluated from a single scan of specially designed phantoms such as the Jaszczak SPECT phantom.* The Jaszczak phantom can be used to provide consistent performance information (see Figure 5.153) for any SPECT (or PET) system.

5.10.10 Performance Assessment and Quality Control for PET

Performance measurement standards for PET have been produced by NEMA (NEMA 1984, 2001b, 2007, 2008) and the EC (Guzzardi et al. 1990). Whilst there are differences between the standards, they all address the main parameters listed in the following sections. The standards address both the intrinsic camera parameters (i.e. spatial resolution,

* See http://www.biodex.com/nuclear-medicine/products/phantoms/jaszczak-aspect-phantom (accessed February 20, 2012).

sensitivity, scatter fraction, randoms and count rate) and the accuracy of correction factors (i.e. corrections for uniformity, scatter, attenuation and count rate non-linearity). A 20 cm diameter cylindrical phantom with a variety of inserts is used for the NEMA measurements, whilst the EC standard employs a more realistic chest phantom including a cardiac insert and a series of small spheres for measurement of partial-volume recovery coefficients. NEMA assessments for a BGO PET camera can be found in Brix et al. (1997). The clinically orientated performance standards (NEMA 2001b) for an ADAC molecular coincidence detection dual-head camera can be found in Sossi et al. (2001). The most important performance parameters are described in the following paragraphs.

The *transaxial resolution* is defined as the FWHM of the image of a line source reconstructed with a ramp filter cut-off at the Nyquist frequency. The measurement is made at the centre of the FOV and offset by various distances. For the offset measurements, both radial and tangential components must be specified, as the radial resolution becomes larger towards the edge of the FOV. Because the scatter distribution is very broad the FWHM measurement is usually changed negligibly by the introduction of scattering medium. If the FWTM is specified (as in the EC specification) then this should be measured in a scattering medium. Because many PET cameras are undersampled in the axial direction, an *axial slice width* rather than an axial resolution is usually specified for 2D imaging, whilst for 3D mode the axial resolution is usually quoted.

The *system sensitivity* can be defined as the true count rate (corrected for scatter and randoms) in response to a uniformly filled cylinder 20 cm in diameter and 20 cm long, and the result is given in units of cps Bq^{-1} mL^{-1}. This measurement is not independent of the volume being imaged and so the average sensitivity per slice, or per cm in the axial direction, is a more meaningful measurement. This measurement is not ideal for cameras with an axial FOV greater than the length of the phantom as the measurement depends on the size and position of the phantom. For this case, a 70 cm long phantom is used. Sensitivity is sometimes quoted as an absolute percentage (i.e. cps $Bq^{-1} \times 100\%$). However, due to the variation in sensitivity over the FOV in 3D mode, this can also be misleading if the source distribution is not specified.

The *scatter fraction* is measured by examining the sinogram of a line source within a 20 cm diameter water-filled cylinder (Figure 5.154). The count profile at each projection angle shows a sharp peak from unscattered events and broad tails that are attributed to scatter. The counts in the tails are integrated and compared with the total counts to obtain the scatter fraction.

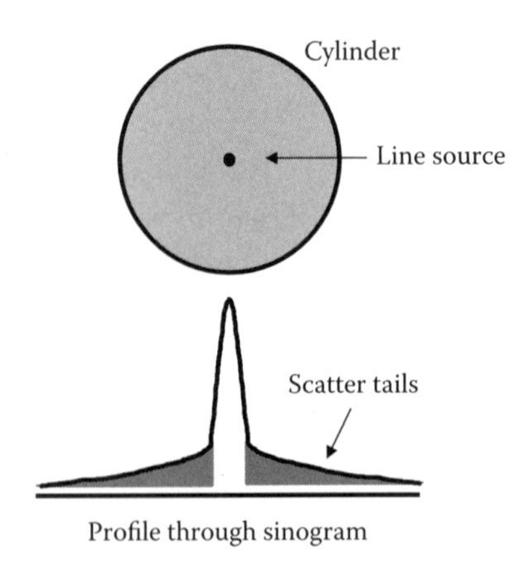

FIGURE 5.154
Schematic illustrating measurement of scatter fraction in PET using a line source placed on the axis of a water-filled cylindrical phantom.

Randoms fraction is characterised by a plot of the random and true count rates as a function of the activity concentration in the phantom.

Slice and volume uniformity can be characterised using essentially the same tests as those applied to gamma-camera images. All data corrections should be applied when this test is made.

The *count-rate performance* is characterised by the activity for which the percentage dead time is 50%. The percentage *dead time* is estimated by comparing the measured count rate with that which would be expected by extrapolating from the count rate measured with only a small activity in the FOV.

The *noise equivalent count rate* (NECR), described in Section 5.9.2.9, is often used to measure the overall count-rate performance of a PET system (Strother et al. 1990). The NECR is equal to the true coincidence rate that would give the observed SNR if there were no randoms and no scatter. Because randoms and scatter reduce the SNR, this 'effective true coincidence rate' is always less than the observed true coincidence rate. Whilst of limited value in comparing between different camera configurations, it is a simple objective measure often used to evaluate protocols and acquisition modes on similar machines.

The establishment of measures of performance that can be compared between cameras, and between different modes of operation, is difficult due to the range of PET camera configurations currently in use (i.e. 2D or 3D acquisition, TOF or non-TOF, scintillator type, etc.). It is also difficult to translate the various performance parameters into an index of how well a particular camera might be expected to perform in a given clinical situation. For this reason performance tests based on the image quality obtained from realistic phantoms and clinical imaging protocols have been developed (NEMA 2001b, 2007). Even then the subject of performance assessment remains controversial and the only definitive way to establish whether a given technology is clinically effective is to compare it against a known gold standard or against a clinical outcome. Such assessments are difficult to perform, so until some broad concensus emerges, physical measurements and images of phantoms are probably the only real option.

The main QC issue for block detectors is that the gain of individual PMTs may drift, resulting in artefacts and a reduced SNR. Routine QC consists of acquiring a daily blank scan and comparing this with the last normalisation scan. If the two differ by more than a predetermined amount then a new normalisation should be performed. If the PMT gains have drifted beyond predetermined limits then the gains themselves must be brought back into the normal operating range prior to normalising.

5.10.11 Summary

Performance assessment and QC of radioisotope imaging systems are important in limiting the effects of detector-related artefacts in the resultant images. Much effort has been put into these areas in application to both the hardware and software used in nuclear medicine.

5.11 Concluding Remarks

In the first edition of this book, we ended the chapter on the physics of radioisotope imaging with a section dedicated to clinical applications. In this edition we have included examples of various clinical applications as appropriate throughout the chapter. The reader can refer to several textbooks on clinical nuclear medicine (e.g. Ell and Gambhir 2004, Cook et al. 2006) for more information about this unique imaging modality.

References

Acton P, Agnew G, Cotton R, Hedges S, McKemey A K, Robbins M, Roy T, Watts S J, Damerell C J S, English R L, Gillman A R, Lintern A L, Su D and Wickens F J 1991 Future potential of charged-coupled –devices as detectors of ionising radiation *Nucl. Instrum. Methods* **305** 504–511.

Adam L-E, Karp J S, Daube-Witherspoon M E and Smith R J 2001 Performance of a whole-body PET scanner using curve-plate NaI(Tl) detectors *J. Nucl. Med.* **42** 1821–1830.

Aloj L and Morelli G 2004 Design, synthesis and preclinical evaluation of radiolabelled peptides for diagnosis and therapy *Curr. Pharm. Des.* **10** 3009–3031.

Anderson D F, Bouclier R, Charpak G and Majewski S 1983 Coupling of a BaF_2 scintillator to a TMAE photocathode and a low-pressure wire chamber *Nucl. Instrum. Methods* **217** 217–223.

Anger H O 1958 Scintillation camera *Rev. Sci. Instrum.* **29** 27–33.

Anger H O 1964 Scintillation camera with multichannel collimators *J. Nucl. Med.* **5** 515–531.

Anger H O 1969 Multiplane tomographic gamma ray scanner In *Medical Radioisotope Scintigraphy* (Vienna, Austria: IAEA) pp. 203–216.

Arano Y 2002 Recent advances in $^{99}Tc^{m}$ radiopharmaceuticals *Ann. Nucl. Med.* **16** 79–93.

Arano Y, Yokoyama A, Furukawa T, Horiuchi K, Yahata T, Saji H, Sakahara H, Nakashima T, Koizumi M, Endo K and Torizuka K 1987 Technetium-99m-labeled monoclonal antibody with preserved immunoreactivity and high *in vivo* stability *J. Nucl. Med.* **28** 1027–1033.

Attix F H and Roesch W C 1966 *Radiation Dosimetry* (New York: Academic Press).

Babich J W 1989 Radionuclide generators In *Radiopharmaceuticals: Using Radioactive Compounds in Pharmaceutics and Medicine* ed. A Theobald (Chichester, UK: Ellis Horwood Limited) pp. 3–18.

Badawi R D, Lodge M A and Marsden P K 1998 Algorithms for calculating detector efficiency normalization coefficients for true coincidences in 3D PET *Phys. Med. Biol.* **43** 189–205.

Bailey D L 1998 Transmission scanning in emission tomography *Eur. J. Nucl. Med.* **25** 774–787.

Bailey D L, Hutton B F and Walker P J 1987 Improved SPECT using simultaneous emission and transmission tomography *J. Nucl. Med.* **28** 844–851.

Bailey D L, Young H, Bloomfield P M, Meikle S R, Glass D, Myers M J, Spinks T J, Watson C C, Luk P, Peters A M and Jones T 1997 ECAT ART – A continuously rotating PET camera: Performance characteristics, initial clinical studies, and installation considerations in a nuclear medicine department *Eur. J. Nucl. Med.* **24** 6–15.

Bakker W H, Krenning E P, Breeman W A, Kooij P P M, Reubi J-C, Koper J W, De Jong M, Lameris J S, Visser T J and Lamberts S W 1991 In vivo use of a radioiodinated somatostatin analogue: Dynamics, metabolism and binding to somatostatin receptor-positive tumours in man *J. Nucl. Med.* **32** 1184–1189.

Baldwin R M 1986 Chemistry of radioiodine *Appl. Radiat. Isot.* **37** 817–821.

Banerjee S, Pillai M R A and Ramamoorthy N 2001 Evolution of Tc-99m in diagnostic radiopharmaceuticals *Semin. Nucl. Med.* **31** 260–277.

Banerjee S R, Schaffer P, Babich J W, Valliant J F and Zubieta J 2005 Design and synthesis of site-directed maleimide bifunctional chelators for technetium and rhenium *Dalton Trans.* **24** 3886–3897.

Barber H B, Apotovsky B A, Augustine F L, Barrett H H, Dereniak E L, Doty F P, Eskin J D, Hamilton W J, Marks D G, Matherson K J, Venzon J E, Woolfenden J M and Young E T 1997 Semiconductor pixel detectors for gamma-ray imaging in nuclear medicine *Nucl. Instrum. Methods Phys. Res., Sect. A* **395** 421–428.

Barr A, Bonaldi L, Carugno G, Charpak G, Iannuzzi D, Nicoletto M, Pepato A and Ventura S 2002 A high-speed, pressurised multiwire gamma camera for dynamic imaging in nuclear medicine *Nucl. Instrum. Methods Phys. Res., Sect. A* **477** 499–504.

Bateman J E, Connolly J F and Stephenson R 1985 High speed quantitative digital beta autoradiography using a multistep avalanche detector and an Apple II microcomputer *Nucl. Instrum. Methods Phys. Res. A* **241** 275–289.

Beck R N 1964 A theory of radioisotope scanning systems In *Medical Radioisotope Scanning* Vol. I (Vienna, Austria: IAEA).

Belhocine T, Shastry A, Driedger A and Urbain J-L 2007 Detection of 99mTc-sestamibi uptake in brown adipose tissue with SPECT-CT *Eur. J. Nucl. Med. Mol. Imaging* **34** 149.

Bellini S, Piacentini M, Cafforio C and Rocca F 1979 Compensation of tissue absorption in emission tomography *IEEE Trans. Acoust. Speech Signal Process.* **27** 213–218.

Bendriem B Casey M, Dahlbom M, Trebossen R, Blohm K, Nutt R and Syrota A 1996 Evaluation of the ECAT EXACT HR+: A new positron camera with 2D/3D acquisition capabilities and nearly isotropic spatial resolution *J. Nucl. Med.* **37** 743.

Berger H J, Malthay R A, Pytlik L M, Gottschalk A and Zaret B 1979 First-pass radionuclide assessment of right and left ventricular performance in patients with cardiac and pulmonary disease *Semin. Nucl. Med.* **9** 275–295.

Bergeron M, Cadorette J, Tétrault M-A, Viscogliosi N, Beaudoin J-F, Selivanov V, Norenberg J, Fontaine R and Lecomte R 2008 Imaging performance of the LabPET small-animal PET scanner *J. Nucl. Med.* **49 (Suppl 1)** 25.

Bergstrom M, Litton J, Eriksson L, Bohm C and Blomquist G 1982 Determination of object contour from projections for attenuation correction in cranial positron emission tomography *J. Comput. Assist. Tomogr.* **6** 365–372.

Beyer T, Townsend D W, Brun T, Kinahan P E, Charron M, Roddy R, Jerin J, Young J, Byars L G and Nutt R 2000 A combined PET/CT scanner for clinical oncology *J. Nucl. Med.* **41** 1369–1379.

Birks J B 1964 *The Theory and Practice of Scintillation Counting* (Oxford, UK: Pergamon).

Bizais Y, Zubal I G, Rowe R W, Bennett G W and Brill A B 1983 Potentiality of D7PHT for dynamic tomography *J. Nucl. Med.* **24** 75.

Blankespoor S C, Wu X, Kalki K, Brown J K, Tang H R, Cann C E and Hasegawa B H 1996 Attenuation correction of SPECT using X-ray CT on an emission- transmission CT system: Myocardial perfusion assessment *IEEE Trans. Nucl. Sci.* **NS-43** 2263–2274.

Bocher M, Balan A, Krausz Y, Shrem Y, Lonn A, Wilk M and Chisin R 2000 Gamma camera-mounted anatomical x-ray tomography: Technology, system characteristics and first images *Eur. J. Nucl. Med.* **27** 619–627.

Boudreau R J and Efange S M 1992 Computer-aided radiopharmaceutical design *Invest. Radiol.* **27** 653–658.

Bowley A R, Taylor C G, Causer D A, Barber D C, Keyes W I, Undrill P E, Corfield J R and Mallard J R 1973 A radioisotope scanner for rectilinear, arc, transverse section and longitudinal section scanning (ASS – the Aberdeen Section Scanner) *Br. J. Radiol.* **46** 262–271.

Boyd M, Ross S C, Dorrens J, Fullerton N E, Tan K W, Zalutsky M R and Mairs R J 2006 Radiation-induced biologic bystander effect elicited in vitro by targeted radiopharmaceuticals labeled with α-, β-, and auger electron–emitting radionuclides *J. Nucl. Med.* **47** 1007–1015.

BRH (Bureau of Radiological Health) 1970 *Radiological Health Handbook* DHEW Publ. no. 2016 (Rockville, MD: BRH).

Brigham E O 1974 *The Fast Fourier Transform* (Englewood Cliffs, NJ: Prentice-Hall).

British Pharmacopoeia 2012 (London, UK: The Stationary Office).

Britton K E and Brown N J G 1971 *Clinical Renography* (London, UK: Lloyd-Luke).

Brix G, Zaers J, Adam L E, Bellemann M E, Ostertag H, Trojan H, Haberkorn U, Doll J, Oberdorfer F and Lorenz W J 1997 Performance evaluation of a whole-body PET scanner using the NEMA protocol *J. Nucl. Med.* **38** 1614–1623.

Brownell G L and Sweet W H 1953 Localization of brain tumours with positron emitters *Nucleonics* **11** 40–45.

Bruland O S, Nilsson S, Fisher D R and Larsen R H 2006 High-linear energy transfer irradiation targeted to skeletal metastases by the *a*-emitter ^{223}Ra: Adjuvant or alternative to conventional modalities *Clin. Cancer Res.* **12** 6250–6257.

Bruyndonckx P, Liu X A, Tavernier S and Zhang SP 1997 Performance study of a 3D small animal PET scanner based on BaF_2 crystals and a photo sensitive wire chamber *Nucl. Instrum. Methods Phys. Res. A* **392** 407–413.

BSI (British Standards Institution) 1998 *Radionuclide Imaging Devices. Characteristics and Test Conditions* British Standard BS EN 61675 (London, UK: BSI).

Buck A K, Nekolla S, Ziegler S, Beer A, Krause B J, Herrmann K, Scheidhauer K, Wester H J, Rummeny E J, Schwaiger M and Drzezga A 2008 SPECT/CT *J. Nucl. Med.* **49** 1305–1319.

Budinger T F, Derenzo S E, Greenberg W L, Gullberg G T and Huesman R H 1978 Quantitative potentials of dynamic emission computed tomography *J. Nucl. Med.* **19** 309–315.

Budinger T F, Derenzo S E, Gullberg G T, Greenberg W L and Huesman R H 1977 Emission computer assisted tomography with single-photon and positron annihilation photon emitters *J. Comput. Assist. Tomogr.* **1** 131–145.

Budinger T F and Gullberg G T 1974 Three-dimensional reconstruction in nuclear medicine emission imaging *IEEE Trans. Nucl. Sci.* **NS-21** 2–20.

Burns H D 1978 Design of radiopharmaceuticals In *The Chemistry of Radiopharmaceuticals* ed. N O Heindel, H D Burns, T Honda and L W Brady (New York: Masson) Chapter 3.

Burns H D, Worley P, Wagner H N Jr, Marzilli L and Risch V 1978 Design of technetium radiopharmaceuticals. Chapter 17 in *The Chemistry of Radiopharmaceuticals* eds. N O Heindel, H D Burns, T Honda and L W Brady (New York: Masson).

Buvat I, Benali H, Todd-Pokropek A and Di Paola R 1994 Scatter correction in scintigraphy: The state of the art *Eur. J. Nucl. Med.* **21** 675–694.

Cabello J, Wells K, Metaxas A, Bailey A, Kitchen I, Clark A, Prydderch M and Turchetta R 2007 Digital autoradiography imaging using CMOS technology: First tritium autoradiography with a back-thinned CMOS detector and comparison of CMOS imaging performance with autoradiography film *IEEE Nucl. Sci. Symp. Conf. Rec.* **5** 3743–3746.

Casey M E and Nutt R 1986 A multicrystal 2-dimensional BGO detector system for positron emission tomography *IEEE Trans. Nucl. Sci.* **NS-33** 460–463.

Celler A, Sitek A, Stoub E, Hawman P, Harrop R and Lyster D 1998 Multiple line source array for SPECT transmission scans: Simulation, phantom and patient studies *J. Nucl. Med.* **39** 2183–2189.

Chang L T 1978 A method for attenuation correction in radionuclide computed tomography *IEEE Trans. Nucl. Sci.* **NS-26** 638–643.

Chappell L L, Ma D, Milenic D E, Garmestani G, Venditto V, Beitzel M P and Brechbiel M W 2003 Synthesis and evaluation of novel bifunctional chelating agents based on 1,4,7,10-tetraazacyclododecane-N,N′,N″,N‴-tetraacetic acid for radiolabelling proteins *Nucl. Med. Biol.* **30** 581–595.

Chatziioannou A and Dahlbom M 1996 Detailed investigation of transmission and emission data smoothing protocols and their effects on emission images *IEEE Trans. Nucl. Sci.* **43** 290–294.

Chen W 2007 Clinical applications of PET in brain tumors *J. Nucl. Med.* **48** 1468–1481.

Cherry S R, Dahlbom M and Hoffman E J 1992 Evaluation of a 3D reconstruction algorithm for multislice PET scanners *Phys. Med. Biol.* **37** 779–790.

Cherry S R, Shao Y, Silverman R W, Siegel S, Meadors K, Mumcuoglu E, Young J, Jones W F, Moyers C, Andreaco M, Paulus M, Binkley D, Nutt R and Phelps M E 1996 MicroPET: A dedicated PET scanner for small animal imaging *J. Nucl. Med.* **37** 334.

Cherry S R, Shao Y, Slates R B, Wilcut E, Chatziioannou A F and Dahlbom M 1999 MicroPET II – Design of a 1 mm resolution PET scanner for small animal imaging *J. Nucl. Med.* **40** 303.

Clack R, Townsend D and Jeavons A 1984 Increased sensitivity and field-of-view for a rotating positron camera *Phys. Med. Biol.* **29** 1421–1431.

Clarke M J and Podbielski J 1987 Medical diagnostic imaging with complexes of $^{99}Tc^{m}$ *Coord. Chem. Rev.* **78** 253–331.

Coates E A 1981 Quantitative structure-activity relationships In *Radiopharmaceuticals: Structure–Activity Relationships* ed. R P Spencer (New York: Grune and Stratton) pp. 8–10.

Coenen H H, Mertens J and Mazière B 2006 *Radioionidation Reactions for Radiopharmaceuticals: Compendium for Effective Synthesis Strategies* (Dordrecht, The Netherlands: Springer).

Coenen H H, Moerlin S M and Stocklin G 1983 No-carrier added radiohalogenation methods with heavy halogens *Radiochim. Acta* **34** 47–68.

Colsher J G 1980 Fully three-dimensional positron emission tomography *Phys. Med. Biol.* **20** 103–115.

Cook G J R, Maisey M N, Britton K E and Chengazi V 2006 *Clinical Nuclear Medicine* 4th edn. (London, UK: Hodder Arnold).

Cormack J, Towson J E C, Flower M A 2004 Radiation protection and dosimetry in clinical practice In *Nuclear Medicine in Clinical Diagnosis and Treatment* Vol II 3rd edn. eds. P J Ell and S S Gambhir (Edinburgh, UK: Churchill Livingstone) pp. 1871–1902.

Cosgriff PS and Sharp PF 1989 Nuclear-medicine software: Safety aspects *Nucl. Med. Commun.* **10** 535–538.

Cotton F A and Wilkinson G 1980 *Advanced Inorganic Chemistry* (New York: Wiley) pp. 71–81.

Couturier O, Supiot S, Degraef-Mougin M, Faivre-Chauvet A, Carlier T, Chatal J-F, Davodeau F and Cherel M 2005 Cancer radioimmunotherapy with alpha-emitting nuclides *Eur. J. Nucl. Med. Mol. Imaging* **32** 601–614.

Cygne 1993 *Product Information Leaflet for ^{123}I-IBZM* (Eindhoven, the Netherlands: Cygne bv).

Daghighian F, Shenderov P, Pentlow K S, Graham M C, Eshaghian B, Melcher C L and Schweitzer J S 1993 Evaluation of cerium-doped lutetium oxyorthosilicate (LSO) scintillation crystal for PET *IEEE Trans. Nucl. Sci.* **NS-40** 1045–1047.

Dahlbom M and Hoffman E J 1988 An evaluation of a two-dimensional array detector for high-resolution PET *IEEE Trans. Med. Imaging* **MI-7** 264–272.

Datz F L, Gullberg G T, Zeng G L, Tung C H, Christian P E, Welch A and Clack R 1994 Application of convergent-beam collimation and simultaneous transmission emission tomography to cardiac single-photon emission computed tomography *Semin. Nucl. Med.* **24** 17–37.

Daube-Witherspoon M E, Zubal I G and Karap J S 2003 Developments in instrumentation for emission computed tomography *Semin. Nucl. Med.* **33** 28–41.

De Ruysscher D and Kirsch C-M 2010 PET scans in radiotherapy planning of lung cancer *Radiother. Oncol.* **96** 335–338.

Defrise M, Kinahan P E, Townsend D W, Michel C, Sibomana M and Newport D F 1997 Exact and approximate rebinning algorithms for 3-D PET data *IEEE Trans. Med. Imaging* **MI-16** 145–158.

Defrise M, Townsend D W and Clack R 1992 FaVoR: A fast reconstruction algorithm for volume imaging in PET *Proc. IEEE Nucl. Sci. Symp. Med. Imaging Conf.*, Santa Fe, NM, November 2–9, 1991 pp. 1919–1923.

Derenzo S E, Moses W W, Huesman R H and Budinger T F 1993 Critical instrumentation issues for resolution smaller than 2 mm, high sensitivity brain PET In *Quantification of Brain Function. Tracer Kinetics and Image Analysis in Brain PET* eds. K Uemura, N A Lassen, T Jones and I Kanno (Amsterdam, the Netherlands: Elsevier Science) pp. 25–40.

Deutsch E, Libson K, Jurisson S and Lindoy L F 1983 Technetium chemistry and technetium radiopharmaceuticals *Prog. Inorg. Chem.* **30** 75–139.

Devaraj A, Cook G J R and Hansell D M 2007 PET/CT in non-small cell lung cancer staging – Promises and problems *Clin. Radiol.* **62** 97–108.

DHSS (Department of Health and Social Security) 1980 *Performance Assessment of Gamma Cameras* part I, Report no. STB 11 (London, UK: DHSS).

DHSS 1982 *Performance Assessment of Gamma Cameras* part II, Report no. STB 13 (London, UK: DHSS).

Diffey B L, Hall F M and Corfield J R 1976 The Tc-99m DTPA dynamic renal scan with deconvolution analysis *J. Nucl. Med.* **17** 352–355.

Dillman L T and Von der Lage F C 1975 MIRD Pamphlet Number 10. *Radionuclide Decay Schemes and Nuclear Parameters for Use in Radiation Dose Estimation* (New York: Society of Nuclear Medicine).

Divoli A, Erlandsson K, Dickson J, Flower M A and Ott R J 2004 Estimation of random coincidences from the prompt PET data *IEEE Nucl. Sci. Sym. Conf. Rec.* **6** 3703–3707.

Divoli A, Flower M A, Erlandsson K, Reader A J, Evans N, Meriaux S, Ott R J, Stephenson R, Bateman J E, Duxbury D M and Spill E J 2005 The PETRRA positron camera: Design, characterisation and results of a physical evaluation *Phys. Med. Biol.* **50** 3971–3988.

Duxbury D M, Ott R J, Flower M A, Erlandsson K, Reader A J, Bateman J E, Stephenson R and Spill E J 1999 Preliminary results from the new large-area PETRRA positron camera *IEEE Trans. Nucl. Sci.* **NS-46** 1050–1054.

Eastman K 2001 *Charged-Coupled Device (CCD) Image Sensors*. Primer MTD/PS-0218 (Rochester, N Y: Eastman Kodak Company).

Eckelman W C, Karesh S M and Reba R C 1975 New compounds: Fatty acids and long chain hydrocarbon derivatives containing a strong chelating agent *J. Pharm. Sci.* **64** 704–706.

Eckelman W C and Volkert W A 1982 In vivo chemistry of ^{99}Tc^{m}-chelates *Int. J. Appl. Radiat. Isot.* **33** 945–951.

EEV 1987 *CCD Imaging* (Chelmsford, UK: e2v Technologies).

Eisen Y, Shor A and Mardor I 1999 CdTe and CdZnTe gamma ray detectors for medical and industrial imaging systems *Nucl. Instrum. Methods Phys. Res. A* **428** 158–170.

Ell P J and Gambhir S S 2004 *Nuclear Medicine in Clinical Diagnosis and Treatment* 3rd edn. (Edinburgh, UK: Churchill Livingstone).

Endo M and Iinuma T A 1984 Software correction of scatter coincidence in positron-CT *Eur. J. Nucl. Med.* **9** 391–396.

Englert D, Roessler N, Jeavons A and Fairless S 1995 Microchannel array detector for quantitative electronic autoradiography *Cell Mol. Biol.* **41** 57–64.

Erlandsson K, Reader A J, Flower M A and Ott R J 1998 A new backprojection and filtering method for PET using all detected events *IEEE Trans. Nucl. Sci.* **NS-45** 1183–1188.

Esser P D, Alderson P O, Mitnick R J and Arliss J J 1984 Angled-collimator SPECT (A-SPECT): An improved approach to cranial single-photon emission tomography *J. Nucl. Med.* **25** 805–809.

Evans N T S, Keyes W I, Smith D, Coleman J, Cumpstey D, Undrill P E, Ettinger K V, Ross K, Norton M Y, Bolton M P, Smith F W and Mallard J R 1986 The Aberdeen Mark II single-photon-emission tomographic scanner: Specification and some clinical applications *Phys. Med. Biol.* **31** 65–78.

Everett D B, Fleming J S, Todds R W and Nightingale J M 1977 Gamma radiation imaging system based on the Compton effect *Proc. IEE* **124** 995–1000.

Farrell M A, McAdams H P, Herndon J E and Patz E F Jr 2000 Non-small cell lung cancer: FDG PET for nodal staging in patients with stage I disease *Radiology* **215** 886–890.

Feldkamp L A, Davis L C and Kress J W 1984 Practical cone-beam algorithm *J. Opt. Soc. Am. A* **1** 612–619.

Ferlin G, Borsato N, Camerani M, Conte N and Zotti D 1983 New perspectives in localising enlarged parathyroids by technetium–thallium subtraction scan *J. Nucl. Med.* **24** 438–441.

Fessler J A 1994 Penalized weighted least-squares image-reconstruction for positron emission tomography *IEEE Trans. Med. Imaging* **MI-13** 290–300.

Fleming J S 1979 A technique for the absolute measurement of activity using a gamma camera and computer *Phys. Med. Biol.* **24** 176–180.

Flower M A, Parker R P, Coles I P, Fox R A and Trott N G 1979 Feasibility of absolute activity measurements using the Cleon emission tomography system *Radiology* **133** 497–500.

Flower M A, Rowe R W and Keyes W I 1980 Sensitivity measurements on single-photon emission tomography systems *Radioakt. Isot. Klin. Forsch.* **14** 451–462.

Flux G, Bardies M, Monsieurs M, Savolainen S, Strand S E and Lassmann M 2006 The impact of PET and SPECT on dosimetry for targeted radionuclide therapy *Z. Med. Phys.* **16** 47–59.

Flyckt S O and Marmonier C 2002 *Photomultiplier Tubes: Principles and Applications* (Brive, France: Photonis).

Freedman G S 1970 Tomography with a gamma camera – Theory *J. Nucl. Med.* **11** 602–604.

Friston K J, Frith C D, Liddle P F and Frackowiak R S J 1991 Comparing functional (PET) images: The assessment of significant change *J. Cereb. Blood Flow Metab.* **11** 690–699.

Garlick P B, Marsden P K, Cave A C, Parkes H G, Slates R, Shao Y, Silverman R W and Cherry S R 1997 PET and NMR dual acquisition (PANDA): Applications to isolated, perfused rat hearts *Nucl. Magn. Res. Biomed.* **10** 138–142.

Gatti E, Rehak P, Longoni A, Kemmer J, Holl P, Klanner R, Lutz G, Wylie A, Goulding F, Luke P N, Madden N W and Walton J 1985 Semiconductor drift chambers *IEEE Trans. Nucl. Sci.* **NS-32** 1204–1208.

Gerber M S, Miller D W, Schlösser P A, Steidley J W and Deutchman A H 1977 Position sensitive gamma ray detectors using resistive charge division readout *IEEE Trans. Nucl. Sci.* **NS-24** 182–187.

Gilbert P 1972 Iterative methods for the three dimensional reconstruction of an object from projections *J. Theor. Biol.* **36** 105–117.

Gottschalk S C, Salem D, Lim C B and Wake R H 1983 SPECT resolution and uniformity improvements by noncircular orbit *J. Nucl. Med.* **24** 822–828.

Gowar J 1993 *Optical Communication Systems* 2nd edn. (Hempstead, NY: Prentice-Hall).

Graham L S and Neil R 1974 *In vivo* quantitation of radioactivity using the Anger camera *Radiology* **112** 441–442.

Grandin C, Wijns W, Melin J A, Bol A, Robert A R, Heyndrickx G R, Michel C and Vanoverschelde J-L J 1995 Delineation of myocardial viability with PET *J. Nucl. Med.* **36** 1543–1552.

Grosu A-L and Weber W A 2010 PET for radiation treatment planning of brain tumours *Radiother. Oncol.* **96** 325–327.

Gu Y, Matteson J L, Skelton R T, Deal A C, Stephan E A, Duttweiler F, Gasaway T M and Levin C S 2011 Study of a high-resolution, 3D positioning cadmium zinc telluride detector for PET *Phys. Med. Biol.* **56** 1563–1584.

Guzzardi R, Jordan K, Spinks T and Knoop B 1990 Performance evaluation of positron emission tomographs *Medical and Public Health Research Programme of the Commission of the European Communities* (Pisa, Italy: EC).

Haie-Meder C, Mazeron R and Magné N 2010 Clinical evidence on PET–CT for radiation therapy planning in cervix and endometrial cancers *Radiother. Oncol.* **96** 351–355.

Hawman P C and Haines E J 1994 The cardiofocal collimator: A variable-focus collimator for cardiac SPECT *Phys. Med. Biol.* **39** 439–450.

Hawman E G and Hsieh J 1986 An astigmatic collimator for high-sensitivity SPECT of the brain *J. Nucl. Med.* **27** 930.

Heindel N D, Burns H D, Schneider R and Foster N I 1981 Principles of rational radiopharmaceutical design In *Radiopharmaceuticals: Structure–Activity Relationships* ed. R P Spencer (New York: Grune and Stratton) pp. 101–128.

Herman G T, Lent A and Rowland S W 1973 ART: Mathematics and applications —A report on the mathematical foundations and on the applicability to real data of the algebraic reconstruction techniques *J. Theor. Biol.* **42** 1–32.

Hider R C and Hall A D 1991 Clinically useful chelators of tripositive elements In *Progress in Medicinal Chemistry* eds. G P Ellis and G B West (Amsterdam, the Netherlands: Elsevier) pp. 40–173.

Hilson A J W, Maisey M N, Brown C B, Ogg C S and Bewick M S 1978 Dynamic renal transplant imaging with Tc-99m DTPA (Sn) supplemented by a transplant perfusion index in the management of renal transplants *J. Nucl. Med.* **19** 994–1000.

Hines H, Kayayan R, Colsher J, Hashimoto D, Schubert R, Fernando J, Simcic V, Vernon P and Sinclair R L 1999 Recommendations for implementing SPECT instrumentation quality control *Eur. J. Nucl. Med.* **26** 527–532.

Hisada K I, Ohba S and Matsudaira M 1967 Isosensitive radioisotope scanning *Radiology* **88** 124–128.

Hoefnagel C A 1991 Radionuclide therapy revisited *Eur. J. Nucl. Med.* **18** 408–431.

Hoffman E J, Huang S C and Phelps M E 1979 Quantitation in positron emission computed tomography: 1. Effect of object size *J. Comput. Assist. Tomogr.* **3** 299–308.

Hoffman E J, Huang S C and Phelps M E and Kuhl D E 1981 Quantitation in positron emission tomography: 4. Effect of accidental coincidences *J. Comput. Assist. Tomogr.* **5** 391–400.

Holman B L, Carvalho P A, Zimmerman R E, Johnson K A, Tumeh S S, Smith A P and Genna S 1990 Brain perfusion SPECT using an annular single crystal camera: Initial clinical experience *J. Nucl. Med.* **31** 1456–1461.

Holte S, Schmidlin P, Linden A, Rosenqvist G and Eriksson L 1990 Iterative image reconstruction for positron emission tomography: A study of convergence and quantitation problems *IEEE Trans. Nucl. Sci.* **NS-37** 629–635.

HPA (Hospital Physicists' Association) 1978 *The Theory, Specification and Testing of Anger-Type Gamma Cameras* Topic Group Report no. 27 (London, UK: HPA).

HPA 1983 *Quality Control of Nuclear Medicine Instrumentation* ed. R F Mould (London, UK: HPA).
Hudson H M and Larkin R S 1994 Accelerated image reconstruction using ordered subsets of projection data *IEEE Trans. Med. Imaging* **MI-13** 601–609.
Huesman R H 1977 The effects of a finite number of projection angles and finite lateral sampling of projections on the propagation of statistical errors in transverse section reconstruction *Phys. Med. Biol.* **22** 511–221.
Hutton B F, Hudson H M and Beekman F J 1997 A clinical perspective of accelerated statistical reconstruction *Eur. J. Nucl. Med.* **24** 797–808.
Hyde R J, Ott R J, Flower M A, Meller S T and Fox R A 1988 A simple method of producing parenchymal renograms using parametric imaging *Clin. Phys. Physiol. Meas.* **9** 255–266.
Ichihara T, Ogawa K, Motomura N, Kubo A and Hashimoto S 1993 Compton scatter compensation using the triple-energy window method for single- and dual-isotope SPECT *J. Nucl. Med.* **34** 2216–2221.
IEC (International Electrotechnical Commission) 1998 *Radionuclide Imaging Devices. Characteristics and Test Conditions* Publication IEC 61675 (Geneva, Switzerland: IEC).
Iida H, Kanno I, Miura S, Murukami M, Takahashi K and Uemura K 1986 A simulation study of a method to reduce positron-annihilation spread distributions using a strong magnetic-field in positron emission tomography. *IEEE Trans. Nucl. Sci.* **NS-33** 597–600.
IPEM (Institute of Physics and Engineering in Medicine) 2003 *Quality Control of Gamma Camera Systems* ed. A Bolster. Report no. 86 (York, UK: IPEM).
IPSM (Institute of Physical Sciences in Medicine) 1992a *Quality Standards in Nuclear Medicine* eds. G C Hart and A H Smith. Report no. 65 (York, UK: IPSM).
IPSM 1992b *Quality Control of Gamma Cameras and Associated Computer Systems* ed. J Hannan. Report no. 66 (York, UK: IPSM).
Iqbal S M, Khalil M E, Lone B A, Gorski R, Blum S and Heller E N 2004 Simple techniques to reduce bowel activity in cardiac SPECT imaging *Nucl. Med. Commun.* **25** 355–359.
Irvine A T, Flower M A, Ott R J, Babich J W, Kabir F and McCready V R 1990 An evaluation of Tc-99m-HM-PAO uptake in cerebral gliomas: A comparison with x-ray CT *Eur. J. Nucl. Med.* **16** 293–298.
Jain D 1999 Technetium-99m labeled myocardial perfusion imaging agents *Semin. Nucl. Med.* **29** 221–236.
Jambhekar S S 1995 Biopharmaceutical properties of drug substances In *Principles of Medicinal Chemistry* eds. W O Foye, T L Lenke and D A Williams (Philadelphia, PA: Williams and Wilkins) pp. 12–24.
Jarritt P H and Acton P D 1996 PET imaging using gamma camera systems: A review *Nucl. Med. Commun.* **17** 758–766.
Jaszczak R J, Chang L T and Murphy P H 1979 Single photon emission computed tomography using multi-slice fan beam collimators *IEEE Trans. Nucl. Sci.* **NS-26** 610–611.
Jaszczak R J, Floyd C E Jr, Greer K L, Coleman R E and Manglos S H 1986a Cone beam collimation for SPECT: Analysis, simulation and image reconstructions using filtered backprojection *Med. Phys.* **13** 484–489.
Jaszczak R J, Floyd C E Jr, Manglos S H, Greer K L and Coleman R E 1986b Cone-beam SPECT: Experimental validation using a conventionally designed converging collimator *J. Nucl. Med.* **27** 930.
Jaszczak R J, Greer K L, Floyd C E Jr, Harris C C and Coleman R E 1984 Improved SPECT quantification using compensation for scattered photons *J. Nucl. Med.* **25** 893–900.
Jaszczak R J, Greer K L, Floyd C E, Manglos S H and Coleman R E 1988 Imaging characteristics of a high-resolution cone beam collimator *IEEE Trans. Nucl. Sci.* **NS-35** 644–648.
Jeavons A P, Chandler R A and Dettmar C A R 1999 A 3D HIDAC-PET camera with sub-millimetre resolution for imaging small animals *IEEE Trans. Nucl. Sci.* **NS-46** 468–473.
Jeavons A, Hood K, Herlin G, Parkman C, Townsend D, Magnanini R Frey P and Donath A 1983 The high-density avalanche chamber for positron emission tomography *IEEE Trans. Nucl. Sci.* **NS-30** 640–645.

Jurisson S, Berning D, Jia W and Ma D S 1993 Coordination compounds in nuclear medicine *Chem. Rev.* **93** 1137–1156.

Kagawa S, Kaneda T, Mikawa T, Banba Y, Toyama Y and Mikami O 1981 Fully ion-implanted p + -n germanium avalanche photodiodes *Appl. Phys. Lett.* **38** 429–431.

Karp J S, Muehllehner G, Geagan M J and Freifelder R 1998 Whole-body PET scanner using curve-plate Na(Tl) detectors *J. Nucl. Med.* **39** 190.

Karp J S, Muehllehner G, Qu H and Yan X H 1995 Singles transmission in volume-imaging PET with a Cs-137 source *Phys. Med. Biol.* **40** 929–944.

Keyes J W 1995 SUV: Standard uptake or silly useless value *J. Nucl. Med.* **36** 1836–1839.

Kimble T, Chou M and Chai BHT 2002 Scintillation properties of LYSO crystals *IEEE Nucl. Sci. Symp. Conf. Rec.* **3** 1434–1437.

Kimura K, Hashikawa K, Etani H, Uehara A, Kozuka T, Moriwaki H, Isaka Y, Matsumoto M, Kamada T and Moriyama H 1990 A new apparatus for brain imaging: Four-head rotating gamma camera single-photon emission computed tomograph [see comments] *J. Nucl. Med.* **31** 603–609.

Kinahan P E and Rogers J G 1989 Analytic 3D image-reconstruction using all detected events *IEEE Trans. Nucl. Sci.* **36** 964–968.

King M A, Long D T and Brill A B 1991 SPECT volume quantitation: Influence of spatial resolution, source size and shape, and voxel size *Med. Phys.* **18** 1016–1024.

King M A, Miller T R, Doherty P W and Schwinger R B 1988a Stationary and nonstationary spatial domain Metz filtering *Nucl. Med. Commun.* **9** 3–13.

King M A, Penney B C and Glick S J 1988b An image-dependent Metz filter for nuclear medicine images *J. Nucl. Med.* **29** 1980–1989.

Knapp F F and Mirzadeh S 1994 The continuing important role of radionuclide generator systems for nuclear medicine *Eur. J. Nucl. Med.* **21** 1151–1165.

Koeppe R A 1996 Tracer kinetics: Principles of compartmental analysis and physiologic modelling In *Nuclear Medicine* eds. R E Henkin, M A Boles, G L Dillehay, J R Halama, S M Karesh, R H Wagner and A M Zimmer (St. Louis, MO: Mosby-Year Book) pp. 292–315.

Kolan H, Li J and Thakur M L 1996 Sandostatin® labeled with Tc-99m: In vitro stability, *in vivo* validity and comparison with In-111 DTPA-Octreotide *Peptide Res.* **9** 144–150.

Krenning E P, Kwekkeboom D J, Bakker W H, Breeman W A P, Kooij P P M, Oei H Y, Van Hagen M, Postema P T E, De Jong M, Reubi J C, Visser T J, Reijs A E M, Hofland L J, Koper J W and Lamberts S W J 1993 Somatostatin receptor scintigraphy with [^{111}In-DTPA-D-Phe^{1}]- and [^{123}I-Tyr^{3}]-octreotide: The Rotterdam experience with more than 1000 patients *Eur. J. Nucl. Med.* **20** 716–731.

Kuhl D E and Edwards R Q 1963 Image separation radioisotope scanning *Radiology* **80** 653–661.

Kuhl D E, Edwards R Q, Ricci A R, Yacob R J, Mich T J and Alavi A 1976 The mark IV system for radionuclide computed tomography of the brain *Radiology* **121** 405–413.

Kuhn A, Surti S, Karp J S, Muehllehner G, Newcomer F M and VanBerg R 2006 Performance assessment of pixelated $LaBr_3$ detector modules for time-of-flight PET *IEEE Trans. Nucl. Sci.* 2006 **NS-53** 1090–1095.

Lacy J L, LeBlanc A D, Babich J W, Bungo M W, Latson L A, Lewis R M, Poliner L R, Jones R H and Johnson P C 1984 A gamma camera for medical applications, using a multiwire proportional counter *J. Nucl. Med.* **25** 1003–1012.

Lambrecht M and Haustermans K 2010 Clinical evidence on PET-CT for radiation therapy planning in gastro-intestinal tumors *Radiother. Oncol.* **96** 339–346.

Larsson S A, Bergstrand G, Bergstedt H, Berg J, Flygare O, Schnell P O, Andersson N and Lagergren C 1984 A special cut-off gamma camera for high-resolution SPECT of the head *J. Nucl. Med.* **25** 1023–1030.

Lear J L 1986 Principles of single and multiple radionuclide autoradiography. Chapter 5 in *Positron Emission Tomography and Autoradiography* eds. M E Phelps, J C Mazziotta and H R Schelbert (New York: Raven).

Lee J, Garmestani K, Wu C, Brechbiel M W, Chang H K, Choi C W, Gansow O A, Carrasquillo J A and Paik C H 1997 *In vitro* and *in vivo* evaluation of structure-stability relationship of ^{111}In- and ^{67}Ga-labelled antibody via 1B4M or C-NOTA chelates *Nucl. Med. Biol.* **24** 225–230.

Lewellen T K 1998 Time-of-flight PET *Semin. Nucl. Med.* **28** 268–275.
Lewellen T K, Miyaoka R S and Swan W L 1999 PET imaging using dual-headed gamma cameras: An update *Nucl. Med. Commun.* **20** 5–12.
Lewis M R, Raubitschek A and Shively J E 1994 A facile, water-soluble method for modification of proteins with DOTA. Use of elevated temperature and optimised pH to achieve high specific activity and high chelate stability in radiolabelled conjugates *Bioconj. Chem.* **5** 565–576.
Liang Z, Turkington T G, Gilland D R, Jaszczak R J and Coleman R E 1992 Simultaneous compensation for attenuation, scatter and detector response for SPECT reconstruction in three dimensions *Phys. Med. Biol.* **37** 587–603.
Lieberman D E (ed.) 1977 *Computer Methods: The Fundamentals of Digital Nuclear Medicine* (St. Louis, MO: C V Mosby).
Liu S and Edwards D S 2001 Bifunctional chelators for therapeutic lanthanide radiopharmaceuticals *Bioconjug. Chem.* **12** 7–34.
Ljungberg M and Strand S-E 1990a Attenuation correction in SPECT based on transmission studies and Monte Carlo simulations of build-up functions *J. Nucl. Med.* **31** 493–500.
Ljungberg M and Strand S-E 1990b Scatter and attenuation correction in SPECT using density maps and Monte Carlo simulated scatter functions *J. Nucl. Med.* **31** 1560–1567.
Maina T, Stolz B, Albert R, Bruns C, Koch P and Macke H 1994 Synthesis, radiochemistry and biological evaluation of a new somatostatin analogue (SDZ 219–387) labelled with technetium-99m *Eur. J. Nucl. Med.* **21** 437–444.
Maisey M 1980 *Nuclear Medicine: A Clinical Introduction* (London, UK: Update).
Maisey M N, Wahl R L and Barrington S F 1999 *Atlas of Clinical Positron Emission Tomography* (London, UK: Arnold).
Mallard J R and Peachey C J 1959 A quantitative automatic body scanner for the localisation of radioisotopes *in vivo Br. J. Radiol.* **32** 652–657.
Marsden P K 2006 Principles and methods. Chapter 1 in *Atlas of Clinical Positron Emission Tomography* eds. S F Barrington, M N Maisey and R L Wahl (London, UK: Edward Arnold).
Marsden P K, Bateman J E, Ott R J and Leach M O 1986 The development of a high efficiency cathode converter for a multiwire proportional chamber positron camera *Med. Phys.* **13** 703–706.
Mather S J and Ellison D 1990 Reduction-mediated technetium-99m labeling of monoclonal antibodies *J. Nucl. Med.* **31** 692–697.
Mayneord W V and Newbery S P 1952 An automatic method of studying the distribution of activity in a source of ionizing radiation *Br. J. Radiol.* **25** 589–596.
McCready V R, Parker R P, Gunnerson E M, Ellis R, Moss E, Gore W G and Bell J 1971 Clinical tests on a prototype semiconductor gamma-camera *Br. J. Radiol.* **44** 58–62.
McKenzie E H, Volkert W A and Holmes R A 1985 Biodistribution of ^{14}C-PnAO in rats *Int. J. Nucl. Med. Biol.* **12** 133–134.
Meijs W E, Haisma H J, Van der Schors R, Wijbrandts R, Van den Oever K, Klok R B, Pinedo H M and Herscheid J D M 1996 A facile method for the labelling of proteins with zirconium *Nucl. Med. Biol.* **23** 439–448.
Meikle S R, Bailey D L, Hooper P K, Eberl S, Hutton B F and Jones W F 1995 Simultaneous emission and transmission measurements for attenuation correction in whole-body PET *J. Nucl. Med.* **36** 1680–1688.
Meikle S R, Hutton B F and Bailey D L 1994 A transmission-dependent method for scatter correction in SPECT *J. Nucl. Med.* **35** 360–367.
Meikle S R, Matthews J C, Cunningham V J, Bailey D L, Livieratos L, Jones T and Price P 1998 Parametric image reconstruction using spectral analysis of PET projection data *Phys. Med. Biol.* **43** 651–666.
Melcher C L and Schweitzer J S 1992 Cerium-doped lutetium oxyorthosilicate: A fast, efficient new scintillator *IEEE Trans. Nucl. Sci.* **NS-39** 502–5
Mercer J R 2007 Molecular imaging agents for clinical positron emission tomography in oncology other than fluorodeoxyglucose (FDG): Applications, limitations and potential *J. Pharm. Pharmaceut. Sci.* **10** 180–202.
Mine P, Santiard J C, Scigocki D Suffert M, Tavernier S and Charpak G 1988 A BaF_2-TMAE detector for positron emission tomography *Nucl. Instrum. Methods Phys. Res. A* **273** 881–885.

Moehrs S, Del Guerra A, Herbert D J and Mandelkern M A 2006 A detector head design for small-animal PET with silicon photomultipliers (SiPM) *Phys. Med. Biol.* **51** 1113–1127.
Moore S C, Doherty M D, Zimmerman R E and Holman B L 1984 Improved performance from modifications to the multidetector SPECT brain scanner *J. Nucl. Med.* **25** 688–691.
Moore S C, Kouris K and Cullum I 1992 Collimator design for single-photon emission tomography *Eur. J. Nucl. Med.* **19** 138–150.
Muehllehner G 1971 A tomographic scintillation camera *Phys. Med. Biol.* **16** 87–96.
Mughabghab S F, Divadeenam M and Holden N E 1981 *Neutron Cross Sections from Neutron Resonance Parameters and Thermal Cross Sections* (New York: Academic Press).
Murase K, Itoh H, Mogami H, Ishine M, Kawamura M, Iio A and Hamamoto K 1987 A comparative study of attenuation correction algorithms in single-photon emission tomography *Eur. J. Nucl. Med.* **13** 55–62.
Murphy P H, Burdine J A and Mayer R A 1975 Converging collimation and a large field-of-view scintillation camera *J. Nucl. Med.* **16** 1152–1157.
Narita Y, Eberl S, Iida H, Hutton B F, Braun M, Nakamura T and Bautovich G 1996 Monte Carlo and experimental evaluation of accuracy and noise properties of two scatter correction methods for SPECT *Phys. Med. Biol.* **41** 2481–2496.
Natterer F 2001 Inversion of the attenuated Radon transform *Inv. Prob.* **17** 113–120.
Neirinckx R D, Canning L R, Piper I M, Nowotnik D P, Pickett R D, Holmes R A, Volkert W A, Forster A M, Weisner P S, Marriott J A and Chaplin S B 1987 Tc-99m *d,l*-HMPAO: A new radiopharmaceutical for SPECT imaging of regional cerebral blood perfusion *J. Nucl. Med.* **28** 191–202.
NEMA (National Electrical Manufacturers' Association) 1984, 2001b, 2007 *Performance Measurements of Positron Emission Tomographs* Standards Publ. no. NU2 (Washington, DC: NEMA).
NEMA 1986, 1994, 2001a *Performance Measurements of Scintillation Cameras* Standards Publ. no. NU1 (Washington, DC: NEMA).
NEMA 2008 *Performance Measurements for Small Animal Positron Emission Tomographs* Standards Publ. no. NU4 (Washington, DC: NEMA).
Ollinger J M 1996 Model-based scatter correction for fully 3D PET *Phys. Med. Biol.* **41** 153–176.
Ott R 2003 Positron emission tomography: A view of the body's function *Contemp. Phys.* **44** 1–15.
Ott R, Evans N, Evans P, Osmond J, Clark A and Turchetta R 2009 Preliminary investigations of active pixel sensors in nuclear medicine imaging *Nucl. Inst. Meth. Phys. Res. A.* **604** 86–88.
Ott R J, Evans N, Harris E, Evans P, Osmond J, Holland A, Prydderch M, Clark A, Crooks J, Halsall R, Key-Charriere M, Martin S and Turchetta R 2006 A CsI-active pixel sensor based detector for gamma ray imaging *IEEE Nucl. Sci. Sym. Conf. Rec.* **5** 2990–2992.
Ott R J, Flower M, Erlandsson K, Reader A, Duxbury D, Bateman J, Stephenson R and Spill E 2002 Performance characteristics of the novel PETRRA positron camera *Nucl. Instrum. Methods Phys. Res. A* **477** 475–479.
Ott R J, Flower M A, Khan O, Kalirai T, Webb S, Leach M O and McCready V R 1983a A comparison between 180° and 360° data reconstruction in single-photon emission computed tomography of the liver and spleen *Br. J. Radiol.* **56** 931–937.
Ott R J, Grey L J, Zivanovic M A, Flower M A, Trott N G, Moshakis V, Coombes R C, Neville A M, Ormerod M G, Westwood J H and McCready V R 1983b The limitations of the dual radionuclide subtraction technique for the external detection of tumours by radioiodine labelled antibodies *Br. J. Radiol.* **56** 101–109.
Ott R, MacDonald J and Wells K 2000 The performance of a CCD digital autoradiography imaging system *Phys. Med. Biol.* **45** 2011–2027.
Owunwanne A, Patel M and Sadek S 1998 *The Handbook of Radiopharmaceuticals* (London, UK: Hodder Arnold) pp. 1–13.
Paganelli G, Zoboli S, Cremonesi M, Macke H R and Chinol M 1999 Receptor-mediated radionuclide therapy with 90Y-DOTA-D-phe1-tyr3-octreotide: Preliminary report in cancer patients *Cancer Biother. Radiopharm.* **14** 477–483.
Paik C H, Murphy P R, Eckelman W C, Volkert W A and Reba R C 1983 Optimization of the DTPA mixed-anhydride reaction with antibodies at low concentration *J. Nucl. Med.* **24** 932–936.

Palmer A J and Taylor D M (ed) 1986 Radiopharmaceuticals labelled with halogen isotopes: including the proceedings of the International Symposium held in Banff, Alberta, Canada, 10–11 September 1985 *Appl. Radiat. Isot.* **37** (8) 645–921 Special issue. (Oxford, UK: Pergamon Press).

Parker R P, Smith P H S and Taylor 1978 *Basic Science of Nuclear Medicine* (Edinburgh, UK: Churchill Livingstone).

Patlak C S, Blasberg R G and Fenstermacher J D 1983 Graphical evaluation of blood-to-brain transfer constants from multiple-time uptake data *J. Cereb. Blood Flow Metab.* **3** 1–7.

Peters A M 1998 Fundamentals of tracer kinetics for radiologists *Br. J. Radiol.* **71** 1116–1129.

Phelps M E 1986 Positron emission tomography: principles and quantitation. In *Positron Emission Tomography and Autoradiography* eds M E Phelps, J C Mazziotta and H R Schelbert (New York: Raven).

Phelps M E 2004 *PET: Molecular Imaging and its Biological Applications* (New York: Springer).

Phelps M E, Huang S C, Hoffman E J, Selin C, Sokoloff L and Kuhl D E 1979 Tomographic measurement of local cerebral glucose metabolic rate in humans with (F-18)-2-fluoro-2-deoxy-D-glucose: Validation of method *Ann. Neurol.* **6** 371–388.

Picchio M, Giovannini E, Crivellaro C, Gianolli L, di Muzio N and Messa C 2010 Clinical evidence on PET/CT for radiation therapy planning in prostate cancer *Radiother. Oncol.* **96** 347–350.

Pichler B, Boning G, Lorenz E, Mirzoyan R, Pimpl W, Schwaiger M and Ziegler S I 1998 Studies with a prototype high resolution PET scanner based on LSO-APD modules *IEEE Trans. Nucl. Sci.* **NS-45** 1298–1302.

Pichler B J, Kolb A, Nägele T and Schlemmer H-P (2010) PET/MRI: paving the way for the next generation of clinical multimodality imaging applications *J. Nucl. Med.* **51** 333–336.

Prasad R, Ratib O and Zaidi H 2011 NEMA NU-04-based performance characteristics of the LabPET-8™ small animal PET scanner *Phys. Med. Biol.* **56** 6649–6664.

Purohit A, Liu S, Ellars C E, Casebier D, Haber S B and Edwards D S 2004 Pyridine-containing 6-hydrazinonicotinamide derivatives as potential bifunctional chelators for $^{99}Tc^{m}$–labelling of small biomolecules *Bioconjug. Chem.* **15** 728–737.

Qi J Y, Leahy R M, Cherry S R, Chatziioannou A and Farquhar T H 1998a High-resolution 3D Bayesian image reconstruction using the microPET small-animal scanner *Phys. Med. Biol.* **43** 1001–1013.

Qi J Y, Leahy R M, Hsu C H, Farquhar T H and Cherry S R 1998b Fully 3D Bayesian image reconstruction for the ECAT EXACT HR+ *IEEE Trans. Nucl. Sci.* **NS-45** 1096–1103.

Reivich M, Alavi A, Wolf A, Fowler J, Russell J, Arnett C, MacGregor R R, Shiue C Y, Atkins H, Anand A, Dann R and Greenberg J H 1985 Glucose metabolic-rate kinetic-model parameter determination in humans: The lumped constants and rate constants for [F-18] fluorodeoxyglucose and [C-11] deoxyglucose *J. Cereb. Blood Flow Metab.* **5** 179–192.

Ritter M A 1995 *Monoclonal Antibodies: Production, Engineering and Clinical Application* eds. M A Ritter and H M Ladyman (Cambridge, UK: Cambridge University Press) pp. 1–8.

Rogers A W 1979 *Techniques of Autoradiography* 3rd edn. (North Holland, the Netherlands: Elsevier).

Rogers W L, Han K S, Jones L W and Beierwaltes W H 1972 Application of a Fresnel zoneplate to gamma-ray imaging *J. Nucl. Med.* **13** 612–615.

Rohren E M, Turkington T G and Coleman R E 2004 Clinical applications of PET in oncology *Radiology* **231** 305–332.

Roselli M, Schlom J, Gansow O A, Brechbiel M W, Mirzadeh S, Pippin C G, Milenic D E and Colcher D 1991 Comparative biodistribution studies of DTPA-derivative bifunctional chelates for radiometal labeled monoclonal antibodies *Nucl. Med. Biol.* **18** 389–394.

Rosenthal M S, Cullom J, Hawkins W, Moore S C, Tsui B M and Yester M 1995 Quantitative SPECT imaging: A review and recommendations by the Focus Committee of the Society of Nuclear Medicine Computer and Instrumentation Council *J. Nucl. Med.* **36** 1489–1513.

Sandler M P, Coleman R E, Patton J A, Wackers F J and Gottschalk A. 2002 *Diagnostic Nuclear Medicine* 4th edn. (Philadelphia, PA: Lippincott Williams & Wilkins).

Schibli R and Schubiger P A 2002 Current use and future potential of organometallic radiopharmaceuticals *Eur. J. Nucl. Med. Mol. Imaging* **29** 1529–1542.

Schmand M, Casey M E, Wienhard K, Eriksson L, Jones W F, Lenox M, Young J W, Baker K, Miller S D, Reed J H, Heiss W D and Nutt R 1999 HRRT a new high resolution LSO-FET research tomograph *J. Nucl. Med.* **40** 306.

Schmand M, Dahlbom M, Eriksson L, Andreaco M, Casey M E, Vagneur K, Phelps M E and Nutt R 1998a Performance of a LSO/NaI(Tl) phoswich detector for a combined PET/SPECT imaging system *J. Nucl. Med.* **39** 24.

Schmand M, Eriksson L, Caset M E, Andreaco M S, Melcher C, Wienhard K, Flugge G and Nutt R 1998b Performance results of a new DOI detector block for a high resolution PET-LSO research tomograph HRRT *IEEE Trans. Nucl. Sci.* **NS-45** 3000–3006.

Schotanus P, Vaneijk C W E and Hollander R W 1988 A BAF2-MWPC gamma-camera for positron emission tomography *Nucl. Instrum. Methods Phys. Res. A* **269** 377–384.

Shankar L K, Hoffman J M, Bacharach S, Graham M M, Karp J, Lammertsma A A, Larson S, Mankoff D A, Siege B A, Van den Abbeele A, Yap J and Sullivan D 2006 Consensus recommendations for the use of ^{18}F-FDG PET as an indicator of therapeutic response in patients in National Cancer Institute trials *J. Nucl. Med.* **47** 1059–1066.

Shao Y P, Cherry S R, Farahani K, Meadors K, Siegel S, Silverman R W and Marsden P K 1997 Simultaneous PET and MR imaging *Phys. Med. Biol.* **42** 1965–1970.

Sharp P F, Gemmell H and Smith F W (eds.) 1998 *Practical Nuclear Medicine* (Oxford, UK: Oxford University Press).

Shepp L A and Logan B F 1974 The Fourier reconstruction of a head section *IEEE Trans. Nucl. Sci.* **NS-21** 21–43.

Short M D 1984 Gamma-camera systems *Nucl. Instrum. Methods Phys. Res.* **221** 142–149.

Singh M and Doria D 1983 An electronically collimated gamma camera for single-photon emission computed tomography. Part 1, Theoretical considerations in design criteria; Part 2, Image reconstruction and preliminary experimental measurements *Med. Phys.* **10** 421–435.

Smith B D 1985 Image reconstruction from cone beam projections: Necessary and sufficient conditions and reconstruction methods *IEEE Trans. Med. Imaging* **MI-4** 14–25.

Sokoloff L, Reivich M, Kennedy C, Des Rosiers M H, Patlak C S, Pettigrew K D, Sakurada O and Shinohara M 1977 The [^{14}C]deoxyglucose method for the measurement of local cerebral glucose utilization: Theory, procedure, and normal values in the conscious and anesthetized albino rat *J. Neurochem.* **28** 897–916.

Sorenson J A and Phelps M E 1980 *Physics in Nuclear Medicine* (New York: Grune and Stratton).

Sossi V, Pointon B, Boudoux C, Cohen P, Hudkins K, Jivan S, Nitzek K, deRosario J, Stevens C and Ruth T J 2001 NEMA NU-2000+ performance measurements on an ADAC MCD camera *IEEE Trans. Nucl. Sci.* **NS-48** 1518–1523.

Spradau T W, Edwards W B, Anderson C J, Welch M J and Katzenellbogen J A 1999 Synthesis and biological evaluation of Tc-99m-cyclopentadienyltricarbonyltechnetium-labeled octreotide *Nucl. Med. Biol.* **26** 1–7.

Srivastava S C and Richards P 1983 Technetium labelled compounds *Radiotracers for Medical Applications* Vol. **I**, ed. G V S Rayudn (Boca Raton, FL: CRC Press).

Stearns C W 1995 Scatter correction method for 3D PET using 2D fitted Gaussian functions *J. Nucl. Med.* **36** 105.

Stickel J R and Cherry S R 2005 High-resolution PET detector design: Modelling components of intrinsic spatial resolution *Phys. Med. Biol.* **50** 179–195.

Stoddart H F and Stoddart H A 1979 A new development in single gamma transaxial tomography: Union Carbide focussed collimator scanner *IEEE Trans. Nucl. Sci.* **NS-26** 2710–2712.

Stokely E M, Sveinsdottir E, Lassen N A and Rommer P 1980 A single-photon dynamic computer assisted tomograph (DCAT) for imaging brain function in multiple cross sections *J. Comput. Assist. Tomogr.* **4** 230–240.

Strobel K, Burger C, Schneider P, Weber M, and Hany T F 2007 MIBG-SPECT/CT-angiography with 3-D reconstruction of an extra-adrenal pheochromocytoma with dissection of an aortic aneurysm *Eur. J. Nucl. Med. Mol. Imaging* **34** 150.

Strother S C, Casey M E and Hoffman E J 1990 Measuring PET scanner sensitivity: Relating countrates to image signal-to-noise ratios using noise equivalent counts *IEEE Trans. Nucl. Sci.* **NS-37** 783–788.
Surti S, Karp J S, Freifelder R and Liu F 2000 Optimizing the performance of a PET detector using discrete GSO crystals on a continuous lightguide *IEEE Trans. Nucl. Sci.* **NS-47** 1030–1036.
Surti S, Kuhn A, Werner M E, Perkins A E, Kolthammer J and Karp J S 2007 Performance of Philips Gemini TF PET/CT scanner with special consideration for its time-of-flight imaging capabilities *J. Nucl. Med.* **48** 471–480.
Tai Y-C, Chatziioannou A F, Yang Y, Silverman R W, Meadors K, Siegel S, Newport D F, Stickel J R and Cherry S R 2003 MicroPET II: Design, development and initial performance of an improved microPET scanner for small-animal imaging *Phys. Med. Biol.* **48** 1519–1537.
Tan P, Bailey D L, Meikle S R, Eberl S, Fulton R R and Hutton B F 1993 A scanning line source for simultaneous emission and transmission measurements in SPECT *J. Nucl. Med.* **34** 1752–1760.
Tanabe S, Zodda J P, Libson K, Deitsch E and Heineman W R 1983 The biological distributions of some technetium MDP components isolated by anion-exchange high-performance liquid-chromatography *Int. J. Appl. Radiat. Isot.* **34** 1585–1592.
Tanaka E 1996 Instrumentation for PET and SPECT studies *Proc. Int. Symp. Tomography Nucl. Med. Vienna 1995* IAEA-SM-**337/38** 19–29 (Vienna, Austria: IAEA).
Todd-Pokropek A (ed.) 1982 The use of computers in nuclear medicine *IEEE Trans. Nucl. Sci.* **NS-29** 1272–1367.
Todd-Pokropek A 1983 Non-circular orbits for the reduction of uniformity artefacts in SPECT *Phys. Med. Biol.* **28** 309–313.
Townsend D, Frey P, Donath A, Clack R, Schorr B and Jeavons A 1984 Volume measurements *in vivo* using positron tomography *Nucl. Instrum. Methods* **221** 105–112.
Townsend D W, Geissbuhler A, Defrise M, Hoffman E J, Spinks T J, Bailey D L, Gilardi M C and Jones T 1991 Fully 3-dimensional reconstruction for a PET camera with retractable septa *IEEE Trans. Med. Imaging* **MI-10** 505–512.
Townsend D, Schorr B, Jeavons A, Clack R, Magnanini R, Frey P, Donath A and Froidevaux A 1983 Image reconstruction for a rotating positron tomograph *IEEE Trans. Nucl. Sci.* **NS-30** 594–600.
Troost E G C, Schinagl D A X, Bussink J, Oyen W J G and Kaanders J H A M 2010 Clinical evidence on PET–CT for radiation therapy planning in head and neck tumours *Radiother. Oncol.* 96 328–334.
Tsui B M W, Frey E C, Zhao X, Lalush D S, Johnston R E and McCartney W H 1994a The importance and implementation of accurate 3D compensation methods for quantitative SPECT *Phys. Med. Biol.* **43** 546–552.
Tsui B M W, Gullberg G T, Edgerton E R, Gilland D R, Perry J R and McCartney W H 1986 Design and clinical utility of a fan beam collimator for SPECT imaging of the head *J. Nucl. Med.* **27** 810–819.
Tsui B M W and Zhao X 1994 Practical iterative reconstruction methods for quantitative cardiac SPECT image-reconstruction *IEEE Trans. Nucl. Sci.* **NS-41** 325–330.
Tsui B M W, Zhao X, Frey E C and McCartney W H 1994b Quantitative single-photon emission computed tomography: Basics and clinical considerations *Semin. Nucl. Med.* **24** 38–65.
Tuy H K 1983 An inversion formula for cone beam reconstruction *SIAM J. Appl. Math.* **43** 546–552.
Vaidyanathan G and Zalutsky M R 1992 1-(meta-[211At]Astatobenzyl)guanidine: Synthesis via astato demetalation and preliminary *in vitro* and *in vivo* evaluation *Bioconjugate Chem.* **3** 499–503.
Van Elmbt L and Walrand S 1993 Simultaneous correction of attenuation and distance-dependent resolution in SPECT: An analytical approach *Phys. Med. Biol.* **38** 1207–1217.
Van Paesschen W, Dupont P, Sunaert S, Goffin K and Van Laere 2007 The use of SPECT and PET in routine clinical practice in epilepsy [Seizure disorders] *Curr. Opin. Neurol.* **20** 194–202.
Vandenbulcke K, Thierens H, De Vos F, Philippé J, Offner F, Janssens A, Apostolidis C, Morgenstern A, Bacher K, de Gelder V, Dierckx R A and Slegers G 2006 *In vitro* screening for synergism of high-linear energy transfer ^{213}Bi-radiotherapy with other therapeutic agents for the treatment of B-cell chronic lymphocytic leukemia *Cancer Biother. Radiopharm.* **21** 364–372.
Verger L, Gentet A C, Gerfault L, Guillemaud R, Mestais C, Monnet O, Montemont G, Petroz G, Rostaing J P and Rustique J 2004 Performance and perspectives of a CdZnTe-based gamma camera for medical imaging *IEEE Trans. Nucl. Sci.* **NS-51** 3111–3117.

Visvikis D, Ott R J, Wells K, Flower M A, Stephenson R, Bateman J E and Connolly J 1997 Performance characterisation of large-area BaF_2-TMAE detectors for use in whole body clinical PET camera *Nucl. Instrum. Methods Phys. Res. A* **392** 414–420

Vogel R A, Kirch D, LeFree M and Steele P 1978 A new method of multiplanar emission tomography using a seven pinhole collimator and an Anger scintillation camera *J. Nucl. Med.* **19** 648–654.

Volkert W A, Goeckeler W F, Ehrhardt G J and Ketring A R 1991 Therapeutic radionuclides: Production and decay property considerations *J. Nucl. Med.* **32** 174–185.

Wagner H N Jr (ed.) 1975 *Nuclear Medicine* (New York: H P Publishing).

Wallis J and Miller T R 1993 Rapidly converging iterative reconstruction algorithms in single-photon emission computed tomography *J. Nucl. Med.* **34** 1793–1800.

Wallis J and Miller T R 1997 An optimised rotator for iterative reconstruction *IEEE Trans. Med. Imaging* **MI-16** 118–123.

Walters T E, Simon W, Chesler D A, Correia J A and Riederer S J 1976 Radionuclide axial tomography with correction for internal absorption *Information Processing in Scintigraphy – Proc. 4th Int. Conf.*, Orsay, France, 1975 eds. C Raynaud and A Todd-Pokropek (French Atomic Energy Authority) pp. 333–342.

Watabe H, Ikoma Y, Kimura Y, Naganawa M and Shidahara M 2006 PET kinetic analysis – Compartmental model *Ann. Nucl. Med.* **20** 583–588.

Webb S 1987 Significance and complexity in medical images: Space variant texture dependent filtering *Proc. 10th Inf. Process. Med. Imaging Conf.*, Utrecht, the Netherlands eds. M A Viergever and C N de Graaf (New York: Plenum).

Webb S, Broderick M and Flower M A 1985a High resolution SPECT using divergent geometry *Br. J. Radiol.* **58** 331–334.

Webb S, Flower M A, Ott R J, Leach M O, Fielding S, Inamdar C, Lowry C and Broderick M D 1986 A review of studies in the physics of imaging by SPECT In *Recent Developments in Medical and Physiological Imaging* eds. R P Clark and M R Goff (London, UK: Taylor and Francis) (*J. Med. Eng. & Tech. Suppl.* 132–146).

Webb S, Long A, Ott R J, Flower M A and Leach M O 1985b Constrained deconvolution of SPECT liver tomograms by direct digital image restoration *Med. Phys.* **12** 53–58.

Webb S, Parker R P, Dance D R and Nicholas A 1978 A computer simulation study for the digital processing of longitudinal tomograms obtained with a zoneplate camera *IEEE Trans. Biomed. Eng.* **BME-25** 146–154.

Welch A, Clack R, Natterer F and Gullberg G T 1997 Toward accurate attenuation correction in SPECT without transmission measurements *IEEE Trans. Med. Imaging* **MI-16** 532–541.

Welch A and Gullberg G T 1997 Implementation of a model-based non-uniform scatter correction scheme for SPECT *IEEE Trans. Med. Imaging* **MI-16** 717–726.

Welch A, Gullberg G T, Christian P E and Datz F L 1994 A comparison of Gd/Tc versus Tc/Tl simultaneous transmission and emission imaging using both single and triple detector fan-beam SPECT systems *IEEE Trans. Nucl. Sci.* **NS-41** 2779–2786.

Welch A, Gullberg G T, Christian P E, Datz F L and Morgan T 1995 A transmission-map based scatter correction technique for SPECT in inhomogeneous-media *Med. Phys.* **22** 1627–1635.

Welch M J and McCarthy T T 2000 The potential role of generator-produced radiopharmaceuticals in clinical PET *J. Nucl. Med.* **41** 315–317.

Welch A, Webb S and Flower M A 1993 Improved cone-beam SPECT via an accurate correction for non-uniform photon attenuation *Phys. Med. Biol.* **38** 909–928.

Wieland D M, Tobes M C and Mangner T J (eds.) 1986 *Analytical and Chromatographic Techniques in Radiopharmaceutical Chemistry* (New York: Springer).

Wieland D M, Wu J-L, Brown L E, Mangner T J, Swanson D P and Beierwaltes W H 1980 Radiolabeled adrenergic neuron-blocking agents: Adrenomedullary imaging with [^{131}I]iodobenzylguanidine *J. Nucl. Med.* **21** 349–353.

Wilbur D S 1992 Radiohalogenation of proteins: An overview of radionuclides, labeling methods, and reagents for conjugate labelling *Bioconj. Chem.* **3** 433–470.

Winchell H S, Baldwin R M and Lin T H 1980 Development of ^{123}I-labeled amines for brain studies: Localization of ^{123}I-iodophenyl alkyl amines in rat brain *J. Nucl. Med.* **21** 940–946.
Wong D F and Pomper M G 2003 Predicting the success of a radiopharmaceutical for *in vivo* imaging of central nervous system neuroreceptor systems *Mol. Imag. Biol.* **5** 350–362.
Wu R K and Siegel J A 1984 Absolute quantitation of radioactivity using the build-up factor *Med. Phys.* **11** 189–192.
Xia W S, Lewitt R M and Edholm P R 1995 Fourier correction for spatially variant collimator blurring in SPECT *IEEE Trans. Med. Imaging* **MI-14** 100–115.
Xu M, Cutler P D and Luk, W K 1996 Adaptive, segmented attenuation correction for whole-body PET imaging *IEEE Trans. Nucl. Sci.* **NS-43** 331–336.
Yanch J C, Flower M A and Webb S 1988 A comparison of de-convolution and windowed subtraction techniques for scatter compensation in SPECT *IEEE Trans. Med. Imaging* **MI-7** 13–20.
Yu S K and Nahmias C 1995 Single-photon transmission measurements in positron tomography using Cs-137 *Phys. Med. Biol.* **40** 1255–1266.
Zito F, Savi A and Fazio F 1993 CERASPECT: A brain-dedicated SPECT system. Performance evaluation and comparison with the rotating gamma-camera *Phys. Med. Biol.* **38** 1433–1442.

6

Diagnostic Ultrasound

J. C. Bamber, N. R. Miller and M. Tristam

CONTENTS

6.1 Introduction ... 353
6.2 Basic Physics ... 354
6.2.1 Wave Propagation and Interactions in Biological Tissues ... 354
6.2.1.1 Speed of Sound ... 354
6.2.1.2 Acoustic Field Parameters ... 357
6.2.1.3 Acoustic Impedance ... 358
6.2.1.4 Intensity ... 358
6.2.1.5 Attenuation ... 359
6.2.1.6 Absorption ... 360
6.2.1.7 Scattering ... 361
6.2.1.8 Non-Linear Propagation ... 367
6.2.2 Movement Effects ... 369
6.2.3 Parameters for Imaging ... 370
6.2.4 Acoustic Radiation Fields ... 371
6.2.4.1 Continuous-Wave and Pulsed Excitation ... 371
6.2.4.2 Effect of Focusing ... 374
6.2.4.3 Effect of Using Short Pulses ... 376
6.2.4.4 Resolution ... 377
6.2.4.5 Effect of Non-Linear Propagation ... 379
6.2.5 Physical Principles and Theory of Image Generation ... 380
6.2.5.1 Pulse-Echo Scanning ... 380
6.2.5.2 Speckle ... 382
6.2.5.3 Speckle Reduction ... 383
6.3 Engineering Principles of Ultrasonic Imaging ... 384
6.3.1 Generation and Reception of Ultrasound: Transducers ... 384
6.3.1.1 Conventional Construction (Single Element) ... 385
6.3.1.2 Multiple-Element Transducers ... 388
6.3.2 Pulse-Echo Techniques (Echography) ... 394
6.3.2.1 Transducer and Frequency ... 394
6.3.2.2 Clock Pulse ... 395
6.3.2.3 Transmitter ... 396
6.3.2.4 Linear RF Amplifier ... 396
6.3.2.5 Time Gain Control ... 396
6.3.2.6 Compression Amplifier ... 398
6.3.2.7 Demodulation ... 398
6.3.2.8 Pre- and Postprocessing ... 398

6.3.2.9 Digitisation and Display ... 398
6.3.2.10 Additional Sections ... 399
6.3.2.11 Scanning and Display Format ... 399
6.3.2.12 Scanning Methods ... 400
6.3.2.13 Real-Time (Rapid) Scanning ... 401
6.3.3 Tissue Harmonic (Non-Linear) Imaging ... 403
6.3.4 Three-Dimensional Imaging ... 405
6.3.4.1 Data Acquisition Using a 1D Array ... 405
6.3.4.2 Data Acquisition Using a 2D Array ... 408
6.3.4.3 Data Processing and Display ... 409
6.3.4.4 Benefits and Limitations ... 411
6.3.5 Doppler Methods ... 412
6.3.5.1 Choice of Frequency (f_0) ... 413
6.3.5.2 Notes on the 'Doppler Equation' ... 413
6.3.5.3 Signal Processing in Doppler Systems ... 416
6.3.5.4 Range–Velocity Compromise ... 423
6.3.5.5 Doppler Display Methods ... 425
6.3.5.6 Tissue Doppler Imaging ... 429
6.3.6 Special Imaging Modes ... 429
6.3.6.1 Microbubble Contrast-Specific Imaging ... 429
6.3.6.2 Elasticity Imaging ... 431
6.4 Acoustic Output and Performance Checks ... 433
6.4.1 Acoustic Output Measurements ... 434
6.4.1.1 Piezoelectric Hydrophone ... 434
6.4.1.2 Hydrophone-Based Measurement Methods ... 438
6.4.1.3 Radiation Force Balance ... 442
6.4.1.4 Calculation of Derated Acoustic Parameters ... 443
6.4.2 Performance Tests and Phantoms for Pulse-Echo Systems ... 443
6.4.2.1 Element Dropout ... 445
6.4.2.2 Low Contrast Penetration ... 445
6.4.2.3 Distance Measurement Accuracy ... 445
6.4.2.4 Axial and Lateral Resolution ... 446
6.4.2.5 Slice Thickness (Elevational Focus) ... 446
6.4.3 Performance Tests and Phantoms for Doppler Systems ... 446
6.4.4 Summary ... 447
6.5 Biological Effects and Safety of Diagnostic Ultrasound ... 448
6.5.1 Biological Effects ... 449
6.5.2 Standards and Guidelines ... 451
6.6 Clinical Applications of Diagnostic Ultrasound ... 454
6.6.1 Obstetrics ... 454
6.6.1.1 First Trimester ... 454
6.6.1.2 Second and Third Trimesters ... 455
6.6.2 Gynaecology ... 456
6.6.3 Abdomen ... 456
6.6.3.1 GI Tract ... 456
6.6.3.2 Liver and Biliary Tree ... 457
6.6.3.3 Pancreas ... 457
6.6.3.4 Spleen and Lymphatics ... 458
6.6.3.5 Kidneys ... 458

6.6.3.6 Bladder 458
6.6.3.7 Prostate 459
6.6.4 Cardiovascular System 459
6.6.4.1 Heart 459
6.6.4.2 Arteries and Veins 460
6.6.5 Superficial Structures 461
6.6.5.1 Musculoskeletal System 461
6.6.5.2 Breast 462
6.6.5.3 Thyroid and Parathyroid 463
6.6.5.4 Male Genital System 463
6.6.6 Eye and Orbit 463
6.6.7 Brain 464
6.6.8 Lungs 465
6.6.9 Skin 465
6.6.10 Paediatric 466
6.6.11 Ultrasound in Invasive Procedures 466
6.6.11.1 Needle Puncture Techniques 466
6.6.11.2 Intra-Operative Applications 467
6.6.11.3 Minimally Invasive Therapeutic Procedures 467
6.6.11.4 External-Beam Radiation Therapy 467
6.7 Broader View of Ultrasonic Imaging 467
6.7.1 Synthetic-Aperture Focusing Technique and Computed Tomography 468
6.7.2 Ultrafast and Zone-Based Imaging 471
6.7.3 Image Optimisation 472
6.7.4 Motion-Vector Imaging 473
6.7.5 Molecular Imaging and Drug Delivery 473
6.7.6 Multiphysics Imaging 475
6.8 Conclusion 476
References 476
Further Reading 485

6.1 Introduction

Our attention is now turned away from those imaging techniques that make use of ionising radiation and to the subject of diagnostic ultrasound. Ultrasound is a form of radiation that, like X-rays, is useful for medical imaging because of a good balance between its penetration of and interaction with the body. Information about the body structures is encoded on the transmitted and scattered radiation, and it is the job of the imaging system to decode it. Unlike X-rays, but by analogy with light, ultrasound experiences refraction and reflection at interfaces between media of different acoustic refractive indices, and it is possible to build focusing systems. However, as we shall see later, the relative dimensions of wavelength and typical focusing apertures are such that wave (rather than geometric) optics is applicable, and phenomena such as diffraction and interference become limiting factors when attempting to build lenses, mirrors or shaped source apertures.

From the point of view of the way in which one makes use of a radiation for imaging, there are also marked differences between ultrasound and both light and X-rays.

Ultrasonic waves propagate sufficiently slowly so that, for the distances travelled in the body, transit times are easily measurable and radar-like pulse-echo methods may be used to create images. Conversely, the propagation speed is fast enough that all of the data needed for a complete image may be gathered and reconstructed rapidly enough, for example, to view moving images of structures in the living heart of a mouse. Coupled with the apparent very low risk of hazard of the examination (see Section 6.5) and the low cost of the equipment (being nearly all electronic), these features have eventually, over the several decades of development of the technique, led to ultrasound being the most frequently used imaging method in diagnostic medicine, apart from plane film X-ray.

Another difference, relative to other forms of radiation, is that ultrasound is a coherent radiation and, as with laser light, pronounced interference effects tend to dominate the images. Unlike even laser light, where the sensor is usually sensitive to optical intensity, conventional ultrasound receivers are amplitude and phase sensitive and are able to generate coherent interference-like fluctuations in apparent received magnitude, due to integration over a non-planar received wave.

Ultrasound is usually defined as sound of frequency above approximately 20 kHz, the limit of human hearing. In the context of 'imaging' techniques, this frequency range covers a considerable variety of applications: from underwater sonar and animal echo location (up to about 300 kHz), through medical diagnosis, therapy and industrial non-destructive testing (0.8–40 MHz), to acoustic microscopy (12 MHz to above 1 GHz). The relationship between frequency, f, and wavelength, λ, is given by

$$c = f\lambda \tag{6.1}$$

where c is medium dependent, but for water and most soft tissues it is ~1500 m s^{-1}. Therefore, typical wavelengths encountered in medical diagnosis range from 1.5 mm at 1 MHz to 0.1 mm at 15 MHz.

Although this chapter cites references to explicit sources of information, it was written using knowledge gained from general reading of a wide range of material. This fact is acknowledged in a non-explicit manner, in the form of the section on Further Reading at the end of the chapter.

6.2 Basic Physics

6.2.1 Wave Propagation and Interactions in Biological Tissues

6.2.1.1 Speed of Sound

To a good approximation, with the exception of bone or cartilage, biological tissues propagate ultrasound as if they were fluids, in that they are not able to support transverse ultrasonic waves to any great extent. Ultrasound, like sound in gases, therefore, propagates largely as a longitudinal pressure wave; the energy of a moving source (piston) is transferred to the medium either as a local compression, when the source pushes on the medium, or as a local stretching (rarefaction) when the source pulls on the medium (see Figure 6.1). In either case,

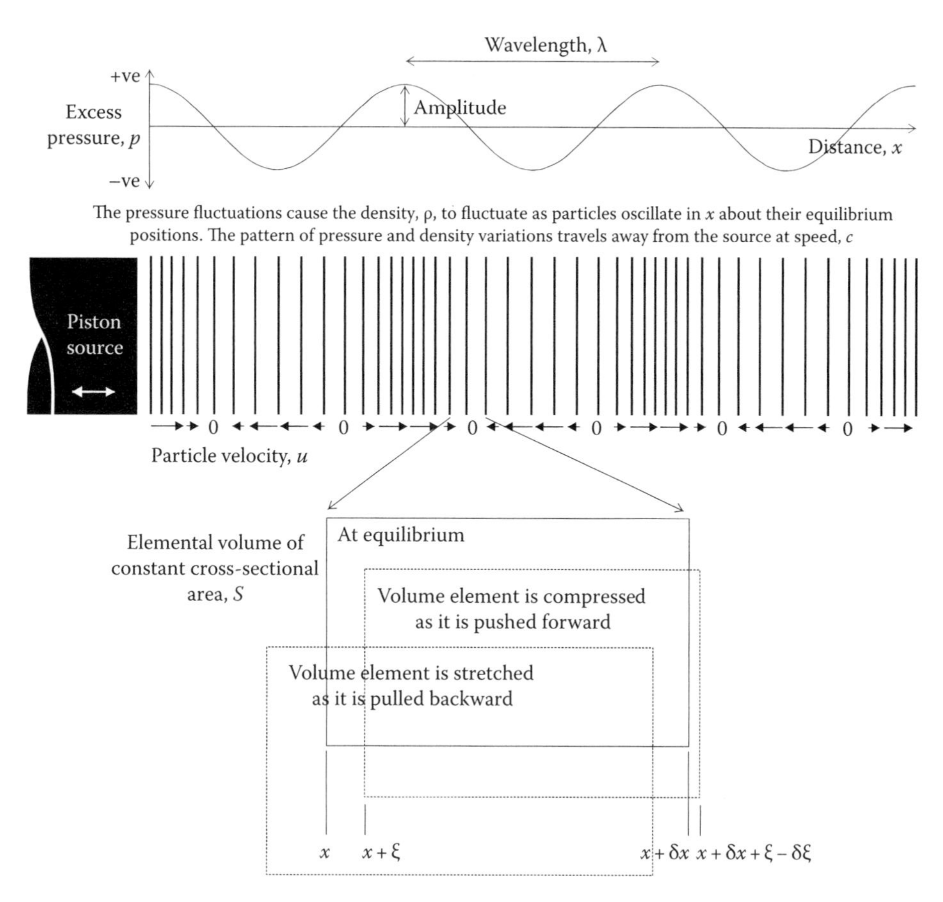

FIGURE 6.1
Schematic illustration of a continuous longitudinal sound wave at one instant. See text for more details.

potential energy is stored elastically in the layer of the medium that is immediately adjacent to the source and is released by motion of the adjacent layer of the medium, which takes time to respond because of the inertia of its mass (i.e. density). This layer then elastically compresses or stretches the next layer, and so on, so that the disturbance of pressure and density about their ambient values travels away from the source. For this to happen the particles of the medium must move forwards at times of compression and backwards at times of rarefaction, although their mean position over time remains fixed. A simple derivation of the equation describing this particle motion can be arrived at by considering a small volume element of the fluid, as illustrated in Figure 6.1, defined by the boundaries at equilibrium of x and $x + \delta x$, and constant cross-sectional area, S. In response to a force applied by either the source or the adjacent layer of the medium, this volume element undergoes motion and deformation. For a compression, it moves forwards by the particle displacement, ξ, and reduces in length, so that the same boundaries become $(x + \xi)$ and $(x + \delta x + \xi - \delta\xi)$. According to the usual approximation of small increments, the variation in force over the incremental distance δx is given by

$$\delta F = \frac{\partial F}{\partial x}\delta x. \tag{6.2}$$

For constant cross-sectional area the fractional volume change per unit length, or volumetric strain of the medium, is given by $\partial\xi/\partial x$. If this elastic response to the stress, or applied pressure F/S, is linear (non-linearity will be discussed in Section 6.2.1.8) then the two are related by the *bulk elastic modulus* of the medium, K, according to Equation 6.3. The bulk elastic modulus is a property of the medium that relates the applied pressure needed to achieve a given fractional change in volume,

$$\frac{F}{S} = K\frac{\partial\xi}{\partial x} \tag{6.3}$$

which can be differentiated with respect to distance and rearranged to give the force per unit length

$$\frac{\partial F}{\partial x} = KS\frac{\partial^2\xi}{\partial x^2}. \tag{6.4}$$

Using Newton's second law, $F = ma$, with $m = \rho_0 S\delta x$ and $a = \partial^2\xi/\partial t^2$, where ρ_0 is the density at equilibrium, gives

$$\delta F = \rho_0 S\,\delta x\frac{\partial^2\xi}{\partial t^2}. \tag{6.5}$$

Substituting for δF from Equation 6.2 into Equation 6.5 and combining with Equation 6.4,

$$KS\frac{\partial^2\xi}{\partial x^2}\delta x = \rho_0 S\,\delta x\frac{\partial^2\xi}{\partial t^2} \tag{6.6}$$

which can be rearranged into the form for a 1D wave equation,

$$\frac{\partial^2\xi}{\partial x^2} = \frac{1}{c^2}\frac{\partial^2\xi}{\partial t^2} \tag{6.7}$$

in which the longitudinal wave speed is

$$c = \sqrt{\frac{K}{\rho_0}}. \tag{6.8}$$

It is, therefore, clear that the speed of the wave depends on the density ρ_0 and the bulk elastic modulus K (strictly speaking, this is the adiabatic bulk modulus). Typical ranges of values for speed of sound are shown in Figure 6.2; further details of these data and their sources may be found in Bamber (1997) and Bamber (2004a). Note that the average speed, and the value often assumed by ultrasonic instrument designers, for soft tissues is about 1540 m s^{-1}, with a total range of ±6%. This variation, which is so small that it is neglected by

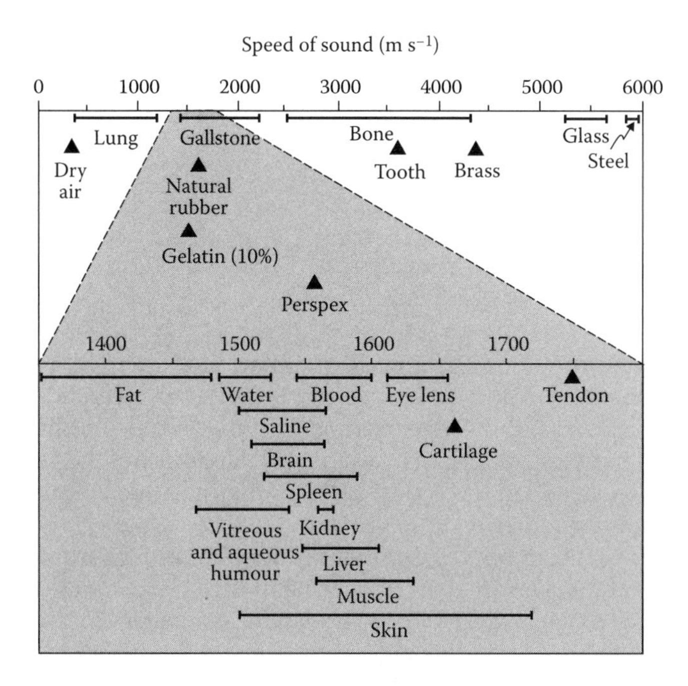

FIGURE 6.2
Ranges of measured values for speed of sound in various biological and non-biological media. The data for soft tissues and biological liquids, which fall within a narrow range, are shown using an expanded scale. (After Bamber, 1986a.)

conventional image reconstruction methods, is thought to be due mainly to fluctuations in bulk elastic modulus rather than density. Practically, c is not a strong function of frequency f (the dispersion is less than 1% over the medical range of frequencies), except in bone. The speed of sound is temperature dependent, the temperature coefficient being positive for non-fatty tissues/organs and negative for fatty tissue. Fat is also the only soft-tissue component that has a speed of sound lower than that for water, except for lung, which has a value close to that for air but dependent on the degree of inflation. For soft tissues other than fat and lung, there is an inverse relationship between speed of sound and water content, but a direct relationship with structural protein (collagen) content – hence, the very high values for tendon and cartilage.

6.2.1.2 Acoustic Field Parameters

Both energy and momentum are propagated by the compressional wave but, as described previously, there is no net transfer of matter unless this is induced as a result of energy lost from the wave. As the wave passes any particular point in the medium, oscillatory changes occur at that locality in a number of wave parameters related as described in the aforementioned derivation. For a sound pulse emerging from a typical diagnostic ultrasound scanner, the peak longitudinal displacement of the medium induced by the wave is only about 8×10^{-8} m. The peak particle velocity is of the order of 0.5 m s^{-1} and the maximum acceleration is a massive 3×10^5 times the acceleration due to gravity. The peak local positive pressure increase would be about 8 atm. All of these values have been calculated using approximate expressions, assuming that at the moment of peak amplitude the pulse is equivalent to a plane sine wave of intensity 2×10^5 W m^{-2}.

The instantaneous wave intensity I (W m^{-2}), which is the energy flowing per unit time per unit area, is related to the oscillating incremental pressure p and particle velocity u via the sum of the kinetic energy density and the potential energy density (J m^{-3}), that is,

$$I = \frac{c}{2}\left(\rho_0 u^2 + \frac{p^2}{\rho_0 c^2}\right). \tag{6.9}$$

Note that (under ideal conditions of a perfectly plane sine wave in an infinite uniform medium), at the moment of peak particle velocity U (i.e. when u reaches a maximum), the pressure is at ambient level and the potential energy term disappears. Similarly, at the time of peak pressure P (maximum p), the displacement is maximal but the velocity is zero, and so too is the kinetic energy term. In other words, the pressure and velocity fluctuations are 90° out of phase with each other.

We shall see later (e.g., Section 6.3.2) that diagnostic systems use very short sound pulses, which are repeated relatively infrequently. Because of this, typical time-averaged diagnostic intensities are very much lower than the corresponding instantaneous peak intensities.

6.2.1.3 Acoustic Impedance

A derived field parameter, analogous to electrical impedance, is the specific acoustic impedance

$$Z_{sp} = \frac{p}{u}, \tag{6.10a}$$

which in general is a complex quantity dependent on the relative phase of p and u, which in turn may be a function of spatial position and is dependent on the type of wave field and propagation conditions. On the other hand, the *characteristic acoustic impedance*,

$$Z = \rho_0 c = (\rho_0 K)^{1/2}, \tag{6.10b}$$

is a property of the medium only (units are kg m^{-2} s^{-1} or rayl). Z is only equal to Z_{sp} for the case of perfect plane waves in a lossless medium. As described in Section 6.2.1.7, the ratio of Z between two adjacent media determines the strength of the echo reflected at their interface. Materials with a high value of Z include bone, perspex, ceramics, metals, etc., while fat, blood, muscle and water have low Z. Air has very low acoustic impedance, explaining the need for coupling gel when imaging the body.

6.2.1.4 Intensity

Intensity can be determined from pressure, as can be seen by substituting for u from Equation 6.10a into Equation 6.9, to obtain

$$I = \frac{p^2}{\rho c} = \frac{p^2}{Z} \tag{6.11}$$

for free-field plane-wave lossless conditions. Calibrated hydrophones (see Section 6.4.1.1) are used to measure instantaneous pressure at a specific point in the ultrasound field, which can then be used to derive intensity. Intensity is often expressed as a ratio, measured in dB:

$$\text{Number of dB} = 10\log_{10}\left(\frac{I_1}{I_2}\right) = 10\log_{10}\left(\frac{A_1}{A_2}\right)^2 = 20\log_{10}\left(\frac{A_1}{A_2}\right) \tag{6.12}$$

where A is any amplitude quantity, such as pressure, P, and is, therefore, proportional to the square root of intensity, as in Equation 6.11.

6.2.1.5 Attenuation

All media attenuate ultrasound, so that the intensity of a plane wave propagating in the x direction decreases exponentially with distance as

$$I_x = I_0\,\mathrm{e}^{-\mu x} \quad \text{or} \quad \mu = -\frac{1}{x}\ln\left(\frac{I_x}{I_0}\right) \tag{6.13}$$

where μ is the intensity attenuation coefficient.

Similarly, for any of the amplitude parameters, P, U, etc. (represented by Q),

$$Q_x = Q_0\,\mathrm{e}^{-\alpha x} \quad \text{or} \quad \alpha = -\frac{1}{x}\ln\left(\frac{Q_x}{Q_0}\right) \tag{6.14}$$

where α is the amplitude attenuation coefficient.

Since $(I_x/I_0) = (Q_x/Q_0)^2$ we see that $\mu = 2\alpha$. The units of μ and α are cm^{-1}, but are usually called nepers cm^{-1} (from the use of the Naperian logarithm in the aforementioned equations). Practically, the intensity and amplitude ratios are often expressed in decibels (dB). Hence, Equation 6.13 becomes

$$\begin{aligned}\mu(\text{dB cm}^{-1}) &= -\frac{1}{x}10\log_{10}\left(\frac{I_x}{I_0}\right) \\ &= -\frac{1}{x}\ln\left(\frac{I_x}{I_0}\right)10\log_{10} e = 4.343\,\mu\ (\text{cm}^{-1}).\end{aligned} \tag{6.15}$$

Similarly,

$$\begin{aligned}\alpha(\text{dB cm}^{-1}) &= -\frac{1}{x}20\log_{10}\left(\frac{Q_x}{Q_0}\right) \\ &= 8.686\alpha\ (\text{cm}^{-1}) = \mu(\text{dB cm}^{-1}).\end{aligned}$$

Thus, when expressed in units of cm^{-1} the intensity attenuation coefficient is twice the amplitude attenuation coefficient, whereas when expressed in units of dB cm^{-1} the two attenuation coefficients are numerically equal to each other.

The attenuation coefficient has contributions from absorption and scattering; thus,

$$\mu = \mu_a + \mu_s \tag{6.16}$$

where

μ_a is the intensity absorption coefficient, and
μ_s is the intensity scattering coefficient.

In practice, if one attempts to measure μ or α in a simple transmission loss experiment, additional losses termed 'diffraction losses' (and sometimes corresponding gains) will be observed due to the diffraction field of the sound source (see Section 6.2.4.1). The relative contributions of μ_a and μ_s to μ are not known for many tissues, although it is believed that for normal liver μ_s might be between 10% and 30% of μ. Substantial departures from this occur for media such as lung, bone and contrast microbubbles. For blood, scattering accounts for less than 1% of the total attenuation.

The attenuation of ultrasound increases with frequency. Many soft tissues of the body attenuate ultrasound to a similar degree and display a nearly linear frequency dependence (see Figure 6.3, further details of which may be found in Bamber 1997 and Bamber 2004b). This gives rise to the ultrasonic instrument designer's rule of thumb for soft tissues, which is

$$\alpha = A\,\mathrm{dB\,cm^{-1}\,MHz^{-1}} \tag{6.17}$$

where, for a wide range of soft tissues, $A \approx 1$. Attenuation decreases as a function of the temperature of soft tissue, up to about 40°C.

6.2.1.6 Absorption

Absorption results in the conversion of the wave energy to heat, and is responsible for the temperature rise made use of in physiotherapy, ultrasound-induced hyperthermia and high-intensity focused ultrasound (HIFU) therapy. There are many mechanisms by which heat conversion may occur, although they are often discussed in terms of three classes. Classical mechanisms, which for tissues are small and involve mainly viscous (frictional) losses, give rise to an f^2 frequency dependence. Molecular relaxation, in which the temperature or pressure fluctuations associated with the wave cause reversible alterations in molecular configuration, are thought to be predominantly responsible for absorption in tissue (except bone and lung), and, because there are likely to be many such mechanisms simultaneously in action, produce a variable frequency dependence close to, or slightly greater than, f^1. Finally, relative motion losses, in which the wave induces a viscous or thermally damped movement of small-scale structural elements of tissue, are thought to be potentially important. A number of such loss mechanisms might also produce a frequency dependence of absorption somewhere between f^1 and f^2. Generally, however, one can say that, for simple solutions of molecules, increasing

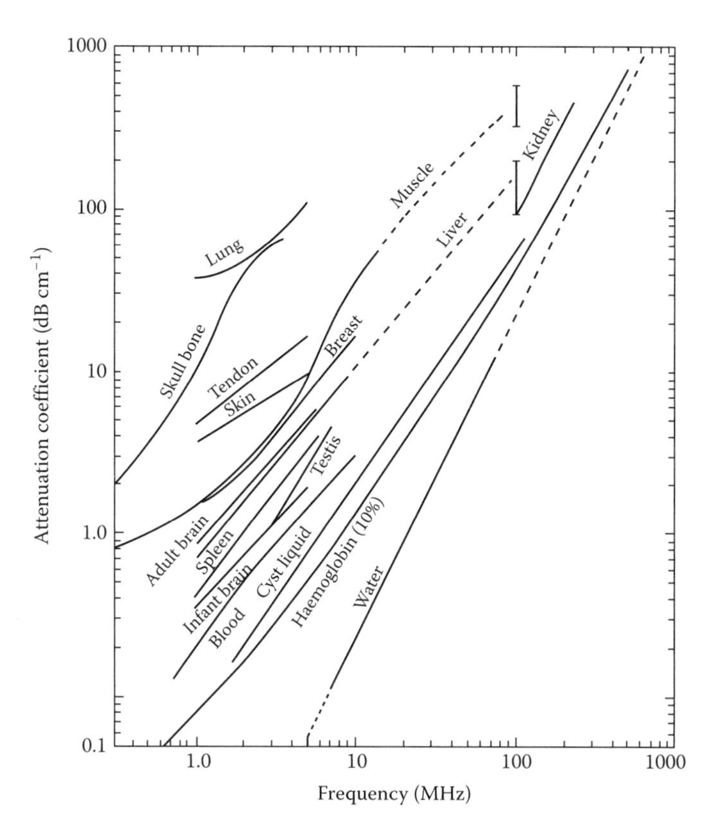

FIGURE 6.3
Illustration of the general trends observed for the variation of the ultrasonic attenuation coefficient (and its frequency dependence) over various biological tissues and solutions. (After Bamber, 1986b.)

molecular complexity results in increasing absorption. For tissues, a higher protein content (especially structural proteins such as collagen), or a lower water content, is associated with greater absorption of sound.

The temperature dependence of absorption is complicated, tending to be increasingly negative at higher frequencies (above 1 or 2 MHz) and positive at low frequencies. Fatty and non-fatty tissues do, however, appear to behave similarly. These changes with temperature are reversible, but when temperature is held above that at which proteins denature (in the region of 43°C), an irreversible increase occurs in the absorption and attenuation coefficients, whatever the frequency.

6.2.1.7 Scattering

Any inhomogeneity in Z (ρ, or c or K) will scatter ultrasound. Scattering structures within tissues range over at least four orders of magnitude in size, from cells (at about 10 μm, or 0.03λ at 5 MHz) to organ boundaries (up to 10 cm, or 300λ at 5 MHz). Different kinds of scattering phenomena occur at different levels of structure. These are classified in Table 6.1 and will now be discussed in more detail.

TABLE 6.1

Types of scattering interaction classified according to the scale of the characteristic dimension d of the scattering structure relative to the wavelength λ of sound for frequencies typical of those used in medical imaging.

Scale of Interaction	Frequency Dependence	Scattering Strength	Examples
$d \gg \lambda$ Geometrical-like, ray theory for reflection and refraction	f^0	Strong	Diaphragm, large vessels, soft tissue/bone, cysts and eye orbit
$d \sim \lambda$ Stochastic phenomena (and diffractive)	Variable	Moderate	Predominates for most structures (and modifies the examples in the other two categories)
$d \ll \lambda$ Rayleigh-like	f^4	Weak	Blood, cells within soft tissues

6.2.1.7.1 *Geometrical-Like Scattering*

At a large-scale boundary, representing the interface between two homogeneous media, the usual law of reflection and Snell's law for refraction apply to predict the direction of the reflected and refracted sound, that is (referring to Figure 6.4),

$$\frac{\sin\theta_i}{\sin\theta_t} = \frac{c_1}{c_2} \quad \text{and} \quad \theta_i = \theta_r. \tag{6.18}$$

The intensity of the reflected sound beam, relative to the incident intensity, is given by the intensity reflection coefficient:

$$R = \frac{I_r}{I_i} = 1 - \frac{I_t}{I_i}\left(\frac{Z_2\cos\theta_i - Z_1\cos\theta_t}{Z_2\cos\theta_i + Z_1\cos\theta_t}\right)^2. \tag{6.19}$$

At normal incidence, this becomes

$$R = \left(\frac{Z_2 - Z_1}{Z_2 + Z_1}\right)^2. \tag{6.20}$$

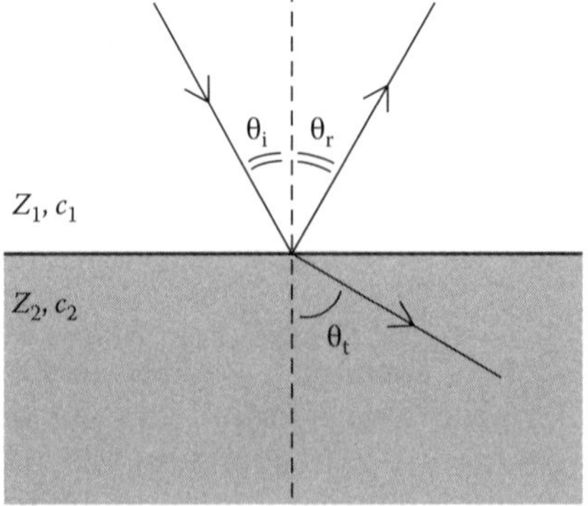

FIGURE 6.4
Geometrical scattering (reflection and refraction) at a plane boundary between two media, of characteristic acoustic impedances Z_1 and Z_2 and speeds of sound c_1 and c_2.

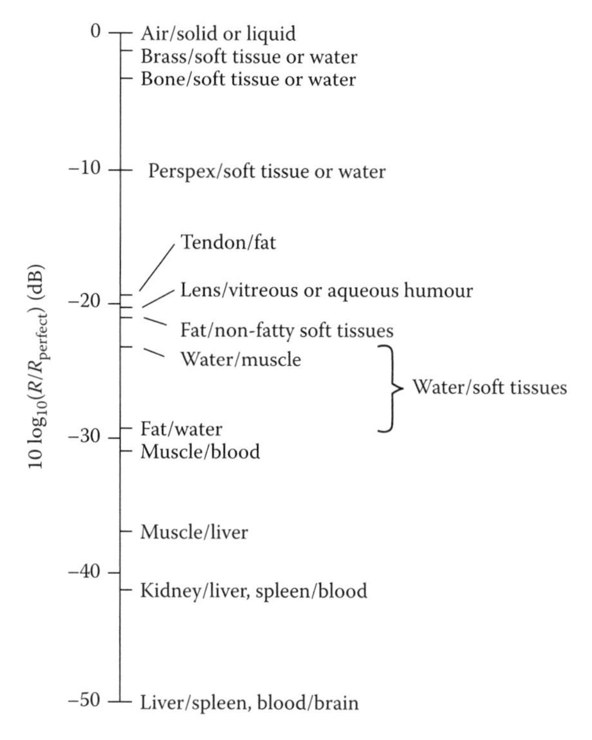

FIGURE 6.5
Calculated reflection coefficients (in decibels relative to a perfect reflector) for sound at normal incidence to a variety of hypothetical boundaries between biological and non-biological media.

There are no absolutely plane, smooth interfaces in the body, but the diaphragm has been noted to behave somewhat like a mirror, causing a second image of structures that are actually in the liver to appear where the lungs are situated. Other structures, such as large round cysts and the eye, produce refractive effects that can be successfully modelled using ray acoustics.

Refraction and reflection (and scattering in general) are determined by inhomogeneities in the speed of sound (c) and the characteristic impedance ($Z = \rho c$). Densities do not vary as much between different types of soft tissue as sound speeds. Therefore, values for R computed from differences in values for speed c, are useful in providing an intuitively helpful measure of the relative magnitude of echoes from various interfaces (see Figure 6.5). As one would expect, interfaces between media that are separated by the greatest difference in speed of sound in Figure 6.2 provide the largest reflection coefficient, and it is easy to see why lung attenuates sound so much, and why it is difficult for ultrasound to penetrate (and visualise beyond) bone or gas (as in lung or sometimes in the gastrointestinal (GI) tract). It might not, however, have been obvious that the range of echo levels from boundaries between different soft tissues and liquids could span at least 30 dB, given the relatively narrow range of sound speeds for these tissues. Echo imaging, therefore, is a technique with an inherently excellent contrast for depicting the boundaries between media with different sound speeds.

6.2.1.7.2 *Rayleigh-Like Scattering*

Examples of this kind of scattering structure are blood (predominantly from red cells) and cells in soft tissues (which contribute to the frequency dependence of scattering in tissues). Contrast agent microbubbles also exhibit Rayleigh scattering at frequencies lower than their resonant frequency. The scattered signal strength is very weak, is proportional to the volume of the scatterer and follows an f^4 frequency dependence. The angular distribution of scattering is fairly uniform but predominantly backwards.

6.2.1.7.3 *Stochastic Scattering and Diffraction Phenomena*

This kind of interaction tends to predominate in internal regions of organs, and modifies reflections from boundaries (i.e. rough surfaces). Scattering in this regime is characterised by a variable frequency dependence and variable angular distribution, although for some tissues (like liver) the average scattering has been measured to be somewhat forwards. Interference of scattered waves gives rise to 20–30 dB fluctuations in measured scattered energy as a function of angle of the scattered wave, orientation and position of the tissue volume and ultrasonic frequency. The fluctuations as a function of position contribute to a form of noise in the echo image known as speckle (discussed further in Section 6.2.5.2).

6.2.1.7.4 *Conventional Ultrasound Imaging*

Conventional ultrasound imaging produces images of the *backscattered amplitude,* (i.e. sound scattered at 180°). The backscattered part of the scattering coefficient is shown as a function of frequency for various biological media in Figure 6.6. Only general trends can be portrayed, as there is large variation between different experimental studies.

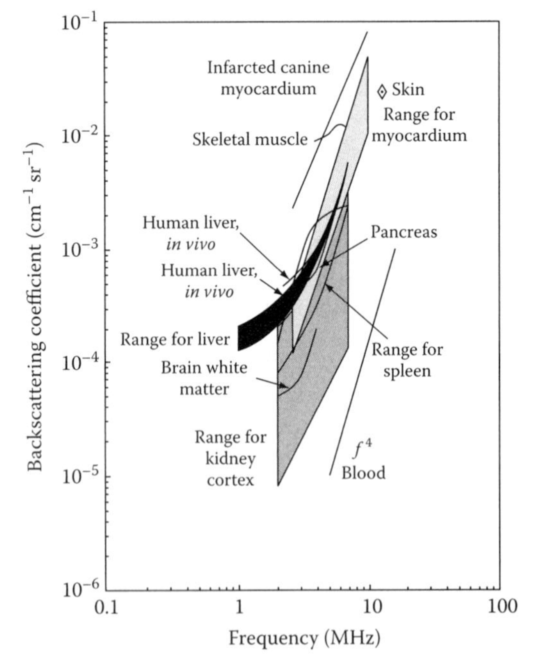

FIGURE 6.6
Frequency dependence of the backscattering coefficient for a range of soft tissues. The data in this figure have been pooled from different sources. The units of the backscattering coefficient are inverse centimetre (distance) and inverse steradian (solid angle). (After Bamber, 1997, which also contains the references to the original sources of data and details of the definition of the backscattering coefficient.)

The backscattering coefficient is influenced by a variety of parameters, including the size of the scattering structures, their acoustic impedance relative to that of the background, their characteristic spacing, their degree of alignment, and their orientation with respect to the sound beam.

6.2.1.7.5 Resonant and Non-Linear Scattering

This is a specialised topic, describing the resonant and non-linear behaviour of gas bubbles within a liquid. It is mainly relevant to contrast-agent microbubbles, which are gas bubbles of diameter 1–8 μm encapsulated within thin shells that may consist of a surfactant, or a mono or bilayer of a phospholipid or a protein. Microbubble contrast agents are used to enhance the scattering from blood. Early contrast agents were filled with air but the most recent use less soluble gases such as perfluorocarbons to increase the time during which they provide useful echo signal and image improvement, which is generally many tens of minutes; they eventually dissolve and the gas is exhaled via the lungs. Gas bubbles in fluids generate relatively strong echoes due to the large difference in acoustic impedance between the fluid and the gas. Moreover, the high compressibility of the gas that they contain and their spherical symmetry make bubbles strongly resonant scatterers. When bubbles are insonified at their resonant frequency, they produce echoes of strength (as measured by a quantity known as the scattering cross section) that is several orders of magnitude larger than it would be without this resonance effect (Figure 6.7a).

The resonant behaviour shown in Figure 6.7a was predicted from a model of gas bubble dynamics. Models of varying degrees of complexity have been proposed, but the simplest approach (and that used to derive Figure 6.7a) is to neglect the shell and to consider the medium as a free gas bubble in an incompressible Newtonian fluid of infinite extent. The pulsation of the bubble in response to an incident pressure field can be then described by the Rayleigh–Plesset equation (Church 1995). The solution to this equation enables calculation of the acoustically induced oscillations of the bubble radius, which allows the bubble then to be treated as a source of sound so that one may calculate the sound pressure scattered by the bubble, which is dependent on frequency. Finally, the scattering cross section is given by the ratio of the acoustic power scattered by the bubble (calculated from the scattered pressure) to the incident acoustic intensity.

Models of increased sophistication include those that take account of the stiffening effects of the shell on bubble dynamics, the principal effect being by an increase in the resonance frequency (e.g. Church 1995, Frinking and de Jong 1998, Hoff et al. 2000). Such models contain terms related to the shell thickness, density, elasticity and viscosity. A further step towards clinical realism is to predict the behaviour of a cloud of bubbles rather than an individual bubble. Church (1995) derived expressions for the frequency dependence of scattering and attenuation of a cloud of encapsulated bubbles in a liquid. The ratio of scattering strength to attenuation can be considered as a measure of the effectiveness of the contrast agent, the ideal agent being one that increases the echo signal without attenuating the sound beam (Bouakaz et al. 1998).

The graph in Figure 6.7a shows only the linear component of the bubble's response. For finite pressure amplitudes, there would also be non-linear (harmonic) components, and while some models have predicted these higher-order terms, they are not sufficiently complex to provide close agreement with observations from *in vitro* experiments. An example of an experimental result is shown in Figure 6.7b. It can be seen that the harmonic components are most visible at the intermediate pressure amplitude of the incident sound field. This is due to the fact that at the highest pressure amplitude, microbubble destruction occurs, which increases the background noise level.

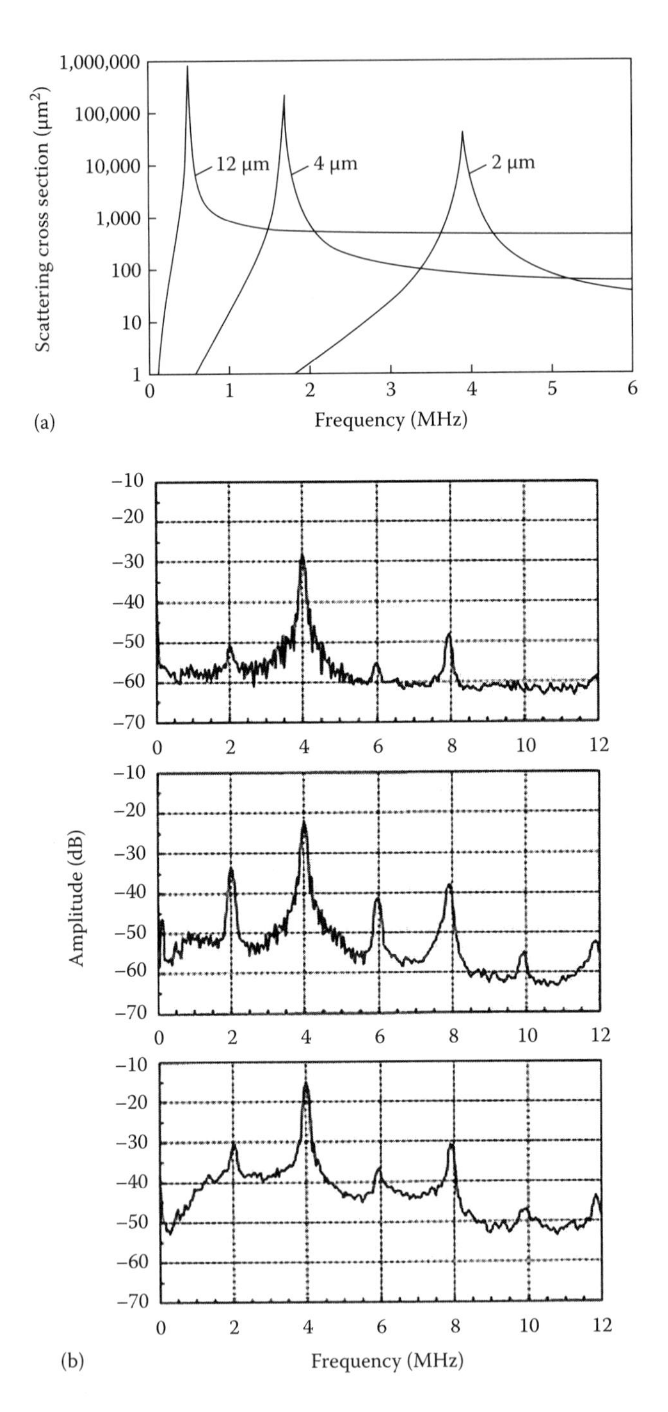

FIGURE 6.7
(a) Predicted resonant behaviour of free gas bubbles in an infinite, incompressible medium. (Reproduced with permission from de Jong et al., 1991.) (b) Experimentally observed frequency dependence of scattering of 4.0 MHz, 32 cycle pulses from Optison™ microbubbles as a function of the pressure amplitude of the incident acoustic wave (top: 0.8 MPa; middle: 1.6 MPa, bottom: 2.4 MPa). The spectra are not corrected for the frequency dependence of the receiving transducer. (Reproduced with permission from Shi and Forsberg, 2000.)

6.2.1.8 Non-Linear Propagation

This subject has become important due to the prominence of harmonic imaging in clinical ultrasound examinations. Everything discussed so far (and much of what is described later) in this chapter assumes that the medium in which the sound wave is travelling responds linearly to the mechanical stresses imposed by the wave. However, this is an approximation, valid only for waves of very small amplitude. All media are, to differing degrees, non-linear, and a consequence of this is that the wave travels at a speed dependent on the local wave amplitude (e.g. particle velocity). The phase speed is highest at the maximum compression and lowest at the peak rarefaction. This causes a progressive change in the wave shape as it propagates further into the medium (Figure 6.8) – the distortion takes the form of a gradual transferral of energy in the fundamental to higher harmonics. The high-frequency components are attenuated more quickly, due to the frequency dependence of the attenuation coefficient, such that far from the source, the wave returns to an attenuated version of its original shape. Figure 6.8 shows the progressive change in the waveform for an infinite sinusoidal plane wave as it propagates through a non-linear medium. Further details of this expected behaviour may be found in Bamber 2004b. The distortion experienced by real diagnostic ultrasound pulses is discussed in various texts (e.g. Barnett and Kossoff 1998, Duck 2000, Martin and Ramnarine 2010) and is illustrated in Figure 6.9b. Unlike the symmetrical saw tooth wave seen in Figure 6.8, the magnitude of the peak compression exceeds that of the peak rarefaction. This is a consequence of the finite size of the source and is a non-linear version of the wave interference that produces the source diffraction field (see Section 6.2.4), as described by Baker (1998).

The distortion of the wave is greater for larger inherent non-linearity of the medium, higher frequencies, lower speed of sound, larger distances (in a non-attenuating medium), higher initial wave amplitude and lower attenuation coefficient. The non-linearity of the medium is described by the parameter B/A, which is calculated from the Taylor series expansion of the relationship between density and acoustic pressure (Duck 2002, Leeman 2004). Examples of reported measurements of B/A for biological and other media are shown in Figure 6.10.

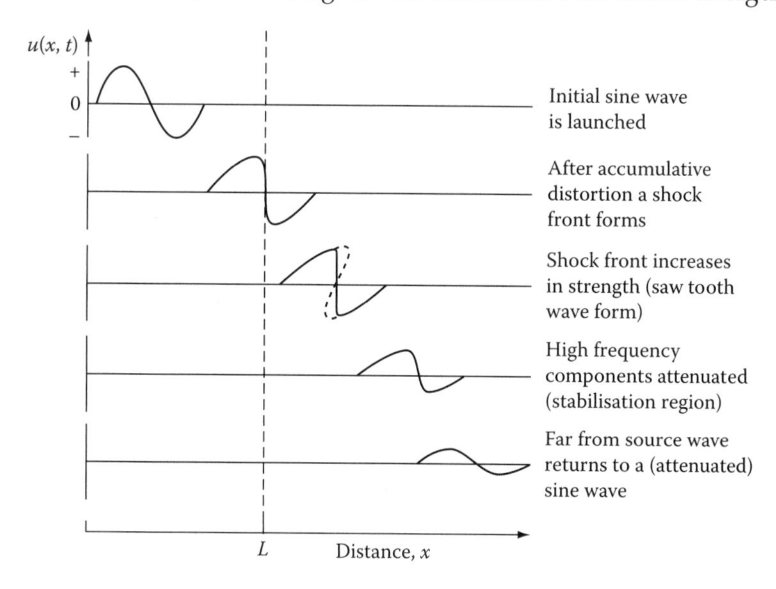

FIGURE 6.8
Change in waveform for an infinite sinusoidal plane wave propagating through a loss-less non-linear medium. L is known as the discontinuity distance which, in the absence of attenuation, occurs when the wavefront develops a step in particle velocity profile known as a shock front. (After Bamber, 1986b.)

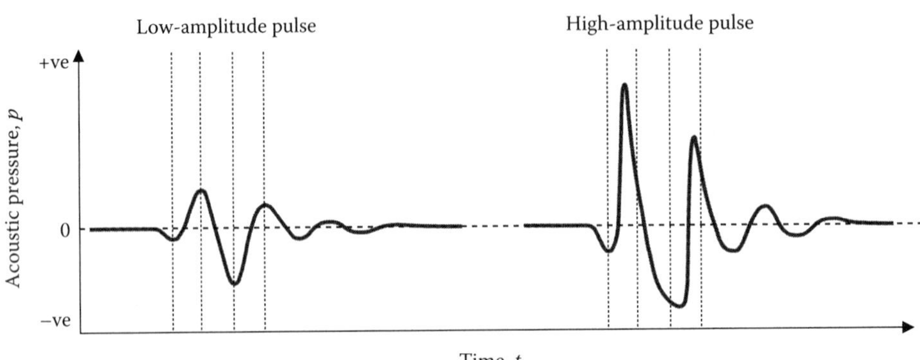

FIGURE 6.9
Acoustic pressure waveforms of the type that would be observed in water for (left) low-amplitude and (right) high-amplitude diagnostic pulses. The dashed vertical lines have been drawn so that the cause of the distortion in the high-amplitude pulse can be seen, viz., relative to arrival times for the low-amplitude pulse: high-pressure regions arrive earlier and low-pressure regions arrive later. (For clarity, the low-amplitude pulse (left) has been drawn much larger, relative to the high-amplitude pulse (right), than it would need to be to produce this relatively undistorted waveform shape. In reality, its peak pressure-amplitude (tens of kPa) would need to be about a hundredth of that of the high-amplitude pulse (greater than a MPa), for such a difference in waveform shape.)

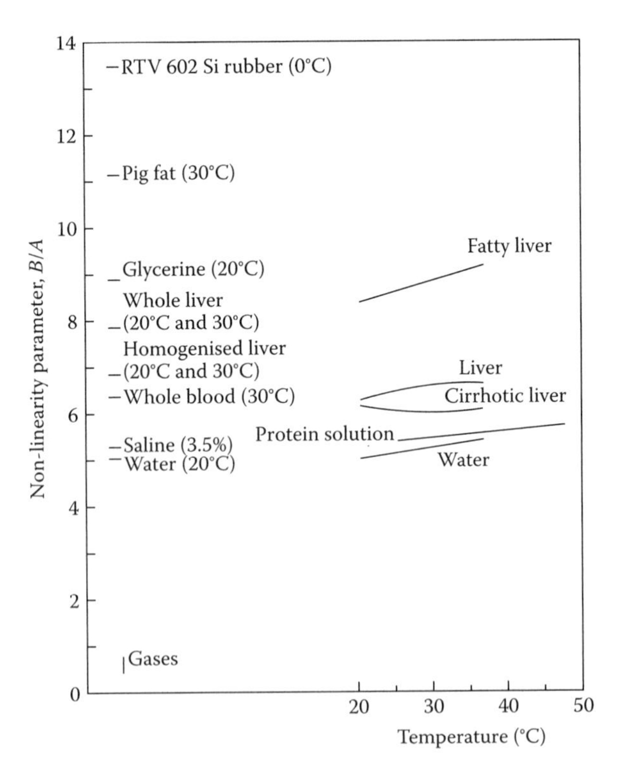

FIGURE 6.10
Parameter of non-linearity, *B/A*, for a variety of media. The temperature dependence of *B/A* for selected biological media is also shown. (After Bamber, 1997, which also contains the references to the original sources of data.)

6.2.2 Movement Effects

Figure 6.11 depicts the situation of a scatterer, such as a red blood cell, moving with velocity of magnitude v and direction θ with respect to the direction of the transmitted sound wave whose amplitude variations are described by

$$q_t = Q_0 \cos(\omega_0 t).$$

If the round-trip distance to the scatterer is $2x$, then the received backscattered wave will be

$$q_r = Q \cos\left[\omega_0\left(t + \frac{2x}{c}\right)\right].$$

However, x, which determines the phase of the wave, is changing at a rate $\pm v \cos\theta$, where the sign depends on whether the scatterer is moving towards or away from the sound source-receiver. If $v \ll c$, a linear relationship between rate of change of phase and scatterer velocity can be assumed (i.e. higher-order effects of the velocity on the measurement of phase can be ignored). Therefore, substituting for $x = \pm tv \cos\theta$, we get

$$q_r = Q \cos\left[\omega_0\left(t \pm \frac{2tv\cos\theta}{c}\right)\right]$$

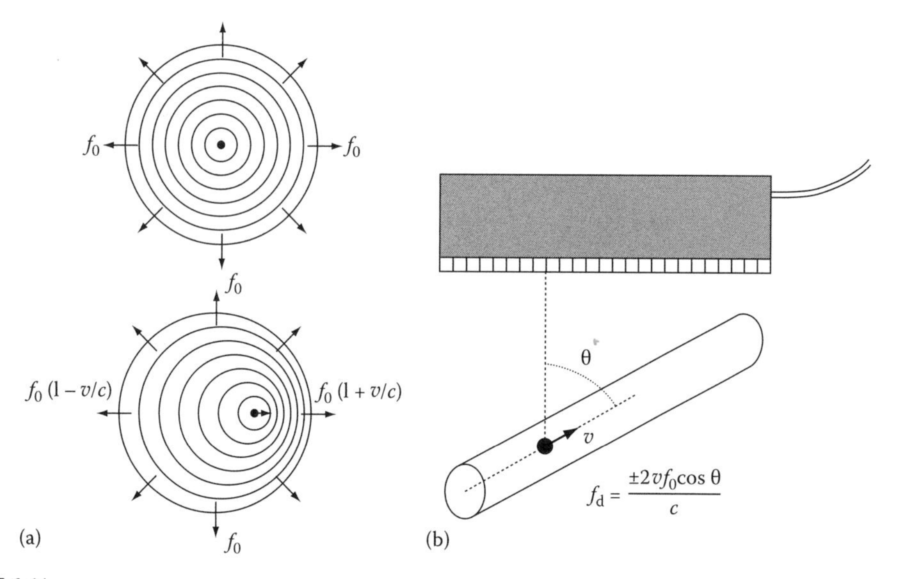

FIGURE 6.11
(a) Graphic illustration of the origin of the Doppler shift due to a moving source. If a source is stationary (upper diagram), then no matter where the observer is positioned the same (source) frequency (rate of arrival of wavefronts) will be detected. If the source is moving however (lower diagram), then it will be in a different position for each cycle of the wave that it emits. As a result, wavefronts will be detected as arriving at a rate that depends on the position of the observer; relative to the source frequency, the detected frequency will be higher when the observer is ahead of the direction of motion, lower when behind the direction of motion and unchanged when orthogonal to the direction of motion. (b) Depiction of a scatterer, such as a red blood cell, moving with velocity of magnitude v and direction θ with respect to the direction of propagation of the sound wave from an ultrasound transducer that is both a transmitter and receiver.

which (if $v \ll c$) is approximately equal to

$$Q\cos\left[\omega_0\left(1 \pm \frac{2v\cos\theta}{c}\right)t\right].$$

Putting this in the form $q_r = Q\cos(\omega t)$ produces

$$2\pi f = 2\pi f_0\left(1 \pm \frac{2v\cos\theta}{c}\right).$$

Therefore, the Doppler shift (or difference frequency) for backscattered sound is given by

$$f_d = f - f_0 = \frac{(\pm 2 f_0 v\cos\theta)}{c}. \tag{6.21}$$

Note that the Doppler shift is double that which arises for a stationary observer and a moving source emitting (rather than reflecting) sound waves. For f_0 in the range 1–10 MHz, and $v\cos\theta$ in the range 0–1 m s^{-1}, f_d is 0–13 kHz, which is within the audio-frequency range. Thus, a common method for subjective interpretation of Doppler-shifted echo signals from moving blood is to detect the Doppler shift, then amplify and listen to it.

Other (i.e. non-Doppler) methods have also been used for measuring and imaging the movement of blood and tissues. These are often based on correlation processing of the echo signals (Eckersley and Bamber 2004). For example, the correlation coefficient computed between two echoes from the same position but at different times is, for small movements of the tissue, inversely related to the distance moved (the constants of the equation being determined by the shape of the radio frequency point spread function described in Sections 6.2.4.3 and 6.2.4.4). Alternatively, a search may be conducted to locate the position of the echo pattern that best correlates with the echo pattern at a known position at some previously known time (see Section 6.3.6.2).

6.2.3 Parameters for Imaging

Of the wave propagation characteristics mentioned earlier, those that fundamentally govern the fate of the sound wave are ρ_0, K, μ_a and B/A. The ideal acoustic imaging system might well aim to produce maps of the spatial distribution of these quantities, plus other information associated with movement. Practically, clinical imaging of ρ_0, K and μ_a is not possible (and imaging B/A is very difficult). Instead, and only to a degree dependent upon the circumstances and access to the part of the body being imaged, one may be able to create images of parameters associated with c, μ and μ_s, which are determined by ρ_0, K and μ_a. Although practical imaging methods are often only able to provide relatively crude (and sometimes qualitative) representations of them, they often represent measures of relatively independent aspects of tissue structure and function.

The most widespread (conventional) ultrasound imaging technique aims to generate images of the backscattered part of the scattering coefficient, μ_s, although these images contain both artefacts and information related to μ, c and perhaps B/A. The second most important class of ultrasound imaging methods uses the Doppler shift to characterise blood flow. Other uses of the aforementioned propagation properties include (1) using knowledge of those properties in the design of contrast agents, (2) observing variation of the properties with temperature, time, etc., so as to make images of additional tissue characteristics, and (3) taking advantage of *B/A* in tissue harmonic imaging (THI), to improve images of backscattering (see Section 6.3.3).

6.2.4 Acoustic Radiation Fields

6.2.4.1 *Continuous-Wave and Pulsed Excitation*

The distribution of acoustic field parameters (intensity, pressure, etc.), as a function of both space (three dimensions) and time (one more dimension), in front of a radiating source of ultrasound, and the corresponding spatial and temporal sensitivity pattern of a receiver, are factors on which the performance of any ultrasonic imaging system is critically dependent. Prediction of acoustic radiation fields is often possible by calculation, particularly for many of the simple shapes of sources and detectors used in practice. Two alternative theoretical approaches have been used.

First, direct application of Huygens' principle to a distribution of elemental point sources of appropriate strength and phase covering a source vibrating with a continuous sinusoidal displacement leads to the Fresnel theory. Analytical solutions exist for the on-axis field variations, and for the field distribution far from the source, for a plane circular transducer (see Figure 6.12), and for other simple shapes such as a line source. For other field positions, and for more complicated transducer shapes, numerical integration of the contributions

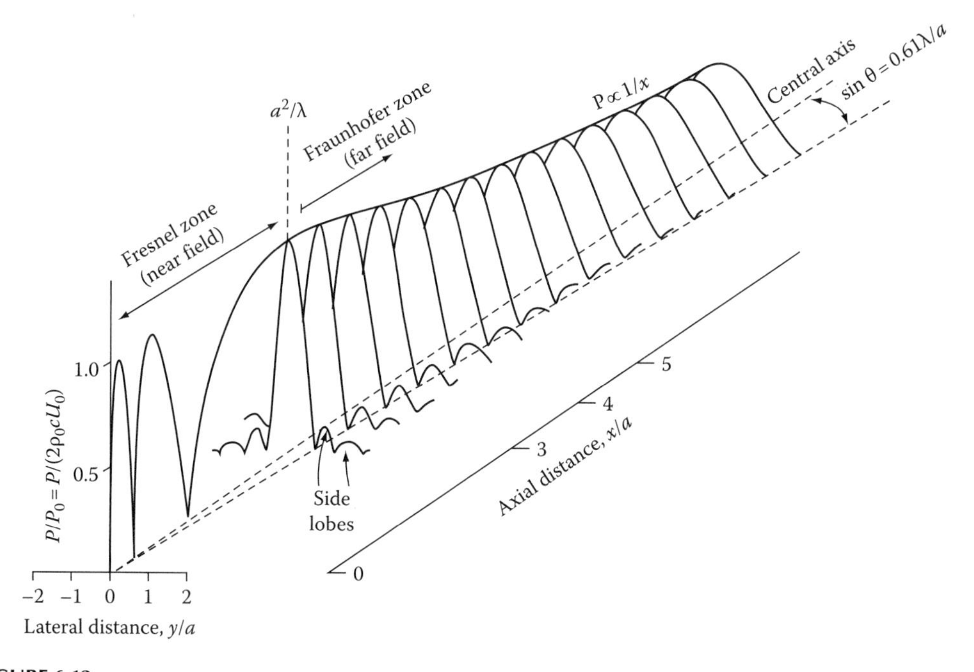

FIGURE 6.12
Fraunhofer and (on-axis) Fresnel solutions for the pressure amplitude in front of a plane circular source of radius a radiating a continuous sine wave.

from each element is necessary. Although the model of a continuous-wave (CW) plane circular source is not often applicable to realistic imaging situations, since both focusing and pulsed excitation are often employed, it does help to understand the general features of a radiation field pattern. One of the main features is the (somewhat arbitrary) division of the radiation field pattern into two regions. One is close to the transducer (often referred to as the Fresnel zone or 'near field'), where pronounced interference maxima and minima occur but where most of the energy is confined within a transducer radius of the central axis. The other is further away from the transducer (often called the Fraunhofer zone or 'far field'), where the wave field is more uniform but tends towards a spherically divergent wave whose amplitude is modulated as shown in the figure. The on-axis variation is such that, starting with a value of zero at infinity, as one moves closer to the source, alternate maxima and minima occur at distances x_m, given by:

$$x_m = \left(\frac{a^2}{m\lambda}\right) - \left(\frac{m\lambda}{4}\right) \tag{6.22}$$

where

a is the radius of the transducer, and
m is an odd integer for maxima and an even integer for minima.

The position of x_1, whose approximate value is a^2/λ for $a \gg \lambda$ (the so-called last axial maximum), is usually regarded as the boundary between the two zones. In the far field, a large proportion of the energy (about 84%) is contained within the main lobe, defined by the first off-axis minimum, which occurs at a divergence angle θ given by

$$\sin\theta \approx \frac{0.61\lambda}{a}. \tag{6.23}$$

The remaining energy in the far field resides within the 'side lobes' indicated in Figures 6.12 and 6.13. Note that the quantity x_1 has no physical meaning if $a < \lambda/2$, in which case $x_1 \leq 0$ and θ is 90°, that is, there is no near field and the transducer behaves much like a point source.

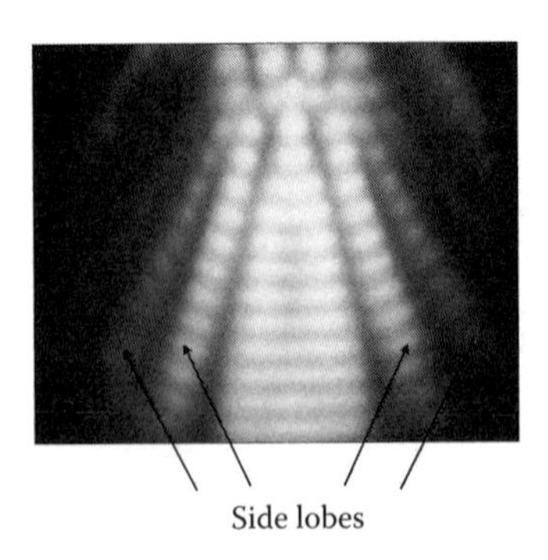

FIGURE 6.13
Schlieren image showing the field pattern from a CW single-element transducer. The banding is a consequence of the individual cycles of the wave that have been frozen in the stroboscopic image. (Courtesy of J. Weight, M. Restori and S. Leeman.)

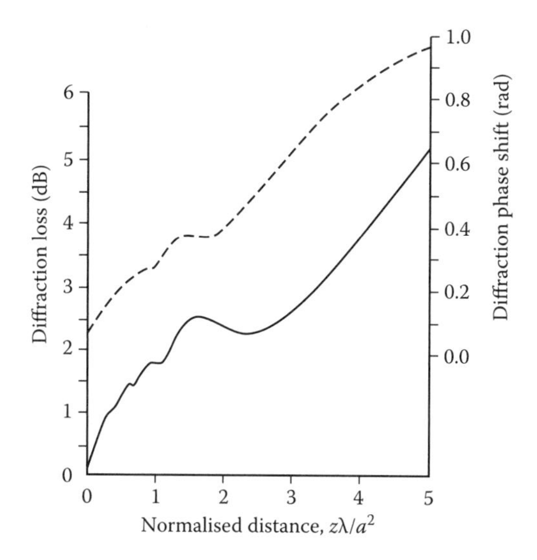

FIGURE 6.14
Diffraction loss (solid line) and phase shift (dashed line) from a circular source of radius a, seen by an identical receiver, as a function of the normalised distance between them. (After Bamber, 1986b.) Data from Seki et al. (1956) and Papadakis (1966) were used to plot the graphs shown here.

It was mentioned previously (see Section 6.2.1.5) that when measuring the acoustical characteristics of a medium (e.g. attenuation coefficient, backscattering coefficient, sound speed and non-linearity parameter), which are defined in terms of plane-wave propagation, a diffraction correction must be applied to obtain values that are independent of the source. For a single-frequency component, the magnitude and phase of the signal received in a lossless medium by a transducer identical to the transmitting transducer varies in a complex manner as a function of the transducer separation. An example of the diffraction loss and diffraction phase shift for a circular transducer is plotted in Figure 6.14. Numerical diffraction corrections may be applied to measurements of wave amplitude, echo amplitude or arrival time as a function of propagation distance, but it is difficult to obtain the required accuracy by calculating or measuring curves such as those shown in Figure 6.14. The approach that is often preferred is to make such measurements in a system designed so that either the medium to be measured is compared with a medium whose properties are known, or the propagation distance in the medium to be measured is arranged to vary without changing the position of the measurement in the diffraction field. This is an extensive and specialised topic that is dealt with elsewhere (e.g. Bamber 2004b).

An alternative to Huygens' interpretation is to adopt Young's method, in which the diffraction field is considered to be the result of superimposing just two waves, rather than the infinite number of spherical wavelets, ideally considered in the classical Fresnel summation. In this case, the two waves considered are (1) a wave emerging with identical spatial extent and phase as the radiating aperture and (2) a wave (which has a special phase variation) spreading out in all directions from the edge of the transducer. The diffraction field is then the result of interference of these two waves. For a plane circular transducer, the two waves are a plane wave and a hemi-toroidal edge wave, as illustrated in Figure 6.15. This approach helps to understand the nature of pulsed acoustic fields, calculated using the impulse response method described in Section 6.2.4.3.

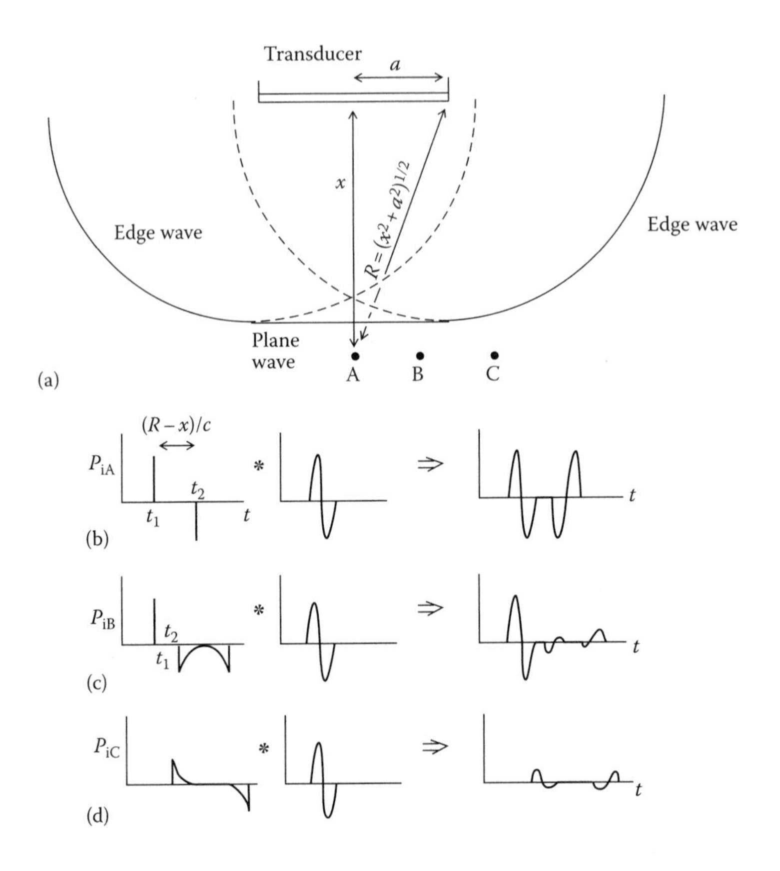

FIGURE 6.15
(a) Plane- and edge-wave components of the acoustic field generated by a plane circular source. The plane wave, and portions of the edge wave that are in phase with the plane wave, are shown as full curves. The broken portions of the edge wave indicate that the wave in this region is 180° out of phase with the plane wave. (b) through (d) The impulse response functions at field points A, B and C, respectively, and the corresponding waveforms resulting from convolution (represented by *) with a single-cycle sine-wave source driving function. This part of the figure is referred to by the text in Section 6.2.4.3.

Many factors can influence the ultrasonic field, and not all can easily be included in a calculation (e.g. very complex transducer shapes, non-uniform mounting and clamping of the piezoelectric element, electrode connection). It is therefore preferable, for a good understanding of the behaviour of a particular device, to be able to make direct measurements of the acoustic field distribution in space and time. There are many methods by which this may be done. The subject is beyond the intended scope of this chapter, but some of them are mentioned in Section 6.4.

6.2.4.2 Effect of Focusing

If the source is focused (as is usually the case in medical imaging), so that a spherically-curved wavefront (rather than the plane one discussed earlier) is launched, the acoustic field distribution is modified in a number of ways, some of which will now be described. The CW on-axis solution is modified such that the position of the last axial maximum of the equivalent plane source (sometimes referred to as the 'natural focus' of the aperture) is always moved to a position (called the true focus) where it is closer to the transducer than either the geometrical focus (given by the radius of curvature of the initial wavefront, R) or the distance x_1 ($=a^2/\lambda$). There is also an increase in the on-axis maximum pressure amplitude at the focus (a 'focusing gain'), accompanied by a decrease in the width of the sound beam at the same position. The shape of the field distribution off-axis, around the position of the focus and beyond, is similar to that in the far field of the plane transducer depicted in Figure 6.12. However, since the wavefront has been made to converge towards the focus, it diverges beyond the focus more rapidly than does the wavefront in the far field of the equivalent plane-wave radiator. This brings about the depth-of-field compromise (which can be overcome using multiple-element transducers, as described in Section 6.3.1.2), that is, the stronger the focusing, the narrower the beam width at the focus but the shorter the depth range over which a narrow beam is maintained. It is often convenient to talk about transducers in terms of their 'strength of focusing' (see Figure 6.16). Strong, medium and weak focusing are sometimes defined, respectively, as $R < 0.25x_1$, $0.25x_1 \leq R < 0.5x_1$ and $R \geq 0.5x_1$. Note that, even if R is made equal to x_1, the position of the true focus becomes $0.6x_1$, and the on-axis intensity at this point is four times as large as that at the natural focus of the equivalent plane source (Figure 6.17). This is sometimes quite a useful focusing design for a single-element transducer, since it results in a moderately narrow but fairly long, parallel beam around the focal region.

Another, often used measure of focusing strength is the 'f-number' of the aperture, n_f, given by the ratio of the radius of curvature to the diameter of the source, $R/2a$. In combination with the wavelength, the f-number provides a convenient method of assessing lateral resolution as defined in Section 6.2.4.4. The first off-axis minimum, at the true focal distance, occurs at a radial distance approximately equal to $0.6\lambda n_f$. About 84% of the total power radiated by the source is contained within the main peak defined by this radius.

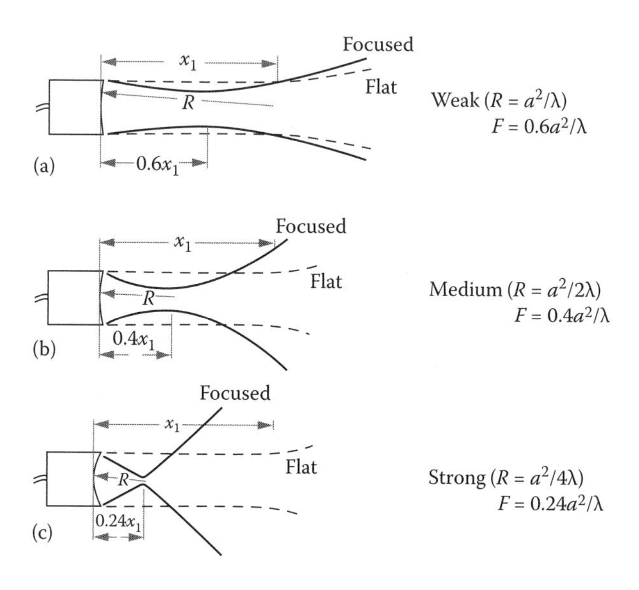

FIGURE 6.16
Beam shapes for (a) weak, (b) medium and (c) strong focusing. The dashed line shows the beam shape of an unfocused (flat) transducer of the same diameter and frequency. The depth at which the beam width is narrowest, known as the true focus, is given by F. (Adapted from Kossoff, 1978, with permission.)

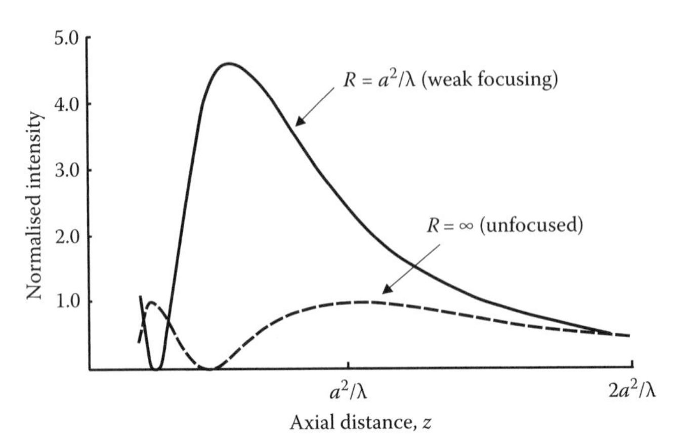

FIGURE 6.17
Axial profile of a weakly focused beam in comparison to that of the equivalent plane-wave (unfocused) source. (Adapted from Kossoff, 1979.)

6.2.4.3 Effect of Using Short Pulses

It is possible to compute the field distribution associated with a sound pulse emerging from a given aperture by adding all of the CW solutions, with weights given by the components of the frequency spectrum of the original pulse (i.e. an approach analogous to modulation transfer function (MTF) characterisation of imaging systems). The impulse response method, however, is more suitable for producing results that permit direct inspection of the temporal shape of the pulse waveform at each point in the field (directly analogous to the point-spread function (PSF) method of characterising imaging systems). Also, it is a two-stage calculation such that, once the 4D pressure impulse response h' of the source has been calculated, one can obtain the acoustic field distribution for any source velocity–time waveform by a convolution:

$$p(x,y,z,t) = \rho_0 u(t) * h'(x,y,z,t). \tag{6.24}$$

The pressure impulse response may be derived from a graphical solution, in which the instantaneous response at a given point (at position x, y, z) is given by the time differential of the value obtained from a surface integration of those source elements which lie on an arc of equal phase, for the situation of a source that undergoes a theoretical unit velocity impulse, that is, a step in position. At any on-axis point (A in Figure 6.15a) the plane-wave impulse will arrive first, followed by the simultaneous arrival of all parts of the antiphase portion of the edge wave. This is depicted in Figure 6.15b. The time separation ($t_2 - t_1$) is given by the difference in acoustic paths from the centre and edges of the transducer ($(R-x)/c$). The rest of Figure 6.15b shows the final pulse waveform at point A, resulting from the convolution of the impulse response with a single-cycle sine wave. Figure 6.15c and d illustrates the same sequence of operations for field points B and C, which are off-axis but inside and outside the geometrical beam, respectively.

From Figure 6.15, it can be seen that very complex pulse shapes can occur, particularly close to the source and off-axis, where (if the driving pulse is short) relative path differences may be sufficient for several pulses to be received when only one was sent out. In practice, however, such pulse splitting is difficult to observe unless a receiver with extremely good time resolution is used. Large changes in pulse structure occur moving down the axis and off-axis, corresponding to the summation of CW solutions for all the

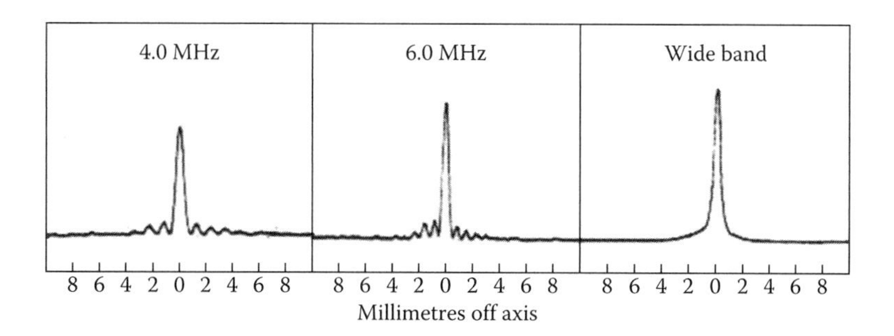

FIGURE 6.18
Lateral beam profiles, measured in the far field of a plane disc-shaped transducer, to show the considerable smoothing resulting from a short (i.e. wide band) pulse (right) when compared with a CW (left or middle, at the frequencies shown). (Courtesy of J. Weight.)

frequency components present in the original pulse; that is, high frequencies are concentrated near the central axis and extend further down the axis before diverging, whereas low-frequency components spread off-axis much sooner. When the same transducer is used to detect the echoes from a point scatterer, similar (reciprocal) properties apply to the time sensitivity of the transducer as a receiver. The result is to introduce still more complexity into the received pulse shape.

It is also possible to explain features of the CW field of Figure 6.12 using the impulse response plus edge-wave descriptions. Take, for example, the response at the on-axis point A (Figure 6.15b). At $x = \infty$ the plane wave and antiphase edge wave overlap and exactly cancel each other. As we move closer to the source, these components gradually separate until, at a distance corresponding to $R - x = \lambda/2$ (in Figure 6.15a), they are a half-wavelength apart and constructively interfere. At closer distances, the other on-axis minima and maxima discussed earlier can be predicted to occur.

Although the diffraction field of the aperture has a complicated effect on the structure of the pulse as it propagates, the use of shorter pulses has a simplifying effect on the overall structure of the sound beam, if it is assessed in terms of some single field parameter such as the peak pressure amplitude within the pulse. This effect can be thought of as being the result of averaging the CW field distributions for each frequency component, which produces considerable smoothing of the far-field side lobes (see Figure 6.18) and the near-field interference structure (not shown). The position of the last axial maximum also becomes blurred, so that it is no longer easy to define a boundary between the near field and the far field. Note, however, that the precise distribution observed depends on the pulse parameter used as a measure of the acoustic field at each point (i.e. peak positive pressure, peak negative pressure, integrated energy in the pulse, etc.).

6.2.4.4 Resolution

There can be no absolute simple definition of the resolution of ultrasonic imaging systems, since in practice too many variables affect the displayed resolution. Even for a particular transducer, there are focusing schemes other than the spherically curved wavefronts discussed earlier. Thus, a resolution measure such as the full width at half-maximum (FWHM), which is commonly used in medical imaging systems (see Chapter 11), will produce a result that depends on the shape of the response function, which may include side lobes, either in a narrow-band system or when they are smoothed due to the use of a short pulse, as mentioned in Sections 6.2.4.1 and 6.2.4.3. Nevertheless, such variations are often ignored. The point response function of ultrasonic imaging systems tends to be highly asymmetrical, the equivalent length of the sound pulse usually being smaller than the beam width. Hence, it is common to talk separately about axial and lateral resolution of pulse-echo systems:

Axial resolution = Half the length of the pulse envelope at some defined level below the peak, multiplied by the sound speed

Lateral resolution = Full width of the beam at some defined level below the maximum

It is also common to use the FWHM criterion, but many other levels are in use, for example, 3, 6 and 10 dB. Note that FWHM refers simply to the received signal, whatever it may be (usually the amplitude of the echo from some small point-like reflector), whereas the method of calculation of the decibel levels depends on whether the distribution is in intensity or in amplitude. This can be a source of confusion if one does not state precisely how the measure of resolution has been calculated.

Pulse length is inversely proportional to pulse bandwidth; hence transducer bandwidth (see Section 6.3.1.1) determines axial resolution. With the assumption that the speed of sound is 1540 m s^{-1}, and using the FWHM resolution criterion, for a pulse that has a Gaussian-shaped envelope with half-power bandwidth Δf (in MHz), the aforementioned definition of axial resolution in mm becomes (Greenleaf 1997)

$$\text{FWHM}_{\text{axial}}\,(\text{mm}) = \frac{1.37}{\Delta f}. \tag{6.25}$$

The lateral resolution is determined by the transducer aperture used to focus the wave. For the spherical focusing system described previously, the lateral resolution at the focus, defined by the FWHM of the intensity distribution (i.e. the 3 dB level), is given by (Fry and Dunn 1962)

$$\text{FWHM}_{\text{sphere}} \approx 1.2\lambda n_{\text{f}}. \tag{6.26}$$

Since the echo amplitude distribution of a transducer is effectively proportional to the square of the one-way pressure amplitude distribution, Equation 6.26 also provides an effective measure of the lateral resolution defined in terms of pulse-echo amplitude.

On-axis, the depth range d_x over which the intensity remains within 3 dB of the maximum is given approximately by

$$d_x \approx 15(1 - 0.01\phi)\,\text{FWHM} \tag{6.27}$$

where $\phi = \sin^{-1}(a/R)$, expressed in degrees. Although d_x does not provide a measure of the change in beam width with depth, it does provide a helpful relative measure of depth of focus.

A very important aspect of resolution in ultrasonic systems is that (at least for the simple beam-forming systems so far discussed) the axial and lateral resolutions are highly spatially variant. The component of this variation due to the diffraction field of the source aperture is present even in a homogeneous, lossless medium. In tissues, additional beam divergence and deviation occur due to scattering phenomena, which include refraction and wavefront aberration due to large- and small-scale variations in sound speed. Occasionally, such effects may cause considerable image degradation as, for example, when imaging across the abdominal wall in the presence of overlying fat and muscle. However, the major distorting effect is often due to frequency-dependent attenuation. This is present for the majority of imaging paths, even when aberrations due to sound speed fluctuation are absent. Selective attenuation of the higher frequencies causes the beam to widen and the pulse to lengthen. Therefore, axial resolution, lateral resolution and SNR all deteriorate with depth. The situation is a complicated one, being influenced by size of aperture, focal length, tissue properties, depth in tissue, initial pulse bandwidth and the mismatch of the speed of sound between the body and any coupling medium. As an example of an empirical result, a 5 MHz transducer with a 6 dB bandwidth extending from 2 to 6 MHz, a focus at 10 cm and an aperture of 2 cm produces a FWHM (at the focus) of 3 mm in water, but 7 mm after propagating 5 cm in breast tissue (see Foster and Hunt 1979). Typically, for such distances of propagation, one observes a degradation in lateral resolution, compared to that observed in water, of a factor of 2.3 for breast and 1.3 for liver.

6.2.4.5 Effect of Non-Linear Propagation

There are a multitude of practical consequences of non-linear propagation, including increased absorption of the wave energy (due to harmonic generation), greater spatial variation of the PSF, and increased dependence of the beam shape both on the field parameter observed (see Section 6.2.1.2) and on the initial pressure amplitude at the source. The effect of non-linear propagation on a beam pattern is illustrated in Figure 6.19.

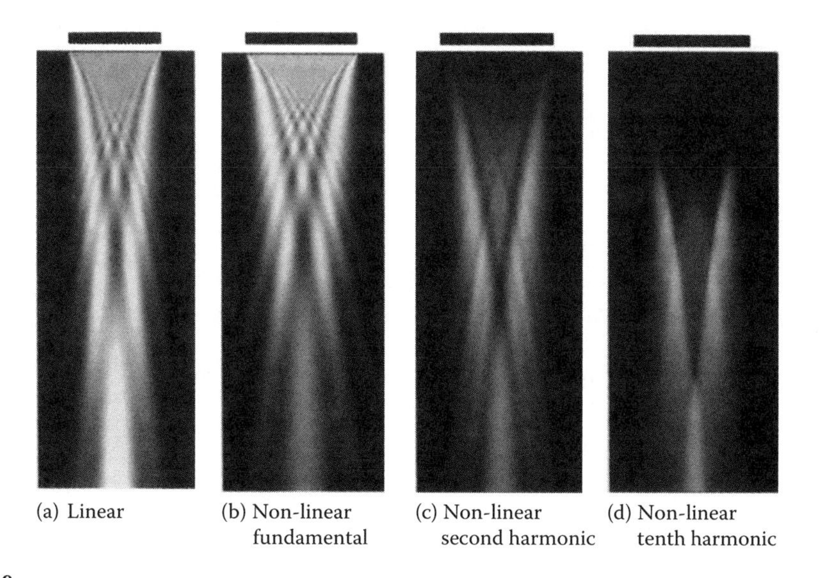

FIGURE 6.19
Fundamental and harmonic field patterns calculated across the diagonal of a 2.25 MHz CW square source (dimension 20 mm), shown to a depth of 200 mm. (a) Low amplitude case, (b) through (d) the fundamental, second-harmonic and tenth-harmonic field patterns for a source pressure of 1 MPa. (Reproduced with permission from Baker, 1998.)

Calculations have been carried out for a square source aperture as if measurements of the pressure field had been made in the plane across the diagonal of the source, because this emphasised diffraction effects for illustrative purposes. It can be seen that for linear propagation (a), the peak amplitude occurs on the acoustic axis as expected (at the bottom of the image) but for the non-linear cases (b) through (d), the peak amplitude occurs in off-axis near-field maxima. Furthermore, at the fundamental frequency, the maxima and minima in the near field occur earlier, and the on-axis average pressure drops sooner, for the non-linear (b) than for the linear (a) case. Finally, the higher the harmonic the longer it takes for the field to build up, the finer the fringe pattern is in the near field and the narrower the main lobe is once it has emerged (c and d). Note that Figure 6.19 represents the case of a CW unfocused source. The non-linear propagation patterns due to more complex (diagnostic) transducers can be difficult to characterise, largely due to the cumulative nature of non-linear effects. Numerical methods have been developed for this purpose (see Duck 2002 for a brief summary) and these may be useful for predicting non-linear effects *in vivo*, which is important for assessing ultrasound safety (see Section 6.5). The non-linear interaction of ultrasound with tissue also has a number of benefits for imaging, as will be described in Section 6.3.3.

6.2.5 Physical Principles and Theory of Image Generation

Chapter 11 covers the subject of formal mathematics of imaging in some detail. This section is primarily for comparison purposes and serves to highlight those aspects of ultrasound imaging which distinguish it from other imaging methods.

6.2.5.1 Pulse-Echo Scanning

The reflected waveform is a convolution of the wave field incident on the tissue structure being imaged and an impulse response associated with the scattering properties of the tissue:

$$I(x,y,z) = h(x,y,z) * T(x,y,z) \tag{6.28}$$

where

$I(x, y, z)$ is the radio-frequency (RF) image (before envelope detection),
$h(x, y, z)$ is the 3D PSF of the ultrasound system, and
$T(x, y, z)$ is the backscattering impulse response of the tissue.

Although the PSF is not separable, it is often considered as such:

$$h(x,y,z) = h_1(x) * h_2(y) * h_3(z) \tag{6.29}$$

where

$h_1(x)$ is the pulse-echo impulse response (beam profile) within the scan plane,
$h_2(y)$ is the axial pulse-echo impulse response of the system (RF pulse shape), and
$h_3(z)$ is the beam profile across the width of the imaged slice.

A considerable amount of signal processing may be applied to $I(x, y, z)$ before it is displayed (as we shall see in Section 6.3.2). Major differences with respect to other areas of imaging occur in that:

(i) the processing is non-linear;
(ii) the PSF is spatially variant (radiation field pattern);
(iii) the PSF depends on the object (speed, attenuation and scattering);
(iv) the radiation is coherent and the detector is phase sensitive (results in interference between different parts of the image);
(v) the PSF has, and therefore so will the image have, negative contributions;
(vi) the PSF is not circularly symmetric; and
(vii) the PSF is not separable (especially in the near field), although it is often assumed to be, as in Equation 6.29.

Equation 6.28 makes many assumptions, for example uniform speed of sound, no attenuation, no multiple scattering, no system noise and a spatially invariant PSF.

For the sake of simplicity, Equation 6.28 will be reduced to 2D hereafter, so as to consider just the 2D plane of the ultrasound image:

$$I(x,y) = h(x,y) * T(x,y). \tag{6.30}$$

If $T(x, y)$ is an impulse (i.e. a single point scatterer), then the image is the PSF of the system. It is instructive to use this situation to see how the different stages of convolution, followed by envelope detection, lead to the image of a point (see Figure 6.20). More usually, we have distributed or extended targets. As more scatterers are introduced, and the same process is repeated, it becomes easy to see how interference occurs and how the speckle patterns mentioned in Section 6.2.5.2 might build up (Figure 6.21).

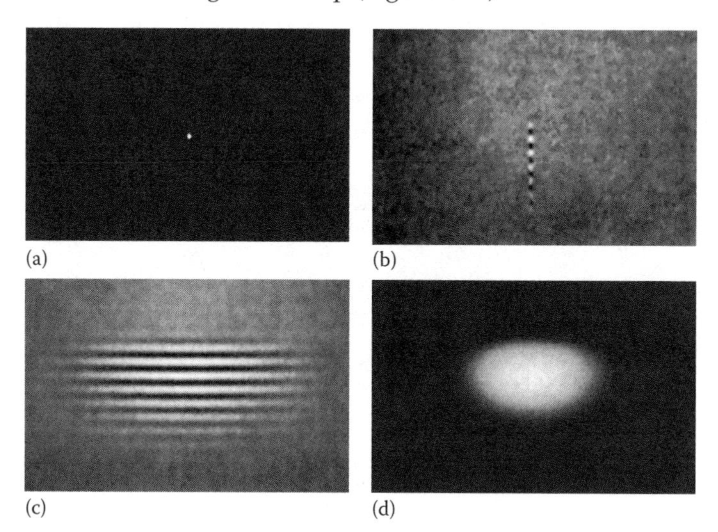

FIGURE 6.20
Use of a digital simulation to illustrate the two stages of convolution, first with the RF pulse (b) and then with the beam profile (c), followed by envelope detection, leading to an image (d) of a single point (a). (From Bamber and Dickinson, 1980.)

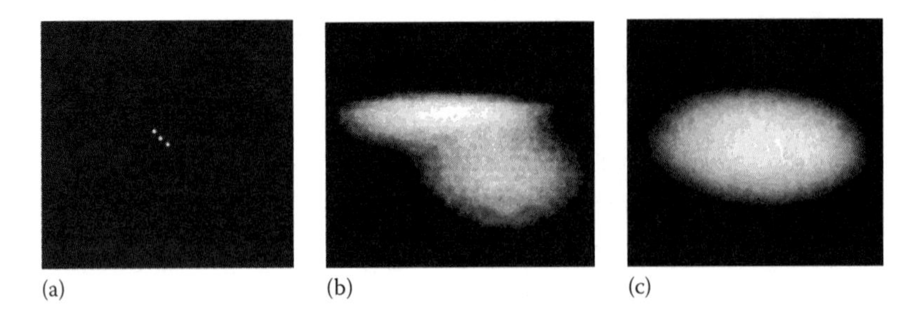

FIGURE 6.21
Use of the same imaging model as in Figure 6.20 to illustrate the generation of artefactual detail, or speckle, (b) from coherent interference between waves from closely spaced scatterers (a). Image (c) simulates the blurring produced by an incoherent imaging system with the same PSF. It shows no artefactual fine detail.

The most difficult part in this model of ultrasonic image formation is that no one yet knows exactly what the tissue backscattering impulse response should look like. Various models exist, but none has proven to be generally applicable. For example, it is popular (though not realistic) to model tissue as a random distribution of point scatterers of a given number per unit volume. Another alternative (which is more instructive) has been to consider that the density and bulk elastic modulus vary continuously in the object. Then one (of many) expression that has been derived for the backscattering impulse response (in the 2D approximation) is:

$$T(x,y) = \frac{1}{4}\frac{\partial^2}{\partial x^2}\left(\frac{\delta\rho(x,y)}{\rho_0} + \frac{\delta K(x,y)}{K_0}\right) \tag{6.31}$$

where
$\delta\rho$ and δK represent small fluctuations in density and bulk elastic modulus about their mean values, and
x is the direction of axial pulse propagation.

The main point to note is that the scattering results from the second spatial derivative of density and bulk elastic modulus fluctuations.

6.2.5.2 *Speckle*

For any uniform region of an object in which many scattering sources exist within a resolution cell (defined by the PSF), the image will have a magnitude that varies apparently randomly with position, due to constructive and destructive interference of the waves scattered from each source (Figure 6.22). This leads to the characteristic speckled effect in ultrasound images and contributes a form of visual noise or clutter, as the spatial frequency of the noise can be similar to that of the image features of interest. Even when electronic noise is low, speckle limits the SNR of the envelope-detected image to a value of approximately 1.9. It also reduces contrast discrimination and effective resolution. The size and shape of the speckle cells are roughly equivalent to the size and shape of the resolution cell. Therefore, smaller speckle cells would be achieved by using shorter pulses, wider bandwidths and narrower beams.

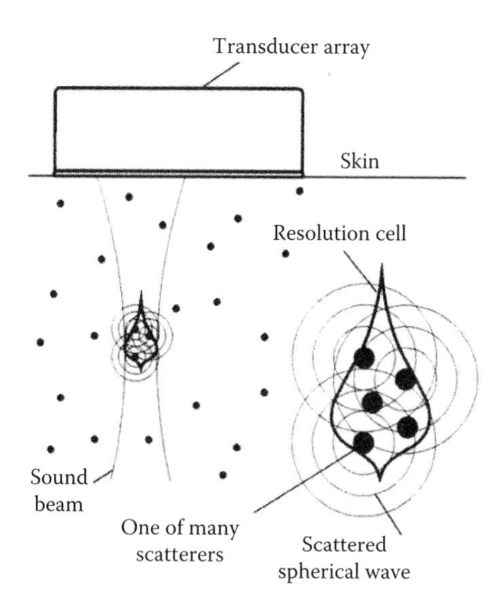

FIGURE 6.22
Mechanism of speckle production by interference between waves received from scatterers at multiple locations with the resolution cell (the volume encompassed by the axial, lateral and elevational resolution distances).

6.2.5.3 Speckle Reduction

Commercial diagnostic ultrasound machines often invoke one or more techniques for reducing speckle. Methods based on a technique known as *compounding* average multiple amplitude demodulated (envelope detected) images of the same object, obtained under different imaging conditions, such that the speckle pattern within uniform regions of the object is different for each image ('decorrelated') but the anatomical detail remains similar across the component images. The result is an averaging of the speckle pattern and a consequent improvement of signal (anatomical structure) to noise (speckle) ratio (SNR) that would be proportional to the square root of the number of images if their speckle patterns were to be completely independent, which usually they are not, and if the structure were to remain the same, which usually it does not. The decorrelated images may be obtained by varying either frequency (*frequency compounding*), relative orientation (*angle compounding*), position (volume projection), or, if the object is moving, time (persistence). Unfortunately, all of these methods lead to some loss of spatial resolution, as well as a reduction in frame rate due to the acquisition of multiple frames. Nevertheless, compounding is a powerful method for image quality improvement and is in widespread use throughout medical ultrasound. All ultrasound imaging involves working with a limited acoustic aperture for forming the sound beam and a limited acoustic bandwidth for generating the acoustic pulse. Frequency compounding involves a deliberate choice to sum a series of subimages, each created with a pulse that has a bandwidth narrower than the available bandwidth. Angle compounding involves a similar choice, to sum a series of subimages each created with an aperture that is smaller than the available aperture. As was noted in Section 6.2.4.4, axial resolution is poorer the narrower the bandwidth, and lateral resolution is poorer the smaller the aperture. In practice, therefore, there is a direct trade-off between SNR improvement from speckle reduction and loss of spatial resolution, and it is desirable for the user of an ultrasound scanner to be able to switch compounding on or off, depending on whether an image optimised for contrast resolution or spatial resolution is desired. Compounding is discussed further in Section 6.3.2.12.

The averaging over position or time mentioned in the previous paragraph is equivalent to filtering of the envelope-detected images. Simple filtering methods (e.g. median filters, Weiner filters) have been proposed, but these suffer from a loss of spatial resolution. A better approach is to use adaptive filtering, whereby the statistical properties of regions of the image are examined and regions are then classified as either mostly signal or mostly noise (speckle). The algorithm then smoothes the image, spatially or temporally, but only in the regions identified as speckle noise (Bamber and Daft 1986). Another type of adaptive method involves detecting regions of explicitly destructive interference, where the phase of the RF echo pattern passes through a discontinuity, and replacing these regions with estimates from surrounding amplitude values (Healey and Leeman 1991). Considerable further development of adaptive speckle reduction algorithms has taken place and they now exist, in various forms, in most commercial ultrasound scanners. In addition to improving contrast resolution, speckle reduction could prove useful for enabling computer-aided image interpretation (e.g. boundary recognition). A more extensive discussion of speckle reduction may be found in Bamber (1993).

6.3 Engineering Principles of Ultrasonic Imaging

In the previous sections it has been shown that there is a range of mechanisms for the interaction of ultrasound with tissues and that this information is encoded on echo signals in a rich but complex manner. The signal and image processing of ultrasound instruments should be designed both to cope with and to take advantage of this richness, for example, by separating different kinds of information, compensating for the interaction between variables, reducing noise and artefacts, and displaying the information such that it is well matched to the human observer's perceptual abilities. Current conventional processing is not ideal, leaving much scope for future development.

The stages of processing in an ultrasonic pulse-echo imaging machine are presented in Figure 6.23. Transducer arrays (described later) allow electronic scanning, focusing and steering of the ultrasound beam, controlled by the transmit beam former, which separately excites each array element within a set. They also give rise to a parallel set of echo signal trains corresponding to a set of array elements. After preamplification and other preprocessing of these signals in parallel-receive channels they will be combined in the process of receive beam-line formation. If a single-element transducer is used rather than an array, both the transmit and the receive beam formers will be missing because their functions are performed by the transducer, which may be focused as described in Section 6.3.1.1. Single-beam-line processing is present in all systems and is, therefore, described in detail in the following (see Section 6.3.2). In some systems, this may be followed by a stage of processing involving multiple-beam-line data sets (e.g. compounding, Section 6.3.2.12). Finally, there will be the stage of scan conversion, which may be preceded or followed by processing to manipulate the image data.

6.3.1 Generation and Reception of Ultrasound: Transducers

The transducer is probably the single-most important component in an ultrasonic imaging system. Its function is to convert applied electrical signals to pressure waves, which propagate through the medium, and to generate an electrical replica of any received acoustic waveform. A well-designed transducer will do this with high fidelity, with good conversion efficiency and with little introduction of noise or other artefacts. Also, it is primarily through transducer design that one has control over the system resolution and its spatial variation.

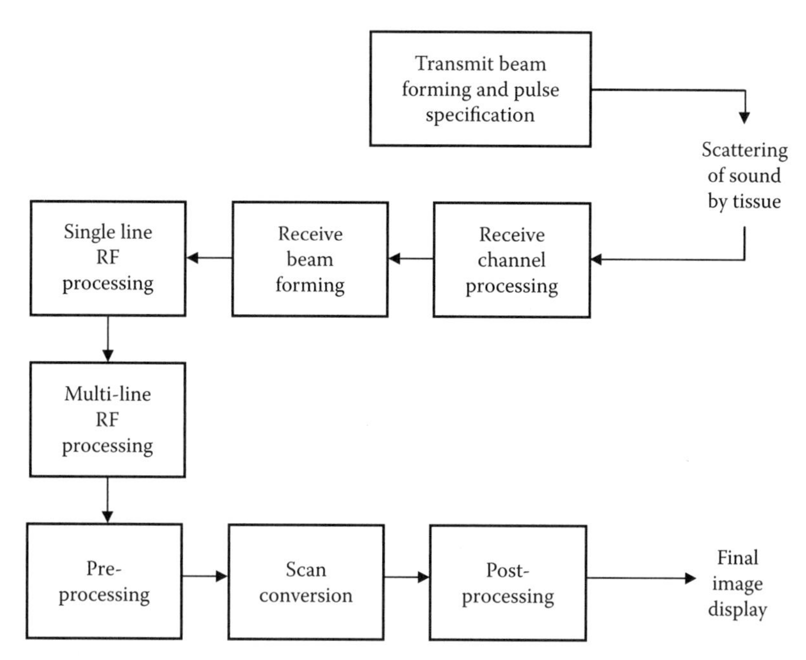

FIGURE 6.23
Stages of processing in ultrasonic image formation. (Adapted from Bamber, 1999a.)

6.3.1.1 Conventional Construction (Single Element)

A good-quality transducer will possess the following design features, as illustrated in Figure 6.24.

6.3.1.1.1 Piezoelectric Element

This is cut and shaped from a piezoelectric ceramic (usually lead zirconate titanate, PZT) or plastic (polyvinylidine difluoride, PVDF). Silver electrodes are deposited on the front and back faces, and the element is permanently polarised across its thickness. It then has the property that any voltage applied across the electrodes produces a proportional

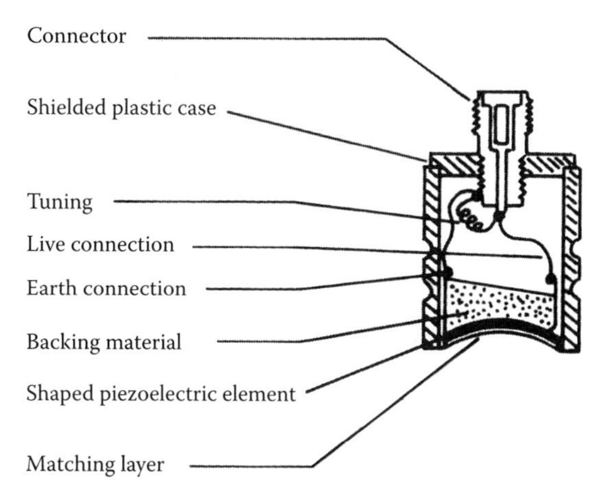

FIGURE 6.24
Typical components in the design of a conventional single-element ultrasonic transducer.

change in thickness and, conversely, pressure applied across its two faces produces a potential difference between the electrodes. The speed of sound in PZT is approximately 4000 m s^{-1}, which gives a fundamental resonance ($\lambda/2$) at frequency f and thickness T related by

$$T\,(\mathrm{mm}) \approx \frac{2}{f(\mathrm{MHz})}. \tag{6.32}$$

For example, at 5 MHz, $T \approx 0.4$ mm.

The characteristic acoustic impedance Z for PZT is approximately 14 times that for water and soft tissue, that is, the equivalent reflection coefficient (calculated from Equation 6.20 as $10 \log_{10} R$), $R_{PZT} \approx -1$ dB, relative to $R_{perfect}$. On the other hand, for PVDF, Z is only about 1.5 times that of water, so that $R_{PVDF} \approx -14$ dB. Thus, PVDF couples its energy to tissue much better than does PZT, and correspondingly has a lower mechanical Q (resonance factor). This makes it a material with a much wider bandwidth and a flatter frequency response than PZT, but it also suffers from a somewhat poorer conversion efficiency and, thus, lower sensitivity. More recently, composite materials have been developed. For example, ceramic elements can be embedded in a polymer matrix to achieve overall acoustic impedance that is lower than that of PZT. There is also a move towards using large single pieces of piezoelectric material (so-called single crystals), for which the degree of polarisation and, hence, conversion efficiency achieved is higher than for ceramics, which are made from sintered powders. Finally, an older technology is now returning, that of capacitive transducers that use electrostatic forces rather than the piezoelectric effect. These may be manufactured as small elements using micromachining techniques on silicon substrates, and possess good conversion efficiency as well as very wide bandwidth and flat frequency response.

Focusing may be applied by a number of methods: (1) a shaped concave (bowl) ceramic as shown in the Figure 6.24; (2) a lens (concave for materials such as epoxy or Perspex; or convex for silicone rubber) plus a plane ceramic disc; (3) a bowl plus a defocusing lens; and (4) overlapping beams from two elements (as in CW Doppler systems).

6.3.1.1.2 Matching Layer

This attempts to overcome the aforementioned acoustic mismatch between the element and the tissue, to increase efficiency of energy transfer. The theory is only strictly applicable to continuous waves, in which case the ideal impedance and thickness of the layer are, respectively,

$$Z_{\mathrm{matching}} = (Z_{\mathrm{element}} \times Z_{\mathrm{tissue}})^{1/2}, \quad \text{and} \quad T_{\mathrm{matching}} = \frac{\lambda}{4}. \tag{6.33}$$

A good example is aluminium powder in Araldite, which has $T \approx 0.14$ mm for the above 5 MHz element. A matching layer is not required for PVDF elements.

6.3.1.1.3 Backing Medium

This is usually required for mechanical support, but should be minimal (preferably air only) if maximum efficiency (high Q) is desired. For pulse-echo imaging, the Q factor needs to be reduced to a value appropriate for short pulses ($Q \approx 1$–3). Both electrical and mechanical

Q contribute to the overall Q. Mechanical damping is partly provided by the $\lambda/4$ layer but also by choosing $Z_{backing}$ as close to $Z_{element}$ as possible. This involves a compromise between wide bandwidth and sensitivity. Typically, tungsten powder in epoxy resin produces a $Z_{backing} \approx ½Z_{element}$. The energy that enters the backing should be absorbed and not reflected back into the element. To accomplish this, plasticised epoxy, embedded fine scatterers and shaped backs have all been used. Note that PVDF has a naturally low mechanical Q and does not generally need special backing to improve the bandwidth.

6.3.1.1.4 Casing

This should be electrically screened and acoustically decoupled from the element (otherwise the dynamic range is reduced due to either acoustic ringing or electrical interference). This either means a plastic case with a screening layer, or a metal case with an acoustic insulator.

6.3.1.1.5 Electrical Tuning

This is often used to filter out low-frequency radial-mode vibrations of the element, and to manipulate the electrical Q for the best compromise between sensitivity and resolution. The capacitance of the transducer element C_t is given by the 'parallel-plate' formula:

$$C_t = \frac{\varepsilon A_t}{T} \tag{6.34}$$

where
 ε is the dielectric constant of the element, and
 A_t is the element area.

In short-pulse applications, a single shunt inductance (as shown in Figure 6.24),

$$L \approx \frac{1}{(2\pi f)^2 C_t}$$

is sometimes used to reduce the electrical Q, but again at the expense of sensitivity. Expressing Equation 6.34 in terms of c and f, for thickness $T = \lambda/2$, and using C_t to calculate the reactance $\chi_t = \omega L$, we obtain

$$\chi_t = \frac{c}{\varepsilon(2\pi f r_t)^2} \tag{6.35}$$

where r_t is the radius of a plane disc $(=(A_t/\pi)^{1/2})$. For PZT-5A, f in MHz and r_t in mm:

$$\chi_t \approx \frac{6.4\times10^3}{(r_t f)^2}. \tag{6.36}$$

For example, if $f = 5$ MHz and $r_t = 10$ mm, then $\chi_t \approx 3\ \Omega$, which is a bad match to the transmitting and receiving electronics if it uses the usual transmission impedances of 50 or 75 Ω. Generally, at above 3–4 MHz, transformer matching will improve this electrical coupling, although with some loss of bandwidth.

6.3.1.2 Multiple-Element Transducers

Single-element transducers, as described earlier, are not often used in modern scanning equipment, although the other basic aspects of design still apply to multiple-element systems. More than one element may be required (in the simplest case) to permit the use of continuous waves (as in a Doppler system), where separate transmitting and receiving elements are required. In pulse-echo imaging, multiple element arrays may be used for electronic (and rapidly changing) beam focusing, translation and steering. In this context, there exists a great variety of array and element shapes and corresponding field patterns.

The general principles of transmit beam forming for focusing, scanning and steering are illustrated schematically, and in two dimensions only, in Figure 6.25. By simultaneously exciting a group of tiny elements, one can, if each element behaves like a Huygens' source, synthesise a plane wavefront, which emerges from the aperture formed by the spatial extent of that group of elements (Figure 6.25a). The sound beam may then be translated from position 1 to position 2 by exciting a different, but overlapping, group of elements.

In order to focus (Figure 6.25c) or steer (Figure 6.25b) the sound beam, one must be able to excite each element via a variable delay. Systems using this technology to create a sector-shaped image by 'phase steering' the sound beam, as described in this section, have come to be known as 'phased-array' systems, the term (and indeed the method) being adopted from the world of radar.

Receive beam formation involves selecting a group of array elements that will form the desired acoustic receiving aperture, applying a variable delay to the signal received on

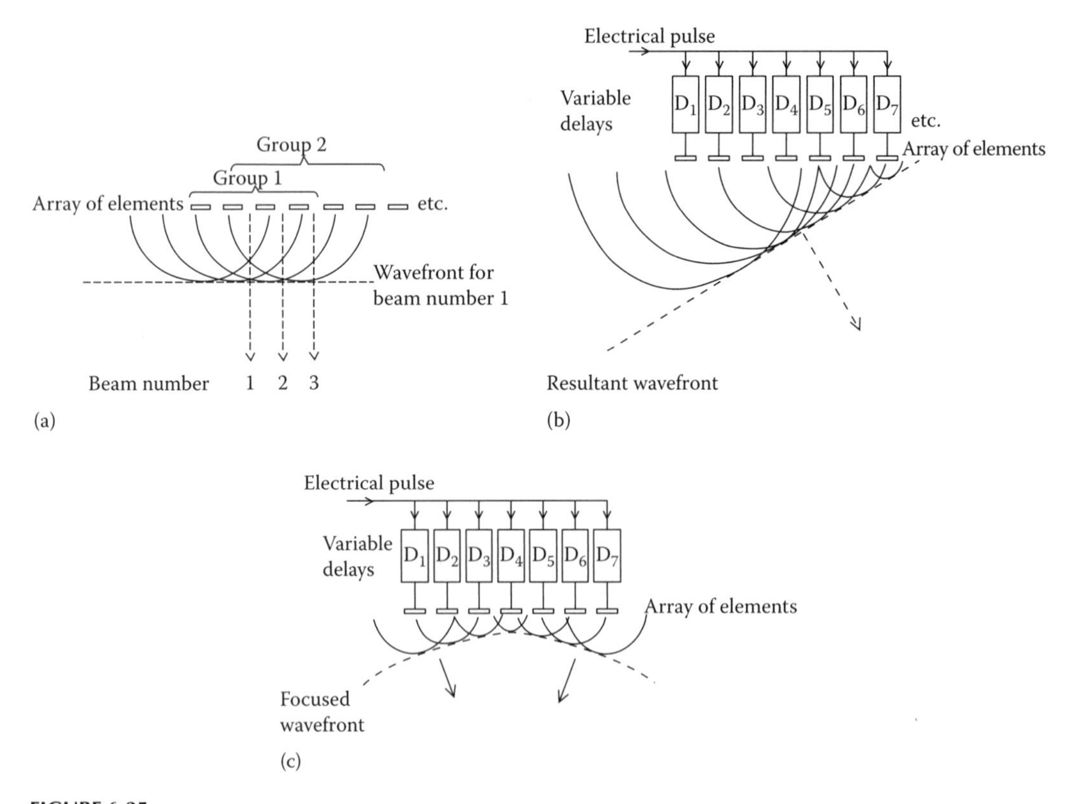

FIGURE 6.25
Schematic, 2D illustration of the principles of beam forming (a through c), focusing (c), lateral scanning (a) and steering (b), using arrays of elemental sound sources.

each element, and summing the signals over all elements in the group. This will form a single receive beam-line. Use of the delays that, on transmit, would have produced a beam steered to a given angle and focused to a given distance (as shown in Figures 6.25b and 6.25c) will generate a receiver that is maximally sensitive to echoes arriving from that steering direction and focal distance. In fact, within the time taken for a complete sequence of echoes to return (as a result of a single transmitted sound pulse), it is possible to adjust the focusing delays continuously so that the system has a receiving directivity pattern that is maximally sensitive to, and therefore focused on, each echo position as it arrives from each and every depth. This approach is known as 'swept (or dynamic) focusing' and, clearly, can only be applied to the received signal. Changes in the focal properties of the transmitted beam can only be made over successive sound pulses, by subjecting each pulse to a different combination of delays (i.e. each transmitted pulse is focused at a different depth). The use of multiple transmit foci therefore results in a reduced rate of data capture (see Section 6.3.2.13) whereas the use of dynamic receive focusing does not.

Generally, one finds combinations of these approaches: focusing plus scanning, focusing plus steering, or all three (where the purpose of lateral scanning is to produce an image, while the purpose of beam steering in addition to lateral scanning is to obtain images at different orientations, e.g. for speckle reduction). To summarise, electronic beam formation using an array is a very powerful method of creating ultrasonic images, providing:

(i) beam translation and steering without physical motion of the transducer;
(ii) variable transmit focusing and dynamic receive focusing;
(iii) arbitrary sequencing of beam-line positions and directions;
(iv) arbitrary sequencing of frames;
(v) easy digital control; and
(vi) possibility of parallel processing, which will reduce the acquisition time and which may be used, for example, to reduce speckle by compounding, increase the number of different types of information that may be simultaneously acquired and displayed, or to increase SNR.

The arbitrary sequencing mentioned in (iii) and (iv) may be used to interleave the acquisition of different types of information (such as B-mode with M-mode, pulsed Doppler or colour Doppler – see following sections) and place them onto one display. The parallel processing in (vi) can be achieved, for example, by transmitting a broad ultrasound beam and using multiple simultaneous receive beam formers, each of which creates a single image line that lies within the geometrical limits of the transmitted beam. In this way multiple image lines may be formed with each transmitted pulse, overcoming the line- and frame-rate limitations imposed by the speed of sound when serial processing of image lines is used (see Sections 6.3.2.2 and 6.3.2.13). The logical eventual extension of this is to form a complete image frame (or volume) with a single ultrasound pulse, using receive beam forming only to create a focused image. This approach is discussed towards the end of the chapter in Sections 6.7.1 and 6.7.2.

The disadvantages of electronic beam formation using an array, which might be overcome in the future, are:

(i) sampling problems (grating lobes, quantisation errors – dealt with later in this section); and
(ii) system cost and complexity.

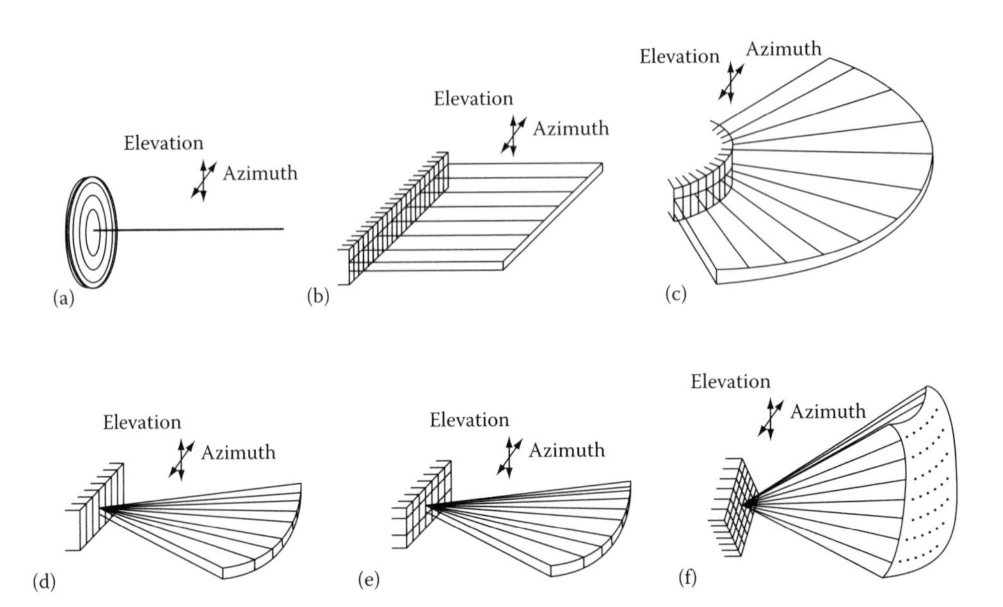

FIGURE 6.26
Six transducer array configurations currently in use: (a) annular array (often prefocused by shaped ceramic or a lens, so that the maximum electronic delay between elements is minimised), (b) linear array (shown with a fixed focus in the plane orthogonal to the scan plane), (c) curvilinear array (a linear array laid down on a curved surface to produce a sector-type image), (d) a phased array (whereby the sector-type image is produced using phase-controlled beam steering) and (e) a 1.5D phased array and (f) a 2D (matrix) phased array. (Schematics (b) through (f) are after Goldberg et al., 2006.)

Typical array configurations are shown in Figure 6.26. The annular array (Figure 6.26a) permits the use of electronic focusing (see also the following paragraph) to improve depth of focus over that of a single-element circular transducer. However, as with single-element transducers, to produce an image, one must steer or translate the beam by mechanically moving the whole transducer assembly. Linear arrays (Figure 6.26b) and curvilinear arrays (Figure 6.26c) allow rapid beam translation and/or steering with no moving parts. The advantage of the curvilinear array is its sector scan format, achieved without phase steering of the beam, which provides a larger field of view than a linear array of equivalent scanning aperture. In situations where a wide field of view is required at depth, but the anatomy of the body does not allow ultrasound to pass through a sufficiently large window (e.g. imaging between the ribs), the transducer of choice is the phased-array transducer (Figure 6.26d). These arrays use only beam steering (not translation) to build up an image; all the transducer elements are involved in transmitting and receiving the sound beam for every line of the image. The disadvantages relative to linear and curvilinear arrays are the narrow field of view near the skin, overall poorer lateral resolution due to the smaller aperture used to form the beam, and a reduced lateral resolution and sensitivity at the edges of the sector field where the steering angle is large (see the following discussion on diffraction gratings to understand why this happens). Furthermore, the requirements for array manufacture, in terms of element pitch (that is, the distance between elements), are more stringent for phased arrays than for linear or curvilinear arrays (see Equation 6.38).

Electronic focusing schemes, which are illustrated in Figures 6.27b to 6.27d, may be used for all arrays. For comparison with the electronic focusing methods, focusing with a single transducer is shown in Figure 6.27a. The transducer may be either curved in a bowl shape, or a flat disc plus a lens. The resulting focus can not be varied but this method has the advantage

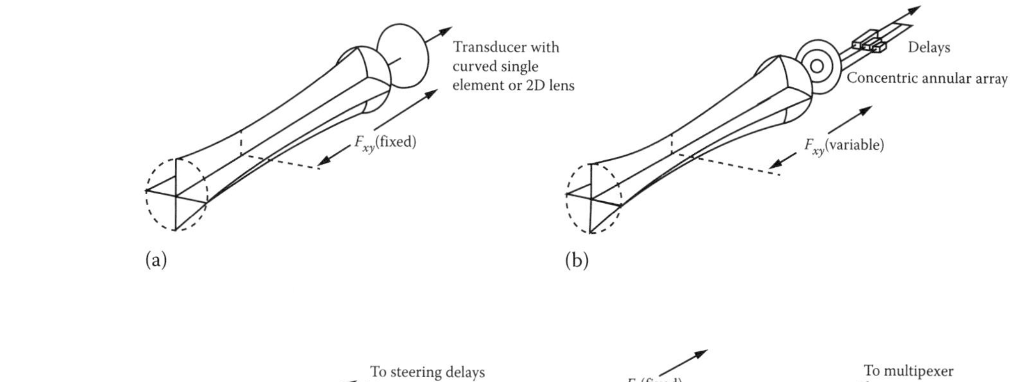

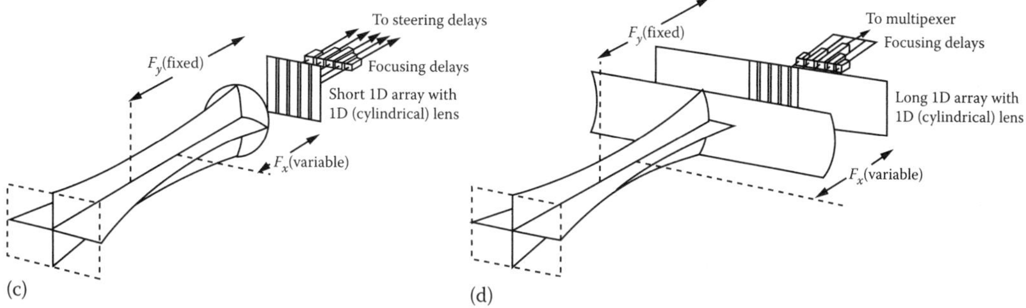

FIGURE 6.27
Ultrasound focusing systems that use 1D arrays. (a) A single transducer that provides a fixed focus, F_{xy} (provided for comparison with the arrays). (b) A concentric array of annular transducer elements that provides an adjustable focus, F_{xy}. (c) A short array of rectangular elements (also known as a phased array) that provides a variable focus, F_x, and a fixed focus, F_y. (d) A long array of rectangular elements (also known as a linear array) that provides a variable focus, F_x, and a fixed focus, F_y. (See text for more details.) (Adapted from Whittingham, 1981.)

of having a single focal length F_{xy} and aperture, producing identical resolutions in the lateral and orthogonal ('elevational') directions. It suffers from the depth-of-focus compromise, mentioned in Section 6.2.4.2, that usually dictates the use of weak focusing. Current use of single transducers for imaging is confined to very-high-frequency work (above ~20 MHz), where there are difficulties associated with manufacturing high-frequency arrays. A concentric array of annular transducer elements (Figure 6.27b) permits focusing delays to be used to provide an adjustable focus to overcome the depth-of-focus compromise, allowing strong focusing to be used. As for the single transducer, lateral and elevational resolutions are identical. For the linear phased (Figure 6.27c), curvilinear (not shown) and linear (Figure 6.27d) arrays, electronic adjustment of the lateral focal depth, F_x, is made possible by the focusing delays shown, but only in the plane of the scan – fixed mechanical (lens) focusing is usually applied in the elevational plane such that there is only one focal depth (F_y) in this direction. The elevational beam width (or 'slice thickness') is limited by the elevational aperture (determined by the element length) and increases considerably beyond the depth F_y. It is generally the poorest component of resolution for these types of array. The linear array provides scope for use of strong lateral focusing (if a large number of elements are used simultaneously) whereas, for the phased array, only medium lateral focusing may be possible. The 1.5-dimensional arrays (Figure 6.26e), which contain several rows of elements (perhaps five or seven), instead of the single row of elements employed in a linear array, have the advantages of a wider elevational aperture and the ability to provide a degree of electronic focusing in the elevational direction, thereby reducing the slice thickness at depth. Two-dimensional arrays, also called matrix arrays (Figure 6.26f), will permit full control of all components of the focus, as do annular arrays, but without loss of electronic scanning capability. Indeed, electronic scanning then becomes possible for 3D imaging (see Section 6.3.4). At the time of writing, 2D arrays are starting to become available in early form (with relatively low element count, suitable for sector scanning).

Construction of arrays is usually accomplished by cutting (using a diamond saw or laser trimmer) one large piece of piezoelectric material, after plating, polarising and mounting. However, the aforementioned capacitive micromachined ultrasound transducers (CMUTs) may be built as large arrays on silicon wafers. Although annular arrays have been built with as many as 64 elements, they generally contain far fewer elements than this (normally less than 12), providing a similar number of overlapping focal zones. Linear and curvilinear arrays have typical apertures of 4 cm or more and may easily contain 512 elements. Phased arrays typically consist of 128 elements fitted into a 1–2 cm row. In all cases, as discussed later, the pitch, p, (or distance) between the elements should be as small as possible and the width, b, of an element should be less than $\lambda/2$ (e.g. 0.3 mm at 5 MHz), if the elements are to behave as omnidirectional sources, so that steering and focusing are possible. These are very difficult to manufacture for high frequencies.

A problem that occurs with multi-element arrays is that they act as diffraction gratings; it is not possible to make the distance between element centres small enough to represent the continuous distribution of Huygens' sources necessary to produce a beam shape identical to that produced by a single-element transducer of the same overall aperture size as a group of elements (see Figure 6.28a). The result is that other sound beams (grating lobes) are generated at various angles to the intended main beam (see Figures 6.28b and 6.28c).

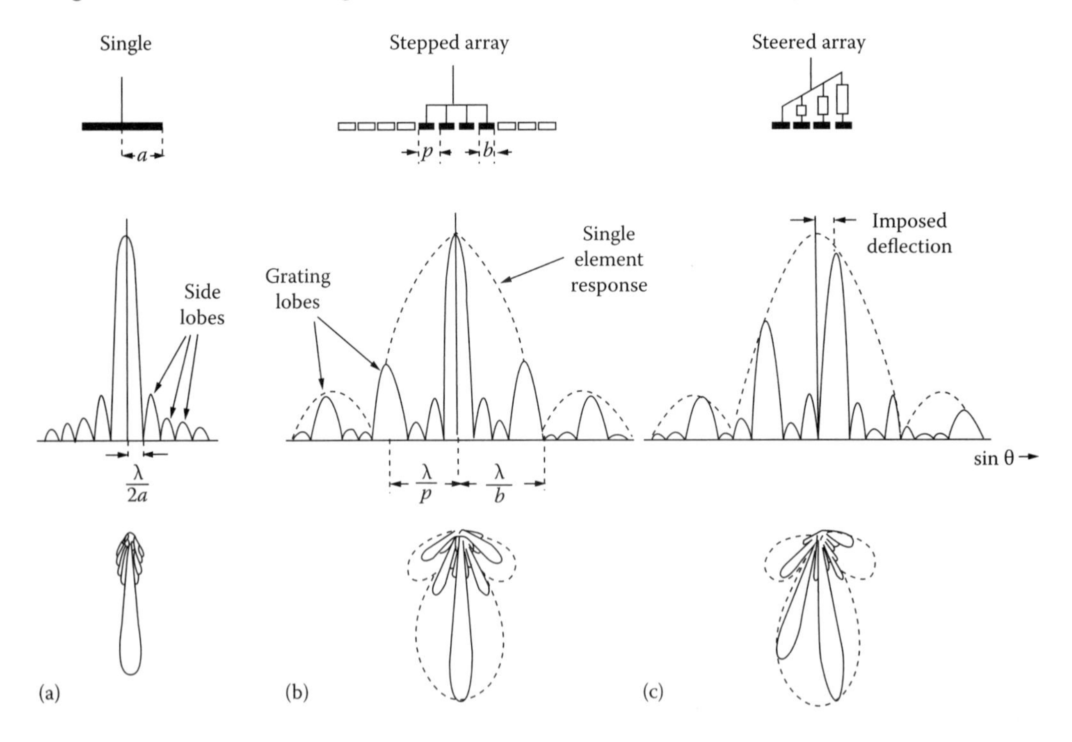

FIGURE 6.28
Beam divergence patterns from (a) a single-element linear transducer, (b) a stepped linear array of the same aperture as (a), and (c) a steered linear array, again of the same aperture. The top row shows the transducer element arrangements, the middle row shows the lateral far-field beam pressure amplitude profile as a function of the sine of the beam direction angle, and the bottom row provides the same information in the form of a polar diagram that shows the loci of the vectors of pressure amplitude and direction. (Adapted from Whittingham, 1981.)

If one considers that the far-field, or Fraunhofer, diffraction pattern is in fact obtained from the Fourier transform of the source aperture, then it will be seen that this phenomenon is equivalent to the sampling theorem (see Chapter 11), that is, sampling of the transmitting or receiving aperture produces repetition of the angular spectrum at angular intervals given by the reciprocal of the sampling distance. In terms of conventional Bragg diffraction theory for the grating, one obtains sound beams in directions θ for

$$\sin\theta = \frac{n\lambda}{p}. \tag{6.37}$$

Each of these replica beams also replicates the side lobe pattern due to the width of acoustic aperture a, now represented by all of the elements in the group used to form the beam.

The primary beam corresponds to $n = 0$. The secondary beam, at $n = 1$, is often referred to as the first 'grating lobe'. If the images are to be free from artefactual echoes, which may occur if a grating lobe interacts with a strongly reflecting structure, the aperture should be sampled finely enough such that no grating lobes exist. This condition is just met if $\theta = \pi/2$ for $n = 1$, which, from Equation 6.37, gives $p = \lambda$.

For steered arrays the problem is actually much more serious, since the sensitivity varies with the steering angle according to the main lobe and side lobe profile of the response of a single element, width b (dotted line in Figures 6.28b and 6.28c). As the main beam is steered off-axis, the grating lobes move closer to the axis, thus increasing in strength as signals from reflectors positioned along the intended line of sight decrease in strength (Figure 6.28c). The maximum usable deflection of the main beam is

$$\sin\theta_{max} \approx \frac{\lambda}{2p} \tag{6.38}$$

which is the angle at which the main lobe and the first grating lobe become equal in magnitude, for example, for $\theta_{max} = \pi/2$, $p = \lambda/2$.

Much of the improvement in image quality in recent years has been due to advances in transducer-array construction and electronic processing to avoid grating lobe artefacts, which appear in the image as echoes, displayed along the steering direction, that infact arise from strong scatterers along the direction of a grating lobe.

Specialised intraluminal and intracavity (or endoscopic) scanners are also available. The use of body cavities in this way enables one to place the ultrasound transducer physically closer to the organ of interest (thus enabling better resolution to be obtained by using higher frequencies) and may permit, in some cases, the examination to be performed with less discomfort, or an organ to be imaged that is otherwise inaccessible because it is obscured (from surface imaging) by gas or bone. Examples of applications are the imaging of the prostate and rectal wall from within the rectum, the uterus and ovaries from the vagina, the bladder wall from the bladder (entered via the urethra) and the stomach wall from within the stomach (entered via the oesophagus). Doppler instruments have also been built to operate with catheter-tip transducers, although this is more of a research activity. Some combination instruments are beginning to appear, incorporating both ultrasound and optical endoscopic imaging devices. Various scanning systems are in use, including linear arrays mounted on one edge of a long cylinder. A scanning system unique to endoscopic application, however, is the 360° mechanically rotating transducer, producing a radar-like 'plan-position' display. Figure 6.29 illustrates one of these devices and provides an example of this form of image display.

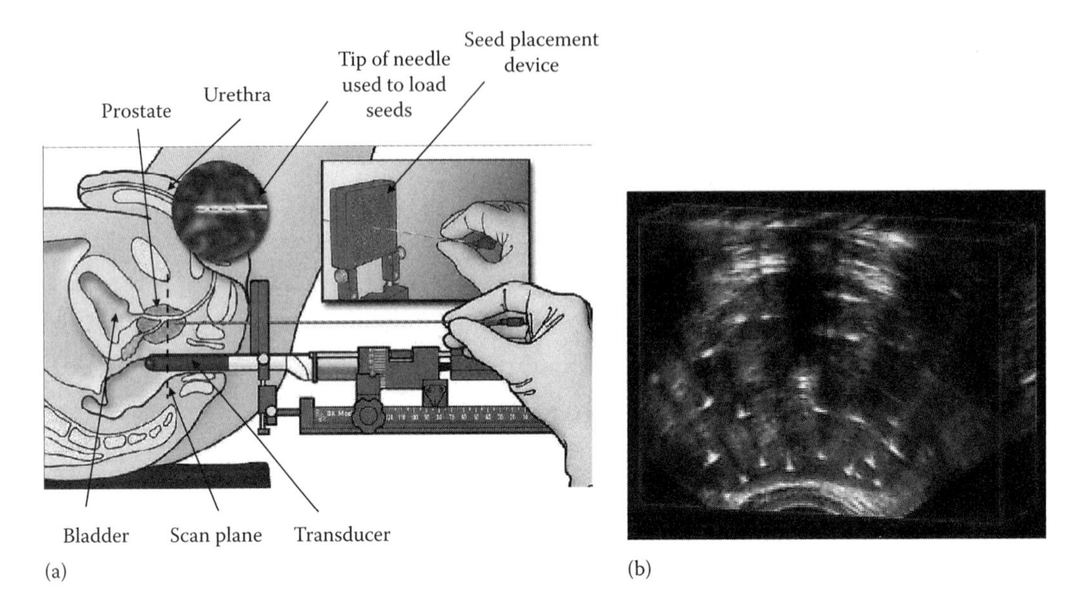

FIGURE 6.29
(a) Diagram of assembly and mechanics of one type of transrectal ultrasound transducer, illustrating its use to guide radioactive seed placement in prostate cancer brachytherapy (see Section 6.6.3) and (b) a slice through a 3D dataset showing the prostate with implanted radioactive seeds (the small, highly echogenic features), obtained with this system. (Diagram adapted from sales brochure of BK Medical, Peabody, MA; grey-scale image courtesy of BK Medical, Peabody, MA.)

6.3.2 Pulse-Echo Techniques (Echography)

Figure 6.30 shows a generalised scheme for signal generation and processing in a pulse-echo imaging system, with a pictorial representation of the signals to be expected at each stage. Referring to Figure 6.23, one can see that transmit pulse generation is also present in Figure 6.30 but that receive channel processing and beam formation (discussed previously) are not. We now discuss, in turn, each of the components in Figure 6.30, noting that on the receive side of the system this figure largely deals with the 'single-line RF processing' stage and later stages in Figure 6.23.

6.3.2.1 Transducer and Frequency

The choice of frequency is a compromise between good resolution and deep penetration. One often finds frequencies in the range 3–5 MHz used to image the liver and other abdominal organs, the uterus and the heart, whereas more superficial structures such as thyroid, carotid artery, breast, testis and various organs in infants would warrant the use of somewhat higher frequencies (5–15 MHz). The eye, being extremely superficial and exhibiting low attenuation, has been imaged with frequencies in the range 10–30 MHz.

The short pulse of sound that emerges from a diagnostic transducer or array is generally no more than 2–3 cycles in length, and is generated (at the transmitter) by applying to the transducer either a momentary voltage step or a brief (e.g. single-cycle) gated sine wave of frequency equal to the resonant frequency of the transducer (see Section 6.3.1).

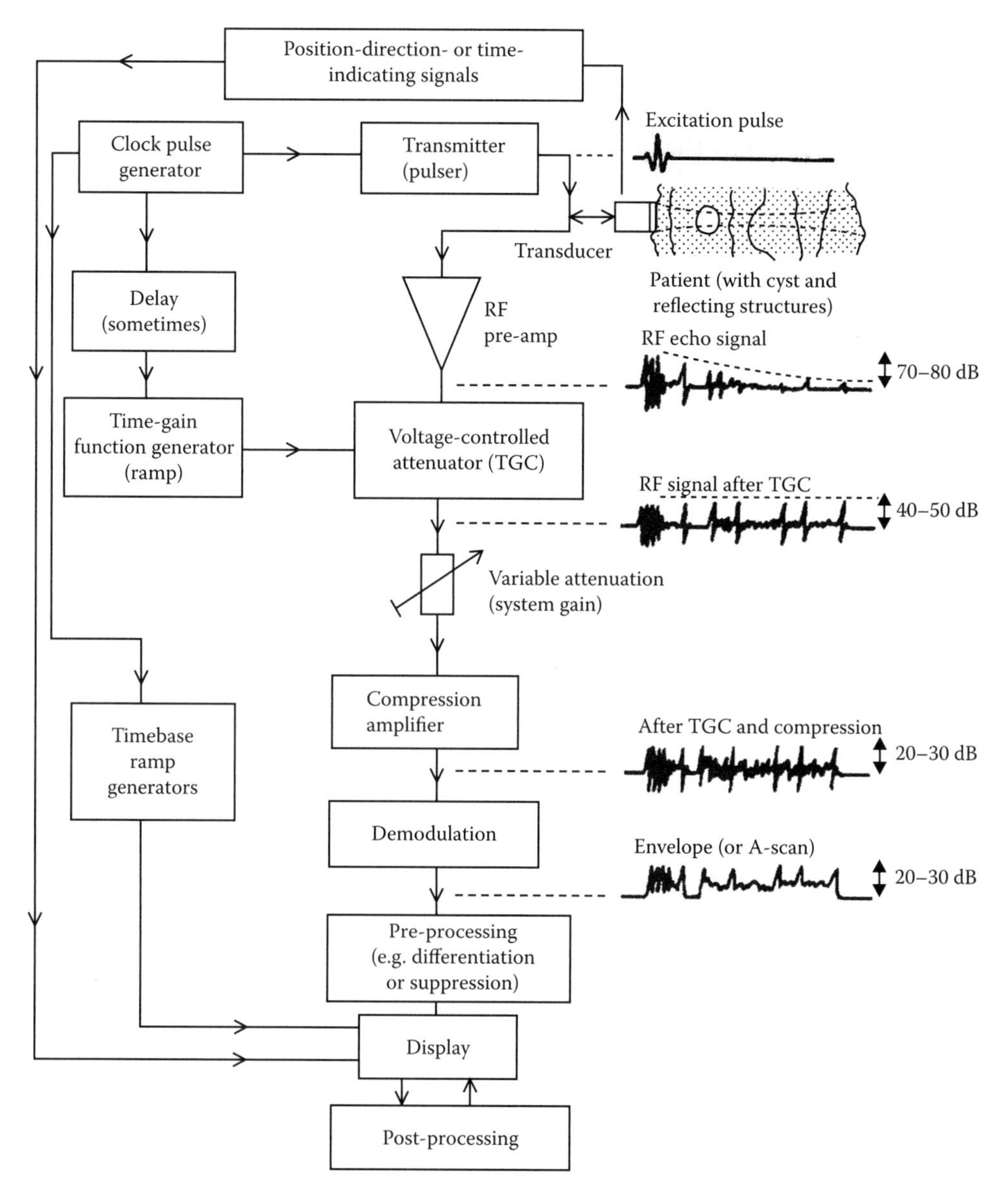

FIGURE 6.30
Block diagram of the essential components in the signal-processing chain of an ultrasound pulse-echo imaging system. On the right of the figure are labelled schematic versions of the kind of signal one might expect to observe at the points indicated, in response to a single transmitted sound pulse. See text for more details.

6.3.2.2 *Clock Pulse*

This triggers the excitation of the transducer and acts, in some circumstances, to synchronise the display. Following the emergence of one sound pulse, a stream of echoes returns, so that all of the echo information for one image line is contained in one clock cycle. High repetition rates are desirable, however, for fast scanning or for following moving structures. The maximum pulse repetition rate, usually referred to as the pulse repetition frequency (PRF_{max}), is limited by the maximum depth (D_{max}) to which one wishes to image, according to

$$PRF_{max}\,(kHz) = \frac{c}{2D_{max}} \approx \frac{1.5}{2D_{max}} \times 10^3 \qquad (6.39)$$

where D_{max} is in millimetres and the factor of 1.5 is the approximate speed of sound (conveniently expressed in mm μs^{-1} for such calculations). If this repetition rate is exceeded, echo ambiguity occurs, since echoes from deep structures associated with a previous pulse become coincident in time with those from more superficial structures for the current pulse. Typically, for abdominal imaging, the PRF is in the region of 1 kHz.

6.3.2.3 Transmitter

The voltage pulse applied to the transducer is usually below 500 V and often in the range 100–200 V. Its shape is equipment dependent, but it must have sufficient frequency components to excite the transducer properly; for example, for frequency components up to 10 MHz, a pulse rise time of less than 25 ns is required.

6.3.2.4 Linear RF Amplifier

This is an important component, since noise generated at this point may well limit the performance of the complete instrument. Clever design is required, since, whilst maintaining low noise and high gain, the input must be protected from the high-voltage pulse generated by the transmitter. Any circuits that do this must have a short recovery time, and the amplifier itself should possess a large dynamic range and good linearity. As indicated in Figure 6.30, the RF echo signal returning from the object may have a large dynamic range (from the noise level to the largest echoes) of some 70–80 dB for soft tissues, and much larger if bone, gallstones and other strongly reflecting structures are included.

6.3.2.5 Time Gain Control

This is provided by a voltage-controlled attenuator. Some form of time-varying function, synchronised with the main clock and triggered via a delay, is used as a control voltage so that the system gain roughly compensates for attenuation of sound in the tissue. The simplest function used is a logarithmic voltage ramp, usually set to compensate for some mean value of attenuation. The effect that this has on the time-varying gain of the system is illustrated schematically in Figure 6.31. The delay period indicated is often adjustable, so that the attenuation compensation does not become active until echoes begin returning from attenuating tissue. This might be used, for example, to scan through a full bladder or through a water bath (e.g. see Section 6.6.5). As a general result of this attenuation correction, the dynamic range of the signal at the output of the time gain control (TGC) section has often been reduced to some 40–50 dB.

An artefact of this type of attenuation correction is that echoes posterior to a structure that attenuates sound more than the average for the surrounding tissue tend not to be amplified enough. This causes the appearance of a shadow behind such structures. Conversely, the simple TGC function will overcompensate for structures that attenuate sound less than average, so that a kind of negative shadow, or 'enhancement', appears on images posterior to such structures (usually fluid filled, such as cysts). Even though these features are rather

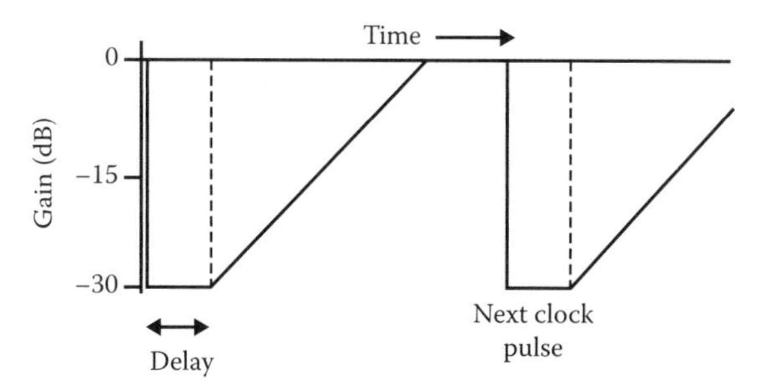

FIGURE 6.31
A simple, but common, form of time-varying gain for a pulse-echo imaging system.

crude artefacts of a simple attenuation correction system, they have, over the years, come to represent important diagnostic features, which often permit experienced diagnosticians to understand more of the nature and consistency of discrete lesions. Figure 6.32 illustrates the possible extreme types of imaging appearance, when a TGC is used, of a discrete structure ('lesion') when the material inside it may have any combination of either a high or low scattering coefficient, or a high or low attenuation coefficient, relative to the surrounding tissue. Various attempts at providing automatic time gain compensation have been tried, but fully automatic systems, which attempt to compensate for the spatially varying attenuation coefficient at all points in the field of view, have failed to gain widespread acceptance. In most systems, the approach taken is to provide a level of automation, where a preset TGC is provided to produce an average uniform image with depth, and a series of sliders allows the user to modify the depth-gain curve by controlling the system gain at depths that correspond to each slider.

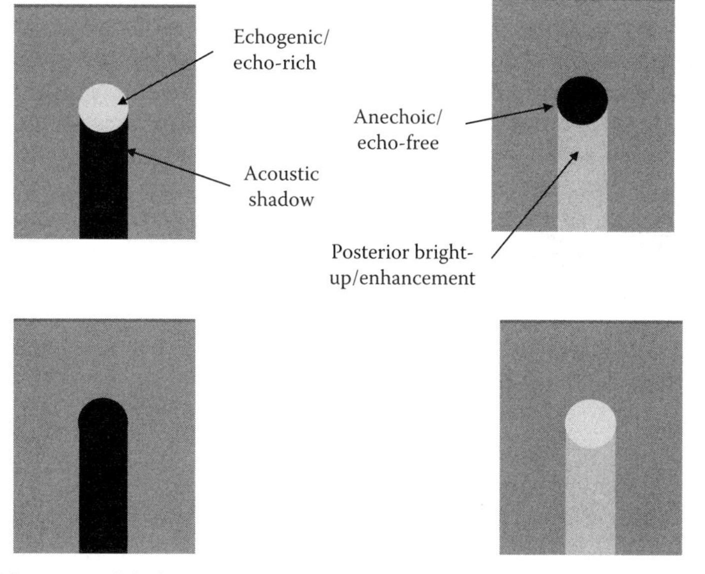

FIGURE 6.32
Schematic illustration of the combinations of B-mode appearances that may occur for discrete lesions of either high or low internal scattering or attenuation, relative to the surrounding tissue, when a TGC is set to produce a background-echo brightness that is uniform with depth.

6.3.2.6 Compression Amplifier

There is a wide range of gain characteristics in use but a general feature is that the gain decreases with increasing input signal. One example is an amplifier with a logarithmic response. This allows the remaining 40–50 dB echo range to be displayed as a grey scale on a screen that might typically have only a 20–30 dB dynamic range. This may be compared with the amplitude windowing technique used to view X-ray CT images, which also possess a wide dynamic range (see Chapter 3). Amplitude windowing is inappropriate in ultrasound imaging because of the speckle modulation.

6.3.2.7 Demodulation

At this point the envelope of the RF echo signal is extracted, for example, by rectification followed by smoothing with a time constant of about 1.5λ, although more accurate techniques of complex demodulation (beyond the scope of this introductory text) are often used. The signal depicted at this point is known as a detected A-scan (representing the amplitude of the echoes). Note that there are two types of demodulation used in ultrasound signal processing, amplitude demodulation as described here for anatomical imaging, and frequency demodulation described in Section 6.3.5.3 for detecting the Doppler shift due to target motion.

6.3.2.8 Pre- and Postprocessing

It is common to preprocess the A-scan echo signal both before and after it is digitised and stored in the display memory (see Section 6.3.2.9). Examples are various forms of edge detection (usually differentiation), and further adjustments to the gain characteristic, such as suppression, which is dynamic-range restriction by rejecting from the display echoes that are below an operator-defined threshold (noise suppression). In modern machines, the term 'suppression' is generally not used, but the same result may be effected by various processing options, often provided for dynamic-range manipulation, or even simply by reducing the system gain until the noise in anechoic regions disappears. Although removing noise from an image in this way may make it aesthetically pleasing, unless the threshold is set extremely carefully, so that the minimum displayed echo amplitude is only just above noise, it will worsen the ambiguity of image interpretation when trying to judge whether a region is free of echoes. Furthermore, postprocessing to display a restricted band of grey levels is of debatable utility, probably because of the presence of the large fluctuations in grey level caused by the coherent speckle phenomenon. These make it difficult to appreciate subtle changes in grey level, and they break up the continuity of regions that would otherwise be uniform and of limited dynamic range. Speckle-reduction methods only partially solve this problem.

6.3.2.9 Digitisation and Display

As the sound beam is scanned across the object being imaged, a sequence of A-scans is generated, each of which provides one line in the final image. This sequence of scan lines must be stored and geometrically reconstructed to form an image, which is achieved by a process known as a scan conversion (see Figure 6.23). Data are stored in digital memory and read out continuously, in conventional television format, but are updated with new scan data by writing to memory pixels in the order, and in the coordinate system, in which the echo line data arrive. Hence, at some stage in the signal processing, the echo voltage signal is digitised. In old systems the signal may be digitised after envelope detection, but in modern systems the signal is digitised at the array-element level after a preamplifier (and possibly a TGC) on each channel, representing the 'channel processing' stage in Figure 6.23, in which case all signal processing becomes digital, including the channel summation for beam formation and all of the signal-processing elements described previously. In the most modern systems, all processing after digitisation is accomplished in software implemented on fast digital-signal processing units. Returning to the subject of scan conversion, the digital echo amplitude data are written into the 2D (or 3D, see Section 6.3.4) display memory along a vector calculated from coordinates of a position-sensing system if the scan is mechanical, or from knowledge of the element selection and beam-steering delay profiles used to generate the sound beam from a transducer array. This involves a coordinate transformation (e.g. from the polar coordinates of a sector scan to the Cartesian coordinates of the display memory), and some interpolation to fill display memory locations that do not have a vector passing through them. Adequate memory is required to avoid aliasing the PSF and to avoid Moiré patterns due to undersampling of the vector lines, and at least eight bits of information are needed at each location to avoid a contoured appearance to the image.

6.3.2.10 Additional Sections

The design of echo signal-processing schemes varies widely from manufacturer to manufacturer, and is often proprietary information. Even within the simplest of (effectively idealised) processing schemes discussed here, other amplification stages would be present, providing an overall gain of between 70 and 100 dB.

6.3.2.11 Scanning and Display Format

There are three commonly used formats for displaying ultrasound echo data, and one rarely used format.

The simplest is the *A-mode* of display, referred to above as the A-scan, which is a 1D display of the envelope-detected-echo amplitude versus time. This is now rarely used as a display method but has applications in eye examinations, when precise measurements of axial length of the eye are required. It is also essential to understand the A-scan since images using the display methods described hereafter are built from multiple A-scans. For adequate spatial sampling within an A-scan at least three display sample points are required per axial resolution distance.

The *M-mode* (also called *TM-mode*) is used for observing changes in the A-scan as a function of time, to provide motion information. The A-scan position is fixed to examine a constant line of interest through the body, and is used to brightness-modulate a display in the vertical direction. A slow time base in the horizontal direction then produces the TM (time-motion) recording. A typical example of an application of M-mode scanning is the examination and quantification of the pattern of movement of heart-valve leaflets. In addition to the adequate spatial sampling down the A-line, M-mode display requires adequate temporal sampling. So long as the echo pattern from the moving structure of interest remains largely unchanged (correlated) then recognition of this pattern allows the motion to be followed. For a random scattering medium (of the type considered in Section 6.2.5.1), the echo pattern will remain correlated so long as the PRF is high enough that the structure does not move laterally or elevationally by more than about a third of the lateral or elevational resolution distance, respectively, between pulses.

If the scatterer itself has spatial correlation, such as a surface normal to the sound beam, then its motion in the axial direction can be tracked even when this temporal sampling condition is substantially violated.

The *B-mode* is by far the most commonly used mode, and is indeed a true image format. A B-scan is what one normally refers to implicitly when talking of ultrasonic pulse-echo imaging. There are many designs of scanning arrangement, but a general feature is that the A-scan is used to brightness-modulate points along a line on an x-y display. The starting point and the direction of the resulting image line vector are determined from the position and direction of the source of the sound beam in the imaging plane. The 2D image is built up by moving the sound beam in the x-y plane and, for each new position and/or direction of the sound beam, writing a new image line vector onto the display. For adequate spatial sampling, at least three adjacent image lines are required per lateral resolution distance.

Finally, in the *C-mode*, or constant-depth scan, a gated portion of the A-scan signal is used in a brightness-modulated display of the projected 2D position of the beam in a plane at a fixed distance from the transducer plane. The method in principle enables a wide-aperture, strongly focused, transducer to be used and the image to be produced in the plane of the focus. The term C-scan is, however, also used to describe a constant-depth plane selected from a 3D dataset obtained using a 3D scanner (see Section 6.3.4), even though the benefits of improved elevational resolution may not (depending on the method of 3D data acquisition) be available. A disadvantage of C-scan planes when viewed in isolation is that out-of-plane attenuating structures may influence the received echo level. For adequate spatial sampling, at least three C-scan display samples are required per lateral resolution distance.

6.3.2.12 Scanning Methods

These are illustrated in Figure 6.33. A major classification exists between simple scans and compound scans; compounding refers to the process of summing multiple simple scans onto the same image (see Section 6.2.5.3). Two types of simple scan are often referred to. The simple linear scan requires a large window of access to the body but gives the best images (because relatively large apertures can be used to form beams with good lateral resolution) with a wider field of view near the skin. The simple sector scan has the advantage that it is easier to move the beam rapidly if the scanning is mechanical, and a much narrower access window may be used, particularly if a phased array is used to scan by steering the beam. However, this limits the beam-forming aperture and hence lateral resolution. The narrower field of view near the skin may be a further disadvantage, but this form of scanning is often the only way to get a wide enough field of view at depth with limited window of access. Typical examples of this

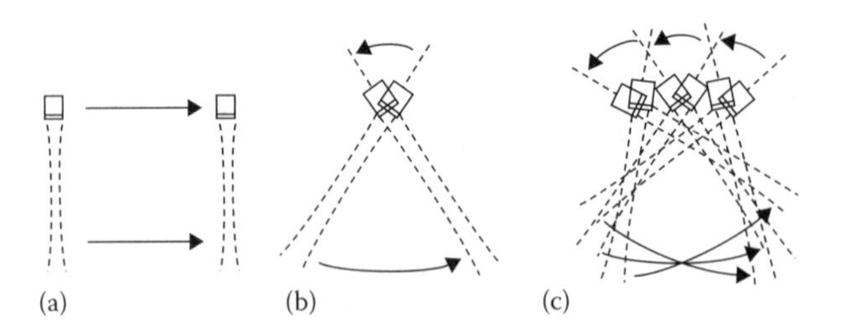

FIGURE 6.33
Some of the scanning methods used in echographic imaging: (a) simple linear scan, (b) simple sector scan and (c) compound scan formed by integrating a number of sector scans.

TABLE 6.2

Advantages and disadvantages of compound and simple scanning, demonstrating the complementary nature of the two approaches.

Method	Advantages	Disadvantages
Compound scan	More complete display of boundaries If a large access window is available, improves effective lateral resolution for single targets (in homogeneous medium) by averaging Reduces speckle by averaging (Section 6.2.5.3) and, therefore, improves image grey-level discrimination	Requires large window of access to the object Reduces frame rate. If access window is fixed (i.e. same for compound and simple scan) compounding will reduce image resolution Registration of individual scans may not be good if inhomogeneities cause refraction or if body moves during scan, which will offset improved speckle averaging
Simple scan	Provides some information related to the diffractive scattering properties of tissues (seen as the texture of the image) Portrays regions of differing attenuation coefficient (and sometimes velocity) by changes in posterior echo level (or position) Takes less time (higher frame rate) Image distortion due to refraction, or body movement, is less of a problem than with compound imaging	Only boundaries or structures that lie normal to the beam are clearly delineated Speckle makes it difficult to perceive low-contrast targets

are imaging of the heart or some parts of the liver through intercostal spaces (between the ribs), or imaging the infant brain through the fontanelle (a space between component bones of the skull).

Compound and simple scanning methods have a number of relative advantages and disadvantages. These features are summarised in Table 6.2. The two methods thus provide complementary information and they are often employed sequentially on the same object.

A further division occurs between contact scanning and water-bath scanning. Again, there are advantages and disadvantages to both approaches, which are summarised in Table 6.3. In contact scanning, the transducer or array surface is placed effectively in contact with the patient's skin, the junction being lubricated with a water-based gel or viscous oil (the main function of which is to provide good acoustic coupling). Water-bath scanning is where the part of the body to be imaged is immersed in a bath of water (or saline or oil), which acts as a propagation delay. For some scanners the transducer is housed within a small self-contained water (or oil) bath. This remains necessary for high-frequency skin imaging, for example, where mechanical scanning may be necessary because arrays of the frequency required are not yet available. Such systems possess some of the advantages of water-bath scanning, for example, the ability to make use of larger-aperture, more tightly focused transducers.

6.3.2.13 Real-Time (Rapid) Scanning

It is important that the image refresh (scanning) rate is fast enough such that one can (1) obtain information about moving structures, (2) eliminate movement artefacts from single frames and (3) permit rapid coverage during a fine search through a large volume of any stationary organ such as the liver (although even organs like the liver are never

TABLE 6.3

Comparison between contact scanning and automated water-bath scanning methods.

Method	Advantages	Disadvantages
Contact scanning	Very versatile (user directs to examine an arbitrary scan plane) Relatively simple to use Possible to be relatively cheap For simple sector scan, only a small access window is required	Large-aperture (strongly focused) transducers not easily applied (especially 3D) Superficial tissue structure may not be imaged well due to recovery time, poor focus and near field May distort tissues while scanning due to contact pressure Requires medically trained operator on many occasions
Automated water-bath scanning	Can move transducer without distorting tissue Can use wide-aperture transducers with strong (3D) focusing Easier and better imaging of superficial structures Easily adapted for compound scanning Suitable for a rapid, automatic and systematic study of an organ by serial tomography May be operated by non-medically trained personnel	Large – may occupy a lot of floor space Tends to be more expensive Complex mechanical systems may be prone to failure Cumbersome for patient handling Tends to require more maintenance by the user Reduced access to some deeper parts of the body Multiple echoes that occur between the skin and the transducer face produce echo confusion and thus may limit the maximum penetration depth

entirely stationary). As with M-mode examination (Section 6.3.2.11), if the echo pattern for a structure of interest remains correlated, its motion may be followed between frames by recognising the pattern. For a random scattering medium (Section 6.2.5.1), such correlation will be maintained so long as the frame rate is high enough for the structures to move by no more than about a third of the elevational resolution distance between frames. For spatially extended scatterers (e.g. blood vessels, boundaries or surfaces) motion may be followed in 1D even when this temporal sampling criterion is violated. One of the reasons why M-mode scanning is still used for the detailed examination and quantification of heart-valve motion is that with conventional sequential beam-formation techniques the frame rate for whole images cannot be as high as that required (of the order of 100 Hz), whereas the M-mode sampling rate may be as fast as the PRF itself. In general, for sequential line-by-line beam formation and scanning, the frame rate is limited by the number of lines (A-scans) in the image, N_{lines}, and the maximum depth to be visualised. Using the same terminology as for Equation 6.39, we have

$$\mathrm{PRF} = N_{\mathrm{lines}} \times \mathrm{frame\,rate}.$$

Hence,

$$\mathrm{max.\,frame\,rate\,(Hz)} = \frac{c}{2D_{\max}N_{\mathrm{lines}}} \approx \frac{1.5\times10^{6}}{2D_{\max}N_{\mathrm{lines}}} \qquad (6.40)$$

where again, D_{max} is measured in millimetres. The frame rate will be lower than this if multiple transmit pulses are used, for example, to enable multiple transmit focal zones or compounding. On some scanners, the frame rate may be higher than would be predicted by Equation 6.40 due to a technique known as parallel beam forming transplant, as was described in Section 6.3.1.2 (see also Section 6.3.4.4).

For the majority of clinical applications, electronic beam forming is used to rapidly translate or steer the sound beam (see Section 6.3.1.2). A further advantage of purely electrical systems is their lack of moving parts and consequent inherent reliability and convenience. Mechanical systems, however, use circular transducers to provide good lateral resolution obtained orthogonal to, as well as within, the scan plane (see Figures 6.27a and 6.27b). This advantage of mechanical systems does not, however, appear to have outweighed the advantages of linear arrays, which have displaced the use of mechanical scanning in all but a few applications such as high-frequency imaging (e.g. in skin, small animal, endoscopic and intravascular imaging), because of the aforementioned difficulty of manufacturing arrays with small closely spaced elements. It is anticipated that the best of all worlds will eventually be obtained by the use of electronic scanning in combination with 2D (matrix) array transducers mentioned earlier (Section 6.3.1.2).

6.3.3 Tissue Harmonic (Non-Linear) Imaging

Non-linear acoustics now play a major role in medical ultrasonics. Tissue harmonic imaging (THI) makes use of signals received at harmonic frequencies of the source frequency to form an image. These harmonics develop in tissue as a consequence of non-linear propagation, as described in Sections 6.2.1.8 and 6.2.4.5. It was shown that an abrupt change from compression to rarefaction (known as an acoustic shock) may be observed when ultrasound propagates at high amplitude in water. Such a dramatic change in pulse shape would not be observed in soft tissue, due to the increased attenuation. The slight distortion experienced by an acoustic pulse that has propagated 10 cm into tissue is shown, neglecting pulse amplitude reduction due to attenuation, in Figure 6.34a.

When compared with the single-frequency peak of the transmitted signal, the pulse at 10 cm has a series of peaks at multiples of the fundamental frequency (Figure 6.34b). The second-order harmonic, which is used in THI, varies with depth, as shown in Figure 6.35. The harmonic

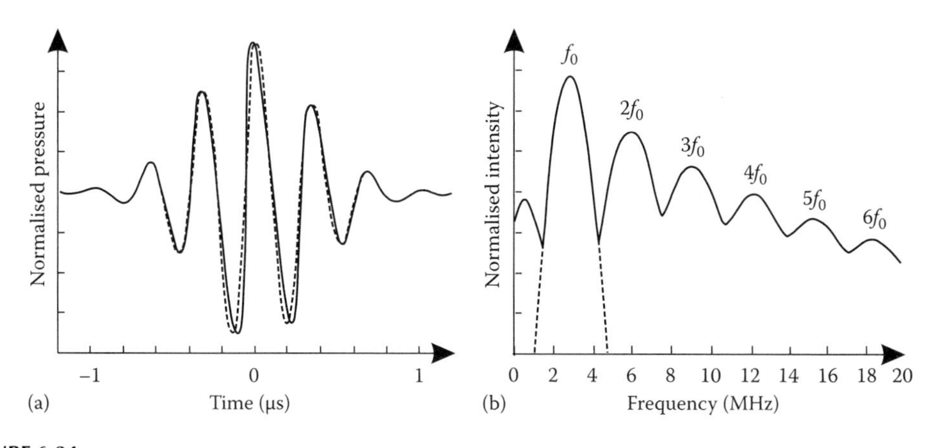

FIGURE 6.34
(a) Time- and (b) frequency-domain representations of an acoustic pulse at the transducer face (dotted line) and after it has propagated through 10 cm of tissue (solid line). (Adapted from Desser et al., 2000.)

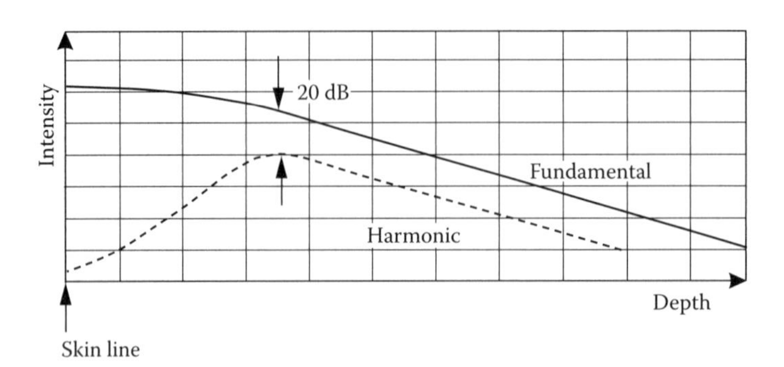

FIGURE 6.35
Depth dependence of the fundamental (solid line) and second-harmonic (dashed line) signals. (Adapted from Desser et al., 2000.)

intensity initially increases, due to the cumulative nature of non-linear effects, but eventually decreases due to frequency-dependent attenuation and energy transfer to higher harmonics. At its maximum amplitude, the intensity of the second harmonic is approximately 20 dB (100 times) less than that of the fundamental. Nevertheless, images formed from this signal have a number of advantages over those formed at the fundamental.

The main benefit of harmonic imaging is often said to be that, as can be seen in Figure 6.35, the harmonic signal intensity is low, close to the face of the transducer. In medical imaging, these depths consist of heterogeneous subcutaneous tissues that produce beam distortion and reverberations due to sound speed differences between fat and muscle layers. Therefore, the degree of distortion and reverberation will be reduced in the harmonic signal when compared to the fundamental. This makes harmonic imaging particularly useful in obese or muscular patients, as it reduces noise and clutter, thereby improving the SNR. A further advantage of the harmonic signal arises from the fact that it is proportional to the square of the fundamental signal (Desser et al. 2000). Consequently, the most significant harmonics are produced along the beam axis, where the fundamental signal is strongest, while harmonic signals due to low-energy side lobes, scattered waves and reverberations are generally too small to be detected. This results in superior lateral resolution and contrast-to-noise ratio. The axial resolution, however, is generally not improved, and may be worsened by THI; the bandwidth reduction needed to separate the harmonic component may lengthen the effective imaging pulse relative to that which uses the full transducer bandwidth. Interestingly, in the presence of wave aberration and reverberations the speckle pattern can possess fine detail smaller that the beam width, giving the (incorrect) impression that an image made with the fundamental has better resolution than the harmonic image. THI is now the default imaging mode for many clinical applications, such as cardiac, breast and abdomen, and some examples comparing the fundamental and second harmonic image are shown in Figure 6.36. Further details of how THI works, and the benefits that it provides, may be found in Whittingham (1999).

Another consequence of non-linear propagation is that the medium can act as an acoustic mixer; overlapping acoustic fields of different frequency may produce a field that, in the region of overlap, also contains the sum and difference frequencies. In this way, narrow, very sharply defined, low-frequency sound beams can be generated, overcoming the diffraction limit of linear acoustics. Commercial medical ultrasound scanners have yet to make use of this transduction method which, in other fields, is known as a parametric array (Westervelt 1963, Berktay 1965, Ritty 2006, Wygant et al. 2009), although similar methods have been employed in the elastographic method known as vibro-acoustography (see Section 6.3.6.2).

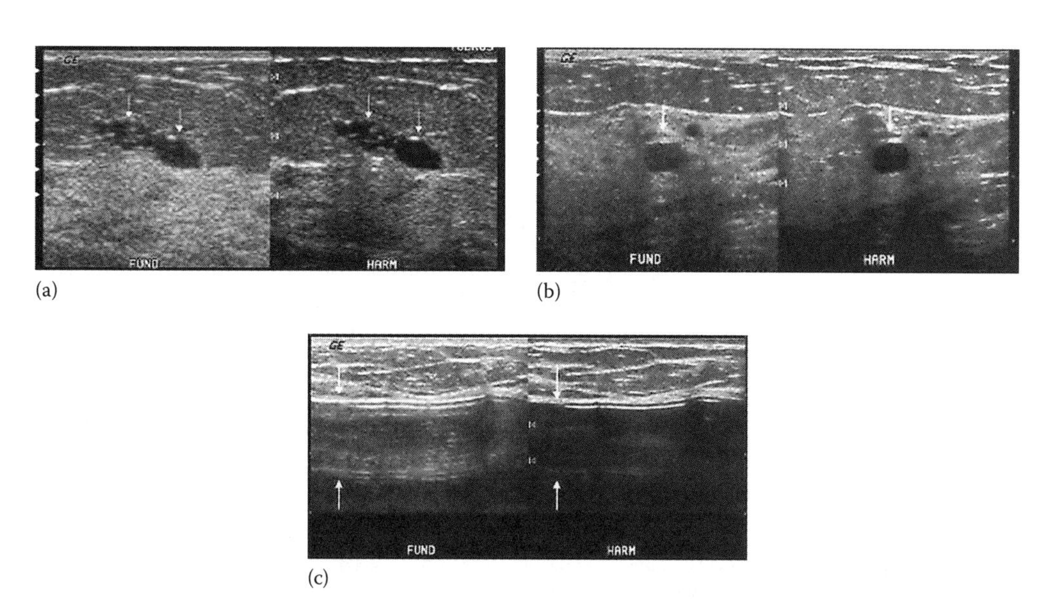

FIGURE 6.36
Examples of the reduction in artefacts and improvement in contrast-to-noise ratio afforded by tissue harmonic imaging (right-hand image of each pair) compared to imaging at the fundamental frequency (left-hand image of each pair). All the images were taken in the breast, showing: (a) A cluster of cysts. Note the reduction in artefactual internal echoes and in clutter throughout the harmonic image. (b) A cyst. This was initially classified as a fibroadenoma (benign solid lesion) on the fundamental image. (c) A silicone implant. Note the reduction in clutter, particularly within the prosthesis. (Reproduced from Rapp and Stavros, 2001, by permission of SAGE Publications, copyright 2001 by SAGE Publications.)

Finally, as Akiyama et al. (2006) have shown, the harmonic frequencies generated as a consequence of non-linear propagation have potential, if a suitably wide-band receiver is available to detect them, to be used in an ultra-broadband implementation of speckle reduction by frequency compounding (Section 6.2.5.3).

6.3.4 Three-Dimensional Imaging

3D imaging is now widely available on high-end ultrasound systems. Although 2D ultrasound is still the primary imaging method for most clinical applications, evidence is beginning to accrue to show that 3D ultrasound has the potential to aid certain diagnostic and therapeutic procedures. The utility of 3D ultrasound is likely to increase further as the image quality and temporal resolution of the technique are improved. 3D data acquisition may use a single-element transducer, an annular array, a 1D array (Section 6.3.4.1) or a 2D array (Section 6.3.4.2). Movement of the transducer in two dimensions is required to acquire 3D echo data using a single-element transducer or an annular array. The following description assumes that the cases of a single-element transducer or an annular array are effectively covered by describing how movement of an array in one dimension can be used to collect 3D data from a 1D array.

6.3.4.1 Data Acquisition Using a 1D Array

Systems using a 1D array acquire a series of 2D images in different planes. To achieve adequate sampling, data should be acquired in at least three scan planes per elevational

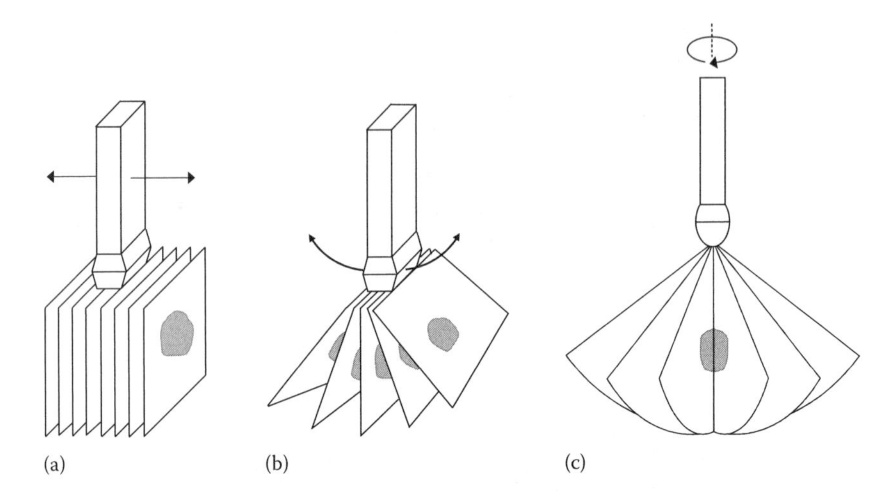

FIGURE 6.37
Examples of methods of moving a 1D or 1.5D array transducer to produce a 3D image: (a) translational, (b) tilting and (c) rotational. (Modified from Downey et al., 2000.)

resolution distance of the ultrasound system. Movement of the transducer can be achieved using either a mechanical or a freehand approach. Freehand 3D imaging can be further subdivided into methods that use a position sensor and those that use image calculation to determine the relative position and orientation of each plane.

6.3.4.1.1 Mechanical

In these systems, the transducer is moved in a predetermined way, either using an external motor or one that is integrated into the probe. For integrated systems, the transducer is contained within an oil-filled or water-filled housing. The transducer is moved in one of the following ways (see Figure 6.37):

(i) Translational – images are acquired parallel to each other. This approach is often used for intraluminal and endocavital imaging (see Section 6.3.1.2), by pulling the transducer back through the lumen or cavity at a constant speed.
(ii) Tilting – images are acquired in a fan shape. This approach is suitable for general applications, especially abdominal and obstetric imaging with a curvilinear array. It is also used with side-firing endocavital probes, for example, in transoesophageal (TOE) and transrectal applications, in which case the probe can be rotated through 360°.
(iii) Rotational – images are acquired in a propeller-like manner. If each image is a sector scan, this sweeps out a conical volume. This approach is used for automated measurement of bladder volume, and has been used in a research context for cardiac imaging with a phased array and for gynaecological applications with an end-firing endocavity transducer (Merz 2007).

The translational approach has the advantage that the spatial sampling between scans does not degrade with distance; for the tilt method, it degrades with depth (although this may be less important than it seems, if the elevational resolution also degrades with depth), while for the rotational method, it degrades with distance from the rotational axis. Nevertheless, the tilt and rotation methods have the practical benefit that the mechanisms can be very compact.

6.3.4.1.2 Freehand Imaging with Position Sensor

In this approach, the operator moves the transducer using one or a combination of the aforementioned types of motion (translation, rotation and tilt), and a position sensor attached to the probe records the position and orientation of each 2D image. The system needs to be calibrated to calculate the three rotation angles and the three translations between the image data and the origin of the position sensor. A new calibration must be performed each time the sensor is reattached (unless it can be reattached in a reproducible way) and when a different probe is used. Calibration is carried out using specially designed phantoms, as described in Mercier et al. (2005). The advantages and disadvantages of freehand 3D ultrasound, relative to mechanical systems, are listed in Table 6.4.

Having acquired the sequence of images, which consists of a set of arbitrarily positioned and orientated 2D planes, the data are generally reconstructed onto a regular 3D grid using an interpolation algorithm. This is the 3D equivalent of the process of scan conversion described earlier (Section 6.3.2.9). Different types of postprocessing may then be applied to enable different types of information to be displayed (see Section 6.3.4.3), including the selection of any desired 2D scan plane for viewing. Some authors recommend direct calculation of the image in the desired 2D scan plane, without reconstructing onto a 3D regular grid (Prager et al. 1999), as this speeds up the visualisation process and also reduces interpolation errors.

The two most-common types of position sensor are electromagnetic and optical. Electromagnetic position sensors consist of a transmitter (to emit a spatially varying magnetic field), a receiver (three orthogonal coils to measure the strength of the three components of the magnetic field) and an electronics unit (to compute the position and orientation of the receiver). The transmitter needs to be placed less than 50 cm from the patient in a stable, fixed position. The receiver is encased in a small plastic cube and is attached to the ultrasound probe. Using commercially available electromagnetic position sensors, the RMS precision in each 3D coordinate is approximately 1–2 mm, while the mean accuracy is better than 0.25 mm (Fenster et al. 2001). A potential problem with electromagnetic sensors is their sensitivity to the environment; the position and orientation data can be corrupted due to interference from other magnetic fields. Optical sensors consist either of active markers (infrared light emitting diodes) or passive markers (retro-reflective markers). Typically, four spherical markers, each of ~1 cm diameter, are attached to the probe. An infrared camera tracks the markers, calculates the position and orientation of the probe, and reports

TABLE 6.4

Advantages and disadvantages of freehand 3D ultrasound, relative to mechanical systems.

Advantages of Freehand 3D	Disadvantages of Freehand 3D
Versatile (any probe can be used for the acquisition, allowing image quality to be optimised)	Imposes constraints on scanning protocol (e.g. requires line of sight or maximum distance between transmitter and sensor)
Large field of view of arbitrary size and shape	Not amenable to 4D (i.e. real-time 3D) imaging
Low cost	Requires more skill and experience (particularly to achieve accurate calibrations)
Less bulky	
Less heavy (ergonomically advantageous)	Prone to image distortion due to variable probe pressure (though correction methods are possible, see Treece et al. 2002)
	Less accurate (for distance, area and volume measurements)

the results to a host computer. The RMS precision of optical position sensors can be less than 0.35 mm, making them more accurate than electromagnetic sensors. However, they are more expensive and require a line of sight between the markers and the camera.

6.3.4.1.3 *Freehand without Position Sensor (Sensorless)*

In this approach, image calculation is used to determine the local transformation (both in-plane and out-of-plane) between each pair of B-scans in the acquired sequence. The in-plane component of the motion consists of two translations (x and y) and a rotation (roll). These can be estimated by conventional 2D image registration techniques. The out-of-plane motion consists of two rotations (tilt and yaw) and one translation (z). Considering just the translation to begin with, this is estimated using a model of how the speckle pattern changes as a function of distance moved in the elevational (z) direction. Various models have been proposed (Tuthill et al. 1998, Prager et al. 2003), but they all make use of the fact that the width of the beam in the elevational direction is large (several mm) compared to the image separation. Therefore, the speckle patterns in neighbouring images are statistically related. In order to account for out-of-plane rotation, the images are subdivided into at least three non-colinear patches and the elevational translation is calculated separately for each patch. A best-fit plane to these values allows calculation of the overall out-of-plane transformation.

The principal advantages of sensorless freehand 3D ultrasound are that it dispenses with the need for extra hardware and it does not impose any constraints on the scanning protocol. The calibration procedure is probe specific, as it requires measurement of the elevational beam width and its variation with depth. However, it only needs to be carried out once per transducer. The drawback of the technique is lower accuracy compared to both sensor-based and mechanical methods. A significant source of error arises from the fact that the models assume speckle statistics, whereas clinical B-mode data also contain coherent scattering. Gee et al. (2006) have proposed a heuristic approach for adjusting the calibration curves to account for local coherent scattering, but further work is required in this area. A second source of error is the inevitable drift due to the cumulative approach of combining transformations at neighbouring positions to build up the scanned volume. The potential clinical applications of sensorless 3D ultrasound are yet to be determined.

6.3.4.2 Data Acquisition Using a 2D Array

Many of the drawbacks associated with freehand and mechanical 3D ultrasound would be overcome by using a 2D (matrix) array to electronically scan the volume. Such arrays have the added advantage of electronic focusing in the elevational plane. However, the development of 2D arrays has been hindered by the complexity and cost of incorporating a large number of elements, and their electrical connections, into an ergonomically shaped transducer. Consequently, there has been considerable research into the viability of using sparse 2D arrays to achieve acceptable image quality (see, e.g. Davidsen et al. 1994, Nikolov and Jensen 2000, Austeng and Holm 2002, Kirkebø and Austeng 2007). Although sparse 2D arrays have proved capable of *in vivo* imaging (Yen and Smith 2002), they are generally thought of as an imperfect or interim solution, as they inevitably have poorer SNR than a fully sampled $\lambda/2$ pitch array. Fully sampled 2D arrays became commercially available in 2002, but at the time of writing, they are limited to small aperture sizes suitable only for cardiac imaging. Further technological advances are required to develop arrays for applications that require larger aperture sizes or higher frequencies, both of which would result in an increase in the number of elements.

6.3.4.3 Data Processing and Display

The most useful method for viewing and interacting with 3D ultrasound data is to allow the user to select for display any image plane arbitrarily orientated and positioned within the volume. This method is known as multiplanar reformatting, or reslicing. The usual approach is to provide simultaneous display of three orthogonal slices through an object (see Figure 6.38).

Rendering is the process of mapping a 3D dataset onto a 2D image for display. The rendering techniques for 3D ultrasound have been copied from other imaging modalities. There are two main techniques that are used to render 3D ultrasound data:

(i) Surface rendering is the process of extracting a surface or boundary from the volume and displaying it as though it is illuminated from behind the viewer. The surface is extracted by segmentation, which may be carried out in one of three ways: manually, using a computer algorithm or using a combined (semi-automated) approach. Segmentation is a challenging task for ultrasound, as artefacts such as speckle and shadowing can reduce the definition of boundaries. Therefore, surface rendering tends to be limited to situations in which the surface of the structure is bounded by liquid, for example, the foetal face and body (Figure 6.39), blood vessels, the bladder, cardiac structures and the eye.

(ii) Volume rendering utilises the whole data volume and produces 2D views by extracting data from lines of sight projected through the volume ('ray casting'). The voxels that intercept each ray are weighted and summed to achieve the desired result in the rendered image. For example, maximum intensity projection displays only the maximum voxel intensity encountered by the ray. In another approach, each pixel

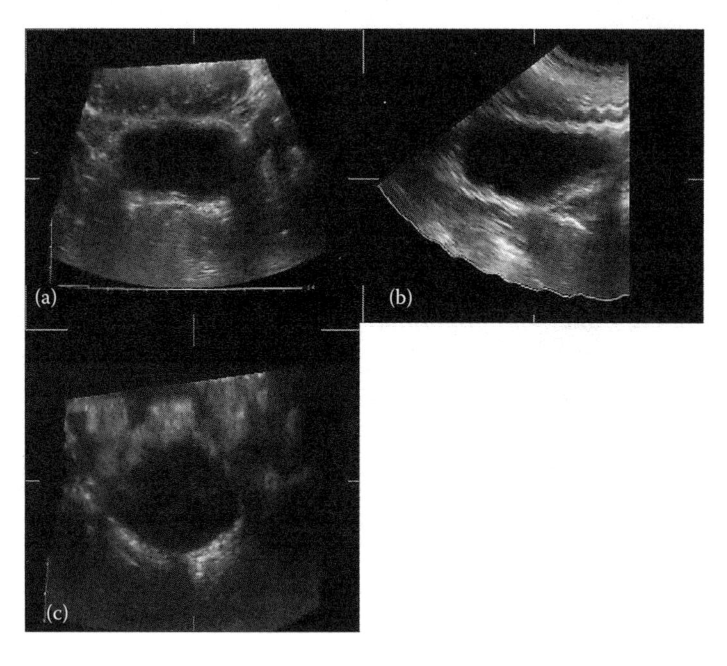

FIGURE 6.38
Example of multiplanar reformatting of a 3D volume of the urinary bladder, showing (a) a transverse view, (b) a longitudinal view and (c) a coronal view (which would not be accessible using conventional 2D ultrasound imaging). The 3D data were acquired by a freehand tilting movement using a Flock-of-Birds position sensor. The jagged edges of the bladder, seen in the longitudinal view, are a consequence of breathing motion.

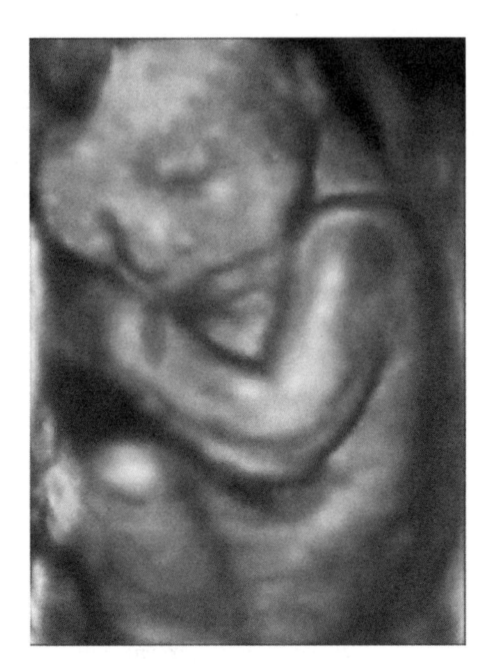

FIGURE 6.39
Example of surface rendering of surface-echo time of arrival, of a 24 week foetus. (Courtesy of Medison Inc., Miami, FL.)

in the 2D view is given the value of the first voxel whose intensity falls between two threshold limits. This tends to extract a surface. Since volume rendering projects the whole volume onto a 2D plane, interpretation of complex images is difficult and the approach is not recommended for B-mode data with subtle contrast between different soft tissues. It is most suited to situations in which 3D anatomical structures that occupy a relatively small percentage of the volume, or anatomical surfaces, are clearly distinguishable, for example, the foetus, tissue/blood interfaces in the heart and large arteries, and vascular structures demonstrated by the technique of power Doppler imaging described in Section 6.3.5.5 (Figure 6.40).

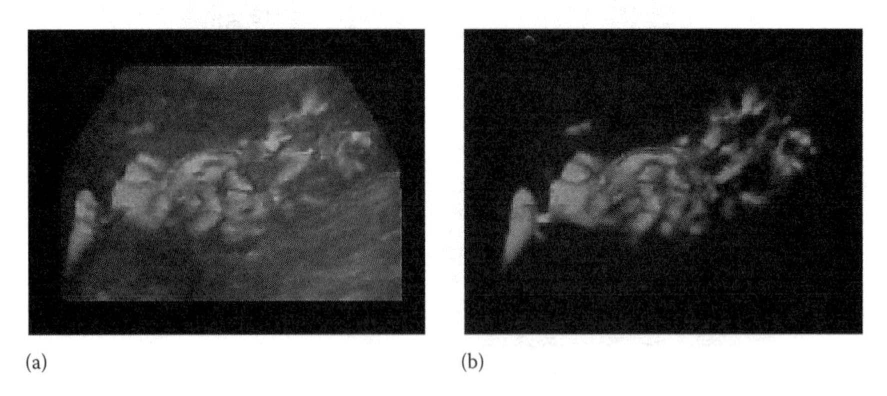

FIGURE 6.40
(See colour insert.) Example of volume rendering (of a 3D power Doppler scan of a normal kidney): (a) maximum intensity projection volume-rendering of grey-scale data, with blood flow superimposed using surface rendering and (b) blood-flow component of the image only.

6.3.4.4 Benefits and Limitations

3D ultrasound imaging has the potential to make a significant impact in a wide variety of clinical settings. Its main advantages are as follows:

(i) It provides access to new information in the form of views of tissue that are inaccessible with 2D ultrasound. For example, there is some evidence to suggest that multiplanar reformatting could be a valuable adjunct to the characterisation of breast masses using ultrasound (Rotten et al. 1999).

(ii) It produces a more repeatable and complete record of the anatomy. This may be useful for planning or guiding surgery, for comparative assessment over a long period of time (e.g. response to therapy), for multi-modality imaging and for tele-medicine.

(iii) It results in more accurate volume measurements than 2D methods. The latter measure distances in two or three planes and use an elliptical assumption to calculate volume.

(iv) It provides views of continuous curved surfaces, such as the foetal limbs and face, cardiac structures and blood vessels.

(v) It has the potential to provide better image quality, by compounding images taken from different views.

To date, the most widespread use of 3D ultrasound imaging has been in obstetrics. Rendering techniques allow foetal malformations, such as cleft palate, to be evaluated with greater precision than is possible with 2D ultrasound (Rotten and Levaillant 2004). Other important applications include ultrasound-guided procedures (e.g. needle biopsy and seed placement in prostate brachytherapy), gynaecology (e.g. diagnosing anomalies of the uterus), image-guided surgery (e.g. correcting for brain shift during tumour resection), echocardiography (e.g. assessing valve geometry), 3D angiography and volume measurement.

Apart from practical issues such as the need for extra equipment and the learning curve associated with acquiring and interpreting the data, the main disadvantage of 3D ultrasound is its reduced temporal resolution. Unlike 2D ultrasound, the temporal resolution may not be sufficient to avoid artefacts due to patient motion (e.g. random motion, respiration or cardiovascular motion). The time for one volume acquisition (when no parallel-processing techniques are used) is given by the 3D equivalent of Equation 6.40:

$$T = \frac{(N_l N_f)2D_{max}}{c} \tag{6.41}$$

where

N_l is the number of scan lines in a single frame,
N_f is the number of frames in a volume, and
D_{max} is the scan depth.

Therefore, for a typical abdominal imaging situation using a fan-shaped acquisition, where there may be 128 scan lines, 65 frames (e.g. to scan through an angle of 65° with a sampling rate of one frame per degree), and an imaging depth of 15 cm, the value of T would be 1.62 s.

A number of different approaches have been taken to improve the temporal resolution of 3D ultrasound. First, commercial systems generally allow the user to increase the volume rate by reducing the spatial sampling and/or the size of the field of view. Another solution that involves a trade-off is parallel-beam receive (Smith et al. 1991), whereby a fairly broad ultrasound beam is emitted and the received signals are fed to several receive beam formers working in parallel. The subject of parallel-beam receive is also discussed in Section 6.7.2. For example, if four beams are acquired simultaneously, it increases the frame rate by a factor of four. As with reducing the spatial sampling, the increased volume rate is achieved at the expense of spatial resolution, although without generating undersampling artefacts that would be present when sampling alone is reduced. A third option is to use respiration or cardiac gating, in which each image in the 3D sequence is captured at the same point in the respiratory or cardiac cycle. However, this approach does not eliminate artefacts due to random motion.

6.3.5 Doppler Methods

The essential principle of Doppler methods, as presented in Section 6.2.2, is to detect the difference in frequency between the transmitted and received ultrasound signal. The principle of Doppler-shift detection is illustrated in Figure 6.41 for the simple case of a single

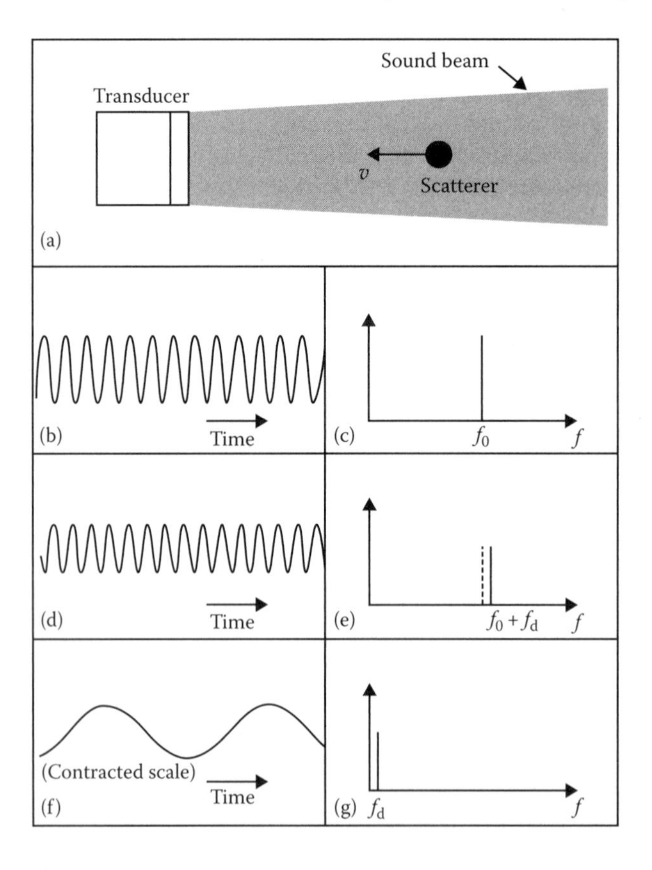

FIGURE 6.41
Simple CW Doppler-shift detection for a single scatterer moving axially towards the receiver (a). The left-hand graphs (b), (d) and (f) show the transmitted, received and Doppler-shift signals, respectively. The right-hand graphs (c), (e) and (g) show the corresponding frequency spectra. (Adapted from Atkinson and Woodcock, 1982, with permission.)

scatterer moving towards a CW receiver. The transmitted signal and its corresponding frequency spectrum are shown in Figure 6.41b and c, respectively. As indicated in Figure 6.41d and e, the signal reflected from the moving scatterer has a higher frequency than the transmitted signal. The purpose of Doppler instrumentation is to detect the Doppler-shift frequency, illustrated in Figure 6.41f and g, by a process known as frequency demodulation (see Section 6.3.5.3).

6.3.5.1 Choice of Frequency (f_0)

As with pulse-echo imaging, the choice of operating frequency for the transmitted signal is governed by a compromise between resolution (but this time we include both spatial and flow velocity resolution) and depth of penetration. However, in the case of Doppler-shift measurement, there is a basis for a quantitative analysis of the optimum operating frequency, since it is known that the scattering of ultrasound from blood has an f^4 dependence. The analysis estimates the blood-echo SNR, SNR_e, as a function of frequency and depth, which leads to an equation for predicting how the frequency f_0 at which SNR_e is maximum varies with depth (Figure 6.42). The analysis depends on many factors, including the frequency dependence of attenuation in the tissue overlying the blood vessel. Many assumptions are made, and, hence, the resulting relationship is an approximation. Nevertheless, it is a useful general guide to the choice of frequency for a particular application.

6.3.5.2 Notes on the 'Doppler Equation'

Equation 6.21 is often referred to as the Doppler equation for (backscattered) ultrasound in medical applications. Before we proceed to a discussion of practical ultrasound Doppler measurement systems, it is useful to note some of the features, limitations and approximations that are associated with this simple equation.

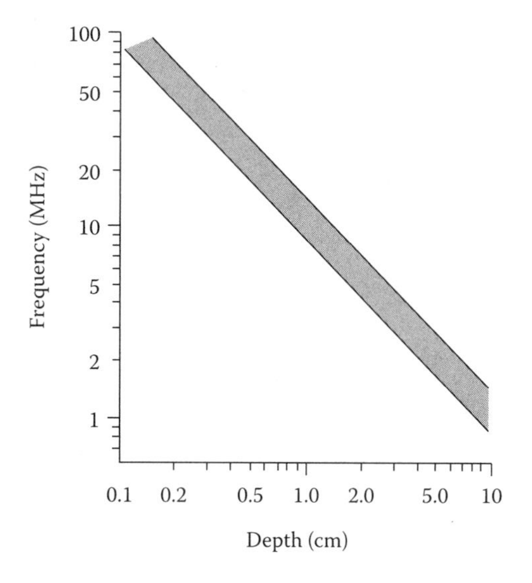

FIGURE 6.42
Estimated range of frequencies, as a function of depth in muscle tissue, that will maximise the SNR for detection of ultrasound scattering from blood (plot of equation f_0 [MHz] = $9/d$ [cm]). (After Baker et al., 1978.)

(i) One needs to know cos θ and c in order to measure v in absolute terms. It is often not easy to determine θ accurately, and this makes accurate determination of flow velocity difficult.

(ii) The information about velocity is encoded by frequency modulation, which helps the detection of weak echoes in the presence of noise. However, this is vital since the echoes from blood are very much smaller than the backscattered signal from nearby slowly moving tissue such as the vascular wall. The latter part of the signal is often called 'clutter', and it is the function of a signal-processing element, known as the 'wall filter', to reject it.

(iii) The simple Doppler equation neglects to take into account the finite width of the sound beam, and predicts that a single scatterer travelling at a constant velocity should produce a line spectrum. In reality, scattering structures that move in a non-axial direction pass across the sound beam and thus their echoes are amplitude modulated, according to the lateral beam profile (see Figure 6.43). This amplitude modulation broadens the Doppler line spectrum. Another way of thinking of this phenomenon (known as transit-time broadening) is to say that, to define v for a given scatterer by measuring f_d, we would need access to the

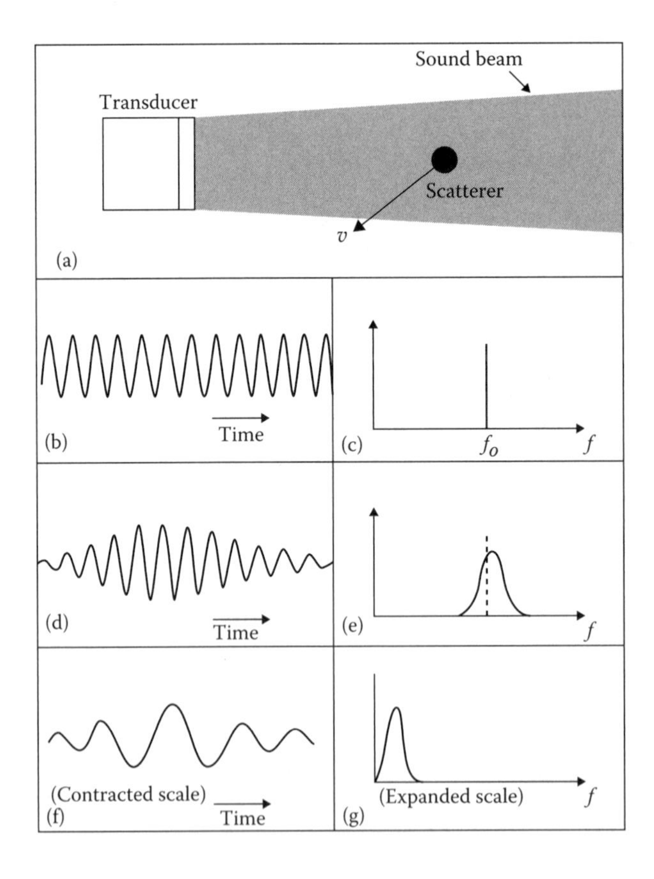

FIGURE 6.43
Transit-time broadening due to the finite width of the ultrasound beam. (a) shows the scatterer moving in a non-axial direction. The left-hand graphs (b, d and f) show the transmitted, received and Doppler-shift signals, respectively. The right-hand graphs (c, e and g) show the corresponding frequency spectra. (Adapted from Atkinson and Woodcock, 1982, with permission.)

backscattered signal for an infinite amount of time, whereas we only possess data lengths of the time taken for the scatterer to traverse the beam, that is, this is a Fourier-domain 'leakage' phenomenon. There exists an uncertainty compromise here, in that if the beam width is reduced to improve lateral resolution, this is at the expense of the frequency (velocity) resolving properties of the Doppler system, and vice versa. The Doppler-shift bandwidth due to transit-time broadening can be derived approximately by using Equation 6.26 for the lateral beam width at the 3 dB level, FWHM $\approx 1.2\lambda n_f$. The time taken for the scatterer to traverse the beam, Δt, is then given by

$$\Delta t = \frac{\text{FWHM}}{v \sin\theta} = \frac{1.2\lambda n_f}{v \sin\theta}. \tag{6.42}$$

The Doppler-shift bandwidth, Δf_d, is the reciprocal of this time period such that

$$\Delta f_d = \frac{v \sin\theta}{1.2\lambda n_f}.$$

Substituting the Doppler equation (6.21) results in

$$\Delta f_d = \frac{c f_d}{2 f_0 \cos\theta} \cdot \frac{\sin\theta}{1.2\lambda n_f} = \frac{f_d \tan\theta}{2.4 n_f}. \tag{6.43}$$

Equation 6.43 was not arrived at by a rigorous derivation and it can be seen to break down at $\theta = 90°$ (although there is no Doppler shift in this case). Nevertheless, it is useful for illustrating the inverse relationship between spatial and frequency (velocity) resolution.

(iv) CW transmission is assumed. This will not be the case for many systems, as we shall see later.

(v) In addition to the transit-time broadening described in (iii), blood vessels consist of multiple scatterers moving at different velocities and in different directions, which further broadens the Doppler-shift spectrum. The spectrum varies with time due to the cardiac cycle, as depicted schematically in Figure 6.44. This 3D distribution contains a great deal of potentially useful information. The ability to observe the distribution of flow velocities within a blood vessel, and time variations to within a relatively small fraction of the cardiac cycle, is a unique feature of ultrasound Doppler techniques. The specific shape of the Doppler-shift spectrum at a given instant is a complicated function of the beam shape, the shape and size of the vessel, and the distribution of flow velocities within the vessel. Much effort has been directed towards understanding these relationships and, given simple shapes for the sound beam (e.g. Gaussian) and blood vessel (cylindrical), it is possible to infer the radial distribution of flow velocities (e.g. parabolic or otherwise) from the measured Doppler spectrum. Much use is made clinically of observations of spectral-shape changes as a function of longitudinal distance along a blood vessel, for detection of flow disturbance induced by lesions and constrictions of the blood vessel. This can be carried out qualitatively, by listening to

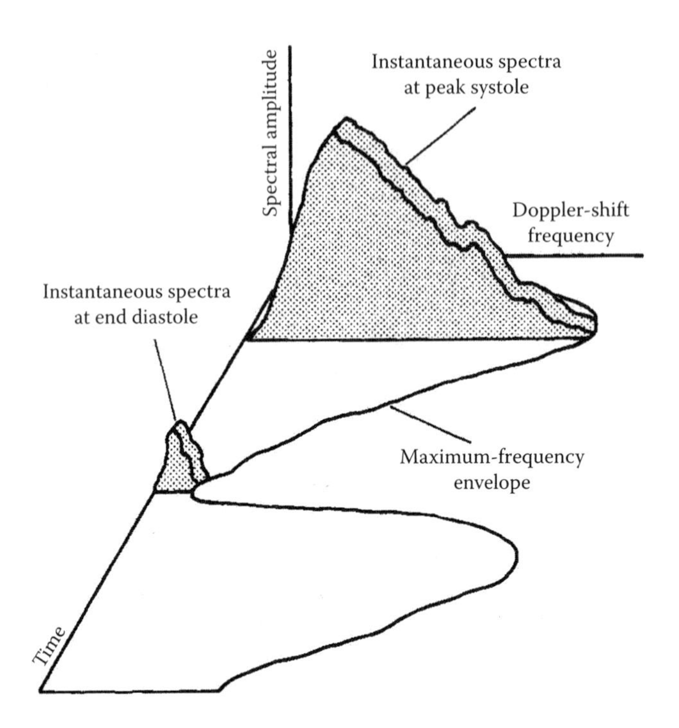

FIGURE 6.44
Schematic illustration of the 3D nature of the Doppler-shift signal from pulsatile flowing blood.

the Doppler-shift audio spectrum, or by quantitative analysis of the spectrum, as described in Section 6.3.5.5. Generally, the disturbed flow results in an increased range of Doppler frequencies, that is, spectral broadening. The time envelope of the maximum Doppler-shift frequency is also useful since its shape is influenced by arterial-wall elasticity and downstream vascular impedance.

6.3.5.3 Signal Processing in Doppler Systems

6.3.5.3.1 CW Flow Detector

In order to extract f_d (positive or negative) from the returned echo signal, which includes the clutter from nearly stationary structures, some signal processing is required. The achievements in this respect are directly related to the sophistication of the system. A number of levels of sophistication may be defined. The simplest is the CW flow detector, illustrated in Figure 6.45. A continuous acoustic wave is transmitted from one element of the transducer, whilst a second element receives the echoes. After amplification, the difference frequencies are extracted by the demodulator, amplified and made audible directly or processed further, prior to some other form of presentation. Many different designs of demodulator exist (e.g. as discussed by Atkinson and Woodcock 1982, Evans and McDicken 2000, Eckersley and Bamber 2004). The simplest mixes the received echo signal with the transmitted reference frequency (producing both the sum and the difference frequency), and then low-pass filters the result to extract only the difference frequency. Such a simple demodulator does not distinguish between flow towards or away from the sound source. An improvement is the heterodyne system, in which the received signal is mixed with an offset reference frequency (f_R), producing a signal where forward flow is represented as frequencies above f_R and reverse flow as frequencies below f_R. Such a signal is not so suitable for audible presentation (because of the constant frequency

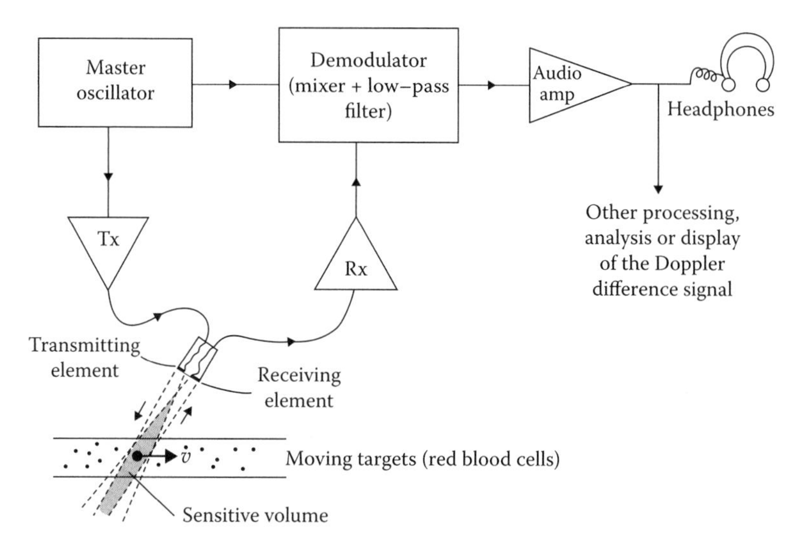

FIGURE 6.45
Block diagram of the essential components in the signal-processing chain of a CW ultrasound Doppler blood-flow detector. Tx = transmitter, Rx = receiver.

for zero flow) and places a severe limit on the maximum measurable reverse-flow velocity. The best demodulation method, and the most widely used, is the so-called quadrature-phase demodulator with 'phase-domain' processing. This provides two audio signals, one for forward flow and the other for reverse flow (sometimes used with stereo audio presentation). The method requires the demodulation to be performed twice in parallel. When implemented with analogue processing, it involves mixing the received signal with the reference signal (a cosine wave) along one channel and with a 90° phase-shifted version of the reference signal (a sine wave) along the other channel, thereby resulting in the direct and quadrature components of the Doppler signal. The two signals are 90° out of phase but otherwise identical. If flow is towards the probe, then the quadrature signal leads the direct signal by 90°, and vice versa. Excellent directional separation is possible with this system.

Methods of further processing (other than listening), analysis and display of the time-varying Doppler-shift signal (Figure 6.44) have, in the past, varied greatly. Real-time spectral analysis (referred to as 'spectral Doppler'), is the preferred method. Spectral analysis breaks down the Doppler signal into frequency components, usually using a digital technique known as the fast Fourier transform (FFT). A Fourier transform is performed serially on small (approximately 5 ms) segments of the digitised Doppler signal. The output is a 2D spectral display known as a 'sonogram' in which the signal amplitude at each frequency is encoded as a grey level and this is plotted as a function of time (abscissa) and frequency shift (ordinate). An example of the sonogram display will be provided when discussing the pulsed-wave flow detector (see Figure 6.53b). Derived characteristics of the instantaneous spectrum (such as the mean frequency, peak frequency or some form of flow disturbance index) may also be computed, and their time variation displayed, usually after obtaining several cardiac cycles' worth of sonogram data. Indices of the flow characteristics may be calculated from more than one property of the spectrum. For example, the pulsatility index (PI) is defined as

$$\mathrm{PI} = \frac{(\mathrm{max} - \mathrm{min})}{\mathrm{mean}} \tag{6.44}$$

where

max is the maximum instantaneous peak velocity during systole,
min is the minimum instantaneous peak velocity during diastole, and
mean is the average velocity, or average peak velocity, over the cardiac cycle.

The flow resistance index, RI, is defined as

$$RI = \frac{(\max - \min)}{\max}. \tag{6.45}$$

An advantage of these indices is that they use relative, rather than absolute velocity parameters, meaning that errors in estimating the flow angle are unimportant.

6.3.5.3.2 Systems that use Pulsed-Wave Flow Detection

As indicated by Figure 6.46a, the CW Doppler system possesses limited resolution in the depth direction and would be unable to distinguish between the signals from two blood vessels lying within the sensitive volume but at different depths (unless, of course, the blood in each of the vessels was flowing in opposite directions). This problem is resolved by use of increased sophistication in signal coding and processing to preserve range information as well as frequency information. The usual method of achieving this is embodied in the pulsed-wave Doppler system, as illustrated in Figures 6.46b and 6.47, which possesses attributes similar to both the pulse-echo imaging and CW Doppler signal-processing schemes already described. So as to aid the description of the pulsed Doppler method, Figure 6.47 also depicts a schematic representation of the signal that might be observed at the various numbered points in the circuit block diagram.

At point 1, the master oscillator generates a sine wave similar to the aforementioned CW system. This is gated under the control of a clock, so that at point 2 one would observe a train of short pulses (a few cycles long) repeated at intervals determined by the PRF, as for echography. These pulses are amplified and used to excite a single-element transducer, which also acts as a receiver. After preamplification, the echo signal, which (at point 3) looks just like the exponentially decaying A-scan prior to TGC correction and envelope

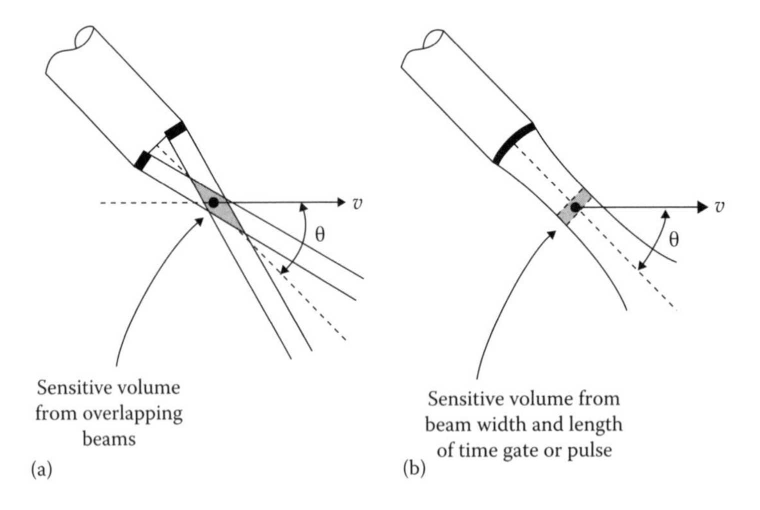

FIGURE 6.46
Two types of Doppler system, the CW (a) and the pulsed wave (b) illustrating the angle θ, which is of significance in the backscatter Doppler equation (6.21).

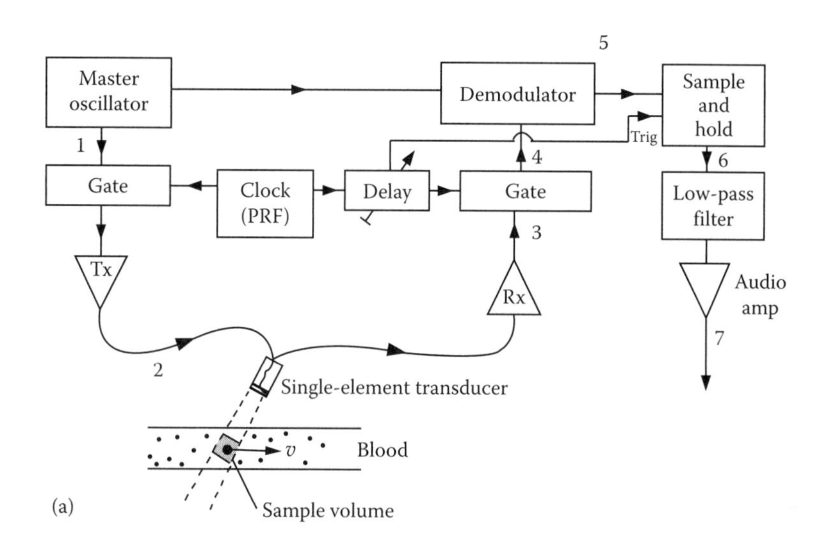

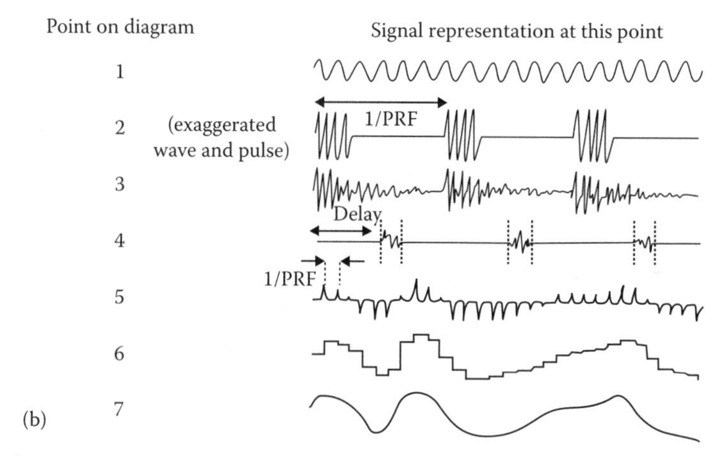

FIGURE 6.47
(a) Block diagram of the essential components in the signal processing for a pulsed-Doppler system and (b) schematic versions of the signal one might expect to observe at the labelled points.

detection, passes through another gate. This gate may be of adjustable length but, in all systems, it is triggered at a time that is adjustable relative to the clock that triggered the transmitted pulse. The gate output, at point 4, is therefore a depth-selected portion of the RF echo signal. Demodulation of this sampled echo signal, by one of the methods described for the CW system (e.g. mixing and smoothing), results in a sampled Doppler-shift signal (point 5), where the sampling frequency is the PRF. A sample-and-hold amplifier, synchronised to the delayed clock, is used to convert this to a continuous, though quantised, Doppler signal, as shown for point 6. Further smoothing then produces the final audio signal at point 7.

The way in which the sampled Doppler-shift signal (i.e. the output of Figure 6.47 at point 5) is built up can be understood by considering the successive echoes received from a plane target that is receding from the transducer at a constant speed (see Figure 6.48). Pulses are transmitted at the PRF at one beam location. For each successive pulse, the target has moved to a greater depth such that the received signal is delayed compared to the

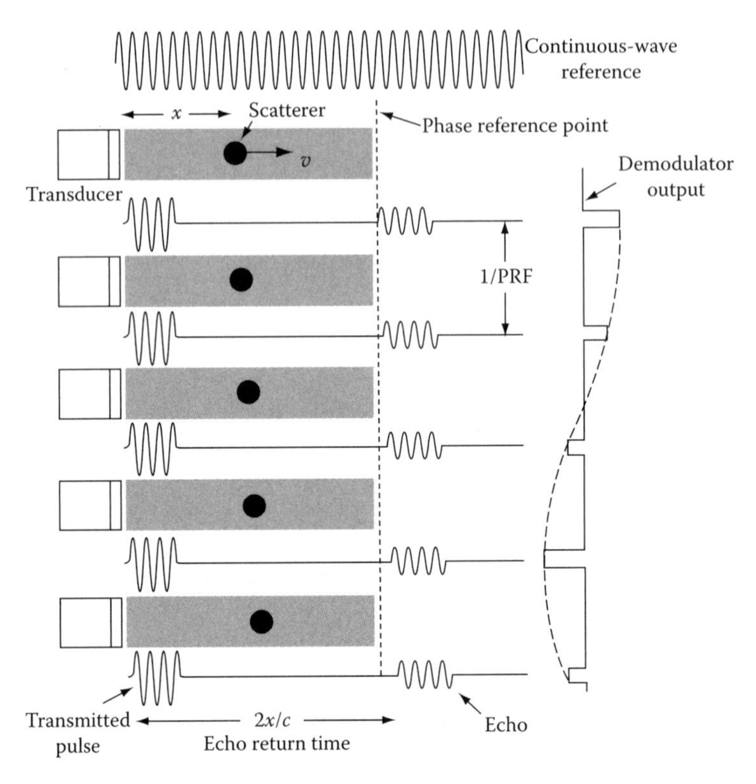

FIGURE 6.48
Build-up of the sampled Doppler-shift signal. (Adapted from Atkinson and Woodcock, 1982, with permission.)

previous pulse. The range gate samples the received signal between two fixed time points. Therefore, the output of the demodulator, which depends on the phase of the received echo signal with respect to the transmitted signal, varies from one pulse to the next. The rate of variation of this relative phase is determined by the speed of the target and is the desired Doppler-shift frequency. The fact that the Doppler signal is reconstructed from a series of successive pulses places a limitation on the maximum velocity that can be measured (see Section 6.3.5.4).

The pulsed Doppler method introduces another Fourier-domain leakage problem, in addition to that described earlier due to the finite width of the sound beam. As a result of the limited pulse length (see Figure 6.49b), the transmitted signal consists of a range of frequencies rather than a single frequency (see Figure 6.49c). This causes further broadening of the Doppler shift spectrum (see Figure 6.49e and 6.49g) and may be considered as an axial transit-time broadening phenomenon, due to the fact that echoes from a scatterer will grow and fade as the scatterer moves past a fixed phase reference point (Figure 6.48). The phenomenon is distinct from that of lateral transit-time broadening due to the finite beam width (shown in Figure 6.43), and in pulsed Doppler both types of transit-time broadening will be present. If the pulse length is reduced to improve range resolution, it is at the expense of frequency resolution, and vice versa. Assuming the scatterers move in the axial direction ($\theta = 0$), the Doppler frequency shift is given by

$$f_{\mathrm{d}} = \frac{2vf_0}{c}.$$

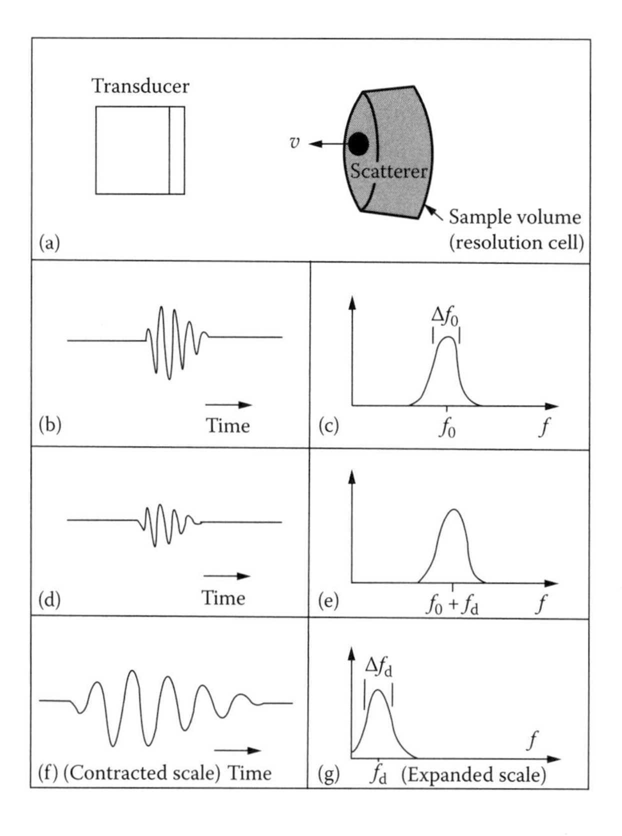

FIGURE 6.49
Axial transit-time broadening due to the finite axial size of the Doppler gate and/or pulse length. (a) Shows the scatterer moving through the sample volume. The left-hand graphs (b, d and f) show the transmitted, received and Doppler-shift signals, respectively. The right-hand graphs (c, e and g) show the corresponding frequency spectra. (Adapted from Atkinson and Woodcock, 1982, with permission.)

The axial resolution, l, is given by $c\tau/2$, where τ is the length of the pulse. Therefore, the time taken for the scatterer to pass a given phase reference point, Δt, is

$$\Delta t = \frac{l}{v} = \frac{c\tau}{2v}.$$

The Doppler-shift bandwidth due to broadening, Δf_{d}, is the reciprocal of this time period such that

$$\Delta f_{\mathrm{d}} = \frac{2v}{c\tau}.$$

Expressing the pulse length as the reciprocal of the bandwidth of the transducer (Δf_0) and then combining with the Doppler equation gives the following equation:

$$\Delta f_{\mathrm{d}} = \frac{2v\Delta f_0}{c} = \frac{f_{\mathrm{d}}\Delta f_0}{f_0} = \frac{f_{\mathrm{d}}}{Q} \tag{6.46}$$

where

Q is the quality factor of the transducer (the inverse fractional bandwidth) and is equal to the number of cycles in the pulse.

Equation 6.46 serves to illustrate the reciprocal relationship between the frequency (velocity) resolution and the axial spatial resolution in pulsed Doppler.

The essential principles of deriving the Doppler signal at one spatial location in pulsed Doppler imaging are summarised in Figure 6.50. The scatterer is moving towards the transducer face in this example, such that successive received echoes occur at progressively shallower depths. The first point to note is that the received echo is amplitude modulated due to both axial and lateral transit-time broadening, the relative contributions depending on the direction of motion and the ratio of the pulse length to the beam width. Frequency demodulation is illustrated here in its simplest form, that of sampling the RF echo at a fixed point in space to obtain a measure of its relative phase with each successive acoustic pulse. This can be thought of as the limit of the process described in Figure 6.48, for a gate size that has been reduced to a single sample (a finite gate size has the practical benefit of providing a better estimate of relative phase in the presence of noise). Direction discrimination is assumed to be performed using quadrature-phase demodulation (see the first paragraph of Section 6.3.5.3). In digital form, and with the idealised single-sample gate that we are using in this illustration, this involves sampling the received signal twice, at slightly different time points (depths), indicated by D (direct) and Q (quadrature) in Figure 6.50. In practice, the signal is sampled over a finite period of time equal to the length of the Doppler gate and the sampling gate is centred at the two different time points. The interval between D and Q is equivalent to a 90° phase difference with respect to the centre frequency of the transmitted signal. The Doppler signal, which is the phase of the echo at a fixed point in space as a function of time, is identical for both D and Q, but there is a phase difference between them. For a scatterer moving towards the transducer, Q leads D, while for a scatterer moving away from the transducer, D leads Q. One can also see that the time-varying phase estimates at both the D and Q positions (which, as stated previously, are amplitude and frequency modulated at rates that will depend on the speed of

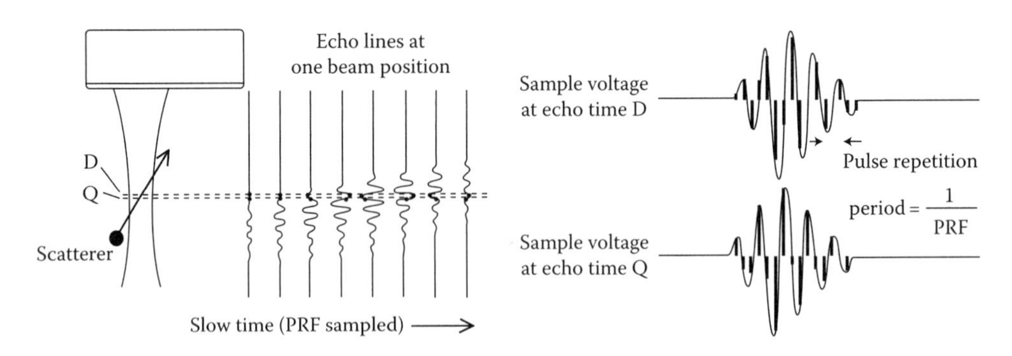

FIGURE 6.50
Essential principles of deriving the Doppler signal at one spatial location in pulsed-Doppler imaging using the simplest implementation of quadrature-phase frequency demodulation, sampling of the RF echo signal at two points with a 90° phase difference between them (see text for description). In order to create a pulsed-Doppler image, the process is repeated for multiple locations down each beam line and for multiple beam lines, followed by further processing at each location to extract a suitable parameter, such as mean Doppler frequency, for display (see Section 6.3.5.5).

the scatterer) are sampled at the PRF, which is fixed. Hence, if the scatterer were to move faster than illustrated here, there would be a risk of undersampling the frequency modulated component (i.e. the Doppler signal). The consequences of such undersampling are discussed in Section 6.3.5.4. In order to create a pulsed Doppler image, the process just described is repeated for a sequence of locations down each beam line and for many beam lines, followed by further processing at each location to extract a summary parameter, such as mean Doppler frequency, for display. The description of pulsed Doppler imaging is continued in Section 6.3.5.5.

6.3.5.4 Range–Velocity Compromise

An important problem arises with all pulsed Doppler systems. As may be seen from Figure 6.51, the quadrature sampling points must be sufficiently frequent to allow accurate reconstruction of the Doppler signal. At the slow flow rate, the sampling is adequate and the measured Doppler signal faithfully reproduces the true Doppler signal. However, at the faster flow rate, the measured Doppler signal (black line) misrepresents the true Doppler signal (grey line), a phenomenon known as aliasing. In addition to underestimating the speed of flow, aliasing in pulsed Doppler misrepresents the flow direction. To avoid aliasing (see Sections 3.9.6 and 11.6), the Doppler-shift frequency, f_d, must be less than or equal to PRF/2. Therefore, from the Doppler equation (6.21), the maximum velocity that can be detected is given by

$$(v\cos\theta)_{\max} = \frac{c\,\mathrm{PRF}}{4f_0}.$$

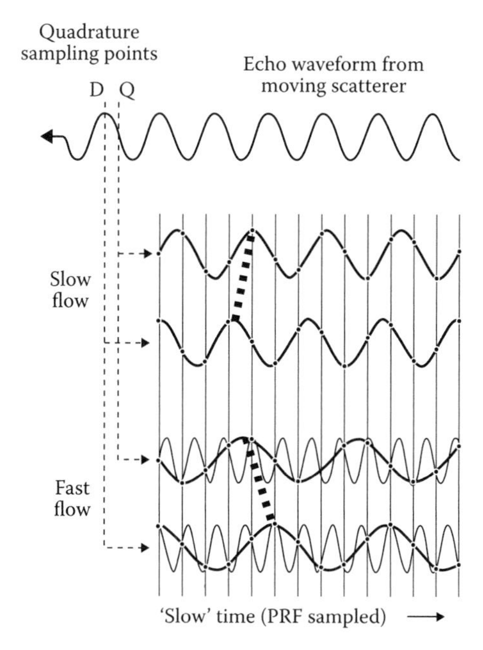

FIGURE 6.51
Aliasing in pulsed-Doppler systems. Note that both the speed and direction of flow are misrepresented.

When this is combined with Equation 6.39 from pulse-echo imaging, we arrive at the 'range–velocity' compromise for the maximum velocity v_{max} that may be unambiguously observed for a given maximum depth of the gate, D_{max}:

$$(v\cos\theta)_{max} = \frac{c^2}{8f_0 D_{max}}$$

that is,

$$v_{max} \approx \frac{2.8\times10^5}{f_0 D_{max}} \tag{6.47}$$

(f_0 in MHz, D_{max} in mm and v_{max} in mm s^{-1}). The rapid rate of decrease in v_{max} with D_{max}, due to the inverse relationship, imposes severe limits on the applicability of pulsed Doppler methods to studying deep-lying blood vessels. Figure 6.52 illustrates the limits of range and velocity, obtained from Equation 6.47, which must be observed if range–velocity ambiguities are not to be present. A common method of maximising system performance at all depths is to link the PRF to the delay that controls the position of the gate in the receiving circuit. This causes other complications in the design, such as the necessity for a sharp cut-off smoothing filter at point 6 in Figure 6.47, whose cut-off frequency is variable and linked to the PRF. Another approach to this problem is mentioned in the following when discussing linked pulsed-Doppler and echographic imaging (Section 6.3.5.5).

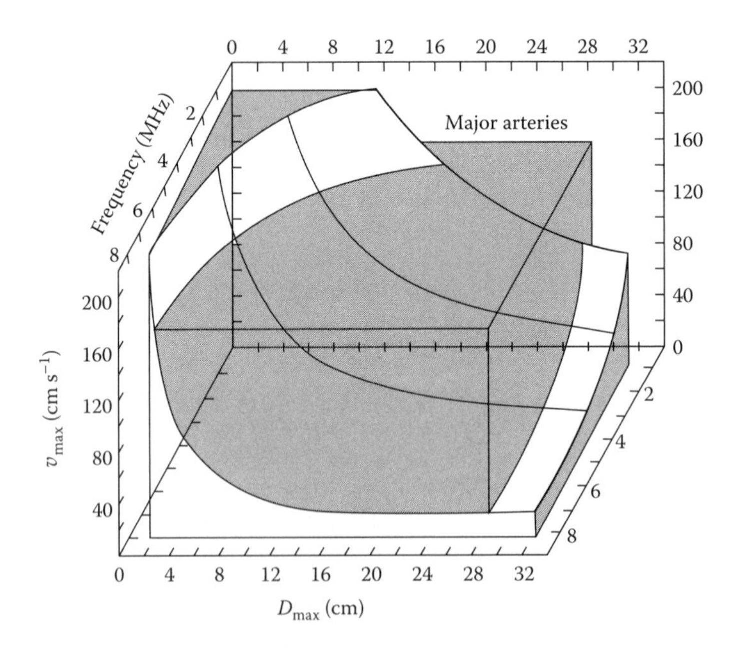

FIGURE 6.52
Maximum range, velocity and frequency limits for pulsed Doppler, compared with depths and flow velocities of major arteries. Points above the depicted surface produce range/velocity ambiguities. The solid box shows the range of depths and frequencies (assuming θ = 0) for the major arteries in the body. Aliasing can often be avoided by control of θ. (Adapted from Baker et al., 1978.)

6.3.5.5 Doppler Display Methods

6.3.5.5.1 Spectral Doppler and Duplex

As with CW Doppler, the preferred method of processing the time-varying Doppler-shift signal is to use real-time spectral analysis. The derived time-varying spectrum can then be displayed as a sonogram. Modern scanners are able to perform pulse-echo imaging and pulsed Doppler apparently simultaneously by interleaving between the methods, giving rise to an operating and display mode known as 'Duplex'. Where colour Doppler imaging is also available (see Section 6.3.5.5.2), all three types of information are displayed together, known as 'Triplex' (see Figure 6.53). The user is able to select the line of site and range gate for the pulsed Doppler from the real-time B-mode image. This approach offers advantages in two ways: the Doppler information can enhance ability to identify correctly vascular anatomy and, more importantly, the imaging information can make the Doppler examination less ambiguous and more quantitative. The increased quantification results from the fact that, when imaging and studying the blood flow in a well-defined vessel (e.g. the carotid artery), the imaging system permits specification of the angle θ in Equation 6.21, so that the Doppler-shift spectrum may be expressed in terms of actual flow velocities (m s^{-1}). The diameter of the vessel is also available, in principle, for volume flow-rate calculation, although, in practice, the errors in estimating the active cross-sectional area by this method may be too large for it to be useful. The range–velocity ambiguity (Equation 6.47) can also now be overcome, allowing the PRF to remain high enough not to alias the Doppler signal even when the Doppler gate is set at a depth that exceeds the limit imposed by Equation 6.47, that is, a pulse is transmitted before the echo from the previous transmit pulse has been received. The scanner displays on-screen markers for all of the possible locations from which the Doppler signal might have arisen. Thus, under the restriction

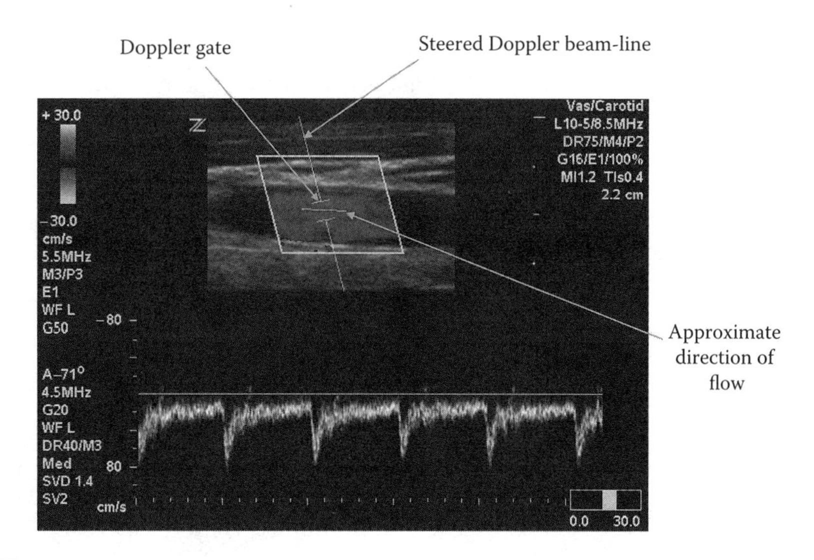

FIGURE 6.53
(See colour insert.) Triplex display of blood flow in the carotid artery. Above: the position and size of the Doppler gate, the steered Doppler beam line and the approximate direction of flow as indicated by the user are superimposed on the combined B-mode and colour Doppler image. Below: the spectral Doppler information is displayed as a sonogram, which depicts spectral power (brightness) as a function of flow velocity and direction (vertical) and time during the cardiac cycle (horizontal). When the user changes the scanner's direction-of-flow indicator, the scanner alters cos θ in Equation 6.21 to change the vertical scale of the sonogram.

that the vascular anatomy in the scan plane is sufficiently simple (e.g. it is known that there is only one significant vessel), the operator is able to use the image, plus knowledge of anatomy, to recognise the correct Doppler gate and correctly position it.

6.3.5.5.2 Colour Doppler Imaging

Flow imaging provides information on blood flow over a 2D tissue section in real time. It requires parallel multi-gated pulsed Doppler (i.e. simultaneous Doppler gates at many depths) as well as scanning of the Doppler beam to different lateral positions. At each line of sight and gate position, a single summary parameter is extracted from the Doppler-shift signal and encoded as a time-varying colour. The most common parameter is the mean Doppler frequency shift, which indicates the mean velocity of flow. Images of this parameter are often called simply 'colour Doppler' images, as distinct from power Doppler images described later, although both methods employ colour to code the flow information and overlay it on a grey-scale image of echo anatomy. Display logic decides when to display flow information (as a colour) and when to display anatomical information (as a grey level), resulting in an image such as that shown in Figure 6.54a. Generally, relative to thresholds that have been set within the scanner and beyond user control, echoes of high amplitude and low Doppler-shift frequency will be displayed as amplitude (grey level) and echoes of low amplitude but high Doppler-shift frequency will be displayed as mean velocity (colour). The Doppler-shift frequency and its sign (direction of flow) are usually encoded as different, easily identifiable aspects of hue, saturation and intensity of the colour scale.

The summary parameter is extracted directly from the Doppler signal after frequency demodulation shown at step 7 of Figure 6.47, or as shown in Figure 6.50, rather than being calculated from the Doppler-shift spectrum in the frequency domain. This approach is necessary to achieve real-time 2D information, as spectral analysis over a 2D region would be too computationally intensive. A further time saving is achieved by using only a small number of repeat transmissions to estimate the flow parameter at a particular position (i.e. approximately 5–10 instead of the 40 or so used in spectral pulsed Doppler). Demodulation is performed using the phase quadrature scheme. The mean Doppler frequency shift is a measure of the average rate of change of phase of the Doppler signal. The instantaneous estimator normally used for this purpose is the autocorrelation

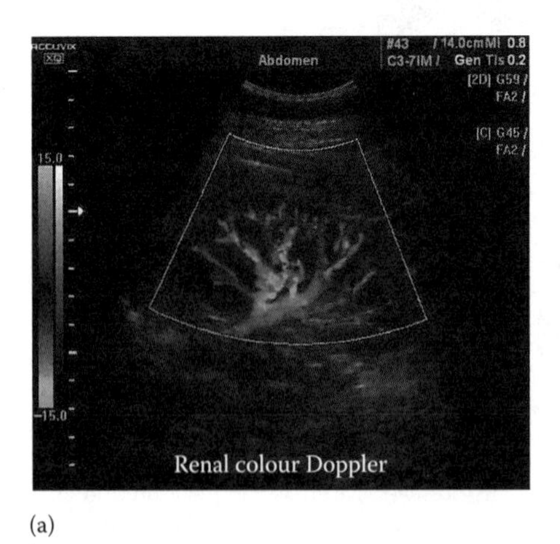

(a)

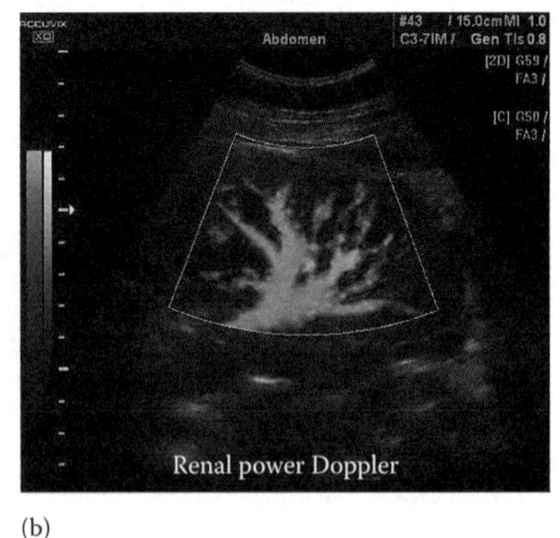

(b)

FIGURE 6.54
(See colour insert.) (a) Colour Doppler and (b) power Doppler images of the same view of a normal human kidney. (Courtesy of Medison Inc., Miami, FL.)

detector, the description of which is beyond the scope of this chapter (see, e.g. p. 315 in Eckersley and Bamber 2004).

There is an alternative method of processing that does not extract the mean velocity from the Doppler signal, but calculates it directly from changes in the arrival time of the RF echoes from one transmit pulse to the next, as determined from the position of the peak in the cross-correlation function between the echo signals for the two pulses. This signal processing is referred to as time-domain correlation (TDC) and it has two advantages, in principle, over Doppler-based processing. First, it can be implemented using shorter pulse durations, which results in superior axial resolution. Second, it can measure higher velocities, as it is not subject to aliasing. It is also the basis for the processing that may be used to estimate tissue displacement in elastography, described in Section 6.3.6.2.

Although colour Doppler images display a parameter related to mean flow speed, the information cannot be considered as truly quantitative. Figure 6.55 shows the Doppler frequency spectra that would be obtained at two different positions along the curved path of a hypothetical single blood vessel. Even if the flow speed were the same at both positions, the Doppler angle is different. Therefore, the mean Doppler frequency shift, which is the parameter displayed in colour Doppler imaging, will be lower for the part of the vessel that is imaged at a larger angle. For some parts of the vessel, the Doppler angle will be close to 90°, resulting in nearly zero Doppler frequency shift and, hence, a dropout in the colour image.

Colour Doppler imaging, that is, flow imaging performed by extracting the mean frequency shift from the sampled Doppler signal, is subject to aliasing of the type illustrated in Figure 6.51. Just as for spectral pulsed Doppler, the consequences are misrepresentation of high flow velocities as low, and an apparent reversal of flow direction. This appears as an erroneous region of colour, which is often confined to the centre of the vessel (where the velocities are highest). As for spectral pulsed Doppler, the maximum unambiguously displayable flow velocity is limited by the maximum scan depth, according to Equation 6.47.

There is also a limit on the smallest flow velocity that can be distinguished from stationary clutter. It was explained previously that for pulsed-Doppler systems, the Doppler signal is reconstructed from a number of repeat transmissions along the same beam position. Therefore, the slower the movement of the scatterer, the fewer the number of cycles that are available to estimate the Doppler-shift frequency, f_d. A minimum of one cycle of the Doppler signal is required to reduce the uncertainty on the estimate of f_d to a value of f_d. If this condition is imposed, the system must pause its scan at each colour Doppler image line position for a time (given by the number of transmit pulses per line divided

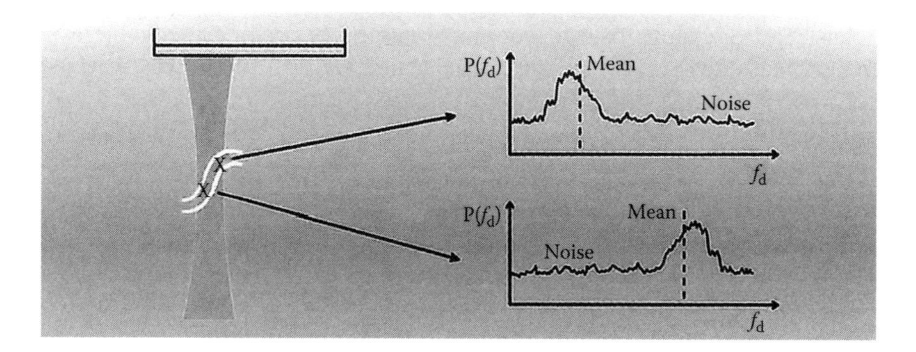

FIGURE 6.55
Doppler spectra as they might exist for two different positions in a tortuous vessel within which the flow speed is constant.

by the PRF) that is equal to the reciprocal of f_d (given by Equation 6.21). Consequently, the minimum velocity that can be detected above noise is

$$v_{min} = \frac{c\,\text{PRF}}{2f_0 N \cos\theta} \tag{6.48}$$

where N is the number of transmissions per line. In practice, at least two to three cycles are generally acquired to improve the estimate of the frequency. If this is combined with the constraint that to avoid aliasing, we require at least two to three samples per cycle of the highest Doppler-shift frequency component, then it becomes clear that the minimum number of repeat transmissions is approximately five to ten, which, as mentioned earlier, is the number typically used in colour Doppler systems.

The requirement to wait until at least one cycle of the lowest Doppler-shift frequency of interest has been obtained, before moving to the next lateral position, leads to a trade-off between temporal and spatial resolution, that is, for a given minimum velocity that one wishes to measure, $(v \cos\theta)_{min}$, the frequency of the scan lines is inversely related to the frame rate:

$$N_{lines/cm} = \frac{2f_0(v\cos\theta)_{min}}{(\text{Frame rate}\cdot\text{Width}\cdot c)}. \tag{6.49}$$

In practice, colour and power Doppler images undergo multi-line processing (Figure 6.23) to reduce noise, in the form of both spatial and temporal averaging (persistence), which may further reduce the effective frame rate.

6.3.5.5.3 *Power Doppler Imaging*

An alternative processing method to the mean-frequency-shift imaging used in 'colour Doppler' is Doppler spectral energy imaging (also known as 'power Doppler'), in which the total area under the Doppler-frequency-shift spectrum is calculated and then displayed as a colour, usually on a colour scale that increases its intensity with increasing spectral energy. As with colour velocity imaging, the parameter of interest is estimated over a 2D region and superimposed on a B-mode image (see Figure 6.54b). The area under the Doppler spectrum is related to the total number of red blood cells in the sampled region, that is, it is a measure of blood volume.

A comparison of the key characteristics of Doppler spectral energy (power Doppler) and Doppler-frequency-shift (colour Doppler) imaging is provided in Table 6.5 and can be understood with reference to Figure 6.55. As explained previously, the mean flow

TABLE 6.5

Comparison of colour Doppler and power Doppler imaging.

Colour Doppler Imaging	Power Doppler Imaging
Related to flow velocity	Related to blood volume
Dependent on θ	Independent of θ
Directional information	No directional information
Independent of attenuation	Requires depth correction
Colour noise covers full range	Noise occupies a fixed low level

velocity varies from one part of a vessel to another, according to the imaging angle. In contrast, the area under the Doppler spectrum is independent of the imaging angle. It is also independent of flow direction, meaning that colour energy imaging does not provide directional information. The two imaging methods differ in their response to signal attenuation. For colour Doppler imaging, provided the ultrasonic SNR is sufficiently high for the Doppler frequency shift to be detectable above Doppler noise, attenuation has no effect, as the mean frequency is independent of the amplitude of the Doppler signal. However, for power Doppler imaging, the strength of the Doppler signal (and, hence, the pixel brightness) decreases with depth. The consequence of attenuation is that power Doppler imaging is not quantitative, despite its angle independence, as there is no means of correcting accurately for the decrease in signal power with depth or for the reduction or complete absence of information in parts of the image that experience shadowing. The final difference mentioned in Table 6.5 is with respect to the appearance of noise within the two types of image. For colour Doppler imaging, the estimated mean frequency that is displayed will be strongly influenced by the frequency corresponding to the largest spectral power, which, for a noisy situation, could be different from the true mean frequency shift. In fact, when the Doppler signal is below Doppler noise, the largest spectral power could be observed for any frequency in the spectrum, meaning that colour noise in colour Doppler imaging covers the full range and is visually quite disturbing. For power Doppler imaging, on the other hand, the displayed brightness just gets weaker, as the Doppler signal gets weaker, until it reaches the noise floor, as is the case for most imaging systems.

6.3.5.6 Tissue Doppler Imaging

Doppler signals that originate from blood flow generally have low amplitude and high velocity. As described previously, a high-pass 'wall filter' is used to eliminate signals due to slowly moving reflectors such as the vessel walls. However, in some applications, it may be of interest to characterise the motion of vessel walls or cardiac structures (such as the myocardium). This can be achieved using an imaging mode known as colour tissue Doppler imaging (TDI), whereby the high-amplitude low-frequency Doppler signals, due to structural components that would be rejected by the wall filter, are isolated and displayed. The 2D distribution of the mean tissue velocity is colour coded and superimposed on a simultaneously acquired B-mode image.

6.3.6 Special Imaging Modes

Having described the basic methods of forming images and making measurements with ultrasound, it is now worth mentioning some of the more sophisticated signal acquisition and processing techniques that have become commonplace on high-end diagnostic ultrasound machines in recent years. Two such 'special imaging modes' that are becoming more widespread are microbubble contrast-specific imaging and elasticity imaging. Each of these is described in the following.

6.3.6.1 Microbubble Contrast-Specific Imaging

This technique refers to images produced from the second harmonic, sub-harmonic, and ultra-harmonic signals scattered from contrast microbubbles, as discussed in Section 6.3.3, and to applications of the fact that microbubbles may be destroyed at high-incident sound pressures.

The practical application of the echo signal enhancement provided by the simple resonant scattering behaviour of microbubbles, described by Figure 6.7a, has been restricted mainly to helping Doppler examination reach deeper and smaller blood vessels. This is because such echoes, whilst providing a useful enhancement of the blood echo signal, are still small relative to the echoes from solid tissues. The Doppler shift from relatively fast flowing blood is, therefore, necessary to discriminate the bubble echoes from solid-tissue echoes. The desire to image slowly moving or stationary blood, and to image stationary microbubbles that may have been 'taken up' by or bound to the cells within a target tissue (see Section 6.7.5) has led to the investigation and application of bubble-echo 'signatures' that improve the discrimination between bubble echoes and background tissue echoes. Research in this area continues, and each new generation of device continues to provide an improvement in microbubble echo signal to background ratio. Detecting the echoes with a very wide bandwidth receiving transducer and then filtering the echo signal to select only the second-harmonic scattering that is illustrated in Figure 6.7b provides an improvement over imaging at the fundamental, but not as much as would be hoped because (as was seen in Sections 6.2.1.8 and 6.2.4.5) propagation through tissue is non-linear and therefore also generates harmonics, this being the basis of THI described in Section 6.3.3. Even more harmonic distortion occurs in a wave propagating through tissue that contains microbubbles, because their presence greatly increases the value of B/A of the tissue (see Section 6.2.1.8). The problem is made worse by the fact that current microbubble preparations contain a distribution of bubble sizes. Although this provides a degree of tolerance to the use of different insonating frequencies, it makes it difficult to take advantage of, for example, the frequency dependence of scattering around the resonance frequency. More advanced microbubble imaging techniques take advantage of the wide-band-harmonic-scattered signals generated at higher pressures (Figure 6.7b) and of the non-linear relationship that exists for microbubbles between the incident and backscattered pressures. Thus, bubble-specific imaging modes tend to employ multiple transmit pulses for a given image line, where each pulse has a different phase and/or amplitude. Combining the received pulses in an appropriate manner and after appropriate processing will (tend to) cancel echoes from tissue scatterers but retain the echoes from the microbubbles. One example of this type of processing is the summation of the stored echo signal received from a first pulse with an inverted echo signal received from a second pulse that is a phase-inverted replica of the first. Another might be where the second pulse is twice the amplitude of the first so that, to obtain cancellation of the tissue echoes by summation, its corresponding echo signal must first be multiplied by −0.5. Combinations of such phase and amplitude modulation techniques are common, using sometimes complex multiple pulse sequences. A problem with multiple-pulse techniques, however, is that they require that the separate pulses are scattered by the same structures; tissue movement may, therefore, cause a reduction in the signal to background contrast. Nevertheless, all of these contrast imaging methods have useful applications. Figure 6.56 illustrates how a liver metastasis may appear highly vascular during an early phase of contrast enhancement but later, when most of the agent has cleared from the blood pool and the only remaining microbubbles in the liver have been held stationary within Kupfer cells that have engulfed them, the mass is clearly identified as devoid of agent because it is not functioning liver tissue.

Bubble-specific imaging methods are also employed in dynamic contrast-enhanced imaging studies in which a bolus of microbubble contrast agent is injected intravenously and kinetic parameters (such as 'time-to-peak enhancement' or 'duration of enhancement') of the Doppler signal power obtained from a vascular region of interest may be extracted from the so-called Doppler time-intensity curve that may be plotted. The relationship between

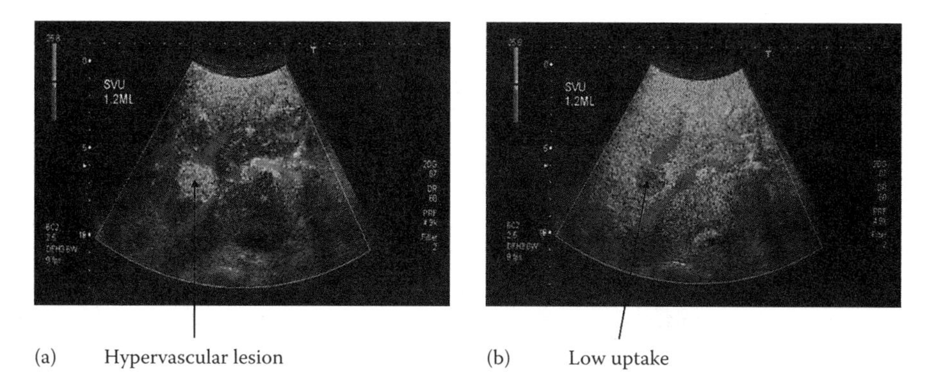

FIGURE 6.56
(See colour insert.) Illustration of bubble-specific imaging; liver metastasis from colo-rectal cancer. After Sonovue™ microbubble injection, the early 'arterial phase' (a) shows the lesion to be hypervascular, but in 'late phase' (b) the same lesion shows low uptake. The type of image display illustrated combines pulse-inversion bubble-specific imaging with colour Doppler. Green is used when the mean Doppler shift is below a threshold (i.e. 'stationary' bubbles) and red or blue (according to flow direction) when the Doppler shift is above the threshold (i.e. 'moving' bubbles). (Images courtesy of D.O. Cosgrove and M. Blomley, and reproduced with permission from Toshiba Medical Systems Visions Magazine.)

such time-intensity curves and the time variation of microbubble concentration may be complicated by factors such as the time variation of the attenuation coefficient and non-linear propagation properties of the intervening perfused tissue. Research continues to determine the value of the method in tumour diagnosis and assessment of therapeutic response.

Finally, as was mentioned in the text referring to Figure 6.7b, microbubble destruction tends to occur at high insonating pressures. This has been taken advantage of in at least two types of microbubble destruction imaging. The first employs the fact that the disappearance of scatterers from those that exist within the resolution cell of the imaging system will cause a decorrelation of the speckle pattern (see Section 6.2.5.2). This decorrelation appears as a random and rapid modulation with time of the amplitude and phase of the echo signal, which can be imaged, for example, by a colour Doppler system as a Doppler noise signal. Disadvantages of such 'agent destruction imaging' are that the contrast signal tends only to be apparent over a limited depth range where the insonating pressure is high enough (usually around the transmit focal depth) and that if one is imaging stationary microbubbles the image can be obtained only once (i.e. repeat scanning of the same tissue volume cannot use the same contrast agent). In the case of flowing blood, however, the narrow region of a single scan plane within which the agent will have been destroyed is quickly replenished with new contrast microbubbles that flow into the scan plane. This has given rise to a 'flash imaging' or 'destruction-replenishment' method, in which a single frame's sequence of pulses of high amplitude is used to destroy the agent in the scan plane at one moment, after which a time sequence of images is obtained using a low-amplitude bubble-specific mode. Local analysis of this time sequence, to extract the rate of recovery of the bubble signal after the initial destructive pulse, can then provide an image with contrast related to local rate of blood flow at the microvascular level and in a manner that is largely independent of flow direction.

6.3.6.2 Elasticity Imaging

In Section 6.3.5.5, time-domain correlation was mentioned as an alternative to Doppler processing for colour flow imaging. Interestingly, this processing method has found greater application as a method for tracking tissue displacements that result from deliberately applied stresses, in what is now known as tissue elasticity imaging, or *elastography* (Gao et al. 1996, Ophir et al. 2000, Greenleaf et al. 2003). As was stated in Section 6.2.1.1, soft tissues propagate ultrasound as if they were fluids, transverse (shear) ultrasonic vibrations being damped over extremely short distances. A more complete description of soft tissue mechanics, however, includes a solid-like behaviour, whereby the tissue may elastically shear (to reversibly change its shape) if a static or slowly varying stress is applied. Elastography involves imaging parameters that are related to the low-frequency shear-elastic modulus that governs this behaviour, as opposed to the high-frequency bulk-elastic modulus that is almost entirely responsible for determining the speed of ultrasound (see Section 6.2.1.1). The shear modulus has been shown to possess potential for excellent tissue discrimination because (like the ultrasound scattering coefficient) it is determined by the large-scale structural organisation of tissue which varies much more between tissues than the molecular composition that largely determines characteristics such as bulk modulus (Sarvazyan and Hill 2004). *Elastograms* may contain information related to a wide range of mechanical characteristics of tissues such as Young's modulus (which equals three times the shear modulus for an incompressible medium), shear viscosity, shear non-linearity, mechanical continuity, anisotropy, Poisson's ratio, and even fluid permeability and filtration. For this reason, and because imaging the elastic properties of tissues can be accomplished with other modalities such as MRI, X-rays or optical imaging, it is rapidly acquiring the status of an imaging modality in its own right. It deserves a thorough treatment but can receive only a brief introduction in a text such as this.

In ultrasound elastography time-varying echo data are processed to extract the spatial and/or temporal variation of a stress-induced tissue displacement or strain. There is a range of signal-processing techniques for extracting the tissue displacements, such as searching for the peak of the cross-correlation between two echo patterns, seeking the time shift that maintains zero phase change, or even measuring the phase change itself which, as was seen in Section 6.3.5.3, is the Doppler method; indeed, there is a strong overlap between elastography and techniques such as TDI described in Section 6.3.5.6. Displaying strain was first suggested by Ophir et al. (1991). This is usually obtained by computing the local average of a spatial gradient of displacement. Applying this to a TDI produces a 'strain rate' image. There are many ways of applying the stress (force), and there are many approaches to extracting properties for image display. The various combinations and refinements of these possibilities provide ample opportunity for ongoing research.

'Freehand' elasticity imaging (Bamber 1999b, Doyley et al. 2001, Bamber et al. 2002, Zhu and Hall 2002) has emerged as an option on some commercial ultrasound systems, and is proving clinically valuable, particularly, for example, in providing additional information to assist breast cancer diagnosis (see Section 6.6.5). The technique derives its name from the use of a hand-held ultrasound transducer to apply a gentle pressure to the surface of the body, or to the use of internal cardiovascular pulsations as the source of stress, while the system displays a real-time elastogram as a grey-scale image alongside the conventional echo image, or the two images are combined using a colour overlay for strain. This form of elastography remains a subjective technique that, as with palpation and ultrasound echo imaging, requires interpretive skills to be learnt. It is hoped that a more quantitative approach will lead to clinically more valuable measures of tissue composition, function or state, with images that are easier to interpret.

One approach to quantitative elastography is to employ a model of the relevant tissue mechanics to convert displacement or strain data to images of tissue properties (e.g. Doyley et al. 2000, Berry et al. 2006), although true quantification of Young's modulus requires knowledge of the stress distribution at the tissue boundaries (Barbone and Bamber 2002). Another is to make use of shear-wave propagation (Sarvazyan et al. 1998). Equation 6.8 may be rewritten for shear waves by replacing c with the shear-wave speed and K by the shear modulus. Under the assumption that the density is a known constant, a measurement of shear-wave speed may then provide a direct measurement of the shear elastic modulus. Because shear waves in soft tissue travel about 1000 times slower than longitudinal waves, the application of a TDC algorithm to map tissue displacements through a sequence of ultrasound echo lines can be used to observe the passage of a shear wave and hence measure its speed, so long as the ultrasound echoes can be generated frequently enough. One commercial device uses such concepts to characterise liver disease; a thumper on the surface of the abdomen launches a shear wave into the body and ultrasound A-scans (at thousands of repetitions per second) are used to measure its speed of propagation in the liver, which is strongly related to degree of fibrosis (Sandrin et al. 2002b, Tatsum et al. 2008). Ultrafast imaging using advanced parallel-receive electronics (see Section 6.7.2) has made it possible to acquire whole echograms at thousands of frames per second, so that it is possible to make full images of the shear modulus using this method (Sandrin et al. 2002a). Alternatively, it is possible to 'slow down' the shear waves, by having two interfering continuous-shear-wave fields generated by vibrating sources, so that even a conventional scanner can be used to follow their motion (Wu et al. 2006).

A final option is to generate the stress deep inside the tissue using ultrasound radiation force. This may be used to create elastograms of tissue either too deep to reach with surface palpation or beyond elastic discontinuities such as fluid collections. In what is effectively an application for a parametric array (see Section 6.3.3), the technique known as vibro-acoustography induces a highly localised low-frequency vibration using two overlapping highly focused ultrasound fields of frequencies that differ in the kHz region, which modulates the radiation force in the region of overlap so that the amplitude of the induced vibratory displacement can be detected either as acoustic emissions using an audio microphone (Fatemi and Greenleaf 1999) or as ultrasound echo displacements using one of the signal-processing methods mentioned earlier (Vappou et al. 2009). Alternatively, a short burst of a single, highly focused ultrasound field may create a small (few tens of microns) displacement impulse within the tissue. Images may then be made of the transient displacement created (Nightingale et al. 2001), the transient strain (Melodelima et al. 2006), or the speed of the shear wave that propagates away from the focal displacement (Bercoff et al. 2004). The first and third of these have been made into successful commercial ultrasound elastography systems.

6.4 Acoustic Output and Performance Checks

These are extensive subjects and only brief descriptions can be given here.

Performance comparison of diagnostic ultrasound equipment (e.g. for the purposes of making a purchasing decision) is difficult to achieve solely by objective physical methods. It is generally necessary for the equipment to be subjectively assessed during a trial period of genuine clinical use. Instead, the current objectives of quality assurance (QA) tests are to ensure that (1) the equipment is safe to use and (2) imaging performance is optimal. Safety checks are achieved by measuring parameters related to the acoustic output of the scanner and transducer combination, as described in Section 6.4.1. Such measurements indicate the likelihood of biological effects in tissue. Imaging performance is assessed by measuring system properties such as sensitivity or spatial resolution and comparing with a baseline measurement. Performance tests for pulse-echo imaging systems are still under development, but some of the more commonly used tests are described in Section 6.4.2. Doppler performance tests, which are at an even more experimental stage, are briefly described in Section 6.4.3.

6.4.1 Acoustic Output Measurements

The primary purpose of these measurements is as a safety check, although acoustic output also provides an indication of imaging performance. Safety checks are generally only performed as an acceptance test (i.e. after initial installation and any equipment upgrades), as the acoustic output is highly unlikely to increase with time.

Modern ultrasound equipment is capable of operating at intensities and pressures that have the potential for causing bioeffects (see Section 6.5). Consequently, the availability of figures for the acoustic output levels has become a prerequisite for sale of diagnostic ultrasound equipment in some countries. For example, the FDA has approved an AIUM/NEMA real-time output display standard (ODS) that incorporates two indicators of possible bioeffects, the thermal index (TI) and the mechanical index (MI) (see Section 6.5). Other parameters commonly used to assess acoustic output are described in Section 6.4.1.2.

A fundamental property of the acoustic field in the context of acoustic output measurements is pressure. Many other parameters of interest, such as acoustic intensity and power, can be derived from the spatio-temporal distribution of the acoustic pressure. The gold standard equipment for measuring pressure is the piezoelectric hydrophone, which is described in the following section (Section 6.4.1.1). The subsequent section (Section 6.4.1.2) describes the experimental and computational techniques for obtaining acoustic field measurements using hydrophones. The third section (Section 6.4.1.3) describes a device for measuring acoustic power directly, the radiation force balance. The final section (Section 6.4.1.4) introduces the concept of 'derating' acoustic measurements made in water to account for tissue attenuation.

6.4.1.1 Piezoelectric Hydrophone

A hydrophone is a miniature sensing element that converts an incident acoustic pressure waveform into an electrical signal. The most common device uses a piezoelectric element. Calibration against an absolute standard provides the relationship between the applied acoustic pressure and the output voltage. This voltage/pressure ratio, which is referred to as the hydrophone's sensitivity, varies as a function of the acoustic frequency of the incident ultrasound wave, in a manner dependent on the structure and design of the hydrophone. Desirable properties of a hydrophone include the following:

(i) Small size to reduce averaging of the spatial peak pressure and to achieve a good directional response. (The latter is important for wide-aperture sources.)

(ii) High sensitivity to measure small pressures and small pressure differences. For piezoelectric hydrophones, smaller active elements are less sensitive, so there is a compromise between sensitivity and spatial resolution.

(iii) Noninterference with the acoustic field being measured.
(iv) Wide bandwidth, much greater than that of the likely range of transmitter frequencies to be measured. Note that high-frequency harmonic components are introduced due to non-linear propagation in the test medium (water).
(v) Linearity (in terms of the relationship between pressure and output voltage).
(vi) A flat frequency response, so that a faithful reproduction of the transmitted acoustic signal is obtained.
(vii) Stability as a function of time, so that the hydrophone calibration remains valid.

Various designs of piezoelectric hydrophone exist, but the membrane design is considered to be the gold standard for acoustic pressure measurements. It is constructed from a thin sheet of unpoled PVDF stretched over a supporting ring (diameter ~100 mm). Electrodes are vacuum deposited on the surface of the film, one electrode on each side of the film, such that they only overlap in a small area (Figure 6.57), which forms the active element of the device. This pressure-sensitive region, which typically has a diameter of 0.3–0.5 mm, determines the spatial resolution of the hydrophone.

Membrane hydrophones have a non-perturbing structure, as, in most cases, the acoustic wave simply passes through the aperture of the ring. Some of the transmitted wave is reflected at the membrane, and this can cause problems when attempting to characterise CW sources (see the following), but the important feature of the device is that the diameter of the supporting ring is large compared to the ultrasonic wavelength. Consequently, the acoustic wave is not diffracted and the fundamental radial resonance mode (that of the internal diameter of the ring) is well below ultrasonic frequencies. Nevertheless, the frequency response of the membrane hydrophone is not completely flat, as there is a resonance associated with the thickness of the film.

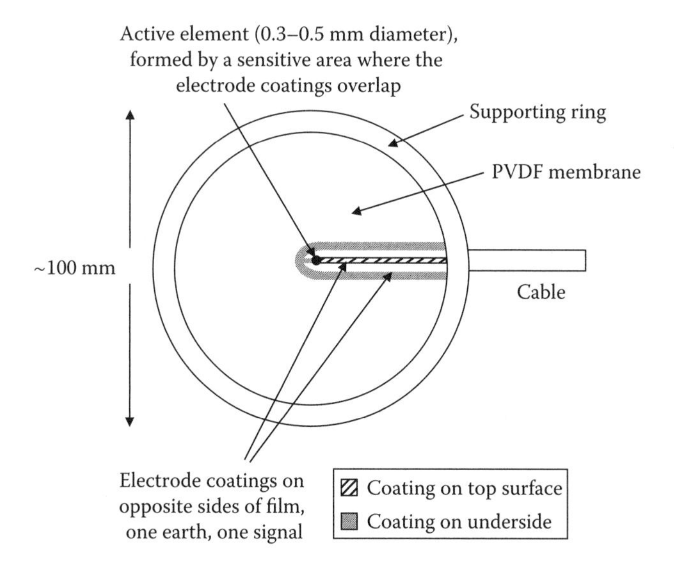

FIGURE 6.57
Schematic diagram of a membrane hydrophone. (Adapted from Preston, 1991, with permission.)

The sensitivity monotonically increases up to the thickness-mode resonance frequency, f_r, given by (Preston 1991):

$$f_r = \frac{c}{2t} \tag{6.50}$$

where
t is the membrane thickness (typically less than 50 μm), and
c is the sound speed in PVDF.

At frequencies above f_r, the sensitivity rapidly decreases. Membrane hydrophones are often used in conjunction with a matching amplifier whose gain decreases with frequency. This cancels out the hydrophone's frequency response so that the combined response is reasonably flat up to the thickness-mode resonance frequency. At the time of writing, commercial devices with the thinnest membranes offer bandwidths up to 50 MHz. There is a compromise between bandwidth and sensitivity, as the sensitivity increases with increasing film thickness.

An alternative design is the needle-type hydrophone, where the active disc-shaped element is supported on the end of a thin-walled metal tube (Figure 6.58). These hydrophones have a diameter that is comparable with the ultrasonic wavelength and as a result, there is significant structure in their frequency response; at low frequencies (<1 MHz), there is diffraction around the probe tip, which causes a decrease in sensitivity. There are also fluctuations in sensitivity at frequencies between about 1 and 6 MHz, due to diffraction phenomena and radial resonance modes (Fay et al. 1994). Finally, as with the membrane hydrophone, there is an increase in sensitivity up to the thickness-mode resonance frequency and a fall-off in sensitivity thereafter. Under certain assumptions, the thickness-mode resonance frequency, f_r, can be estimated from (Fay et al. 1994):

$$f_r = \frac{c}{4t} \tag{6.51}$$

where
t is the thickness of the PVDF film, and
c is the sound speed in PVDF.

Despite the fact that the usefulness of the needle hydrophone is limited by its non-uniform frequency response, it may be more appropriate than a membrane hydrophone

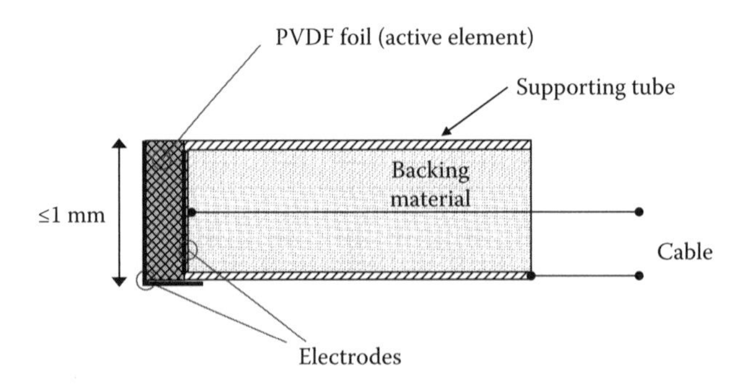

FIGURE 6.58
Schematic diagram of a needle hydrophone.

in certain situations. For example, needle hydrophones can be constructed with smaller active elements (down to 0.04 mm), which may be useful in applications requiring superior spatial resolution. Furthermore, needle hydrophones can be used *in vivo* and in confined spaces. Finally, the needle design is more suitable for CW measurements because it does not disturb the ultrasound field. In contrast, the large supporting ring of the membrane hydrophone causes reflections between the transducer face and the hydrophone, which can interfere with the direct-bath beam. The amplitude of the reflected beam increases with increasing frequency. A potential solution is to tilt the membrane hydrophone to destroy the reflections, but knowledge of the directional response is then required to calculate the acoustic pressure.

In order to assess whether there is a statistically significant difference between acoustic output measurements made at different times or by different people (e.g. the scanner manufacturer and the end user), the uncertainty on the measurements must be estimated. A detailed discussion is beyond the scope of this review, but three of the most important sources of uncertainty will be briefly discussed. The first source is the equipment; there will be uncertainty on the calibration of the hydrophone and its preamplifier as well as uncertainty on the output voltage recorded by the oscilloscope or digitiser. The equipment manufacturers should specify these sources of uncertainty. Two other sources of error that can be considerable are (1) spatial averaging and (2) structure in the frequency response of the hydrophone-amplifier combination. These issues will now be discussed in turn.

Safety indices usually require measurement of the maximum acoustic pressure or intensity in the ultrasonic field, known as the spatial peak. If the active element of the hydrophone is too large, the acoustic pressure is averaged over the sensitive area, leading to an underestimation of the spatial peak value. To avoid this effect (known as spatial averaging), the hydrophone element should be several times smaller than the −6 dB beam width. Since it is not a trivial matter to measure beam width, AIUM/NEMA (1992) provide an alternative method of determining the appropriate hydrophone size. They specify that the effective diameter, d_e, should be less than or approximately equal to $\lambda/2$, where λ is the acoustic wavelength. The effective diameter of a hydrophone is larger than its actual diameter and is a function of the acoustic frequency of the incident wave. It is calculated from measurements of the hydrophone's directional response (Smith 1989). As the frequency increases, the effective diameter decreases, levelling out as it approaches the physical diameter of the element.

For high-frequency transducers, it may not be possible to use a hydrophone that is small enough to reduce the effects of spatial averaging to an acceptable level (e.g. <5% measurement error). Some authors have suggested algorithms to correct for the effects of spatial averaging (e.g. Smith 1989). However, these algorithms are of limited accuracy and are not routinely applied.

The third significant source of error on hydrophone measurements is due to the conversion from voltage to pressure. The standard approach is to divide the measured voltage by a single value for the sensitivity, namely, the sensitivity at the transducer's centre frequency:

$$p(t) = \frac{V(t)}{M(f_{\text{aw}})} \tag{6.52}$$

where

M is the sensitivity, and

f_{aw} is the acoustic working frequency, a measure of centre frequency (see Section 6.4.1.2).

However, this conversion is only strictly valid for a hydrophone-amplifier combination that has a completely flat frequency response. Furthermore, it ignores the fact that the transfer function of the hydrophone is in fact complex valued (i.e. there is a phase response, which may be non-linear). The consequences of ignoring these factors are that the peak positive pressure may be overestimated by up to 50% (Wilkens and Koch 2004). A more accurate approach would be to convert voltage to pressure by deconvolution, to take account of the broadband, complex-valued frequency response. Deconvolution methods are likely to become standard practice in hydrophone measurements in the future.

6.4.1.2 Hydrophone-Based Measurement Methods

Having chosen the appropriate hydrophone, careful attention must be paid to the measurement set-up to ensure meaningful and accurate measurements. The first consideration is the scanner settings. In most cases, the objective of the safety test is to determine the maximum values of the output parameters for the machine-plus-probe under investigation, or at least the maximum value that can be achieved with the range of scanner settings in common clinical use. Front-panel settings that can affect acoustic output include (but may not be limited to) the scanning mode (e.g. B-mode, spectral Doppler, colour Doppler), the number and depth of the focal zones, the output power control, the size and position of the imaged region or Doppler gate, and the use of special modes (e.g. coded excitation, harmonic imaging).

Apart from the output power control, it is difficult to predict the effect of changing an individual scanner setting because manufacturers often design the machine to change more than one parameter at the same time. Therefore, it can be time consuming to find the settings that maximise output, unless the data capture and analysis system provide instantaneous read-out of the desired acoustic parameter. The process can be speeded up somewhat by using the on-screen indices (see Section 6.5.2) to provide a rough indication of acoustic output. However, it would be imprudent to rely on these displayed indices completely, especially when the aim of the test is to verify their accuracy. Another possible short cut is to use a radiation force balance (see Section 6.4.1.3) to determine the conditions that deliver maximum power (Dixon and Duck 1998).

Hydrophone measurements should be made in a large tank of degassed (and depending on the hydrophone, deionised) water. The tank should be lined with absorbing materials, such as dimpled rubber, to reduce reverberations. Since both the ultrasound beam and the hydrophone have a directional response, they must be carefully aligned to ensure that the maximum acoustic signal is detected. They should be securely clamped and the hydrophone mount should have 5 degrees of freedom (three axes of translation, plus rotation and tilt). Computer-controlled hydrophone positioning is a useful capability, as it allows rapid characterisation of the spatial distribution of the acoustic field. The output of the hydrophone-amplifier combination is recorded as a voltage-time waveform on an oscilloscope or digitiser and then converted to a pressure-time waveform using the hydrophone's calibration. Averaging a large number of pulses (over time) improves the SNR and, hence, the precision of the derived acoustic parameters.

As may be appreciated by reviewing Sections 6.2.1, 6.2.4, and 6.3.3, complete and accurate specification of acoustic output is far from being a simple matter. In CW systems, one may measure the spatial distribution of either intensity or pressure. The intensity may be integrated over the beam area to provide the total power output of the system. For pulsed systems, one must consider both the instantaneous and the time-averaged versions of these quantities. Finally, the non-linear characteristics of the test medium (water) will mean that the pulse shape and beam shape are amplitude dependent, and that measurement of different acoustic-field parameters (e.g. intensity, peak positive pressure, peak negative pressure) may result in different observed spatial distributions. The following paragraphs describe the most common acoustic parameters derived from the quantities pressure, intensity and power for a pulse-echo system. The derivations assume that the system is unscanned (e.g. single-element transducer, M-mode or pulsed Doppler), so that the hydrophone receives identical pulses at the pulse repetition rate. The final part of this section briefly discusses the modifications that need be made to the measurement and analysis methods to deal with scanned systems (e.g. B-mode).

6.4.1.2.1 Pressure Parameters

A typical acoustic waveform from a pulse-echo system is shown in Figure 6.59a. Two parameters that can be derived directly from this pulse are p_+, the peak positive pressure, and p_-, the peak negative pressure. The latter is a particularly important parameter, as it is widely accepted to relate to the potential for tissue damage by means of cavitation (see Section 6.5 and Equation 6.64).

In order to derive the pressure waveform, the ultrasonic frequency must be known (see Equation 6.52). Therefore, it is first necessary to calculate the Fourier transform of the pulse (Figure 6.59b), so as determine the acoustic working frequency, f_{aw}. This is defined as the arithmetic mean of the frequencies at which the amplitude of the pressure spectrum becomes 3 dB lower than the peak.

One of the most important quantities that can be derived from the pressure waveform is the time integral of the square of the instantaneous acoustic pressure, shown in Figure 6.59c. This integral allows the pulse duration, t_d, to be calculated as

$$t_d = 1.25(t_2 - t_1) \tag{6.53}$$

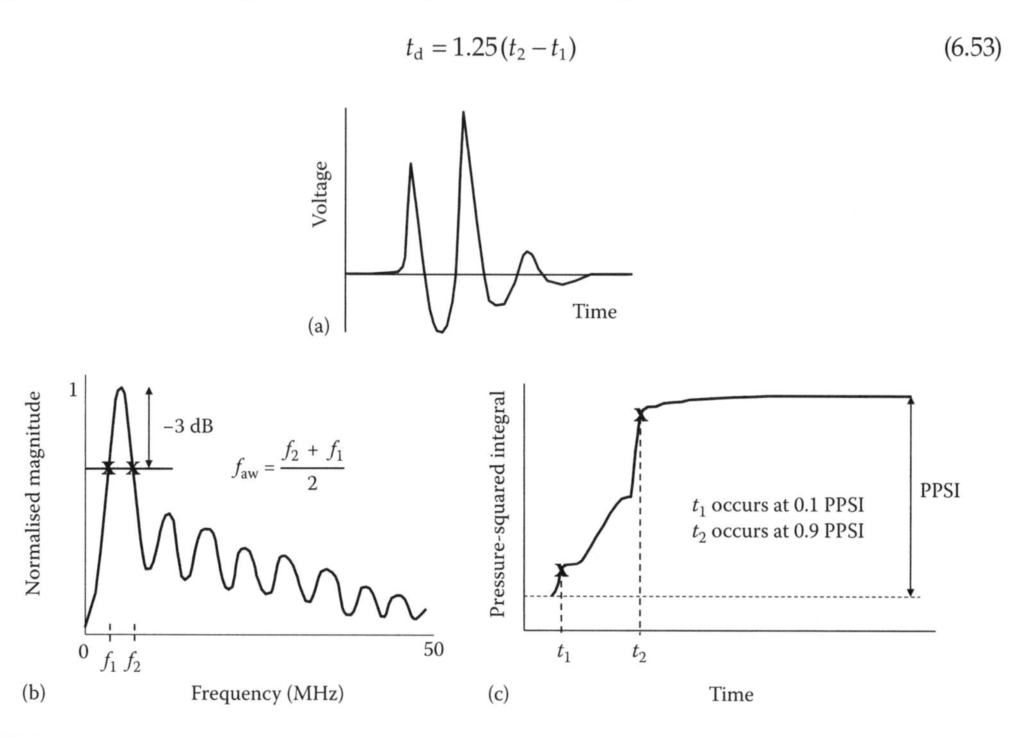

FIGURE 6.59
A typical acoustic waveform (averaged over many pulses) from a pulse-echo system (a), its Fourier transform (b) and the time integral of the square of the pressure wave (c). (Adapted from Preston, 1991, with permission.)

where t_1 and t_2, respectively, equal the time taken for the integral of the squared pressure to reach 10% and 90% of its final value. The final value of the integral (see Figure 6.59c) is referred to as the pulse-pressure-squared integral (PPSI). The reader should be aware that the PPSI is referred to in the literature variously, for example, p_i, p_{ii}, PII, and pulsed-intensity integral. It is often used as a summary measure for further calculations. For example, beam-shape information is generally obtained by plotting PPSI as a function of the position of the hydrophone within the field. Furthermore, most of the intensity parameters (see Section 6.4.1.2.2) are calculated from PPSI. The reason for this is that, as previously mentioned, non-linear propagation results in different spatial distributions being observed for different acoustic parameters. Therefore, the spatial maxima of the pressure parameters can occur at different depths. The peak value of p_+ usually occurs at a greater distance from the transducer than the peak value of p_-. The maximum value of PPSI occurs at a distance between the two, such that the plane containing PPSI is the best choice for a summary measure.

6.4.1.2.2 Intensity Parameters

Under the assumption of plane progressive waves, the instantaneous intensity, i, can be calculated from the instantaneous pressure, p, as follows:

$$i = \frac{p^2}{\rho c} \tag{6.54}$$

where ρ and c are respectively the density and sound speed in the water. These two properties are temperature dependent, so the temperature of the water needs be measured so that the correct values are used. A consequence of the plane-wave assumption is that significant errors can occur close to the face of the transducer. An intensity parameter that can be determined directly from Equation 6.54 is the spatial-peak temporal-peak intensity, I_{SPTP}. This is calculated by using the spatial peak value of p_+ or p_- (whichever is larger) as the instantaneous pressure. It is worth noting that the term 'instantaneous intensity' has no real meaning, as the measurement of intensity requires a time window (see Leighton 2007). However, values of I_{TP} are sometimes reported, perhaps as it is the intensity parameter that is most likely to be relevant to cavitation hazard. The remaining intensity parameters are used more frequently and are time-averaged values. The time-averaged intensity, I, is given by

$$I = \frac{\overline{p^2}}{\rho c} \tag{6.55}$$

where p^2 is the square of the instantaneous acoustic pressure and overlining indicates the temporal average value over a specific time period. Two time periods are generally used, giving rise to two further intensity parameters.

The spatial-peak pulse-average intensity, I_{SPPA}, is the spatial peak value of I_{PA}, which is the intensity averaged over the duration of the pulse:

$$I_{PA} = \frac{\text{PPSI}}{\rho c t_d}. \tag{6.56}$$

The spatial-peak temporal-average intensity, I_{SPTA}, is the spatial-peak value of I_{TA}, which is the intensity averaged over the time when the sound is both off and on. When all pulses are identical, it is defined as

$$I_{TA} = \frac{\text{PPSI}}{\rho c}\text{PRF}. \tag{6.57}$$

For a scanning beam, I_{TA} needs to include contributions due to neighbouring scan lines, that is, scan lines that do not coincide with the axis of measurement, but still produce a measurable signal at the hydrophone.

Although it is more common to report spatial-peak values for the intensity parameters, the temporal-average intensity is sometimes averaged over the beam area to calculate the spatial-average temporal-average intensity, I_{SATA}. For an infinite number of measurements, this would be determined by integrating I_{TA} over the beam area and dividing by the beam area. In practice, hydrophone measurements are taken at a large number of equally spaced sample points and I_{SATA} is determined as follows:

$$I_{SATA} = \frac{(\Sigma\, I_{TA} \text{ for all points at which } I_{TA} > 0.25\, I_{SPTA})}{\text{Number of points}}. \tag{6.58}$$

Beam plotting with a hydrophone is a time-consuming procedure. It is worth noting that a number of devices exist that allow direct visualisation of the entire acoustic field. These include the hydrophone array and optical methods, such as a Schlieren system. The latter enables visualisation of the change in refractive index of the water, induced as the density fluctuates with the passage of the acoustic wave.

6.4.1.2.3 *Power Parameters*

Power is an important acoustic output parameter, as it appears in all the formulae for the TI and MI. It can be measured by spatial integration of the temporal-average intensity, I_{TA}, over the beam area (i.e. over all points that produce a measurable hydrophone signal). A detailed description of the measurement procedure can be found in AIUM/NEMA 2004. However, power measurement with a hydrophone is a time-consuming procedure and the precision of the measurement is likely to be poor relative to that achieved with a radiation force balance (see Section 6.4.1.3).

6.4.1.2.4 *Scanned Systems*

For an unscanned system, all pulses are identical, so the oscilloscope or digitiser can be triggered from the received hydrophone signal. For scanned modes, the pulse changes with time, as each beam is transmitted from a different position or at a different angle. Measurement of I_{SPTP} and I_{SPPA} is still possible, as the threshold level for data acquisition can be increased until only the largest (i.e. on-axis) signal is captured. However, this approach is not useful for temporally averaged values, where contributions from off-axis scan lines need to be measured, and yet the signals from these lines will be smaller than the on-axis value. Similarly, measurement of the beam width is not possible because as the hydrophone is moved off the central position, signals from off-centre scan lines will exceed the desired signal (i.e. due to the central scan line).

The easiest way of selecting the scan line of interest in these situations is to use the electrical signals sent by the scanner signifying the start-of-frame and start-of-line events. A delayed-timebase oscilloscope or logic-counter unit is required to select the *n*th start-of-line pulse after the start-of-frame pulse (see Preston 1991). However, the manufacturer's cooperation is usually required to access these signals. Alternatively, if the parameter of interest is I_{TA}, this can be obtained by connecting the hydrophone to an RF power meter to measure the temporal average power, and then converting power to acoustic intensity (see Whittingham et al. 2004). Given that I_{SPTA} is an important output parameter, it is worth pointing out that for linear arrays, the maximum value is likely to occur around the depth of the elevational focus. For phased and sector arrays, the greatest value of I_{SPTA} tends to occur close to the probe because all the transmitted beams overlap in this region. For more detail, see Whittingham (2007).

A further difference between unscanned and scanned modes is in the measurement of acoustic power. To determine the total power in the array using hydrophone measurements, the power determined from one beam (e.g. the central scan line) needs to be multiplied by the number of beams in the array.

Beyond discrete scanning modes, another level of complexity is combined modes, such as simultaneous B- and M-mode or simultaneous B-mode and pulsed Doppler (known as Duplex). One could also include in this category multiple focal zones achieved with multiple firings. The triggering techniques for dealing with these systems are beyond the scope of this review, but two essential principles must be applied. First, it must be possible to isolate the repeating acoustic signal for each mode separately and treat it as a discrete mode. Second, when determining the values of I_{SPTA} or total power, the contributions from all modes (or focal zones) must be included.

6.4.1.3 Radiation Force Balance

If only the total output power of the diagnostic ultrasound system is required, then a device known as a radiation force balance may be used. This is essentially a very sensitive balance that measures the steady-state force exerted by the sound beam on a target intercepting the field. The theory underlying the phenomenon is complex, but under the assumptions of incident plane waves and a target size sufficient to intercept the entire beam, it results in a relatively simple practical equation for the time-averaged spatially integrated power, *W*:

$$W = \frac{Fc}{h} \tag{6.59}$$

where

c is the speed of sound in the coupling medium (degassed, and usually deionised, water),
F is the force on the target measured by the balance, and
h is a factor that depends on the target geometry and equals unity for a totally absorbing target or for a reflecting target orientated at 45° to the incident beam.

The plane-wave assumption has been shown to be sufficiently accurate for many diagnostic fields (Beissner 1993), but power measurements from highly focused or divergent fields may not be accurate. In these cases, it may be possible to modify Equation 6.59 to account for the range of angles of incidence on the target (Preston 1991, p. 76).

Various designs of power balance exist, but the most common approach is to use a servo to apply a restoring force to the target in order to keep it stationary. The targets are specially designed and are usually reflectors as it is easier to manufacture a perfect reflector than a perfect absorber. However, in principle, absorbing targets are to be preferred because they are more accurate for non-plane waves. The typical radiation force per unit power is 0.69 μN mW^{-1} and the time averaged power output of diagnostic ultrasound equipment tends to be in the range 1–50 mW. Therefore, these devices need to be very sensitive for use with diagnostic beams.

For scanning systems, it is recommended that the power be measured in a static mode such as M-mode (AIUM/NEMA 2004). If the power measured in M-mode is denoted by W_m, then the power in B-mode can be calculated as:

$$W_B = N \cdot \left(\frac{W_m}{\mathrm{PRF}_m} \right) \tag{6.60}$$

where

N is the number of scan lines, and
PRF_m is the pulse repetition frequency in M-mode.

6.4.1.4 Calculation of Derated Acoustic Parameters

Acoustic output measurements are made in water, which is a non-attenuating medium. Thus, it is common to reduce (or 'derate') the measured values to account for the attenuation in the body. The attenuation coefficient is taken to be 0.3 dB cm^{-1} MHz^{-1}, which is lower than the value for soft tissue (see Section 6.2.1.5), but allows for the possibility that a significant proportion of the propagation path consists of amniotic fluid or urine. Derated values are calculated according to:

$$A_{.3} = A \exp\left(\frac{-0.3 fz}{4.343} \right) \tag{6.61}$$

where

A is the measured value of the acoustic parameter (either intensity or power),
$A_{.3}$ is the derated value,
f is frequency (in MHz), and
z is the distance from the transducer (in cm).

To derate a pressure value, the factor 4.343 would need to be doubled to 8.686. Examination of Equation 6.61 reveals that the peak value of the derated parameter may occur closer to the transducer face than the peak value measured in water, so it is advisable to calculate the derated values in real time when searching for the spatial peak of a particular parameter.

6.4.2 Performance Tests and Phantoms for Pulse-Echo Systems

Properties related to image quality are generally assessed by obtaining an image of a specially constructed object. Such objects, as in other areas of medical imaging, are known as test phantoms. No single phantom is able to provide all the information of interest, but at some time or other all of the characteristics listed in Table 6.6 have been assessed using a

TABLE 6.6

Performance characteristics of ultrasound pulse-echo imaging equipment that have been assessed using imaging phantoms.

Element dropout	Elevational resolution (also called slice thickness, orthogonal to scan plane)
Image uniformity	Penetration depth (low and high contrast)
Dead zone (also called dead time, paralysis and ringdown, due to the length of the excitation voltage pulse)	System noise
Registration error	Dynamic range
Linearity/distortion (also called geometric uniformity)	Shape of compression curve
Image aspect ratio	System artefacts
Vertical and horizontal distance accuracy (and electronic callipers)	Low-contrast detectability
	Cyst (anechoic object) imaging
Area and/or volume measurement accuracy	Signal-processing options
Axial resolution	Movement-related parameters
Lateral resolution (within scan plane)	

Source: Adapted from the list given by McCarty, 1986.

phantom (although many are as relative measures). Phantoms also find limited application in areas that have little to do with performance assessment. Examples are in the training of personnel in the use of ultrasound equipment for imaging a particular part of the body (e.g. breast phantoms), and in physics research (e.g. for testing and optimising tissue-characterisation procedures and computer algorithms).

A large number of test phantom designs have been proposed by manufacturers and researchers. The two most popular designs have been the AIUM (American Institute of Ultrasound in Medicine) standard test object and the tissue-mimicking phantom. The AIUM phantom can be used to test high-contrast resolution and geometrical accuracy. It consists of a grid of wires in a water–alcohol mixture (used to obtain the correct speed of sound). Tissue-mimicking phantoms are generally described as 'multipurpose', as they are designed to test a large number of scanner characteristics. These phantoms usually consist of nylon fibres ($\leq$0.1 mm diameter to avoid reverberations), cyst-like structures and solid masses in a gelatine background loaded with small scattering beads. The background tissue-mimicking material (TMM) attempts to simulate the scattering from liver parenchyma. Commercially available TMMs tend to be gel based (e.g. Gammex-RMI, CIRS), but urethane phantoms (e.g. ATS) are also available and are generally cheaper and more durable. A disadvantage of urethane phantoms is that they have a sound speed of 1430–1450 m s^{-1} at room temperature, which is lower than the value assumed by ultrasound scanners (1540 m s^{-1}). Although some manufacturers adjust the positions of the wires such that the phantom may be used to test distance measurement accuracy (this test is described later), it is not possible to make adjustments that are simultaneously valid for all types of multi-element transducer (Goldstein 2000). Furthermore, phantoms with acoustic velocities different from 1540 m s^{-1} cause beam defocusing, which means that they cannot be used to provide absolute measures of resolution (Chen and Zagzebski 2004). However, they are still useful for assessing changes in resolution over time. It is worth noting that these problems are not peculiar to urethane phantoms, as the speed of sound in most TMMs can vary somewhat from their quoted value of 1540 m s^{-1}, either as a function of time (due to dehydration) or as a function of temperature. Therefore, it is important that phantoms are kept in a temperature-controlled environment and that they are replaced when they begin to deteriorate.

In practice, only a small subset of the characteristics listed in Table 6.6 is tested in most centres. This is for a variety of reasons, including time constraints, lack of appropriate equipment and test methods, and lack of an evidence base that faults will be detected before being noticed by the end user (Dudley et al. 2001). The following paragraphs describe the characteristics most commonly tested. New or repaired equipment should be tested to establish baseline data. Subsequently, performance tests should be performed every 6–12 months and after any software or hardware upgrade. Corrective action should be taken (e.g. contacting the manufacturer) if the test result falls outside the chosen tolerance. Most test results are highly dependent on the equipment settings (e.g. gain, output power, dynamic range, TGC, etc.), so it is essential to reproduce the same scanner settings each time.

Prior to performing the specific tests detailed hereafter, a physical and mechanical inspection of the scanner and its transducers is recommended to reveal any obvious signs of damage. Another important step is to check that the monitor contrast and brightness settings are correct and that the ambient lighting is optimised. An internally stored greyscale test pattern can be used to set up the monitor at the initial baseline test and to check the monitor's performance at subsequent tests (see Goodsitt et al. 1998). The fidelity of hard-copy photographs should also be checked, particularly if these are being used to record the test results.

6.4.2.1 *Element Dropout*

This can be assessed by running a thin piece of wire, such as a paper clip, across the face of the transducer. A reduction or loss of brightness at a particular position is indicative of element failure or degradation. Often, element dropout is visible in an image taken in air, which is a test that can be easily performed by the user. There are also commercially available devices for testing transducers (e.g. Sonora Medical Systems).

6.4.2.2 *Low Contrast Penetration*

This is essentially a combined measure of the system's sensitivity and noise. It can be assessed using a TMM by estimating the depth at which the speckle pattern first disappears into noise.

6.4.2.3 *Distance Measurement Accuracy*

In this test, the distance between two targets in the phantom is measured using the scanner's callipers (electronic markers) and compared with the known target separation. Transducer pressure must be minimised to avoid distorting the target positions. The phantom must have a sound speed of 1540 m s^{-1} for this test. If there is any doubt, it may be preferable to use targets in water and then apply a correction to the measured distance. For a linear-array transducer, only the vertical distance needs to be corrected according to:

$$d_{\text{corr}} = d_{\text{meas}} \left(\frac{c_{\text{water}}}{1540} \right) \tag{6.62}$$

where

d_{meas} is the distance measured by the callipers,
d_{corr} is the corrected value of the measured distance, and
c_{water} is the speed of sound in water at the temperature of the measurement.

For a curvilinear or phased array, both the vertical and horizontal distance measurements need to be corrected. However, this requires knowledge of both the transducer geometry and of the exact positions of the targets within the imaged region (Goldstein 2000). Therefore, a better approach might be to use water whose sound speed has been corrected to 1540 m s^{-1}, for example, by the addition of glycerol (approximately 10% by weight).

6.4.2.4 *Axial and Lateral Resolution*

Resolution can be defined as the smallest separation between two point targets such that they can just be resolved. Multi-purpose phantoms often contain groups of resolution targets in which each pair of adjacent targets is separated by a different distance. The aim is to identify the closest pair that can still be resolved. However, apart from the fact that this method is quantised, it is difficult to manufacture phantoms whose targets are sufficiently closely spaced to challenge the highest frequency transducers. An alternative method is to measure the axial and lateral dimension of the echo caused by the wire targets. Due to the small distances involved in such measurements (particularly the axial dimension), computerised analysis is recommended to reduce the uncertainty (Gibson et al. 2001).

6.4.2.5 *Slice Thickness (Elevational Focus)*

The recommended method for measuring the width of the beam in the elevational direction (i.e. orthogonal to the scan plane) is to image across a diffuse scattering plane inclined at 45° to the ultrasound beam. The vertical dimension of the imaged bar indicates the slice thickness (Skolnick 1991). If such a phantom is not available, then a qualitative assessment of slice thickness can be made by imaging the long axis of an anechoic cylinder and observing the echo level inside the object (known as 'cyst fill-in').

In order to reduce the subjectivity of the performance tests just described (in Sections 6.4.2.2 onwards), computerised image analysis can be used to extract the parameters of interest (e.g. Rownd et al. 1997, Gibson et al. 2001). For example, resolution can be measured by computing the FWHM or full width at tenth maximum of a profile through a wire target. A further reduction in uncertainty can then be achieved by taking a number of independent scans of the object and averaging the parameter of interest over all scans.

6.4.3 Performance Tests and Phantoms for Doppler Systems

Two main designs of test object have been used for assessment of Doppler performance: the string phantom and the flow phantom. String phantoms consist of a filament (e.g. silk, nylon or rubber) that is moved by a motor with a known velocity. The string and motor are immersed in speed-of-sound-corrected degassed water. An example of a commercially available string phantom is that produced by CIRS. The motor speed can be varied between 10 and 200 cm s^{-1} and a variety of waveforms can be applied, including sine waves, square waves, and clinically realistic waveforms. Flow phantoms consist of a TMM containing tubes of varying diameter through which blood-mimicking fluid is pumped at a known speed. Careful design of flow phantoms is required to ensure laminar and parabolic flow. They also require careful calibration to ensure accurate knowledge of the flow velocity. ATS Laboratories produce a flow phantom in which the flow channels are embedded horizontally in a background of urethane rubber. The surfaces of the phantom are angled to achieve a known angle between the imaging beam and the test fluid.

The main advantage of the flow phantom is that both the background material and the velocity profiles are more realistic than those of the string phantom. The disadvantages are that it does not provide a single velocity (to assess intrinsic spectral broadening) and it is more expensive and difficult to construct, especially in terms of velocity calibration. Neither phantom is ideal, although a possible solution is to construct a flow phantom that also contains a vibrating filament. There is considerable scope for further development of Doppler test objects.

A third type of phantom that has been used in Doppler assessment, although not as extensively as the string and flow phantoms, is acoustic signal injection (Sheldon and Duggan 1987). This technique has also been proposed for performance testing of pulse-echo ultrasound (Duggan and Sik 1983), but it has been more widely implemented for Doppler tests. The principle, which is possibly unique to ultrasound, involves the use of electronically synthesised test signals. The signal may be produced entirely by a separate transducer or, alternatively, the transmitted pulse from the scanner may be captured by a receiving transducer, mixed with an acoustic test signal, and transmitted back to the scanner ('acoustic reinjection'). The transducer of the injection device is coupled to the transducer under test via a Perspex block, which can cause problems for curved arrays. There are also problems associated with the complex beam-forming methods used on modern ultrasound scanners (Li et al. 1998).

Doppler QA tests have been most frequently performed on spectral Doppler, although recently, colour Doppler assessment has gained popularity. At the time of writing, tissue Doppler tests remain a matter of research. The following is a list of the spectral and colour Doppler performance tests that have been most widely implemented. As with pulse-echo imaging, it is essential to use identical scanner settings for serial tests on a particular system. For a detailed description of Doppler QA tests, see Hoskins et al. (1994), Thijssen et al. (2002) and di Nallo et al. (2006).

1. Accuracy of maximum velocity estimation and intrinsic spectral broadening (spectral).
2. Range-gate registration (spectral): compare the position of the range gate for which the Doppler signal is maximised with the known position of the pipe or string.
3. Sample-volume dimensions (spectral): the size of the sample volume is measured for the minimum range gate setting, which is usually 1 mm.
4. Vessel angle measurement accuracy (spectral): the angle measured using the scanner is compared with the known angle of the pipe or string.
5. Accuracy of waveform indices (spectral): using a clinically realistic test waveform, the scanner's calculation of parameters such as the PI can be verified.
6. Depth of penetration (spectral and colour): determine the maximum depth at which flow is distinguishable from background Doppler noise.
7. Velocity direction discrimination (spectral and colour): this is a 'pass or fail' test.
8. Minimum and maximum detectable velocity (spectral and colour): the maximum value is the value just before aliasing occurs.

6.4.4 Summary

Safety checks play an important role in equipment acceptance. There is reasonable international agreement over the important acoustic output parameters (pressure, intensity and power) and how they should be measured (the piezoelectric hydrophone and the radiation force balance). It is worth noting that other measurement devices have been proposed and that these might become more widely used in the future. Of particular interest is the optical-fibre hydrophone, where the tip of an optical fibre is coated with a material whose refractive index changes in response to an applied pressure (Morris et al. 2006). The principal advantage of this approach, relative to the piezoelectric sensor, is that sensitivity is not dependent on element size so it should be possible to achieve high sensitivity with elements as small as ~10 μm. Alternatives to the radiation force balance are also desirable, particularly for diagnostic fields, as these produce very small radiation forces. Zeqiri et al. (2004) have proposed a thermal method based on the pyroelectric effect.

In comparison to safety checks, performance tests and test methods are poorly developed. Only a small subset of the proposed tests (e.g. sensitivity, noise and element dropout) has been shown to be capable of detecting faults with the equipment (Dudley et al. 2001). An evidence base needs to be established to justify other tests such as spatial resolution. It is likely that computerised image analysis will greatly benefit this endeavour. New test methods and phantoms are required for all scanning conditions, but particularly for Doppler and for the newer technologies (e.g. 3D, harmonic imaging, elasticity imaging and high-frequency probes).

If constancy checks are poorly developed, then there is even more work to be done to devise tests that produce absolute measures of performance. Such tests would be useful for informing scanner purchase and for deciding when a scanner should be replaced. It is interesting to note that, in the United Kingdom, only one clinical application (breast imaging) has a set of formal requirements for the technical performance of the scanner (Medical Devices Agency 1999). Part of the problem is fundamental in the sense that the complexity of ultrasound interaction with tissue (e.g. frequency-dependent attenuation, non-linear propagation and reverberations) makes it difficult to design phantoms that are indicative of clinical performance. However, setting aside this more challenging problem, it has proven difficult even to devise tests and test methods that can distinguish between state-of-the art and antiquated systems. Recently, a number of new techniques have been proposed for this purpose. For example, Pye et al. (2004) have patented a pipe phantom and measurement method that produces an overall figure of merit known as the Resolution Integral. It is a combined measure of two of the most important indices of performance: depth of penetration and spatial resolution. Browne et al. (2004) have proposed a sensitivity performance index for colour Doppler that combines the lowest detectable velocity, vessel size, and penetration depth. This metric was found to correlate well with the cost and age of a series of scanners. However, further work is required to validate such approaches and to make them readily available.

6.5 Biological Effects and Safety of Diagnostic Ultrasound

The fact that ultrasound techniques have gained such an important place in diagnostic medicine is due, in large part, to their presumed lack of hazard to the patient. Indeed, ultrasound has been in extensive clinical use in obstetrics, the application where there would be greatest cause for concern over a potential biological effect, since the 1960s. Since that time, numerous investigations have been undertaken in an endeavour to detect adverse effects. No confirmed deleterious effect to the foetus or the mother caused by ultrasound at diagnostic intensities has been reported. It is extremely unlikely that any causal association would not by now have been detected.

Nevertheless, there is continuing awareness that lack of evidence of a hazard is not proof of its nonexistence. Indeed, at levels of acoustic power higher than those currently used diagnostically, biological effects are well known, understood to some extent, and made use of in medical therapy.

Ultrasonic diagnostic equipment manufacturers are constantly striving, in a competitive market, to improve the quality of the information provided by their systems. This may sometimes be achieved by increasing the acoustic power output of the device and is definitely associated with increased spatial peak intensities due to improved focusing methods. Before standards can be defined for safe equipment output levels and measurement methods, it is necessary to gain an understanding of the potential biological effects, the mechanisms by which they occur, the degree to which they may constitute a hazard, and the appropriate dose–effect relationships. This subject is still a matter of much debate and research.

6.5.1 Biological Effects

There are two main categories of mechanism by which ultrasound has the potential to cause biological effects in tissue: thermal and nonthermal (mechanical). Thermal mechanisms arise because of molecular relaxation, viscous losses (due to relative motion) and heat conduction, all of which contribute to acoustic absorption, which in turn produces a local temperature rise. Elevated maternal or foetal body temperatures have been shown to result in birth defects in the offspring of a wide range of mammalian species. The risk of tissue damage depends on both the temperature rise and the exposure time. A number of authors have reviewed the mammalian data and, in general, the following conclusions have been reached:

1. A temperature rise of less than approximately 1.5°C can be considered to present no hazard to human tissue, even if maintained indefinitely.
2. The lower bound for adverse effects is a dose equal to a 6°C temperature rise maintained for 1 min (Miller and Ziskin 1989).
3. There is a non-linear relationship between temperature rise and the minimum exposure time needed to produce adverse effects. To achieve the same bioeffects as those caused by a 6°C temperature rise for 1 min, the exposure time must increase by a factor of 4 for every 1°C decrease in temperature rise (e.g. for a temperature rise of 5°C, the necessary exposure time is 4 min).

There is a potential for thermal effects with modern diagnostic ultrasound equipment, if used imprudently. The greatest risk is when imaging bone, due to its high absorption coefficient. From calculations and *in vivo* animal experiments, it has been shown that temperature rises in excess of 4°C can be achieved in soft tissue adjacent to bone (Church and Miller 2007).

Non-thermal bioeffects occur due to mechanical processes that may be classed either as cavitational or noncavitational. Cavitation is the name given to a range of complex phenomena that involve the vibration, growth, and collapse of gas bubbles. When an existing bubble is exposed to ultrasonic pressure, it alternately expands and contracts at the frequency of the ultrasound field, a phenomenon known as non-inertial cavitation. A possible consequence of non-inertial cavitation is rectified diffusion, whereby the equilibrium radius of the bubble grows with time. This occurs because the surface area of the bubble is larger during the expansion phase than during compression, so more gas diffuses into the bubble when the pressure inside is low than diffuses out when the pressure is high.

Non-inertial cavitation may arise at any pressure amplitude, provided gas bubbles are present, but it generally requires long pulse lengths for its effects to be observed (Church and Miller 2007). A second type of behaviour, known as inertial cavitation, occurs when the peak negative pressure is increased beyond a certain threshold and the bubble experiences a large expansion followed by a rapid, violent collapse. Inertial cavitation can occur over very short pulse durations, but the threshold pressure increases with ultrasonic frequency and is inversely related to bubble radius (Apfel and Holland 1991). Therefore, in the absence of exogenous microbubbles, inertial cavitation is unlikely to occur *in vivo*, even at the highest pressure levels available on commercial systems (Carstensen et al. 2000). A possible exception is the intestine, since it contains a wide range of bubble sizes, some of which may be small enough to act as cavitation nuclei (Carstensen et al. 2000).

Cavitation can produce a number of different physical effects in tissue (Humphrey 2007). Non-inertial cavitation results in enhanced absorption (hence, an enhanced temperature rise), as well as microstreaming, whereby eddies of flow are set up in the fluid surrounding the bubble. High shear stresses are produced in the region of strong velocity gradients near the bubble surface, which have the potential to rupture cell membranes. Inertial cavitation results in extremely high temperatures and pressures at the moment of collapse of the bubble. This can cause a number of harmful effects, including cell rupture, free radical production and the formation of microjets of liquid.

Apart from cavitation, an ultrasonic field can generate mechanical effects in other ways. The two most significant non-cavitational phenomena are thought to be radiation pressure and the related effect of acoustic streaming. Radiation pressure (as used for output power measurement described in Section 6.4.1.3) refers to the fact that an acoustic wave transfers momentum to the propagation medium, either due to energy loss by absorption or scattering, or a change in sound speed, or as a consequence of non-linear propagation. This produces a force (usually) in the direction of wave propagation, which, in principle, could set tissue structures or fluids in motion. Although there are no known harmful effects of radiation pressure at diagnostic intensities (Church and Miller 2007), the phenomenon has been associated with a number of biological responses, including increased foetal activity (O'Brien 2007). There is perhaps greater cause for concern over acoustic streaming (also known as macrostreaming), which refers to the radiation pressure-induced bulk movement of fluid. If the flowing liquid impinges on a biological membrane, it can change the membrane permeability, which in turn can affect the cell behaviour (Church and Miller 2007). Macrostreaming is only likely to occur where the ultrasound beam passes through a fluid path, for example, in obstetric scanning.

Despite the wide range of potential mechanisms discussed in the previous paragraphs, the risk of a significant non-thermal bioeffect occurring *in vivo*, under diagnostic exposure conditions, is thought to be extremely low. A large number of laboratory experiments have been undertaken, in which *in vitro* and *in vivo* mammalian tissues have been irradiated at diagnostic exposure levels, and only a small number of physical effects have been observed. In the absence of contrast agents, the most widely reported effect is capillary bleeding at tissue/air interfaces in the lungs of *in vivo* mammals (see, e.g. Dalecki et al. 1997). Further research is required to understand the underlying mechanism, but there is evidence to suggest that it is not a cavitational phenomenon (Carstensen et al. 2000, O'Brien et al. 2000), despite the fact that the lung only becomes susceptible to this type of damage when it is filled with air. There is a theoretical risk that current diagnostic-ultrasound equipment can be made to operate so as to cause haemorrhaging on the surface of neonatal or infant lung. However, the extent of the damage and the ability of the lung to heal are such that the long-term implications of the injury may not be serious (Baggs et al. 1996, O'Brien 2007).

Foetal lungs are not thought to be at risk, as they do not contain air. Once contrast agents are introduced into the body, the potential for bioeffects due to inertial cavitation markedly increases (Carstensen et al. 2000, O'Brien 2007). Adverse effects that have been reported include haemolysis, haemorrhaging and cardiac arrhythmogenesis.

A large number of *in vitro* experiments have aimed to discover whether non-thermal mechanisms can have adverse effects on cells and their components. Unfortunately, *in vitro* experiments involving cell systems are prone to artefacts if not very carefully executed, and the conditions of exposure may permit mechanisms of interaction that would not occur *in vivo* (Miller et al. 1996). Nevertheless, such experiments are important for understanding the basic mechanisms of action. The most widely studied *in vitro* effect is cell lysis, whereby the cell membrane is ruptured and the cell is destroyed. This process is thought to occur primarily as a result of inertial cavitation and it, therefore, exhibits a similar dependence on the exposure parameters; cell death increases with an increase in ultrasound intensity or pulse length, but decreases with increasing frequency. Cells that are not destroyed by lysis can still be damaged by cavitational processes. A number of different effects have been reported, including membrane transport effects, alterations in cell growth, and DNA damage (Miller et al. 1996). The latter, in particular, has been fairly extensively investigated and the vast majority of studies have reported negative results. However, some subtle effects (such as single-strand breaks) have been reproducibly observed using exposure levels high enough to cause cavitation (Miller et al. 1996). The relevance of these results to *in vivo* ultrasound examinations is not currently known, although the presence of inertial cavitation does appear to be required, implying that situations in which microbubble contrast agents are used could pose the greatest risk.

The previous paragraphs described specific physical effects that have been observed in the laboratory, by irradiating *in vivo* and *in vitro* mammalian tissues at diagnostic exposure levels. Unfortunately, it is difficult to extrapolate from these results to assess the likely risk *in vivo*. There is a serious lack of long-term, prospective epidemiological studies of the kind that would be required to detect some subtle stochastic effect to the foetus. Moreover, as time passes, the opportunities for carrying out such studies are reduced, because the widespread use of ultrasound in obstetrics makes it difficult to set up a control group of pregnant women who will not be scanned. A review of the epidemiological evidence to date (Salvesen 2007) reveals that there are no confirmed associations between ultrasound during pregnancy and subsequent harmful effects on childhood development. However, the data for these studies were collected over a period when the acoustic output of diagnostic ultrasound devices was substantially lower than it is today. It is perhaps more instructive to consider the evidence from recent *in vivo* experiments of ultrasonic exposure of mammalian embryos *in utero*. These studies indicate that, in the absence of a thermal effect, the risk of embryonic loss, congenital malformations or neurobehavioral effects is insignificant at diagnostic exposure levels (Jensh and Brent 1999).

6.5.2 Standards and Guidelines

In different parts of the world, government bodies place demands on manufacturers wishing to legally market their ultrasound scanners. This legislation is guided by international standards, most notably those published by the International Electrotechnical Commission (IEC). The IEC standard relating to diagnostic ultrasound equipment (IEC 2001) sets no upper limit on acoustic output and instead stipulates that two indices related to exposure and safety must be continually updated on the screen in response to any change in the scanner settings. These indices were introduced in the ODS published jointly by the

American Institute of Ultrasound in Medicine and the National Electrical Manufacturers Association (AIUM/NEMA 1992). Mathematical expressions were derived to quantify the thermal and mechanical hazards from in-water measurements of acoustic intensity or power. These result in values, known as the thermal index (TI) and mechanical index (MI), which are displayed on the scanner alongside the ultrasound image or Doppler spectrum. The methods for determination of the safety indices are published in the ODS, as well as in an IEC standard (IEC 2006).

The TI is a guide to the potential for tissue heating. It is a dimensionless quantity, defined as:

$$\mathrm{TI} = \frac{W_0}{W_{\mathrm{deg}}} \tag{6.63}$$

where

W_0 is the acoustic power of the ultrasound system, and

W_{deg} is the acoustic power required to elevate the temperature by 1°C under very specific conditions.

Six formulae have been derived to calculate TI in six different situations, each associated with a different value of W_{deg}. The first class of conditions, which gives rise to three different thermal indices, concerns the tissue type. The soft-tissue thermal index (TIS) assumes a uniform homogeneous medium with an attenuation coefficient of 0.3 dB cm^{-1} MHz^{-1}. The bone-at-focus thermal index (TIB) includes a layer of bone within the soft-tissue model, situated at the depth that maximises the temperature rise. The absorption and attenuation of the bone are not specified. Instead, the formula assumes that half the incident power is absorbed in the bone layer. Finally, the cranial bone thermal index (TIC) assumes that bone is in direct contact with the transducer face. TIS is the appropriate index if the path is principally made up of soft tissue or fluid (e.g. first trimester exam, abdominal exam). TIB is appropriate if bone is in the beam (e.g. second or third trimester exam). TIC is appropriate if bone is within about 1 cm of the transducer (e.g. transcranial imaging). For each of the three tissue types, two different formulae have been derived. The first assumes unscanned exposure conditions (appropriate for M-mode or pulsed Doppler), while the second assumes scanned conditions (appropriate for B-mode or colour-flow imaging). Although the TI has been defined to be without units, it is intended to give a rough indication of the likely maximum (i.e. worst case) temperature rise in °C that might be produced after several minutes of keeping the probe stationary. However, due to the many assumptions used in its calculation, the TI can underestimate the *in situ* temperature elevation by a factor of more than 2 (Jago et al. 1999).

The MI is a guide to the potential for mechanical effects due to acoustically driven gas bubbles. The formula is:

$$\mathrm{MI} = \frac{p_-}{\sqrt{f_{\mathrm{aw}}}} \tag{6.64}$$

where

f_{aw} is the acoustic working frequency (centre frequency), and

p_- is the derated value of the peak negative pressure at the position of the maximum derated PPSI (see Section 6.4.1.2).

The formula for MI is derived from the theoretical analysis of the threshold for inertial cavitation, presented by Apfel and Holland (1991).

Various organisations, such as the British Medical Ultrasound Society (BMUS), have published documents advising the user on the course of action to be taken for different values of the safety indices. These specific recommendations will not be discussed here. Instead, some general remarks will be made concerning responsible use of the indices. The most important point is that MI and TI should be kept as low as reasonably achievable (known as the ALARA principle) without compromising the diagnostic value of the examination. Similarly, the total examination time should be kept as short as possible and the image should be frozen when real-time information is not being used. Extra vigilance should be exercised for special circumstances, for example, when scanning sensitive tissues (such as the foetus or the eye), when the patient has pre-existing temperature elevation, or when ultrasound contrast agents are used. Temperature rises are likely to be greatest at bone surfaces and adjacent soft tissues, so care should be taken when scanning foetal structures close to bone such as the brain or spinal cord. The AIUM strongly discourages the nonmedical use of ultrasound for psychosocial or entertainment purposes, for example, when the sole purpose is to obtain a picture of the foetus or to determine the foetal gender (AIUM Statement on Prudent Use in Obstetrics 2007).

The safety indices have been subject to much criticism, both of a technical and practical nature (Duck 2007). For example, they are based on simplified models that make inappropriate assumptions, they do not inform the user of the depth at which the maximum hazard occurs (making it difficult to judge the risk to sensitive tissues), and they provide a poor fit to experimentally measured bioeffects (Church and O'Brien 2007). Other strategies are continually being proposed and it is possible that the safety indices and their measurement methods will change in the future. Some authors have suggested an entirely different approach. For example, Shaw et al. (1999) proposed the use of a thermal test object to provide a direct estimate of the *in situ* temperature rise. The test object consisted of a thin-film thermocouple sandwiched between two discs of TMM. The geometry and properties of the material were chosen to represent homogeneous soft tissue, but other designs modelling different clinical situations are possible.

Although the introduction of the safety indices has passed much of the responsibility onto the operator, there are still exposure limits imposed on manufacturers. Those imposed by the U.S. Food and Drug Administration (FDA) have greatest practical consequence for most people, as diagnostic ultrasound equipment is designed and manufactured largely for the U.S. market. Manufacturers complying with track 3 of the 510(k) regulations must specify whether the scanner is intended for ophthalmic or non-ophthalmic applications and must meet the following upper limits on ultrasound exposure, which are imposed by the FDA for market clearance of diagnostic ultrasound equipment (Duck 2007):

Non-ophthalmic applications	derated $I_{SPTA} \leq 720\,\mathrm{mW\,cm^{-2}}$, derated $I_{SPPA} \leq 190\,\mathrm{W\,cm^{-2}}$, MI ≤ 1.9, TI ≤ 6.0 (values of TI above 6.0 are allowed but require justification).
Ophthalmic applications	derated $I_{SPTA} \leq 50\,\mathrm{mW\,cm^{-2}}$, MI ≤ 0.23, TI ≤ 1.0, where the limit on the derated I_{SPPA} was not specified.

The derated values quoted for I_{SPTA} and I_{SPPA} are calculated as described in Section 6.4.1.4. At least one of the two quantities MI and I_{SPPA} must be below the specified limit. A further requirement is that manufacturers must display the two safety indices according to the IEC standard (IEC 2001).

Although the IEC standard (IEC 2001) does not set an upper limit on acoustic output, it does impose the following limits on the temperature and temperature rise at the surface of the transducer (Duck 2007):

Temperature	≤50 °C in air and 43 °C on tissue whether for external or internal use
Temperature rise	≤27 °C in air, 10 °C on tissue for external use, and 6 °C on tissue for internal use.

Note that the maximum temperature rise is higher for extracorporeal probes than for endoprobes because in the latter case, the adjacent tissue is already at a temperature of 37°C and, furthermore, the heat cannot be dissipated by convection or radiation. At present, transducer self-heating provides a practical limitation on ultrasound exposure, in the sense that further increases in output power are not possible without the transducer becoming too hot to use (Duck 2007). This issue will need to be addressed in future transducer designs.

6.6 Clinical Applications of Diagnostic Ultrasound

Ultrasound accounts for about one in four of all imaging procedures worldwide (Wells 2006). Ultrasound imaging and Doppler methods have found vast application throughout the breadth of medical diagnostic investigations. It is impossible, in this brief survey, to do justice to the extent to which they are used, or to their value. The approach taken is to provide examples of major uses.

6.6.1 Obstetrics

Ultrasound plays a major role in the management of both normal and complicated pregnancies. It is used to monitor foetal development and to diagnose maternal and foetal disorders.

6.6.1.1 First Trimester

The main applications are as follows:

(i) Confirmation of pregnancy – this can be demonstrated even a few days after the missed menstruation.
(ii) Assessment of gestational age – this is initially based on the size and structure of the gestational sac. After 6–8 weeks, the crown-rump length is the most reliable index.
(iii) Multiple pregnancy – this can be assessed with confidence by the end of the first trimester.
(iv) Differential diagnosis of vaginal bleeding – this can occur due to missed abortion (e.g. embryonic demise or blighted ovum) or abnormal pregnancy (e.g. ectopic pregnancy).
(v) Screening – measurements may be made of the thickness of the so-called nuchal translucency, where (at 11 – 14 weeks) a collection of lymph fluid thickens an echo-poor appearance of the soft tissues at the back of the neck (the nuchal fold) of the foetus. An increased thickness may be associated with chromosomal,

cardiac or structural disorders. Ultrasound is also used to guide chorionic villus sampling (CVS), in which cells are taken from the placenta. Genetic disorders such as Down's syndrome can be diagnosed.

(vi) Detection of major structural abnormalities – ultrasound can detect some structural abnormalities, for example, of the central nervous system and urinary tract.

6.6.1.2 Second and Third Trimesters

Ultrasound is used in the following examinations:

(i) Evaluation of foetal maturity and growth in such standardised measurements as biparietal diameter, crown-rump length, abdomen circumference and femur length.

(ii) Placental localisation and diagnosis of placenta previa.

(iii) Imaging of foetal structural abnormalities such as achondroplasia, cleft palate, microcephaly, hydrocephaly, spina bifida, renal agenesis, urinary and bowel obstruction, skeletal and cardiac anomalies and malignant tumours.

(iv) Ultrasound-guided diagnostic procedures – ultrasound can be used to guide sampling of amniotic fluid (amniocentesis) and foetal blood (percutaneous umbilical blood sampling). These procedures are used to diagnose genetic abnormalities and blood disorders.

(v) Assessment of foetal movement – foetal activity such as pseudo-breathing, trunk movement and response to stimuli (e.g. vibration) is characteristic of a healthy foetus and is decreased in a hypoxic foetus.

(vi) Ultrasound-guided therapies, for example, removal of tumours, placement of vesico-amniotic shunts in cases of bladder neck obstruction, blood transfusion.

In Figure 6.60, two examples of B-scans of the normal foetus at different ages are shown.

In most of the aforementioned applications, examinations are performed transabdominally using real-time B-mode, equipped with sector or linear-array probes operating at frequencies of 3.5–5 MHz. For early or suspected pregnancy, transvaginal scanning is often preferred, as it enables the use of higher frequencies (5–7 MHz), which provide

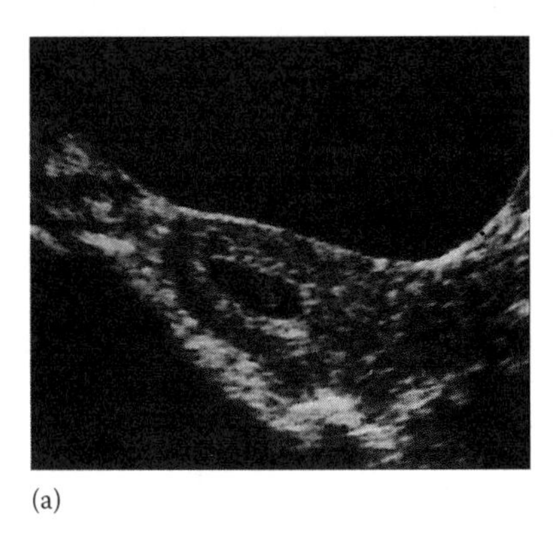

(a)

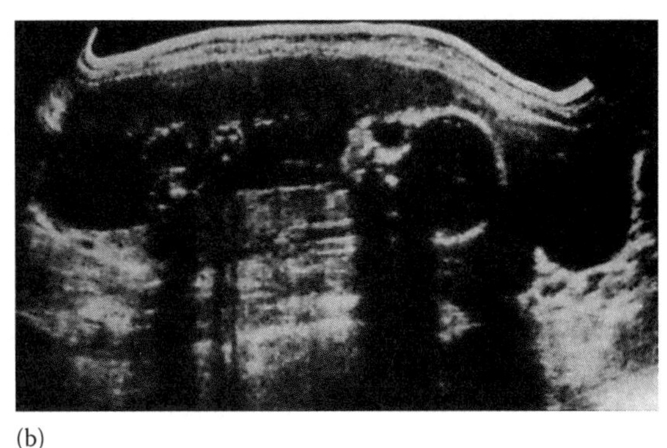

(b)

FIGURE 6.60
Image of a normal foetus at 5 weeks (a) taken with an electronic real-time scanner and another at 22 weeks (b) taken (circa 1975) with a contact manual B-scan system. (Courtesy of S.G. Schoenberger and D.O. Cosgrove, respectively.)

superior resolution. M-mode scans are sometimes used in the assessment of the foetal heart, while Doppler methods find use in measurements of placental and foetal blood flow. CW Doppler is used for foetal-heart monitoring during labour. 3D ultrasound (see Section 6.3.4) is used for volume measurement of foetal organs and for better visualisation of surface foetal abnormalities, such as cleft palate.

6.6.2 Gynaecology

Ultrasound is the preferred imaging modality for studying the female pelvis. The following are the most common applications:

(i) Assessment of congenital uterine anomalies – abnormal size and shape can cause infertility or failed pregnancy.
(ii) Differential diagnosis of abnormal vaginal bleeding or acute pelvic pain – such symptoms are usually due to a benign condition, for example, adenomyosis, uterine fibroids or ovarian torsion.
(iii) Detection, diagnosis and follow-up of benign and malignant masses of the uterus, cervix, ovaries and Fallopian tubes.
(iv) Study and management of infertility – ultrasound is used to assess the integrity of the reproductive tract, to detect pathological changes that may be contributing to infertility, and to monitor cyclic changes of pelvic organs. Ultrasound also plays a role in the guidance of infertility treatment (e.g. ultrasound-guided oocyte retrieval and monitoring ovarian follicle size following drug treatment).
(v) Ultrasound-guided procedures, for example, abscess drainage.
(vi) Intrauterine devices – ultrasonic examination is useful in demonstrating the presence and position of the contraceptive coil.

In gynaecological applications, real-time B-mode, pulsed-wave, colour and power Doppler are used. The examination techniques are either transabdominal scanning, using the full bladder as an acoustic window, or transvaginal scanning. Transvaginal ultrasound can be combined with a technique known as sonohysterography, in which a catheter is passed through the cervix into the uterine cavity and sterile saline is gently injected. This provides excellent delineation of the endometrial cavity, facilitating the diagnosis of morphological and structural anomalies.

6.6.3 Abdomen

In the upper abdomen, the most useful scanning method is real-time B-mode, with a transabdominal sector or linear array, operating at frequencies of 3.5–5 MHz. In the pelvis, and for parts of the GI tract, intracavity probes are often preferred. Doppler methods are used to assess abdominal arteries and veins and to evaluate tumours. Contrast agents also play an important role for these purposes, particularly in the liver, pancreas and kidneys.

6.6.3.1 GI Tract

Stomach and bowel are filled with varying amounts of gas, fluid and solid faecal material, which render transabdominal ultrasonic scanning difficult. Therefore, the method of choice is usually endoscopic ultrasound, performed either through the mouth (for the upper GI tract) or the rectum (for the lower GI tract). The frequency range is typically 5–12 MHz, although recently, miniprobes have been developed with diameters of 2–3 mm and frequencies up to 30 MHz. These small flexible catheters can be passed through a standard endoscope such that regular (optical) endoscopy and ultrasound imaging can be performed at the same time. The most common uses of ultrasound in the GI tract are:

(i) Differential diagnosis, staging, and follow-up of lesions in the oesophagus, stomach, intestines and rectum.
(ii) Assessment and management of inflammatory bowel diseases.
(iii) Detection of oesophageal varices.
(iv) Ultrasound-guided procedures, for example, fine-needle aspiration (FNA) and core biopsy, drainage of cysts and abscesses and creation of digestive anastomosis.

6.6.3.2 Liver and Biliary Tree

Liver examinations are usually performed transabdominally, using an intercostal approach, but endoscopic ultrasound (transgastric or transduodenal) is useful for assessing the biliary tree. It is also possible to pass miniprobes into the biliary system for intraductal scanning. The most common applications of ultrasound imaging of the liver are:

(i) Assessment of liver size and shape – hepatomegaly often accompanies liver disease, while change of shape (e.g. an increased roundness in ascites) might also be indicative of pathology.
(ii) Focal disorders – diagnosis of primary and secondary malignant tumours, abscesses and haematomas. Contrast-enhanced ultrasound is particularly useful for this purpose.
(iii) Diffuse diseases – diagnosis of fatty change, cirrhosis and, rarely, malignant infiltration.
(iv) Detection of gallstones and inflammation of the gall bladder.
(v) Diagnosis of obstructive jaundice.
(vi) Ultrasound-guided procedures, for example, FNA, minimally invasive thermal therapies.

6.6.3.3 Pancreas

The superior image quality has caused endoscopic ultrasound (transgastric or transduodenal) to replace transabdominal ultrasound as the method of choice when evaluating the pancreas, provided skilled and experienced operators are available. The most common applications of pancreatic imaging are:

(i) Diagnosis of pancreatic carcinoma (including guidance of FNA).
(ii) Differential diagnosis and management of benign and pre-malignant pancreatic cysts.
(iii) Diagnosis and management of pancreatitis (acute and chronic) and pancreatic trauma.
(iv) Ultrasound-guided procedures, for example, FNA, celiac plexus block for the management of pancreatic pain, cyst fluid analysis and drainage of pseudocysts.

6.6.3.4 *Spleen and Lymphatics*

The spleen is usually assessed transabdominally, although endoscopic (transgastric) ultrasound is also used, especially for ultrasound-guided procedures. The most common applications are:

(i) Assessment of spleen size – splenomegaly often accompanies inflammatory disease and it is also common in many malignancies, as a manifestation of the immune response to a tumour.
(ii) Management of spleen trauma.
(iii) Assessment of focal lesions – the role of ultrasound is generally limited to differentiating between solid and cystic lesions.
(iv) Diagnosis of accessory spleen – contrast-enhanced ultrasound is useful for this purpose.
(v) Detection and follow-up of lymphadenopathy in staging of malignant diseases.
(vi) Ultrasound-guided procedures, for example, FNA and abscess drainage.

6.6.3.5 *Kidneys*

Transabdominal ultrasound plays a major role in assessing the kidneys and is often the first imaging technique to be employed in patients presenting with renal failure, haematuria or proteinuria. Contrast agents have improved the value of renal ultrasound, particularly for the diagnosis of renal masses and for the assessment of renal vasculature and perfusion. Some of the most common applications are:

(i) Detection of congenital anomalies (agenesis, hypoplasia, etc.).
(ii) Diagnosis of urinary obstruction.
(iii) Diagnosis of renal cysts (e.g. polycystic disease) and tumours.
(iv) Management of renal trauma.
(v) Assessment of complications (rejection) of renal transplants.
(vi) Detection of kidney stones and of hydronephrosis.
(vii) Detection and quantification of abnormal renal vasculature, especially renal artery stenosis.
(viii) Ultrasound-guided procedures, for example, FNA and minimally invasive thermal therapies.

6.6.3.6 *Bladder*

The full bladder is readily examined by contact scanning through the abdominal wall, using a real-time sector probe, operating at frequencies of 3.5–5 MHz. Endoscopic ultrasound via the urethra and occasionally the rectum, at 7 MHz and above, is superior to transabdominal scanning for staging bladder carcinoma. Transvaginal ultrasound may be used for women. The most common applications of bladder ultrasound imaging are:

(i) Detection of congenital anomalies (e.g. diverticula) and pathological changes (e.g. trabeculation).
(ii) Assessment of size, shape and volume (capacity and residual, post-void urine volume) – these factors are indicative of pathology in several disorders, for example, neurological disturbances.

(iii) Diagnosis and staging of bladder tumours.
(iv) Diagnosis of bladder stones and clots.

6.6.3.7 Prostate

Due to the superior image quality, particularly in obese patients, transrectal imaging has replaced transabdominal imaging as the method of choice for visualising the prostate gland. The most common applications are:

(i) Differential diagnosis of tumours – the value of ultrasound in prostate cancer detection has increased dramatically in recent years, partly due to contrast-enhanced ultrasound (including contrast-enhanced targeted biopsies) and elastography.
(ii) Diagnosis of prostatitis/BPH.
(iii) Investigating male infertility – this is usually due to a congenital anomaly or the presence of pathology that is causing ductal obstruction.
(iv) Ultrasound-guided procedures, for example, biopsy, minimally invasive thermal therapies and brachytherapy.

6.6.4 Cardiovascular System

6.6.4.1 Heart

In contrast with the majority of abdominal applications, a very common scanning technique in echocardiography is the M-scan. In fact, the two techniques, B-mode and M-mode, are complementary: although M-mode is ideally suited to investigation of valve and wall motion, information can be obtained only from along lines of sight defined by the ultrasonic beam. A better method for examining the spatial relationships within the heart is a 2D scan, which, on the other hand, provides less quantitative information about motion. Modern echocardiography permits examination of anatomy in motion by utilising both techniques simultaneously. The need for this scan flexibility, and to scan through small acoustic windows (either intercostally or subcostally), has resulted in phased-array sector scanning (2.5–5.0 MHz) being the most commonly employed technology. For some cardiac structures, superior image quality can be achieved using TOE imaging (5–10 MHz). TOE probes also have the advantage that the facility to image in two or more planes can be readily incorporated. Haemodynamic effects of structural and functional abnormalities are assessed in Doppler measurements. Echocardiography tends to make rapid use of advances in technology and has found application for harmonic imaging, 3D imaging, contrast-enhanced ultrasound, tissue Doppler imaging and strain-rate imaging.

An example of a cardiac echogram, together with a corresponding M-mode, is shown in Figure 6.61.

Some of the most common applications are:

(i) Assessment of left-ventricular function, in terms of wall thickness and movement, diastolic and systolic volume, ejection fraction, jet velocity, etc.
(ii) Valve disorders – assessment of cardiac murmurs, stenosis and regurgitation of all four cardiac valves and follow-up of prosthetic valves.
(iii) Pericardial disease – diagnosis of pericardial effusion and guidance of pericardiocentesis (drainage of pericardial fluid).
(iv) Diagnosis and management of infective endocarditis.

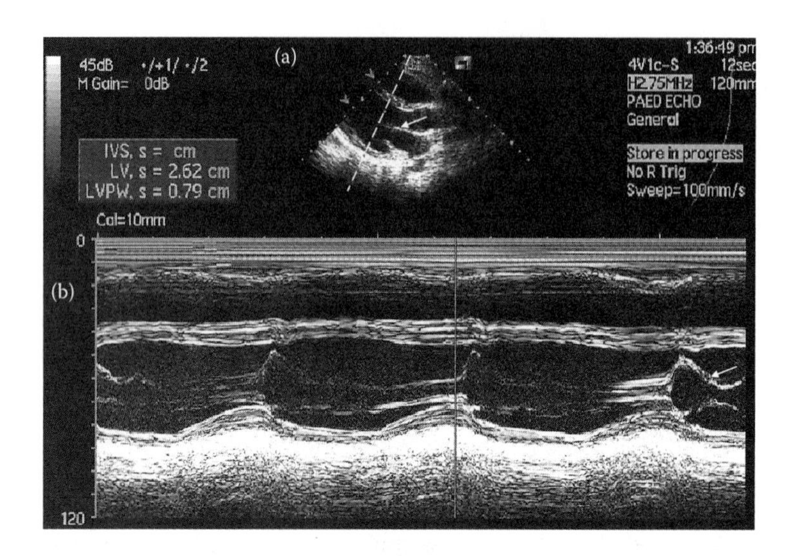

FIGURE 6.61
A single frame (a) from a sequence of real-time sector scans of a normal heart, indicating the line of site (dashed line) used for an M-mode recording (b) of mitral valve motion. The white arrow (in both (a) and (b)) indicates the mitral valve. (Courtesy of N. Joshi.)

(v) Atrial fibrillation – assessing structural cause and assessing risk of thromboembolism.
(vi) Evaluation of congenital heart disease (largely a problem of infancy and childhood).
(vii) Diagnosis of cardiac tumours – these are rare, but often benign, and early diagnosis can lead to effective surgery.
(viii) Stress echocardiography – this is a comparative evaluation of cardiac function at rest and under exercise- or drug-induced stress. It is used to diagnose coronary artery disease, to assess response to treatment, etc.
(ix) Ultrasound-guided procedures, for example, endomyocardial biopsy, cardioversion and pericardiocentesis.

6.6.4.2 Arteries and Veins

Doppler techniques, often in conjunction with ultrasonic imaging methods, are used in clinical investigation of the vascular system. Carotid and peripheral vascular scanning is performed at frequencies of 7–13 MHz, while abdominal vessels are scanned at 3.5–5 MHz. Intravascular probes operating at frequencies up to 30 MHz can be used to study pathology in the vessel wall. These tend to be mechanically scanned transducers that rotate through 360°, providing 3D visualisation of vascular structures and their related pathologies. The main fields of application are:

(i) Cerebrovascular disease – ultrasound provides an accurate method of screening for carotid artery disease; detection of signs premonitory to stroke might justify surgical intervention.
(ii) Investigation of lower-limb arterial disease (detection and measurement of stenoses and aneurysms).
(iii) Diagnosis of venous disorders in both the upper and lower extremities (e.g. thrombosis, varicose veins and chronic venous insufficiency).

(iv) Imaging of large abdominal vessels (e.g. to assist identification of vascular anatomy or to diagnose and evaluate abdominal aortic diseases, such as aortic dissection and aneurysm).
(v) Detection and evaluation of atherosclerotic plaque – elastography has shown promise in this context.
(vi) Pre-operative vein mapping and post-operative graft surveillance in arterial bypass surgery.
(vii) Evaluation of dialysis access grafts to verify that they are functioning properly.
(viii) Ultrasound-guided procedures, for example, foam injections to treat varicose veins, thrombin injection to treat pseudoaneurysms.

Figure 6.53 provides an example of a B-scan of a vessel, together with corresponding Doppler signal.

6.6.5 Superficial Structures

In ultrasonography of superficial structures (and peripheral vessels), that is, musculoskeletal system, breast, thyroid and male genitalia, the most useful scanning methods are real-time B-mode and colour-Doppler imaging. These are performed with a linear array, operating at frequencies of 5–15 MHz. The method of examination is either contact scanning or a water-bath technique, where a polythene bag filled with water is placed between the skin and the probe, or the organ (breast, scrotum) hangs in a water bath in which the probe is also placed. Stand-off pads may also be used.

6.6.5.1 Musculoskeletal System

Ultrasound represents a cost-effective alternative to MRI for assessing a variety of musculoskeletal disorders of the shoulder, elbow, wrist, hip, knee and ankle. The two most common causes of musculoskeletal disorders are rheumatological conditions and trauma or injury (in particular, sports or work related). Soft-tissue involvement in arthritis can be detected by ultrasound long before the appearance of bone erosion, making it helpful for early diagnosis. Some of the uses of ultrasound are as follows:

(i) Evaluation of soft-tissue masses – ultrasound can distinguish between solid and cystic lesions.
(ii) Injuries and abnormalities in the muscle and fascia, for example, assessment of muscle tears or contusions, evaluation of changes in muscle size and architecture due to neuromuscular diseases, thickening of fascia (e.g. in plantar fasciitis).
(iii) Diagnosis of bleeding or other fluid collections within the muscles, bursae and joints.
(iv) Detection of tendon pathologies, for example, tendonitis, tenosynovitis and tears.
(v) Detection of vascular pathologies, for example, aneurysm, haemangioma.
(vi) Abnormalities of the outer surface of the bone, for example, discontinuities or fractures.
(vii) Ultrasound-guided procedures, for example, biopsy of soft-tissue masses, needle aspiration of fluid in the joints, injection of local anaesthetic or steroid.

6.6.5.2 Breast

In the breast, ultrasound plays an important role (complementary to palpation and mammography) in the diagnosis of inflammation, cysts, fibrocystic disease, and malignant and benign tumours. In particular, in differential diagnosis of breast carcinoma in the symptomatic breast, it helps to establish whether a palpable mass is solid and, if so, whether it exhibits features that suggest malignancy (illustrated in Figure 6.62). B-scans are used for imaging breast structure and also for assessing mobility and elasticity, based on real-time observation during palpation. Doppler examinations and contrast-enhanced ultrasound are used to assess tumour vasculature. Recently, new ultrasonic imaging methods, such as 3D (Section 6.3.4) and elasticity imaging (Section 6.3.6.2), have demonstrated improvements in diagnostic accuracy. Furthermore, it has been shown that high-frequency ultrasound has the potential to accurately predict the presence of metastatic disease in axillary lymph nodes.

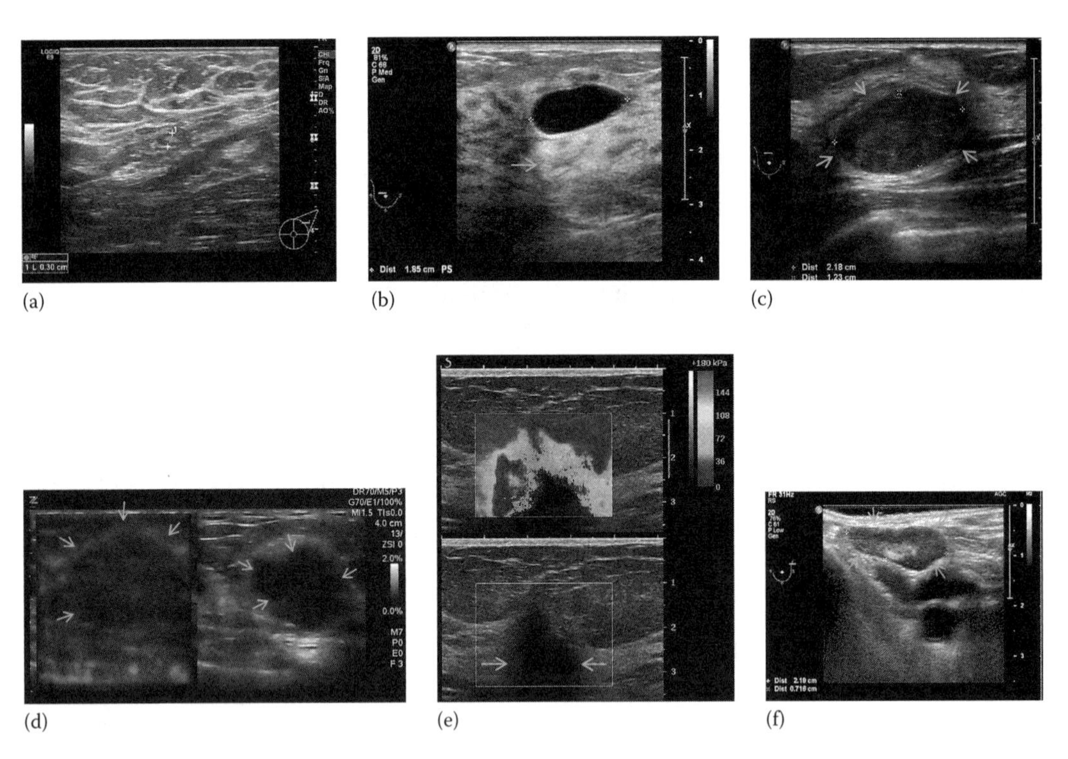

FIGURE 6.62
(See colour insert.) Single B-scan frames from real-time studies of the breast, showing (a) normal breast gland (calipers indicate lymph node), (b) cyst (note the sharp borders, lack of internal echoes, posterior enhancement indicated by arrows and calipers used to measure the maximum dimension of the cyst (1.85cm)), (c) a fibroadenoma indicated by arrows (well defined regular border, uniform echoes, lateral edge shadow and some posterior enhancement), (d) a B-scan (right) of a large carcinoma (irregular fuzzy border, heterogeneous internal echoes and slight enhancement) and a freehand strain image (left) of the same tumour (relatively stiff and heterogeneous with irregular fuzzy borders and stiff tissue that extends beyond the borders defined by the echo-poor regions in the B-scan) (see elastography, Section 6.3.6.2), (e) B-scan (below) of a highly suspicious breast mass and a colour coded Young's modulus image (above) obtained from shear-wave speed (see Section 6.3.6.2) (maximum tumour stiffness > 160 kPa at the tumour margin and a central strong acoustic shadow (indicated by arrows) providing very little shear-wave signal), and (f) an enlarged lymph node indicated by arrows. (Courtesy of S. Allen, L. Hastings, and P. Smith.)

Ultrasound is of particular value in the young or dense breast, during pregnancy, and in evaluating the radiographically evident but non-palpable mass. Ultrasound guidance of FNA, core biopsy, draining cysts and surgical-guide-wire placement is also of great value, particularly for non-palpable lesions.

6.6.5.3 *Thyroid and Parathyroid*

In the thyroid, ultrasound provides structural information, which complements radionuclide functional studies. The chief contribution of ultrasonic examinations is differentiation of cystic from solid nodules. The following are the main applications:

(i) Demonstration of cysts and tumours (adenoma, carcinoma), especially those lesions which appear cold on radioisotope scans.
(ii) Diagnosis of multinodular goitre and diffuse disease.
(iii) Parathyroid tumours.
(iv) Thyroid volume measurement for calculating dose in nuclear medicine treatment of benign thyroid disease.
(v) Ultrasound-guided FNA of thyroid nodules.

6.6.5.4 *Male Genital System*

Ultrasound imaging, in particular colour and power Doppler, is the procedure of choice in most scrotal and penile pathologies. The main applications are as follows:

(i) Detection and assessment of scrotal masses, including cysts, tumours, abscesses, haemorrhage, etc. It allows differentiation of cystic and solid masses; most solid masses located inside the testicles are malignant or cancerous.
(ii) Diagnosis of testicular torsion.
(iii) Management of inflammation – ultrasound is used to evaluate the degree and extent of inflammation and to monitor the effects of therapy.
(iv) Detection of absent or undescended testicle.
(v) Study of male infertility (e.g. screening for varicocele, measuring testicular volume and detecting anomalies of the epididymis and vas deferens).
(vi) Diagnosis of hydrocele testis.
(vii) Diagnosis and management of testicular injuries due to blunt scrotal trauma.
(viii) Confirmation of the diagnosis of inguinal scrotal hernia.
(ix) Ultrasound-guided biopsy of non-palpable scrotal masses.

6.6.6 Eye and Orbit

Ophthalmic ultrasonography has become an essential diagnostic tool. Because of the small size of ocular and orbital structures, high resolution is required without the demand for deep penetration. Equipment is designed specifically for ophthalmic use, with a typical frequency range of 13–20 MHz, although frequencies up to 50 MHz are used for some applications (e.g. biometry). The eye is scanned either in direct contact or using a water bath. A-scans are used to provide precise information about the location of reflecting interfaces and to make measurements of the axial length of the eye. Relatively cheap semi-automatic

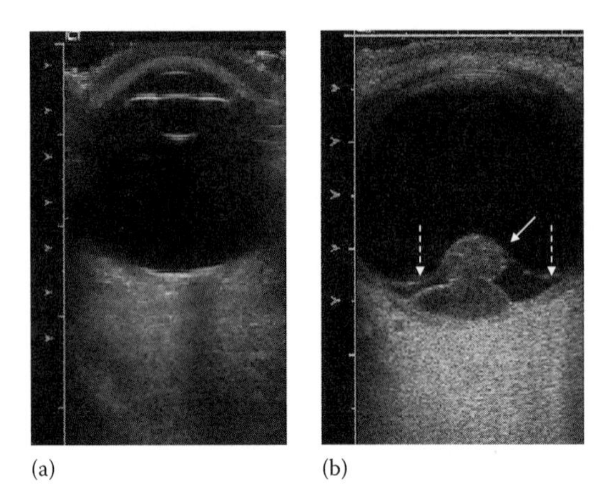

FIGURE 6.63
(a) Central transverse B-mode image of a normal eye. (b) Transverse B-Mode image of a 'Collar-stud'-shaped malignant melanoma (solid arrow) with an overlying retinal detachment (dashed arrows). (Courtesy of M. Restori.)

A-scan instruments are available for this purpose. Real-time B-scans provide topographical and motional information and, in conjunction with Doppler techniques, represent the primary diagnostic method. Some of the applications are:

(i) Detection of foreign bodies, trauma, mucocele and haemorrhage.
(ii) Diagnosis of intra-ocular and orbital tumours.
(iii) Diagnosis and assessment of retinal detachment (illustrated in Figure 6.63).
(iv) Diagnosis of pathology of the orbital vessels, for example, cavernous haemangioma and dilated orbital veins.
(v) Biometry – the required power of intra-ocular or contact lenses can be calculated.
(vi) Pre-operative assessment in patients with pathologies (such as dense cataract) that make the eye difficult to image with optical methods.
(vii) Follow-up of orbital disease.

6.6.7 Brain

Echographic visualisation of adult-brain anatomy is made difficult by the attenuating and refracting properties of the skull, although recent studies have shown that several brain disorders can be depicted (e.g. Parkinson's disease). In infants, however, an acoustic window (the anterior fontanelle) allows scanning during the early months of life and has become an important method of diagnosing various conditions (see Section 6.6.10). Ultrasound is also used for intra-operative guidance of neurosurgery in adults, although not as commonly as MRI and CT. The main use of ultrasound in the adult brain is transcranial Doppler, which allows flow velocity and direction to be measured in the proximal portions of large intracranial arteries. Some applications of this technique are:

(i) Detection and evaluation of cerebral vasospasm following subarachnoid haemorrhage.
(ii) Confirmation of a clinical diagnosis of brain death.
(iii) Detection of haemodynamic and embolic events that may result in perioperative stroke during and after carotid endarterectomy.
(iv) Evaluation of occlusive lesions of intracranial arteries.

6.6.8 Lungs

Ultrasound does not normally penetrate the lung, which is filled with air, so diagnostic information is limited to the chest wall, the pleura, and lesions that extend to the pleural surface. In addition, there are certain pathological processes that create an acoustic window, allowing visualisation of lung parenchyma distal to the window. Examples include pleural effusion (when fluid accumulates within the pleural cavity) and atelectasis (when all or part of the lung collapses). Transthoracic lung imaging is carried out using linear, convex and sector transducers with frequencies in the range 3.5–10 MHz. An alternative to transthoracic imaging is endobronchial imaging, which allows visualisation of the tracheobronchial wall and the immediate surrounding structures. The technique is most effective when performed using balloon-tipped miniprobes in the frequency range 12–20 MHz. Endobronchial ultrasound (EBUS) has demonstrated usefulness for lung cancer staging in terms of depth of tumour invasion in the bronchial wall and lymph node involvement. However, it is a skilled procedure that has been slow to gain acceptance, perhaps due to the length and cost of the training. Some of the uses of ultrasound in the lungs are:

(i) Assessment and differential diagnosis of tumours of the pleura, chest wall and peripheral lung. EBUS has also been shown to be useful for assessing central pulmonary lesions.
(ii) Confirmation of pleural effusion and estimation of the quantity of effusion.
(iii) Detection of pleural thickening.
(iv) Diagnosis and monitoring of pneumothorax.
(v) Diagnosis of pulmonary embolism.
(vi) Evaluation of the regional extension of lymph-node involvement in lung cancer.
(vii) Ultrasound-guided procedures, for example, FNA or core biopsy of tumours, thoracocentesis. EBUS has been used to guide transbronchial needle aspiration of lymph nodes.

6.6.9 Skin

Ultrasound imaging of the skin has traditionally been performed using B-mode grey-scale imaging at frequencies of around 20 MHz. Recently, colour- and power-Doppler techniques have gained importance and ultrasound biomicroscopy systems have been developed with typical frequencies between 40 and 60 MHz. Systems are available that can produce cross-sectional images and 3D reconstructions. The main applications of ultrasound imaging in the skin are:

(i) Monitoring the effects of therapy – ultrasound measurement of features such as skin (or tumour) thickness and the degree of vascularity can help assess response to therapy in skin tumours and in inflammatory diseases such as scleroderma or psoriasis.
(ii) Pre-operative evaluation of skin tumours (depth and localisation of margins).
(iii) Differential diagnosis of skin lesions – the specificity of ultrasound is currently low, so pathological confirmation is required. Colour Doppler improves specificity, particularly for lesions with vascular pathology, such as angiomas and pseudoaneurysms.

6.6.10 Paediatric

Ultrasound is widely used in the paediatric setting, as it is a non-invasive, well-tolerated and low-hazard procedure. The choice of transducer and scanning mode is determined by the body site, and, in general, the information provided in the previous sections is relevant. However, specially designed transducers with higher operating frequencies are used for babies and young children. Some common applications of paediatric ultrasound are:

(i) Neonatal neurology – the anterior fontanelle is used as an acoustic window. Disorders such as cerebral haemorrhage in premature babies, hydrocephalus and congenital malformations can be diagnosed.
(ii) Heart – Doppler techniques provide quantitative measures of flow and function.
(iii) GI tract – ultrasound is helpful in the diagnosis of various disorders including pyloric stenosis, intussusception, malrotation, bowel obstruction, bowel wall thickening and appendicitis.
(iv) Liver and gallbladder – for example, the assessment of cirrhosis, liver tumours, portal hypertension and gallstones.
(v) Adrenal gland – diagnosis of adrenal haemorrhage and tumours (the most common being neuroblastoma).
(vi) Kidneys – detection of congenital anomalies and hydronephrosis, assessment and follow-up of urinary tract infections, assessment of renal tumours (the most common being Wilms' tumour).
(vii) Bladder – detection of congenital anomalies, assessment of bladder volume (both pre- and post-voiding), assessment of bladder-wall thickening.
(viii) Female pelvis – detection of congenital anomalies of the uterus and ovaries, assessment of tumours, detection of ovarian cysts and ovarian torsion, diagnosis of precocious puberty.
(ix) Male genitalia – detection of undescended testis, hydrocele, epididymitis, torsion, varicocele and testicular tumours. Assessment of scrotal trauma.
(x) Musculoskeletal – assessment of soft-tissue masses, cystic lesions, haematomas, osteomyelitis, injuries, trauma and hip effusion.

6.6.11 Ultrasound in Invasive Procedures

Although one of the main advantages of ultrasonography is the non-invasive character of examinations, ultrasound also finds diagnostic applications in some invasive procedures, such as needle puncture and intra-operative scanning.

6.6.11.1 Needle Puncture Techniques

The ability of ultrasonic imaging to visualise the depth and position of internal tissue structures makes it a useful tool in guiding the insertion of biopsy needles. Tips of needles penetrating the tissue will be seen as bright reflections and can be guided precisely to the position desired for obtaining material for cytological and histopathological evaluation and for drainage of abscesses. The procedure, carried out under local anaesthesia, involves minimal hazard and discomfort to the patient.

Puncture techniques are used in the diagnosis of lesions of many organs, including breast, liver, pancreas, kidney, retroperitoneum and prostate. Collection of fluid, as in amniocentesis, and aspiration of cysts (e.g. in liver, breast, etc.) is easily performed under control of ultrasound. For non-palpable breast malignancies, echography may be useful in guiding the placement of wires with hooking tips, which aid the location of a mass during surgery.

Specially designed transducer heads, equipped with needle holders, are sometimes used in these techniques.

6.6.11.2 *Intra-Operative Applications*

Internal lesions can be visualised by ultrasonic imaging pre-operatively, but the intra-operative use of ultrasound in direct contact with an organ gives better resolution and provides the surgeon with additional information, aiding the process of decision making. Some of the applications are:

(i) Neurosurgery – localisation of tumours, controlled biopsy and abscess aspiration.
(ii) Abdominal surgery – pre-operative examination of liver, pancreas and kidney (tumours), and search for stones in kidneys and the common bile duct.

Intra-operative transducers are of special design, satisfying the requirements of smaller size and easy maintenance of sterility.

6.6.11.3 *Minimally Invasive Therapeutic Procedures*

These include RF ablation, HIFU, ethanol injection, cryoablation and microwave ablation. Ultrasound has a role to play in all aspects:

(i) Planning (identification of the location and spatial extent of the pathology to be treated).
(ii) Guidance (prediction of the location and spatial extent of the treated region before it is formed).
(iii) Monitoring (visualising the size, shape and location of the treated region), both immediately after therapy and long term.

6.6.11.4 *External-Beam Radiation Therapy*

Ultrasound is used to verify the size, shape and position of the organ or tumour prior to treatment. Research is also looking at tracking organ motion during therapy and feeding this information back into the therapy system.

6.7 Broader View of Ultrasonic Imaging

In order to conclude this review of ultrasonic imaging there is value in looking briefly at how the aforementioned methods relate to alternative imaging approaches and emerging techniques. Six interrelated topics will be mentioned in this context: synthetic-aperture imaging, ultrafast and zone-based imaging, image optimisation, molecular imaging, motion-vector imaging and multiphysics imaging.

6.7.1 Synthetic-Aperture Focusing Technique and Computed Tomography

The process of forming an image by focusing and pulse-echo scanning described in Sections 6.3.1.2 and 6.3.2.12 appears, at first glance, to be fundamentally different from the manner in which X-ray or NMR tomograms, for example, are reconstructed. It is, however, a special (hardware) case of image reconstruction that originated initially with mechanical scanning and remained necessary when only single-channel receive processing (Section 6.3.2) was available, but does not make maximum use of all the potentially available information. More general approaches exist, which, by spatially sampling the amplitude *and phase* of many wavefronts that have travelled through the same object via many different paths, can treat the ultrasonic echo signals as projection data for digital image reconstruction. If a sufficient number of such measurements (i.e. in three dimensions and as a function of wavelength over a wide bandwidth) could be made, and if a sufficiently good model were available for the physics of the ultrasound scattering interactions, it would, in principle, be possible to reconstruct images of the distribution of density and compressibility. Much theoretical work has been published on this subject, but quantitative reconstruction of density and compressibility is not something that one is likely to see used for practical purposes in the very near future. Here we consider some methods that fall short of this ideal, but provide techniques for imaging a variety of the acoustic characteristics of tissues mentioned in Section 6.2.3. The principal disadvantage of all of these methods used to be the time needed to collect the projection information and to reconstruct the image. Recently, however, some of them have started to find their way into commercial medical instrumentation. As capacity for construction of transducer arrays with large numbers of elements, parallel channel data acquisition, fast data transfer and fast digital processing increases this trend is likely to continue, probably to the point where all ultrasound scanners will eventually employ such methods.

Figure 6.64 illustrates some of the great variety of approaches to digital ultrasound image reconstruction. The simplest is probably the so-called time-of-flight (TOF) method for reconstructing the distribution of speed of sound in the object. Consider the case of speed of sound and attenuation coefficient distributions, $c(x, y)$ and $\alpha(x, y)$, in the imaging plane within an object surrounded by water of speed of sound c_0 and negligible attenuation coefficient (Figure 6.64a). A sound pulse is transmitted in the y' direction and travels a distance D to be received after a measured time t. If, in the region of the water only, $t = t_0$, then the shift parameter τ is defined as $\tau = t_0 - t$. From the figure,

$$t_0 = \int_0^D \left(\frac{1}{c_0}\right) dy' = \frac{D}{c_0}. \tag{6.65}$$

Scanning both the transmitter and the receiver in the x' direction (at angle θ to the x-axis) produces a 'shift projection' at orientation θ:

$$\tau(x', \theta) = \left(\frac{D}{c_0}\right) - \int_0^D \left[\frac{1}{c(x, y)}\right] dy'. \tag{6.66}$$

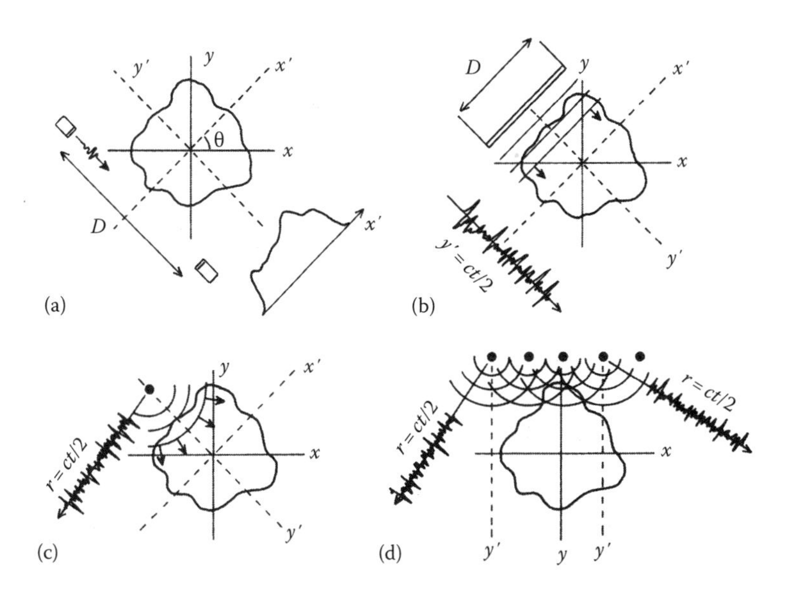

FIGURE 6.64
Schematic illustration of coordinate systems, transducer arrangements and lines of projection data, for ultrasound computed tomography image reconstruction: (a) speed of sound or attenuation-coefficient reconstruction from transmission projections; (b) coherent plane-wave backscatter reconstruction; (c) coherent cylindrical-wave backscatter reconstruction and (d) limited(linear)-aperture cylindrical-wave coherent backscatter reconstruction, also called synthetic aperture imaging.

The problem of reconstruction is then to estimate $c(x, y)$ from a number of shift projections at various angles θ. The various methods available for doing this are described in Sections 3.6–3.8.

By measuring signal magnitude, instead of TOF, the same approach may be used to reconstruct images of the attenuation coefficient $\alpha(x, y)$. The projection would be described by

$$A(x',\theta) = A_0 \exp\left(-\int_0^D \alpha(x,y)\,dy'\right) \tag{6.67}$$

where A_0 is a constant (the magnitude of the transmitted wave). The need to be able to pass sound completely through the object is a major problem for transmission methods. Applications are limited to accessible organs containing no bone, such as the female breast. Indeed, the distribution of fat in the breast is imaged extremely well by speed of sound images, which tend to complement the attenuation and conventional pulse-echo images, as far as the nature of the information displayed is concerned. Other difficulties with transmission reconstruction methods are as follows: (1) the resolution tends to be poor, being limited by the width of the transmitted sound beam and the receiver sensitivity pattern; (2) sound refraction causes image degradation by invalidating backprojection along straight lines (although iterative methods starting with a first estimate of the sound speed image may be employed); and (3) phase cancellation at the receiver tends to produce measurement errors, adding object-dependent noise to the attenuation images (mainly at edges) and making them qualitative rather than quantitative.

Referring to Figure 6.64b, the echo signal from a large, plane-wave source/receiver can be viewed (neglecting attenuation, amongst others things) as a projection of line integrals across the width of the wavefront, that is, the projection itself now occurs as a function of y', scaled in terms of time and speed of sound:

$$\dot{A}(y',\theta) = A_0 \int_0^D \alpha_{bs}(x,y)\,dx' \qquad (6.68)$$

where $\alpha_{bs}(x, y)$ is the image to be reconstructed, which can be referred to as the amplitude backscattering coefficient of the tissue (the intensity backscattering coefficient was discussed in Section 6.2.1.7). Note that, unlike in Equation 6.67, A is now the signal amplitude and may be positive or negative.

Linear backprojection of such coherent projection data is essentially a digital method of creating a focusing system having a completely circular aperture. To avoid corruption of the projection data by phase cancellation, it is necessary for the object to lie in the far field of the source, meaning that large distances, small objects or low frequencies (or all) must be used. This approach is seldom used. More commonly, 2D point sources (i.e. a line source) are used to satisfy far-field conditions by generating cylindrical waves, as in Figure 6.64c. Reconstruction, however, is no longer possible by backprojection along straight lines. Curved lines, corresponding to the shape of the wavefront, must be used (as shown in the figure). The advantage of this system arises with limited-angle reconstruction, where new projection data may be obtained by linearly shifting the reference frame, instead of by the usual method of rotating it. As illustrated in Figure 6.64d, with ultrasound this can be achieved by making use of linear-array technology, using each element of the array as a source for gathering a backscatter projection. Digital reconstruction by backprojection along the paths of the wavefronts will then form an image in which the whole array has been used to focus for each and every point in the image. This method of focusing, and extensions of it, has been used in many fields of remote imaging where the phase of the wave may be measured as well as its magnitude. Examples of applications occur in underwater acoustics, satellite and airborne radar imaging, and geophysics (seismology). It passes by various names, including holography, diffraction tomography, *synthetic-aperture* imaging and *synthetic-aperture focusing technique* (SAFT). Note that in comparison with the conventional line-by-line imaging methods discussed in Sections 6.3.1.2 and 6.3.2.12, the SAFT method accomplishes focusing and scanning in one image reconstruction computation. The most extensive discussion of SAFT methods may, arguably, be found in the radar imaging literature (e.g. Soumekh 1999). The mathematics of focusing in these systems may be formulated as a matrix inversion problem, and may be accomplished in the spatial or frequency domains, as with CT. Some focusing algorithms, such as those in the field known as seismic migration or wave field extrapolation (Berkhout et al. 1982), have been designed to focus through media that have an inhomogeneous sound propagation speed, particularly those used in seismology, where a layered medium may be a reasonable model and iterative methods can be used to determine the speed of sound for optimum focusing for each successive layer (see image optimisation, Section 6.7.3). In practice, three different methods for synthetic aperture data acquisition have been conceived: the 'plane-wave' method, the 'zero-offset' method and the 'complete-dataset' method (Ridder et al. 1982). The zero-offset method is in fact that illustrated in Figure 6.64d, where each element only receives echoes that have arisen from the pulse that it emitted. Airborne and satellite radar is limited to using this method because at any moment during the trajectory of the aircraft or satellite only one source-receiver exists (although there are 'spotlight' methods in which a short phase array is transported along the length of the synthetic aperture and for each source-receiver position is used to impart directionality to the wave and to steer this towards an object region of interest, whatever the source position). The data are, therefore, acquired sequentially, element by element. In the complete dataset method, each element is again used in turn to transmit a pulse, but in this case the backscattered wave for each pulse is sampled on reception by all possible elements. Finally, in the plane-wave method all elements are excited simultaneously to produce a single transmit pulse that propagates to cover the entire field of view, and the backscattered wave for this pulse is sampled on reception by all possible elements. If a multi-channel receiver is available with as many channels as there are transducer elements, and if data transfer and processing bandwidth are not limitations, then there would appear to be no advantage to the zero-offset method as compared with the complete dataset method, the latter providing improved resolution and SNR compared with the former. The plane-wave method, and variations of it, has the advantage that imaging may be accomplished at a very high frame rate (at some cost of spatial resolution), as discussed in the following.

6.7.2 Ultrafast and Zone-Based Imaging

Given the availability of an independent (parallel) receive signal-processing channel for each transducer element in the focusing aperture, the aforementioned plane-wave insonation SAFT method allows all of the data to be acquired for reconstruction of a complete 2D image (or a 3D image if a 2D array and sufficient channels and processing capability are available) from a single sound pulse, whereas the other two methods require a separate transmit pulse for each element, reducing data acquisition speed by the number of elements. As a result, the imaging frame (or volume) rate for the plane-wave method may be equal to the PRF, which amounts to thousands of frames (or volumes) per second. This method has been termed ultrafast imaging. It would be extremely useful for real-time observation of fast events such as the motion of a heart valve, and is essential for observation and tracking of the motion of a shear wave in the shear-wave elastography technique described in Section 6.3.6.2. The PSF of images formed with the plane-wave method, which focuses only on receive, essentially follows the square root of the function that would be obtained if the same PSF were applied for each object point on both transmit and receive. Such transmit focusing is never fully achieved in conventional line-by-line imaging, even with a large number of transmit foci, although transmit focusing also improves signal strength, due to focusing gain (Section 6.2.4.2) and due to improved coherence of backscatter waves at the receiving array (Mallart and Fink 1994, Bamber et al. 2000). Some of the lost resolution, signal strength and image contrast-to-noise ratio may be recovered by trading for frame (or volume) rate, by dividing the imaged field into lateral zones that are insonated sequentially (Mo et al. 2007). The frame rate will, however, be reduced by the number of transmit zones employed. The effect of a transmit focus may also be achieved synthetically in the full plane-wave method, or the zone method, by transmitting multiple waves each phased to steer to a particular angle. If the resulting reconstructed B-mode images were incoherently summed, this would be equivalent to spatial compounding for speckle reduction (see Section 6.2.5.3). Coherent summation of the RF backscatter, however, is equivalent to focusing on transmit (Montaldo et al. 2009). Here the frame (or volume) rate would be reduced by the number of transmit wave angles employed.

Ultrafast and zone-based methods are derived from techniques that appeared in the early 1980s; Cauvard and Hartemann (1982) built a system that employed a surface acoustic-wave device to implement 20 parallel-processing channels to achieve a frame rate of 800 Hz, albeit with very limited image quality, and the 'explososcan' (Shattuck et al. 1984) employed transmit beams that encompassed four receive lines formed with parallel-receive channels using a single transmitted sound pulse. Such parallel-receive beam forming, to the level of 4:1, became an option on most top-of-the-range commercial ultrasound scanners from the 1990s. The high data rates achieved using parallel-receive processing need not always be employed to display images at high frame rates. It is possible, for example, to average images to reduce noise and thus improve imaging depth, or to produce an image at standard frame rates while reducing patient exposure, or to increase the field of view without sacrificing frame rate. High-quality ultrafast imaging, at frame capture rates of many kHz, appeared in the early 2000s, followed by commercial systems employing zone-based and ultrafast methods. Such methods, in combination with the 2D array transducers (Section 6.3.4.2) will eventually provide the means for high-volume-rate 3D data acquisition, needed for artefact-free volumetric imaging of the heart, and time-consuming quantitative measurements such as motion-vector imaging described in Section 6.7.4.

6.7.3 Image Optimisation

The conventional approaches to focusing and echo localisation described in Sections 6.3.1.2 and 6.3.2.12 assume that there is a single, known value for the speed of sound along the tissue paths from all points on the transducer to each point being imaged. In reality, this value is not known, which means that, for electronic focusing and steering, the delay profiles applied across the transducer aperture may not generate the intended direction and shape of wavefront, resulting in image distortion due to incorrect beam-steering angles and defocusing due to incorrect wavefront curvature. Similar distortion and blurring of the PSF would occur in the aforementioned SAFT methods, due to an incorrect time-distance scaling for RF echo backprojection. Scanners that work on the assumption that the average sound speed in the tissue is 1540 m s^{-1} do not usually produce significant image degradation, although they may do so when there is a substantial amount of fat in the acoustic path, such as may happen in breast or abdominal imaging. For this situation some ultrasound scanner manufacturers offer an imaging mode for which a lower value of assumed sound speed is used by the scanner. Manufacturers may also offer an automated image optimisation mode, which may work by forming a range of trial images, each using a different assumed sound speed, and an image metric such as spatial frequency content is used to choose the sound speed that produces the sharpest image (Napolitano et al. 2006).

As was mentioned in Section 6.2.4.4, there may be further distortion and defocusing from wavefront aberration due to variations in sound speed across the sound beam. In SAFT this is equivalent to the backprojection needing to be along irregular, rather than smoothly curved, circular paths. Although THI goes some way towards alleviating this problem, as explained in Section 6.3.3, it is by no means a complete solution. This is a difficult problem, the solution to which will require the 2D arrays mentioned earlier because aberration occurs in 3D. One approach has been to invent algorithms that attempt to find the set of irregular delay profiles that, when applied to the elements prior to beam forming, produces the best image. Measures such as speckle brightness, speckle sharpness, image contrast, and cross-channel signal correlation and coherence have all been employed (and shown to be generally equivalent) to automatically adjust the profiles and select the optimum set of delays (Flax and O'Donnell 1988, Nock et al. 1989, Liu and Waag 1994, 1998, Rigby et al. 2000, Ivancevich et al. 2006b, Li and Robinson 2008). An implementation for general use has not yet emerged although worthwhile improvements in ultrasound beam and image quality have been obtained, including through sections of human skull. Difficulties have included the lack of uniqueness to the solution, the presence of frequency-dependent absorption, the limited apertures available, and the need to unravel aberration in transmission and reception, the former leading to a change in speckle target volume size which itself alters echo coherence (Mallart and Fink 1994, Bamber et al. 2000). If an echo from a known point scatterer (a so-called beacon signal) can be identified in the imaged field then this separates the transmission and reception problems and the correction methods improve their performance considerably. Microbubbles may provide such signals (Seo et al. 2005, Ivancevich et al. 2006a). In a technique known as time reversal imaging (Tanter et al. 2000), the unfocused phase profile across all elements for a beacon echo is reversed in time and use on transmit to create a well-focused beam. Another approach, which may employ the concept of time-reversal, is to use independent information, such as from X-ray CT or MR images, to form a map of likely sound speeds and attenuation coefficients, with which to model the propagation and hence derive corrected delay profiles (Hynynen and Sun 1999, Marquet et al. 2009). It has also been shown to be possible to adjust transmit delay profiles to maximise the generation of radiation force as measured by tissue displacement in elastography (Urban et al. 2007). This continues be an active research area.

6.7.4 Motion-Vector Imaging

Current commercial Doppler measurement and imaging systems measure and display only the component of velocity or displacement in the direction of the sound beam. The same is true of 1D RF echo-correlation tracking methods. With 2D or 3D echo-correlation tracking it is possible to measure the lateral or elevational components of motion, but the precision obtained (limited by beam width) is more than ten times worse than that (a small fraction of a wavelength) in the axial direction. Methods are needed to measure displacement and velocity vectors with high precision for all components. This would make it possible, for example, to obtain quantitative Doppler blood flow measurement, to take full advantage of the potential of elastographic methods, and to accurately track tissues moving in 3D for real-time guidance of treatments such as radiotherapy or HIFU. Substantial research progress has been made towards solving this problem using various techniques, including the use of multiple phase-steered beams to obtain independent estimates of the axial component of motion for different beam angles (Dunmire et al. 2000, Steel et al. 2004, Chen and Varghese 2009, Rao and Varghese 2009), the use of a separate beam to estimate the direction of blood flow by seeking a Doppler angle of 90° using the symmetry obtained in the Doppler spectrum at this angle (Ricci et al. 2009), detection of the lateral gradient of time-varying phase across the aperture of the receiving array (Nitta and Shiina 1998), and the generation of transversely modulated sound fields that have fine structure in the lateral direction (Jensen 2001, Sumi et al. 2008, Hansen et al. 2009). To take advantage of such methods for full vector motion imaging in real time it will be necessary to employ well-sampled large-aperture 2D arrays with ultrafast echo acquisition, and high-bandwidth data transfer and processing.

6.7.5 Molecular Imaging and Drug Delivery

Molecular imaging is the name now given to any imaging method that shows contrast for the bio-distribution and local concentration of molecules that can be used to remotely observe or measure biochemical processes, or that are highly specific indicators of things such as disease, subgroups of disease, cell movement, cell proliferation, cell death, invasion, gene expression, expression of a drug target, drug delivery, gene delivery, prognosis, effects of therapy and response to therapy (Weissleder and Mahmood 2001, Fujibayashi et al. 2002, Rollo 2003, Cherry 2004, Jaffer and Weissleder 2005, Ottobrini et al. 2006, Debbage and Jaschke 2008). Some imaging methods possess intrinsically high specificity for endogenous molecular contrast; optical absorption spectroscopy, for example, may be sensitive to blood oxygen concentration. As can be seen from Section 6.2.1, endogenous ultrasound contrast could exist for collagen, fat and water, for example, but the molecular specificity would be poor (although if temperature can be deliberately altered then fat may be uniquely identified). For many applications, across all imaging modalities, molecular imaging 'probes' have been developed. This is viewed as one of the most promising areas of medical imaging, and research to develop new molecular imaging probes continues at a fast pace. Such probes are comprised of two parts, a chemically reactive part that will bind to and may report an interaction with the biochemical target of interest, and a signal-generating component that allows the probe to be localised by an imaging method. Thus, immunochemical and some conventional histological stains may be classed as molecular probes, imaged, for example, with transmission or fluorescence microscopy. Current practically useful ultrasound probes consist of microbubbles, with or without attached targeting molecules (Liang and Blomley 2003, Klibanov 2007, Kiessling et al. 2009). Microbubbles alone may be specific to being taken up by functioning cells of the reticular-endothelial system. Their application to image functioning regions of the liver, as described in Section 6.3.6.1, may, therefore, be regarded as molecular imaging (see Figure 6.56). In general, however, microbubbles need to be targeted, or 'functionalised', to use them as molecular imaging probes. In a research context this has been shown to be highly successful for a number of molecular targets, including, for example, integrins, p-selectin, vascular endothelial growth factor receptors and prostate specific membrane antigen. A current limitation of microbubbles for molecular imaging however, is that their size prevents them from reaching targets that are extravascular, although this may eventually be overcome using phase change agents that start as nanodroplets of liquid that under ultrasound activation are induced to vaporise into small microbubbles (Asami et al. 2007). It is also difficult to distinguish between stationary bound microbubbles and those that are still circulating in the blood, although current research aims to solve this problem. One approach may be to employ the method of combining a microbubble-specific imaging mode with a thresholded Doppler measurement of microbubble velocity, as illustrated in Figure 6.56. Another is to use a variant of the flash imaging mode described in Section 6.3.6.1; once a steady state is thought to have been reached, for bubbles binding to their targets, a high-amplitude sound pulse is emitted to destroy all bubbles within a region, and the difference between the integrated bubble echo strength before destruction and after reperfusion has occurred is regarded as the molecular target signal strength (Rychak et al. 2007). For intravascular applications, microbubble imaging by ultrasound has a number of useful features, including very high sensitivity to small quantities of the probe (a single microbubble may be detectable *in vivo*), high spatial resolution, high frame rate, low cost and wide availability in the clinic, and the molecular target information is obtained simultaneously and in registration with anatomic and physiologic (blood flow) information. Finally, a potentially unique advantage of microbubbles as molecular imaging probes is their ability to carry a payload of drugs or genes to the target site and, with imaging confirmation of local dose, to (1) release the payload using a bubble-destroying sound pulse and (2) assist the entry of the drug or gene into the cell by means of a phenomenon known as sonoporation, in which transient increases in cell membrane permeability are brought about by the action of the sound field enhanced by the presence of microbubbles (Liang and Blomley 2003, Ferrara et al. 2007, Bohmer et al. 2009).

6.7.6 Multiphysics Imaging

For the purposes of the present discussion, it is convenient to define the term, multiphysics imaging, when referring to any imaging method that cannot be described without invoking a combination of theories taken from what have traditionally been regarded as separate branches of physics. Ultrasound elastography (Section 6.3.6.2) is an example of this; to understand what determines the performance of an elastographic imaging system, it is necessary to bring together ultrasound imaging theory with the theory of rheology. The ability of ultrasound to introduce a localised mechanical or thermal disturbance in tissue, which may be detected in other images, makes it possible to invent many multiphysics imaging methods that involve ultrasound. In addition to the tissue elasticity imaging methods based on acoustic radiation force described in Section 6.3.6.2, backscatter temperature imaging (Seip and Ebbini 1995, Miller et al. 2002) uses the echo-tracking and strain-imaging techniques of elastography to visualise localised changes in sound speed induced by reversible heating with focused ultrasound. This may be extended to temperature-change visualisation using other modalities (Meaney et al. 2008), or applied in a reverse sense by heating with light or microwaves to create an ultrasound-derived image that shows optical or microwave absorption at the electromagnetic wavelength employed (Shi et al. 2003, Mokhtari-Dizaji et al. 2007, Kawakami et al. 2008). In acousto-optic imaging, ultrasound pulses are used to modulate light that is present at the time and spatial position of the sound pulse. If only the light that exhibits this modulation is used to reconstruct a transmission image for example, then an optical attenuation image may be formed with similar spatial resolution to that of an ultrasound scan (Draudt et al. 2009, Sandhu et al. 2009). In some multiphysics imaging methods tissue and contrast agents may be induced to emit sound waves that can then be imaged using receive-only ultrasound beam forming. 'Hall-effect' imaging is one example where, in principle, contrast related to local electrical conductivity may be accessible from sound emitted when a pulsed electrical current is passed through the tissue in the presence of a strong magnetic field (Wen 2000). A particularly promising method of this kind is optoacoustic (also called photoacoustic) imaging, where tissue is illuminated with a short (many nanoseconds) light pulse at a chosen wavelength and the rapid thermal expansion from the heat generated by local absorption produces ultrasonic emissions (Wang 2009). Microwave excitation may also be used, in which case the method may be termed thermoacoustic imaging. A single pulse of electromagnetic radiation may provide enough information for an ultrasound array with a high-channel-count parallel-receive system to reconstruct an image using SAFT techniques of the spatial distribution of optical absorption (Jaeger and Frenz 2009). Frame rates of 15 Hz have been achieved with this type of approach (Fournelle et al. 2008). The frame rate would in general be limited by considerations of the safety of tissue heating rather than the speed of sound which would permit thousands of frames per second (or volumes per second if a 2D array were employed). This imaging technique provides resolutions typical of ultrasound but with endogenous optical molecular imaging capability and potential for *in vivo* absorption spectroscopy, albeit to a depth limited by light penetration but this can be several centimetres. Exogenous molecular imaging probes exist for optoacoustic imaging that could make the method even more powerful (e.g. Copland et al. 2004, Kim et al. 2007, Mallidi et al. 2007, Li et al. 2008, Song et al. 2009, Xiang et al. 2009). These include gold and carbon nanoparticles that are easily functionalised and may be small enough (e.g. 40 nm) to reach extravascular targets. Magnetic nanoparticles, which may also be functionalised for molecular targeting, making it possible to conceive of yet another class of multiphysics imaging methods, such as magneto-motive ultrasound imaging (Mehrmohammadi et al. 2007). In this, the nanoparticles are induced to move using a time-varying magnetic field, and their motion, indicating their location, is detected using ultrasound methods. Interestingly, in addition to its use in localising magnetic nanoparticles as a molecular imaging probe, this bi-physics imaging method may be extended to become a tri-physics technique by analysing the pattern of motion to derive tissue-elasticity information in the micro-vicinity of the particle. Finally, the nanodroplets mentioned in Section 6.7.5 may be used in this way too; their sudden vaporisation causes an acoustic emission, the characteristics of which provide tissue elasticity information on a microscopic scale (Asami et al. 2008). Many of these multiphysics imaging methods are at an embryonic stage of development, requiring considerable further research and development before their clinical potential is understood, but all the indications point to a promising future for the subject and to the extensive involvement of ultrasound in multiphysics imaging.

6.8 Conclusion

At various points in this chapter, reference was made to the extensive ongoing research aimed at increasing knowledge and understanding of the interaction of ultrasound with tissues, and the further development of medical ultrasound methodology and instrumentation. Ultrasound continues to provide new opportunities for extracting diagnostic and other useful information from the human body, and remains a rapidly evolving medical imaging modality. This is especially demonstrated by the five topics covered in Section 6.7. Currently, our lack of technical ability (e.g. in areas such as construction of fully effective systems using 2D arrays, complete with the ability to rapidly extract and process data from them) prevents full advantage from being taken of existing knowledge of the physical limits of what can, in principle, be done to extend the nature and quality of information extracted from the body using ultrasound. These technical restrictions are likely to be overcome with time. It is also likely that we shall continue to invent new ways of improving the information and of extracting different kinds of information. Some limits may remain, such as the difficulty of imaging through bone or gas, but even these areas are under investigation.

References

AIUM Statement on Prudent Use in Obstetrics 2007. This is available from the American Institute of Ultrasound in Medicine, on the Internet at: http://www.aium.org/publications/viewStatement.aspx?id=33. Accessed on 4th January 2012.

AIUM/NEMA 1992 Revision 2 2004 standard for real time display of thermal and mechanical acoustic output indices on diagnostic ultrasound equipment UD 3-2004 (AIUM/NEMA: Laurel, MD).

AIUM/NEMA 2004 Acoustic output measurement standard for diagnostic ultrasound equipment (AIUM/NEMA: Laurel, MD).

Akiyama I, Saito S and Ohya A 2006 Development of an ultra-broadband ultrasonic imaging system: Prototype mechanical sector device *J. Med. Ultrason.* **33** 71–76.

Apfel R E and Holland C K 1991 Gauging the likelihood of cavitation from short-pulse, low-duty cycle diagnostic ultrasound *Ultrasound Med. Biol.* **17** 179–185.

Asami R, Azuma T, Yoshikawa H and Kawabata K I 2007 Tumor enhanced imaging and treatment with targeted phase-change nano particles pp. 1973–1976 in: *IEEE Ultrasonics Symposium Proceedings,* New York, October 28–31, 2007.

Asami R, Ikeda T, Azuma T, Yoshikawa H and Kawabata K-I 2008 Measuring viscoelastic properties with in-situ ultrasonically induced microbubbles pp. 658–661 in: *IEEE Ultrasonics Symposium Proceedings,* Beijing, China, November 2–5, 2008.

Atkinson P A and Woodcock J P 1982 *Doppler Ultrasound and Its Use in Clinical Measurement* (London, UK: Academic Press).

Austeng A and Holm S 2002 Sparse 2-D arrays for 3-D phased array imaging—Design methods *IEEE Trans. Ultrason. Ferroelectr. Freq. Control* **49** 1073–1086.

Baggs R, Penney D P, Cox C, Child S Z, Raeman C H, Dalecki D and Carstensen E L 1996 Thresholds for ultrasonically induced lung hemorrhage in neonatal swine *Ultrasound Med. Biol.* **22** 119–128.

Baker A C 1998 Nonlinear effects in ultrasound propagation Chapter 2 pp. 23–38 in: Duck F A, Baker A C and Starritt H C (eds.) *Ultrasound in Medicine* (IOP Publishing: Bristol, UK).

Baker D, Forster F K and Daigle R E 1978 Doppler principles and techniques in: Fry F J (ed.) *Ultrasound, Its Applications in Medicine and Biology* (Elsevier: New York) part I, Chapter 3.

Bamber J C 1986a Speed of sound. Chapter 6 pp. 200–224 in: Hill CR (ed.) *Physical Principles of Medical Ultrasonics,* 1st Edn (Ellis Horwood: Chichester, UK).

Bamber J C 1986b Attenuation and absorption. Chapter 5 pp. 118–199 in: Hill CR (ed.) *Physical Principles of Medical Ultrasonics,* 1st Edn (Ellis Horwood: Chichester, UK).

Bamber J C 1993 Speckle reduction. pp. 55–67 in: Wells P N T (ed.) *Advances in Ultrasonic Techniques and Instrumentation* (Churchill Livingstone: New York).

Bamber J C 1997 Acoustical characteristics of biological media. pp. 1703–1726 in: Crocker M J (ed.) *Encyclopedia of Acoustics* (John Wiley: New York).

Bamber J C 1999a Medical ultrasonic signal and image processing *Insight* **41** 14–15.

Bamber J C 1999b Ultrasound elasticity imaging: Definition and technology *Eur. Radiol.* **9** S327–S330.

Bamber J C 2004a Speed of sound Chapter 5 pp. 167–190 in: Hill C R, Bamber J C and ter Haar G R (eds.) *Physical Principles of Medical Ultrasonics* (John Wiley: Chichester, UK).

Bamber J C 2004b Attenuation and absorption Chapter 4 pp. 93–166 in: Hill C R, Bamber J C, ter Haar G R (eds.) *Physical Principles of Medical Ultrasonics* 2nd Edn (John Wiley: Chichester, UK).

Bamber J C, Barbone P E, Bush N L, Cosgrove D O, Doyley M M, Fuechsel F G, Meaney P M, Miller N R, Shiina T and Tranquart F 2002 Progress in freehand elastography of the breast *IEICE Trans. Inf. Syst.* **85-D(1)** 5–14.

Bamber J C and Daft C 1986 Adaptive filtering for reduction of speckle in pulse-echo images *Ultrasonics* **24** 41–44.

Bamber J C and Dickinson R J 1980 Ultrasonic B-scanning: A computer simulation *Phys. Med. Biol.* **25** 463–479.

Bamber J C, Mucci R A and Orofino D P 2000 Spatial coherence and beamformer gain pp. 43–48 in: Lee H (ed.) *Acoustical Imaging* Vol. 24 (Kluwer Academic: New York).

Barbone P E and Bamber J C 2002 Quantitative elasticity imaging: What can and can not be inferred from strain images *Phys. Med. Biol.* **47** 2147–2164.

Barnett S B and Kossoff G 1998 *Safety of Diagnostic Ultrasound: Progress in Obstetric and Gynecological Sonography Series.* (Parthenon: Carnforth, UK).

Beissner K 1993 Radiation force and force balances pp. 127–142 in: Ziskin M C and Lewin P A (eds.) *Ultrasonic Exposimetry* (CRC Press: Boca Raton, FL).

Bercoff J, Tanter M and Fink M 2004 Supersonic shear imaging: A new technique for soft tissue elasticity mapping *IEEE Trans. Ultrason. Ferroelectr. Freq. Control* **51** 396–409.

Berkhout A J, Ridder J and van de Wal L F 1982 Acoustic imaging by wave field extrapolation part I: Theoretical considerations pp. 513–540 in: Alai P and Metherell F (eds.) *Acoustical Imaging* **10** (Plenum Press: New York).

Berktay H O 1965 Possible exploitation of nonlinear acoustics in underwater transmitting application *J. Sound Vib.* **2** 435–461.

Berry G P, Bamber J C, Armstrong C G, Miller N R and Barbone P E 2006 Towards an acoustic model-based poroelastic imaging method: I. Theoretical foundation *Ultrasound Med. Biol.* **32** 547–567.

Bohmer M R, Klibanov A L, Tiemann K, Hall C S, Gruell H and Steinbach O C 2009 Ultrasound triggered image-guided drug delivery *Eur. J. Radiol.* **70** 242–253.

Bouakaz A, De Jong N and Cachard C 1998 Standard properties of ultrasound contrast agents *Ultrasound Med. Biol.* **24** 469–472.

Browne J E, Watson A J, Hoskins P R and Elliott A T 2004 Validation of a sensitivity performance index test protocol and evaluation of colour Doppler sensitivity for a range of ultrasound scanners *Ultrasound Med. Biol.* **30** 1475–1483.

Carstensen E L, Gracewski S and Dalecki D 2000 The search for cavitation in vivo *Ultrasound Med. Biol.* **26** 1377–1385.

Cauvard P and Hartemann P 1982 Ultrafast acoustical imaging with surface acoustic wave components pp. 17–22 in: Alai P and Metherell F (eds.) *Acoustical Imaging* **10** (Plenum Press: New York).

Chen H and Varghese T 2009 Multilevel hybrid 2D strain imaging algorithm for ultrasound sector/phased arrays *Med. Phys.* **36** 2098–2106.

Chen Q and Zagzebski J A 2004 Simulation study of effects of speed of sound and attenuation on ultrasound lateral resolution *Ultrasound Med Biol.* **30** 1297–1306.

Cherry S R 2004 In vivo molecular and genomic imaging: New challenges for imaging physics *Phys. Med. Biol.* **49** R13–R48.

Church C C 1995 The effects of an elastic solid surface layer on the radial pulsations of gas bubbles *The J. Acoust. Soc. Am.* **97** 1510–1521.

Church C C and Miller M W 2007 Quantification of risk from fetal exposure to diagnostic ultrasound *Prog. Biophys. Mol. Biol.* **93** 331–353.

Church C C and O'Brien W D Jr. 2007 Evaluation of the threshold for lung hemorrhage by diagnostic ultrasound and a proposed new safety index *Ultrasound Med. Biol.* **33** 810–818.

Copland J A, Eghtedari M, Popov V L, Kotov N, Mamedova N, Motamedi M and Oraevsky A A 2004 Bioconjugated gold nanoparticles as a molecular based contrast agent: Implications for imaging of deep tumors using optoacoustic tomography *Mol. Imag. Biol.* **6** 341–349.

Cosgrove D and Blomley M 2003 Contrast harmonic imaging (CHI) some case studies *Toshiba Med. Syst. J. (Visions Magazine)* **3** 59–63.

Dalecki D, Child S Z, Raeman C H, Cox C and Carstensen E L 1997 Ultrasonically induced lung hemorrhage in young swine *Ultrasound Med. Biol.* **23** 777–781.

Davidsen R E, Jensen J A and Smith S W 1994 Two-dimensional random arrays for real time volumetric imaging *Ultrason. Imaging* **16** 143–163.

de Jong N, Tencate F J, Lancee C T, Roelandt J R T C and Bom N 1991 Principles and recent developments in ultrasound contrast agents *Ultrasonics* **29** 324–330.

Debbage P and Jaschke W 2008 Molecular imaging with nanoparticles: Giant roles for dwarf actors *Histochem. Cell. Biol.* **130** 845–875.

Desser T S, Jedrzejewicz T M S and Bradley C 2000 Native tissue harmonic imaging: Basic principles and clinical applications *Ultrasound Q.* **16** 40–48.

Di Nallo A M, Strigari L and Benassi M 2006 A possible quality control protocol for Doppler ultrasound for organizational time optimization *J. Exp. Clin. Cancer Res.* **25** 373–381.

Dixon K L and Duck F A 1998 Determining the conditions for measurement of spatial-peak temporal-averaged intensity in scanned ultrasound beams *Br. J. Radiol.* **71** 968–971.

Downey D B, Fenster A and Williams J C 2000 An investigation of the clinical utility of 3-dimensional ultrasound *Radiographics* **20** 559–571.

Doyley M M, Bamber J C, Fuechsel F and Bush N L 2001 A freehand elastographic imaging approach for clinical breast imaging: System development and performance evaluation *Ultrasound Med. Biol.* **27** 1347–1357.

Doyley M M, Meaney P M and Bamber J C 2000 Evaluation of an iterative reconstruction method for quantitative elastography *Phys. Med. Biol.* **45** 1521–1540.

Draudt A, Lai P X, Roy R A, Murray T W and Cleveland R O 2009 Detection of HIFU lesions in excised tissue using acousto-optic imaging pp. 270–274 in: Ebbini E S (ed.) *8th International Symposium on Therapeutic Ultrasound*, September 10–13, 2008, *AIP Conference Proceedings*, Minneapolis, MN, (The American Institute of Physics: Melville, NY).

Duck F A 2000 The propagation of ultrasound through tissue Chapter 2 in: ter Haar G R and Duck F A (eds.) *The Safe Use of Ultrasound in Medical Diagnosis* (The British Institute of Radiology: London, UK).

Duck F A 2002 Nonlinear acoustics in diagnostic ultrasound *Ultrasound Med. Biol.* **28** 1–18.

Duck F A 2007 Medical and non-medical protection standards for ultrasound and infrasound *Prog. Biophys. Mol. Biol.* **93** 176–191.

Dudley N J, Griffith K, Houldsworth G, Holloway M and Dunn M A 2001 A review of two alternative ultrasound quality assurance programmes *Eur. J. Ultrasound* **12** 233–245.

Duggan T C and Sik M J 1983 Assessment of ultrasound scanners by acoustic signal injection pp. 179–184 in: Lerski R A and P Morley P (eds.) *Ultrasound'82* (Pergamon: Oxford, UK).

Dunmire B, Beach K W, Labs K H, Plett M and Strandness D E 2000 Cross-beam vector Doppler ultrasound for angle-independent velocity measurements *Ultrasound Med. Biol.* **26** 1213–1235.

Eckersley R J and Bamber J C 2004 Methodology for imaging time-dependent phenomena Chapter 10 pp. 304–335 in: Hill C R, Bamber J C and ter Haar G R (eds.) *Physical Principles of Medical Ultrasonics* 2nd Edn (John Wiley: Chichester, UK).

Evans D H and McDicken W N 2000 *Doppler Ultrasound: Physics, Instrumentation and Signal Processing* 2nd Edn 0-471-97001-8 (John Wiley & Sons: Chichester, UK).

Fatemi M and Greenleaf J F 1999 Vibro-acoustography: An imaging modality based on ultrasound-stimulated acoustic emission *Proc. Natl Acad. Sci. USA* **96** 6603–6608.

Fay B, Ludwig G, Lankjaer C and Lewin P A 1994 Frequency response of PVDF needle-type hydrophones *Ultrasound Med. Biol.* **20** 361–366.

Fenster A, Downey D and Cardinal N 2001 Review: 3-dimensional ultrasound imaging *Phys. Med. Biol.* **46** R67–R99.

Ferrara K, Pollard R and Borden M 2007 Ultrasound microbubble contrast agents: Fundamentals and application to gene and drug delivery *Annu. Rev. Biomed. Eng.* **9** 415–417.

Flax S W and O'Donnell M 1988 Phase-aberration correction using signals from point reflectors and diffuse scatterers: Basic principles *IEEE Trans. Ultrason. Ferroelectr. Freq. Control* **35** 758–757.

Foster F S and Hunt J W 1979 Transmission of ultrasound beams through human tissue—Focusing and attenuation studies *Ultrasound Med. Biol.* **5** 257–268.

Fournelle M, Maass K, Fonfara H, Welsch H-J, Moses M, Hewener H, Günther C, Frosini S, Masotti L and Lemor R 2008 A combined platform for b-mode and real-time optoacoustic imaging based on raw data acquisition *J. Acoust. Soc. Am.* **123** 3641.

Frinking P J A and de Jong N 1998 Acoustic modeling of shell-encapsulated gas bubbles *Ultrasound Med. Biol.* **24** 523–533.

Fry W J and Dunn F 1962 Ultrasound analysis and experimental methods in biological research pp. 261–394 in: Nastuk W L (ed.) *Physical Techniques in Biological Research* (Academic Press: New York).

Fujibayashi Y, Furukawa T, Takamatsu S and Yonekura Y 2002 Molecular imaging: An old and new field connecting basic science and clinical medicine *J. Cell Biochem. Suppl.* **39** 85–89.

Gao L, Parker K J, Lerner R M and Levinson S F 1996 Imaging of the elastic properties of tissue—A review *Ultrasound Med. Biol.* **22** 959–977.

Gee A H, Housden R J, Hassenpflug P, Treece G M and Prager R W 2006 Sensorless freehand 3D ultrasound in real tissue: Speckle decorrelation without fully developed speckle *Med. Image Anal.* **10** 137–149.

Gibson N M, Dudley N J and Griffith K A 2001 Computerised ultrasound quality control testing system *Ultrasound Med. Biol.* **27** 1697–1711.

Goldberg R L, Smith S W, Mottley J G and Ferrara K W 2006 Ultrasound in Bronzino J D (ed.) *Medical Devices and Systems Biomedical Engineering Handbook* 3rd Edn (Taylor & Francis: Boca Raton, FL).

Goldstein A 2000 The effect of acoustic velocity on phantom measurements *Ultrasound Med. Biol.* **26** 1133–1443.

Goodsitt M M, Carson P L, Witt S, Hykes D L and Kofler J M 1998 Real-time B-mode ultrasound quality control test procedures; Report of AAPM Ultrasound Task Group No. 1. *Med. Phys.* **25** 1385–1406.

Greenleaf J F 1997 Acoustical medical imaging instrumentation Chapter144 pp. 1751–1760 in: Crocker M J (ed.) *Encyclopedia of Acoustics* (John Wiley: New York).

Greenleaf J F, Fatemi M and Insana M 2003 Selected methods for imaging elastic properties of biological tissues *Annu. Rev. Biomed. Eng.* **5** 57–78.

Hansen K L, Udesen J, Thomsen C, Jensen J A and Nielsen M B 2009 In vivo validation of a blood vector velocity estimator with MR angiography *IEEE Trans. Ultrason. Ferroelectr. Freq. Control* **56** 91–100.

Healey A J and Leeman S 1991 A complex z-transform technique for speckle reduction pp. 1109–1112 in: McAvoy B R (ed.) *Ultrasonics Symposium Proceedings* Lake Buena Vista, FL, December 8–11, 1991.

Hill C R and Bamber J C 2004 Methodology for clinical investigation Chapter 9 pp. 255–302 in: Hill C R, Bamber J C and ter Haar G R (eds.) *Physical Principles of Medical Ultrasonics* 2nd Edn (John Wiley: Chichester, UK).

Hoff L, Sontum P C and Hovem J M 2000 Oscillations of polymeric microbubbles: Effect of the encapsulating shell *J. Acoust. Soc. Am.* **107** 2272–2280.

Hoskins P R, Sherriff S B and Evans J A 1994 (ed.) Testing of Doppler ultrasound equipment IPSM Report No. 70 (IPSM: York, UK).

Humphrey V F 2007 Ultrasound and matter—Physical interactions *Prog. Biophys. Mol. Biol.* **93** 195–211.

Hynynen K and Sun J 1999 Trans-skull ultrasound therapy: The feasibility of using image-derived skull thickness information to correct the phase distortion *IEEE Trans. Ultrason. Ferroelectr. Freq. Control* **46** 752–755.

IEC 2001 IEC 60601 part 2–37: *Medical Electrical Equipment: Particular Requirements for the Safety of Ultrasound Diagnostic and Monitoring Equipment 2001 and Amendment 1, 2005* (IEC: Geneva, Switzerland).

IEC 2006 IEC 62359 *Ultrasonics—Field Characterization—Test Methods for the Determination of Thermal and Mechanical Indices Related to Medical Diagnostic Ultrasound Fields* (IEC: Geneva, Switzerland).

Ivancevich N M, Dahl J D, Light E D, Nicoletto H A, Scism M, Laskowitz D T, Trahey G E and Smith S W 2006a Phase aberration correction on a 3D ultrasound scanner using RF speckle from moving targets pp. 120–123 in: *Ultrasonics Symposium Proceedings,* Vancouver, Canada, October 2–6, 2006.

Ivancevich N M, Dahl J J, Trahey G E and Smith S W 2006b Phase-aberration correction with a 3-D ultrasound scanner: Feasibility study *IEEE Trans. Ultras. Ferroelectr. Freq. Control* **53** 1432–1439.

Jaeger M and Frenz M 2009 Combined ultrasound and photoacoustic system for real-time high-contrast imaging using a linear array transducer p. 289 in: Wang L V (ed.) *Photoacoustic Imaging and Spectroscopy* (CRC Press: Boca Raton, FL).

Jaffer F A and Weissleder R 2005 Molecular imaging in the clinical arena *J. Am. Med. Assoc.* **293** 855–862.

Jago J R, Henderson J, Whittingham T A and Mitchell G 1999 A comparison of AIUM/NEMA thermal indices with calculated temperature rises for a simple third trimester pregnancy tissue model *Ultrasound Med. Biol.* **25** 623–628.

Jensen J A 2001 A new estimator for vector velocity estimation *IEEE Trans. Ultrason. Ferroelectr. Freq. Control* **48** 886–894.

Jensh R P and Brent R L 1999 Intrauterine effects of ultrasound: Animal studies *Teratology* **59** 240–251.

Kawakami S, Nakamura N, Mukaiyama T, Ishibasi S, Wada K, Matsuyama T, Matsunaka T, Kono K and Horinaka H 2008 Spectroscopic imaging of nano-particle distribution in biological tissue using optically assisted ultrasonic velocity-change detection pp. 1302–1305 in: *IEEE Ultrasonics Symposium Proceedings*, Beijing, China, November 2–5, 2008.

Kiessling F, Huppert J and Palmowski M 2009 Functional and molecular ultrasound imaging: Concepts and contrast agents *Curr. Med. Chem.* **16** 627–642.

Kim G, Huang S W, Day K C, O'Donnell M, Agayan R R, Day M A, Kopelman R and Ashkenazi S 2007 Indocyanine-green-embedded PEBBLEs as a contrast agent for photoacoustic imaging *J. Biomed. Opt.* **12** Article Number: 044020.

Kirkebø J E and Austeng A 2007 Improved beamforming using curved sparse 2D arrays in ultrasound *Ultrasonics* **46** 119–128.

Klibanov A L 2007 Ultrasound molecular imaging with targeted microbubble contrast agents *J. Nucl. Cardiol.* **14** 876–884.

Kossoff G 1978 The transducer pp. 25–30 in: de Vlieger M, Holmes J H, Kratochwil A, Kazner E, Kraus R, Kossoff G, Poujol J and Stransdness D E (eds.) *Handbook of Clinical Ultrasound* (Wiley: New York).

Kossoff G 1979 Analysis of focusing action of spherically curved transducers *Ultrasound Med. Biol.* **5** 359–365.

Leeman S J 2004 Basic acoustic theory Chapter1 pp. 1–40 in: Hill C R, Bamber J C and ter Haar G R (eds.) *Physical Principles of Medical Ultrasonics* 2nd Edn (John Wiley: Chichester, UK).

Leighton T G 2007 Review: What is ultrasound? *Prog. Biophys. Mol. Biol.* **93** 3–83.

Li S F, Hoskins P R, Anderson T and McDicken W N 1998 An acoustic injection test object for colour flow imaging systems *Ultrasound Med. Biol.* **24** 161–164.

Li Y and Robinson B 2008 The cross algorithm for phase-aberration correction in medical ultrasound images formed with two-dimensional arrays *IEEE Trans. Ultrason. Ferroelectr. Freq. Control* **55** 588–601.

Li P C, Wang C R C, Shieh D B, Wei C W, Liao C K, Poe C, Jhan S, Ding A A and Wu Y N 2008 In vivo photoacoustic molecular imaging with simultaneous multiple selective targeting using antibody-conjugated gold nanorods *Opt. Express* **16** 18605–18615.

Liang H-D and Blomley M J K 2003 The role of ultrasound in molecular imaging *Br. J. Radiol.* **76** S140–S150.

Liu D L and Waag R C 1994 Time-shift compensation of ultrasonic pulse focus degradation using least-mean-square error-estimates of arrival time *J. Acoust. Soc. Am.* **95** 542–555.

Liu D L D and Waag R C 1998 Estimation and correction of ultrasonic wavefront distortion using pulse-echo data received in a two-dimensional aperture *IEEE Trans. Ultrason. Ferroelectr. Freq. Control* **45** 473–490.

Mallart R and Fink M 1994 Adaptive focusing in scattering media through sound speed inhomogeneities: The van Cittert-Zernike approach and focusing criterion *J. Acoust. Soc. Am.* **96** 3721–3732.

Mallidi S, Larson T, Aaron J, Sokolov K and Emelianov S 2007 Molecular specific optoacoustic imaging with plasmonic nanoparticles *Opt. Express* **15** 6583–6588.

Marquet F, Pernot M, Aubry J F, Montaldo G, Marsac L, Tanter M and Fink M 2009 Non-invasive transcranial ultrasound therapy based on a 3D CT scan: Protocol validation and in vitro results *Phys. Med. Biol.* **54** 2597–2613.

Martin K and Ramnarine K V 2010 Physics Chapter 2 pp. 7–22 in: Hoskins P R, Martin K and Thrush A (eds.) *Diagnostic Ultrasound Physics and Equipment*, 2nd Edn. (Cambridge University Press: Cambridge, UK).

McCarty K 1986 *Quality Assurance in Medical Imaging* IOP Short Meetings Series no. 2 (Institute of Physics: Bristol, UK) pp. 77–98.

Meaney P M, Tian Z, Fanning M W, Geimer S D and Paulsen K D 2008 Microwave thermal imaging of scanned focused ultrasound heating: Phantom results *Int. J. Hyperthermia* **24** 523–536.

Medical Devices Agency 1999 *Evaluation Report MDA/98/52* (HMSO: London, UK).

Mehrmohammadi M, Oh J, Ma L, Yantsen E, Larson T, Mallidi S, Park S, Johnston K P, Sokolov K, Milner T and Emelianov S 2007 Imaging of iron oxide nanoparticles using magneto-motive ultrasound pp. 652–655 in: *IEEE Ultrasonics Symposium Proceedings*, New York, October 28–31, 2007 (IEEE: New York).

Melodelima D, Bamber J C, Duck F, Shipley J and Xu L 2006 Elastography for breast cancer diagnosis using radiation force: System development and performance evaluation *Ultrasound Med. Biol.* **32** 387–396.

Mercier L, Langø T, Lindseth F and Collins L D 2005 A review of calibration techniques for freehand 3-D ultrasound systems *Ultrasound Med. Biol.* **31** 143–165.

Merz E 2007 Ultrasound in obstetrics and gynecology Vol 2 p. 222 in: *Gynecology* 2nd Edn (Thieme: New York).

Miller N R, Bamber J C and Meaney P M 2002 Fundamental limitations of noninvasive temperature imaging by means of ultrasound echo strain estimation *Ultrasound Med. Biol.* **28** 1319–1333.

Miller M W, Miller D L and Brayman A A 1996 A review of in vitro bioeffects of inertial ultrasonic cavitation from a mechanistic perspective *Ultrasound Med. Biol.* **22** 1131–1154.

Miller M W and Ziskin M C 1989 Biological consequences of hyperthermia *Ultrasound Med. Biol.* **15** 707–722.

Mo L Y L, DeBusschere D, Bai W, Napoilan D, Irish A, Marschall S, McLaughlin G W, Yang Z, Carson P L and Fowlkes I B 2007 Compact ultrasound scanner with built-in raw data acquisition capabilities pp. 2259–2262 in: *IEEE Ultrasonics Symposium Proceedings*, New York, October 28–31, 2007 (IEEE: New York).

Mokhtari-Dizaji M, Gorjiara T and Ghanaati H 2007 Assessment of pixel shift in ultrasound images due to local temperature changes during the laser interstitial thermotherapy of liver: In vitro study *Ultrasound Med. Biol.* **33** 934–940.

Montaldo G, Tanter M, Bercoff J, Benech N and Fink M 2009 Coherent plane-wave compounding for very high frame rate ultrasonography and transient elastography *IEEE Trans. Ultrason. Ferroelectr. Freq. Control* **56** 489–506.

Morris P, Hurrell A and Beard P (2006) Development of a 50 MHz Fabry-Perot type fibre-optic hydrophone for the characterisation of medical ultrasound fields *Proc. Inst. Acoust.* **28** 717–725.

Napolitano D, Chou C H, McLaughlin G, Ji T L, Mo L, DeBusschere D and Steins R 2006 Sound speed correction in ultrasound imaging *Ultrasonics* **44(Suppl 1)** E43–E46.

Nightingale K R, Palmeri M L, Nightingale R W and Trahey G E 2001 On the feasibility of remote palpation using acoustic radiation force *J. Acoust. Soc. Am.* **110** 625–634.

Nikolov S I and Jensen J A 2000 Application of different spatial sampling patterns for sparse array transducer design *Ultrasonics* **37** 667–671.

Nitta N and Shiina T 1998 Real-time three-dimensional velocity vector measurement using the weighted phase gradient method *Jpn. J. Appl. Phys.* **37** 3058–3063.

Nock L, Trahey G E and Smith S W 1989 Phase aberration correction in medical ultrasound using speckle brightness as a quality factor *J. Acoust. Soc. Am.* **85** 1819–1833.

O'Brien W D Jr. 2007 Ultrasound—Biophysics mechanisms. *Progr. Biophys. Mol. Biol.* **93** 212–255.

O'Brien W D Jr., Frizzell L A, Weigel R M and Zachary J F 2000 Ultrasound-induced lung hemorrhage is not caused by inertial cavitation *J. Acoust. Soc. Am.* **108** 1290–1297.

Ophir J, Cespedes I, Ponnekanti H, Yazdi Y and Li X 1991 Elastography: A quantitative method for imaging the elasticity of biological tissues *Ultrason. Imaging* **13** 111–134.

Ophir J, Garra B, Kallel F, Konofagou E, Krouskop T, Righetti R and Varghese T 2000 Elastographic imaging *Ultrasound Med. Biol.* **26(Suppl 1)** S23–S29.

Ottobrini L, Ciana P, Biserni A, Lucignani G and Maggi A 2006 Molecular imaging: A new way to study molecular processes in vivo *Mol. Cell Endocrinol.* **246** 69–75.

Papadakis E P 1966 Ultrasonic diffraction loss and phase change in anisotropic materials. *J. Acoust. Soc. Amer.* **40** 863–876.

Prager R W, Gee A and Berman L 1999 Stradx: Real-time acquisition and visualization of freehand three-dimensional ultrasound *Med. Image Anal.* **3** 129–140.

Prager R W, Gee A H, Treece G M, Cash C J C and Berman L H 2003 Sensorless freehand 3-D ultrasound using regression of the echo intensity *Ultrasound Med. Biol.* **29** 437–446.
Preston R C (ed.) 1991 *Output Measurements for Medical ultrasound* (Springer: Berlin, Germany).
Pye S D, Ellis B and MacGillivray T J 2004 Medical ultrasound: A new metric of performance for grey-scale imaging *J. Phys.: Conf. Ser.* **1** 187–192.
Rao M and Varghese T 2009 Estimation of the optimal maximum beam angle and angular increment for normal and shear strain estimation *IEEE Trans. Biomed. Eng.* **56** 760–769.
Rapp C L and Stavros A T 2001 Coded harmonics in breast ultrasound: Does it make a difference? *J. Diag. Med. Sonography* **17** 22–28.
Ricci S, Diciotti S, Francalanci L and Tortoli P 2009 Accuracy and reproducibility of a novel dual-beam vector Doppler method *Ultrasound Med. Biol.* **35** 829–838.
Ridder J, Berkhout A J and van de Wal L F 1982 Acoustic imaging by wave field extrapolation part II: Practical aspects pp. 541–565 in: Alai P and Metherell F (eds.) *Acoustical Imaging Vol. 10* (Plenum Press: New York).
Rigby K W, Chalek C L, Haider B H, Lewandowski R S, O'Donnell M, Smith L S and Wildes D G 2000 Improved in vivo abdominal image quality using real-time estimation and correction of wavefront arrival time errors pp. 1645–1653 in: Schneider S C, Levy M and McAvoy B R (eds.) *IEEE Ultrasonics Symposium Proceedings*, San Juan, Puerto Rico, October 22–25, 2000 (IEEE: New York).
Ritty A 2006 Directional loudspeaker using a parametric array *Acta Polytechnica* **46** 47–48.
Rollo F D 2003 Molecular imaging: An overview and clinical applications *Radiol. Manage.* **25** 28–32.
Rotten D and Levaillant J M 2004 Two- and three-dimensional sonographic assessment of the fetal face. 2. Analysis of cleft lip, alveolus and palate *Ultrasound Obstet. Gynecol.* **24** 402–411.
Rotten D, Levaillant J M and Zerat L 1999 Analysis of normal breast tissue and of solid breast masses using three-dimensional ultrasound mammography *Ultrasound Obstet. Gynecol.* **14** 114–124.
Rownd J, Madsen E L, Zagzebski J A, Frank G R and Dong F 1997 Phantoms and automated system for testing the resolution of ultrasound scanners *Ultrasound Med. Biol.* **23** 245–260.
Rychak J J, Graba J, Cheung A M Y, Mystry B S, Lindner J R, Kerbel R S and Foster F S 2007 Microultrasound molecular imaging of vascular endothelial growth factor receptor 2 in a mouse model of tumor angiogenesis *Mol. Imaging* **6** 289–296.
Salvesen K A 2007 Epidemiological prenatal ultrasound studies *Prog. Biophys. Mol. Biol.* **93** 295–300.
Sandhu J S, Schmidt R A and La Riviere P J 2009 Full-field acoustomammography using an acousto-optic sensor *Med. Phys.* **36** 2324–2327.
Sandrin L, Tanter M, Catheline S and Fink M 2002a Shear modulus imaging with 2D transient elastography *IEEE Trans. Ultrason. Ferroelectr. Freq. Control* **49** 426–435.
Sandrin L, Tanter M, Gennisson J L, Catheline S and Fink M 2002b Shear elasticity probe for soft tissues with 1-D transient elastography *IEEE Trans. Ultrason. Ferroelectr. Freq. Control* **49** 436–446.
Sarvazyan A P and Hill C R 2004 Physical chemistry of the ultrasound-tissue interaction pp. 223–235 in: Hill C R, Bamber J C, ter Haar G R (eds.) *Physical Principles of Medical Ultrasonics*, 2nd Edn (John Wiley: Chichester, UK).
Sarvazyan A P, Rudenko O V, Swanson S D, Fowlkes J B and Emelianov S Y 1998 Shear wave elasticity imaging: A new ultrasonic technology of medical diagnostics *Ultrasound Med. Biol.* **24** 1419–1435.
Seip R and Ebbini E S 1995 Noninvasive estimation of tissue temperature response to heating fields using diagnostic ultrasound *IEEE Trans. Biomed. Eng.* **42** 828–839.
Seki H, Granato A and Truell R 1956 Diffraction effects in the ultrasonic field of a piston source and their importance in the accurate measurement of attenuation. *J. Acoust. Soc. Amer.* **28** 230–238.
Seo J, Choi J J, Fowlkes J B, O'Donnell M and Cain C A 2005 Aberration correction by nonlinear beam mixing: Generation of a pseudo point sound source *IEEE Trans. Ultrason. Ferroelectr. Freq. Control* **52** 1970–1980.
Shattuck D P, Weinshenker M D, Smith S W and von Ramm O T 1984 Explososcan: A parallel processing technique for high speed ultrasound imaging with linear phased arrays *J. Acoust. Soc. Am.* **75** 1273–1282.

Shaw A, Bond A D, Pay N M and Preston R C 1999 A proposed standard thermal test object for medical ultrasound *Ultrasound Med. Biol.* **25** 121–132.

Sheldon C D and Duggen T C 1987 Low-cost Doppler signal simulator *Med. Biol. Eng. Comput.* **25** 226–228.

Shi W T and Forsberg F 2000 Ultrasonic characterization of the nonlinear properties of contrast microbubbles *Ultrasound Med. Biol.* **26** 93–104.

Shi Y, Witte R S, Milas S M, Neiss J H, Chen X C, Cain C A and O'Donnell M 2003 Ultrasonic thermal imaging of microwave absorption pp. 224–227 in: *IEEE Ultrasonics Symposium Proceedings*, Honolulu, Hawaii, October 5–8, 2003 (IEEE: New York).

Skolnick M L 1991 Estimation of ultrasound beam width in the elevation (section thickness) plane *Radiology* **180** 286–288.

Smith R A 1989 Are hydrophones of diameter 0.5 mm small enough to characterise diagnostic ultrasound equipment? *Phys. Med. Biol.* **34** 1593–1607.

Smith S W, Pavy H E and von Ramm O T 1991 High-speed ultrasound volumetric imaging system—Part I: Transducer design and beam steering *IEEE Trans. Ultrason. Ferroelectr. Freq. Control* **38** 100–108.

Song K H, Kim C, Maslov K and Wang L V 2009 Noninvasive in vivo spectroscopic nanorod-contrast photoacoustic mapping of sentinel lymph nodes *Eur. J. Radiol.* **70** 227–231.

Soumekh M 1999 *Synthetic Aperture Radar Signal Processing with MATLAB Algorithms* (John Wiley & Sons: New York).

Steel R, Ramnarine K V, Criton A, Davidson F, Allan P L, Humphries N, Routh H F, Fish P J and Hoskins P R 2004 Angle-dependence and reproducibility of dual-beam vector Doppler ultrasound in the common carotid arteries of normal volunteers *Ultrasound Med. Biol.* **30** 271–276.

Sumi C, Noro T and Tanuma A 2008 Effective lateral modulations with applications to shear modulus reconstruction using displacement vector measurement *IEEE Trans. Ultrason. Ferroelectr. Freq. Control* **55** 2607–2625.

Tanter M, Thomas J L and Fink M 2000 Time reversal and the inverse filter *J. Acoust. Soc. Am.* **108** 223–234.

Tatsum C, Kudo M, Ueshima K, Kitai S, Takahashi S, Inoue T, Minami Y, Chung H, Maekawa K, Fujimoto K, Akiko T and Takeshi M 2008 Noninvasive evaluation of hepatic fibrosis using serum fibrotic markers, transient elastography (fibroscan) and real-time tissue elastography *Intervirology* **51(Suppl 1)** 27–33.

Thijssen J M, van Wijk M C and Cuypers M H M 2002 Performance testing of medical echo/Doppler equipment *Eur. J. Ultrasound* **15** 151–164.

Treece G M, Prager R W, Gee A H and Berman L H 2002 Correction of probe pressure artifacts in freehand 3D ultrasound *Med. Image Anal.* **6** 199–215.

Tuthill T A, Krücker J F, Fowlkes J B and Carson P L 1998 Automated three-dimensional US frame positioning computed from elevational speckle decorrelation *Radiology* **209** 575–582.

Urban M W, Bernal M and Greenleaf J F 2007 Phase aberration correction using ultrasound radiation force and vibrometry optimization *IEEE Trans. Ultrason. Ferroelectr. Freq. Control* **54** 1142–1153.

Vappou J, Maleke C and Konofagou E E 2009 Quantitative viscoelastic parameters measured by harmonic motion imaging *Phys. Med. Biol.* **54** 3579–3594.

Wang L V 2009 *Photoacoustic Imaging and Spectroscopy* (CRC Press: Boca Raton, FL).

Weissleder R and Mahmood U 2001 Molecular imaging *Radiology* **219** 316–333.

Wells P N T 2006 Ultrasound imaging *Phys. Med. Biol.* **51** R83–R98.

Wen H 2000 Feasibility of biomedical applications of Hall effect imaging *Ultrason. Imaging* **22** 123–136.

Westervelt P J 1963 Parametric acoustic array *J. Acoust. Soc. Am.* **35** 535–537.

Whittingham T A 1981 Real-time ultrasonic scanning. pp 153–166 in: Moores B M, Parker R M and Pullan B R (eds.) *Physical Aspects of Medical Imaging* (John Wiley and Sons: Chichester, UK).

Whittingham T A 1999 Tissue harmonic imaging *Eur. Radiol.* **9** (Suppl. 3) S323–S326.

Whittingham T A 2007 Medical diagnostic applications and sources *Prog. Biophys. Mol. Biol.* **93** 84–110.

Whittingham T A, Mitchell G, Tong J and Feeney M 2004 Towards a portable system for the measurement of thermal and mechanical indices *J. Phys. Conf. Ser.* **1** 64–71.

Wilkens V and Koch C 2004 Improvement of hydrophone measurements on diagnostic ultrasound machines using broadband complex-valued calibration data *J. Phys. Conf. Ser.* **1** 50–55.
Wu Z, Hoyt K, Rubens D J and Parker K J 2006 Sonoelastographic imaging of interference patters for estimation of shear velocity distribution in biomaterials *J. Acoust. Soc. Am.* **120** 535–545.
Wygant I O, Kupnik M, Windsor J C, Wright W M, Wochner M S, Yaralioglu G G, Hamilton M F and Khuri-Yakub B T 2009 50 kHz Capacitive micromachined ultrasonic transducers for generation of highly directional sound with parametric arrays *IEEE Trans. Ultrason. Ferroelectr. Freq. Control* **56** 193–203.
Xiang L Z, Yuan Y, Xing D, Ou Z M, Yang S H and Zhou F F 2009 Photoacoustic molecular imaging with antibody-functionalized single-walled carbon nanotubes for early diagnosis of tumor *J. Biomed. Opt.* **14** Article Number 021008.
Yen J T and Smith S W 2002 Real-time rectilinear volumetric imaging using a periodic array *Ultrasound Med. Biol.* **28** 923–931.
Zeqiri B, Shaw A, Gélat P N, Bell D and Sutton Y C 2004 A novel device for determining ultrasonic power *J. Phys. Conf. Ser.* **1** 105–110.
Zhu Y and Hall T J 2002 A modified block matching method for real-time freehand strain imaging *Ultrason. Imaging* **24** 161–176.

Further Reading

General Acoustics and Medical Diagnostic Ultrasound

Alty J, Hoey E, Wolstenhulme S and Weston M 2006 *Practical Ultrasound: An Illustrated Guide* (Royal Society of Medicine Press: London, UK).
Beyer R T 1997 *Nonlinear Acoustics* Reprint edition (Acoustical Society of America: Melville, NY).
Beyer R T 1999 *Sounds of Our Times: Two Hundred Years of Acoustics* (Springer Verlag: New York).
Blackstock D T 2000 *Fundamentals of Physical Acoustics* (John Wiley & Sons: New York).
Blake G M, Wahner H W and Fogelman I 1999 Technical principles of ultrasound, and commercial ultrasound instruments in: *The Evaluation of Osteoporosis: Dual Energy X-Ray Absorptiometry and Ultrasound in Clinical Practice* (Martin Dunitz: London, UK).
Bushong S C 1999 *Diagnostic Ultrasound (Essentials of Medical Imaging)* (McGraw-Hill: New York).
Cobbold R S C 2006 *Foundations of Biomedical Ultrasound* (Oxford University Press: Oxford, UK).
Crocker M J (ed.) 1997 *Encyclopedia of Acoustics* (John Wiley: New York).
Dowsett D J, Kenny P A and Johnston R E 1998 Ultrasound principles, and ultrasound imaging Chapter 17 and 18 pp. 413–465 in: *The Physics of Diagnostic Imaging* (Chapman & Hall: London, UK).
Duck F A, Baker A C and Starritt H C 1998 *Ultrasound in Medicine* (IOP Publishing: Bristol, UK).
Dunn F and O'Brien W D Jr. (eds.) 1977 *Ultrasonic Biophysics Benchmark Papers in Acoustics* Vol 7 (Dowden, Hutchinson and Ross: Stroudsburg, PA).
Fish P 1990 *Physics and Instrumentation of Diagnostic Medical Ultrasound* (Wiley: Chichester, UK).
Gooberman G L 1968 *Ultrasonics: Theory and Application* (English Universities Press: London, UK).
Greenleaf J F (ed.) 1986 *Tissue Characterisation with Ultrasound* (CRC Press: Boca Raton, FL).
Greenleaf J F 1997 Acoustical medical imaging instrumentation pp. 1751–1760 in: Crocker M J (ed.) *Encyclopedia of Acoustics* (John Wiley: New York).
Hedrick W R, Hykes D L and Starchman D E 2005 *Ultrasound Physics and Instrumentation* 4th Edn (Elsevier Mosby: St. Louis, MO).
Hill C R, Bamber J C and ter Haar G R (eds.) 2004 *Physical Principles of Medical Ultrasonics* 2nd Edn (John Wiley: Chichester, UK).
Hoskins P R, Martin K and Thrush A (eds.) 2010 *Diagnostic Ultrasound: Physics and Equipment*, 2nd Edn (Cambridge University Press: Cambridge, UK).

Hussey M 1985 *Basic Physics and Technology of Medical Diagnostic Ultrasound* (Macmillan: London, UK).
Kinsler L E, Austin R F, Copens A B and Sanders J V 1982 *Fundamentals of Acoustics* (Wiley: New York).
McDicken W N 1981 *Diagnostic Ultrasonics. Principles and Use of Instruments* (Wiley: New York).
Raichel D R 2000 *The Science and Applications of Acoustics* (Springer Verlag: New York).
Rossing T D and Fletcher N H 2004 *Principles of Vibration and Sound* (Springer-Verlag: New York).
Shung K K and Cloutier G 1997 Medical diagnosis with acoustics pp. 1739–1750 in: Crocker M J (ed.) *Encyclopedia of Acoustics* (John Wiley: New York).
Szabo T 2004 *Diagnostic Ultrasound Imaging: Inside Out* (Academic Press: Burlington, VT).
Taylor K J W, Burns P N and Wells P N T (eds.) 1988 *Clinical Applications of Doppler Ultrasound* (Raven: New York).
Wells P N T 1977 *Biomedical Ultrasonics* (Academic Press: London, UK).
Wells P N T (ed.) 1993 *Advances in Ultrasound Techniques and Instrumentation* (Churchill Livingstone: New York).
Woodcock J P 1979 Ultrasonics In: *Medical Physics Handbook* (Adam Hilger: Bristol, UK).
Zagzebski J A 1996 *The Essentials of Ultrasound Physics* (Mosby-Year Book Inc.: St. Louis, MO).

Microbubble Contrast Agents

Forsberg F, Goldberg B B and Raichlen J S 2001 *Ultrasound Contrast Agents* (Martin Dunitz: London, UK).
Hoff L 2001 *Acoustic Characterization of Contrast Agents for Medical Ultrasound Imaging* (Kluwer Academic: Dordrecht, the Netherlands).
Leighton T G 1994 *The Acoustic Bubble* (Academic Press: San Diego, CA).
Lencioni R 2006 *Enhancing the Role of Ultrasound with Contrast Agents* (Springer: Milan, Italy).
Nanda N C, Schlief R and Goldberg B B 1997 *Advances in Echo Imaging Using Contrast Enhancement* (Kluwer Academic Publishers: Dordrecht, the Netherlands).
Quaia E 2004 *Contrast Media in Ultrasonography: Basic Principles and Clinical Applications* (Springer: Berlin, Germany).

Ultrasound Biological Effects and Safety

AIUM 1993 *Bioeffects and Safety of Diagnostic Ultrasound* (AIUM: Laurel, MD).
Carstensen E L 1997 Biological effects of ultrasound pp. 1727–1737 in: Crocker M J (ed.) *Encyclopedia of Acoustics* (John Wiley: New York).
ter Haar G and Duck F A (eds.) 2000 *The Safe Use of Ultrasound in Medical Diagnosis* (The British Institute of Radiology: London, UK).

7

Spatially Localised Magnetic Resonance

S. J. Doran and M. O. Leach

CONTENTS

7.1 Introduction 489
7.2 Historical Development of Nuclear Magnetic Resonance 490
7.3 Principles of Nuclear Magnetic Resonance 493
  7.3.1 Classical Description of the Nuclear Magnetic Moment 493
  7.3.2 Vector Model and Interaction with a Radiofrequency Field 495
  7.3.3 Rotating Reference Frame 496
  7.3.4 Relaxation and the Bloch Equations 499
  7.3.5 Elementary Quantum-Mechanical Description 501
  7.3.6 Statistical Distribution of Spin States 503
  7.3.7 Signal Detection 504
7.4 Basic Magnetic Resonance Pulse Sequences 507
  7.4.1 Basic Classifications 507
  7.4.2 Response to a Single RF Pulse: Saturation- and Inversion-Recovery 508
  7.4.3 Spin Echo 509
  7.4.4 General Response to Two or More RF Pulses: Stimulated Echoes 511
7.5 Relaxation Processes and Their Measurement 512
  7.5.1 Longitudinal Relaxation, T_1 512
  7.5.2 Transverse Relaxation, T_2 513
  7.5.3 T_2^* 514
  7.5.4 $T_{1\rho}$ and Spin Locking 515
  7.5.5 Information Available from Relaxation Times 516
  7.5.6 Measurement of Longitudinal Relaxation Times 516
  7.5.7 Measurement of Transverse Relaxation Times and Diffusion 518
7.6 Nuclear Magnetic Resonance Image Acquisition and Reconstruction 520
  7.6.1 Goal of an MRI Acquisition 520
  7.6.2 Magnetic Field Gradients 520
  7.6.3 Selective Excitation 522
  7.6.4 Introduction to Common Imaging Pulse Sequences 524
  7.6.5 Projection Reconstruction 526
  7.6.6 2D Fourier Imaging 527
  7.6.7 Mathematical Description of 2D Fourier Imaging 528
  7.6.8 *k*-Space and Image Resolution 530
  7.6.9 Single-Point Imaging: An Alternative Explanation of Phase Encoding 532
  7.6.10 Signal-to-Noise Ratio of MR Images 534
  7.6.11 Relaxation Contrast in Images 535
  7.6.12 Spoiling 537

7.6.13 Review of the Main Classes of Imaging Pulse Sequence 537
7.6.13.1 Rapid Gradient-Echo Sequences 537
7.6.13.2 Echo-Planar and Spiral Imaging 538
7.6.13.3 Spin-Echo Sequences 541
7.6.13.4 Interleaved and Hybrid Sequences 541
7.6.13.5 Multi-Slice vs. 'True-3D' Sequences for Volume Imaging 542
7.6.14 Advanced Techniques 543
7.6.14.1 Acquisition Strategies in *k*-Space 543
7.6.14.2 Gating 545
7.6.14.3 Parallel Imaging 546
7.6.15 Other Imaging Methods with Historical Interest 548
7.6.15.1 Point Methods 548
7.6.15.2 Line-Scanning Methods 548
7.7 Image Contrast Mechanisms and Quantitative Imaging 549
7.7.1 General Principles of Quantitative Imaging 549
7.7.2 Relaxation Times and Dynamic Contrast-Enhanced MRI 550
7.7.3 Flow 552
7.7.4 Diffusion and Perfusion 555
7.7.5 Susceptibility and Functional MRI 558
7.7.6 Ultra-Short T_E Imaging, Magnetisation Transfer and Polarisation Transfer 561
7.7.7 Chemical-Shift Imaging 564
7.7.8 Current-Density Imaging 564
7.7.9 Contrast due to Long-Range Dipolar Fields and Multiple-Quantum Imaging 565
7.7.10 Imaging of Nuclei Other than Hydrogen 565
7.8 Image Artefacts 566
7.8.1 Artefacts Caused by the Instrumentation 567
7.8.2 Artefacts Relating to Experimental Design 571
7.8.3 Artefacts Generated by the Sample 571
7.9 Spatially Localised MR Spectroscopy 576
7.9.1 Introduction 576
7.9.2 Chemical-Shift Spectroscopy 577
7.9.3 1D Spectroscopy Data Processing 581
7.9.4 Localised Spectroscopy 582
7.9.4.1 Localisation Using Slice Selective Techniques 583
7.9.4.2 Image-Selected *In Vivo* Spectroscopy 584
7.9.4.3 Point-Resolved Spectroscopy 585
7.9.4.4 Stimulated Echo Acquisition Mode 585
7.9.4.5 Spectroscopic Imaging or Chemical-Shift Imaging 586
7.9.4.6 Choice of Localisation Method 587
7.9.5 Advanced Topics 587
7.9.6 *In Vivo* and Clinical Applications 588
7.10 MRI Technology 592
7.10.1 Equipment 592
7.10.1.1 Main Magnet 594
7.10.1.2 Magnetic Field Gradients 597
7.10.1.3 Radiofrequency System 597
7.10.1.4 Computer Systems and Pulse Programmer 600

7.10.2 Microscopy and Pre-Clinical Systems 600
7.10.3 Siting Considerations 600
7.11 MR Safety 601
7.11.1 Biological Effects and Hazards of MR 602
7.11.2 Static $\boldsymbol{B}_0$ Magnetic Fields 602
7.11.3 Time-Varying Magnetic Fields 603
7.11.4 Radiofrequency $\boldsymbol{B}_1$ Fields 604
7.11.5 Gradient-Induced Noise 604
7.11.6 Safety Considerations in MR Facilities 605
7.12 Applications of Magnetic Resonance in Medicine 607
7.12.1 Cancer 607
7.12.2 Neurology 607
7.12.3 Cardiology 608
7.12.4 Interventional Procedures and Monitoring of Therapy 609
7.12.5 Other Medical Applications 609
7.12.6 Molecular and Cellular Imaging 609
7.12.7 Genetic Phenotyping for Pre-Clinical Imaging 610
7.13 Conclusions 610
References 611

7.1 Introduction

Having discussed transmission and emission imaging with ionising radiation, and imaging with ultrasound, the use of magnetic resonance (MR) to measure and image tissue properties is now considered.

This chapter describes the physics associated with nuclear MR (NMR), explaining how NMR methods are used, together with the necessary instrumentation, to obtain images whose intensities are a function of proton spin density, relaxation times, water diffusion coefficients and other sample properties. Advanced imaging techniques and related topics are described. Methods of measuring the spatial distribution of other nuclei and metabolites using MR imaging (MRI) and MR spectroscopy (MRS) are discussed. Clinical applications of these techniques are introduced. The current convention of describing clinical applications of NMR as 'MR' is adopted, whilst remembering that electron spin resonance (ESR) techniques also have an important role in biomedical research.

MRI has become a major diagnostic tool in medical diagnosis and management and has a powerful role in clinical and pre-clinical research. In the years since the first edition of this book was produced, clinical MRI has truly come of age, and the technology is now mature. Nevertheless, despite the wide availability and range of applications, the modality continues to develop rapidly. Spatially localised MRS is now available on many clinical MR instruments and is contributing to clinical examinations and research. The technique has benefited from improvements in hardware and the continuing automation of set-up and measurement.

In common with cross-sectional X-ray tomography (Chapter 3), MR can produce three-dimensional (3D) image data sets providing detailed visualisation of the anatomy. However, rather than generating a map of electron density, related principally to physical density, MR images provide information relating to the number density of hydrogen atoms, principally in water and fat (commonly termed the *proton density*), and also to a whole host of other properties of the sample. This makes MRI arguably the most flexible of medical-imaging modalities.

The freedom of hydrogen-containing molecules to rotate is reflected in the *relaxation times* of a tissue. By appropriate manipulation of the sample's nuclear magnetisation, the relative contrast of different anatomical structures can be varied, and quantitative measurements of tissue relaxation times and other properties can be obtained. In addition to anatomical information, the MR image can also provide functional information more akin to nuclear medicine (Chapter 5) and ultrasound (Chapter 6), by using paramagnetically labelled tracers to provide selectively increased contrast and by providing direct measurements of blood flow, tissue perfusion and water diffusion. Imaging sequences that measure brain activity indirectly by its effect on blood flow and blood oxygenation are of major importance in neurological research. Nuclei other than hydrogen can also be measured by imaging or spectroscopy. ^{31}P MRS provides a method of assessing tissue energy metabolism noninvasively. Both ^{31}P and ^{1}H spectroscopies allow assessment of lipid metabolism, with ^{1}H spectroscopy having particular application in assessing neurological conditions. Measurements on ^{19}F allow the metabolism of exogenous fluorine-containing compounds, such as drugs, to be monitored. Hyperpolarisation of ^{129}Xe and ^{3}He provides powerful tracers and allows the examination of lung spaces.

This chapter is written at a level appropriate to postgraduate courses in imaging and medical physics. It also contains information relevant to the operation of MR units and for those conducting postgraduate and more advanced research. References to the primary literature and other general reference works are given later.

The historical development of NMR is described briefly in Section 7.2. The authoritative reference for this topic is Volume 1 of the *Encyclopedia of NMR* (Grant and Harris, 1996, Ch. 12), which contains an excellent article by Becker et al. (1996), together with personal reminiscences from over a hundred of the early researchers in the field. NMR theory is discussed in Section 7.3. More advanced texts on the theory of NMR include Andrew (1958), Slichter (1978), Abragam (1983), Ernst et al. (1987), Goldman (1991), Freeman (1997), Kimmich (1997) and Levitt (2001). A number of textbooks, such as Krane and Halliday (1988) on nuclear physics, Bleaney (1989) on electricity and magnetism and Kittel (2005) on solid-state physics, have useful sections.

Basic pulse sequences are described in Section 7.4 and the phenomena and measurement of relaxation in Section 7.5. The principles of MR image acquisition, outlined in Section 7.6, are covered in detail by Mansfield and Morris (1982b), Morris (1986), Callaghan (1991), Haacke et al. (1999) and Liang and Lauterbur (2000) among others. In Section 7.7, we discuss contrast mechanisms other than relaxation and describe the methods available for quantitative imaging. Artefacts may have a deleterious effect on the results of imaging experiments: these are categorised and explained in Section 7.8. Spatially localised MRS is described in Section 7.9, including an introduction and a short discussion of *in vivo* applications. Useful advanced reading on NMR includes Farrar and Becker (1971), Dwek (1973), James (1975), Shaw (1976), Fukushima and Roeder (1981), Gadian (1995) and Gunther (1995). Section 7.10 is concerned with instrumentation, and Section 7.11 discusses the potential hazards and the safety of NMR. Finally, Section 7.12 gives a brief overview of the application of MRI in medicine. For a fuller account of the latest developments, readers are referred to the comprehensive set of review articles listed in this section.

7.2 Historical Development of Nuclear Magnetic Resonance

Examination of the fine structure of atomic spectra led to the discovery that a magnetic field caused each spectral line to be split into a triplet, with spacing proportional to the strength of the applied magnetic field: the Zeeman effect (Zeeman 1897). The demonstration of quantised electron magnetic moments and spin was made in 1921, when a beam of silver ions was passed through an inhomogeneous magnetic field, in experiments by Gerlach and Stern (1921). Similar techniques were employed to measure *nuclear* magnetic moments in atoms with no electronic magnetic moment (H_2, D_2, Na, Cl), indirectly by measuring the influence of the nuclear moment on the electronic magnetic moment. The first suggestion that nuclear susceptibility should show a resonance effect for radiofrequency (RF) fields, dependent on the value of an imposed static field, was made by Gorter (1936), who attempted unsuccessfully to measure the resonant absorption by calorimetric methods. During the late 1930s, Rabi's laboratory had been pioneering particle beam methods, and a remark by Gorter on a visit to the laboratory led Rabi to introduce 'a new method of measuring the nuclear magnetic moment' (Rabi et al., 1938). This employed RF irradiation of the ion beam, in the presence of a uniform magnetic field. As the uniform field was swept, the beam current dropped markedly when the system passed through its resonance, providing what is generally regarded as the first demonstration of NMR. Stern and Rabi were awarded the Physics Nobel Prizes of 1943 and 1944 for their pioneering work. Measurements in bulk materials, however, required different techniques to be developed. Nuclear magnetism in solid hydrogen, where the low temperature results in a large polarisation, was detected, via its susceptibility effect, by Laserew and Schnubikow (1937), with electronic magnetism being demonstrated in 1944 (Zavoisky, 1945). The magnetic moment of the neutron was first measured by Alvarez and Bloch (1940).

The years immediately after World War II proved extremely fruitful, and many of the seminal publications in NMR come from this period. NMR of hydrogen in solid materials close to room temperature was discovered independently by Bloch et al. (1946) in paraffin wax and by Purcell et al. (1946) in water. Because the two teams used somewhat different terminology, it was at first unclear that what were being observed were two different manifestations of the same phenomenon. Bloch and Purcell were jointly awarded the Nobel Prize for Physics in 1952 for their discoveries. The first *in vivo* observations were performed surprisingly quickly after these initial experiments. At Stanford, Bloch placed his finger in the RF coil of the spectrometer and observed the hydrogen signal. At Harvard in 1948, Purcell and Ramsey both inserted their heads within an RF coil into the centre of the Harvard University 2 T cyclotron.

The theory of NMR relaxation was proposed by Bloembergen et al. (1948), and this was soon followed by the discovery of the chemical shift (Dickinson, 1950; Proctor and Yu, 1950) and a related phenomenon, the Knight shift (Knight, 1949), allowing nuclei in different chemical environments to be identified as a result of the small change in resonant frequency caused by the electron cloud of the molecule. The 1950s also saw the discovery of the spin echo (SE) by Hahn (1950), on whose work Carr and Purcell built with their analysis of the effects of diffusion on the NMR signal (Carr and Purcell, 1954).

A major step forwards was made by Ernst and Anderson (1966) who showed that the swept-field method of *continuous wave (CW)* NMR could be replaced by the application of short, intense pulses of RF. This not only paved the way for the development of modern high-resolution NMR spectroscopy as a versatile tool for studying the chemistry and structure of solids and liquids but is also a vital ingredient of imaging.

Notwithstanding the impressive analytical capabilities of NMR, the major biochemical and medical interest has arisen from the possibilities of making non-invasive measurements in living tissue. Initial measurements of ^{31}P in intact blood cells were carried out by Moon and Richards (1973). Measurements of frog sartorius muscle followed, but early experiments were limited by the small bore of the available magnets. Developments in magnet technology permitted phosphorus studies to be extended, initially to small animals and later, with the advent of wide-bore high-field magnets, to studies of humans. Proton, fluorine and carbon spectroscopy can also be performed on these instruments. As *in vivo* NMR spectroscopy of animals and humans has become possible, methods of obtaining spatially localised signals from a well-defined region of tissue have been developed.

In parallel with the development of spectroscopic techniques, methods of imaging the distribution of hydrogen (protons) in tissue evolved. These techniques again depended on the spatial localisation of the NMR signal, although in this case with a much higher spatial resolution. A major impetus for this work was the demonstration by Damadian (1971) of a systematic difference in the T_1 relaxation properties of cancerous and normal tissue and, in 1972, he filed a patent for an 'apparatus and method for detecting cancer in tissue'. Whilst the originality of the idea proposed in his patent (granted in 1974) was evident, no mention is made of magnetic field gradients, now seen as essential for image formation. In 1973, the principle of utilising the shift in resonant frequency resulting from the imposition of a magnetic-field gradient was first proposed by Lauterbur (1973) and then by Mansfield and Grannell (1973). Again, the language used by these two publications is very different. The early images that were formed were limited to small objects, but the first whole-body image was published in 1977 by Damadian et al. (1977). In 2003, the Nobel Prize for Physiology and Medicine was awarded jointly to Lauterbur and Mansfield but not Damadian, who, to this day, remains a controversial figure, having vigorously protested at his exclusion from the award. Work on spatially resolved measurements in one dimension (1D) using a gradient, presented by Carr in his PhD thesis (Carr, 1952), also received less acknowledgement than, in retrospect, it might have deserved. This situation is mirrored by the controversies associated with the award of the Nobel Prize for X-ray CT when inevitably some were excluded.

With the concept of imaging introduced, the race was on to produce results, and the remainder of the 1970s saw a succession of 'firsts'. In 1974, the first image of a laboratory animal was produced by Lauterbur (1974); Hinshaw et al. in Nottingham devised the so-called 'sensitive-point' method of imaging (Hinshaw, 1974); Garroway, Grannel and Mansfield invented slice selection (Garroway et al., 1974), now ubiquitous in MRI. In 1975, Kumar, Welti and Ernst introduced Fourier imaging (Kumar et al., 1975). Although modified in one key respect by Edelstein and co-workers to give 'spin-warp' imaging (Edelstein et al., 1980), Fourier methods are now dominant over the original back-projection technique of Lauterbur. The year 1976 saw the first *in vivo* image of a mouse (Damadian et al., 1976), whilst in 1977, the first human images were produced: a hand from Andrew et al. (1977), a finger cross section from Mansfield's group (Mansfield and Maudsley, 1977), a wrist from Hinshaw et al. (1977) and a cross section through the chest from Damadian, acquired using his 'focused NMR' or FONAR technique (Damadian et al., 1977). Another milestone was reached when, in 1978, Clow and Young obtained the first MR head images (Clow and Young, 1978). See also Chapter 1 for the general historical context.

The clinical potential of MRI was rapidly realised, but in the early days, investigators were severely limited by the available technology, particularly by the requirement to create a homogeneous magnetic field over a large enough region for whole-body imaging. Nevertheless, pioneers such as Young and Bydder at the Hammersmith Hospital in London gradually won over their medical colleagues. The story of the 1980s was largely one of developing the equipment required, such as high-field superconducting magnets, and the commercialisation that was to turn MR scanners from speculative research tools into routine items of hospital diagnostic equipment.

These early results were followed rapidly by technical and commercial developments, producing a variety of techniques that allowed proton images to be acquired and provided information on spin density and T_1 and T_2 relaxation times. During the late 1980s and early 1990s, two important areas of development were fast imaging and quantitative imaging.

In 1986, Haase et al. introduced the Fast Low-Angle SHot (FLASH) method, which for the first time allowed snapshot imaging (Haase et al., 1986). Hennig used successive 180° pulses to acquire multiple echoes from a single excitation. This method, known as Rapid Acquisition with Relaxation Enhancement (RARE), allowed a reduction in imaging time by a factor equal to the number of echoes (Hennig et al., 1986). At about the same time, major improvements in gradient performance using the concept of active shielding (Mansfield and Chapman, 1986) started to make feasible the acquisition of data using a technique known as Echo-Planar Imaging (EPI), first suggested by Mansfield (1977).

A number of papers in which images were made sensitive to water diffusion (Merboldt et al., 1985; Taylor and Bushell, 1985; Le Bihan et al., 1986) showed that contrast mechanisms other than relaxation had potential uses in medicine. Diffusion imaging has gone on to have major applications in the management of stroke. Methods of quantifying diffusion and other parameters, such as flow, perfusion and magnetisation transfer, were developed. These have provided a means of exploring a wide variety of sample properties (e.g. tissue temperature and oxygenation, or blood vessel permeability) that can be related to quantities measurable via MRI. Targeted MR contrast agents provide more sensitive identification of abnormality and facilitate assessment of organ and tumour function. In 1990 and 1991, experiments involving susceptibility contrast demonstrated how activated regions of the brain could be imaged (Ogawa et al., 1990; Belliveau et al., 1991). Taken together with developments in brain fibre tractography, achieved via diffusion tensor imaging (DTI), this has had profound implications, not just for the development of MRI as a discipline, but for neuroscience and society in general.

The 1990s were characterised by the increasing standardisation of MRI equipment and a vast increase in the number of systems installed worldwide. With continuing high levels of investment by the main scanner manufacturers, technical developments were rapid. In addition, the number of MRI researchers, both in hospitals and universities, mushroomed. Two consequences have flowed from this: MRI has fragmented into a number of subspecialities (cardiology, brain mapping, cancer, etc.), and the level of sophistication of imaging protocols and reconstruction techniques has increased enormously. The continual demand from radiologists for ever faster imaging means that we have now reached the stage where it is the human body itself that limits the speed with which RF and gradient pulses may be applied – beyond the current regulatory limits, unacceptable RF heating and peripheral nerve stimulation occur. Physicists have responded creatively, and the major hardware development of the early 2000s was the introduction of *partially parallel acquisition*, dominated by the SMASH and SENSE methods (Sodickson and Manning, 1997; Pruessmann et al., 1999) and variants thereof. In these techniques, as described in Section 7.6.14.3, the differing spatial sensitivity patterns of multiple receiver coils are exploited to replace a fraction of the phase-encoding steps.

Despite the rapid rate of development in recent years, MR is still developing rapidly, and many further advances are likely.

7.3 Principles of Nuclear Magnetic Resonance

7.3.1 Classical Description of the Nuclear Magnetic Moment

The behaviour of the net nuclear magnetic moment of materials can usually be sufficiently described by using the classical theory of electromagnetism. For the moment, consider an isolated proton. This possesses a charge $+e$ and has spin angular momentum $\boldsymbol{I}$. An electric charge circulating in a conducting loop produces a magnetic field through the loop normal to the plane of current rotation. The charge on a proton can be considered as being distributed and rotating about a central axis as a result of the angular momentum. This gives rise to a magnetic field and dipole moment **μ** parallel (for a proton) to the angular-momentum vector and normal to the plane of charge circulation, as shown in Figure 7.1a, that is,

$$\boldsymbol{\mu} = \gamma \boldsymbol{I}, \tag{7.1}$$

where γ is a constant known as the *gyromagnetic ratio*, whose value is characteristic of a given nucleus and measured in SI units of rad s^{-1} T^{-1}. (Note that there is some inconsistency both within the NMR literature and between different specialities as to this definition and in the units of γ. Some authors call this the *magnetogyric ratio* and describe the gyromagnetic ratio as its reciprocal. Others treat it as synonymous with the Landé g-factor described in the following. Sometimes the gyromagnetic ratio is quoted in Hz T^{-1}, in which case there is a factor of 2π difference. Also seen is the unit s^{-1} T^{-1}, which is ambiguous.)

For the classical model of the proton, as described earlier,

$$\gamma_{\text{classical}} = \frac{e}{2m_p}, \tag{7.2}$$

where m_p is the proton mass. Quantum mechanics modifies this by inserting an extra factor, known as the Landé g-factor, which in the case of the proton is approximately 5.59, such that $\gamma = 2.675 \times 10^8$ rad s^{-1} T^{-1} or 42.58 MHz T^{-1}.

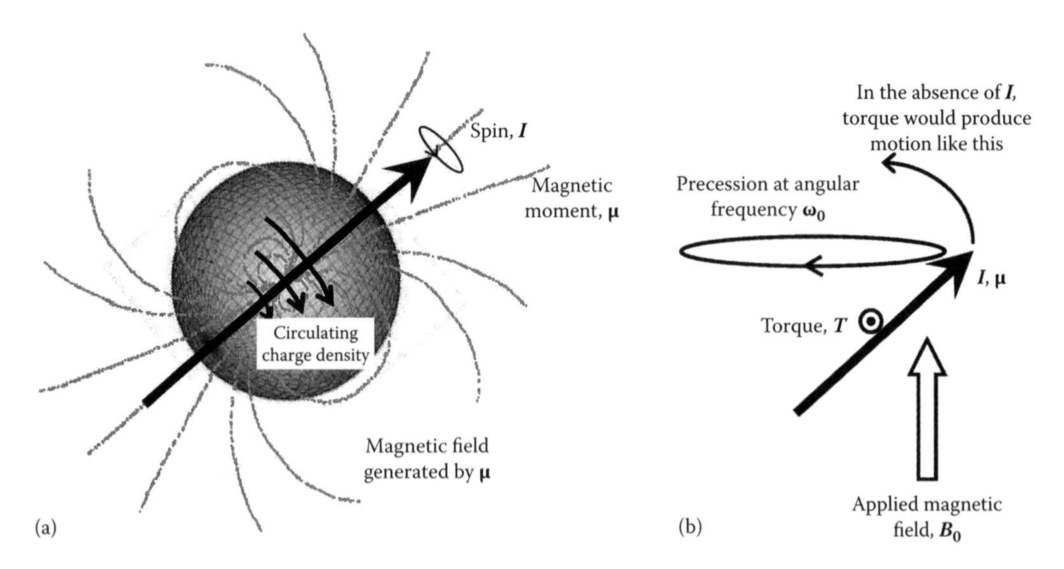

FIGURE 7.1
(a) Representation of a nucleus with spin angular momentum $\boldsymbol{I}$. The circulating charge density gives rise to a magnetic moment **μ**, which produces a dipolar magnetic field, just like a current-carrying loop of wire. (b) In the presence of an applied magnetic field $\boldsymbol{B}_0$, there is a torque $\boldsymbol{T}$ on the magnetic moment. If the dipole were not associated with an angular momentum, it would behave like a compass needle and align itself with $\boldsymbol{B}_0$. However, the existence of the angular momentum leads to precession about $\boldsymbol{B}_0$ at the Larmor angular frequency $\boldsymbol{\omega}_0$.

An external magnetic flux density $\boldsymbol{B}_0$ will exert a couple $\boldsymbol{C}$ on a magnetic dipole moment, causing the angular momentum to change at a rate equal to the torque or couple, that is,

$$\boldsymbol{C} = \boldsymbol{\mu} \times \boldsymbol{B}_0 = \frac{d\boldsymbol{I}}{dt}. \tag{7.3}$$

Substituting for $\boldsymbol{I}$ from Equation 7.1 gives

$$\frac{d\boldsymbol{\mu}}{dt} = \gamma\boldsymbol{\mu} \times \boldsymbol{B}_0. \tag{7.4}$$

Equation 7.4 is the Larmor equation, which describes the precession of $\boldsymbol{\mu}$ about $\boldsymbol{B}_0$ with angular velocity

$$\boldsymbol{\omega}_0 = -\gamma\boldsymbol{B}_0, \tag{7.5}$$

where $\boldsymbol{\omega}_0$ is known as the *Larmor angular frequency*.

The angle between $\boldsymbol{\mu}$ and $\boldsymbol{B}_0$ is not changed by this precession and depends only on the orientation of $\boldsymbol{\mu}$ when $\boldsymbol{B}_0$ was established (which, as we shall see later, is determined by quantum mechanics). Figure 7.1b shows the effect of the couple acting on the magnetic moment.

7.3.2 Vector Model and Interaction with a Radiofrequency Field

In a sample comprising many nuclei, the magnetisation $\boldsymbol{M}$ is defined by

$$\boldsymbol{M} = \sum_{\substack{\text{all spins } i \\ \text{in volume } V}} \frac{\boldsymbol{\mu}_i}{V}, \tag{7.6}$$

and this summation is illustrated schematically in Figure 7.2 for the case of thermal equilibrium. Classical theory is inadequate to describe fully the behaviour of individual

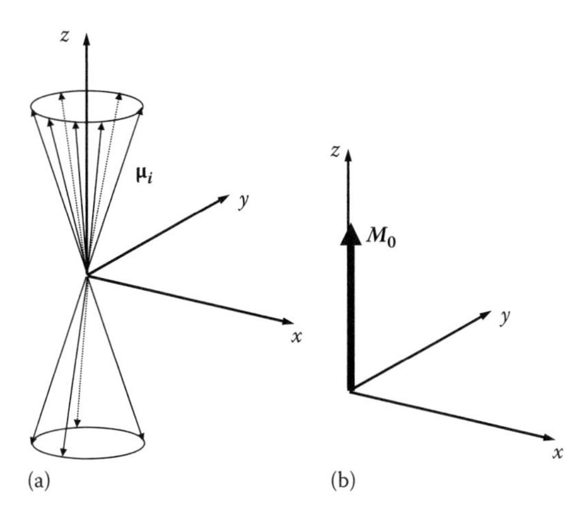

FIGURE 7.2
(a) At thermal equilibrium, there is an excess of spins with z components parallel to the main field $\boldsymbol{B}_0$ (conventionally along $+z$). (b) Summing all the elementary magnetic moments $\boldsymbol{\mu}_i$ leads to the net equilibrium magnetisation vector $\boldsymbol{M}_0$.

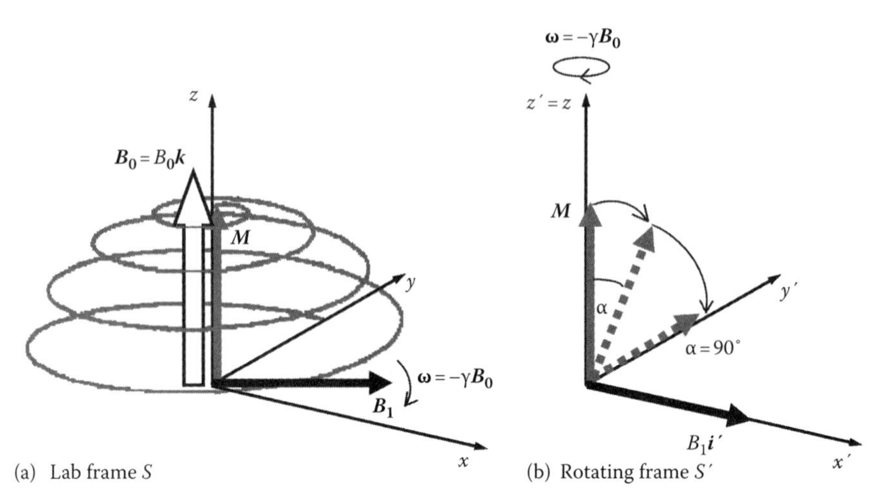

FIGURE 7.3
The effect of a radiofrequency magnetic field $\boldsymbol{B}_1$ on longitudinal magnetisation $\boldsymbol{M}$, seen in (a) the laboratory reference frame S and (b) the rotating frame of reference S'.

spins, because of the quantum nature of nuclear properties, but for a large ensemble of nuclei, a classical description in terms of $\boldsymbol{M}$ is normally adequate. This is known as the *vector model* of NMR and is the way in which Bloch first described the phenomenon. In a magnetic flux density $\boldsymbol{B}_0$, the net magnetic moment at equilibrium will be aligned with $\boldsymbol{B}_0$, which is conventionally taken to lie along the z axis. Only the xy or *transverse* component of $\boldsymbol{M}$ gives rise to a measurable signal in the standard NMR experiment and therefore at equilibrium, $\boldsymbol{M}$, which is then equal to $M_0\mathbf{k}$, cannot be measured directly. In order to measure $\boldsymbol{M}$, it must be tilted away from the $\boldsymbol{B}_0$ direction, to produce a measurable component in the xy plane. This is the basis of NMR measurements. If a magnetic field of flux density $\boldsymbol{B}_1$, oriented in the xy plane and rotating at the Larmor angular frequency $\boldsymbol{\omega}_0$, is applied, $\boldsymbol{M}$ will experience a second torque. The combination of the $\boldsymbol{B}_0$ and $\boldsymbol{B}_1$ fields rotates $\boldsymbol{M}$ into the xy plane along the spiral trajectory shown in Figure 7.3a.

By substituting (7.6) into (7.4), the motion of $\boldsymbol{M}$ under the influence of an arbitrary applied field with flux density $\boldsymbol{B}$ can be described, leading to the expression

$$\frac{d\boldsymbol{M}}{dt} = \gamma \boldsymbol{M} \times \boldsymbol{B}. \tag{7.7}$$

7.3.3 Rotating Reference Frame

The behaviour of $\boldsymbol{M}$ can often be more simply described by considering a reference frame (S') rotating with angular velocity $\boldsymbol{\omega}$ with respect to the laboratory frame (S). $\boldsymbol{\omega}$ is a vector, implying that the angular velocity may be about an arbitrary axis, but, in elementary NMR theory, it is common to restrict the discussion to frames of reference rotating about the z axis. The motion of $\boldsymbol{M}$ in the rotating frame $D\boldsymbol{M}/Dt$ can be related to its motion in the laboratory frame $\mathrm{d}\boldsymbol{M}/\mathrm{d}t$ as follows. A point $\boldsymbol{r}$ fixed in S' will move a distance

$$\delta \boldsymbol{r} = \omega \delta t \, r \sin\theta \, \hat{e} = \boldsymbol{\omega} \times \boldsymbol{r} \delta t \tag{7.8}$$

in time δt, with respect to the laboratory frame, where θ is the angle of $\boldsymbol{r}$ to the z axis and $\hat{e}$ is the unit vector perpendicular to $\boldsymbol{r}$ and the z axis. Hence, the velocity of $\boldsymbol{r}$ is

$$\lim_{\delta t \to 0} \frac{\delta \boldsymbol{r}}{\delta t} = \frac{d\boldsymbol{r}}{dt} = \boldsymbol{\omega} \times \boldsymbol{r}. \tag{7.9}$$

If $\boldsymbol{r}$ is not fixed but has a velocity $D\boldsymbol{r}/Dt$ in the S', then by the addition of the velocities

$$\frac{d\boldsymbol{r}}{dt} = \frac{D\boldsymbol{r}}{Dt} + \boldsymbol{\omega} \times \boldsymbol{r}. \tag{7.10}$$

This result is discussed in more detail by Kibble and Berkshire (2004, pp. 106–108). It simply says that the motion of a point in the laboratory frame is the motion of S' relative to S plus the motion within S'. If we think not of a point that is moving, but rather the rate of change of its associated vector, it is easy to see how this relates to changes in the direction and magnitude of our magnetisation $\boldsymbol{M}$, as seen in the two different frames:

$$\frac{d\boldsymbol{M}}{dt} = \frac{D\boldsymbol{M}}{Dt} + \boldsymbol{\omega} \times \boldsymbol{M}. \tag{7.11}$$

Substituting from Equation 7.7, we obtain

$$D\boldsymbol{M}/Dt = \gamma \boldsymbol{M} \times \boldsymbol{B}_0 - \boldsymbol{\omega} \times \boldsymbol{M} \tag{7.12}$$

$$= \gamma \boldsymbol{M} \times \boldsymbol{B}_0 + \boldsymbol{M} \times \boldsymbol{\omega} \tag{7.13}$$

$$= \gamma \boldsymbol{M} \times \left(\boldsymbol{B}_0 + \frac{\boldsymbol{\omega}}{\gamma} \right). \tag{7.14}$$

This is equivalent to Equation 7.4 with $\boldsymbol{B}_0$ replaced by the effective field $(\boldsymbol{B}_0 + \boldsymbol{\omega}/\gamma)$, which is the laboratory flux density with a 'fictitious flux density' $\boldsymbol{B}_f = -\boldsymbol{\omega}/\gamma$ subtracted from it. Equation 7.14 takes on the particularly simple form $D\boldsymbol{M}/Dt = 0$ at the so-called on-resonance condition, where $\boldsymbol{\omega} = \boldsymbol{\omega}_0 = -\gamma \boldsymbol{B}_0$.

In the rotating frame, $\boldsymbol{M}$ can be expressed as components $M_{x'}\boldsymbol{i}' + M_{y'}\boldsymbol{j}' + M_{z'}\boldsymbol{k}'$, where $\boldsymbol{i}'$, $\boldsymbol{j}'$ and $\boldsymbol{k}'$ are unit vectors in the direction of the x', y' and z' axes which rotate at $\boldsymbol{\omega}$, with respect to the laboratory-frame axes x, y and z. If a circularly polarised RF field of angular frequency ω (in the laboratory frame) and with magnetic flux density $\boldsymbol{B}_1$ is applied in the plane perpendicular to $\boldsymbol{B}_0$, then $\boldsymbol{B}_1$ will rotate, as shown in Figure 7.3a. In the rotating frame, it will have a fixed orientation, as shown in Figure 7.3b. In the absence of additional information, the direction of $\boldsymbol{B}_1$ in the $x'y'$ plane of the rotating frame is undefined, depending on the exact time at which the x' axis is defined as being coincident with the x axis of the laboratory frame. (On an MR scanner, this is normally adjustable electronically, both for transmission and reception, allowing one to vary the phase angle of the transmit pulse and 'phase' the signal.)

The effective flux density in the rotating frame is

$$\boldsymbol{B}_{\text{eff}} = \boldsymbol{B}_0 + \frac{\boldsymbol{\omega}}{\gamma} + \boldsymbol{B}_1, \tag{7.15}$$

which, for the case of $\boldsymbol{\omega}$ parallel to $\boldsymbol{B}_0$, can be resolved into components

$$\boldsymbol{B}_{\text{eff}} = \left(B_0 + \frac{\omega}{\gamma} \right) \boldsymbol{k}' + B_1 \boldsymbol{i}'. \tag{7.16}$$

The effect of $\boldsymbol{B}_1$ is thus to act as an additional couple on $\boldsymbol{M}$ and then, on resonance,

$$\frac{D\boldsymbol{M}}{Dt} = \gamma \boldsymbol{M} \times B_1 \boldsymbol{i}', \tag{7.17}$$

which has the effect of rotating $\boldsymbol{M}$ about the direction $\boldsymbol{B}_1$ (the x' direction) with an angular frequency $\omega_1 = -\gamma B_1$. When the RF irradiation corresponding to $\boldsymbol{B}_1$ is applied for a time interval that is short by comparison with the sample's relaxation times – see Section 7.5 – we call this an RF *pulse* (as distinguished from continuous wave NMR). If the pulse is applied for time τ, $\boldsymbol{M}$ will rotate through an angle $\alpha = \gamma B_1 \tau$. Figure 7.3b shows the rotation of $\boldsymbol{M}$ during an RF pulse; note in particular the convention for positive rotation angles generally employed. When $\alpha = \pi/2$, $\boldsymbol{M}$ will be directed along the $+y'$ axis and will have been subjected to a 90° pulse. A 180° pulse will rotate $\boldsymbol{M}$ along the $-z$ axis, antiparallel to $\boldsymbol{B}_0$.

Precession about x' occurs only during the RF pulse, and α depends upon both B_1 and τ, so either can be varied to give the desired angle. As α also depends upon the gyromagnetic ratio γ, the RF power or duration of pulse required to obtain a given tilt angle α will depend upon the nucleus under observation. For instance, phosphorus will require some 2.5 times the RF power that hydrogen requires to tip $\boldsymbol{M}$ through a given angle, and this can be important when considering RF power deposition and NMR safety.

In much NMR equipment, a plane-polarised RF field is applied, as this is technically easier and sometimes the only possible method. A plane-polarised field with flux density $2B_1\cos(\omega t)$ can be considered as the superposition of two circularly polarised fields with flux density $\boldsymbol{B}_1$, rotating in opposite directions. The field rotating in the same sense as the precession will interact with $\boldsymbol{M}$, whilst the influence of the counter-rotating field, the Bloch–Siegert effect (Abragam, 1983, Ch. 2), is negligible. The linear or plane-polarised field is, however, less efficient, as the power deposited is twice that required. For this reason, most clinical MR systems have been designed to transmit circularly polarised signals – see Section 7.10.1.

Now consider the general case in which the rotating frame of reference (where $\boldsymbol{B}_1$ is fixed) moves with an angular velocity $\boldsymbol{\omega}$ different from the Larmor angular frequency $\boldsymbol{\omega}_0$. From Equation 7.16,

$$\boldsymbol{B}_{\text{eff}} = \left(B_0 + \frac{\omega}{\gamma} \right) \boldsymbol{k}' + B_1 \boldsymbol{i}' = \frac{-\boldsymbol{\omega}_{\text{eff}}}{\gamma}, \tag{7.18}$$

where $\boldsymbol{\omega}_{\text{eff}}$ is the angular velocity of the Larmor precession of $\boldsymbol{M}$ about $\boldsymbol{B}_{\text{eff}}$. It will be seen from Figure 7.4 that the magnitude of the effective B field and the effective angular velocity are, respectively,

$$|\boldsymbol{B}_{\text{eff}}| = \left[(B_0 - B_f)^2 + B_1^2 \right]^{1/2}; \quad |\boldsymbol{\omega}_{\text{eff}}| = \left[(\omega_0 - \omega)^2 + \omega_1^2 \right]^{1/2}. \tag{7.19}$$

If the angle between $\boldsymbol{B}_{\text{eff}}$ and $\boldsymbol{B}_0$ is θ, then

$$\tan\theta = \frac{\omega_1}{\omega_0 - \omega}. \tag{7.20}$$

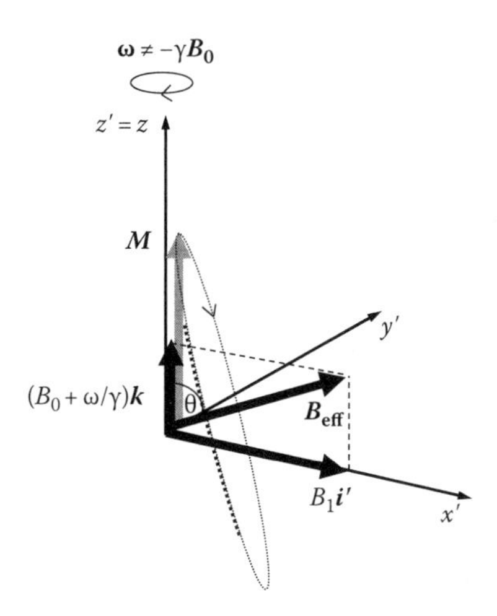

FIGURE 7.4
Precession of the magnetisation about the effective magnetic field B_{eff} in a frame of reference rotating at a frequency different from the Larmor frequency.

From this, it can be seen that M will only be significantly reoriented if θ becomes large, which occurs when $|\omega - \omega_0|$ becomes close to $|\omega_1|$, the latter usually being very much smaller than ω_0. Thus, B_1 must have a frequency close to the resonant frequency to have a significant effect.

An adiabatic rapid-passage experiment is one in which the angular frequency of the B_1 field is swept from a value far below resonance to a value far above. From Equation 7.20, we see that this corresponds to θ ranging from 0° to 180°. If B_1 is sufficiently large and the sweep is rapid then, effectively, no relaxation (see Section 7.5) takes place whilst the B_1 field is applied. The term *adiabatic* means that no energy is exchanged with the surroundings, that is, no transitions between spin states occur. At the same time, the nuclear precession about B_{eff} is always fast compared with the rate of change of θ, and this is a requirement for adiabaticity. This means that the value of $|M|$ is unchanged during the experiment whilst the magnetisation always has the same relationship with B_{eff}. Adiabatic techniques are particularly useful for providing pulses with accurate flip angles (e.g. 180° inversions) in situations where the B_1 field is inhomogeneous.

7.3.4 Relaxation and the Bloch Equations

Bloch (1946) proposed a set of equations that, for most purposes, accurately describe the behaviour of the nuclear magnetic moment of a sample of weakly interacting spins, such as those in a liquid sample. In a homogeneous field with flux density B, the so-called free-precession part of the equation of motion is given by Equation 7.7. However, interactions between a spin and its neighbours (Figure 7.5), which we discuss in more detail in Section 7.5, lead to additional terms in the equation. The phenomenon is known as *relaxation* and is described empirically by the following equations.

In a static field of flux density $B = B_0 k$, the return of the z component of magnetisation, M_z, to its equilibrium value, M_0, following a stimulus such as an RF pulse, can be described by

$$\frac{dM_z}{dt} = -\frac{(M_z - M_0)}{T_1}, \tag{7.21}$$

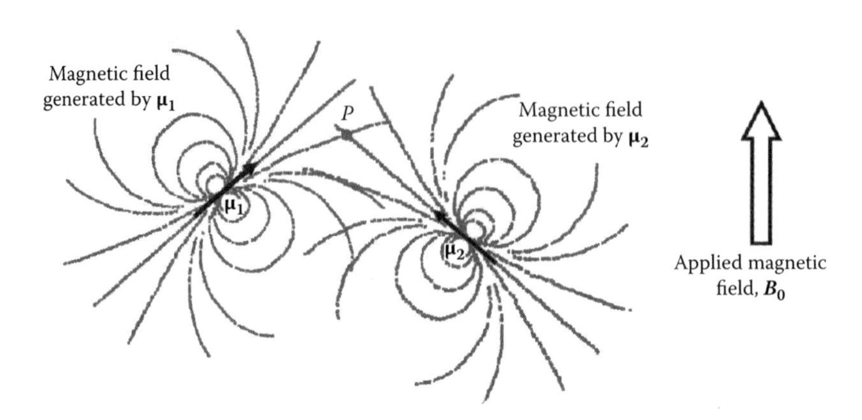

FIGURE 7.5
Illustration of the dipolar interaction. A nucleus at point P experiences not only the static magnetic field $\boldsymbol{B}_0$ and any oscillating $\boldsymbol{B}_1$ field applied but also time-varying magnetic fields created by the magnetic moments $\boldsymbol{\mu}_i$ of all the surrounding nuclei, of which two are shown in this figure. In general, the $\boldsymbol{\mu}_i$ will be precessing and the nuclei to which they are attached will be undergoing translation and molecular rotation.

where T_1 is the longitudinal relaxation time. If the nuclear magnetisation has a component perpendicular to the z direction, this transverse magnetisation will decay due to interactions with local spins, with the rate of change being described by

$$\frac{dM_x}{dt} = -\frac{M_x}{T_2}; \quad \frac{dM_y}{dt} = -\frac{M_y}{T_2}, \tag{7.22}$$

where T_2 is the transverse relaxation time. In theoretical treatments, the transverse magnetisation is sometimes written as a complex number $M_+ = M_x + iM_y$, which allows the equation of transverse relaxation to be written in the compact form

$$\frac{dM_+}{dt} = -\frac{M_+}{T_2}. \tag{7.23}$$

This emphasises the fact that the transverse magnetisation is independent of any particular choice of x and y axes in the transverse plane.

If it is assumed that the change in $\boldsymbol{M}$ due to relaxation can be superimposed on the motion of the free spins under the influence of a static field and a much smaller RF field, the behaviour of the magnetisation can be described by

$$\frac{d\boldsymbol{M}}{dt} = \gamma \boldsymbol{M} \times \boldsymbol{B} - \frac{(M_x\boldsymbol{i} + M_y\boldsymbol{j})}{T_2} - \frac{(M_z - M_0)\boldsymbol{k}}{T_1}, \tag{7.24}$$

where $\boldsymbol{i}$, $\boldsymbol{j}$ and $\boldsymbol{k}$ are unit vectors in the laboratory frame. As in Section 7.3.1, the behaviour of $\boldsymbol{M}$ in the rotating frame can be described by

$$\frac{d\boldsymbol{M}}{dt} = \gamma \boldsymbol{M} \times \boldsymbol{B}_{\text{eff}} - \frac{(M_x\boldsymbol{i}' + M_y\boldsymbol{j}')}{T_2} - \frac{(M_z - M_0)\boldsymbol{k}'}{T_1}, \tag{7.25}$$

with $\boldsymbol{i}'$, $\boldsymbol{j}'$ and $\boldsymbol{k}' = \boldsymbol{k}$ being unit vectors in the rotating frame. By expanding the cross-product and making use of Equation 7.18 for $\boldsymbol{B}_{\mathbf{eff}}$, Equation 7.25 can be rewritten to provide expressions for the evolution of the three orthogonal components of $\boldsymbol{M}$ in the rotating frame:

$$\begin{aligned} \frac{dM_{x'}}{dt} &= -\frac{M_{x'}}{T_2} + \Delta\omega M_{y'} \\ \frac{dM_{y'}}{dt} &= -\Delta\omega M_{x'} - \frac{M_{y'}}{T_2} - \omega_1 M_{z'} \\ \frac{dM_{z'}}{dt} &= \omega_1 M_{y'} - \frac{(M_{z'} - M_0)}{T_1}. \end{aligned} \tag{7.26}$$

Here, $\Delta\omega = \omega - \omega_0$, where ω is the angular frequency of the rotating reference frame, ω_0 is the Larmor angular frequency and $\omega_1 = -\gamma B_1$. This system of coupled first-order differential equations can be solved under a variety of boundary conditions, often numerically using a simple Runge–Kutta algorithm (Press et al., 2002, Ch. 16), to describe the time-domain (also known as '*k*-space' – see Section 7.6.8) behaviour of the NMR signal following RF irradiation. Alternatively, it can be instructive to work directly with Equation 7.25 using small time steps and finite rotations.

7.3.5 Elementary Quantum-Mechanical Description

The previous text is an approximate description, useful when considering the net effect on the assembly of spins as a whole. For an individual nucleus, a quantum-mechanical description is required. Consider a nucleus with spin angular momentum $\boldsymbol{I}$ and magnetic moment $\boldsymbol{\mu}$. The magnitude of the nuclear angular momentum is

$$|\boldsymbol{I}| = \hbar[I(I+1)]^{1/2}, \tag{7.27}$$

where I is the nuclear spin quantum number and $\hbar = h/2\pi$, where h is Planck's constant. Because the quantum-mechanical operators for x, y and z angular momentum do not commute, the direction of $\boldsymbol{I}$ cannot be established. If an axis of quantisation z is defined by applying an external magnetic field with flux density $\boldsymbol{B}_0$, the z component of $\boldsymbol{I}$ will be $m_z\hbar$, where m_z is the magnetic quantum number, which may take $(2I + 1)$ possible values $-I, -(I - 1), \ldots, +(I - 1), +I$.

For the proton and a number of other nuclei of interest, such as ^{31}P and ^{19}F, $I = 1/2$, and, therefore, m_z can take values $\pm 1/2$, which correspond to so-called 'spin-up' and 'spin-down' states. However, for other more complex nuclei than hydrogen, m_z may have more than two values. For example, $I = 3/2$ for ^{23}Na, giving $m_z = -3/2, -1/2, 1/2$ and $3/2$, whilst for deuterium (^2H), $I = 1$, leading to three spin levels $m_z = -1, 0$ and 1. Nuclei with $I > 1/2$ have nuclear quadrupole moments that can produce splitting of the resonance lines or line-broadening effects. The possible orientations of $\boldsymbol{I}$ for a spin 1 nucleus are shown in Figure 7.6a, and it can be seen that $\boldsymbol{I}$ is oriented at a semi-angle β to the z axis such that

$$\cos\beta = \frac{m_z\hbar}{|\boldsymbol{I}|}. \tag{7.28}$$

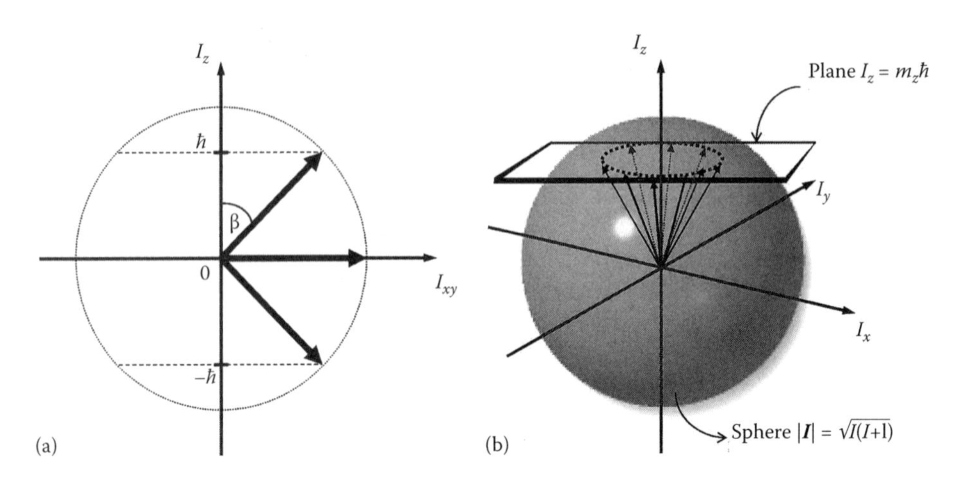

FIGURE 7.6
(a) Schematic diagram of the possible spin states of a spin 1 nucleus, such as deuterium. (b) The situation is really 3D and the spin $\boldsymbol{I}$ may take any value on the locus of intersection of the plane $I_z = m_z\hbar$ and the sphere $|\boldsymbol{I}| = \sqrt{I(I+1)}$, that is, the spin vector lies on a cone, with the azimuthal angle undefined.

The angular momentum is, in fact, a 3D vector, as is made clear by Figure 7.6b. The locus of all possible orientations for $\boldsymbol{I}$ is given by the intersection of the spherical shell described by Equation 7.27 and an appropriate plane of constant I_z. It will thus be seen that the azimuthal angle is undetermined and the spin vector lies on a cone. For the remainder of the theoretical development, we will consider exclusively spin 1/2 particles, such as the ^{1}H nucleus, with $m_z = \pm 1/2$ and thus just two states.

The energy of these states is

$$E = -\boldsymbol{\mu} \cdot \boldsymbol{B}_0 = -\gamma\hbar m_z B_0, \tag{7.29}$$

and their energy separation is $\Delta E = \gamma\hbar|\boldsymbol{B}_0|$. The quantity $\gamma\hbar$ is known as the *nuclear magneton*. Transitions can take place from the lower energy spin-up state (nuclear moment parallel to $\boldsymbol{B}_0$) to the higher energy spin-down state (nuclear moment antiparallel to $\boldsymbol{B}_0$) if a quantum of energy $\gamma\hbar|\boldsymbol{B}_0|$ is absorbed – see Figure 7.7a. Using the de Broglie equation $E = \hbar\omega$,

$$|\omega_0| = \gamma|\boldsymbol{B}_0|, \tag{7.30}$$

and, thus, energy at the Larmor angular frequency ω_0 must be supplied.

As the magnetic moment of the proton is allowed to occupy only two states, spin up and spin down, it might not be immediately apparent how the experimentally observed (and important) transverse component of nuclear magnetisation can arise. The behaviour of the proton magnetic moment in quantum mechanics is described in terms of the probability of the proton occupying a given state, and this occupancy can vary from observation to observation. Moreover, although the x and y components of angular momentum are not eigenvalues of the spin-up and spin-down states (due to the non-commutation of the operators for I_x and I_y with I_z), it can nevertheless be shown that the *expectation value* taken over the wavefunction of a free spin is consistent with the classical equation (Abragam, 1983, Ch. 2). Since the relationship is linear, large assemblies of independent nuclei will also behave classically, and so the net magnetic moment can

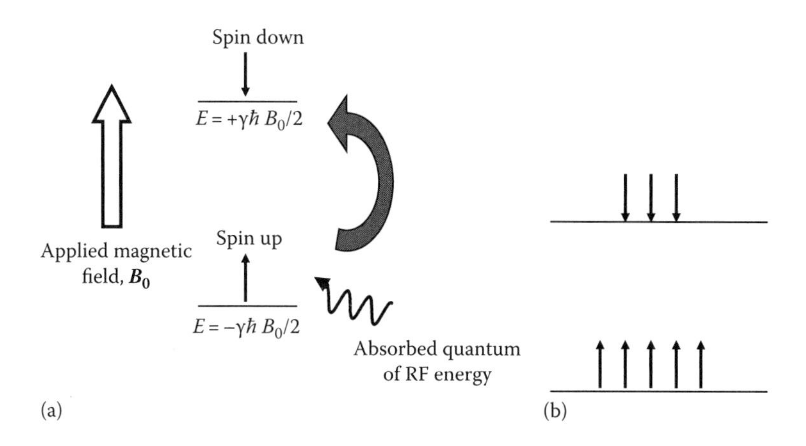

FIGURE 7.7
(a) Energy levels for a spin 1/2 nucleus, separated by energy gap $E = \gamma\hbar B_0 = \hbar\omega_0$, demonstrating how a quantum of radiofrequency energy oscillating at the Larmor frequency ω_0 undergoes resonant absorption, causing a 'spin flip'. (b) Schematic representation of the difference in population between the spin-up and spin-down energy levels. NB: In reality, the population difference is only a few in every million spins.

be described classically, provided there is negligible interaction between spins. A full quantum-mechanical treatment of the problem uses the *density matrix* formalism – see, for example, Ch. 4 of Goldman (1991).

7.3.6 Statistical Distribution of Spin States

In the absence of a magnetic field, all the spin states are degenerate, so spin-up and spin-down states cannot be defined. If a magnetic field with flux density $\boldsymbol{B}_0$ is applied, spins will be distributed between the two states, and, after a period dependent on the T_1 relaxation time (see Section 7.5), the population of the two levels will reach equilibrium, with the spin-up state being more favoured (Figure 7.7b).

The ratio of the populations in the two states is described by the Boltzmann distribution

$$\frac{n_\uparrow}{n_\downarrow} = \exp\left(\frac{\Delta E}{kT_s}\right), \tag{7.31}$$

where

$n_\uparrow$ is the number of nuclei in the spin-up ($m_z = +1/2$) state,
$n_\downarrow$ is the number of nuclei in the spin-down ($m_z = -1/2$) state,
k is the Boltzmann constant, and
T_s is the absolute temperature of the spin system (which is equal to the so-called lattice temperature when the sample is at equilibrium).

When a hydrogen sample is at equilibrium at room temperature (293 K) in a field of flux density 1.5 T, Equation 7.31 yields approximately 1.00001 for the population ratio. This is equivalent to a fractional excess of protons in the spin-up state (also known as the polarisation – see Section 7.7.6) of about 5×10^{-6}. In combination with the very low energy of the radiation quanta involved, this explains the relatively low sensitivity of NMR, compared with, for instance, radioisotope measurements, where under some circumstances a single decay can be measured.

The concept of spin temperature is sometimes useful. If the population of spins is disturbed, for instance, by supplying energy, $n_{\downarrow}$ will increase, changing the population difference between the states. As ΔE is fixed, for Equation 7.31 to hold, T_s (the temperature of the spin system) must increase. Energy loss is characterised by T_s returning to its equilibrium value.

Unlike γ radiation, where energy loss via γ ray emission is a spontaneous process, energy loss in NMR from the high- to the low-energy state occurs almost exclusively via *stimulated* emission. This probability of *spontaneous* emission is extremely low (Hoult and Ginsberg, 2001) and is given by

$$p = \frac{\mu_0 \omega_0^3 \gamma^2 \hbar}{6\pi c^3}, \tag{7.32}$$

where c is the speed of light *in vacuo*. For a proton in a magnetic field of flux density 1.5 T, this corresponds to a mean lifetime of approximately 10^{24} s, some 10^7 times the age of the universe! Stimulated emission occurs as a result of coupling with a magnetic field fluctuating at the Larmor frequency. Non-coherent coupling with a radiation field in thermal equilibrium is again a very weak energy-loss mechanism, and the major energy-loss mechanism is coupling with the lattice. This is further discussed in the section on relaxation (Section 7.5).

7.3.7 Signal Detection

In the laboratory frame of reference, transverse magnetisation that has been excited by an RF field rotates at the Larmor frequency. Thus,

$$M_x = M_0 \sin\theta \cdot \cos(\omega_0 t + \phi_0); \quad M_y = M_0 \sin\theta \cdot \sin(\omega_0 t + \phi_0); \quad M_z = M_0 \cos\theta. \tag{7.33}$$

ϕ_0 is an *arbitrary constant*, corresponding to the *phase* of our detector system (Section 7.10.1). We can set it to anything we like by modifying the electronics of the detector. As indicated by Equation 7.23, the transverse magnetisation decays via T_2 relaxation, yielding

$$M_+ = M_0 \exp(i\phi_0)\exp(i\omega_0 t)\exp\left(\frac{-t}{T_2}\right). \tag{7.34}$$

The reason we are able to detect a signal is because this rotating magnetisation induces an EMF in the detector (Faraday's law) – see Figure 7.8. Although this EMF is purely real, the technique of *heterodyne detection* allows us to extract from the signal, which is oscillating at the Larmor frequency, both real and imaginary components, corresponding to the two components of the transverse magnetisation. The final time-domain signal output is of the form

$$S(t) = S_0 \exp(i\phi_0)\exp(i[\omega_0 - \omega_{\text{dem}}]t)\exp\left(\frac{-t}{T_2}\right), \tag{7.35}$$

where the constant S_0 includes all effects of proton density, detection efficiency and amplification, as described in Hoult and Richards (1976), and ω_{dem} is the spectrometer demodulation angular frequency. This oscillating and decaying signal is known as the *free induction decay* (FID). ('Free' indicates that no applied RF field is perturbing the evolution of $\boldsymbol{M}$ in the static magnetic field $\boldsymbol{B}_0$ and is contrasted with 'forced' precession when an RF pulse is applied.)

It is a common misconception that, by analogy with optical spectroscopy, the NMR signal arises because the probe somehow 'catches' RF photons arising from the downward transitions

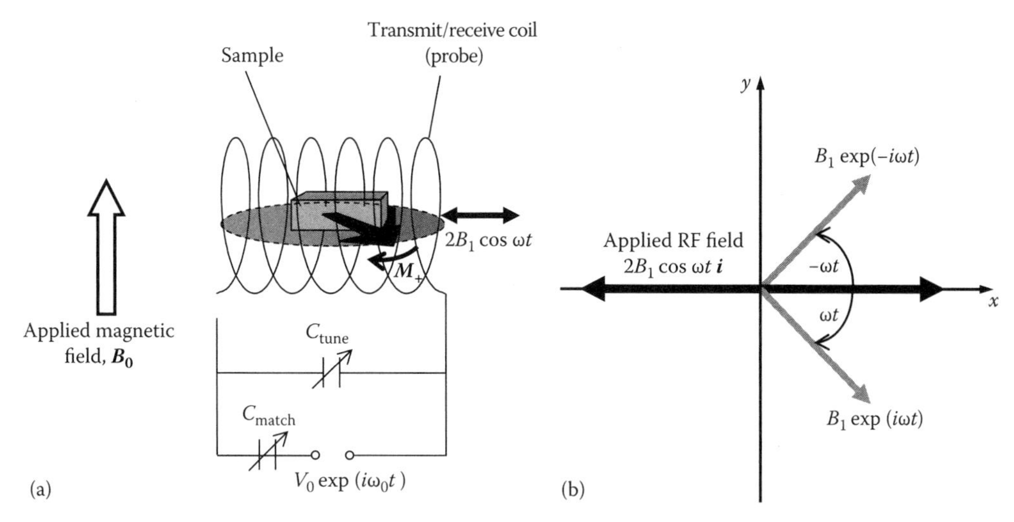

FIGURE 7.8
In (a), a solenoidal coil – see Section 7.10.1.3 for details of other designs – creates a linearly oscillating RF field of amplitude $2B_1 \cos \omega t$ in a direction perpendicular to the main magnetic field $\boldsymbol{B}_0$. This RF field can be decomposed into two counter-rotating components, as shown in (b), of which the clockwise component, rotating near the Larmor angular frequency ω_0, excites magnetisation into the transverse plane. This transverse magnetisation $\boldsymbol{M}_+$ rotates at ω_0, and the resultant time-varying magnetic field induces an EMF $V_0 \exp(i\omega_0 t)$ in the receiver coil.

of spins, in the same way that RF quanta originating in the probe promote spins from the low- to the high-energy state. In fact, there are two terms in the classical expression for the magnetic field produced by the rotating magnetisation: the 'near' and 'far' fields. In a typical case analysed by Hoult and Ginsberg (2001), the far-field (or radiative) term in the expression, which is associated with RF photons, accounts for only about 0.2% of the signal. The EMF induced in the probe comes predominantly from the non-propagating near field, and if one wishes to discuss this in terms of photons, one must resort to the language of virtual photons and quantum electrodynamics.

Although the signal initially acquired from the transverse component of magnetisation is a time-varying signal, it is the frequency-domain representation of the signal, obtained following Fourier transformation (FT), that is of most interest in both imaging (being proportional to displacement; see Section 7.6) and spectroscopy (giving chemical shift; see Section 7.9.2).

Following quadrature detection and FT of a free induction decay, we obtain, by simple integration of Equation 7.35,

$$\begin{aligned}
\tilde{S}(\omega) &= \int_{-\infty}^{\infty} S(t)\exp(-i\omega t)dt \\
&= S_0 \exp(i\phi_0)\int_0^{\infty} \exp\left(i\left[\omega_0 - \omega_{\text{dem}} - \omega + \frac{i}{T_2}\right]t\right)dt \\
&= S_0 \exp(i\phi_0)\cdot\frac{T_2^2}{1+(\omega_0-\omega_{\text{dem}}-\omega)^2 T_2^2}\cdot\left[\frac{1}{T_2}+i(\omega_0-\omega_{\text{dem}}-\omega)\right] \\
&= S_0 \exp(i\phi_0)\cdot\frac{T_2^2}{1+(-\Delta\omega-\omega_{\text{dem}})^2 T_2^2}\cdot\left[\frac{1}{T_2}-i(\Delta\omega-\omega_{\text{dem}})\right],
\end{aligned} \tag{7.36}$$

where $\Delta\omega = \omega - \omega_0$, as in Equation 7.26. Notice that the limits of integration are changed to $(0, \infty)$ in the second line of the aforementioned equation, because the NMR signal is zero before the RF pulse, assumed to occur at $t = 0$. If the system is 'phased' appropriately (corresponding to $\phi_0 = 0$), then the real part of the Fourier transform is a Lorentzian function with FWHM $2\omega_{1/2}$. The two components of the Fourier transform are sometimes known as the 'absorption' and 'dispersion' mode signals, these terms being more frequently encountered in older texts, particularly those that discuss CW NMR spectroscopy. Prior to the widespread availability of rapid computation and pulsed FT scanners, spectra were commonly acquired by sweeping the transmission frequency of a steady, rather than pulsed, excitation and observing the absorption of RF energy – an extension of the way in which Purcell et al. first discovered NMR (Purcell et al., 1946). The steady state (SS) may easily be found by setting the LHS of Equation 7.26 to zero and solving the coupled differential equations simultaneously. If one works in a regime where $\omega_1^2 T_1 T_2 \ll 1$ (i.e. B_1 is a negligible perturbation of the system and $M_{z'} \approx M_0$), then the solution is

$$M_{x'\text{SS}} = \frac{M_0\omega_1 T_2^2}{1+(\Delta\omega)^2 T_2^2}\cdot(-\Delta\omega); \quad M_{y'\text{SS}} = \frac{M_0\omega_1 T_2^2}{1+(\Delta\omega)^2 T_2^2}\cdot\left(-\frac{1}{T_2}\right), \tag{7.37}$$

provided the rate of scanning of ω is sufficiently slow – see Goldman (1991, Ch. 1). The terms

$$u = S_{\text{disp}}(\Delta\omega) = M_{x'\text{SS}}; \quad v = S_{\text{abs}}(\Delta\omega) = -\frac{\gamma}{|\gamma|}M_{y'\text{SS}}, \tag{7.38}$$

introduced by Bloch, are seen to correspond to the imaginary and real parts, respectively, of the spectrum in Equation 7.36, for the case $\omega_{\text{dem}} = 0$.

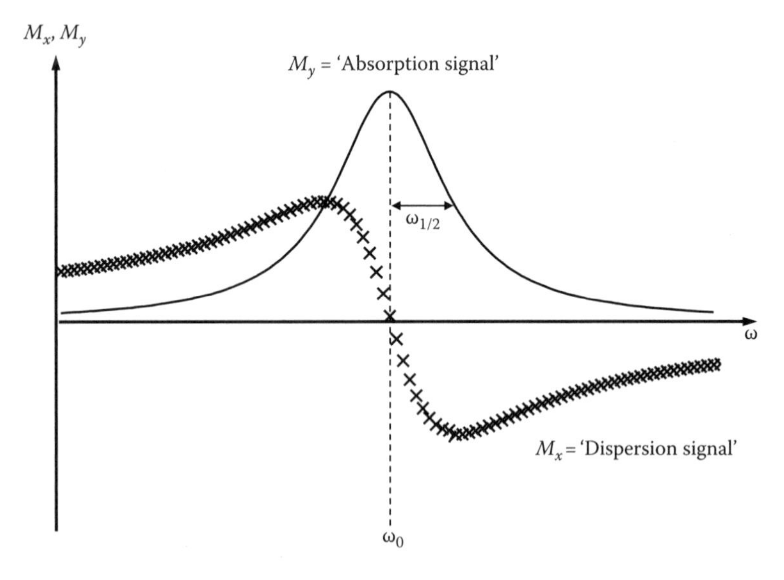

FIGURE 7.9
Form of the absorption and dispersion mode parts of the CW NMR signal – see main text for further details.

The two signals are shown in Figure 7.9. The dispersion mode signal u is 90° out of phase with the absorption signal and is an odd, as opposed to an even, function of $\Delta\omega$, going to 0 at ω_0. Often the modulus of the two signals is taken to give the absolute signal, with a consequent loss of phase information. Where there is significant saturation, the Lorentzian shape is broadened.

7.4 Basic Magnetic Resonance Pulse Sequences

The preceding sections have shown how an NMR signal can be generated by manipulating $\boldsymbol{M}$ with suitable RF pulses, and much of MRI may be regarded as simply the working out of the consequences of the Bloch equation (7.25). This section describes a number of basic sequences involving RF pulses that provide the foundation for much of spectroscopy and imaging. Discussion of pulse sequences involving gradients is deferred until Section 7.6.4.

7.4.1 Basic Classifications

The neophyte in NMR is presented with a panoply of MRI and MRS sequences that might at first sight seem bewildering. However, despite the apparent complexity, there are actually only four basic operations of an MRI scanner:

- *transmission of an RF pulse* – this might be to one of a number of different coils (see Section 7.6.14.3) or on the decoupler channel;
- *reception of the NMR signal;*
- *application of a gradient pulse* on one or more of the x, y and z physical gradient coils;
- *doing nothing* – the precise use of delays in MR sequences to allow evolution of spins to occur is of cardinal importance.

An MR *pulse sequence* is simply a time-ordered set of operations. The same term is also commonly used for the computer code executed by the scanner to implement the sequence. It is often represented by a *sequence diagram*, a visual description of the sequence with an agreed notation.

The four basic operations may be combined to form *modules* that achieve certain aims. Although attempts to classify the range of MR sequences (Boyle et al., 2006) on the basis of the modules that they contain are often hampered by the fact that a given set of pulses, gradients and delays often performs more than one function, one may nevertheless make the following broad divisions:

- modules that *manipulate longitudinal magnetisation* – these often have the aim of modifying image contrast by subsequent transfer of this magnetisation into the transverse plane for observation;

- modules that *manipulate transverse magnetisation* – this includes *creation* of transverse magnetisation via a suitable RF pulse; *destruction* (or 'spoiling') of coherent magnetisation in the transverse plane via either RF or gradient pulses; *coherent movement* of magnetisation around the transverse plane via RF pulses (often regarded as a discrete and instantaneous flip through a finite angle) and *dephasing/rephasing*, the continuous movement of magnetisation around the transverse plane using magnetic field gradients.

7.4.2 Response to a Single RF Pulse: Saturation- and Inversion-Recovery

Section 7.3.3 describes how an RF pulse at the Larmor frequency tips magnetisation from the z' axis to the $x'y'$ plane, whilst the signal obtained is given in Section 7.3.7. Notice that in Equation 7.35, the initial amplitude of the FID S_0 depends on both the pulse angle and the z magnetisation immediately prior to the pulse, that is, $S_0 \propto M_z(0_-) \sin\theta$. If the system is at thermal equilibrium prior to the pulse, then $M_z(0_-) = M_0$, but in many MR sequences this is not the case.

The maximum FID signal is obtained when $\theta = 90°$, and this type of sequence, known as *saturation-recovery* (SR), is a very common mode of measurement in spectroscopy. The term 'saturation' indicates that the populations of the two energy levels are equal. Provided a perfect 90° pulse is used throughout the sample, $M_z(0_+)$, that is, the z component of magnetisation immediately after the pulse, should be zero everywhere.

Following the 90° pulse, an ideal FID will decay with an exponential envelope, as given by Equation 7.35, whilst M_z will recover with relaxation time T_1. Solving Equation 7.21 using an integrating factor yields

$$M_z(t) = M_0 + [M_z(0_+) - M_0]\exp\left(\frac{-t}{T_1}\right), \tag{7.39}$$

which, for the particular case of SR, becomes

$$M_z(t) = M_0\left[1 - \exp\left(\frac{-t}{T_1}\right)\right]. \tag{7.40}$$

In a practical NMR sequence, the basic 90° pulse will often be repeated many times, with a characteristic repetition time T_R. If T_R is less than about $5T_1$, this will result in a reduction in M_z compared with M_0, as a result of partial saturation. As shown in Figure 7.10a, the response to the very first pulse is as given in Equation 7.40, but thereafter, the 'input' magnetisation for all subsequent pulses is

$$M_z(0_-) = M_0\left[1 - \exp\left(\frac{-T_R}{T_1}\right)\right]. \tag{7.41}$$

In an imaging context, this type of manipulation of the z magnetisation leads to *T_1-weighted contrast*.

In an *inversion-recovery* (IR) *sequence*, the z component of magnetisation, M_z, is inverted to lie along the $-z$ axis, by using a 180° RF pulse. This process is shown in Figure 7.10b. If an accurate 180° pulse is used, there should be no xy component of $\mathbf{M}$. Hence, this inverted magnetisation can be observed only by interrogating M_z with a second pulse (often 90°) at

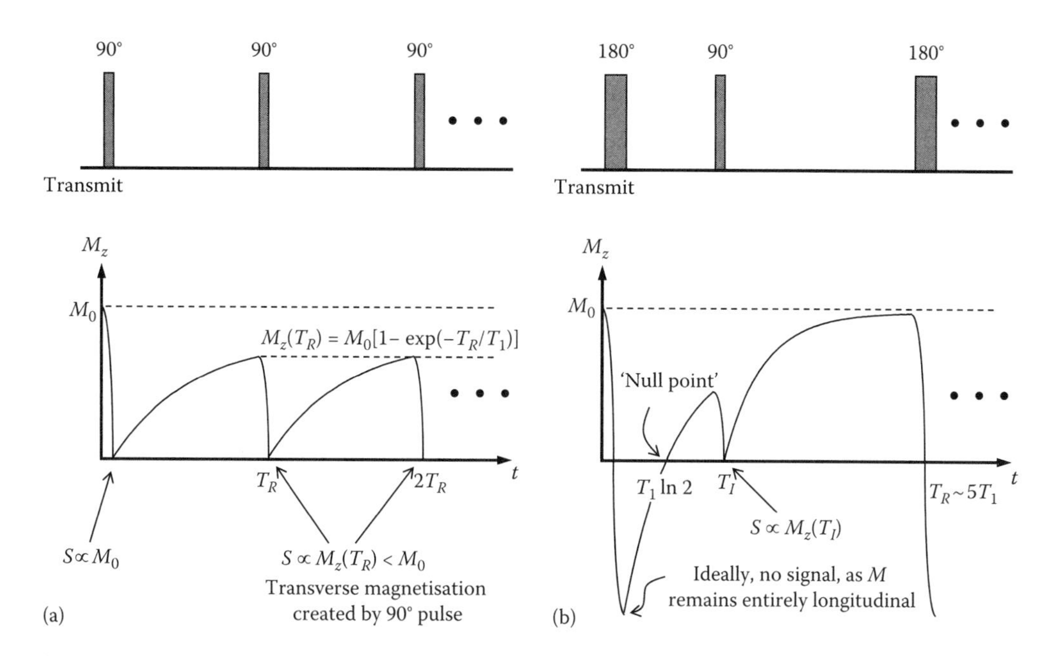

FIGURE 7.10
Schematic diagrams of (a) the basic SR pulse sequence and (b) the IR sequence, demonstrating the evolution of the longitudinal magnetisation M_z, as described in the main text.

a time T_I later. T_I is known as the *inversion time*, and by varying it, one is able to investigate the return to equilibrium:

$$M_z(T_I) = M_0\left[1 - 2\exp\left(\frac{-T_I}{T_1}\right)\right], \tag{7.42}$$

which is simply Equation 7.39 with $M_z(0_+) = -M_0$. Notice that M_z must pass through zero and then continue until $M_z = M_0$. It, therefore, takes longer to return to any given M_z value than for SR, and a longer repetition time T_R is required between sequences (although the difference between SR and IR decreases as T_R becomes longer, becoming insignificant for repetition times of order $5T_1$). A particularly interesting value of T_I is the 'null point', $T_I = \ln 2\ T_1$. If one appends a spatially resolved acquisition onto the end of the IR module, then one can create images in which all regions of the sample with a particular T_1 are nulled. This technique is widely used in medical imaging for the suppression of signals from fat.

7.4.3 Spin Echo

This sequence is shown in Figure 7.11. As in the SR sequence, a 90° pulse interrogates the z axis magnetisation. After an interval $T_E/2$ (T_E is the echo time), a 180° pulse is applied that 'refocuses' the xy magnetisation, producing an echo centred at time T_E after the 90° pulse. In the rotating frame (Figure 7.12), spins at flux density B_0, and thus rotating at Larmor frequency ω_0, will appear stationary. Due to inhomogeneities in the magnetic field – see Section 7.5.3 – some spins will experience a higher field and rotate faster in the lab frame, while others will see a lower field and precess more slowly. The latter ('slow spins') will gain phase compared to the spins at ω_0 and hence appear to move anticlockwise in the rotating frame, whilst the 'fast spins' at higher fields (i.e. larger negative values of $\omega = -\gamma B$)

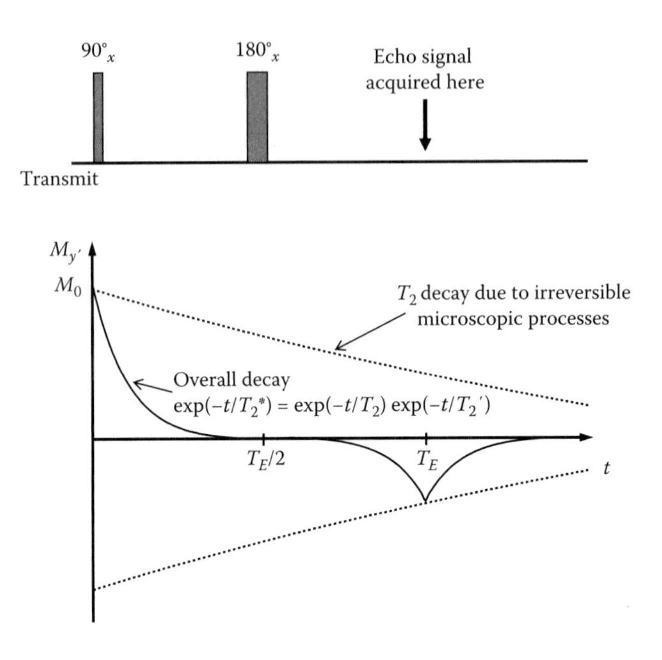

FIGURE 7.11
Schematic diagrams of the basic SE pulse sequence and the response of the transverse magnetisation. Note that the Meiboom–Gill modification, in which the $180°_x$ pulse is replaced by a $180°_{y'}$, would lead to an echo along the positive M_y axis.

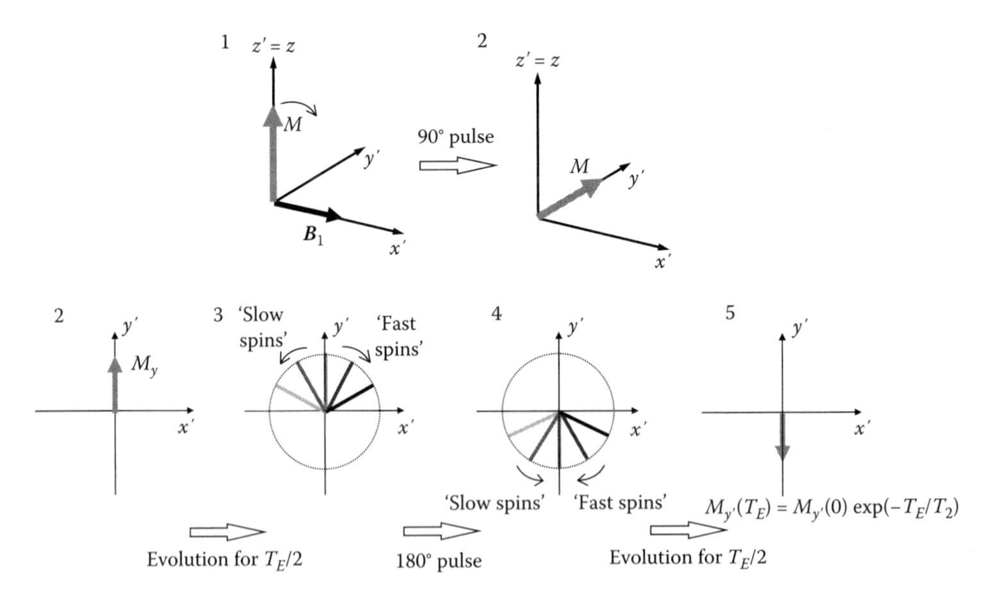

FIGURE 7.12
Step-by-step explanation of the process of formation of an SE: (1) Magnetisation initially at equilibrium along the z axis is excited by the 90° pulse into the transverse plane (2) (both a 3D view and a plan view looking down on the transverse plane are shown). (3) During the evolution period spins precessing at different rates dephase, losing or gaining phase with respect to spins stationary in the rotating frame, that is, precessing at angular frequency ω_0 in the laboratory frame. (4) The 180° pulse flips the spins about the x' axis, so that spins that were lagging are now leading and vice versa. However, the *relative* directions of motion in the rotating frame remain the same and so all the spins come back into phase after a second evolution period (5).

will move clockwise in the rotating frame. Overall, there will be a *dephasing* of the xy signal. After being flipped through 180° about the x' axis, each spin will still be subject to the same field, and its motion will still be in the same direction as before. However, *relative to the other spins*, the direction of motion will be opposite to that before the refocusing pulse. As a whole, the spins will now be converging on the $-y'$ axis to produce an echo.

A useful analogy is to consider a group of runners, some fast, some slow and some of average speed. At time $t = 0$, they start running forwards from the start line for a time $T_E/2$, by which point they are spread out. At time $T_E/2$, an instruction is given for the runners to reverse their direction. They retrace their steps at the same speeds as before, and at $t = T_E$, the runners are all level again back at the start line.

The SE pulse sequence is the basis for many of the current imaging sequences. Further echoes can be obtained by repeating the 180° pulse, and several different multi-echo sequences are discussed in more detail in Section 7.6.15. As with the SR and IR, the whole sequence can be repeated with repetition time T_R.

7.4.4 General Response to Two or More RF Pulses: Stimulated Echoes

The pictorial explanation outlined earlier does not tell the whole story. In the presence of magnetic field inhomogeneities that give rise to dephasing, *any set of two pulses* (except two successive 180°s) will yield a spin echo. Furthermore, adding additional pulses leads to initially unexpected effects, and a more advanced treatment is needed to understand these rich multi-echo phenomena. Generating the desired echoes, suppressing the ones that are not wanted and controlling the uniformity of amplitude of the overall echo train form an important part of modern sequence design.

Ignoring the effects of diffusion, whose explanation we defer to Section 7.5.7, a pulse sequence of the form $\theta_1 - \tau - \theta_2$ (where θ_1 and θ_2 represent two pulses with flip angles θ_1 and θ_2 and τ is a delay) will give a single spin echo with amplitude

$$A = M_0 \sin\theta_1 \sin^2\left(\frac{\theta_2}{2}\right)\exp\left(\frac{-t}{T_2}\right), \tag{7.43}$$

whilst the three-pulse sequence θ_1–τ_1–θ_2–τ_2–θ_3 gives five echoes, all with different amplitudes, which are functions of θ_1, θ_2, θ_3, τ_1, τ_2, T_1 and T_2 and diffusion. Depending on the relative time intervals and pulse angles, some of these echoes may fall on top of one another or not be present.

The echo occurring at $t = \tau_2 + 2\tau_1$ after the initial pulse is of a particularly interesting type and is known as a *stimulated echo*. It arises from a portion of the magnetisation that has 'visited' the transverse plane and been dephased there during the first delay; has been flipped back up onto the longitudinal axis by θ_2, relaxing with time-constant T_1 during delay τ_2 and, finally, has been flipped back down to the transverse plane by θ_3. The echo appears at a time τ_1 after the third pulse. It is as if, despite the bulk z magnetisation vector not being associated with a phase, as is the case for transverse magnetisation, the isochromats (Section 7.5.3) have somehow 'remembered' the phase that they had at the point of being flipped up to the longitudinal axis. The magnetisation behaves as if it has simply taken a 'time out' from transverse relaxation for the period τ_2 and then refocused as normal to give an echo. This 'z-storage' process proves extremely useful in situations where one wishes to insert a long delay into a sequence (e.g. to allow diffusion evolution), but where the T_2 of the sample is short.

A full explanation for these phenomena is beyond the scope of this chapter. Descriptions in the primary literature are given by Woessner (1961) and Kaiser et al. (1974), whilst Ch. 4

of Liang and Lauterbur (2000) provides an excellent exposition with full derivations. Analysing the effects in terms of the Bloch equations rapidly becomes inconvenient for large numbers of pulses. From N pulses, it is theoretically possible to generate up to

$$N_{\max} = \frac{1}{2}\left[\frac{3^N - 1}{2} - N\right] \tag{7.44}$$

different echoes (Liang and Lauterbur, 2000, p. 130). A formalism known as the *extended phase graph*, pioneered by Hennig (1988), expresses the results in a compact matrix notation and, in recent years, has become an indispensable tool in analysing multi-echo imaging sequences.

7.5 Relaxation Processes and Their Measurement

7.5.1 Longitudinal Relaxation, T_1

Longitudinal relaxation is often known as T_1 or 'spin-lattice relaxation'. The term 'lattice' has its origins in the early studies of NMR, many of which were performed in crystalline solids, where the entire lattice was described quantum-mechanically by a single wavefunction. Most medical MRI, by contrast, is performed on samples with a large concentration of water that is relatively mobile ('weakly bound'), and this simplifies greatly the theoretical treatment of the relaxation phenomena. Many of the relevant couplings in the quantum-mechanical models are considerably reduced and may disappear altogether to a first approximation. In this context, 'lattice' implies a 'thermal bath', of well-defined temperature, to which each of the spin systems is independently coupled.

Spin-lattice relaxation describes the transfer of energy to or from the spin system (by exchanging it with the lattice), resulting in changes to the longitudinal or z component of magnetisation. For spin 1/2 nuclei, fluctuating magnetic fields that can interact with the nuclear dipole moment provide the required stimulus. For spins having an electric quadrupole moment ($I > 1/2$), electric-field gradients must also be considered. The major mechanism, in the absence of paramagnetic nuclei with unpaired electrons, is the magnetic field generated by the dipole moments of neighbouring molecules such as water (see Figure 7.5). The magnetic fields due to these dipole moments will have components in the xy plane, some of which will vibrate at ω_0 and can therefore stimulate emission or absorption of energy. The de-excitation of spin states results in a reduction in the spin temperature. The rate at which energy is lost indicates how closely the spins are coupled to the lattice or how tightly the spins are bound in the liquid.

A mathematical description of relaxation was first proposed by Bloembergen et al. (1948). Given its somewhat extensive nature, only the briefest of summaries will be presented here. For further details, see Abragam (1983, Ch. VIII), Ernst et al. (1987, p. 51f), McConnell (1987, p. 37ff) and Goldman (1991, Ch. 9) – not for the faint-hearted! Chapters 15 and 16 of Levitt (2001) give an excellent review of the physical principles, together with clear diagrams, but without going into the quantum mechanics in depth.

In characterising the random molecular tumbling and translation that gives rise to relaxation, a key parameter is the so-called *correlation time* τ_c, which describes the correlation of a particle's position and orientation at any instant, with those at time t later.

Abragam notes: 'a crude but convenient assumption that we will often make and in certain cases be able to justify, is that … the correlation function … can be represented by $\exp(-t/\tau_c)$'. A molecule undergoing rapid motion will have a short correlation time and for small molecules such as glucose in solution, $\tau_c \sim 10^{-12}$ to 10^{-10} s. Correlation times are discussed in more detail in James (1975). The temperature dependence of τ_c can often be described by a form of Arrhenius equation

$$\tau_c \propto \exp\left(\frac{-E_a}{kT_{\text{lattice}}}\right), \tag{7.45}$$

where

E_a is a relevant activation energy,
T_{lattice} is the lattice temperature, and
k is Boltzmann constant.

Often T_1 has a temperature dependence of this form.

The Fourier transform of the correlation function is known as the *spectral density function* $J(\omega)$. This describes the amplitude of the spectral component of the time-varying field oscillating at angular frequency ω. Crudely, we may say that longitudinal relaxation is caused by the transverse component of flux density arising from molecular motion at the Larmor frequency and write $1/T_1 \propto B_{xy}^2 J(\omega_0)$. However, a more careful treatment yields

$$\frac{1}{T_1} \propto \gamma^4 \hbar^2 I(I+1)[J(\omega_0) + 4J(2\omega_0)], \tag{7.46}$$

and the same analysis proves theoretically that T_1 relaxation (at least for homogeneous samples) is a single-exponential process, something that appears as an empirical observation in the Bloch equations. Note the dependence of the T_1 relaxation rate on fluctuations at twice the Larmor frequency, which might not intuitively be expected. Interestingly, spin-state changes can also be induced by causing vibrations at ω_0, for instance, by applying an ultrasound field. Given the form of the correlation function, it will readily be seen that the spectral density is a Lorentzian function (cf. Section 7.3.7):

$$J(\omega) \propto \frac{\tau_c}{1 + \omega^2 \tau_c^2}. \tag{7.47}$$

The presence of paramagnetic species, with unpaired electrons, and therefore with magnetic moments a factor of about 1000 greater than the nuclear magnetic moment, leads to enhanced relaxation rates. This effect is proportional to r^{-6}, where r is the distance between the nucleus and the paramagnetic ion. Relaxation effects due to paramagnetic agents are considered in detail by Dwek (1973) and described by the Solomon–Bloembergen equations (Solomon, 1955; Bloembergen, 1957). The use of paramagnetic spin-enhancement agents in imaging is discussed in Section 7.7.2.

7.5.2 Transverse Relaxation, T_2

Transverse relaxation (also known as T_2 or spin-spin relaxation) is the loss of magnetisation from the xy plane, as a result of a loss of phase coherence between the precessing magnetic moments. T_2 relaxation is the result of two types of process: indirect exchanges via the lattice (flips of more than one spin, involving no *net* transfer of energy to or from the lattice, but a loss of phase information) and small differences in precession frequency between spins. The randomly varying magnetic fields referred to in the previous section usually give rise to a *z* component in addition to the *xy* component already discussed. This has the effect of slightly altering the effective field B_0 (and hence Larmor frequency) experienced by the nucleus, and it is characterised by the $J(0)$ component of the spectral density function. The equivalent of Equation 7.46 for T_2 relaxation is

$$\frac{1}{T_2} \propto \gamma^4 \hbar^2 I(I+1)[3J(0)+5J(\omega_0)+2J(2\omega_0)]. \tag{7.48}$$

T_2 is always less than or equal to T_1, and, in tissue samples, T_2 relaxation is usually considerably faster than T_1 relaxation, as indicated by the extra $J(0)$ term in the previous equation. However, when $\omega_0\tau_c \ll 1$ (i.e. very short correlation times, often known as the 'extreme motional narrowing limit'), the T_2 relaxation time approaches T_1; in other words, the dephasing processes are not significant when compared with the lattice-mediated relaxation processes. This occurs for rapid motion of small molecules in a dilute solution, and, thus, water doped with the commonly used paramagnetic agent gadolinium often has T_2 values similar to the T_1 values of the solution. In order to mimic the relaxation behaviour of tissues, it is common to incorporate these solutions into agarose gels (Walker et al., 1989), which allows T_1 to be manipulated semi-independently from T_2.

7.5.3 T_2^*

Where single-exponential T_2 relaxation is the only significant process producing a loss of transverse magnetisation, we obtain the Lorentzian line shape of Equation 7.36, with linewidth

$$FWHM = \frac{2}{T_2}. \tag{7.49}$$

(The commonly encountered variant $FWHM = 1/\pi T_2$ applies if we use frequency on the horizontal axis rather than angular frequency.)

In practice, factors other than T_2 relaxation also cause a loss of transverse magnetisation, leading to a shorter decay time and hence a broader spectral line. In particular, inhomogeneities in the main magnetic field cause spins in different regions of the sample to have different Larmor frequencies. $\mathbf{B}_0$ inhomogeneities have two main causes:

- imperfections in the field produced by the main magnet. The development of high-quality superconducting magnets during the 1980s was a key advance.
- variations in the magnetic susceptibility of the sample. These may be on a large scale (e.g. the air in the nasal sinuses) or microscopic (e.g. blood of different oxygenation levels in capillaries). The resulting changes in precession frequency and additional signal decay are either problematic (e.g. giving rise to distortion in echo-planar images of regions around the sinuses—see Section 7.8.3) or a useful source of contrast (e.g. in functional MRI (fMRI), Section 7.7.5).

Notionally, we can divide our sample into 'packets' or *isochromats* that contain a large enough number of spins not to be subject to quantum fluctuations (and hence obey the macroscopic Bloch equations), yet a small enough number to be highly spatially localised and hence associated with a single precession frequency. The net transverse magnetisation is given by the vector sum of all the isochromats, which individually decay according to T_2. As time passes in an inhomogeneous field, the isochromats dephase, and the net magnetisation decreases faster than T_2 – see Section 7.4.3.

There is no particular reason for the inhomogeneities in $\mathbf{B}_0$ to be distributed in such a way as to give an exponential decrease of signal with time. Moreover, many real FIDs (particularly before shimming – see Section 7.10.1) are manifestly nonexponential. Despite this, it still proves convenient to associate a characteristic decay time (T_2') with the decay caused by field inhomogeneities, according to

$$\frac{1}{T_2^*} = \frac{1}{T_2} + \frac{1}{T_2'}, \tag{7.50}$$

where $\exp(-T_2^*)$ describes the overall decay. Note that the often-repeated formula

$$\frac{1}{T_2^*} = \frac{1}{T_2} + \gamma\Delta B_0 \tag{7.51}$$

is the result of: (a) *assuming* the special case of a Lorentzian distribution for the number density $n(\omega)$ of isochromats having precession angular frequency ω, such that the signal is exponential with time constant T_2'

$$n(\omega) \propto \frac{1/T_2'}{1/T_2'^2 + (\omega - \omega_0)^2} \xleftrightarrow{\text{FT}} S(t) \propto \exp\left(\frac{-|t|}{T_2'}\right); \tag{7.52}$$

then (b) noting that $\omega - \omega_0 = -\gamma(B - B_0)$; and finally (c) deciding arbitrarily that an appropriate measure of the magnetic field inhomogeneity is $\Delta B_0 = B_{HH} - B_0$, the field offset at which the Lorentzian $n(\omega)$ curve falls to half its height.

7.5.4 $T_{1\rho}$ and Spin Locking

A further relaxation time encountered occasionally in the literature is the *longitudinal relaxation time in the rotating frame*, or $T_{1\rho}$. Consider the pulse sequence $90°_x$ – SL_y, in which SL_y is a pulse (often of relatively long duration) along the y' axis of the rotating frame, known as a *spin-locking* pulse. As normal, the $90°_x$ flips magnetisation from the z' to the y' axis. Suppose now that we have a distribution of spin precession frequencies that would normally give rise to T_2' relaxation (or alternatively a number of species with different chemical shifts (Section 7.9.2)). Consider the *effective* magnetic field in the rotating frame. Provided $|\omega_1| \gg |\gamma\Delta B_0|$, this field still points close to the x' axis, and precession occurs around this field, rather than about the z' axis. Hence, the magnetisation is locked in place pointing closely along y', rather than dephasing.

Under these conditions, it can be shown that, instead of relaxation with time constant T_2, as in Equation 7.26, we get

$$\frac{\partial M_{y'}}{\partial t} = -\frac{M_{y'}}{T_{1\rho}},$$

where $T_{1\rho}$ is related to but separate from T_1. For further details, see Kimmich (1997, Ch. 10).

7.5.5 Information Available from Relaxation Times

Given the microscopic origins of the relaxation phenomena, NMR relaxation parameters are extremely sensitive indicators of the *physico-chemical micro-environment* in which the nuclei of interest are situated. This environment comprises many elements, which include the following:

- bonding, motional hindrance and chemical exchange on a molecular length scale and, particularly, in the context of medical imaging, the degree to which water is bound or adsorbed onto surfaces in a tissue;
- variable surface relaxation parameters on the micron length scale of porous media;
- local variations in magnetic field, due to susceptibility differences in the sample (both gross morphology and tissue microstructure) and static field inhomogeneities;
- physical quantities such as temperature, viscosity, chemical concentration, oxygen tension and radiation dose, which may vary on macroscopic length scales.

For the scientist, the broad range of phenomena potentially accessible via measurement of relaxation times is simultaneously a great strength of relaxometry and its biggest weakness: an advantage because the number of questions that may be addressed is huge, but a disadvantage because many of the factors are inextricably intertwined.

7.5.6 Measurement of Longitudinal Relaxation Times

The evolution of longitudinal magnetisation after a saturation or inversion pulse is described in Section 7.4.2. Adding a 90° pulse to the end of either of these sequences gives an FID that allows the partially relaxed longitudinal magnetisation to be sampled. By recording the initial amplitude of the FID for a number of different values of t in Equation 7.40 or T_I in Equation 7.42, one can plot a *magnetisation recovery curve*. This is preferably analysed using a non-linear Levenberg–Marquardt fit – see Press et al. (2002, Ch. 15) – to the function

$$\hat{M}_z(t) = \hat{M}_0\left[1-\hat{\beta}\exp\left(\frac{-t}{\hat{T}_1}\right)\right], \tag{7.53}$$

where $\hat{M}_0$, $\hat{\beta}$ and $\hat{T}_1$ are adjustable model parameters. The reason for estimating $\hat{\beta}$ from the data rather than inserting 1 for SR and 2 for IR is that, with non-homogeneous probes (Section 7.10.1) or selective excitations that have non-square slice profiles (Section 7.6.3), one cannot guarantee that a flip angle will be exactly 90° or 180°.

If one wishes to obtain longitudinal relaxation time data in an imaging context (i.e. an individual T_1 value for each voxel), then one replaces the final 90° with an appropriate imaging module. This will result in a *stack* of images, one for each delay value. The fit in Equation 7.53 is then repeated for each voxel in turn, leading to a *parametric image* for each of $\hat{M}_0$, $\hat{\beta}$ and $\hat{T}_1$. Notice that the imaging module might be either single-shot, in which case the SR or IR module is used only once, or multishot. In the latter case, the T_1 contrast must be prepared by a fresh application of the SR or IR module before each repetition of the imaging module. The speed of the overall T_1 mapping experiment will clearly depend on the number of repetitions of the underlying imaging module required to acquire the image data—see Section 7.6.13.4. In all sequential measurements, there is obviously scope for errors to arise due to electronic drift in receiver gain and probe tuning (the latter leading to both incorrect values of pulse amplitude and reception efficiency). However, these issues have largely been overcome on modern clinical systems.

Many studies have been performed concerning the most efficient method of measuring T_1. If only a *single* relaxation time is being measured, then it can be shown (via so-called Cramér–Rao lower bound theory) that the precision of the parameter estimates is optimised by choosing just two values of t, one where the magnetisation is fully relaxed and one shorter time that can be obtained via the theory (Spandonis et al., 2004). However, it is important to note that, although T_1 is defined only for exponential processes, the relaxation for any given bulk sample or single voxel may *not* be accurately described by a single exponential function. This will be evident only if the recovery of the longitudinal magnetisation is measured at a number of different times. Multi-exponential behaviour occurs typically because of microscopic variations in composition and is often described, for biological systems, in terms of different tissue 'compartments'. Porous media also generally have complex relaxation time distributions, because of the effect of surface relaxation (Brownstein and Tarr, 1979).

Both SR and IR have advantages for different reasons. The dynamic range of the IR experiment is double that of SR, leading to more precise results for a given set of time points. However, it is necessary to wait for the magnetisation to recover fully between successive time points, thus extending the acquisition time. This is an important consideration, both for samples with intrinsically long T_1s and for imaging studies where the IR module must be repeated for multiple phase-encoding steps. By contrast, an SR acquisition requires no recovery delay between time points (providing either T_2 is short or an adequate means exists for spoiling transverse magnetisation). In practice, neither IR nor SR is the optimal solution, with a better performance being obtained by the so-called *fast IR* module $(180° - \tau_i - 90° - W)_n$, where W is a waiting time less than $5T_1$ (Doran et al., 1992).

An alternative to these methods is to make use of low flip-angle pulses, which are an intrinsic part of many rapid imaging sequences (Section 7.6.13.1). It is well known that, provided there is adequate spoiling of transverse magnetisation between pulses, the steady-state response to a train of pulses of flip angle θ is

$$S = S_0 \sin\theta \frac{1-\exp(-T_R/T_1)}{1-\cos\theta\cdot\exp(-T_R/T_1)}, \tag{7.54}$$

where S_0 is the maximum signal, obtained from a single 90° pulse. Figure 7.13 plots this expression for different values of the ratio T_R/T_1 and flip angle. If one acquires two images with known θ and T_R, then their ratio may be used to estimate T_1, with the unknown S_0 cancelling out. For small values of θ, $S \propto \theta$, thus allowing very simple calculation, whilst for the remainder of the monotonically increasing region, it is sufficient to generate a look-up table of ratios and T_1s.

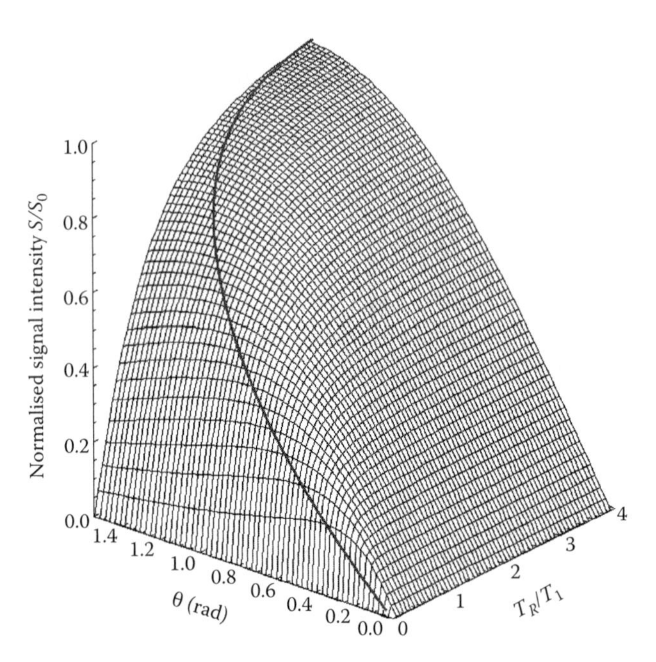

FIGURE 7.13
Surface plot of the theoretical expression for the signal intensity from a GE sequence, normalised to the signal from a single 90° pulse, S_0. The plot shows that the signal is always increased by making T_R longer, although this effect is insignificant for T_R greater than about $3T_1$. For any given value of the ratio T_R/T_1, setting the flip angle to the Ernst angle, $\theta_E = \cos^{-1}(-T_R/T_1)$, provides the maximum signal, as illustrated by the curved line, which tracks the maximum value on the θ axis for all T_R/T_1.

Wang et al. (1987) described a method for extracting the T_1 value analytically from two or more such experiments. Equation 7.54 may be rewritten as

$$\frac{S}{\sin\theta} = \exp\left(\frac{-T_R}{T_1}\right)\cdot\frac{S}{\tan\theta} + S_0\left[1 - \exp\left(\frac{-T_R}{T_1}\right)\right], \tag{7.55}$$

which is of the form $y = mx + c$, thus allowing T_1 to be extracted from the gradient m. This method must, however, be used with some caution, as inhomogeneities in the $\boldsymbol{B}_1$ field produced by the probe can cause errors.

7.5.7 Measurement of Transverse Relaxation Times and Diffusion

As the envelope of the FID decays with T_2^* rather than T_2, the T_2 relaxation time cannot be measured from the shape of the FID or SE. The method usually adopted is to use an SE sequence that gives rise to multiple echoes. During formation of the SE by rephasing the transverse magnetism, the effect of field inhomogeneities is removed, providing that protons do not move into regions of different field strength during the course of the acquisition. Figure 7.11 showed a simple 90° – τ – 180° SE sequence. In practice, diffusion of the spins, first investigated by Hahn (1950) and Carr and Purcell (1954), also occurs, causing the nuclei to experience different magnetic fields in an inhomogeneous field (often in the form of a field gradient G). The expression for the amplitude of the SE at time $T_E = 2\tau$, therefore, contains a term to describe the effects of diffusion,

$$M_{y'}(T_E) = -M_{y'}(0)\exp\left(-\frac{T_E}{T_2} - \frac{\gamma^2 G^2 D T_E^3}{12}\right), \tag{7.56}$$

where D is the self-diffusion coefficient of the spins, which for proton imaging will be predominantly water molecules. It will be evident that the second term in the exponential function becomes rapidly more important as T_E increases.

To obtain T_2 and $M_{y'}(0)$, one must carry out a number of measurements using different values of T_E. However, the dependence on T_E^3 as well as T_E complicates the analysis of such experiments consisting of multiple Hahn echoes, and there is also a time penalty in having to repeat the sequence, particularly in an imaging context.

These problems were overcome by the development of the Carr–Purcell (CP) SE train (Carr and Purcell, 1954), which, by repeating 180° pulses at intervals of T_E, generates a train of echoes of alternating polarities – see Figure 7.14. In this sequence, the 90° and 180° pulses are all applied along the x' axis, and the sequence can be described as $90°_x - T_E/2 - (180°_x - T_E)_n$, where the y' magnetisation for the nth echo is ideally described by

$$M_{y'}(nT_E) = (-1)^n M_{y'}(0)\exp\left(\frac{-nT_E}{T_2} - \gamma^2 G^2 D n T_E^3\right). \tag{7.57}$$

Notice the reduction by a factor of n^2 in the value of the second term in the exponential, compared with Equation 7.56. This reduces significantly the sensitivity of the CP sequence to diffusion. A curve can be plotted for the many echo values obtained, allowing a T_2 measurement to be carried out within a single pulse train, in a time of order $5T_2$. However, the success of this method is critically dependent on an accurate 180° pulse length and amplitude. As the $180°_x$ pulse is always applied in the same sense about the x' axis, it will cause a *cumulative* error in the projection of M onto the xy plane if the rotation angle is incorrect.

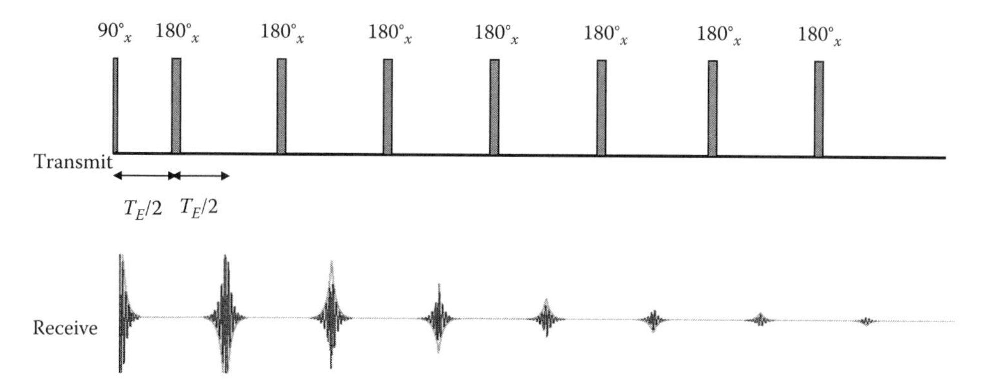

FIGURE 7.14
The CP sequence used for obtaining multiple T_2-weighted echoes: After an initial 90° pulse to excite transverse magnetisation, a regular train of refocusing 180° pulses, separated by time T_E, is transmitted. A simulated signal, typical of what might be received when the spectrometer is slightly off resonance, is shown in black. The grey line is the ideal (demodulated and phased) signal, corresponding to $M_{y'}$. Note the alternating polarity. In the CPMG modification, the 180° pulses are along the y' direction of the rotating frame of reference and all the echoes have the same polarity.

The method is therefore sensitive to inhomogeneities in $\boldsymbol{B}_1$ and often results in a measured value of T_2 that is too short.

The sequence was modified by Meiboom and Gill (1958) to give the CPMG sequence in which all of the 180° pulses are shifted 90° in phase with respect to the initial 90° pulse. In other words, they are oriented along the $\pm y'$ axis. This results in all of the echoes being formed along y' regardless of the exact tip angle thus avoiding cumulative errors – cf. the discussion on spin locking in Section 7.5.4.

The simplest way of calculating T_2 on a pixel-by-pixel basis is to acquire two images with equal T_R but differing T_E (see Section 7.6.11), either by repeating a standard SE sequence or by using a double-echo sequence. However, this has limited accuracy because of the diffusion problem mentioned earlier, whilst the corresponding multi-echo CPMG imaging sequence can be obtained in a practicable clinical acquisition time and is the preferred option. The resulting stack of images is preferably analysed via a non-linear fit, in much the same way as described earlier for T_1. Important points to bear in mind when fitting the data are as follows:

- The imaging signal-to-noise ratio (SNR) is generally lower than for bulk samples and so one must be careful not to include late-echo data from rapidly relaxing regions that simply introduce additional noise into the fit.
- Static magnetic inhomogeneities and imperfections in the 180° pulses (particularly non-rectangular slice profiles) can cause systematic errors – see Majumdar et al. (1986a, 1986b).

The apparent disadvantage of a sensitivity to diffusion in Equation 7.56 turns out to be of great importance, because it can be used to calculate the diffusion coefficient of the spin species under study. This is discussed further in Section 7.7.4.

7.6 Nuclear Magnetic Resonance Image Acquisition and Reconstruction

7.6.1 Goal of an MRI Acquisition

The goal of an MRI experiment is to obtain information about the spatial distribution of the nucleus under study. Mathematically speaking, we want to evaluate the *proton density* function $\rho(\mathbf{r}) = \rho(x, y, z)$. (In later sections, we will modify this goal to include measurement of relaxation, diffusion and other contrast generating mechanisms in addition to $\rho(\mathbf{r})$). Note that x, y and z are coordinates referred to the isocentre of the magnet (i.e. the point at which the fields generated by the gradient coils are all zero). These should be distinguished from the (x, y, z) and (x', y', z') of previous sections, which referred to axes centred on an individual nucleus.

By far and away, the most commonly studied nucleus is ^{1}H, present in the water and fat that make up the majority of body tissues. Imaging of other nuclei, such as ^{19}F, is possible, but many species that might potentially be of interest are present in concentrations that are too low to provide images with adequate SNR.

7.6.2 Magnetic Field Gradients

We saw in Section 7.3.3 that the resonant frequency of protons is proportional to the applied magnetic field. The key idea in MRI is that if one makes *B vary with position, then different locations in the sample will give signals that can be distinguished from one another by their frequencies.* One could make B vary according to some arbitrary function, but in almost all medically related MRI, we are interested only in linearly varying magnetic fields, that is, uniform *field gradients.*

Thus, if a small magnetic field gradient along the x axis is added to the main (static) field $\boldsymbol{B}_0$, the resonant frequency of protons will change with x displacement according to

$$\omega(x) = -\gamma B_z(x) = -\gamma(B_0 + xG_x). \tag{7.58}$$

Notice here that the direction of the magnetic field causing the precession always lies along z, but the direction of *change* of that field is along x. We may equally consider a gradient along either the y or z axis, or, indeed, along any arbitrary direction. In the general case, one creates the gradient by passing currents of appropriate amplitudes through all three physical gradient coils – see Section 7.10.1 – and the result is expressed as a vector, with the variation in angular frequency given by

$$\omega(\boldsymbol{r}) = -\gamma B_z(\boldsymbol{r}) = -\gamma(B_0 + \boldsymbol{r}\cdot\boldsymbol{G}). \tag{7.59}$$

We will find it convenient during the rest of this section to work with the *demodulated* angular frequency (Section 7.3.7), and, hence, we will *redefine* ω. Arranging our demodulation so that ω is zero at the origin of the coordinate system (magnet isocentre) corresponds simply to the case

$$\omega(\boldsymbol{r}) := \omega(\boldsymbol{r}) - (-\gamma B_0) = -\gamma\boldsymbol{r}\cdot\boldsymbol{G}. \tag{7.60}$$

By changing the direction of the gradient, we can look at the distribution of spins in different directions. *Magnetic field gradients map space onto frequency* (Figure 7.15). This is known as *frequency encoding.* Any sample other than a single point will give rise to an RF signal containing a number of different frequencies, that is it will possess a frequency spectrum and the amplitude at each point of the spectrum will be related to the number of protons at a particular position. Thus, in the same way that raw data from a high-resolution chemical acquisition are Fourier transformed to provide an NMR spectrum, so data acquired in

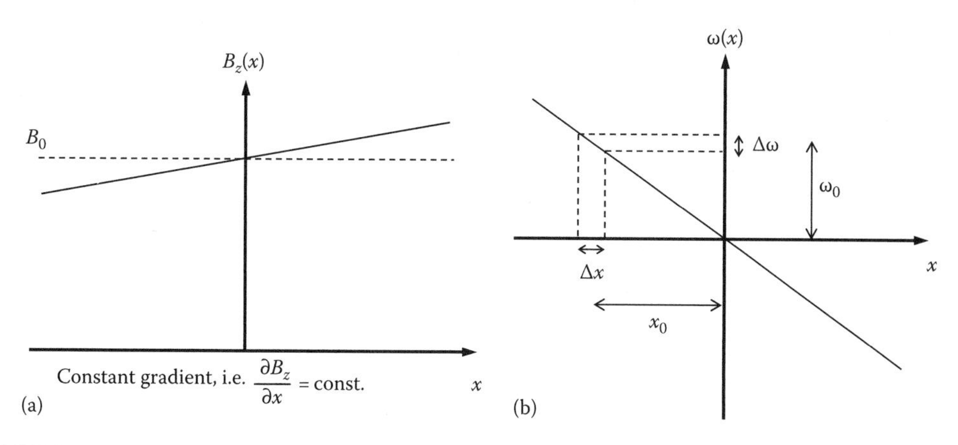

FIGURE 7.15
By creating a linear variation (constant gradient) in magnetic field across a sample, as shown in (a), different spatial positions along a given axis (e.g. x) are mapped onto different angular frequencies of NMR signal (b), via the relation $\omega = -\gamma B$. If the sample is excited by an RF pulse consisting of a range of angular frequencies $\Delta\omega$, then a slice perpendicular to the gradient axis is selected with a range of coordinates Δx, with the slice position x_0 being determined by the transmitter frequency offset ω_0.

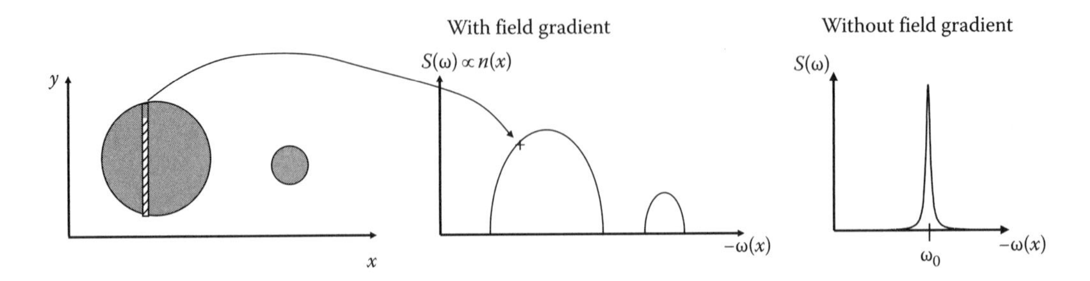

FIGURE 7.16
Spatial discrimination along a single direction in MRI is analogous to the creation of a chemical spectrum in which the angular frequency axis is a function of position. The magnitude of the spectrum is proportional to the number of spins $n(\omega(x))$ at a particular x position. This implies an integration (or 'projection') along the y and z (not shown) directions, and this type of 1D representation is often known as a *projection*. Note the minus sign on the horizontal axis of the right-hand graph, which comes about from the relation $\omega = -\gamma B$. Without a magnetic field gradient, all the spins would resonate with angular frequencies close to $\omega_0 = -\gamma B_0$ (any spread being due to magnetic field inhomogeneities), as indicated in the right-hand graph.

the presence of a gradient can be converted by FT into a 'positional spectrum' or 1D image. This concept is formalised in Section 7.6.4 and illustrated in Figure 7.16.

7.6.3 Selective Excitation

It is often convenient to assume that RF pulses are infinitely short or 'hard'. The Fourier transform of such a delta-function pulse is a constant, indicating that spins at all precession frequencies are excited equally, that is, covering all B_0 offsets. (In practice, the lower limit of pulse length on a liquid state machine might be of order 1 μs, leading to excitation over a bandwidth of order 1 MHz.) By contrast, if a sample is irradiated with a longer RF pulse composed of a narrow band of frequencies – often known as a *selective* or 'soft' pulse – then only spins in a slice resonating at frequencies within that band are excited. This is particularly important in imaging: if one applies the pulse in conjunction with a gradient, then a 'slice' of the sample, perpendicular to the gradient axis, is excited or 'selected' – see Figure 7.15. Until the advent of ultra-high-speed CT scanners allowing multi-planar reformatting of 3D data sets, the ability of MRI to obtain images with arbitrary slice orientation was a major advantage of the modality. The ability to align the imaging plane with a particular feature of interest is still of great importance in clinical MRI.

Without loss of generality, let the slice selection direction be z. For *small flip angles*, the Bloch equations form a linear system, and, in this approximation, the angle through which the magnetisation at a particular z coordinate precesses is proportional to the component of the pulse with angular frequency $\omega(z) = -\gamma z G_z$. It can be shown that, for a pulse applied during the time period $[-t_p/2, t_p/2]$, the transverse magnetisation generated is

$$M_+[\omega(z)] = i\gamma M_0 \exp\left[\frac{i\omega(z)t_p}{2}\right] \int_{-t_p/2}^{t_p/2} B_1(t)\exp[-i\omega(z)t]dt. \tag{7.61}$$

The profile of the slice, thus, depends upon the Fourier transform of the RF pulse. With our convention that ω is the demodulated angular frequency, note that the correct interpretation of B_1 in Equation 7.61 is that it represents the variation of the RF field in the rotating frame, that is, the *envelope* of the RF pulse as seen in the lab frame. By varying the angular frequency offset of the transmitted RF pulse, the position of the slice may be moved along z – a straightforward application of the Fourier shift theorem – whilst varying the pulse length or the gradient strength allows the slice thickness to be changed. Typical pulse shapes used are based on sinc or Gaussian envelopes, whose Fourier transforms have attractive properties. This FT approximation is only a starting point, however. For larger flip angles, the Bloch equations become non-linear, and the design of pulses to excite arbitrarily specified profiles of transverse magnetisation is a non-trivial exercise.

The degree of saturation occurring, a function of T_1, also plays a key role – see the discussion of steady-state NMR in Section 7.3.7 – as do chemical-shift differences between different regions in the sample (e.g. fat-water) and the effects of $\boldsymbol{B}_1$ fields that are not uniform throughout the volume of the probe. The larger the number of pulses that are applied to a sample, the more important it becomes to understand and control the slice profile. Multiple SE experiments can be very susceptible to flip angles that vary throughout the imaging slice – see Section 7.4.4 and Majumdar et al. (1986a). Since the final image intensity is related to the integral over the whole slice, these problems can be difficult to diagnose and correct.

Notice in Equation 7.61 that the transverse magnetisation generated acquires a position-dependent phase $\exp[i\omega(z)t_p/2]$. The need to rephase this contribution explains why MRI sequences have a 'negative gradient lobe' in the slice-selection direction after the end of the pulse application – see Figure 7.17.

Chapter 3 of Callaghan (1991) and Chapter 16 of Haacke et al. (1999) have excellent descriptions of selective excitation.

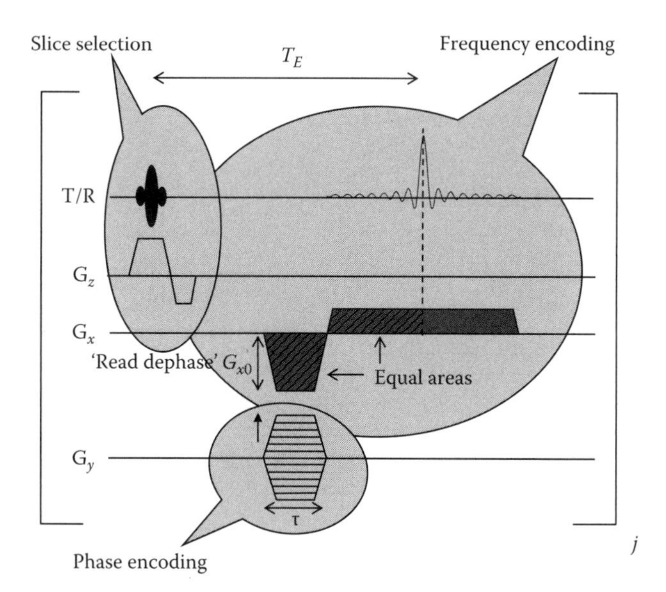

FIGURE 7.17
Schematic diagram of a GE sequence, introducing symbols used in NMR pulse sequence diagrams. The four long horizontal lines labelled T/R (transmit/receive), G_z, G_x and G_y (gradients) indicate the different output channels of the NMR spectrometer. The shapes on these axes indicate a time-ordered (but not, generally, to scale) sequence of 'pulses' on these channels. Trapezoidal shapes indicate gradient pulses for which the 'ramp-up' and 'ramp-down' of the gradient coils are not instantaneous. The first symbol on the T/R line represents a frequency-selective RF pulse, and the grey line shaped like a sinc function is the acquisition of an echo signal. The horizontal lines on the double trapezoid of the phase-encoding gradient represent a gradient whose amplitude changes stepwise from negative to positive on each successive iteration of the sequence, indexed by j.

7.6.4 Introduction to Common Imaging Pulse Sequences

In the following sections, we discuss in detail the physics and mathematics underlying MR image formation and review systematically the major classes of imaging pulse sequence, together with more technical aspects of the subject. Before that, we present an overview of a number of methods commonly encountered in the clinic.

The simplest type of image acquisition uses the *gradient-echo* (GE) pulse sequence of Figure 7.17 – see also Sections 7.6.6, 7.6.7 and 7.6.13.1. In 2D or multislice (Section 7.6.13.5) measurements, a selective RF pulse excites magnetisation in a single slice along the z direction. A so-called *read gradient* provides frequency encoding (Section 7.6.2), and, thus, the frequency spectrum of the acquired echo corresponds to the distribution of proton spins in the chosen slice along the x direction. This excitation-echo sequence is repeated a large number of times, and, at each repetition, a *phase-encoding gradient*, orthogonal to both the read and slice directions, is stepped to a different amplitude. This phase-encoding procedure, an explanation of whose operation will be deferred to Section 7.6.7, provides spatial localisation along y. GE pulse sequences provide T_1- and T_2^*-weighted image contrast, which can be varied in a flexible manner by altering the repetition time T_R, the echo time T_E and the RF flip angle θ – see Section 7.6.11. The time taken to acquire a GE image depends on the desired contrast, resolution and SNR, but is typically in the region of 1 s to 5 min for a single slice. It is also common practice to obtain 'true-3D' data sets (Section 7.6.13.5) using GE sequences, and, on modern equipment, a high-resolution 3D volume may be acquired in as little as a few minutes.

A number of applications, such as functional imaging (Section 7.7.5), require T_2^*-weighted images to be obtained more quickly than is possible using simple GE sequences. By rapidly reversing the read gradient, as illustrated in the *EPI* sequence of Figure 7.18, a long train of GEs is formed. Each of these echoes is phase encoded differently by an appropriate

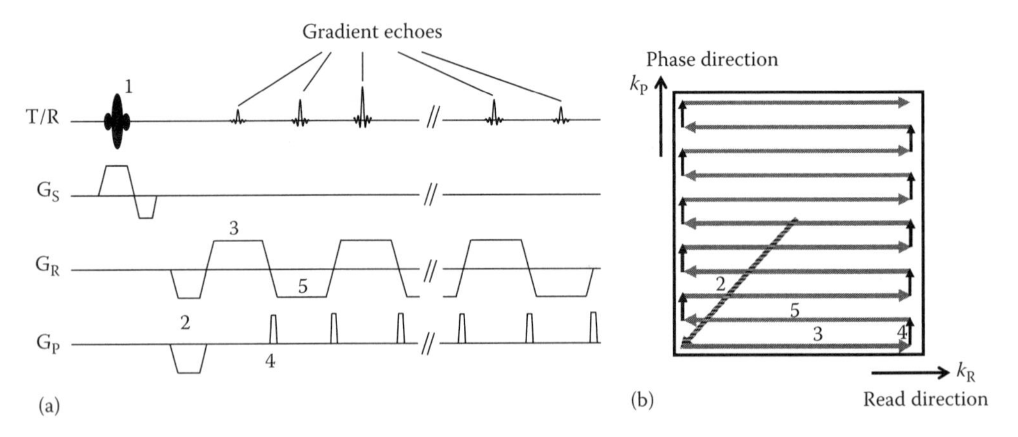

FIGURE 7.18
(a) Schematic diagram of an EPI sequence and (b) corresponding path traversed in k-space by the magnetisation. Magnetisation is excited into the transverse plane by slice-selective pulse **1**. Initial dephasing lobes **2** move the position in k-space to its extreme negative value in both phase and read directions. Signal read-out occurs during a positive gradient lobe **3**, as the position in k-space moves to the right. Between successive read-out periods, a phase-encoding blip **4** moves the k-space position up one line, before a reverse read-out occurs under the influence of negative read-gradient lobe **5**. For a full explanation of the various features of the sequence, see Section 7.6.13.2 of the main text. Note that (in contrast to Figure 7.17) we use the notation G_S, G_R and G_P (the so-called 'logical' gradients in the slice, read and phase directions) instead of G_z, G_x and G_y (the 'physical' gradients). This emphasises the fact that it is the *function* rather than spatial direction that is of primary importance. In practice, MRI is often used with oblique slicing, which means that each of G_S, G_R and G_P is physically applied by currents through two or more of the physical gradient coils.

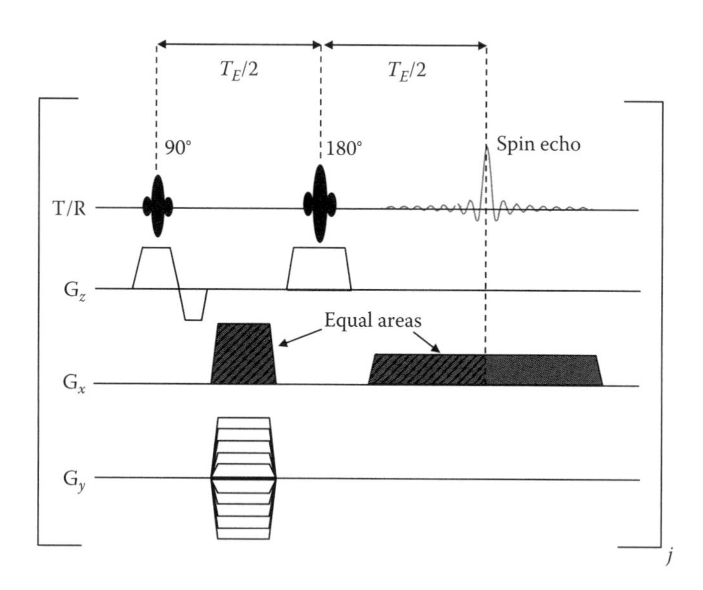

FIGURE 7.19
Schematic diagram of an SE pulse sequence. The sequence functions similarly to that of Figure 7.17, but with the insertion of a frequency-selective 180° pulse, which causes inhomogeneities in the $\boldsymbol{B}_0$ field to be refocused. Note the reversal in direction of the read-dephase gradient that this entails.

manipulation of the phase-encoding gradient, and this allows complete 2D images to be acquired in as little as a few tens of ms. Rapid imaging using EPI is described in more detail in Section 7.6.13.2.

Where T_2, rather than T_2^*, contrast is required, *spin*-echo, rather than gradient-echo, sequences are used (Figure 7.19). As in GE methods, transverse magnetisation is generated with an initial selective RF pulse. However, instead of acquiring the GE immediately, a second RF pulse is applied to refocus isochromats that have dephased under the influence of magnetic field inhomogeneities (Section 7.4.3). SE sequences tend to use longer echo times (up to around 100 ms) than GEs, and, consequently, the overall acquisition time is increased.

The RARE or fast SE (FSE) sequence combines the attraction of T_2 weighting with a rapid scan time and is illustrated in Figure 7.20. This sequence is the imaging equivalent of the CPMG sequence described earlier (Section 7.4.3): after an initial 90° pulse, a series of 180° pulses gives rise to an echo train. Each of these echoes is separately phase encoded. The sequence may be run either in a single-shot mode, in which case, enough echoes are acquired from one 90° excitation pulse to provide all the data for an image, or, more commonly, in an interleaved mode (Section 7.6.13.4). In the former case, image acquisition time is typically between 100 and 500 ms; images suffer from a degree of blurring in the phase-encoding direction due to T_2 signal decay during the course of the acquisition. In the interleaved version, a number of echoes (the echo-train length, ETL) are acquired at each 90° excitation and the whole sequence repeated, typically of order 8 times, to achieve the desired resolution in the phase-encoding direction. A certain degree of re-learning is required by radiologists to relate the T_2-weighted contrast seen with RARE sequences to that in traditional SE images.

A more complex form of image contrast, involving a mixed T_1 and T_2 weighting, is found using sequences based on the steady-state free precession (SSFP) principle – see Section 7.6.13.1. These are superficially very similar to the GE described earlier, but have a much higher SNR; transverse magnetisation used to create imaging echoes is not lost, but instead

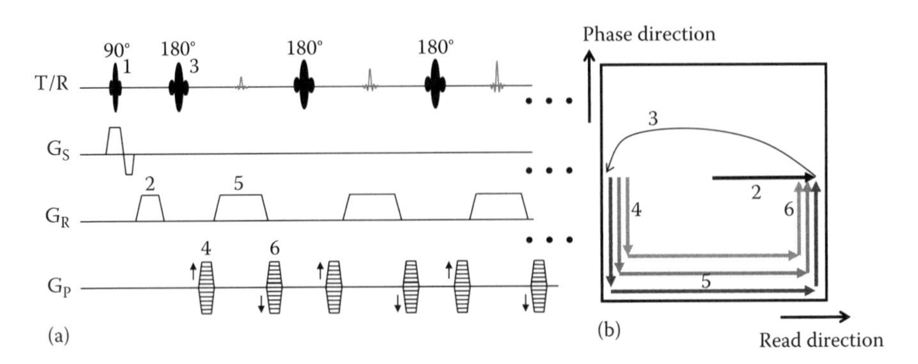

FIGURE 7.20
(a) Schematic diagram of the first three phase-encoding steps of a TSE sequence and (b) corresponding path traversed in *k*-space by the magnetisation. Magnetisation is excited into the transverse plane by slice-selective 90° pulse **1**. Initial read-dephasing lobes **2** move the position in *k*-space to its extreme positive value in the read direction, and the subsequent 180° pulse **3** flips this signal to the extreme negative value while simultaneously refocusing any dephasing caused by magnetic field inhomogeneities. A phase-encoding gradient **4** moves the position in *k*-space to the extreme negative in the phase-encoding direction. Signal read-out occurs during a positive gradient lobe **5**, as the position in *k*-space moves to the right. A so-called phase-encode rewind gradient **6** returns the magnetisation to its state immediately after **2**, ideally modified only by T_2 relaxation. A repetition of the previous steps with different values of the phase-gradient allows for acquisition of all the lines in *k*-space, either in a single shot or, more commonly, in an interleaved fashion. For a full explanation of the various features of the sequence, see Section 7.6.13.3 of the main text.

'recycled' onto the longitudinal axis. SSFP images are widely used in cardiac imaging, and a cine sequence, depicting heart motion throughout the cardiac cycle, can now be acquired in less than 10 s, well within a single breath-hold, even for most sick patients.

The sequences described earlier are all generally implemented as 2D Fourier methods (Section 7.6.6ff). However, we start our detailed explanation of imaging with the method used by Lauterbur to acquire the very first MR images (Lauterbur, 1973).

7.6.5 Projection Reconstruction

In projection reconstruction imaging, the first step is to select a slice. Again, without loss of generality, we will assume a 'transverse' or 'axial' slice is obtained by applying a z gradient. Having selected the plane of interest, the next task is to resolve x and y displacements within that plane. For an SR sequence, reading the FID out in the presence of an x gradient will relate each frequency present in the FID to a particular x displacement. Thus, the Fourier transform of the FID will be a projection of the transverse magnetisation onto the x axis, as shown in Figure 7.16:

$$\rho(x) = \int_{z_0-\frac{\Delta z}{2}}^{z_0+\frac{\Delta z}{2}} \int_{-\infty}^{+\infty} \rho(x,y,z)dy\, dz. \tag{7.62}$$

This projection is conceptually similar to those obtained in transmission CT imaging (Chapter 3) or in SPECT (Section 5.8). If the x and y gradients are combined during read-out of the signal, projections at other angles can be obtained – see Section 7.6.2. A projection

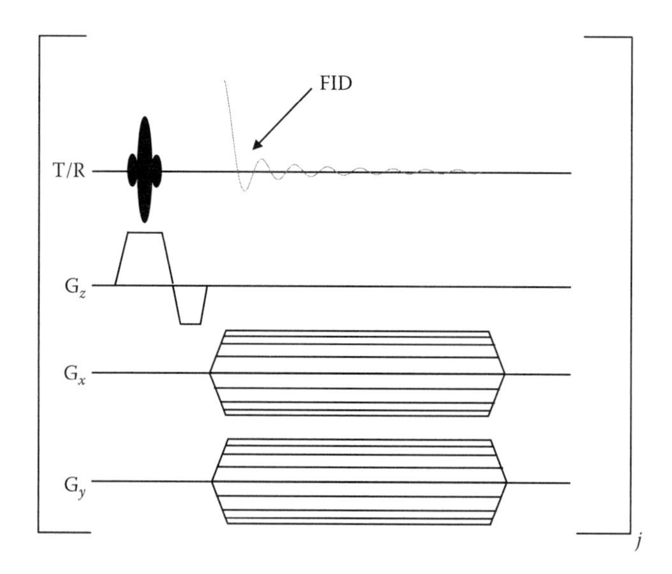

FIGURE 7.21
Pulse sequence diagram for the projection reconstruction MRI acquisition sequence. The unevenly spaced horizontal lines for the G_x and G_y phase-encoding gradients represent gradient values that are stepped sinusoidally.

at angle ϕ is represented mathematically by an analogue of Equation 7.62 in which x and y are replaced by

$$x_\phi = x\cos\phi + y\sin\phi$$
$$y_\phi = -x\sin\phi + y\cos\phi. \tag{7.63}$$

By cyclically repeating the sequence of selective excitations of the plane of interest in the presence of a z gradient and then reading out the FID in the presence of a combination of x and y gradients that are changed for each measurement, a series of projections of the plane can be built up, allowing an image to be reconstructed by convolution and back-projection (see Section 3.6). A schematic diagram of the pulse sequence that achieves this task is shown in Figure 7.21. Alternatively, the FIDS can be directly projected into Fourier space. The 2D data set composed of all of the FIDs is multiplied by a suitable filter, before a 2D discrete Fourier transform (DFT) is carried out to obtain the real-space image. Often, a *gridding* procedure will be used to place the data onto a regular array in Fourier space, suitable for a fast Fourier transform (FFT). Note that, from a signal-processing point of view, projection information is oversampled at the centre of Fourier space.

7.6.6 2D Fourier Imaging

Fourier zeugmatography was introduced by Kumar et al. (1975). Their original pulse sequence is shown in Figure 7.22. Instead of obtaining projections of the spin density at different angles as with projection imaging, the read-out gradient is held constant, and the other two dimensions are obtained by *phase encoding*. The FID is allowed to evolve under the influence first of a z gradient, then of a y gradient applied for variable periods t_z and t_y. These gradients introduce phase shifts that are functions of z and y displacements, due

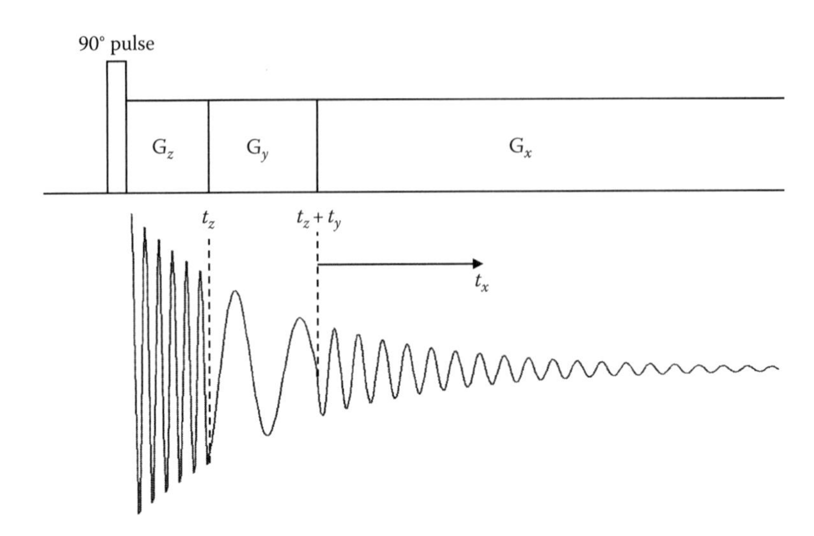

FIGURE 7.22
Method of 3D phase-encoding originally proposed by Kumar et al. (1975). As noted in the main text, this proposal has a number of inherent disadvantages and was superseded by the spin-warp method of Edelstein et al. (1980). (Redrawn from Kumar et al., 1975, figure 1.)

to the linear change in precession frequency. After a period equal to the maximum phase-encoding period, an x gradient is applied, so that the remainder of the FID is frequency encoded in the x direction, and digitisation is performed during this period. A set of FIDS is obtained and stored in a 'time space' (often called Fourier space or k-space – see Section 7.6.8) in ascending order of phase-encoding period. A 3D Fourier transform is carried out to produce the frequency and therefore the spatial distribution in the x, y and z directions.

This original concept is associated with several problems. Firstly, it is not possible to choose negative times for the durations of the gradients. Sampling the negative portions of k-space is crucial to obtain a complete image data set. Furthermore, changing the length of the phase-encoding periods t_z and t_y means that each acquired FID is differently weighted by T_2. This introduces a filter in the time domain and is equivalent to convolving the real-space image with a Lorentzian function. When the 3D sequence was first introduced, the length of acquisition was prohibitive, as were the data storage and processing requirements.

These problems are all overcome by a technique known as spin-warp imaging (Edelstein et al., 1980), which is similar to the GE sequence of Figure 7.17. In this case, a slice-selective pulse is introduced along z, as in the projection reconstruction technique mentioned earlier, such that only one phase-encoding gradient is required. Different phases are obtained by changing the *amplitude* of the y gradient, rather than by changing its *duration*. The phase-encoding period remains constant, leading to a fixed T_2 decay, whilst negative gradient amplitudes (obtained by reversing the current direction through the coil) allow easy access to the negative portions of k-space.

7.6.7 Mathematical Description of 2D Fourier Imaging

In this section, we explain how a simple GE imaging sequence like that in Figure 7.17 gives rise to a signal that is the 2D Fourier transform of the proton density.

Immediately after the selective RF pulse, there are gradients in both *read* (x) and *phase-encode* (y) directions. This means that the spins have a precession angular frequency that depends on both their x and y coordinates:

$$\omega(y) = -\gamma(xG_{x0} + yG_y), \tag{7.64}$$

where G_{x0} is defined in Figure 7.17. Thus, during the period before the acquisition starts, the spins acquire a *phase* which depends on both their x and y positions:

$$\phi = \phi_{\text{read-dephase}}(x) + \phi_{\text{phase-encode}}(y) = -\gamma x G_{x0}\tau - \gamma y G_y \tau. \tag{7.65}$$

During the acquisition itself, the spins are precessing, with an angular frequency that depends *only on* x:

$$\omega(x) = -\gamma x G_x. \tag{7.66}$$

An isochromat at the point (x, y), thus, gives the following contribution to the total NMR signal:

$$dS(t) = \exp[i\phi_{\text{phase-encode}}(y)] \cdot \exp[i\phi_{\text{read-dephase}}(x) + i\omega(x)t] \cdot \rho(x,y)dx\, dy, \tag{7.67}$$

where

$$\rho(x,y) = \int_{z_0 - \frac{\Delta z}{2}}^{z_0 + \frac{\Delta z}{2}} \rho(x,y,z)dz. \tag{7.68}$$

In the second exponential term in Equation 7.67, notice how $\phi_{\text{read-dephase}}(x)$ is independent of time and can be absorbed into the other x term. If we make a suitable redefinition of our time variable $t' = t - t_0$, where t_0 represents the time taken for the read gradient to reverse exactly the effect of the read-dephase gradient lobe, we can express the signal as a combination of a single *frequency-encoding term* in x and a *phase-encoding term* entirely in y:

$$dS(t) = \exp[i\phi_{\text{phase-encode}}(y)] \cdot \exp[i\omega(x)t'] \cdot \rho(x,y)dx\, dy. \tag{7.69}$$

At the point where the read gradient G_x has exactly reversed the dephasing caused by G_{x0}, that is at $t' = 0$, a so-called *gradient echo* forms. If the phase encoding is very weak, this echo signal will be observable as a visible increase in the magnitude of the time-domain data, but for large G_y values, the re-phasing by the read gradient is masked by the dephasing caused by the G_y. When the phase-encoding gradient is zero, the peak magnetisation for the GE is given by $M_0 \exp(-t/T_2^*)$, as compared with $M_0 \exp(-t/T_2)$ for the SE discussed earlier. For the sake of simplicity in this section, we suppress the relaxation dependence in the equations for S – see Section 7.6.11.

We obtain the total signal by integrating over the whole sample:

$$\begin{aligned} S(t', \tau) &= \int_{-\infty}^{\infty}\int_{-\infty}^{\infty} \exp[-i\phi_{\text{phase-encode}}(y)] \cdot \exp[-i\omega(x)t'] \cdot \rho(x,y)dx\, dy \\ &= \int_{-\infty}^{\infty}\int_{-\infty}^{\infty} \exp[-i\gamma y G_y \tau] \cdot \exp[-i\gamma x G_x t'] \cdot \rho(x,y)dx\, dy. \end{aligned} \tag{7.70}$$

For a single acquisition period, we obtain data at a whole range of t' values, but a single τ. We repeat the sequence a number of times to get data for different values of $\phi_{\text{phase-encode}}(y)$ by stepping the phase-encoding gradient G_y.

Notice that Equation 7.70 is symmetrical with respect to x and y. We can emphasise this even more by defining so-called k-space variables:

$$k_x = \gamma G_x t' \quad \text{and} \quad k_y = \gamma G_y \tau. \tag{7.71}$$

Inserting these variables transforms Equation 7.70 into a formula for $S(k_x/\gamma G_x, k_y/\gamma G_y)$. It is more convenient to work with a function S', which has the same values as S, but for function arguments scaled by γG_x and γG_y. When performing a 2D digital FT, this mathematical subtlety is rarely of any consequence:

$$S'(k_x, k_y) = \int_{-\infty}^{\infty}\int_{-\infty}^{\infty} \exp(-ixk_x)\cdot\exp(-iyk_y)\cdot\rho(x,y)\,dx\,dy. \tag{7.72}$$

This is a *two-dimensional* (2D) Fourier transform.

7.6.8 *k*-Space and Image Resolution

The concept of a reciprocal Fourier space is found in many branches of physics. A particularly relevant comparison is the representation of particle beams in quantum mechanics in terms of either position (***r***) or momentum (***k***) space. The application to X-ray scattering yields equations that are almost identical to Equation 7.72, a consideration which evidently lay behind Mansfield's earliest ideas on NMR imaging (Mansfield and Grannell, 1973).

Interpreting the MRI experiment in terms of k-space leads to a richer understanding of the imaging process, in which the original projection reconstruction technique is seen to be merely as special case. The idea of 'filling' this space with measured data gives rise to the concept of a k-space *trajectory*, fundamental to the design of all modern sequences. The order in which k-space is filled defines the *filtering* that is applied to the image, affecting both spatial resolution and contrast – see Section 7.6.14.1. Spatial resolution itself and artefacts such as aliasing appear as a natural result of the theory of discrete Fourier transforms.

Consider the 2D k-space shown in Figure 7.23. The method of Sections 7.6.6 and 7.6.7 gives a straightforward recipe for acquiring an image: execute the imaging sequence in such a way as to acquire data for every point (k_x, k_y), then arrange the data appropriately and take the 2D Fourier transform. The only problem is, there are an infinite number of possible values of (k_x, k_y) and so it would take an infinite amount of time to acquire the image and infinite computer disk space to store it. In practice, S' is sampled for a *discrete* set of values of the phase-encoding gradient G_y and a discrete set of time points t' during the acquisition phase. The data are processed by using a DFT, with the acquired image matrix size often being a multiple of two to take advantage of the FFT algorithm – see Press et al. (2002, Ch. 12).

A finite grid of $S'(k_x, k_y)$, leads to finite resolution for $\rho(x, y)$. The Nyquist sampling theorem, which states that the highest frequency present should be sampled a minimum of twice per cycle, provides a simple relation between the step length in k-space – which translates

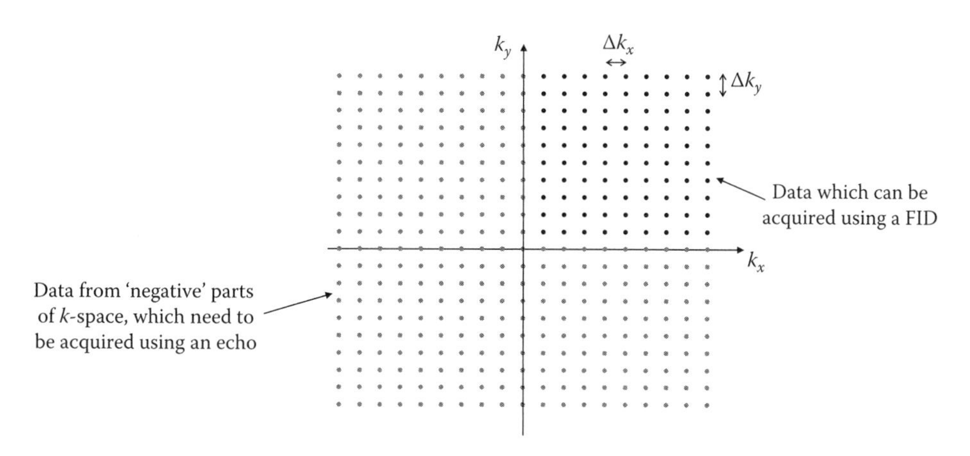

FIGURE 7.23
Representation of k-space, demonstrating the idea of acquiring data at a finite set of locations in order to obtain an image.

into either a y gradient phase-encoding step ΔG_y or time step $\Delta t'$ via formula (7.71) – and the *field of view* (FOV) of the image:

$$\mathrm{FOV}_x = \frac{2\pi}{\Delta k_x} = \frac{2\pi}{\gamma G_x \Delta t'}; \quad \mathrm{FOV}_y = \frac{2\pi}{\Delta k_y} = \frac{2\pi}{\gamma \Delta G_y \tau}. \tag{7.73}$$

If, in addition, we know that there are (N_x, N_y) samples in x and y, such that the total expanse of k-space covered in the two directions is $K_x = N_x\, \Delta k_x$ and $K_y = N_y\, \Delta k_y$, then

$$\Delta x = \frac{\mathrm{FOV}_x}{N_x} = \frac{2\pi}{K_x}; \quad \Delta y = \frac{\mathrm{FOV}_y}{N_y} = \frac{2\pi}{K_y}. \tag{7.74}$$

A common measure is the bandwidth per pixel of a given imaging experiment $1/(N_x\, \Delta t')$, which is inversely proportional to the duration of each ADC sampling point, as mentioned earlier. Historically, it was common for gradients to be expressed in Hz cm^{-1} rather than the more modern units of mT m^{-1}. With the former convention, the spatial resolution is simply the bandwidth per pixel divided by the gradient strength.

The resolution obtained in the frequency direction can be increased by sampling an increased number of points, at the same time as increasing the read gradient. There is no time penalty in this operation, but it is not a case of 'getting something for nothing', because SNR is proportional to $1/\sqrt{\Delta t'}$ and hence 'oversampling' by a factor of 2 reduces SNR in a way that can be recouped only by averaging adjacent points in the read direction. In the phase direction, resolution is dictated by the number of phase-encoding steps used, and by halving the number of steps, the total acquisition time can be halved, at a cost either of a loss of resolution or generation of an aliasing artefact (Sections 7.8.2 and 7.6.14.3). 'Zero padding' the raw data set prior to FT (Bartholdi and Ernst, 1973) can be a useful technique in both read and phase directions. Although no new information is added and true spatial resolution is not increased despite a larger matrix size, perceived image quality is often substantially improved. A further scan time reduction can be achieved by 'partial-Fourier' imaging (Margosian, 1985, 1987), where the Hermitian symmetry of the complex time-domain data is taken into account by only acquiring the positive phase steps, together with a few negative steps to allow for accurate phase correction. This can also reduce the acquisition time by a factor of approximately 2, but at the expense of SNR, which is similarly reduced.

7.6.9 Single-Point Imaging: An Alternative Explanation of Phase Encoding

The descriptions of the frequency spectrum of an object in Section 7.6.2 and phase-encoded 2D imaging in Section 7.6.7 might lead one to suppose that there is some difference between frequency and phase encoding. Via description of a sequence known as *single-point imaging*, this section explains how the two are, in fact, one and the same, and how the GE sequence of Figure 7.17 evolves naturally from a sequence in which all dimensions are phase encoded.

Consider Figure 7.24. A 90° RF pulse is applied, followed by three gradients at the same time. Then, a few ms later, a single datapoint is acquired. The square brackets and the indices p, q and r represent a set of *three nested loops*, just like in a computer programme. The x gradient changes with loop index p, the y gradient with index q and the z gradient with r. The sequence is thus repeated for all possible combinations of p, q and r.

In order to understand how this sequence produces an image, we start by working out what the magnetic field is during the period τ:

$$B_z(p,q,r) = B_0 + xG_x(p) + yG_y(q) + zG_z(r). \tag{7.75}$$

As before, we will ignore the B_0, since we will be working with demodulated signals. This means that during the phase-encoding period, the isochromat at point (x, y, z) will be precessing at angular frequency

$$\omega(x,y,z;p,q,r) = -\gamma[xG_x(p) + yG_y(q) + zG_z(r)]. \tag{7.76}$$

After time τ, isochromats in the sample will thus have acquired a set of different phases $\phi = \omega\tau$, dependent both on position and on the values p, q and r. Each isochromat

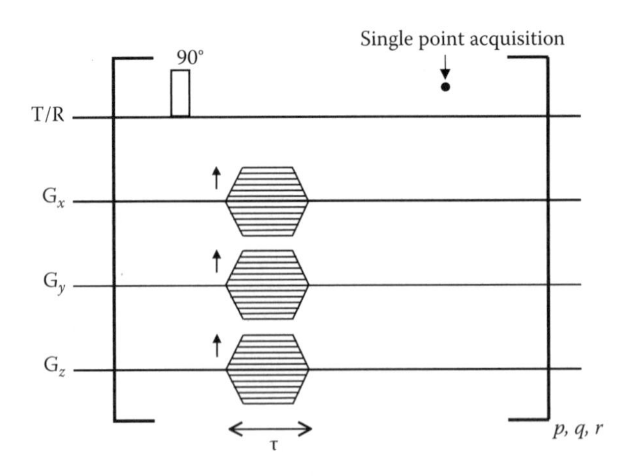

FIGURE 7.24
A pulse sequence for a 'single point' imaging sequence that demonstrates the principle of phase-encoding. For further details, see Section 7.6.9.

will contribute a small quantity of magnetisation towards the signal at the point of acquisition AQ:

$$dS(p,q,r;\ x,y,z) = [\rho(x,y,z)dx\ dy\ dz]\exp[i\phi(p,q,r;\ x,y,z)]. \tag{7.77}$$

If we reinsert the explicit expression for the phase ϕ and integrate over the whole sample, we obtain

$$S(p,q,r) = \int_{\text{sample volume}} \rho(x,y,z)\exp\{-i\gamma[xG_x(p) + yG_y(q) + zG_z(r)]\tau\}\ dx\ dy\ dz$$

$$= \int_{\text{sample volume}} \rho(\boldsymbol{r})\exp\{-i\boldsymbol{k}(p,q,r)\cdot\boldsymbol{r}\}d^3\boldsymbol{r}, \tag{7.78}$$

where $\boldsymbol{k} = (k_x, k_y, k_z) = \gamma\boldsymbol{G}\tau$, by analogy with Equation 7.71. Notice that the signal we get depends on the loop indices p, q and r, because each time we go round the loop, the gradients will be different and hence so will ϕ.

Noting that there is a direct one-to-one correspondence between a given loop variable and the corresponding k-space variable, we define a function $S'(k_x, k_y, k_z)$, as in Section 7.6.7, such that we now have a 3D Fourier transform. The Fourier inversion theorem tells us that if we can measure the value of S' for all possible different values of (k_x, k_y, k_z), then we can calculate the function $\rho(x, y, z)$ using

$$\rho(x,y,z) = \frac{1}{(2\pi)^3}\int_{\text{all } k\text{-space}} S'(k_x,k_y,k_z)\exp[+i(xk_x + yk_y + zk_z)]dk_x dk_y dk_z, \tag{7.79}$$

or, more simply,

$$\rho(\boldsymbol{r}) = \frac{1}{(2\pi)^3}\int S'(\boldsymbol{k})\exp(-i\boldsymbol{k}\cdot\boldsymbol{r})d^3\boldsymbol{k}. \tag{7.80}$$

In practice, the preceding imaging sequence would be impractical for use in patients, because it would take so long – a repetition of the excitation and acquisition for every point in the 3D k-space, with three nested loops. We can remove one of these (say the r loop) using selective excitation. To obtain a further saving, we note that it is very wasteful to acquire just *one* point every time we go round the sequence. Instead of regarding the G_x gradient of Figure 7.17 as a *frequency*-encoding gradient, consider it as a *phase*-encoding gradient that is left on, allowing us to collect multiple points, each of which has a different phase encoding.

During the period τ, the analogue of Equation 7.76 is

$$\omega(x,y;q) = -\gamma[xG_{x0} + yG_y(q)], \tag{7.81}$$

and the phase acquired is $\phi = \gamma\tau$, as before. When the acquisition starts, the x gradient changes to G_x. Points are acquired at time intervals Δt, and the phase continues to evolve. The pth point of the acquisition will have phase

$$\phi(x,y;p,q) = -\gamma[xG_{x0} + yG_y(q)]\tau - \gamma xG_x\cdot p\Delta t. \tag{7.82}$$

Following closely our earlier derivation, we can write

$$dS(p,q;x,y) = \left\{ \int_{z_0-\Delta z/2}^{z_0+\Delta z/2} \rho(x,y,z_0)dz \right\} dy\, dx \exp[i\phi(x,y)], \tag{7.83}$$

remembering that we are looking only at the slice at position z_0, with thickness Δz. We now redefine our variable k_x as follows:

$$k_x = \gamma G_{x0}\tau + \gamma G_x \cdot p\Delta t. \tag{7.84}$$

With this new definition, our integrated signal is exactly what we had in Equation 7.72 of Section 7.6.7, namely,

$$S(p,q) = S'(k_x,k_y) = \int_{\text{2D slice}} \rho(x,y)\exp[-i(xk_x + yk_y)]dx\, dy. \tag{7.85}$$

What we have achieved here is to allow the acquisition of *successive points* to substitute for repeating a phase-encoding loop, the p loop in Figure 7.24. We have thus explained formally how frequency encoding is simply a time-saving variant of phase encoding.

7.6.10 Signal-to-Noise Ratio of MR Images

A complete description of the SNR of NMR measurements is complex, depending on a large number of factors including the coil design, loading and signal amplification, the RF detection method used (quadrature or single phase), operating field strength and nucleus to be observed, imaging method and pulse sequence used, T_1 and T_2 relaxation times, T_R and T_E, image resolution and slice thickness required. Many of these areas are considered elsewhere in this chapter and are discussed in more depth by a number of authors. Hoult and Richards (1976), Hoult and Lauterbur (1979) and Edelstein et al. (1986) have all considered the system SNR, with emphasis on the crucial first stage, the detection probe. The authors provide analytical formulae on the basis of various assumptions concerning coil geometry. Using quadrature detection gives a gain of $\sqrt{2}$ over single-phase detection (rarely used nowadays), and the use of circularly polarised receiver coils (Chen et al., 1983) can provide a further $\sqrt{2}$ increase in SNR – see Section 7.10.1.

Translating the time-domain SNR (peak signal divided by RMS noise, S_0/σ_t) to the corresponding image SNR again involves a number of assumptions. Callaghan (1991, Ch. 4) derives formulae based on a circular object of radius N_s pixels in an $N \times N$ FOV. He obtains the relation

$$\frac{I_0}{\sigma_I} = \frac{4}{\pi} \cdot \frac{N}{N_s^2} \cdot \frac{S_0}{\sigma_t}, \tag{7.86}$$

where

I_0 is the peak image signal, and
σ_I is the image noise standard deviation.

Now $S_0/\sigma_t \propto V_s$, the excited sample volume, and in a single-slice imaging sequence, $V_s = (N\Delta x)^2\Delta z$. If we further assume that the image fills the FOV, such that $N_s \approx N$, then

$$\frac{I_0}{\sigma_I} \propto N(\Delta x)^2 \Delta z. \tag{7.87}$$

It will be seen that as the voxel size decreases, the SNR goes down very rapidly. To some extent, this can be offset by *averaging*. If the scan is repeated n_{av} times and the data accumulated, then the signals add coherently and the noise incoherently, such that SNR $\propto n_{av}^{1/2}$. However, because of the square-root dependence, we have a 'law of diminishing returns' and, as the voxels become smaller, the number of averages required to reach an acceptable SNR rises very sharply. Callaghan (1991) describes a 'brick wall' of sensitivity-related resolution at around a few microns. For further general information on the treatment of noise in images, see Liang and Lauterbur (2000).

In general, SNR improves with increasing field strength in an approximately linear fashion, in line with Equation 7.31, where $\Delta E \propto B_0$. This is why NMR microscopes are based around high-field magnets. However, this benefit is partially offset by increased T_1 at high fields, often requiring a longer T_R, and changing contrast between different tissues (Section 7.6.11). The imaging method and pulse sequence have major effects on the image SNR, and these effects were considered for basic pulse sequences by Brunner and Ernst (1979), Mansfield and Morris (1982a) and Morris (1986). The SNR of more modern sequences needs to be addressed on a case-by-case basis as each is developed. Changes in contrast are also important, as ultimately it is the *contrast*-to-noise that determines whether one will be able to distinguish between two tissues in the presence of noise. This has been considered by Young et al. (1987) for several pulse sequences.

7.6.11 Relaxation Contrast in Images

If all that MRI were able to do was to map the distribution of nuclei, then it would be of limited utility in medicine. The concentration of mobile ^{1}H nuclei is broadly similar in many tissues, and, hence, maps of the proton density ρ often show relatively little contrast. Instead, clinical imaging exploits the fact that NMR relaxation times are extremely sensitive to the physico-chemical environment of the nuclear spins (Section 7.5.5).

Including the effect of T_2^* relaxation explicitly in Equation 7.85 yields

$$S'(k_x, k_y) = \int_{\text{2D slice}} \rho(x, y)\exp[-T_E/T_2^*(x, y)]\exp[-i(xk_x + yk_y)]dx\, dy, \tag{7.88}$$

provided that the read-out period is much shorter than T_E. Since the term including T_2^* is a purely spatial function, with no contribution from k_x or k_y, we may combine it with the proton density to give

$$S'(k_x, k_y) = \int_{\text{2D slice}} \rho_{T_2^*}(x, y)\exp[-i(xk_x + yk_y)]dx\, dy. \tag{7.89}$$

Performing our standard 2D FT image acquisition on a sample with finite T_2^*, thus, leads to an image of $\rho_{T_2^*}(x,y)$, rather than $\rho(x, y)$, and we call this a T_2^*-weighted image. If one requires T_2, rather than T_2^* weighting, then an SE, rather than GE, imaging sequence is used (Section 7.6.13.3).

We have already seen in Equations 7.41 and 7.54 the response of the z magnetisation to a set of RF pulses repeated with time interval T_R. These formulae apply equally to simple GE imaging, *provided that residual transverse magnetisation does not persist between consecutive phase-encoding steps.* In the general case, the signal intensity is

$$I_{GE}(x,y) = \rho(x,y)\sin\theta \cdot \frac{1-\exp(-T_R/T_1)}{1-\cos\theta\exp(-T_R/T_1)}\exp(-T_E/T_2^*). \tag{7.90}$$

Image contrast can be manipulated in a very straightforward, yet flexible manner by changing T_E and T_R, as illustrated in Figures 7.25 and 7.26. With a knowledge of the expected

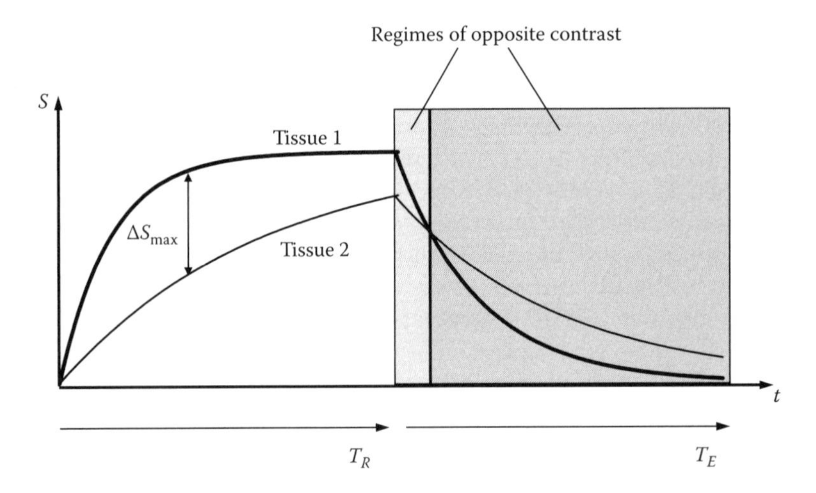

FIGURE 7.25
Illustration of the origin of relaxation contrast in MR images. Relative contrast between tissues of different T_1 and T_2 can be manipulated and sometimes even reversed by an appropriate choice of repetition time T_R and echo time T_E.

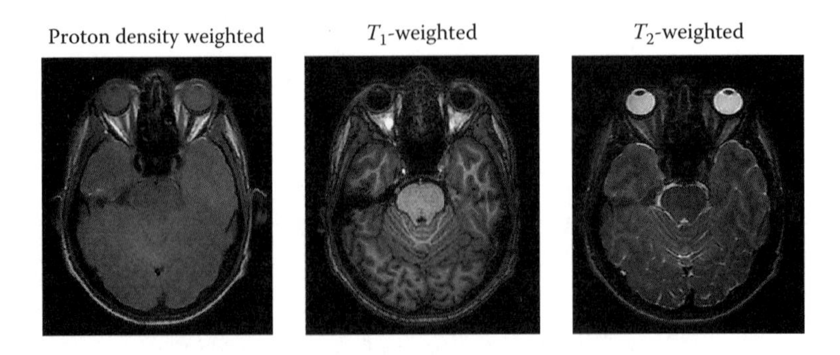

FIGURE 7.26
Example of relaxation time contrast. Similar slices of the brain of a volunteer are imaged with proton density, T_1 and T_2 weightings, revealing different features.

relaxation times for specific organs, excellent soft tissue discrimination may be achieved. This is a major advantage of MRI over other imaging modalities.

For further information on basic relaxation contrast, see the very clear explanation in Chapter 7 of Liang and Lauterbur (2000).

7.6.12 Spoiling

We have noted on a number of occasions that the analysis of some imaging sequences is simplified if we can assume that the transverse magnetisation is zero at the time of application of an excitation pulse. If $T_R \gg T_2$, then this condition is satisfied by default. However, in many situations, this is not the case, and magnetisation that may, at first sight, appear to have 'gone away', having decayed with $\exp(-t/T_2^*)$, may reappear 'unexpectedly' after being refocused by a later pulse in the sequence. This magnetisation is usually unwanted and will often cause image artefacts.

In order to remove these contributions, a variety of techniques have been proposed, and these go by the generic name *spoiling*. Both RF pulses and gradients may be used, but the technical details are beyond the scope of this chapter. See the detailed discussion in Haacke et al. (1999, Ch. 18) for further details.

7.6.13 Review of the Main Classes of Imaging Pulse Sequence

7.6.13.1 Rapid Gradient-Echo Sequences

GE sequences may be divided into two general classes: *spoiled* and *coherent*. MR vendors have a variety of different commercial names for these generic sequences. For further details, see Boyle et al. (2006).

The spoiled sequence described earlier is variously known as FLASH (Haase et al., 1986), T1-FFE, RAGE and SPGR and is one of the main clinical 'workhorses'. A wide range of different T_E and T_R is possible. For high T_2^* contrast, T_E will tend to be of the order of tens of ms, imposing a lower limit on the acquisition time of the sequence, since $T_R > T_E$. For snapshot imaging, T_R might be as little as a few ms, such that a 2D image is acquired in a few hundred ms or an entire 3D image in just a few seconds. In this case, the image will be largely proton-density weighted, as neither relaxation mechanism has a time to cause significant signal change between excitations, and will have a relatively poor SNR. T_2 effects aside, the largest signal for any given T_1 species will be obtained when T_R and the pulse flip angle θ are matched according to

$$\cos\theta_E = \exp\left(\frac{-T_R}{T_1}\right), \tag{7.91}$$

where θ_E is known as the Ernst angle and the expression derived by minimising Equation 7.90 with respect to θ – see, Figure 7.13.

Coherent GE sequences are GE sequences with short T_Rs (such that the condition $T_R \gg T_2$ does not apply), in which the transverse magnetisation is not spoiled between excitation pulses. These are often known generically as SSFP sequences – although, technically, spoiled sequences are also acquired in the steady state. As implied by the discussions of Section 7.4.4, the signal arises out of a complex set of magnetisation pathways. Since the pulses are regularly spaced, so are the echoes. Many different echoes are superimposed on top of each other, and, after a large number of pulses, an equilibrium state is obtained, in which signal contrast depends on the ratio of transverse and longitudinal relaxation times (as well as flip angle

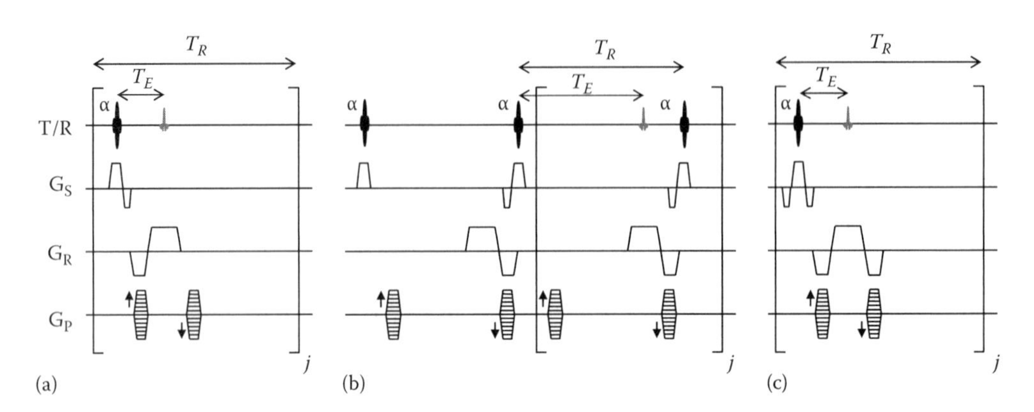

FIGURE 7.27
Pulse-sequence diagrams for the various coherent GE sequences described in the main text: (a) FISP, (b) PSIF and (c) true FISP. In (b), two additional RF pulses are shown outside the loop to illustrate the echo origin of the PSIF signal (magnetisation excited by the first pulse is refocused by the second), as opposed to the FID signal in FISP. It must be stressed, however, that by the time an equilibrium state is reached, the signals of all three techniques contain information from many refocusing pathways. α is the flip angle of the elementary pulses.

and diffusion). Three variants of the sequence are in common use. FISP (synonyms GRASS and FFE) uses the so-called FID signal just after the RF pulse and gives image weighting determined largely by T_1/T_2^*. Figure 7.27a shows the sequence diagram. The sequence for PSIF (reversed FISP, CE-GRASS, CE-FFE-T_2) in Figure 7.27b is simply the mirror image, in which the 'echo' portion of the signal is sampled, and leads to T_1/T_2 contrast.

A sequence that has gained in popularity, as improved gradient technology has made it feasible on clinical systems, is 'balanced' SSFP (True-FISP, FIESTA, balanced FFE) – see Figure 7.27c – in which the effect of all three gradients is 'rewound', so that the state of the magnetisation prior to each pulse is (ideally) exactly the same as for the preceding pulse. Scheffler and Lehnhardt (2003) state, 'If T_1 and T_2 are similar (i.e. for CSF or fat) the optimal flip angle is around 70°–90°, and the maximum possible signal approaches 50% of M_0! This is a very remarkable feature of balanced SSFP, and there exists no other type of sequence that is able to continuously acquire 50% of the total available spin polarisation M0! Based on its totally coherent steady-state magnetisation, balanced SSFP thus offers the highest possible SNR per unit time of all known sequences'.

Despite its excellent SNR performance, SSFP sequences have several disadvantages. Firstly, a tissue having both a long T_1 and T_2 will give the same contrast as one with short T_1 and T_2, since only the ratio is of importance. Secondly, the sequence is very sensitive to 'off-resonance' artefacts (due either to poor shimming or chemical shift), which can lead to severe banding in images.

There is much literature on the subject, and the interested reader is referred in the first instance to Haacke et al. (1999, Ch. 18), Liang and Lauterbur (2000, Ch. 9), Scheffler and Lehnhardt (2003) and Boyle et al. (2006).

7.6.13.2 Echo-Planar and Spiral Imaging

The sequences that we have looked at so far have all been *multi-shot* sequences: each line of *k*-space is acquired after a separate excitation pulse. The earliest ultra-rapid sequence, EPI, introduced in 1977 by Mansfield (1977) but not implemented commercially until the 1990s because of its technical demands, is an example of a *single*-shot sequence. The basic pulse sequence for the 2D form of the measurement is shown in Figure 7.18a. A selective excitation is applied to define a plane and produce a transverse component of magnetisation. This is then dephased and rephased repeatedly by a strong read gradient alternating in polarity. This has the effect of producing a train of GEs, each of which contains information about the spin density along the read direction. At the end of each read-out period, a short 'blipped' gradient in the third dimension provides phase encoding.

It is instructive to look at the trajectory of the magnetisation in *k*-space, shown in Figure 7.18b. Under the influence of a gradient $\boldsymbol{G}(t)$, our location in *k*-space evolves according to

$$\boldsymbol{k}(t) = \int_0^t \gamma \boldsymbol{G}(t) dt. \tag{7.92}$$

Thus, the initial phase-encode and read-dephase gradients move us to the bottom left of the diagram. During the first acquisition period, signal is acquired moving left to right, and then, the blipped phase-encode gradient moves us up to the next line of *k*-space. We 'fill' this row of our raw data matrix by travelling backwards, right to left, under the influence of a negative read gradient. Every time we pass the centre of *k*-space, an echo forms. As the net integral of the phase-encode gradient decreases (moving upwards towards the centre), the echoes become larger and then get smaller again as the combined effect of all the phase-encoding blips heads towards the maximum positive value of k_p.

T_2^* relaxation occurs continuously throughout the extended data acquisition period. Although it is possible to include an additional 180° pulse to give 'spin-echo' EPI, the archetypal single-shot EPI sequence contains just one pulse, and so there is no RF refocusing of dephasing caused by magnetic field inhomogeneities. The $\exp(-t/T_2^*)$ decay acts as a *k*-space filter – see Section 7.6.14.1 – and produces images that are, broadly speaking, T_2^* weighted. This contrast has proved ideal in fMRI, whose goal is to detect signal changes caused by the susceptibility differences between oxygenated and deoxygenated blood in the brain. It is also very straightforward to incorporate a pair of Stejskal–Tanner gradients (Section 7.7.4), making EPI a popular method for measuring diffusion coefficients *in vivo*.

Single-slice EPI is extremely fast, with imaging acquisition times down to as little as 10 ms, and can easily be combined with a variety of contrast preparation modules (see Section 7.4.1). Until the advent of parallel imaging strategies (Section 7.6.14.3), however, the matrix size in single-shot mode was relatively limited, because of the need to finish acquiring all the data within a time significantly less than $5T_2^*$. Other disadvantages of the technique are a sensitivity to susceptibility variations and poor shim settings, which manifest themselves as distortion and signal 'dropout' artefacts (Figure 7.28a), and the possibility of significant 'N/2 Nyquist ghosts' (Figure 7.28b), caused by subtle mismatches between the acquired echoes read out left-to-right and the alternate echoes read out right-to-left.

It will be noted that the trajectory of EPI is Cartesian, filling *k*-space in a regular grid. The 'sharp corners' of the ideal trajectory do not fit well with the physical requirements for turning gradients off and on – imagine trying to drive a car rapidly along a track this shape and consider the strain on the braking system of such instantaneous accelerations – and for many years, EPI was 'slew-rate limited'. A number of alternative *k*-space trajectories have been proposed, each of which has the aim of providing a complete coverage of *k*-space with uniformly dense sampling. One of the longest standing of these is *spiral* imaging. Here, the read and phase gradients are oscillated together smoothly in such a way that their integral yields a *k*-vector that executes a spiral pattern, as shown in Figure 7.29. The technique is a

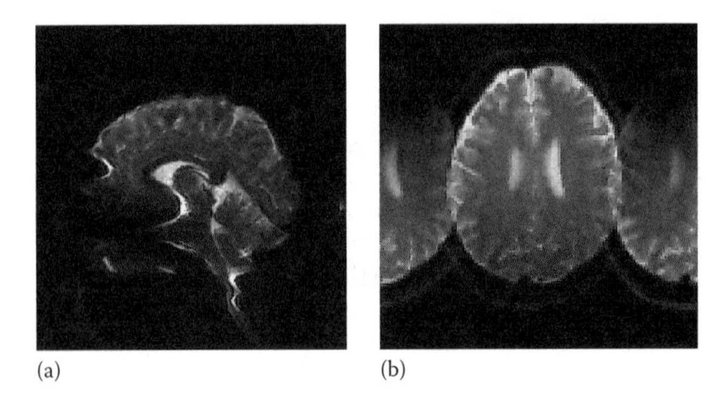

FIGURE 7.28
Extreme examples of imaging artefacts common in EPI: (a) image distortion and signal dropout in a sagittal view of the author's brain (note, in particular, the distortion around the eyes and mouth, which is related to the presence of susceptibility discontinuities at the air cavities) and (b) Nyquist ghost artefacts, caused by mismatches between read-outs going left to right and right to left in k-space. (Data for (b) from Lee et al., 2002.)

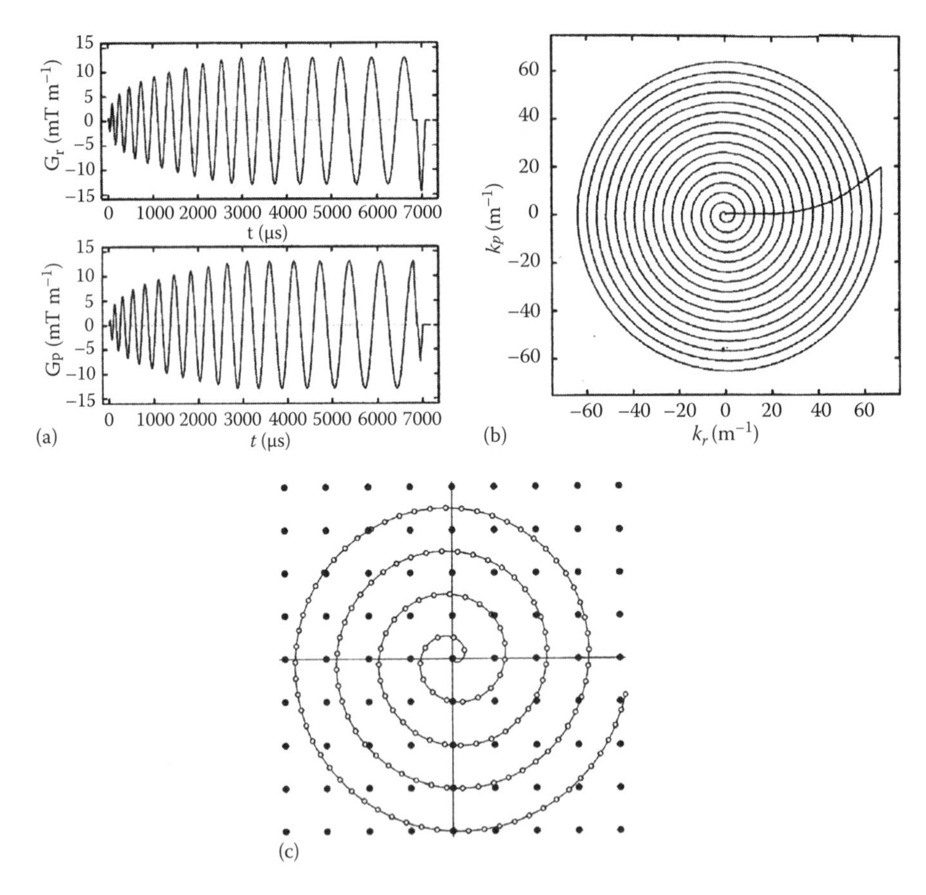

FIGURE 7.29
(a) Pulse sequence for a spiral imaging scan, (b) corresponding trajectory in k-space and (c) comparison of the sampling of datapoints in a traditional Cartesian raster scan and a spiral scan. (Figures taken from Block and Frahm, 2005.)

very efficient way of covering *k*-space and is very robust against motion (including flow), which can cause problems in EPI. However, spiral scanning is sensitive to off-resonance effects (e.g. poor shim or fat-water chemical shift), which cause image blurring and require more sophisticated image reconstruction algorithms. It has seen a much less widespread uptake on clinical scanners than has EPI (Block and Frahm, 2005).

7.6.13.3 Spin-Echo Sequences

The basic SE sequence of Figure 7.19 is very similar to the GE shown in Figure 7.17 and analysed mathematically in Section 7.6.7. An additional 180° pulse is introduced a time $T_E/2$ after the original excitation and refocuses the effects of magnetic field inhomogeneities at the point of the echo, leading to T_2, rather than T_2^* contrast. The size and position of the two read gradient lobes are such that their effect is rephased to form a GE that is coincident with the SE. Notice that the read-dephase gradient is reversed in sign to account for the fact that all isochromats have their phase reversed by the 180° pulse.

In the same way that a multi-shot GE sequence can be made into single-shot EPI by adding extra gradient refocusing lobes, so an SE sequence can generate multiple echoes by adding additional 180° pulses. The imaging equivalent of the CPMG sequence (Section 7.5.7) is generically known as single-shot FSE. Two-dimensional slices may be acquired in a few hundred ms and have good SNR. However, T_2 decay during the acquisition acts to filter the *k*-space data, leading to image blurring in the phase direction on FT. These problems are mitigated either by acquiring only half of the *k*-space data and taking advantage of the Hermitian symmetry of the *k*-space data for a real object – the so-called HASTE sequence – or by using parallel imaging (Section 7.6.14.3). A further disadvantage of the basic implementation of the FSE method is that the large number of 180° pulses leads to a high power deposition in the patient, particularly if many slices are acquired. A number of sophisticated techniques based on the extended phase graph algorithm of Section 7.4.4 have recently been developed that not only reduce the power deposited but also provide echo trains of controllable amplitude – see Weigel and Hennig (2006) for further details.

7.6.13.4 Interleaved and Hybrid Sequences

Looking at the pulse sequence diagrams and *k*-space trajectories for the standard GE and EPI experiments in previous sections, it should be clear that these methods represent two extremes of a family of sequences. Requiring a separate excitation for each phase-encoding step can be slow, because time is required both for the RF pulse to be played out and for the recovery of longitudinal magnetisation between 'shots'. On the other hand, the resolution and contrast of single-shot EPI sequences are inherently limited by T_2^*. A compromise solution is the *interleaved EPI* sequence, in which the image is split into a number of shots, each with an excitation pulse at the start, followed by the acquisition of a number (often known as the ETL) of GEs. Some typical *k*-space trajectories of interleaved sequences are shown in Figure 7.30 and compared with single-shot and multi-shot counterparts. The optimum number of points in the read direction, ETL and number of excitations are determined on an individual basis for each application and depend on the capabilities of the particular scanner used.

The notion of an interleaved sequence is very common. The interleaved version of the SE imaging sequence is known variously as RARE, FSE or turbo SE (TSE). Spiral sequences may also be interleaved, and these have been used with some success in cardiac imaging (Kerr et al., 1997; Nayak et al., 2001).

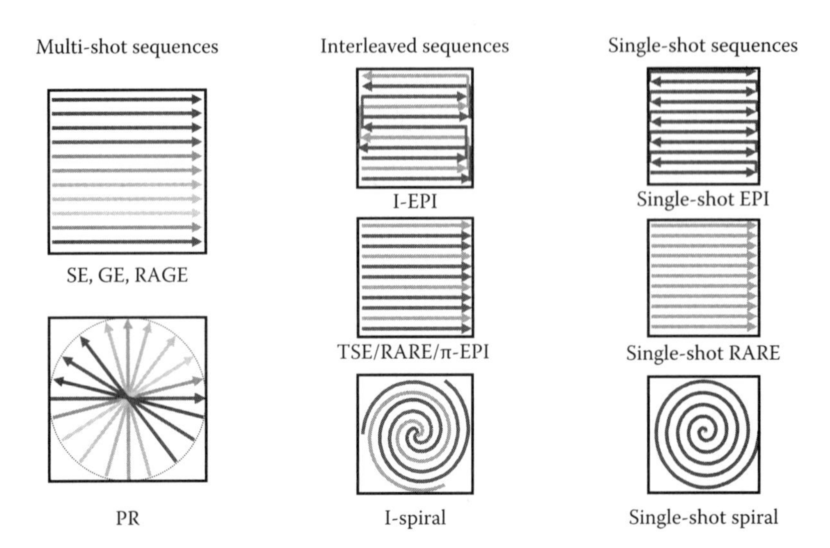

FIGURE 7.30
(See colour insert.) Schematic comparison of the *k*-space trajectories in multi-shot, interleaved and single-shot sequences. Different coloured lines represent data acquired on different shots.

Occasionally, it proves useful to take even further the idea of combining attractive features of different sequence types. GRASE (Oshio and Feinberg, 1991) is a *hybrid* sequence in which an initial excitation pulse is followed by a series of refocusing pulses as in RARE. However, instead of obtaining a single read-out corresponding to each RF echo, the read gradient is reversed repeatedly as in EPI. The method is a compromise between the speed of EPI and the robustness to magnetic field inhomogeneity of RARE.

7.6.13.5 Multi-Slice vs. 'True-3D' Sequences for Volume Imaging

Most clinical imaging examinations require information to be obtained from a volume rather than a single plane. Often this is achieved by the acquisition of a set of planes, which may or may not be contiguous. This is the most common method of acquiring volumetric data in NMR, and, depending on the repetition time and the number of echoes obtained, different slices can be examined by varying the slice-select pulse during the repetition period. In this *multi-slice* imaging mode, each slice individually experiences the full repetition time, but a number of slices can be examined within a single T_R period. This means that there is no time penalty to examining multiple slices rather than just one, provided that they can all be accommodated within the repetition time. As the slice profile is not exactly rectangular, with non-negligible excitation taking place in the 'wings', it is normal to use an interlaced sequence to examine alternate slices. Then, after a degree of relaxation has occurred, the intervening slices are measured to give a contiguous volume overall. This reduces the degree of suppression of adjacent slices due to overlap of the slice profile.

By contrast, it is also possible to obtain 3D information directly via phase encoding, which has an SNR advantage, since signal comes from the entire volume of interest, not just a set of slices. The most common method for this purpose is the 3D FT method, in which the 2D method of Section 7.6.7 is modified by using a non-selective excitation and adding a second phase gradient in the z direction. If a cube of side n pixels is to be imaged, this means that the total acquisition time will be equal to the product of n^2 and the repetition time rather than the product of n and the repetition time for the multi-slice 2D FT method. The image can then be reconstructed by carrying out a 3D FT. Often, the method is combined with selective excitation to allow one or more thick 'slabs' to be subdivided into a set of thin phase-encoded slices.

The dependence of imaging time on n^2 meant that, until the mid-1990s when high-performance gradient systems became standard in commercial systems, there was little interest in using true 3D FT methods for normal imaging. In recent years, there has been a rapid growth in the use of such sequences in areas such as cardiology and angiography, both of which require 3D data with isotropic high resolution, and dynamic contrast MRI, which requires rapid acquisition with 3D coverage. With a 2 ms repetition time and a parallel acquisition acceleration factor of 2, for example, a 128 × 128 × 64 voxel image can be acquired in around 8 s, whilst high-resolution isotropic imaging is achievable within 1 min. Note that, as described in Section 7.6.13.1, image contrast may be sacrificed for speed in such sequences.

7.6.14 Advanced Techniques

7.6.14.1 Acquisition Strategies in k-Space

From the previous sections, it will be seen that, in rapid imaging, each different line of *k*-space is potentially acquired with a different relaxation weighting: progressive T_2 attenuation in RARE, T_2^* attenuation in EPI and a complex oscillatory weighting if data are acquired during the approach to M_z equilibrium of an SSFP sequence. These 'envelope' functions are multiplied by the 'ideal' data in the time domain, and, thus, their Fourier transforms are convolved with the ideal image (convolution theorem). For example, signal decay in the time domain causes blurring along the phase-encoding direction of the image domain, in much the same way that T_2 decay in a spectroscopy experiment causes a broader line shape via Equation 7.49. Interleaved sequences need particularly careful attention, since the envelope functions are not smooth and their point-spread functions give rise to ringing artefacts (Liang and Lauterbur, 2000, p. 294ff).

With all this in mind, one must consider carefully the *order* of acquisition of phase-encoding steps. By suitable manipulation of the phase gradient, one can change this order at will. For most images, the largest fraction of signal power (and hence the most significant contribution to the final SNR of the image) is acquired in the *centre* of *k*-space and so it often makes sense to acquire these data early on in an echo train where the signal available is high. One very common pattern, known as *centric phase encoding*, is to acquire the central line first and then work gradually outwards, on alternating sides of the origin, as in Figure 7.31.

It is often stated that 'image contrast is determined by the weighting at the centre of *k*-space'. This will determine the time at which we choose to form the central echo and hence the order of the phase-encode steps. Whilst the statement is broadly correct for most applications, it is also somewhat simplistic. The inner part of *k*-space is associated with low-resolution features of the image, whilst the high *k*-space parts of the data are related to the image detail. If we are keen to obtain quantitative measures for the contrast of high-resolution structures in the image, then the experimental design must take into account the relaxation weighting of the outer part of *k*-space, too.

An acquisition taking advantage of these Fourier properties of the signal is the so-called *keyhole* method, sometimes used in dynamic contrast experiments consisting of a large number of sequential acquisitions of the same slice. If an assumption is made that the

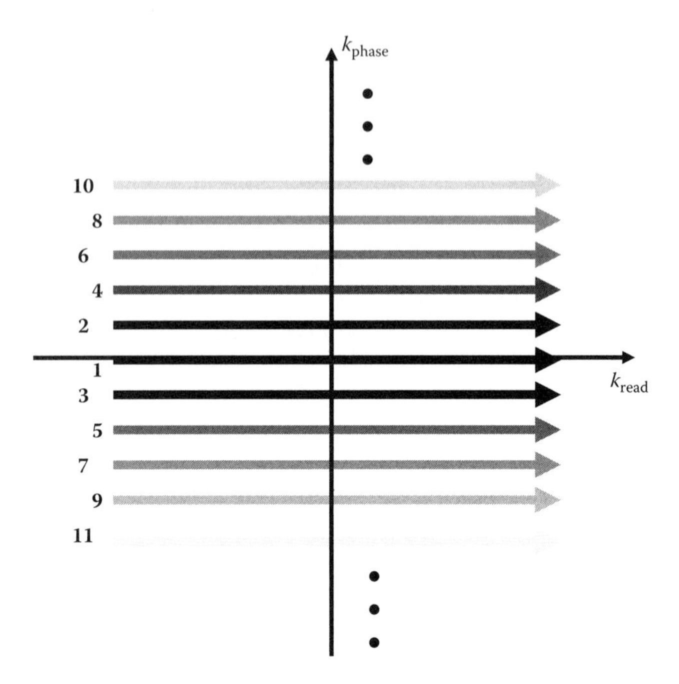

FIGURE 7.31
Acquisition order of data in centric phase encoding. The arrows become lighter as the lines move outwards from the centre of k-space indicating the decrease in signal amplitude due, say, to T_2 decay in a single-shot sequence.

changes of interest take place in structures several pixels across (as might be the case in a large, homogeneous tumour), then it is primarily the central regions of k-space that provide the relevant contrast information. Thus, one may acquire a high-resolution image at the start of an experiment and *update only the central phase-encode steps* during the time course of contrast agent uptake – see Figure 7.32a. All images are reconstructed at the full resolution, but share data from two acquisitions, with the high-resolution information coming from the pre-contrast image. This allows improved temporal resolution at the cost of potential inconsistency in k-space data and lack of fidelity. Caution is needed in applying the technique (Oesterle et al., 2000), as erroneous results may be obtained even though the high-temporal-resolution images themselves may seem to be artefact free.

A further method involving the sharing of k-space data between multiple image acquisitions is the *sliding window* (Kerr et al., 1997), which does allow for the high-resolution features of k-space to be updated, whilst still increasing imaging speed for successive images of the same slice. In its simplest form, images are acquired sequentially, and phase-encoding steps from consecutive data sets are combined to form intermediate images – see Figure 7.32b. More sophisticated techniques use an interleaved acquisition mode, as described in Section 7.6.13.4, but, rather than simply obtaining the data for a single image in n interleaves and performing a reconstruction, acquisition takes place continuously, with the n interleaves being repeated in a cyclic fashion. Then, to reconstruct an image corresponding to the time point t, one uses the n interleaves acquired nearest in time to t. More sophisticated algorithms combine keyhole and sliding window principles by reacquiring the central lines of k-space with every interleave, such that the dominant image contrast is better recovered (D'Arcy et al., 2002).

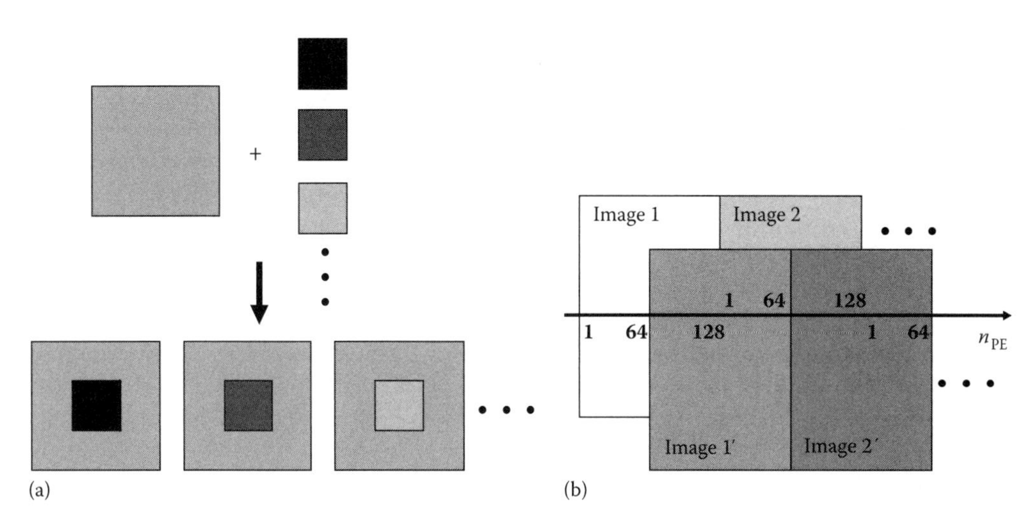

FIGURE 7.32
Two methods for increasing data acquisition speed for multiple images. (a) 'Keyhole imaging' involves the acquisition of one complete (full *k*-space) data set, followed by a number of faster partial data sets, in which only the central portion of *k*-space is obtained. Reconstructed images are obtained by combining the outer portion of the original *k*-space data with the dynamically updated central region. (b) The sliding window approach illustrated for images with 128 phase-encoding (PE) steps: In a traditional technique, images 1, 2, 3,... are acquired sequentially. However, we can create additional, temporally interleaved (but not independent), images 1′, 2′, 3′, ... by using PE steps 65–128 from Image 1 and PE steps 1–64 of Image 2 to form Image 1′ and so on.

In recent years, acquisition strategies have become increasingly sophisticated, as exemplified by the so-called *k–t* BLAST technique (Tsao et al., 2003). This is an adaptive method that uses prior information from training data to improve significantly on the artefact performance of sliding window. A detailed explanation is beyond the scope of this chapter.

7.6.14.2 Gating

Although methods such as EPI are now able to acquire images very rapidly, they are not instantaneous. Moreover, most MRI techniques are susceptible to *motion artefacts*, for example, blurring or ghosting. Since many of the structures that one is interested in clinically move as part of the respiratory or cardiac cycles, various techniques have been developed to address the problem. The simplest of these is long-term averaging: one accepts the fact that motion will occur, but acquires enough repetitions of the data set for the artefacts generated to average themselves out, leaving an image that reflects the average position of the organ concerned.

Gating is an alternative approach in which the MR acquisition is synchronised with the output of an appropriate monitor such that acquisition occurs at the same point of every cardiac or respiratory cycle. Originally, a single phase-encode step would have been acquired at each cycle, but this approach is too lengthy to be practical. An obvious extension is to acquire the same phase-encoding step n times during each cycle. Then, at the end of the sequence, n separate images are reconstructed, corresponding to n different *phases* of the cycle. By contrast, in the 'retrospective' gating technique, phase-encode steps are acquired asynchronously for a pre-determined number of complete cycles of the phase-encode gradient. At the time of reconstruction, the phase-encoding steps are 'binned' into appropriate phases of the cardiac cycle on the basis of a co-acquired ECG trace.

7.6.14.3 Parallel Imaging

Possibly, the most significant development in MRI data acquisition during the early 2000s was the introduction of parallel imaging on clinical instruments. Parallel imaging makes use of data acquired simultaneously using two or more receiver coils whose sensitivity profiles overlap. The basic premise is that the spatially non-uniform sensitivities of the coils provide information that can be used to aid in localising the origin of signals. This information makes redundant a fraction of the phase-encoding steps, allowing images to be acquired more quickly. The beauty of the technique is that it applies to any imaging sequence, and speedup factors typically in the regions 2–6 may be obtained with negligible rewriting of the underlying pulse programmes themselves. Although additional hardware is required in the form of multiple receiver coils and analogue-to-digital converters (ADCs), this type of instrumentation was being introduced into clinical scanners in any case during the 1990s, because of the SNR benefits to be gained from using phased-array receiver coils (Section 7.10.1).

Two broad approaches may be taken to parallel imaging. Simultaneous acquisition of spatial harmonics (SMASH) (Sodickson and Manning, 1997) and its derivatives consider the signal in the time domain, whilst sensitivity encoding (SENSE) (Pruessmann et al., 1999) processes data in the spatial domain after FT. Owing to the limited space available in a chapter of this nature, we will provide only a brief explanation and describe only the SENSE reconstruction.

Consider a 2D-Fourier-transform MRI experiment – the actual pulse sequence used is arbitrary – that acquires data on a Cartesian grid. From Equations 7.73 and 7.74 of Section 7.6.8, the resolution of an imaging sequence is determined by the maximum sampling extent in k-space ($\pm K/2$), whilst the FOV of an image is determined by the k-space sampling interval (Δk). If one takes a standard k-space matrix for a 2D image and acquires only every other line in k-space, the resolution of the data set is unaltered, because K does not change. However, the FOV is reduced by a factor of 2, because Δk has been doubled – see Figure 7.33. This leads to an aliasing artefact (Section 7.8.2), with each pixel in the Fourier-transformed image being the superposition of signals from *two* points in the object.

With image data from just a single coil, this aliasing problem may not be resolved, but the differing sensitivity profiles of the two receive coils allow us to express the signal for an arbitrary pixel, at point (x, y) along the phase-encoding direction as

$$\begin{aligned} I_1(x,y) &= C_1(x,y)I(x,y) + C_1\left(x, y+\frac{FOV}{2}\right)I\left(x, y+\frac{FOV}{2}\right) \\ I_2(x,y) &= C_2(x,y)I(x,y) + C_2\left(x, y+\frac{FOV}{2}\right)I\left(x, y+\frac{FOV}{2}\right), \end{aligned} \tag{7.93}$$

where

$I_{1,2}$ are the intensities of the measured images from the two coils,
I is the ideal image intensity that we wish to calculate, and
$C_{1,2}$ are the coil sensitivity functions, which are different.

Typical data are shown in Figure 7.34. Thus, we have two simultaneous equations for two unknowns: $I(y)$ and $I(y + FOV/2)$. The aliasing is removed by solving these equations for all points y in the phase-encoding direction and repeating the whole procedure for all positions x in the perpendicular direction.

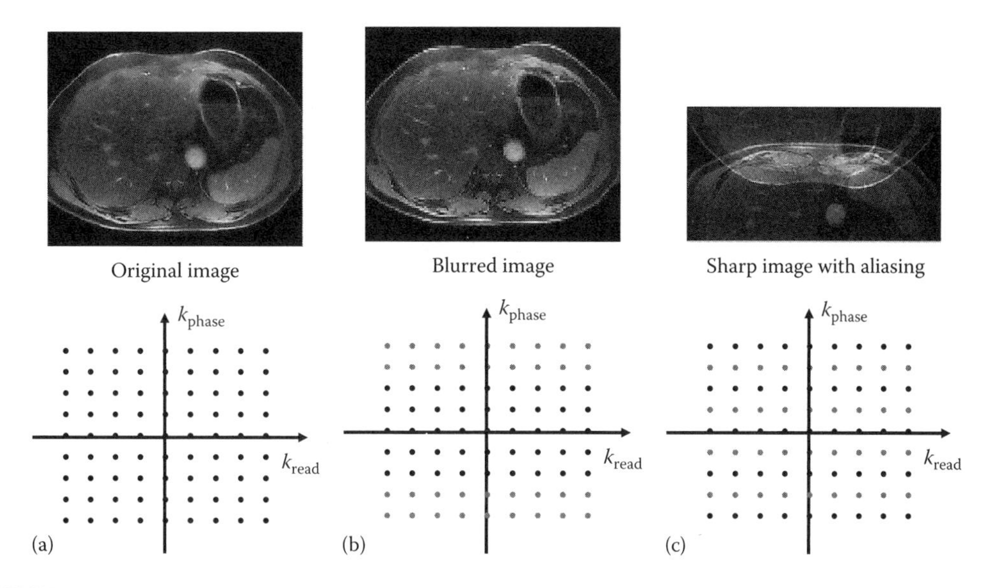

FIGURE 7.33
Illustration of the effects on an image of reducing the amount of k-space data acquired: (a) original image data set; (b) the quantity of data has been reduced by truncating the outermost phase-encode steps and (c) maintaining the original span in k-space preserves image resolution, but increasing the gap between sampling points reduces the field-of-view and leads to aliasing.

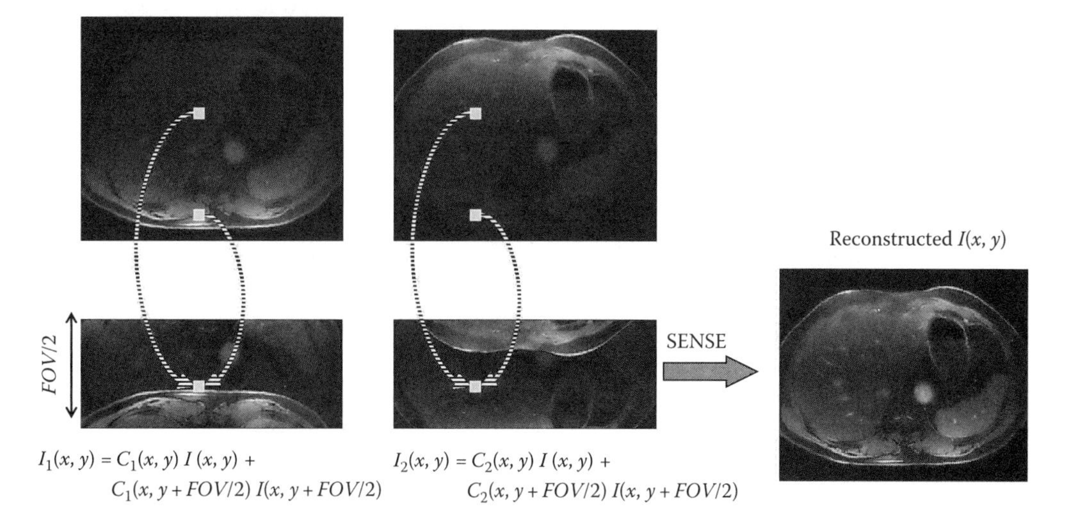

FIGURE 7.34
Principle of the SENSE reconstruction technique. Images on the top row are those that would be obtained with full phase-encoding for each of two RF coils. The two images on the bottom row (left) are those that are actually obtained in a SENSE acquisition with acceleration factor 2. Provided the coil sensitivity profiles C_1 and C_2 are known, the unaliased image (right) may be reconstructed. For full details, see the main text in Section 7.6.14.3.

The preceding analysis had originally been proposed by Ra and Rim (1993) for an arbitrary number of coils, but the major achievement of the landmark paper by Pruessmann et al. (1999) was to generalise the formalism to an arbitrary k-space trajectory and to describe in detail the contribution of noise to the image reconstruction process. The noise level in a SENSE image is not constant over the entire FOV and is, moreover, correlated between different pixels. The so-called geometry factor, g, which describes the noise propagation, is strongly dependent on the relative orientations and overlap of the different receiver coils. In general, noise in SENSE reconstructions rises with increasing acceleration and tends to provide a limit to the speedup factors that are achievable.

7.6.15 Other Imaging Methods with Historical Interest

7.6.15.1 Point Methods

These methods are designed to obtain a signal from only one location in the sample. Compared with plane- and volume-imaging methods, they are therefore inefficient and of low sensitivity, but they are of value in certain applications.

The sensitive-point method, developed by Hinshaw (1974), makes use of three orthogonal gradient fields which oscillate in such a way that only one sensitive point is subjected to a non-time-varying gradient. The position of this point can be moved electronically through an object to build up an image. The method can also be used to select the signal from a particular region of interest, for instance, in measuring T_1, or carrying out ^{31}P spectroscopy.

Two other methods depend on modifying the $\boldsymbol{B}_0$ field such that only a small region is homogeneous. Most of the sample will be in an inhomogeneous field, giving broad spectral components, whereas the signal from the homogeneous region will have a narrow line shape. (Techniques exist to remove the non-homogeneous component to the signal, or a selective pulse can be used to affect only the sample within the homogeneous region.) In field-focused NMR (FONAR) (Damadian et al., 1976, 1977), the patient was moved across the sensitive point. A similar technique, topical MR (Gordon et al., 1980), developed for use in spectroscopy, used static-field-profiling coils to destroy the homogeneity of the magnetic field other than at a central point. Acquisition of spectroscopic data from a single voxel will be discussed further in Section 7.9.3.

7.6.15.2 Line-Scanning Methods

The initial problem in line scanning is how to ensure that a signal is only observed from an individual column of spins. This is technically more difficult than ensuring that the signal comes from a plane. Two types of method are available: the use of two consecutive pulses with slice gradients in different directions or the use of a single '2D pulse' (Hardy and Cline, 1989; Pauly et al., 1989). The general theory of 2D selective pulses is beyond the scope of this chapter, and this class of measurements will not be discussed further here.

Several variants of the first approach have been suggested. One may saturate selectively all but one plane and then excite an orthogonal plane, producing a line of excited magnetisation at the intersection of the two planes (Mansfield and Maudsley, 1976; Maudsley, 1980). A method (volume-selective excitation) based on two selective 45° pulses applied in the presence of a linear gradient and a 90° broadband pulse of opposite phase was developed by Aue et al. (1984). This method retains the equilibrium magnetisation in the plane perpendicular to an applied gradient and can be repeated to select a line or an isolated voxel, leading to applications in selective spectroscopy as well as in imaging. Once magnetism in a line has been selected, spatial information along the line can be obtained by using a read-out gradient in the manner described for projection imaging.

Focused selective excitation, proposed by Hutchison (1977), makes use of a different method to define the selected line. A selective 180° pulse in the presence of a z gradient is followed by a selective 90° pulse in the presence of a y gradient, producing spins in the selected line that are 180° out of phase with the signal in the remainder of the sample. A second acquisition is then carried out in the absence of the selective 180° pulse. If the results from the two experiments are then subtracted, the signals from the line will add and the signals from the remainder of the sample will cancel out. The method is, however, particularly sensitive to motion artefacts. A related SE technique was suggested by Maudsley (1980). A selective 90° pulse is applied in the presence of a z gradient, and this is followed by a selective 180° pulse along the y' axis applied in the presence of a y gradient, which refocuses only those spins lying along the line of intersection of the two planes.

A variant of the sensitive-point method, known as the multiple-sensitive-point method, uses two oscillating field gradients to define the line, read-out along the line occurring in the presence of a linear gradient (Andrew et al., 1977).

7.7 Image Contrast Mechanisms and Quantitative Imaging

7.7.1 General Principles of Quantitative Imaging

As has been seen in previous sections, the contrast in MR images is influenced by a wide variety of factors. We may usefully divide these into two groups:

- properties whose quantitative values are directly measurable by MRI (e.g. diffusion, flow velocity, proton density);
- properties whose values are obtained indirectly, via some form of model, from quantitative MR measurements – examples here are temperature, obtained from NMR phase shifts, diffusion values or T_1s; brain activation, quantified by analysing statistically T_2^*-related signal changes and radiation dose, obtained from the T_2 of a suitable radiosensitive phantom and a calibration relation.

In order to eliminate the undesired factors contributing to the intensity of image voxels, it is common to acquire two or more images to form an 'image stack', as illustrated in Figure 7.35. The pulse sequence is designed such that each image is weighted differently by the parameter of interest, but identically (as far as possible) by all other sample properties. We may describe the intensity of the voxel at (x, y, z) in the ith image by

$$I(x,y,z;i) = I(\{\text{properties of voxel}(x,y,z)\}, \{\text{experimental parameters}\}_i). \tag{7.94}$$

Processing takes place by extracting corresponding voxels from each image in the stack and fitting a model to them. The fit output may consist of a number of separate parameters; these are stored in position (x, y, z) in a set of *parameter images* that are built up voxel by voxel. Other examples of parametric imaging can be found in other imaging modalities (see, e.g. Section 5.7).

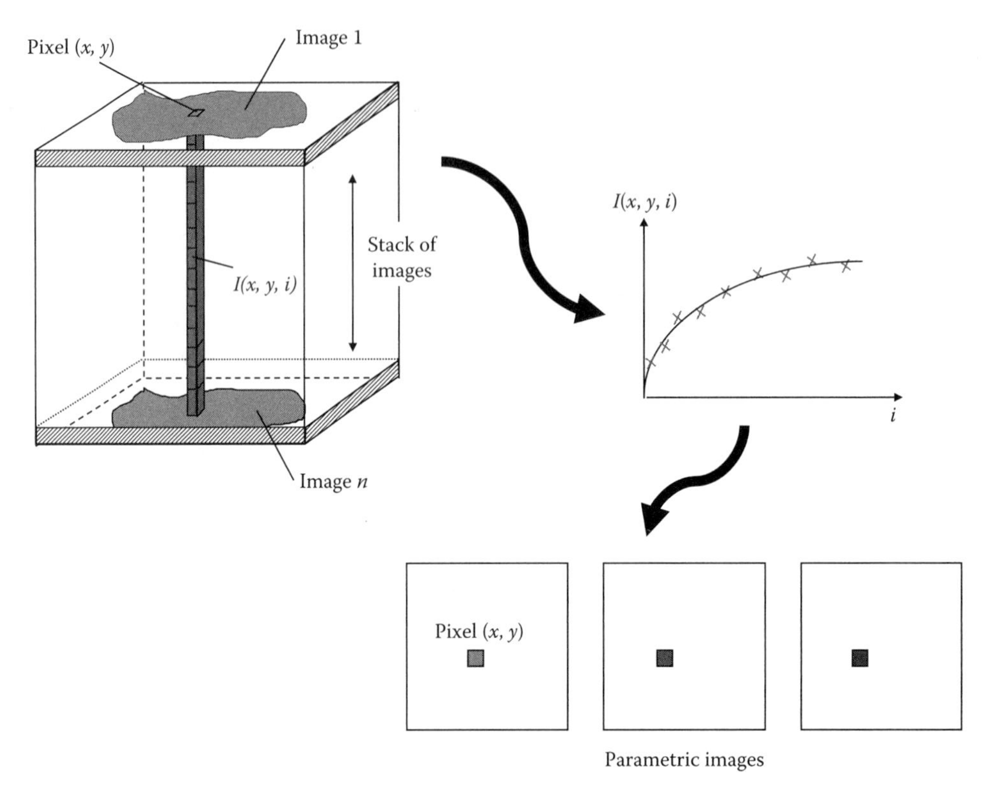

FIGURE 7.35
(See colour insert.) Principle of quantitative imaging: A stack of images is acquired, each weighted differently by the parameter to be measured. A model fit is performed for each pixel (x, y), and the results are placed in the output parametric images.

7.7.2 Relaxation Times and Dynamic Contrast-Enhanced MRI

One of the ways of modifying image contrast is via the administration of exogenous contrast agents. The most common mechanism of action is a variation in the relaxation properties of the tissue targeted, and, almost invariably, this involves the *shortening* of T_1, T_2 or T_2^* (often a combination of these). If a T_1-weighted sequence is used, then a positive enhancement is observed where the agent accumulates, whilst T_2- or T_2^*-weighted sequences yield negative contrast. Agents may be classified according to their biodistribution (e.g. freely diffusible tracers, blood pool agents and targeted organ-specific agents), and Table 7.1 gives a comprehensive list of compounds.

Dynamic contrast-enhanced MRI (DCE-MRI) (Jackson et al., 2003) is of particular utility in the diagnosis of cancer, because malignant tumours generally demonstrate more rapid and higher levels of contrast enhancement than is seen in normal tissue. This reflects the greater underlying vasculaturisation and higher endothelial permeability of the tumour tissues. Both T_1- and $T_2^{(*)}$-based techniques are widely used: broadly speaking, T_1 provides information on vascular permeability (Tofts, 1997; Collins and Padhani, 2004), whilst $T_2^{(*)}$ gives estimates of perfusion, blood volume and potentially vessel size (Calamante, 2003).

Although some useful measures, such as IAUGC (initial area under the gadolinium curve), are not dependent on any particular pharmacokinetic model, the analysis of DCE-MRI data usually involves the fitting of models that attempt to simplify a complex

TABLE 7.1

MRI contrast agents.

Short Name	Generic Name	Trade Name	Enhancement Pattern
Extracellular fluid (ECF) space agents			
Gd-DTPA	Gadopentetate dimeglumine	Magnevist/ [Magnograf]	Positive
Gd-DOTA	Gadoterate meglumine	Dotarem/[Artirem]	Positive
Gd-DTPA-BMA	Gadodiamide injection	Omniscan	Positive
Gd-HP-DO3A	Gadoteridol injection	ProHance	Positive
Gd-DTPA-BMEA	Gadoversetamide	Optimark	Positive
Gd-DO3A-butrol	Gadobutrol	Gadovist	Positive
Gd-BOPTA	Gadobenate dimeglumine	MultiHance	Positive
Targeted/organ-specific agents			
Liver agents			
Mn-DPDP	Mangafodipir trisodium	Teslascan	Positive
Gd-EOB-DTPA	Gadoxetic acid	Primovist	Positive
Gd-BOPTA	Gadobenate dimeglumine	MultiHance	Positive
AMI-25	Ferumoxides (SPIO)	Endorem/Feridex	Negative
SH U 555 A	Ferucarbotran (SPIO)	Resovist/Cliavist	Negative
Other targets			
Gadofluorine-M	—	—	Positive (lymph nodes)
AMI-227	Ferumoxtran (USPIO)	Sinerem/Combidex	Positive or negative (lymph nodes)
AMI-25	Ferumoxides (SPIO)	Endorem/Feridex	Negative (lymph nodes)
EP-2104R	—	—	Positive (visualisation of blood clots)
P947	—	—	Positive (visualisation of matrix Metallo-proteinases, MMPs)
Gd-DTPA mesoporphyrin (gadophrin)	—	—	Positive (myocardium, necrosis)
Blood pool agents			
NC-100150**	PEG-feron (USPIO)	Clariscan	Positive
SH U 555 C	Ferucarbotran (USPIO)	Supravist	Positive
MS-325	Gadofosveset	Formerly Angiomark; Vasovist	Positive
Code 7228	Ferumoxytol		Positive
Gadomer-17	—	—	Positive
Gadofluorine-M	—	—	Positive
P792	Gadomelitol	Vistarem	Positive
AMI-227	Ferumoxtran (USPIO)	Sinerem/Combidex	Positive or negative
GdBOPTA	Gadobenate dimeglumine	MultiHance	Positive

(*continued*)

TABLE 7.1 (continued)

MRI contrast agents.

Short Name	Generic Name	Trade Name	Enhancement Pattern
Enteral agents (orally or rectally administered)			
Gd-DTPA	Gadopentetate dimeglumine	Magnevist enteral	Positive
—	Ferric ammonium citrate	Ferriseltz	Positive
—	Manganese chloride	LumenHance	Positive
—	Manganese-loaded zeolite	Gadolite	Positive
OMP	Ferristene	Abdoscan	Negative
AMI-121	Ferumoxsil (SPIO)	Lumirem/Gastromark	Negative
PFOB	Perfluoro-octylbromide	Imagent-GI	Negative
—	Barium sulphate suspensions	—	Negative
—	Clays	—	Negative
Ventilation agents			
Perfluorinated gases	—	—	Positive
Hyperpolarised ^{3}He, ^{129}Xe	—	—	Positive

Source: Information taken from http://www.emrf.org/New%20Site/FAQs/FAQs%20MRI%20Contrast%20Agents.htm (accessed February 12, 2009).

biological process in terms of a number of *compartments*, each of which contains a well-defined concentration of the pharmacokinetic tracer (contrast agent). One uses the image data to map quantitatively the evolution of contrast agent concentration in each voxel and tries to match the model predictions to what is observed. Adjustable parameters typically include the volume fractions of each compartment and the inter-compartment transfer rate constants. The choice of the optimum model under a given set of circumstances is nontrivial, and a considerable effort has been made over recent years to standardise methodology and notation between the various groups working in this area (Tofts et al., 1999). MRI-derived parameters are now becoming widely accepted as measures of outcome in clinical trials (Leach et al., 2005).

7.7.3 Flow

It is convenient to describe macroscopic motion of the nuclear spins in terms of two classes: motion that is broadly *coherent* within a voxel (e.g. blood flowing along a large vessel) and motion that is *incoherent* (e.g. molecular diffusion). Perfusion may be considered as a hybrid of these two cases: molecules flow coherently, but along a myriad of randomly oriented capillaries within a single imaging voxel, mimicking diffusive motion, but on a different length scale.

Consider the effect on the NMR signal of spin motion through a gradient applied along x. The Larmor frequency is $\omega_0 = -\gamma(B_0 + xG_x)$, and, in the rotating frame of reference, the phase evolution during the interval $[t_1, t_2]$ is

$$\Delta\phi = -\gamma \int_{t_1}^{t_2} G_x x(t)dt = -\gamma \int_{t_1}^{t_2} G_x \left[x_0 + vt + \frac{1}{2}at^2 + \cdots \right] dt, \tag{7.95}$$

where v and a are the x components of velocity and acceleration of the spins initially at x_0. Where the motion is coherent, all the spins in a given voxel have a similar value of $\Delta\phi$, and the value of v may be extracted from Equation 7.95 on the assumption that a and higher-order time derivatives of x are zero. By applying gradients along different axes and repeating the image acquisition a number of times, we are able to map the 3D velocity field with high spatial resolution. An important clinical research application of this *phase contrast imaging* is angiography in areas of complex flow, such as the carotid bifurcation or the heart.

The presence of flowing spins may also be measured via the so-called *time-of-flight* effect, using one of several different methods. In the most common implementation, a spoiled GE sequence is repeated rapidly, thus saturating spins within the imaging slice. In voxels where there is significant flow perpendicular to the imaging plane, fresh (unsaturated) spins continually replace those that have been excited by previous pulses – see Figure 7.36a. Thus, the majority of the imaging slice is dark, but bright circles (or ellipses) mark the intersections of the imaging plane with cylindrical vessels, as shown in Figure 7.36b. Multiple slices are acquired, and a pseudo-3D image is reconstructed, usually via *maximum intensity projection* (MIP) (Figure 7.36c and d).

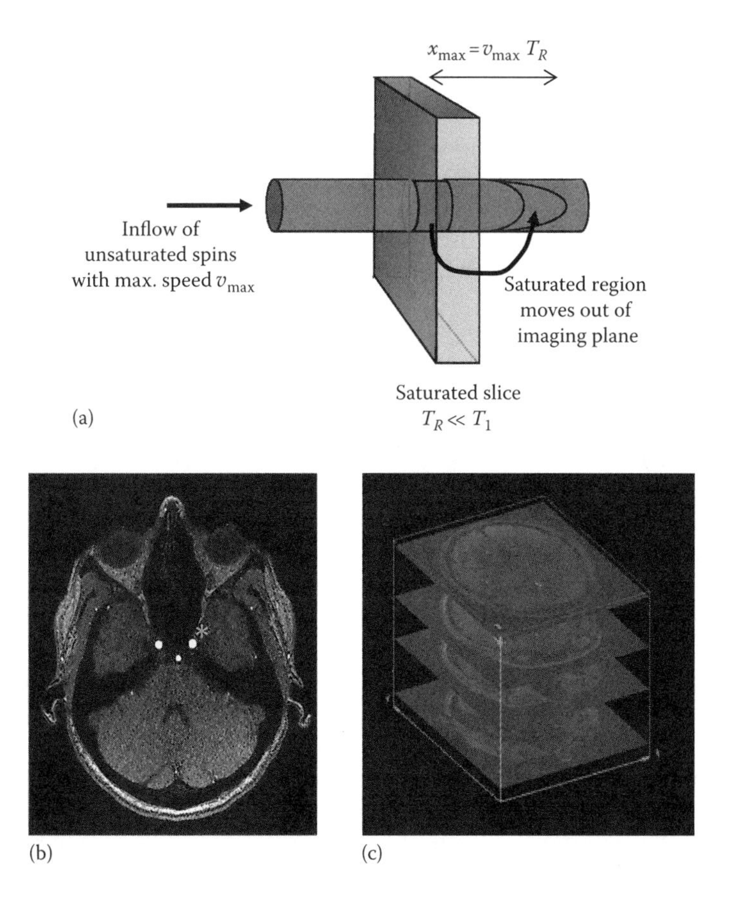

FIGURE 7.36
(See colour insert.) (a) Principle of time-of-flight MR angiography. (b) Sample angiography image of a cross section in the head. High vessel contrast is achieved by saturation of stationary spins and inflow of fully relaxed spins. (c) A 3D image is acquired consisting of a stack of highly contrasted slices.

(continued)

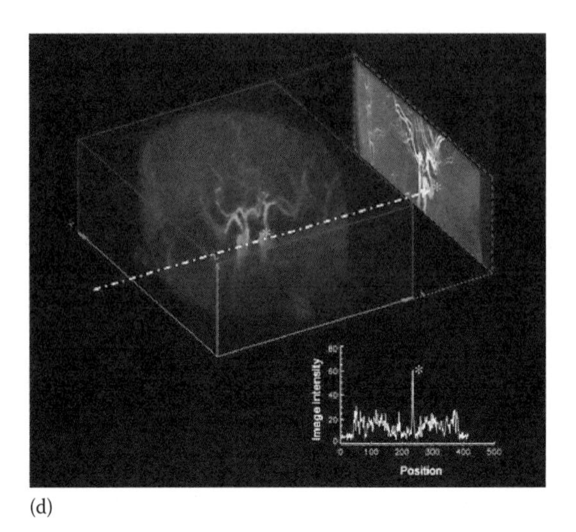

(d)

FIGURE 7.36 (continued)
(See colour insert.) (d) A maximum intensity projection (MIP) image is formed from projections taken through the 3D data set, with the inset graph showing the signal intensity along the dotted line. The orange asterisks shown in (b) and (d) represent corresponding points in the different representations of the data.

An alternative technique is to 'label' magnetisation in one slice and then image the appearance of this magnetisation in a second slice positioned some distance distal to the first. An SE sequence can be used in which the first slice receives the 90° pulse and the second slice then receives a 180° refocusing pulse. Only blood that has received both pulses and therefore flowed between the two planes gives a signal. The velocity of the blood may, in principle, be related to the two pulses and the distance between the planes.

Since the late 1990s and the advent of scanners with high-performance gradients and parallel imaging, both of the techniques mentioned previously have been largely superseded by dynamic contrast angiography. A bolus of contrast agent is injected and ultrarapid 3D images are acquired during the first pass of the contrast agent through the vasculature. The contrast agent lowers the T_1 of the blood so much that, even at short T_R values, for which background tissues are completely saturated, a strong signal is observed from the blood.

Further details on angiography and methods of measuring flow in MRI may be found in Laub et al. (1998) and Haacke et al. (1999, Ch. 24), and examples of MIP images are shown in Figure 7.37.

Two other types of MRI experiment that exploit motion are worthy of note. In *spin-tagging* experiments, a specialised RF pulse scheme is used to excite tissue whilst leaving thin and regularly repeating 'stripes' of unexcited material. As the stripes are associated with the tissue rather than the corresponding image location, they will move if this part of the tissue moves. Such a technique is often used to measure strain in images of the heart. *MR elastography* (cf. US elastography, Section 6.3.6.2) is used to create images whose contrast depends on the local amplitude of an elastic wave passing through the sample. By synchronising the timing of a motion-sensitive MRI sequence with the forced mechanical vibration of the object being examined, displacement amplitudes as small as 1 μm may be measured. With various assumptions, the elastic properties of tissue may be mapped. This provides a quantitative assessment of many breast, thyroid, prostate and abdominal pathologies, which have historically been identified by palpation.

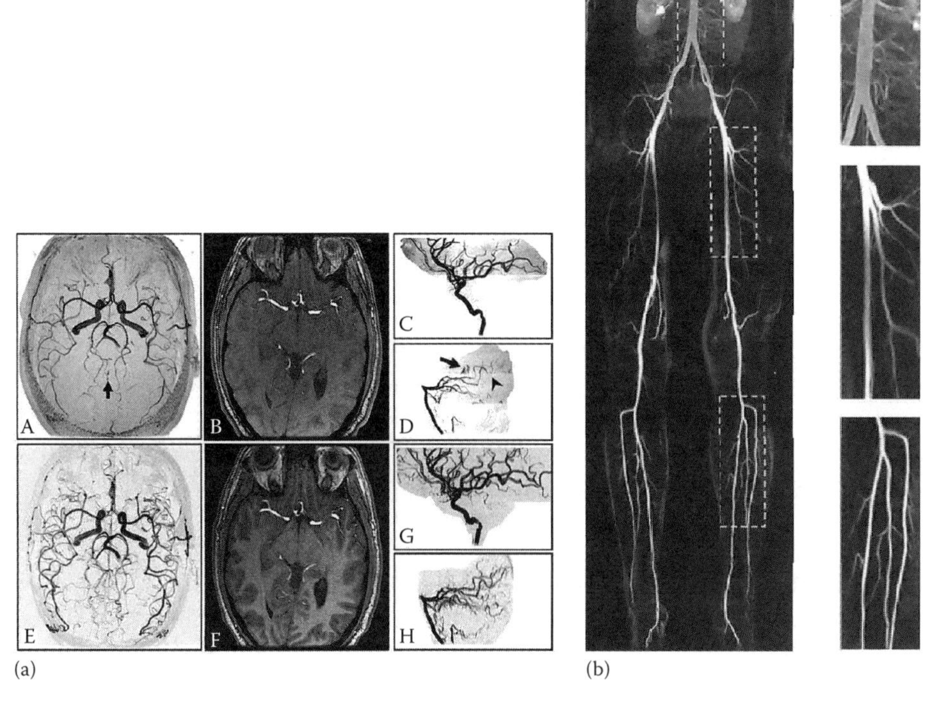

FIGURE 7.37
Examples of MR angiography. (a) Cranial. (Data from Tan et al., 2010.) (b) Peripheral, with enlargement of specified regions. (Data from Zenge et al., 2006.)

7.7.4 Diffusion and Perfusion

Random molecular diffusion means that Equation 7.95 leads to a distribution of phases $\Delta\phi$. If one sums the effect of all the isochromats contributing to the signal, one finds that the net magnetisation decays exponentially for a constant applied field gradient – see Section 7.5.7. When measuring diffusion, it turns out to be more convenient to 'concentrate' the diffusion sensitisation gradient into a module that can be decoupled from the rest of the sequence, rather than apply a constant gradient. A widely used pattern is the *bipolar* or *pulsed-field gradient* scheme suggested by Stejskal and Tanner (1965), and a schematic diagram of their pulse sequence appears in Figure 7.38. Pulsed-field gradients may also be used without the 180° RF pulse, in which case the sign of the second lobe of the gradient must be reversed. The added gradients have no effect on any spins that remain in the same place, since the dephasing effect of the first gradient is exactly balanced by the rephasing of the second, and the echo is attenuated only by $T_2^{(*)}$. By contrast, for diffusing spins, the amplitude of the SE in Figure 7.38 is given by

$$\ln\frac{S(T_E)}{S(0)} = -\frac{T_E}{T_2} - \gamma^2 G^2 D \delta^2 \left(\Delta - \frac{\delta}{3}\right). \tag{7.96}$$

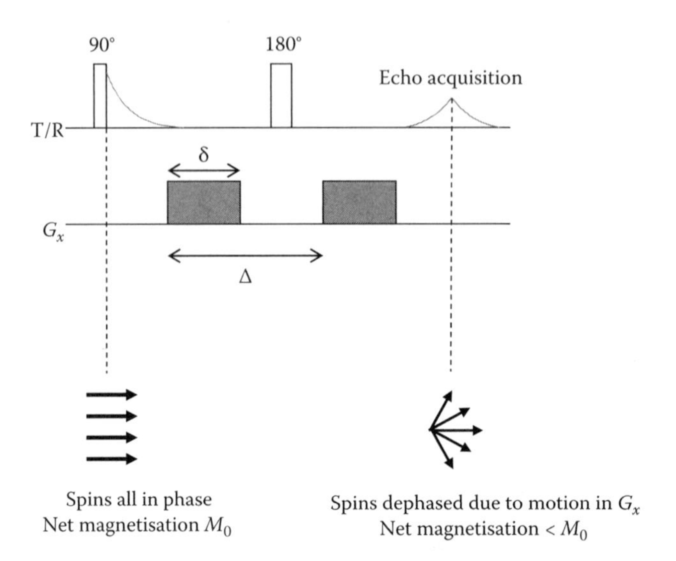

FIGURE 7.38
Basic pulsed-field gradient SE (PGSE) pulse sequence.

Notice how the diffusion attenuation is independent of the echo time. This makes the Stejskal–Tanner module ideal for incorporation into an imaging sequence. It is common to write the image intensity as $I(x, y) = I_0(x, y) \exp[-b\, D(x, y)]$, where the so-called *b*-value incorporates the influence of the pulse sequence parameters. Although the diffusion effect of the imaging gradients themselves is normally much lower than that of the Stejskal–Tanner gradients – their duration and amplitude being smaller – their effect cannot be ignored. Ahn and Cho (1989) present an elegant proof showing that, for a general sequence of gradient pulses, the *b*-value is

$$b(t) = \gamma^2 \int_0^t \left[\int_0^{t'} G(t'')dt'' \right]^2 dt' = \int_0^t [k(t')]^2 dt'. \tag{7.97}$$

In biological tissues, diffusion MRI measures not *D*, the free diffusion coefficient of water (given by the Stokes–Einstein equation $D = kT/6\pi\eta r$), but a quantity known as the *apparent* diffusion coefficient (ADC). This is usually a factor of 2–5 smaller than *D* and reflects *restriction* of the diffusive motion of water molecules by the cellular microstructure. It is the information on this aspect of the tissue micro-environment that gives MR diffusion-weighted imaging (DWI) its great utility in modern diagnosis and research. If two or more DW images are available, a quantitative calculation of ADC(x, y) (normally expressed in the units $mm^2\ s^{-1}$) is possible. However, the sequence will often not be applied clinically in a quantitative mode, but rather as a means of generating contrast for distinguishing between different types of tissue.

Given the need to detect small diffusion-related displacements, the influence of gross sample motion (both involuntary patient movement and physiological motion associated with the respiratory and the cardiac cycles) is very destructive in an imaging context, where the measurement takes place over an extended period. The introduction of so-called *navigator-echo* techniques (Anderson and Gore, 1994) was a major advance in this area.

Initial DWI applications were intracranial. The technical issues surrounding respiratory motion, compounded by issues of susceptibility and chemical-shift artefacts, have led to a limited literature on body DW-MRI. However, this situation is rapidly changing.

Since the first demonstration of DWI by Taylor and Bushell (1985), the technique has become increasingly important in a number of areas of medical diagnosis and research:

- In ischaemic stroke, a cascade of events is set in train following the interruption of the blood supply. One of the results is a swelling of cells (cytotoxic oedema) with water flowing in as the 'pump' maintaining homeostasis fails. As the extra-cellular space becomes smaller, restriction increases, thus decreasing the ADC. Within minutes of arterial occlusion, a corresponding increase in the DWI signal intensity may be observed, and this provides the earliest indication of infarct size. See Figure 7.39a and b for an example.

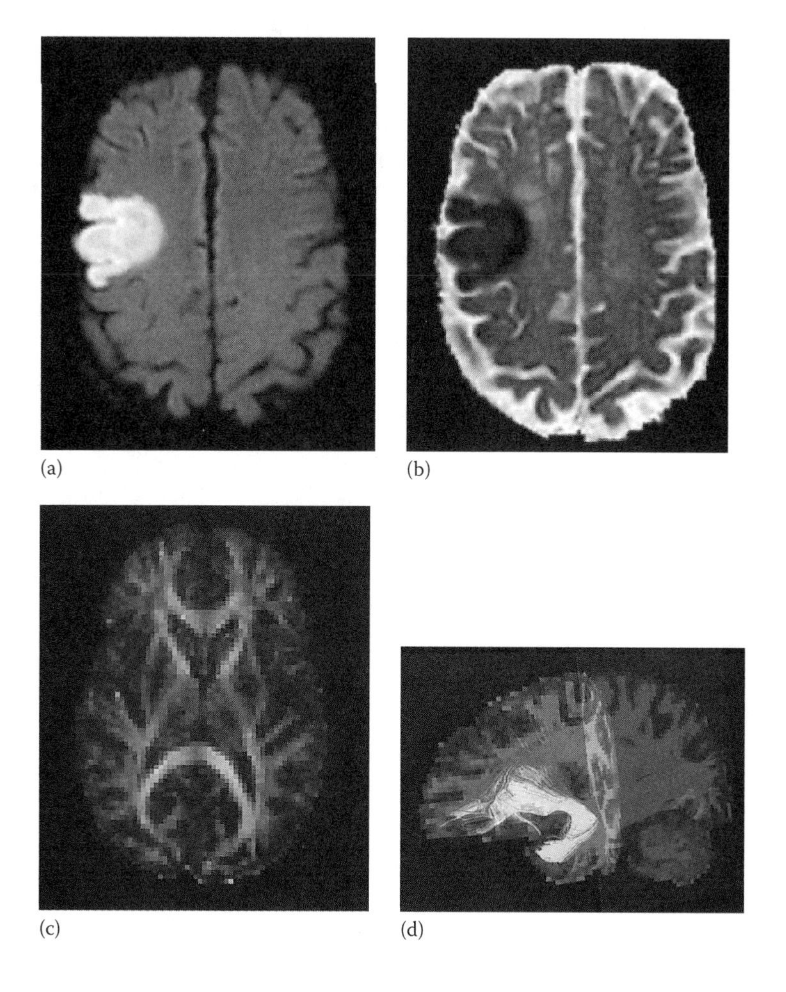

FIGURE 7.39
(See colour insert.) Examples of MR diffusion imaging. (a) Diffusion-weighted image of a patient with an acute stroke. (b) Calculated ADC image. (Data for (a) and (b) from Pipe and Zwart, 2006.) (c) Transverse DTI colour map: red, green and blue components of each colour represent fibres running along right-left, anterior-posterior and superior-inferior axes, respectively. (d) White matter tracts identified via 'diffusion streamlines'. (Data for (c) and (d) are courtesy of the Autism Research Centre, University of Cambridge, Cambridge, UK.)

- In DWI of cancer, a diagnostic benefit was found in differentiating purulent brain processes from cystic or necrotic brain tumours, giving a positive predictive value of 93% and a negative predictive value of 91% (Chang et al., 2002; Guzman et al., 2002). Several studies have now verified histologically the strong correlation between the measured ADC of gliomas and tumour cellularity (Sugahara et al., 1999). Early research in pelvic, abdominal and breast MRI shows promise for diagnosis, whilst there have been indications that DW-MRI may have a role to play in early assessment of response to chemo- and radiation therapy (Dzik-Jurasz et al., 2002). The field has recently been reviewed by Koh and Collins (2007).
- It is well known that many tissues (e.g. cardiac muscle tissue and white matter tracts) consist of arrangements of elongated cells with a preferred alignment. Since water diffuses more readily along the cells than across the membranes, the quantitative values of ADC measured in DWI are strongly dependent on the orientation of the diffusion-encoding gradient. Such *anisotropic diffusion* is studied systematically by repeating the DWI experiment with a number of different diffusion gradient directions. The result is a spatial mapping of the so-called *diffusion tensor.* Clinical applications of DTI have developed very rapidly over recent years (see Figure 7.39c), supported by technical improvements in rapid imaging capabilities. By considering the principal eigenvector of the tensor to be representative of the local fibre orientation and following it from one pixel to another, fibre tracking is now capable of producing extremely high-quality maps of connectivity, as demonstrated by Wakana et al. (2004) and illustrated in Figure 7.39d. These classify different regions of white matter in the brain, allow a visualisation of the relation of different tracts with each other and with the grey matter that they connect, and monitor their disruption by pathological processes. Le Bihan (2003) reviewed the neuroscience applications of DTI, whilst a special double issue of the journal *NMR in Biomedicine* (Le Bihan and van Zijl, 2002) discussed the topic in detail.

7.7.5 Susceptibility and Functional MRI

As has previously been described in Section 7.5.3, inhomogeneities in the main magnetic field lead to additional signal decay, characterised by the T_2' relaxation time defined in Equation 7.50. Insofar as these inhomogeneities arise from within the sample, four different types of behaviour may be distinguished: diamagnetism, paramagnetism, superparamagnetism and ferromagnetism.

It is found experimentally that materials acquire a magnetic moment when placed in a magnetic field, and, for many materials, the induced magnetisation is directly proportional to the applied *field strength* $\boldsymbol{H}$ (measured in A m^{-1}). The constant of proportionality is known as the *susceptibility* χ. In SI units, the *magnetic flux density* $\boldsymbol{B}$ (measured in tesla T) is equal to the sum of $\boldsymbol{H}$ and $\boldsymbol{M}$ multiplied by the constant scale factor μ_0, which is known as the permeability of free space. That is,

$$\boldsymbol{M} = \chi \boldsymbol{H}; \quad \boldsymbol{B} = \mu_0(\boldsymbol{H} + \boldsymbol{M}) = \mu_0 \boldsymbol{H}(1 + \chi). \tag{7.98}$$

A further common notation is that of relative magnetic permeability, with μ_r defined by $\boldsymbol{B} = \mu_0\mu_r\boldsymbol{H}$, so that $\chi = \mu_r - 1$. (It should be noticed that a number of older references, e.g. Abragam (1983), do not use these conventions, writing $\boldsymbol{\omega}_0 = -\gamma \boldsymbol{H}_0$, instead of $\boldsymbol{\omega}_0 = -\gamma \boldsymbol{B}_0$.)

Materials are classified according to the sign and size of χ. In diamagnetic substances (e.g. water), the magnetisation is induced in the opposite direction to the main field $\boldsymbol{H}$ and so χ is negative, typically of order -10^{-6}. The effect is weak, because the electrons in such materials are paired, leading to no net magnetic moment, and the diamagnetic effect arises as a result of Lenz's law.

Paramagnetic materials, by contrast, usually have unpaired electrons; individual molecules have magnetic moments even in the absence of an external field. Normally, these dipoles are randomly oriented, but when a magnetic field is applied, they become aligned, producing a net magnetisation, in a similar fashion to that described in Section 7.3.1. Values of χ are positive and typically in the range 10^{-5}–10^{-3}. Gadolinium-based MR contrast agents, for example, Magnevist® (gadopentetate dimeglumine, Gd DTPA), are strongly paramagnetic.

Ferromagnetism is the phenomenon by which certain substances such as iron become magnetised when placed in a magnetic field and remain magnetised when the field is switched off. Ferromagnetic materials are characterised by an $\boldsymbol{M}$ vs. χ relation that is non-linear and exhibits hysteresis. Extremely large (typically greater than 100), positive values of χ are observed. Above the so-called Curie temperature, the alignment of microscopic magnetic moments that creates this strong magnetisation breaks down, and the materials behave like paramagnets.

Superparamagnetism occurs when a ferromagnetic material is composed of very small crystallites (1–10 nm). In this case, even when the temperature is below the Curie temperature, thermal energy is great enough to change the direction of magnetisation of an entire crystallite. Individual crystallites have magnetic moments pointing in different directions when there is no external field and the consequent magnetisation averages to zero. Thus, the material behaves like a paramagnet, except that instead of each individual atom being independently influenced by an external magnetic field, the magnetic moment of the entire crystallite tends to align with the magnetic field. The last decade has seen the introduction of a number of superparamagnetic contrast agents, for example, Sinerem® (ferumoxtran, AMI-227) and Endorem® (ferumoxide, AMI-25).

In general, one does not attempt to measure quantitatively absolute values of susceptibility via MRI. This would have to be done by solving a rather complex inverse problem, based on the measurable effects. Variations in susceptibility perturb the local magnetic field $\boldsymbol{B}(\boldsymbol{r})$ from the $\boldsymbol{B}_0$ field applied by the main magnet. This inhomogeneous field leads either to signal attenuation via T_2' (including complete signal 'dropout' if the perturbation is severe) or to severe image distortion when using certain techniques such as EPI – see Section 7.8.3. Mapping of susceptibility variations in the context of correcting distortion artefacts was discussed by Jezzard and Balaban (1995).

Far from being merely a nuisance, susceptibility variations are the origin of possibly the most far-reaching application of MRI to be developed since the first edition of this book: fMRI. The magnetic properties of blood vary depending on its oxygen content: oxygenated haemoglobin is diamagnetic, while deoxygenated haemoglobin is paramagnetic and thus has a shorter T_2^*. Ogawa et al. (1990) demonstrated that this phenomenon could be used to obtain blood oxygenation level dependent (BOLD) image contrast from vessels in the brain. Then, after an initial experiment in which brain activation during visual stimulation was measured using susceptibility contrast generated by the administration of an exogenous contrast agent (Belliveau et al., 1991), BOLD contrast in humans was demonstrated in 1992 (Kwong et al., 1992).

What is observed in fMRI is not the direct activation of neurons, but rather the increased blood flow that accompanies it. Nerve impulses travel from neuron to neuron via the release of neurotransmitters at synapses, and the process of recycling these neurotransmitters back into the cell requires energy. This is provided, with a latency of some 4–6 s after the onset of the task, by the haemodynamic response. An initial dip in signal, caused by the deoxygenation of blood close to the activated area, is followed by sustained signal increase, as the region is over-supplied by fresh, oxygenated blood (longer T_2^*). A sequence such as EPI that gives T_2^* weighting produces images with BOLD contrast. The difference in image intensity between oxygenated and deoxygenated conditions is in the region of 3%–5% and depends on B_0, being larger at higher field strengths.

The potential of this technique for mapping human brain activity noninvasively was realised extremely rapidly. From a few publications in 1992, growth was explosive with many thousands of papers published annually at the time of writing. The technique has revolutionised the way in which the brain is studied. Because MRI involves no ionising radiation, studies may be performed repeatedly on healthy volunteers as well as patients. By designing appropriate stimuli, neurophysiologists have been able to elucidate the regions of the brain associated with perception, recognition, interpretation, movement, memory, learning, pain and a whole host of other functions in normal subjects. Diseases of many types have been studied, including psychiatric conditions such as schizophrenia and compulsive disorders for which objective measurement was hitherto unimaginable. fMRI is now used routinely prior to surgery to map out the position of eloquent structures. Monitoring the rehabilitation after acute stroke has demonstrated the plasticity of the brain, with new areas being recruited to take over functions from damaged regions. It has even been shown that parts of the brain may change in size in response to high usage in the same way that muscles enlarge in response to exercise. See Figure 7.40 for an illustration of the type of data produced by fMRI.

For further details on many of the subjects discussed earlier, see Jezzard et al. (2003). Although fMRI is primarily associated with the brain, the BOLD effect also has applications in other types of tissue (e.g. for cancer studies). There are a number of other interesting applications of susceptibility-induced contrast, for example, the incorporation of

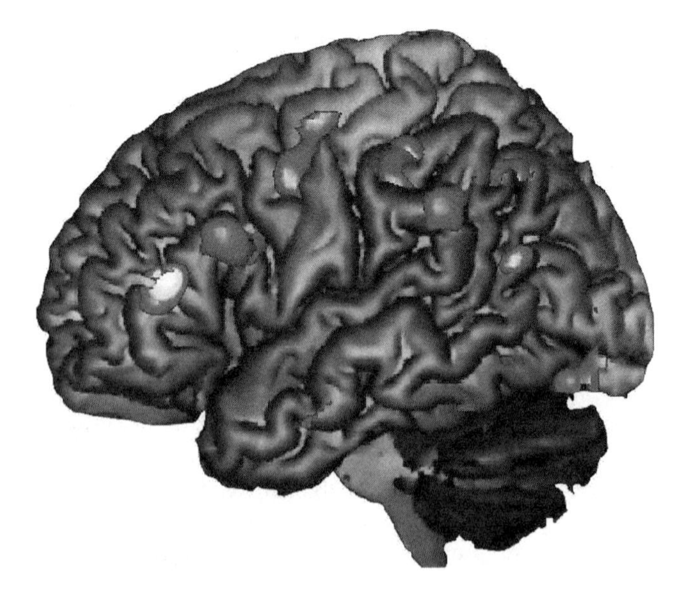

FIGURE 7.40
(See colour insert.) Example of a fMRI data set superimposed on a 3D surface rendering of a high-resolution anatomical data set. (Data courtesy of the Autism Research Centre, University of Cambridge, Cambridge, UK.)

superparamagnetic particles into single cells for stem-cell tracking (Bulte and Kraitchman, 2004; Modo et al., 2005) and the development of gene-targeted contrast agents for molecular imaging (King, 2002; Weissleder and Mahmood, 2001).

7.7.6 Ultra-Short T_E Imaging, Magnetisation Transfer and Polarisation Transfer

Clinical MRI forms images using signals obtained primarily from hydrogen nuclei that are 'mobile' or 'weakly bound', such as those in tissue water. This is because the T_2 values for strongly bound protons are very short (microseconds, rather than milliseconds) and there is not enough time for imaging pulse sequences to run before the signal decays. Hardware developments have dramatically pushed back the frontiers, with the advent of so-called *ultra-short* T_E sequences, which provide previously unseen contrast in cortical bone, cartilage, tendons and ligaments (Robson and Bydder, 2006). See Figure 7.41 for an example.

Contrast may also be generated by the mechanism of *magnetisation transfer* from bound to mobile protons in close proximity (Henkelman et al., 2001). Figure 7.42 illustrates how mobile protons with long T_2 values correspond to narrow spectral lines, whilst the short T_2 values associated with bound protons correspond to wide spectral lines. Crucially, this applies equally well to the excitation of the spins as to the signal read-out. Thus, if one transmits with an off-resonance RF pulse, it will saturate spins in the bound pool, while leaving the mobile spins fully relaxed. If one then obtains an image of the mobile component of the sample, the signal intensity will be lowered if processes are occurring that transfer magnetisation between the two species. Thus, the technique is sensitive to chemical exchange and may provide valuable information about tissue micro-environment.

As described in Section 7.3.6, the cause of the low SNR in MRI is the very small difference between $n_\uparrow$ and $n_\downarrow$. The *nuclear polarisation* $P = (n_\uparrow - n_\downarrow)/(n_\uparrow + n_\downarrow)$ for hydrogen at room temperature and 1.5 T is $\sim 5 \times 10^{-6}$. For certain applications, one may image species whose polarisation can be artificially enhanced to a non-thermal state, leading to a huge increase in signal (often of order 10^5). The most common examples in biomedical MRI are the noble gases ^{3}He and ^{129}Xe, which can be *hyperpolarised* to $P > 0.1$. During the early

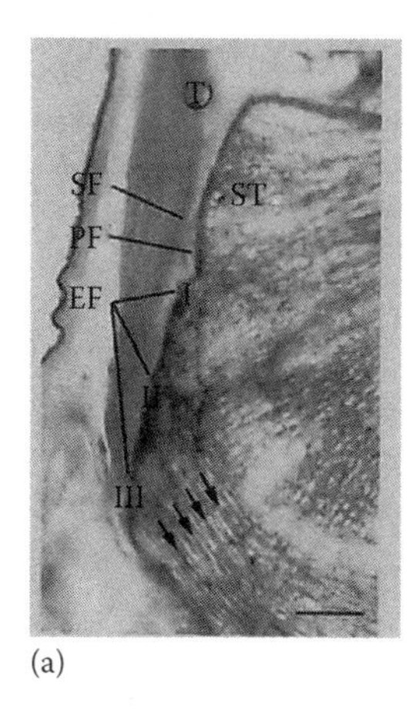

(a)

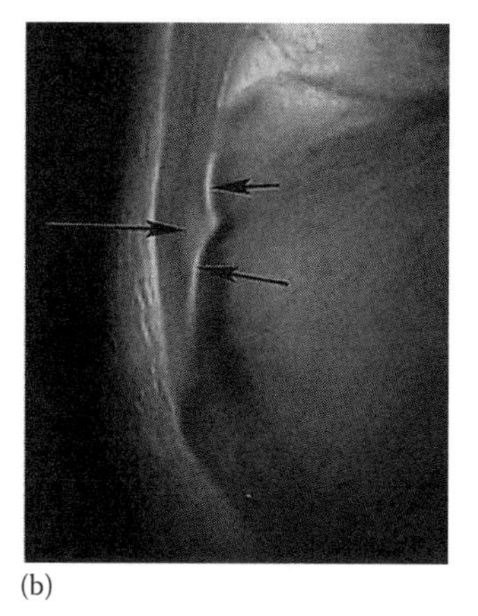

(b)

FIGURE 7.41
(See colour insert.) Achilles tendon. Anatomical specimen (a) and UTE image (T_R/T_E 500/0.08 ms) (b). The fibrocartilage in the tendon has a high signal (see arrows in (b)). The enthesis fibrocartilage (EF), periosteal fibrocartilage (PF) and sesomoid fibrocartilage (SF) are shown with the tendon (T) opposite the superior tuberosity (ST). Arrows in (a) show the direction of the trabecular bone. (Data from Robson and Bydder, 2006.)

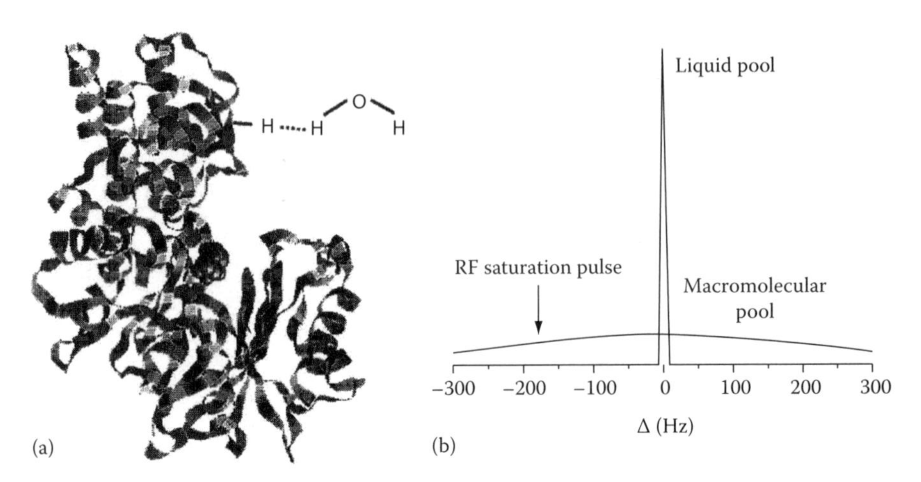

FIGURE 7.42
Principle of magnetisation transfer. (a) Magnetisation can be transferred between protons in a macromolecule and mobile water protons via chemical exchange and other effects. (b) The macromolecular spins, being more tightly bound, have a broader NMR lineshape and can thus be saturated by transmitting off resonance. (Data from Henkelman et al., 2001.)

years of the twenty-first century, hyperpolarised helium was widely used for imaging the air spaces in the lungs (Moller et al., 2002; van Beek et al., 2004) and it is still seen as an extremely promising technique (Fain et al., 2010), as illustrated in Figure 7.43a. However, to a certain extent, the method has become a victim of its own success, leading to significant difficulties in obtaining supplies of this rare gas. The enormous chemical-shift range of xenon in different environments makes it an extremely sensitive probe (see Cherubini and Bifone, 2003; Figures 7.43b and 7.43c). Hyperpolarised xenon is highly soluble, and it may be used to enhance the signal from adjacent hydrogen or carbon nuclei, via the nuclear Overhauser effect (NOE) (Navon et al., 1996). Significant progress has been made in imaging hyperpolarised ^{13}C (Golman et al., 2003; Golman and Petersson, 2006), and this holds out the tantalising prospect of translating ^{13}C spectroscopic techniques, ubiquitous in analytical chemistry, into the imaging arena.

A number of techniques are now available for enhancing the polarisation of NMR nuclei. The simplest to understand is the so-called 'brute force' approach. Noting from Equation 7.31 that the nuclear polarisation is approximately proportional to B and inversely to T, one subjects samples to as a high a 'pre-polarising' magnetic field and as low a temperature as is feasible. The sample is then warmed up and transported to the (lower field) magnet in which the NMR experiment is performed. It is found experimentally that the T_1 of the species involved is long enough for magnetisation to 'survive' extraction from the cryostat, providing data acquisition takes place within a few minutes of the pre-polarisation. Signal gains of order several hundred are achievable via this route. A non-cryogenic alternative, but with the advantage of not requiring transfer of the polarised material, is rapid field-cycling MRI (Matter et al., 2006). True hyperpolarisation is achieved in noble gases via *optical pumping* methods. Two routes are available: direct optical pumping via metastable atoms (Colegrove et al., 1963), for which polarisations of up to ~80% may be achieved in ^{3}He, and indirect optical pumping, which proceeds via spin exchange with an optically pumped alkali metal vapour (Walker and Happer, 1997; Appelt et al., 1998). In each case, the hyperpolarised gas may be stored for some time, often in a low magnetic field. Samples of ^{3}He may have extremely long

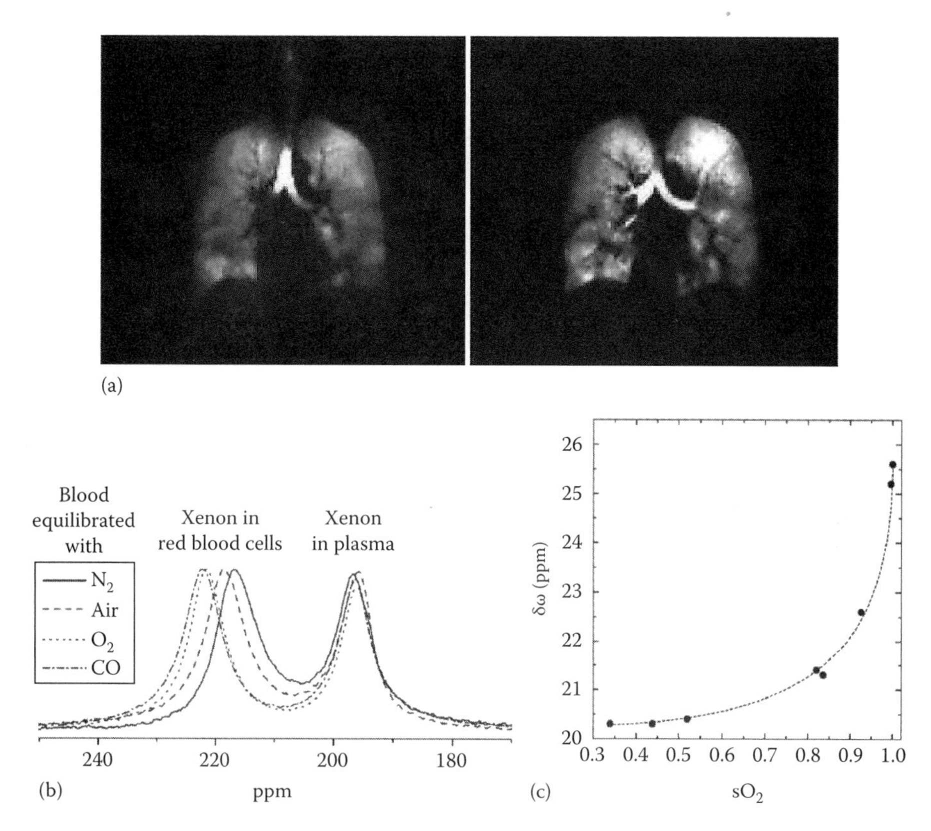

FIGURE 7.43
Examples of the use of hyperpolarised gases in MR. (a) Two adjacent slice images of the lungs ventilated with hyperpolarised ^{3}He. (Data from Lee et al., 2006.) (b) Schematic diagram illustrating the large range of chemical shift of xenon in red blood cells and plasma under various conditions. (c) Separation $\delta\omega$ of the peaks corresponding to the extra- and intracellular Xe NMR signal is a sensitive indicator of blood oxygenation level sO_2. (Data for (b) and (c) from Wolber et al., 2000.)

T_1 values (hours rather than seconds) depending on the surface relaxation properties of the container, and this has even allowed successful shipping of the product between countries by plane. Four main techniques have so far been used to hyperpolarise ^{13}C: dynamic nuclear polarisation (Wolber et al., 2004), parahydrogen-induced polarisation (Golman et al., 2001), thermal mixing and the spin polarisation-induced nuclear Overhauser effect (Cherubini et al., 2003). The arrival of dynamic nuclear polarisation has proved particularly exciting (Kurhanewicz et al., 2011) and offers the potential of increasing the sensitivity of measurements of molecules containing ^{13}C or ^{15}N by hyperpolarising them, increasing the SNR by a factor of perhaps 10,000, albeit an advantage that decays with T_1 relaxation.

ESR is a well-established method of detecting free radicals but one that is not well adapted for high-resolution imaging. The PEDRI technique (Lurie et al., 1998) combines ESR with an MRI read-out. Electron spins in the free-radical target are irradiated in a low magnetic field, which is then rapidly increased to one suitable for MRI. Spin angular momentum is transferred from the electron to the nuclear spins, thus leading to contrast in the NMR image that reflects the density of the free radicals. ESR may also be used for measuring the oxygenation of tissue, which is potentially of interest in cancer applications.

7.7.7 Chemical-Shift Imaging

Chemical-shift imaging (CSI) is the acquisition of images showing separately the distribution of different metabolites or compounds of the nucleus of interest. This is most commonly performed in fat/water proton imaging (see Figure 7.44), where only two major components are present, and allows chemical-shift artefacts to be corrected and the differing distribution of water and fat to be mapped. A method proposed by Dixon (1984) acquires two SE images in which the position of the 180° RF pulse is changed slightly with respect to the gradients, such that, in the first image, the water and fat signals are in phase and, in the second, they are in antiphase. The sum of the two images gives a water-only image, whilst their difference yields a fat-only image. An alternative approach used in chemical-shift-selective (CHESS) imaging (Rosen et al., 1984; Haase et al., 1985) is to saturate one component with a frequency-selective 90° pulse and then use a spoiler gradient (Section 7.6.12) to destroy transverse magnetisation, prior to executing the usual imaging sequence. This has the advantage that a fat or water image can be obtained in only one acquisition. Both the methods mentioned earlier encounter problems if the $\boldsymbol{B}_0$ homogeneity is not high, when it can prove difficult to obtain satisfactory suppression over the whole image.

A more detailed discussion of spectroscopic imaging may be found in Section 7.9.4.

7.7.8 Current-Density Imaging

The imaging modality magnetoencephalography uses detectors positioned outside the head to measure electric currents deep inside the brain by making exquisitely sensitive measurements of the tiny (~10^{-14} T) magnetic fields they generate. It has been suggested that direct 3D imaging of these currents, of major importance in the location of epileptic seizures, might be possible using MRI because of the local changes in Larmor frequency they create (Scott et al., 1992; Huang et al., 2006). Several phantom studies have been successful, but detection of the effect *in vivo* remains elusive. Although work by Pell et al. (2006) gives some grounds for optimism, it may remain forever below the sensitivity limits of the MRI experiment, except insofar as non-physiological currents are applied artificially (Joy et al., 1999).

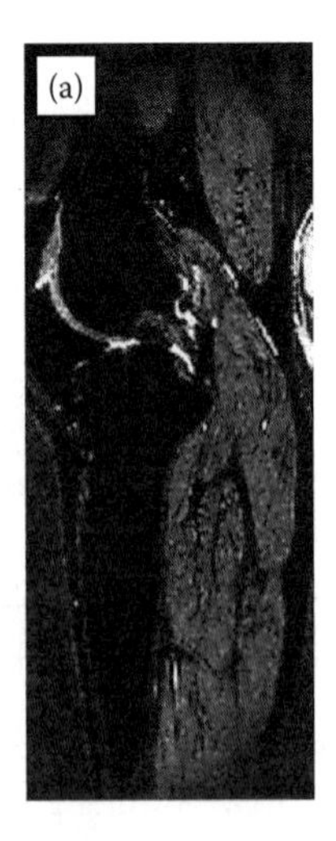

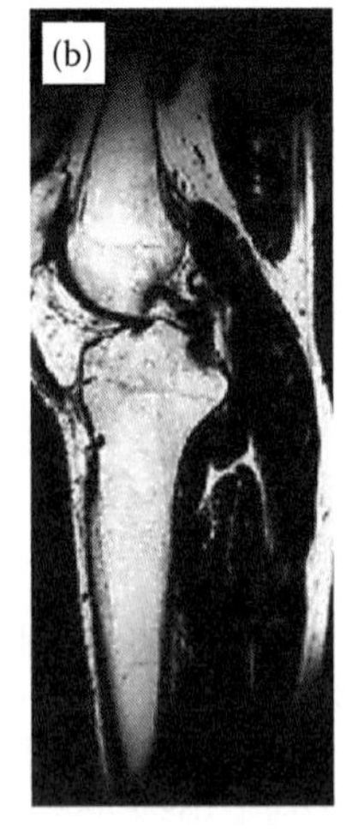

FIGURE 7.44
Examples of fat-water selective imaging. (a) Water-only image of a volunteer knee. (b) Fat-only image. (Data from Hargreaves et al., 2006.)

7.7.9 Contrast due to Long-Range Dipolar Fields and Multiple-Quantum Imaging

One of the implicit assumptions in the detailed quantum mechanics of NMR (not studied in this chapter) is that one may ignore the long-range dipolar interactions between NMR spins. The dipolar magnetic field at one nuclear spin as a result of its neighbour is proportional to the term $(3\cos^2\theta - 1)/r^3$, where θ is the angle between the main magnetic field $\boldsymbol{B}_0$ and the line joining the two spins, and r is the distance between them. Although the magnitude of this term decreases rapidly with the separation of the nuclei, the number of participating nuclei at distance r from the site of interest rises according to r^2 and so their contribution cannot be ignored. The reason that this topic was little studied in the early days of MRI was that, in a magnetically homogeneous medium, this term averages to zero and spins see no long-range effect (although the random *local* effects are significant and give rise to relaxation – see Section 7.5).

In the latter part of the 1990s, a number of papers demonstrated that if the long-range order of the nuclear spins is perturbed by the application of an appropriate gradient, this fundamental assumption is no longer valid (Lee et al., 1996). The CRAZED sequence, introduced by Richter et al. (1995), is a particularly simple set of RF pulses and gradients, yet it displays behaviour that cannot be explained with standard classical and quantum-mechanical arguments, which ignore the long-range dipolar interaction. Gutteridge et al. (2002) have demonstrated that a useful consequence of these effects is the ability to measure the absolute value of the sample equilibrium magnetisation M_0 in a way that is independent of any reference calibration. The contrast arising when an imaging module is appended to CRAZED is potentially of great interest, because long-range dipolar coupling is one of the few NMR contrast mechanisms to be related to structure on a *mesoscopic* length scale (rather than microscopic like T_2). Moreover, the *correlation distance* of the sequence (i.e. the length scale of sample inhomogeneities which have an effect on contrast) can be 'tuned' within the range 10 μm to 1 mm simply by changing the gradient amplitude in the CRAZED module. It has been suggested that this type of imaging may, in future, play an important role in many fields, from studies of porous media to fMRI (Richter et al., 2000). However, at the time of writing, 'the jury is still out': there are significant difficulties in interpretation of the results (Charles-Edwards et al., 2004) and in relating the observed changes in signal intensity back to underlying sample structure.

A significant body of work has also been performed in the area of *multiple-quantum* imaging. Whilst it has a number of clear applications in elucidating tissue structure – see, for example, Navon et al. (2001) – the theory of multiple-quantum coherences and the associated imaging methods is beyond the scope of this introductory review.

7.7.10 Imaging of Nuclei Other than Hydrogen

In the clinic, the dominance of ^{1}H imaging is overwhelming. This is due primarily to (i) the large concentration of mobile hydrogen-containing compounds within the body, (ii) a high natural abundance of the ^{1}H isotope, (iii) the large gyromagnetic ratio of hydrogen and (iv) broadly speaking, favourable T_1 and T_2 relaxation times. NMR is a technique with relatively low sensitivity, and the SNR of images is a constant issue – see Section 7.6.10. It thus proves impossible to obtain images of adequate quality from a number of species that would potentially be of interest, for example, ^{13}C with an atomic concentration in the body only a tenth that of hydrogen, a natural abundance of only 1.1% and a gyromagnetic ratio only a quarter that of hydrogen, together with a prohibitively long T_1 relaxation time.

Whilst techniques such as hyperpolarisation (Section 7.7.6) have started to change the outlook for carbon, much research has already been performed on other nuclei. ^{31}P spectroscopy studies have demonstrated over many years the potential for measuring the high-energy metabolites, adenosine triphosphate (ATP) and phosphocreatine (PCr), with altered levels detected in a number of clinical conditions. Sodium is often elevated in tumours, whilst its transport across membranes is also vital for cell function, and energetic failure leads to altered sodium concentrations that can be detected by ^{23}Na NMR. Imaging of both of these species has been successfully demonstrated on commercial clinical scanners, albeit with relatively low spatial resolution (Lee et al., 2000; Romanzetti et al., 2006). Moreover, multiple quantum ^{23}Na imaging, together with deuterium MRI, gives us a powerful tool for looking at microscopic structure in tissues (Navon et al., 2001). ^{39}K is also quadrupolar and can be used to measure potassium behaviour in the body. ^{19}F is an excellent candidate nucleus for imaging, as it has a high gyromagnetic ratio and 100% natural abundance. It is not present naturally in the body and is therefore an ideal exogenous tracer. It has been administered preclinically, for example, in the form of perfluorocarbon emulsions for measuring the oxygenation status of tumours (Robinson and Griffiths, 2004), and incorporated into amyloidophilic compounds for labelling the types of plaque found in Alzheimer's disease (Higuchi et al., 2005). ^{17}O has been used for direct monitoring of the cerebral metabolic rate of oxygen consumption, with potential applications in functional imaging (Zhu et al., 2002; Ugurbil et al., 2003). Finally, the potential applications of a number of less commonly imaged nuclei (^{7}Li, ^{10}B/^{11}B, ^{17}O, ^{59}Co, ^{87}Rb and ^{133}Cs) were extensively reviewed in a special issue of *NMR in Biomedicine* (Kupriyanov, 2005 and other articles in the same issue). Spectroscopic studies have been reported using ^{14}N and ^{15}N (e.g. Shen et al., 1998), although the latter is a low-abundance and low-sensitivity nucleus.

7.8 Image Artefacts

In common with most imaging techniques, NMR images often contain *artefacts*, which we define here as intensity values in the measured image that do not correspond accurately to features in the object being studied. These might, for instance, be features that are recorded in the correct place but with the wrong intensity, features that are recorded in the wrong place due to image distortions or spurious features that do not relate to any portion of the real sample. Even in a well-adjusted machine, there are a number of effects that will inevitably give rise to artefacts. There is also a very wide range of effects that can occur due to the failure or poor adjustment of equipment.

In MRI, artefacts are generated whenever the data contain information that is encoded in a way that is *a priori* unexpected on the part of the reconstruction algorithm. The algorithm then assumes the wrong things about the data and hence processes it 'incorrectly'. For example, a very common assumption in Fourier imaging is that the raw data are related to the required image by Equation 7.72, or in 3D,

$$S(\boldsymbol{k}) = \int \rho(\boldsymbol{r}) \exp(i\boldsymbol{k} \cdot \boldsymbol{r}) d^3\boldsymbol{r}. \tag{7.99}$$

(Note the S' of Equation 7.72 is dropped for simplicity in what follows, and we use the ideal continuous version of what is, in fact, a discrete data set.) In practice, this signal is often corrupted, by addition or multiplication of various additional terms. If one knew before performing the processing what the relevant terms were, an appropriately modified method of analysis might possibly be employed. Unfortunately, the exact nature of the data corruption is often unknown.

In the sections that follow, the types of artefact are classified by their origin: problems due to the scanner or its settings; problems related to the design of the NMR experiment and problems caused by the sample. For additional details, see the excellent section on image artefacts in Liang and Lauterbur (2000, Ch. 8) and the review article by Henkelman and Bronskill (1987). A number of websites contain comprehensive galleries of MR image artefacts, with some, such as Beckman Institute (1998), also including raw, time-domain MRI data.

7.8.1 Artefacts Caused by the Instrumentation

Most MRI systems are surrounded by some form of Faraday cage, which prevents RF interference in the images. This interference takes the form of an additional term in Equation 7.99:

$$S(\boldsymbol{k}) = \int \rho(\boldsymbol{r}) \exp(i\boldsymbol{k}\cdot\boldsymbol{r}) d^3\boldsymbol{r} + A(k_y, k_z) \exp(i\omega_{\text{int}} t), \qquad (7.100)$$

where

ω_{int} is the angular frequency of the spurious radio signal, and
t is typically related to $\boldsymbol{k}$ by $t = k_x/\gamma G_x$ – cf. Equation 7.71.

Fourier transformation of $S(\boldsymbol{k})$ yields the sum of the desired signal together with an additional term at a constant position in the x (frequency-encoding) direction. Since the repetition of the phase-encoding steps is asynchronous with the RF interference, the term $A(k_y, k_z)$ is effectively random, and, thus, correction of the effect is impossible. The reconstruction artefact takes the form of a 'zipper' running along the phase-encoding direction of the image, as illustrated in Figure 7.45. Generally speaking, the artefact is not too troublesome, as it occupies only a single column of pixels. Moreover, RF screening is a straightforward, if somewhat expensive, procedure clinically.

A special case of Equation 7.100 with $\omega_{\text{int}} = 0$ and constant A occurs if the receiver electronics give an output signal that has a small offset. In this case, the additional term Fourier

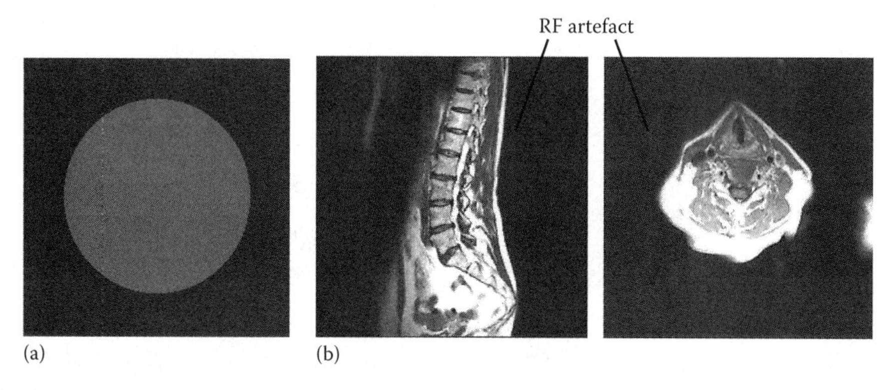

FIGURE 7.45
Examples of 'zipper' artefacts caused by RF interference. In this and a number of subsequent illustrations, we present both experimental data and images in which the clinical effect of such artefacts is simulated using the mathematical operations described in the main text. (a) Simulated artefact obtained by applying a sinusoidal signal of amplitude 0.0005 of the time-domain data maximum and phase randomised between phase-encoding steps. (b) Clinical images windowed to reveal the presence of a weak RF artefact. A more obvious example of the effect in a phantom image can be found at http://chickscope.beckman.uiuc.edu/roosts/carl/artifacts.html (accessed on February 27, 2012).

transforms to a delta function centred at the origin, which manifests itself as a single bright spot in the centre of the image. Whilst most systems now automatically correct for this effect, it should be realised that it is intrinsically impossible to distinguish the artefact from the (admittedly highly unlikely) case of a sample that *genuinely does have* a bright peak at this location.

The converse of this type of artefact is the case of an isolated glitch or spike in the time-domain data, usually due to some equipment fault (and now rarely seen on well-maintained clinical scanners). Since the Fourier transform of a delta function is a sinusoidal oscillation, the effect on the image will be a set of light and dark bands at an angle and frequency dependent on the position of the noise spike in k-space, as illustrated in Figure 7.46a. Although apparently alarming, this type of artefact is very easy to rectify, by changing a single value in the time-domain data. More difficult to correct, however, is a related type of banding artefact, as shown in Figure 7.46b, which occurs when a spurious spin or stimulated echo is refocused in an unwanted position during the data acquisition window.

When acquiring images, a modern clinical scanner performs extensive calibration work that is generally unseen by the end user. One of the important tasks is to set the receiver gain such that the data 'fill' the ADC, that is, make optimal use of the number of bits available for digitising each data point. Occasionally, this algorithm used will fail, leading to an echo of the form shown in Figure 7.47a, which 'over-ranges' or 'saturates' the detector. Thus,

$$S(\boldsymbol{k}) = \text{Lesser of}\left[\int \rho(\boldsymbol{r})\exp(i\boldsymbol{k}\cdot\boldsymbol{r})d^3\boldsymbol{r},\quad S_{\max}\right]. \tag{7.101}$$

After FT, the image tends to appear dim and blurry, often with a halo, as shown in Figures 7.47b and 7.47c. No correction is possible, since data have been lost, and the image must be re-acquired with the correct receiver gain.

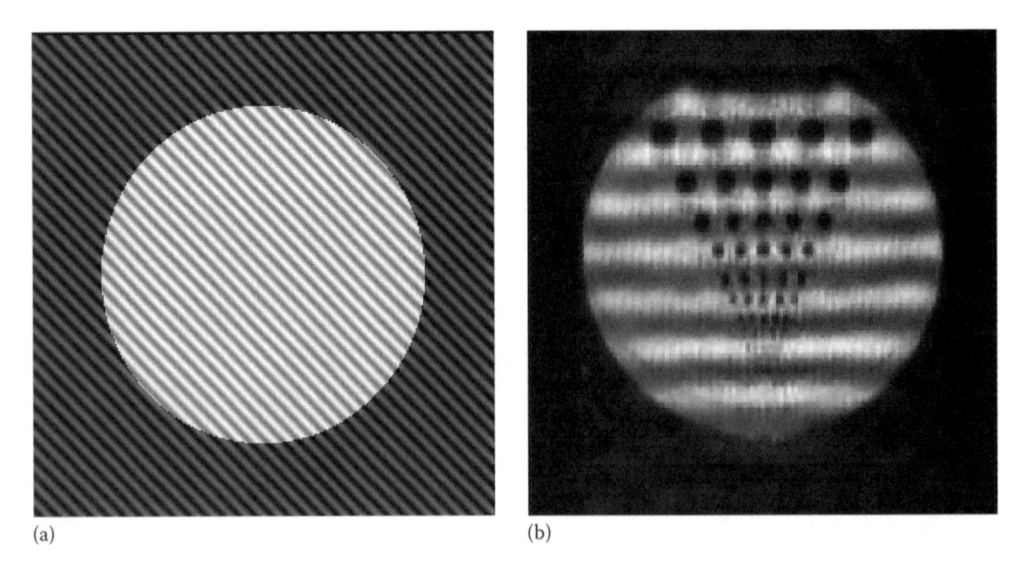

FIGURE 7.46
Examples of 'stripe' artefacts caused by RF interference. (a) 'Data glitch' artefact simulated by setting a single point (100,100) in the 256 × 256 data set to the data maximum. Note that the angulation, width and spacing of the stripes will depend on the point at which the glitch occurs. (b) Experimentally measured echo overlap artefact, in this case caused by the generation of an unwanted stimulated echo in a multiecho sequence. (Data for (b) from Doran et al., 2005a.)

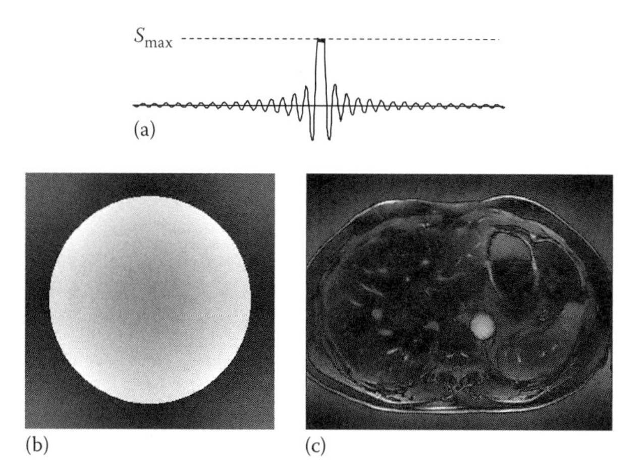

FIGURE 7.47
The 'data clipping' artefact. (a) Illustration of the cause of the artefact. (b) Image corresponding to a simulated time-domain data set of a uniform phantom clipped at 0.2 of the peak modulus value in the original data: the image shows subtle intensity variations. (c) Image corresponding to a liver data set with the modulus of the time-domain data clipped at 0.05 of the peak modulus value in the original data: the image shows significant loss of original signal intensity and a 'halo' artefact.

Occasionally, the amplifiers for the real and imaginary channels of the data might become unbalanced. This leads to a so-called quadrature artefact, as shown in Figure 7.48. On some occasions the ghost will be shifted diagonally (as illustrated at Beckman Institute 1998), which distinguishes this from most other sources of ghost artefact.

As discussed in Section 7.10.1, the RF probe often produces a $\boldsymbol{B}_1$ field that is spatially inhomogeneous, and this relates directly to the amplitude with which signals from different spatial locations are received. The MRI signal equation becomes

$$S(\boldsymbol{k}) = \int C(\boldsymbol{r})\rho(\boldsymbol{r})\exp(i\boldsymbol{k}\cdot\boldsymbol{r})d^3\boldsymbol{r}. \tag{7.102}$$

Since the spatial sensitivity profile of the coil $C(\boldsymbol{r})$ is independent of $\boldsymbol{k}$, the resulting Fourier transform is just a weighted image (cf. Section 7.6.11). Although the drop-off of

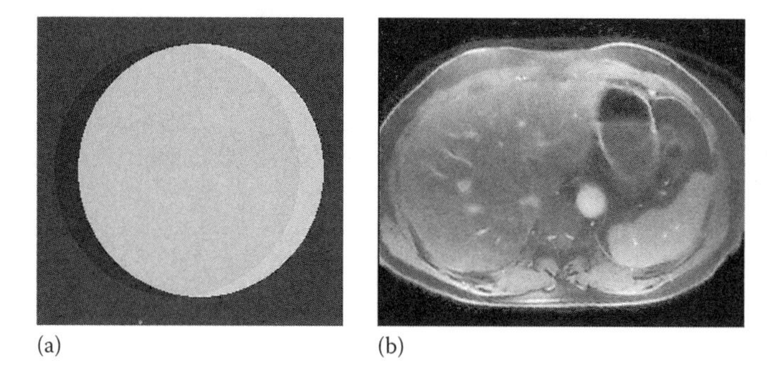

FIGURE 7.48
Examples of 'quadrature ghost' artefacts. (a) Simulation data for uniform phantom (shifted off centre) with real channel multiplied by 0.8 to simulated unbalanced amplifiers on the two receiver channels. (b) Liver data set with the imaginary channel multiplied by 0.8. Note that, in both cases, the image has been windowed to visualise the ghost more easily.

signal illustrated in Figure 7.34 causes a major loss of SNR, the effect will generally not be regarded as an artefact *per se*, as it is a design feature of the probe. If $C(\mathbf{r})$ is known in advance, then it can be used productively for parallel imaging – see Section 7.6.14.3. Occasionally, however, as shown in the 'RF Inhomogeneity' section of Beckman Institute (1988), unexpected signal dropout may reveal a technical problem with the probe.

Implicit in the formulae earlier is the assumption that all the magnetic field gradients are constant (i.e. the change in B_z is linear with distance) throughout the sample volume imaged. In practice, this is not true. Modern gradient coils are a sophisticated compromise between the competing requirements of linearity, low inductance (important to achieve fast switching rates) and shielding (Section 7.8.3).

Figure 7.49a shows the departure from linearity of a typical clinical gradient coil, whilst Figures 7.49b and 7.49c demonstrate the distortion that this causes for a linearity test phantom. Doran et al. (2005b) have demonstrated how these distortions can be mapped (Figure 7.49d) and corrected (Figure 7.49e). These spatial distortions are a particularly important consideration if MRI data are to be used for stereotactic biopsy or radiotherapy treatment planning.

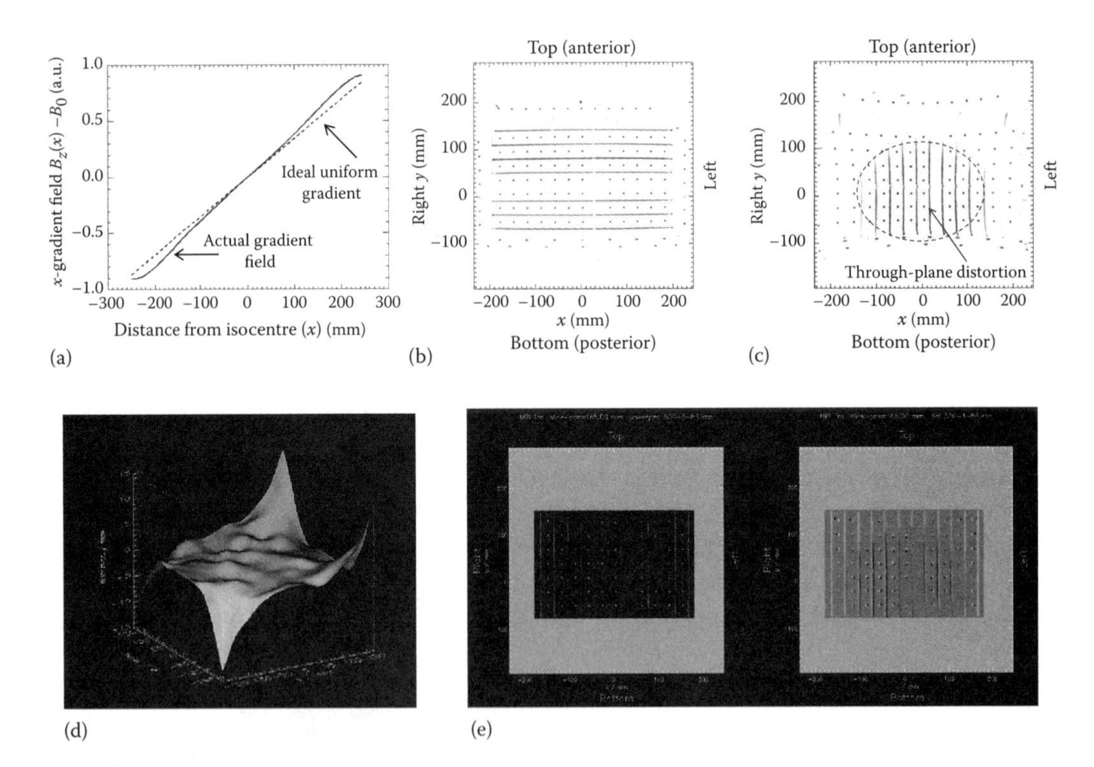

FIGURE 7.49
Explanation of gradient distortion. (a) A real, physical gradient coil does not produce an ideal uniform gradient away from the magnet isocentre (data simulated on the basis of known spherical harmonic decomposition of clinical gradient coil along a line at $y = 200$ mm, $z = 165$ mm, corresponding to the heavily distorted top row of points in (c)). (b) Scan of linearity test object in a transverse plane through the magnet isocentre, demonstrating little distortion. (c) Same acquisition but viewing a slice 165 mm from isocentre (through-plane distortion is clearly seen). (d) Experimentally measured x distortion for a coronal plane at $y = 135$ mm. (e) Distortion-corrected version of the slice in (c) and subtraction image, showing the movement of points needed. Continuous blocks of light grey indicate parts of the image that cannot be corrected either because the distortion data were unobtainable or because the image data were distorted out of field of view. (Data from Doran et al., 2005b.)

7.8.2 Artefacts Relating to Experimental Design

From Section 7.6.8, it is seen that image resolution and FOV are directly related to *k*-space sampling. If the time-domain signal is undersampled, higher frequencies may give rise to *aliasing artefacts*, by being 'folded back' into the low-frequency part of the spectrum. See Liang and Lauterbur (2000, p. 255ff) and Figure 7.34.

When resolution is improved, allowing a part of the image to be zoomed by increasing the gradient strength without increasing the total number of datapoints, some of the sample often extends outside of the FOV. This would normally lead to aliasing. However, in the read direction, one has the option to *filter* the RF signal prior to digitisation, using analogue RF electronics. The frequency response of this filter should be adjusted to match the digitisation step used. Irrespective of aliasing, it is important to apply an appropriate filter to avoid the introduction of noise into the image – recall that the RMS thermal noise voltage is proportional to the square root of the frequency interval observed in the Johnson–Nyquist formula.

A problem with limiting the extent of sampling in *k*-space by having a finite number of phase-encoding steps is that the Fourier data are often 'brutally' truncated. This may be expressed mathematically by multiplying the ideal k-space data by a 'top-hat' window function. By the convolution theorem, the resulting image will be the convolution of the desired image and the Fourier transform of the window function. The latter is a rapidly oscillating sinc function, and, thus, the image obtained from truncated *k*-space data exhibits oscillatory behaviour known as *Gibbs ringing*. Examples of Gibbs ringing artefacts are shown in Figure 7.50; in clinical practice, one tends to observe ringing most commonly with high-intensity features such as abdominal fat. Full details of the mathematics involved are given in Liang and Lauterbur (2000, p. 251ff). The deleterious effects of Gibbs ringing can be mitigated, at the cost of a somewhat reduced spatial resolution, by applying an *apodising filter* prior to FT.

7.8.3 Artefacts Generated by the Sample

In Fourier imaging, major artefacts arise due to respiratory motion and to blood flow in the body. Flowing blood usually has a high contrast with respect to surrounding tissue because of the 'inflow effect' (Section 7.7.3). This in itself is an advantage, in clearly demarcating

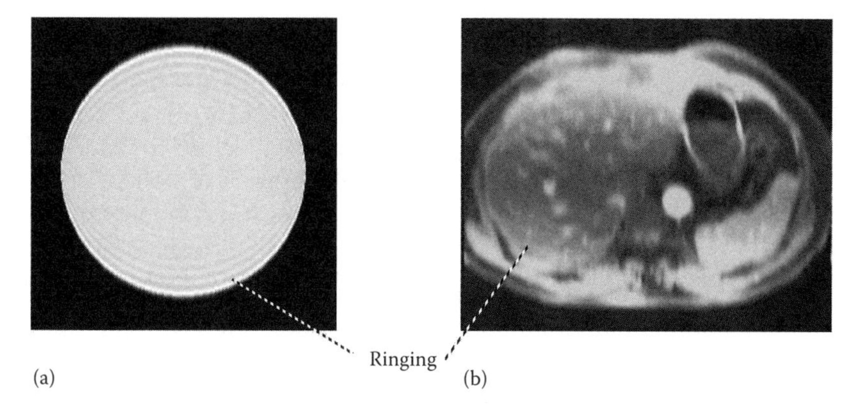

FIGURE 7.50
Examples of Gibbs ringing artefacts caused by abrupt truncation of the *k*-space data. (a) Uniform phantom data from 256 × 256 matrix with only the central 64 phase-encode lines retained. (b) Liver data set with only the central 31 phase-encode lines of the *k*-space data retained.

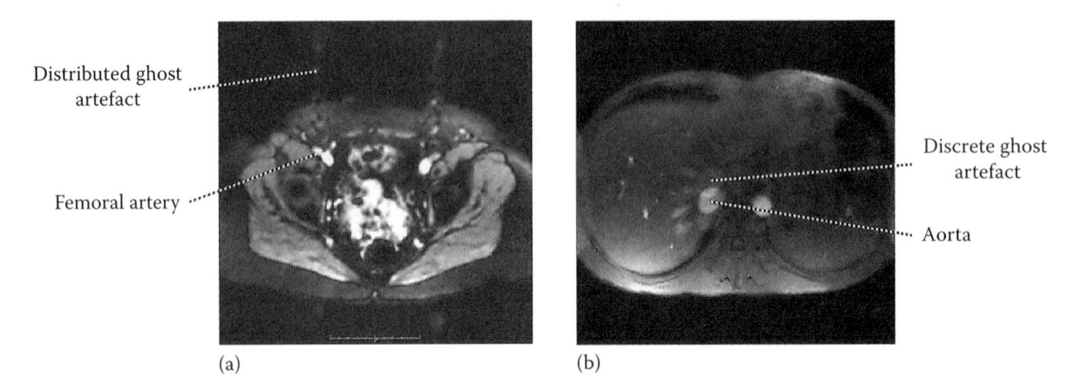

FIGURE 7.51
Examples of flow-related artefacts. (a) Fat-suppressed image of the pelvis showing an artefact distributed along the phase-encoding direction, due to flow in the femoral artery. (b) Abdominal image containing discrete ghosts of the aorta and other vessels. The exact appearance of the artefacts depends on the relative values of flow velocity and repetition time.

vessels and also in providing a means of assessing blood flow. However, the motion of the blood also results in the blood acquiring a different phase shift compared to the surrounding tissue in the presence of the phase-encoding gradients. This incorrect phase information results in signal from the blood being wrongly positioned in the reconstructed image. Generally fast-flowing blood gives rise to a disturbance across the entire image in the phase-encoding direction, in line with the vessel giving rise to the artefact – see Figure 7.51a – but sometimes discrete ghost images may be formed, as in Figure 7.51b. The high intensity of the blood means that the artefact also has a high intensity, and this effect is particularly pronounced at high fields and in fast-imaging sequences. The artefact can be reduced in conventional sequences by ECG triggering (see Section 7.6.14.2), although the degree of artefact suppression will depend on the degree to which the ECG trace represents local flow-rate changes. If the orientations of the phase- and frequency-encoding gradients are exchanged, the orientation of the artefact will be rotated through 90°, allowing pathology of interest to be observed. With fast imaging, a more productive approach is to saturate either regions adjacent to the slice, or the whole sample, to reduce the intensity of the signal from flowing blood.

Respiratory motion also results in signals being misplaced in the phase-encoding direction, and again the effect is particularly a problem when the moving tissue is of high intensity. The major artefact is due to abdominal and superficial fat (Figure 7.52), which manifests as ghosts of the body contour or internal fat displaced by a distance related to the period of the motion and the repetition time of the sequence. One commonly used method to reduce this effect is signal averaging, which, to a certain extent, averages out the effects of motion. Under some circumstances, however, respiratory-gated acquisitions (Section 7.6.14.2) or fat saturation (Section 7.4.2 and Section 7.7.7) are more appropriate.

Mathematically speaking, the origin of the artefact is an extra term in the equation for the time-domain signal:

$$S(\boldsymbol{k}) = \int \rho(\boldsymbol{r}) \exp[i\boldsymbol{k} \cdot \boldsymbol{r} + i\Delta\phi(\boldsymbol{k})] d^3\boldsymbol{r}. \tag{7.103}$$

Analysis of the effects and generation of artefact models is nontrivial, but nevertheless tractable (Liang and Lauterbur, 2000, p. 260ff) for simple cases. A variety of motion compensation

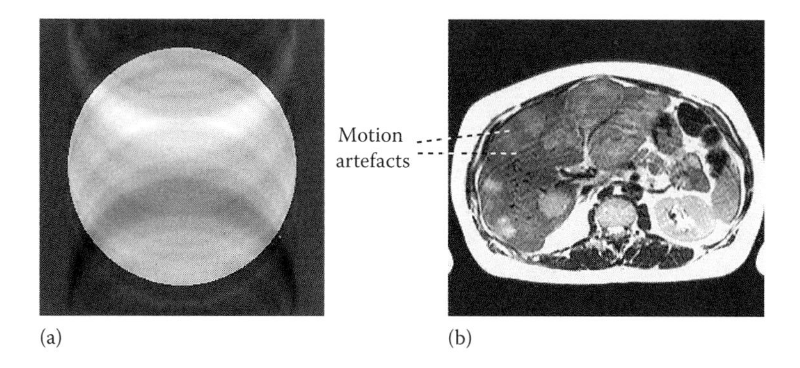

FIGURE 7.52
Examples of motion artefacts. (a) Simulation of severe ghosting (uniform phantom with random phase variation over range $\pm\pi/2$). (b) Clinical image of the abdomen, showing internal ghosts from highly intense fat signal.

strategies are available (e.g. gradient moment nulling, navigator echoes and analysis in (k, t) space), but these are beyond the scope of this chapter.

In clinical imaging, the *chemical-shift artefact* relates to tissues composed of both fat and water. The two species contain protons that are in different chemical environments and thus resonate at slightly different frequencies. If frequency is used to define position, then water and fat protons from the *same* location in the patient will appear in the image slightly displaced from one another. The artefact occurs to a detectable degree only in the read direction. Consider the angular frequencies of precession of the water and fat spins (ω_w and ω_f, respectively) in the presence of a read gradient. These are described by Equation 7.58, but with the electron shielding parameter σ – see Section 7.9.2 – explicitly included:

$$\omega_w = -\gamma(B_0 + xG_x)(1-\sigma_w) \quad \text{and} \quad \omega_f = -\gamma(B_0 + xG_x)(1-\sigma_f). \tag{7.104}$$

Suppose we set our demodulation angular frequency such that water spins are on resonance at the magnet isocentre. The demodulated angular frequencies (cf. Equation 7.60) are

$$\omega_w = -\gamma x G_x(1-\sigma_w) \quad \text{and} \quad \omega_f = -\gamma[B_0(\sigma_w - \sigma_f) + xG_x(1-\sigma_f)]. \tag{7.105}$$

Since σ_w and σ_f are both of order 10^{-6}, the terms $xG_x(1 - \sigma_w)$ and $xG_x(1 - \sigma_f)$ are both negligibly different from xG_x. However, since $B_0 \gg xG_x$, the term $B_0(\sigma_w - \sigma_f)$ cannot be ignored, and this is what gives rise to the chemical-shift artefact. The phases developed by the two species in the read-out period are thus

$$\phi_w = -\gamma\, xG_x t \quad \text{and} \quad \phi_f = -\gamma[B_0(\sigma_w - \sigma_f) + xG_x]t. \tag{7.106}$$

Why do we see no corresponding shift in the phase-encoding direction? Superficially, the situation would appear the same, with the only modification to Equation 7.105 being that xG_x is replaced by yG_y. However, the crucial difference is in how phase is accumulated in spin-warp imaging: for the phase-encoding direction, it is the *gradient* G_y that is changed not the encoding time τ – see Section 7.6.7. Thus, while the phase differences between two successive points in the read direction are

$$\Delta\phi_w = -\gamma\, xG_x\Delta t \quad \text{and} \quad \Delta\phi_f = -\gamma\,[B_0(\sigma_w - \sigma_f) + xG_x]\Delta t, \tag{7.107}$$

between successive phase-encoding steps, we have $\Delta\phi = -\gamma y\Delta G_y\tau$ for *both* species, with the term $-\gamma[B_0(\sigma_w - \sigma_f)\tau]$ being a constant for all phase-encoding steps.

We may regard our image as the superposition of two separate images, one for water and one for fat, that are *misaligned* by an amount Δx_{CS} in the read direction:

$$S(\boldsymbol{k}) = \int [\rho_{\text{water}}(\boldsymbol{r})\exp(i\boldsymbol{k}\cdot\boldsymbol{r}) + \rho_{\text{fat}}(\boldsymbol{r})\exp(i\boldsymbol{k}\cdot[\boldsymbol{r} + \Delta x_{CS}\boldsymbol{i}])]\,d^3\boldsymbol{r}, \tag{7.108}$$

where $\boldsymbol{i}$ is the unit vector in the x direction. See Figure 7.53a for an example. The difference in precession angular frequency between fat and water is $\gamma B_0\delta$, where δ is the fractional chemical shift (i.e. the value in ppm (parts per million) divided by 10^6). The relevant scale factor to convert this to a distance is the read gradient G_x and so

$$\Delta x_{CS} = \frac{B_0\delta}{G_x}. \tag{7.109}$$

It is often of more interest to know the size of the chemical-shift artefact in pixel units, for which we make use of Equations 7.73 and 7.74:

$$\text{Chemical shift in pixels} = \frac{\Delta x_{CS}}{\Delta x} = \frac{B_0\delta N\gamma\Delta t}{2\pi}. \tag{7.110}$$

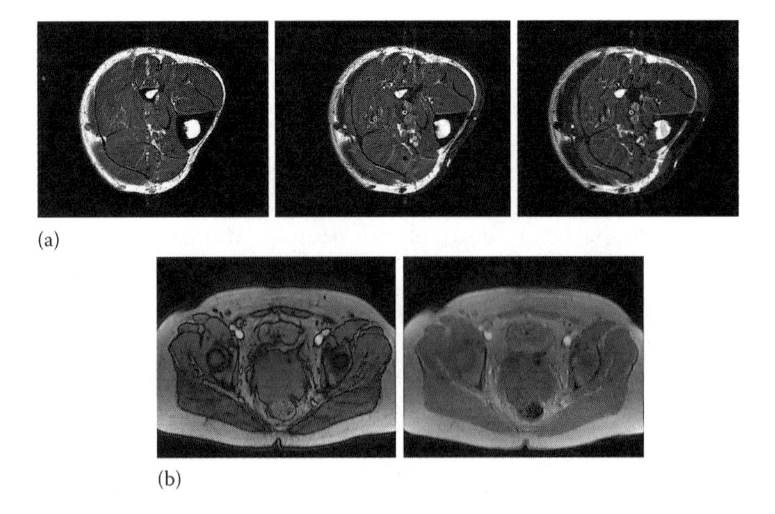

FIGURE 7.53
Effects related to the difference in chemical shift between water and fat. (a) 'Chemical shift artefact', as described in the main text. The three images were acquired on the same 1.5 T scanner with progressively decreasing acquisition bandwidths (sampling rates) of 62.5, 7.81 and 3.91 kHz and correspondingly increasing chemical shifts of 2, 15 and 30 pixels. (Data from Servoss and Hornak, 2011.) (b) Clinical images of the pelvis showing (left) dark boundary artefact at 1.5 T with T_E = 2.3 ms and (right) same slice and pulse sequence, but with T_E = 4.9 ms.

Numerically, δ is about 3.5 ppm leading to shifts around a pixel for an image bandwidth of 50 kHz ($\Delta t = 2\,\mu$s) at 1.5 T. The effect is more pronounced on high-field systems, but it can be reduced by using a stronger read gradient, resulting in greater bandwidth (shorter Δt). The artefact is particularly noticeable where there are large water/fat interfaces, for instance, around the bladder, or at interfaces between muscle and fat. The effect causes a dark rim on one side of the object where a signal void occurs and a high signal on the other side where the water and fat signals overlap. The artefact can be removed altogether by carrying out CSI to separate the water and fat components (Section 7.7.7). Although there is a great deal of lipid in the brain, it does not normally contribute to proton images, presumably as the protons are more rigidly bound than in fat, and thus the chemical-shift artefact is not a problem in the brain.

A second issue related to the difference in precession between water and fat is the relative *phases* of the two echo signals. If, as mentioned earlier, the water spins are on resonance, then at $t = T_E$, the fat spins will be out of phase by an angle $\Delta\phi = \Delta\omega T_E$. Vector addition of the fat and water signals, thus, leads to the possibility of signal cancellation. At 1.5 T, an echo time of $T_E \approx 2.3$ ms gives $\Delta\phi = \pi$. Any voxels containing a *mixture* of fat and water will appear dark, as shown in Figure 7.53b. Typically, mixed voxels are at the interface between tissues, and so this artefact is sometimes known as the *black boundary artefact*. See also the discussion of the Dixon method in Section 7.7.7.

A phenomenon closely related to the chemical-shift artefact is distortion caused by magnetic susceptibility differences in the sample (Section 7.7.5). Equations 7.108 through 7.110 are also applicable in this case, with the understanding that $B_0\delta$ should be replaced by $B(\mathbf{r}) - B_0$. Successful correction of image distortion in the read direction may be obtained using an algorithm first proposed by Chang and Fitzpatrick (1992) and developed further by Reinsberg et al. (2005). Metallic items in the body, even if nonferromagnetic, can disrupt the local field and spoil the images – see Figure 7.54. There is an important distinction between materials that are MR *safe* (i.e. there is no risk of them being attracted into the magnet) and those that are truly MR *compatible* (i.e. do not lead to degradation of the image).

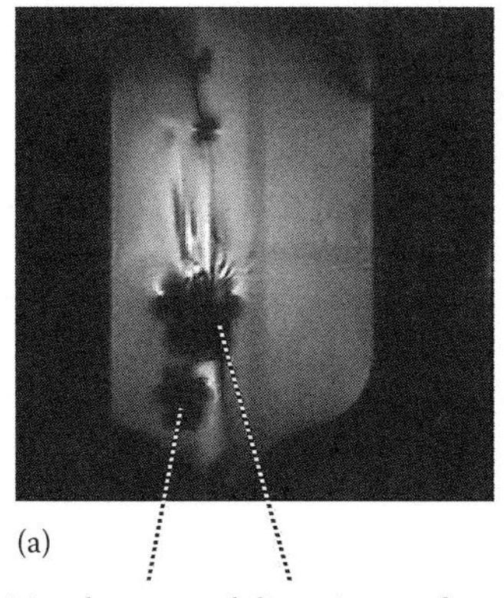

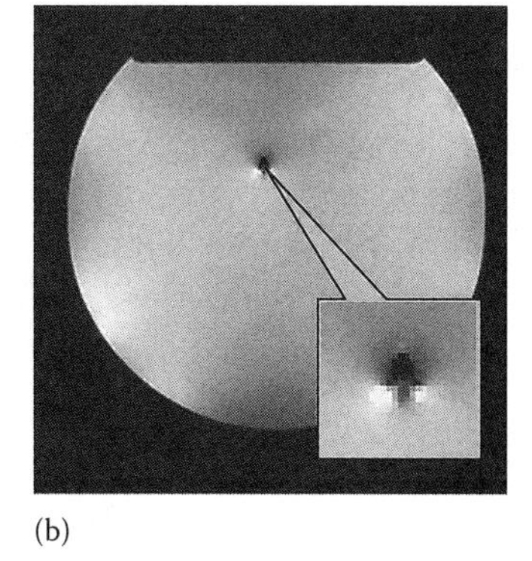

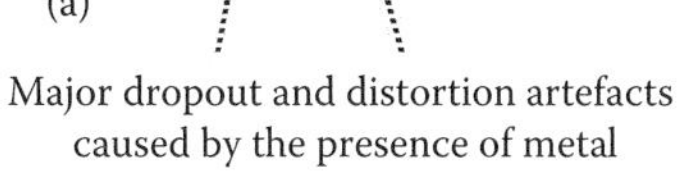

FIGURE 7.54
Examples of MR distortion artefacts induced by metallic objects in the FOV. (a) Massive dropout and distortion artefacts throughout image caused by an MR-safe (but not MR-compatible) metallic radiotherapy applicator device. (b) Classic 'arrowhead' artefact caused by the presence of an isolated cylindrical metallic object cutting the FOV – this pattern can be deduced theoretically.

EPI sequences are particularly prone to susceptibility distortion. Since the entire image is acquired during a single read period, the different phase-encoding lines occur via *progressive* evolution of the spins rather than separate encoding periods each of length τ, as discussed earlier. This means that distortion *does* occur in the phase direction for EPI. Moreover, since the gap between the acquisitions of successive phase-encoding lines is of order $N\,\Delta t$, the distortion, as calculated with the analogue of Equation 7.110, is far *worse* in the phase direction – see Figure 7.28a. Current strategies for overcoming this type of distortion include the acquisition of phase maps to correct the data (Jezzard and Balaban, 1995) or shortening the read-out period using parallel imaging (Section 7.6.14.3). Where susceptibility effects are severe and the main magnetic field becomes significantly inhomogeneous at the level of an individual pixel, so-called signal dropout occurs leading to dark regions in the image and unrecoverable loss of information.

Metallic objects and/or conductive structures in the apparatus also allow the circulation of *eddy currents*. These are generated, via Lenz's law, in response to pulsing the magnetic-field gradients. These circulating currents modify the field generated by the magnetic-field-gradient coils, leading to local distortions and often a reduction in signal intensity in the image. Nowadays, gradient coils are designed to be *shielded* so that no gradient field extends outside them. This avoids the induction of eddy currents in the conductive structures of the main magnet (Turner, 1993).

7.9 Spatially Localised MR Spectroscopy

7.9.1 Introduction

As applied in biomedicine, MRS provides a means of measuring and discriminating a range of molecules of biological significance, due to the distinct chemical shift, or resonance frequency, exhibited by different molecules. Early developments in the field led to analytical spectrometers that have since become widespread tools in chemistry, materials and protein sciences.

In the 1970s, chemical-shift differences were exploited to measure non-invasively the behaviour of phosphorus-containing metabolites concerned with cellular bioenergetics. PCr, inorganic phosphate and nucleotide triphosphates (the major cellular energy currency) could be observed, thus enabling energy metabolism and pH to be measured in intact tissues and perfused organs. Initial studies used standard, analytical, vertical-bore NMR systems (Hoult et al., 1974; Gadian et al., 1976), but appreciation of the potential of this technique led to the creation of specially engineered horizontal-bore superconducting magnets, at what was then considered a high field (circa 2 T) and which could accommodate small animals for *in vivo* spectroscopy (Gordon et al., 1980). After encouraging results, measurements on human limbs were made in this type of magnet (Chance et al., 1981; Radda et al., 1982), providing important insights into tissue biochemistry. Larger systems followed, allowing neonates (Delpy et al., 1982; Cady et al., 1983) and adult humans (Oberhaensli et al., 1986; Blackledge et al., 1987) to be scanned. These developments were distinct from the parallel developments in MRI, and early MRS systems had limited means of localising signal. Combined MRI and MRS systems with a field strength suitable for MRS were not introduced until the mid-1980s, but on such clinical equipment, the MRS facilities were still only experimental and by no means routinely used in the clinic (Leach, 2006).

Pre-clinical *in vivo* spectroscopy can now be performed on a wide range of experimental equipment, with 200 MHz (4.7T) to 400 MHz (9.4T) being common for experimental horizontal-bore systems, but measurements in perfused organs and small rodents being possible on higher field vertical-bore systems. These can be supplemented by *ex vivo* measurements such as magic-angle spinning (MAS) – see Section 7.9.6 and Griffiths and Griffin (1993) – and measurements of fluids, extracts or perfused cell systems. Pre-clinical MRS can be associated with MRI studies, and specialist systems to measure multiple animals simultaneously, particularly to aid understanding of gene function, have been developed. Clinical measurements – see, for example, Nelson (2003) – have been performed principally at 1.5T, with a number of centres utilising higher field systems between 3 and 8 T. At the time of writing, several ultra-high-field systems (e.g. 9.4 T) are being planned or installed.

While standard receive-only coils can often be used for ^1H studies, measurements of other nuclei will generally require specialist RF coils, which may be multifrequency if decoupling (Section 7.9.2) or complex multi-nuclear studies are required (unless the body coil is used for decoupling). In general, such coils are not provided with the clinical scanner or directly by the scanner manufacturer (although they may be sold as an option by some vendors). They must either be sourced separately from one of a number of specialist companies or built in house. Facilities for tuning coils (if not 'spoilt-Q' designs) and for measuring power deposition to ensure compliance with safety requirements are necessary. Pre-clinical coils can often be supplied by the manufacturer of the scanner, but spectroscopy experiments may still require considerable local investment in development.

While initial MRS measurements were concerned mainly with measuring phosphorus-containing metabolites, using ^{31}P MRS, early work also exploited ^{13}C, and exogenous agents using ^{19}F. As the technology developed, it became possible to produce good quality spectra of metabolites containing ^1H, providing increased sensitivity and allowing access to a wider range of molecules of biological importance. Several other nuclei, including ^{23}Na, have also been employed in spectroscopic studies.

As this book is concerned primarily with imaging, the current section will provide only the necessary background material to explain NMR spectroscopy and will concentrate primarily on localisation methods, a number of which are similar to methods used in NMR imaging measurements. For more advanced topics, the interested reader is encouraged to consult specialist textbooks, such as Abraham et al. (1988), Bovey et al. (1988), Goldman (1991) and Freeman (1997).

7.9.2 Chemical-Shift Spectroscopy

The relationship between resonant frequency and applied magnetic field for a particular isotope was discussed in Section 7.3. An isolated atom is subject only to the externally imposed magnetic field. In a molecule, however, atoms are shielded by the electron cloud of the molecule, which produces a small additional magnetic field that is characteristic of the atom's position in a given molecule. This results in a chemical shift (δ) with respect to a fixed reference angular frequency (ω_{ref}), given by

$$\delta = \frac{\omega_{ref} - \omega}{\omega_{ref}} \times 10^6, \tag{7.111}$$

where

ω is the Larmor precession angular frequency of the atom of interest in a particular molecule, and

δ is expressed in terms of ppm.

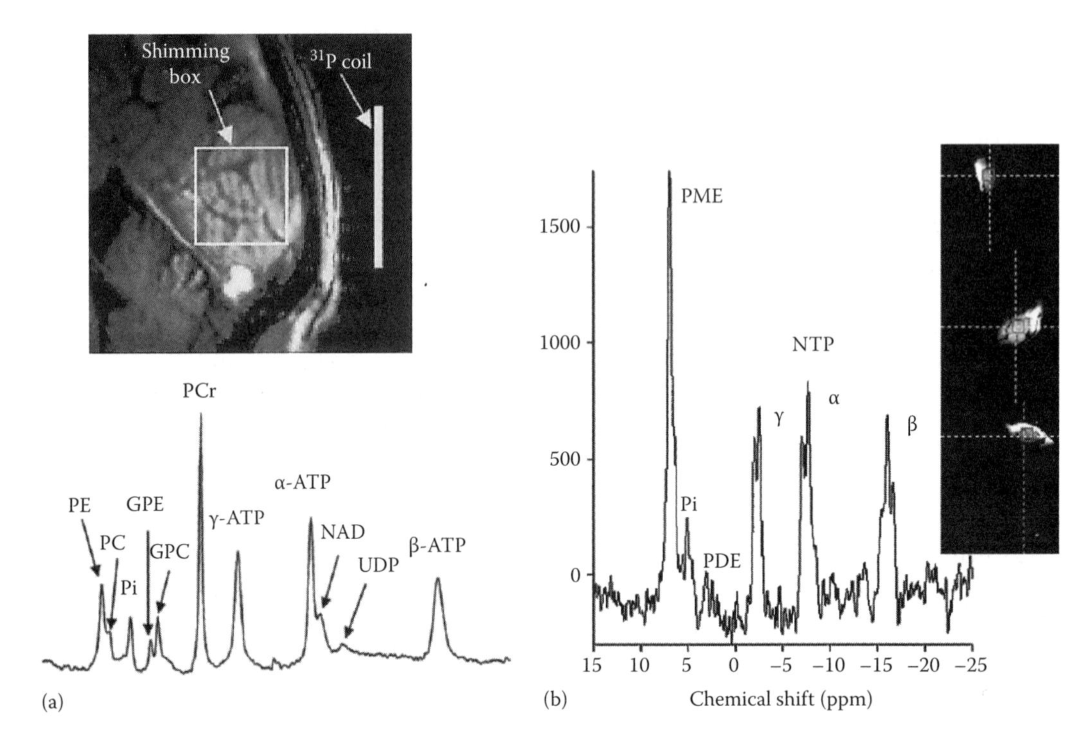

FIGURE 7.55
Examples of ^{31}P spectra acquired *in vivo*. (a) Localiser image and corresponding spectrum acquired from the primary human visual cortex at 7 T (128 signal averages, $T_R = 3$ s, total acquisition time = 6.4 min). Note the excellent SNR and spatial resolution, as well as the absence of splitting of the three ATP peaks, since the J-coupling has been removed by the particular acquisition method. (Data from Lei et al., 2003.) (b) Localiser image and spectrum from a non-Hodgkin's lymphoma in the groin of a patient, measured at 1.5 T. Note the absence of a PCr peak. (Data from Payne et al., 2006.) PE, phosphoethanolamine; PC, phosphocholine (PE and PC are grouped together as PMEs in the 1.5 T spectrum); Pi, inorganic phosphate; GPE, glycerophosphoethanolamine; GPC, glycerophosphocholine (GPE and GPC are grouped together as PDEs in the 1.5 T spectrum); PCr, phosphocreatine; ATP, adenosine triphosphate; NAD, nicotinamide adenine dinucleotides; UDP, uridine diphospho sugar (provisional peak assignment); NTP, nucleoside triphosphate.

This is a useful measure, because whilst the separation of characteristic peaks, expressed in terms of angular frequency, increases with the strength of the applied magnetic field, the chemical shift is field independent. Figure 7.55 (Lei et al., 2003; Payne et al., 2006) shows two example of *in vivo* ^{31}P spectra, and the various peaks are described in more detail in Section 7.9.5.

The dispersion of chemical shifts for different isotopes varies greatly, and Table 7.2, compiled using data from Callaghan (1991) and Hoffman (2009), shows typical chemical-shift ranges for a number of isotopes. The chemical shift for a particular molecule is due to the sum of shielding effects from local diamagnetism, the diamagnetism and paramagnetism of neighbouring atoms, and intra-atomic currents, which in turn will depend on the electronegativity of the various elements involved. In many molecules, the diamagnetic effects dominate. These are due to small electron currents induced perpendicular to $\boldsymbol{B}_0$, which in turn generate a small induction $\boldsymbol{B}_{opp}$ that opposes $\boldsymbol{B}_0$. The other effects may generate inductions that either oppose or reinforce $\boldsymbol{B}_0$. The total induction experienced by the nucleus, $\boldsymbol{B}_{eff}$, will then be the vector sum of $\boldsymbol{B}_0$ and all these shielding inductions. In more advanced analysis, the interactions are expressed in terms of the *chemical-shift tensor*, σ,

TABLE 7.2

Properties of NMR nuclei of medical interest.

		Gyromagnetic Ratio					
Nucleus	**Spin I**	**γ/(10^8 rad s^{-1} T^{-1})**	**($\gamma/2\pi$)/ (MHz T^{-1})**	**Natural Abundance[a]/ %**	**Typical *In Vivo* Concentration[b]/ mM**	**Sensitivity per Nucleus Relative to ^1H[c]**	**Approx. Chemical Shift Range[d]/ ppm**
^1H	1/2	2.675	42.57	99.985	100,000	1	13
^2H	1	0.41	6.54	0.015	15	9.65×10^{-3}	13
^3He	1/2	2.038	32.43	1.34×10^{-3}	0	0.442	58
^{13}C	1/2	0.673	10.71	1.108	200	1.59×10^{-2}	200
^{14}N	1	0.193	3.08	99.63	1,800	1.01×10^{-3}	900
^{15}N	1/2	−0.271	−4.31	0.37	6	1.04×10^{-3}	900
^{17}O	5/2	−0.363	−5.77	0.037	15	2.91×10^{-2}	1160
^{19}F	1/2	2.517	40.05	100	2	0.83	700
^{23}Na	3/2	0.708	11.26	100	80	9.27×10^{-2}	72
^{31}P	1/2	1.083	17.23	100	350	6.63×10^{-3}	430
^{129}Xe	1/2	0.744	11.84	26.44	0	2.16×10^{-2}	5300

Source: Compiled from multiple sources—Callahan (1991), Emsley (1997) and Hoffman (2009).

[a] Natural abundance refers to the global occurrence of different isotopes of the element.

[b] Figures are derived from the overall elemental composition of the human body (Emsley, 1997). The concentration of the nucleus in any given metabolite may differ and so the concentrations observed by MRS may be greater or much lower. Values for different isotopes have been multiplied by the natural abundance, so this column gives the isotopic concentration.

[c] Relative sensitivity is calculated using the factor $\gamma^3 I(I+1)$. To obtain an estimate of the overall sensitivity of an experiment, the concentration relative to water and the relative efficiency of the NMR probe and other features of the experiment must also be taken into account.

[d] Data from Callaghan (1991) and Hoffman (2009).

but in liquid-state NMR, molecular rotations produce such a rapid variation of the tensor that the spins 'respond' only to the time-averaged value of the Hamiltonian, leading to an angular frequency

$$\omega = -\gamma B_0(1-\sigma), \tag{7.112}$$

where σ is often known as the *shielding parameter*. The chemical-shift characteristics of atoms in many molecules for a number of nuclei have been tabulated and, in many cases, calculated theoretically. Several factors may give rise to the shift of a peak from its characteristic position: these include local pH, the solvent used, in some circumstances the temperature, and the presence of paramagnetic or ferromagnetic agents.

Another important feature of some NMR spectra is *spin–spin* splitting, which causes a single peak to be split into a *multiplet*. This is a result of the so-called *indirect interaction* (or J-coupling) between the magnetic moment of that nucleus and the magnetic moment of another neighbouring spin. The multiplet structure (doublet, triplet, etc.) can be predicted from the molecular configuration. Good explanations of this phenomenon may be found in both Goldman (1991) and Freeman (1997). Figure 7.55b shows an example in which splitting of the three ATP peaks is caused by the presence of the two other phosphorus atoms in the molecule (*homonuclear coupling*). However, in other molecules, the splitting may be the result of *heteronuclear coupling* where a ^{31}P or ^{13}C atom is affected by nearby ^{1}H atoms. The spin-spin coupling constant J describes the magnitude of the splitting.

J-coupling does not require an external magnetic field and is therefore independent of the applied field. The coupling constant is defined in Hertz (Hz). If rapid exchange of some of the atoms concerned occurs, this can lead to expected splitting not being observed or to the multiplets being broadened. The splitting can be removed by a technique known as *decoupling,* where a second strong RF field is used to saturate the resonance(s) of the nuclei that are causing the splitting. This reduces the multiplet to a singlet and can be important to increase the conspicuity of a given spectral line or to 'de-clutter' a crowded spectrum. Both homonuclear and heteronuclear decoupling are possible, depending on whether or not the nuclei causing the splitting of a given spectral line are of the same element as the nucleus being observed.

As discussed in Section 7.5.3, the linewidth of a singlet resonance is inversely related to the T_2 relaxation time by Equation 7.49: FWHM = $2/T_2$, or $1/\pi T_2$ if we are working in Hz. In an inhomogeneous magnetic field, we must use the T_2^* relaxation time, which takes into account both the natural T_2 and the effect of field variations. In spectroscopy, *ex vivo* samples are generally prepared in small NMR tubes, allowing the latter effects to be minimised, but for *in vivo* studies, the effects of local susceptibility often dominate and ensure that the local field inhomogeneity is still an important component.

In general, in NMR spectroscopy, it is desirable to operate at as high a field as possible, to increase the frequency separation of different spectral lines and to maximise the SNR available. In order fully to realise the benefits of the higher field, it is necessary to retain or improve on the field homogeneity at higher fields. Although this is practicable in small samples, particularly where these are spun to reduce the effects of susceptibility differences within the sample, the local susceptibility effects may prevent a significant gain in resolution at high fields *in vivo.* See Section 7.10.1.1 for further details.

In order to excite an appropriate range of spectral lines at differing frequencies, it is necessary to use a pulse with sufficient bandwidth to cover the chemical-shift range. For an RF pulse of length τ, the frequency bandwidth is $\Delta f = 1/\tau$. In receiving the NMR signal, it is similarly important to ensure that the equipment is appropriately set up. As with imaging, a hardware or digital filter is applied to limit the bandwidth of the received signal and therefore to prevent noise aliasing back into the spectrum. These filters should cover an appropriate bandwidth, and the sampling rate of the ADC should be adjusted to sample the complete bandwidth. If the dwell time (Δt) of the ADC is equal to the time between sampling points, the maximum frequency sampled with quadrature detection is

$$f_{\max} = \frac{1}{\Delta t}, \tag{7.113}$$

and the number of spectral points will again be limited to the number of sampling points. Thus, if one were to sample for a total of 500 ms with 1024 samples, Δt = 500/1024 ms, and, therefore, $f_{\max} = 1/\Delta t \approx 2\,\text{kHz}$. This should be the maximum frequency present in the FID or else data will be undersampled and aliasing will occur. The frequency resolution is another important factor, which is equal to the acquired bandwidth divided by the number of sample points. This resolution should be adequate to discriminate between spectral lines of interest. Both of these quantities have direct imaging analogues in Equations 7.73 and 7.74.

7.9.3 1D Spectroscopy Data Processing

A number of data processing operations are commonly carried out when converting the acquired FID to its corresponding 1D spectrum. It is important to recognise that none of these provide any additional information content to the spectrum as a whole – this is fixed at the time of acquisition – but they do enhance substantially the presentation of the spectrum and make it significantly easier for the eventual human observer to extract the features of interest.

An ideal FID may be described as a set of damped sinusoids, plus a noise term, viz.,

$$S_k = \sum_{m=1}^{M} A_m \exp\left(i[\omega_m t_k + \phi_m] - \frac{t_k}{T_{2m}} \right) + n_k, \tag{7.114}$$

where A_m, ω_m, ϕ_m and T_{2m} are the amplitude, precessional angular frequency, phase and T_2 of the mth spectral component, whilst S_k is the signal at the kth datapoint, which is sampled at $t_k = k\Delta t$, and n_k is the noise. A direct approach to interpreting spectra is explicitly to fit to this functional form and obtain the various parameters in the aforementioned equation. Since the time and frequency domains contain equivalent data, one may equally well model the spectrum, as a set of M Lorentzian functions. However, spectrum modelling can be a time-consuming and laborious process, and it is more common to use methods such as the following to enhance the visual appearance of the spectrum, with the aim of optimising the perceived spectral resolution and SNR, as well as the removal of unwanted artefacts.

Zero filling (padding): By analogy with Equation 7.74, the spectral resolution is

$$\Delta\omega = \frac{2\pi}{T} = \frac{2\pi}{N\Delta t} \quad \text{or} \quad \Delta f = \frac{1}{T} = \frac{1}{N\Delta t}, \tag{7.115}$$

and thus, in order to resolve closely separated peaks, one must acquire a large number of datapoints. However, given that the signal decreases exponentially, whilst the noise level stays the same, it is clear that a value of N will be reached when all the additional points collected contain to all intents and purposes just noise. An apparent increase in resolution may be gained by replacing this noise with zeroes, or simply padding the FID by adding zeroes at the end to increase its length. Of course, no new information is generated by the process and what is obtained is merely a sinc interpolation of the frequency-domain data. However, Bartholdi and Ernst (1973) noted early on that, if one is interested in *phased data* (see the following), then zero padding a complex array to twice its original length ensures that, after FT, all the information content from *both* channels of the original FID appears in the real channel of the spectrum. This results in a genuine gain of $\sqrt{2}$ in SNR over what would be obtained by looking at the real channel of the non-zero-padded FT.

Apodisation is the process of multiplying the time-domain data by a 'window' or 'envelope' function and has three main goals. (i) *Reduction of ringing artefacts*: By analogy with the imaging artefacts of Figure 7.50, spectra may exhibit rapid oscillation around peaks (convolution with a sinc function). This occurs if the time-domain data are truncated abruptly by the end of the sampling window, before the FID has decayed to zero. By multiplying by an appropriately decaying function (e.g. a negative exponential), the time-domain data may be brought smoothly to zero by the end of the acquisition, and ringing is eliminated in the spectrum. (ii) *Improvement of SNR:* Consider the random noise in a spectrum. Relative to the data of interest, it has a strong component that fluctuates rapidly and is thus associated with high values of the reciprocal Fourier variable t. This is another way of saying that the data acquired at the end of the acquisition window contribute mainly noise, as we saw earlier. Premultiplication of the FID by a decaying function suppresses this source of noise. However, beauty is very much in the eye of the beholder, because in both (i) and (ii), the 'improvement' in the spectrum comes at the cost of increasing the spectral linewidth. This may blur peaks unacceptably. A so-called matched filter of form $\exp(-t/T_2)$ is often used, as this leads to optimum SNR, but it results in a doubling of the peak width. (iii) *Improvement of spectral resolution*: By performing the converse operation (i.e. multiplication by an *in*creasing function) on data sets with intrinsically high SNR, one may *over*-emphasise the parts of the time-domain data that lead to high-resolution features in the spectra. This helps to sharpen peaks and make them more conspicuous. It is analogous to edge-enhancement filtering in image analysis.

Phase correction: Examination of Figure 7.9 shows that the imaginary (dispersion) component of the Lorentzian function is much broader than the real (absorption) part. This means that there is a strong incentive to 'phase' the spectra, rather than to add the two components in quadrature, as is done routinely in magnitude images. Phasing involves finding the values ϕ_m in Equation 7.114. In general, ϕ_m consists of both a constant *zero-order* phase shift, which is related to the receiver electronics and is the same for all peaks, and a second term that is angular frequency dependent. This *first-order* or linear phase shift is often related to delays in the start of acquisition after a pulse. During the delay, nuclei in the different species precess through a phase angle equal to the product of the angular frequency offset and the delay time.

Baseline suppression: From standard FT theory, we know that low-frequency components in a spectrum (i.e. the 'baseline') are related to the first few points in the time-domain data. Unfortunately, these points are often the most susceptible to measurement error, for various instrumental reasons concerning the turning-off of the RF pulse and the turning-on and stabilisation of the receiver. Considerable efforts are therefore made to recover the correct values of these points and hence to produce a flat baseline.

Solvent suppression: It is often the case that the compound of interest to the spectroscopist is present at much lower concentration than that of the reagent in which the sample is dissolved. If the solvent contains NMR-active nuclei of the same element as is being observed, then the desired signal may be swamped. A number of possibilities are available for eliminating this problem: (i) modify the solvent by isotopic substitution (e.g. deuterate a hydrogen-containing solvent); (ii) saturate the solvent resonance during acquisition by selective irradiation prior to the wider bandwidth pulse used to obtain the spectrum; (iii) excite all angular frequencies except that of the solvent resonance; (iv) make use of any differential relaxation properties of the solvent and the compound being studied (e.g. invert all the spins and acquire the spectrum as the solvent passes through its null point); and (v) computationally, one might proceed by creating a model of the solvent peak of interest and then subtracting this from the spectral data.

For more information on all these techniques, see Chapter 3 of Hoch and Stern (1996).

7.9.4 Localised Spectroscopy

Whilst the initial *in vivo* applications of MRS were measurements of perfused organs with no localisation, the introduction of systems capable of accommodating whole animals led to a need to localise spatially the origin of the measured signals. Two early approaches were topical MR, based on the use of special B_0 field profiling coils that destroyed homogeneity outside a selected region (Gordon et al., 1980, 1982), and the use of *surface coils* for reception. A surface coil has a small sensitive volume, allowing signal to be obtained from a region of interest by appropriate physical positioning of the coil (Ackerman et al., 1980). This type of RF probe is still in frequent use, providing the benefit of high SNR for regions close to the coil, a feature also now widely used in parallel imaging and phased-array coils (Section 7.6.14.3).

MRS is generally measuring metabolites present at millimolar concentrations in the body, compared with water and fat at nearly 60 M as observed in imaging. Thus, signal is much lower and has to be obtained from a larger region to provide sufficient SNR. Typically, these volumes may be of 8–27 cm^3 for ^{31}P or 0.2–1.0 cm^3 for ^1H. Surface coils are often used in spectroscopy for both transmission and reception. This is most common for nuclei other than ^1H; for the latter, the conventional body coil can be used to transmit. Excitation by the surface coil provides further localisation albeit with a spatially inhomogeneous profile. For experiments on superficial tissues, or subcutaneously implanted tumours, surface coils have thus often been used with no additional means of spatial selection, but the signals obtained can be contaminated with signal from other nearby tissues, and it is often advisable to use one of the gradient-based methods of signal localisation described later.

The spectroscopy measurement involves a number of phases. First, the subject is positioned with appropriate coils to transmit and receive signal. This may follow on from a first stage of imaging the patient, or imaging may be achieved with the spectroscopy coil in place. Coils are tuned or adjusted to optimise performance (this may be automatic or unnecessary). A region for localisation may be selected. Shimming then needs to be performed (see Section 7.10.1.1), and this may be optimised for the selected region. The results of the shimming need to be inspected to ensure that the linewidth is sufficiently good. One or more spectroscopy acquisitions are then performed, having selected and set up for the appropriate region. For ^1H spectroscopy, the sequences will include pulses or techniques to eliminate the very large water signal, as described earlier. If possible, the performance of this should be checked. A ^1H examination may include measurements at different echo times to optimise conditions for long and short T_2 metabolites. The chosen echo time may also affect the relative intensity of multiplet spectral lines, due to J-coupling and the evolution of the J-coupling in the time period. This should be taken account of when designing the measurements and interpreting the results. Signals need to be spatially reconstructed (if spectroscopic imaging (SI) was used) and Fourier transformed. Phasing is performed manually or automatically, followed by any quantification required.

7.9.4.1 Localisation Using Slice Selective Techniques

The basic principles underlying slice selection have been introduced in Section 7.6.3. By irradiating the sample with a narrow band of frequencies, whilst applying an appropriate gradient, spins in a selected slice at any orientation can be excited. For spin excitation, a nominal flip angle of 90° is commonly applied, to obtain maximum SNR in the spectrum. However, the exact flip angle depends on the shape and hence slice profile, of the RF pulse used, together with the spatial $\boldsymbol{B}_1$ field profile of the RF coil. Slice-selective 180° inversion or refocusing pulses can be applied to an individual slice. Similar techniques can be used to define suppression slices in which the signal is either excited then destroyed with spoiler gradients or nulled by the use of slice-selective noise pulses.

7.9.4.2 Image-Selected In Vivo Spectroscopy

Image-selected *in vivo* spectroscopy (ISIS) (Ordidge et al., 1986) is a method of preparing and reading out a set of slice-selected signals whose summation provides only information from a localised region (usually cuboidal). A sequence of (typically three) selective inversions is performed, in which each successive pulse acts on an orthogonal slice by virtue of an applied gradient along a different axis – see Figure 7.56a. This is followed

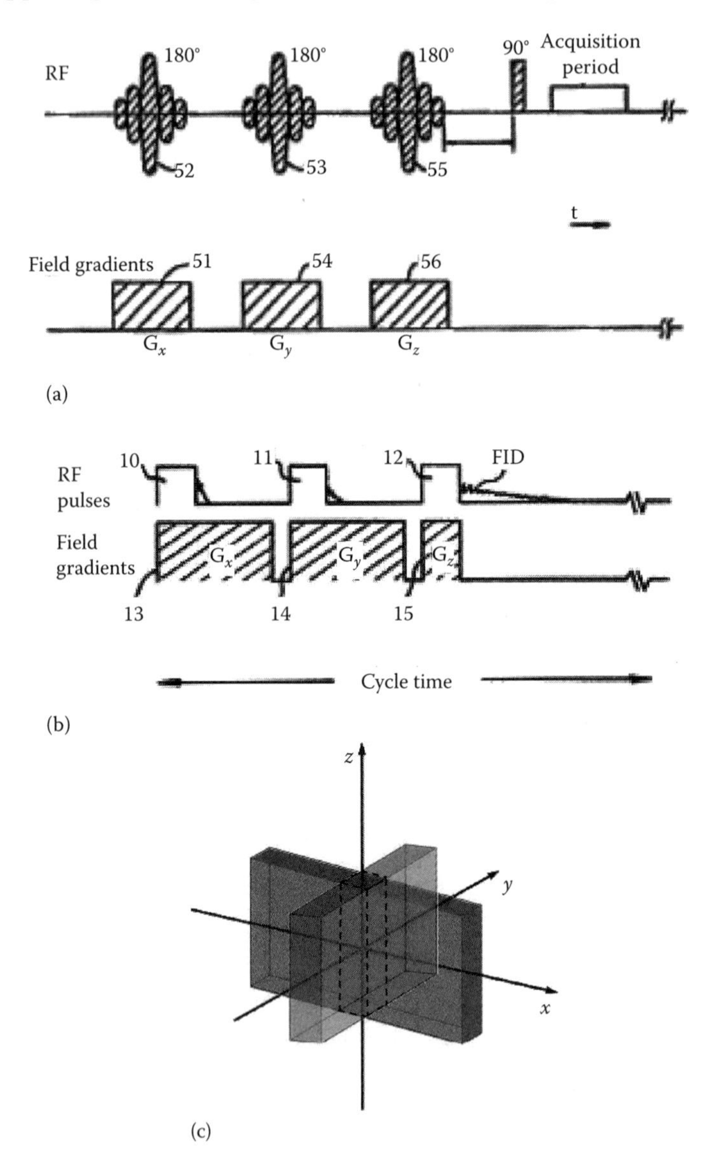

FIGURE 7.56
(a) Modified version of original drawing of ISIS localisation pulse sequence. (Modified from Ordidge, R. J., U.S. Patent 4714883.) (b) Modified version of original drawing of PRESS sequence. (Modified from Ordidge, R. J., U.S. Patent 4531094.) (Note that RF pulses labelled 10, 11 and 12 may be arbitrary, but optimum signal is obtained with the sequence 90°–180°–180°, in which case an echo, rather than an FID, is formed.) (c) Principle of selecting a region via intersecting orthogonal planes. Note that, for clarity, the effect of only an x and a y pulse has been shown, leading to a column of spins being selected. With the addition of a third plane, perpendicular to z, a cuboidal voxel is selected.

by a 90° read-out pulse and acquisition of the FID. By repeating this for the 8 possible on-off permutations of the selective inversions, and using an appropriate summation scheme for the 8 FIDs so obtained, the signal from outside of the intersection of the slices can be cancelled, leaving only the signal from the selected voxel. In the ideal case, this represents the full signal (8 averages) from that region.

This technique has been widely used for ^{31}P measurements. The collection of an FID (as opposed to an echo) allows metabolites with the short T_2 relaxation times typical of ^{31}P metabolites to be observed. The signal-averaging process implicit in the method is an advantage for this low-signal nucleus. For work with surface coils, further modifications have been required to achieve good performance. The non-uniformity of the surface coil leads to a spatially variable flip angle, so an accurate inversion is not obtained, and the read-out pulse varies with distance from the coil. These problems can be overcome by using adiabatic inversion pulses – see Section 7.3.3 – such as the hyperbolic secant, which, by frequency sweeping, provide an accurate inversion and good slice profile above a threshold $\boldsymbol{B}_1$ field. An adiabatic read-out pulse such as the BIR4 can provide a uniform flip angle above a threshold $\boldsymbol{B}_1$ field, aiding signal quantification. A modified version of ISIS allowing selection of more complex regions has also been developed (Sharp and Leach, 1992).

7.9.4.3 Point-Resolved Spectroscopy

Point-resolved spectroscopy (PRESS) (Ordidge et al., 1985; Bottomley, 1987) employs similar methods of slice selection to ISIS, in the sense that a cuboid is defined by the intersection of three orthogonal planes, but achieves 3D localisation with a single excitation and read-out, requiring no summation of spectra. In this approach, a slice is first selected by a combination of a 90° pulse and appropriate gradient. Then, in the manner of an SE sequence, a 180° pulse is applied but with a slice-selective gradient orthogonal to the first; see Figure 7.56b. Only the magnetisation from a *column* of spins is refocused, as shown in Figure 7.56c. Application of a further selective 180° pulse in a direction orthogonal to the first leads to the selection of the required cuboid. Magnetisation from outside of the selected cuboid is suppressed by spoiler gradients applied before and after the refocusing pulses. The signal is read out as an echo following the second 180° pulse. This technique is generally employed for ^{1}H spectroscopy, where the echo time set by the two 180° pulses is not a limitation, as the T_2 of many of the metabolites of interest is long enough to avoid significant signal loss. The body coil is generally used for transmission, avoiding the problems of inhomogeneous $\boldsymbol{B}_1$ fields that would affect the refocusing pulse. Standard adiabatic pulses cannot be used, as for ISIS, as they do not refocus. With the ideal experiment, the entire signal available (i.e. the equivalent of a single 90° pulse) is obtained from the volume of interest.

7.9.4.4 Stimulated Echo Acquisition Mode

Stimulated echo acquisition mode (STEAM) (Frahm et al., 1987) uses similar slice selective pulses but exploits the phenomenon of the stimulated echo (Section 7.4.4). The sequence starts with a 90° excitation as with PRESS, but follows this with a further slice-selective 90° pulse for an orthogonal slice. This has the effect of flipping 50% of the signal from the intersected column from the *xy* plane to lie along the *z* axis, where it is stored. A further selective 90° pulse with a gradient in the third orthogonal direction returns the magnetisation from the intersection of the three slices back into the *xy* plane, rephasing it as an echo. Spoiler gradients eliminate unwanted magnetisation. The sequence has two advantages over PRESS: (i) pulse profiles for 90° pulses tend to be better than for 180° standard pulses, leading to a better definition of the volume of interest, and (ii) shorter effective echo times can be obtained, as a result of the storage period not being subject to T_2 relaxation. This can be significant for metabolites with short T_2 values. Whilst primarily used to observe ^{1}H-containing metabolites, the technique has also been demonstrated for ^{31}P. The disadvantage of STEAM is that signal is obtained from only 50% of the magnetisation.

7.9.4.5 Spectroscopic Imaging or Chemical-Shift Imaging

Two-dimensional Fourier imaging and its commonly used variant, spin-warp imaging, were introduced in Section 7.6.6. This approach uses slice selection, followed by phase encoding and frequency encoding for the read-out. The technique can be generalised to a 3D imaging technique by using a further phase-encoding gradient in the slice-selective direction, to encode a series of partitions. The presence of the frequency-encoding read-out gradient prevents any intrinsic chemical-shift information from being recovered. SI is a development of this technique. Two-dimensional SI retains the slice selection, but then employs phase-encoding in *both* in-plane directions. An FID is then obtained in the absence of a frequency-encoding gradient, allowing the chemical-shift information to be obtained. Three-dimensional SI eliminates the initial slice selection, but typically applies a non-selective 90° pulse, followed by phase encoding all three spatial directions. After read-out, the subsequent FID is processed using a 4D Fourier transform, leading to a 3D set of spectra. Figure 7.57 (Lei et al., 2003) gives an example of a 3D CSI data set.

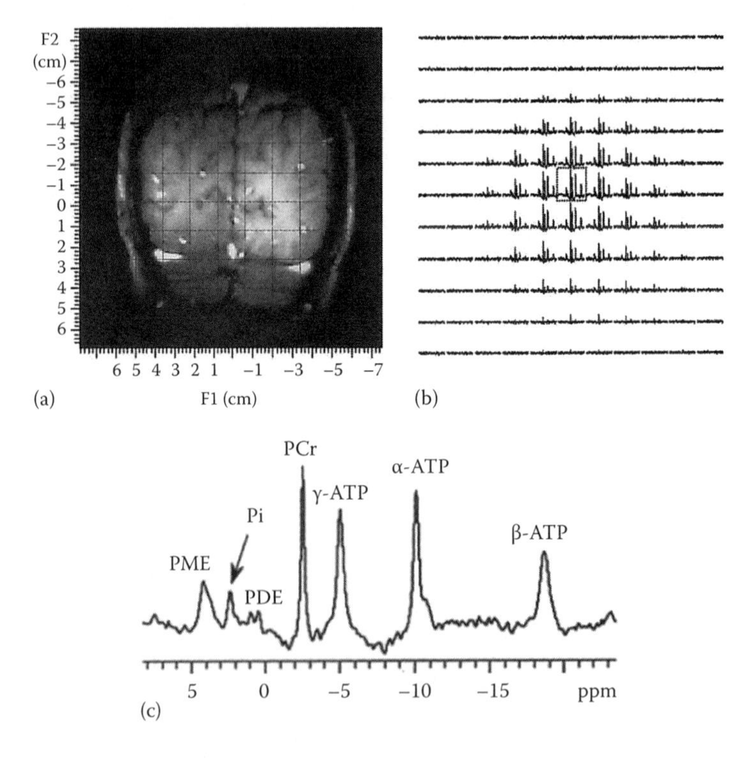

FIGURE 7.57
(a) T_1-weighted coronal localiser image of the brain at 7 T. (b) One plane of an 11 × 11 × 11 voxel 3D ^{31}P CSI data set acquired at the locations specified in (a), with voxel size 7.5 mL and acquisition time 7.85 min. (c) Expansion of spectrum from dotted voxel in (b). Peak labelling as in Figure 7.55. (Data from Lei et al., 2003.)

The technique preserves the entire signal, apart from that lost due to T_2^* dephasing, and also simultaneously samples a large volume, leading to a higher SNR. However, it has a longer measurement time than single-voxel techniques as many phase encoding steps are required. For ^{31}P measurements, it is usual to use a 3D acquisition. For ^{1}H measurements, 2D acquisitions are common, and the sequence may be used in combination with PRESS or STEAM to define a selected volume, within which SI is applied.

The number of phase-encoding steps in each direction is usually small, due both to time constraints and to the need to obtain large voxels for SNR reasons. This can result in a poor and complex profile for each voxel. The potential for so-called Fourier bleed of signal from adjacent voxels into the voxel of interest is a concern, as well as aliasing of signal from more distant voxels. For this reason, combination with a single-voxel selection technique to select a region of interest avoiding problematic tissues (such as fat) and/or the use of suppression slabs is common to improve the quality of an examination. Collection of spectroscopic signals over a large area allows metabolite maps to be produced, where the distribution of a single metabolite, or several shown in different colours, provides a new diagnostic tool. Care has to be taken to ensure that the spectral quantification is reliable, and this is particularly the case where the process is automatic and uses software that is not well characterised.

7.9.4.6 Choice of Localisation Method

There are pros and cons of all the major localisation methods. Usually, they are used as part of an overall investigation that also includes imaging examinations. Although the latter are used to aid the localisation of the spectroscopic study, they also take some of the available time, reducing that available for acquisition of the spectrum. In some cases, it may also be necessary to use different coils for spectroscopy compared with imaging, to improve sensitivity, or allow a different nucleus to be examined. If time is short, a single-voxel approach may be best. If on the other hand the location of the region of interest is not known, or a tissue needs to be surveyed, SI may be more appropriate. In both cases, a knowledge of the way in which voxel size and position are defined is valuable (this can vary between manufacturers), and it is advisable to perform QA checks to ensure the reported positioning is correct. Motion can degrade both the shim and the spectroscopic results. All localisation methods are affected by motion but in different ways. PRESS appears to be better than STEAM in the presence of motion, but has a poorer slice profile. ISIS is relatively robust, but loses signal. Slice-selective methods are affected by a chemical-shift artefact, which shifts the localisation volume for each metabolite as a function of frequency. This is particularly a problem for nuclei with large dispersion such as ^{31}P and ^{19}F. In the past, eddy currents caused by rapid gradient switching have caused poor line shape, particularly in short echo-time spectra. Advances in instrumentation have greatly reduced this effect, but it remains a possible cause of problems and should be borne in mind when investigating poor results.

7.9.5 Advanced Topics

Spectroscopy is a developing area, as improved instrumentation increases the capability of instrumentation. The spectroscopy techniques discussed so far yield 1D spectra, where the amplitude present at different frequencies is plotted. High-resolution spectroscopy (as used in *ex vivo* chemical analysis) commonly uses a range of so-called 2D spectroscopy techniques. Here, the dimensions referred to are spectral rather than spatial. One dimension is commonly resonance frequency, as mentioned earlier, whilst the other may be one of a number of NMR phenomena. The most common approach is *correlation spectroscopy* (COSY), which reports on scalar coupling processes. This correlates signal from nuclei up to three chemical bonds apart, giving insight into molecular structure. This approach has been used to help identify coupled metabolites in complex ^{1}H *in vivo* spectra of patients with leukaemia (Prescot et al., 2005). Total correlation spectroscopy (TOCSY) shows all the protons in a spin system, not just those within three bonds. NOESY uses the NOE, whose strength is related to inter-spin distance, to show the relative distance of spins apart. There are many more techniques, including multinuclear approaches, and some of these are likely to find applications for *in vivo* measurements. See Ch. 6–8 of Goldman (1991) for a more in-depth explanation of multidimensional spectroscopy.

The presence of multiplets in a spectrum complicates its appearance and can limit sensitivity for detecting some nuclei. In liquid state NMR, it is common to decouple the J-coupling between ^{1}H and ^{13}C and ^{1}H and ^{31}P (Section 7.9.2). This is performed by irradiating the ^{1}H nuclei whilst acquiring the ^{31}P or ^{13}C signals. This process can also produce NOE enhancement of the ^{31}P or ^{13}C signal. A variety of decoupling approaches are available, and a common approach is known as WALTZ (Shaka et al., 1983a, 1983b). Coils capable of operating suitably at the two frequencies are also required, and it is important to ensure that the total RF power deposition is within safety limits.

Magic-angle spinning is a solid-state NMR technique that can be applied to *ex vivo* tissue samples (Griffiths and Griffin, 1993). The dipolar interaction — see Sections 7.3.4, 7.5.1, 7.5.2 and 7.7.9—is responsible for a significant fraction of the observed signal decay (and hence line broadening) in solids and 'soft solids' such as tissues. It is described quantum-mechanically by a so-called Hamiltonian, which contains a term of form $3\cos^2\theta - 1$. This term vanishes at the magic angle of 54.7° ($\cos^{-1} 1/\sqrt{3}$), resulting in a longer T_2 equivalent to much narrower spectral lines, and a consequent improvement in spectral resolution for tissue samples. This technique uses a high-resolution spectrometer with a specially attached spinning rotor inclined at the magic angle. The technique is used to match spectral lines seen *in vivo* with those from tissue samples from the same patients. It may also provide a metabolic adjunct to histology to aid diagnosis and assessment of prognosis in cancer. Imaging at the magic angle has been used *in vivo* (without spinning) in human imaging studies to improve visualisation of structured tissues such as cartilage (Bydder et al., 2007). See also Section 7.7.6.

7.9.6 *In Vivo* and Clinical Applications

Since the first edition of this book, there have been major advances in the instrumentation for spectroscopy that have brought it to routine clinical use in a number of applications. However, its major role remains in pre-clinical and clinical research where it plays an important role in furthering our understanding of diseases and their treatment, particularly where metabolism is important. In this section, an overview of applications is given to illustrate the uses of MRS. For more extensive discussion of applications, several references provide more detail (Gadian, 1995; Matson and Wiener, 1999; Alger et al., 2006).

^{31}P MRS (Payne et al., 2006) provides a valuable means of monitoring the cellular energy balance, due to the use of high-energy phosphate bonds to store energy. A ^{31}P spectrum obtained from human brain is shown in Figure 7.55a, with detailed peak assignments in the figure caption. The main features of the spectrum are as follows:

- the three peaks due to nucleotide triphosphates, predominantly ATP, which is the major source of energy in the body;
- phosphocreatine (PCr), which provides a readily accessible supply of energy, useful for meeting short-term energy demands, for instance, in muscle contraction, but is not present in liver;

- inorganic phosphate (Pi), which is the product of the breakdown of ATP to produce adenosine diphosphate (ADP) and energy;
- the phosphomonoester (PME) peak(s) and the phosphodiester (PDE) peak(s).

By measuring the chemical shift of the Pi peak with respect to a reference frequency (usually that of PCr), the local pH can be determined. For instance, in the absence of an adequate oxygen supply, the Krebs cycle cannot provide ATP as a source of energy, and it must be derived less efficiently via anaerobic glycolysis. This produces lactic acid as an end product, which gives rise to more acidic conditions and therefore to a fall in pH. This is reflected by a shift of the Pi peak towards the PCr peak. If adequate spectral resolution is available, it may be possible to see a separation of the Pi peak into components from intracellular and extracellular compartments. *In vivo* spectra may represent a large population of differing cell types under varying degrees of anoxia, and the Pi peak may be broadened due to this distribution.

The PME peak is representative of PMEs, such as phosphocholine (PC) and phosphoethanolamine (PE), as well as adenosine monophosphate (AMP) and sugar phosphates such as glucose-6-phosphate and glycerol-3-phosphate. Some of the constituents of the PME peak act as precursors of lipid production. The PDE peak is due to compounds such as glycerophosphoethanolamine (GPE) and glycerophosphocholine (GPC), products of lipid breakdown.

By making quantitative measurements of the areas under these different peaks and taking account of factors that may affect their relative heights (such as T_1 suppression effects, the bandwidth of the irradiating pulse and the analogue filter), it is possible to observe relative changes in the concentration of these metabolites during the course of an investigation. *In vivo* ^{31}P measurements usually suffer from the lack of an internal standard, so results are generally presented as ratios of metabolites. However, with careful (and generally complex) calibration steps, the absolute concentrations of different compounds can sometimes also be assessed.

As will be apparent from the ^{31}P spectra of Figure 7.55, different organs present characteristic spectra indicative of their differing metabolism. For example, large PME and PDE peaks in liver are related to its role in producing and storing glucose. In the brain, the relatively large PME and PDE peaks may be related to the high lipid content of the brain, whilst, as shown by Figure 7.55b, elevated PME levels are also characteristic of tumours. Monitoring the PME peak can provide an early indicator of response to therapy.

Phosphorus spectroscopy studies *in vivo* have been used to investigate the effect of exercise on metabolism (Taylor et al., 1983) and metabolic myopathies such as McArdle's syndrome (Ross et al., 1981). The method has also proved valuable in the study of mitochondrial myopathies (Radda et al., 1982) and in studies of muscular weakness and dystrophies (Arnold et al., 1984). An important application has been in the study of birth asphyxia in premature babies (Cady et al., 1983). ^{31}P MRS remains an important tool for evaluating energy metabolism in skeletal muscle, cardiac muscle, brain and in organs such as the kidneys. For a broad review of NMR spectroscopy in cerebral bioenergetics, see Chen and Zhu (2005). Recently, decoupling techniques have been employed to separate PC and PE in the PME peak, and GPC and GPE in the PDE peak, potentially providing greater insight into the biochemical changes resulting from cancer treatment. In pre-clinical investigations in cells and in xenograft models, ^{31}P MRS is widely employed, and recently, changes in PC and GPC have been shown to report on the inhibition of specific pathways in cells (Beloueche-Babari et al., 2010). Some drugs contain phosphorus, and a number of studies have been performed reporting on the distribution and metabolism of drugs such as ifosfamide and cyclophosphamide (Mancini et al., 2003; Payne et al., 2000). In human studies, the relatively low sensitivity at 1.5 T has been a limitation. However, with the growing availability of higher field magnets, it is likely that there will be increasing use of ^{31}P MRS in metabolic investigations in a range of diseases.

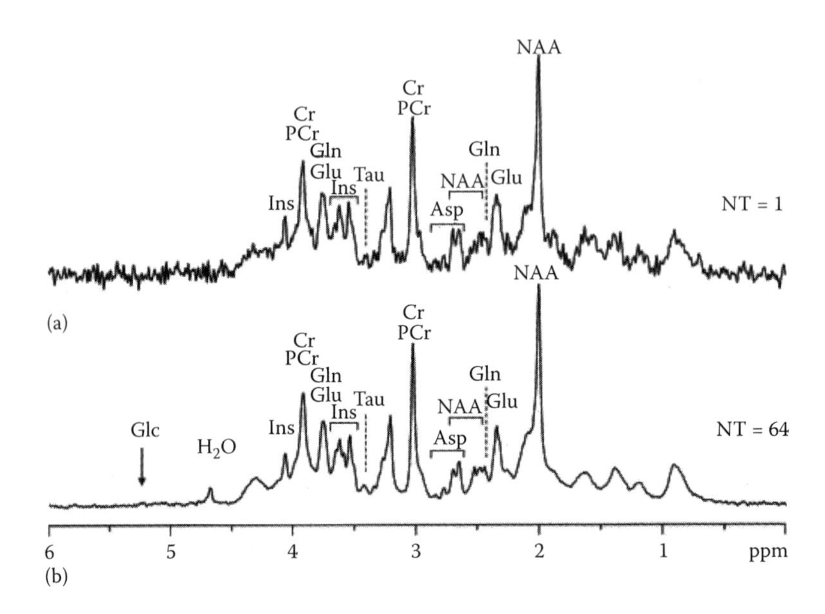

FIGURE 7.58
^{1}H NMR spectrum of the human brain at 7T acquired over the occipital lobe, largely over grey matter, using STEAM (T_E = 6ms, T_R = 5s, VOI = 8mL) for (a) single shot and (b) 64 averages. (Data from Ugurbil et al., 2003.)

^{1}H spectroscopy is technically more demanding than ^{31}P MRS due to the smaller chemical-shift range and, therefore, higher resolution requirements. The dominating effect of the water resonance can be a problem, whilst extra-cranially contamination of spectra by fat can occur. Eliminating this requires good shimming (Section 7.10.1.1), together with effective background suppression (Section 7.9.3). On the plus side, the higher sensitivity of ^{1}H MRS allows smaller voxels to be interrogated than is the case with ^{31}P.

Figure 7.58 shows a high-quality ^{1}H spectrum from human brain obtained at 7T (Ugurbil et al., 2003), whilst typical clinical spectra, acquired at 1.5T, are presented in Figure 7.59 (Murphy et al., 2004). Note how two echo times are often used, to allow a degree of *spectral editing*. All components of the spectra appear in the short T_E spectrum (20ms) of Figure 7.59a, while the choice of 135ms for the spectrum of Figure 7.59b not only removes short T_2

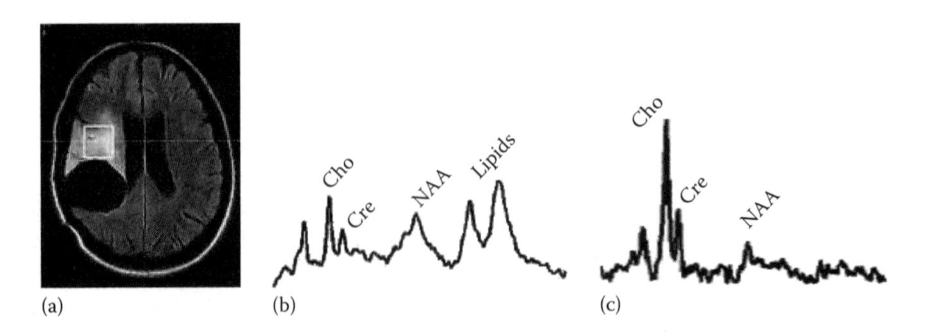

FIGURE 7.59
(a) Localiser image of a low-grade glioma patient showing spectroscopy voxel. (b) Short echo-time ^{1}H spectrum (T_E = 20ms), acquired using STEAM. (c) Long echo-time (T_E = 135ms). (Modified from Murphy et al., 2004.)

components but also substantially reduces the amplitude of the two J-coupled lipid peaks, which are in antiphase at the start of the acquisition and cancel. This allows weak underlying signals such as lactate to be seen. Major components of the spectra include the following:

- total creatine (Cr) – concerned with energy metabolism and often used as an internal reference (ratios of peak areas with respect to the Cr peak are often taken);
- total choline (Cho) – a compound of the vitamin B group; choline and its accompanying metabolites are needed for three main physiological purposes: structural integrity and signalling roles for cell membranes, cholinergic neurotransmission (acetylcholine synthesis) and as a major source for methyl groups in other syntheses;
- *N*-acetyl aspartate (NAA) – a mitochondrial product found exclusively in neurons and thus a marker for neuronal viability, of particular relevance in the study of degenerative diseases;
- glutamate and glutamine (Glu) – excitatory neurotransmitter and its precursor;
- myo-inositol (mI) – an osmolyte involved in volume regulation and lipid metabolism, commonly used as a marker for glial cells;
- lactate (Lac) – concerned with energy metabolism (see also discussion on phosphorous spectroscopy at the start of this section); this compound has a very low concentration in normal brain tissue but is elevated in ischaemia and demyelination (multiple sclerosis (MS));
- lipids and other macromolecules.

A number of other metabolites may also be seen, depending on the tissue of interest. These can include alanine (Al), aspartate (Asp), citrate (Cit), gamma-aminobutyric acid (GABA), CH_2 and CH_3. Some of these peaks are overlapping and can require sophisticated approaches to separate them. An excellent review of the NMR properties and major biochemical roles of all major ^{1}H MRS-visible brain metabolites is given by Govindaraju et al. (2000).

^{1}H MRS provides a metabolic signature of an area of tissue. It can therefore be used to identify areas of abnormal tissue that may not be clear, or may have produced no structural change, on MRI. MRS can identify disturbance in normal metabolism that may confirm or clarify imaging findings, or help in degenerative diseases that may not have shown detectable macroscopic changes in morphology. Examples include identifying tumours in the brain and prostate, identifying the focus of seizures in epilepsy, evaluating MS plaques, evaluation of stroke and myocardial lipid changes resulting from ischaemia, and investigation of neurodegenerative and psychiatric diseases. The evaluation of brain tumours is an important area, where there is growing evidence that MRS can effectively discriminate different types of tumour, as well as necrosis. MRS may also demonstrate areas of tumour that are not evident on MRI, for example, where breach of the blood–brain barrier has not occurred, aiding in the planning of treatment and identification of recurrence. This is leading to use in the diagnosis and management of paediatric and adult brain tumours, and in planning treatments such as radiotherapy and surgery. In the prostate (for which an endorectal coil is generally used), MRS shows promise in confirming the presence of areas of disease. MRS can be used to evaluate response to treatment, as the height of the choline peak appears to be a sensitive indicator. Further details on these areas can be found in Alger et al. (2006) and the references therein.

^{19}F is a nucleus that is not normally present in the body, but has been tracked usefully as an exogenous label. Several drugs contain fluorine, such as the widely used 5-fluorouracil anti-cancer agent, and these may be employed directly, or generated *in vivo* by metabolic processes from a less toxic pro-drug. By performing ^{19}F MRS, it has been possible to evaluate the metabolism and distribution of the drug, showing that retention, or half-life, can predict response to treatment, as can the presence of some fluorinated metabolites (Wolf et al., 2000; Hamstra et al., 2004). Spectroscopic imaging has been used to confirm biliary recirculation.

Another valuable nucleus is ^{13}C, which, although having low natural abundance and low NMR sensitivity, can still be seen in a number of molecules at physiological concentrations. Owing to the cost of isotopically enriched carbon, ^{13}C-radiolabelled compounds in human studies are rare, although this is a very important technique in pre-clinical investigations. The large chemical-shift range and the ability to detect a number of hydrocarbon compounds of interest will continue to support study of this nucleus. Decoupling techniques are also possible in ^{13}C spectroscopy, but these techniques require substantial RF power, and care is required to ensure that such sequences do not exceed RF power deposition guidelines (see Section 7.9.1). The use of hyperpolarisation techniques such as dynamic nuclear polarisation (Section 7.7.6) has significantly increased the utility of ^{13}C spectroscopy in recent years (Kurhanewicz et al., 2011).

A number of other nuclei can be observed, and for more information on these, see Section 7.7.10.

7.10 MRI Technology

Since the first edition of this book was published, MRI has moved from a novel scanning method, only recently introduced into the clinical mainstream, to a mature imaging modality with well over 10,000 installations worldwide. Despite this, the development of imaging hardware still continues apace. Some areas have reached a natural plateau – for example, the strength and switching speed of gradient coils are now limited largely by human physiology and the need to avoid neural stimulation. On the other hand, some aspects of the hardware have recently undergone major changes, chief among which is the RF receiver chain, driven by the potential of parallel imaging (Section 7.6.14.3).

A notable feature of recent years is the major investment that vendors have been making in software. The scanners themselves have become easier to use, allowing radiographers routinely to perform increasingly sophisticated examination protocols in a shorter time. Postprocessing of data has also changed significantly. As occurred a generation ago with the hardware, techniques that were previously the exclusive province of academic research laboratories are now reaching the clinic. This development goes hand in hand with the widespread introduction of picture archiving and communications systems (PACS), radiology information systems (RIS) and the digital imaging and communications in medicine (DICOM) standard for radiology workflow and image exchange (see Chapter 14). As physicists, one of our major challenges is now to ensure the integrity of an increasingly complex data-processing chain. The task becomes more and more difficult as MRI scanners take a larger number of decisions automatically during acquisition and as access to the truly 'raw' data is further restricted by the manufacturers.

7.10.1 Equipment

Figure 7.60 illustrates schematically the apparatus required for an MRI scanner, and the following sections describe the main sub-systems in further detail. For specific

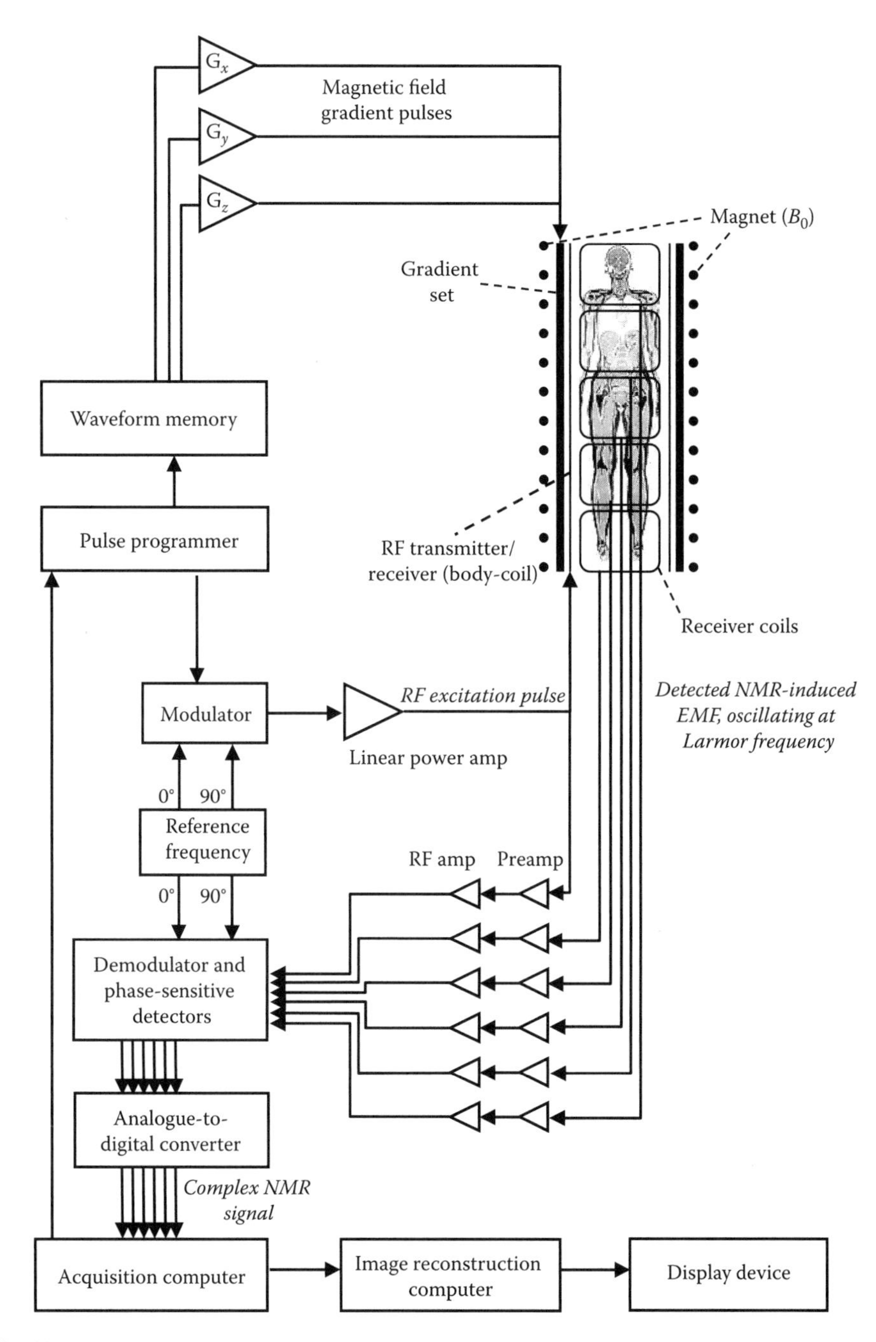

FIGURE 7.60
Schematic diagram of a clinical MRI scanner. See main text for further explanation.

technical details of recent generations of clinical scanner from the major vendors, the reader is referred to the excellent set of reports issued by the NHS Purchasing and Supply Agency and compiled by MagNET, part of the independent evaluation programme supported by UK Government's Centre for Evidence-based Purchasing (CEP) (Price et al., 2005, 2006a, 2006b, 2006c).

7.10.1.1 Main Magnet

The main magnet, which produces the B_0 field, represents the largest capital cost. Whilst the majority of clinical scanners now employ superconducting solenoidal designs (Figure 7.61a), a variety of different types of magnet are sold, many of which occupy niche markets, such as interventional MRI or low-cost MRI of extremities. Huge improvements have been made over recent years in the available field strength (currently up to around 10 T for human and over 20 T for animal systems), in magnetic field homogeneity (which is related to T_2^*) and cryogen utilisation.

In the early days of MRI, many research systems and some commercial equipment used resistive electromagnets. These require a continuous power supply, generate a significant amount of heat and require cooling, either with rapidly circulating air or water. An important consideration is the time taken for such systems to reach thermal stability and the maintenance of this state. These magnet designs are not suitable for generating fields greater than approximately 0.3 T. Nowadays, low-field systems (typically 0.2–0.4 T) are often based around 'C-shaped' permanent magnets (Figure 7.61b), which allow good access to the patient from the side. Competing with this type of scanner in the interventional market is the so-called double-doughnut design, a 0.5 T superconducting system consisting of two solenoidal coils with a gap between (Figure 7.61c). The advantage of this geometry is that the surgeon may operate from above the patient as well as to the side. Both systems represent a major advance from the traditional 'tunnel' design of the solenoidal magnet, although here too, the trend has been for shorter, more open systems provoking fewer patient refusals on the grounds of claustrophobia. Despite these advances, surgical access with all the configurations is still somewhat restricted, limiting the type of procedure that may be performed with MR image guidance.

As described in Section 7.3.6, the available signal in NMR depends on the strength of the main magnetic field. Low-field systems suffer from the disadvantage of a poor SNR, and the majority of clinical imagers now use 1.5 T cylindrical magnets with a clear bore diameter of approximately 60–90 cm and the axis of symmetry lying along the patient's body. (Although these are widely described as 'solenoids', the coil geometry is, in fact, more complicated, since a true solenoid would need to be unacceptably long to achieve the desired homogeneity of magnetic field.) There has been a trend in recent years to increase this field strength further, and 3 T is now a very popular choice. However, a number of problems detract from the improved SNR and spectral resolution available at higher field: T_1 values increase with field and so the experimental times required to obtain equivalent

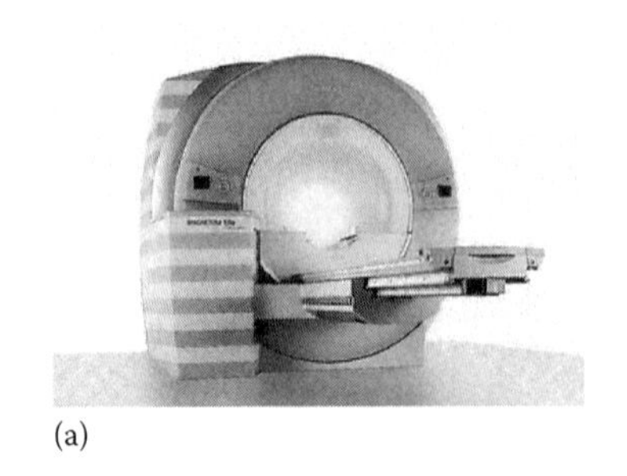

(a)

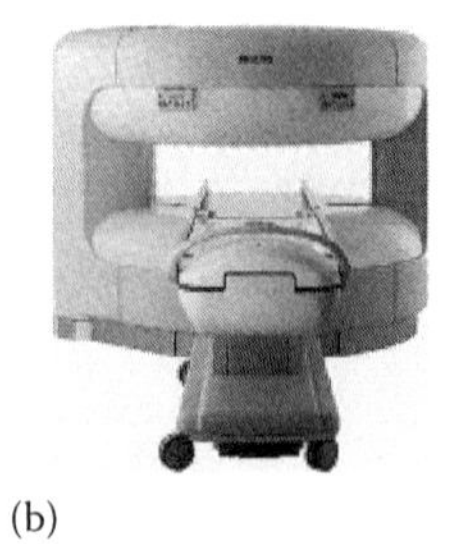

(b)

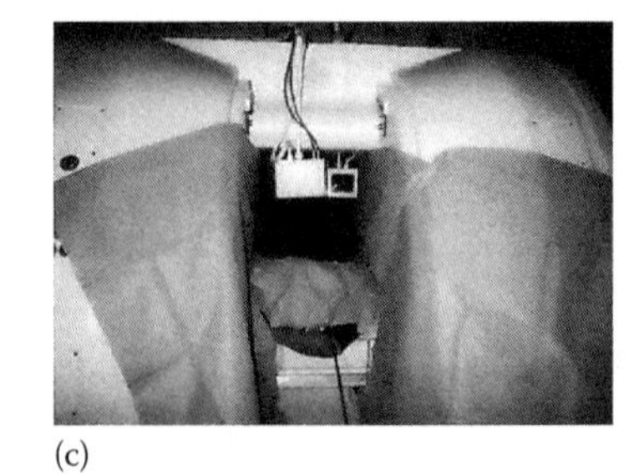

(c)

FIGURE 7.61
Examples of different types of clinical magnet. (a) Siemens Magnetom Trio 3 T cylindrical solenoid. (Courtesy of Siemens Healthcare.) (b) Philips Panorama HFO 1.0 T 'open' system. (Courtesy of Philips Healthcare.) (c) GE Signa 'double-doughnut' 0.5 T magnet. (Data from Gedroyc, 2000.) Both (b) and (c) may be used for intra-operative imaging.

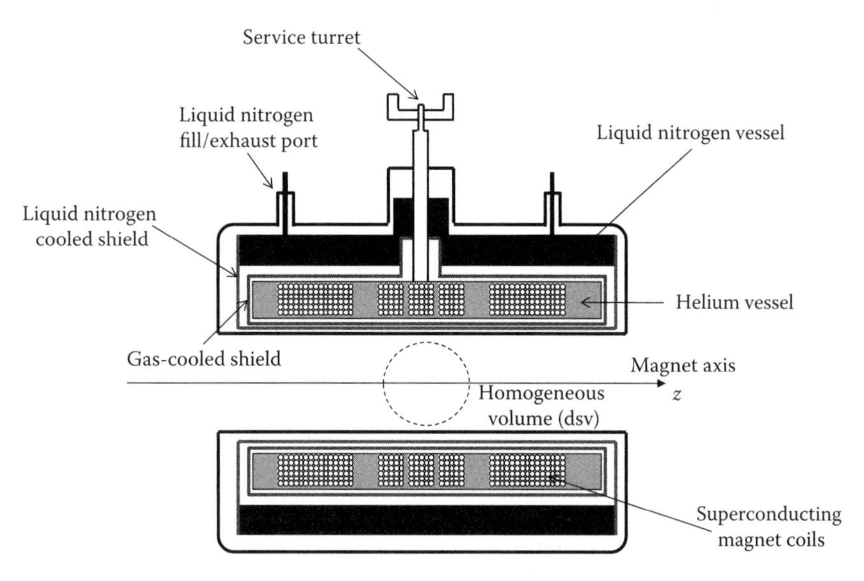

FIGURE 7.62
Simplified schematic of the vertical cross section through a superconducting magnet.

image contrast rise, the magnitude of susceptibility and chemical-shift artefacts goes up and there is poor penetration of RF into the body. The last of these effects leads to a non-uniformity of flip angles in the volume of interest and makes it difficult to use sequences that depend on precise 180° pulses (such as IR fat saturation or CPMG pulse trains).

The superconducting magnet technology on which the systems are based is now mature, and a typical superconducting magnet is shown schematically in Figure 7.62a. The magnet windings, typically made from strands of niobium-tin or niobium-titanium alloy embedded in copper wire (Figure 7.62b), are immersed in liquid helium, which provides a stable temperature of approximately 4.2 K, in order to maintain the superconducting state. Considerable effort is directed at minimising the heat losses from the helium vessel. The magnet is normally constructed with a number of different evacuated spaces, filled with insulation and helium-gas-cooled heat shields. In early cryostats, a liquid-nitrogen jacket surrounded the helium vessel, but in recent years, it has become more common for an active refrigeration unit to be attached to the magnet. By these means, the helium boil-off rate can be reduced significantly, to the extent that some modern systems no longer need any cryogen refills during an installed lifetime of approximately 10 years.

It is a general requirement of all MRI and spectroscopy procedures to obtain as homogeneous a magnetic field as possible in order to maximise the period over which signal may be obtained (Section 7.5.3). This is achieved by a process known as *shimming* and is achieved either via pieces of iron inserted around the magnet bore (passive shims) or via current-carrying coils (active shims). The latter may be further subdivided into superconducting cryoshims, whose values remain fixed after installation, and room temperature shims, whose values are altered as necessary. It can be shown that arbitrary variations in magnetic field can be decomposed into the sum of a set of *spherical harmonics*:

$$B(r,\theta,\phi) = \sum_{n=0}^{\infty}\sum_{m=0}^{n}\left(\frac{r}{r_0}\right)^n P_{nm}(\cos\theta)[C_{nm}\cos m\phi + D_{nm}\sin m\phi] \quad (7.116)$$

where

$B(r, \theta, \phi)$ is the magnetic field expressed in spherical polar coordinates,
r is the distance from isocentre,
θ is the angle from the z axis,
ϕ is the azimuthal angle,
r_0 is an appropriate scaling length,
P_{nm} are the associated Legendre polynomial functions, and
C_{nm} and D_{nm} are constants that quantify the contribution of each harmonic term to the total magnetic field.

The process is similar in many ways to Fourier analysis, in which functions are decomposed into a set of sinusoidal basis functions. The advantage here is that each spherical harmonic term can be physically realised by a *shim coil* of a particular shape. These are commonly known by names such as x, y, z (first-order shims, commonly using the gradient coils); xy, yz, zx, $x^2 - y^2$ (second-order shims) and x^3, y^3, xz^2, xz^3, etc. (high-order shims). By passing different currents through the shim coils (corresponding to the terms C_{nm} and D_{nm} in Equation 7.116), one may, in principle, null the effect of arbitrary field inhomogeneities. In practice, only a finite number of terms of the expansion can be built into the shim coil set and so it is not possible to compensate for all local variations. In the early days of NMR, shimming was carried out manually by optimising the visual appearance of the FID or the linewidth of the Fourier transform. The success depended to a large part on operator experience and varied considerably from measurement to measurement. Modern systems use automated and objective shimming routines, which are often able to perform local shimming that depends on the volume of interest for a particular imaging measurement and takes account of the susceptibility variations in the sample. Some advanced systems are even able to alter shim settings 'on the fly' to optimise imaging on a per-slice basis (De Graaf et al., 2003). The ability to compensate easily for any distortions produced and hence the ease of shimming depend, to a certain degree, on the symmetry of the sample, which affects the field distribution, and so susceptibility-matched bolus materials are often used. In whole-body magnets, the subject is, of course, large and will have a pronounced effect on the magnetic-field distribution.

Sample susceptibility is an important effect, and, in certain cases, where tissues of very different composition lie close together, this may give rise to large changes in local susceptibility that cannot be compensated for. In imaging, these cause severe artefacts, such as that shown in Figure 7.28a, whilst in spectroscopy, the consequent line-broadening will dominate the natural linewidth and blur details of interest. These effects may be particularly pronounced in experimental tumours in animals, where the tumour causes a large deviation from cylindrical symmetry, or in tumours that have large necrotic regions.

The homogeneity of the main magnetic field is now typically around 0.2–0.4 ppm for a modern 1.5 T clinical magnet, quoted as the RMS deviation of the field over a 40 cm diameter spherical volume (dsv) at isocentre. Substituting this value directly into Equation 7.51 leads to an estimate for T_2' of approximately 10 ms, with this value increasing if a small area is locally shimmed. By comparison, a state-of-the-art high-resolution spectrometer might achieve a homogeneity in the region several parts per billion.

The shielding of the static magnetic field is another important consideration. As will be discussed later, the extent of this magnetic field is a major consideration in siting and designing facilities for NMR, and it is important to reduce the 'footprint' of a clinical system as much as possible. Whilst RF electromagnetic signals may be effectively screened using a Faraday cage, passive screening of static magnetic fields is a heavy engineering task and requires large quantities of soft-iron-containing materials, either as a room surrounding the system or in the form of large yokes around the magnet. An alternative arrangement, adopted by most commercial manufacturers, is to use 'active' shielding: extra superconducting coils are built into the system, whose aim is to reduce the field outside the scanner.

7.10.1.2 Magnetic Field Gradients

The individual coils of wire that produce the magnetic field gradients are wound on a former and fixed into place (normally by encasing them in some form of epoxy resin). This ensemble, which may also include some shim coils – see previous section – is known as the *gradient set*. The gradient set, together with its associated controllers and current amplifiers, produces the controlled variation in magnetic field required for spatial localisation. Here, the key performance parameters are the maximum gradient (currently around 40 mT m^{-1} for a typical clinical system), the slew rate (i.e. the speed at which the gradient may be changed from one value to another, typically 200 mT m^{-1} s^{-1} on a state-of-the-art clinical system) and uniformity of the gradient throughout the imaging region. These parameters tend to be competing: for example, a gradient coil producing a larger change in field per unit current tends to have more windings and hence a higher self-inductance, which reduces the slew rate for a given applied voltage. As discussed earlier, gradient performance is unlikely to see major improvements in the future, because current systems have already reached the limit of what is safe to apply on human subjects before peripheral nerve stimulation occurs.

A major problem with earlier generations of gradient design was that they created a time-varying magnetic field *outside* the coils. This led to the induction of *eddy currents* in the magnet and in conductive surfaces of the magnet cold shields and bore. These currents, in turn, gave rise to additional magnetic fields within the imaging volume that disrupted the imaging process. A major innovation in the late 1980s was the introduction of *active gradient screening* that revolutionised the performance of gradient systems and made possible all of the rapid imaging we take for granted today (Turner, 1993). Figure 7.63 shows a schematic diagram of the Maxwell–Golay gradient set, an elementary design used for many years, together with examples of the complex windings on more modern gradient sets.

7.10.1.3 Radiofrequency System

The RF system consists of the following:

- a frequency *synthesiser* to provide signals at the base NMR frequency;
- a system for *modulating* and *gating* this signal, in order to obtain both non-selective and selective RF pulses of appropriate bandwidth;
- an RF *power amplifier*;
- a set of RF *probes*, appropriate for transmitting the pulses to and receiving NMR signals from different parts of the body;
- a *phase-sensitive detector* to convert the single-channel oscillating voltage obtained from the probe into a two-channel signal corresponding to the real and imaginary parts of the magnetisation;

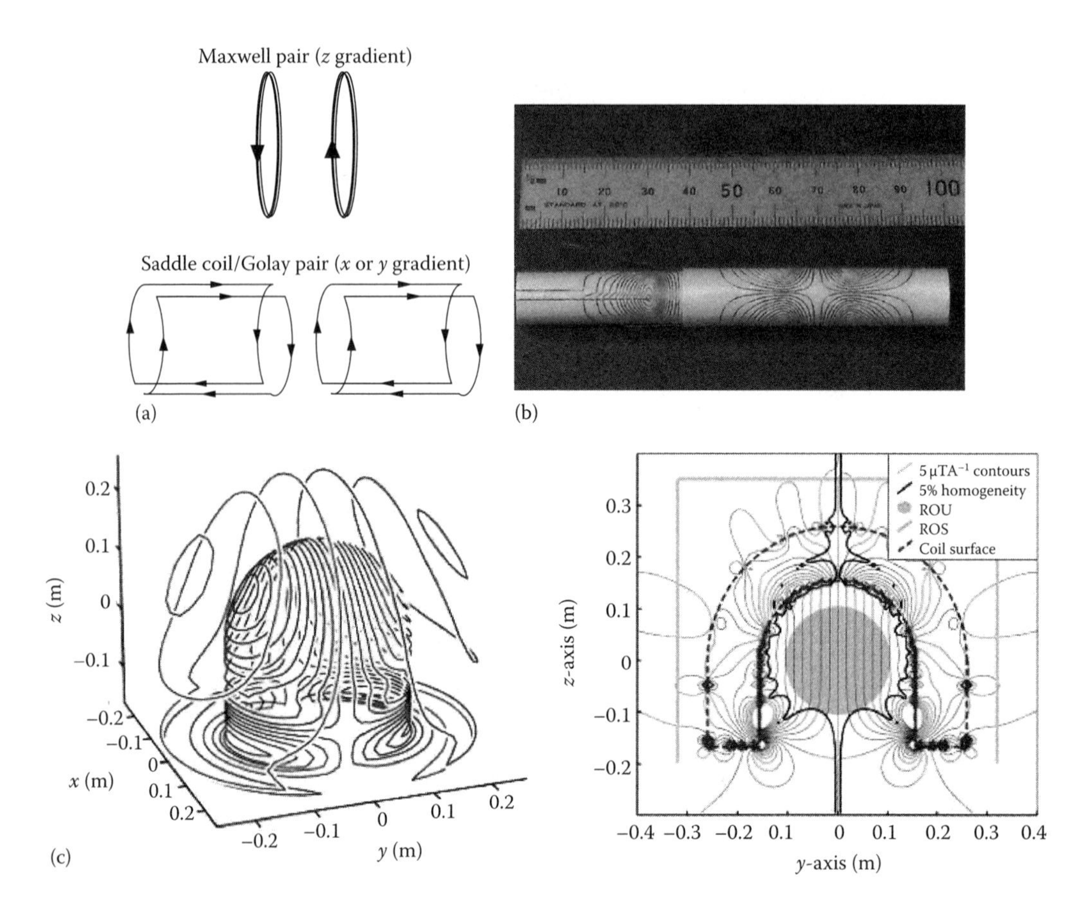

FIGURE 7.63
(See colour insert.) (a) Schematic of wire paths in an elementary Maxwell–Golay gradient set. (b) More complex wire paths in a dual-layer gradient set. (Data for (b) from Leggett et al., 2003.) (c) Design for *y*-gradient of ultra-efficient, highly asymmetrical head gradient coil. (Data for (c) from Poole and Bowtell, 2007.)

- a *demodulator* to transform signals oscillating at the NMR frequency to signals oscillating at lower frequencies (typically up to a maximum of a few hundred kHz) suitable for digitisation;
- *pre-amplification* and *amplification* stages;
- an analogue-to-digital converter.

Increasingly, users of clinical systems require that the RF chain be capable of use with a variety of nuclei. This means that the synthesiser and power amplifier have to be capable of a broadband response. By contrast, individual probes are designed specifically for different nuclei, with the exception of those for nuclei whose resonant frequencies are in close proximity, such as ^{1}H and ^{19}F, which can sometimes be made to dual tune.

The NMR probe (or 'coil') is a critical component of the whole system. Some probes are responsible for both transmitting and receiving the RF signal, whereas others can be transmit-only or receive-only. The optimum requirements for these two functions differ. For imaging, a transmit coil giving a homogenous B_1 field is generally a primary requirement. As described in Section 7.3.2, the B_1 field is perpendicular to the main magnetic field, and this determines the designs of probes that are appropriate. In systems involving small samples,

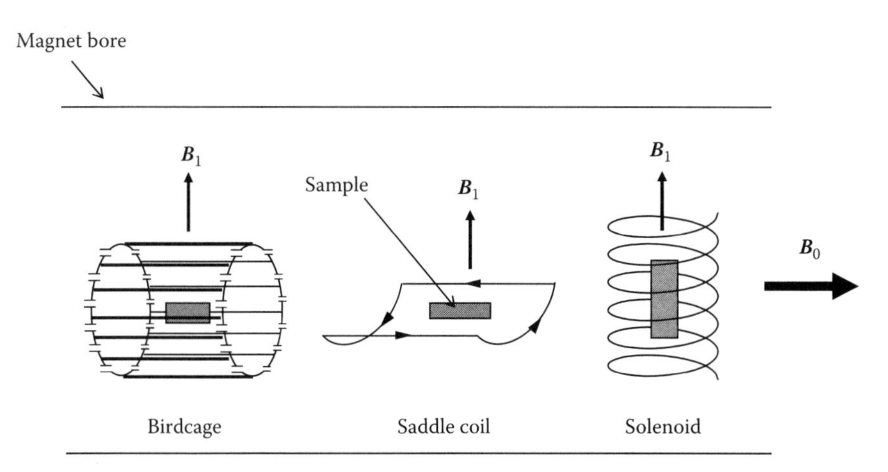

FIGURE 7.64
Elementary designs of RF probe, together with the sample position and orientation relative to the magnet bore.

for which insertion can be transverse to the B_0 axis, a solenoidal coil can be used, but in most clinical systems, this is usually not practical and a birdcage geometry is often used (Hayes et al., 1985). The large 'body coil' is frequently used for transmission and a separate coil for reception. Where two different coils are used, it is important that they are decoupled, so that large signals are not induced in the receiver coil during the transmission phase, possibly damaging sensitive equipment. Similarly, if the same coil is used for both transmission and reception, then the receiver must be appropriately gated so that it is off during the transmit period. It is thus not practicable to measure the FID until after the end of the RF pulse.

The coil will normally form part of a tuned circuit, which must be matched to the 50 Ω output impedance of the RF transmitter to prevent power being reflected back to the transmitter – see Figure 7.64a. An important characteristic of the coil is its Q or *quality factor.* This is a property of a damped oscillator (resonator) that relates the energy stored to the energy dissipated per cycle. For high values of Q, an equivalent definition as the ratio of the resonant frequency to the bandwidth is often used in NMR. Further details may be found in any good electromagnetism textbook, for example, Bleaney (1989). For transmission, this should be relatively low, to ensure that the coil does not continue to 'ring' after the end of the pulse. By contrast, during reception, Q should be as high as possible (subject to the requirements of a high reception bandwidth for fast sequences) to give the greatest sensitivity. It may therefore be necessary to include an active switch that provides for low Q during transmission and high Q during reception.

In the early days of MRI, many successful coils were of relatively simple design, often requiring only a few turns of wire (forming the inductive part of the tuned circuit), together with appropriate capacitors to bring the resonance frequency to the appropriate value. A few simple designs are illustrated in Figure 7.64. By contrast, in modern scanners, the probes are extremely sophisticated, and the RF system is the area that has seen the largest change since the early 2000s. Until the 1990s, a single probe was generally used. The introduction of *phased-array* coils allowed higher SNR with greater coverage and paved the way for the introduction of *parallel imaging*, Section 7.6.14.3. The Siemens Tim® system, for example, allows the patient to be covered with up to 102 separate, but integrated, receiver coils, allowing whole-body 3D coverage via a moving-table technique. Up to 32 receiver channels are simultaneously acquired, and the results combined to provide significant image speedup factors.

The key properties of the amplification chain are a high gain and low noise, with the noise level in the preamplifier being particularly crucial. Typically, the noise figure NF = $\log_{10}(SNR_{out}/SNR_{in})$ is between 1 and 2.5. It is also important that the total gain of the system can be varied, to match the final signal to the range of the ADC.

7.10.1.4 Computer Systems and Pulse Programmer

Typically, a modern MRI scanner will use two computers: a host computer, whose job is to synchronise the operation of all the other pieces of hardware in the system, and a specialised image-reconstruction computer, optimised to obtain the fastest image-rendering performance. Associated with these two systems will be some form of image database, which may be hosted separately. The host computer will communicate with the outside world via a set of protocols such as DICOM. At the time of writing, the current fashion is to use PCs, running either Linux or Microsoft Windows operating systems, although for some time during the early development of the field, non-standard or bespoke operating systems were used. In the same way, there has been a trend away from proprietary, vendor-specific coding systems for creating NMR pulse programmes to widely used languages such as C++. These pulse programmes are compiled into executables by the control PC. At runtime, the PC passes gradient and RF shape data to waveform memories and instructions to a specialist pulse-programmer unit, which coordinates the various lower-level devices.

7.10.2 Microscopy and Pre-Clinical Systems

Whilst the basic principles of NMR microscopy (which, in many cases, is synonymous with pre-clinical animal imaging) are the same as those of standard MRI, several features require further comment. Firstly, as has previously been stated (Section 7.6.10), the SNR of the imaging experiment drops rapidly with the voxel size. Three strategies are used to compensate for this: moving to higher fields, now over 20 T (Fu et al., 2005); designing specialist microcoils to obtain the best possible filling factor for the probe (Glover and Mansfield, 2002) and extending the imaging time to many hours, which restricts the utility of the technique for *in vivo* imaging. Secondly, the applied magnetic field gradients need to be large, and this may require the development of novel gradient hardware (Seeber et al., 2000) or the use of unusual techniques such as stray-field imaging (STRAFI) (McDonald and Newling, 1998).

Glover and Mansfield (2002) discuss the limits of microscopy under various circumstances, showing how the achievable resolution is limited by diffusion, T_2 and SNR. Via a set of examples of recent state-of-the-art measurements, they explain that whilst the resolution of MR microscopy is never likely to be better than 1 μm, a more realistic range of image resolution available is 4–100 μm.

For further details on the physical basis of NMR microscopy, see Callaghan (1991).

7.10.3 Siting Considerations

Siting an NMR system raises a number of difficulties not normally encountered in the design of a clinical installation. Not only is the magnetic assembly itself often massive, weighing tens of tonnes, but also the field produced may cause a number of problems. Much of the equipment common in hospitals and research institutes is highly sensitive to the presence of magnetic fields. This includes most devices in which charged particles are accelerated over a distance, for example, CRT monitors, radioactive counting equipment,

X-ray equipment with image-intensifier systems, electron microscopes, linear accelerators for radiotherapy treatment, some CT scanners and gamma cameras. If these devices are static, the effects may be correctable, but modern systems are often mobile, or at least rotate. If the stray magnetic field from the NMR magnet is of the same order as the Earth's magnetic field (~10^{-5} T), then the effects are likely to lie within the built-in adjustment range of the equipment. However, should a ramping magnet (i.e. one whose field is regularly changed) be used, then a single or very infrequent recalibration of equipment will not suffice. Commercial manufacturers will provide field maps of their magnets, and these are essential when considering a proposed location for the installation.

Whilst the magnet will certainly have an impact on its surrounding environment, the effect of the environment on the magnetic field and the operation of the NMR system also has to be considered. The principal effects are those due to ferrous materials, and these can be divided into moving and static steel. The effect of static ferrous objects can usually be allowed for, provided that they are not too massive or too close to the magnet, by shimming the magnet after installation. The demands of a system to be used for spectroscopy are much higher than those for an imaging system, and this is particularly so when one considers the second class of objects, moving ferrous items. These need to be very carefully controlled where they have an effect on the system, as it is not possible to correct for changes occurring during the course of a measurement. The magnet needs to be sited well away from lifts, lorries and other such heavy equipment. Other factors that need to be taken into consideration include power cables close to the magnet, particularly those at low voltages where high currents may be present. Access must also be provided for cryogen replenishment, as well as venting in the case of a magnet quench (rapid boil-off of helium following a loss of superconductivity in the main coil).

Safety is an important design criterion, and it is a legal responsibility to ensure that members of the public may not be exposed to magnetic fields above statutory limits (typically 0.5 mT). It is also important to ensure that access to the magnet room can be carefully controlled, particularly outside normal working hours. It is advisable to provide a means of screening personnel for the presence of metallic items and of checking whether tools used by maintenance staff are ferromagnetic.

7.11 MR Safety

There are two areas of MR safety to consider in operating an MR facility. First, there are the biological effects of the MR exposure on staff and patients in terms of effects in tissues, organs or the body as a whole, where the possibility of either acute or long-term effects needs to be considered. The second category is the prevention of accidents that may result from the potential hazards of the high magnetic field and the presence of cryogens.

The Medicines and Healthcare products Regulatory Agency (MHRA) has produced a valuable publication that summarises evidence, guidelines and legislation relating to the clinical use of MR (MHRA, 2007). It was first published in 1993 and updated in 2002 and 2007. The International Commission on Non-Ionising Radiation Protection (ICNIRP) has published guidance on general exposure to static magnetic fields (Matthes, 1994) and time-varying electromagnetic fields (Ahlbom et al., 1998). A statement on MR clinical exposure was also published (Matthes, 2004). In the United Kingdom, the National Radiological Protection Board (NRPB, now the Health Protection Agency (HPA)) issued guidance on the protection of patients and volunteers (Saunders, 1991) and has published a number of reviews of the biological effects of magnetic and electromagnetic fields (McKinlay et al., 2004). The World Health Organisation (WHO, 2006) published a review of the possible health effects of exposure to static electric and magnetic fields. The use of MR systems is covered by an international standard that defines how exposure to patients is controlled (IEC, 2006). A European Directive, the Physical Agent (EMF) Directive, has been enacted and will be applied in all member states. This regulates the exposure of staff, applying limits to time-varying gradient and electromagnetic field exposure (EU, 2004). The potential ramifications of this directive for MRI caused great concern among the MRI community and are examined in Keevil et al. (2005) and Riches et al. (2007). Unusually, implementation of the directive was postponed to 2012 and, at the time of writing, there was still considerable uncertainty as to the final outcome of the various representations to the EU authorities. In the United States, the Food and Drug Administration (FDA) provides guidance on safety parameters for MR and receives advice from various sources, including the Blue Ribbon Panel on MR Safety of the American College of Radiology (Kanal et al., 2007).

In the United Kingdom, there are separate limits for exposure of patients, volunteers and staff to time-varying magnetic fields (switched gradients and RF pulses). The limits for patients and volunteers have several levels: the *normal mode* of operation (below the NRPB uncontrolled level, equating to the IEC normal level); the *controlled mode* of operation (above the NRPB uncontrolled level and below the NRPB upper level, equating to the IEC level one), with scanning requiring patient monitoring; and the *research/experimental mode* (exposure exceeds the NRPB upper level, equating to IEC level 2), with scanning requiring ethical approval and patient monitoring.

7.11.1 Biological Effects and Hazards of MR

MR equipment and measurements involve exposure to three different types of electromagnetic field. These are the main $\boldsymbol{B}_0$ (or static) magnetic field, the time-varying magnetic fields produced by the gradients, which have variable time dependence, and the RF $\boldsymbol{B}_1$ field produced from the transmitting RF coils. Both patients and, to a lesser degree, staff and maintenance engineers are exposed to these different fields. The three types of fields are considered in Sections 7.11.2–7.11.4. At the exposure levels encountered in clinical MR measurements, there is the potential for acute short-term effects due to RF power heating, or to currents induced by the switched gradient fields producing nerve stimulation. Exposure to the static magnetic field in high-field units can also cause dizziness and potentially some slowing of reactions.

No clear evidence of any deleterious long-term effects has been reported, and nor have mechanisms been proposed by which such damage could occur. Several authors have considered the physiological effects and long-term hazards that might arise from MR exposure (Budinger, 1979, 1981; Saunders and Smith, 1984).

7.11.2 Static $\boldsymbol{B}_0$ Magnetic Fields

Clinical MR systems have static magnetic fields with flux densities that operate in the range 150 mT to 3 T, with some systems for clinical research ranging up to about 10 T. A large number of studies have been carried out directed at providing evidence of physiological or mutagenic effects from exposure to static magnetic fields. The evidence obtained is contradictory, and no conclusive evidence of long-term effects is available for systems operating in the clinical range. One effect that is evident in magnets is the potential difference generated by charges flowing in blood in the presence of a magnetic field. This can be observed on an ECG as an additional flow-potential peak. Although this induced potential can be observed, it appears to have no effect on the heart rate. Saunders (1982) and Saunders and Orr (1983) have calculated the peak flow potential generated on the aorta wall, assuming a peak blood velocity of 0.63 m s^{-1} and an aortic diameter of 0.025 m to be 16 mV T^{-1}. Thus, 2.5 T would create flow potentials of about 40 mV. This is the depolarisation threshold for individual cardiac muscle fibres. This potential would, of course, be spread across very many individual cells, and thus, for each cell, it would be well below the cell depolarisation threshold.

The NRPB (Saunders, 1991), therefore, recommended that exposure to the static magnetic field for patients or volunteers in the normal mode should not exceed 2.5 T to the head and trunk or 4.0 T to the limbs. The IEC/ICNIRP value currently limits normal exposure to less than 2.0 T. In the controlled mode, the head and trunk may be exposed up to 4.0 T (NRPB and IEC/ICNIRP). Pregnant women should not be routinely exposed outside of the normal mode. Limits for staff are based on the guidelines detailed earlier. Limits have been set to avoid a number of perceptual effects that can arise from movement in a magnetic field, potentially causing short-term effects such as disorientation or dizziness. They have recommended that staff operating the equipment should not be exposed for prolonged periods to more than a time-weighted average of 0.2 T to the whole body, with an upper limit of 2.0 T to the body and 5.0 T to the limbs. The European Directive places no limit on occupational exposure to static magnetic fields.

7.11.3 Time-Varying Magnetic Fields

Time-varying magnetic fields, caused by the switched gradients on imaging systems, will induce currents in conductive pathways in the body. Saunders and Orr (1983) have estimated the tissue current density at which a number of effects in humans occur. These have generally been derived by supplying a current flowing between electrodes placed on the skin surface. Phosphenes, which are the sensations of flashes of light, can be induced with a tissue current density of the order of 1–10 A m^{-2} at 60 Hz. This range of current densities also includes the thresholds for the voluntary release or let-go of a grip contact, thoracic tetanisation (which inhibits breathing) and ventricular fibrillation, with cardiac arrest and inhibition of respiratory centres being thought to occur at greater currents than those producing ventricular fibrillation. The thresholds for these effects are very dependent on both frequency and pulse lengths. Saunders and Orr (1983) also calculated that the current density per unit magnetic field per unit time produced in inductive loops in peripheral head and trunk tissue will be of the order of 10 mA m^{-2}, which is very much less than the threshold values described earlier.

Based on this evidence, the NRPB recommended that for the normal mode, for periods of magnetic flux density change exceeding 120 μs, exposures should be restricted to less than 20 T s^{-1} (RMS). Because the effects tend to fall as frequency increases, they have recommended that, for periods of change less than 120 μs, the relationship $dB/dt \leq 2.4 \times 10^{-3}/t$ T s^{-1} should be observed, where dB/dt is the peak value of the rate of change of magnetic flux density in any part of the body and t is the duration of the change of magnetic field. IEC/ICNIRP normal operating levels allow operation up to 80% of the threshold of the directly determined mean threshold for peripheral nerve stimulation, defined as the onset of sensation. Within this range, operation can fall within the NRPB controlled mode, bringing a requirement for patient monitoring.

For operation in the controlled mode, for periods of change less than 3 ms, the extent of the rate of change of the field strength must satisfy the relationship $dB/dt < 60 \times 10^{-3}/t$ T s^{-1}, to avoid cardiac muscle stimulation (NRPB guidance). The IEC/ICNIRP controlled operating mode encompasses exposures up to 100% of the threshold of the directly determined mean threshold for peripheral nerve stimulation, defined as the onset of sensation.

7.11.4 Radiofrequency B_1 Fields

Power deposition from RF radiation is well known to produce local heating if sufficient power is absorbed in tissue. RF burns are the most common injury to patients undergoing MR examinations. As is apparent from studies of RF hyperthermia, most tissues in the body have a high capacity to dissipate deposited heat. However, certain tissues, such as the eye and the testes, have a low blood flow and so cannot readily dissipate heat. The lens of the eye and the testes are therefore particularly susceptible tissues. Exposure limits have been drawn up on the basis that any significant rise in the temperature of the sensitive tissues of the body should be avoided. Acceptable exposures should not result in a rise of body temperature of more than 0.5°C (normal mode) or 1°C (controlled mode) with the maximum temperature in the head or foetus limited to 38°C, the trunk limited to 39°C and the limbs limited to 40°C for both normal and controlled exposure. Where a local exposure is involved, these limits apply for the appropriate region of the body in any mass of tissue not exceeding 10 g.

In practice, these limits are met by ensuring that the mean specific absorption rate (SAR) is limited. Software in the MR machine will usually implement appropriate limits, although these are most likely to be based on the IEC standard, which may differ in some respects from NRPB guidance. The NRPB imposes limits whereby account is taken of the overall length of exposure, and the time taken to cause a temperature rise. So short exposures of less than 15 min are allowed twice the power deposition of longer exposures (longer than 30 min), with a sliding scale in between. The uncontrolled power deposition limit to the whole body for less than 15 min is 2.0 W kg^{-1} and is 1.0 W kg^{-1} for longer than 30 min. The upper controlled level whole body limits are twice these values. Higher values have been set for the head/foetus, trunk and limbs. These limits also need to take account of ambient relative humidity and temperature, which can reduce the ability to dissipate heat.

The RF power limit is perhaps the most important limit in the practical operation of an MR facility. If a receiver coil is not properly decoupled from the transmitter coil, a large local RF field-focusing effect can occur, giving much increased local power deposition. It is also important to note that RF power deposition in loops of wire inadvertently placed on the body, for instance, due to ECG leads, can also give rise to local heating in the cables. Whilst the manufacturers take precautions to ensure that the pulse sequences and equipment they supply will not deliver too great an RF power to tissues, using sophisticated power monitoring equipment to limit the power applied, it is important for the user to be aware of the possibility of malfunction of components and to take urgent action if a patient complains of heating.

7.11.5 Gradient-Induced Noise

Vibration of the gradient coils in the main magnetic field, due to the switched currents, is a source of audio-frequency acoustic noise. In modern machines, the level of noise can cause discomfort and potentially damage. These effects increase with the strength of the gradients and the $\boldsymbol{B}_0$ magnetic field. It is recommended that patients are supplied with hearing protection when a noise level of 80 dB(A) is exceeded. This is best provided by headphones, which allow continued communication with the operator and the playing of music to put patients more at ease in a potentially stressful situation.

7.11.6 Safety Considerations in MR Facilities

Cardiac pacemakers may be adversely affected by strong magnetic fields as well as by time-varying magnetic fields. Most modern pacemakers usually run in a demand mode, only providing a stimulus when the heart is not functioning correctly. In order to prevent failure in the presence of RF interference, the units are usually designed to operate in automatic pacing mode in the presence of electromagnetic fields of sufficiently high strength. Often, they also have a provision to be switched to this mode by switching an internal reed switch with a small hand-held magnet. Normally, then, anyone with a cardiac pacemaker approaching the magnet would only suffer the effect of their pacemaker switching to automatic pacing mode and producing a higher than normal pulse rate. However, should the electromagnetic interference mimic the profile of a naturally detectable cardiac signal, this could falsely inhibit the pacemaker or cause false synchronisation if the pacemaker has failed to switch to automatic mode. There is also some possibility that induced voltages in the pacemaker leads could cause direct stimulation of the heart, and this may be a potential cause of death. Thus, it is necessary to indicate that the hazard exists by placing appropriate signs around an MR facility. Patients and visitors must be screened for the presence of cardiac pacemakers, and unscreened persons must be restricted from having access to areas where the field strength exceeds 0.5 mT.

A major hazard in operating an NMR facility is the risk of ferromagnetic materials inadvertently being brought close to the magnet and then acting as projectiles in the presence of the magnetic field. This is a problem with both large ferromagnetic objects, which may partially crush anyone caught between them and the magnet, and smaller items such as scissors or small tools, which will travel with considerable velocity close to the magnet and can, again, inflict serious damage on staff close to the magnet or to patients within the magnet. Particular care, therefore, has to be taken to exclude all ferromagnetic objects from the vicinity of the magnet and to devise procedures for ensuring that such objects cannot be introduced. This is particularly a problem as staff become more familiar with the system and may therefore be less alert to the potential hazards. Oxygen cylinders and other items associated with anaesthesia are a particular concern (Chaljub et al., 2001), and a 6-year-old boy was killed at a U.S. hospital in 2001 by a ferromagnetic cylinder that was mistakenly brought into the MRI suite and subsequently attracted violently into the bore of the magnet (Archibold, 2001). Poorly trained staff, technical failures and poor communications procedures were all cited as reasons for the tragedy.

These considerations also affect equipment required for maintenance of the system, replenishment of cryogens and resuscitation. Perhaps one of the major risks is to staff, for instance, cleaning staff, entering the building outside of normal working hours with conventional cleaning equipment, and not having been familiarised with the risks. Thus, it is necessary to take precautions to prevent staff, service personnel and patients entering the magnet room during normal operation with any ferromagnetic items, and also to prevent any unauthorised persons entering the magnet room when the facility is unattended. This may, perhaps, best be done by ensuring that the magnet facility is locked when the building is unattended, with a key that is not available to normal domestic personnel. During normal working hours, anyone entering the magnet room should be screened, with provision also to check tools that are supposedly non-ferromagnetic for ferromagnetic content using a strong magnet. The problems are most severe with a large high-field superconducting magnet, but precautions must also be taken with lower-field resistive magnets. Some magnets are 'self-shielded', meaning that field outside the magnet casing (and hence the projectile risk) is substantially reduced.

With a superconducting magnet, there is also the risk that a major incident would necessitate an emergency shutdown of the magnet, requiring the magnetic field to be quenched. This releases the energy stored in the magnetic field as heat and boils off much of the liquid helium. It obviously involves considerable expense and may also risk damaging the superconducting coils of the magnet. During an emergency quench, the field is typically reduced to 50% in 10 s and 99% in 30 s. There has been some concern that the rate of collapse during a quench could cause field induced currents *in vivo* sufficient to cause cardiac arrest. However, this problem was evaluated in a deliberate emergency quench of a magnet system (Bore et al., 1985); no adverse effects due to the quench were observed.

It is particularly important that all staff working close to the magnet and also all patients are screened for the presence of ferromagnetic plates in the skull, post-operative clips or other implanted metal. New et al. (1983) have studied a variety of metal surgical implants and found that, of these, a number suffered sufficient forces and torques to produce a risk of haemorrhage or injury by displacement. Evidently, great care is necessary in screening patients for the presence of such objects. If there is any risk that ferromagnetic clips were used during any previous surgery, very great care must be exercised in deciding whether it is safe to examine that patient in an MR system. Patients with shrapnel injuries or fragments of steel embedded after industrial accidents should also be excluded, as they may be at risk of previously benign objects moving and causing harm. Local heating can occur in some non-ferromagnetic implants, and patients should be warned to be aware of this possibility so that the examination can be stopped immediately if this occurs. Some non-ferromagnetic implants will cause local image distortions. The best method of ascertaining any risks is a detailed discussion with each patient, with the use of a questionnaire indicating the possible operations or experiences that may give rise to hazards, followed by an X-ray examination if there is any doubt.

The use of cryogens presents particular hazards in superconducting facilities. If handled improperly, liquid helium and nitrogen can cause severe burning, and soft tissues are rendered brittle if they approach these low temperatures. Thus, great care is required during filling of the cryostat, particularly where there is any risk of the filling lines fracturing. Most facilities and magnets are designed to vent exhaust cryogen gases from the room, usually by means of a large-diameter vent pipe that is connected not only to the main vents from the magnet but also to rupture discs, so that, in the event of a quench or any other large release of gases, these will not enter the magnet room but will be released to the atmosphere. Alternatively (Bore and Timms, 1984), the magnet room can be designed so that the low-density helium released during a quench would not occupy the lower 2 m of the room. In these circumstances, it is also necessary to provide the room with windows or panels designed to open on overpressure, and this would not be practicable in an RF-screened room. If there is a possibility that, due to large spillages or leaks of cryogen, a significant proportion of oxygen in the atmosphere of the room could be displaced, some warning system should be incorporated in the design.

The response to fire alarms in an MR building also presents particular problems, as the fire brigade are usually clad in a considerable quantity of steel. Thus, should they need to enter the magnet room close to a high-field magnet, it is necessary for them to remove all steel, or for the magnet to be quenched. Thus, it may be advisable in designing a facility to include provision external to the room for quenching the magnet. However, it is also important to ensure that the fire brigade are familiar with the risks that they would encounter on entering the magnet room, and are also well aware of the costs and implications of quenching the magnet. The situation is best handled by having a member of staff familiar with the system on call, but it is also necessary to provide for emergencies where no expert member of staff is available.

7.12 Applications of Magnetic Resonance in Medicine

It is well beyond the scope of a chapter devoted to the physics of MRI to discuss in detail all the many applications in medicine of the techniques discussed earlier. Nevertheless, since the first edition of this book, many new targets for imaging have been identified, and the more important of these are described briefly in the following sections.

7.12.1 Cancer

MRI and MRS play a major role in all stages of the treatment of cancer, from research on the fundamental biochemistry, through diagnosis (which involves the identification of a lesion, characterisation of tumours as benign or malignant and determination of their extent), to treatment planning, image guidance during surgery and monitoring of the outcome of therapy, assessment of residual disease and finally detecting relapse in treated or distant tissues.

Key imaging targets are the pelvic organs, breast, liver and brain, reflecting the prevalence of cancers of these organs. However, MRI is used routinely to diagnose cancers in all parts of the body and also in experimental animal models of disease.

Early imaging studies were based on intrinsic differences in relaxation times between tumours and healthy tissue. However, better understanding of the tumour micro-environment and improved hardware have dramatically increased the range of techniques available. These include dynamic contrast imaging (Section 7.7.2), ideal for monitoring the treatment efficacy of *anti-angiogenic* drugs, which have been developed specifically to target the tumour micro-vasculature; targeted contrast agents; spectroscopy (^{1}H, ^{19}F, ^{23}Na, ^{31}P) for monitoring tumour and drug metabolism; DWI to probe the response to treatment and DTI to determine tumour infiltration into white matter tracts (Section 7.7.4); BOLD imaging (Section 7.7.5) for both tumour oxygen utilisation and fMRI in surgical planning; correlation of histopathology with micro-MRI of biopsy samples; elastography (Section 7.7.3); monitoring of thermotherapy treatments; and surgical guidance in biopsy.

In many cases, cancer diagnosis requires the development of *biomarkers*, that is, markers with properties measurable via MRI that can be correlated with changes taking place at a microscopic level. It is believed that properties such as diffusion, or the identification of drug metabolites, may be earlier indicators of the success or failure of treatment than gross morphological changes in the tumour.

7.12.2 Neurology

MRI is capable of detecting a variety of different neural disorders. In addition to cancer, it can be useful to classify these into diseases of the white matter and degenerative disorders. White matter diseases, perhaps the best known of which is MS, are readily detected via MRI, but poorly understood. Lesions are conspicuous on standard T_2-weighted scans and change over time, but their presence correlates very poorly with clinical symptoms of the disease. The three major characteristics of MS (inflammation, destruction of the myelin sheaths surrounding axons and axonal injury) may all be detected via MRI, using a combination of DCE-MRI (Section 7.7.2), cellular imaging of macrophages, whole-brain volumetry, magnetisation transfer (Section 7.7.6), diffusion imaging (Section 7.7.4) and spectroscopic imaging (Section 7.9).

Although Alzheimer's disease (Petrella et al., 2003) is the most widely known neurodegenerative disease, a number of others are similarly characterised by the loss of nerve cells in the brain. It is often difficult to diagnose these diseases radiologically, because of the lack of distinctive lesions or obvious changes in brain morphology and also because of the significant overlap with the symptoms of normal ageing. MRI in these conditions is important primarily to exclude other conditions such as tumour or stroke. Nevertheless, at a research level, some progress is being made in using imaging (e.g. whole-brain volumetry) and spectroscopy (in particular, levels of the compound NAA) to investigate the mechanisms of brain damage in these diseases and develop biomarkers to assess treatment outcomes.

The use of MRI for the assessment of brain function and stroke has previously been discussed in Sections 7.7.4 and 7.7.5.

7.12.3 Cardiology

Capturing sharp 3D images of a moving target with the time resolution necessary to see individual phases of the heart cycle is a daunting task. Cardiac MRI is, arguably, the field of diagnosis that has seen the greatest impact since the 1990s of the phenomenal increase in acquisition speed of MR scanners. This has been brought about via the introduction of fast, high-contrast sequences such as True-FISP (Section 7.6.13.1), gradient modulation strategies (e.g. spiral imaging, Section 7.6.13.2), parallel imaging techniques (Section 7.6.14.3) and improvements in gradient performance.

Respiratory motion creates significant difficulties, and, when cardiac imaging was first introduced, it was necessary to gate the acquisition with both the cardiac and respiratory cycles (Section 7.6.14.2), leading to extremely long acquisition times. Three-dimensional images may now be acquired within a single breath-hold and so the motion artefacts introduced by respiration are removed. Nevertheless, cardiac patients are precisely the group that find holding their breath difficult, and so free-breathing strategies are sometimes used (usually in conjunction with navigator techniques) despite the difficulties this may lead to in co-registering slices acquired at different times.

Standard morphological imaging and cine imaging, employing various mechanisms of contrast to distinguish between different tissues, enable measurement of ventricular volumes and visualisation of valve operation, wall thickening, congenital abnormalities and aortic or pulmonary anomalies. Flow imaging (Section 7.7.3) may be used to quantify blood velocities in the chambers of the heart, allowing the ejection fraction (a key diagnostic variable) to be calculated by integrating over the whole cycle. Velocity-sensitive MRI has the potential to replace colour Doppler ultrasound (see Section 6.3.5) in diagnosing problems such as mitral valve regurgitation. In conjunction with computational fluid dynamics simulations, it may predict pressure gradients across stenoses and regions of turbulence. Angiography techniques are employed for 3D visualisation of the coronary arteries and the evaluation of bypass grafts. Cardiac tagging studies allow one to quantify wall motion, strain and twisting in the heart muscle. The orientation of the muscle fibres themselves, despite their sub-voxel size, can be measured via diffusion tensor mapping (Section 7.7.4). Perfusion imaging via dynamic susceptibility contrast is used to measure regional variations in blood supply to the myocardium and thus to assess the areas of tissue compromised by ischaemic heart disease, with a vastly better spatial resolution than the alternative technique of $^{99}Tc^{m}$ SPECT. ^{23}Na MRI (Section 7.7.10) can be used to gain an insight into sodium homeostasis in myocytes and thus may allow discrimination between viable and non-viable tissue after a heart attack, whilst specially developed Gd contrast agents may be used to similar effect with ^{1}H imaging.

7.12.4 Interventional Procedures and Monitoring of Therapy

The design of MR systems for use in the operating theatre has already been discussed in Section 7.10.1.1. A second technological challenge is the manufacture of MR-compatible surgical tools. There is a distinction to be drawn between 'MR-safe' materials, such as 'non-magnetic stainless steel', which are not attracted into the magnet bore but still perturb the local homogeneity of the magnetic field, causing massive disruption to the resulting images, and 'MR-compatible' materials, in whose presence images can successfully be acquired. Titanium and some plastics are suitable in this regard, but cannot be made as sharp or stiff as stainless steel and may not be so easy to sterilise.

Applications of interventional MRI include the following: *biopsy*, where one wishes to both localise accurately the tissue to be sampled and visualise structures to avoid en route to the target; *surgery of brain and skull base tumours*, where access is limited and where there are many critical areas that must not be damaged; and *endovascular treatments*, in which the progress of a catheter or device must be monitored as it passes from the point of insertion to the site of action or placement. Here, MRI has the potential to replace X-ray fluoroscopy (see Section 2.8.1), which provides a dynamically updated view of the surgery, but at the expense of a large patient dose of ionising radiation.

Unique to MRI is the ability to make images of *temperature distribution* within a sample (Rieke and Butts Pauly, 2008). This allows us to monitor the delivery of clinical hyperthermia treatments, for example, the ablation of liver tumours by laser or thermal catheters or a high-intensity focused ultrasound beam (Jolesz and McDannold, 2008). By a feedback-controlled coupling of the scanner to the delivery system, very precise manipulation of the tissue temperature may be achieved.

7.12.5 Other Medical Applications

Numerous other specialist application areas of MRI exist, the description of which lies beyond the scope of this summary. These include MRI of the musculoskeletal system (with important applications in surgical planning for the lumbar spine, for trauma patients and for the knee and other joints; the study of the ligaments, tendons, cartilage and of arthritis); imaging of the gastro-intestinal tract; MRI of the kidneys and urinary system; and imaging of the foetus *in utero* and placenta. MRI is also becoming increasingly important for drug development. Whilst this report has focused on the significant role of quantitative imaging, the importance of exquisitely detailed morphological information must never be underestimated.

7.12.6 Molecular and Cellular Imaging

In an era when the human genome has been sequenced and in which we are rapidly learning the biological function of the individual genes that make it up, it is predicted that medical therapy will move rapidly towards a model where drugs are purpose-designed to act on specific genetic/molecular targets. One might imagine that it would be impossible to image such phenomena, since they occur at concentrations and at length scales far below the resolution limits of MRI. However, localisation of the point of action of a drug to better than the 0.1 mm available on today's MRI scanners is clinically unnecessary. To visualise the effects of a molecularly directed therapy we simply (!) need a probe with (i) a high affinity and specificity for the target molecule, enzyme or receptor; (ii) a delivery route to the tissue of interest; and, crucially, (iii) the ability to *amplify* nano-molar changes in concentration to signals detectable at the level of an imaging voxel. The most commonly used amplification mechanism in

MRI is relaxation: small quantities of ferromagnetic or paramagnetic materials have an effect on the NMR signal over a large region. Thus, the signal we detect comes not from the relatively few target molecules but from the much larger pool of surrounding water molecules.

Probes may be divided into two classes: agents that influence water relaxation merely by an increase in concentration and 'smart' agents that change their conformation on interaction with the target. An example of the former is a so-called polymerised vesicle containing the relaxation agent Gd^{3+}, which is conjugated to an antibody targeting the compound $\alpha_v\beta_3$ integrin found in many angiogenesis contexts (Winter et al., 2003). By contrast, the 'Egad' complex contains a β-galactose sub-unit that prevents access of water molecules to the paramagnetic Gd^{3+} ion trapped inside it. When the enzyme β-galactosidase is present, the ring is cleaved and the relaxation efficiency of the Gd^{3+} is vastly increased (Louie et al., 2000). This mechanism has a potentially wide application, because expression of β-galactosidase is widely used as a marker to report successful gene transfection. A ball-park figure for the number of Gd complex units required per cell to obtain successful visualisation on an image is $\sim$4–5 × 10^7 and initial studies appear to show that this will be achievable *in vivo*. Finally, so-called PARACEST contrast agents (Zhang et al., 2003) with markedly different chemical-shift properties allow the possibility of 'turning on or off' contrast selectively using the principles of chemical exchange and magnetisation transfer.

7.12.7 Genetic Phenotyping for Pre-Clinical Imaging

It is known that humans share a very large fraction of the genome with other mammals, and, thus, there is a strong likelihood that human gene function may be inferred from studies on transgenic mice. To this end, a systematic programme is underway to produce heritable mutations in every gene and study the phenotype of each resulting mouse.

High-resolution ~ $(100\,\mu m)^3$ 3D MRI is a valuable screening approach. It has a number of advantages compared with traditional histological methods: (i) it is nondestructive, allowing serial studies; (ii) the data do not exhibit the geometric distortion that can be caused by stresses or dehydration during histological sectioning; (iii) small organs can be isolated and imaged at higher resolution, then co-registered with the main data set; (iv) the data acquisition procedure is easier to automate than physical sectioning of the animal – important, since it has been estimated that some 20,000 mouse strains will need to be examined just to characterise the effects of modifying dominant genes; and (v) MRI is inherently 3D and digital. Not only does this allow the researcher to slice the data set in any arbitrary orientation to view features of interest, but via image registration and pattern matching software, the data for a transgenic phenotype might be compared automatically with a control.

Over recent years, this type of research has led to growing databases of phenotype images and atlases with rapid dissemination of data via web-based archives (Kovacevic et al., 2005; Chuang et al., 2011).

7.13 Conclusions

This chapter has provided an introduction to the physics, technology and applications of MRI. It is clear that the scope of MRI is enormous and spans the physical and medical sciences. Here, we have merely scratched the surface of the subject, providing a starting point for further study.

References

Abragam A 1983 *The Principles of Nuclear Magnetism* (Oxford, UK: Clarendon Press).

Abraham R J, Fisher J and Loftus P 1988 *Introduction to NMR Spectroscopy* (Chichester, UK: John Wiley & Sons).

Ackerman J J H, Grove T H, Wong G G, Gadian D G and Radda G K 1980 Mapping of metabolites in whole animals by P-31 NMR using surface coils *Nature* **283** 167–170.

Ahlbom A, Bergqvist U, Bernhardt J H, Cesarini J P, Court L A, Grandolfo M, Hietanen M, McKinlay A F, Repacholi M H, Sliney D H, Stolwijk J A J, Swicord M L, Szabo L D, Taki M, Tenforde T S, Jammet H P and Matthes R 1998 Guidelines for limiting exposure to time-varying electric, magnetic, and electromagnetic fields (up to 300 GHz) *Health Physics* **74** 494–522.

Ahn C B and Cho Z H 1989 A generalized formulation of diffusion effects in μm resolution nuclear magnetic resonance imaging *Medical Physics* **16** 22–28.

Alger J R, Von Kienlen M, Ernst T, McLean M, Kreis R, Bluml S, Cecil K, Cha S, Vigneron D B, Lazeyras F, Bizzi A, Barker P B, Waldman A D, Renshaw P F, Cozzone P J and Gullapalli R P 2006 MR spectroscopy in clinical practice (educational course, 16 presentations). *Proceeding of the International Society for Magnetic Resonance in Medicine* (Seattle, WA: ISMRM).

Alvarez L W and Bloch F 1940 A quantitative determination of the neutron moment in absolute nuclear magnetons *Physical Review* **57** 111.

Anderson A W and Gore J C 1994 Analysis and correction of motion artifacts in diffusion-weighted imaging *Magnetic Resonance in Medicine* **32** 379–387.

Andrew E R 1958 *Nuclear Magnetic Resonance* (Cambridge, UK: Cambridge University Press).

Andrew E R, Bottomley P A, Hinshaw W S, Holland G N, Moore W S and Simaroj C 1977 NMR images by multiple sensitive point method—Application to larger biological-systems *Physics in Medicine and Biology* **22** 971–974.

Appelt S, Baranga A B, Erickson C J, Romalis M V, Young A R and Happer W 1998 Theory of spin-exchange optical pumping of He-3 and Xe-129 *Physical Review A* **58** 1412–1439.

Archibold R C 2001 Hospital details failures leading to M.R.I. fatality *New York Times* August 22.

Arnold D L, Radda G K, Bore P J, Styles P and Taylor D J 1984 Excessive intracellular acidosis of skeletal muscle on exercise in a patient with a post-viral exhaustion fatigue synrome—a P-31 nuclear magnetic resonance study *Lancet* **1** 1367–1369.

Aue W P, Muller S, Cross T A and Seelig J 1984 Volume-selective excitation—A novel approach to topical NMR *Journal of Magnetic Resonance* **56** 350–354.

Bartholdi E and Ernst R R 1973 Fourier spectroscopy and the causality principle *Journal of Magnetic Resonance* **11** 9–19.

Becker E D, Fisk C and Khetrapal C L 1996 The development of NMR In Grant D M and Harris R K (eds.) *Encyclopedia of Nuclear Magnetic Resonance* (Chichester, UK: Wiley).

Beckman Institute 1998 *MRI Artifact Gallery* [Online] Available at http://chickscope.beckman.uiuc.edu/roosts/carl/artifacts.html (accessed on May 3, 2007).

Belliveau J W, Kennedy D N, McKinstry R C, Buchbinder B R, Weisskoff R M, Cohen M S, Vevea J M, Brady T J and Rosen B R 1991 Functional mapping of the human visual-cortex by magnetic-resonance-imaging *Science* **254** 716–719.

Beloueche-Babari M, Chung Y L, Al-Saffar N M S, Falck-Miniotis M and Leach M O 2010 Metabolic assessment of the action of targeted cancer therapeutics using magnetic resonance spectroscopy *British Journal of Cancer* **102** 1–7.

Blackledge M J, Rajagopalan B, Oberhaensli R D, Bolas N M, Styles P and Radda G K 1987 Quantitative studies of human cardiac metabolism by P-31 rotating-frame NMR *Proceedings of the National Academy of Sciences of the United States of America* **84** 4283–4287.

Bleaney B I 1989 *Electricity and Magnetism* (Oxford, UK: Oxford University Press).

Bloch F 1946 Nuclear induction *Physical Review* **70** 460.

Block K T and Frahm J 2005 Spiral imaging: A critical appraisal *Journal of Magnetic Resonance Imaging* **21** 657–668.

Bloch F, Hansen W W and Packard M 1946 Nuclear induction *Physical Review* **69** 127.

Bloembergen N 1957 Proton relaxation times in paramagnetic solutions *Journal of Chemical Physics* **27** 572–573.

Bloembergen N, Purcell E M and Pound R V 1948 Relaxation effects in nuclear magnetic resonance absorption *Physical Review* **73** 679.

Bore P J, Galloway G J, Styles P, Radda G K, Flynn G and Pitts P 1985 Are quenches dangerous? In *Proceedings of the Fourth Annual Meeting of the Society of Magnetic Resonance*, London, UK (Berkeley, CA: SMRS) pp. 914–915.

Bore P J and Timms W E 1984 The installation of high-field NMR equipment in a hospital environment *Magnetic Resonance in Medicine* **1** 387–395.

Bottomley P A 1987 Spatial localization in NMR spectroscopy *in vivo Annals of the New York Academy of Sciences* **508** 333–348.

Bovey F A, Mirau P A and Gutowsky H S 1988 *Nuclear Magnetic Resonance Spectroscopy* 2nd edn. (Orlando, FL: Academic Press).

Boyle G E, Ahern M, Cooke J, Sheehy N P and Meaney J F 2006 An interactive taxonomy of MR imaging sequences *Radiographics* **26** e24.

Brownstein K R and Tarr C E 1979 Importance of classical diffusion in NMR studies of water in biological cells *Physical Review A* **19** 2446.

Brunner P and Ernst R R 1979 Sensitivity and performance time in NMR imaging *Journal of Magnetic Resonance* **33** 83–106.

Budinger T F 1979 Thresholds for physiological effects due to RF- and magnetic-fields used in NMR imaging *IEEE Transactions on Nuclear Science* **26** 2821–2825.

Budinger T F 1981 Nuclear magnetic resonance (NMR) *in vivo* studies—Known thresholds for health-effects *Journal of Computer Assisted Tomography* **5** 800–811.

Bulte J W M and Kraitchman D L 2004 Iron oxide MR contrast agents for molecular and cellular imaging *NMR in Biomedicine* **17** 484–499.

Bydder M, Rahal A, Fullerton G D and Bydder G M 2007 The magic angle effect: A source of artifact, determinant of image contrast, and technique for imaging *Journal of Magnetic Resonance Imaging* **25** 290–300.

Cady E B, Dawson M J, Hope P L, Tofts P S, Costello A M D, Delpy D T, Reynolds E O R and Wilkie D R 1983 Non-invasive investigation of cerebral metabolism in newborn-infants by phosphorus nuclear magnetic resonance spectroscopy *Lancet* **1** 1059–1062.

Calamante F 2003 Quantification of dynamic susceptibility contrast t_2^* MRI in oncology In Jackson A, Buckley D L and Parker G J M (eds.) *Dynamic Contrast-Enhanced Magnetic Resonance Imaging in Oncology* (Berlin, Germany: Springer).

Callaghan P T 1991 *Principles of Nuclear Magnetic Resonance Microscopy* (Oxford, UK: Clarendon).

Carr H Y 1952 Free precession techniques in nuclear magnetic resonance. PhD thesis, Harvard University, Cambridge, MA.

Carr H Y and Purcell E M 1954 Effects of diffusion on free precession in nuclear magnetic resonance experiments *Physical Review* **94** 630.

Chaljub G, Kramer L A, Johnson R F, Johnson R F, Singh H and Crow W N 2001 Projectile cylinder accidents resulting from the presence of ferromagnetic nitrous oxide or oxygen tanks in the MR suite *American Journal of Roentgenology* **177** 27–30.

Chance B, Eleff S, Leigh J S, Sokolow D and Sapega A 1981 Mitochondrial regulation of phosphocreatine inorganic phosphate ratios in exercising human muscle—a gated P-31 NMR study *Proceedings of the National Academy of Sciences of the United States of America—Biological Sciences* **78** 6714–6718.

Chang H and Fitzpatrick J M 1992 A technique for accurate magnetic resonance imaging in the presence of field inhomogeneities *IEEE Transactions on Medical Imaging* **11** 319–329.

Chang S C, Lai P H, Chen W L, Weng H H, Ho J T, Wang J S, Chang C Y, Pan H B and Yang C F 2002 Diffusion-weighted MRI features of brain abscess and cystic or necrotic brain tumors—Comparison with conventional MRI *Clinical Imaging* **26** 227–236.

Charles-Edwards G D, Payne G S, Leach M O and Bifone A 2004 Effects of residual single-quantum coherences in intermolecular multiple-quantum coherence studies *Journal of Magnetic Resonance* **166** 215–227.
Chen C N, Hoult D I and Sank V J 1983 Quadrature detection coils—A further square-root 2 improvement in sensitivity *Journal of Magnetic Resonance* **54** 324–327.
Chen W and Zhu X H 2005 Dynamic study of cerebral bioenergetics and brain function using in vivo multinuclear MRS approaches *Concepts in Magnetic Resonance Part A* **27A** 84–121.
Cherubini A and Bifone A 2003 Hyperpolarised xenon in biology *Progress in Nuclear Magnetic Resonance Spectroscopy* **42** 1–30.
Cherubini A, Payne G S, Leach M O and Bifone A 2003 Hyperpolarising C-13 for NMR studies using laser-polarised Xe-129: SPINOE vs thermal mixing *Chemical Physics Letters* **371** 640–644
Chuang N, Mori S, Yamamoto A, Jiang H, Ye X, Xu X, Richards L J, Nathans J, Miller M I, Toga A W, Sidman R L and Zhang J 2011 An MRI-based atlas and database of the developing mouse brain *NeuroImage* **54** 80–89.
Clow H and Young I 1978 Britain's brains produce first NMR scans *New Scientist* **80** 588.
Colegrove F D, Schearer L D and Walters G K 1963 Polarization of He-3 gas by optical pumping *Physical Review* **132** 2561.
Collins D J and Padhani A R 2004 Dynamic magnetic resonance imaging of tumor perfusion *IEEE Engineering in Medicine and Biology Magazine* **23** 65–83.
D'arcy J A, Collins D J, Rowland I J, Padhani A R and Leach M O 2002 Applications of sliding window reconstruction with Cartesian sampling for dynamic contrast enhanced MRI *NMR in Biomedicine* **15** 174–183.
Damadian R 1971 Tumor detection by nuclear magnetic resonance *Science* **171** 1151–1153.
Damadian R, Goldsmith M and Minkoff L 1977 NMR in cancer.16. Fonar image of live human-body *Physiological Chemistry and Physics* **9** 97–100.
Damadian R, Minkoff L, Goldsmith M, Stanford M and Koutcher J 1976 Field focusing nuclear magnetic-resonance (fonar)—Visualization of a tumor in a live animal *Science* **194** 1430–1432.
De Graaf R A, Brown P B, McIntyre S, Rothman D L and Nixon T W 2003 Dynamic shim updating (DSU) for multislice signal acquisition *Magnetic Resonance in Medicine* **49** 409–416.
Delpy D T, Gordon R E, Hope P L, Parker D, Reynolds E O R, Shaw D and Whitehead M D 1982 Non-invasive investigation of cerebral ischemia by phosphorus nuclear magnetic resonance *Pediatrics* **70** 310–313.
Dickinson W C 1950 Dependence of the F^{19} nuclear resonance position on chemical compound *Physical Review* **77** 736.
Dixon W T 1984 Simple proton spectroscopic imaging *Radiology* **153** 189–194.
Doran S J, Attard J J, Roberts T P L, Carpenter T A and Hall L D 1992 Consideration of random errors in the quantitative imaging of NMR relaxation *Journal of Magnetic Resonance* **100** 101–122.
Doran S J Bourgeois M E and Leach M O 2005a Burst imaging – can it ever be useful in the clinic? *Concepts in Magnetic Resonance A* **26** 11–34.
Doran S J, Charles-Edwards L, Reinsberg S A and Leach M O 2005b A complete distortion correction for MR images: I. Gradient warp correction *Physics in Medicine and Biology* **50** 1343–1361.
Dwek R A 1973 *Nuclear Magnetic Resonance (NMR) in Biochemistry: Applications to Enzyme Systems* (Oxford, UK: Clarendon Press).
Dzik-Jurasz A, Domenig C, George M, Wolber J, Padhani A, Brown G and Doran S 2002 Diffusion MRI for prediction of response of rectal cancer to chemoradiation *Lancet* **360** 307–308.
Edelstein W A, Glover G H, Hardy C J and Redington R W 1986 The intrinsic signal-to-noise ratio in NMR imaging *Magnetic Resonance in Medicine* **3** 604–618.
Edelstein W A, Hutchison J M S, Johnson, G and Redpath T 1980 Spin warp NMR imaging and applications to human whole-body imaging *Physics in Medicine and Biology* **25** 751–756.
Emsley J 1997 *The Elements* (Oxford, UK: Oxford University Press).
Ernst R R and Anderson W A 1966 Application of Fourier transform spectroscopy to magnetic resonance *Review of Scientific Instruments* **37** 93–102.

Ernst R R, Bodenhausen G and Wokaun A 1987 *Principles of Nuclear Magnetic Resonance in One and Two Dimensions* (Oxford, UK: Oxford University Press).

EU 2004 Directive 2004/40/EC of the European Parliament and of the council on the minimum health and safety requirements regarding the exposure of workers to the risks arising from physical agents (electromagnetic fields) (18th individual Directive within the meaning of Article 16(1) of Directive 89/391/EEC). (Brussels: EU).

Fain S, Schiebler M L, McCormack D G and Parraga G 2010 Imaging of lung function using hyperpolarized helium-3 magnetic resonance imaging: Review of current and emerging translational methods and applications *Journal of Magnetic Resonance Imaging* **32** 1408.

Farrar T C and Becker E D 1971 *Pulse and Fourier Transform NMR: Introduction to Theory and Methods* (New York: Academic Press).

Frahm J, Merboldt K D and Hanicke W 1987 Localized proton spectroscopy using stimulated echoes *Journal of Magnetic Resonance* **72** 502–508.

Freeman R 1997 *Spin Choreography: Basic Steps in High Resolution NMR* (Oxford, UK: Spektrum).

Fu R, Brey W W, Shetty K, Gor'kov P, Saha S, Long J R, Grant S C, Chekmenev E Y, Hu J, Gan Z, Sharma M, Zhang F, Logan T M, Bruschweller R, Edison A, Blue A, Dixon I R, Markiewicz W D and Cross T A 2005 Ultra-wide bore 900 MHz high-resolution NMR at the national high magnetic field laboratory *Journal of Magnetic Resonance* **177** 1–8.

Fukushima E and Roeder S B W 1981 *Experimental Pulse NMR: A Nuts and Bolts Approach* (Boulder, CO: Westview Press).

Gadian D G 1995 *NMR and Its Applications to Living Systems* (Oxford, UK: Oxford University Press).

Gadian D G, Hoult D I, Radda G K, Seeley P J, Chance B and Barlow C 1976 Phosphorus nuclear magnetic resonance studies on normoxic and ischemic cardiac tissue *Proceedings of the National Academy of Sciences of the United States of America* **73** 4446–4448.

Garroway A N, Grannell P K and Mansfield P 1974 Image-formation in NMR by a selective irradiative process *Journal of Physics C—Solid State Physics* **7** L457–L462.

Gedroyc W 2000 Interventional magnetic resonance imaging *British Journal of Urology International* **86** (Suppl. 1), 174–180.

Gerlach W and Stern O 1921 Der experimentelle nachweis des magnetischen moments des silberatoms *Zeitschrift für Physik* **8** 110–111.

Glover P and Mansfield P 2002 Limits to magnetic resonance microscopy *Reports on Progress in Physics* **65** 1489–1511.

Goldman M 1991 *Quantum Description of High-Resolution NMR in Liquids* (Oxford, UK: Clarendon Press).

Golman K, Ardenaer-Larsen J H, Petersson J S, Mansson S and Leunbach I 2003 Molecular imaging with endogenous substances *Proceedings of the National Academy of Sciences of the United States of America* **100** 10435–10439.

Golman K, Axelsson O, Johannesson H, Mansson S, Olofsson C and Petersson J S 2001 Parahydrogen-induced polarization in imaging: Subsecond C-13 angiography *Magnetic Resonance in Medicine* **46** 1–5.

Golman K and Petersson J S 2006 Metabolic imaging and other applications of hyperpolarized C-13 *Academic Radiology* **13** 932–942.

Gordon R E, Hanley P E and Shaw D 1982 Topical magnetic resonance *Progress in Nuclear Magnetic Resonance Spectroscopy* **15** 1–47.

Gordon R E, Hanley P E, Shaw D, Gadian D G, Radda G K, Styles P, Bore P J and Chan L 1980 Localization of metabolites in animals using P-31 topical magnetic resonance *Nature* **287** 736–738.

Gorter C J 1936 Paramagnetic relaxation *Physica* **3** 503–514.

Govindaraju V, Young K and Maudsley A A 2000 Proton NMR chemical shifts and coupling constants for brain metabolites *NMR in Biomedicine* **13** 129–153.

Grant D M and Harris R K 1996 *Encyclopedia of Nuclear Magnetic Resonance* (New York: Wiley), pp. xviii, 861.

Griffiths J M and Griffin R G 1993 Nuclear magnetic resonance methods for measuring dipolar couplings in rotating solids *Analytica Chimica Acta* **283** 1081–1101.

Gunther H 1995 *NMR Spectroscopy: Basic Principles, Concepts, and Applications in Chemistry* (Chichester, UK: John Wiley & Sons).

Gutteridge S, Ramanathan C and Bowtell R 2002 Mapping the absolute value of M-0 using dipolar field effects *Magnetic Resonance in Medicine* **47** 871–879.

Guzman R, Barth A, Lovblad K O, El-Koussy M, Weis J, Schroth G and Seiler R W 2002 Use of diffusion-weighted magnetic resonance imaging in differentiating purulent brain processes from cystic brain tumors *Journal of Neurosurgery* **97** 1101–1107.

Haacke E M, Brown R W, Thompson M R and Venkatesan R 1999 *Magnetic Resonance Imaging: Physical Principles and Sequence Design* (New York: John Wiley & Sons).

Haase A, Frahm J, Hanicke W and Matthaei D 1985 H-1-NMR chemical-shift selective (CHESS) imaging *Physics in Medicine and Biology* **30** 341–344.

Haase A, Frahm J, Matthaei D, Hanicke W and Merboldt K D 1986 Flash imaging—Rapid NMR imaging using low flip-angle pulses *Journal of Magnetic Resonance* **67** 258–266.

Hahn E L 1950 Spin echoes *Physical Review* **80** 580.

Hamstra D A, Lee K C, Tychewicz J M, Schepkin V D, Moffat B A, Chen M, Dornfeld K J, Lawrence T S, Chenevert T L, Ross B D, Gelovani J T and Rehemtulla A 2004 The use of F-19 spectroscopy and diffusion-weighted MRI to evaluate differences in gene-dependent enzyme prodrug therapies *Molecular Therapy* **10** 916–928.

Hardy C J and Cline H E 1989 Spatial localization in 2 dimensions using NMR designer pulses *Journal of Magnetic Resonance* **82** 647–654.

Hargreaves B A, Bangerter N K, Shimakawa A, Vasanawala S S, Brittain J H and Nishimura D G 2006 Dual-acquisition phase-sensitive fat-water separation using balanced steady-state free precession *Magnetic Resonance Imaging* **24** 113–122.

Hayes C E, Edelstein W A, Schenck J F, Mueller O M and Eash M 1985 An efficient, highly homogeneous radiofrequency coil for whole-body NMR imaging at 1.5 T *Journal of Magnetic Resonance* **63** 622–628.

Henkelman R M and Bronskill M J 1987 Artifacts in magnetic resonance imaging *Reviews of Magnetic Resonance in Medicine* **2** 1–126.

Henkelman R M, Stanisz G J and Graham S J 2001 Magnetization transfer in MRI: A review *NMR in Biomedicine* **14** 57–64.

Hennig J 1988 Multiecho imaging sequences with low refocusing flip angles *Journal of Magnetic Resonance* **78** 397–407.

Hennig J, Nauerth A and Friedburg H 1986 RARE imaging—A fast imaging method for clinical MR *Magnetic Resonance in Medicine* **3** 823–833.

Higuchi M, Iwata N, Matsuba Y, Sato K, Sasamoto K and Saido T C 2005 F-19 and H-1 MRI detection of amyloid beta plaques in vivo *Nature Neuroscience* **8** 527–533.

Hinshaw W S 1974 Spin mapping—Application of moving gradients to NMR *Physics Letters A* **48A** 87–88.

Hinshaw W S, Bottomley P A and Holland G N 1977 Radiographic thin-section image of human wrist by nuclear magnetic-resonance *Nature* **270** 722–723.

Hoch J C and Stern A 1996 *NMR Data Processing* (New York: John Wiley & Sons).

Hoffman 2009 *Multinuclear NMR* [Online] Available at http://chem.ch.huji.ac.il/nmr/techniques/1d/multi.html (accessed on June 23, 2009).

Hoult D I, Busby S J W, Gadian D G, Radda G K, Richards R E and Seeley P J 1974 Observation of tissue metabolites using P-31 nuclear magnetic resonance *Nature* **252** 285–287.

Hoult D I and Ginsberg N S 2001 The quantum origins of the free induction decay signal and spin noise *Journal of Magnetic Resonance* **148** 182–199.

Hoult D I and Lauterbur P C 1979 The sensitivity of the zeugmatographic experiment involving human samples *Journal of Magnetic Resonance* **34** 425–433.

Hoult D I and Richards R E 1976a Signal-to-noise ratio of nuclear magnetic-resonance experiment *Journal of Magnetic Resonance* **24** 71–85.

Hoult D I and Richards R E 1976b The signal-to-noise ratio of the nuclear magnetic resonance experiment *Journal of Magnetic Resonance* **24** 71–85.

Huang R, Posnansky O, Celik A, Oros-Peusquens A M, Ermer V, Irkens M, Wegener H P and Shah N J 2006 Measurement of weak electric currents in copper wire phantoms using MRI: Influence of susceptibility enhancement *Magnetic Resonance Materials in Physics Biology and Medicine* **19** 124–133.
Hutchison J M S 1977 Imaging by nuclear magnetic resonance In Hay G A (ed.) *7th L. H. Gray Conference 1976: Medical Images: Formation, Perception and Measurement*, Leeds, West Yorkshire, UK, pp. 135–141.
IEC 2006. Medical electrical equipment—Part 2-33: Particular requirements for the safety of magnetic resonance equipment for medical diagnosis. (Geneva, Switzerland: IEC).
Jackson A, Buckley D L and Parker G J M (eds.) 2003 *Dynamic Contrast-Enhanced Magnetic Resonance Imaging in Oncology* (Berlin, Germany: Springer).
James T L 1975 *Nuclear Magnetic Resonance in Biochemistry: Principles and Applications* (New York: Academic Press).
Jezzard P and Balaban R S 1995 Correction for geometric distortion in echo-planar images from B_0 field variations *Magnetic Resonance in Medicine* **34** 65–73.
Jezzard P, Matthew P M and Smith S M (eds.) 2003 *Functional Magnetic Resonance Imaging: An Introduction to Methods* (Oxford, UK: Oxford University Press).
Jolesz F A and McDannold N 2008. Current status and future potential of MRI-guided focused ultrasound surgery *Journal of Magnetic Resonance Imaging* **27** 391–399.
Joy M L G, Lebedev V P and Gati J S 1999 Imaging of current density and current pathways in rabbit brain during transcranial electrostimulation *IEEE Transactions on Biomedical Engineering* **46** 1139–1149.
Kaiser R, Barthold E and Ernst R R 1974 Diffusion and field-gradient effects in NMR Fourier spectroscopy *Journal of Chemical Physics* **60** 2966–2979.
Kanal E, Barkovich A J, Bell C, Borgstede J P, Bradley W G, Froelich J W, Gilk T, Gimbel J R, Gosbee J, Kuhni-Kaminski E, Lester J W, Nyenhuis J, Parag Y, Schaefer D J, Sebek-Scoumis E A, Weinreb J, Zaremba L A, Wilcox P, Lucey L and Sass N 2007 ACR guidance document for safe MR practices: 2007 *American Journal of Roentgenology* **188** 1447–1474.
Keevil S F, Gedroyc W, Gowland P, Hill D L G, Leach M O, Ludman C N, McLeish K, McRobbie D W, Razavi R S and Young I R 2005 Electromagnetic field exposure limitation and the future of MRI *British Journal of Radiology* **78** 973.
Kerr A B, Pauly J M, Hu B S, Li K C, Hardy C J, Meyer C H, Macovski A and Nishimura D G 1997 Real-time interactive MRI on a conventional scanner *Magnetic Resonance in Medicine* **38** 355–367.
Kibble T W B and Berkshire F H 2004 *Classical Mechanics* (London, UK: Imperial College Press).
Kimmich R 1997 *NMR: Tomography, Diffusometry, Relaxometry* (Berlin, Germany: Springer).
King C P L (ed.) 2002 Molecular Imaging *Journal of Magnetic Resonance Imaging* **16**(4) Special Issue, pp. 333–483.
Kittel C 2005 *Introduction to Solid State Physics* (Hoboken, NJ: Wiley).
Knight W D 1949 Nuclear magnetic resonance shift in metals *Physical Review* **76** 1259.
Koh D-M and Collins D J 2007 Diffusion-weighted MRI in the body: Applications and challenges in oncology *American Journal of Roentgenology* **188** 1622–1635.
Kovacevic N, Henderson J T, Chan E, Lifshitz N, Bishop J, Evans A C, Henkelman R M and Chen X J 2005 A three-dimensional MRI atlas of the mouse brain with estimates of the average and variability *Cerebral Cortex* **15** 639–645.
Krane K S and Halliday D 1988 *Introductory Nuclear Physics* (New York: John Wiley & Sons).
Kumar A, Welti D and Ernst R R 1975 NMR Fourier zeugmatography *Journal of Magnetic Resonance* **18** 69–83.
Kupriyanov V 2005 Special issue: Biomedical applications of MRS/I of uncommon nuclei *NMR in Biomedicine* **18** 65–66.
Kurhanewicz J, Vigneron D B, Brindle K, Chekmenev E Y, Comment A, Cunningham C H, Deberardinis R J, Green G G, Leach M O, Rajan S S, Rizi R R, Ross B D, Warren W S and Malloy C R 2011 Analysis of cancer metabolism by imaging hyperpolarized nuclei: Prospects for translation to clinical research *Neoplasia* **13** 81–97.

Kwong K K, Belliveau J W, Chesler D A, Goldberg I E, Weisskoff R M, Poncelet B P, Kennedy D N, Hoppel B E, Cohen M S, Turner R, Cheng H M, Brady T J and Rosen B R 1992 Dynamic magnetic-resonance-imaging of human brain activity during primary sensory stimulation *Proceedings of the National Academy of Sciences of the United States of America* **89** 5675–5679.

Laserew G and Schnubikow L W 1937 *Physik Zeits. Sowjetunion* **11** 445.

Laub G, Gaa J and Drobnitzky M 1998 Magnetic resonance angiography techniques *Electromedica*

Lauterbur P C 1973 Image formation by induced local interactions—Examples employing nuclear magnetic-resonance *Nature* **242** 190–191.

Lauterbur P C 1974 Magnetic-resonance zeugmatography *Pure and Applied Chemistry* **40** 149–157.

Leach M O 2006 Magnetic resonance spectroscopy (MRS) in the investigation of cancer at The Royal Marsden Hospital and The Institute of Cancer Research *Physics in Medicine and Biology* **51** R61–R82.

Leach M O, Brindle K M, Evelhoch J L, Griffiths J R, Horsman M R, Jackson A, Jayson G C, Judson I R, Knopp M V, Maxwell R J, McIntyre D, Padhani A R, Price P, Rathbone R, Rustin G J, Tofts P S, Tozer G M, Vennart W, Waterton J C, Williams S R and Workmanw P 2005 The assessment of antiangiogenic and antivascular therapies in early-stage clinical trials using magnetic resonance imaging: Issues and recommendations *British Journal of Cancer* **92** 1599–1610.

Le Bihan D 2003 Looking into the functional architecture of the brain with diffusion MRI *Nature Reviews Neuroscience* **4** 469–480.

Le Bihan D, Breton E, Lallemand D, Grenier P, Cabanis E and Lavaljeantet M 1986 MR imaging of intravoxel incoherent motions—Application to diffusion and perfusion in neurologic disorders *Radiology* **161** 401–407.

Le Bihan D and Van Zijl P 2002 From the diffusion coefficient to the diffusion tensor *NMR in Biomedicine* **15** 431–434.

Lee K J, Barber D C, Paley M N, Wilkinson I D, Papadakis N G and Griffiths P D 2002 Image-based EPI ghost correction using an algorithm based on projection onto convex sets (POCS) *Magnetic Resonance in Medicine* **47** 812–817.

Lee R F, Giaquinto R, Constantinides C, Souza S, Weiss R G and Bottomley P A 2000 A broadband phased-array system for direct phosphorus and sodium metabolic MRI on a clinical scanner *Magnetic Resonance in Medicine* **43** 269–277.

Lee R F, Johnson G, Grossman R I, Stoeckel B, Trampel R and McGuinness G 2006 Advantages of parallel imaging in conjunction with hyperpolarized helium—A new approach to MRI of the lung *Magnetic Resonance in Medicine* **55** 1132–1141.

Lee S, Richter W, Vathyam S and Warren W S 1996 Quantum treatment of the effects of dipole-dipole interactions in liquid nuclear magnetic resonance *Journal of Chemical Physics* **105** 874–900.

Leggett J, Crozier S, Blackband S, Beck B and Bowtell R 2003 Multilayer transverse gradient coil design *Concepts in Magnetic Resonance B* **16** 38–46.

Lei H, Zhu X H, Zhang X L, Ugurbil K and Chen W 2003 In vivo P-31 magnetic resonance spectroscopy of human brain at 7 T: An initial experience *Magnetic Resonance in Medicine* **49** 199–205.

Levitt M H 2001 *Spin Dynamics: Basic Principles of NMR Spectroscopy* (New York: John Wiley & Sons).

Liang Z-P and Lauterbur P C 2000 *Principles of Magnetic Resonance Imaging: A Signal Processing Perspective* (Bellingham, WA: IEEE Press).

Louie A Y, Huber M M, Ahrens E T, Rothbacher U, Moats R, Jacobs R E, Fraser S E and Meade T J 2000 In vivo visualization of gene expression using magnetic resonance imaging *Nature Biotechnology* **18** 321–325.

Lurie D J, Foster M A, Yeung D and Hutchison J M S 1998 Design, construction and use of a large-sample field-cycled PEDRI imager *Physics in Medicine and Biology* **43** 1877–1886.

Majumdar S, Orphanoudakis S C, Gmitro A, Odonnell M and Gore J C 1986a Errors in the measurements of T2 using multiple-echo MRI techniques. 1. Effects of radiofrequency pulse imperfections *Magnetic Resonance in Medicine* **3** 397–417.

Majumdar S, Orphanoudakis S C, Gmitro A, Odonnell M and Gore J C 1986b Errors in the measurements of T2 using multiple-echo MRI techniques. 2. Effects of static-field inhomogeneity *Magnetic Resonance in Medicine* **3** 562–574.
Mancini L, Payne G S, Dzik-Jurasz A S K and Leach M O 2003 Ifosfamide pharmacokinetics and hepatobiliary uptake in vivo investigated using single- and double-resonance P-31 MRS *Magnetic Resonance in Medicine* **50** 249–255.
Mansfield P 1977 Multi-planar image-formation using NMR spin echoes *Journal of Physics C—Solid State Physics* **10** L55–L58.
Mansfield P and Chapman B 1986 Active magnetic screening of coils for static and time-dependent magnetic-field generation in NMR imaging *Journal of Physics E—Scientific Instruments* **19** 540–545.
Mansfield P and Grannell P K 1973 NMR diffraction in solids *Journal of Physics C—Solid State Physics* **6** L422–L426.
Mansfield P and Maudsley A A 1976 Line scan proton spin imaging in biological structures by NMR *Physics in Medicine and Biology* **21** 847–852.
Mansfield P and Maudsley A A 1977 Medical imaging by NMR *British Journal of Radiology* **50** 188–194.
Mansfield P and Morris P G 1982a *NMR Imaging in Biomedicine* (London, UK: Academic Press).
Mansfield P and Morris P G 1982b *NMR Imaging in Biomedicine: Supplement 2, Advances in Magnetic Resonance* (Orlando, FL: Academic Press).
Margosian P 1985 Faster MR imaging with half the data In *Proceedings of the 4th Annual Meeting of the Society of Magnetic Resonance in Medicine*, 1985, London, UK, pp. 1024–1025.
Margosian P 1987 MR images from a quarter of the data: Combination of half-Fourier methods with a linear recursive data extrapolation In *Proceedings of the 6th Annual Meeting of the Society of Magnetic Resonance in Medicine*, New York, p. 375.
Matson G B and Wiener M W 1999 Spectroscopy In Stark D D and Bradley W G J (eds.) *Magnetic Resonance Imaging* (St Louis, MO: Mosby, year book.
Matter N I, Scott G C, Grafendorfer T, Macovski A and Conolly S M 2006 Rapid polarizing field cycling in magnetic resonance imaging *IEEE Transactions on Medical Imaging* **25** 84–93.
Matthes R 1994 Guidelines on limits of exposure to static magnetic fields *Health Physics* **66** 100–106.
Matthes R 2004 Medical magnetic resonance (MR) procedures: Protection of patients *Health Physics* **87** 197–216.
Maudsley A A 1980 Multiple-line-scanning spin-density imaging *Journal of Magnetic Resonance* **41** 112–126.
McConnell J 1987 *The Theory of Nuclear Magnetic Relaxation in Liquids* (Cambridge, UK: Cambridge University Press).
McDonald P J and Newling B 1998 Stray field magnetic resonance imaging *Reports on Progress in Physics* **61** 1441–1493.
McKinlay A F, Allen S G, Cox R, Dimbylow P J, Mann S M, Muirhead C R, Saunders R D, Sienkiewicz Z J, Stather J W and Wainwright P R 2004 Review of the scientific evidence for limiting exposure to electromagnetic fields (0–300 GHz) *Doc. NRPB* **15** 197–204.
Meiboom S and Gill D 1958 Modified spin-echo method for measuring nuclear relaxation times *Review of Scientific Instruments* **29** 688–691.
Merboldt K D, Hanicke W and Frahm J 1985 Self-diffusion NMR imaging using stimulated echoes *Journal of Magnetic Resonance* **64** 479–486.
MHRA (Medicine and Healthcare products Regulation Agency) 2007 *Safety Guidelines for Magnetic Resonance Imaging Equipment in Clinical Use* – DB 2007(03), available from http://www.mhra.gov.uk/Publications/Safetyguidance/DeviceBulletins/CON2033018 (accessed on February 27, 2012).
Modo M, Hoehn M and Bulte J W M 2005 Cellular MR imaging *Molecular Imaging* **4** 143–164.
Moller H E, Chen X J, Saam B, Hagspiel K D, Johnson G A, Altes T A, De Lange E E and Kauczor H U 2002 MRI of the lungs using hyperpolarized noble gases *Magnetic Resonance in Medicine* **47** 1029–1051.
Moon R B and Richards J H 1973 Determination of intracellular pH by P-31 magnetic-resonance *Journal of Biological Chemistry* **248** 7276–7278.

Morris P G 1986 *Nuclear Magnetic Resonance Imaging in Medicine and Biology* (Oxford, UK: Clarendon Press).

Murphy P S, Viviers L, Abson C, Rowland I J, Brada M, Leach M O and Dzik-Jurasz A S K 2004 Monitoring temozolomide treatment of low-grade glioma with proton magnetic resonance spectroscopy *British Journal of Cancer* **90** 781–786.

Navon G, Shinar H, Eliav U and Seo Y 2001 Multiquantum filters and order in tissues *NMR in Biomedicine* **14** 112–132.

Navon G, Song Y Q, Room T, Appelt S, Taylor R E and Pines A 1996 Enhancement of solution NMR and MRI with laser-polarized xenon *Science* **271** 1848–1851.

Nayak K S, Pauly J M, Yang P C, Hu B S, Meyer C H and Nishimura D G 2001 Real-time interactive coronary MRA *Magnetic Resonance in Medicine* **46** 430–435.

Nelson S J 2003 Multivoxel magnetic resonance spectroscopy of brain tumors *Molecular Cancer Therapeutics* **2** 497–507.

New P F J, Rosen B R, Brady T J, Buonanno F S, Kistler J P, Burt C T, Hinshaw W S, Newhouse J H, Pohost G M and Taveras J M 1983 Potential hazards and artifacts of ferromagnetic and nonferromagnetic surgical and dental materials and devices in nuclear magnetic resonance imaging *Radiology* **147** 139–148.

Oberhaensli R D, Galloway G J, Taylor D J, Bore P J and Radda G K 1986 Assessment of human-liver metabolism by P-31 magnetic resonance spectroscopy *British Journal of Radiology* **59** 695–699.

Oesterle C, Strohschein R, Kohler M, Schnell M and Hennig J 2000 Benefits and pitfalls of keyhole imaging, especially in first-pass perfusion studies *Journal of Magnetic Resonance Imaging* **11** 312–323.

Ogawa S, Lee T M, Kay A R and Tank D W 1990 Brain magnetic-resonance-imaging with contrast dependent on blood oxygenation *Proceedings of the National Academy of Sciences of the United States of America* **87** 9868–9872.

Ordidge R J, Bendall M R, Gordon R E and Connelly A 1985 Volume detection for *in vivo* spectroscopy In Govil C, Khetrapal C L and Saran A (eds.) *Magnetic Resonance in Biology and Medicine* (New Delhi: Tata-McGraw-Hill).

Ordidge R J, Connelly A and Lohman J A B 1986 Image-selected in vivo spectroscopy (ISIS)—A new technique for spatially selective NMR-spectroscopy *Journal of Magnetic Resonance* **66** 283–294.

Oshio K and Feinberg D A 1991 GRASE (gradient-echo and spin-echo) imaging—A novel fast MRI technique *Magnetic Resonance in Medicine* **20** 344–349.

Pauly J, Nishimura D and Macovski A 1989 A *k*-space analysis of small-tip-angle excitation *Journal of Magnetic Resonance* **81** 43–56.

Payne G S, Al-Saffar N and Leach M O 2006 Phosporous magnetic resonance spectroscopy on biopsy and *in vivo* In Webb G (ed.) *Modern Magnetic Resonance* (London, UK: Springer).

Payne G S, Pinkerton C R, Bouffet E and Leach M O 2000 Initial measurements of ifosfamide and cyclophosphamide in patients using P-31 MRS: Pulse-and-acquire, decoupling, and polarization transfer *Magnetic Resonance in Medicine* **44** 180–184.

Pell G S, Abbott D F, Fleming S W, Prichard J W and Jackson G D 2006 Further steps toward direct magnetic resonance (MR) imaging detection of neural action currents: Optimization of MR sensitivity to transient and weak currents in a conductor *Magnetic Resonance in Medicine* **55** 1038–1046.

Petrella J R, Coleman R E and Doraiswamy P M 2003 Neuroimaging and early diagnosis of Alzheimer disease: A look to the future *Radiology* **226** 315–336.

Pipe J G and Zwart N 2006 Turboprop: Improved PROPELLER imaging *Magnetic Resonance in Medicine* **55** 380–385.

Poole M and Bowtell R 2007 Novel gradient coils designed using a boundary element method *Concepts in Magnetic Resonance B* **31** 162–175.

Prescot A P, Dzik-Jurasz A S K, Leach M O, Sirohi L, Powles R and Collins D J 2005 Localized COSY and DQF-COSY H-1-MRS sequences for investigating human tibial bone marrow in vivo and initial application to patients with acute leukemia *Journal of Magnetic Resonance Imaging* **22** 541–548.

Press W H, Teukolsky S A, Vetterling W T and Flannery B P 2002 *Numerical Recipes in C: The Art of Scientific Computing*, 2nd edn. (Cambridge, UK: Cambridge University Press).
Price D, Delakis I and Renaud C 2005 1.5T Issue 6 *Comparison Reports* NHS Purchasing and Supply Agency, Centre for Evidence Based Purchasing, Cheshire, UK, Report Number 06005.
Price D, Delakis I and Renaud C 2006a 0.3T to 0.4T Open MRI Systems Issue 6 *Comparison Reports* MagNET/NHS Purchasing and Supply Agency, Centre for Evidence Based Purchasing, Cheshire, UK, Report Number 06032.
Price D, Delakis I and Renaud C 2006b 0.6T to 1.0T Open MRI Systems Issue 6 *Comparison Reports* MagNET/NHS Purchasing and Supply Agency (Centre for Evidence Based Purchasing, Cheshire, UK), Report Number 06030.
Price D, Delakis I and Renaud C 2006c 3T Systems Issue 3 *Comparison Reports* MagNET/NHS Purchasing and Supply Agency, Centre for Evidence Based Purchasing, Cheshire, UK, Report Number 06006.
Proctor W G and Yu F C 1950 The dependence of a nuclear magnetic resonance frequency upon chemical compound *Physical Review* **77** 717.
Pruessmann K P, Weiger M, Scheidegger M B and Boesiger P 1999 SENSE: Sensitivity encoding for fast MRI *Magnetic Resonance in Medicine* **42** 952–962.
Purcell E M, Torrey H C and Pound R V 1946 Resonance absorption by nuclear magnetic moments in a solid *Physical Review* **69** 37–38.
Ra J B and Rim C Y 1993 Fast imaging using subencoding data sets from multiple detectors *Magnetic Resonance in Medicine* **30** 142–145.
Rabi I I, Zacharias J R, Millman S and Kusch P 1938 A new method of measuring nuclear magnetic moment *Physical Review* **53** 318.
Radda G K, Bore P J, Gadian D G, Ross B D, Styles P, Taylor D J and Morganhughes J 1982 P-31 NMR examination of 2 patients with NADH-COQ reductase deficiency *Nature* **295** 608–609.
Reinsberg S A, Doran S J, Charles-Edwards E M and Leach M O 2005 A complete distortion correction for MR images: II. Rectification of static-field inhomogeneities by similarity-based profile mapping *Physics in Medicine and Biology* **50** 2651–2661.
Riches S F, Collins D J, Scuffham J W and Leach M O 2007 EU directive 2004/40: Field measurements of a 1.5T clinical MR scanner *British Journal of Radiology* **80** 483–487.
Richter W, Lee S H, Warren W S and He Q H 1995 Imaging with intermolecular multiple-quantum coherences in solution nuclear-magnetic-resonance *Science* **267** 654–657.
Richter W, Richter M, Warren W S, Merkle H, Andersen P, Adriany G and Ugurbil K 2000 Functional magnetic resonance imaging with intermolecular multiple-quantum coherences *Magnetic Resonance Imaging* **18** 489–494.
Rieke V and Butts Pauly K 2008 MR thermometry *Journal of Magnetic Resonance Imaging* **27** 390.
Robinson S P and Griffiths J R 2004 Current issues in the utility of F-19 nuclear magnetic resonance methodologies for the assessment of tumour hypoxia *Philosophical Transactions of the Royal Society of London Series B—Biological Sciences* **359** 987–996.
Robson M D and Bydder G M 2006 Clinical ultrashort echo time imaging of bone and other connective tissues *NMR in Biomedicine* **19** 765–780.
Romanzetti S, Halse M, Kaffanke J, Zilles K, Balcom B J and Shah N J 2006 A comparison of three sprite techniques for the quantitative 3D imaging of the Na-23 spin density on a 4T whole-body machine *Journal of Magnetic Resonance* **179** 64–72.
Rosen B R, Wedeen V J and Brady T J 1984 Selective saturation NMR imaging *Journal of Computer Assisted Tomography* **8** 813–818.
Ross B D, Radda G K, Gadian D G, Rocker G, Esiri M and Falconersmith J 1981 Examination of a case of suspected McArdle's syndrome by P-31 nuclear magnetic-resonance *New England Journal of Medicine* **304** 1338–1342.
Saunders R D 1982 The biological hazards of NMR In *Proceedings of the First International Symposium on NMR Imaging* Bowmen Gray School of Medicine, Winston-Salem, NC, pp. 65–71.

Saunders R D 1991 Principles for the protection of patients and volunteers during clinical magnetic resonance diagnostic procedures. Limits on patient and volunteer exposure during clinical magnetic resonance diagnostic procedures: Recommendations for the practical application of the board's statement *Doc. NRPB* **2**(1).

Saunders R D and Orr J S 1983 Biologic effects of NMR In Partain C L, James A E, Rollo F D and Price R R (eds.) *Magnetic Resonance Imaging* (Philadelphia, PA: W.B. Saunders).

Saunders R D and Smith H 1984 Safety aspects of NMR clinical imaging *British Medical Bulletin* **40** 148–154.

Scheffler K and Lehnhardt S 2003 Principles and applications of balanced SSFP techniques *European Radiology* **13** 2409–2418.

Scott G C, Joy M L G, Armstrong R L and Henkelman R M 1992 Sensitivity of magnetic-resonance current-density imaging *Journal of Magnetic Resonance* **97** 235–254.

Seeber D A, Hoftiezer J H, Daniel W B, Rutgers M A and Pennington C H 2000 Triaxial magnetic field gradient system for microcoil magnetic resonance imaging *Review of Scientific Instruments* **71** 4263–4272.

Servoss T G and Hornak J P 2011 Converting the chemical shift artifact to a spectral image *Concepts in Magnetic Resonance A* **38** 107–116.

Shaka A J, Keeler J and Freeman R 1983a Evaluation of a new broad-band decoupling sequence—WALTZ-16 *Journal of Magnetic Resonance* **53** 313–340.

Shaka A J, Keeler J, Frenkiel T and Freeman R 1983b An improved sequence for broad-band decoupling—WALTZ-16 *Journal of Magnetic Resonance* **52** 335–338.

Sharp J C and Leach M O 1992 Rapid localization of concave volumes by conformal NMR-spectroscopy *Magnetic Resonance in Medicine* **23** 386–393.

Shaw D 1976 *Fourier Transform NMR Spectroscopy* (Amsterdam, the Netherlands: Elsevier Scientific).

Shen J, Sibson N R, Cline G, Behar K L, Rothman D L and Shulman R G 1998 N-15-NMR spectroscopy studies of ammonia transport and glutamine synthesis in the hyperammonemic rat brain *Developmental Neuroscience* **20** 434–443.

Slichter C P 1978 *Principles of Magnetic Resonance* (Berlin, Germany: Springer).

Sodickson D K and Manning W J 1997 Simultaneous acquisition of spatial harmonics (SMASH): Fast imaging with radiofrequency coil arrays *Magnetic Resonance in Medicine* **38** 591–603.

Solomon I 1955 Relaxation processes in a system of two spins *Physical Review* **99** 559.

Spandonis Y, Heese F P and Hall L D 2004 High resolution MRI relaxation measurements of water in the articular cartilage of the meniscectomized rat knee at 4.7 T *Magnetic Resonance Imaging* **22** 943–951.

Stejskal E O and Tanner J E 1965 Spin diffusion measurements: Spin echoes in the presence of a time-dependent field gradient *Journal of Chemical Physics* **41** 288–292.

Sugahara T, Korogi Y, Kochi M, Ikushima I, Shigematu Y, Hirai T, Okuda T, Liang L X, Ge Y L, Komohara Y, Ushio Y and Takahashi M 1999 Usefulness of diffusion-weighted MRI with echo-planar technique in the evaluation of cellularity in gliomas *Journal of Magnetic Resonance Imaging* **9** 53–60.

Tan E T, Huston J, Campeau N G and Reiderer S J 2010 Fast inversion recovery magnetic resonance angiography of the intracranial arteries *Magnetic Resonance in Medicine* **63** 1648–1658.

Taylor D J, Bore P J, Styles P, Gadian D G and Radda G K 1983 Bioenergetics of intact human muscle. A ^{31}P nuclear magnetic resonance study *Molecular Biology and Medicine* **1** 77–94.

Taylor D G and Bushell M C 1985 The spatial-mapping of translational diffusion-coefficients by the NMR imaging technique *Physics in Medicine and Biology* **30** 345–349.

Tofts P S 1997 Modeling tracer kinetics in dynamic Gd-DTPA MR imaging *Journal of Magnetic Resonance Imaging* **7** 91–101.

Tofts P S, Brix G, Buckley D L, Evelhoch J L, Henderson E, Knopp M, Larsson H B W, Lee T Y, Mayr N A, Parker G J M, Port R E, Taylor J and Weisskoff R M 1999 Estimating kinetic parameters from dynamic contrast-enhanced T_1-weighted MRI of a diffusable tracer: Standardized quantities and symbols *Journal of Magnetic Resonance Imaging* **10** 223–232.

Tsao J, Boesiger P and Pruessmann K P 2003 k-t BLAST and k-t SENSE: Dynamic MRI with high frame rate exploiting spatiotemporal correlations *Magnetic Resonance in Medicine* **50** 1031–1042.

Turner R 1993 Gradient coil design: A review of methods *Magnetic Resonance Imaging* **11** 903–920.

Ugurbil K, Adriany G, Andersen P, Chen W, Garwood M, Gruetter R, Henry P G, Kim S G, Lieu H, Tkac I, Vaughan T, Van De Moortele P F, Yacoub E and Zhu X H 2003 Ultrahigh field magnetic resonance imaging and spectroscopy *Magnetic Resonance Imaging* **21** 1263–1281.

Van Beek E J R, Wild J M, Kauczor H U, Schreiber W, Mugler J P and De Lange E E 2004 Functional MRI of the lung using hyperpolarized 3-helium gas *Journal of Magnetic Resonance Imaging* **20** 540–554.

Wakana S, Jiang H Y, Nagae-Poetscher L M, Van Zijl P C M and Mori S 2004 Fiber tract-based atlas of human white matter anatomy *Radiology* **230** 77–87.

Walker P M, Balmer C, Ablett S and Lerski R A 1989 A test material for tissue characterization and system calibration in MRI *Physics in Medicine and Biology* **34** 5–22.

Walker T G and Happer W 1997 Spin-exchange optical pumping of noble-gas nuclei *Reviews of Modern Physics* **69** 629–642.

Wang H Z, Riederer S J and Lee J N 1987 Optimizing the precision in T1 relaxation estimation using limited flip angles *Magnetic Resonance in Medicine* **5** 399–416.

Weigel M and Hennig J 2006 Contrast behavior and relaxation effects of conventional and hyperecho-turbo spin echo sequences at 1.5 and 3T *Magnetic Resonance in Medicine* **55** 826–835.

Weissleder R and Mahmood U 2001 Molecular imaging *Radiology* **219** 316–333.

WHO 2006 Static fields *Environmental Health Criteria Monographs* (Geneva, Switzerland: WHO Press).

Winter P M, Morawski A M, Caruthers S D, Fuhrhop R W, Zhang H, Williams T A, Allen J S, Lacy E K, Robertson J D, Lanza G M and Wickline S A 2003 Molecular imaging of angiogenesis in early-stage atherosclerosis with αvβ3-integrin-targeted nanoparticles *Circulation* **108** 2270–2274.

Woessner D E 1961 Effects of diffusion in nuclear magnetic resonance spin-echo experiments *Journal of Chemical Physics* **34** 2057–2061.

Wolber J, Cherubini A, Leach M O and Bifone A 2000 Hyperpolarized Xe-129 NMR as a probe for blood oxygenation *Magnetic Resonance in Medicine* **43** 491–496.

Wolber J, Ellner F, Fridlund B, Gram A, Johannesson H, Hansson G, Hansson L H, Lerche M H, Mansson S, Servin R, Thaning M, Golman K and Ardenkjaer-Larsen J H 2004 Generating highly polarized nuclear spins in solution using dynamic nuclear polarization *Nuclear Instruments and Methods in Physics Research Section A: Accelerators, Spectrometers, Detectors and Associated Equipment* **526** 173–181.

Wolf W, Presant C A and Waluch V 2000 F-19-MRS studies of fluorinated drugs in humans *Advanced Drug Delivery Reviews* **41** 55–74.

Young I R, Khenia S, Thomas D G T, Davis C H, Gadian D G, Cox I J, Ross B D and Bydder G M 1987 Clinical magnetic-susceptibility mapping of the brain *Journal of Computer Assisted Tomography* **11** 2–6.

Zavoisky E K 1945 Paramagnetic relaxation of liquid solutions for perpendicular fields *Journal of Physics USSR* **9** 245.

Zeeman P 1897 The effect of magnetisation on the nature of light emitted by a substance *Nature* **55** 347.

Zenge M O, Vogt F M, Brauck K, Jökel M, Barkhausen J, Kannengiesser S, Ladd M E and Quick H H 2006 High-resolution continuously acquired peripheral MR angiography featuring partial parallel imaging GRAPPA *Magnetic Resonance in Medicine* **56** 859–865.

Zhang S R, Merritt M, Woessner D E, Lenkinski R E and Sherry A D 2003 Paracest agents: Modulating MRI contrast via water proton exchange *Accounts of Chemical Research* **36** 783–790.

Zhu X H, Zhang Y, Tian R X, Lei H, Zhang N Y, Zhang X L, Merkle H, Ugurbil K and Chen W 2002 Development of O-17 NMR approach for fast imaging of cerebral metabolic rate of oxygen in rat brain at high field *Proceedings of the National Academy of Sciences of the United States of America* **99** 13194–13199.

8

Physical Aspects of Infrared Imaging

C. H. Jones

CONTENTS

8.1 Introduction 623
8.2 Infrared Photography 624
8.3 Infrared Imaging 624
8.3.1 Thermal Radiation 625
8.3.2 Single-Photon Detectors 626
8.3.3 Thermographic Scanning Systems 628
8.3.4 Focal-Plane Staring Arrays 630
8.3.5 Pyroelectric Imaging Systems 634
8.3.6 Temperature Measurement 634
8.3.7 Clinical Thermography: Physiological Factors 636
8.3.8 Clinical Thermography: Applications 638
8.3.8.1 Assessment of Inflammatory Conditions 638
8.3.8.2 Investigations of Vascular Disorders 639
8.3.8.3 Metabolic Studies 640
8.3.8.4 Assessment of Pain and Trauma 640
8.3.8.5 Oncological Investigations 641
8.3.8.6 Physiological Investigations 641
8.4 Liquid-Crystal Thermography 642
8.5 Microwave Thermography 643
8.6 Future Developments 644
References 644

8.1 Introduction

The human body may be regarded as having a thermally regulated inner core where heat production is centred and where the temperature is maintained within fine limits (typically 37°C ± 1°C). The temperature of superficial tissues is more variable and is influenced by several factors including proximity to the core, subcutaneous heat production, blood perfusion, thermal properties of body tissues and radiant heat transfer from the skin surface. Blood perfusion has a major role in modifying the skin temperature. Many disease processes and physical conditions affect superficial tissue blood flow, and, consequently, various means have been developed to record and measure surface temperatures. This chapter is concerned principally with the use of IR thermal imaging but the related subjects of IR photography and microwave radiometry will also be described.

8.2 Infrared Photography

Clinical IR photography depends upon the spectral transmission and reflection characteristics of tissue and blood in the wavelength range 0.7–0.9 μm. The technique requires the subject to be photographed in diffuse, shadow-less illumination. A lens filter is used to absorb radiation of wavelengths less than 0.7 μm. Penetration and reflection are maximal in the red end of the visible spectrum, where the radiation penetrates the superficial layers of skin and tissue up to a depth of 2.5 mm and is then reflected out again. The spectral transmission of radiation through blood depends upon the relative concentration of reduced and oxygenated haemoglobin: the latter transmits radiation of wavelengths greater than 0.6 μm, whereas reduced haemoglobin transmits radiation significantly when the wavelength is greater than 0.7 μm.

These properties allow images to be recorded on IR-sensitive film and will show blood vessels (veins) that lie within 2.5 mm of the skin. IR film can be replaced by a silicon-target vidicon tube which is sensitive up to 1.2 μm to allow real-time viewing of vascular patterns (Jones 1982).

Colour IR photography can also be used and is useful for recording pigmented lesions. The film consists of three image layers of emulsion that are sensitive to green, red and infrared radiation, so the image records false colour as well as tone differences.

8.3 Infrared Imaging

Thermal imaging is dependent upon the detection of radiant emission from the skin surface. A typical imaging device consists of a system for collecting radiation from a well-defined field of view and a detector that transduces the radiation focused onto it into an electrical signal. The use of thermal imaging in medicine falls into three periods. From 1960 to about 1975, imaging was principally with scanning systems that used single IR detectors. The period 1975–1995 saw the commercial development of the SPRITE detector, linear arrays and 2D arrays. From 1993 focal-plane array (FPA) detectors became available.

There are two principal classes of IR detector: thermal detectors, such as thermocouples and semiconductor bolometers, which rely upon the temperature rise in the absorbing element, and photon detectors. Thermal detectors are less sensitive than photon detectors and have longer time constants. In the main, medical thermography has employed photon detector systems.

Because the single detector element senses the entire image in a short time it needs to react very quickly – typically, in a few microseconds. An advantage of this kind of detector is that there is only one detector involved so that there is a high degree of uniformity of response over the whole image. Photon detectors are wavelength dependent and detect only a small proportion of the total radiation emitted from the object's surface.

The measurement of surface temperature by IR imaging techniques requires knowledge of the emissive properties of the surface being examined over the range of wavelengths to which the detector is sensitive. The emissivity of human skin approaches unity and, providing the skin is clean and devoid of perspiration, emissivity problems in clinical thermography are usually of secondary importance.

From 1958 to 1985, medical thermography employed devices that depended upon single detectors of either indium antimonide or cadmium mercury telluride. The imaging device consisted of an optical system for collecting radiation and focusing it onto a detector cooled with liquid nitrogen. The build-up of the thermal image was achieved by means of an optical scanning system and a processing unit to display the image.

More recently, the technology used for manufacturing FPA photon detectors has been applied to produce microbolometers and pyroelectric and quantum-well-type FPA detectors (Kutas 1995). FPA systems do not require a scanning system but employ a lens to focus the IR image onto a detector array, much like the CCD detector in a cam recorder. A detector array suitable for high-quality imaging consists typically of 640 × 320 pixels but arrays of 1024 × 1024 pixels have also been made.

8.3.1 Thermal Radiation

The Stefan–Boltzmann law for total radiation emitted from a perfectly black body is given by

$$R(T) = \sigma A T^4 \tag{8.1}$$

where

$R(T)$ is the total power radiated into a hemisphere,
σ is the Stefan–Boltzmann constant (5.67×10^{-8} W m^{-2} K^{-4}),
A is the effective radiating area of the body, and
T is the absolute temperature of the radiating surface.

For a thermal black body, the emissivity (ε) is unity. In practice, a thermal body is not only radiating energy but is also being irradiated by surrounding sources: it is the net radiant exchange that is important. For surfaces that are not full radiators,

$$R(T) = \varepsilon(T)\sigma A T^4 \tag{8.2}$$

and the rate of heat loss between two surfaces at T_1 and T_2 is

$$W = A\sigma\left[\varepsilon(T_1)T_1^4 - \varepsilon(T_2)T_2^4\right]. \tag{8.3}$$

This radiant heat loss forms the basis of thermal imaging. Although $\varepsilon(T)$ is a slowly varying function of T, at body temperatures it can be considered to be constant.

If the emissivities of the skin and surroundings are taken as unity and if the ambient temperature is 20°C, the heat loss from a 300 × 300 mm^2 surface area at 32°C will be about 6.5 W.

The spectral distribution of emitted radiation from a surface depends upon its temperature and is represented accurately by Planck's formula. Skin temperature is normally maintained within the range 25°C–35°C, and over this range emission occurs between 2 and 50 μm with maximum emission around 10 μm. Steketee (1973a) has measured the emissivity of living tissue between 1 and 14 μm at normal incidence: for white skin, black skin and burnt skin he found that the emissivity was independent of wavelength and equal to 0.98 ± 0.01. The emissivity of a surface is affected principally by surface structure, and consequently surface contaminants (such as talcum powder and cosmetics) can alter the emissive properties of skin and affect the apparent surface temperature.

If the surface is at T_0 and has an emissivity ε_0, the radiant power emitted from the surface will be proportional to $\varepsilon_0 T_0^4$. If T_b is the temperature of a black body emitting the same radiant power, then

$$T_b^4 = \varepsilon_0 T_0^4 \quad \text{and} \quad \varepsilon_0 = \left(\frac{T_b}{T_0}\right)^{1/4} \tag{8.4}$$

so a 1% change in emissivity will cause an apparent reduction in T of about 0.75°C. In the case of photon detectors, emissivity changes affect temperature measurements in a complex way, which depends upon the spectral sensitivity of the detector and the temperature of the skin surface.

Electromagnetic theory for electrical insulators shows that maximum emission always occurs at normal incidence to a skin surface. Watmough et al. (1970) and Clark (1976) have predicted the variation of emissivity with angle of view for a smooth surface and have shown that significant errors occur if view angles of greater than 60° are used. Clinically, this is not a problem except, maybe, in female-breast examinations.

8.3.2 Single-Photon Detectors

These are semiconductor-type devices in which photon absorption results in the freeing of bound electrons or charge carriers in proportion to the intensity of the incident radiation. One of the earliest and most frequently used detectors was indium antimonide (InSb). The width of the bandgap energy is 0.18 eV at room temperature, corresponding to a cut-off of 7 μm: on cooling to 77 K, the bandgap energy increases to 0.23 eV, giving rise to a cut-off wavelength of 5.5 μm. The search for detectors with energy gaps small enough to allow detection of radiation beyond 10 μm led to the formation of mixed crystal detectors such as cadmium mercury telluride (CdHgTe) (often abbreviated to CMT). CdHgTe has been used frequently for medical imaging in the 8–15 μm band. Lead telluride (PbTe) has also been used.

Figure 8.1 illustrates the spectral detectivities of these detectors as well as for typical thermal detectors, some of which were used in early imagers. Photon detectors may be used in either photoconductive or photovoltaic modes. In the photoconductive mode, the conductivity change due to incident photon flux is used to cause a corresponding voltage change across a resistor. Although photoconductive detectors are easier to couple to the required preamplifiers, the heat generated by bias current can cause cooling problems. When many detector elements are used close together, photovoltaic detectors are to be preferred. Spectral response and detectivity depend on the mode of operation. For InSb and CMT detectors, both modes have been used for medical imaging. To make the detector sensitive enough to resolve small temperature differences and to overcome thermal noise within the detector, cryogenic cooling to −196°C is used. Initially this was achieved with liquid nitrogen or by Joule–Thomson cooling using air at high pressure. Subsequently, it became more usual to employ a Stirling-cycle cooling engine. This has the advantage of being small but it has the disadvantage of generating heat and some devices have an inconvenient cool-down time of a few minutes. Although detector cooling is not usually a major difficulty for imaging in a clinical environment, the need to use imaging devices in the field for military and security purposes and also in firefighting situations has brought about the development of detectors that do not need to be cooled.

The parameters by which photon IR detectors are usually specified are responsivity, noise, detectivity, cut-off wavelength and time constant. Responsivity (R) is the ratio of output voltage to radiant input power, expressed in volts per watt. Because the output

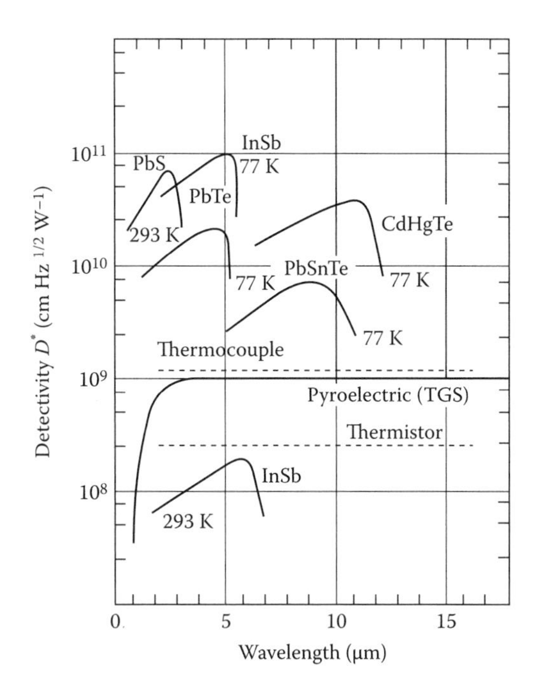

FIGURE 8.1
Variation of D^* (normalised detectivity of IR detector) with wavelength for various detectors. (From Jones and Carnochan, 1986.)

voltage due to incident IR radiation is a very small fraction (about 10^{-5}) of the DC bias voltage across the detector, the responsivity is measured by exposing the detector to chopped radiation from a calibrated source (frequently 500 K), and measuring the alternating voltage component at the chopping frequency. Responsivities (R, 500 K) are typically 10^4–10^5 V W^{-1} for CMT detectors operated at −196°C. Noise, together with responsivity, determines a detector's ability to detect small input signals. This is specified in volts per hertz at one or a number of frequencies or as a noise spectrum. Detectivity (D) is given by:

$$D = 1/\text{NEP} \tag{8.5}$$

where NEP is the noise equivalent power, which is the RMS value of the sinusoidally modulated radiant power falling upon a detector that will give rise to an RMS signal voltage (V_s) equal to the rms noise voltage (V_n) from the detector. For many photon detectors, the NEP is directly proportional to the square root of the area of the detector, and it becomes appropriate to use a normalised detectivity D^* given by

$$D^* = DA_d^{1/2}(\Delta f)^{1/2} = \frac{V_s}{V_n}\frac{[A_d(\Delta f)]^{1/2}}{W} \tag{8.6}$$

where

A_d is the area of detector,
Δf is the frequency bandwidth of the measuring system, and
W is the radiation power incident on the detector (RMS value in watts).

As D^* varies with the wavelength of the radiation and the frequency at which the noise is measured, these are stated in parentheses: $D^*(5\,\mu m, 800\,Hz, 1)$ means that the specified value was measured at 5 μm wavelength with the noise measured at 800 Hz and that the measurement has been normalised to unit bandwidth. D^* may also be measured at a specific black-body temperature, and this is used in place of the wavelength. As a figure of merit, D^* enables a theoretical maximum detectivity to be calculated that would apply when performance is limited only by noise due to fluctuation of the background radiation.

The detector time constant Γ is the time between incident radiation being cut off and the output of the detector falling by 63%; typical time constants range from a fraction of a microsecond for CMT detectors to a few microseconds for InSb detectors.

8.3.3 Thermographic Scanning Systems

The detector forms part of an imaging system whose performance is usually specified in terms of temperature resolution, angular resolution and field of view. Temperature resolution is a measure of the smallest temperature difference in the scene that the imager can resolve. It depends on the efficiency of the optical system, the responsivity and noise of the detector, and the SNR of the signal-processing circuitry. Temperature resolution can be expressed in two ways: noise equivalent temperature difference (NETD), which is the temperature difference for which the SNR at the input to the display is unity, and minimum resolvable temperature difference (MRTD), which is the smallest temperature difference that is discernible on the display. Most medical thermography has been carried out with systems that have MRTD between 0.1 and 0.3 K. Angular resolution is typically 1–3 mrad but can be as small as 0.5 mrad. Table 8.1 summarises the principal parameters

TABLE 8.1

Details of thermographic scanning systems.

Component/Parameter	Range of Values	
Detector		
Single element	InSb (3–5.5 μm), 77 K	
	CdHgTe (8–15 μm), 77 K	
	CdHgTe (3–5 μm), 195 K	
	PbTe (3–5 μm), 77 K	
	PbSnTe (8–12 μm), 77 K	
Linear arrays	×10	InSb
	SPRITE	CdHgTe (CMT)
	×512	PbTe
Matrix	6 × 8	InSb
	1 × 8	SPRITE CMT
Pyroelectric	TGS	
	Vidicon TGS, DTGS (room temp.)	
Spatial resolution	0.5–3 mrad	
Number of horizontal lines	90–625	
Elements per line	100–600	
NETD	0.05°C–0.4°C at 30°C	
MRTD	0.1°C–0.5°C at 30°C	
Number of frames per second	0.5–50	
Angular field of view	5° × 5° to 60° × 40°	
Temperature ranges	1°C–50°C	

of systems that have been used for medical imaging prior to the development of FPA detectors. A comprehensive analysis of the factors influencing the design and construction of first-generation thermographic systems used in clinical imaging has been given by Lawson (1979).

In an imaging system capable of transmitting and focusing IR radiation, the scene is viewed by an optical lens. A high value of IR refractive index is advantageous in lens design, but materials that have high refractive indices tend to have low transmittances. For example, for 2 μm radiation, germanium has a refractive index of about four and a transmittance of 47%. This high reflective loss may be eliminated by antireflection coatings, which raise the transmittance to as high as 95%–97% for a given wavelength interval. Germanium and silicon are used frequently for IR optical components. Advances in IR technology resulted in improvements in the resolution of imaging systems and designers of imaging systems are now hard pressed to develop matching optical systems. This has led to a search for achromatic IR optical materials and the use of chalcogenide glasses. These selenium-based glasses combine well with other IR optical materials (such as germanium, zinc selenide and gallium arsenide) and provide high-resolution optical material operating in the 3–5 and 8–12 μm atmospheric windows.

First- and second-generation scanning systems employed configurations of lenses, rotating prisms, rocking mirrors or rotating multisided mirror drums. The precise design was dependent on commercial features and the functional purpose of the imager. Single-element detectors have the advantage of simplicity, both electronically and mechanically. There are two advantages in replacing a single detector by a multi-element array of n similar detectors. First, there is a decrease in noise level since the signal increases in proportion to n whereas noise increases in proportion to $n^{1/2}$. Second, the higher scan speeds obtainable with array detectors make these instruments important for investigating rapid temperature changes. In a clinical context, high-resolution real-time imaging allows precise focusing on the skin surface and continuous observation of thermal changes so that transient or dynamic studies can be made on patients.

A significant development in imaging technology was the SPRITE (signal processing in the element) detector. The SPRITE CMT detector carries out the delay and add functions within the element, so that a single detector replaces a linear array of detectors. In a conventional in-line array, the signal from each IR-sensitive element is preamplified and then added to the signal that is generated in the adjacent element. In the SPRITE, the individual elements are replaced by a single IR strip mounted on a sapphire substrate. It requires only one amplifier channel and has optimum gain at high speeds. An eight-element SPRITE detector is equivalent in performance to an array of at least 64 discrete elements but requires far fewer connections (Mullard 1982, Figure 8.2). By arranging the detectors

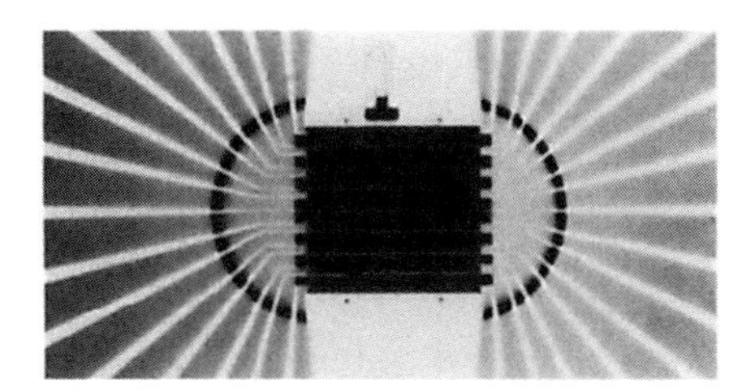

FIGURE 8.2
Eight-element SPRITE CMT array. (Courtesy of Mullard Ltd., London, UK.)

FIGURE 8.3
AGEMA Thermovision 870 Camera with germanium lens and CMT detector. (Courtesy of AGEMA Infrared Systems Ltd., Secaucus, NJ.)

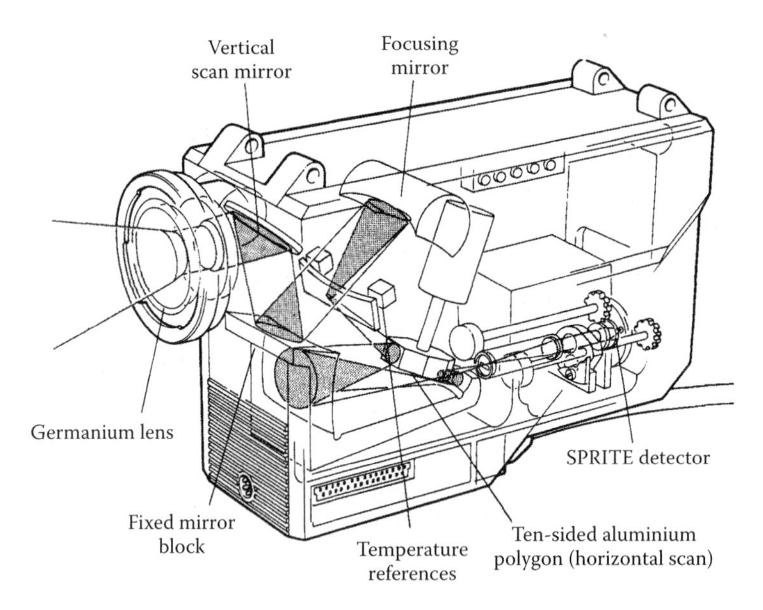

FIGURE 8.4
Schematic of AGEMA Thermovision 870 Camera. (Courtesy of AGEMA Infrared Systems Ltd., Secaucus, NJ.)

in a stack, outputs can be stored in parallel in-line registers and serially combined to a TV-compatible display rate. Figures 8.3 and 8.4 illustrate the resulting compactness of the AGEMA Thermovision 870 Camera which has a SPRITE CMT detector.

8.3.4 Focal-Plane Staring Arrays

Since 1993 much of the earlier technology employing single detectors, linear arrays, and SPRITE detector arrays has been replaced by the development of staring-array detectors. These third-generation cameras offer higher temperature-resolution images in real time.

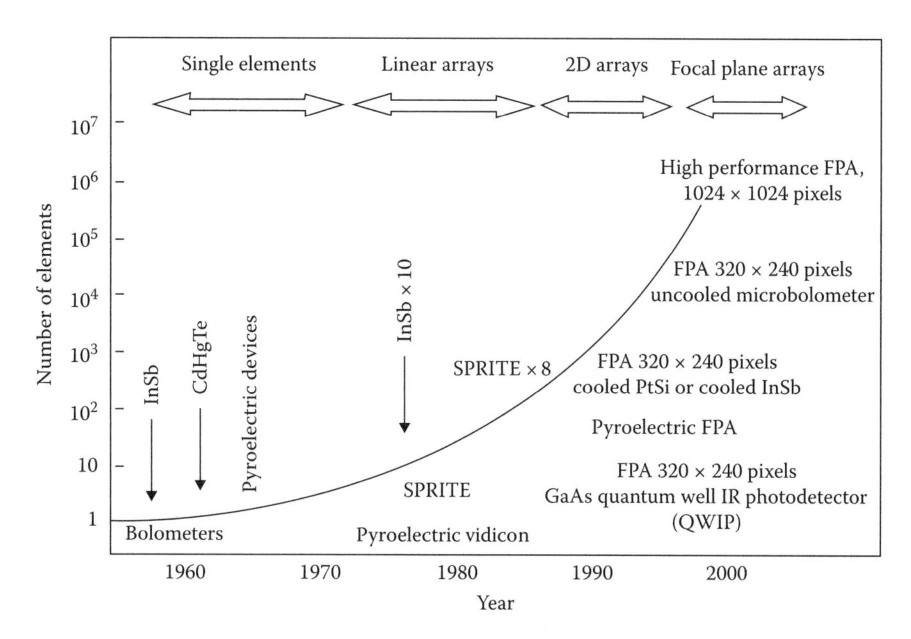

FIGURE 8.5
Forty-year development of IR detectors.

The absence of a scanning mechanism means that all FPA solid-state cameras are very compact and quiet in use. The scene is viewed through a lens and, when appropriate, a filter can be incorporated to view IR above a specific wavelength. Figure 8.5 illustrates detector development over the 40 year period 1960–2000.

FPA arrays have been constructed from a number of different materials including InSb, PtSi, CdHgTe and InGaAs. Complex scanning systems are no longer required and the inherent simplicity of the staring-array detector together with advances in micro-cooler technology has resulted in the manufacture of very compact staring-array systems with high performance (see Table 8.2).

In a scanning system, the detector or each pixel of the detector only sees the object for a very short time and this reduces the amount of energy collected. In FPA systems, the scanned detector is replaced by an array of detector cells, one for each pixel 'staring' constantly at the object being imaged. In order to increase the number of detectors inside the sensor vacuum Dewar, most quantum detector arrays operate in photovoltaic mode. The detectors can be fabricated on substrate as p–n junctions using integrated circuit (IC)

TABLE 8.2

Details of thermographic FPA systems.

FPA Detector	No. of Elements	Operating Temperature	Spectral Range (μm)	NETD (°C)
InSb	256 × 256	77 K	3–5.5	0.025
PtSi	512 × 512	77 K	3–5	0.2
Uncooled microbolometer vanadium oxide	320 × 240	Room temperature	7.6–13	0.08
Quantum-well infrared photodetector GaAs	340 × 240	77 K	8–9	0.02

Note: Cell size is typically 30 × 30 μm² and spatial resolution of all systems is 1 mrad.

techniques with very high packing density. There are two ways of constructing FPA detectors. Monolithic detectors are easier and cheaper to construct because both the IR-sensitive material and the signal transmission paths are on the same layer. In contrast, in the case of hybrid FPAs, the detector is on one layer and the signal and processing circuitry is on another layer. The advent of micron and sub-micron silicon technology has led to the manufacture of complex signal-conditioning electronics and multiplexers integrated onto a silicon chip. This in turn is incorporated directly behind the IR detector within the vacuum encapsulation. The problem of nonuniformity of detector response in FPAs is addressed by using digital signalling electronics and computing technology to match all channels.

Uncooled FPA detectors based on the principle of the bolometer have also been developed. These devices consist of a sensitive area whose electrical resistance has a strong dependence on its temperature. Absorption of the incident thermal radiation changes the temperature of the sensitive area and the change in the measured electrical resistance results in a signal proportional to that radiation. The disadvantage of this type of 'thermal' detector is that they react relatively slowly compared with the response of photon detectors. Nevertheless, such detectors respond fast enough to work well in FPA systems where response requirements are in the millisecond range.

Lindstrom (1998) described the construction of the AGEMA 570 imaging system based on a 320 × 240 bolometer detector array. The AGEMA 570 hand-held system is shown in Figure 8.6. The manufacturing process includes the micromachining of silicon in combination with silicon IC technology. The distance between two adjacent cells is approximately 50 μm. To improve thermal isolation, each bolometer sensing area comprises a thin-film layer of vanadium oxide which acts as a temperature-dependent resistor. The vanadium oxide is deposited on a free-standing bridge structure of dielectric material supported on

FIGURE 8.6
Cut-away view of the hand-held AGEMA 570 uncooled FPA radiometer. (Courtesy of FLIR Systems Inc., Wilsonville, OR.)

two very thin legs. The low thermal mass achieved by this design results in high sensitivity and a lower time constant. The isolation of each bolometer pixel minimises detector crosstalk and smearing of the resultant thermal image. Further improvement is achieved by encapsulating the detector in a vacuum. The temperature coefficient of resistance of vanadium oxide is such that a temperature difference of 1°C results in a 2% change in resistance. If the imaging system is to be capable of measuring a temperature difference as low as 0.1 K then the temperature of the detector itself must be carefully controlled. This is achieved in the AGEMA 570 system by means of a solid-state thermoelectric stabilising device built into the detector to keep it at a fixed point temperature close to room temperature.

The signals from the 76,800 detector pixels are all A/D converted and multiplexed to form one single 14-bit digital output stream of data from a CMOS-based digital read-out IC. By using 14-bit digital image processing, high linearity, and a wide dynamic range, a good thermal resolution over a broad range of temperatures is achieved. An illustration of a FPA single cell is shown in Figure 8.7.

Staring-array technology has also been applied to quantum-well-type IR photodetectors (QWIPs). These devices are constructed to have a quantum well with only two energy states, the ground state and the first excited state. The excited state is arranged to be near the top of the well so that it can detect light photons. By alternating layers of the well material (such as GaAs) and the potential barrier, it becomes possible to control the characteristics of the QWIP so that it will respond to a particular wavelength of radiation. Typically QWIP detectors are designed to detect radiation in the 8–9 μm range.

FPA-based systems requiring detector cooling have also been developed. In these cameras the matrix of detector cells is fashioned from InSb, or from platinum silicide (PtSi), and must be cooled to 80 K for optimal use as a thermal imaging device. PtSi is a semiconductor made by ion implantation of platinum into a silicon surface, forming a Schottky diode structure and is relatively simple to manufacture. It has reliable long-term stability but it has low quantum efficiency. It is sensitive to radiation in the 1–5 μm range. Cooling of FPA detectors is usually accomplished by a Stirling cooler. Typically, matrices used in clinical imaging are either 320 × 240 pixels or 640 × 320 pixels but, for research purposes, detectors with arrays of 512 × 512 pixels and 1024 × 1024 pixels have been developed.

Additional features have been incorporated into the latest imaging systems. Images can be stored on disc and then displayed and analysed on a PC with data-analysis software.

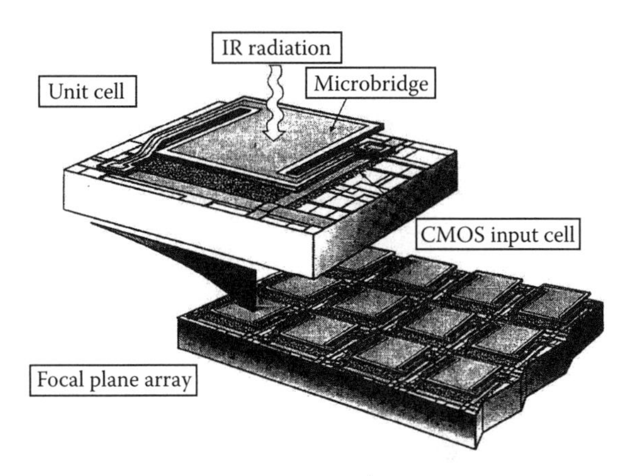

FIGURE 8.7
Schematic of a FPA single cell. (Courtesy of FLIR Systems, Wilsonville, OR.)

One very useful facility incorporated in some systems is an analysis package for recording multiple thermal images and simultaneous visual images.

The image quality of FPA detectors being used clinically is superior to that of earlier scanning systems; image capture and image processing are easier and faster. Clearly, the use of an uncooled device for dynamic imaging of patients in a ward or clinic is advantageous. This is demonstrated well by studies that have been made at airports when a large number of passengers have been examined for unsuspected temperature increases due to infectious diseases such as Avian influenza and severe acute respiratory syndrome (SARS).

8.3.5 Pyroelectric Imaging Systems

The pyroelectric effect is exhibited by certain ferromagnetic crystals such as barium titanate and triglycine sulphate (TGS). When exposed to a change in radiance, these materials behave as capacitors on which electrical charge appears. The magnitude of the effect depends on the rate of temperature change in the detector, and so the sensor does not respond to a steady flux of radiation. Pyroelectric detection has been developed as a cheaper alternative to photon-detector-based systems. Although pyroelectric detectors were originally incorporated into systems that employed mechanical scanning devices to build up a thermal image, most value for clinical work came from the development of pyroelectric vidicon camera tubes. In pyroelectric systems the scene is panned or modulated by a rotating disc and the IR radiation enters the vidicon tube by means of a germanium IR-transmitting lens (8–14 μm), which focuses the image of the thermal scene onto a thin disc of TGS pyroelectric material (20 mm diameter, 30 μm thick). On the front of the TGS target there is an electrically conducting layer of material chosen to be a good absorber of thermal radiation. The target is scanned in a TV raster by the electron beam of the vidicon tube and the image displayed on a TV monitor (Watton et al. 1974).

The latest pyroelectric imagers are based upon FPAs using the pyroelectric effect in ceramic barium-strontium titanate. Multi-pixel arrays having an NETD of 0.5°C have been developed largely for industrial use and surveillance purposes.

8.3.6 Temperature Measurement

Medical investigations require quantification of the thermal scene. Some systems have adjustable, built-in, temperature reference sources, and the detector output signal is compared with that from the reference source at a known temperature. Generally, it is more reliable to use an independent temperature reference located in the FOV. The image display of typical thermographic equipment is equipped with an isotherm function by which a signal is superimposed upon the grey-tone picture so that all surface areas with the same temperature are presented as saturated white (or in a specific colour if the image is displayed in colour mode). The isotherm can be adjusted as required within the temperature range of the thermogram (Figure 8.8).

Temperature measurement of the skin by an IR camera must allow for the following:

(i) radiation emitted by the skin as a result of its surface temperature (T_0) and its emissivity (ε);

(ii) radiation reflected by the skin into the measuring system as a result of the ambient temperature of the surroundings (T_a), which will be proportional to the reflectance (ρ) of the skin; and

(iii) radiation transmitted through the skin as a result of the temperature (T_d) at a depth within the body of the subject.

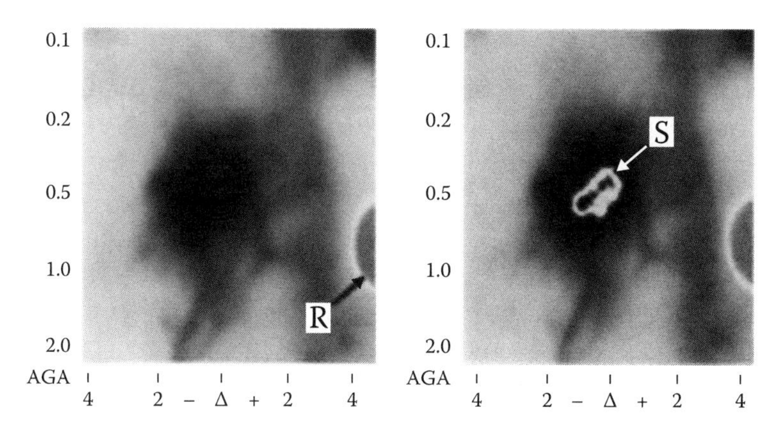

FIGURE 8.8
AGA Thermovision images (white cold – black hot; white to black 10°C) of skin tumour showing a maximum temperature of 34.9°C indicated by isotherm S. The reference temperature R is 32°C.

In practice, the skin is nontransparent for the wavelengths detected, and the transmitted contribution can be neglected. The signal S resulting from a small quantity of the total radiation emitted by an object irradiating the detector can be written as

$$S = \varepsilon f(T_0) + \rho f(T_a) \tag{8.7}$$

where
$\varepsilon f(T_0)$ is the emitted radiation, which is a function of the object's surface temperature, and
$\rho f(T_a)$ is the reflected radiation, which is a function of the ambient temperature.

Since $\rho = 1 - \varepsilon$, then

$$S = \varepsilon f(T_0) + (1 - \varepsilon) f(T_a). \tag{8.8}$$

The camera signal versus black-body temperature characteristic curve

$$S_b = f(T_b) \tag{8.9}$$

depends upon the detector and system optics but is usually exponential in form. Skin radiates with a spectral distribution similar to a black body and, if ε is the surface emissivity, the corresponding characteristic response curve for the detector will be

$$S_0 = \varepsilon f(T_0). \tag{8.10}$$

Calibration curves are usually provided by manufacturers for various aperture settings in isotherm units (or signal voltages). Since skin emissivity is very high, the calibration curve for unit emissivity may be used to correlate isotherm-level differences with temperature–magnitude differences. The isotherm facility is used to measure the temperature of an area relative to the temperature of a reference standard. The isotherm can be adjusted first to match the reference and then to match the area of interest. With the aid of the appropriate calibration curve, the difference between these two measurements can be used to read off the unknown temperature.

8.3.7 Clinical Thermography: Physiological Factors

As a homeotherm, man maintains a stable deep-body temperature within relatively narrow limits (37°C ± 1°C), even though the environmental temperature may fluctuate widely. The minimal metabolic rate for resting man is about 45 W m^{-2}, which for the average surface area of 1.8 m^2 gives a rate of 81 W per man (Mount 1979). Metabolic rates increase with activity, with a consequential temperature rise. Furthermore, the core temperature is not constant throughout the day: it shows a 24 h variation, with maximum body temperature occurring in the late afternoon and minimum temperature at 4.00 a.m., the difference being about 0.7°C. Blood perfusion has an important role in maintaining deep-body temperature, even though the environmental temperature might alter. The rate of blood flow in the skin is the factor that has the chief influence on the internal conductance of the body: the higher the blood flow and the conductance, the greater is the rate of transfer of metabolic heat from the tissues to the skin for a given temperature difference. For these reasons, thermal imaging is best carried out in a constant ambient temperature. Blood flowing through veins coursing just below the skin plays an important part in controlling heat transfer. The thermal gradient from within the patient to the skin surface covers a large range, and gradients of 0.05°C–0.5°C mm^{-1} have been measured. It might appear that, when there is a large thermal gradient, IR temperature-measuring techniques could be inaccurate because of the effect of radiation that emanates from deeper tissues being transmitted through the skin. Steketee (1973b) has shown that this is not the case for IR radiation of wavelength greater than 1.2 μm.

The surface temperature distribution of a person depends upon age, sex and obesity as well as on metabolism and topography. Superimposed on a general pattern will be a highly specific pattern due to the thermal effects of warm blood flowing through subcutaneous venous networks. Surface temperature gradients will be affected by these vessels providing they are within 7 mm or so of the skin surface (Jones and Draper 1970); the deeper the vessel is, the smaller the surface temperature gradient and the wider the half-height of the thermal profile.

Temperature increments associated with clinical conditions are typically 2°C–3°C in magnitude (Table 8.3), and prominent patterns due to the effect of underlying vasculature (Figure 8.9) have profiles with widths at half-height of 1–3 cm and gradients of 1°C–2.5°C cm^{-1}. To image these profiles requires equipment of high resolution. Figure 8.10, which shows the distribution of tiny droplets of perspiration over sweat pores on the surface of a finger, illustrates the image quality that high-resolution systems are capable of producing.

There are two main methods of thermographic investigation:

1. Examinations made in a constant ambient temperature (within 19°C–26°C, depending upon the investigation); this type of study requires patient equilibrium for about 15 min prior to data collection.
2. Examinations on patients undergoing thermal stress; usually this involves the application of a hot or cold load onto an area of the patient's skin or getting the patient to place a hand or foot into a bath of cold (or hot) water for a short time, after which thermal patterns are recorded sequentially.

Most studies are based on the observation of temperature differences (or temperature changes from a known baseline) rather than the measurement of temperature magnitudes. A knowledge of the normal baseline pattern is a fundamental requirement.

TABLE 8.3

Details of clinical investigations.

Thermographic Investigation	Typical Temperature Changes, ΔT(°C)	Notes
Malignant disease		
Breast	1.5–4	Tumours with large ΔT tend to have a poor prognosis
Malignant melanoma	2–4	
Soft-tissue tumours	2–5	Vascular patterns are anarchic
Bone tumours	2–5	
Skin lesions	0.5–3	Skin infiltration causes increased ΔT
Vascular studies		
DVT	>1.5	Vascular patterns are significant
Scrotal varicocele	>1 (>32)	Examination under stress
	>1.5 (32)	Presence of varicocele indicated by ΔT; effect on infertility greater if $T > 32$°C
Locomotor-disease studies		
Rheumatic diseases	2	Thermographic index useful for monitoring thermal changes
Trauma		
Burns	1.5, also –3	Higher-degree burn has larger drop in ΔT
Pain	2, also –1	
Frostbite		Pain is sometimes associated with hypothermia
Therapy studies		
Radiotherapy	2	Radiotherapy skin reactions increase ΔT
Chemotherapy	~1.5	
Hormone therapy	~1.5	

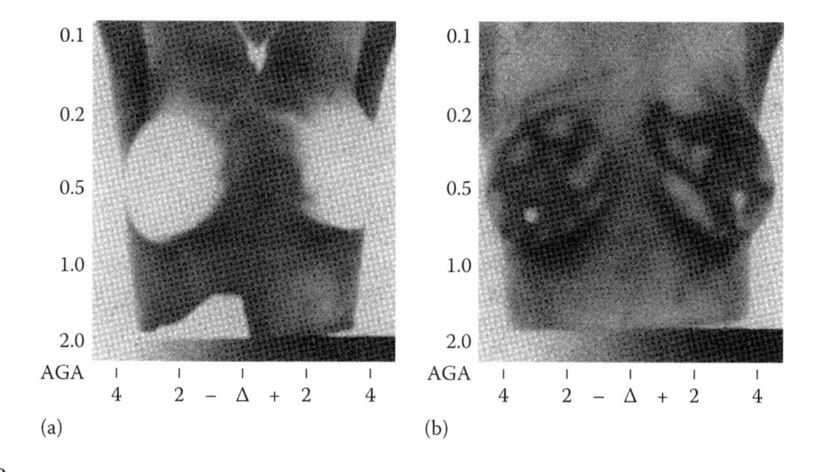

FIGURE 8.9
AGA Thermovision images (white cold – black hot; white to black 10°C) showing examples of the appearance of vasculature. (a) Patient with avascular thermal pattern over breasts due to mammoplasty. (b) Patient with prominent vascular pattern associated with lactation.

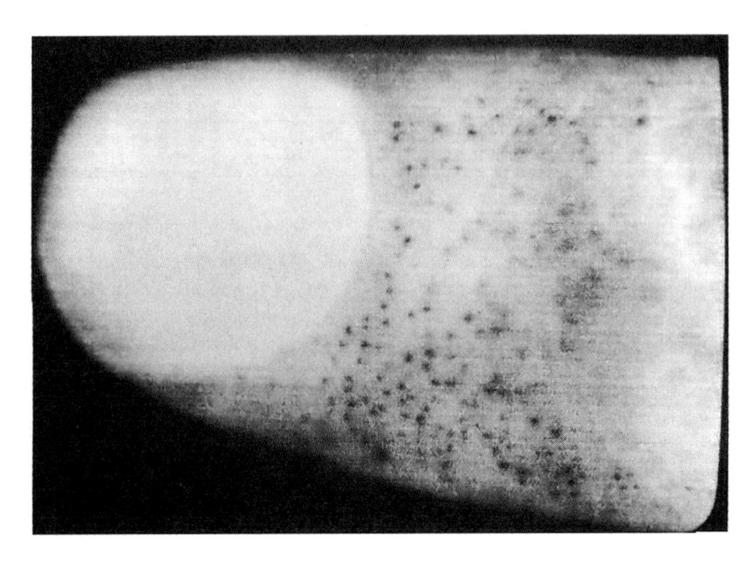

FIGURE 8.10
High-resolution thermogram of finger showing tiny droplets of perspiration over sweat pores. Image recorded with Barr and Stroud IR18 (Mark 2 Camera). (Courtesy of Barr and Stroud, Glasgow, UK.)

8.3.8 Clinical Thermography: Applications

Medical thermography has a large bibliography (Jones 1983, Ring and Phillips 1984, Jones 1987, Ammer and Ring 1995, Ring and Ammer 2000, Ammer and Ring 2006, Ring 2010), and the technique has been used to investigate a wide variety of clinical conditions. Most important among these are the following, and we now consider briefly each of these in turn:

(i) the assessment of inflammatory conditions such as rheumatoid arthritis;
(ii) vascular-disorder studies including;
  (a) the assessment of deep vein thrombosis (DVT),
  (b) the localisation of varicosities,
  (c) the investigation of vascular disturbance syndromes, and
  (d) the assessment of arterial disease;
(iii) metabolic studies;
(iv) the assessment of pain and trauma;
(v) oncological investigations; and
(vi) physiological studies.

8.3.8.1 Assessment of Inflammatory Conditions

Arthritis is frequently a chronic inflammatory lesion that results in overperfusion of tissue and a consequential increase in skin temperature. Thermography is able to distinguish between deep-seated inflammation and more cutaneous involvement. Furthermore, it is useful for evaluating and monitoring the effects of drug and physical therapy. By standardising conditions and cooling peripheral joints so that the skin is within a specific temperature range (26°C–32°C for the lower limbs and 28°C–34°C for the joints of upper limbs), it is possible to quantify the thermal pattern in the form of a thermographic index on a scale from 1.0 to 6.0, in which healthy subjects are usually found to be less than 2.5 and inflammatory joints are raised to 6.0 (Collins et al. 1974). This quantitative analysis is a very effective means of assessing the efficacy of anti-inflammatory drugs used in the treatment of rheumatic conditions.

8.3.8.2 Investigations of Vascular Disorders

8.3.8.2.1 Deep Vein Thrombosis

Venography and ultrasound are the most commonly used investigations for investigating DVT. Venography is invasive, time consuming and labour intensive. It also exposes the patient to the risks of contrast allergy. Doppler ultrasound avoids the risks of venography but is operator dependent and can be time consuming. An increase in limb temperature is one of the clinical signs of DVT, and thermography can be used to image the suspected limb. The thermal activity is thought to be caused by the local release of vasoactive chemicals associated with the formation of the venous thrombosis, which cause an increase in the resting blood flow (Cooke 1978). The thermal test is based on the observation of delayed cooling of the affected limb. The patient is examined in a resting position and the limbs cooled, usually with a fan. Recent calf-vein thrombosis produces a diffuse increase in temperature of about 2°C.

As a result of a prospective study of 1000 patients with suspected DVT, Harding (1998) concluded that the ideal investigative pathway in clinically suspected DVT should be initial thermography, proceeding to Doppler ultrasound only when thermography is positive, and finally proceeding to venography, after negative or equivocal duplex Doppler ultrasound.

8.3.8.2.2 Localisation of Varicosities

The localisation of incompetent perforating veins in the leg prior to surgery may be found thermographically. The limb is first cooled with a wet towel and fan and the veins are drained by raising the leg, to an angle of 30°–40° with the patient lying supine. A tourniquet is applied around the upper third of the thigh to occlude superficial veins and the patient then stands and exercises the leg. The sites of incompetent perforating veins are identified by areas of 'rapid rewarming' below the level of the tourniquet (Patil et al. 1970).

Varicosities occurring in the scrotum can alter testicular temperatures. It is known that the testes are normally kept at about 5°C lower than core-body temperature: if this normal temperature differential is abolished, spermatogenesis is depressed. The incidence of subfertility in man is high and the cause is often not known, but the ligation of varicocele in subfertile men has been shown to result in improvements in fertility, with pregnancies in up to 55% of partners (Dubin and Amelar 1977). The presence of a varicocele is not always clinically evident. Even after surgical ligation, residual dilated veins can affect testicular temperature adversely. Thermography has been used extensively to determine the thermal effect and extent of clinical varicocele, to investigate infertile men or subfertile men who might have unsuspected varicocele, to examine patients who have had corrective surgery and to determine whether any residual veins are of significance after ligation of the varicocele. In a cool environment of 20°C, the surface temperature of the normal scrotum is 32°C, whereas a varicocele can increase this to 34°C–35°C (Vlaisavljevic 1984).

8.3.8.2.3 Vascular Disturbance Syndromes

Patients with Raynaud's disease and associated disorders have cold and poorly perfused extremities. The viability of therapeutic intervention will depend upon whether the vasculature has the potential for increased blood flow. This may be assessed by subjecting the hands (or feet) to hot or cold stress tests and comparing the thermal recovery of the skin temperature with that of healthy limbs (Clark and Goff 1986). Howell et al. (1997) have measured thermographically the temperature of toes in Raynaud's phenomenon. These authors concluded that the baseline mean toe temperature and medial lateral toe temperature difference are good diagnostic indicators of this condition in the feet. Cold challenge can enhance temperature differentials since these recover markedly better over a 10 min period in healthy individuals than in Raynaud's patients.

Temperature changes also occur in Paget's disease. This is a disease of bone that results in increased blood flow through bone tissue which, in turn, can increase the bone temperature and overlying skin.

8.3.8.2.4 Assessment of Arterial Disease

Peripheral arterial disease can cause ischaemia, necessitating amputation of an affected limb. Tissue viability of the limb has to be assessed to determine the optimal level of amputation; this will depend upon skin blood flow, perfusion pressure and arterial pressure gradients. Skin-temperature studies that show hypothermal patterns can indicate the presence of arterial stenosis (Spence et al. 1981). The non-invasive nature of thermal imaging allows the method to be used even on patients who are in extreme pain.

8.3.8.3 Metabolic Studies

Skin temperature is influenced by the proximity of the skin and superficial tissues to the body core and the effects of subcutaneous heat production, blood perfusion and the thermal properties of the tissues themselves. Subcutaneous fat modifies surface temperature patterns, as does muscular exercise. The complex interplay of these factors limits the role of IR imaging in metabolic investigations to the study of the most superficial parts of the body surface. For example, in the case of newborn infants, it has been postulated that the tissue over the nape of the neck and interscapular region consists of brown adipose tissue, which plays an important role in heat production. Thermal imaging has been used to study this heat distribution directly after birth (Rylander 1972). The presence of brown adipose tissue in man has also been investigated by this means: it has been observed that metabolic stimulation by adrenaline (ephedrine) produces an increase in skin temperature in the neck and upper back.

Thermal imaging has also been used in the study of metabolic parameters in diabetes mellitus (Jiang et al. 2002).

8.3.8.4 Assessment of Pain and Trauma

The localisation of temperature changes due to spinal-root-compression syndromes, impaired sympathetic function in peripheral nerve injuries and chronic-pain syndromes depends largely upon the finding that in healthy subjects thermal patterns are symmetrical. Asymmetrical heat production at dermatomes and myotomes can be identified thermographically (Wexler 1979). Temperature changes are probably related to reflex sympathetic vasoconstriction within affected extremity dermatomes and to metabolic changes or muscular spasm in corresponding paraspinal myotomes.

Thermography has also found a place in physical medicine in the assessment of such conditions as 'frozen shoulder' (Ammer et al. 1998). Frozen shoulder is usually characterised by a lowering of skin temperature, which is probably due to decreased muscular activity resulting from a decreased range of motion.

Thermography can be used to assess tissue damage caused by a burn or frostbite. The treatment of a burn depends upon the depth of injury and the surface area affected. Whereas a first-degree burn shows a skin erythema, a third-degree burn is deeper and shows a complete absence of circulation. Identification of a second-degree burn is sometimes difficult, and temperature measurements are used to assist with this assessment. Third-degree burns have been found to be on average 3°C colder than surrounding normal skin. In conditions of chronic stress (such as bed sores, poorly fitting prosthetic devices), thermal imaging can be used to assess irritated tissue prior to frank breakdown.

8.3.8.5 **Oncological Investigations**

Thermal imaging has been used as an adjunct in the diagnosis of malignant disease, to assess tumour prognosis and to monitor the efficacy of therapy. Malignant tumours tend to be warmer than benign tumours due to increased metabolism and, more importantly, due to vascular changes surrounding the tumour. It has been observed that surface temperature patterns are accentuated and temperature differences increased by cooling the skin surface in an ambient 20°C. This procedure reduces blood flow in the skin and subcutaneous tissues and, since blood flow through tumour vasculature is less well controlled than through normal vasculature, the effects of cooling the skin surface are less effective over the tumour than over normal tissue. Imaging has been used to determine the extent of skin lesions and to differentiate between benign and malignant pigmented lesions. The method has been used by many investigators as an aid in the diagnosis of malignant breast disease, but it lacks sensitivity and specificity. It has been advocated as a means of identifying and screening 'high-risk groups' and for selecting patients for further investigation by mammography (Gautherie and Gros 1980). Breast tumours that cause large temperature changes (more than 2.5°C) tend to have a poor prognosis (Jones et al. 1975).

The treatment of malignant disease by radiotherapy, chemotherapy or hormone therapy can be monitored thermographically. Serial temperature measurements indicating temperature drops of 1°C or more are usually consistent with tumour regression.

8.3.8.6 **Physiological Investigations**

In the physiology of cutaneous perfusion, Anbar et al. (1997) used dynamic area telethermometry. In this technique, 256 images at 66 IR images per second were captured with a 256 × 256 FPA gallium-arsenide quantum-well photon-detector camera. The authors studied different areas of skin on the human forearm and observed temperature modulation about 1 Hz with an amplitude of about 0.03 K due to the cardiac cycle. Their findings demonstrated the potential use of high-speed imaging in peripheral haemodynamic studies.

Other workers have shown that imaging is useful in experimental respiratory function studies (Perks et al. 1983), and for studying hyperthermia (Jones and Carnochan 1986).

Improvements in examination techniques and the advantages of digital imaging have improved the reliability of thermal imaging significantly in the study of diseases such as diabetes mellitus and coronary heart disease (Marcinkowska-Gapinska and Kowal 2006, Ring 2010).

The development of FPA systems allowed thermography to be used for public screening such as for SARS and pandemic influenza. Outgoing or incoming travellers have been screened at airports to identify individual travellers with a raised body temperature caused by SARS or Avian influenza. In practice, real-time screening of facial temperature patterns is used to identify individuals for further investigation. Adequate digital data can be captured within 2 s. The positive threshold temperature is usually taken to be 38°C.

The control of infectious diseases is an international problem and various international organisations are involved in establishing international standards for temperature screening of the public (Ring 2006). Cooled FPA detector systems give good thermal and spatial resolution but suffer from the limited length of time the cooling system can operate. Uncooled camera systems are more suited for continuous use over long time periods but are best used in conjunction with an external reference source. A further problem is the lack of published data relating facial temperature patterns and deep body temperatures. This problem is accentuated by possible ethnic differences caused by facial topography.

8.4 Liquid-Crystal Thermography

Thermochromic liquid crystals are a class of compound that exhibit colour–temperature sensitivity. They are used encapsulated in pseudo-solid powders and incorporated in a thin film with a black background to protect the crystals from chemical and biological contamination. The reflective properties of cholesteryl nonanoate and chiral nematic liquid crystals are temperature dependent: when viewed on a black background, the scattering effects within the material give rise to iridescent colours, the dominant wavelength being influenced by small changes in temperature. The absolute temperature range of response of each plastic film depends upon the liquid-crystal constituents, but is typically about 3°C–4°C. When used clinically, the plate should be placed in uniform contact with the skin surface, but this is not always possible. The response time varies according to plate thickness (which ranges from 0.06 to 0.3 mm) and is typically 20–40 s. To reduce this time, plates may be warmed by body contact prior to clinical use. The method provides an inexpensive method of recording temperature distributions but it has some disadvantages:

1. The thermal mass of the plate can affect the temperature of the surface of the skin.
2. Alterations to the heat exchange between the skin and its surroundings alter radiant heat transfer from the skin surface.
3. The pressure of the plate on the skin can change vascular circulation and, consequently, affect the convection of body heat to the skin surface.
4. The presence of fat, oil or other artefacts between the skin and the plate can alter the temperature distribution and the colour of the liquid-crystal screen.
5. The heat radiation of the surroundings can change the temperature and colour of the screen.
6. It is often not possible to obtain uniform contact over the whole of the surface being examined.

In spite of these limitations, liquid-crystal plate thermography is an effective way of investigating specific clinical problems especially where precise temperature measurement is not required.

Kalodiki et al. (1992) used liquid-crystal thermography (LCT), duplex scanning and venography to examine 100 patients with suspected DVT. LCT was found to have a negative predictive value of 97% if performed within 1 week of the symptoms. The detector was placed over the calves, with the patient prone on a bed having first elevated and exposed the legs to room temperature for 10 min. After recording the temperature patterns of both legs, the procedure was repeated with the patient supine and with legs elevated 30°, with the LCT detectors placed on the shins and thighs. The presence of a homogeneous area with a temperature difference of 0.8°C or more in the symptomatic limb, compared with the same area of the opposite asymptomatic limb was considered to be a positive clinical result. The thermogram was considered to be negative when the temperature distribution was similar in both limbs. The authors concluded that the combination of LCT with duplex scanning contributes not only towards an accurate diagnosis but also the method is cost effective.

Kalodiki et al. (1995) examined 200 patients with varicose veins preoperatively with the aid of LCT after an exercise consisting of 20 'tip toes'. Distinct patterns on the LCT plate were identified and corresponding incompetent perforating veins were identified and marked on the patients' limbs prior to surgical investigation. It was concluded that LCT was an accurate, simple and relatively cheap preoperative means of identifying incompetent perforating veins.

8.5 Microwave Thermography

In contrast to IR thermography, which records surface temperature, microwave radiometry is capable of measuring radiation intensities, which mirror subcutaneous and deeper body temperatures. This is because human tissues are partially transparent to microwaves: for example, radiation of 10 μm wavelength is absorbed in 1 mm of skin, whereas radiation of 10 cm wavelength can travel through several centimetres of tissue. However, the intensity of microwave radiation emitted by the human body at 37°C is some 10^{-8} that of IR radiation. However, specially designed radiometers placed in contact with the skin surface can be used to detect this radiation. Since the body tissue is partially transparent to this radiation, the method can be used to estimate body temperature at depths of a few centimetres.

The receiver probes have apertures about 1 cm × 2 cm and are filled with dielectric material of appropriate permittivity for matching when placed in contact with the skin surface. Contact probes of this type, sensitive to 1.3 and 3.3 GHz radiation, have been used as Dicke-switch-type radiometers in diagnostic investigations. However, the technique is limited by inadequate spatial resolution (approximately 1 cm) and lack of thermal discrimination with tissue depth. The significance of the signal measured is frequently difficult to determine because:

1. The transmittance of microwave radiation through tissue depends strongly on the relative absorption properties of high and low water content tissue, for example, for 3 GHz radiation, the penetration depth (e^{-1}) is about 5 cm in fat and about 8 mm in muscle or skin.
2. Multiple reflections in parallel tissue planes within the body result in interference effects.
3. Other errors can be caused by impedance mismatch at the antenna interface.
4. The radiometer signal depends on the temperature distribution within the tissue volume coupled to the antenna.

In spite of these physical limitations, the potential advantage of being able to measure temperatures below the skin surface led to investigations by a number of groups. Surface probes sensitive to 1.3 GHz (23 cm wavelength) and 3.3 GHz (9.1 cm wavelength) radiation have been used experimentally (Myers et al. 1979, Land et al. 1986). Jones and Dodhia (1989) used a 3 GHz probe to determine the effective depth of measurement when placed in contact with tissue and bone, in which known temperature gradients had been imposed. For temperature gradients typical of those found clinically, it was found that the effective depth of measurement was between 4 and 10 mm according to the temperature gradient and whether the bulk of material was tissue, fat or bone. The authors also reported the use of microwave radiometry for measuring testicular temperatures in infertile men. They found that the difference between the infrared and microwave measurements was 2°C indicating that the testicular temperature was less than 34°C which corresponded well with measurements made by other workers who had inserted needle probes into the testicle.

Thermal imaging has been achieved with 30 and 68 GHz radiations, but at these frequencies only radiation originating in the most superficial layers of body tissue is detected.

8.6 Future Developments

The development of uncooled detectors capable of real-time imaging represents a major step forward in reducing the cost and complexity of IR imaging systems for clinical work. However, IR lenses are still expensive and remain an obstacle to any further significant reduction in the cost of imaging systems. If cheaper optical components were to become available, then it might be envisaged that uncooled hand-held FPA thermal imaging systems could be afforded by most hospitals.

The compactness and high sensitivity of the latest FPA cameras opens up a wide range of investigations in medicine. The use of thermography in screening travellers at airports for SARS and Avian influenza emphasises the need for ongoing research to establish normal facial temperature patterns and how these are related to deep body temperatures.

It has been suggested that the new generation of imagers will be useful for culture testing, cardiac surgery, skin grafting, micro-vascular surgery and diabetic care (Kutas 1995). The detection of temperature changes as low as 0.02 K suggests that imaging might become suitable for metabolic studies. In the clinic, imaging could be used to identify patients who are hypersensitive to ionising radiation exposure and to monitor radiotherapy skin reactions. Other foreseeable applications will be in the field of sports injuries, DVT studies on long haul airline passengers and topical skin reactions resulting from drugs.

References

Ammer K, Engelbert B, Hamerle S, Kern E, Solar S and Kuchar K 1998 Thermography of the painful shoulder *Eur. J. Thermol.* **8** 93–100.

Ammer K and Ring E F J (eds.) 1995 *The Thermal Image in Medicine and Biology* (Vienna, Austria: Uhlen-Verlag).

Ammer K and Ring E F 2006 Standard procedures for infrared imaging in medicine In *Medical Devices and Systems* (*The Biomedical Engineering Handbook*) 3rd edn. ed. J D Bronzino (Boca Raton, FL: Taylor & Francis) Chapter 36 pp. 1–15.

Anbar M, Grenn M W, Marino M T, Milescu L and Zamani K 1997 Fast dynamic area telethermometry (DAT) of the human forearm with a Ga/As quantum well infrared focal plane array camera *Eur. J. Thermol.* **7** 105–118.
Clark J A 1976 Effects of surface emissivity and viewing angle on errors in thermography *Acta Thermogr.* **1** 138–141.
Clark R P and Goff M R 1986 Dynamic thermography in vasospastic diseases In *Recent Developments in Medical and Physiological Imaging* Suppl. to *J. Med. Eng. Technol.* eds. R P Clark and M R Goff (London, UK: Taylor & Francis) pp. 95–101.
Collins A J, Ring E F J, Cosh J A and Bacon P A 1974 Quantitation of thermography in arthritis using multi-isotherm analysis 1. The thermographic index *Ann. Rheum. Dis.* **33** 113–115.
Cooke E D 1978 *The Fundamentals of Thermographic Diagnosis of Deep Vein Thrombosis* Suppl. **1** to *Acta Thermogr.* (Padova, Italy: Bertoncello Artigrafiche).
Dubin L and Amelar R 1977 Varicocelectomy: 986 cases in a twelve-year study *Urology* **10** 446–449.
Gautherie M and Gros C M 1980 Breast thermography and cancer risk prediction *Cancer* **45** 51–56.
Harding J R 1998 Thermal imaging in the investigation of deep vein thrombosis (DVT) *Eur. J. Thermol.* **8** 7–12.
Howell K, Kennedy L F, Smith R E and Black C M 1997 Temperature of the toes in Raynaud's phenomenon measured using infra-red thermography *Eur. J. Thermol.* **7** 132–137.
Jiang G, Shang Z and Zhang M 2002 Metabolism parameter analysis of diabetics on thermography *Eng. Med. Biol. 24th Annual Conference and Meeting of the Biomedical Engineering Society* **3** 2226–2227.
Jones C H 1982 Review article: Methods of breast imaging *Phys. Med. Biol.* **27** 463–499.
Jones C H 1983 *Thermal Imaging With Non-Ionising Radiations* ed. D F Jackson (Glasgow, UK: Surrey University Press) pp. 151–216.
Jones C H 1987 Medical thermography *IEE Proc. A* **134** 225–235.
Jones C H and Carnochan P 1986 Infrared thermography and liquid crystal plate thermography In *Physical Techniques in Clinical Hyperthermia* eds. J W Hand and R J James (Letchworth, UK: Research Studies Press) pp. 507–547.
Jones C H and Dodhia P 1989 Microwave radiometry: Heterogeneous phantom studies *J. Photo. Science Proc.* **37** 161–163.
Jones C H and Draper J W 1970 A comparison of infrared photography and thermography in the detection of mammary carcinoma *Br. J. Radiol.* **43** 507–516.
Jones C H, Greening W P, Davey J B, McKinna J A and Greeves V J 1975 Thermography of the female breast: A five year study in relation to the detection and prognosis of cancer *Br. J. Radiol.* **48** 532–538.
Kalodiki E, Calahoras L, Geroulakos G and Nicolaides A N 1995 Liquid crystal thermography and duplex scanning in the preoperative marking of varicose veins *Phlebology* **10** 110–114.
Kalodiki E, Marston R, Volteas N, Leon M, Labropoulos N, Fisher C M, Christopoulos D, Touquet R and Nicolaides A N 1992 The combination of liquid crystal thermography and duplex scanning in the diagnosis of deep vein thrombosis *Eur. J. Vasc. Surg.* **6** 311–316.
Kutas M 1995 Staring focal plane array for medical thermal imaging In *The Thermal Image in Medicine and Biology* eds. K Ammer and E F J Ring (Vienna, Austria: Uhlen-Verlag).
Land D V, Fraser S M and Shaw R D 1986 A review of clinical experience of microwave thermography In *Recent Advances in Medical and Physiological Imaging* Suppl. to *J. Med. Eng. Technol.* eds. R P Clark and M R Goff (London, UK: Taylor & Francis) pp. 109–113.
Lawson W D 1979 Thermal imaging In *Electronic Imaging* eds. T P McLean and P Schagen (New York: Academic Press) pp. 325–364.
Lindstrom K 1998 Private communication.
Marcinkowska-Gapinska A and Kowal P 2006 Blood fluidity and thermography in patients with diabetes mellitus and coronary heart disease in comparison to healthy subjects *Clin. Hemorheol. Microcir.* **35** 473–479.
Mount L E 1979 *Adaptation to Thermal Environment – Man and His Productive Animals* (London, UK: Edward Arnold) p. 146.

Mullard 1982 Electronic components and applications **4**, no. 4, Mullard Technical Publ. M82-0099.
Myers P C, Sadowsky N L and Barrett A H 1979 Microwave thermography: Principles, methods and clinical applications *J. Microwave Power* **14** 105–113.
Patil K D, Williams J R and Lloyd-Williams K 1970 Localization of incompetent perforating veins in the leg *Br. Med. J.* **1** 195–197.
Perks W H, Sopwith T, Brown D, Jones C H and Green M 1983 Effects of temperature on Vitalograph spirometer readings *Thorax* **38** 592–594.
Ring E F J 2006 Thermography for human temperature screening – International Standards Organisation Project *Thermol. Int.* **16** 110.
Ring E F 2010 Thermal imaging today and its relevance to diabetes *J. Diabetes Sci. Technol.* **4** 857–862.
Ring E F and Ammer K 2000 The technique of infrared imaging in medicine *Thermol. Int.* **10** 7–14.
Ring E F J and Phillips B (eds.) 1984 *Recent Advances in Medical Thermology* (New York: Plenum).
Rylander E 1972 Age dependent reactions of rectal and skin temperatures of infants during exposure to cold *Acta Paediatr. Scand.* **61** 597–605.
Spence V A, Walker W F, Troup I M and Murdoch G 1981 Amputation of the ischemic limb: Selection of the optimum site by thermography *Angiology* **32** 155–169.
Steketee J 1973a Spectral emissivity of skin and pericardium *Phys. Med. Biol.* **18** 686–694.
Steketee J 1973b The effect of transmission on temperature measurements of human skin *Phys. Med. Biol.* **18** 726–729.
Vlaisavljevic V 1984 Thermographic characteristics of the scrotum in the infertile male In *Recent Advances in Medical Thermology* eds. E F J Ring and B Phillips (New York: Plenum) pp. 415–420.
Watmough D J, Fowler P W and Oliver R 1970 The thermal scanning of a curved isothermal surface: Implications for clinical thermography *Phys. Med. Biol.* **15** 1–8.
Watton R, Smith C, Harper B and Wreathall W M 1974 Performance of the pyroelectric vidicon for thermal imaging in the 8–14 micron band *IEEE Trans. Electron Devices* **ED-21** 462–469.
Wexler C E 1979 Lumbar, thoracic and cervical thermography *J. Neurol. Orthop. Surg.* **1** 37–41.

9

Imaging of Tissue Electrical Impedance

B. H. Brown and S. Webb

CONTENTS

9.1 Electrical Behaviour of Tissue 647
9.2 Measurement of Tissue Parameters 648
  9.2.1 Resistivity 648
  9.2.2 Permittivity 649
9.3 Tissue Impedance Imaging 650
  9.3.1 Forward Problem 651
  9.3.2 Inverse Problem 651
9.4 Clinical Applications of Electrical Impedance Tomography 656
  9.4.1 Pulmonary Imaging 656
  9.4.2 Gastric Imaging 658
  9.4.3 Cardiac Imaging 658
  9.4.4 Breast Imaging 659
9.5 Conclusions 659
References 660

9.1 Electrical Behaviour of Tissue

Considerable effort has been expended during the past 30 years to investigate the exciting possibility of making images of the spatial variation of the electrical properties of biological tissues. Biological tissue exhibits at least two important passive electrical properties. Firstly, it comprises free charge carriers and may thus be considered an electrical conductor with an associated conductivity. Different tissues have associated characteristic conductivities and hence images of electrical conductivity may resolve structure and even be indicative of pathology. Secondly, tissues also contain bound charges leading to dielectric effects. It should be possible to form images of relative electrical permittivity that should also relate to tissue structure.

Both the electrical conduction and displacement currents, which arise when a potential gradient is applied to tissue, are frequency dependent. However, the conduction current is only very weakly dependent on frequency and may be thought of as essentially constant over the six decades of frequency between 10 Hz and 10 MHz. The displacement current exhibits strong frequency dependence. Below 1 kHz, the conduction current may be some 3 orders of magnitude greater than the displacement current, and even at 100 kHz the conduction current is still an order of magnitude greater for many biological tissues. If an image of a biological structure could be made at several frequencies, then it would be expected that the structure would have a characteristic impedance spectrum. However, the spectral measurements may not be easy to make, and, at very high frequencies (more than about 1 MHz), stray wiring and body capacitances become a serious experimental problem. The concept of producing spatial images of electrical conductivity (or the inverse, resistivity) is attractive but technically difficult. However, considerable research progress has been made and there are currently at least two commercial machines based on these principles. There is also a large body of research that has led to several reviews and special issues of journals (Bayford and Woo 2006, Brown 1983, 2001a, 2001b, Holder 2000, 2002, 2005, Isaacson et al. 2003).

When low-frequency currents are applied to the skin via a pair of electrodes, a threshold of sensation may be found, which increases with frequency by some two decades between 10 Hz and 10 kHz. The dominant electrical mechanism for sensation also changes with frequency and there are three identifiable causes of sensation. At low frequency (<10 Hz), the sensation arises from local electrolysis. At higher frequencies (>10 Hz), electrolysis appears to be reversible, and neural stimulation is the dominant mechanism. The body is electrically most sensitive at around 50 Hz, explaining the potential danger of domestic electric shock. At much higher frequencies (>10 kHz), the dominant biological mechanism is heating of the tissue. As a result of the inherent delay for a nerve to respond to a stimulus because of the low propagation speed of an impulse along a nerve, alternating currents of some tens of kilohertz may be used to probe the body without any danger to heart, nerves or muscles.

9.2 Measurement of Tissue Parameters

9.2.1 Resistivity

For electrical resistivity to be a useful parameter to attempt to image, it is necessary to know whether the resistivity of biological material is linear with electric field strength; there is experimental evidence to support linearity to within 5% in the region 100 Hz to 100 kHz. There is also a large body of experimental data on the electrical resistivity of biological tissue, but there are many discrepancies in the values reported in the literature, which is not surprising when one considers the difficulty of making measurements *in vivo*. Much of this difficulty arises from electrode contact impedance. Current can be passed between two electrodes attached to tissue and the resulting potential measured and used to calculate an impedance. However, this impedance will be dominated by the impedance of the electrode/tissue interface and not by the characteristic impedance of the tissue itself. In order to reduce the effect of electrode impedance four electrodes are required. Current is injected between one pair of electrodes and the resulting potential is measured between the remaining pair of electrodes.

At low frequencies, cell membranes are almost perfect insulators and electric current flows around the cell boundaries. Hence, tissue can be considered as resistive at these frequencies. For example, at room temperature, blood has a resistivity of about 1.5 Ω m, liver 3–6 Ω m, neural tissue 5.8 Ω m (brain), 2.8 Ω m (grey matter), 6.8 Ω m (white matter), bone 40–150 Ω m and skeletal muscle 1–23 Ω m (anisotropic resistivity). For comparison, the resistivity of sea water is less than 1 Ω m. These values arise from the different concentrations and mobilities of electrolytes in the individual organs and tissues. The value for liver depends on the blood content. Detailed tables of resistivities for biological tissue are provided in the reviews by Barber and Brown (1984), Duck (1990), Foster and Schwan

TABLE 9.1

Typical values of resistivity and X-ray linear attenuation coefficient for five biological tissues.

Tissue	Resistivity (Ω m)	X-Ray Attenuation Coefficient (m^{-1})
Bone	150.0	35.0
Muscle	3.0	20.4
Blood	1.6	20.4
Fat	15.0	18.5
Cerebrospinal fluid	0.65	20.0

Source: Barber et al., 1983.

(1989) and Gabriel et al. (1996a, 1996b) including full references to the sources of these and other measurements. It should be possible to differentiate tissues *in vivo*, and indeed it is noteworthy that electrical resistivity can vary greatly between two tissues whose X-ray linear attenuation coefficients may be somewhat similar and whose appearance in X-ray CT would then be more difficult to discriminate. For example, Table 9.1 shows how five biological tissues compare. Muscle and fat have similar X-ray linear attenuation coefficients at X-ray energies typical of CT scanners, but differ by about a factor of 5 in resistivity. This is a good example of the general truism that new imaging modalities tend to complement rather than replace existing ones, since they generally rely on different physical mechanisms.

In view of the ratios of resistivity of blood, fat and muscle, it is necessary to remember that, when the volume of blood changes in the region being imaged, the resistivity of blood-filled tissues also changes. It is possible to evaluate the magnitude of this effect in some cases by noting the (possibly anisotropic) expansion of (say) a limb, whose volume will be greater during systole than during diastole, and using the elementary formula $R = \rho L/A$ relating the resistance R to the length L, area A and resistivity ρ of a cylinder. Such expansion leads to non-trivial changes in electrical resistance but the resistivity of the tissue is also seen to change. The resistivity of body fluids *in vitro* can also be readily obtained using laboratory conductivity cells, but accurate measurements *in vivo* are considerably more difficult to perform.

9.2.2 Permittivity

Tissue has an impedance that falls with increasing frequency. This can be handled mathematically by considering both the conductivity and permittivity of the tissue. An alternative is to consider the conductivity as a complex parameter σ^*, given by

$$\sigma^* = \sigma + j\omega\varepsilon_0\varepsilon_r \tag{9.1}$$

where

ω is the frequency,
ε_0 is the permittivity of free space, and
ε_r is the relative permittivity.

There has been a large volume of work in recent years on the frequency-dependent properties of tissue. Much of this considers tissue as composed of dispersions that can be

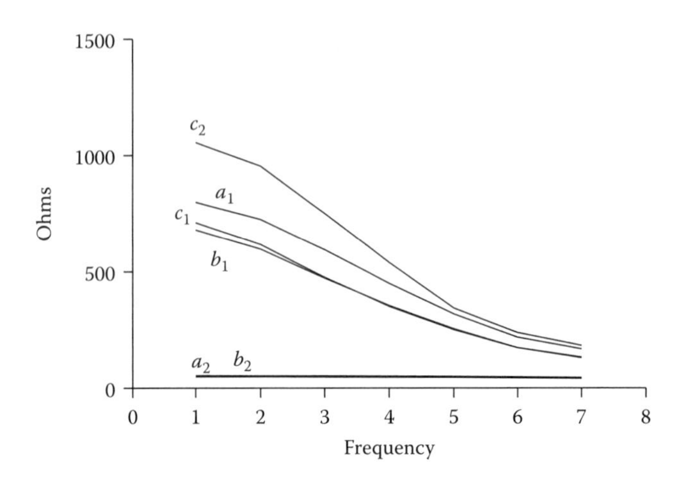

FIGURE 9.1
Graph showing the effect of freezing on plant tissue. The impedance of three courgettes (zucchini) was measured at seven frequencies in binary steps from 10 kHz (point 1) to 640 kHz (point 7). The initial measurements, a_1, b_1 and c_1, all gave low-frequency readings of about 750 Ω. In all three cases, the impedance fell with increasing frequency because of the impedance presented by the cell membranes at low frequencies. Twenty four hours later, the measurement c_2 made on the vegetable that had been kept in a refrigerator showed an increase in low-frequency impedance because of some dehydration. However, the readings a_2 and b_2 made on the two vegetables that had been frozen and then thawed showed a dramatic fall in impedance and a total absence of cellular structure.

associated with the structure of the tissue. A very commonly used model for each dispersion is that described by the Cole equation (Cole and Cole 1941). The reader is referred to Brown et al. (1999) for a detailed treatment of this area of work.

The importance of considering the frequency-dependent nature of tissue impedance is illustrated in Figure 9.1. This shows the results of making measurements of impedance spectra on vegetables before and after freezing. It can be seen that the frequency spectra depend almost totally upon the integrity of the cell membranes within the tissue. When the cell membranes are destroyed by freezing, the conductivity of the vegetable is increased and there is very little change of impedance with frequency.

9.3 Tissue Impedance Imaging

The aim of electrical impedance tomography (EIT) is to produce images of the distribution of electrical impedance within sections of the human body. For simple conductors we can speak of either the resistance or the conductance, where conductance is the inverse of resistance and vice versa. If the conductor is complex we speak of the impedance and its inverse the admittance. The specific resistance of a material is the resistance between opposite faces of a unit cube of the material. This is usually called the resistivity with its inverse the conductivity. Most theoretical formulations of EIT are in terms of admittance rather than impedance and it is common to use the term conductivity rather than admittivity, it being understood that, in this context, conductivity is complex, that is, it has both a real and an imaginary part.

9.3.1 Forward Problem

For volume conductors the relationship between the current vector **J** within the conductor, the (complex) conductivity σ and the electric field vector **E** is given by

$$\mathbf{J} = \sigma \mathbf{E}. \tag{9.2}$$

If the material is isotropic, then **E** and **J** are parallel. If the material is not isotropic, then this is generally not the case. We will assume isotropic conductivity, although some tissues are anisotropic. Every point within a conducting object through which electric current is flowing has an associated electrical potential ϕ. **E** is the local gradient of potential, usually written as $\nabla\phi$. If there are no sources or sinks of current within the object, then the divergence of **J**, written $\nabla\mathbf{J}$, is zero. So we can write

$$\nabla(\sigma\nabla\phi) = 0. \tag{9.3}$$

We can place electrodes on the surface of the object and apply a pattern of current to the object through these electrodes. The only constraint is that the total current passed into the object equals the total current extracted from the object. Suppose we apply a known current pattern through an object with conductivity distribution σ. ϕ within the object must satisfy the aforementioned equation. If we know the conductivity distribution σ, it is possible to calculate the distribution of ϕ within the object and on the surface of the object for a given current pattern applied to the surface of the object. This is known as *the forward problem*. For some simple cases it is possible to do this analytically. However, for most cases, this is not possible and a numerical solution must be used. The technique most commonly used to do this is the finite element method (FEM). The object is divided into small volume elements and it is usually assumed that the conductivity value within each element is uniform. It is then possible to solve numerically the last equation and calculate the potential distribution within the object and on the surface of the object. The principal reason for doing this is to predict the surface voltage measurements obtained for a real object whose internal conductivity distribution is known. The accuracy with which this can be done will depend on the number of volume elements and their shape and distribution, how accurately the boundary shape of the object is known and how well the electrodes are modelled.

9.3.2 Inverse Problem

We cannot in general measure ϕ within the volume of the object but we can measure it on the surface of the object. Since ϕ on the surface of the object is determined by the distribution of conductivity within the object and the pattern of current applied to the surface of the object, an obvious question is: Can we use knowledge of the applied current pattern and the measured surface voltage values to determine the distribution of conductivity within the object? This is known as *the inverse problem*. Provided the voltage is known everywhere on the surface and a sufficient number of different current patterns are applied the answer is yes, provided the conductivity is isotropic. For the anisotropic case the answer is no. The best that can be achieved is that the conductivity can be isolated to a class of possible conductivity distributions. This is a rather esoteric result which in practice is usually ignored.

The aforementioned arguments have implicitly assumed that the current is DC. In practice, we want to use AC because DC can produce polarisation effects at the electrodes which can corrupt measurements and can stimulate responses in the nervous system. Accurate measurement of alternating voltages is easier than DC voltages and in any case we want to see how conductivity changes with frequency. Provided the wavelength of the AC (i.e. the velocity of light divided by the frequency) within the tissue is much larger than the size of the object, we can treat alternating currents as though they are DC currents. This is called the quasi-static case. The frequencies used in EIT are usually less than 1 MHz and the quasi-static assumption is a reasonable one.

The major difficulty to be overcome in constructing maps of tissue impedance is that caused by the divergence of the electrical field lines when a potential is applied between two electrodes attached to the tissue. Although the curvilinear field lines can be predicted for uniform tissue, for the real case of unknown inhomogeneous tissue even this becomes impossible. The measurement made at a recording electrode does not relate directly to the bioelectrical properties along a straight line to the source electrode. It is simply not true that, if a potential gradient is applied and an exit-current measurement made, the summed electrical resistance along the path between the electrodes is the ratio of the potential difference to the current. If this difficulty did not arise, one might imagine making measurements of the straight-line integral resistivity in many directions and employing identical reconstruction techniques to those used in X-ray computed tomography (CT) and single-photon emission computed tomography (SPECT) (see Chapters 3 and 5). If we consider for the moment the electrical analogue of a planar X-radiograph, two experimental possibilities arise. Either one could constrain the electrical field lines to become straight by applying guard potentials to the (small) recording electrodes or one could, with unguarded electrodes, attempt to backproject along curved isopotentials to cope with the complicated field pattern. The latter is more difficult. Considering the former, one could imagine applying a large pair of electrodes to a slab of biological tissue and making measurements of exit current. In such an arrangement, the field lines are relatively straight (for uniform tissue) away from the periphery of the electrodes, and only bend at the edges. If one now imagined that the recording electrode were divided up into a matrix of small recording electrodes, the electric field lines remain relatively straight between each and the driving electrode (for uniform tissue), and straight-line integral resistivity measurements may be made. In this experimental arrangement of essentially guarded electrodes, the straight-line assumption holds exactly only if the tissue is uniform, and for electrically inhomogeneous material this becomes only a first-order approximation. Henderson and Webster (1978) reported measurements, using an impedance camera, of integral thoracic impedance by such a technique but, in addition to poor spatial resolution, image distortion was inevitable due to electrical inhomogeneity (Figure 9.2). A similar technique has been adopted by a number of workers and is currently under investigation for imaging of the breast (Choi et al. 2007).

Bates et al. (1980) argued that conventional CT approaches to impedance imaging are subject to fundamental limitations, which make the solutions nonunique. Nonetheless, an alternative image reconstruction technique, known initially as applied potential tomography (APT) but now as EIT, was extensively developed in Sheffield, United Kingdom (Barber and Brown 1984); this relies for its physical basis on a departure from the measurement of exit current in favour of the measurement of the potential distribution arising from an impressed current.

When a current is applied to two points on the surface of a conducting biological material, a potential distribution is established within the material. If this is sampled by making measurements of peripheral potential, the ratio of such measurements to what would be the expected peripheral potentials if the medium were uniform can be backprojected along the (uniform material) curved isopotentials to create an image. Repeating this

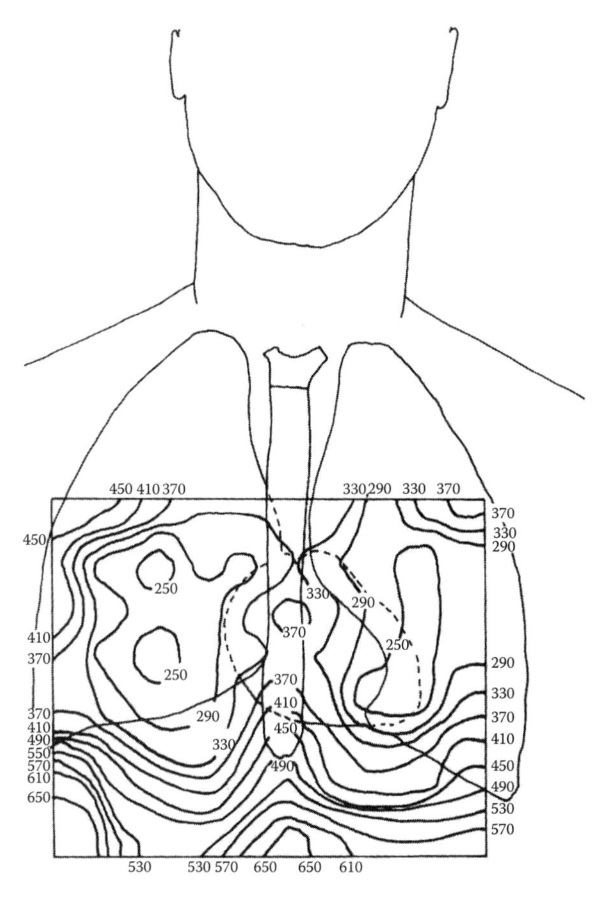

FIGURE 9.2
Iso-admittance contour map of a human thorax from Henderson and Webster's impedance camera. (Reprinted from Henderson and Webster, 1978, Copyright 1978 IEEE.)

procedure for different locations of the stimulating electrodes leads to a form of CT. Imagine a circular planar distribution of electrical conductivity (or equivalently an irregularly shaped biological specimen immersed in a circular saline bath) around which are arranged N electrodes. When current is applied across a pair of electrodes at each end of a diagonal of the system, $(N - 4)$ measurements of peripheral potential gradient may be made. (No measurements may be made connecting to the driven electrodes because these would include the electrode impedances.) If for the moment we imagine the electrical conductivity to be uniform, then an exact solution to Laplace's equation for the electrical potential distribution exists, predicting the peripheral potential gradients between the electrodes. Indeed, for symmetrically placed electrodes and assuming a cylindrical object for measurement, the electrodes may be diagrammatically joined by curved equipotentials defining bands of electrical conductivity within the circle (see Figure 9.3). The departure of the measured peripheral potential gradients from these calculated gradients may then be interpreted as arising from departures of the integral conductivity within these bands from uniform values as a result of the spatial dependence of electrical conductivity. Where the band conductivity between equipotentials is lower, the peripheral potential gradient will be higher, and vice versa. This explains the philosophy behind backprojection of the

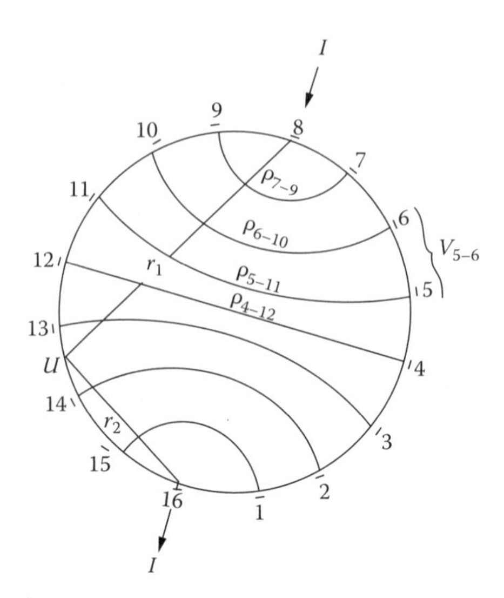

FIGURE 9.3
Isopotentials for a circular conducting system in which current is applied across diagonally opposite electrodes. (From Brown, 1983.)

measured-to-calculated peripheral potential gradient along curved equipotentials, which should then generate the relative distribution of electrical conductivity, in the limit of many backprojections for every pair of peripheral electrodes. The system of backprojection along such curved isopotentials is difficult to handle mathematically. The techniques described in Chapters 3 and 5 for X-ray CT and SPECT, respectively, do not apply.

If there are N electrodes, there are $N(N-1)/2$ possible pairs through which the current may be driven, and for each of these configurations one might naively think N measurements can be made, leading to a total of $N^2(N-1)/2$ measurements. These are not all independent, however, and considerations of reciprocity lead to the result that there are only $N(N-3)/2$ independent measurements. We see that for a typical experimental arrangement of 16 electrodes, only 104 independent measurements can be made, and one could argue that the reconstructed pixel values are only independent if the matrix size is 10×10 or smaller. Note that, in the practical realisation, the current injection is not restricted to diagonally opposed electrodes, but instead is often applied to adjacent electrodes. Use of this configuration appears to provide the best resolution, although not the highest sensitivity (Barber and Brown 1986). The isopotentials corresponding to this experimental arrangement are shown in Figure 9.4, and the shaded area is that 'receiving' backprojected data from the measurement across electrodes 5 and 6.

There are several factors that work in favour of this method of image reconstruction. With good electronic design, the measurement of peripheral potentials can be accurate to within 1% over a dynamic range of 40 dB. No electrode guarding is necessary and the measurements are not affected by the electrode contact impedance. We recall that the latter was the major impediment to image reconstruction using measured exit currents.

The reconstructed image, being a backprojection, is blurred (as is the case for all CT backprojections), and it would be desirable to deconvolve the point response function of the system (see Chapter 11). Using a detailed analysis of the reconstruction algorithm, Barber and Brown (1986, 1987) showed that the point-source response function is not spatially invariant. It proves possible, however, to convert the filtration problem

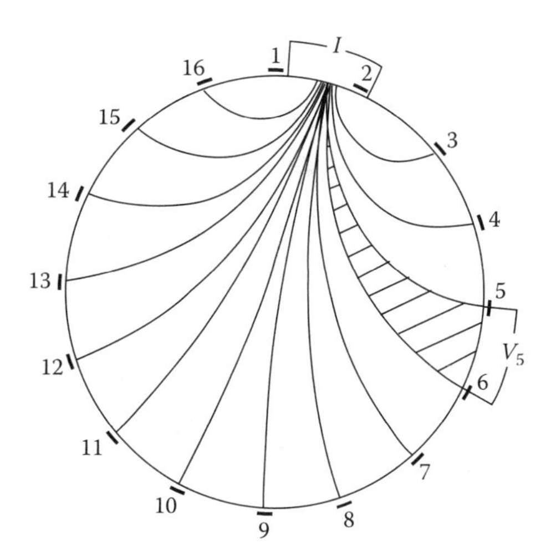

FIGURE 9.4
Isopotentials of a practical EIT system for uniform conductivity. (From Brown et al., 1986.)

to a position-independent one by a coordinate transformation. After deconvolution, the reverse transformation is applied to the result. Additionally, the method is only accurate to first order because the experimental data are backprojected along the equipotentials corresponding to a uniform distribution of conductivity rather than the true non-uniform distribution, giving rise to the measured peripheral potentials, and in which the isopotentials are significantly different from in the uniform case. Iterative reconstruction schemes have been devised in which the first-order isopotentials are recomputed from the first-order backprojected distribution of conductivity and used to recompute difference projections prior to re-backprojection along the new first-order isopotentials. However, there is evidence that use of the true isopotentials does not necessarily lead to improved image accuracy (Avis et al. 1992).

Distinguishing features of EIT are the possibility of very rapid data collection (about 0.1 s or less per slice), modest computing requirements, relatively low cost and no known significant hazards. Against these may be set the limitations described of poor spatial resolution and technical difficulties arising from the oversimplified assumptions concerning the distribution of the isopotentials. Brown et al. (1985) estimated that the best resolution that EIT might expect to achieve is of the order 1.5% of the reconstructed field diameter (with 128 electrodes), but presently the experimental limit has been 5%–10% of the reconstructed-field diameter. When the number of drive electrodes is increased, the practical problems of data collection rise, and Brown and Seagar (1985) have provided a detailed analysis of the factors involved. The data collection may be serial or parallel. The currents must, of course, be applied serially, but the recording electrodes can be read in parallel provided an arrangement employing one amplifier for each electrode is used. In the system described by Brown (1983), the data collection is in fact serial but still takes less than a second.

There is a large literature on image reconstruction algorithms (see Holder 1992 and Molinari et al. 2002) and the previous description should only be seen as an introduction to the inverse problem.

The ability to collect data rapidly from complex arrays of electrodes has enabled 3D imaging to be developed. Metherall et al. (1996) describe a system using 64 electrodes placed in four planes around the thorax and the reconstruction of dynamic images of both lung ventilation and perfusion. Figure 9.5 shows a set of images collected using this system.

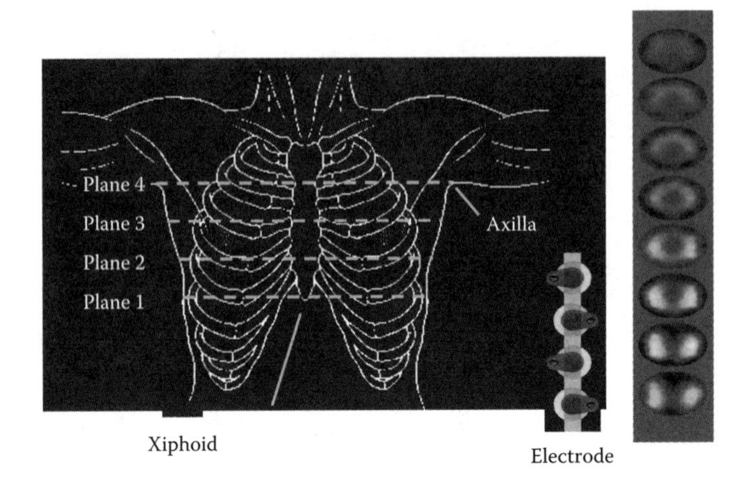

FIGURE 9.5
(See colour insert.) Four planes of 16 electrodes were used to reconstruct dynamic images at eight levels showing lung resistivity at inspiration with respect to expiration. (Courtesy of P. Metherall.)

Multi-frequency imaging was mentioned as a possibility in Section 9.2. A number of workers have now implemented this and data have been collected at many frequencies, usually within the range 10 kHz to 1 MHz. By using data collected at a single frequency as a reference, images can be reconstructed to show the variations of tissue impedance with frequency. By this means a tissue impedance spectrum can be produced for each pixel in the image and this spectrum can be used to characterise tissue (see Brown et al. 1995, Hampshire et al. 1995).

9.4 Clinical Applications of Electrical Impedance Tomography

It must be stated at the outset that EIT is not currently at the same state of development as most of the imaging modalities described in this volume. Whilst commercial equipment is now available, clinical applications of EIT are still mainly in a research setting. In this section, we therefore report on those proposed applications for which there is significant evidence of clinical viability. There are currently well over 30 groups worldwide working on EIT and there have been European Union Concerted Action programmes in this field. At the time of writing the best reviews of the clinical utility of EIT are those of Holder and Brown (1993), Boone et al. (1997) and Holder (2005).

9.4.1 Pulmonary Imaging

Most clinical implementations of EIT utilise dynamic rather than static images. A dynamic image is produced by comparing a reference EIT dataset with a new dataset from the same subject in order to image changes in resistivity. With typical data collection times of 0.1 s or less, changes in impedance during the cardiac cycle are amenable to imaging. During dynamic imaging, static structures do not appear in the images, although they may contribute to image artefacts (Brown et al. 1985). For example, the ventilation of normal lungs can

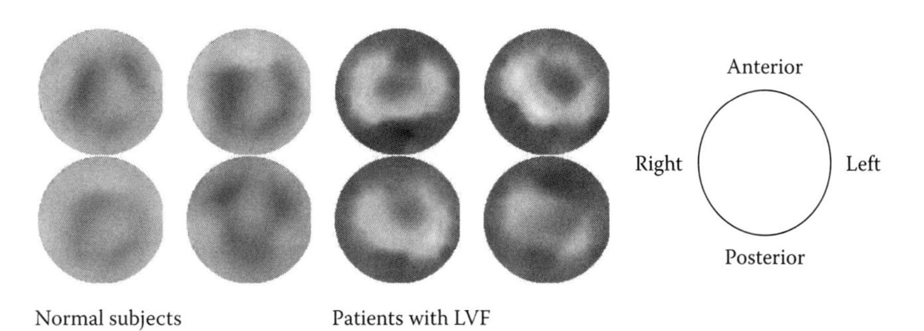

FIGURE 9.6
(See colour insert.) EIT images of the thorax for four normal subjects and four patients admitted to coronary care with lung water associated with left ventricular failure (LVF). The images show conductivity with respect to a mean normal thorax. The increased conductivity in the lungs of the patients with pulmonary oedema is clear.

be seen very easily on a dynamic EIT image as the lungs are more resistive (c. 23 Ω m) when inflated than when deflated (c. 7 Ω m) due to current flow around the aerated alveoli.

It is possible to image pulmonary oedema (fluid in the lungs) in this way using a mean normal EIT dataset as a reference. Fluid appears as a region of reduced resistivity within the lungs. Pulmonary oedema is a consequence of many diseases and, although extensive pulmonary oedema is usually readily diagnosed by clinical and radiological means (chest X-ray), any technique that allowed quantification of the amount of water present would be useful, for example in monitoring therapy. In Figure 9.6, some images from patients admitted to coronary care with LVF are compared with some normal images (see Noble et al. 1999). The increased lung conductivity associated with lung water is very clearly visible. It may also be interesting to look at differences in resistivity between simple pulmonary oedema, that is, a fluid overload, and oedema associated with, for example, pulmonary infection. In particular, this would be useful if it allowed a more definite discrimination between pulmonary oedema and some of the unusual pulmonary infections that can appear to be very similar on a chest radiograph (Cherryman 1987). Normal lung comprises 80% air and only 5% fluid (Brown et al. 1985), and it has been estimated that volumes of fluid as small as 10 mL should be able to be imaged. Campbell et al. (1994) looked at the changes in EIT images during the aspiration of pleural effusion and the infusion of normal saline. Newell et al. (1996) showed changes in the lung region of animals during oedema induced by oleic acid. Noble et al. (2000) measured the changes in humans following the administration of diuretics.

Changes in pulmonary resistivity during inspiration for six normal volunteers were mapped using EIT techniques by Harris et al. (1987). An impedance index derived from the images was shown to vary linearly with the volume of air inspired up to 5 L if the subject sits upright. The function of right and left lungs may be observed separately. In particular, it was demonstrated that for a normal volunteer lying in the lateral decubitus position (on his/her right-hand side), the upper (left) lung increases in impedance more rapidly than the right for a low inspiration volume, possibly due to a reduction in the compliance of the lower lung under gravity. Harris et al. (1987) also provide images of a patient with emphysematous bulla in the right lung, the defect being conspicuous by non-uniform changes in impedance over the affected lung. The differential impedance in the lung between inspiration and expiration has been used via impedance plethysmography to provide a gating signal for a gamma camera (see Section 5.3.2) in order to reduce respiratory artefacts by collecting data during only part of the respiratory cycle (Heller et al. 1984). This is claimed to be more appropriate than spirometry for extremely ill patients. The use of EIT for pulmonary investigation has been systematically investigated in both animal and human situations and it has been shown that EIT can show regional differences in resistivity that are in good agreement with CT scans on the same patients (Hahn et al. 2006). The use of EIT for pulmonary monitoring of artificially ventilated patients has also been investigated. It appears that EIT might be used to control the application of positive end-expiratory pressure (PEEP) for maximising lung ventilation (Hahn et al. 1995) and this area of application is the subject of considerable current research.

9.4.2 Gastric Imaging

Measurement of the rate of emptying of the stomach after eating is an important test in the management of certain gastric disorders, for example, pyloric stenosis. The most widely used method for assessment of gastric stenosis is gamma scintigraphy: the patient consumes a radiolabelled test substance, and the rate of gamma emission is measured in the region of the stomach. In infants, and in adults who require frequent investigation, the radiation dose is potentially dangerous. One may, in these cases, resort to the rather uncomfortable intubation methods. For example, in dye-dilution studies, the patient swallows a nasogastric tube, through which small amounts of a substance with intense colouring can be delivered. The nasogastric tube is then used to withdraw samples of stomach contents. The volume of the stomach at any given time may be found from the ratio of the amount of dye delivered to its concentration in the gastric sample. Many patients find the intubation uncomfortable, and there is a small risk of aspiration of the stomach contents. Clearly, some alternative to these techniques is required.

The assessment of gastric emptying is one of the best validated applications of EIT. Mangnall et al. (1987) obtained significant correlations between EIT and gamma scintigraphy. A number of groups compared the volume of fluid introduced into the stomach with the corresponding EIT images (e.g. Devane 1993, Nour 1992). A further validation technique is to investigate whether EIT can distinguish between healthy individuals and those with known gastric disease. Lamont et al. (1988) measured the times for the stomach volume to decrease by half in 18 infants, after a conductive drink. Infants with pyloric stenosis showed longer emptying times (mean 46 min) than those without (mean 12 min). The apparent resistivity of the stomach region is influenced not only by its size, but by its contents. Baxter et al. (1988) investigated the relationship between the mean resistivity in the EIT image and the concentration of fixed volumes of hydrochloric acid introduced into the stomach. Their results suggested that the presence of stomach acid would significantly affect the resistivity. Acid secretion is typically suppressed using a drug such as cimetidine, although this may not be ethical in all cases, for example, studies in young children.

9.4.3 Cardiac Imaging

Changes in resistivity take place in the thorax during the cardiac cycle; these are due to the redistribution of blood between the heart, great vessels and lungs. Such changes are small compared to those due to ventilation. Cardiac gating, that is, synchronisation of the data acquisition to the cardiac cycle, is normally used to separate the cardiac and ventilation information in the EIT recording. Cardiac gating was first used in EIT by Eyüboglu et al. (1987) for imaging the thorax. In this technique, the R-wave of the ECG starts the acquisition of EIT data, typically at about 20 frames per second. Each set of frames is averaged to produce an image of resistivity during the cardiac cycle. The authors tentatively suggested that there were separate features in the images corresponding to the descending aorta, right atrium, left and right ventricles and left and right lungs. McArdle et al. (1993)

compared EIT images of perfusion in the thorax with M-mode ECG. They also reduced ventricular stroke volume using glyceryl trinitrate. In contrast to earlier studies, they concluded that EIT could only resolve one atrial and one ventricular component in the region of the heart, and that even these might overlap. If this were the case, the true resistivity changes in the region of the heart would be underestimated, since a decreasing resistivity in the atria might 'cancel' an increasing resistivity in the aorta. Furthermore, the heart moves in the chest during the respiratory cycle, and this would further blur the distinction between different parts of the heart. The relative sizes of the cardiac and ventilation components make it necessary to acquire a large number of data acquisition cycles. Leathard et al. (1993) proposed that EIT could be used to monitor blood flow rate in the heart and lungs using concentrated (1.8%) saline as a tracer, rather than by averaging. This technique, if successful, would avoid the ventricular and atrial signals 'cancelling', as was thought to take place in cardio-synchronous imaging. In one subject, a resistivity decrease of about 4% was observed in the chest after injection of 20 mL of saline into the arm.

A more promising application is the imaging of cardiac-related changes in the lungs. As blood perfuses the lungs there is an associated change of a few percentage points in lung conductivity because blood has a relatively high resistivity. Smit et al. (2003) have shown that EIT imaging of pulmonary perfusion is sufficiently reproducible for it to be considered as a clinical tool.

9.4.4 Breast Imaging

Carcinoma of the breast is a significant cause of death in the Western world; prospects for the sufferer are much better if the disease is detected at an early stage. While X-ray mammography (see Section 2.11.2) is a useful technique for detecting the condition, it is too expensive to use routinely (e.g. by general practitioners), and can itself increase the risk of the disease if overused. EIT may be a much more practicable way to apply routine screening for breast cancer. There is known to be a significant difference in impedance between normal breast tissues and tumours (e.g. Jossinet et al. 1981, Jossinet and Risacher 1996). This application demands a method for producing images of absolute (i.e. an image produced from a single EIT dataset as opposed to imaging the change in impedance between datasets recorded at two different times) resistivity as we cannot image during the development of a tumour. Multiple-frequency imaging, in which an image of the difference in impedance between two datasets is recorded at different frequencies, may be a successful route to achieving this.

A major problem with imaging tumours is that they are usually highly anisotropic and hence they present a problem for EIT which depends upon the passage of current from many directions in order to produce an image. The approach of Choi et al. (2007) in which current is passed across the tissues partially avoids this problem. An even more successful approach might be to make contact with the ductal tissue of the breast and make local electrical impedance spectroscopy measurements.

9.5 Conclusions

The pace of research on EIT does not appear to be diminishing even though clinical applications are not yet clear. In addition to the areas considered previously, other areas are under investigation. For example, during cerebral ischaemia, the impedance of brain tissue increases by up to 100% (Holder 1992, Romsauerova et al. 2006). This might allow for a non-invasive method for assessing the severity of stroke. Holder obtained images of cerebral ischaemia in anaesthetised rats, with both intracranial and extracranial electrodes.

One of the most active areas of technical innovation concerns the development of methods to make absolute as opposed to relative or dynamic measurements of tissue conductivity. A major obstacle to clinical application is that absolute measurements are required if normal ranges are to be established. For example, dynamic EIT might be used to track the progress of pulmonary oedema in a patient, but absolute measurement is required if an initial diagnosis of the presence of oedema is required or the return to a normal range for the lung tissue is to be established. To enable absolute measurements to be made, the boundary shape of the body is required so that an accurate forward model can be produced. Computing resources have recently improved to the point where it is possible to make sufficiently accurate forward models using either finite element or finite difference techniques. Full absolute *in vivo* imaging has not yet been achieved but techniques to make absolute measurements of organ conductivity *in vivo* have been developed and used to measure lung resistivity (Babaeizadeh et al. 2007, Brown et al. 2002a, 2002b).

A fundamentally different approach to the imaging of tissue conductivity is the technique of magnetic induction tomography (MIT) which has been developed during the past 15 years (see Griffiths H. Chapter 8, pp. 213–233 in Holder 2005). MIT applies a magnetic field using an excitation coil to induce eddy currents in tissue. The magnetic field then produced by the eddy currents is detected by sensing coils. No direct contact with the body is required. A number of systems have been reported, including hybrid systems using both coils and electrodes. This is a very active area for research but the fundamental problem that needs to be overcome is the achievement of an adequate signal-to-noise ratio using currents that are within the safety standards for electromedical equipment.

Whereas EIT can never be expected to compete with X-ray CT and nuclear magnetic resonance (NMR) for high-resolution anatomical imaging, it may offer a new route to rapid dynamic measurement of physiological function. In this respect, it is more reasonable to make comparisons between what is offered and measured by EIT and the imaging of physiological function using radiolabelled pharmaceuticals and the gamma camera. It is concluded that currently EIT shows much clinical promise, and it can rival existing alternative techniques for selected applications, but it retains significant disadvantages that probably destine its arrival in the clinic to be an evolving process rather than an acclaimed overnight innovation.

References

Avis N J, Barber D C, Brown B H and Kiber M A 1992 Back-projection distortions in applied potential tomography images due to non-uniform reference conductivity distributions *Clin. Phys. Physiol. Meas.* **13 Suppl. A** 113–117.

Babaeizadeh S, Brooks D H and Isaacson D 2007 3-D Electrical impedance tomography for piecewise constant domains with known internal boundaries *IEEE Trans. BME* **54** 2–10.

Barber D C and Brown B H 1984 Applied potential tomography *J. Phys. E: Sci. Instrum.* **17** 723–733.

Barber D C and Brown B H 1986 Recent developments in applied potential tomography—APT In *Information Processing in Medical Imaging* ed. S L Bacharach (the Hague, the Netherlands: Nijhoff).

Barber D C and Brown B H 1987 Construction of electrical resistivity images for medical diagnosis *SPIE Conference*, Long Beach, CA, Paper 767 04.

Barber D C, Brown B H and Freeston I L 1983 Imaging spatial distributions of resistivity using applied potential tomography *Electron. Lett.* **19** 933–935.

Bates R H T, McKinnon G C and Seagar A D 1980 A limitation on systems for imaging electrical conductivity distributions *IEEE Trans. Biomed. Eng.* **BME-27** 418–420.

Baxter A J, Mangnall Y E, Loj E J, Brown B H, Barber D C, Johnson A G and Read N W 1988 Evaluation of applied potential tomography as a new non-invasive gastric secretion test *Gut* **30** 1730–1735.

Bayford R and Woo E J (Eds.) 2006 Special issue: *Sixth Conference on Biomedical Applications of Electrical Impedance Tomography Phys. Meas.* **271** 1–280.

Boone K, Barber D C and Brown B H 1997 Review: Imaging with electricity: Report of the European concerted action on impedance tomography *J. Med. Eng. Tech.* **21** 201–232.

Brown B H 1983 Tissue impedance methods In *Imaging with Non-Ionising Radiations* ed. D F Jackson (Guildford, UK: Surrey University Press).

Brown B H 2001a Measurement of the electrical properties of tissue—New developments in impedance imaging and spectroscopy *IEICE Trans. Inf. Syst.* **E85-D** 2–5.

Brown B H 2001b Medical impedance tomography and process impedance tomography: A brief review *Measurement Science and Technology* **12** 991–996.

Brown B H, Barber D C, Harris N and Seagar A D 1986 Applied potential tomography: A review of possible clinical applications of dynamic impedance imaging. In *Recent Developments in Medical and Physiological Imaging* Suppl. to *J. Med. Eng. Technol.* eds. R P Clark and M R Goff (London, UK: Taylor & Francis) pp. 8–15.

Brown B H, Barber D C and Seagar A D 1985 Applied potential tomography: Possible clinical applications *Clin. Phys. Physiol. Meas.* **6** 109–121.

Brown B H, Leathard A D, Lu L, Wang W, and Hampshire A 1995 Measured and expected Cole parameters from electrical impedance tomographic spectroscopy images of the human thorax. *Physiol. Meas.* **16 A** 57–67.

Brown B H, Primhak R A, Smallwood R H, Milnes P, Narracott A J and Jackson M J 2002a Neonatal lungs—Can absolute lung resistivity be determined non-invasively? *Med. Biol. Eng.* **40** 388–394.

Brown B H, Primhak R A, Smallwood R H, Milnes P, Narracott A J and Jackson M J 2002b Neonatal lungs—Maturational changes in lung resistivity spectra *Med. Biol. Eng.* **40** 506–511.

Brown B H and Seagar A D 1985 Applied potential tomography—Data collection problems *Proceedings of IEE Conference on Electric and Magnetic Fields in Medicine and Biology*, London, UK Conf. Publ. no. 257 pp. 79–82.

Brown B H, Smallwood R H, Barber D C, Hose D R and Lawford P V 1999 Non-ionizing electromagnetic radiation: Tissue absorption and safety issues In *Medical Physics and Biomedical Engineering* (Bristol, UK: Institute of Physics Publishing) ISBN 0 7503 0368 9 Chapter 8 pp. 224–257.

Campbell J H, Harris N D, Zhang F, Brown B H and Morice A H 1994 Clinical applications of electrical impedance tomography in the monitoring of changes in intrathoracic fluid volumes *Physiol. Meas.* **15 Suppl. 2A** 217–222.

Cherryman G 1987 Personal communication.

Choi M H, Hao T J, Isaacson D, Saulnier G J and Newell J C 2007 A reconstruction algorithm for breast cancer imaging with electrical impedance tomography in mammography geometry *IEEE Trans. BME* **54** 700–710.

Cole H S and Cole R H 1941 Dispersion and absorption in dielectrics *J. Chem. Phys.* **9** 341–351.

Devane S P 1993 Application of EIT to gastric emptying in infants: Validation against residual volume method. In *Clinical and Physiological Applications of Electrical Impedance Tomography* ed. D S Holder (London, UK: UCL Press).

Duck F A 1990 *Physical Properties of Tissue* (London, UK: Academic Press).

Eyuboglu B M, Brown B H, Barber D C and Seagar A D 1987 Localisation of cardiac related impedance changes in the thorax *Clin. Phys. Physiol. Meas.* **8 Suppl. A** 167–173.
Foster K R and Schwan H P 1989 Dielectric properties of tissues and biological materials: A critical review. *Crit. Rev. Biomed. Eng.* **17** 25–104.
Gabriel C, Gabriel S and Corhout E 1996a The dielectric properties of biological tissues, I, Literature survey *Phys. Med. Biol.* **41** 2231–2249.
Gabriel S, Lau R W and Gabriel C, 1996b The dielectric properties of biological tissue, II, Measurements in the frequency range 10 Hz to 20 GHz *Phys. Med. Biol.* **41** 2251–2269.
Hahn G, Just A, Dudykevych T, Frerichs I, Hinz J, Quintel M and Hellige G 2006 Imaging pathologic pulmonary air and fluid accumulation by functional and absolute EIT *Phys. Meas.* **27** S187–S198.
Hahn G, Sipinkova I, Baisch F and Hellige G 1995 Changes in the thoracic impedance distribution under different ventilatory conditions *Physiol. Meas.* **16 3A** 161–173.
Hampshire A R, Smallwood R H, Brown B H and Primhak R A 1995 Multi-frequency and parametric EIT images of neonatal lungs *Phys. Meas.* **16 A** 175–189.
Harris N D, Sugget A J, Barber D C and Brown B H 1987 Applications of applied potential tomography (APT) in respiratory medicine *Clin. Phys. Physiol. Meas.* **8 Suppl. A** 155–165.
Heller S L, Scharf S C, Hardaff R and Blaufox M D 1984 Cinematic display of respiratory organ motion with impedance techniques *J. Nucl. Med.* **25** 1127–1131.
Henderson R P and Webster J G 1978 An impedance camera for spatially specific measurements of the thorax *IEEE Trans. Biomed. Eng.* **BME-25** 250–253.
Holder D S 1992 Electrical impedance tomography with cortical or scalp electrodes during global cerebral ischaemia in the anaesthetised rat *Clin. Phys. Physiol. Meas.* **13** 87–98.
Holder D S (ed.) 2000 Biomedical applications of electrical impedance tomography *Phys. Meas.* **21** 1–207.
Holder D S (ed.) 2002 Biomedical applications of electrical impedance tomography *Phys. Meas.* **23** 95–243.
Holder D S (ed.) 2005 *Electrical Impedance Tomography, Methods, History and Applications* (Bristol, UK: Institute of Physics Publishing), pp. 1–456.
Holder D S and Brown B H 1993 Biomedical application of EIT: A critical review. In *Clinical and Physiological Applications of EIT* ed. D S Holder (London, UK: UCL Press).
Isaacson D, Mueller J and Siltanen S (eds.) Biomedical applications of electrical impedance tomography 2003 *Phys. Meas.* **24** 237–637.
Jossinet J J, Fourcade C and Schmitt M 1981 A study for breast imaging with a circular array of impedance electrodes In *Proceedings of the 5th International Conference on Electrical Bioimpedance*, Tokyo, Japan pp. 83–86.
Jossinet J and Risacher F 1996 The variability of resistivity in human breast tissue *Med. Biol. Eng. Comp.* **34** 346–350.
Lamont G L, Wright J W, Evans D F and Kapila L A 1988 An evaluation of applied potential tomography in the diagnosis of infantile hypertrophic pyloric stenosis *Clin. Phys. Physiol. Meas.* **9A** 65–69.
Leathard A D, Caldicott L, Brown B H, Sinton A M, McArdle F J, Smith R W M and Barber D C 1993 Cardiovascular imaging of injected saline. In *Clinical and Physiological Applications of Electrical Impedance Tomography* ed. D S Holder (London, UK: UCL Press).
Mangnall Y F, Baxter A J, Avill R, Bird N C, Brown B H, Barber D C, Seagar A D, Johnson A G and Read N W 1987 Applied potential tomography: A new non-invasive technique for assessing gastric function *Clin. Phys. Physiol. Meas.* **8 Suppl. A** 119–129.
McArdle F J, Turley A, Hussain A, Hawley K and Brown B H 1993 An in-vivo examination of cardiac impedance changes imaged by cardiosynchronous averaging. In *Clinical and Physiological Applications of Electrical Impedance Tomography* ed. D S Holder (London, UK: UCL Press).
Metherall P, Barber D C, Smallwood R H and Brown B H 1996 Three-dimensional electrical impedance tomography *Nature* **380** 509–512.

Molinari M, Cox SJ, Blott B H and Daniell G J 2002 Comparison of algorithms for non-linear 3D electrical tomography reconstruction *Phys. Meas.* **23** 95–104.

Newell J C, Edic P M, Ren X, Larson-Wiseman J L and Danyleiko M D 1996 Assessment of acute pulmonary edema in dogs by electrical impedance imaging *IEEE Trans. BME* **43** 133–139.

Noble T J, Harris N D, Morice A H, Milnes P and Brown B H 2000 Diuretic induced changes in lung water assessed by electrical impedance tomography *Phys. Meas.* **21** 155–163.

Noble T J, Morice A H, Channer K S, Milnes P, Harris N D and Brown B H 1999 Monitoring patients with left ventricular failure by electrical impedance tomography *Eur. J. Heart Fail.* **1** 379–384.

Nour S 1992 *Measurement of Gastric Emptying in Infants Using APT* PhD dissertation (Sheffield, UK: University of Sheffield).

Romsauerova A, McEwan A, Horesh L, Yerworth R, Bayford R H and Holder D S 2006 Multi-frequency electrical impedance tomography (EIT) of the adult human head: Initial findings in brain tumours, arteriovenous malformations and chronic stroke, development of an analysis method and calibration *Phys. Meas.* **27** S147–S162.

Smit H J, Handoko M L, Noordegraaf Vonk A, Faes Th J C, Postmus P E, de Vries P M J M and Boonstra A 2003 Electrical impedance tomography to measure pulmonary perfusion: Is the reproducibility high enough for clinical practice? *Phys. Meas.* **24** 491–499.

10

Optical Imaging

J. C. Hebden

CONTENTS

10.1 Introduction 665
10.2 Origin of Optical Contrast 666
  10.2.1 Refraction 666
  10.2.2 Scatter 666
  10.2.3 Absorption 667
10.3 Optical Coherence Tomography 668
  10.3.1 Principles of Optical Coherence Tomography 668
  10.3.2 Applications of Optical Coherence Tomography 670
10.4 Diffuse Optical Imaging 671
  10.4.1 Breast Transillumination 671
  10.4.2 Optical Topography 673
  10.4.3 Optical Tomography 675
  10.4.4 Image Reconstruction 676
10.5 Photoacoustic and Acousto-Optic Imaging 678
  10.5.1 Photoacoustic Imaging 678
  10.5.2 Acousto-Optic Imaging 679
10.6 Other Advances in Optical Imaging 680
  10.6.1 Exogenous Contrast Agents and Molecular Imaging 680
  10.6.2 Multimodality Imaging 680
References 681

10.1 Introduction

All of us are exposed to optical (i.e. visible and near-infrared [NIR]) radiation from the sun and other sources throughout our lives. Assuming our eyes are shielded from excessive intensity, and our skin is protected from the ultraviolet content of sunlight, we accept this exposure in the knowledge that it is perfectly safe. Unlike X-rays, optical photons are not sufficiently energetic to produce ionisation, and unless light is concentrated to such a high degree that it causes burning to the skin, optical radiation offers no significant hazard. It is therefore apparent why optical imaging is an attractive concept. Isolated cases of the use of transmitted light as a means of diagnosing disease were reported as far back as the early nineteenth century (Bright 1831, Curling 1843), and the first attempts to use light to identify cancer in the female breast by shining light through it were described by Cutler (1929). The potential of optical methods as a diagnostic indicator of tissue function has been widely known since Jöbsis (1977) first demonstrated that transmittance measurements of NIR radiation could be used to monitor the degree of oxygenation of certain metabolites. This led to the development and increasingly widespread use of clinical NIR spectroscopy, which offers a safe, non-invasive means of monitoring blood flow and oxygenation at the bedside without the use of radioisotopes or other exogenous contrast agents. Inevitably there has been considerable interest in the possibility of extending this global monitoring technique into imaging, and a variety of potential clinical applications have been explored.

The principal challenge underlying imaging the human body with light is to identify contrast between different tissue types while minimising the blurring effects of scatter. As a consequence of scatter, a collimated beam of light loses all directionality after transmission through more than a few millimetres of tissue. Optical images generated using light which has passed through several centimetres of tissue exhibit poor spatial resolution, particularly for features located well below the surface. Since the 1980s there has been a rapid growth of interest in the diagnostic potential of optical radiation, and the field known as biomedical optics has flourished. Progress in optical imaging has been greatly assisted by advances in NIR laser and detector technology, as well as by the availability of fast inexpensive computers. In this chapter, three broad areas of optical imaging of the living human body are examined: optical coherence tomography (OCT), diffuse optical imaging (DOI) and photoacoustic imaging. This will follow a brief description of the basic interactions between light and tissue, and the mechanisms of image contrast at optical wavelengths.

10.2 Origin of Optical Contrast

The interaction between optical radiation and tissue can be characterised by three distinct mechanisms: refraction, scatter and absorption, which are described separately in the following.

10.2.1 Refraction

Refraction is the change in the direction of a beam of light due to change in the local speed of light, and is characterised by the refractive index, n. The range of refractive indices of tissues at optical wavelengths is small, being confined almost entirely between those of water (~1.33) and adipose tissue (~1.55). Consequently refractive index is usually considered to be a constant on a macroscopic scale. Nevertheless, microscopic variation in refractive index is the origin of scatter which strongly influences all forms of optical imaging, and gives rise to the reflections which produce contrast in OCT (Section 10.3). Detection of the very small changes in refractive index and the changes in scatter which result from them has been proposed as a means of monitoring blood glucose content, although studies have indicated that the effect on external transmittance measurements is very small (Kohl et al. 1995).

10.2.2 Scatter

The dominant cause of scatter in soft tissues is the chaotic variation in refractive index at a cellular and sub-cellular scale. This occurs at the cell membrane boundaries with extracellular fluid, and at boundaries between intracellular fluid and internal organelles such as mitochondria, ribosomes, fat globules, glycogen and secretory globules. The probability of a photon being scattered per unit length of travel is characterised by the scatter coefficient μ_s. Scatter is by far the most dominant tissue-photon interaction at NIR wavelengths, with scattering coefficients within the range of 10–100 mm^{-1} (Cheong et al. 1990). In order to fully describe scatter of light in tissue, it is necessary to consider the probability of a photon being scattered in a given direction at each interaction. The probability of a photon incident along a unit vector $\hat{\mathbf{s}}'$ being scattered into a direction $\hat{\mathbf{s}}'$ is described by the phase function $f(\hat{\mathbf{s}}, \hat{\mathbf{s}}')$. For a random medium it can be assumed that this probability is independent of $\hat{\mathbf{s}}$ and only depends on the angle between the incident and scattered directions, θ. Thus the phase function can be conveniently expressed as a function of the scalar product of the unit vectors in the initial and final directions, which is equal to the cosine of the scattering angle $\cos(\theta)$. The anisotropy in the probability distribution is commonly characterised in terms of the mean cosine of the scattering angle, g. In biological tissues, scatter occurs principally in a forward direction, corresponding to an anisotropy in the range $0.69 \leq g \leq 0.99$ (Cheong et al. 1990). Despite the forward scatter, typical values of scattering coefficient ensure that light travelling through more than a few millimetres of tissue loses all of its original directionality, and can be treated as effectively isotropically distributed. Thus it is often convenient to express the scatter by tissues in terms of a transport scatter coefficient:

$$\mu_s' = \mu_s(1-g) \tag{10.1}$$

which represents the effective number of isotropic scatters per unit length and is a fundamental parameter in diffusion theory (see Section 10.4.4). Typically μ_s' is of the order of 1 mm^{-1} in most soft tissues. Scatter is also a potential source of contrast for DOI techniques (Section 10.4). In principle, variation in scatter coefficient may offer a means of imaging myelination in the brain, and of discriminating between types of tissue in the breast.

10.2.3 Absorption

Human tissues contain a variety of substances whose absorption spectra at NIR wavelengths are well defined, and which are present in sufficient quantities to contribute significant attenuation to measurements of transmitted light. The probability of a photon being absorbed per unit length is described by the absorption coefficient, μ_a. For a specific absorbing substance,

$$\mu_a = c \cdot \varepsilon \tag{10.2}$$

where

c is the molar concentration of the absorber, and
ε is known as the specific absorption coefficient (mm^{-1} mol^{-1}).

The concentration of some absorbers, such as water, melanin and bilirubin, remains virtually constant with time. However, some absorbing compounds, such as oxygenated haemoglobin (HbO_2), deoxyhaemoglobin (Hb) and oxidised cytochrome oxidase (CtOx), have concentrations in tissue which are strongly linked to tissue oxygenation and metabolism. Since water is so prevalent in tissue, constituting around 80% of the human brain for example (Woodward and White 1986), increasingly dominant absorption by water at longer wavelengths limits spectroscopic and imaging studies to less than about 1000 nm. The lower limit on wavelength for *in vivo* NIR studies is dictated by the overwhelming absorption of haemoglobin below about 650 nm. The haemoglobin compounds are the dominant and most clinically interesting

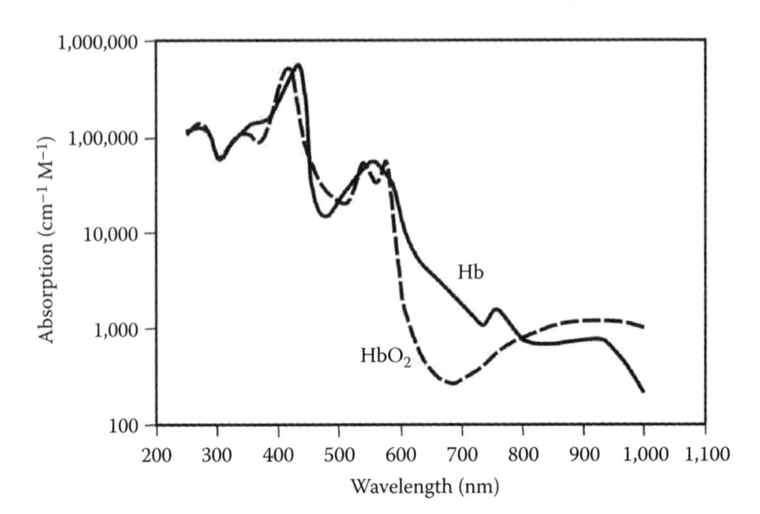

FIGURE 10.1
The specific absorption coefficients of oxyhaemoglobin and deoxyhaemoglobin in the 250–1000 nm wavelength range.

chromophores in the 650–1000 nm wavelength range, and their specific absorption coefficients are illustrated in Figure 10.1. The difference in absorption of the two compounds plays a role in the familiar observation that arterial blood (containing around 98% HbO_2) appears bright red whereas venous (deoxygenated) blood has a more bluish appearance. Simultaneous monitoring of the transmittance at two different wavelengths enables changes in attenuation to be converted into changes in concentrations of oxyhaemoglobin [HbO_2] and deoxyhaemoglobin [Hb]. Increasing the number of wavelengths enables changes in concentration of other chromophores such as cytochrome to be also taken into account.

10.3 Optical Coherence Tomography

OCT was first developed in the early 1990s. The principles of OCT are analogous to those of diagnostic ultrasound (Chapter 6) except they involve measuring reflections of light rather than sound, and that the range-gating is achieved using the optical interference of the probing beam with a reference beam rather than by measurement of time-of-flight. Although scattering limits the depth of penetration in biological tissues to a millimetre or so, OCT can provide images of tissue pathology *in vivo* with a spatial resolution of a few micrometers. The dominant application of OCT is currently ophthalmology, although it promises to have a broad range of clinical applications involving the non-invasive assessment of tissue (Section 10.3.2). The principles of OCT are briefly summarised in the following, and further details can be found in several excellent reviews (Fercher et al. 2003, Fujimoto 2003, Zysk et al. 2007).

10.3.1 Principles of Optical Coherence Tomography

OCT is conventionally performed using a Michelson-type interferometer and a low-coherence source, as illustrated in Figure 10.2. Light from the source is split into two equal beams, and

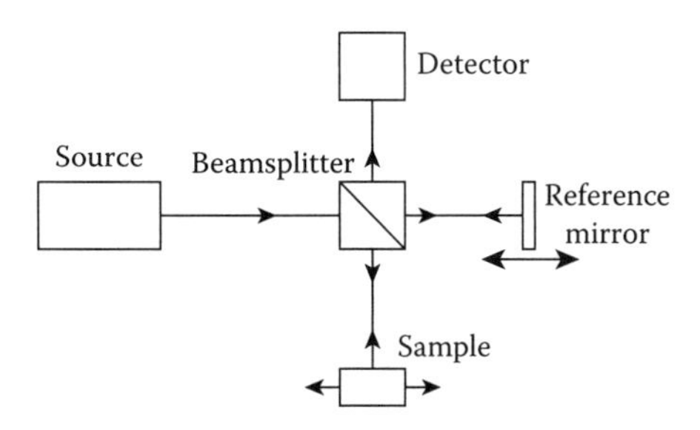

FIGURE 10.2
OCT using a Michelson-type interferometer and a low-coherence source.

sent along the sample arm and reference arm of the interferometer. Reflections off the sample and off the reference mirror are then recombined and sent to the detector. Every interface within the sample where there is a change in refractive index will reflect a portion of the incident beam. Interference fringes are produced if the optical path lengths in the two arms of the interferometer coincide within the coherence length l_c of the source, which is equal to

$$l_c = \frac{2\ln 2}{\pi} \cdot \frac{\lambda^2}{\Delta\lambda} \cong \frac{\lambda^2}{\Delta\lambda} \tag{10.3}$$

where
λ is the mean wavelength of the source, and
$\Delta\lambda$ is the spectral bandwidth (assuming a Gaussian spectral distribution).

Superluminescent diodes are commonly used sources for OCT, which have a typical coherence length of around 10 μm. The amplitudes of the interference fringes depend on the reflectance/backscatter occurring at a specific depth within the tissue, and OCT systems scan over a range of depths by varying the optical pathlength in the reference arm. Two-dimensional cross-sectional images can be generated in real time by rapidly scanning the beam laterally. Alternatively, scanning can be performed in two lateral directions to obtain 3D images, or 2D images corresponding to a constant depth. While transverse spatial resolution depends on the spot size, axial resolution depends on the source coherence length. OCT has the ability to acquire high resolution images of biological tissue because of its facility to isolate a coherent reflected signal in the sample beam from the incoherent scattered light. Very high dynamic range and sensitivity (>100 dB) can be achieved by using heterodyne detection schemes. The maximum optical power to which biological tissue can be safely exposed limits the maximum penetration depth to around 1–3 mm, depending on wavelength and type of tissue.

Technological advances have steadily yielded improvements in OCT spatial resolution, spectral resolution and scanning speed. Doppler OCT has been developed in order to detect fluid and blood flow as slow as 10 μm s^{-1} (Barton et al. 1998). Fourier-domain OCT has been introduced as a means of replacing the mechanical depth scan (translation of the reference mirror in Figure 10.2) with a spectrometric measurement (Fercher et al. 2003). Although the technique requires a more sophisticated and expensive detector, it achieves very much faster scan times. Faster imaging with standard OCT has also been achieved using linear or 2D array detectors, allowing a whole line or area to be scanned at once.

10.3.2 Applications of Optical Coherence Tomography

OCT is often described as 'optical biopsy', where real-time high-resolution cross-sectional images of carefully targeted tissues can be generated *in vivo* without the need for excision. As reviewed by Fujimoto (2003) and Fercher et al. (2003), OCT has already made a substantial impact in ophthalmology and has demonstrated the potential to contribute towards a very broad range of other clinical applications, including arterial imaging, cancer diagnosis, gastroenterology, dermatology and dentistry. The first commercial OCT instrument for ophthalmic applications was developed by Carl Zeiss Meditec (formally Humphrey Systems) in 1996. Subsequent generations of their system have been introduced as OCT technology has become widespread in ophthalmology. OCT has already demonstrated the ability to detect retinal diseases such as glaucoma, macular oedema, central serous chorioretinopathy and age-related macular degeneration. OCT has the potential to detect the early stages of such diseases where morphological changes to the retina normally precede physical symptoms and permanent loss of vision. Retinal imaging using OCT is illustrated in Figure 10.3, which shows a 3D scan across the optic nerve head of a healthy volunteer (from Hitzenberger et al. 2003). The scan field size is 10° × 10°, and data were acquired to a depth of 1.05 mm. Since OCT imaging can be performed at the end of a narrow optical fibre, it is particularly powerful when applied to tissues which are normally inaccessible to conventional microscopes, such as via endoscopes (Li et al. 2000a), catheters (Das et al. 2001), laparoscopes (Boppart et al. 1999) and needles (Li et al. 2000b). However, intravascular imaging (such as for the identification of plaques) requires balloon occlusion or flushing with saline to displace the highly scattering blood which would otherwise obscure the vessel walls. Cancer diagnosis represents another major potential application of OCT, where differentiation is still largely dependent on observation of changes in architectural morphology. However, the most important architectural indicators of cancer are generally considered to occur at the sub-cellular level, requiring further improvements in spatial resolution. Meanwhile, considerable effort is also being devoted to improving the distinction between tissues based on spectral-dependent or birefringent properties.

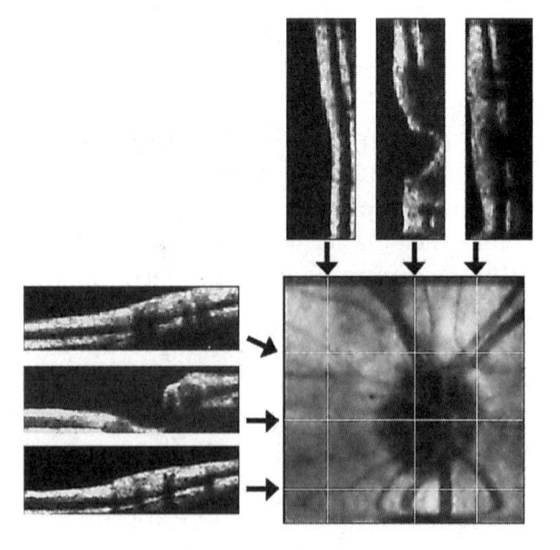

FIGURE 10.3
A 3D OCT scan of the optic nerve head of a healthy volunteer. Bottom right is a projection image. Above and to the left are cross-sectional scans along the white lines indicated in the projection image. (Reproduced with permission from Hitzenberger et al., 2003.)

10.4 Diffuse Optical Imaging

A parallel beam of light incident on tissue is rendered diffuse (i.e. becomes isotropically distributed) after penetrating a distance of about $1/\mu_s'$. Light which has penetrated much more than this distance (a few millimetres) has lost all initial coherence, polarisation and directionality. The formation of images using such highly scattered light is known as DOI. The simplest example of DOI is transillumination, which involves viewing or otherwise forming an image when the tissue is illuminated from behind. This method has mainly been applied to imaging the female breast, and the development of breast transillumination is described in Section 10.4.1. Since the mid-1990s two additional DOI approaches have emerged, known as optical topography and optical tomography (see Figure 10.4). Although the distinction between them is somewhat arbitrary, the term optical topography is generally reserved to describe techniques which provide maps of changes in optical properties (usually absorption) close to the surface, with little or no depth resolution, while optical tomography involves reconstructing an image representing the 3D volume of tissue or a transverse slice. The distinction evolved because of the differences in the complexity of the imaging problem and of the associated instrumentation. The two techniques are briefly described in separate sections in the following, and further details are available in reviews by Boas et al. (2001), Gibson et al. (2005) and Durduran et al. (2010).

10.4.1 Breast Transillumination

The first report of a systematic attempt to image patients using optical radiation was that of Cutler (1929) who transilluminated the female breast with a bright light source in a darkened room. His studies revealed that light propagating through tissue experiences considerable multiple scattering, which causes features below the surface to appear extremely blurred. Cutler discovered the dominant role played by blood concentration in the degree of opacity of a given tissue, and found that solid tumours often appeared to be opaque, although it was not possible to differentiate between benign and malignant tumours. Despite some optimism from Cutler regarding the potential of transillumination for certain classes of breast pathology, the method exhibited very poor sensitivity and specificity resulting from low inherent spatial resolution. Although a small number

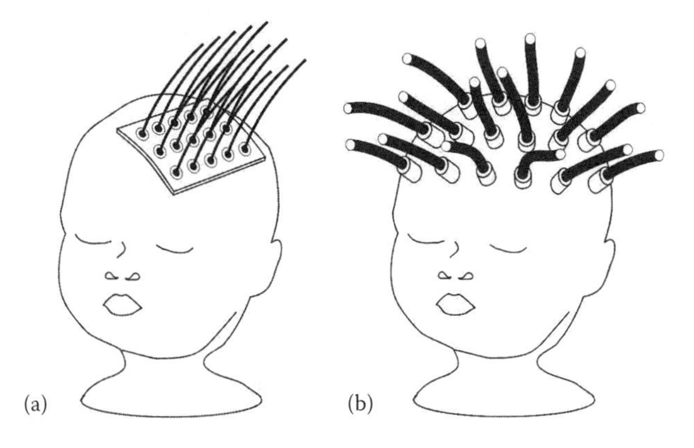

FIGURE 10.4
DOI using arrangements of sources and detectors for performing (a) optical topography and (b) optical tomography of the brain.

of further papers were published on the subject in the 1930s, the technique did not catch on. In fact there was only sporadic interest shown over the following 40 years, until Gros et al. (1972) reported the use of a water-cooled probe with a high-intensity white-light source. This device provided sufficient transmitted intensity to record the images on colour film. These researchers were also responsible for coining the term 'diaphanography' to describe transillumination of the breast, derived from the Greek words *dia* (through) and *phanes* (showing). Nevertheless, the new technology still exhibited poor sensitivity (Di Maggio et al. 1975). It then occurred to some researchers that, because of the very high absorption by blood, the vast majority of the optical radiation emitted by a white light source was achieving little except heat the tissue. Consequently Ohlsson et al. (1980) employed NIR sources and IR-sensitive film, which considerably improved their ability to identify carcinoma, and a 95% accuracy was claimed. This development, in conjunction with the employment of video cameras and recorders, led to a minor explosion of interest. The study which probably generated the greatest optimism during the early 1980s was that of Carlsen (1982), who reported results obtained using a commercial instrument designed and marketed by Spectrascan, Inc. This device recorded transmission images of the breast at two (red and NIR) wavelength regions simultaneously, and displayed their relative intensity on a monitor. Carlsen claimed a sensitivity better than 90%, and an overall performance comparable to X-ray mammography. Although some further initial evaluations of the Spectrascan device were reasonably positive (Bartrum and Crow 1984, Marshall et al. 1984), subsequent detailed comparisons between transillumination and X-ray mammography were generally unfavourable. Geslien et al. (1985) reported a sensitivity of only 58% (compared to 97% for X-ray mammography), and deduced that the performance was inadequate for screening. Even some tumours larger than 1 cm in diameter were not detectable. An equally thorough and no less negative assessment of breast transillumination was reported shortly afterwards by Monsees et al. (1987a, 1987b).

Subsequent development of transillumination (and other forms of DOI) focused on increasing the amount of information available from light transmitted across the tissue by increasing the number of wavelengths, and measuring either the times-of-flight of photons travelling through the tissue or the modulation and phase delay of light detected from sources modulated at MHz frequencies. Devices that perform such measurements are known as time-domain and frequency-domain systems respectively whereas those that measure intensity only are known as continuous wave (CW) systems. (Note that frequency-domain measurements performed over a sufficiently broad range of frequencies provides equivalent information to a time-of-flight measurement.) In the mid-1990s two major companies in Germany announced development of breast imaging systems based on frequency domain measurements. Prototypes constructed by Carl Zeiss (Kaschke et al. 1994, Franceschini et al. 1997) and by Siemens (Götz et al. 1998) both involved rectilinear scanning of a source and detector over opposite surfaces of a compressed breast, resulting in single-projection images at multiple NIR wavelengths. Scan times were minutes rather than seconds, and compression more moderate than that used for X-ray mammography. Unfortunately the performance of both systems during quite extensive trials fell below that required of a method of screening for breast cancer. Later, DOBI Medical International announced their ComfortScan® System which records CW measurements of red light transmitted across the compressed breast (Bartoňková et al. 2005). Transillumination of the compressed breast continues to be explored using frequency-domain (Culver et al. 2003, Pera et al. 2003, Zhang et al. 2005) and time-domain (Grosenick et al. 2005, Taroni et al. 2005) systems, although this approach has yet to emerge as a major clinical or research tool. Optical imaging of the uncompressed breast is described in Section 10.4.3.

10.4.2 Optical Topography

The technique known as optical topography evolved as means of providing NIR spectroscopy measurements at multiple sites on the scalp in order to localise the origin of haemodynamic signals in the brain, and the earliest attempts at optical topography were indeed performed using single-channel NIR spectrometers (Hoshi and Tamura 1993, Gratton et al. 1995). In general, optical topography involves acquiring measurements of diffusely reflected light at small (<3 cm) source-detector separations over an area of tissue simultaneously or in rapid succession (Figure 10.4a). Measured signals are relatively high and therefore may be acquired quickly, enabling images to be displayed in real-time at a rate of a few Hertz or faster. Optical topography was developed principally as a means of mapping haemodynamic activity in the cerebral cortex, such as an evoked response to sensory stimuli. Thus the technique is analogous to electroencephalography (EEG), which maps electrical activity of the brain. By making some assumptions about the background optical properties of the investigated medium, measurements of changes in attenuation of signals at two or more wavelengths can be converted into estimates of changes in [HbO_2] and [Hb].

Purpose-built optical topography instruments consist of an array of discrete sources and detectors coupled to the head via optical fibres. While each detector records signals continuously in parallel, sources are either illuminated sequentially, or simultaneously with the intensity of each modulated at a distinct frequency. Frequency encoding allows the detected signal from each source to be isolated using lock-in amplifiers or via a Fourier transform (Everdell et al. 2005). The fibres are usually attached to a flexible net worn by the subject, or to a pad which is held in direct contact with the scalp. The first commercial optical-topography systems were developed by Hitachi Medical Corporation (Japan). Their ETG-7000 system, for example, was designed for whole-head coverage and consists of 40 frequency-encoded laser-diode sources and 40 parallel APDs (Figure 10.5a) (Koizumi et al. 2003). Hitachi researchers have published various brain studies including changes in [HbO_2] and [Hb] in response to motor (Yamashita et al. 2001) and visual (Taga et al. 2003) stimulation, language processing (Watanabe et al. 1998a) and epilepsy (Watanabe et al. 1998b). Figure 10.5b shows activation maps obtained using the Hitachi ETG-100 system and a probe placed on the left and right sides of an adult head during a left-hand finger-tapping exercise (Sato et al. 2006). The maps exhibit significant increases in [HbO_2] on the contralateral side of the head. Shimadzu Corporation (Japan) also developed a commercial system known as the OMM-2000, which was used by Kusaka et al. (2004) to study the evoked response to visual stimulation in infants.

The majority of instruments developed for optical topography, including those marketed by Hitachi and Shimadzu, measure only the attenuation of source intensity. However, Arridge and Lionheart (1998) have shown that intensity measurements alone are insufficient to distinguish changes in absorption from those in scatter. To facilitate such distinction, some optical topography systems have been developed which are based on either time-domain (Quaresima et al. 2005) or frequency-domain (Danen et al. 1998) technology. A commercial frequency-domain optical-topography system developed by ISS Inc. (Champaign, IL) uses up to 64 sources modulated at MHz frequencies and employs up to 8 detectors which measure the modulation depth and phase delay of the detected signal (Choi et al. 2004). This system, called the Imagent, was designed specifically to monitor fast changes in optical properties in the brain associated with neuronal activity as well as the much slower changes due to haemodynamic activity. The neuronal signal exhibits a latency of around 50–150 ms (Franceschini and Boas 2004).

A broad variety of methods have been employed to generate images from data acquired using optical topography systems. Most are based on backprojection of the measured

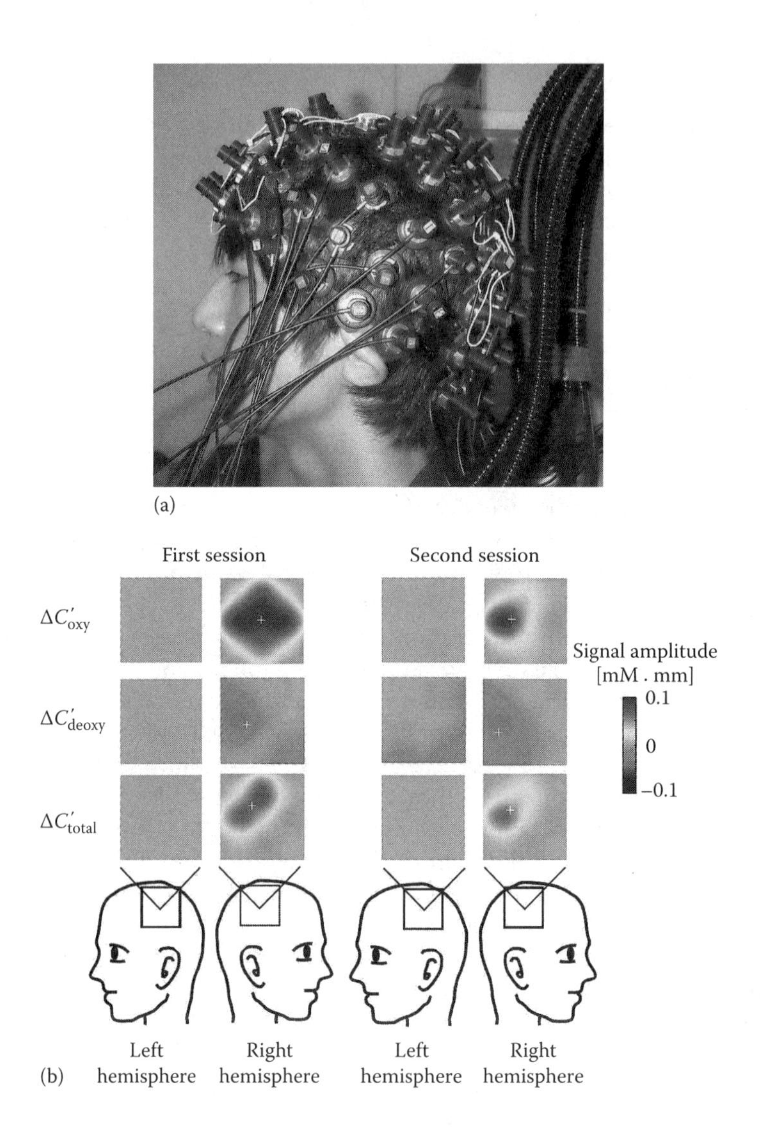

FIGURE 10.5
(See colour insert.) (a) The Hitachi ETG-7000 whole-head optical topography system. (Reproduced with permission from Koizumi et al., 2003.) (b) Topographic maps of activation on the left and right sides of the adult brain during a left-hand finger-tapping exercise, obtained using the Hitachi ETG-100 system. The maps show changes in concentrations of oxy-($\Delta C'_{oxy}$), deoxy-($\Delta C'_{deoxy}$), and total haemoglobin ($\Delta C'_{total}$). All changes occur on the contralateral side of the brain. (Reproduced with permission from Sato et al., 2006.)

changes in attenuation along a line between neighbouring source-detector pairs, and interpolating to produce a simple 2D map (e.g. Franceschini et al. 2000). More sophisticated algorithms have been employed which permit limited control of the depth at which the technique is most sensitive (e.g. Hintz et al. 2001), while it is also possible to produce 3D reconstructions of the optical properties directly beneath the probe using yet more complex reconstruction techniques (Bluestone et al. 2001) described in more detail in Sections 10.4.3 and 10.4.4. The penetration depth of optical topography is typically 1–2 cm, although it varies according to the distribution of optical fibres on the surface and the sensitivity of the detectors. A facility to display changes in [Hb] and [HbO_2] with temporal resolution of a few tens of milliseconds is commonly achieved, which offers a significant advantage over other functional imaging techniques such as fMRI and PET. However, the spatial resolution is typically no better than about 10 mm, and deteriorates with increasing depth below the surface. Further information on the principles and applications of optical topography can be found in papers by Franceschini and Boas (2004) and Gratton et al. (2005).

10.4.3 Optical Tomography

Optical tomography involves generating 3D or cross-sectional images using measurements of light transmitted across large thicknesses of tissue, such as across a breast or infant head (Figure 10.4b). By measuring light transmitted between sources and detectors spread over as much of the available surface of the tissue as possible, optical tomography attempts to avoid the overlapping of internal features exhibited in images obtained using the simple transillumination approach described in Section 10.4.1. However, measuring light which is strongly attenuated requires powerful sources (consistent with patient safety) and sensitive detectors, and inevitably optical tomography has a much lower temporal resolution than optical topography. Optical tomography requires a method to reconstruct the 3D distribution of optical properties or 2D cross-section, and such methods are discussed in Section 10.4.4.

Since the mid-1990s, commercial prototypes which use 3D optical tomography to image the breast have been developed by Philips Research Laboratories (the Netherlands) (Colak et al. 1999), Imaging Diagnostic Systems Inc. (Fort Lauderdale, FL) (Grable and Rohler 1997), ART Inc. (Winnipeg, Manitoba, Canada) (Intes 2005) and NIRx Medical Technologies (Glen Head, NY) (Schmitz et al. 2005). All four systems require the patient to lie on a bed with her breast suspended within a cavity surrounded by sources and detectors. The ART SoftScan® system has the potential advantages offered by time-domain technology, while the other three are based on CW measurements. The NIRx DYNOT device has the unique capacity to scan both breasts simultaneously. Further devices which perform 3D optical tomography of the breast based on CW (Intes et al. 2003), frequency- (Dehghani et al. 2003, Choe et al. 2005) and time-domain (Yates et al. 2005) technology have been developed. Overall, optical tomography images of the breast typically demonstrate a strong heterogeneity, and significant sensitivity to many types of abnormal breast lesions. However, spatial resolution remains low (1–2 cm) and a specificity sufficient for cancer screening remains elusive.

Optical tomography has also been applied to imaging of the brain, and of the newborn infant brain in particular. The first successful demonstration of optical tomography of an infant brain was performed by researchers at Stanford University, who employed an imaging system which measures photon flight times between points on the circumference of the infant head (Hintz et al. 1998, Benaron et al. 2000). Scans performed on infants at a variety of gestational ages successfully identified intracranial haemorrhage (Hintz et al. 1998, 1999), and focal regions of low oxygenation after acute stroke (Benaron et al. 2000). A major drawback was the prolonged scan times (up to several hours), which is a consequence of the low source power and the use of a single electronic detector. This and other technical limitations were largely overcome by a group at University College London using a 32-channel time-resolved system (Schmidt et al. 2000). Their system was used to reconstruct 3D images of the whole infant brain in scans lasting just a few minutes. Images revealed incidence of intraventricular haemorrhage (Hebden et al. 2002) and changes in blood volume and oxygenation induced by small alterations to ventilator settings (Hebden et al. 2004). A further study on a small cohort of pre-term babies led to the first 3D optical images of the entire neonatal head during motor evoked response (Gibson et al. 2006), as illustrated in Figure 10.6.

The attenuation of light across much larger heads prevents any optical system from imaging the centre of the adult brain, although partial tomographic reconstruction of

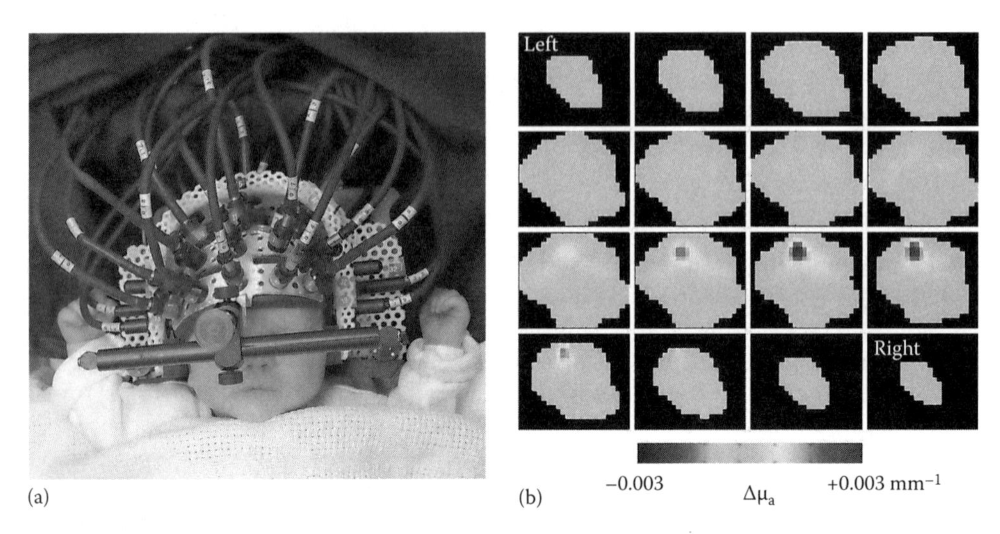

FIGURE 10.6
(See colour insert.) 3D optical imaging of evoked response in the newborn infant brain: (a) a helmet attaching an array of optical fibres to the infant head; (b) sagittal slices across a 3D image of the absorption change occurring on the right side of the brain produced by passive movement of the left arm.

the adult cortex is feasible. For example, the Stanford researchers successfully imaged localised contralateral oxygenation increases in the motor cortex of a healthy adult during hand movement (Benaron et al. 2000). A more sophisticated approach to 3D tomographic image reconstruction of the adult brain has been demonstrated using the NIRx DYNOT system by Bluestone et al. (2001) who imaged haemodynamic changes within the frontal regions of the adult brain during the Valsalva manoeuvre.

10.4.4 Image Reconstruction

The fundamental task of optical tomography is to estimate 3D distributions of optical properties within the investigated volume of tissue from a set of measurements made of light propagating between pairs of points on the surface. In X-ray CT (Chapter 3), where the majority of detected radiation has not been scattered, the Radon transform is applicable and reconstruction can be achieved using backprojection methods. Although various *ad hoc* backprojection methods have been proposed for optical tomography (Benaron et al. 1994, Walker et al. 1997), they are not applicable to arbitrary-shaped volumes. Consequently is it necessary to utilise a model of the propagation of light within the tissue which allows a set of predicted measurements on the surface to be derived for any given distribution of internal optical properties. Optical tomography then involves 'inverting' the model in order to recover the optical properties for a given set of measurements. This approach, which was first formulated for DOI by Singer et al. (1990) and Arridge et al. (1991), requires three components: the so-called forward-model; the definition of an objective function to be minimised, based on the error between model predictions and experimental data; and a scheme for adjusting the parameters of the forward-model to achieve the minimisation.

A broad variety of models of photon migration in tissue have been explored and are summarised in Gibson et al. (2005). The model most commonly used for DOI image reconstruction is the diffusion equation, which is derived from the radiative transport (or Boltzmann transport) equation (RTE). The RTE itself is in many ways an ideal model. Although it does not accommodate wave effects (such as polarisation), the RTE is applicable to a wide range of media. However, analytical solutions to the RTE are only available for very simple geometries, and numerical solutions are highly computationally intensive. The diffusion equation, which is much more amenable to both analytical and numerical solution, may be derived from the RTE by making the assumptions that scatter dominates over absorption and that scattering is isotropic (Arridge 1999). Fortunately the assumption that $\mu_s \gg \mu_a$ is reasonable for most biological tissues, while isotropic scattering simply implies that scatter is characterised by μ_s' alone rather than by μ_s and g.

Analytical solutions to the diffusion equation exist for some regular geometries (Arridge et al. 1992), but numerical techniques are required if complex geometries are to be modelled. The finite element method (FEM) has become the method of choice for most DOI applications, although the finite difference method (FDM), the finite volume method (FVM) and the boundary element method (BEM) have also been used. The FEM involves dividing the reconstructed volume into a fine mesh, and the objective of the reconstruction process is to derive the optical properties assigned to each element in the mesh.

Optical tomography is an inverse problem which is non-linear, severely ill-posed and generally underdetermined (Arridge 1999). Consequently finding reliable and unique solutions for a given set of data involves considerable theoretical and practical difficulties, and various compromises and approximations are unavoidable. A common approximation is to linearise the inverse problem, as will be illustrated later. Suppose the forward problem is represented by the following equation:

$$y = F(x) \tag{10.4}$$

where

y is the data,
x is the set of internal optical properties, and
F is an operator representing the model.

The inverse problem may therefore be expressed as

$$x = F^{-1}(y). \tag{10.5}$$

Now suppose Equation 10.4 is expanded about a fixed value x_0 in a Taylor series as follows:

$$y = y_0 + \frac{dF(x_0)}{dx} \cdot (x - x_0) + \frac{d^2F(x_0)}{dx^2} \cdot (x - x_0)^2 + \cdots \tag{10.6}$$

where $y_0 = F(x_0)$. If the true optical properties x are close to an initial estimate x_0 and the measured data y are close to the modelled data generated from that estimate, y_0, then the forward problem can be linearised by neglecting the higher terms. This may be expressed as

$$\Delta y = \mathbf{J}\Delta x \tag{10.7}$$

where

$\Delta y = y - y_0$,
$\Delta x = x - x_0$, and
$\mathbf{J}$ is the matrix form of the first derivative of $F(x_0)$, known as the Jacobian.

Thus image reconstruction involves inverting the matrix J. Regularisation is required to stabilise inversion against ill-conditioning caused by the ill-posedness of the inverse problem. While various regularisation techniques are available, the Tikhonov method is commonly employed as follows:

$$\Delta x = \mathbf{J}^{\mathrm{T}}(\mathbf{J}\mathbf{J}^{\mathrm{T}} + \lambda\mathbf{I})^{-1}\Delta y \tag{10.8}$$

where
I is the identity matrix, and
λ is a regularisation parameter.

The matrix **J** can be calculated from the forward-model based on a 'best guess' of the distributions of optical properties. However, selecting λ is often critical. Various objective methods for choosing λ have been proposed, although selection is often based on what gives the most 'acceptable' image. This linear approach has been implemented successfully for many DOI studies, particularly where measurements are obtained before and after a small localised change in properties (e.g. studies of evoked response). Nevertheless, the approach is clearly inappropriate if the measured data do not represent a difference between two similar states, or if the 'best guess' is insufficiently representative of the true optical properties. In this case, a non-linear approach must be used, which involves iteratively updating the estimate of **J** by replacing the 'best guess' with the distributions derived from the previous iteration. Iteration continues until the difference between the measured data and the simulated data generated by the forward-model reaches an acceptable minimum. Convergence towards the correct solution should be assisted by including prior information whenever possible. A comprehensive review of the theory of optical tomography image reconstruction is presented by Arridge (1999), and other summaries of this field have been provided by Boas et al (2001), Gibson et al. (2005) and Durduran et al. (2010).

10.5 Photoacoustic and Acousto-Optic Imaging

10.5.1 Photoacoustic Imaging

When a medium is illuminated by a beam of light, an acoustic wave can be generated by a process known as the photoacoustic effect. If the light is concentrated into a very short (nanosecond) pulse of light, regions within the medium which strongly absorb the light will undergo a sudden thermoelastic expansion, and emit broadband ultrasonic waves which can be detected by ultrasound detectors at the surface (Kruger 1994). Using the basic principles of pulse-echo ultrasound imaging, measurements of the time delay between the irradiation of the tissue and the detection of the acoustic waves can be used to generate 3D images of the regions where the ultrasound originated. At NIR wavelengths, where the overall absorption by tissue is relatively low, light can irradiate large volumes providing a penetration depth of several centimetres. Photoacoustic imaging provides a unique combination of the sub-millimetre spatial resolution associated with high-frequency diagnostic ultrasound with the contrast and sensitivity to physiological parameters which are characteristic of optical techniques.

Photoacoustic imaging systems can be categorised into two detection modes: forward-mode, where the light source and ultrasound detector are on opposite surfaces of the tissue; and backward mode, where the source and detector are on the same side. So far, forward-mode has been the most widely explored, although it has the obvious limitations that the tissue volume must be accessible from two opposite sides and the tissue must not be too attenuating.

A forward-mode breast imaging system described by Ermilov et al. (2006) uses a curved 128-element piezoelectric transducer array and laser pulses at 1064 nm. Images of a slice across the breast are obtained with a repetition rate of up to 1 Hz. Another system, known as a photoacoustic mammoscope, generates forward-mode photoacoustic images of the breast compressed between two plates built into a hospital bed (Manohar et al. 2005). This also employs laser pulses at 1064 nm and a 590-element array. By rotating the source and detectors around the tissue, photoacoustic measurements can be used to reconstruct cross-sectional images tomographically. For example, Wang et al. (2003) employed a circular scanning arrangement to generate cross-sectional images of the blood vessels within a living rat brain. Pulses at a wavelength of 532 nm were used, and a single ultrasound transducer was rotated around the head. Later, by acquiring photoacoustic signals at two optical wavelengths, Wang et al. (2006) generated photoacoustic images of the rat brain which display relative values of [Hb] and [HbO_2] and consequently of blood oxygen saturation. Using a similar tomographic approach, researchers with OptoSonics Inc. (Oriental, NC) acquired photoacoustic cross-sectional images of the head and upper thorax of a mouse (Kruger et al. 2003). Their system used a conventional linear transducer array and a 1064 nm laser source. OptoSonics have also pioneered a related technique, known as thermoacoustic imaging, which involves irradiating the tissue with microwaves instead of NIR light (Kruger et al. 2002). Microwaves can provide greater penetration than NIR light, although water rather than haemoglobin is the principal absorber.

The first human *in vivo* photoacoustic images, of blood vessels in the palm of the hand were presented by Zhang et al. (2006). They used a backward-mode scanning system based on a conventional ultrasound transducer which is mechanically scanned over the surface of the tissue. They also demonstrate vessel-by-vessel mapping of blood oxygenation in blood vessels by combining photoacoustic data at four optical wavelengths. Meanwhile a fast backward-mode photoacoustic imaging system based on optical scanning of a Fabry Perot sensor has been demonstrated for *in vivo* imaging by Zhang et al. (2009). Further information about photoacoustic imaging can be found in Wang (2009).

10.5.2 Acousto-Optic Imaging

Acousto-optic imaging (also known as ultrasound modulated optical imaging) involves directing one or more beams of ultrasound into tissue in order to modulate the physical properties of the tissue at a specific site. The idea is that a measurement of light diffusely transmitted throughout the volume of tissue during the ultrasound irradiation will exhibit a modulated component (known as a virtual source) which is sensitive to the perturbation of the local optical properties at the selected site. A common approach involves measuring the modulation of laser speckle occurring at the surface of the tissue. The stronger the optical absorption at the focus of the ultrasound beam, the weaker the modulation of the speckle. By measuring the speckle modulation as the focus of an ultrasound beam is scanned across the tissue, an image can be reconstructed which has a spatial resolution typical of ultrasound, but contrast dependent on optical properties. Like photoacoustic imaging, the method can be applied in either forward or backward modes. The technique has already been demonstrated on thick samples of biological tissues (Lévêque-Fort 2000, Li and Wang 2002, Atlan et al. 2005).

10.6 Other Advances in Optical Imaging

10.6.1 Exogenous Contrast Agents and Molecular Imaging

The distinction between healthy and diseased tissues using optical methods can be significantly enhanced by intravenous administration of contrast agents. Agents can be synthesised to be tumour specific and to therefore highlight diseased tissues directly. Alternatively, agents may enhance the natural contrast of blood, which improves visibility of lesions associated with angiogenesis. So far, the only contrast agent which has received widespread approval for clinical studies is indocyanine green dye (ICG), which has already been employed for DOI of the breast (Intes et al. 2003) and photoacoustic imaging (Ku and Wang 2005). When ICG is injected into the blood stream, it binds immediately and totally to blood proteins, primarily albumin. This binding ensures that ICG is confined almost entirely to the vascular compartment except for incidences of abnormal blood capillaries with high permeability, as in the case of tumour hypervascularity. After 10 min following injection, only a small fraction is detectable in the blood, and the ICG is removed almost entirely by the liver, and eliminated via the bile.

There has been a massive growth of interest in the development of new molecular markers which can highlight specific pathological tissues, and they are likely to have a significant impact on the future role of optical imaging. Among the types of agents being explored for optical imaging, fluorescent markers are of particular interest (Hawrysz and Sevick-Muraca 2000, Ntziachristos et al. 2003). Fluorescence occurs when a molecule absorbs light corresponding to a transitional energy level, and becomes activated into a singlet state from which it can decay, releasing light of lower energy (higher wavelength). The fluorescence lifetime, which can be measured using time- or frequency-domain instruments, provides an additional measure with which to distinguish tissues and generate images (Tadrous et al. 2003). Changes in fluorescent lifetimes can arise from the different molecular environments at which the agent becomes bound. A tumour-specific fluorescent contrast agent can provide a so-called beacon which enables relatively simple linear image reconstruction methods to be applied. However, if fluorescent agents are not exclusively localised within a tissue volume of interest, time or frequency measurements become essential in order to isolate the much weaker signals arising from deeper tissues of interest from the fluorescence emitted near to the surface. Sufficiently short fluorescent lifetimes (<1 ns) are also necessary for this separation to be achievable. Optical imaging using contrast agents is clearly a very promising area for future development, particularly in the field of drug evaluation in animals. However, it should be noted that probing deep tissues using fluorescence imaging suffers from the inevitable severe attenuation of light in both directions due to intrinsic absorption, as well as due to the low quantum efficiencies of fluorophores.

10.6.2 Multimodality Imaging

Optical imaging techniques offer tremendous advantages in terms of physiologically related contrast, high temporal resolution, patient safety and the use of low-cost portable instrumentation. However, optical imaging suffers a severe trade-off between spatial resolution and penetration depth. Consequently there has been a growing interest in combining the positive attributes of optical systems with conventional forms of diagnostic imaging, enabling the functional data derived optically to be correlated with high-resolution anatomical information (Pogue et al. 2010). Another very important potential benefit of such combinations is the facility to gain anatomical structure to be used as prior information in the image reconstruction process and thereby achieve better optical images (Schweiger and Arridge 1999, Carpenter et al. 2007). One of the first examples was simultaneous time-domain DOI and MRI of the compressed breast (Ntziachristos et al. 2000). Optical imaging is particularly suitable for combining with MRI since coupling the instrument to the patients can be achieved using long optical fibres, which have no effect on the magnetic field. A combination of optical imaging with ultrasound has also been achieved (Chen et al. 2001), and a significant benefit of structural prior information on the optical image reconstruction was demonstrated. Finally, Zhang et al. (2005) have performed sequential DOI and tomographic X-ray imaging of the compressed breast, which promises to deliver better diagnostic information than that achieved using the two techniques separately.

References

Arridge S R 1999 Optical tomography in medical imaging *Inverse Prob.* **15** R41–R93.

Arridge S R, Cope M and Delpy D T 1992 The theoretical basis for the determination of optical pathlengths in tissue: Temporal and frequency analysis *Phys. Med. Biol.* **37** 1531–1560.

Arridge S R and Lionheart W R B 1998 Non-uniqueness in diffusion-based optical tomography *Opt. Lett.* **23** 882–884.

Arridge S R, van der Zee P, Cope M and Delpy D T 1991 Reconstruction methods for infrared absorption imaging *Time-Resolved Spectroscopy and Imaging of Tissues* ed. B Chance *SPIE* **1431** 204–215.

Atlan M, Forget B C, Ramaz F, Boccara A C and Gross M 2005 Pulsed acousto-optic imaging in dynamic scattering media with heterodyne parallel speckle detection *Opt. Lett.* **30** 1360–1362.

Barton J K, Welch A J and Izatt J A 1998 Investigating pulsed dye laser-blood vessel interaction with color Doppler optical coherence tomography *Opt. Express* **3** 251–256.

Bartoňková H, Standara M and Schneiderová M 2005 The results of DOBI examinations in Masaryk Memorial Cancer Institute *Klinická Onkologie* **18** 149–151.

Bartrum R J and Crow H C 1984 Transillumination lightscanning to diagnose breast cancer: A feasibility study *Am. J. Radiol.* **142** 409–414.

Benaron D A, Hintz S R, Villringer A, Boas D, Kleinschmidt A, Frahm J, Hirth C et al. 2000 Noninvasive functional imaging of human brain using light *J. Cerebral Blood Flow Metab.* **20** 469–477.

Benaron D A, Ho D C, Spilman S, Van Houten J P and Stevenson D K 1994 Tomographic time-of-flight optical imaging device *Adv. Exp. Med. Biol.* **361** 207–214.

Bluestone A Y, Abdoulaev G, Schmitz C, Barbour R and Hielscher A 2001 Three-dimensional optical tomography of hemodynamics in the human head *Opt. Express* **9** 272–286.

Boas D A, Brooks D H, Miller E L, DiMarzio C A, Kilmer M, Gaudette R J and Quan Zhang 2001 Imaging the body with diffuse optical tomography *IEEE Signal Process. Mag.* **18** 57–75.

Boppart S A, Goodman A, Libus J, Pitris C, Jesser C A, Brezinski M E and Fujimoto J G 1999 High resolution imaging of endometriosis and ovarian carcinoma with optical coherence tomography: Feasibility for laparoscopic-based imaging *Br. J. Obstet. Gynaecol.* **106** 1071–1077.

Bright R 1831 Reports of medical cases selected with a view of illustrating the symptoms and cure of diseases by a reference to morbid anatomy *Diseases of the Brain and Nervous System* Vol. 2 (Longman: London, UK) p. 431.

Carlsen E N 1982 Transillumination lightscanning *Diagn. Imaging* **3** 28–33.

Carpenter C M, Pogue B W, Jiang S, Dehghani H, Wang X, Paulsen K D, Wells W A et al. 2007 Image-guided optical spectroscopy provides molecular-specific information in vivo: MRI-guided spectroscopy of breast cancer haemoglobin, water, and scatterer size *Opt. Lett.* **32** 933–935.

Chen N G, Guo P, Yan S, Piao D and Zhu Q 2001 Simultaneous near-infrared diffusive light and ultrasound imaging *Appl. Opt.* **40** 6367–6380.

Cheong W-F, Prahl S and Welch A J 1990 A review of the optical properties of biological tissues *IEEE J. Quant. Electron.* **26** 2166–2185.

Choe R, Corlu A, Lee K, Durduran T, Konecky S D, Grosicka-Koptyra M, Arridge S R et al. 2005 Diffuse optical tomography of breast cancer during neoadjuvant chemotherapy: A case study with comparison to MRI *Med. Phys.* **32** 1128–1139.

Choi J, Wolf M, Wolf U, Polzonetti C, Safonova L P, Gupta R, Michalos A et al. 2004 Noninvasive determination of the optical properties of adult brain: Near-infrared spectroscopy approach *J. Biomed. Opt.* **9** 221–229.

Colak S B, van der Mark M B, Hooft G W, Hoogenraad J H, van der Linden E S and Kuijpers F A 1999 Clinical optical tomography and NIR spectroscopy for breast cancer detection *IEEE J. Sel. Top. Quant. Electron.* **5** 1143–1158.

Culver J P, Choe R, Holboke M J, Zubkov L, Durduran T, Slemp A, Ntziachristos V, Chance B and Yodh A G 2003 Three-dimensional diffuse optical tomography in the parallel plane transmission geometry: Evaluation of a hybrid frequency domain/continuous wave clinical system for breast imaging *Med. Phys.* **30** 235–247.

Curling T B 1843 *A Practical Treatise on the Diseases of the Testis and of the Spermatic Cord and Scrotum* (Samual Highley: London, UK) pp. 125–181.

Cutler M 1929 Transillumination as an aid in the diagnosis of breast lesions *Surg. Gynecol. Obstet.* **48** 721–728.

Danen R M, Wang Y, Liz X D, Thayer W S and Yodh A G 1998 Regional imager for low resolution functional imaging of the brain with diffusing near-infrared light *Photochem. Photobiol.* **67** 33–40.

Das A, Sivak M V Jr., Chak A, Wong R C K, Westphal V, Rollins A M, Willis J, Isenberg G and Izatt J A 2001 High-resolution endoscopic imaging of the GI tract: A comparative study of optical coherence tomography versus high-frequency catheter probe EUS *Gastrointest. Endosc.* **54** 219–224.

Dehghani H, Pogue B W, Poplack S P and Paulsen K D 2003 Multiwavelength three-dimensional near-infrared tomography of the breast: Initial simulation, phantom and clinical results *Appl. Opt.* **42** 135–145.

Di Maggio C, Di Bello, Pescarini L and Dus R 1975 La valeur de la diaphanoscopie dans le diagnostic des lesions due sein *J. Radiol.* **56** 627–628.

Durduran T, Choe R, Baker W B and Yodh A G 2010 Diffuse optics for tissue monitoring and tomography *Rep. Prog. Phys.* **73** 076701.

Ermilov S A, Conjusteau A, Mehta K, Lacewell R, Henrichs P M and Oraevsky A A 2006 128-channel laser optoacoustic imaging system (LOIS-128) for breast cancer diagnostics *Photons Plus Ultrasound: Imaging and Sensing 2006: The Seventh Conference on Biomedical Thermoacoustics, Optoacoustics, and Acousto-optics* eds. A A Oraevsky and L V Wang *SPIE* **6086** 608609.

Everdell N L, Gibson A P, Tullis I D C, Vaithianathan T, Hebden J C and Delpy D T 2005 A frequency multiplexed near infrared topography system for imaging functional activation in the brain *Rev. Sci. Instrum.* **76** 093705.

Fercher A F, Drexler W, Hitzenberger C K and Lasser T 2003 Optical coherence tomography – Principles and applications *Rep. Prog. Phys.* **66** 239–303.

Franceschini M A and Boas D A 2004 Noninvasive measurement of neuronal activity with NIR optical imaging *Neuroimage* **21** 372–386.

Franceschini M A, Moisia K T, Fantini S, Gaida G, Gratton E, Jess H, Manlulin W W, Seeber M, Schlal P M and Kaskuke M 1997 Frequency-domain techniques enhance optical mammography: Initial clinical results *Proc. Natl Acad. Sci. USA* **94** 6468–6473.

Franceschini M A, Toronov V, M. Filiaci M, Enrico Gratton E and Sergio Fantini S 2000 On-line optical imaging of the human brain with 160-ms temporal resolution *Opt. Express* **6** 49–57.

Fujimoto J G 2003 Optical coherence tomography for ultrahigh resolution in-vivo imaging *Nat. Biotechnol.* **21** 1361–1367.

Geslien G E, Fisher J R and DeLaney C 1985 Transillumination in breast cancer detection: Screening failures and potential *Am. J. Radiol.* **144** 619–622.

Gibson A P, Austin T, Everdell N L, Schweiger M, Arridge S R, Meek J H, Wyatt J S, Delpy D T and Hebden J C 2006 Three-dimensional whole-head optical tomography of passive motor evoked responses in the neonate *Neuroimage* **30** 521–528.

Gibson A P, Hebden J C and Arridge S R 2005 Recent advances in diffuse optical imaging *Phys. Med. Biol.* **50** R1–R43.

Götz L, Heywang-Köbrunner S H, Schütz O and Siebold H 1998 Optische mammography an präoperativen patientinnen *Akt. Radiol.* **8** 31–33.

Grable R J and Rohler D P 1997 Optical tomography breast imaging *Optical Tomography and Spectroscopy of Tissue: Theory, Instrumentation, Model, and Human Studies II* eds. B Chance and R R Alfano *SPIE* **2979** 197.

Gratton G, Corballis P M, Cho E, Fabiani M and Hood D C 1995 Shades of gray matter: Noninvasive optical images of human brain responses during visual stimulation *Psychophysiology* **32** 505–509.

Gratton E, Toronov V, Wolf U, Wolf M and Webb A 2005 Measurement of brain activity by near-infrared light *J. Biomed. Opt.* **10** 011008.

Gros C M, Quenneville Y and Hummel Y 1972 Diaphanologie mammaire *J. Radiol. Electrol. Med. Nucl.* **53**, 297–306.

Grosenick D, Moesta K T, Möller M, Mucke J, Wabnitz H, Gebauer B, Stroszczynski C, Wassermann B, Schlag P M and Rinneberg 2005 Time-domain scanning optical mammography: I, Recording and assessment of mammograms of 154 patients *Phys. Med. Biol.* **50** 2429–2449.

Hawrysz D J and Sevick-Muraca E M 2000 Developments toward diagnostic breast cancer imaging using near-infrared optical measurements and fluorescent contrast agents *Neoplasia* **2** 388–417.

Hebden J C, Gibson A, Austin T, Yusof R M, Everdell N, Delpy D T, Arridge S R, Meek J H and Wyatt J S 2004 Imaging changes in blood volume and oxygenation in the newborn infant brain using three-dimensional optical tomography *Phys. Med. Biol.* **49** 1117–1130.

Hebden J C, Gibson A, Yusof R M, Everdell N, Hillman E M C, Delpy D T, Arridge S R, Austin T, Meek J H and Wyatt J S 2002 Three-dimensional optical tomography of the premature infant brain *Phys. Med. Biol.* **47** 4155–4166.

Hintz S R, Benaron D A, van Houten J P, Duckworth J L, Liu F W H, Spilman S D, Stevenson D K and Cheong W-F 1998 Stationary headband for clinical time-of-flight optical imaging at the bedside *Photochem. Photobiol.* **68** 361–369.

Hintz S R, Benaron D A, Siegel A M, Zourabian A, Stevenson D K and Boas D A 2001 Bedside functional imaging of the premature infant brain during passive motor activation *J. Perinat. Med.* **29** 335–343.

Hintz S R, Cheong W-F, van Houten J P, Stevenson D K and Benaron D A 1999 Bedside imaging of intracranial hemorrhage in the neonate using light: Comparison with ultrasound, computed tomography, and magnetic resonance imaging *Pediatr. Res.* **45** 54–59.

Hitzenberger C K, Trost P, Lo P-W and Zhou Q 2003 Three-dimensional imaging of the human retina by high-speed optical coherence tomography *Opt. Express* **11** 2753–2761.

Hoshi Y and Tamura M 1993 Dynamic multichannel near-infrared optical imaging of human brain activity *J. Appl. Physiol.* **75** 1842–1846.

Intes X 2005 Time-domain optical mammography SoftScan: Initial results *Acad. Radiol.* **12** 934–947.

Intes X, Ripoll J, Chen Y, Nioka S, Yodh A G and Chance B 2003 In-vivo continuous-wave optical breast imaging enhanced with indocyanine green *Med. Phys.* **30** 1039–1047.

Jöbsis F F 1977 Noninvasive infrared monitoring of cerebral and myocardial oxygen sufficiency and circulatory parameters *Science* **198** 1264–1267.

Kaschke M, Jess H, Gaida G, Kaltenbach J M and Wrobel W 1994 Transillumination imaging of tissues by phase modulation techniques *Proc. OSA Advances in Optical Imaging and Photon Migration* **21** ed. R R Alfano (Optical Society of America: Washington, DC) pp. 88–92.

Kohl M, Essenpreis M and Cope M 1995 The influence of glucose concentration upon the transport of light in tissue-simulating phantoms *Phys. Med. Biol.* **40** 1267–1287.

Koizumi H, Yamamoto T, Maki A, Yamashita Y, Sato H, Kawaguchi H and Ichikawa N 2003 Optical topography: Practical problems and new applications *Appl. Opt.* **42** 3054–3062.
Kruger R A 1994 Photoacoustic ultrasound *Med. Phys.* **21** 127–131.
Kruger R A, Kiser W L Jr, Reinecke D R and Kruger G A 2003 Thermoacoustic computed tomography using a conventional linear transducer array *Med. Phys.* **30** 856–860.
Kruger R A, Stantz K and Kiser W L 2002 Thermoacoustic CT of the breast *Medical Imaging 2002: Physics of Medical Imaging* eds. L E Antonuk and M J Yaffe *SPIE* **4682** 521–525.
Ku G and Wang L V 2005 Deeply penetrating photoacoustic tomography in biological tissues enhanced with an optical contrast agent *Opt. Lett.* **30** 507–509.
Kusaka T, Kawada K, Okubo K, Nagano K, Namba M, Okada H, Imai T, Isobe K and Itoh S 2004 Noninvasive optical imaging in the visual cortex in young infants *Human Brain Mapping* **22** 122–132.
Lévêque-Fort S 2000 Three-dimensional acousto-optic imaging in biological tissues with parallel signal processing *Appl. Opt.* **40** 1029–1036.
Li X D, Boppart S A, Van Dam J, Mashimo H, Mutinga M, Drexler W, Klein M, Pitris C, Krinsky M L, Brezinski M E and Fujimoto J G 2000a Optical coherence tomography: Advanced technology for the endoscopic imaging of Barrett's esophagus *Endoscopy* **32** 921–930.
Li X D, Chudoba C, Ko T, Pitris C and Fujimoto J G 2000b Imaging needle for optical coherence tomography *Opt. Lett.* **25** 1520–1522.
Li J and Wang L V 2002 Methods for parallel-detection-based ultrasound-modulated optical tomography *Appl. Opt.* **41** 2079–2084.
Manohar S, Kharine A, van Hespen J C G, Steenbergen W and van Leeuwen T G 2005 The Twente photoacoustic mammoscope: System overview and performance *Phys. Med. Biol.* **50** 2543–2557.
Marshall V, Williams D C and Smith K D 1984 Diaphanography as a means of detecting breast cancer *Radiology* **150** 339–343.
Monsees B, Destouet J M and Gersell D 1987a Light scan evaluation of nonpalpable breast lesions *Radiology* **163** 467–470.
Monsees B, Destouet J M and Totty W G 1987b Light scanning versus mammography in breast cancer detection *Radiology* **163** 463–465.
Ntziachristos V, Bremer C and Weissleder R 2003 Fluorescence imaging with near-infrared light: New technological advances that enable in-vivo molecular imaging *Eur. Radiol.* **13** 195–208.
Ntziachristos V, Yodh A G, Schnall M and Chance B 2000 Concurrent MRI and diffuse optical tomography of the breast after indocyanine green enhancement *Proc. Natl Acad. Sci. USA* **97** 2767–2772.
Ohlsson B, Gundersen J and Nilsson D M 1980 Diaphanography: A method for evaluation of the female breast *World J. Surg.* **4** 701–707.
Pera V E, Heffer E L, Siebold H, Schutz O, Heywang-Kobrunner S, Gotz L, Heinig A and Fantini S 2003 Spatial second-derivative image processing: An application to optical mammography to enhance the detection of breast tumours *J. Biomed. Opt.* **8** 517–524.
Pogue B W, Leblond F, Krishnaswamy V and Paulsen K D 2010 Radiologic and near-infrared optical spectroscopic imaging: Where is the synergy? *Am. J. Roent.* **195** 312–332.
Quaresima V, Ferrari M, Torricelli A, Spinelli L, Pifferi A and Cubeddu R 2005 Bilateral prefrontal cortex oxygenation responses to a verbal fluency task: A multichannel time-resolved near-infrared topography study *J. Biomed. Opt.* **10** 011012
Sato H, Kiguchi M, Maki A, Fuchino Y, Obata A, Yoro T and Koizumi H 2006 Within-subject reproducibility of near-infrared spectroscopy signals in sensorimotor activation after 6 months *J. Biomed. Opt.* **11** 014021
Schmidt F E W, Fry M E, Hillman E M C, Hebden J C and Delpy D T 2000 A 32-channel time-resolved instrument for medical optical tomography *Rev. Sci. Instrum.* **71** 256–265.
Schmitz C H, Klemer D P, Hardin R, Katz M S, Pei Y, Graber H L, Levin M B, Levina R D, Franco N A, Solomon W B and Barbour R L 2005 Design and implementation of dynamic near-infrared optical tomographic imaging instrumentation for simultaneous dual-breast measurements *Appl. Opt.* **44** 2140–2153.

Schweiger M and Arridge S R 1999 Optical tomographic reconstruction in a complex head model using a priori region boundary information *Phys. Med. Biol.* **44** 2703–2721.
Singer J R, Grünbaum F A, Kohn P and Zubelli J P 1990 Image reconstruction of the interior of bodies that diffuse radiation *Science* **248** 990–993.
Tadrous P J, Siegel J, French P M W, Shousha S, Lalani E-N and Stamp G W H 2003 Fluorescence lifetime imaging of unstained tissues: early results in human breast cancer *J. Pathol.* **199** 309–317.
Taga G, Asakawa K, Maki A, Konishi Y and Koizumi H 2003 Brain imaging in awake infants by near-infrared optical topography *Proc. Natl Acad. Sci. USA* **100** 10722–10727.
Taroni P, Torricelli A, Spinelli L, Pifferi A, Arpaia F, Danesini G and Cubeddu R 2005 Time-resolved optical mammography between 637 and 985 nm: Clinical study on the detection and identification of breast lesions *Phys. Med. Biol.* **50** 2469–2488.
Walker S A, Fantini S and Gratton E 1997 Image reconstruction by backprojection from frequency-domain optical measurements in highly scattering media *Appl. Opt.* **36** 170–179.
Wang L V (ed.) 2009 *Photoacoustic Imaging and Spectroscopy* (CRC Press, Taylor & Francis Group: Boca Raton, FL).
Wang X, Pang Y, Ku G, Xie X, Stoica G and Wang L V 2003 Noninvasive laser-induced photoacoustic tomography for structural and functional in-vivo imaging of the brain *Nat. Biotechnol.* **21** 803–806.
Wang X, Xie X, Ku G, Wang L V and Stoica G 2006 Noninvasive imaging of haemoglobin concentration and oxygenation in the rat brain using high-resolution photoacoustic tomography *J. Biomed. Opt* **11** 024015.
Watanabe E, Maki A, Kawaguchi F, Takashiro K, Yamashita Y, Koizumi H and Mayanagi Y 1998a Non-invasive assessment of language dominance with near-infrared spectroscopic mapping *Neurosci. Lett.* **256** 49–52.
Watanabe E, Maki A, Kawaguchi F, Yamashita Y, Koizumi H and Mayanagi Y 1998b Noninvasive cerebral blood volume measurement during seizures using multichannel near infrared spectroscopic topography *J. Epilepsy* **11** 335–340.
Woodward H Q and White D R 1986 The composition of body tissues *Br. J. Radiol.* **59** 1209–1219.
Yamashita Y, Maki A and Koizumi H 2001 Wavelength dependence of the precision of noninvasive optical measurement of oxy-, deoxy-, and total-hemoglobin concentration *Med. Phys.* **28** 1108–1114.
Yates T, Hebden J C, Gibson A, Everdell N, Arridge S R and Douek M 2005 Optical tomography of the breast using a multi-channel time-resolved imager *Phys. Med. Biol.* **50** 2503–2517.
Zhang Q, Brukilacchio T J, Li A, Stott J J, Chaves T, Hillman E, Wu T et al. 2005 Coregistered tomographic x-ray and optical breast imaging: Initial results *J. Biomed. Opt.* **10** 024033.
Zhang E Z, Laufer J G, Pedley R B and Beard P C 2009 In vivo high-resolution 3D photoacoustic imaging of superficial vascular anatomy *Phys. Med. Biol.* **54** 1035–1046.
Zhang H F, Maslov K, Stoica G and Wang L V 2006 Functional photoacoustic microscopy for high-resolution and non-invasive in-vivo imaging *Nat. Biotechnol.* **24** 848–851.
Zysk A M, Oldenburg A L and Marks D L 2007 Optical coherence tomography : A review of clinical development from bench to bedside *J. Biomed. Opt.* **12** 051403

11

Mathematics of Image Formation and Image Processing

S. Webb

CONTENTS

11.1 Concept of Object and Image ... 687
11.2 Relationship between Object and Image ... 689
11.3 General Image Processing Problem ... 692
11.4 Discrete Fourier Representation and the Models for Imaging Systems ... 694
11.5 General Theory of Image Restoration ... 697
  11.5.1 Least-Squares Image Processing ... 699
  11.5.2 Constrained Deconvolution ... 700
  11.5.3 Maximum-Entropy Deconvolution ... 701
11.6 Image Sampling ... 702
11.7 Iterative Image Processing ... 705
  11.7.1 Spatial Frequency Analysis of the Iterative Method ... 706
11.A Appendix ... 708
References ... 711

11.1 Concept of Object and Image

Why is it interesting or necessary to study the mathematics of image formation, and can it really be possible that the many imaging modalities that we have met so far can be represented by the same set of equations? In this chapter, we shall see that, in general, it is indeed possible to write down a small set of equations that are sufficiently general to be applicable to many imaging situations. This property endows the equations of image formation with a certain majesty, and they become powerful tools for coming to grips with the underlying principles of imaging. As such, they also provide the basis for understanding why and how images are imperfect, and they suggest techniques for image processing. In this chapter, the words 'image processing' are used to imply the *processing* of images for the removal of degradations that have arisen during their formation. (Elsewhere in this book, these same words have often been used more generally to imply the *creation* of images from some form of raw or recorded data. For example, the theory of reconstruction of tomographic images from projection data has been referred to as image processing in earlier chapters.)

Here we describe as the *object* that property which is distributed in multidimensional space and which we require to measure. We describe as the *image* that measurement estimate of the object (also, in general, multidimensional) which has been made and which is regarded as best representing the object distribution. It has become somewhat conventional to represent the object mathematically by the symbol f and the image by the symbol g. We can write straightaway that a perfect imaging system is one in which, at all locations in the space, $f = g$. Almost without exception, no medical imaging system achieves this and, in general, it is necessary to conduct experiments to parameterise the relationship between f and g.

In general, both f and g are 3D when they describe the distribution of some time-stationary property *in vivo*. However, we have already seen that many medical imaging modalities are tomographic and that the complete 3D distribution of a property is substituted by a contiguous sequence of 2D descriptors. This achievement of tomography has not only revolutionized our understanding of the true 3D distribution of some property, removing artefactual interference or cross talk, but has also led to a most useful simplification of the mathematical description of imaging. Hence, in this chapter, we shall restrict the mathematics to two spatial dimensions, with the understanding that this will be applicable both to distributions that are genuinely 2D and to tomographic images. We shall make use of the terms 'image plane' and 'object plane' freely, implying total containment of information within these planes.

To perceive the beautiful generality of the mathematics that follows, some examples are given here to illustrate the conceptual statement in the previous paragraph. Stepping aside from medicine for one moment, we might imagine the simplest optical imaging method of taking a photograph (image plane) of a scene painted on a 2D canvas (object plane). Introducing a coordinate scheme, the painting is $f(\alpha, \beta)$ and the photograph is $g(x, y)$. In this example, both f and g might be thought of as having the same dimensions, since they are both the spatial distribution of colour or light intensity in a plane. The photograph was taken by quanta of light reflected from the painting through the lens system of the camera and onto the photographic film. Such image formation is, of course, degraded by distortions and the detector response, and hence f is not equal to g. If we had a description of the degradation, image processing might be performed to compensate for the degradation in image formation, as we shall see from what will follow.

As a second example, let us recall that, from a series of 2D images recorded by a rotating gamma camera at sequential orientations around a patient, it is possible to generate a series of 2D images g of the distribution f of the radiopharmaceutical *in vivo* by the process of single-photon emission computed tomography (SPECT) (Chapter 5). In this example, the diagnostician is led to believe that the images being viewed are actually at the location of interest in the patient. There is a suspension of disbelief, of course, because what is being observed (the image g) is a 2D distribution of reconstructed pixel values, whereas the clinician is thinking of this as a 2D distribution of the uptake of activity within the patient. The image g is thus a *representation* of the object f, which has different dimensions but is in more or less the same spatial location. The aim of SPECT is, of course, that g be related to f in a linear quantitative way. Interestingly, the gamma camera, which was the means of recording the data, does not feature at all in the mathematical description relating image to object! Indeed, it is quite incidental that the data used to reconstruct SPECT tomograms are themselves images. We recall, for instance, that the recorded data for a multiwire proportional chamber (MWPC) positron camera (see Chapter 5) were no more than a string of coordinate values, images only appearing after backprojection.

It is thus apparent that we should remove ourselves from the restriction of thinking that images of an object are created at a different place from the source distribution and with the same dimensions as the object itself. Indeed, it is more appropriate to think of the image and object as being physically coincident, sometimes with differing dimensions, but related to each other by the processes characterising the imaging modality.

11.2 Relationship between Object and Image

Against this background, we proceed to consider 2D distributions, knowing these to incorporate tomographic images. We shall represent the object distribution by $f(\alpha, \beta)$, entirely contained within the object plane, and the image distribution by $g(x, y)$, entirely contained within the image plane. In general, there is no perfect 1:1 correspondence between the information at a particular location (α', β') and the corresponding location (x', y'). 'Information' can, in principle, disperse *to all* image locations *from each* object location. For any useful imaging modality, however, the principal contribution to each (α', β') will come from a particular single location (x', y'). Other neighbouring points will contribute smaller amounts of information, the contribution decreasing rapidly away from the principal contributing location (x', y'). This is known as the neighbourhood principle, and simply recognises that the image of some point in object space may be dependent on the object point and on points in an infinite neighbourhood surrounding the object point. Later we shall see that this concept can be described quantitatively by use of the point source response function (PSRF), which will be narrow for a 'good' imaging system in which the aforementioned neighbourhood is confined to a small (certainly not infinite!) region. By way of illustration, imagine a gamma camera fitted with a parallel-hole collimator viewing a point source of activity in a scattering medium. One particular image pixel (that corresponding to the direct line-of-sight view) will record maximum counts, but adjacent pixels will register some recorded events, with the number of events decreasing rapidly with distance from the maximum-valued pixel. This is known as the point process of image formation. Conversely, imagine reversing all the photon paths from a single camera pixel back towards a plane parallel to the face of the camera. Most paths will intersect the direct line-of-sight object pixel but paths will also branch out to a myriad of other pixels because of imperfect collimation and photon scatter.

What is this physical link between object and image space? Conceptually, it is the transport of information. The image plane is built up from a knowledge of what is in the object plane brought by whatever are the appropriate 'coding messengers' of the imaging modality. Thus a photograph is created by optical photon transport. A distribution of acoustic scattering centres is mapped to a B-scan brightness image by scattered longitudinal ultrasound waves. A gamma camera records a measure of the uptake of a radiopharmaceutical by counting the emitted γ-ray photons. A transmission x-radiograph is a brightness image related to the linear attenuation of X-ray photons that traverse the object space to form the image. A thermogram is a map of pixel intensity corresponding to an *in vivo* temperature distribution, the messengers carrying the information being infrared photons. Because the smallest unit of radiant energy transport is non-negative (implying that there is either something to measure or there is not), then the image distribution must also be non-negative (formal ultrasonographers may disagree with this statement). Mathematically we can write

$$f(\alpha,\beta) \geq 0 \quad \text{and} \quad g(x,y) \geq 0. \tag{11.1}$$

Measuring the total information in an object and image distribution is also physically meaningful, and so it is reasonable to assume that f and g are integrable functions, since integration corresponds to measuring totals.

Now postulate a function $h(x, y, \alpha, \beta)$ that describes the spatial dependence of the point process. For a point process in which the object is only non-zero at (α', β'), the recorded image is

$$g'(x,y) = h(x,y,\alpha',\beta',f'(\alpha',\beta')). \tag{11.2}$$

Here the express dependence on the magnitude of the point-object signal has been made clear by including f as a fifth argument of the function h. Now imagine a second signal at the same location giving rise to recorded image

$$g''(x,y) = h(x,y,\alpha',\beta',f''(\alpha',\beta')). \tag{11.3}$$

Since the superposition principle states that radiant energies are additive

$$g'(x,y)+g''(x,y) = h(x,y,\alpha',\beta',f'(\alpha',\beta'))+h(x,y,\alpha',\beta',f''(\alpha',\beta')). \tag{11.4}$$

Equation 11.4 is non-linear superposition (since Equation 11.2 is non-linear), that is, additive components in the object plane do not lead to additive measurements in the image plane. If the imaging system is linear, then Equation 11.2 becomes

$$g'(x,y) = h(x,y,\alpha',\beta')\,f'(\alpha',\beta') \tag{11.5}$$

and Equation 11.4 becomes

$$g'(x,y)+g''(x,y) = h(x,y,\alpha',\beta')[f'(\alpha',\beta')+f''(\alpha',\beta')]. \tag{11.6}$$

From this, we see that additive components in the object plane lead to additive measurements in the image plane via a single use of the transformation function h. Mathematically, this is such an important simplification that, as we shall see, linearity is often assumed to first order even when it is not strictly true.

It is now possible to invoke the concept of a distribution as the superposition of a finite set of points to extend the aforementioned ideas to generate the general equations that link the spaces. For a non-linear imaging system,

$$g(x,y) = \iint h(x,y,\alpha,\beta,f(\alpha,\beta))\,\mathrm{d}\alpha\,\mathrm{d}\beta \tag{11.7}$$

and for a linear system

$$g(x,y) = \iint h(x,y,\alpha,\beta)\,f(\alpha,\beta)\,\mathrm{d}\alpha\,\mathrm{d}\beta. \tag{11.8}$$

The function h, which we have been using to relate f to g, is called the PSRF. In Equations 11.7 and 11.8 the PSRF h is a function of all four spatial coordinates and is consequently referred to as a space-variant PSRF (or SVPSRF). These are the most general descriptions of imaging that it is possible to write down.

If, however, the imaging system is such that the point process is the same for all locations of the point in the object plane, then the system is said to be spatially invariant and the PSRF is called a space-invariant PSRF (or SIPSRF). In these special circumstances, the function h depends only on the *difference* coordinates $(x - \alpha, y - \beta)$, because the PSRF depends only on the relative distance between points in the object plane and points in the image plane. Another description is to refer to this situation as isoplanatic imaging. This concept is quite separate from any consideration of linearity, and so we may write down that for a space-invariant imaging system that is non-linear

$$g(x,y) = \iint h(x-\alpha, y-\beta, f(\alpha,\beta))\,\mathrm{d}\alpha\,\mathrm{d}\beta \tag{11.9}$$

whereas for a linear space-invariant imaging system

$$g(x,y) = \iint h(x-\alpha, y-\beta)\, f(\alpha,\beta)\,\mathrm{d}\alpha\,\mathrm{d}\beta. \tag{11.10}$$

This is recognised as the familiar *convolution integral*. An image is the convolution of the object distribution with the PSRF. It is this function h that carries the information from object to image space and embodies all the geometric 'infidelities' of the imaging modality. It is therefore not surprising that attempting to remove these degradations is often referred to as *deconvolution*. Strictly, this only applies to image processing when the modality is space-invariant, although the word 'deconvolution' has slipped into common use as a general substitute for the words 'image processing'. This casual use will not be made here. We shall return at length to the subject of deconvolution later in this chapter.

A final simplification of the general image equations arises if the behaviour in orthogonal directions is unconnected. By this it is meant that the 2D PSRF can be constructed by multiplying together two 1D PSRF. For a space-variant system

$$h(x,y,\alpha,\beta) = h'(x,\alpha)h''(y,\beta) \tag{11.11}$$

and for a space-invariant modality

$$h(x,y,\alpha,\beta) = h'(x-\alpha)h''(y-\beta). \tag{11.12}$$

This property is known as separability. Finally, then, we give the imaging equation for a linear, space-invariant, separable modality as

$$g(x,y) = \int h'(x-\alpha) f(\alpha,\beta)\,\mathrm{d}\alpha \int h''(y-\beta) f(\alpha,\beta)\,\mathrm{d}\beta \tag{11.13}$$

which is the product of two orthogonally separate convolution integrals.

11.3 General Image Processing Problem

Against this background, it is easy to appreciate that by a suspension of disbelief, when we observe an image, we imagine we are looking at an object distribution. That, after all, was the rationale for creating the image. We are not! We are inspecting an image that is a *fair representation* of the object distribution. It has already been seen that this may not even have the same dimensions (i.e. be the same physical quantity) as the object in which we are interested. Built into the measured image g are all the characteristics and imaging imperfections of the imaging modality in question. To what we might call first order, this does not matter. If the instrument has been constructed to a high specification, we would expect that 'fair' means 'good' and that a more or less 1:1 spatial correspondence exists between the spaces, and that the measured quantity is more or less linear with the object distribution that it represents. However, when we become a little more critical, we lose this suspension of disbelief, and notice the imperfections introduced by the imaging process. In some cases, what we are inspecting clearly cannot be real, and we want to rid the image of these imperfections. Using a simple analogy, if an optical photograph shows a familiar face with four eyes, we can be pretty sure the camera or object has moved during the exposure! Generally, degradations introduced in medical imaging are unfortunately more subtle and consequently more difficult to remove. For example, a SPECT tomogram taken through a small point source will produce an image whose width is much larger than the source. We immediately recognise that degradations are often synonymous with loss of resolution.

This state of affairs should not be interpreted that 'something has gone wrong'. Imaging degradations are present because of the underlying physical laws governing the image formation process. *They have to be there* and it would be 'wrong' if they were not. It is for this reason that considerable theoretical and experimental effort is expended to describe properly the physical processes of image formation. In attempting image processing, one is thus using the known performance of a system (usually its point response) to cheat nature and produce an image that the system could not otherwise possibly generate!

Let us take a naive approach to the problem of image processing and also concentrate on a linear space-invariant model. We have already found that, in this case, a convolution integral (Equation 11.10) links the object f to the image g. In simpler notation, Equation 11.10 may be written:

$$g(x,y) = h(x,y) * f(x,y) \tag{11.14}$$

where $*$ represents convolution. Notice that in Equation 11.14 x and y are used simply to represent whatever are the local 2D coordinates in the appropriate space. When we rewrite Equation 11.10 in Fourier space, the convolution becomes a simple multiplication by the familiar theorem. Using upper-case letters to represent Fourier variables, we can write

$$G(u,v) = H(u,v)F(u,v) \tag{11.15}$$

where (u, v) are spatial frequencies corresponding to the x and y directions.

(*Note.* The remainder of this chapter makes extensive use of Fourier transforms. To follow the arguments it is, however, only necessary to be familiar with the *concept* of the Fourier transform (as defined earlier) and with the theorem that states that convolutions in one space correspond to multiplications in the reciprocal space. Readers who wish to implement the techniques described would, of course, require a greater understanding. The theory necessary to follow the rest of this chapter is summarised in the appendix (Section 11.A), and the book *Radiological Imaging* by Barrett and Swindell (1981) is a good starting point for understanding the Fourier transform.)

Before considering the implications of Equation 11.15, let us step aside to review what is meant by the Fourier representation of the object and the image. The linking equations are

$$F(u,v) = \iint f(x,y)\exp[-2\pi i(ux+vy)]\,\mathrm{d}x\,\mathrm{d}y \tag{11.16}$$

and

$$f(x,y) = \iint F(u,v)\exp[+2\pi i(ux+vy)]\,\mathrm{d}u\,\mathrm{d}v. \tag{11.17}$$

Equation 11.16 shows how the object f may be split up into its component spatial frequencies, and Equation 11.17 shows how these components can be recombined to yield again the original object f. The Fourier representation F contains the same information as f, but in a different form. The same set of statements can be made about the image g and its Fourier representation G. In verbal terms, we can now interpret the information in the Fourier representation. An image that is 'sharp' contains spatial frequencies that are higher than those in a 'more blurred' image. We recognise sharpness as relating to resolution, and so we would expect that the description of the image in spatial-frequency terms contains information pertaining to the available resolution.

More importantly, returning to Equation 11.15, it is clear that the 2D Fourier transform H of the PSRF h gives the fraction of the component of the object distribution at spatial frequency (u, v) that is transferred to the image distribution at the same spatial frequency. H regulates the transfer of information at each spatial frequency and is often called the modulation transfer function (MTF) (see also Chapter 2). If the imaging system introduced no loss of spatial-frequency information (i.e. $H = 1$ for all spatial frequencies up to infinity), then this would imply that $G = F$ and, in turn, $g = f$ by definition. That is, the image would be a perfect representation of the object. Remembering that the Fourier transform of unity is a delta function, this corresponds to an infinitely narrow PSRF, which, in turn, means a precise 1:1 correspondence between object space and image space. We recognise the impossibility of this situation in practice. For all imaging systems, the PSRF has a finite width, and this corresponds to a falling-off in magnitude of the MTF with increasing spatial frequency. One can define mathematically the resolution of a system in terms of either the width of the PSRF at (say) half-maximum or correspondingly the width of the MTF at half-maximum, there being a reciprocal relationship between the two.

It is thus clear that, for all real imaging systems, there is a loss of spatial-frequency information at high spatial frequencies, this being parameterised by the MTF. What are the implications for image processing? *They are dire!* Returning to Equation 11.15, we might naively imagine that we can recover the object distribution from the measured image by direct deconvolution. Inverting Equation 11.15 gives

$$F(u,v) = \frac{G(u,v)}{H(u,v)}$$

and hence

$$f(x,y)=\iint\left[\frac{G(u,v)}{H(u,v)}\right]\exp[+2\pi i(ux+vy)]\,du\,dv. \tag{11.18}$$

Directly invoking Equation 11.18 is a foolish procedure. The reason is that the MTF H will decrease in magnitude with increasing spatial frequency, and there will exist a cut-off frequency beyond which it has zero magnitude. Even before this cut-off is reached, the magnitude will become small. Hence, it is clear that, because at high spatial frequencies the divisor of G becomes vanishingly small or zero, the integral becomes dominated by the larger terms generated at these frequencies. Moreover, it is precisely at such high spatial frequencies that the image (and hence G) becomes dominated by noise. Direct deconvolution thus leads to unacceptable noise amplification. We shall return to this difficulty later in this chapter when a resolution of the problem is proposed. Before that, however, we need to give more thought to the nature of the image formation process and its mathematical representation.

To conclude this section, let us recall that the experimental technique of imaging a point source can yield the MTF. That is, when $f = 1$ we have directly from Equation 11.15 that $G = H$ or $g = h$. The image measured *is* the PSRF and its 2D Fourier transform *is* the MTF. The PSRF of many imaging systems in nuclear medicine, for example, is a Gaussian function, and since the Fourier transform of a Gaussian is another Gaussian, then so also is the MTF. High spatial frequencies are lost and the system behaves as a low-pass filter.

11.4 Discrete Fourier Representation and the Models for Imaging Systems

Returning to Equation 11.7, we have

$$g(x,y)=\iint h(x,y,\alpha,\beta,f(\alpha,\beta))\,d\alpha\,d\beta.$$

This can be regarded as an operator equation in which the function $H\{\,\}$ operates on the object to give the image, that is,

$$g=H\{f\}. \tag{11.19}$$

Here $H\{\,\}$ implies a real-space operation on whatever is within the braces $\{\,\}$. In the special circumstances that the imaging system is linear and spatially invariant, we have (Equation 11.10)

$$g(x,y)=\iint h(x-\alpha,y-\beta)f(\alpha,\beta)\,d\alpha\,d\beta.$$

Equation 11.10 is a Fredholm integral with a 2D kernel, and a model of this type is referred to as 'continuous-continuous', in that both the object and the image space are depicted as a continuous distribution of values. More realistically, the image plane is usually discrete, comprising a matrix of sensors that sample the image. In medical imaging, images are usually made and stored as discrete matrices of numbers in picture elements (pixels). Also, within the conceptual generalisation introduced in Section 11.1, the tomographic image, which is thought of as being physically located within the object distribution (patient), is also generally discrete or digital. Under these circumstances, the Fredholm integral becomes

$$g_{i,j} = \iint h_{i,j}(\alpha,\beta) f(\alpha,\beta)\, d\alpha\, d\beta. \tag{11.20}$$

The PSRF depends on the discrete image-space variables i and j and on the continuous object-space variables α and β. This is referred to as the 'continuous-discrete' model. *It is the closest representation of what really occurs.*

All biological systems are, of course, continuous on the spatial scale that we are considering, rather than discrete, and yet routinely we inspect images (which are discrete) that purport to represent objects, which by inference are therefore being regarded as if they were discrete. We have become very familiar with this, the greatest suspension of disbelief. Not only are we not really looking at the distribution of the parameter of interest—we are looking at its image—we are also imagining the representation to be discrete. Nevertheless, this is extremely convenient and is, in practice, what corresponds with our requirements to store digital images and display them as matrices. If we therefore imagine the object to comprise digital pixels also, we can write down the 'discrete-discrete' model of imaging as

$$g_{i,j} = \sum_{k=1}^{N} \sum_{l=1}^{N} h_{i,j,k,l} f_{k,l}. \tag{11.21}$$

This is unreal, but is customary practice in many medical imaging modalities. Equation 11.21 is familiar as a matrix multiplication in which the object f and image g are $N \times N$ 2D matrices and the PSRF is an $N^2 \times N^2$ 2D matrix. In view of our expectation that imaging systems will have 'good' resolution and that, to first order, there is a more or less 1:1 correspondence between locations in object and image spaces, the matrix h will be very sparse, that is, the majority of terms will be zero. It turns out that this can be a useful property in image processing. Imagine, however, the enormous task of storing the digital form of h. If $N = 256$, as is common, then h has close to half a million terms. It is also easy to see the enormous computational complexity of attempting digital deconvolution by real-space matrix multiplication techniques! We already know that direct deconvolution will have other physical disadvantages. The size problem alone is a very important incentive to attempt image processing in the Fourier-space, rather than the real-space, domain.

Let us therefore return to the Fourier representation and discuss the representation when the object and image are regarded as discrete matrices. We have already seen that the advantage of Fourier-space representation is that the behaviour of each frequency component in object space can be traced through to image space via the MTF. All convolutions disappear and are replaced by simpler multiplications. The MTF carries within it the resolution of the system in a manner that is easier to understand than the somewhat arbitrary real-space definitions of resolution.

For a discrete distribution f, the integrals in Equation 11.16 become replaced by discrete summations, and we have

$$F(u,v) = \left(\frac{1}{N}\right) \sum_{x=0}^{N-1} \sum_{y=0}^{N-1} f(x,y) \exp\left[\frac{-2\pi i(ux+vy)}{N}\right] \tag{11.22}$$

where it is implicit that x, y, u, and v are discrete variables representing sample points in the object and image space.

Evaluating Equation 11.22 on a digital computer is relatively simple even for large matrix sizes N. The computation is separable because of the form of the exponential and an 'intermediate transform' can first be taken in the x direction to generate $F(u, y)$. This can then be transformed by a series of 1D transforms in the orthogonal y direction to yield $F(u, v)$. Thus an N^2 2D transform breaks down into $2N$ 1D transforms. The development of the Cooley-Tukey algorithm (which has become loosely known as the fast Fourier transform or FFT) has revolutionised the computation of Fourier transforms. These no longer increase linearly with increasing size N but rather increase in computational time required roughly as $N^2 \ln N$. Many different forms of the algorithm exist and each user has a favourite.

Reversing Equation 11.22 gives

$$f(x,y) = \left(\frac{1}{N}\right)\sum_{u=0}^{N-1}\sum_{v=0}^{N-1} F(u,v)\exp\left[\frac{+2\pi i(ux+vy)}{N}\right] \tag{11.23}$$

and shows how a discrete distribution is constructed from its discrete Fourier representation. From this equation, it is very easy to introduce the idea of digital image processing. Imagine that some function $T(u, v)$ is inserted on the right-hand side to give

$$\hat{f}(x,y) = \left(\frac{1}{N}\right)\sum_{u=0}^{N-1}\sum_{v=0}^{N-1} F(u,v)T(u,v)\exp\left[\frac{+2\pi i(ux+vy)}{N}\right]. \tag{11.24}$$

The function acts as a frequency modulator or filter, since it multiplies into the distribution F. By choosing different forms of T, a set of filtered images $\hat{f}$ can be generated from f by first operating Equation 11.22 followed by Equation 11.24 (see also Section 5.5.3). Indeed, direct deconvolution is just a special form of Equation 11.24 when the function T is chosen to be $1/H$ and the operation is performed on G, the resulting equation

$$\hat{f}(x,y) = \left(\frac{1}{N}\right)\sum_{u=0}^{N-1}\sum_{v=0}^{N-1}\left[\frac{G(u,v)}{H(u,v)}\right]\exp\left[\frac{+2\pi i(ux+vy)}{N}\right] \tag{11.25}$$

being the discrete version of Equation 11.18. Once again beware the dangers!

Before leaving the consideration of imaging models, let us briefly return to the form of the PSRF. If the imaging system were perfect, then, in the 'discrete-discrete' notation, we have

$$h_{i,j,k,l} = \delta(i-k,\ j-l) \tag{11.26}$$

and hence, from Equation 11.21, $g_{i,j} = f_{i,j}$ and the image perfectly maps the object. When this is not so (which is always), the degradation is coded into h, but the nearer diagonal (or more sparse) is h, the better is the representation. In general, the degradation is not known *a priori*, being a characteristic of the instrument and possibly also a function of the imaging conditions. At worst, it may be object dependent. Hence, in general, experimental work has to establish the form of the PSRF. For medical imaging modalities, this is often possible by imaging a point source. The theory described, however, would equally apply, say, to an optical telescope, and in this case isoplanatic areas of the sky would be required to contain isolated point sources of optical emission in order to determine the PSRF. The theory presented also applies to processing the images from TV cameras carried by space platforms, and, in this case, known degradations have often been accepted in the design of such instruments knowing that they can be largely removed after transmission to Earth of the imperfect images. The reason this is done is to save on transmission bandwidth.

11.5 General Theory of Image Restoration

Imaging can be considered as an operation, and in operator format we have written (Equation 11.19)

$$g = H\{f\}.$$

The image restoration problem is then to find the inverse operator S such that the operation

$$S\{H\{f\}\} = f \tag{11.27}$$

recovers f or at least a best estimate $\hat{f}$ of f. Several difficulties may arise:

1. It may be that S does not exist. In this case, the problem is said to be singular and the problem is, of course, insoluble.
2. If S exists, it may not be unique. Two different inverse operators could yield the same result. This is not really a problem except that it disguises the true nature of the degradation.
3. Even if S exists and is unique, it may be ill conditioned. By ill conditioning we mean that a trivial perturbation in g could lead to non-trivial perturbations in the estimate of f, that is,

$$S\{g + \varepsilon\} = f + Y \quad \text{where } Y \gg \varepsilon. \tag{11.28}$$

In practice, most medical imaging systems conform to the third situation, and non-trivial perturbations (noise) in the image g give artefactual signals in the estimate of f which swamp the true signal.

In reality, it has been wrong to represent the imaging process by Equation 11.8, which gives the false impression that object and image space are connected only by geometrical transforms. In practice, images are contaminated by a variety of noise processes; these differ from modality to modality and have been described in earlier chapters. That being the case, the correct mathematical description of imaging for a linear system is

$$g(x,y) = \iint h(x,y,\alpha,\beta) f(\alpha,\beta)\,\mathrm{d}\alpha\,\mathrm{d}\beta + n(x,y) \tag{11.29}$$

where $n(x, y)$ represents the distribution of noise in the image. There is *no unique solution* to this equation. The reason is as follows. For every element in the *ensemble* of the noise process and for every corresponding image element, there is an element for f (always assuming S exists, is unique and is well conditioned). Since, for every realisation of the noise process, the restored element f is different, it makes no sense to talk about *the* restored image. The fact must be faced that image restoration depends on finding the 'best'-fitting solution to the imaging equation with some *a priori* constraints. We are now in a position to quantify noise amplification simply in mathematical terms.

As a further simplification, let us invoke lexicographic stacking to portray the functions f and g as 1D vector strings of data and the operator $H\{\ \}$ (in the sense of Equation 11.19) as a 2D stacked matrix. In order to make this distinction, the 1D vectors will be set in bold italic type (e.g. $\boldsymbol{g}$) and the 2D matrix will be set in bold sans serif type (e.g. H). Equation 11.29 now becomes

$$\boldsymbol{g} = \mathsf{H}\boldsymbol{f} + \boldsymbol{n}. \tag{11.30}$$

Restated, the problem is that, from a measured image $\boldsymbol{g}$, there is an infinity of solutions for $\boldsymbol{f}$ if the noise term $\boldsymbol{n}$ is non-zero (as it always is). Suppose we operate on both sides of Equation 11.30 with the operator H^{-1}, which is the inverse operator; we obtain

$$\mathsf{H}^{-1}\boldsymbol{g} = \mathsf{H}^{-1}\mathsf{H}\boldsymbol{f} + \mathsf{H}^{-1}\boldsymbol{n}. \tag{11.31}$$

If we define a measure of the true object ($\hat{\boldsymbol{f}}$) as $\mathsf{H}^{-1}\boldsymbol{g}$ then we find

$$\hat{\boldsymbol{f}} = \boldsymbol{f} + \mathsf{H}^{-1}\boldsymbol{n}. \tag{11.32}$$

This equation may be interpreted that the processed image data, which give a measure of the true object, are the sum of the true object and a term that is the noise amplification. If the operator H is singular, clearly we cannot even proceed this far, but if it is 'merely' ill conditioned, then the second term in Equation 11.32 will dominate the first and effectively invalidate the technique. The degree of ill conditioning is sometimes represented by an 'imaging Reynolds number' given by the square root of the ratio of the largest to the smallest eigenvalue of the composite matrix operator $\tilde{\mathsf{H}}\mathsf{H}$ (here the superscript tilde ~ represents transpose).

It is then clear that it may not be too wise to perform direct deconvolution of the measured data. Some more mathematics gives an expression for the degradation of SNR in direct deconvolution. Representing the Euclidean norm of a vector by $|\boldsymbol{g}|\left(=\sum_i g_i^2\right)$ from Equation 11.30

$$|\boldsymbol{g}| = |\mathsf{H}\boldsymbol{f}| + |\boldsymbol{n}| \tag{11.33}$$

but by the conservation of energy implied by the theorem of non-negativity discussed earlier, by which information cannot become 'lost' in being transferred from object to image space,

$$|\mathsf{H}\boldsymbol{f}| = |\boldsymbol{f}| \tag{11.34}$$

and so the SNR in the observed image $\boldsymbol{g}$ is

$$\mathrm{SNR}_{\mathrm{im}} = \frac{|\boldsymbol{f}|}{|\boldsymbol{n}|}. \tag{11.35}$$

Now performing the same operations on Equation 11.32 we have

$$|\hat{f}| = |f| + |\mathbf{H}^{-1}n| \tag{11.36}$$

and so the signal-to-noise ratio in the estimate of the object is

$$\mathrm{SNR_{ob}} = \frac{|f|}{|\mathbf{H}^{-1}n|}. \tag{11.37}$$

Now although $\mathbf{H}$ is energy-preserving, there is no reason why $\mathbf{H}^{-1}$ should be, and we thus have

$$\mathrm{SNR_{ob}} = |\mathbf{H}^{-1}|^{-1}\,\mathrm{SNR_{im}}. \tag{11.38}$$

Now the 'magnification factor' $|\mathbf{H}^{-1}|$ could be of many orders of magnitude, the noise amplification destroying the aims of the image processing. Hence we may make the statement that *for all real medical imaging systems, which are necessarily ill conditioned, it is usual to seek some solution to the image processing problem that optimises some feature.* We shall now consider such optimisations.

11.5.1 Least-Squares Image Processing

Let us suppose that the image formation process may be represented by Equation 11.30. Now seek a solution in which the norm of the noise is minimised. That is, the solution is in this sense 'least squares'. The norm of the noise vector may be represented by the outer product $\tilde{n}\cdot n$. From Equation 11.30,

$$\tilde{n}\cdot n = (\widetilde{g - \mathbf{H}\hat{f}})(g - \mathbf{H}\hat{f}) \tag{11.39}$$

where $\hat{f}$ is now the best estimate, under these conditions, of the object. All we have to do now is solve for this estimate. Differentiating with respect to the elements f_k of $\hat{f}$, putting $d_k = \partial\hat{f}/\partial f_k$ and setting $\partial\tilde{n}\cdot n/\partial\hat{f} = 0$, we obtain

$$0 = -d_k\tilde{\mathbf{H}}g - \tilde{g}\,\mathbf{H}d_k + d_k\,\tilde{\mathbf{H}}\mathbf{H}\hat{f} + \hat{f}\,\tilde{\mathbf{H}}\mathbf{H}d_k. \tag{11.40}$$

This equation is satisfied by

$$\tilde{\mathbf{H}}\mathbf{H}\hat{f} = \tilde{\mathbf{H}}g \tag{11.41}$$

because this, plus its transpose, sum to zero in Equation 11.40. In turn, this reduces to the simpler form

$$\hat{f} = (\tilde{\mathbf{H}}\mathbf{H})^{-1}\tilde{\mathbf{H}}g$$

or

$$\hat{f} = \mathbf{H}^{-1}g. \tag{11.42}$$

This is surprising! The result seems to be advocating precisely what we have been counselling against, namely, direct deconvolution. Equation 11.42 is just the same as if we had struck out the noise term as if it did not exist. It is just the direct reversal of Equation 11.19. What we are being told is that a least-squares solution minimising the norm of $\boldsymbol{n}$ is equivalent to just this. As such, we have unfortunately been once again thwarted. The result is of very little use in view of the previous discussion. Not all is lost, however, as the method is suggestive of one further refinement, which will prove to be most useful.

11.5.2 Constrained Deconvolution

Suppose now we choose $\hat{f}$ such that some linear operator acting on f, namely, $\mathbf{Q}(f)$, is minimised, subject to the constraint that the norm of the noise is fixed. This is a Lagrangian problem, which is equivalent to minimising the function

$$|\mathbf{Q}f| + \Gamma |g - \mathbf{H}f| \tag{11.43}$$

where Γ is a Lagrange multiplier. Writing out the Euclidean norm in full as the outer product of a vector with its transpose, we need to minimise

$$\tilde{\hat{f}}\tilde{\mathbf{Q}}\mathbf{Q}\hat{f} + \Gamma(\widetilde{g - \mathbf{H}\hat{f}})(g - \mathbf{H}\hat{f}). \tag{11.44}$$

Differentiating with respect to $\tilde{\hat{f}}$ and setting the derivative to zero, we obtain

$$d_k\tilde{\mathbf{Q}}\mathbf{Q}\hat{f} + \tilde{\hat{f}}\tilde{\mathbf{Q}}\mathbf{Q}d_k - \Gamma d_k\tilde{\mathbf{H}}g - \Gamma\tilde{g}\mathbf{H}d_k + \Gamma d_k\tilde{\mathbf{H}}\mathbf{H}\hat{f} + \Gamma\tilde{\hat{f}}\tilde{\mathbf{H}}\mathbf{H}d_k = 0 \tag{11.45}$$

with d_k as in Section 11.5.1. This equation is satisfied by

$$\tilde{\mathbf{Q}}\mathbf{Q}\hat{f} - \Gamma\tilde{\mathbf{H}}g + \Gamma\tilde{\mathbf{H}}\mathbf{H}\hat{f} = 0 \tag{11.46}$$

since this, plus its transpose, sum to zero to give Equation 11.45. Simplifying

$$(\tilde{\mathbf{Q}}\mathbf{Q} + \Gamma\tilde{\mathbf{H}}\mathbf{H})\hat{f} = \Gamma\tilde{\mathbf{H}}g \tag{11.47}$$

or

$$\hat{f} = (\tau\tilde{\mathbf{Q}}\mathbf{Q} + \tilde{\mathbf{H}}\mathbf{H})^{-1}\tilde{\mathbf{H}}g \tag{11.48}$$

with $\tau = \Gamma^{-1}$. This is the most general solution to the image processing problem because it contains the free operator $\mathbf{Q}$. In order to implement the technique, it is necessary to consider some specific forms for $\mathbf{Q}$. First, let us recover a trivial result by setting $\tau = 0$. We then obtain

$$\hat{f} = (\tilde{\mathbf{H}}\mathbf{H})^{-1}\tilde{\mathbf{H}}g \tag{11.49}$$

or (remembering to reverse terms in taking the inverse of the product)

$$\hat{f} = \mathbf{H}^{-1}g. \tag{11.50}$$

This is the familiar least-squares solution (or direct deconvolution) appearing again. This is not surprising because setting τ to zero was equivalent to letting the minimisation of the Euclidean norm of the noise term dominate the Lagrange minimisation. In mathematical terms, Equation 11.44 is under these conditions reduced to Equation 11.39.

Now let the operator $\mathbf{Q}$ be the identity operator $\mathbf{I}$. Equation 11.48 becomes

$$\hat{f} = (\tau\mathbf{I} + \tilde{\mathbf{H}}\mathbf{H})^{-1}\tilde{\mathbf{H}}g. \tag{11.51}$$

Choosing this form of $\mathbf{Q}$ is such that the restored picture itself has a minimum norm because in Equation 11.43 $|f|$ has been minimised. Equation 11.51 is a very important result for image processing and has been described by several names including 'pseudo-inverse filter', 'constrained deconvolution' and 'maximum-entropy deconvolution'. Essentially, the reason why this equation leads to successful image processing is that, for those frequencies where the MTF becomes vanishingly small or even zero, the term $\tau\mathbf{I}$ dominates the denominator in Equation 11.51 and avoids noise amplification. It is therefore clear that the value of τ, which remains a free parameter in general, has to be chosen taking due account of the shape of the decline of the MTF with increasing frequency. When τ itself is considered to be frequency dependent, the filter is a Wiener filter. If the MTF is real (corresponding to a symmetric PSRF), then the outer product can be replaced by the modulus and the transpose by the complex conjugate (denoted by an asterisk), giving the neater form for the equation as

$$\hat{f} = \frac{\mathbf{H}^* g}{\tau + |\mathbf{H}^2|}. \tag{11.52}$$

11.5.3 Maximum-Entropy Deconvolution

The entropy in a scene f can be defined as

$$-f \ln f \tag{11.53}$$

by analogy with entropy in information theory or statistical mechanics. If we maximise this expression, this is equivalent to yielding a solution to the image processing problem which makes the fewest presuppositions concerning the form of the processed result. The analysis is similar to that detailed in the previous two sections, except that it is a little more complicated, and here the result only is given that

$$\hat{f} = \exp[-\mathbf{1} - 2\Gamma\tilde{\mathbf{H}}(g - \mathbf{H}\hat{f})]. \tag{11.54}$$

This is a transcendental equation because $\hat{f}$ appears on both sides of the equation. Suppose, however, the result is expanded to just the first term in the Taylor series for the exponential. Then we obtain

$$\hat{f} = -\mathbf{1} - 2\Gamma\tilde{\mathbf{H}}(g - \mathbf{H}\hat{f}) + \mathbf{1}. \tag{11.55}$$

Rearranging the result and cancelling terms gives

$$(\mathbf{I} - 2\Gamma\tilde{\mathbf{H}}\mathbf{H})\hat{f} = -2\Gamma\tilde{\mathbf{H}}g \tag{11.56}$$

or

$$\hat{f} = (\kappa\mathbf{I} + \tilde{\mathbf{H}}\mathbf{H})^{-1}\tilde{\mathbf{H}}g \tag{11.57}$$

with κ replacing $-1/2\Gamma$.

Equation 11.57 is none other than the equation for constrained deconvolution 11.51 and it is now clear from where its other name derives.

Necessarily, the preceding discussion is a somewhat simplified summary of some of the mathematics of image formation and image processing, presented to illustrate the essential concepts rather than provide a comprehensive review. The subject is treated in depth in the book by Andrews and Hunt (1977) and readers are referred there for further reading.

11.6 Image Sampling

In Section 11.4 the concept of discretely sampled space was introduced for both object and image. The formal imaging equation was interpreted as a matrix multiplication and the implications for the Fourier-space representation were considered. It was appreciated that a form of self-deception was in operation when an observer studied a discrete representation of an image and interpreted it as the object distribution of some continuously distributed property. In this section, we further discuss the relationship between a continuous function and its sampled form. The description is given in terms of image space but the same considerations apply in the domain of the object. Let us also confine the discussion to images that are sampled uniformly at spacing δx and δy and let us require that the image being sampled is band-limited at Ω_x and Ω_y. By band limiting we mean that the image does not comprise any spatial frequencies greater than Ω_x and Ω_y. Sampling the continuous image g to form the discrete sampled image g_S is equivalent to multiplication by a 2D Dirac delta function or Shah function $d(x, y)$, that is,

$$g_S(i\delta x, j\delta y) = \sum_i \sum_j g(x,y)\delta(x - i\delta x, y - j\delta y). \tag{11.58}$$

Let us now inspect the implications for Fourier space and we will discover that this leads to a very important statement concerning the relationship between the sampling intervals and the band-limiting frequencies. Equation 11.16 can be used to give the Fourier transform $G(u, v)$ of g and with the same notation we write $G_S(u, v)$ as the Fourier transform of the sampled image g_S. Let us write $D(u, v)$ as the Fourier transform of the 2D Dirac sampling comb $d(x, y)$. This also turns out to be another 2D Dirac comb in frequency space, namely,

$$D(u,v) = (\delta x \delta y)^{-1} \sum_i \sum_j \delta\left(u - \frac{i}{\delta x}, v - \frac{j}{\delta y}\right). \tag{11.59}$$

Equation 11.58 is a multiplicative expression in real space, that is,

$$g_S(i\delta x, j\delta y) = g(x,y)d(x,y) \tag{11.60}$$

and hence the corresponding Fourier-space relationship involves a convolution

$$G_S(u,v) = G(u,v) * D(u,v) \tag{11.61}$$

or in full

$$G_S(u,v) = (\delta x\,\delta y)^{-1} \sum_i \sum_j G\left(u - \frac{i}{\delta x}, v - \frac{j}{\delta y}\right). \tag{11.62}$$

Equation 11.62 implies that the transform $G(u, v)$ is *replicated* on an infinite regularly spaced grid. Wherever a value $G(u', v')$ occurs for the continuous transform, it reappears at all locations spaced apart from (u', v') by $(\delta x^{-1}, \delta y^{-1})$. This is best appreciated diagrammatically and Figure 11.1 shows a figurative form of the transform $G(u, v)$ of the continuous image g. Figure 11.2 shows the corresponding transform $G_S(u, v)$ of the sampled image g_S. This diagram

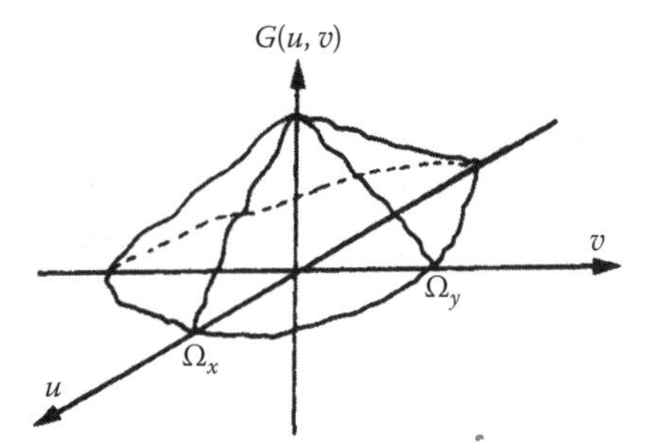

FIGURE 11.1
Schematic 2D Fourier transform $G(u, v)$ of continuous image g.

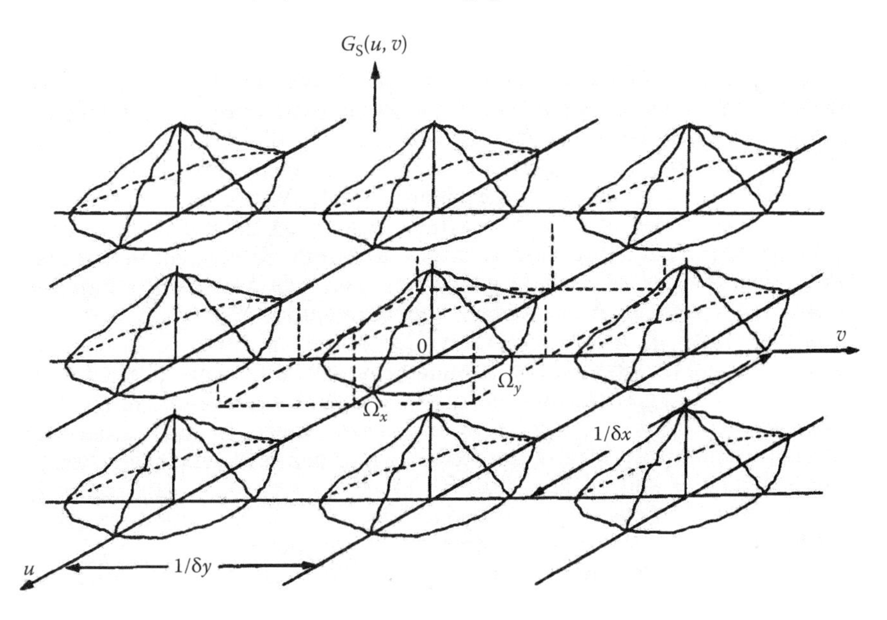

FIGURE 11.2
Schematic 2D Fourier transform $G_S(u, v)$ of sampled image g_S.

enables a clear interpretation of the relationship between the sampling interval δx and the cut-off frequency Ω_x and the corresponding orthogonal pair δy and Ω_y. It is required that

$$2\Omega_x \le (\delta x)^{-1} \quad \text{and} \quad 2\Omega_y \le (\delta y)^{-1}. \tag{11.63}$$

If these conditions are not satisfied, then the sampled function $G_S(u, v)$ will self-interfere and contributions at a particular frequency may arise from a sample and its neighbour. Such overlapping in frequency space is called aliasing, and it is an unwanted phenomenon. Rewriting Equation 11.63 we arrive at the rule that must govern unaliased sampling of an image, namely,

$$\delta x \le \frac{1}{2\Omega_x} \quad \text{and} \quad \delta y \le \frac{1}{2\Omega_y}. \tag{11.64}$$

These are known as the Nyquist sampling criteria.

Hence, for a given image, band-limited at Ω_x and Ω_y, there is a maximum size limitation on the sampling intervals in real space. There is, of course, no minimum limit except that determined by the physical limitations of the technique. As the sampling intervals are made smaller and smaller, so the 'Fourier ghost images' 'move' away from the central transform. Of course, in the limit as the sampling intervals become vanishingly small, all the 'ghost transforms' have congregated at infinity! This is then entirely consistent with the statement that a continuous function has only a single-valued transform. It is as if the situation of a variable sampling interval corresponds to the Fourier images being on a rubber sheet that is being continually stretched.

This view of the relationship between a continuous image and its sampled form generates a technique for recovery of a continuous distribution from a sampled image. If it can be arranged that a 'box' isolates just the central part of the Fourier transform and sets to zero all the replicated part of the transform, then inverse-transforming the function so generated back to real space gives a continuous image. Such a Fourier box filter is the 'zero-one' filter shown as broken lines (base only shown) in Figure 11.2. This filter (let us call it $P(u, v)$) is multiplied in Fourier space. Thus the operation is equivalent in real space to convolving the sampled distribution with the Fourier transform $p(x, y)$ of $P(u, v)$, that is,

$$g(x,y) = g_S(i\delta x, j\delta y) * p(i\delta x, j\delta y). \tag{11.65}$$

The real-space transform of a 'zero-one' box function is a 2D sinc function (see Section 11.A.4):

$$p(i\delta x, j\delta y) = \mathrm{sinc}[2\Omega_x(x - i\delta x)]\,\mathrm{sinc}[2\Omega_y(y - j\delta y)]. \tag{11.66}$$

The operation of recovery in real space can be thought of as an interpolation. This is also useful if the image that one requires to view does not contain pixels of the right size to match the visual acuity of the eye (see Chapter 13). In short, inspection of an image may be obscured by the obvious presence of large pixels. The theory presented here can be used to redisplay an image at a finer resolution than that at which it could be acquired. Note that it is not a unique interpolation because there are a number of ways in which one could choose the box function in frequency space. Also, unless the convolution were taken to an infinite number of terms, Equation 11.65 never quite generates the correct interpolated result. Additionally, our discussion has glossed over the fact that images are not normally sampled at a finite number of *points* but are rather binned into finite-size *pixels*. In this sense, the sampling is not truly a 2D Dirac comb but a comb in which the 'teeth' have a finite size and in fact fill up all the space with only infinitesimally thin 'inter-teeth' gaps. The corresponding full analysis would be more complex and for this reason the simpler analysis has been presented.

In conclusion, the previous analysis has demonstrated the Shannon sampling theorem that a band-limited function is fully specified by samples spaced at intervals not exceeding $1/2\Omega$.

11.7 Iterative Image Processing

For a linear space-invariant image processing system, it has been seen that the image is related to the object distribution via the convolution integral (Equation 11.10)

$$g(x,y) = \int\int h(x-\alpha, y-\beta) f(\alpha,\beta)\,\mathrm{d}\alpha\,\mathrm{d}\beta$$

and several methods of inverting this equation have been discussed. A neat iterative solution exists, due to Iinuma and Nagai (1967), which, for a finite number of iterations, falls short of deconvolution but in the infinite limit approaches unconstrained deconvolution. The advantage lies in the user's ability to terminate iteration interactively before noise amplification becomes too great. To illustrate the method, consider the 1D form of Equation 11.10

$$g(x) = \int h(x-\alpha) f(\alpha)\,\mathrm{d}\alpha \tag{11.67}$$

or equivalently

$$g(x) = \int h(\phi) f(x-\phi)\,\mathrm{d}\phi. \tag{11.68}$$

First, make a 'zeroth approximation' to the object distribution, f_0, which is simply the observed image distribution g. When this is substituted for f on the RHS of Equation 11.68, a zeroth-order approximation to the image (or pseudo-image), g_0, is generated:

$$g_0(x) = \int h(\phi) f_0(x-\phi)\,\mathrm{d}\phi. \tag{11.69}$$

The first-order iteration for the object distribution f_1 is then defined as the sum of the zeroth-order f_0 and the difference between the measurement g and the pseudo-image g_0, that is,

$$f_1(x) = f_0(x) + \left(g(x) - \int h(\phi) f_0(x-\phi)\,\mathrm{d}\phi \right). \tag{11.70}$$

By continuing the same reasoning up to the nth order,

$$f_n(x) = f_{n-1}(x) + \left(g(x) - \int h(\phi) f_{n-1}(x-\phi)\,\mathrm{d}\phi \right). \tag{11.71}$$

The underlying reasoning supporting this particular iterative scheme is as follows. If the imaging modality were noiseless, the term in large parentheses in Equation 11.71 would gradually tend to zero as n increases and $f_n(x)$ would tend to the true object distribution $f(x)$, which is the inversion of Equation 11.68. This is consequent on realising that the nth-order pseudo-image

$$g_n(x) = \int h(\phi) f_{n-1}(x-\phi)\,\mathrm{d}\phi$$

gradually tends to $g(x)$ as n increases.

In practice, noise in the imaging modality precludes proceeding to high orders of iteration and the scheme is concluded when the value of $|g_n(x) - g(x)|$ falls below some prearranged value. Now generalising for a 2D imaging system gives

$$f_n(x,y) = f_{n-1}(x,y) + \left(g(x,y) - \iint (h(\phi,\theta) f_{n-1}(x-\phi, y-\theta)\,\mathrm{d}\phi\,\mathrm{d}\theta \right). \quad (11.72)$$

11.7.1 Spatial Frequency Analysis of the Iterative Method

An analysis of this method in Fourier space gives further insight into why it succeeds and why in the limit as n tends to infinity it approximates unconstrained deconvolution. The first iteration is given by Equation 11.70 and the second yields

$$f_2(x) = f_1(x) + \left(g(x) - \int h(\phi) f_1(x-\phi)\,\mathrm{d}\phi \right). \quad (11.73)$$

Rewriting Equation 11.70 in Fourier space and recalling that a real-space convolution becomes a multiplication in frequency space, we obtain

$$F_1(u) = F_0(u) + [G(u) - H(u)F_0(u)] \quad (11.74)$$

and since $f_0 = g$, $F_0 = G$ and we may group terms to obtain

$$F_1(u) = F_0(u)K_1(u) \quad \text{where } K_1(u) = 2 - H(u). \quad (11.75)$$

Doing the same for Equation 11.73 we find

$$F_2(u) = F_1(u) + [F_0(u) - H(u)F_1(u)] = F_0(u)K_2(u)$$

where

$$K_2(u) = 3 - 3H(u) + H^2(u). \quad (11.76)$$

If this conversion to Fourier space is continued up to the nth iteration

$$F_n(u) = F_0(u)K_n(u) \quad (11.77)$$

where

$$K_n(u) = 1 + [1-H(u)] + [1-H(u)]^2 + \cdots + [1-H(u)]^{n-1} + [1-H(u)]^n. \quad (11.78)$$

Equation 11.78 is a geometric series with first term 1 and common ratio $1 - H(u)$. Its sum, to the nth term is thus

$$K_n(u) = \frac{1-[1-H(u)]^n}{1-[1-H(u)]} = \frac{1}{H(u)}\{1-[1-H(u)]^n\}. \tag{11.79}$$

Combining Equations 11.77 and 11.79, remembering $F_0 = G$, we get

$$F_n(u) = \frac{G(u)}{H(u)}\{1-[1-H(u)]^n\}. \tag{11.80}$$

As the normalised MTF is always less than 1, as $n \to \infty$, the term in braces in (11.80) tends to 1 and Equation 11.80 reduces to pure unconstrained deconvolution (Equation 11.18). For any other finite n, Equation 11.80 represents a truncated deconvolution. The analogy is further clarified by noting that unconstrained deconvolution involves the multiplication of the transform of the image, $G = F_0$, by the function $1/H(u)$. Now

$$\frac{1}{H(u)} = [1-S(u)]^{-1} \quad \text{with } S(u) = 1 - H(u) \tag{11.81}$$

and expanding binomially to an infinite number of terms

$$\frac{1}{H(u)} = 1 + S(u) + S^2(u) + S^3(u) + \cdots + S^n(u) + \cdots (\text{to } \infty). \tag{11.82}$$

Thus unconstrained deconvolution is

$$F_\infty(u) = F_0(u)K_\infty(u) \tag{11.83}$$

where

$$K_\infty(u) = 1 + [1-H(u)] + [1-H(u)]^2 + \cdots + [1-H(u)]^n + \cdots (\text{to } \infty). \tag{11.84}$$

We now observe the important difference between the pairs of Equations 11.77 and 11.78 and Equations 11.83 and 11.84. In the former, the series has a finite number n of terms, while the latter has an infinite number of terms. Hence finite iteration corresponds to *truncating* the deconvolution. It is thus possible to perform image processing that can approach *arbitrarily* close to deconvolution without reaching the limit and incurring the penalties described earlier in the chapter.

Inspecting Equations 11.77 and 11.78 more closely, it is clear that they represent a frequency-dependent modulation with the precise dependence governed by the number of iterations and the shape of the MTF. K_n achieves its largest value $(n + 1)$ at and above those frequencies for which $H(u) = 0$. For any finite number of iterations $K_n(u) < H(u)^{-1}$ at *all* frequencies u. It is for precisely this reason that noise amplification otherwise associated with small values of the MTF is not so severe. There is, however, a computational penalty associated with the implementation in real space, namely, that n iterations necessitate n discrete convolutions (see Equation 11.71) and convolution is a time-consuming computational process. This technique was originally applied by Iinuma and Nagai (1967) to improve the images of radioactive distributions from a rectilinear scanner, and further suggestions were made for overcoming this computational problem. The method is included here because it is completely general and could be applied to images from other medical imaging modalities. The discussion also nicely completes our review of the physical limitations of processing images whose PSRF is linear and space-invariant. All the clinical examples from the first edition have been deleted and instead the next chapter will complement this one by reviewing the theory of Medical Image Processing in a wider context.

11.A Appendix

11.A.1 Fourier Transform

In this appendix the equations that define the Fourier transform are introduced and the convolution theorem, which is vital to understanding the basis of image deconvolution, is proved.

It is shown in elementary mathematical texts that, provided a function $f(x)$ has certain basic properties, it can be expanded in a trigonometric series

$$f(x) = a_0 + \sum_{n=1}^{\infty} a_n \cos(nx) + \sum_{n=1}^{\infty} b_n \sin(nx). \tag{11.A.1}$$

For this to be possible, $f(x)$ must be periodic. The coefficients of the series are unique and the series itself is differentiable. If the function $f(x)$ is odd, then the series has only sine terms, and if the function $f(x)$ is even, then the series has only cosine terms. It is not true that all trigonometric series are Fourier series and some functions have finite rather than infinitely extending Fourier series. As an example, the function

$$f(x) = \begin{cases} 0 & \text{for} -\pi \le x < 0 \\ 0.5 & \text{when } x = 0 \\ 1 & \text{for } 0 < x \le \pi \end{cases} \tag{11.A.2}$$

can be represented by the infinite series:

$$f(x) = \frac{1}{2} + \left(\frac{2}{\pi}\right)\left\{\sin x + \frac{1}{3}\sin(3x) + \frac{1}{5}\sin(5x) + \cdots + [(1-(-1)^n)/2n]\sin(nx) + \cdots\right\}. \tag{11.A.3}$$

Now making use of the complex exponential $\exp(i\phi) = \cos\phi + i\sin\phi$ where $i = \sqrt{(-1)}$, a more general Fourier series may be defined as

$$f(x) = \sum_{n=-\infty}^{\infty} z_n \exp\left(\frac{2\pi i n x}{X}\right)$$

where

$$z_n = \left(\frac{1}{X}\right)\int f(x)\exp\left(\frac{-2\pi inx}{X}\right)\mathrm{d}x. \tag{11.A.4}$$

By expanding the complex exponential, it is easy to relate the coefficients of this series to those of the trigonometric series.

It is now a small step to develop the Fourier transform. First *define* the function $F(u)$ as

$$F(u) = \int f(x')\exp(-2\pi iux')\mathrm{d}x'. \tag{11.A.5}$$

Now evaluate the integral

$$I = \int F(u)\exp(2\pi iux)\mathrm{d}u.$$

Substituting from Equation 11.A.5 for $F(u)$ gives

$$\begin{aligned} I &= \int\int f(x')\exp[2\pi iu(x-x')]\mathrm{d}u\,\mathrm{d}x' \\ &= \int f(x')\mathrm{d}x'\int \exp[2\pi iu(x-x')]\mathrm{d}u \\ &= \int f(x')\delta(x-x')\mathrm{d}x' \end{aligned}$$

by the definition of the delta function:

$$\delta(x-x') = \int \exp[2\pi iu(x-x')]\mathrm{d}u.$$

Hence

$$I = \int F(u)\exp(2\pi iux)\mathrm{d}u = f(x). \tag{11.A.6}$$

Equations 11.A.5 and 11.A.6 define the 1D Fourier pair. Equations 11.16 and 11.17 are the 2D extensions and can be similarly derived in view of the orthogonal separability of the Fourier transform.

11.A.2 Convolution Theorem

It is now a simple matter to develop the convolution theorem. Consider a 1D version of Equation 11.14

$$g(x) = \int h(x-x')f(x')\mathrm{d}x'.$$

Taking the 1D Fourier transform (via Equation 11.A.5)

$$G(u) = \int g(x)\exp(-2\pi iux)\,\mathrm{d}x \tag{11.A.7}$$

and substituting from the previous equation for $g(x)$ gives

$$\begin{aligned} G(u) &= \int\int h(x-x')f(x')\exp(-2\pi iux)\,\mathrm{d}x\,\mathrm{d}x' \\ &= \int\int h(x'')f(x')\exp[-2\pi iu(x'+x'')]\,\mathrm{d}x'\,\mathrm{d}x'' \\ &= \int h(x'')\exp(-2\pi iux'')\,\mathrm{d}x''\int f(x')\exp(-2\pi iux')\,\mathrm{d}x' \\ &= H(u)F(u). \end{aligned} \tag{11.A.8}$$

This is the 1D form of Equation 11.15. That is, a convolution in real space has become a multiplication in Fourier space. It is an easy matter to prove the converse that a convolution in Fourier space becomes a multiplication in real space.

11.A.3 Autocorrelation Function and the Power Spectrum

We show that the Fourier transform of the autocorrelation function is the power spectrum of the image. The autocorrelation function is

$$c(X) = \left(\frac{1}{J}\right)\int f^*(x)f(x+X)\,\mathrm{d}x \tag{11.A.9}$$

for a real-space shift X, where the asterisk represents complex conjugate and where

$$J = \int [f(x)]^2\,\mathrm{d}x.$$

The Fourier transform of the autocorrelation function is $C(u)$ where

$$\begin{aligned} C(u) &= \int c(X)\exp(-2\pi iuX)\,\mathrm{d}X \\ &= \left(\frac{1}{J}\right)\int\int f^*(x)f(x+X)\exp(-2\pi iuX)\,\mathrm{d}X\,\mathrm{d}x \\ &= \left(\frac{1}{J}\right)\int f^*(x)\exp(2\pi iux)\,\mathrm{d}x\int f(x+X)\exp[-2\pi iu(x+X)]\,\mathrm{d}X \\ &= \frac{F^*(u)F(u)}{\int [F(u)]^2\mathrm{d}u} \end{aligned} \tag{11.A.10}$$

where in the second integral $(x + X)$ is a dummy and the denominator J can be replaced by the corresponding Fourier-space integral by Parseval's theorem that

$$\int [f(x)]^2 \mathrm{d}x = \int [F(u)]^2 \mathrm{d}u.$$

Equation 11.A.10 is the normalised power spectrum and the proof is complete.

11.A.4 Interpolation Function

In Section 11.6 on image sampling it was stated that the zero-one box function that isolates the central order part of the discrete Fourier transform of a sampled image is a sinc function. We prove this here for the 1D case as an example of the calculation of a Fourier transform.

A 1D box function is specified by

$$f(x) = \begin{cases} b & \text{for } |x| \leq a/2 \\ 0 & \text{elsewhere.} \end{cases} \quad (11.A.11)$$

From Equation 11.A.5

$$\begin{aligned} F(u) &= \int_{-\infty}^{\infty} f(x) \exp(-2\pi i u x) \mathrm{d}x \\ &= \int_{-a/2}^{a/2} b \exp(-2\pi i u x) \mathrm{d}x \\ &= \frac{b[\exp(-2\pi i u x)]_{-a/2}^{a/2}}{(-2\pi i u)} \\ &= \frac{b[-2i \sin(\pi a u)]}{(-2i\pi u)} \\ &= \frac{ab \sin(\pi a u)}{\pi a u} \\ &= ab\, \mathrm{sinc}(au). \end{aligned} \quad (11.A.12)$$

The generalisation to the 2D box function follows the same line of argument.

References

Andrews H C and Hunt B R 1977 *Digital Image Restoration* (Englewood Cliffs, NJ: Prentice-Hall).

Barrett H H and Swindell W 1981 *Radiological Imaging: The Theory of Image Formation, Detection and Processing* Vol. 1 (New York: Academic Press).

Iinuma T A and Nagai T 1967 Image restoration in radioisotope imaging systems *Phys. Med. Biol.* **12** 501–509.

12

Medical Image Processing

J. Suckling

CONTENTS

12.1 Introduction 714
12.1.1 Technicalities and Notation 715
12.2 Overview of Image Processing 715
12.2.1 Image-Generation Systems 715
12.2.2 Image Compression 717
12.2.3 Image Restoration 717
12.2.4 Image Enhancement 717
12.2.5 Image Analysis 717
12.2.6 Image Segmentation 717
12.2.7 Image Registration 718
12.2.8 Image Understanding 718
12.3 Image Enhancement 718
12.3.1 Contrast Enhancement 718
12.3.1.1 Thresholding 719
12.3.1.2 Pseudocolour 719
12.3.2 Histogram Equalisation 719
12.3.3 Image Subtraction 719
12.3.4 Spatial Filtering 719
12.4 Image Analysis 720
12.4.1 Texture Features 720
12.4.1.1 Grey-Level Histogram Features 720
12.4.1.2 Edge Density 722
12.4.1.3 Fractal Dimension 722
12.4.2 Edges and Boundaries 722
12.4.2.1 First-Order Operators 722
12.4.2.2 Second-Order Operators 723
12.4.3 Shape and Structure 723
12.4.3.1 Geometric Features 723
12.4.3.2 Object Moments 724
12.4.3.3 Object Skeleton 725
12.5 Linear Scale-Space 726
12.5.1 Irreducible Invariants 727
12.5.2 Composite Invariants 729
12.5.3 Line Detection 730

12.6 Image Classification 731
12.6.1 Supervised Classifiers 731
12.6.1.1 Bayesian Classifier 731
12.6.1.2 Linear Discriminant Functions 732
12.6.1.3 Multilayer Neural Networks 733
12.6.1.4 *K*-Nearest Neighbour Classifier 733
12.6.2 Unsupervised Classifiers 733
12.6.2.1 'c'-Means Classifier 733
12.6.2.2 Self-Organising Artificial Neural Networks 734
12.7 Summary 735
References 736

12.1 Introduction

Digital image data have transformed radiological practice permitting the introduction of computational approaches, and in particular image processing, as an aid to (or claims of even replacement for) the radiologist. Automation and reliability accompanying the use of computers is recognised as beneficial in a number of clinical settings with particular potential for health screening programmes where large numbers of cases need to be accurately examined with rapid turnaround. This scenario also occurs in epidemiological imaging studies. To demonstrate significant atypical or pathological differences (which are often small), statistical power lies in using large cohorts. Automation with image processing then becomes attractive in terms of both reliability and tractability.

Generally, image-processing solutions are engineered from a broad literature of techniques, the construction of systems issuing primarily from an understanding of the problem at hand and the interjection of accumulated knowledge and experience. There is no clear 'first principles' approach; indeed there are always several equivalent techniques to consider, and trial and error forms a large part of development. The reader is reminded that this text is merely an introduction to the subject including outlines of some useful algorithms. Its intention is to encourage authorship of a few systems and initiate an exploration into this area. Medical image processing is a part of a larger community of image processing and machine-vision research with systems found in astronomy, remote sensing, robot vision, geophysics and many others. There is a large overlap between these areas and much fertile ground is to be found in collaborative efforts.

Image processing is a practical discipline and hence there are a few prerequisites for those interested in applying its ideas. First, it is necessary to write code in programming languages like FORTRAN or C/C++. Software toolkits are commercially available (or through shareware and freeware licensing) providing modules from which an author can compose a system through a graphical interface or scripted 'macro' language. Although these toolkits are ideal for rapid prototyping, a working knowledge of structured programming is still needed and, of course, some idea of a module's effect. Furthermore, image processing draws heavily on statistical (Altman 1991, Everitt and Dunn 1991) and numerical (Press et al. 1992, Ciarlet and Lions 1998) analysis. Many functions of calculus, linear algebra, inference and function minimisation are taken for granted.

12.1.1 Technicalities and Notation

Digital images are arrays of individual pixels ('picture elements'), or more generally voxels ('volume elements'), that is, pixels with finite thickness. Arrays can be two-dimensional (2D), stacked 2D slices or fully three-dimensional (3D). Stacked 2D images may have gaps between the acquired slices and are not therefore true volumetric representations. Even if 2D stacked images are spatially contiguous the slices may be sequentially collected, for example, with subject movement between acquisitions resulting in misregistration between anatomy on adjacent slices. Within slices voxels are close-packed.

Many image-generation systems now produce images in colour. Representing colour is done in a number of equivalent ways all based on mixtures of three primary colours (Wyszecki and Styles 1982). Perhaps the most conceptually accessible representation is RGBA in which a colour voxel is described by three intensities of monochromatic primary sources at 700.0 nm (red, R), 546.1 nm (green, G) and 435.8 nm (blue, B), which translate directly to the intensities of the cathode-ray tubes of a monitor. The fourth component (A) is the voxel transparency and is important for simple algebraic operations (Sangwine and Horne 1998). From a computational perspective RGBA is represented at each voxel by four bytes. With 8 bits per byte, 256 intensities of each component are possible resulting in 16.5 million unique colours.

Mathematically, an image array is an ordered set of voxels, each represented by an integer or floating point value: $I = \{I_i\}$, where $i = 1,\ldots,N$, the total number of voxels. A binary image, $B = \{B_i\}$, has the further constraint that voxels may only take one of two values: zero or non-zero; the latter assumed to be one unless otherwise indicated. Objects within a scene described by a binary image are represented by $O = \{O_i : i \in object\} \subset B$. For compactness algorithmic aspects are described herein assuming a 2D image. When describing the local spatial environment of a voxel, it is convenient to refer to a cartesian coordinate system, explicitly indicated by the use of two indices in square parentheses: $I = \{I[i,j]\}$, $B = \{B[i,j]\}$ and $O = \{O[i,j]\}$. Image features are calculated voxelwise and are represented by a vector, $\mathbf{F}_i$, with N_F elements. The vector may contain the original voxel values as well as co-registered data from other modalities.

12.2 Overview of Image Processing

A chart describing the processing flow of a generic image-processing solution is shown in Figure 12.1 and forms the basis for this discussion. Not all the stages are covered here and the discussion is limited to those involving calculations on the image data.

12.2.1 Image-Generation Systems

Data from most imaging modalities (e.g. X-ray CT, MRI and PET) emerge in digital form after reconstruction with a suitable algorithm (Figure 12.2a). Many computer file structures are designed to contain patient and scan information in addition to the image array. There are many different formats, each with their own merits with some developed for integration within databases. Alternatively, images may initially be recorded on an analogue medium (most notably planar X-ray film) and subsequently digitised with a laser microdensitometer or CCD camera. However, the resulting image is not an exact representation of the original but now includes peculiarities of the digitisation, such as geometric distortions or non-linear signal responses, requiring correction or accommodation in later processing.

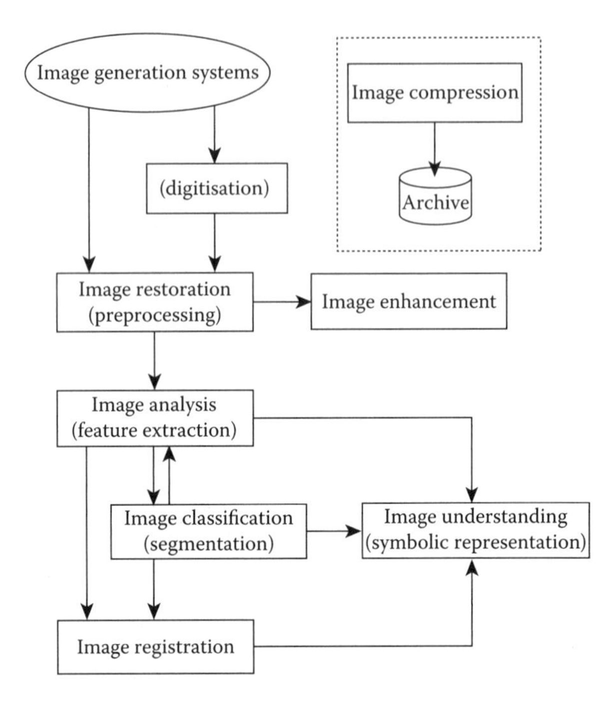

FIGURE 12.1
Overview of the processing flow of an image-processing system.

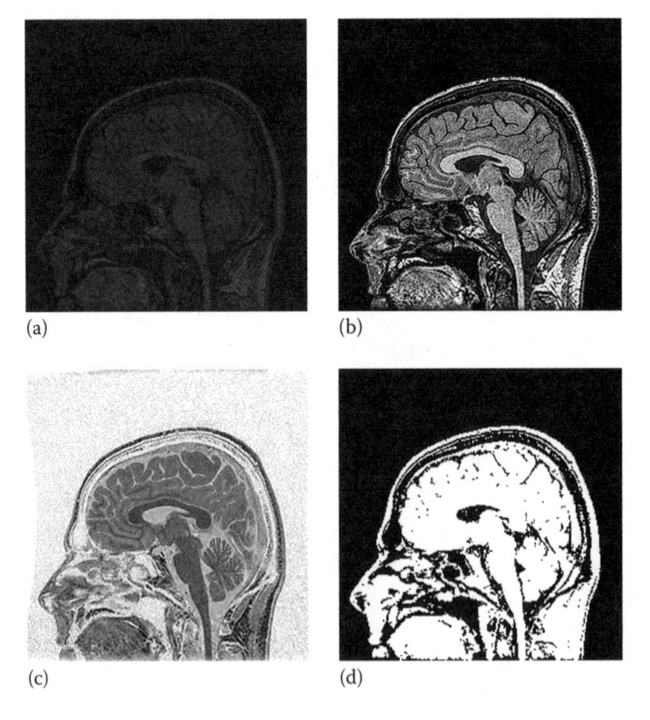

FIGURE 12.2
Examples of image enhancement techniques: (a) sagittal MRI of the human head used as the source image for all subsequent examples, (b) contrast enhanced, (c) inverted and (d) thresholded.

12.2.2 Image Compression

Improved spatial and grey-level resolution, 3D imaging and colour information are making ever greater demands on computer storage and, moreover, on data transmission rates or network 'bandwidth,' upon which emerging technologies such as teleradiology are dependent. For storage and transmission to be as economical as possible compression is used to represent images with the minimum number of bytes. The techniques employed broadly fall into two categories: unlossy and lossy (Chen and Bovik 1990, Netraveli and Haskell 1995). Unlossy compression dictates that an image compressed and uncompressed is an exact reproduction of the original. Lossy compression allows some information to be lost in this transformation. For medical images unlossy techniques typically reduce file sizes by a factor of 2; lossy techniques by a factor of 20.

12.2.3 Image Restoration

Both image-generation systems and digitisation devices introduce distortions of geometry or image intensity (see Chapter 11). Removal of these effects can be achieved through modelling with analytic expressions (such as the optical density–dose relationship of screen-film planar X-ray) or calibration experiments (Karssemeijer 1993). More complex effects require approximations or numerical techniques (e.g. artefacts arising in MRI) (Li et al. 1995, Sled et al. 1998). Correcting these effects ('restoring the image') may not be necessary for successful image processing, but doing so removes the need to customise techniques to particular image properties.

12.2.4 Image Enhancement

The ability to interactively alter the appearance of an image on a computer monitor is a standard feature of any image-display software. Changing apparent image contrast and brightness, or thresholding to remove background regions are simple tasks which can greatly improve the performance of a human operator in detection and interpretation of an image. These techniques do not necessarily generate any additional information from the original data, but merely alter the appearance of the image on a monitor.

12.2.5 Image Analysis

An image scene is composed of 'features' such as edges, lines, textures, and so on. Quantifying these features forms the foundation of image-processing systems. Used directly, features characterise predefined regions, useful for serial measurements, but more often they are classified into the tissues the voxel represents in order to delineate (segment) objects in the scene. There are several different methods for calculating the same features and it is a case of matching the method to properties of the acquisition (e.g. noise, spatial and grey-level resolution).

12.2.6 Image Segmentation

Segmentation connects features (and thus voxels) to form objects and subsequently labelling them as real-world bones, vessels, organs, their parenchyma, and so on is a key aspect of the process. Features of the objects themselves are often of interest and image analysis may again be invoked.

There are two approaches to segmentation. The first aggregates features by classification to form regions. This is bottom-up processing and will be considered in detail later. The alternative—top-down processing—requires specification of target objects which are located and then scaled and distorted to best correspond to local features (Jain 1989).

12.2.7 Image Registration

It is often useful to have images aligned, or registered, so that the mapping of a voxel in one image to the corresponding voxel in another is known. The diversity of information from different medical-imaging techniques has made this an area of rapidly increasing importance (Barber 1992, Glasbey and Mardia 1998, Little and Hawkes 1997, Woods et al. 1993). Furthermore, time-series data from sequentially acquired images are analysed assuming voxels represent the same anatomical location. Patient movement, voluntary or involuntary, is inevitable and registration is essential (Bullmore et al. 1996, Prokopowicz and Cooper 1995).

12.2.8 Image Understanding

To understand a scene is to interpret the elements identified within it. Image and object features, co-registered images and 'world-knowledge' (such as the history of a patient or parameters of the acquisition hardware) are integrated so that a statement about the scene can be constructed: 'the tumour is located near X' or 'The organ has an asymmetrical structure'. Decisions may then be taken (Taylor 1995).

12.3 Image Enhancement

12.3.1 Contrast Enhancement

When an image is displayed, a voxel value, I_i, is usually linearly mapped onto a scale of monitor brightness: $v = \lambda \cdot I_i$. Thus large values of λ give high contrast over a small range of grey levels. This can be generalised so that an operator has interactive control over both λ and the range of the values mapped, using 'width', W, and 'level', L, controls:

$$
\begin{aligned}
I_i < L - W/2 &\Rightarrow v = 0 \\
L - W/2 \le I_i < L + W/2 &\Rightarrow v = (I_i - (L - W/2)) \cdot v_{\max}/W \\
I_i \ge L + W/2 &\Rightarrow v = v_{\max}
\end{aligned}
\tag{12.1}
$$

where $v_{\max}$ is the maximum display brightness. This is a frequently used viewing tool and is particularly effective when the voxel values within a region of interest have a small range relative to $I_{\max}$. With W set to cover this range and L as the approximate mean of the values in the region, contrast is enhanced (Figure 12.2b). Other mappings include multi-linear (multiple L and W), exponential and the logarithmic function. The latter is useful when the grey-level range across the image is extremely large. Reversing the mapping (i.e. $-\lambda$) gives the negative of the original image (Figure 12.2c), often improving apparent contrast.

12.3.1.1 Thresholding

This is a special case of contrast enhancement. If $W = 0$, then v steps from 0 to v_{max} at L and the output is a binary image. Thresholding appears throughout image processing and not only for image display. Regions representing an object often have similar grey levels or features and an appropriate threshold setting can remove extraneous voxels leaving only those of interest: a simple segmentation method (Figure 12.2d).

12.3.1.2 Pseudocolour

The mapping of voxel values is not restricted to changes in monitor brightness alone. Arbitrary colour scales can be constructed and applied to the image, for example, red-through-yellow (the 'hotwire' spectrum) or indeed any scale of colours, which may contain discontinuous jumps, for example, green-to-red, designed to highlight particular regions.

12.3.2 Histogram Equalisation

The voxel histogram of an image is the frequency spectrum of its voxel values. The contrast enhancement algorithms mentioned earlier can be applied by altering the voxel values through manipulation of the parameters of the histogram, rather than just by the grey-level mapping (Jain 1989, Lindley 1991). Additionally, histogram equalisation is a non-interactive transformation which equalises the number of voxel values at each grey level. This is achieved by using the cumulative histogram, scaled to the brightness range desired, as the mapping from voxel value to grey level. Histogram equalisation is useful when displaying images containing large ranges of values.

12.3.3 Image Subtraction

Images acquired before and after the administration of a contrast agent can be subtracted to remove a distracting background and reveal structures of interest. This technique is prevalent in X-ray angiography (Katzen 1995) and enhancement of pathology in neuroanatomical MR (Curati et al. 1996, Lee et al. 1996). Note that this technique assumes the two images are in perfect alignment and that voxel values of non-enhancing tissues remain constant.

12.3.4 Spatial Filtering

The output of a spatial-filtering operation at a voxel depends on its neighbourhood. Implementation is through convolution with single or multiple masks:

$$I^*[i,j] = \sum_{k=-n/2}^{k=+n/2} \sum_{l=-m/2}^{l=+m/2} N[k,l] \cdot I[i-k, j-l] \qquad (12.2)$$

where $N[k,l]$ is a mask defined over a $m \times n$ neighbourhood. This operation is denoted as

$$I^* = N * I$$

There are numerous convolution masks for image analysis, but only a few are of significance in the context of enhancement. The mean of the region is obtained by assigning $1/(m \times n)$ to each $N[k,l]$. This has the effect of smoothing the data and 'suppressing' noise. Increasing m and n increases the amount of smoothing but also increases processing time. The median is also useful when the image contains especially large or small isolated voxel values and the mode more suited for images containing only a few values (e.g. segmented images).

12.4 Image Analysis

12.4.1 Texture Features

Textures like rough, smooth, gnarled, rippled and so on are qualitative terms for well-known properties. The local neighbourhood of a voxel is used to quantify texture and a few popular techniques are described in what follows. They are important features for segmentation especially when objects in a scene have indistinct borders.

12.4.1.1 Grey-Level Histogram Features

The simplest texture descriptors are the voxel values themselves. The shape of the local-neighbourhood voxel histogram has been used to reveal objects embedded in a complex background. For example, if the histogram is bimodal it may intimate the presence of a 'high-density' object. If a threshold value is set at the minima between the two modes, only the object voxels are retained (Davies and Dance 1990).

Other measures from the local grey-level histogram can be extracted, for example, from the central moments, μ_α, and absolute central moments, μ^*_α:

$$\mu_\alpha = \sum_{v=I_{\min}}^{I_{\max}} (v-\bar{v})^\alpha p(v) \tag{12.3}$$

$$\mu^*_\alpha = \sum_{v=I_{\min}}^{I_{\max}} |(v-\bar{v})|^\alpha p(v) \tag{12.4}$$

where $p(v)$ is the voxel probability distribution, that is, the number of voxels with value v divided by the total number of voxels in the region and $I_{\min}$ and $I_{\max}$ are the minimum and maximum voxel values in the neighbourhood, respectively. Then, μ^*_1 is the dispersion (Figure 12.3a), μ_2 is the variance (Figure 12.3b), μ_3 is the skewness (Figure 12.3c) and $\mu_4 - 3$ is the kurtosis (Figure 12.3d).

The second-order histogram is the frequency of grey-level v_2 occurring at an orientation θ and a distance d from v_1. The associated probability distribution, $q_{\theta,d}(v_1, v_2)$, is the histogram

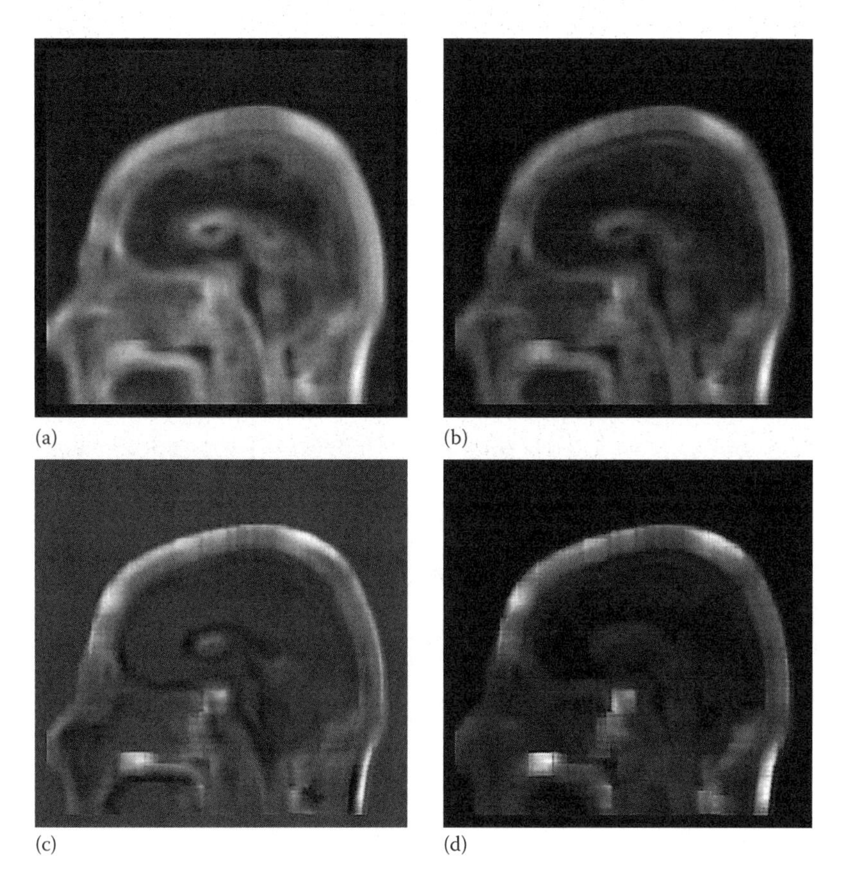

FIGURE 12.3
Examples of textures from the grey-level histogram: (a) dispersion, (b) variance, (c) skewness and (d) kurtosis.

divided by the number of voxels within the neighbourhood. $q_{\theta,d}(v_1, v_2)$ is sometimes known as the co-occurrence matrix. Parameters from the matrix (Haralick et al. 1973) include the following:

$$\text{Angular second moment:} \quad \sum_{v_1=I_{\min}}^{v_1=I_{\max}} \sum_{v_2=I_{\min}}^{v_2=I_{\max}} \{q_{\theta,d}(v_1,v_2)\}^2 \tag{12.5}$$

$$\text{Contrast:} \quad \sum_{v_1=I_{\min}}^{v_1=I_{\max}-1} v^2 \sum_{v_1=I_{\min}}^{v_1=I_{\max}} \sum_{v_2=I_{\min}}^{v_2=I_{\max}} \{q_{\theta,d}(v_1,v_2)\}, |v_1 - v_2| = v \tag{12.6}$$

$$\text{Correlation:} \quad \frac{\sum_{v_1=I_{\min}}^{v_1=I_{\max}} \sum_{v_2=I_{\min}}^{v_2=I_{\max}} v_1 \cdot v_2 \cdot q_{\theta,d}(v_1,v_2) - \bar{q}_{v_1}, \bar{q}_{v_2}}{\sigma(q_{v_1}) \cdot \sigma(q_{v_2})} \tag{12.7}$$

where $\bar{q}_{v1}$ and $\sigma(q_{v1})$ are the mean and standard deviation of the 'marginal' probability matrix $q_{v1} = \sum_{v_2=I_{\min}}^{v_2=I_{\max}} q_{\theta,d}(v_1,v_2)$ and similarly for v_2.

12.4.1.2 Edge Density

The number of ridges or edges per unit area estimates the roughness of the texture. A number of methods exist for edge and line detection and will be developed later. The properties of edge direction and magnitude in a moving-neighbourhood processing scheme infer measures of larger-scale structures in the scene. For example, if the neighbourhood histogram of edge directions is peaked (has small variance), then the neighbourhood may be within a region of stratified appearance.

12.4.1.3 Fractal Dimension

Natural formations such as living tissues are not easily described using Euclidean geometry. Their shape and structure cannot be generated from simple formulae characterised by a few sizes such as radii or edge length. Instead, this form is known as a fractal and is described by recursive relationships, which are independent of scale (Barnsley 1988, Mandelbrot 1975). Euclidean shapes have integer dimension. In contrast, a fractal has a fractal dimension, D, which is continuous with an upper limit of the dimensionality of the space in which it is 'embedded'. Thus, $D = 1$ for a line embedded in the 2D plane. If the line becomes more complex and convoluted in appearance it begins to increasingly 'fill' the plane and $D > 1$. Eventually, if the line twists and turns to the point when it can be considered as filling the entire plane, then $D = 2$.

To find an approximate measure of the fractal dimension, we can exploit the expression (Mandelbrot 1975) relating D to the apparent length, $l(\sigma)$, of a line measured at scale, σ:

$$D = -\frac{\log\{l(\sigma)\}}{\log\{\sigma\}}. \tag{12.8}$$

Practically, the binary image matrix containing the line is resampled with contiguous regions of size $\sigma \times \sigma$ and the number of regions the line intersects, $l(\sigma)$, is calculated. Repeated for several values of σ, a plot of $\log\{l(\sigma)\}$ vs. $\log\{\sigma\}$ is a straight line for a fractal, and $-D$ is its slope. The fractal dimension for a surface texture is the 3D equivalent of this procedure. More computationally expedient methods for estimating D are available (Penn and Loew 1997).

12.4.2 Edges and Boundaries

12.4.2.1 First-Order Operators

Edge-detection algorithms which test for large values of the first-order spatial derivative are generally implemented through two orthogonal convolution masks (spatial filters): H^1 and H^2. Common 3×3 ($m \times n$) voxel masks are Sobel and Prewitt (Table 12.1) and generate almost identical output. Implementation is via two intermediate images, E^1 and E^2, generated by convolution of the masks with the original image, I, that is,

$$E^1 = H^1 * I \tag{12.9}$$

$$E^2 = H^2 * I. \tag{12.10}$$

TABLE 12.1

First-order derivatives (Sobel and Prewitt) 3 × 3 voxel and second-order derivative (Laplacian) 3 × 3 voxel edge-detection spatial masks.

Operator	H^1				H^2	
Sobel	−1	0	1	−1	−2	−1
	−2	0	2	0	0	0
	−1	0	1	1	2	1
Prewitt	−1	0	1	−1	−1	−1
	−1	0	1	0	0	0
	−1	0	1	1	1	1
Laplacian			0	−1	0	
			−1	4	−1	
			0	−1	0	

The magnitude, M_i, and direction, θ_i, of the edge gradient are then obtained at each voxel:

$$M_i = \sqrt{(E_i^1)^2 + (E_i^2)^2} \tag{12.11}$$

$$\theta_i = \tan^{-1}\left\{\frac{E_i^1}{E_i^2}\right\}. \tag{12.12}$$

12.4.2.2 Second-Order Operators

Operators which estimate the second-order derivative, for example, the Laplacian, are useful where edges are smooth and long range. They are, however, more sensitive to noise. Here, a single mask is convolved (Section 12.3.4) with the image and the zero-crossing points identified. The 3 × 3 voxel masks are given in Table 12.1 and example outputs are shown in Figure 12.4.

Both first- and second-order operators have poor performance when detecting lines, rather than edges. First-order operators generate low values at the centre of the line and, moreover, do not give any output for lines of one voxel thickness. Second-order operators give maxima at the line centre but information about direction is not readily available. Compass masks (Jain 1989) constitute a set of second-order derivative operators which are sensitive to predefined angles. Each is applied in turn to the original image and the filter with the highest output determines the direction of any line or edge feature at the centre of the mask. Improvements in angular resolution are made by increasing the filter bank, requiring increased mask size; this is unattractive from a computational viewpoint.

12.4.3 Shape and Structure

Measures of shape and structure are derived from binary images generated by segmentation. These features describe properties of located objects and not the voxels that form them.

12.4.3.1 Geometric Features

Geometric features are effective descriptors of gross morphology. Area, perimeter, and minimum and maximum radii from the centre of mass to the boundary are well-known

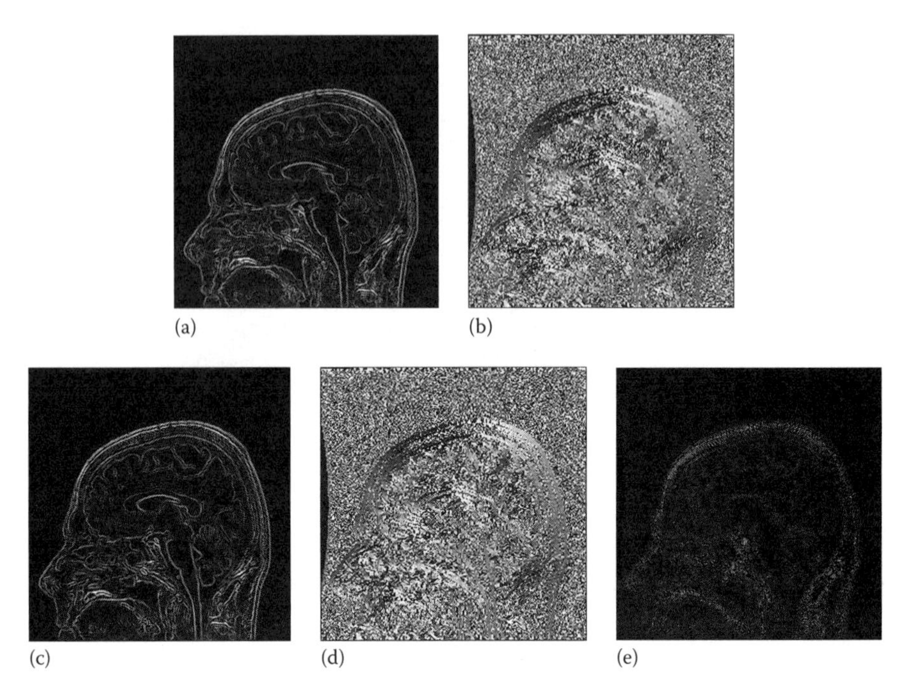

FIGURE 12.4
Examples of edge detection: (a) Sobel filter magnitude, (b) Sobel filter angle, (c) Prewitt filter magnitude, (d) Prewitt filter angle and (e) Laplacian second-order filter. The curved edge to images (b) and (d) are due to a slight edge non-linearity of the imaging coil. This is a common feature of the hardware and is only revealed when the background is non-zero.

properties that are simply calculated (Jain 1989). One other notable parameter is compactness, (perimeter)^2/4π(area), which has a minimum value of 1 for a circle (Parker et al. 1995).

12.4.3.2 Object Moments

The central moments of an object, *O*, in a binary image are defined by

$$\mu_{p,q} = \sum_{\text{all}\,i}\sum_{\text{all}\,j}[i-i']^p[j-j']^q \quad i,j \in O \tag{12.13}$$

where $[i',j']$ is the centre of mass of *O*. This double sequence of moments uniquely determines *O*. The second-order moments ($p + q = 2$) represent the moments of inertia and are combined to calculate the angle of the axis of the least moment of inertia, ϕ:

$$\phi = 0.5\tan^{-1}\left[\frac{2\mu_{1,1}}{\mu_{2,0}-\mu_{0,2}}\right]. \tag{12.14}$$

The bounding rectangle is the smallest enclosing the object and is aligned with ϕ. By performing the coordinate rotation,

$$i^* = j\cos\phi + i\sin\phi \tag{12.15}$$

$$j^* = -j\sin\phi + i\cos\phi \tag{12.16}$$

on each boundary point and recording minimum and maximum values: $i^*_{\min}$, $j^*_{\min}$ and $i^*_{\max}$, $j^*_{\max}$, respectively. The lengths of the rectangle are $i^*_{\max} - i^*_{\min}$ and $j^*_{\max} - j^*_{\min}$ and the ratio of the sides forms a further value. Features derived from moments form a succinct and comprehensive representation for the manipulation and comparison of objects (Ullman 1989).

12.4.3.3 Object Skeleton

The skeleton of an object is a compact representation of its form, composed of lines or curves from which the original object can be reconstructed. Features derived from the skeleton include: the number of line segments, total length of the skeleton, the number of nodes (line intersections) and other, more complex, measures taken from graph theory (Rosen 1988). There are several methods to obtain the skeleton and two are considered here: the distance transform (DT) and morphological processing.

The DT (Borgefors 1986, 1996) operates on a binary mask image, B (Figure 12.5a), which is composed of a background of zero value and the boundary of the object, represented by voxels of infinite (or some suitably large) value. The output generated at each voxel is its closest distance to the object boundary. An algorithm for calculating the DT uses two 'half-masks', C^n, which are applied sequentially: one (C^1) operating from the top to the bottom of the image raster fashion, the other (C^2) in the opposite direction. The output of the voxel on which the mask has its origin is the minimum of the sums of the image and mask values:

$$B[i,j] = \min(B[i+k, j+l] + C^n[k,l]) \quad \text{for each } [k,l] \in C^n. \tag{12.17}$$

An important aspect of the implementation is that, unlike the spatial masks, the output is inserted into the original matrix as the operation proceeds. There are many masks (C^n) to choose from, but a good approximation to the exact distance is given by a 5 × 5 voxel chamfer mask (Table 12.2). The structure function, or skeleton, is the set of points whose DT is locally maximum (Borgefors 1986) (Figure 12.5b).

Morphological processing is based on the application of two operations on the object, O, in a binary image:

$$Erode(O,S) = \{i : S_i \subset O\} \tag{12.18}$$

$$Dilate(O,S) = \{i : S_i \cap O \neq \phi\} \tag{12.19}$$

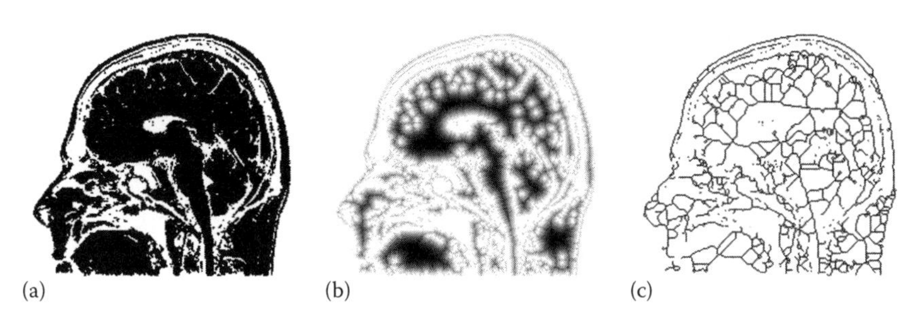

FIGURE 12.5
Examples of object skeleton: (a) binary mask, (b) distance transform and (c) skeleton by erosion.

TABLE 12.2

Half-masks of the 5 × 5 voxel distance transform (DT) chamfer mask.

C^1						C^2				
	11		11					0	5	
11	7	5	7	11		11	7	5	7	11
	5	0					11		11	

Note: The element '0' denotes the mask origin.

where S is a 'structure element' and S_i is the translation of the element to location i. These basic operations are used singularly to expand or contract the boundary of an object but can be used in combination for a variety of operations (Schmitt and Mattoli 1994, Serra 1982). In this context we are concerned only with obtaining the skeleton of an object and in this respect one combination is of particular interest: opening, *Open*(*O*,*S*), is an erosion followed by a dilation (its opposite, closing *Close*(*O*,*S*), is a dilation followed by an erosion). The object structure is obtained by recursive operation:

$$Struct^n(O) = Erode^n(O,S) \setminus Open(Erode^n(O,S)) \tag{12.20}$$

where

'\' denotes a set subtraction, and

n denotes the number of erosions repeated until, at $n = n_{max}$, no voxels remain representing the object.

The skeleton of the object (Figure 12.5c) is then

$$Struct(O) = \bigcup_{n=1}^{n=n_{max}} Struct^n(I). \tag{12.21}$$

12.5 Linear Scale-Space

Several of the image analysis operators thus far discussed are variant under translation or rotation of the image. An example of this is 'spurious resolution' (Koenderink 1984). If an image is resampled with contiguous sub-regions then details that were not previously apparent emerge, dependent on the location of the sub-regions relative to the scene. Linear scale-space addresses these problems with an axiomatic framework for the derivation of image features (Florack et al. 1994, Lindeberg and ter Haar Romeny 1994).

The mathematical formulation of linear scale-space defines the set of image operators from a complete well-behaved solution which satisfies the postulates of:

1. Linearity – no model or *a priori* information is assumed.
2. Spatial homogeneity and isotropy – operators are invariant to translations or orientations of the image, respectively.
3. Spatial scale invariance – operators have no preference for any particular scale. However, digital images have implicit bounds to the spatial scales they represent. The sampling of the scene determines the lower (inner) scale and the field of view the upper (outer) scale.

From these postulates it can be shown that the operator to connect one scale with another must satisfy the linear diffusion equation. The Gaussian kernel and its spatial derivatives are solutions and form a complete set of image operators, or invariants (Florack et al. 1992, Koenderink and van Doorn 1992).

12.5.1 Irreducible Invariants

For a given image dimensionality and an upper limit to the differential order, the number of irreducible invariants, upon which all other constructed operators depend, is limited. For 2D images and up to second-order derivatives, there are five irreducible invariants. For higher order differentials there is no simple method for determining the complete irreducible set (Florack et al. 1992).

To computationally generate linear scale-space operators, we begin with the Gaussian kernel of variance σ^2 in Δ dimensions:

$$G(\boldsymbol{x},\sigma) = \frac{1}{(2\pi\sigma^2)^{\Delta/2}} \exp\left(\frac{-|\boldsymbol{x}|^2}{2\sigma^2}\right) \tag{12.22}$$

where $\boldsymbol{x} = \{x_1, x_2, \ldots, x_\Delta\}$, are the spatial coordinates. A useful expression for the differential of a Gaussian to order $\boldsymbol{n} = \{n_1, n_2, \ldots, n_\Delta\}$ is with Hermite polynomials, $H_n(x)$ (ter Haar Romeny et al. 1994):

$$\partial_n G = \frac{\partial^{n_1+\cdots+n_\Delta}}{\partial x_1^{n_1} \ldots \partial x_\Delta^{n_\Delta}} = (-1)^{\|\boldsymbol{n}\|} \frac{1}{\sigma^{\|\boldsymbol{n}\|}} \left\{ \prod_{k=1}^{\Delta} H_{n_k}\left(\frac{x_k}{\sigma}\right) \right\} \cdot G(\boldsymbol{x},\sigma) \tag{12.23}$$

where $||\boldsymbol{n}||$ is the norm of $\boldsymbol{n}$ and Hermite polynomials have the recursive relationship:

$$\begin{aligned} n=0:&\quad H_n(x) = 1, \\ n=1:&\quad H_n(x) = x, \\ n>1:&\quad H_n(x) = xH_{n-1}(x) - (n-1)H_{n-2}(x). \end{aligned} \tag{12.24}$$

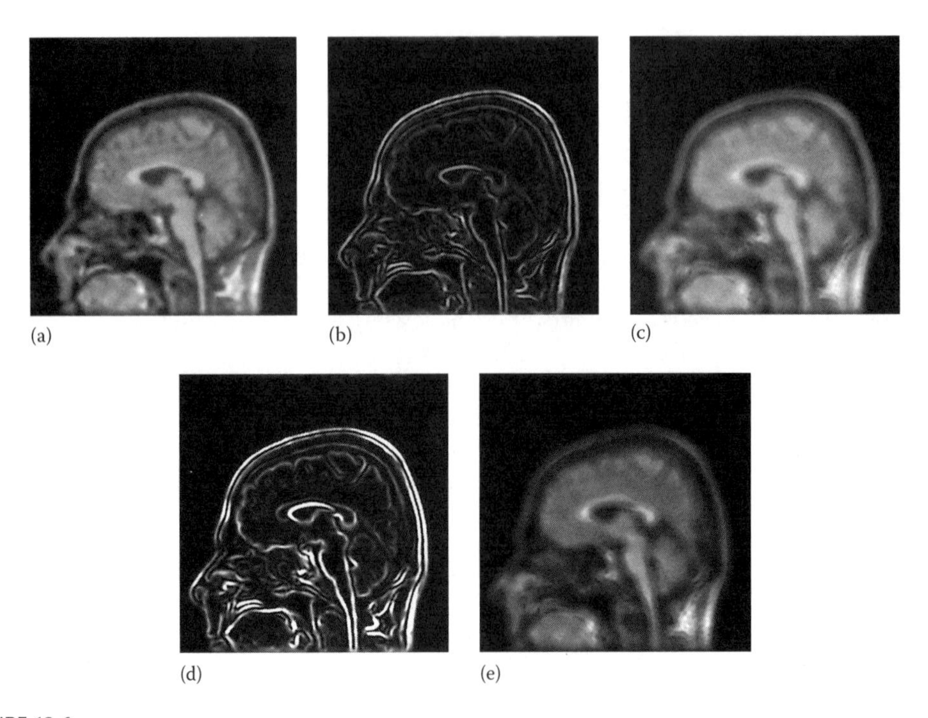

FIGURE 12.6
Linear scale-space irreducible invariants for a 2D image at σ = 2 voxels: (a) I, (b) $I_{x_1}^2 + I_{x_2}^2$, (c) $I_{x_1 \cdot x_1} + I_{x_2 \cdot x_2}$, (d) $I_{x_2}^2 I_{x_1 \cdot x_1} + I_{x_1}^2 I_{x_2 \cdot x_2} + 2I_{x_1} I_{x_2} I_{x_1 \cdot x_2}$ and (e) $I_{x_1 \cdot x_1}^2 + I_{x_2 \cdot x_2}^2 + 2I_{x_1 \cdot x_2}^2$. See text for notational details.

The five irreducible invariants at scale σ, for 2D images up to second order are (Figure 12.6):

1. $I^*(\sigma) = I$, the original voxel values convolved with a Gaussian of scale σ.
2. $I^*(\sigma) = I_{x_1}^2 + I_{x_2}^2$, the squared, first-order gradient, where I_{x_k} denotes convolution of the image with the first-order Gaussian operator in spatial direction x_k.
3. $I^*(\sigma) = I_{x_1 \cdot x_1} + I_{x_2 \cdot x_2}$. (12.25)
4. $I^*(\sigma) = I_{x_2}^2 I_{x_1 \cdot x_1} + I_{x_1}^2 I_{x_2 \cdot x_2} + 2I_{x_1} I_{x_2} I_{x_1 \cdot x_2}$.
5. $I^*(\sigma) = I_{x_1 \cdot x_1}^2 + I_{x_2 \cdot x_2}^2 + 2I_{x_1 \cdot x_2}^2$.

Features are generated by convolution of the image with the invariants. In practice, only a limited number of scales can be used and these are chosen to match the size of the structures of interest. Application of the invariants is via the convolution theorem, which replaces the spatial-domain operation of spatial filtering (Section 12.3.4) with a voxel-by-voxel multiplication after transformation into the Fourier domain, that is,

$$I^* = V * I = FT^{-1}\{FT\{V\} \cdot FT\{I\}\} \qquad (12.26)$$

where
FT denotes the Fourier transform (most often implemented as the fast Fourier transform (Press et al. 1992)), and
V is the invariant.

12.5.2 Composite Invariants

Many interesting operators can be generated from combinations of irreducible invariants (Florack et al. 1992, Koenderink and van Doorn 1992), and only a few are included in this text.

The isophote curvature (Figure 12.7a) is the local deviation from the tangent of a contour of constant voxel values. This is a feature containing information about the angle of curvature of boundaries:

$$I^*(\sigma) = \frac{2I_{x_1}I_{x_2}I_{x_1 \cdot x_2} - I_{x_1}^2 I_{x_2 \cdot x_2} - I_{x_2}^2 I_{x_1 \cdot x_1}}{(I_{x_1}^2 + I_{x_2}^2)^{3/2}}. \quad (12.27)$$

A related feature is the flow-line curvature (Figure 12.7b),

$$I^*(\sigma) = \frac{(I_{x_1}^2 - I_{x_2}^2)I_{x_1 \cdot x_2} + I_{x_1}I_{x_2}(I_{x_2 \cdot x_2} - I_{x_1 \cdot x_1})}{(I_{x_1}^2 + I_{x_2}^2)^{3/2}}. \quad (12.28)$$

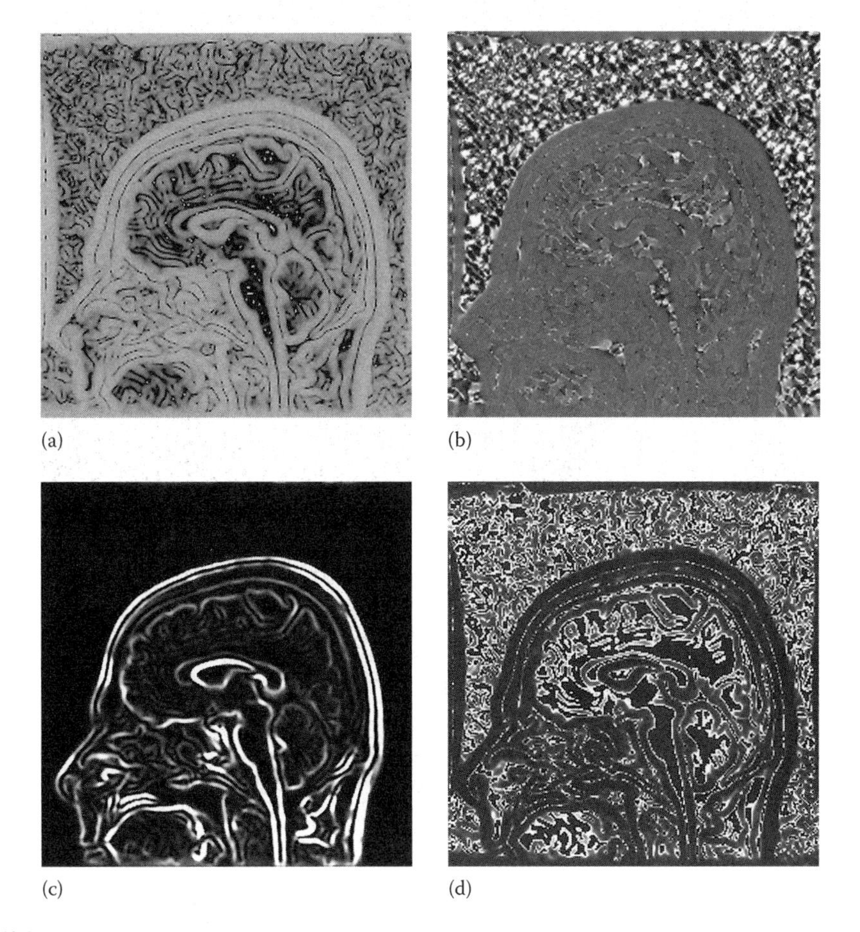

FIGURE 12.7
Examples of linear scale-space composite invariants at $\sigma = 2$ voxels: (a) isophote curvature, (b) flow-line curvature, (c) junction and (d) umbilicity.

Together they are useful for detecting curves that form ridges in the image. Finding the intersection of ridges can be done with a so-called junction invariant (Figure 12.7c), with the points of intersection at local maxima:

$$I^*(\sigma) = \left| I_{x_2}^2 I_{x_1 \cdot x_1} - 2 I_{x_1} I_{x_2} I_{x_1 \cdot x_2} + I_{x_1}^2 I_{x_2 \cdot x_2} \right|. \tag{12.29}$$

Blob-like structures are of interest in medical images as they may represent lesions or regions of localised distortion. The umbilicity (Figure 12.7d) is an invariant which gives high positive output for blobs of high or low grey levels and negative output at saddle-like structures:

$$I^*(\sigma) = \frac{2 I_{x_2 \cdot x_2} I_{x_1 \cdot x_1}}{I_{x_2 \cdot x_2}^2 + I_{x_1 \cdot x_1}^2}. \tag{12.30}$$

Whether regions specifically contain high or low grey levels can be obtained with the Laplacian, which is positive in regions of low values and vice versa.

12.5.3 Line Detection

First-order operators for line and edge detection are implemented by convolution with two orthogonal first-order Gaussian derivatives (Florack et al. 1992, Koenderink and van Doorn 1992). Edge magnitude, M_i, and direction, θ_i, are obtained at each voxel as with spatial masks (Section 12.4.2; Figure 12.4). However, there are no valid results given at the centre of the line and, significantly, the method is insensitive to lines of one voxel width.

An invariant which overcomes these limitations uses the result of Koenderink and van Doorn (1992). The output, $I^*(\theta,\sigma) = \{I_i^*(\theta,\sigma)\}$, from convolution of the image with the second-order Gaussian derivative at an arbitrary direction, θ, can be expressed as a combination of the outputs from convolution with derivatives at $\theta = k\pi/3$ ($k = 0,1,2$):

$$I_i^*(\theta,\sigma) = \frac{1}{3} \begin{bmatrix} (1+2\cos(2\theta)) \cdot I_i^*(0,\sigma) \\ +\left(1-\cos(2\theta)+\sqrt{3}\sin(2\theta)\right) \cdot I_i^*\left(\frac{\pi}{3},\sigma\right) \\ +\left(1-\cos(2\theta)-\sqrt{3}\sin(2\theta)\right) \cdot I_i^*\left(\frac{2\pi}{3},\sigma\right) \end{bmatrix}. \tag{12.31}$$

Now, the direction of the perpendicular to a line passing through voxel, i, is given by the derivative orientation which causes $I_i^*(\theta,\sigma)$ to be maximised. This is found by solving $dI^*(\theta,\sigma)/d\theta = 0$ (Karssemeijer and te Brake 1996):

$$\theta_{\pm} = \frac{1}{2}\left[\sqrt{3}\,\frac{I^*((2\pi/3),\sigma) - I^*((\pi/3),\sigma)}{I^*((\pi/3),\sigma) + I^*((2\pi/3),\sigma) - 2I^*(0,\sigma)} \pm \pi\right] \tag{12.32}$$

and choosing the orientation as the larger of $I_i^*(\theta_+,\sigma)$ and $I_i^*(\theta_-,\sigma)$. Note that lines with positive output correspond to lines in the image of negative contrast (valleys) and negative output corresponds to positive contrast (ridges). The scales chosen are those estimated to be most appropriate to the structures of interest.

12.6 Image Classification

From image analysis we now have a vector of features for each voxel, describing its local environment. To make representations of objects we need to group or classify voxels, thereby segmenting the objects from the scene.

If a feature classifier is 'supervised', it includes *a priori* information about the object classes contained in the scene. These data are voxels labelled with the 'true' classification and collectively form the training set. They are used to calculate the optimum parameters of the classifier which minimise misclassification of the training set. Once complete, the parameters are fixed and the classifier is used to segment previously unencountered data. An expert human operator usually selects data for the training set and the classifier is thus not entirely automated.

With 'unsupervised' classifiers no training set is required. Instead natural groupings, or clusters, of the data in feature space (the domain constructed from features) are automatically defined. These algorithms work well if clusters are well separated in feature space but performance declines as they merge: supervised classifiers are then more suitable.

12.6.1 Supervised Classifiers

At each voxel location the N_F image features are represented by the set of vectors, $F = \{\mathbf{F}_i\}$. With a supervised classifier an additional N_T training vectors, $T_c = \{\mathbf{T}_{cl}\}$, with $l = 1,\ldots,N_T$ are required for each class, $c = 1,\ldots,N_C$ where N_C is the number of classes. Training data can either be obtained from a selected set of images, the training set, or from the image to be segmented, that is, $T_c \subset F$. Usually it is left to a human operator to select T_c as the vectors corresponding to the image voxels which best represent the appearance of each class.

12.6.1.1 Bayesian Classifier

Initially, this classifier (Duda and Hart 1973, Hartigan 1983) requires the *a priori* probabilities of the classes, $P(c = 1,\ldots,N_C)$. That is, in the absence of any measured features, what are the probabilities of a voxel belonging to each class? Naturally,

$$P(c) > 1 \quad c = 1,\ldots,N_C \tag{12.33}$$

$$\sum_{k=1}^{k=N_C} P(c = k) = 1. \tag{12.34}$$

Training data for the Bayesian classifier are in the form of the class-conditional probability distributions, $p(c|\mathbf{T}_{cl})$, of the training-set feature vectors for each class, in other words, the probability distributions of $\mathbf{T}_{cl}$ for each class, c. Each voxel is classified by posing the question; 'Given a feature vector $\mathbf{F}_i$ which class does it belong to?' For the answer we need the *a posteriori* probabilities, $p(\mathbf{F}_i|c)$ for each class, given by Bayes' rule:

$$p(\mathbf{F}_i \mid c) = \frac{p(c \mid \mathbf{T}_{cl})P(c)}{\sum_{k=1}^{k=N_C} p(\mathbf{T}_{cl} \mid c = k)P(c = k)}. \tag{12.35}$$

The value of $p(\boldsymbol{F}_i|c)$ is calculated for each class, and the voxel assigned to the class with the largest probability.

Practically, we need estimates of $P(c)$ and $p(c|\boldsymbol{T}_{cl})$. The *a priori* probabilities can be obtained from existing literature or estimated from tissue volume ratios in manually segmented images. If there are a sufficient number of T_c, then $p(c|\boldsymbol{T}_{cl})$ can be determined non-parametrically directly from these data, interpolating as necessary. If however, as is often the case, the amount of training data is small, then a parameterised function is regressed onto the data; for example, the multivariate Gaussian distribution with non-equal variances:

$$p(c\,|\,\boldsymbol{T}_{cl}) = \frac{1}{2\pi\,|\,V_c\,|^{1/2}} \exp\left\{-\frac{1}{2}(\boldsymbol{T}_{cl} - \bar{\boldsymbol{T}}_c)^{\mathrm{T}} V_c^{-1}(\boldsymbol{T}_{cl} - \bar{\boldsymbol{T}}_c)\right\}, \quad c = 1,\ldots,N_C \tag{12.36}$$

where

$\bar{\boldsymbol{T}}_c$ is the centroid of the distribution, and
V_c is the $N_F \times N_F$ covariance matrix.

A numerical procedure (Press et al. 1992) can estimate the $N_F + N_F(N_F + 1)/2$ parameters of the distribution.

As an aside, N_F is often increased to improve cluster separation in feature space. However, for fixed N_T the number density of training points in feature space decreases as its dimensionality increases and large variances in the derived parameters result.

12.6.1.2 Linear Discriminant Functions

The perceptron (Block 1962, Hinton 1992) is a gradient descent optimisation method (Press et al. 1992) credited with being the first practical artificial neural network, demonstrated in an early computer designed to recognise alphabetic characters (Rosenblatt 1958). From two-class training data, $T = \{\boldsymbol{T}_l\} = \{T_{fl}\}; f = 1,\ldots,N_F$, and $l = 1,\ldots,N_T$, the perceptron estimates a linear function which partitions, or discriminates, feature space into regions corresponding to each class. Conceptually, it consists of a single node with one output and N_F inputs. These inputs are weighted by the vector $\boldsymbol{\omega} = (\omega_0, \omega_1, \ldots, \omega_{N_F})^{\mathrm{T}}$, and define the offset from the origin and normal of the discriminant function, U_l:

$$U_l = \sum_{f=1}^{f=N_F} \omega_f T_{fl} + \omega_0 \tag{12.37}$$

To find $\boldsymbol{\omega}$, the training data are presented to the perceptron in turn and the output O_l, a function of U_l, is calculated and the weights updated using the learning law:

$$\omega_f^{\text{new}} = \omega_f^{\text{old}} + \eta(c_l - O_l)T_{fl} \tag{12.38}$$

where $c_l = \{0, 1\}$ is the class of the training data vector, $\boldsymbol{T}_l$. The output function of the perceptron is the step function:

$$O_l = \begin{cases} 0 & \text{if } U_l < 0 \\ 1 & \text{if } U_l \geq 0. \end{cases} \tag{12.39}$$

The learning rate, η, determines the effect of each iteration and decreases in value as training proceeds. Convergence requires around 50–100 passes through the training set at which point the change in $\boldsymbol{\omega}$ from one pass to the next is within some tolerance value. The derived $\boldsymbol{\omega}$ is adopted and used to classify unlabelled data, with O_l indicating to which class it belongs.

For problems of more than two classes, nodes are connected in a single layer structure. Training data are now labelled such that one node finds the function to discriminate between class '0' and the other combined classes, and the second node discriminates between class '1' and the other combined classes and so on. Note that there is no interaction between the nodes themselves, the discriminant functions are optimised independently of each other.

12.6.1.3 Multilayer Neural Networks

The perceptron, or indeed any linear discriminant function, is ineffective when classes are not linearly separable in feature space. Generalising neural networks to non-linear functions is done by introducing multiple layers of nodes linked via weighted connections between layers, although there are none between nodes within a layer (Hassoun 1995, Hinton 1992). The network weights do not carry the easily interpretable meaning that they have with the perceptron, and trained networks are used as classifiers without knowing the analytic form of the discriminant function (Touretzky and Pomerleau 1989).

Backpropagation (Hassoun 1995, Jones and Hoskins 1987) is a common learning law and is an implementation of gradient descent optimisation. During training, the difference between the network output when presented with a training-set vector and the known classification is used to update the network weights, such that on subsequent presentations of the same vector the output more closely matches the classification.

12.6.1.4 K-Nearest Neighbour Classifier

If N_T is small (from a few tens to a few hundreds) fitting a distribution or optimising a discriminant function can result in high rates of misclassification, as the variances of the fitted parameters are potentially very large.

The *KNN* classifier (Duda and Hart 1973) is a non-parametric approach. An unclassified vector is compared to every training-set vector by calculating the distance between them. A vote is taken based on the classification of the closest K feature-space neighbours: the unlabelled vector is ascribed to the class with the largest vote. The value of K is selected in the range 5–50 depending on the total number of training points. The training set for this classifier must maintain the number of points representing each class in proportion to that expected during operation of the classifier. In other words, the proportions of training data between classes are analogous to the *a priori* probabilities of the Bayesian classifier.

12.6.2 Unsupervised Classifiers

Obtaining a training set is not always possible. For example, a scene representing a time series of acquired images may change on a frame-by-frame basis. A more automated algorithm is required which can identify natural groupings, or clusters, of voxels in feature space.

12.6.2.1 'c'-Means Classifier

This unsupervised classifier (Bezedek 1981) locates clusters based on a similarity measure, $s(\boldsymbol{F}_i, \boldsymbol{F}_j)$, between vectors in feature space. Partitioning of feature space is iteratively achieved, minimising the variance of vectors about a mean within each partition.

Initially assume that there are N_C (≡ 'c' in the name) classes. Associated with each feature vector, $\boldsymbol{F}_i$, is a binary class membership, C_{ic}, which is 1 if it is classified as class c and zero otherwise; thus, $C_{ik} \in \{0,1\}$ and class membership is mutually exclusive. An iterative algorithm is constructed which minimises the within-class sum of square distances:

1. Randomly initialise $\{C_{ic}\}$, maintaining mutual exclusivity of classes.
2. Create a copy of classifications: $\{C^*_{ik}\}$.
3. Calculate the class centroids, $\bar{\boldsymbol{F}}_c$:

$$\bar{\boldsymbol{F}}_c = \frac{\sum_{i=1}^{i=N} C_{ic}\boldsymbol{F}_i}{\sum_{i=1}^{i=N} C_{ic}} \quad c = 1,\ldots,N_C. \tag{12.40}$$

4. Recalculate the new membership values, $\{C_{ic}\}$:

$$C_{ic} = \min_c \left\lfloor s(\boldsymbol{F}_i, \bar{\boldsymbol{F}}_c) \right\rfloor \quad i = 1,\ldots,N; \quad c = 1,\ldots,N_C. \tag{12.41}$$

5. Repeat steps 2–4 until $\left\| s(C_{ic}, C^*_{ic}) \right\| < \varepsilon(\text{typically} \sim 0.001)$, $c = 1,\ldots,N_C$.

The choice of $s(\boldsymbol{F}_i, \boldsymbol{F}_j)$ need not be Euclidean distance. Although rapidly calculated this measure is easily biased if one feature is much larger than the others. Normalisation through the Z-transform (subtract the mean and divide by the variance) is not particularly helpful if, as is hoped, the features form a multimodal distribution. Other measures are available (Jain 1989):

$$\text{Dot product:} \quad s(\boldsymbol{F}_i, \boldsymbol{F}_j) = \langle \boldsymbol{F}_i, \boldsymbol{F}_j \rangle = \left| \boldsymbol{F}_i \right| \left| \boldsymbol{F}_j \cos(\boldsymbol{F}_i, \boldsymbol{F}_j) \right| \tag{12.42}$$

$$\text{Correlation:} \quad s(\boldsymbol{F}_i, \boldsymbol{F}_j) = \rho(\boldsymbol{F}_i, \boldsymbol{F}_j) = \frac{\langle \boldsymbol{F}_i, \boldsymbol{F}_j \rangle}{\sqrt{\langle \boldsymbol{F}_i, \boldsymbol{F}_j \rangle \langle \boldsymbol{F}_i, \boldsymbol{F}_j \rangle}}. \tag{12.43}$$

If N_C is not known then we can begin with a large value of N_C and merge clusters after each iteration of the algorithm if the distance between cluster centres falls below a predefined threshold.

12.6.2.2 Self-Organising Artificial Neural Networks

The self-organising network (Kohonen 1989, 1990) is a topologically connected surface containing N_S nodes (typically varying from a few tens to several hundreds). At each iteration an unlabeled feature vector, $\boldsymbol{F}_i$, is presented to every node in the surface each of which has a corresponding variable weight vector, $\boldsymbol{R}_n$, $n = 1,\ldots,N_S$. A distance measure is made between the $\boldsymbol{F}_i$ and $\boldsymbol{R}_n$ and the node with the closest matching weight vector (the 'winner') is updated, that is, decreasing the distance between them. The neighbourhood of nodes surrounding the winner is also similarly updated, but to a lesser extent. This is competitive learning. Eventually, after many passes through the entire dataset, the weight vectors converge and regions of the network become sensitive to specific clusters of feature vectors. Formalising the algorithm:

1. Define the topology of the network. Nodes can be arranged in any ordered way, on a rectangular or hexagonal matrix, for example, with connections between neighbours. Nodes at the edges of the network are 'wrapped' so that the topological surface formed is continuous; a rectangular matrix thus becomes a toroidal surface.
2. Randomly initialise $\{\boldsymbol{R}_n\}$.
3. Create a copy of the weights, $\{\boldsymbol{R}_n^*\}$.
4. For each feature vector in the dataset find the winning neuron, $\boldsymbol{R}_n^+$, based on

$$\boldsymbol{R}_n^+ = \min_n(|s(\boldsymbol{F}_i, \boldsymbol{R}_n)|). \tag{12.44}$$

5. Update $\boldsymbol{R}_n^+$ and the neighbourhood, R+, around it:

$$\boldsymbol{R}_n^{+\text{new}} = \boldsymbol{R}_n^{+\text{old}} + h^+\eta s(\boldsymbol{F}_i, \boldsymbol{R}_n^{+\text{old}}) \quad n \in \text{R+} \tag{12.45}$$

   where
   $0 \le \eta \le 1$ is the learning rate, its value decreasing with an increasing number of iterations, and
   h^+ is a function which describes the updating of the neighbourhood.

   An effective rule is $h^+ = 1, j \in$ R+; $h^+ = 0$, otherwise. Like the learning rate the size of R+ also decreases as the iterations proceed and robust classification is observed (Kohonen 1990). Initially, large regions are updated giving a coarse segmentation in the first few passes through the data. As both η and R+ are reduced, the classification becomes increasingly finer.
6. Repeat steps 3–5. If, after a pass through the dataset $\| s(\boldsymbol{R}_n, \boldsymbol{R}_n^*) \| < \varepsilon$, the reference vectors have converged. Feature vectors are finally classified with the converged network by recording the node which responds most strongly to it on presentation.

12.7 Summary

In this chapter it has only been possible to present a brief excursion into medical image processing. Some of the techniques included are commonly discussed in basic texts on image processing; others are perhaps only encountered in scientific journals. However, the reader should now be armed with enough information to confidently begin authorship of an image-processing system.

From the outset consideration should be given to the chain of processing steps that form the overall algorithm bearing in mind what information is required from it: a segmented image as aid to object recognition, calculation of object volume or surface area, the location of a particular object and so on. The choice of features will depend on the image properties (noise, contrast and spatial resolution) and on the nature of the objects to be located. For example, texture may be more appropriate in planar X-rays where pixels represent more than one tissue and objects appear more diffuse, embedded in a structured background.

In turn, the selection of a classifier depends on the distribution of vectors in feature space. If features form known or easily modelled distributions, the Bayesian classifier is implicated. Clearly distinct clusters may be linearly separable and easily classified with a perceptron and more complex feature spaces, say where clusters are not simply connected, might require a neural network with one or more hidden layers. These algorithmic elements are selected and chained through intuition, experimentation and experience.

References

Altman D G. 1991 *Practical Statistics for Medical Research.* (Oxford, UK: Blackwell).

Barber D C. 1992 Registration of low resolution medical images. *Phys. Med. Biol.* **37** 1485–1498.

Barnsley M. 1988 *Fractals Everywhere.* (San Diego, CA: Academic Press).

Bezedek J. 1981 *Pattern Recognition with Fuzzy Objective Function Algorithms.* (New York: Plenum Press).

Block H D. 1962 The Perceptron: A model for brain functioning. *Rev. Mod. Phys.* **34** 123–135.

Borgefors G. 1986 Distance transformations in digital images. *Comput. Vis. Graph. Image Process.* **34** 344–371.

Borgefors G. 1996 On digital distance transforms in three-dimensions. *Comput. Vis. Image Und.* **64** 368–376.

Bullmore E, Brammer M, Williams S C R, Rabe-Hesketh S, Janot N, David A, Mellers J, Howard R and Sham P. 1996 Statistical methods of estimation and inference for functional MR image analysis. *Magn. Reson. Med.* **35** 261–277.

Chen D and Bovik A C. 1990 Visual pattern image coding. *IEEE Trans. Commun.* **38** 2137–2146.

Ciarlet P G and Lions J L. 1998 *Handbook of Numerical Analysis.* (Amsterdam, the Netherlands: Elsevier Science).

Curati W L, Williams E J, Oatridge A, Hajnal J V, Saeed N and Bydder G M. 1996 Use of subvoxel registration and subtraction to improve demonstration of contrast enhancement in MRI of the brain. *Neuroradiology* **38** 717–723.

Davies D H and Dance D R. 1990 Automated computer detection of clustered calcifications in digital mammograms. *Phys. Med. Biol.* **35** 1111–1118.

Duda R and Hart P. 1973 *Pattern Classification and Scene Analysis.* (New York: Wiley).

Everitt B S and Dunn G. 1991 *Applied Multivariate Data Analysis.* (London, UK: Edward Arnold).

Florack L M J, ter Haar Romeny B M, Koenderink J J and Viergever M A. 1992 Scale and the differential structure of images. *Image Vision Comput.* **10** 376–388.

Florack L M J, ter Haar Romeny B M, Koenderink J J and Viergever M A. 1994 Linear scale-space. *J. Math. Imaging Vis.* **4** 325–351.

Glasbey C A and Mardia K V. 1998 A review of image-warping methods. *J. Appl. Stat.* **25** 155–171.

ter Haar Romeny B M, Florack L M J, Salden A H and Viergever M A. 1994 Higher order differential structures of images. *Image Vision Comput.* **12** 317–325.

Haralick R M, Shanmugam K and Its'hak D. 1973 Textural features for image classification. *IEEE Trans. Syst. Man Cybern.* **3** 610–621.

Hartigan J A. 1983 *Bayes Theory.* (Berlin, Germany: Springer-Verlag).

Hassoun M H. 1995 *Fundamentals of Artificial Neural Networks.* (Cambridge, MA: MIT Press).

Hinton G E. 1992 How neural networks learn from experience. *Sci. Am.* **267** 144–151.

Jain A K. 1989 *Fundamentals of Digital Image Processing.* (Englewood Cliffs, NJ: Prentice-Hall).

Jones W and Hoskins J. 1987 Backpropagation: A generalised delta learning rule. *Byte* **12** 155–162 October.

Karssemeijer N. 1993 Adaptive noise equalisation and image analysis in mammography. *Information Processing in Medical Imaging. Lecture Notes in Computer Science.* (New York: Springer-Verlag) Vol. **687**, pp. 472–486.

Karssemeijer N and te Brake G M. 1996 Detection of stellate distortions in mammograms. *IEEE Trans. Med. Imaging* **15** 611–619.

Katzen B T. 1995 Current status of digital angiography in vascular imaging. *Radiol. Clin. North Am.* **33** 1–14.

Koenderink J J. 1984 The structure of images. *Biol. Cybern.* **50** 363–370.

Koenderink J J and van Doorn A J. 1992 Generic neighbourhood operators. *Trans. Pattern Anal. Mach. Intell.* **14** 597–605.

Kohonen T. 1989 *Self-Organization and Associative Memory.* (Berlin, Germany: Springer-Verlag).

Kohonen T. 1990 The self-organising map. *Proc. IEEE* **78** 1464–1480.

Lee V S, Flyer M A, Weinreb J C, Krinsky G A and Rofsky N M. 1996 Image subtraction in gadolinium-enhanced MR imaging. *Am. J. Roent.* **167** 1427–1432.

Li S, Williams G D, Frisk T A, Arnold B W and Smith M B. 1995 A computer simulation of the static magnetic field distribution in the human head. *Magn. Reson. Med.* **34** 268–275.

Lindeberg T and ter Haar Romeny B M. 1994 *Geometry-Driven Diffusion in Computer Vision.* (Dordrecht, the Netherlands: Kluwer Academic).

Lindley C A. 1991 *Practical Image Processing in C: Acquisition, Manipulation and Storage.* (Boston, MA: John Wiley & Sons).

Little J A and Hawkes D J. 1997 The registration of multiple medical images acquired from a single subject: Why, how, what next? *Stat. Methods Med. Res.* **6** 239–265.

Mandelbrot B. 1975 *The Fractal Geometry of Nature.* (New York: W H Freeman).

Netraveli A N and Haskell B G. 1995 *Digital Pictures: Representation, Compression and Standards. Applications of Communication Theory.* (New York: Plenum Press).

Parker J, Dance D R, Davies D H, Yeoman L J, Michell M J and Humpreys S. 1995 Classification of ductal carcinoma in situ by image analysis of calcifications from digital mammograms. *Br. J. Radiol.* **68** 150–159.

Penn A I and Loew M H. 1997 Estimating fractal dimension with fractal interpolation function models. *IEEE Trans. Med. Imaging* **16** 930–937.

Press W H, Teukolsky S A, Vetterlin W T and Flannery B P. 1992 *Numerical Recipes in C.* (Cambridge, UK: Cambridge University Press).

Prokopowicz P N and Cooper P R. 1995 The dynamic retina—Contrast and motion detection for active vision. *Int. J. Comput. Vision* **16** 191–204.

Rosen K H. 1988 *Discrete Mathematics and Its Applications.* (New York: Random House).

Rosenblatt F. 1958 The perceptron: A probabilistic model for information storage and organization in the brain. *Psychol. Rev.* **65** 386–408.

Sangwine S J and Horne R E N. 1998 *The Colour Image Processing Handbook.* (Boston, MA: Kluwer Academic).

Schmitt M and Mattoli J. 1994 *Morphologie Mathematique.* (Paris, France: Masson).

Serra J. 1982 *Image Analysis and Mathematical Morphology.* (New York: Academic Press).

Sled J G, Zijdenbos A P and Evans A C. 1998 A non-parametric method for automatic correction of intensity nonuniformity in MRI data. *IEEE Trans. Med. Imag.* **17** 87–97.

Taylor P. 1995 Computer aids for decision-making in diagnostic radiology—A literature review. *Br. J. Radiol.* **68** 945–957.

Touretzky D and Pomerleau D. 1989 What's hidden in the hidden layers? *Byte* **14** 227–233.

Ullman S. 1989 Aligning pictorial descriptions: An approach to object recognition. *Cognition* **32** 193–254.

Woods R P, Mazziotta J C and Cherry S R. 1993 MRI-PET registration with an automated algorithm. *J. Comput. Assist. Tomogr.* **17** 536–546.

Wyszecki G and Styles W S. 1982 *Color Science: Concepts and Methods, Quantitative Data and Formulae.* (New York: John Wiley & Sons).

13

Perception and Interpretation of Images

C. R. Hill

CONTENTS

13.1 Introduction 739
13.2 Eye and Brain as a Stage in an Imaging System 741
13.3 Spatial and Contrast Resolution 742
13.4 Perception of Moving Images 747
13.5 Quantitative Measures of Investigative Performance 749
References 753

13.1 Introduction

Medical imaging occupies a position part way between science and art. Both activities are attempts to transmit to the eye and brain of observers some more or less abstract impression of an object of interest and, in so doing, somehow to influence their state of mind. The pure artist will be aiming for an emotional interpretation of the impression, whilst the medical imager will want to use it to attain an objective judgment about the object, perhaps concerned with the relationships or abnormalities of its anatomy. Thus, even the most technically remarkable medical images might reasonably be likened to a Picasso, but should never be thought of as the real thing. Figure 13.1 illustrates this point.

As outlined in the following text, part of the function of a well-designed medical-imaging system is to extract certain features from the real object and present them to observers in a form that is well matched to their particular perceptual faculties. The conceptual and mathematical relationships between the object and image spaces and frames of reference have already been discussed in Chapter 11. In that chapter, the image in space was considered the physical end point, and no attempt was made to examine how such an image conveys its information to the brain of the observer. In this chapter we proceed to examine this latter question.

Whether in art or science, the ultimate goal of the medical image maker can only be reached through the operation of the particular properties of two sequential channels: the visual faculties of the observers and their mental processes. Both of these are remarkably powerful, and far from fully understood, but both have important characteristics and limitations. The main purpose of this chapter is to give a brief account of the properties and limitations of the human visual perception process, which constitutes a final physical stage and is describable by its own 'optical' or 'modulation transfer function' (MTF), in an imaging system. The mental or interpretative faculties of an observer will only be touched on briefly.

All images are to some extent noisy, in the sense that there are statistical errors in the image data. For many forms of imaging, the magnitude of such noise is small in relation

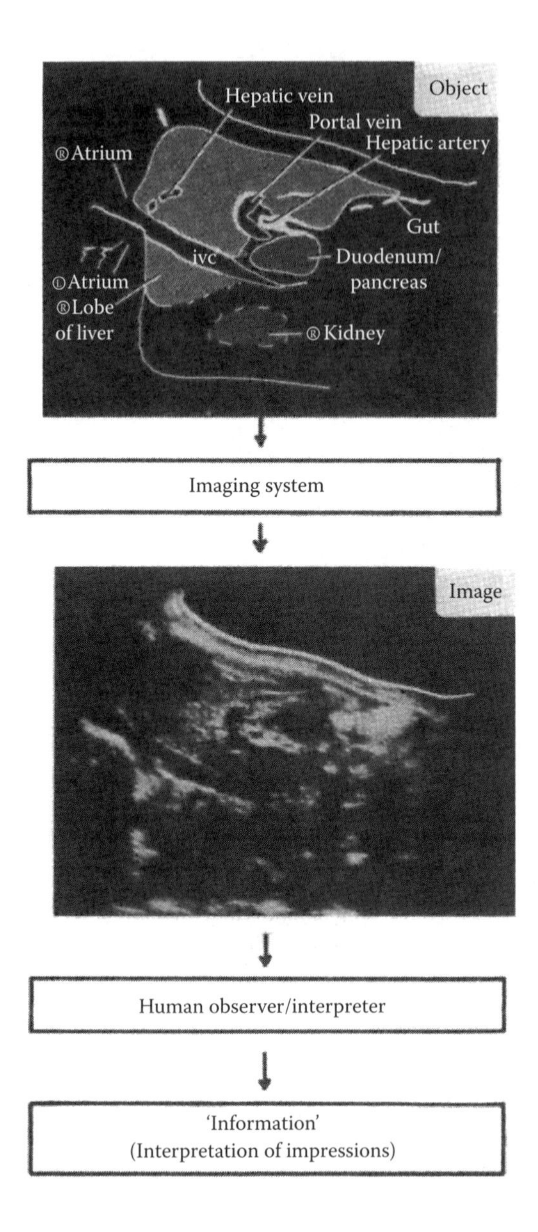

FIGURE 13.1
A representation of the 'medical-imaging' process. An 'object' (in this case, a thin section through the body) has a highly complex (and time-variant) physical–chemical structure. An imaging system transfers (generally by a distorted and non-linear process) a very limited subset of the original data (modified by noise and other artefacts) to form a light amplitude (and/or colour) modulation pattern in 'image space'. This new data set is in turn further filtered (spatio-temporally) by an observer and its contents are interpreted in the context of his/her experience and expectations, to provide some information about the hitherto unknown (or partly known) object. (Anatomical section and image reproduced by permission from Cosgrove and McCready, 1982.)

to the true data ('signal') and the perception process is not appreciably affected. Almost all forms of medical-imaging procedures, however, are subject to a specific constraint on the magnitude of the signal: it is proportional to the radiation exposure of, and hence to some actual or hypothetical damage to, a patient. This is a central consideration in almost all medical imaging and related procedures and is taken up more fully in Chapter 16.

A mathematical treatment of the relationship between SNR and dose was derived in Section 2.4.3 in the context of diagnostic radiology with X-rays. Many medical-imaging procedures are thus characterised by low SNRs, and the ability of the human eye and brain to function effectively under such conditions comes to be of crucial importance. Some relevant aspects of visual performance and also of problems of image interpretation under the conditions of low SNR will be discussed.

The scope of this chapter is necessarily limited and it is not possible to deal, for example, with the important topic of the perception of colour. However, this and other detailed aspects of the subject are well covered in more specialised texts. A stimulating introduction to methods of image analysis and the essential characteristics of human vision is given in a book by Pearson (1975), which is oriented particularly to the engineering of systems for communication of image data. More detailed and comprehensive treatments of the visual perception process have been provided by Cornsweet (1970) and by Haber and Hershenson (1973), whilst engineering aspects, and particularly the implications of noise in imaging systems, are dealt with by Biberman (1973) and Overington (1976). Part of the material of the present chapter is drawn from earlier reviews (Hill et al. 1991, Hill and Bamber 2004).

13.2 Eye and Brain as a Stage in an Imaging System

Elsewhere in this book, we discuss the quantitative description of imaging systems that is necessary to document their behaviour in handling image data. In this context, the eye–brain combination constitutes the final component of such a system, and it will be useful here to recall some of the relevant quantitative measures of spatial transfer characteristics.

The classical, and apparently most straightforward, approach to expressing the spatial properties of an imaging system, in transforming a stimulus in an object space to a signal in the corresponding image space, is to determine the spatial distribution of the image signal that results from a point-object stimulus. Such a distribution is termed a point-source-response function (PSRF) (or, in some older texts, the point spread function – PSF), and a related function, corresponding to an infinitely thin line object, is the line-source-response function (LRSF). The spatial properties of the PSRF were discussed in Section 11.2, where it was shown that, when the imaging system is space- and time-invariant and linear, the object is convolved with the PSRF to form the image distribution.

In dealing with any but the simplest imaging systems, the preceding formulation proves unsatisfactory, particularly since the procedure for computing the combined effect, on the total imaging process, of the set of PSRFs contributed by each stage of the process entails mathematically a series of convolutions for linear space-invariant systems. Thus, it becomes both simpler mathematically and computationally, and more intuitively enlightening, to deal with the situation in the frequency domain (see Sections 2.4.2 and 11.3). The one-dimensional Fourier transform of the LSRF is the optical transfer function (OTF), which is in general a complex quantity whose modulus is termed the modulation transfer function (MTF). The MTF for an imaging system, or for any stage of it, is thus the ratio of the amplitudes of the imaged and original set of spatial sine waves, corresponding to the object that is being imaged, plotted as a function of the sine-wave frequency. The MTF for a system made up of a series of stages is now the product of the MTFs of the individual stages.

This approach to the analysis of imaging systems has proved to be very powerful, and is dealt with in detail in other texts, such as that by Pearson (1975). It is important to

TABLE 13.1

Typical luminance levels.

	Luminance (cd/m^2)
White paper in sunlight	30,000
Highlights of bright CRT display[a]	1,000
Comfortable reading	30
Dark region of low-level CRT display[a]	0.1
White paper in moonlight	0.03

[a] In any one CRT display the ratio of maximum to minimum luminance is seldom more than 100:1, and is typically less than this.

note, however, that its applicability in an exact sense is limited by a number of important conditions, some of which are not met in certain medical-imaging systems. Included in these restrictions are that the processes should be linear and space-time-invariant (i.e. the MTF should not vary either with time or over the surface being imaged), and also that they should be non-negative. This latter restriction, which implies that an image function should not possess negative values, can be met in imaging systems using energy-sensitive detectors of incoherent radiation, for example, X-rays or radioisotope γ emissions (see Section 11.2, and also Metz and Doi 1979), but not in those using field-stress sensors of coherent radiation, which is the case in most ultrasonic systems.

Analytically, a fundamental difference between the use of OTF and MTF is that only the former recognises and preserves phase information. As discussed in the context of Chapter 6, this proves, in practice, to be a vital consideration in the understanding of coherent-radiation imaging systems.

As a preliminary to describing, in a suitably quantitative fashion, the behaviour of human vision, it is necessary to remind ourselves of the conventions used in measuring and reporting the display amplitude of a particular element of an image. This is termed the luminance (*L*) of the image element and is defined (in an observer-independent manner) in terms of the rate of emission, per unit area of the image, of visible light of specified spectral range and shape. Luminance is normally expressed in units of candela per square metre (cd/m^2), although some older texts use the millilambert ($1\,mL = 3.183\,cd/m^2$). The term 'brightness' (*B*), whilst physically equivalent to luminance, is by definition observer-dependent (Pearson 1975). Some examples of typical luminance values found in images and generally in the environment are given in Table 13.1.

An immediate point to note from this table is the very wide (more than 10^6 and probably 10^8 in luminance, or approximately 80 dB) dynamic range of the visual system. This is the range between those luminance values that are so low as to be indistinguishable from noise background and those beyond which any increase in luminance no longer results in an increased perceived response.

13.3 Spatial and Contrast Resolution

Both the sensitivity and resolution characteristics of vision are closely related to the structure of the human retina, in which the sensitive elements include both 'cones' and the relatively more sensitive 'rods' (which constitute highly miniaturised—with dimensions in micrometres—photomultipliers that provide useful electrical output in response to the input of a single photon). Many of the remarkably powerful properties of the human and mammalian visual faculties are related to the manner in which the outputs of these sets of detector elements can be combined, according to the need, to achieve what is in effect a programmable, hard-wired pre-processing system, capable of functions such as edge detection, movement detection and SNR enhancement. Detailed treatment of this subject is beyond the scope of this chapter. It is worth recalling that the mammalian visual system, which evolved essentially as a response mechanism for survival (animals requiring to detect a potential predator), is driven by individual photons. As discussed in Section 13.2, the mechanism has evolved to be responsive even to very small fluxes of photons. The energy of a single photon is only able to disturb a single atom or molecule. Since a nerve pulse involves the movement of millions of atoms or ions, the visual system demands to be a highly efficient low-noise photomultiplier, whose mechanism is still imprecisely understood.

The central region of the retina, the fovea, which subtends (at the focal plane of the lens) an angle of between 1° and 2°, is lined almost entirely with closely packed cones, whilst more peripheral regions include both cones and rods. Within the fovea, the spacing of cones is sufficiently close (about 10 μm) to enable grating resolution to be achieved up to about 60 cycles/degree. The behaviour at lower spatial frequencies, for two display luminances (see Table 13.1), is illustrated in Figure 13.2.

The existence of the peak in sensitivity (at about one and three cycles per degree for the two luminance levels chosen here) illustrates the remarkable and important fact that the human eye–brain system is specifically adapted for the perception of sharp boundaries. It is for this reason that processes such as edge enhancement, grey-level quantisation and colour mapping can prove to be effective features in the design of an imaging system. At high light levels, the colour spectral response of the visual system has its peak near that

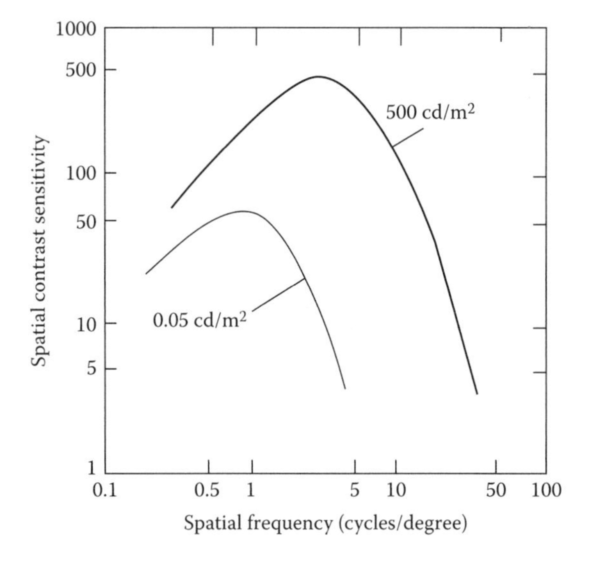

FIGURE 13.2
Typical contrast sensitivity of the eye for sine-wave gratings. Evidently, the perception of fine detail is dependent on luminance level; this has practical implications, for example, on the choice between positive and negative modulation for the display of particular types of image. (After Pearson, 1975.)

of the sun's radiation but shifts towards the blue at lower light levels, thus matching the shifted spectral content of light scattered from the sky under night and twilight conditions.

Other practical implications of the behaviour illustrated in Figure 13.2 are that there will be a limit to the degree of fine detail that can be perceived (even if it is present in an image), that very gradual boundaries (e.g. a diffusely infiltrating border of a tumour) may easily be missed unless processing measures are taken to enhance them and that spatial frequencies roughly in the one to five cycles per degree range will be maximally perceived. This latter feature may clearly be beneficial or the reverse according to whether the structure and magnification of the image are such that the detail in question is anatomical or artefactual (e.g. due to a raster pattern or even to coherent speckle; see the following text).

There is a further important practical message here: one should be careful, in viewing images and in designing image-display arrangements, to ensure a viewing geometry that will enable optimum perception of important image detail whilst maximally suppressing the perception of noise and other artefacts. Practical appreciation of this point may be gained by viewing the scan in Figure 13.1 at a range of viewing distances and also with a magnifying lens.

The value of about 60 cycles/degree, quoted earlier for optically achievable grating resolution, represents a limit set by the anatomical structure of the foveal region of the retina and only holds for high levels of illumination and low levels of image noise. Thus, degradation in acuity to below this anatomical limit can be seen as arising in two separate ways: from the statistics of the visual averaging process that becomes necessary in the absence of adequate illumination, and from the limitations imposed by image noise. These two factors will be considered here in turn, together with their relevance to the question, which is of central importance in several branches of medical imaging, of contrast resolution – the ability to discriminate between neighbouring regions of differing image brightness.

The definition and measurement of contrast resolution, even in the absence of significant image noise, has been a matter for considerable research and, for a detailed account of the various factors involved, reference should be made to one of the specialised texts, for example, Chapter 5 of Haber and Hershenson (1973). Generally, experiments have been carried out in which observers have been presented with a large screen, illuminated in two adjacent segments at different uniform-luminance levels (L, $L + \Delta L$), and have been tested for their ability to detect a difference in those levels. Contrast resolution threshold $\Delta L/L$ is then determined from the difference in luminance ΔL that is just perceivable (reported in 50% of observations) at a given luminance, L. The ratio $\Delta L/L$, sometimes termed the Weber ratio, varies considerably with the level of light falling on the retina, in the manner indicated in Figure 13.3. As the data in this figure indicate, the human eye is capable under ideal conditions (bright illumination and a sharp boundary between two semi-infinite object areas) of discriminating between grey levels separated by as little as 1%. In practical situations, performance is commonly much reduced as a result of four particular factors: the use of suboptimal illuminance, the absence of sharp boundaries (the significance of which has been discussed earlier in relation to Figure 13.2), the limited size of the target area for discrimination, and the presence of image noise and 'clutter'. The significance of these last two factors will now be considered.

In a major series of experiments, Blackwell (1946) demonstrated the manner in which the ability of a human observer to detect the presence of a circular target against a contrasting background depends on the degree of contrast, the level of illumination and the angular size of the target. For this purpose, contrast C is defined as $C = (B_S - B_0)/B_0$, where B_0 is the brightness of the background and B_S is the brightness of the stimulus (target), and $B_S > B_0$, otherwise $C = (B_0 - B_S)/B_0$. This dependence is summarised in Figure 13.3.

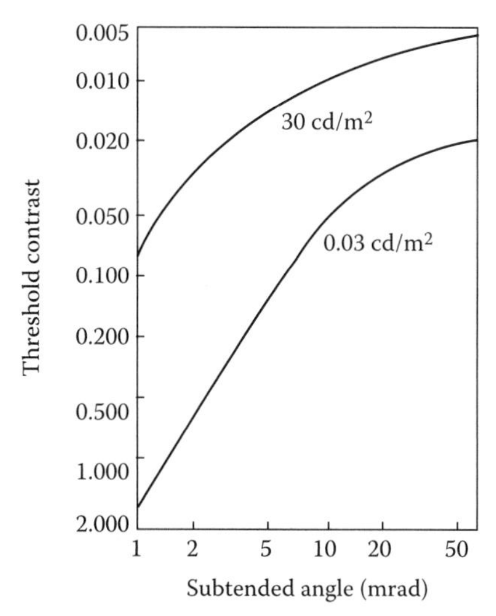

FIGURE 13.3
Dependence of threshold contrast $\Delta L/L$ (the 'Weber ratio') on the size of an observed circular disc object, for two levels of background luminance, with zero noise and a 6s viewing time. The effect of added noise and/or of shorter viewing time will generally be to increase threshold contrast relative to the levels indicated here. (From Blackwell, 1946.)

From these data, it will be seen in particular that, for a given brightness level, there is an inverse relationship between the linear size of a target and the degree of contrast necessary for its discrimination. Although this is sometimes stated to be an inverse linear relationship, such an approximation is clearly only valid over a limited range of target size.

The influence of the noise content of an image on its perceptibility has been the subject of a great deal of study, particularly in relation to electro-optical imaging and photographic grain noise. An often-quoted result is that to detect a single pixel differing from its N neighbours by a contrast C ('$C = 0.01$' means 1% contrast), the total number of photons that must be detected is k^2N/C^2 where $k \approx 5$ (Rose 1973). For example, to detect a 1% single-pixel contrast in a 256×256 digital image would require the detection of 16.38×10^9 photons. In a typical 64×64 gamma-camera image with 10^6 detected photons, the smallest detectable contrast on this basis would be 32%, although this would reduce if a smaller value of k were adopted, corresponding to decreasing the confidence in the detected contrast. It should, however, be remembered that a signal-to-noise value of 5 is not always required if the object to be detected is not simply a single pixel and has some correlation, and also if it is acceptable not to be completely certain that the detected event is not spurious (see also Webb 1987).

In the present context, it is convenient to extend the concept of noise to include that of 'clutter', a term that has the more general connotation of an unwanted signal. A distinction between noise and clutter is that noise is generally incoherent in nature, whereas clutter may exhibit a degree of coherence in relation to the wanted signal. Examples of such coherent 'clutter' in medical images are the reconstruction artefacts sometimes seen in X-ray CT and NMR images and the reverberation artefacts in some ultrasound images (see Chapter 6).

An example of the use of threshold contrast analysis, illustrated in Figure 13.4, is investigation of the effect of coherent speckle noise on the minimum detectable size of an abnormality in an ultrasonic image. This computer modelling study showed that complete removal of speckle noise led to a corresponding factor of 10–15 reduction in the linear dimension of a just detectable abnormality, of any given contrast, in the image.

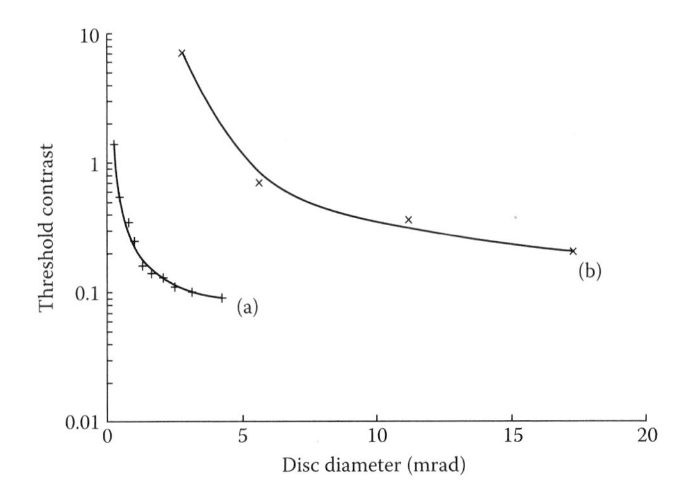

FIGURE 13.4
Experimental contrast-detail curves comparing the following: (a) uniform discs and backgrounds (+++); (b) discs and backgrounds both modulated by simulated echographic speckle (×××). Speckle cell size (FWHM of the auto-correlation function) is approximately 0.55 (axial) × 1.4 (lateral) milliradian. (From Hill et al., 1991.)

Apart from the general level of noise, a factor of major importance is its effective spatial-frequency distribution. Quantitatively, this dependence can be investigated by measuring the degree of modulation that is necessary for visual detectability in a sine-wave bar pattern (raster) as a function of spatial frequency. In the target recognition field, this is sometimes referred to as the demand modulation function (DMF) or 'noise required modulation' (NRM).

The motive for introducing this concept is to provide a somewhat quantitative basis for predicting the effect, on an overall imaging-and-perception process, of noise or clutter having particular spatial-frequency characteristics. A practical example where this concept might usefully be applied in medical imaging is the problem of coping with the phenomenon of coherent speckle in ultrasound images (Chapter 6). Here, the designer has some control over the spatial frequency of the resultant noise, and may indeed ultimately be able to eliminate its ill effects (cf. Figure 13.4). Figure 13.5 illustrates graphically, for this example, how the relationship between the system MTF and the noise-induced degradation of visual perception (expressed as NRM) can lead to a quantitative predictor of the relative system performance for different designs.

In experiments in which varying amounts of grain noise were deliberately added to optical images, it was found that an observer, if given the choice, will tend to adjust the viewing magnification in a manner that keeps constant the spatial-frequency spectrum of the perceived image. This again emphasises the importance, in practical system design, of either providing an optimum set magnification or of allowing for operator selection. A useful discussion of the application of the MTF concept to the problem of image noise is given by Halmshaw (1981).

The concept of operator performance is clearly important in relation to medical imaging, since an image here is generally produced in order to assist in a particular task of detection or recognition of an abnormality. In this connection, therefore, it is of interest to note that, in optical imagery, three stages of perception of an object are formally categorised as 'detection' (i.e. a decision is made as to whether some, as yet unspecified, abnormality is present),

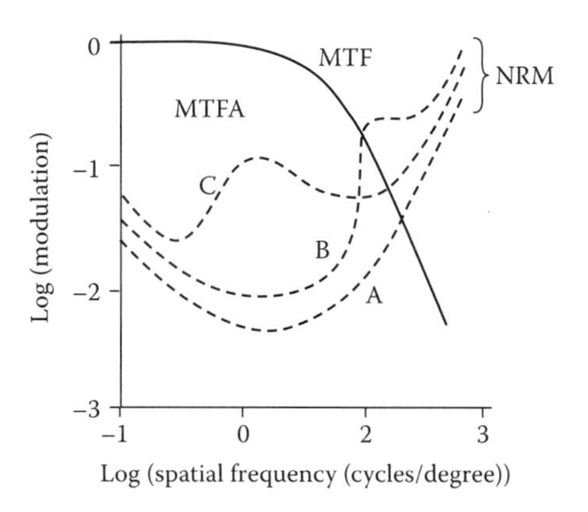

FIGURE 13.5
Illustrative relationship between the modulation transfer function (MTF) of an imaging system and the noise required modulation (NRM) functions for visual perception of image detail under various noise conditions: curve A, low noise; B and C, noise or clutter with spatial spectra having peaks at about 10 and 1 cycles/degree, respectively. Note that curve A is an alternative representation of the behaviour illustrated in Figure 13.2. The area between the MTF and appropriate NRM curves is termed the modulation transfer function area (MTFA) and is considered to be a useful measure of overall performance. The examples given here might arise practically in ultrasonic imaging with coherent speckle noise (Chapter 6) and indicate the advantage (increase in MTFA) of shifting the noise spectrum upwards (C to B) towards frequencies where perceptual ability and MTF are falling off. (Adapted from Biberman, 1973.)

'recognition' (i.e. features such as size and shape of an abnormality are quantified) and 'identification' (i.e. decisions are made as to likely disease patterns that correspond to the recognised, detected abnormalities). Furthermore, these different degrees of perception have been related empirically to the detectability, using the same imaging conditions and degree of modulation, of bar patterns of given spatial frequency. Thus, it is found (in work on non-medical-imaging applications) that 'detection', 'recognition' and 'identification' occur for target angular widths $\alpha \approx 1/v$, $4/v$ and $6.5/v$, respectively, where v is the period of the highest detectable spatial frequency (cycles per unit angle). An alternative way of framing this statement would appear to be (if, for simplicity, one assumes that two image samples can be taken from each modulation period – the so-called Nyquist rate) that the linear dimensions of targets that are 'detectable', 'recognisable' and 'identifiable' will be 2, 8 and 13 resolution cell widths, respectively. This somewhat simplistic statement should, however, be qualified by noting that, in practice, performance of such perceptual tasks will be influenced by a number of other factors, including the time available for the task and also the *a priori* expectation of finding a particular target in a particular region of a 'scene'. It would be of interest to investigate the possible extension of the preceding approach for relating operator performance to imaging system parameters in some particular areas of medical imaging. In particular, it may have relevance to questions of the following type: 'What is the smallest size of tumour that we can expect to see with the XYZ equipment?'

13.4 Perception of Moving Images

Useful information about human anatomy and pathology can be derived from studies of movement, or variation in time. It is unfortunate in this respect that a number of medical-imaging modalities are severely restricted, through the combined effects of noise and the requirement to avoid excessive radiation damage, in the rate at which image data can be generated, but dynamic or 'real-time' imaging is nevertheless possible in some situations, notably with ultrasound but also in some X-ray, nuclear medicine and MRI procedures.

It is, therefore, important to be aware of the time-related factors that affect human visual perception. The discussion here will be limited to the behaviour of the immediate visual process, but it should also be borne in mind that the processes of pattern and feature recognition that take place in the higher levels of the brain will in general be rate-limited and therefore time-dependent.

It appears that the human visual process has evolved in a manner that is responsive to the specific perception of movement. Quantitatively, this can be measured and expressed, analogously to the spatial response, as a temporal-frequency response or 'flicker sensitivity' (Pearson 1975). If a small, uniform-luminance source is caused to fluctuate sinusoidally in luminance about a mean value L, the resulting stimulus will be

$$L + \Delta L \cos(\pi f t)$$

where

ΔL is the peak amplitude of the fluctuation, and
f is its frequency.

If the value of ΔL that produces a threshold sensation of flicker is determined experimentally as a function of f, a flicker sensitivity or temporal contrast sensitivity can be derived as the ratio $L/\Delta L$.

The typical response of the human eye under representative display viewing conditions is illustrated in Figure 13.6. From this, it will be seen that, at low luminance levels, the eye behaves as though it were integrating with a time constant of around 0.2 s and is maximally sensitive to static objects. By contrast, at higher luminance, the system discriminates quite strongly against relatively static aspects of a scene and is maximally sensitive at frequencies of around 8 Hz. The criteria for flicker-free viewing will evidently entail that operation

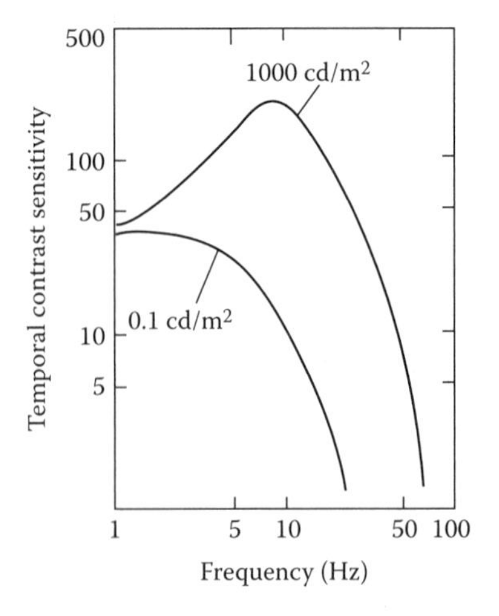

FIGURE 13.6
Typical flicker sensitivity of the eye for two values of retinal illuminance. (After Pearson, 1975.)

must be on the negative slope of the response curve and that relative brightness fluctuation must be below some value set by the flicker frequency. The limitation will be somewhat less stringent at moderate than at high luminance. The time constant of the human visual system appears to be a consequence again of evolution for survival. Since signals from potential predators are arriving in discrete quanta, there is a dilemma between the need to integrate for sufficiently long that the image is not quantum-limited (and possibly useless) and the need to react quickly for survival! The human nervous/muscular response time is about 0.1 s and the visual system has evolved to match this, with a maximal sensitivity around 8 Hz at high illumination.

13.5 Quantitative Measures of Investigative Performance

Whilst medical imaging is only one subdivision of mankind's endeavour to 'investigate' his environment, it reflects a fairly general need to have quantitative measures of how well the investigation is proceeding. Certainly, in the medical-imaging context, it is worth giving some thought to the nature and meaning of 'investigation'. Normally one will be trying to answer a fairly specific question (e.g. does the patient have breast cancer?); one will already have expectations as to the result (e.g. from knowledge of a palpable lump, the patient's age and family history), and one may never know the 'true' answer and therefore whether one's interpretation of the images was correct (e.g. because the patient does not go to surgery, or biopsy is unsatisfactory, etc.). Thus, useful measures of investigative performance will have to be based on the incremental improvement in knowledge achieved and on the comparison with the currently best-available approximation to a true answer (sometimes referred to as a 'gold standard'). It will be clear also that we are dealing with a process of decision making.

Several different kinds of decisions are involved in the total procedure of image interpretation. The first is that of 'detection': a decision as to whether an abnormality is present. Beyond this, however, there is 'localisation' (where is the abnormality?) and 'classification' (what sort of abnormality is it?). Of these, the detection process has been most fully discussed, and seems to be best understood, although both localisation and classification can usefully be considered as modifications of the detection process.

It is clearly important to be able to assess the quality of diagnostic decisions that result from a particular imaging (or similar) procedure, and a certain formalism has been developed for this purpose, particularly in relation to decisions on detection. This starts from the assumption that a 'true' answer exists to the question whether an abnormality is present, and compares this with the answer given by the imaging procedure. In this way, one can construct the statistical decision matrix shown in Table 13.2, covering the four possible situations.

TABLE 13.2

Statistical decision matrix, showing the four possible situations when testing for the presence of an abnormality.

	Actual Disease Is Present	No Actual Disease
Test result positive	True positive (TP)	False positive (FP)
Test result negative	False negative (FN)	True negative (TN)

On the basis of this formalism, it is common to use the following terms to indicate the quality of a diagnostic test:

$$\text{Sensitivity (or true positive fraction, TPF)} = \frac{\text{Number of correct positive assessments}}{\text{Number of truly positive cases}} = \frac{\text{TP}}{\text{TP}+\text{FN}} \tag{13.1}$$

$$\text{Specificity (or true negative fraction, TNF)} = \frac{\text{Number of correct negative assessments}}{\text{Number of truly negative cases}} = \frac{\text{TN}}{\text{TN}+\text{FP}} \tag{13.2}$$

$$\text{Accuracy} = \frac{\text{Number of correct assessments}}{\text{Total number of cases}} = \frac{(\text{TP}+\text{TN})}{(\text{TP}+\text{TN}+\text{FP}+\text{FN})}. \tag{13.3}$$

At this stage it is necessary to amplify some points in the earlier discussion of the objectivity of this kind of analysis. In the first place, particularly if decisions are being made by a human observer (or even by a programmed machine), one must accept the likelihood of bias, deliberate or otherwise. There will always be a certain expectation value for the ratio between normal and abnormal cases, and the assumption of an inappropriate ratio will tend to bias the results. More importantly, however, diagnosticians may be strongly influenced in their decision making by knowledge of the consequence of a particular decision. Consider, for example, the hypothetical situations of a diagnostician wishing first of all to screen a population of apparently healthy women for possible signs of breast cancer, and, secondly, to examine a woman with a suspected breast lesion, in support of a decision on whether major surgery should be undertaken. In the first case, the statistical expectation of an abnormality will be very low, but the consequences of a high false-positive rate will be the relatively mild one of an excessive number of patients undergoing further examinations. Thus, there will be a valid bias towards achieving high sensitivity at the cost of decreased specificity. In the second case, the expectation of an abnormality will be much higher, but the diagnostician will need to be able to convey to the referring surgeon the degree of confidence that can be attached to the eventual assessment.

Another important qualification to note again is that (as Oscar Wilde remarks) 'truth is rarely pure and never simple': the assumption, implicit in the preceding matrix, that one can always expect a 'true' assessment of an abnormality is unrealistic. The best that one can usually hope for in the way of a 'definitive diagnosis' is a report on histopathology following surgery, biopsy, or post-mortem examination. Even this is not always forthcoming, and, when it is, can be subject to considerable uncertainty.

It is thus clear that a simple measure such as sensitivity, specificity, or even accuracy will not be an objective indicator of the quality of decisions available from a particular imaging test procedure. Such an indicator is provided rather better by the so-called receiver operating characteristic (ROC). This is constructed as a plot of TPF against false positive fraction

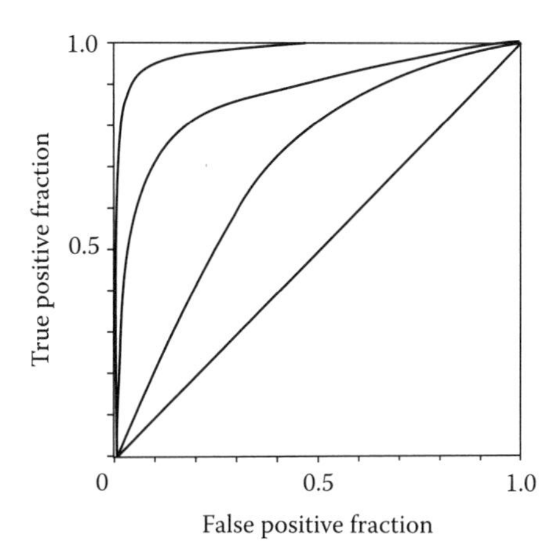

FIGURE 13.7
'Theoretical' examples of receiver operating characteristics (ROC). Any signal (or test) that generates an ROC curve wholly above and to the left of another is the more detectable.

(FPF), the individual points on the curve being obtained by repeating the test on a number of occasions with different degrees of bias (or decision threshold) as to the expectation of a positive result. A 'theoretical' set of such ROC curves is illustrated in Figure 13.7. The different curves in the figure are indicative of differing quality in the decision process: the diagonal straight line would be the result of a totally uninformative test, whilst lines approaching closest to the FPF = 0 and TPF = 1 axes are those corresponding to the best performance (Green and Swets 1966, Todd-Pokropek 1981). A practical example of the closely related 'location ROC' (LROC) curves obtained in perceptual tests on radioisotope images is given in Figure 13.8.

ROC analysis can be used both to compare the detectability of different kinds of abnormality and to compare performance (in the sense of facilitating detection decisions) either of imaging systems or of their operators, or of both. It can, however, be a very

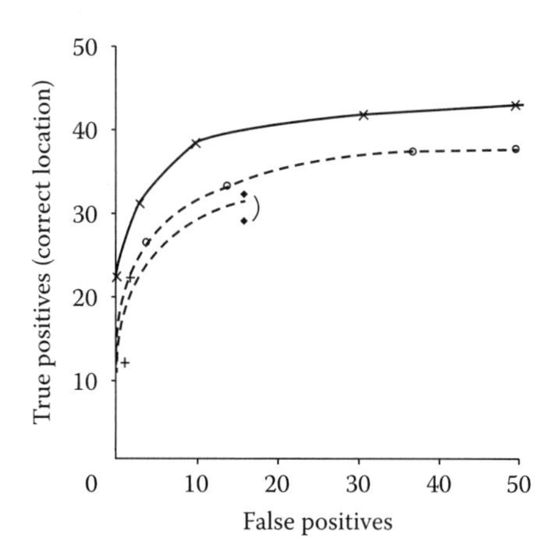

FIGURE 13.8
Measured ROC curves for an investigation of the effectiveness of different processing and display procedures (in use at three different medical diagnostic 'centres') for identifying both the presence and correct location of a simulated brain 'lesion' that had been inserted mathematically into a normal brain radioisotope scan. Curves of this type have been termed 'location ROC' or LROC curves. (From Houston et al., 1979.)

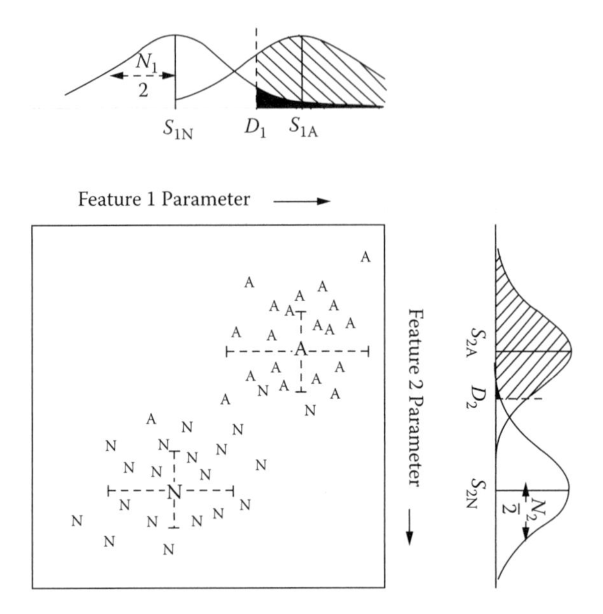

FIGURE 13.9
Illustration of 'noise'-limited feature separation. The signals for the two different features are characterised by their separations (e.g. S_{1A}–S_{1N}) and the widths of their noise spectra (e.g. N_1). Imposing a decision threshold (e.g. D_1) defines true positive fraction (hatched area) and false positive fraction (solid area). Corresponding distributions in two-dimensional feature space are also shown.

time-consuming procedure, and it is useful to note that equally good results can be produced by a faster rating procedure (Chesters 1982). In this, the observers are required to quantify their degree of certainty about the presence of the abnormality, for example, within the categories 0%–20%, 20%–40%, etc.

It is sometimes useful to consider the decision-making procedure in terms of a model in which the separation of an abnormal signal from a normal signal is a noise-limited process (Figure 13.9). In this, the 'signals' may be any of a large number of potentially quantifiable features of an image (e.g. grey level, smoothness of a lesion boundary), and the noise may originate from the observer, the imaging system, and/or variations in the object itself. In this model, one can see rather clearly the effect of placing a decision threshold at some particular value on the scale of signal magnitude. This model also illustrates the fact that decisions (whether observer- or machine-implemented) will often be made on the basis of a number of separate features, for each of which there will be a degree of noise-limited signal separation. In this situation, the decision will be made (whether in the mind of the observer, or in a computer) within the format of a multidimensional feature space (shown as two-dimensional in the figure). A more comprehensive treatment of the subject of image assessment and decision making in diagnostic imaging is given by Goodenough (1977).

As will have been seen in earlier chapters of this book, a concept of substantial and growing interest is that of parametric imaging: the process of deriving, from discrete regions of an image, data that can provide objective characterisation of features of the imaged tissue volume – X-ray absorption coefficient, blood flow velocity, etc. Here, again, questions arise about quality of performance of particular imaging systems and techniques – for example: 'How well does imaging system A obtain useful data of type X, and how does it compare objectively in this regard with system B?'

Answers to such questions will entail at least three measures of performance:

- How good is the system/technique in discriminating between small differences in an object (tissue) characteristic?
- With what spatial resolution does it achieve this?
- How much time does this take?

As discussed earlier, in Section 13.3, these will be interrelated in some way: the data needed to achieve a given degree of tissue discrimination will have to be acquired over finite measures of both space and time. In discussions of this issue, Hill and Bamber (Hill et al. 1990, Hill and Bamber 2004) derived an expression for a 'figure of merit' for any given procedure. Quantitatively, it is supposed that the procedure for differentiation is based on numerical measures, P, of some underlying physical property or parameter of the interrogated tissue, and that, for each value P_i, the technique will generate a corresponding signal of value S_i. The expression for figure of merit is then derived as the contrast-to-noise ratio for a given fractional change in 'property value', P, between two tissue conditions:

$$\Gamma = \frac{\mathrm{d}S}{\mathrm{d}P}\frac{P_i N^{0.5}}{\sigma} \tag{13.4}$$

where
P_i is the property value which we need to distinguish from an adjacent value,
N is the number of uncorrelated measurements performed, and
σ is the square root of the sum of the squares of individual variances in the measurement procedure that contribute to system noise.

This expression (originally derived in application to ultrasonic imaging) is a special example of the general formula for statistical variance (cf. 'Student's *t*-test'), and is indeed identical in form to that derived independently in the context of NMR imaging (MRI) by Edelstein et al. (1983). It should be an interesting exercise to explore further its practical validity and usefulness in a range of medical-imaging contexts.

References

Biberman L M (ed.) 1973 *Perception of Displayed Information* (New York: Plenum).
Blackwell H R 1946 Contrast thresholds of the human eye *J. Opt. Soc. Am.* **36** 624–643.
Chesters M S 1982 Perception and evaluation of images In *Scientific Basis of Medical Imaging* ed. P N T Wells (Edinburgh, UK: Churchill Livingstone) pp. 237–280.
Cornsweet T N 1970 *Visual Perception* (New York: Academic Press).
Cosgrove D O and McCready V R 1982 *Ultrasound Imaging of Liver and Spleen* (New York: Wiley).
Edelstein W A, Bottomley P A, Hart H R and Smith L S 1983 Signal, noise and contrast in nuclear magnetic resonance (NMR) imaging *J. Comput. Assist. Tomogr.* **7** 391–401.
Goodenough D J 1977 Assessment of image quality of diagnostic imaging systems In *Medical Images: Formation, Perception and Measurement* ed. G A Hay (New York: Wiley) pp. 263–277.

Green D M and Swets J A 1966 *Signal Detection Theory and Psychophysics* (New York: Wiley).
Haber R N and Hershenson M 1973 *The Psychology of Visual Perception* (New York: Holt, Rinehart and Winston).
Halmshaw R 1981 Basic theory of the imaging process in the context of industrial radiology In *Physical Aspects of Medical Imaging* eds. B M Moores, R P Parker, and B R Pullan (Chichester, UK: Wiley) pp. 17–37.
Hill C R and Bamber J C 2004 Methodology for clinical investigation In *Physical Principles of Medical Ultrasonics* (2nd edn.) eds. C R Hill, J C Bamber and G R ter Haar, Chapter 9 (Chichester, UK: Wiley) pp. 255–302.
Hill C R, Bamber J C and Cosgrove D O 1990 Performance criteria for quantitative ultrasonology and image parameterisation *Clin. Phys. Physiol. Meas.* **11 Suppl. A** 57–73.
Hill C R, Bamber J C, Crawford D C, Lowe H J and Webb S 1991 What might echography learn from image science? *Ultrasound Med. Biol.* **17** 559–575.
Houston A S, Sharp P F, Tofts P S and Diffey B L 1979 A multi-centre comparison of computer-assisted image processing and display methods in scintigraphy *Phys. Med. Biol.* **24** 547–558.
Metz C E and Doi K 1979 Transfer function analysis of radiographic imaging systems *Phys. Med. Biol.* **24** 1079–1106.
Overington I 1976 *Vision and Acquisition* (London, UK: Pentech).
Pearson D E 1975 *Transmission and Display of Pictorial Information* (London, UK: Pentech).
Rose A 1973 *Vision—Human and Electronic* (New York: Plenum) p. 12.
Todd-Pokropek A 1981 ROC analysis *Physical Aspects of Medical Imaging* eds. B M Moores, R P Parker and B R Pullan (Chichester, UK: Wiley) pp. 71–94.
Webb S 1987 Significance and complexity in medical images: Space variant, texture dependent filtering *Proc. 10th IPMI Conf. Utrecht* eds. M Viergrever and C N de Graaf (New York: Plenum).

14

Computer Requirements of Imaging Systems

G. D. Flux, S. Sassi and R. E. Bentley

CONTENTS

14.1 Introduction 755
14.2 Computing Systems 757
14.2.1 Operating Systems 758
14.2.2 Monitors 759
14.3 Generation and Transfer of Images 759
14.3.1 File Formats 761
14.3.2 Picture Archiving and Communication Systems 762
14.3.3 Internet and Intranet 763
14.3.4 Teleradiology 763
14.3.5 Standards 764
14.4 Conclusion 764
References 764

14.1 Introduction

The exponential increase in processing speed and memory capacity for portable and fixed digital data-storage devices has played a pivotal role in the development of digital imaging systems, as shown in Figure 14.1. This trend is expected to continue with the potential for significant acceleration due to developments in the nanotechnology field leading to large improvement in performance of highly miniaturised integrated circuits and electronic components. It is hard to believe that G. E. Moore's (Intel's cofounder) prediction in 1965 of an exponential growth in the number of transistors per integrated circuit (Moore 1965) would hold for so long and will be true for almost every measure of the capabilities of digital electronic devices. For medical imaging, this has meant a complete transformation from totally analogue imaging using analogue capture and display of imaging data to an almost completely digital environment. This of course has led to significant improvement in the quality of imaging services provided by health care institutions. In order to fully harness and utilise these technologies, medical physicists have played an increasingly central role in the provision of these services. In a trend similar to the increase in the capabilities of digital electronic devices, the role of medical physicists has expanded from that of being mainly a radiation physicist to include that of IT expert, computer programmer and imaging scientist.

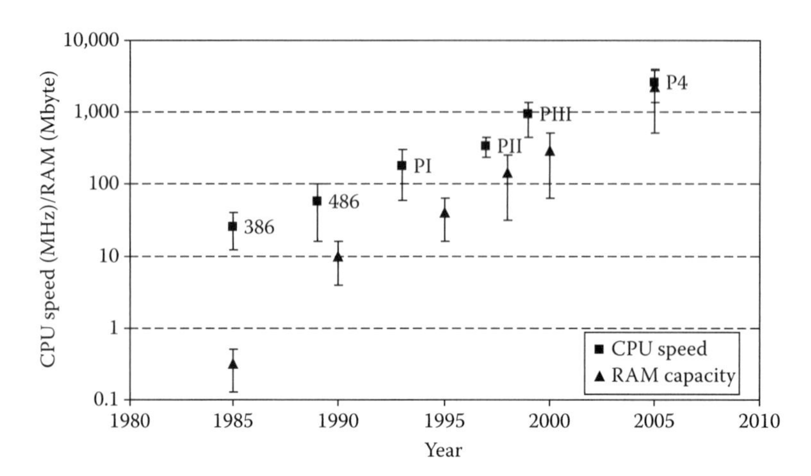

FIGURE 14.1
Increase in typical CPU speed and RAM capacity for desktop PCs. Since the first edition of this book the CPU has increased by 2 orders of magnitude and the RAM capacity has increased by 3 orders of magnitude. The trend looks set to continue with the advent of multiple processors and advances in computer architecture.

Computers play an ever-expanding role in our lives at home and at work. Almost every machine one may operate or use will have a computer on board, and this applies equally to medical imaging equipment. Most medical imaging systems have more than one computer to control various functions of the machine. Probably the best example of that are multimodality systems, where dedicated computers, often with standard user interfaces, are used to control the patient couch, the multimodality gantries, the detection systems, and independent image acquisition, processing and data-storage systems. Often image acquisition, storage, processing, analysis and clinical reporting are carried out on separate computers, sometimes provided by different vendors. This, in addition to the relatively recent introduction of local and national area patient administration systems (PAS) and picture archiving and communication systems (PACS), makes data integrity, safety, compatibility and conformance issues an increasingly essential part of medical imaging systems' specifications. Even though many vendors still use proprietary data formats to process images and transfer data between the acquisition and processing workstation, almost all imaging systems will declare some level of compliance with agreed international standards. These may include DICOM, HL7, IHE and ISO protocols. Despite that, in practice medical physicists have to test the integrity of data transfer between various networks and image processing, storage and display systems often discovering new conflicts, distortions and mislabelling of transferred data. All these factors make it essential for a medical physicist to learn about the various computer systems and international standards and protocols that govern their interactions.

The emergence of multimodality imaging and the rapid growth of sub-second volumetric scanning in CT resulted in an ever-growing number of datasets acquired during each acquisition. In order to fully harness the new technologies, a plethora of advanced image processing and quantitative packages have been developed, but these highly demanding applications are often run on specialised workstations.

In this chapter we provide a resumé of various computer systems used to acquire, process and display digital imaging data, and provide an introduction to some of the international standards and protocols applied to medical imaging.

14.2 Computing Systems

Despite the current trend by medical imaging systems' vendors to migrate their acquisition and processing systems to windows-based instead of Unix-based systems, a range of computing platforms are currently being used in medical imaging departments. These include personal computers (PCs) such as IBM PCs (and clones) and Apple's Mac systems to workstations produced by Sun Microsystems, Silicon Graphics, DEC and Hewlett-Packard. The relative merits of each continue to be debated with each system having staunch supporters. In recent years, PC systems appear to be winning the battle with PCs entering into more and more areas that used to be the exclusive domain of workstations with workstation systems being limited to a number of highly specialised server-type applications. A principal difference between personal computers and workstations is the type of microprocessor instruction set architecture used. PCs mainly employ the CISC (complex instruction set computer) architecture, while most workstations are mainly based on the RISC (reduced instruction set computer) architecture.

The RISC design was originally developed by a team of researchers at the University of California. In 1985 Sun Microsystems developed its well-known SPARC (Scalable Processor Architecture), which was based on the RISC architecture, and later promoted as the SPARC processor by SPARC International. This architecture was fully open and nonproprietary which led to its use in most workstation systems. Due to the relatively slow memory-access times by central processing units (CPU) in older computing systems, the RISC architecture's use of less complex instruction sets to enable a large number of operations to be carried out by the microprocessor with relatively few calls to the memory made this architecture very efficient. The CISC architecture uses more complex instruction sets, resulting in a significant reduction in the number of instruction sets and thus quicker execution of the operations.

Increasingly the two architectures are becoming less separated as RISC microprocessors become capable of handling more complex instruction sets and CISC microprocessors are able to handle techniques such as pipelining (whereby multiple operations may be carried out simultaneously) which was once the reserve of RISC architecture. Until recently, workstations have been viewed as a necessary minimum to handle the processing and display of image data due to their superior computing power and their high-quality graphics capabilities. However, the rapid increase in the home PC market fuelled by a rapid decrease in PC prices has generated improvements resulting in performance and graphics capabilities that compare well with workstations. The advent of Pentium chips by Intel in particular has played a vital role in revolutionising the PC market.

The exponential increase in the volume of medical imaging data and the ever-increasing complexity of imaging systems' configurations has placed exceptional demands on the performance of microprocessors used for these computationally intensive applications. In addition to increasing clock speed, processor performance enhancement has been achieved by employing innovative architectural solutions such as the SIMD (single instruction, multiple datastream) instruction set and its more recent extensions SSE (streaming SIMD extensions), SSE2 and SSE3. These instruction sets speed up applications by enabling basic computational operations to be performed on multiple data elements in parallel using single instructions. Almost all current desktop and notebook computers and the majority of servers and workstation processors are based on the X86 instruction set architecture. This architecture is basically a set of assembly-language instructions created to carry out the basic operations that a computer CPU performs. The basic set of instructions has evolved and a number of extra instructions have been added to it that are processor specific. For example, the MMX instruction and its SSE extensions were introduced by Intel in their Pentium line of microprocessors, while AMD has introduced the 3DNow instruction set to work with floating-point numbers. The X86 platform is supported by a wide range of computer operating systems including MS windows, Apple Mac OS, Solaris, Linux and MS-DOS, and it is implemented on almost all current PC processors produced by Intel, AMD, Cyrix and many others.

These developments, allied with unparalleled increases in CPU speed, have enabled manipulation of large scientific data in very short times. At the time of writing, Pentium-based processors are widely used for scientific computing applications including image reconstruction and mathematical simulations. This allowed these types of machines to seriously challenge the dominance of the RISC-based servers using SPARC architecture such as those produced by the Sun Microsystems of the workstation domain. However, the latter systems still claim superior graphics-handling capabilities due to their ultra port architecture (UPA) interconnect bus architecture, which was introduced with Sun Microsystems' family of high-speed UltraSPARC processors, and an open architecture operating system (UNIX) that is more suited to graphics and large datasets as well as a processor that uses an instruction set designed to accelerate 2D and 3D graphics.

14.2.1 Operating Systems

The computer's operating system (OS) is a programme that controls every task a computer performs. It manages hardware and software resources and organises communications between them. All desktop and notebook computers and most hand-held devices including mobile phones have operating systems. Microsoft windows (MS corporation), Apple Mac OS (Apple Corporation) and UNIX (Bell Laboratories) are the most commonly used operating systems. Choice of the operating system is crucial to the stability, flexibility and upgradability of the system's hardware and software.

Computer operating systems used in medical imaging can be categorised by the type of user interface as either command line or graphical user interface (GUI). Command-line operating systems such as UNIX and MS-DOS require command line entry of all user instructions, while GUI operating systems use menus, icons, tool bars and windows to perform user instructions and operations. A GUI OS is, therefore, much easier to learn and use than a command-line OS.

Until recently most high-end workstations adopted a version of the UNIX operating systems usually with command-language interpreters in the form of a shell such as the C-shell and its various derivates (e.g. Tenex C shell (tcsh), Korn shell (ksh), Hamilton C shell and Bourne-Again shell (bash)). This adds extra functionality and flexibility to the user interface using the command line. However, the advent of graphical applications that run on UNIX operating systems such as X-Windows has allowed windows-based user interfaces similar to the GUI OS to be run on what are primarily command-line systems.

This blurring of the demarcation between the different categories was further entrenched with the introduction of Windows NT by Microsoft to provide all the important features of UNIX OS including improved data security, improved large-scale memory management and support for workgroup networking, while guaranteeing compatibility with most windows applications.

Operating systems that are more suited to multiple-user access, such as Windows NT, Linux and even Solaris are also available for PC servers. Direct comparison of different systems is not trivial. Benchmark tests are available from certain institutions, for example, SPEC (Standard Performance Evaluation Corporation) and TPC (Transaction Processing performance Council), and many national and international user groups endeavour to establish, collect and provide benchmark test results. Information gained from these tests can be very useful although these comparisons are only valid for the specific application and set-up used for a particular test and should be used cautiously.

Both personal computers and workstations may be connected in a local area network (LAN) that allows maximum use to be made of computing resources within a department. Thus priority may be given to interactive use of the system during the day with the system running overnight on, for example, Monte Carlo calculations. Networks also require servers to manage resources such as file storage, database management, printing and the handling of network traffic.

14.2.2 Monitors

Display of medical images has also witnessed a massive revolution with the widespread use of flat panel display (FPD) almost completely replacing the 50 year-old cathode ray tube (CRT) for visual display units (VDUs).

FPD, mainly in the form of liquid crystal display (LCD) screens, come to life when light from behind the screen is shone through the screen's matrix of tiny coloured liquid crystal cells. Signals control each cell, allowing varying amounts of colour through, and a picture is built up pixel by pixel. The screen consists of a liquid-crystal solution between two polarising sheets. An electric current passed through the solution causes the crystals to align, obscuring the passage of light. LCDs can be either active matrix, whereby thin film transistors (TFT) control the pixels, or passive matrix, whereby LCD elements are situated at the intersection of a grid of wires. Active-matrix TFT systems provide better resolution and naturally are more expensive than passive-matrix systems. Research continues into both technologies. LCD monitors generally have a high contrast ratio even though their brightness level is lower than that of CRT monitors. However, CRT monitors are still used on legacy systems and still have an important role in severe ambient conditions including surgical theatres, where very bright lights are required, and where LCD monitors may not be suitable.

Active-matrix LCD (TFT) touch-screen monitors are gaining wide use on many medical imaging devices to enable fast system set-up and acquisition protocol selection. Less common is the electroluminescent display (ELD), whereby a film of phosphorescent substance, sandwiched between plates of horizontal and vertical wires, glows when a current is passed through an intersection. Gas plasma displays work on a similar principle, but use a neon gas between plates. However, only monochrome display is possible with this technology and it requires 50% more power than LCD display.

Even though CRT monitors have a wider dynamic range than most LCD systems, they are more prone to geometric distortion and temporal blurring caused by image lag. It is worth mentioning that standard TV reads out alternate lines in two fields, using so-called interlaced readout, whereas radiology reporting systems used to display medical images often use progressive (line by line) readout.

14.3 Generation and Transfer of Images

Medical imaging systems are often connected to a LAN comprising a number of computer platforms with different methods for mapping registers to memory locations that employ different data-storage depths. A system that allows a code to run on, and transfer data between, platforms with different internal representations was created. This system is often referred to as endianness, which is basically a system (structure) for the order in which bytes in multi-byte numbers should be stored, using either most significant first (Big-Endian) or least significant first (Little-Endian). This system is an important consideration in network programming, since two computers with different byte orders may be communicating. Failure to account for varying endianness when writing code for mixed platforms can lead to image artefacts or bugs that can be difficult to detect and resolve. Luckily, most networked computers support the TCP/IP networking protocol which includes a definition of all required functions to convert data from one endianness system to the other and allows for data storage in an endianness-independent system termed 'network byte order'.

The X86 family and Digital Equipment Corporation architectures (PDP-11, VAX, Alpha) are representatives of Little-Endian, while the Sun SPARC, IBM 360/370, and Motorola 68000 and 88000 architectures are Big-Endians. Still, other architectures such as PowerPC, MIPS and Intel's 64 IA-64 are Bi-Endian, that is, they are capable of operating in either Big-Endian or Little-Endian mode. The system's Endian format is often included in the header of the binary data file.

Although some medical systems still use 8 bit digitisation systems, most modern medical imaging systems use 10 and 12 bit systems. High-resolution (16 bit) imaging systems are primarily used in research applications such as electron and fluorescence microscopy. Data will, in general, be in digital form, which means that the operations of image capture and digitisation associated with visual image formation are normally avoided. Less frequently, images and other data are digitised from analogue media such as X-ray films although it is expected that all radiological data will be digital at the point of capture in the future.

Data acquisition, processing and storage are often carried out on independent computer systems. Acquired data are frequently transferred from the acquisition station to processing stations during or at the end of the acquisition process. Processed data transfers may be purely local, possibly between different computerised devices in the same room, or may involve transmission over long distances for viewing and assessment.

In some cases, a straightforward link between two specific computers is all that is required, and the hardware and software to support this is relatively simple. In general, the requirement is not to be restricted in this way to a single end-to-end connection but to have access to a fully switched network that will allow unrestricted transmission within the network. A LAN should provide the capability to connect from any device to any other, if necessary passing through a series of intermediate stages (routing nodes) to provide for automatic continuation after interruption and to provide an alternative route when the primary route is unavailable. The network must be able to operate between computers of different types and different computer operating systems. It must provide the means for conversion of codes and data formats and must also offer a means of error detection and the facility for retransmission when an error is found.

A distinction is made between LAN and wide area network (WAN), although there is no clear definition of these terms. In general, a LAN, at least in the United Kingdom, does not involve the crossing of public property and is generally limited to about 3 km in extent. A WAN, on the other hand, can extend to the whole world.

In a LAN, higher transmission rates can be used. The two best-known LAN systems are the IBM Token Ring, with a data rate of 4 or 16 Mbit s^{-1}, and Ethernet, which usually operates with a 100 Mbit s^{-1} backbone, although Gigabit Ethernet is now available. These systems have high overheads for such things as message addressing so that, even in a lightly loaded system with no contention between competing users, the actual throughput may be less than 50% of the specified figure. When there is contention between several applications at the same time, the degradation in performance can be very marked. Operating at only 10% of the rated figure, a full CT image set, that may contain 35 slices, each consisting of a 512 × 512 matrix, and will be 16 bits deep, can be transferred over Ethernet in 15 s or so, which becomes feasible for clinical work. It should be realised, however, that transmission in this time assumes that the workload of the two communicating computers is such that their Ethernet interfaces are given sufficient resources to sustain this speed.

As a result of the necessary involvement of phone service providers and relevant authorities, a WAN is often restricted in transmission speed. Integrated Services Digital Network (ISDN) is an international communications standard that works at 64,000 bits s^{-1}. In this case a simple 512 × 512 image, 8 bits deep, will take 35 s to transfer. The CT array mentioned would take 40 min to transfer although for data downloading it is possible to run two lines simultaneously. Broadband ISDN (B-ISDN) is available that can support data rates of 1.5 Mbit s^{-1}, although it is not currently widely used. Mega stream circuits have become more commonplace that run at 8 Mbit s^{-1} in the United Kingdom, and asynchronous transfer mode (ATM) links are available that run in theory from 52 to 622 Mbit s^{-1}, although in practice they may be half of that speed. ATM links are currently a strong contender to be the preferred choice for both LANs and WANs for the transfer of medical image data (Huang et al. 1997). An ATM link transfers packets or cells of a fixed size, but unlike TCP/IP (transmission control protocol/Internet protocol) fixes a channel between two points before transfer begins. This prevents allowances to be made for network traffic. In general figures quoted for network speeds can only be used as a guide. Whilst a 10-fold improvement in the aforementioned figures may be obtained by using data-compression techniques, procedures for error correction and retransmission may reduce the speed of transfer.

14.3.1 File Formats

The use of proprietary file formats and different transfer protocols has proved a significant hindrance to the transfer, analysis and storage of medical imaging data. Until recently each manufacturer used their own proprietary image format, which was invariably incompatible with other vendors' systems.

Recent years have seen greater convergence of format standardisation with the adoption of the DICOM standard by all major medical equipment manufacturers. This facilitated the seamless integration of imaging data across various imaging modalities and vendor systems and encouraged the growth of stand-alone image processing, analysis and storage systems. One example of this is the Interfile image format for nuclear medicine data which arose from the European Cost-B2 project (Todd-Pokpropek et al. 1992). The most common general format that has emerged is DICOM (digital imaging and communications in medicine), which was produced as version 3 of an ACR-NEMA (American College of Radiologists/National Electrical Standards Manufacturers Association) standard aimed at solving the problem of incompatibility (Bidgood and Horii 1992, Ratib et al. 1994, Bidgood et al. 1997).

The DICOM standard is more than simply a file format for medical image data. It specifies a number of 'service classes' such as 'print management', including the use of referenced look-up tables as well as printer queue management; 'query/retrieve' which enables the access of data based on search criteria such as a patient name and 'storage' which aids transfer and storage of image data. 'Information objects' are also defined that are associated with a specific image modality or type. Thus service-object pairs (SOP) may be specified that combine a service with a particular modality (e.g. the transfer of a CT image set). Service class users (SCU) are devices that require some particular service classes to be available for operations on data they hold, and service class providers (SCP) are able to provide a number of these services. An example would be a CT scanner acting as an SCU that requires the use of a printer as an SCP to print images. Integration within PACS systems is aided by service classes such as patient management (involving e.g. admission and discharge) and study management (enabling the tracking of a patient's progress). Manufacturers can specify whether their product supports a 'storage class' for a given modality as a provider or user, thereby enabling the purchaser to determine the compatibility of equipment. Conformance statements specify 'proposed presentation contexts' – a list of services the device will require – and 'proposed acceptance contexts' – a list of services that may be provided. The use of DICOM continues to expand, as it is more readily accepted as a solution to problems generated by proprietary systems. The development of the DICOM standard is central to the digital revolution in health care provision at local, regional, national and international levels. However, at the time of writing, procedures for testing DICOM compliance are not specified but left to the individual manufacturer and in practice there is a danger that the user will be left to solve problems of non-compliance. These problems are often highlighted when institutions start implementing new PACS systems.

14.3.2 Picture Archiving and Communication Systems

The falling cost of mass storage and processing speed, the growth of networking (both within LANs and via the World Wide Web) and international standardisation of image file formats have led to an emergence of a more widespread use of computers. One example of this is the development and subsequent wide implementation of PACS within hospitals (Bick and Lenzen 1999).

These systems enable patient data to be stored digitally and recalled at any time or place within a particular institution. Full implementation of a PACS obviates the need for hard-copy media (e.g. X-ray films) with the inherent problems that may be caused by this means of storage. A PACS system enables ready, simultaneous access of patient data to be viewed simultaneously at different locations, minimising the potential for data to be lost in transit. It is essential for PACS to be integrated with the hospital information system (HIS), radiology information system (RIS) and electronic patient records (EPR) to enable full electronic storage, access to all patient data, and seamless matching of images with reports and patient attendance data for optimal management of short- and long-term data storage. This may be best achieved using ANSI-specified Health Level 7 standards, designed for the storage and transfer of clinical data between hospital departments.

PACS typically consists of four subsystems:

1. An image acquisition subsystem for image capture, requiring video frame grabbers and often film digitisers and document scanners.
2. A virtual image communications network, which can be dedicated or may use existing LANs. However, rapid access involving large amounts of data necessitates network backbones supporting data transfer rates of at least 100 Mbits s^{-1}.

3. An image storage subsystem enabling mass storage. This usually consists of both low-capacity short-term storage enabling rapid access to data relevant to current in- and outpatients, and high-capacity storage for archiving with database management. Integration with the HIS and RIS enables automatic image transfer from long-term to short-term storage in preparation for its recall. Storage capacity may be increased by a factor of 2–3 if lossless compression is employed and by as much as a factor of 40 if lossy compression is used.
4. An output and display subsystem. Ideally, image data will be displayed simultaneously with patient information. Quality of image display must be comparable with that obtained from film. For a large system, VDUs may account for 50% of the total cost. Minimum screen resolution will depend on the data being examined but for radiological review of X-ray images a matrix size of 2048 × 2048 pixels will be required. The choice of VDU should depend on clinical use and type of radiological examinations. Nuclear medicine, PET and ultrasound images require colour display, while computed radiography (CR), digital radiography (DR), computed tomography (CT) and magnetic resonance (MR) image data require high-resolution grey-scale monitors. Medical-imaging-grade monitors are usually more expensive than standard PC monitors.

14.3.3 Internet and Intranet

Recent years have seen a boom in the use of the Internet for both home and business use with about 2.0 billion Internet users worldwide in 2010, registering an increase of 445% in the total number of Internet users over a 10 year period from the year 2000 (Internet World Stats 2011). There are two important implications for medical imaging resulting from this. Firstly, the information accessible to medical centres (and patients) has increased to an extent that the wealth of information is difficult to assimilate, affecting both clinical and research protocols. Secondly, the possibility now exists to process and view image data remotely in a manner that was never before possible. The use of Java, a high-level platform-independent programming language developed by Sun Microsystems to be compatible with the World Wide Web, enables applications (called Java applets) to be downloaded from a Web server and run on a computer by Java-compatible web browsers (Phung and Wallis 1997, Mikolajczyk et al. 1998, Truong et al. 1998).

Increasing use is being made by intranets (networks accessible only within an organisation). These are particularly attractive to medical imaging departments as they may be set up shielded from the Internet, for security, and can use TCP/IP communications protocols, such as telnet and ftp, to access data remotely or to transfer data between systems. TCP/IP communications can also be used to transfer anonymised imaging data safely between various centres.

14.3.4 Teleradiology

The growth of the Internet, video conferencing and information technology has led to a marked increase in telemedicine (also called telehealth) and teleradiology. There are numerous cases of scan data being acquired in remote regions and sent to specialist centres for examination and diagnosis. Combined email, fax and satellite video conferencing can also be used in conjunction with the Internet to allow 'teleconsulting'. A teleradiology set-up consists basically of an image capture and sending station, possibly including a film digitiser, a transmission network and a receiving and viewing station. As with all network issues there is an inevitable trade-off between image resolution and transmission speed. Whilst standards such as DICOM may be used to facilitate teleradiology, there are as yet no international standards specifying transmission and format protocols. The American College of Radiologists (ACR) has, however, defined broad standards for the use of teleradiology, and a European project (EUROMED) has addressed the issue of standards using the Internet and the World Wide Web (Marsh 1998). These include the goals of teleradiology and the personnel that should be involved as well as the required resolution for individual modalities and image processing facilities such as magnification and grey-scale inversion.

14.3.5 Standards

A number of national and international standards have been produced by bodies such as the British Standards Institution (BSI) standard (BS EN 60601-1:2006) which is widely adopted in Europe, the Middle East and Africa, and the Pan-American International Electrotechnical Commission (IEC) standard (medical imaging equipment publication 60601-1). These standards' primary concern is with the safety and essential performance of any medical electrical equipment intended to be used for patient care including monitoring, diagnosis and treatment of patients. Computers used in medical care and in particular image display monitors should comply with these standards. As a result all medical imaging grade computing and VDUs are much more expensive than their equivalent standard equipment. It is important that any equipment used for patient care has the required certification and complies with the relevant standards.

14.4 Conclusion

Medical physicists have become increasingly relied upon for management of and advice on the use of computers. Whilst this may prove to be an onerous responsibility, it should also be recognised that the IT revolution presents greater opportunities for both research and for clinical impact due to the information that may be obtained from advanced image processing and analysis. The ability to distribute and share medical data and knowledge not just within a local area or a region but increasingly globally will hopefully lead to greater quality and equality in health care provisions and more international collaboration.

References

Bick U and Lenzen H 1999 PACS: The silent revolution *Eur. Radiol.* **9** 1152–1160.

Bidgood W D and Horii S C 1992 Introduction to the ACR-NEMA DICOM standard *Radiographics* **12** 345–355.

Bidgood W D, Horii S C, Prior F W and Van Syckle D E 1997 Understanding and using DICOM, the data interchange standard for biomedical imaging *J. Am. Med. Inform. Assoc.* **4** 199–212.

Huang H K, Wong A W K and Zhu X 1997 Performance of asynchronous transfer mode (ATM) local area and wide area networks for medical imaging transmission in clinical environment *Comput. Med. Imaging Graph.* **21** 165–173.

Internet World Stats 2011 http://www.internetworldstats.com/stats.htm, Miniwatts Marketing Group Copyright 2000–2011. Chapter 14 p. 765. Accessed on December 22, 2011.

Marsh A 1998 EUROMED—A 21st century WWW-based telemedical information society *Lect. Notes Comput. Sci.* **1401** 54–63.

Mikolajczyk K, Szabatin M, Rudnicki P, Grodzki M and Burger C 1998 A JAVA environment for medical image data analysis: Initial application for brain PET quantitation *Med. Inform.* **23** 207–214.

Moore G E 1965 Cramming more components onto integrated circuits *Electronics* **38** 114–117.

Phung N X and Wallis J W 1997 An Internet-based, interactive nuclear medicine image display system implemented in the Java programming language *J. Nucl. Med.* **38** 210P–211P.

Ratib O, Hoehn H, Girard C and Parisot C 1994 Papyrus-3.0—DICOM-compatible file format *Med. Inform.* **19** 171–178.

Todd-Pokropek A, Cradduck T D and Deconinck F 1992 A file format for the exchange of nuclear-medicine image data—A specification of interfile version 3.3 *Nucl. Med. Commun.* **13** 673–699.

Truong D C, Huang S C, Hoh C, Vu D, Gambhir S S and Phelps M E 1998 A Java/internet based platform independent system for nuclear medicine computing *J. Nucl. Med.* **39** 278P.

15

Multimodality Imaging

G. D. Flux

CONTENTS

15.1 Introduction ... 767
15.2 Image Transformation ... 768
15.3 Classification Schemes ... 770
15.3.1 Point Matching ... 770
15.3.2 Line or Surface Matching ... 770
15.3.3 Volume Matching ... 771
15.3.3.1 Sum of Absolute Differences ... 771
15.3.3.2 Voxel Ratios ... 772
15.3.3.3 Correlation Coefficient ... 772
15.3.3.4 Mutual Information ... 772
15.4 Non-Rigid Registration ... 773
15.5 Multimodality Image Display ... 773
15.6 Clinical Applications ... 774
15.7 Registration Accuracy ... 777
15.8 Discussion ... 778
15.9 Conclusions ... 778
References ... 778

15.1 Introduction

Each imaging modality provides unique information and it is usual for a patient to have a number of scans in the course of diagnosis and treatment. Multimodality imaging (MMI) addresses the possibility of obtaining information from a combination of image data that may not be available from consideration of each image in isolation. The most obvious use of MMI is visualisation, particularly to localise functional image data by placing it within anatomical context. Whilst this is useful for diagnostic purposes, MMI is increasingly being applied to treatment planning. Image registration (sometimes called co-registration or image fusion) is used to align image sets to achieve a spatial correspondence that enables direct qualitative or quantitative comparison. Registration may be performed on images of the same patient obtained from different modalities, on time-sequential image sets obtained from the same modality, or to align image data with standardised atlases of anatomy.

In some cases image data from different modalities are acquired such that the images are aligned. This can be enabled by ensuring that patient set-up parameters are identical for all scans, for example, by using stereotactic frames (Gill et al. 1991, Lemieux et al. 1994) or masks (Miura et al. 1988, Meltzer et al. 1990) to maintain the patient in the same position relative to the scanner and by acquiring data into similar matrix dimensions. This approach is becoming increasingly common with the advent of hybrid scanners such as PET/CT (see Section 5.9.3.9) and SPECT/CT (see Section 5.8.3.5). However, where this is not possible, image data must be transformed and resampled to achieve registration.

15.2 Image Transformation

A number of classifications of image transformation are possible although essentially image transformations may be considered as either rigid body, whereby spatial coordinates between two image sets are linked only by identical rotation, scaling and translation parameters, or elastic, whereby individual elements within an image set are transformed according to different parameters. Rigid-body registration is used more commonly due to its relative simplicity and to avoid the potential pitfalls of warping image data unrealistically to achieve spatial correspondence. However, it must be noted that with few exceptions (e.g. the skull), the human body is not rigid and relative geometry of a patient's anatomy will change from one scan to the next, as well as during the course of a scan.

Rigid-body registration requires the determination of nine transformation parameters that spatially link any two image sets – scaling, translation and rotation in three dimensions. Mathematically this is best considered in terms of matrix algebra since a digital image set may be represented as a set of 'points' in 3D. Two 3D point sets p_i' and p_i'', $i = 1, 2, \ldots,$ N (for N points) are related by:

$$p_i' = \boldsymbol{R}(\boldsymbol{S}p_i'') + \boldsymbol{T} + \zeta_i \quad (15.1)$$

where

$\boldsymbol{S}$ is a diagonal (3 × 3) scaling matrix,
$\boldsymbol{R}$ is a (3 × 3) rotation matrix,
$\boldsymbol{T}$ is a (3 × 1) translation vector, and
ζ is a (3 × 1) 'noise' or 'uncertainty' vector required where points may not be precisely localised or for point sets not exactly correlated via a rigid-body fit.

Thus, if the scaling transformation is directly applied, that is:

$$p_i = \boldsymbol{S}p_i'' \quad (15.2)$$

it is then necessary to determine the rotation and translation parameters to minimise

$$\sum_{i=1}^{N} |\zeta_i|^2 = \sum_{i=1}^{N} |p_i' - (\boldsymbol{R}p_i + \boldsymbol{T})|^2. \quad (15.3)$$

For the purposes of digital image manipulation the transformations are best regarded as matrix operations. In this case the points p_i' and p_i'' are represented by column vectors:

$$p' = \begin{pmatrix} x' \\ y' \\ z' \end{pmatrix} \tag{15.4}$$

$$p'' = \begin{pmatrix} x'' \\ y'' \\ z'' \end{pmatrix} \tag{15.5}$$

the scaling operator by a single diagonal matrix:

$$\boldsymbol{S} = \begin{pmatrix} s_x & 0 & 0 \\ 0 & s_y & 0 \\ 0 & 0 & s_z \end{pmatrix} \tag{15.6}$$

the rotation operator by separate 3 × 3 matrices for rotation about each orthogonal axis:

$$\begin{aligned} \boldsymbol{R}_x &= \begin{pmatrix} 1 & 0 & 0 \\ 0 & \cos\theta_x & \sin\theta_x \\ 0 & -\sin\theta_x & \cos\theta_x \end{pmatrix} \\ \boldsymbol{R}_y &= \begin{pmatrix} \cos\theta_y & 0 & \sin\theta_y \\ 0 & 1 & 0 \\ -\sin\theta_y & 0 & \cos\theta_y \end{pmatrix} \\ \boldsymbol{R}_z &= \begin{pmatrix} \cos\theta_z & \sin\theta_z & 0 \\ -\sin\theta_z & \cos\theta_z & 0 \\ 1 & 0 & 0 \end{pmatrix} \end{aligned} \tag{15.7}$$

and the translation operator by a column vector:

$$\boldsymbol{T} = \begin{pmatrix} t_x \\ t_y \\ t_z \end{pmatrix}. \tag{15.8}$$

A 4-vector description of this transformation is also possible (Webb 1993). Of these degrees of freedom, rotation and translation parameters must usually be determined from the images themselves. Scaling on the other hand can invariably be performed according to the known image dimensions that will have been defined according to data acquisition and reconstruction parameters.

15.3 Classification Schemes

A number of classification schemes have been defined for image registration, based on various criteria (Van den Elsen et al. 1993, Maintz and Viergever 1998, Hill et al. 2001, Zitova and Flusser 2003). One approach is to consider the parameter that defines the cost function that is to be minimised to achieve registration. It is evident that in the absence of external devices to define the location of image data, any two image sets to be registered must contain some information common to both. This information can consist of anatomically derived features or can be extracted from voxel values.

15.3.1 Point Matching

The most basic, and perhaps the most obvious feature to consider as a basis for registration is a landmark that can be considered as a point (Hill et al. 1991, Papavasileiou et al. 2001a). To achieve rigid-body registration a minimum of three non-coplanar points must be defined, although the accuracy of registration may be increased significantly by considering a larger number. Points may consist of landmarks obtained from anatomical features, which can include intersections of structures. CT and MRI data provide sufficient anatomical detail to enable corresponding anatomical points to be defined although functional data do not offer this possibility with the same accuracy. Nevertheless, a point may be defined in PET or SPECT data by calculation of the centre of a mass which may be a defined region of interest. A further possibility is to fix external ('fiducial') markers to the patient. A variety of external markers have been designed (see, for example, figure 5 in Papavasileiou et al. 2001a) to aid localisation of image data. Many are available as commercial products although it should be noted that such markers will still be imaged as a region of interest from which a single point must be defined (Van den Elsen and Viergever 1991, Malison et al. 1993, Wang et al. 1994, Papavasileiou et al. 2001b). The solution to matching two sets of points in spatial correspondence, known as the 'Procrustes problem' was elegantly solved by Arun et al. (1987) using singular value decomposition.

15.3.2 Line or Surface Matching

As an extension to matching a series of points, lines or surfaces may also be used. Lines may be obtained from external frames (Lemieux et al. 1994) but can also be defined from internal anatomy, for example, by considering principal axes within a given volume of interest (Alpert et al. 1990). An advantage of this approach is that it is less necessary for there to be a large overlapping volume in the image sets to be registered. Surface matching has been used extensively for medical image registration, as this may be applied to anatomical and functional image data. Head and neck data in particular are suitable for this approach and a number of algorithms have been developed for the purpose. One typical variation is the so-called head-hat algorithm, whereby anatomical surfaces may be generated on two sets of image data by outlining a series of tomographic slices (Kessler et al. 1985, Pelizzari et al. 1991). One set of outlines (the 'head') is tiled to create a virtual solid surface and a series of points are generated from the second set of outlines to create the 'hat'. The sum of distances from each point to the 'head' is then minimised by iteratively moving the 'hat' until a best fit is found using

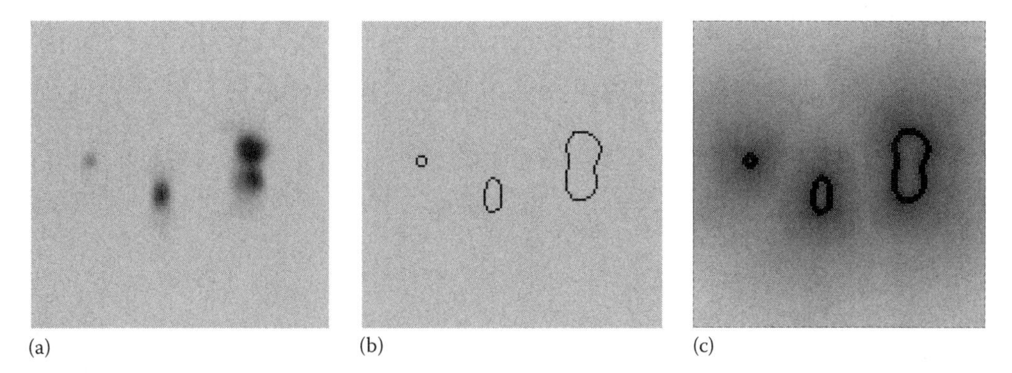

FIGURE 15.1
Distance transform based on chamfer distances. (a) Original SPECT slice of shoulder metastases from differentiated thyroid cancer treated with I-131 NaI. (b) Contour image. (c) Distance transform. See Figure 15.4 for corresponding absorbed dose map and CT. (Images courtesy of P. Papavasileiou.)

a minimisation algorithm such as the Powell method (Powell 1964). Although mostly applied to head and neck data, the approach has also been used with externally applied fiducial bands placed around the thorax (Scott et al. 1995) and in principle the technique may be applied to functional as well as to anatomical data (Sjögreen et al. 2001). An extension to this method is the use of chamfer matching. For this technique a distance transform is performed on an outlined contour from one image to create a grey-scale representation whereby the intensity of each voxel is proportional to its distance from the nearest contour voxel (Figure 15.1). Overlaying the corresponding contour from the image to be matched onto this distance transform enables the determination of a cost function consisting of the sum of intensities of all voxels from the distance transform that correspond to those occupied by the match a contour. This cost function is then minimised (Borgefors 1988, Van den Elsen et al. 1995).

15.3.3 Volume Matching

Cost functions involving voxel-value-derived properties are increasingly being used as parameters to generate registration transformations. There are two main advantages to these techniques. Primarily, they tend to operate well when matching data from different modalities and do not require external devices to be fixed to the patient. Further, they require less prospective user interaction, in that points, lines or surfaces do not need to be defined or delineated. A number of algorithms have been developed in this area. Four of these are described in Sections 15.3.3.1 to 15.3.3.4.

15.3.3.1 Sum of Absolute Differences

An intuitive cost function to apply to image registration, particularly for functional image data, is the sum of the absolute differences (SAD) between two sets of voxel values in images $A(x,y,z)$ and $B(x,y,z)$ (Hoh et al. 1993, Eberl et al. 1996). The cost function is given by:

$$\mathrm{SAD} = \sum_{z}\sum_{y}\sum_{x}\left|A(x,y,z) - B(x,y,z)\right|. \tag{15.9}$$

It is evident that for two perfectly matched image sets of the same modality corresponding voxels will have the same value. Variations will therefore occur only in volumes in which there are anatomical or functional differences. In the limited cases of the same modality where little change is expected, minimisation of the sum of the absolute differences between corresponding voxels is therefore a reasonable basis for image registration. Registration accuracies of 2.1 mm have been achieved in Tl-201 myocardial perfusion studies using this technique (Eberl et al. 1996).

15.3.3.2 Voxel Ratios

A technique particularly applicable to sequential functional image data is that of matching the ratios of voxel values. In the absence of a redistribution of activity in a SPECT or PET scan, successive images will differ only by the effective decay of the radiopharmaceutical (see Section 5.4.4). In this case each voxel within a given image set will be related to corresponding voxels in any previous or later image set by a constant that should be identical for each voxel. A ratio volume is created by dividing the voxel intensities of one set by the other and image sets are most perfectly aligned when the standard deviation of the voxels within this VOI is minimised. For this technique it is important to mask the VOI so that background voxels are not included in the calculation (Woods et al. 1992). An adaptation of this method has been applied to PET and MRI brain scans (Woods et al. 1993).

15.3.3.3 Correlation Coefficient

Maximisation of the correlation coefficient can be used to register image data where it may be assumed that a linear relationship exists between the two data sets to be registered. The correlation coefficient is defined as:

$$CC = \frac{\sum_z \sum_y \sum_x \left[A(x,y,z) - \bar{A}\right]\left[B(x,y,z) - \bar{B}\right]}{\left\{\sum_z \sum_y \sum_x \left[A(x,y,z) - \bar{A}\right]^2 \sum_z \sum_y \sum_x \left[B(x,y,z) - \bar{B}\right]^2\right\}^{1/2}} \quad (15.10)$$

where $\bar{A}$ and $\bar{B}$ are the mean intensity values over the overlap volume of scans A and B, respectively. This has been used to successfully register PET cardiac scans (Bacharach et al. 1993).

15.3.3.4 Mutual Information

In recent years the application of mutual information to medical image registration has become for many cases the technique of choice (Maes et al. 1997, Studholme et al. 1997). Based on information theory, the method relies on aligning image data such that the overlap of information is maximised. An advantage of this method is that the information does not have to be of the same type, so that cross-modality registration can be performed relatively robustly. If the values of two image sets are a and b, and the joint probability distribution function is $p(a,b)$, the mutual information of the two scans is given by:

$$\mathrm{MI}(A,B) = \sum_{a,b} p(a,b)\log\left(\frac{p(a,b)}{p(a)\cdot p(b)}\right) \tag{15.11}$$

where $p(a)$ and $p(b)$ are the marginal probability distribution functions of A and B, respectively.

15.4 Non-Rigid Registration

The application of image registration to medical images must take into account the fact that, with the exception of some bony structures, anatomy is not rigid. Rigid-body registration will lead to local errors that must be assessed for the registration to be considered valid. Ideally, registration of such volumes should be elastic to take account of movement occurring between scans either on different modalities or on the same modality at different times. A number of elastic registration techniques, which are also initially dependent on the identification of common features in the two image sets to be registered, have been developed for this purpose. The inevitable pitfall of adopting this approach is that in principle any two sets of image data may be aligned, even in cases where the images are entirely disparate. This can be a particular drawback if a global transformation is applied, for example, by applying a global higher order polynomial, and techniques have been developed to deal with this problem. A relatively common application is to apply cubic B splines to a grid of defined landmarks in both sets of images, so that local restrictions are imposed (Kybic and Unser 2000), although various other methods of elastic registration have been developed that use, for example, wavelets or radial basis functions (Holden 2008). It is likely that with the rapid increase in clinical hybrid imaging systems, rigid-body registration will become less necessary as image data are inherently aligned. However, the need for elastic registration will continue to exist to enable study of anatomical and functional differences in longitudinal studies and to account for the effects of motion on different imaging modalities.

15.5 Multimodality Image Display

The question of how best to display registered image data has not been resolved and is largely dependent on the data itself and on personal preference. Many methods have been devised to combine both 2D- and 3D-registered image data for optimum display (Viergever et al. 1997). For simultaneous display of registered nuclear medicine and CT or MRI data image overlay is frequently used. This usually employs either a 'chessboard' effect (whereby the two images will occupy the 'black' or 'white' squares, respectively) or a colourwash effect which can be achieved by the display of voxel intensities according to one image but hues according the other. Pseudo-3D effects can also be produced. One example is a stereoscopic technique which involves displaying the two images in red and blue colour scales, whereupon the user wears spectacles with red and blue lenses (Condon 1991). Quarantelli et al. (1999) used a method of display for registered PET and MRI data involving frequency encoding, whereby the Fourier transform of the merged image is obtained by summing the low frequencies of the PET image and the high frequencies of the MR image. They showed that these various display techniques can be beneficial to interpretation, as well as being visually appealing. However, despite numerous possibilities for the display of registered image data, it is not uncommon for images to be simply displayed as adjacent 2D slices with a linked cursor. See Figure 5.133 where a linked cursor has been used in the display of whole-body PET/CT images.

15.6 Clinical Applications

There are a wide range of clinical applications for image registration, involving all medical imaging modalities. Primarily, MMI was developed to enable direct visualisation of corresponding anatomical or functional image data in the same patient. The success of image registration in clinical practice has led to the development of hybrid scanners which are becoming increasingly prevalent. However, more sophisticated uses of MMI have been developed as tools to aid in the diagnosis and treatment of patients.

Localisation and visualisation of patient data can pose significant problems to the clinician when reporting images acquired from different modalities. In particular, direct correspondence of abnormal features as seen, for example, on functional and anatomical image data are not easily assessed from image data that are not aligned. Hybrid PET/CT and SPECT/CT scanners have been developed to solve this problem. A particular application of hybrid imaging is to localise small foci of radiopharmaceutical uptake to determine whether they represent abnormalities (Figure 15.2). Similarly, an abnormal feature seen on CT or MRI can be judged to be malignant or benign according to its functional uptake as viewed, for example, on an FDG scan. An interesting use of registration is to compare a patient scan with an anatomical atlas depicting a standardised anatomy to identify abnormalities. This requires the use of elastic image transformations.

In addition to visualisation and localisation to aid diagnosis, image registration and MMI are increasingly being applied to treatment planning. For example, MMI has been used for over 20 years to provide information to help with planning for surgical procedures. More recently, applications have been towards radiotherapy treatment planning with either external-beam radiotherapy or with radiopharmaceuticals. The superior soft-tissue contrast of MRI data can aid target-volume delineation for radiotherapy treatment planning although the issue of MRI distortion must be addressed. Registration of MRI and CT data, particularly for the head and neck, can be achieved with extreme accuracy using surface matching techniques. In recent years treatment planning for external-beam radiotherapy has begun to incorporate functional image data to optimise the delivery of radiation by minimising doses delivered to healthy normal tissues (Newbold et al. 2006, Lavrenkov et al. 2007, MacManus and Hicks 2008). Both SPECT and PET data have been employed for this purpose (Figure 15.3).

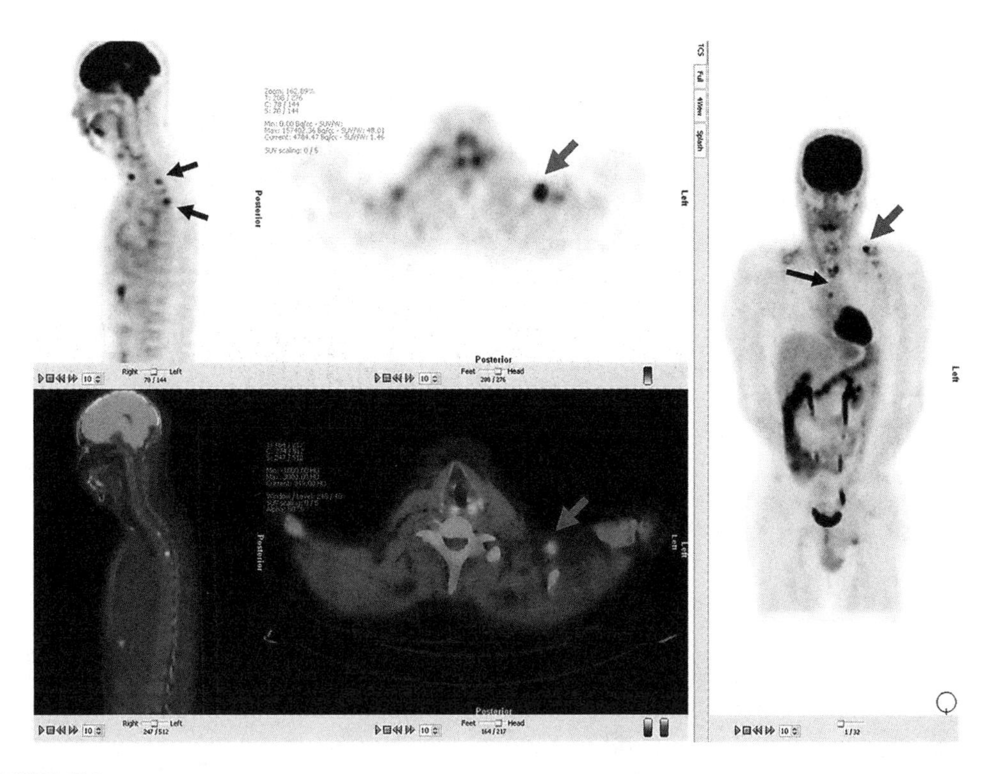

FIGURE 15.2
(See colour insert.) A 27 year old male had a total thyroidectomy and left-sided neck dissection for a papillary thyroid carcinoma. A few years later he presented with abnormally raised thyroglobulin level, but negative whole-body planar imaging I-131 (not shown). PET/CT showed evidence of thoracic bony metastases (black arrows) and a focal area of intense FDG uptake within the left supraclavicular fossa (red arrows), corresponding to a nodal lesion on CT, consistent with a nodal metastasis instead of a left clavicular bony deposit. (Courtesy of S. Chua.)

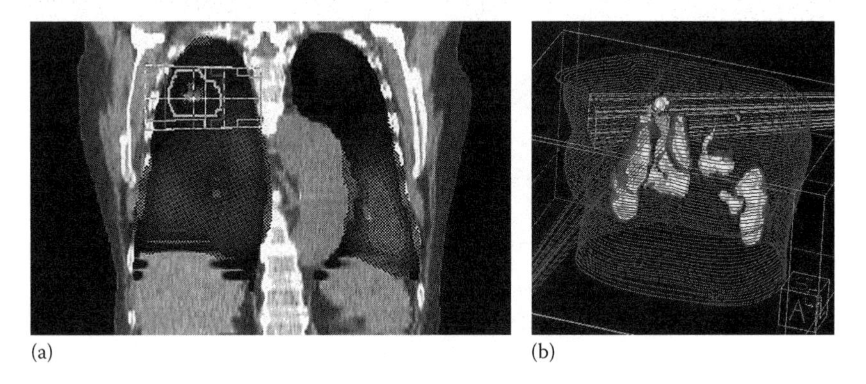

FIGURE 15.3
(See colour insert.) (a) A coronal CT slice with a planning target volume outlined in blue. (b) The registered and overlaid lung SPECT image, which indicates that the top of the contralateral lung is non-functioning and therefore can be irradiated if necessary. (Images courtesy of M. Partridge.)

An example of different methods of image registration is the application of MMI to dosimetry for targeted radionuclide therapy (Flux et al. 2006). In this case a series of gamma-camera scans is acquired following the administration of a therapeutic radiopharmaceutical, and the absorbed dose is calculated from the activity-time curve derived from these scans. Traditional approaches to dosimetry have concentrated on only the mean absorbed dose over a region of interest. However, image registration allows the sequential scans to be registered, so that the absorbed dose may be calculated on a voxel-by-voxel basis (Sgouros et al. 1990, Flux et al. 1997). This produces an absorbed dose distribution, from which isodose contours and dose-volume histograms may be obtained. Registration of gamma-camera scans with CT serves the dual purpose of providing an attenuation map for accurate SPECT reconstruction and localisation of the absorbed dose distribution (Figure 15.4).

Other clinical applications for image registration and MMI have included image-guided surgery performed both pre-operatively and intra-operatively (Roche et al. 2000, McDonald et al. 2007, Nottmeier and Crosby 2007), registration of cardiac data to normal templates (Slomka et al. 2001) and neurology (Stokking et al. 1999). An example of the

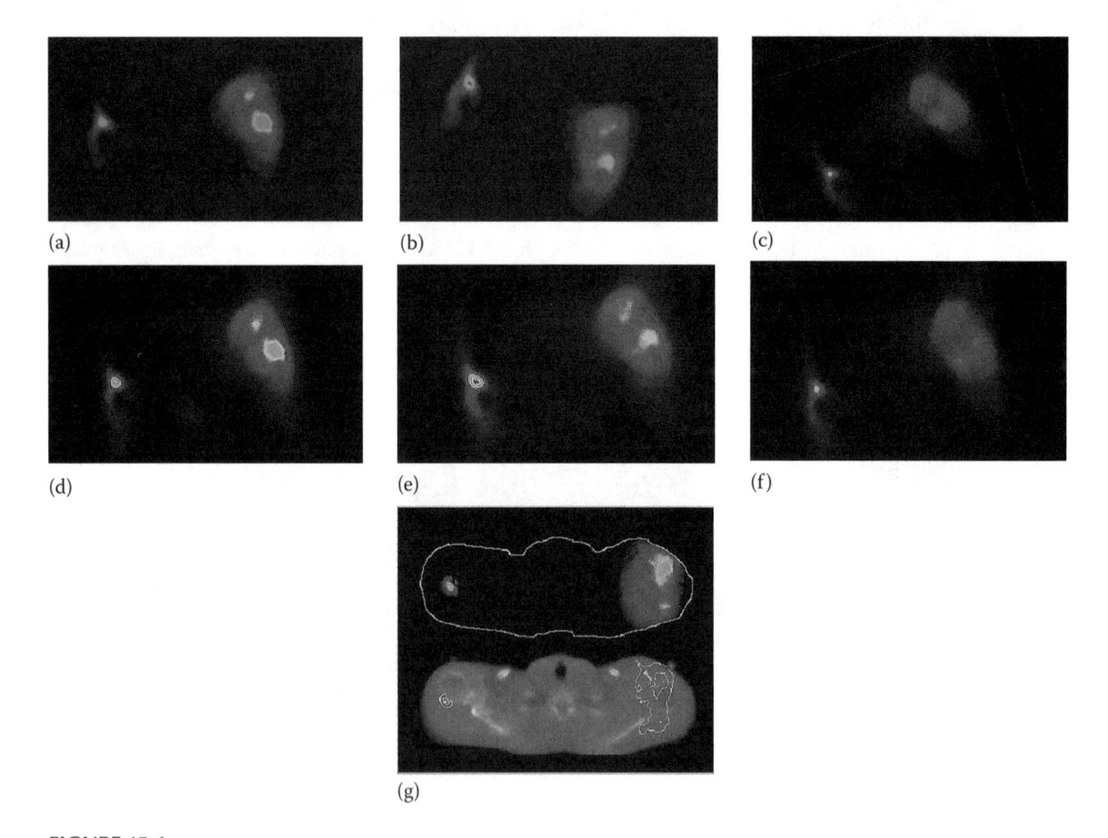

FIGURE 15.4
(See colour insert.) The application of image registration to dosimetry for targeted radionuclide therapy: (a)–(c) show successive SPECT scans taken on days 2, 3 and 6 following administration of 3000 MBq NaI for thyroid metastases; (d)–(f) show these same scans following registration using the mutual information algorithm; and (g) shows (at the top) the resultant absorbed dose distribution and (at the bottom) the isodose contours (ranging from 33 to 70 Gy) overlaid onto the CT scan that has been registered using external markers.

use of registration of functional and anatomical data for purposes other than localisation is that of using CT to provide an attenuation map for the reconstruction of SPECT data (Fleming 1989).

15.7 Registration Accuracy

The aim of image registration is to align image sets such that there is a one-to-one spatial correspondence between each element of the images. There are a number of potential sources of error in this process. Primarily, these are as follows:

1. Definition of features to be used to determine the cost function. Inaccuracies will inevitably occur from the definition of points that are invariably determined from extended features, or from lines or surfaces obtained from either manual delineation or automatic processes such as thresholding. Cost functions based on voxel properties are also dependent on parameters such as data acquisition parameters and reconstruction algorithms, and particularly on the degree of masking employed.
2. The accuracy and robustness of the algorithm used. This can be tested on real and virtual phantom data although these factors, along with their ease of use and speed of operation, do not necessarily translate to patient data.
3. Image transformation. Re-sampling of image data during transformation necessitates trilinear or shape-based interpolation of voxel values which will affect relative voxel values, so that even identical image data cannot be registered precisely if a transformation is involved.
4. Differences in anatomy or imaged parameters. Physiological changes occurring between imaging studies, including anatomical distortion, will mean that a precise registration is not possible except in limited cases with elastic matching which is particularly difficult to assess. Further, the fact that different modalities image different parameters also brings into question the meaning of registration accuracy. Essentially, perfect registration is possible only in two identical data sets, in which case registration is not required.

The fundamental problem of determining the accuracy of image registration is that the optimal method of assessing that accuracy, that is, the correspondence between similarity measures defined in any two image sets, is the method that should be used to determine the image transformation itself. However, it is possible to compare a given method of registration to one that may be more established to assess variations in transformation parameters (Alpert et al. 1990, Acton et al. 1997, Maes et al. 1997), or to apply markers only to assess their correspondence following image registration (Black et al. 1996, Eberl et al. 1996, Flynn et al. 1999). Ultimately, qualitative assessment is often used to judge the accuracy of any registration process and it is recommended that the results of registration are assessed visually before being used. Manual registration, if only to perform minor adjustments to end results, is not uncommon. Further issues to consider when performing registration include definition of the volume to be registered. Whilst it is common to register entire data sets, superior results may be obtained from consideration of isolated regions of interest that will then not be affected by extraneous data. Local, rather than global registration can be performed for a number of sites within the same patient.

15.8 Discussion

Registration is not entirely limited to only two sets of images. Sometimes it is desirable to register a number of data sets acquired over time to monitor progression of disease or of function. For accurate radionuclide therapy dosimetry activity-time curves must be determined from a series of SPECT scans acquired following the administration of a radiopharmaceutical. In such cases decisions must be made as to whether all scans are registered to one base (i.e. reference) scan or whether each scan must be registered to the preceding scan. One approach in such cases is to apply a 4D-registration algorithm, whereby all image sets are registered simultaneously (Acton et al. 1997, Papavasileiou et al. 2007).

The trend of image registration since its inception in the early 1980s is towards automatic methods that facilitate easy clinical implementation. The development of MMI and widespread application of image registration helped to provide the impetus for the introduction of hybrid scanners which are becoming standard equipment in medical-imaging centres. Whilst registration of disparate image data is feasible, the workload in practice precludes its routine application, and yet dual-modality scanners have been shown to be both practical and beneficial. PET/CT in particular has been demonstrated to provide information that aids diagnosis beyond that of PET alone (Hany et al. 2002, Israel et al. 2004). SPECT/CT scanners are becoming commonplace (Hasegawa et al. 2002, Buck et al. 2008), and it is likely that PET/MRI scanners will be introduced to the clinic in the near future (Mackewn et al. 2005, Pichler et al. 2008).

15.9 Conclusions

Since the initial development of image registration in the 1980s, a large body of work has been produced in this field, aided by the rapid increase in accessible computer power, the merging of imaging modalities, and the increase in clinical practice of multidisciplinary teams. Whilst the advent of hybrid scanners increasingly renders routine registration of image data from different modalities for visualisation redundant, novel uses of MMI continue to be developed. As yet, there are no standardised protocols for image registration. The technique chosen for any given clinical application should be selected carefully according to the purpose of the registration and the data available. However, commercial and freeware software packages are becoming increasingly available, bringing MMI into more routine clinical use.

References

Acton P D, Pilowsky L S, Suckling J, Brammer M J and Ell P J 1997 Registration of dynamic dopamine D-2 receptor images using principal component analysis *Eur. J. Nucl. Med.* **24** 1405–1412.

Alpert N M, Bradshaw J F, Kennedy D and Correia J A 1990 The principal axes transformation—A method for image registration *J. Nucl. Med.* **31** 1717–1722.

Arun K S, Huang T S and Blostein S D 1987 Least squares fitting of 2 3-D point sets. *IEEE Trans. Pattern Anal. Mach. Intell.* **9** 699–700.

Bacharach S L, Douglas M A, Carson R E, Kalkowski P J, Freedman N M T, Perronefilardi P and Bonow R O 1993 3-Dimensional registration of cardiac positron emission tomography attenuation scans *J. Nucl. Med.* **34** 311–321.

Black K J, Videen T O and Perlmutter J S 1996 A metric for testing the accuracy of cross-modality image registration: Validation and application *J. Comput. Assist. Tomogr.* **20** 855–861.

Borgefors G 1988 Hierarchical chamfer matching—A parametric edge matching algorithm *IEEE Trans. Pattern Anal.* **10** 849–865.

Buck A K, Nekolla S, Ziegler S, Beer A, Krause B J, Herrmann K, Scheidhauer K, Wester H J, Rummeny E J, Schwaiger M and Drzezga A 2008 SPECT/CT *J. Nucl. Med.* **49** 1305–1319.

Condon B R 1991 Multimodality image combination—5 techniques for simultaneous MR-SPECT display *Comput. Med. Imaging Graph.* **15** 311–318.

Eberl S, Kanno I, Fulton R R, Ryan A, Hutton B F and Fulham M J 1996 Automated interstudy image registration technique for SPECT and PET *J. Nucl. Med.* **37** 137–145.

Fleming J S 1989 A technique for using CT images in attenuation correction and quantification in SPECT *Nucl. Med. Commun.* **10** 83–97.

Flux G, Bardies M, Monsieurs M, Savolainen S, Strand S-E and Lassmann M 2006 The impact of PET and SPECT on dosimetry for targeted radionuclide therapy *Z. Med. Phys.* **16** 47–59.

Flux G D, Webb S, Ott R J, Chittenden S J and Thomas R 1997 Three-dimensional dosimetry for intralesional radionuclide therapy using mathematical modeling and multimodality imaging *J. Nucl. Med.* **38** 1059–1066.

Flynn A A, Green A J, Boxer G, Pedley R B and Begent R H J 1999 A Comparison of image registration techniques for the correlation of radiolabelled antibody distribution with tumour morphology *Phys. Med. Biol.* **44** N151–N159.

Gill S S, Thomas D G T, Warrington A P and Brada M 1991 Relocatable frame for stereotaxic external beam radiotherapy *Int. J. Radiat. Oncol. Biol. Phys.* **20** 599–603.

Hany T F, Steinert H C, Goerres G W, Buck A and Von Schulthess G K 2002 PET diagnostic accuracy: Improvement with in-line PET-CT system: Initial results *Radiology* **225** 575–581.

Hasegawa B H, Iwata K, Wong K H, Wu M C, Da Silva A J, Tang H R, Barber W C, Hwang A H and Sakdinawat A E 2002 Dual-modality imaging of function and physiology *Acad. Radiol.* **9** 1305–1321.

Hill D L G, Batchelor P G, Holden M and Hawkes D J 2001 Medical image registration *Phys. Med. Biol.* **46** R1–R45.

Hill D L G, Hawkes D J, Crossman J E, Gleeson M J, Cox T C S, Bracey E E C M, Strong A J and Graves P 1991 Registration of MR and CT images for skull base surgery using point-like anatomical features *Br. J. Radiol.* 64 1030–1035.

Hoh C K, Dahlbom M, Harris G, Choi Y, Hawkins R A, Phelps M E and Maddahi J 1993 Automated iterative 3-dimensional registration of positron emission tomography images *J. Nucl. Med.* **34** 2009–2018.

Holden M 2008 A review of geometric transformations for non-rigid body registration *IEEE Trans Med Imaging* **27** 111–128.

Israel O, Mor M, Guralnik L, Hermoni N, Gaitim D, Bar-Shalom R, Kaidar Z and Epelbaum R 2004 Is F-18-FDG PET/CT useful for imaging and management of patients with suspected occult recurrence of cancer? *J. Nucl. Med.* **45** 2045–2051.

Kessler M L, Chen G T Y and Pitluck S 1985 Computer techniques for 3-D image spatial correlation *Med. Phys.* **12** 520.

Kybic J and Unser M 2000 Multidimensional elastic registration of images using splines *2000 International Conference on Image Processing*, Vol II, *Proceedings*, Vancouver, British Columbia, Canada, pp. 455–458.

Lavrenkov K, Christian J A, Partridge M, Niotsikou E, Cook G, Parker M, Bedford J L and Brada M 2007 A potential to reduce pulmonary toxicity: The use of perfusion SPECT with IMRT for functional lung avoidance in radiotherapy of non-small cell lung cancer *Radiother. Oncol.* **83** 156–162.

Lemieux L, Kitchen N D, Hughes S W and Thomas D G T 1994 Voxel-based localization in frame-based and frameless stereotaxy and its accuracy *Med. Phys.* **21** 1301–1310.

Mackewn J E, Strul D, Hallett W A, Halsted P, Page R A, Keevil S F, Williams S C R, Cherry S R and Marsden P K 2005 Design and development of an MR-compatible PET scanner for imaging small animals *IEEE Trans. Nucl. Sci.* **52** 1376–1380.

MacManus M and Hicks R J 2008 The Use of Positron Emission Tomography (PET) in the staging/evaluation, treatment, and follow-up of patients with Lung cancer: A critical review *Int. J. Radiat. Oncol.* **72** 1298–1306.

Maes F, Collignon A, Vandermeulen D, Marchal G and Suetens P 1997 Multimodality image registration by maximization of mutual information *IEEE Trans. Med. Imaging* **16** 187–198.

Maintz J B A and Viergever M A 1998 A survey of medical image registration *Med. Image Anal.* **2** 1–36.

Malison R T, Miller E G, Greene R, McCarthy G, Charney D S and Innis R B 1993 Computer-assisted coregistration of multislice SPECT and MR brain images by fixed external fiducials *J. Comput. Assist. Tomogr.* **17** 952–960.

McDonald C P, Brownhill J R, King G J W, Johnson J A and Peters T M 2007 A comparison of registration techniques for computer- and image-assisted elbow surgery *Comput. Aided Surg.* **12** 208–214.

Meltzer C C, Bryan R N, Holcomb H H, Kimball A W, Mayberg H S, Sadzot B, Leal J P, Wagner H N and Frost J J 1990 Anatomical localization for PET using MR imaging *J. Comput. Assist. Tomogr.* **14** 418–426.

Miura S, Kanno I, Iida H, Murakami M, Takahashi K, Sasaki H, Inugami A, Shishido F, Ogawa T and Uemura K 1988 Anatomical adjustments in brain positron emission tomography using CT images *J. Comput. Assist. Tomogr.* **12** 363–367.

Newbold K, Partridge M, Cook G, Sohaib A, Charles-Edwards E, Rhys-Evans P, Harrington K and Nutting C 2006 Advanced imaging applied to radiotherapy planning in head and neck cancer: A clinical review *Br. J. Radiol.* **79** 554–561.

Nottmeier E W and Crosby T L 2007 Timing of paired points and surface matching registration in three-dimensional (3D) image-guided spinal surgery *J. Spinal Disord. Tech.* **20** 268–270.

Papavasileiou P, Divoli A, Hatziioannou K and Flux G D 2007 A generalized 4D image registration scheme for targeted radionuclide therapy dosimetry *Cancer Biother. Radiopharm.* **22** 160–165.

Papavasileiou P, Flux G D, Flower M A and Guy M J 2001a An automated technique for SPECT marker-based image registration in radionuclide therapy *Phys. Med. Biol.* **46** 2085–2097.

Papavasileiou P, Flux G D, Flower M A and Guy M J 2001b Automated CT marker segmentation for image registration in radionuclide therapy *Phys. Med. Biol.* **46** N269–N279.

Pelizzari C A, Tan K K, Levin D N, Chen G T Y and Balter J 1991 Interactive 3D patient—Image registration *Lect. Notes Comput. Sci.* **511** 132–141.

Pichler B J, Wehrl H F and Kolb A and Judenhofer M S 2008 Positron emission tomography/magnetic resonance imaging: The next generation of multimodality imaging? *Semin. Nucl. Med.* **38** 199–208.

Powell M J D 1964 An efficient method for finding the minimum of a function of several variables without calculating derivatives *Comput. J.* **7** 155–162.

Quarantelli M, Alfano B, Larobina M, Tedeschi E, Brunetti A, Covelli E M, Ciarmiello A, Mainolfi C and Salvetora M 1999 Frequency encoding for simultaneous display of multimodality images *J. Nucl. Med.* **40** 442–447.

Roche A, Pennec X, Rudolph M, Auer D P, Malandain G, Ourselin S, Auer L M and Ayach N 2000 Generalized correlation ratio for rigid registration of 3D ultrasound with MR images *Medical Image Computing and Computer-Assisted Intervention—MICCAI*, Pittsburgh, PA, Vol. 1935 2000 pp. 567–577.

Scott A M, Macapinlac H, Zhang J J, Daghighian F, Montemayor N, Kalaigian H, Sgouros G et al. 1995 Image registration of SPECT and CT images using an external fiduciary band and 3-dimensional surface fitting in metastatic thyroid-cancer *J. Nucl. Med.* **36** 100–103.

Sgouros G, Barest G, Thekkumthala J, Chui C, Mohan R, Bigler R E and Zanzonico P B 1990 Treatment planning for internal radionuclide therapy—3-dimensional dosimetry for nonuniformly distributed radionuclides *J. Nucl. Med.* **31** 1884–1891.

Sjögreen K, Ljungberg M, Wingardh K, Erlandsson K and Strand S-E 2001 Registration of emission and transmission whole-body scintillation-camera images. *J. Nucl. Med.* **42** 1563–1570.

Slomka P J, Radau P, Hurwitz G A and Dey D 2001 Automated three-dimensional quantification of myocardial perfusion and brain SPECT *Comput. Med. Imaging Graph.* **25** 153–164.

Stokking R, Studholme C, Spencer S S, Corsi M, Avery R A, Morano G N, Siebyl J P and Zubal I G 1999 Multimodality 3D visualization of SPECT difference and MR images in epilepsy *J. Nucl. Med.* **40** 1270.

Studholme C, Hill D L G and Hawkes D J 1997 Automated three-dimensional registration of magnetic resonance and positron emission tomography brain images by multiresolution optimization of voxel similarity measures *Med. Phys.* **24** 25–35.

Van den Elsen P A, Maintz J B A, Pol E J D and Viergever M A 1995 Automatic registration of CT and MR brain images using correlation of geometrical features *IEEE Trans. Med. Imaging* **14** 384–396.

Van den Elsen P A, Pol E J D and Viergever M A. 1993 Medical image matching—A review with classification *IEEE Eng. Med. Biol. Mag.* **12** 26–39.

Van den Elsen P A and Viergever M A 1991 Marker guided registration of electromagnetic dipole data with tomographic-images *Lect. Notes Comput. Sci.* **511** 142–153.

Viergever M A, Maintz J B A and Stokking R 1997 Integration of functional and anatomical brain images *Biophys. Chem.* **68** 207–219.

Wang C, Pahl J J and Hogue R E 1994 A method for co-registering 3-dimensional multimodality brain images *Comput. Methods Programs Biomed.* **44** 131–140.

Webb S 1993 *The Physics of Tthree-Dimensional Radiation Therapy* 1st edn. (Bristol, UK: IOPP).

Woods R P, Cherry S R and Mazziotta J C 1992 Rapid automated algorithm for aligning and reslicing PET images *J. Comput. Assist. Tomogr.* **16** 620–633.

Woods R P, Mazziotta J C and Cherry S R 1993 MRI-PET registration with automated algorithm *J. Comput. Assist. Tomogr.* **17** 536–546.

Zitova B and Flusser J 2003 Image registration methods: A survey *Image Vision Comput.* **21** 977–1000.

16

Epilogue

S. Webb and C. R. Hill

CONTENTS

16.1 Introduction 783
16.2 Attributes and Relative Roles of Imaging Modalities 784
16.2.1 Anatomical versus Functional Imaging 785
16.2.2 Time Factors 785
16.2.3 Single and Multiple Images 788
16.2.4 Image Processing 788
16.2.5 Resources: Personnel and Space 788
16.2.6 Applications of Imaging 789
16.2.7 Spatial Resolution and Sensitivity 789
16.2.8 Work in Progress 789
References 790

16.1 Introduction

This book has described the physics of a variety of approaches to medical imaging. Each chapter has inevitably concentrated attention on a particular approach, but, in the real world, the different approaches complement each other in the solution of different problems. This chapter tries to put the situation in perspective.

Even when the physics of medical imaging has been neatly parcelled into discrete sections, as we have done in this book, it must be said that these segregations are somewhat artificial. They are convenient for our purposes; they represent how a teaching course on medical imaging might be structured, but the divisions are woolly at the edges since so many common features and problems are shared between the imaging modalities. The earlier chapters presented the major imaging modalities in turn and the later chapters considered some thematic aspects.

To end our account, we might note that the very act of assembling a number of physical imaging techniques under that one label invites *comparison*. Indeed, many conferences have been held with just that intention, whilst at others comparing imaging modalities quickly becomes the topic of discussion. There are a number of reasons why this is important. First, it is essential for the person making a diagnosis to request the appropriate images and, preferably, in some optimum order. For the patient, this is vital also, since the aim of the clinician is to reach a clear diagnosis by the quickest, most accurate, least inconvenient, least harmful and least painful way. The clinical questions being asked thus clearly dictate the choices of imaging method. The choices and order of their execution are generally quite different depending on the nature of these questions and many books (e.g. Preston et al. 1979, Sodee and Verdon 1979,

Simeone 1984, Newhouse 1988, Hohne et al. 1990, Krestel 1990, Cho et al. 1993) have addressed the problem. See also several reviews in *Physics in Medicine and Biology* **51** (13) (2006) the 50th Anniversary Issue. A second reason for comparing imaging modalities is that they rely on different physical principles. What may be a quite impossible task for one method may be solvable by another. The corollary for the physicist is that it may be worth abandoning a line of physical investigation that is becoming progressively more difficult and less rewarding in favour of a quite different approach. For example, we have seen how no amount of struggling with planar X-rays can image deep-seated small-scale disease, whereas this is a comparatively easy problem for X-ray CT or imaging with ultrasound. Thirdly, with only limited financial resources for health care, the questions of what imaging facilities to provide, how many of each and where they should be located are of paramount importance (WHO 1983, 1985). Those charged with making these managerial decisions require detailed knowledge of a comparative nature. Naively, one might imagine that decisions could be based on the known incidences of requesting imaging examinations of various kinds. This would, however, ignore the feedback process whereby the real clinical demand has often only become apparent after the provision of a facility. Demand provision would also ignore the possibility of catering for rarer, but potentially more important, investigations. It is very tempting for the physicist to try to duck out of this question and leave the burden of choice elsewhere. This is wrong, since the physicist can uniquely assess the imaging potential, complementing the clinician who can translate this into terms of assessing clinical benefit.

The manufacturer also plays an important role, since it has been commonly observed that however suitable a laboratory imaging tool may be for some purpose, its full potential is rarely realised until a critical mass of people are all working with it. This has been particularly true of X-ray CT, ultrasound, radioisotope imaging and MRI.

Since the first edition of this book, the use of 3D and 4D medical images, particularly in planning, delivering and monitoring the outcome of *radiation therapy*, has grown enormously. Sections of the books by Webb (2000, 2004) chart a more detailed account than is provided here in Chapters 4 and 16. Imaging also plays an important role in the planning and monitoring of radionuclide and focussed ultrasound therapies, and in bio-robotic surgery.

Even though we have still not strayed far from aspects of imaging to which an understanding of physical principles is the bedrock, we are in dangerous waters. There are aspects of these comparative considerations that are very subjective and which by their nature fuel discussion. It is almost certain that even the physicists contributing to this volume would find it difficult to be in complete agreement on the role of a particular imaging method. There are too many feedback processes involved, including the biases of their own particular interests and experiences, the relative novelty of the technique in question, and indeed the perception of the relative interest in the method by others. Against this background, it is tempting to confess to no consensus and leave the readers to decide for themselves on the roles and relative importance of the subjects covered.

16.2 Attributes and Relative Roles of Imaging Modalities

We note that some aspects of the situation are not unlike the problem of consumer choice and, in this spirit, we provide below a table of fairly objective comparative information. The word 'fairly' is meant to imply that even the data in this table are not entirely devoid of argument! To do the job properly, we would have to provide data for a vast number of different investigations even within a particular modality. Table 16.1 lists some of the questions one may need to answer in considering using or providing some investigative tool. Many of these questions have been addressed in more detail in earlier chapters.

16.2.1 Anatomical versus Functional Imaging

In Table 16.1, we highlight some important attributes that help to determine the roles of the imaging modalities. They divide into two classes: those which essentially give anatomical information and those showing the function of biological tissues. This division breaks down if the anatomical method can generate images fast enough to show changes in anatomy with time (which is then functional information, e.g. B-scanning the heart). It is also common for B-scan imaging to be combined with Doppler functional measurements and/or parametric functional images in one machine. Those imaging modalities which only show function can sometimes be difficult to interpret in the absence of anatomical landmarks. Techniques for geometrically registering images from different modalities were proposed in the 1980s (Gerlot and Bizais 1987) and are now in common use (see review by Webb [1993] and Chapter 15). Some of the methods that yield anatomical information can also yield functional information, generally by the administration of contrast media. For example, blood flow can be measured by computed tomography (Flower et al. 1985), and radio-opaque barium compounds can be used to image the gastrointestinal tract dynamically with X-rays. By varying the pulse sequence of MRI, the images may be interpreted as functional rather than anatomical images (see Chapter 7).

16.2.2 Time Factors

Turning to the time taken for an imaging procedure, it is possible to identify at least three separate components. There is the time it takes to acquire the basic imaging data (which may not actually be an image), the time it takes to reduce these data to recognisable images and finally the time the procedure takes as experienced by the patient. The latter will include the time required to prepare for the examination, the intervals during the examination when data are not being acquired (e.g. time taken resetting geometry) and the further 'dead' time after the data are acquired, but before the patient actually leaves the examination room. It is also important to note wide differences in the time needed to 'read' images. There is at least one order-of-magnitude variation in the time experienced by the patient. X-radiography generally requires the shortest attendance (say 5 min) whilst complex procedures such as dynamic PET and contrast-enhanced dynamic MRI can take over an hour.

There are several orders-of-magnitude variations in data acquisition time. X-radiography, B-scan ultrasonography, optical imaging and EIT are essentially instantaneous data-capture procedures. Conversely, most nuclear medicine and MRI studies require of the order of 20 min or longer to acquire data. We have seen that (in the trade-off for limiting radiation dose) these procedures are noise limited.

For some investigations, any processing of the data before the images are available can be relatively straightforward and rapid (e.g. gamma-camera planar scintigraphy, B-scan ultrasonography and digital radiography). Others, notably those involving image reconstruction, particularly SPECT and PET, can take a few minutes. How many minutes obviously depends on whether the associated computers utilise fast arithmetic units or array processors and on the skill of optimising the reconstruction programming.

Since the publication of the first edition of this book, most image capture has become digital, rather than analogue. Two consequences are apparent. Firstly, such data can be captured

TABLE 16.1

Comparison of imaging modalities.

	Conventional Planar X-Ray	Xeroradiograph	Digital Radiograph	Conventional X-Ray Tomogram	X-Ray CT Study	Gamma-Camera Image	Rectilinear NM Scan	Gamma-Camera SPECT Study	PET Study	US B-Scan and Doppler Imaging	MR Image	MRS Image	Thermogram	Electrical Impedance Tomogram	Optical Image
Anatomy (A) or function (F) measured by basic technique	A	A	A	A	A	F	F	F	F	A + F	A + F	F	F	A + F	A
Data acquisition time	<1 s	<1 s	<1 s	Few seconds	<1 min	5 min	20 min	20 min	20 min	1 s	10 min	10 min	≪1 s	≪1 s	<1 s
Data reconstruction or processing time	2 min[a]	2 min	<1 s	2 min[a]	<1 min	<1 s	Few seconds	1 min	2 min	<1 s	2 min	<1 s	≪1 s	Few minutes	<1 s
Patient handling time	5 min	5 min	5 min	5 min	30 min	10 min[b]	30 min[b]	30 min[b]	30 min[b]	10 min	1 h	1 h	10 min	Few minutes	10 min
Single (S) or multiple (M) images generated	S	S	S	S	M	S	S	M	M	M	M	M	M	S	M
Does imaging require a computer?[c]	N[d]	N	Y	N[d]	Y	Y	Y	Y	Y	Y	Y	Y	N	Y	N
Can images be further processed easily?[c]	N[d]	N	Y	N[d]	Y	Y	Y	Y	Y	Y	Y	Y	N	Y	N
Minimum number of staff needed to use equipment conveniently	1	1	1	1	2	1	1	1	2	1	2	3	1	1	1

Space requirement[e]	S	S	L	S	>1	S	L	L	>1	S	SB	SB	S	S	S
Is a medical doctor needed for imaging?[c]	N	N	N	N	N	N	N	N	N	Y/N	N	N	N	N	Y/N
How common is equipment? (1(rare)→5(very common))	5	1	4	1	5	4	1	4	4	5	5	5	1	1	1
How many body sites can be studied? (1(few)→5(many))	5	2	2	3	5	5	5	4	4	4	5	4	3	2	1
Spatial resolution (1(poor)→5(good))	5	5	4	5	4	3	2	2	4	4	4	3	5	1	2
Sensitivity (1(poor)→5(good))	2	2	2	2	4	2	2	3	4	3	4	2	4	5	3
Inconvenience or unpleasantness to patient (1(little or none)→5(great))	1	3	1	2	3	2	2	3	3	2	4	5	2	2	1
Could the technique be a first-choice investigation?[c]	Y	Y	Y	N	Y	Y	Y	N	Y	Y	Y	Y	Y	N	Y
Number of measured parameters (1(few)→5(many))	1	1	2	1	2	1	1	1	3	2	4	4	1	3	2

[a] For film processing but 'instantaneous' if digital capture.
[b] Excluding injection time and wait time.
[c] N, no; Y, yes.
[d] Yes if digital capture.
[e] S, small room; >1, more than one room; L, large room; SB, special building.

faster, thus bypassing 'processing steps' such as film development. Secondly, such data are inherently immediately available and in a suitable form for further digital processing.

16.2.3 Single and Multiple Images

A further way of classifying the modalities into two groups is by multiplicity (M) or singularity (S) of images generated. Again, we have chosen to refer to common practice for a study. The division is also a little arbitrary, since all the modalities labelled 'M' could, of course, be forced to yield just one image; yet they rarely do so, since full advantage is taken of their ability to generate stacking two-dimensional images, simulating three-dimensional investigation. Equally, those labelled 'S' are generally requested to form two or more separate images from, for example, orthogonal viewing directions.

Some tomographic scanners always generate multislice data (e.g. gamma-camera-based SPECT) whilst others, although in principle able to make just a single tomogram (e.g. X-ray CT, B-scan ultrasonography and MRI), rarely do so, full studies generally being carried out. It should be noted that the estimates for times in Table 16.1 are for full multislice studies rather than single-plane imaging.

16.2.4 Image Processing

The majority of the imaging methods require a computer. This was, of course, a contributing factor to the bunching of developments into the 1970s and 1980s, as discussed in Chapter 1. We should not, however, forget that classical planar diagnostic radiology with X-rays does not necessarily have this requirement and its use probably outnumbers that of other investigative tools, although it could be claimed ultrasound has become the most-used modality. Meanwhile, digital radiology is about to overtake film-based radiology. A glance down the appropriate column in Table 16.1 shows that this, historically the first modality, still retains many attractive features. Other investigations generally complement X-radiology rather than substitute for it. Since images generated by a computer are digital, it follows, as already mentioned, that they are in a form immediately amenable to further processing. The lack of this ability for radiographs on film or electrostatic plate is a major drawback and is ultimately leading to their demise, as was prophesied in the 1980s by Craig and Glass (1985).

In this volume, we have discussed the problems of data compatibility, data compression and storage and data transmission. These are necessarily rapid growth areas because of the vast quantities of data contained in (particularly tomographic) images.

16.2.5 Resources: Personnel and Space

We see from Table 16.1 that most imaging modalities can be operated by just one person. In two instances, B-scan ultrasonography and optical imaging, it is often desirable that this should be a medical doctor, since the information is gathered in real time and the course of action during the investigation may depend on new clinical questions raised at the time. This is a particularly important factor to note in relation to ultrasound, where many of the investigations entail a high degree of interactivity between the investigator, patient and machine. For the others, it is sufficient to arrange a reporting session some time after the images have been taken. That said, an efficient department will often seek a quick medical opinion on images before the patient departs, in case a further study is requested. When this is not possible (e.g. for those modalities requiring complex postprocessing), the patient may need to be recalled after the reporting session.

Some of the least common modalities require a great deal of space, although this is not always the case. The commitment of physical space and the capital and revenue financial costs of equipment are clearly factors in determining the number of devices that can be installed. As we intimated earlier, they are not, however, the only deciding factors.

The imaging modalities divide into two roughly equal classes depending on whether the technique would be a first-choice investigative tool, although clearly this is problem dependent. There is a good correlation that those techniques which are not first-choice tools are also rarer, as one might expect. Surprisingly, perhaps the reverse does not seem to be true. There are several first-choice investigative techniques that are fairly uncommon, for example, EIT and optical imaging. Other reasoning has prevailed. Since it is quite impossible to give more than a brief generalisation here, the reader is referred elsewhere (e.g. Sodee and Verdon 1979) for an approach to this problem.

16.2.6 Applications of Imaging

There is a range of applicability by body site for the different imaging modalities. Almost every part of the body can be radiographed, scanned by X-ray CT or MRI or imaged with a gamma camera, although how successful this will be does vary by body site. Some probes, for example, ultrasound and certain ECT techniques, are not universally applicable, but nevertheless enjoy a wide range of applicability. Other imaging modalities, for example, xeroradiography, EIT and optical imaging, are very specialised tools.

16.2.7 Spatial Resolution and Sensitivity

Earlier chapters have shown a wide range of spatial resolution and sensitivity. X-radiology has excellent spatial resolution (less than 1 mm) but is not very sensitive to changes in the parameters being imaged. X-ray CT retains good resolution (1–2 mm) whilst showing an enormous increase in sensitivity. Ultrasound also can combine substantially sub-millimetre resolution with good sensitivity, particularly for high-frequency (short-range) applications, such as for skin pathology and intraluminal investigations. Many MR imagers have a comparable resolution to X-ray CT. Digital radiology is pixel based and pixel size, whilst dependent on the properties of the digital detector, is ultimately noise and thus dose limited. Gamma cameras have much poorer extrinsic resolution (5–15 mm) and resolution for clinical SPECT can be as poor as 20 mm or more, but 10 mm at best. Since PET makes use of annihilation coincidence detection rather than physical collimation, its spatial resolution is much better (5–10 mm for clinical systems) than SPECT. EIT currently brings up the tail at around 30 mm spatial resolution. Yet the latter runs away with the prize for best sensitivity (to small changes in water content for this modality).

Most imaging tools essentially display one property of biological tissue. Some modalities promise more, but it is only MRI and PET that seem to have the flexibility to investigate many properties. X-ray CT and digital radiology can give the composition of tissues, but struggle to do so; ultrasound can investigate and quantify a variety of elastic and dynamic properties, not least those related to vascular behaviour, and EIT using multiple frequencies also promises to image more than one electrical property of tissue.

16.2.8 Work in Progress

Finally, having tried to look at imaging modalities in this loose common framework, we return to recognising that many of these are 'apples and pears' comparisons. Medical imaging methods build into a formidable arsenal of diagnostic tools, generally complementing each other rather than replacing existing techniques. The search for less hazardous, less invasive investigations goes on. With new technology not only comes better diagnosis and patient management, but imaging that contributes to man's fundamental understanding of human biology. It is just a century since the only way to see inside the human body was literally by eye at surgery (Hill 2009). Anatomical and functional information is now available on the spatial scale of millimetres. Imaging at the cellular level is a developing field that we have not covered in detail in this book. Most 'clinical imaging' relies on information that is macroscopic by cellular dimensions. The story of medical imaging has not yet reached its final chapter.

References

Cho Z H, Jones J P and Singh M 1993 *Foundations of Medical Imaging* (New York: Wiley).

Craig J O M C and Glass H I 1985 The creation of a filmless/digital hospital: The St Mary's experience *Br. J. Radiol.* **58** 803.

Flower M A, Husband J E and Parker R P 1985 A preliminary investigation of dynamic transmission computed tomography for measurements of arterial flow and tumour perfusion *Br. J. Radiol.* **58** 983–988.

Gerlot P and Bizais Y 1987 Image registration: A review and a strategy for medical applications *Proc. 10th IPMI Meeting, Utrecht* eds. M Viergever and C N de Graaf (New York: Plenum).

Hill C R 2009 Early days of scanning: Pioneers and sleepwalkers. *Radiography* **15** e15–e22.

Hohne K H, Fuchs H and Pizer S M 1990 *3D Imaging in Medicine* (Berlin, Germany: Springer Verlag).

Krestel E 1990 *Imaging Systems for Medical Diagnostics* (Berlin, Germany: Siemens).

Newhouse V 1988 *Progress in Medical Imaging* (New York: Springer).

Preston K, Taylor K J W, Johnson S A and Ayers W R (eds.) 1979 *Medical Imaging Techniques: A Comparison* (New York: Plenum).

Simeone J F (ed.) 1984 *Co-ordinated Diagnostic Imaging* (New York: Churchill Livingstone).

Sodee D B and Verdon T A (ed.) 1979 Correlations in diagnostic imaging *Nuclear Medicine, Ultrasound and Computed Tomography in Medical Practice* (New York: Appleton-Century-Crofts).

Webb S 1993 *The Physics of Three-Dimensional Radiation Therapy, Conformal Radiotherapy, Radiosurgery and Treatment Planning* (Bristol, UK: IOPP).

Webb S 2000 *Intensity-Modulated Radiation Therapy* (Bristol, UK: IOPP).

Webb S 2004 *Contemporary IMRT* (Bristol, UK: IOPP).

WHO (World Health Organisation) 1983 A rational approach to radiodiagnostic investigations Technical Report Series 689 (Geneva, Switzerland: World Health Organisation).

WHO 1985 Future use of new imaging technologies in developing countries Technical Report Series 723 (Geneva, Switzerland: World Health Organisation).

Index

A

Abdominal ultrasonic imaging, 456–459
Aberdeen Section Scanner, 191
Accelerators, 205
Accidental coincidences, 299
Accuracy in imaging, 777
Acoustic impedance, 358
Acoustic microscopy, 354
Acoustic output measurements
  derated parameters, 443
  hydrophone-based measurement (*see* Hydrophone measurement)
  piezoelectric hydrophone (*see* Piezoelectric hydrophone)
  radiation force balance, 442–443
Acoustic radiation fields
  continuous-wave and pulsed excitation
    diffraction loss and phase shift, 373
    Fraunhofer and Fresnel solutions, 371–372
    plane and edge-wave components, 373–374
    single-element transducer, 372
    Young's method, 373
  focusing effect, 374–376
  non-linear propagation, 379–380
  resolution, 377–379
  short pulses, 376–377
Acousto-optic imaging, 679
Active-matrix array (AMA)
  fluorescent screens, 66
  selenium photoconductors, 66–68
Activity–time curve, 231, 234–235, 251, 254, 256–260, 310, 317
Adiabatic rapid-passage experiment, 499
Advanced single-slice rebinning (ASSR), 132–133
AGEMA Thermovision 870 Camera, 630
Air-gap techniques, 40–41
Algorithms
  adaptive multiple plane reconstruction (AMPR), 133, 142
  adaptive speckle reduction, 384
  attenuation correction, 281
  automatic exposure, 114
  cone-beam reconstruction, 131–133, 141–142
    FDK-type, 131–132, 141
    Feldkamp algorithm, 160
  distance transform (DT), 725
  edge detection, 253, 722
  fast Fourier transform (FFT), 233, 530, 696
  FaVoR algorithm, 311
  focusing, 470
  Fourier rebinning (FORE), 312
  head-hat, 770
  histogram equalisation, 68, 719
  image reconstruction, 134, 137, 268, 287, 296, 308, 541, 566, 654–655, 676–678, 777
    2D filtered backprojection, 122–128, 277–278, 310–311
    3D PET, 309–313
  interpolation, 143, 147, 407
  iterative reconstruction, 128–130, 141, 278–281, 284, 312–313
    ML–EM, 280, 312
    OSEM, 231, 281, 287, 309, 311–312
    penalised weighted least squares, 279
  motion correction, 120
  reprojection, 311
  Runge–Kutta, 501
  segmentation, 313, 409
  single-slice interpolation, 140
  single-slice rebinning, 132–133, 311
  spiral interpolation, 140–141
  volume-matching or image registration, 771–773, 776–778
Aliasing, 138–139, 423, 704
Alpha decay, 209–210
Analogue hard copy, 59
Analogue image receptors
  direct-exposure X-ray film (*see* Direct-exposure X-ray film)
  image intensifiers (*see* Image intensifiers)
  screen–film combinations (*see* Screen–film combination)
Analogue-to-digital converters (ADC), 59, 62, 228–229, 531, 546, 568, 580, 600
Anger camera, 5, 168, 177, 304; *see also* Gamma camera
Anger logic, 186, 189–190, 303
Angular sampling requirements, 136
Anisotropic diffusion, 558
Annihilation coincidence detection (ACD), 294–295
Annular array, 390, 392

Apodisation, 581–582
Apparent diffusion coefficient (ADC), 556–558
Applied potential tomography (APT), 652
Array processor, 117, 785
Artefacts; *see also* NMR, image artefacts
  aliasing, 136, 138–139, 530–531, 546, 571
  beam-hardening, 115, 138, 141
  black boundary, 575
  circular (or nonuniformity), 160, 269–270, 281, 333
  equipment-related, 84, 139, 141, 147, 160, 191, 313–314, 335, 444, 760
  flow-related artefacts, 571–572
  grating lobe, 393
  motion, 79, 139, 244, 271, 288, 309, 401, 411–412, 545, 549
  partial-volume, 136–138, 286
  patient-related, 288, 290–291
  reconstruction, 108, 110, 131, 745
  respiration, 251, 657
  shadowing, 396–397, 409
  speckle, 382, 409, 744
  spoke or star, 277
  streaking, 138–139, 270, 309, 312
  subtraction, 244–245
  truncation, 271
Artificial radionuclide, 4
A-scan, 399
Asynchronous transfer mode (ATM) links, 761
Attenuation
  PET, 297
  SPECT, 281–283
  ultrasound, 359–360
  X-ray, 18–20
Auger electrons, 19
Autocorrelation function, 710–711
Automatic brightness control (ABC), 70
Automatic exposure control (ACE), 69–70, 83
Autoradiography, 176, 178, 196, 198–200, 318
Axial (ultrasound) resolution, 446

B

Backing medium, 386–387
Baseline suppression, 582
β-emission, 210–211
Biologically relevant radionuclides, 221
Biomarkers, 607
Bloch equations, 499–501
Broadband ISDN (B-ISDN), 761
Brute force approach, 562
B-scan, 400
Bubble-specific imaging, 431
Bucky factor, 39

C

Cavitation, 449–450
Centric phase encoding, 543
Charged coupled device (CCD), 55, 59–60, 64–66, 73, 75, 176, 199–200
Chemical-shift spectroscopy
  decoupling, 580
  heteronuclear coupling, 580
  ^{31}P spectra, 578
  properties of, 578–579
  shielding parameter, 579
  spin–spin splitting, 579
Chemical-shift tensor, 578–579
Clinical applications
  adrenal gland, 466
  Alzheimer's disease, 566, 607
  angiography
    DSA, 2–3, 60, 72
    MRI, 543, 553–555
  anoxia, 589
  anti-angiogenic drugs, 607
  arthritis, 638
  bladder, 154, 458–459
  biopsy and needle-puncture techniques, 466–467, 609
  brain
    MRI, 553, 607–608
    PET, 322
    SPECT, 291–292
    surgery, 609
    ultrasound, 464–465
  breast
    digital mammography, 73–75
    EIT, 659
    transillumination, 671–672
    ultrasound, 462–463
  cancer, 155, 292, 319, 321, 607
  chemotherapy, 153
  deep vein thrombosis, 639
  endovascular treatments, 609
  eye and orbit, 463–464
  gastric (EIT), 658
  GI tract, 456–457
  gynaecology, 456
  heart
    EIT, 658–659
    first-pass studies, 255–256
    LVEF, 251, 253
    MRI, 543, 553, 608

multigated acquisition, 251–254
PET, 322
SPECT, 288–291
ultrasound, 459–461
intraoperative, 467
kidneys
nuclear medicine, 245–246, 256–261, 292
ultrasound, 458
liver and biliary tree, 292, 457
lungs
EIT, 656–658
ultrasound, 465
male genital system, 463
musculoskeletal system, 461, 609
obstetrics, 454–456
pancreas, 457
paediatric, 466
prostate, 155, 158, 459
radiotherapy
IGRT, 155
MMI, 774–775
PET, 321–322
targeted radionuclide therapy, 292–293, 776
ultrasound, 467
Raynaud's disease, 639–640
skeleton
nuclear medicine, 240–242, 291
thermography, 640
skin, 465
spleen and lymphatics, 458
thyroid and parathyroid, 241–243, 463
vascular disorders, 639–640
whole-body, 241
Clinical thermography
inflammatory conditions assessment, 638–639
metabolic studies, 640
oncological investigations, 641
pain and trauma, 640–641
physiological factors, 636–638
physiological investigations, 641–642
vascular disorders
arterial disease, 640
deep vein thrombosis, 639
varicosities localisation, 639
vascular disturbance syndromes, 639–640
Clutter, 414
C-mode, 400
Coherent flow, 552
Collimators, 169, 177–178
511-keV, 275
astigmatic, 271
coded apertures, 264
cone-beam, 183–184, 271, 273, 279
converging multi-hole, 183–184, 264, 274
diverging, 264
fan-beam, 179, 183–184, 272–273, 279
focused, 182, 191–192, 265, 274–275
multileaf, 156
multi-slice, 192
parallel-hole, 179–182, 241–242, 266–267, 270, 273, 275, 689
pinhole, 179, 182–183, 241–242, 264
rotating, 274–275
scanning X-ray beams, 41
slant-hole, 264, 271
variable-focus, 271
X-ray tube, 28
X-ray CT, 113, 115
Complex instruction set computer (CISC), 757
Compression amplifier, 398
Computed radiography, 60–64
Computed tomography (CT); *see also* X-ray computed tomography
discovery, 3–4
scanner (*see* CT scanner)
Computer requirements
computing systems
complex instruction set computer, 757
monitors, 759
operating system, 758–759
reduced instruction set computer, 757
streaming SIMD extensions, 757–758
Sun Microsystems, 758
desktop PC, 756
image generation and transfer
ATM links, 761
file formats, 761–762
integrated services digital network, 761
internet and intranet, 763
LAN, 759–761
picture archiving and communication systems, 756, 762–763
standards, 764
teleradiology, 763–764
Cone-beam reconstruction algorithm
approximate cone-beam reconstruction techniques
adaptive multiple plane reconstruction algorithm, 133
advanced single-slice rebinning algorithm, 132–133
FDK-type algorithms, 131–132
single-slice rebinning algorithm, 132

exact reconstruction algorithms, 131
image quality, 141–142
Cone-beam X-ray computed tomography, 159–161
Constant-potential generator, 32
Continuous wave (CW) method, 491
Contrast media, 70–72
Conventional ultrasound imaging, 364–365
Conversion electron, 211
Convolution integral, 691
Co-registration, *see* Image registration
Correlation distance, 565
Correlation spectroscopy (COSY), 587–588
Correlation time, 512
C-scan, 400
CT dose index (CTDI), 134–136, 145
CT scanner
design
dual-source, 112
electron-beam CT scanner, 109
multi-slice CT, 110–112
spiral CT scanner, 109–110
future aspects
cardiac imaging, 143
CT angiography, 142–143
CT fluoroscopy, 143
multi-planar reformatting, 142
radiation dose, 143
virtual endoscope, 143
operation and components
beam filtration and collimation, 115
computer system, 117–118
data acquisition system, 115–117
gantry, 112–113
image reconstruction and display, 118–119
patient couch, 117
X-ray generation, 113–114
performance assessment (*see* Performance assessment (of X-ray CT))
Current-density imaging, 564

D

Deconvolution, 691, 693–696, 698, 700–702, 705–708
in EIT, 654–655
in PET, 315
in scintigraphy, 259, 261
in SPECT, 264, 283–284
in ultrasound imaging, 438
in X-ray CT, 127–128
Demand modulation function (DMF), 746
Demodulator, 598
Dental radiography, 14, 26, 29, 33, 76, 83
Detective quantum efficiency (DQE), 25
Detector dose indicator (DDI), 83
Detectors, *see* Radiation detectors
Diffusion and perfusion
bipolar/pulsed-field gradient scheme, 555
Stejskal–Tanner module, 556
Stokes–Einstein equation, 556
Diffusion tensor, 558
Digital chest imaging, 75
Digital fluoroscopy, 59–60
Digital image receptors
active-matrix array
fluorescent screens, 66
selenium photoconductors, 66–68
advantages, 68–69
bad pixels, 58
charge-coupled devices, 65
computed radiography
colour and luminescence centres, 60–61
DQE for, 63–64
europium activated barium fluorohalide compounds, 60
image plate construction, 61–62
laser beam, 62
MTF, 62–63
photostimulable phosphor, 60–61
screen–film systems, 62
digital fluoroscopy, 59–60
flat-panel receptors, 66
phosphor plates, 64–65
photon counting detectors, 68
pixel depth, 59
Digital imaging and communications in medicine (DICOM), 761–762
Digital spot imager (DSI), 160
Digital tomosynthesis, 75–76
Direct digital radiography (DDR), 83
Direct-exposure X-ray film
film base, 42
film blackening, 43
film gamma, 45
film latitude, 45
grain sensitisation process, 44
H and D curves, 44
Nutting's law, 43
photographic process, 43
Discrete Fourier representation
discrete–discrete model, 695–697
Fredholm integral, 694–695
operator equation, 694
Doppler methods
choice of frequency, 413

CW flow detector
phase-domain processing, 417
pulsatility index, 417–418
quadrature-phase demodulator, 417
real-time spectral analysis, 417
resistance index, 418
signal processing components, 416–417
Doppler display method
colour imaging, 426–428
power imaging, 428–429
spectral Doppler and Duplex, 425–426
Doppler equation
3D nature, Doppler-shift signal, 415–416
amplitude modulation, 414
cos θ and *c* measurement, 414
CW transmission, 415
Doppler-shift bandwidth, 415
Fourier-domain leakage phenomenon, 415
frequency modulation, 414
transit-time broadening, 414–415
Doppler-shift detection principle, 412–413
pulsed-wave flow detection
components, 419
Doppler frequency shift, 420
Doppler-shift signal, 419–420
gate and/or pulse length, 420–421
phase reference point, 421
signal derivation principles, 422–423
types, 418
range-velocity, 423–424
tissue Doppler imaging, 429
Dual-energy imaging, 73
Dual-isotope imaging, 243, 245, 283
Dual-source CT scanner, 112
Dynamic planar scintigraphy
cardiac imaging
first-pass studies, 255–256
multigated acquisition (*see* Multigated acquisition)
principles, 250–251
renal imaging
$^{99}Tc^{m}$-MAG3, 259
activity–time curve, 256–260
compartments, 260
excretion, 256, 259–260
GFR, 256, 259
graphical analysis, 259–260
kidney anatomy, 256, 257
mean transit time, 261
multiple temporal images, 256
percentage outflow curves, 261
perfusion phase, 256–258
renograms, 256–260
ROI analysis, 258–259

E

Echography
clock pulse, 395–396
compression amplifier, 398
demodulation, 398
digitisation and display, 398–399
linear RF amplifier, 396
pre- and postprocessing, 398
real-time scanning, 401–403
scanning and display format, 399–400
scanning methods, 400–402
signal generation and processing, 394–395
time gain control, 396–397
transducer and frequency, 394
transmitter, 396
Effects
biological, 449–451, 602
Bloch–Siegert, 498
focusing, 374–376
heel, 30
movement, 369–370
partial-volume, 286–287, 315–316
pharmacological/toxicological, 201
Electrical impedance tomography (EIT), 8
clinical applications
breast imaging, 659
cardiac imaging, 658–659
gastric imaging, 658
pulmonary imaging, 656–658
conductance, 650
forward problem, 651
inverse problem
AC and DC voltage, 651–652
electrical conductivity, 652–654
electrical field lines, 652
image reconstruction, 654–656
isopotentials, 654–655
multi-frequency imaging, 656
Electrical tuning, 387
Electroluminescent display (ELD), 759
Electron-beam CT scanner, 109
Electronic gamma camera, 4–5
Electro-optical imaging, 745
Emission computed tomography (ECT), 170
Enhancement
contrast, 58, 71–72, 86, 160, 430, 457–459, 462, 550, 718–719
correlated signal enhancement (CSE), 190

edge, 58, 68, 231–232, 582
image (*see* Image enhancement)
Equalisation radiography, 76
Extended phase graph, 512
External-beam radiation therapy, 467

F

Film transistors (TFT), 759
Filters
adaptive (or speckle-reduction), 384
aluminium, 32, 80
apodising, 571
attenuating or flat, 78, 115
backprojection, 131
beam-shaping ('bow-tie'), 115, 135, 144
box, 704
Butterworth, 232–233
Colsher, 311
convolution masks, 722
copper, 32
discrete, 124
edge-enhancement, 68, 231–232, 582
Fourier-space (or frequency-space), 232–233
Hanning, 126, 232–233, 286
k-space, 539
Laplacian, 724
lens, 624, 631
low-pass, 106, 416–417, 419, 694
median, 58, 384
molybdenum, 31–32
nine-point spatial smoothing, 231
noise reduction or smoothing, 68, 118, 268, 285, 287, 424, 580
Prewitt, 724
pseudo-inverse, 701
Ramachandran–Lakshminarayanan, 125
ramp, 131, 285–286, 311, 334
reconstruction, 118, 141, 278
rhodium K-edge, 32
sharp, 118
Shepp and Logan, 126–127, 286
silver, 32
Sobel, 724
spatial-frequency, 232
temporal smoothing, 231–232
trough, 76
wall, 414, 429
Wiener, 384, 701
z-filter, 120–121, 133, 140–141
^{18}F-labelled fluorodeoxyglucose (^{18}FDG), 294
radiopharmaceuticals, 318–319
standard acquisition protocols, 309
Flat panel display (FPD), 759
Flow-relatedartefacts, 571–572
Focal-plane array (FPA) detectors, 625
2D Fourier imaging
3D phase-encoding, 527–528
mathematical description
gradient echoforms, 529
k-space variables, 530
phase-encoding, 529
spin-warp imaging, 528
Fourier transform, 708–709
Fraunhofer zone, 372
Free induction decay (FID), 504–505, 515–516, 526–528, 580–582, 584–586
Frequency encoding, 521
Fresnel zone, 372
Full width at half maximum (FWHM)
energy resolution, 171–172, 185, 329
MRI, 506, 514, 580
spatial resolution
PET, 194, 295–296, 303, 308, 315, 334
scintigraphy, 179, 183–184, 196, 248, 326–329
SPECT, 274, 315
ultrasound imaging, 377–379, 446, 746
X-radiology, 22
Full width at tenth maximum (FWTM), 196, 295, 326, 328, 334, 446

G

Gamma camera
capacitor network/resistive-coupled network, 186
components, 178–179
converging multi-hole collimators, 183–184
detected count rate (n_d), 188
developments
correlated signal enhancement, 190
modern cameras improvements, 191
PMT gain, 188–189
spatial non-linearity, 188
stability and uniformity, 191
variable energy response, 189
very-high-speed digitisation, 190
energy signal, 186
hexagonal array, PMTs, 185
image-formatting system, 187
light-emission decay constant, 184
output signals, paralysable and non-paralysable detectors, 187–188
parallel-hole collimator
diameter and length, 180

geometrical efficiency, 181
geometrical spatial resolution, 179–180
near-normal incident photons, 179
performance characteristics, 181
point spread function, 179
septal thickness, 182
system spatial resolution, 180–181
types and dimensions, 179–180
pinhole collimator, 182–183
pulse-height analyser, 187
single-channel analysers, 187
spectral response, bialkali photomultiplier, 184–185
Gamma-ray photons, 4
Geiger–Müller tube, 4
Geometric unsharpness, 34–36
Gibbs ringing artefacts, 571
Glomerular filtration rate (GFR), 256, 259
Gradient-echo (GE) pulse, 523–524, 529
Graphical user interface (GUI), 758
Grid-biased X-ray tubes, 28
Grid cut-off, 39
Grid ratio, 38
Gross tumour volume (GTV), 154
Gyromagnetic ratio, 494

H

Halogen radionuclides, 221
Half-value layer, 79–82, 87, 144, 212, 249, 297
Heterodyne detection, 504
Heteronuclear coupling, 580
Homonuclear coupling, 579
Hydrophone-based measurements
intensity parameters, 440–441
power parameters, 441
pressure parameters, 439–440
scanned system, 441–442
scanner settings, 438
spatial distributions, 438–439
voltage–time waveform, 438

I

Image analysis
edges and boundaries
first-order operators, 722–723
second-order operators, 723–724
shape and structure
geometric features, 723–724
object moments, 724–725
object skeleton, 725–726
texture features
edge density, 722
fractal dimension, 722
grey-level histogram, 720–721
Image classification
supervised classifiers
Bayesian classifier, 731–732
K-nearest neighbour classifier, 733
linear discriminant functions, 732–733
multilayer neural networks, 733
unsupervised classifiers
c-Means classifier, 733–734
self-organising artificial neural networks, 734–735
Image display
MMI, 773–774
radioisotope imaging, 235–236
SPECT, 287
ultrasound imaging
A-mode, 399
B-mode, 400
C-mode, 400
Doppler, 425–429
echography, 398–399
M-mode, 399–400
three-dimensional, 409–410
X-ray CT, 118–119
Image enhancement
contrast enhancement
mapping, 718
pseudocolour, 719
thresholding, 719
histogram equalisation, 719
image subtraction, 719
spatial-filtering operation, 719–720
Image formation and processing; *see also* Iterative image processing
autocorrelation function, 710–711
convolution theorem, 709–710
discrete Fourier representation
discrete–discrete model, 695–697
Fredholm integral, 694–695
operator equation, 694
Fourier transform, 708–709
image restoration
constrained deconvolution, 700–701
inverse operator, 697
least-squares image processing, 699–700
linear system, mathematical description, 697–698
magnification factor, 699

maximum-entropy deconvolution, 701–702
signal-to-noise ratio, 698–699
image sampling
2D Fourier transform, 703
discrete sampled image, 702
Fourier-space relationship, 702–703
image size, 704–705
intervals, 704
Shannon sampling theorem, 705
unaliased image sampling, 704
zero-one box function, 704
interpolation function, 711
object and image
2D distribution, 688
3D distribution, 688
linear space-invariant imaging system, 691
non-linear space-invariant imaging system, 691
optical photon transport, 689
spatial dependence, 690
superposition principle, 690
power spectrum, 710–711
problems in
direct deconvolution, 693–694
Fourier representation, 693
Gaussian function, 694
image formation, 692
modulation transfer function, 693
Image fusion, *see* Image registration
Image-guided radiation therapy (IGRT), 155
Image intensifiers
caesium iodide screen, 56
flux gain, 55
Kell factor, 58
photocathode, 56
photoelectrons, 55
pickup tube, 55
plumbicon camera, 57
spot film, 57
Image processing; *see also* Image analysis
compression, 717
enhancement, 717 (*see also* Image enhancement)
image analysis, 717
image-generation systems, 715–716
linear scale-space (*see* Linear scale-space)
restoration, 717
segmentation, 717–718
technicalities and notation, 715
Image quality in X-ray CT
aliasing artefacts, 138–139
angular sampling requirements, 136
beam-hardening artefacts, 138
cone-beam reconstruction algorithm effect on, 141–142
equipment-related artefacts, 139
iterative reconstruction algorithm effect on, 141
motion artefacts, 139
parameters
low-contrast resolution, 134
noise, 134
radiation dose, 134–136
relation between, 136–137
spatial resolution, 134
partial-volume artefacts, 137–138
spiral interpolation algorithm effect on
effective tube current scan time product, 141
multi-slice scanning, 140–141
sensitivity profile dependence, 140
single-slice interpolation algorithms, 140
spatial resolution, 140
Image registration, 767
Image perception and interpretation
dynamic imaging, 748
eye and brain imaging system, 741–742
flicker sensitivity, 748
physical–chemical structure, 739–740
potential predators, 748
quantitative measures
definitive diagnosis, 750
diagnostic decisions, 749
figure of merit, 753
noise-limited process, 752
parametric imaging, 752 (*see also* Parametric imaging)
receiver operating characteristics, 750–751
statistical decision matrix, 749
sequential channels, 739
SNR, 741
spatial and contrast resolution
clutter, 744
coherent speckle noise, 745
cones and rods, 742–743
DMF, 746
image noise, 744
low-noise photomultiplier, 743
luminance levels, 744
NRM, 746
operator performance, 746–747
sine-wave gratings, 744

single photon, 743
single pixel, 745
visual averaging process, 744
visual perception process, 741
Image transformation, 768–769
Imaging modalities, attributes and roles
anatomical *vs.* functional imaging, 785
image processing, 786–788
imaging applications, 789
personnel and space, 788–789
single and multiple images, 788
spatial resolution and sensitivity, 789
time factors, 785, 788
Imaging pulse sequences for magnetic resonance imaging (MRI)
echo-planar and spiral imaging
initial phase-encode and read-dephase gradients, 539
multi-shot sequences, 538
pulse sequence for, 539–540
single-shot EPI sequence, 539
interleaved and hybrid sequences, 541–542
multi-slice *vs.* true-3D sequences, 542–543
rapid gradient-echo sequences, 537–538
spin-echo sequences, 541
Incoherent flow, 552
Inertial cavitation, 450
Infrared (IR) imaging
clinical thermography (*see* Clinical thermography)
focal-plane staring arrays
AGEMA 570 imaging system, 632–633
detector construction, 632
image quality, 634
quantum-well-type IR photodetector, 633
scanning system, 631
single cell FPA, 633
uncooled FPA detectors, 632
future aspects, 644
IR detector, 624–625
IR photography, 624
liquid-crystal thermography, 642–643
microwave thermography, 643–644
pyroelectric imaging systems, 634
single-photon detectors
cadmium mercury telluride, 626
detectivity, 627–628
indium antimonide, 626
responsivity, 626–627
spectral detectivities, 626–627
temperature measurement, 634–635
thermal radiation
radiant heat loss, 625
skin and surface emissivity, 625–626
Stefan–Boltzmann law, 625
thermographic scanning systems
first- and second-generation scanning systems, 629
IR optical components, 629
IR refractive index, 629
minimum resolvable temperature difference, 628
principal parameters, 628–629
SPRITE CMT detector, 629–630
Integrated Services Digital Network (ISDN), 761
Intensity-modulated radiotherapy (IMRT), 157–158
International Electrotechnical Commission standard, 451–452
Internet and intranet, 763
Inversion-recovery (IR) sequence, 508–509
Ionisation chamber detector, 69
Isochromats, 515
Iterative image processing
noise in imaging modality, 706
object distribution
convolution integral, 705
first-order iteration, 705–706
nth-order pseudo-image, 705–706
zeroth approximation, 705
spatial frequency analysis
Fourier space, 706
MTF, 707
time-consuming computational process, 707–708
truncated deconvolution, 707
unconstrained deconvolution, 707
Iterative reconstruction methods; *see also* Reconstruction from projections
additive immediate correction, 129–130
algorithms, 129
effect on image quality, 141
ill-posed problems, 128
principle, 128–129

K

Kell factor, 58
Keyhole method, 543–544
K-shell binding energy (K-edge), 30
kVCT systems, 162

L

Landé g-factor, 494
Larmor angular frequency, 495
Left-ventricular ejection fraction (LVEF), 251, 253
Light-beam diaphragm, 82
Limulus amoebocyte lysate (LAL) assay, 224
Linear scale-space
  composite invariants
    flow-line curvature, 729
    isophote curvature, 729
    junction invariant, 730
    umbilicity, 730
  irreducible invariants, 727–728
  line detection, 730
  postulates of, 727
Line source response function (LSRF), 741
Line-spread function (LSF), 326–328
Liquid crystal display (LCD) screens, 759
Liquid-crystal thermography, 642–643
Location ROC (LROC), 750

M

Magnetic resonance imaging (MRI)
  advanced techniques
    gating, 545
    k-space, acquisition strategy, 543–545
    parallel imaging, 546–548
  applications of
    cancer, 607
    cardiology, 608
    genetic phenotyping, 610
    interventional procedures and monitoring, 609
    molecular and cellular imaging, 609–610
    musculoskeletal system, 609
    neurology, 607–608
  brute force approach, 562
  chemical-shift imaging, 564
  classifications of, 507–508
  current-density imaging, 564
  diffusion and perfusion
    bipolar/pulsed-field gradient scheme, 555
    Stejskal–Tanner module, 556
    Stokes–Einstein equation, 556
  discovery, 6
  dynamic contrast-enhanced MRI, 550–552
  ESR, 563
  flow, 552–555
  image acquisition and reconstruction
    2D Fourier imaging (*see* 2D Fourier imaging)
    goal of, 520
    imaging pulse sequences, 524–527
    k-space and image resolution, 530–532
    magnetic field gradients, 520–522
    projection reconstruction, 526–527
    relaxation contrast, 535–537
    selective excitation, 522–523
    signal-to-noise ratio, MR images, 534–535
    single-point imaging, 532–534
    spoiling, 537
  image artefacts
    arrowhead, 575
    black boundary artefact, 574–575
    bloodflow, 571–572
    chemical shift, 538, 557, 564, 573–574, 587, 595
    demodulation angular frequency, 573
    distortion and signal dropout, 539–540, 559, 575
    experimental design, 571
    Gibbs ringing, 571
    instrumentation, 567–570, 595
    MR distortion, 575
    Nyquist ghosts, 539–540
    phase-encoding direction, 573–574
    respiratory motion, 572–573, 608
    ringing, 543, 581
    time-domain signal, 572
  line-scanning methods, 548–549
  long-range dipolar fields, 565
  magnetisation transfer, 561
  MRI technology
    computer systems and pulse programmer, 600
    equipment, 592–593
    magnetic field gradients, 597
    main magnet, 593–597
    microscopy and pre-clinical systems, 600
    radiofrequency system, 597–600
    siting, 600–601
  MR safety
    biological effects and hazards, 602
    gradient-induced noise, 604
    radiofrequency B_1 fields, 604
    safety considerations, 605–606
    static B_0 magnetic fields, 602–603
    time-varying magnetic fields, 603–604
  MR spectroscopy (*see* Magnetic resonance spectroscopy (MRS))
  multiple-quantum imaging, 565
  nuclear polarisation, 561
  nuclei imaging, hydrogen, 565–566
  parallel imaging, 599
  partially parallel acquisition, 493

point method, 548
principles of
Bloch equations, 499–501
classical theory of electromagnetism, 493–494
elementary quantum-mechanical description, 501–503
gyromagnetic ratio, 494
Landé g-factor, 494
Larmor angular frequency, 495
radiofrequency field interaction, 495–496
relaxation, 499–501
rotating reference frame, 496–499
signal detection, 504–507
statistical distribution, spin states, 503–504
vector model, 495–496
pulse sequences for MRI (*see* Imaging pulse sequences for MRI)
quantitative imaging, 549–550
radiofrequency, 491
relaxation processes
information from, 516
longitudinal relaxation, T_1, 512–513
measurement of, 516–518
$T_{1\rho}$ and spin locking, 515–516
transverse relaxation, T_2, 513–514
transverse relaxation times and diffusion, 518–520
relaxation times, 550–552
rotating reference frame
adiabatic rapid-passage experiment, 499
Bloch–Siegert effect, 498
effective flux density, 497–498
Larmor precession, 498
plane-polarised field, 498
resonance condition, 497
saturation- and inversion-recovery, 508–509
spin echo, 509–511
stimulated echoes, 511–512
susceptibility and functional MRI, 558–561
Magnetic resonance spectroscopy (MRS)
chemical shift, 575
1D spectroscopy data processing
apodisation, 581–582
baseline suppression, 582
phase correction, 582
solvent suppression, 582
zero filling (padding), 581
^{19}F, 591–592
^{1}H, 590–591
image-selected *in vivo* spectroscopy, 584–585
localisation method, 587
localised spectroscopy, 582–583
^{31}P, 588–589
point-resolved spectroscopy, 585
radiotherapy planning, 155
slice selective techniques, 583
spectroscopic imaging/chemical-shift imaging, 586–587
stimulated echo acquisition mode, 585–586
TOCSY, 588
WALTZ, 588
Magnetisation recovery curve, 516
Magnetisation transfer, 561
Magnetogyric ratio, 494
Matching layer, 386
Maximum intensity projection (MIP), 287, 553–554
Megavoltage computed tomography (MVCT), 161–162
Metallic radionuclides, 221
Microwave thermography, 643–644
Modulation transfer function (MTF), 693–695, 701, 707, 739, 741–742, 746–747
scintigraphy, 326
ultrasound imaging, 376
X-radiology 22–23, 47–50, 54, 56, 62–63, 73–75
X-ray CT, 134, 146
Molybdenum, 28–29
Monte-Carlo-based model, 26
Motion-vector imaging, 473
MR elastography, 554
Multi-crystal scanners
Aberdeen Section Scanner, 191
crystal/collimator systems, 191
MSPTSs, 191–192
multi-crystal Cleon 710 single-photon tomographic brain imager, 191–192
SPECT and PET systems, 192–193
Multi-crystal single photon tomographic scanners (MSPTSs), 191–192
Multi-crystal positron emission tomography (PET), 193, 303–304
Multigated acquisition
amplitude and phase image, 254
heart chambers function, 251–252
left anterior oblique view, 251, 253
list-mode acquisition, 252
LVEF, 251, 253
regional wall motion measurement, 254
ROI analysis, 253
R-wave, ECG, 251–252
Multimodality imaging (MMI)
clinical application
abnormality identification, 774–775

image-guided surgery, 776–777
radiotherapy, 774–775
targeted radionuclide therapy, 776
display, 773–774
image registration, 767
image transformation, 768–769
line/surface matching, 770–771
non-rigid registration, 773
point matching, 770
registration accuracy, 777
volume matching
correlation coefficient, 772
cost functions, 771
mutual information, 772–773
SAD, 771–772
voxel ratios, 772
voxel-value-derived properties, 771
Multiphysics imaging, 475–476
Multiple-element transducer
applications, 393
array configuration, 390
array construction, 392
beam divergence patterns, 392–393
beam formation, 388–389
disadvantages, 389
focusing systems, 390–391
intraluminal and intracavity scanners, 393
swept/dynamic focusing, 389
transrectal ultrasound, 393–394
Multiple-quantum imaging, 565
Multi-slice CT, 110–112
Multi-wire proportional chamber (MWPC) detectors
3D FBP reconstruction method, 194
drift and detection region, 195–196
first-pass cardiac imaging, 196
hybrid system, 194
interleaved wire planes and thin lead foil converters, 193
intrinsic resolution, 193, 196
lead channel-plate photon converter, 193–194
PETRRA, 194–195
photoelectrons, 193
quantitative macroscopic autoradiography, 196–197
temporal and spatial resolution, 194
TMAE, 194–195

N

Navigator-echo technique, 556
Needle hydrophone, 436–437
Needle puncture techniques, 466–467
Noise equivalent count rate (NECR), 335
Noise equivalent quanta (NEQ), 25
Noise power spectrum, 23, 52–53
Noise required modulation (NRM), 746
Non-diagnostic computed tomography (NDCT), 159
Non-linear clock, 8–9
Non-screen film, 42
Non-standard computed tomography scanners, 159–161
Nuclear magnetic resonance (NMR), *see* Magnetic resonance imaging (MRI); Magnetic resonance spectroscopy (MRS)
Nuclear medicine, 4–5, 168, 223, 269, 335; *see also* Radioisotope imaging
Nuclear polarisation, 561
Nutting's law, 43
Nyquist sampling theorem, 22, 530
Nyquist sampling criteria, 124, 704
Nyquist sampling frequency, 63, 73–74, 90, 334

O

Optical imaging
acousto-optic imaging, 679
contrast
absorption, 667–668
refraction, 666
scatter, 666–667
diffuse optical imaging
breast transillumination, 671–672
image reconstruction, 676–678
optical tomography, 675–676
optical topography, 673–675
exogenous contrast agents and molecular imaging, 680
multimodality imaging, 680–681
optical coherence tomography
applications, 670
principles of, 668–669
photoacoustic imaging, 678–679
Optical pumping methods, 562
Optical transfer function (OTF), 741–742

P

Paget's disease, 640
Parametric imaging, 752
MRI, 517, 549–550
radioisotope imaging, 234, 251, 254, 261, 317
Partially parallel acquisition, 493

Performance assessment
PET
axial slice width, 334
count-rate performance, 335
NECR, 335
randoms fraction, 335
recovery coefficient, 315, 334
scatter fraction, 334
slice and volume uniformity, 335
standards, 333–334
system sensitivity, 334
transaxial resolution, 334
radioisotope imaging
count-rate performance, 330
energy resolution, 329
flood-field uniformity, 324–326
multiple-energy-window spatial registration, 331
plane sensitivity, 330
recommendations, gamma cameras, 323–324
shield leakage, 331–332
spatial distortion, 329–330
spatial resolution, 326–329
SPECT
axial slice thickness, 333
centre of rotation, 322
contrast, 333
recovery coefficient, 287, 315, 333
transaxial spatial resolution, 333
uniformity, 333
ultrasound imaging (echography)
axial and lateral resolution, 446
distance measurement, 445–446
element dropout, 445
low contrast penetration, 445
performance characteristics, 443–444
phantoms, 444
slice thickness, 446
X-ray CT
CT dosimetry, 145
helical scanning, 146–147
image quality, 145–146
scan localisation, 144–145
tube and generator tests, 144
X-ray unit, 143
X-radiology
beam perpendicularity, 82
exposure time, 79
focal-spot size, 80–82
half-value layer and filtration, 80–81
light-beam diaphragm, 82
practical considerations, 82–83
tube output, 79–80
tube voltage, 78–79
X-ray field alignment, 82
Perfusion index (PI), 258
Peripheral arterial disease, 640
Phantoms
anthropomorphic, 160–161
chest, 334
dosimetry, 134–135, 145, 147
for pulse-echo systems, 443–446
for Doppler systems, 446–447
head, 3
Hoffman brain, 195, 235–236
Jaszczak phantom, 333
Leeds test objects, 85–88
linearity, 189, 329, 570
low-contrast resolution, 134, 146–147
spatial resolution, 160, 326–328
tissue-mimicking, 444
uniform, 134, 145, 333–334, 569–571, 573
Phase contrast imaging, 553, 558
Phase correction, 582
Phased-array coils, 599
Phase-encoding gradient, 524
Phase-sensitive detector, 597
Phosphor plates, 64–65
Photoacoustic imaging, 678–679
Photodiode (PD), 174
Photoelectric interaction, 19–20
Photographic grain noise, 745
Photomultiplier detectors, 69, 171–174
Photon counting detectors, 68
Photon interactions, 15
Compton, 170–171
gamma-ray mass attenuation coefficients, 170
linear attenuation coefficient, 18–19
photoelectric, 18–19, 170
photon energy range, 18
transmission of monoenergetic photons, 18
Phototimer detectors, 69
Physico–chemical micro-environment, 516
Picture archiving and communication system (PACS), 762–763
Piezoelectric hydrophone; *see also* Hydrophone-based measurement
acoustic pressure, 437
amplifier combination, 438
deconvolution methods, 438
membrane design, 435
needle, 436–437
properties, 434–435
sensitivity, 437

spatial peak, 437
thickness-mode resonance frequency, 436
Pinhole/slit techniques, 81
Point-resolved spectroscopy, 585
Point source response function (PSRF), 689, 691, 741
Point spread function (PSF), 179
Positron emission tomography (PET), 155
2D image reconstruction, 310
3D image reconstruction algorithms
2D projections, 310–311
Colsher filter, 311
cross-plane rays, 310
FaVoR algorithm, 311
Fourier rebinning algorithm, 312
single-slice rebinning, 311
annihilation photons, 168–169
camera technology developments
new light detectors, 307
new scintillators, 306–307
PET/MR, 309
small animal PET systems, 308
time of flight, 307–308
clinical data acquisition
dynamic acquisition modes, 309–310
standard acquisition protocols, ^{18}FDG, 309
data correction and image quantification
attenuation correction, 313–314
dead-time correction, 315
normalisation, 313
partial-volume effect, 315–316
quantification/calibration, 315
randoms correction, 314
scatter correction, 314–315
detectors and cameras
block detector, 302–303
dedicated PET NaI camera, 304
detector requirements, 301–302
modified dual-head gamma camera, 305
PET/CT systems, 306
PET/SPECT cameras, 305–306
research tomographs, 305
standard positron camera configuration, 303–304
wire-chamber systems, 305
diagnostic and research applications
cancer staging, 321
cardiology, 322
clinical research, 322–323
neurology, 322
oncology, 319
radiotherapy treatment planning, 322
therapy response, 321–322
discovery, 6
image quantification
analysis techniques, 317
compartmental modelling, 316–317
standard uptake values, 316
iterative and model-based image reconstruction algorithms, 312–313
performance assessment (*see* Performance assessment (for PET))
principles
2D and 3D sensitivity, 299–300
attenuation and scatter, 297–298
coincidence detection, gamma rays, 294–295
coincidence event types, 297
dead time and count-rate losses, 300
gamma-ray detection efficiency, 296
noise equivalent counts, 300–301
spatial resolution, 295–296
true and random coincidences, 298–299
radionuclides, 293–294
radiopharmaceuticals
^{18}FDG, 318–319
properties, 317–318
and radioisotopes production, 318
radioligands, 319–320
tracers and applications, 319–320
radiotracers, 293–294
Power Doppler imaging, 428–429
Proton density function, 520
Pulsatility index (PI), 417–418
Pulsed-field gradient scheme, 555
Pulse-height analyser (PHA), 187
Pyroelectric imaging systems, 634

Q

Quality factor, 599
Quality control
imaging equipment (*see also* Performance assessment)
radioisotope imaging, 323–332
PET, 333–335
SPECT, 332–333
X-radiology, 76–91
X-ray CT, 143–147
radiopharmaceuticals
apyrogenicity, 224
biological purity testing, 224
chemical impurities, 226
colloidal/particulate radiopharmaceuticals, 227

endotoxins, 224–225
pharmacopoeias, 223–224
pH measurement, 227
radiochemical purity, 225–226
radionuclidic purity, 225
sterility, 224
Quantum-well-type IR photodetectors (QWIPs), 633

R

Radiation detectors
autoradiography, 176
CCDs (*see* Charge coupled device (CCD))
gas, 69, 175–176,
MWPC, 193–197
photon counting, 68
scintillation (*see* Scintillation detectors)
semiconductor (*see* Semiconductor detectors)
Radioactive decay
alpha decay, 209–210
Auger electron, 212
β-emission, 210–211
characteristic X-rays emission, 211–212
conversion electron, 211
decay scheme, 209–210
electron capture, 211
γ emission, 211
internal conversion, 211
positron emission, 211
radioactivity, 201
Radioisotope imaging
computers role
applications, 228
data acquisition, 228–229
data processing and image reconstruction
Butterworth filter, 233
complex filtering operations, 232
2D projection data, 231
dynamic datasets, 235
edge enhancement, 231–232
Fourier-space filter, 233
frequency-space filters, 233
Hanning filter, 232–233
image filtering techniques, 231
nine-point spatial smoothing filter, 231
numerical and curve data, 230, 234
procedures for, 230–231
real-space operations, 234
ROIs, 234–235
smoothing and filtering, 230
spatial-frequency filters, 232
data storage, 236–237
image display and manipulation, 235–236
interface electronics, 227–228
online data correction, 230
system control, 237–238
dual-isotope imaging (*see* Dual-isotope imaging)
dynamic planar scintigraphy (*see* Dynamic planar scintigraphy)
ECT, 170
equipment
gamma camera (*see* Gamma camera)
history, 177–178
multi-crystal scanners (*see* Multi-crystal scanners)
MWPC detectors (*see* Multi-wire proportional chamber detectors)
performance assessment (*see* Performance assessment (for radioisotope imaging))
quality control (*see* Quality control)
semiconductor detectors (*see* Semiconductor detectors)
nuclear medicine (*see* Nuclear medicine)
PET (*see* Positron emission tomography)
radiation detectors (*see* Radiation detectors)
radioactive decay (*see* Radioactive decay)
radionuclide (*see* Radionuclide)
radiopharmaceuticals (*see* Radiopharmaceuticals)
single-photon imaging, 168–169
SPECT (*see* Single-photon emission computed tomography)
static planar scintigraphy (*see* Static planar scintigraphy)
Radionuclide
advantages, 200
effective half-life, 212
^{111}In and ^{67}Ga, 213
mass of tracer, 201
pharmacological/toxicological effect, 201
physical characteristics, 212
production
charged-particle bombardment, 205–206
neutron capture, 203–204
nuclear fission, 204–205
proton-to-neutron ratio, 202–203
radionuclide generator, 206–209
radioactive decay (*see* Radioactive decay)
single-photon-emitting radioisotopes, 212
Radio-opaque catheters, 156
Radiopharmaceuticals, 4

biological distribution
absorption properties, 219
administration, distribution and excretion routes, 219–220
hydrophilic compounds, 221
hydrophobicity, 220
intracellular components, 221
lipid solubility, 220
passive and specialised transport systems, 220
renal excretion, 221
urinary and biliary excretion, 221
chemistry
anatomical/morphological changes, 213
halogens, 216–217
indium, 218
iodine, 217–218
octreotide, 218–219
radioactive compound, 213
subpharmacological, 214
technetium, 214–216
quality control (*see* quality control (for radio pharmaceuticals))
single-photon emitters, 221–223
Read gradient, 524
Receiver operating characteristic (ROC), 749
Reconstruction from projections
2D filtered backprojection, 122–124, 277–278
continuous and discrete versions, 127
discrete backprojection, 124–125
fan-beam geometry, 127–128
filtered back projection method, 127, 277–278
FORTRAN code, 125–126
Nyquist condition, 124
projection data, 124
Whittaker–Shannon sampling theorem, 124
iterative methods (*see also* Iterative reconstruction methods)
optical tomography, 676–678
PET, 312–313
SPECT, 278–281
X-ray CT, 128–130
Reduced instruction set computer (RISC), 757
Regions of interest (ROIs), 234–235
Resistance index (RI), 418
Resonant frequency, 364–366, 386, 394, 436, 491–492, 499, 520–521, 598–599
Rhodium, 28
Rigid-body registration, 768

S

Saturation- and inversion-recovery, 508–509
Saturation-recovery (SR), 508
Scattering
Compton scattering, 170–172, 176, 198, 282–283, 297, 301
IR imaging, 642
optical imaging, 666–671, 673, 677
PET, 297–302, 305, 308, 310, 313–315, 334–335
scintigraphy, 170, 187, 197–199, 225, 229, 247–248, 250, 329–330
SPECT, 268, 276, 278, 280, 282–284, 287–288
ultrasound
conventionl ultrasound imaging, 364–365
geometrical-like scattering, 362–363
Rayleigh-like scattering, 363–364
resonant and non-linear, 365–366
stochastic and diffraction phenomena, 364
structure, 361–362
X-radiology, 15–22, 32, 56, 61–62, 73–75
anti-scatter device, 15–17, 37–40
scatter-to-primary ratio, 36–37, 41, 70
X-ray CT, 99, 101, 115
Scintillation detectors
electron-multiplying dynodes, 173
energy resolution, 171–172
light-emission characteristics, 171
mono-energetic gamma ray, 171–172
multidetector-based imaging systems, 172
NaI(Tl)-based scintillation counters, 173
photocathode, 173–174
photodiode, 174
physical properties, inorganic scintillators, 171
position-sensitive PMTs, 174, 307
refractive index, 171, 185
relative photon-energy absorbed fraction, 172
silicon photomultiplier, 174–175
Screen–film combinations; *see also* Direct-exposure X-ray film
contrast, 49–50
double-emulsion systems, 48
film granularity, 51
fluorescent screens, 46
light-fluorescent photons, 46
noise power, 52
phosphor thickness, 47
Poisson distribution, 50
radiographic noise, 52

receptors, 46
sensitivity, 48–50
Selenium photoconductors, 66–68
Semiconductor detectors
avalanche photodiodes, 174
charged coupled device (*see* Charged couple device (CCD))
charge-splitting system, 197
coded-aperture systems, 197
Compton gamma camera, 197–198
double-sided microstrip detector, 199
electron–hole (e–h) pair, 61, 67, 68, 197
germanium gamma camera, 197
photodiodes, 66, 79, 174
pixel detector, 199
position-sensitive devices, 200
scattering material, 198–199
silicon wafers, 198
Shielding parameter, 579
Shimming, 595–596
Signal processing in the element (SPRITE) CMT detector, 629–630
Signal-to-noise ratio (SNR), 23
images perception and interpretation, 741
MR images, 534–535
Silicon-based flat-panel detector, 73
Silicon photomultiplier (SiPM), 174–175
Simultaneous transmission/emission protocols (STEP), 272–273
Single-channel analysers (SCAs), 187
Single element transducer
backing medium, 386–387
casing, 387
design, 385
electrical tuning, 387
matching layer, 386
piezoelectric element, 385–386
Single instruction, multiple datastream (SIMD) instruction set, 757
Single-photon emission computed tomography (SPECT), 155
advantages, 262
clinical applications, 288–293
data correction methods
distance-dependent PRF compensation, 284–285
misaligned projection data, 281
motion correction, 284–285
nonuniformity, 281
partial-volume effect, 286–287
photon attenuation, 281–283
photon scatter, 283–284
statistical noise compensation, 285–286
data processing and display optimisation, 287
discovery, 5–6
dual-modality SPECT/CT, 275–277
dual-modality SPECT/PET, 274–275
image reconstruction methods
filtered backprojection, 277–278
iterative reconstruction techniques, 278–281
limited-angle emission tomography, 264–266
longitudinal and transaxial tomography, 262–263
performance assessment (*see* Performance assessment (for SPECT))
^{99}Tc^{m}-HMPAO images, 262–263, 292
transaxial, sagittal and coronal slices orientation, 263–264
transaxial SPECT
attenuated Radon transform, 266
basic principles, 266–267
camera-based systems advantages, 269
data acquisition, 267
data sampling requirements, 268
fan- and cone-beam, 274
gamma camera, 269–271
hardware developments, 271–274
image reconstruction, 268
physiological and metabolic process image, 266
problems and solutions, 267–268
special-purpose systems, 274–275
system design, 269
X-ray linear attenuation coefficients, 266
tri-modality SPECT/PET/CT systems, 277
Sinogram, 229, 267, 281, 284, 309–315, 334
Si-strip detector, 75
Skull base tumours, 609
Slit-scanning system, 41
Solid-state detector, 69
Solvent suppression, 582
Space-invariant PSRF (SIPSRF), 691
Space-variant PSRF (SVPSRF), 691
Spectral density function, 513
Spectral editing, 590
Spherical harmonics, 595–596
Spin echo
dephasing, 511
process of formation, 509–510
schematic diagram, 509–510
Spin–spin splitting, 579
Spin-tagging experiments, 554
Spiral CT scanner, 109–110
data interpolation

multi-slice scanners, 120–121
single-slice scanners, 119–120
image quality
effective tube current scan time product, 141
multi-slice scanning, 140–141
sensitivity profile dependence, 140
single-slice interpolation algorithms, 140
spatial resolution, 140
Spot film, 57
Spot imaging, mammography, 75
Static planar scintigraphy
data-acquisition parameters
collimator and photopeak window, 240
count density and film exposure, 243
image quality optimisation, 240
patient positioning, 243
time after injection, 241
dual-isotope imaging, 243–245
high-energy photon emitters, 249–250
quantification
anterior and posterior views counts, 246–247
build-up factors, 248
depth corrections, photon attenuation, 245–246
divided renal function study, 245–246
geometric mean, 247–248
object radioactivity distribution, 245
relative transmission, γ-rays, 245
single planar views, 2D superposition, 244
tissue and attenuation correction, 245
requirements
double-headed gamma-camera system, 238–239
electromechanical gantry, 238–239
modern large-FOV gamma camera, 238
multi-formatted storage, X-ray film, 239–340
patient couch, 239
whole-body imaging, 248–249
Stefan–Boltzmann law, 625
Stereotactic biopsy control, 75
Stimulated echo, 511
Stimulated echo acquisition mode, 585–586
Streaming SIMD extensions (SSE), 757–758
Subtraction imaging 719
DSA (*see* Digital subtraction angiography)
fluorography, 86
MRI, 570
radioisotope imaging, 231, 236, 243–245, 283
X-radiology, 69, 72–73
Sum of absolute differences (SAD), 771–772
Sun Microsystems, 758
Synthetic-aperture focusing technique (SAFT)
complete-dataset method, 470–471
image reconstruction, 468
linear backprojection, 470
plane-wave method, 470–471
TOF method, 468–469
transmission reconstruction methods, 469
zero-offset method, 470–471

T

Targeted radionuclide therapy, 776
Teleradiology, 763–764
Temperature distribution, 609
Tetrakis(dimethylamino)ethylene (TMAE), 194
Theorems
central-section, 107, 122, 127, 130
convolution, 123, 543, 571, 692–693, 709–710, 728
Fourier inversion theorem, 533
Fourier shift theorem, 523
Nyquist sampling, 22, 530
Parseval's, 711
sampling, 22, 124, 136, 310, 393, 705
Whittaker–Shannon sampling, 124, 705
Three-dimensional imaging, ultrasound
1D array data acquisition
advantages and disadvantages, 407
elevational resolution distance, 405–406
freehand imaging with position sensor, 407–408
mechanical, 406
sensorless, 408
2D array data acquisition, 408
advantages and disadvantages, 411–412
data processing and display, 409–410
limitations, 411–412
Time-of-flight (TOF), 302, 307–308, 335, 468–469, 553
Tissue Doppler imaging, 429
Tissue electrical impedance
electrical behaviour, tissue
displacement currents, 647–648
electrical conduction, 647–648
low-frequency currents, 648
passive electrical properties, 647
imaging (*see* Electrical impedance tomography)
tissue parameter measurement
permittivity, 649–650
resistivity, 648–649

Tissue harmonic imaging (THI)
acoustic pulse, 403
advantage, 404
clinical applications, 404–405
depth dependence, 403–404
parametric array, 404
Total correlation spectroscopy (TOCSY), 588
Transducers
conventional construction
backing medium, 386–387
casing, 387
design, 385
electrical tuning, 387
matching layer, 386
piezoelectric element, 385–386
function, 384
multiple-element
applications, 393
array configuration, 390
array construction, 392
beam divergence patterns, 392–393
beam formation, 388–389
disadvantages, 389
focusing systems, 390–391
intraluminal and intracavity scanners, 393
swept/dynamic focusing, 389
transrectal ultrasound, 393–394
Transmitter, 396
Tube voltage, 78–79
Tungsten, 28–30

U

Ultra port architecture (UPA), 758
Ultrasound imaging
1D industrial flaw detector, 7
2D ultrasound scan, 6–7
3D imaging (*see* Three-dimensional imaging, ultrasound)
acoustic output measurements
derated parameters, 443
hydrophone-based measurement (*see* Hydrophone measurement)
piezoelectric hydrophone (*see* Piezoelectric hydrophone)
radiation force balance, 442–443
acoustic radiation fields (*see* Acoustic radiation fields)
biological effects
non-thermal, 449–451
thermal, 449
clinical application (*see* Clinical application)
Doppler methods (*see* Doppler methods)
Doppler system, 446–447
echography (*see* Echography)
elasticity imaging, 431–433
image generation
pulse-echo scanning, 380–382
speckle production, 382–383
speckle reduction, 383–384
image optimisation, 472–473
imaging parameters, 370–371
microbubble contrast-specific imaging, 429–431
molecular imaging and drug delivery, 473–474
motion-vector imaging, 473
movement effects, 369–370
multiphysics imaging, 475–476
performance assessment (*see* Performance assessment (for ultrasound imaging))
pulse-echo principle, 6
SAFT
complete-dataset method, 470–471
image reconstruction, 468
linear backprojection, 470
plane-wave method, 470–471
TOF method, 468–469
transmission reconstruction methods, 469
zero-offset method, 470–471
standards and guidelines
IEC standard, 451–452
mechanical index, 452–453
safety indices, 453
temperature limitation, 454
thermal index, 452
THI (*see* Tissue harmonic imaging)
transducers (*see* Transducers)
ultrafast and zone-based imaging, 471–472
ultrasonic image formation, 384–385
wave propagation and interactions
absorption, 360–361
acoustic field parameters, 357–358
acoustic impedance, 358
attenuation, 359–360
intensity, 358–359
non-linear propagation, 367–368
scattering (*see* Scattering)
speed of sound, 354–357
UNIX operating systems, 758

V

Varicosities, 639
Vascular disturbance syndromes, 639–640
Venography, 639

W

Wiener spectrum, 52

X

Xenon detectors, 115–116
Xeroradiographic techniques, 7
X-ray computed tomography, 3
  2D image reconstruction, 310 (*see also* Reconstruction from projections)
  cone-beam reconstruction, 130
    approximate reconstruction algorithms, 131–133
    exact reconstruction algorithms, 131
  image quality (*see* Image quality)
  in treatment planning
    bladder cancer, 154
    Ca prostate treatment, 155
    chemotherapy, 153
    conformal therapy, 156–157
    cranio-caudal direction, 156, 158
    cross-sectional plane, 153
    dose distribution, 156
    GTV, 154
    ICRU Reports, 158
    image registration method, 156
    immobilisation technique, 157–158
    lung cancer, 155
    lymph-node, 155, 158
    vs. MRI, 154–155
    MVCT, 161–162
    non-standard CT scanners, 159–161
    prostate cancer, 155, 158
    skin tattoos, 157–158
    thermoluminescent dosimetry, 157
    topogram, 156
    vacuum-moulded polystyrene bag, 158
  iterative reconstruction methods
    additive immediate correction, 129–130
    algorithms, 129
    ill-posed problems, 128
    principle, 128–129
  scanner (*see* CT scanner)
  sectional imaging
    central-section theorem, 107
    CT image, 108–109
    Fourier transform, 106–107
    line integrals, 103–105
    object distribution, 107
    projection data, 106
    projection sets, 105–106
    reconstruction, 2D Fourier methods, 108
    source–detector geometries, 100–103
X-ray film; *see* Direct-exposure X-ray film, *and* Screen–film combinations
X-ray imaging (or X-radiology)
  applications
    chest imaging, 75
    digital mammography, 73–75
    equalisation radiography, 76
    tomosynthesis, 75–76
  automatic exposure control, 69–70, 83
  components, 15
  contrast, 20–22
    media, 70–72
  conventional tomography, 83–84
  digital image receptors (*see* Digital image receptors)
  digital radiographic systems, 89–91
  digital subtraction imaging, 72
    digital subtraction angiography, 72
    dual-energy imaging, 73
  discovery, 2, 14
  dynamic range, 27
  fluorography systems, 86–87
  fluoroscopy systems, 84–86
  image display monitors, 91
  image formation
    energy absorption efficiency, 17
    primary and secondary photon, 16–17
    scatter distribution function, 17
    scatter-to-primary radiation ratio, 17
  image receptors (*see* Analogue image receptors)
  mammographic systems, 87–89
  noise, 50–54
  noise and dose
    DQE, 25
    glandular dose, 25
    high-dose interventional procedures, 26
    ICRP risk factors, 26
    ionising radiation, 25
    NEQ, 25
    per capita radiation dose, 27
    photon counting detector, 24
    quantum mottle, 23
    quantum noise, 23
    SNR, 23
    statistical fluctuations, 23
    surface dose, 24
  photon detector, 15
  photon interaction (*see* Photon interactions)
  quality assurance, 77
  quality control tests (*see* Performance assessment (for X-radiology))

radiographic image, 15
scatter removal
air-gap techniques, 40–41
anti-scatter grid, 37–40
scanning beams, 41–42
scatter-to-primary ratio, 36–37
stereotactic biopsy control, 75
tube output, 79–80
tube voltage, 78–79
unsharpness, 22–23
X-ray field alignment, 82
X-ray tubes
anode, 28–30
cathode, 28
geometric unsharpness, 34–36
high voltage generation, 32–33
x-ray spectrum, 30–32
X-rays, 2

Z

Zero filling, 581
Zeugmatography, 527–528
Zooming, 271